The ARRL
Operating
Manual

Fifth Edition

Edited by:
Steve Ford, WB8IMY

Assistant Editor:
Paul Danzer, N1II

Published by:
The American Radio Relay
League

Contents

Foreword

The fifth edition of the *ARRL Operating Manual* reflects the many changes that have taken place in Amateur Radio during the last five years. As the amateur population has increased, the needs and interests of average hams have shifted. For example, VHF and UHF operating are in the limelight, as are amateur satellites and new HF digital communication modes.

You'll find that the *Operating Manual* is your most valuable resource when you're navigating these new waters. Between these covers, you'll discover the answers to whatever operating-oriented questions you may have. And if the exact answer is not apparent, you'll find helpful signposts to guide your search in the right direction.

Of course, the book overflows with the kind of information we should all know, but always forget. How do I use a repeater autopatch? Where is the CW subband on 20 meters? How do I use a QSL bureau? What is grayline propagation and how do I make it work for me?

It's this kind of depth that makes the *Operating Manual* essential for all amateurs—whether you've been licensed for 6 days or 60 years. Make sure this book is close to your radio. You'll be reaching for it often!

Please let us know how we can make future editions even better. A handy feedback form is at the back for your convenience.

David Sumner, K1ZZ
Executive Vice President
March 1995

Introduction

WHAT'S NEW IN THE FIFTH EDITION

Whenever we release a new edition of any book, the first question is, "What's new?" In the fifth edition of the *ARRL Operating Manual*, the answer is, "A lot!"

Some chapters have been updated with new information; other chapters have been rewritten completely. Let's hit just a few of the highlights . . .

Chapter 1—Curt Phillips, KD4YU, brings his expertise to the subject of shortwave listening. He's rewritten the entire chapter from top to bottom.

Chapter 3—Put together for the new ham.

Chapter 4—Jerry Hall, K1TD, and Chuck Hutchinson, K8CH, have revised the antenna-orientation data in this chapter for increased accuracy.

Chapter 5—We're entering the low ebb of the sunspot cycle and HF DXing has changed accordingly. Bob Locher, W9KNI, has reedited this chapter with new DX realities in mind.

Chapter 7—Computers have made a huge impact on contesting. Learn the latest!

Chapter 8—Award requirements for several countries have changed substantially (the Federal Republic of Germany, for example). Other countries have added new awards.

Chapter 9—A new chapter with a new name: HF Digital Communications. This section covers everything from RTTY to G-TOR.

Chapter 10—Stan Horzepa, WA1LOU, has revised his chapter to reflect some of the most exciting changes in packet radio such as *APRS*, the Automatic Packet Reporting System.

Chapter 13—New satellites are in orbit and others are on the way. Read about them here!

Chapter 14—Richard Regent, K9GDF, has revamped his chapter to showcase modern emergency communication procedures.

In short, the fifth edition of the *ARRL Operating Manual* is more than just a new cover, albeit a beautiful one. (Thanks Sue Fagan!) The commodity we offer in this book is *information*, something that's becoming harder to obtain fresh off the press. I hope you enjoy it!—*WB8IMY*

The ARRL at Your Service

ARRL Headquarters is open from 8 AM to 5 PM Eastern Time, Monday through Friday, except holidays. Our address is: 225 Main St, Newington, CT 06111-1494. You can call us at 860-594-0200, or fax us at 860-594-0259.

If you have a question, try one of these Headquarters departments . . .

	Contact	Telephone	Electronic Mail
QST Delivery	Circulation Desk	860-594-0338	circulation@arrl.org
Publication Orders	Sales Desk	860-594-0250	pubsales@arrl.org
Regulatory Info	Norm Bliss	860-594-0323	reginfo@arrl.org
Exams	VEC	860-594-0300	vec@arrl.org
Educational Materials	Educational Activities	860-594-0301	ead@arrl.org
Contests	Billy Lunt	860-594-0252	contest@arrl.org
Technical questions	ARRL Lab	860-594-0214	tis@arrl.org
Awards	Eileen Sapko	860-594-0288	awards@arrl.org
DXCC	Bill Moore	860-594-0234	dxcc@arrl.org
Advertising	Brad Thomas	860-594-0207	ads@arrl.org
Media Relations	Jennifer Hagy	860-594-0328	newsmedia@arrl.org
QSL Service	Joe Carcia	860-594-0274	buro@arrl.org
Scholarships	Mary Carcia	860-594-0230	foundation@arrl.org
Emergency Communication	Rick Palm	860-594-0261	k1ce@arrl.org

You can send e-mail to any ARRL Headquarters employee if you know their name or call sign. The second half of every Headquarters e-mail address is **@arrl.org**. To create the first half, simply use the person's call sign. If you don't know their call sign, use the first letter of their first name, followed by their complete last name. For example, to send a message to John Hennessee, KJ4KB, Regulatory Information Specialist, you could address it to: **jhennessee@arrl.org** or **kj4kb@arrl.org**.

If all else fails, send e-mail to **hq@arrl.org** and it will be routed to the right people or departments.

ARRL on the On-Line Services

We maintain accounts on these major on-line services and check for mail several times daily.

Service	ARRL Address
CompuServe	70007,3373
America Online	HQARRL1
GEnie	ARRL
Prodigy	PTYS02A
MCI	215-5052

Downloadable files for the new ham, upgrader, instructor or disabled ham are featured in the various on-line ham radio file libraries.

ARRL BBS

The ARRL *Hiram* Bulletin Board System is as close as your telephone. Hiram offers more than a thousand software files for your enjoyment. You can also use Hiram to send messages to anyone at Headquarters—or to other hams who frequent the BBS. Hiram accepts up to four simultaneous connections at rates from 1200 to 28,800 baud.
Fire up your modem and call 860-594-0306.

Technical Information Server

If you have Internet e-mail capability, you can tap into the ARRL Technical Information Server, otherwise known as the *Info Server*. To have user instructions and a handy index sent to you automatically, simply address an e-mail message to: **info@arrl.org**
Subject: **Info Request**
In the body of your message enter:

> **HELP**
> **SEND INDEX**
> **QUIT**

ARRL on the World Wide Web

You'll also find the ARRL on the World Wide Web at:

> **http://www.arrl.org/**

At the ARRL Web page you'll find the latest W1AW bulletins, a hamfest calendar, exam schedules, an on-line ARRL Publications Catalog and much more. We're always adding new features to our Web page, so check it often!

Shortwave Listening

CURT PHILLIPS, KD4YU
6200 VICKY DR
RALEIGH, NC 27603

S hortwave listening (also known as *world band* listening) has experienced a resurgence of interest in the last several years. This is due to the increased availability of moderately priced shortwave receivers. In addition, there has been a widening realization that the unique information available on the shortwave bands cannot be ignored.

The shortwave bands are often considered to comprise the frequencies between 3 and 30 MHz. For hobbyist purposes, the monitoring of frequencies between 100 kHz and 30 MHz can be considered *shortwave listening*, notwithstanding that the AM broadcast band is in the *medium wave* spectrum and frequencies below that are technically considered *long wave*. The ham bands within the shortwave bands are also referred to as the high frequency or *HF* bands, as opposed to the very high frequency (VHF) and ultra high frequency (UHF) bands, which are above 30 MHz.

There's a good chance that you're a licensed ham, or you're considering becoming one. But shortwave listening is just . . . listening. As an Amateur Radio operator, you can transmit RF, not simply receive it. So what does shortwave listening have to do with ham radio? Isn't shortwave listening a step backward?

The short answer is no. The diverse, worldwide communication that you have access to on the shortwave bands provides a wealth of possibilities that *supplement* the choices provided by Amateur Radio. What shortwave listening can offer you depends in part on when you joined our hobby—and how. Let me explain.

In the "old" days, new recruits entered the hobby through direct exposure to the shortwave bands. Before they took their exams, many new hams spent hours listening to shortwave as they honed their CW skills. The code-practice sessions provided by ARRL flagship station W1AW, and other CW activity on the shortwave bands, were the prime sources of practice fodder. Bear in mind that this was before the days of cassette tape machines and personal computers.

After they were on the air, their shortwave experience was limited. For a period of time from about the mid-1950s until the late 1970s, ham transceivers tuned the Amateur Radio portions of the shortwave spectrum *only*. Why? The technology of ham radio equipment during that period was such that optimum performance was most easily obtained by rigs that covered a small portion of the radio spectrum. So, hams who may have started with a general-coverage receiver quickly "downgraded" to a ham-band-only receiver.

Today, the most popular method of entry to Amateur Radio is the Technician license. This license doesn't require learning the Morse code, or offer operating privileges below 30 MHz. For these hams, shortwave listening offers a chance to become familiar with HF operation.

So, Technicians have a new world to explore and veterans can reacquaint themselves with the fun they may have missed. But for hams of any vintage, there are an amazing number of intriguing signals that can be heard on shortwave. Indeed, since no license is required to listen, everyone in the family can enjoy the diverse offerings of the shortwave bands.

WHAT'S OUT THERE?

Unique Perspectives on the News

As this is being written (early 1995), the United States has troops in Haiti and naval support ships around the island. Boatloads of Cuban expatriates have been flooding Florida and are being relocated to Guantanamo Bay. A couple of months ago, Iraq massed troops near Kuwait and the US military was mobilized to meet this threat.

The shortwave listener has a ringside seat to events such as these. The military frequencies are full of traffic as troops and equipment are moved. The Voice of America broadcasts American views of the world situation. Opposing viewpoints are heard from Radio Iraq International and Radio Habana Cuba. The shortwave stations of other nations weigh in with their analysis. Transmissions on the "Mystic Star" frequencies from government aircraft can be heard as VIPs shuttle about on their diplomatic missions.

Even for events occurring within the United States, shortwave radio can often provide information before it is released by the traditional media. When President Ronald Reagan was shot, listeners monitoring the radio traffic between the White House and Vice President George Bush were privy to reports on the President's condition long before they were broadcast on the major networks.

In our turbulent world there are always hot spots where shortwave radio provides direct access to the information. The broadcast news of the major networks tells you what *they* think is important, filtered through their perspective and edited to fit into a brief newscast. The shortwave bands bring you news direct from the source, and in as much detail as they care to provide.

During the Russian coup in August 1991 and the armed uprising of old-line Soviets in the legislature during September 1993, those monitoring Radio Moscow got quite a different perspective on the internal situation in Russia than those confined to the popular media. In both cases, it was evident that the prodemocratic forces maintained control of Russia's international broadcasting apparatus, portending a favorable outcome for them. If you listened only to the US media, however, you heard a very negative assessment of their chances.

After the end of the Cold War, leaders in several Eastern Bloc countries told of how the Voice of America, Radio Free Europe, Radio Liberty and other Western shortwave broadcasts had helped them overcome the news blackouts and censorship that had been common in their countries.

The ability to listen directly to broadcasts from countries around the world can encourage students to have greater interest in history, geography and foreign languages. Therefore, shortwave listening is an excellent way to both encourage the younger generation in their educational pursuits *and* get them interested in radio as a hobby.

But it's not just wars and military actions where the shortwave bands offer special news access. In December 1986, shortwave listeners were treated to the exciting communications supporting the flight of the *Voyager*, the first nonstop, nonrefueled flight around the world.

Radio Dramas

One art form that might have died without the support of the international shortwave broadcasters is the radio drama. On these stations, radio dramas are not only surviving but thriving.

The BBC's two weekly offerings, "Play of the Week" and "Thirty-Minute Drama" are excellent, the latter having been chosen one of the Ten Best Shows in 1994 by a major shortwave publication.

The religious show "Unshackled" (broadcast on HCJB and several other stations) is the longest-running shortwave radio drama. It sounds about the same today as it did when I first heard it almost 30 years ago. I suspect it sounds about the same as it did on its inaugural broadcast! "Unshackled" retains an "old-timely" aura (swelling organ music, a terse announcer), so listening to it is to be time-warped back to the heyday of radio dramas in the 1930s (or so my parents tell me!). "Unshackled" is produced by Pacific Garden Mission, located (as you would expect) in Chicago.

Music

On shortwave you can hear African rhythms, the sitar of India and other exotic music direct from the source. National shortwave broadcasts present a range of regional music that is not available via any other media, along with commentaries and information on its composition and history.

Language Lessons

With the prevalence of the English language in the world, almost every international broadcaster offers English-language broadcasts. But, if you are fluent in another tongue, the shortwave bands will give you ample opportunity to exercise your skill. For the would-be fluent, many stations offer introductory courses on their native language.

DXing

The activities we discussed so far focus on the *content* of

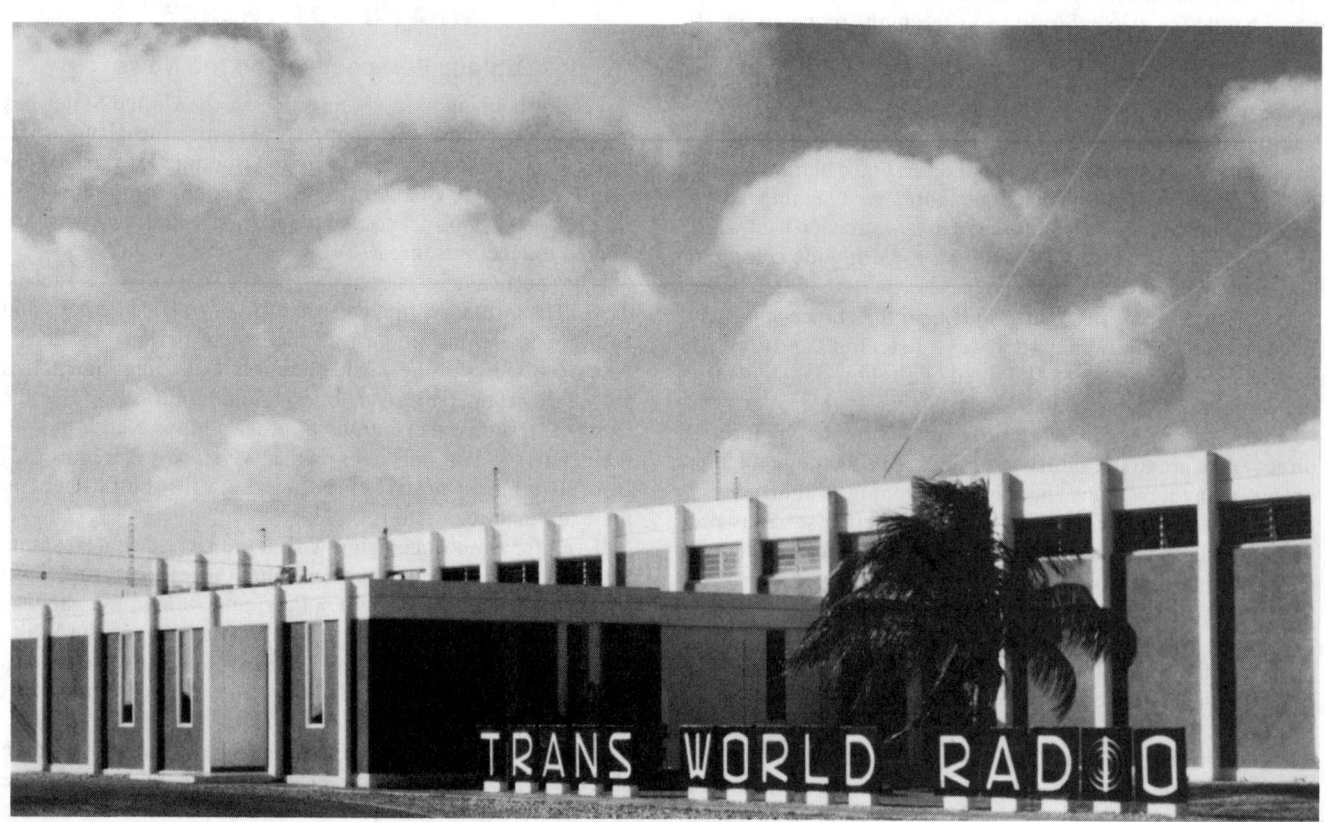

The Bonaire, Netherlands Antilles, transmitting site of Trans World Radio. From this location, their signal booms throughout the Americas. (*photo courtesy of Trans World Radio*)

shortwave signals. But there is a sector of the shortwave listening hobby that concentrates primarily on the technical challenge of receiving hard-to-hear stations. Shortwave DXers pride themselves on obtaining official verification of their reception (via QSL cards), and competing for various awards.

The Ionosphere

Someone accustomed to the line-of-sight nature of the UHF/VHF bands (and who may have forgotten the HF theory they studied for their exam!) might wonder how shortwave signals make it around the world. To answer this question we need to briefly examine the nature of the ionosphere.

The ionosphere is a layer of the Earth's atmosphere approximately 80 to 300 miles above Earth's surface. To oversimplify for the sake of brevity, at the VHF region and above, radio waves will most often pass through the ionosphere. At frequencies below 30 MHz, radio waves are often refracted (bent) back to the Earth's surface. It is this "bouncing" of radio signals off the ionosphere (sometimes via several bounces) that allows shortwave signals to travel (propagate) around the world.

The ability of radio waves to be refracted over long distances by the ionosphere is affected by the time of day, time of year and the number of sunspots. Speaking very generally, higher frequencies (closer to 30 MHz) work best during the daylight hours and during the summer. During late evenings and in the winter, the lower frequencies (close to 3 MHz and even below) begin to peak in efficiency. This phenomenon can be observed on the medium wave AM broadcast band (0.5 MHz-1.6 MHz). During the day you hear only local stations. In the evening, distant stations can be clearly heard. During the summer, however, even evening reception of distant AM broadcast stations is often difficult because of high noise levels created by thunderstorms and other weather phenomena.

These daylight/darkness propagation patterns are amplified by the sunspot cycle. During the peak of the 11-year sunspot cycle, the high frequencies are at their best for long-distance communication. At sunspot minimums, the capabilities of the lower frequencies become more prominent. As this book went to press (1995), the current sunspot cycle was nearly at rock bottom. The higher shortwave frequencies will not peak again for long-distance communication until nearly the end of the century.

Table 1-1
Shortwave Frequency Guide

All frequencies shown in kHz

Frequencies	Service	Comments	Frequencies	Service	Comments
535—1705	AM Broadcast	Standard North America AM	12050—12330	Fixed Service	
1705—1800	Fixed Service	Land/Mobile/Marine	12330—13200	Maritime Mobile	Ship/Shore
1800—2000	Amateur 160 Meters		13200—13360	Aeronautical Mobile	Transoceanic Flights
2000—2107	Maritime Mobile		13360—13600	Fixed Service	
2107—2170	Fixed Service	Land/Mobile/Marine	13600—13800	Shortwave Broadcast	New WARC Allocation
2170—2194	Land Mobile Service		13800—14000	Fixed Service	
2194—2300	Fixed Service		14000—14350	Amateur 20 Meters	
2300—2495	Shortwave Broadcast	120 Meters	14350—14995	Fixed Service	
2495—2505	Time Standard		15010—15100	Aeronautical Mobile	Transoceanic Flights
2505—2850	Fixed Service	Land/Mobile/Marine	15100—15600	Shortwave Broadcast	19 Meters
2850—3155	Aeronautical Mobile	Transoceanic Flights	15600—16460	Fixed Service	
3155—3200	Fixed Service		16460—17360	Maritime Mobile	Ship/Shore
3200—3400	Shortwave Broadcast	90 Meters	17360—17550	Fixed Service	
3400—3500	Aeronautical Mobile	Transoceanic Flights	17550—17900	Shortwave Broadcast	16 Meters
3500—4000	Amateur 80/75 Meters		17900—18030	Aeronautical Mobile	Transoceanic Flights
3900—4000	Shortwave Broadcast	75 Meters, Not in Region 2	18030—18780	Fixed Service	
4000—4000	Time Standard		18780—18900	Maritime Mobile	Ship/Shore
4000—4063	Fixed Service		18900—19680	Fixed Service	
4063—4438	Maritime Mobile	Ship/Shore	19680—19800	Maritime Mobile	Ship/Shore
4438—4650	Fixed Service		19800—21000	Fixed Service	
4650—4750	Aeronautical Mobile	Transoceanic Flights	21000—21450	Amateur 15 Meters	
4750—5060	Shortwave Broadcast	60 Meters	21450—21850	Shortwave Broadcast	13 Meters
5005—5450	Fixed Service		21850—22000	Aeronautical Mobile	
5450—5730	Aeronautical Mobile	Transoceanic Flights	22000—22720	Maritime Mobile	Ship/Shore
5730—5950	Fixed Service		22720—23200	Fixed Service	
5950—6200	Shortwave Broadcast	49 Meters	23200—23350	Aeronautical Mobile	
6200—6525	Maritime Mobile	Ship/Shore	23350—24990	Fixed Service	
6525—6765	Aeronautical Mobile	Transoceanic Flights	24890—24990	Amateur 12 Meters	Shared with Fixed Service
6765—7000	Fixed Service		25010—25330	Petroleum Industry	
7000—7300	Amateur 40 Meters		25330—25600	Government Frequency	
7100—7300	Shortwave Broadcast	41 Meters, Not in Region 2	25600—26100	Shortwave Broadcast	11 Meters
7300—8195	Fixed Service		26100—26480	Land Mobile Service	
8195—8815	Maritime Mobile	Ship/Shore	26480—26950	Government	
8815—9040	Aeronautical Mobile	Transoceanic Flights	26950—26960	International Fixed Service	
9040—9500	Fixed Service		26960—27410	Citizen's Band	Channels start at 26965 kHz
9500—9900	Shortwave Broadcast	31 Meters	27410—27540	Land Mobile Service	
9775—9995	Fixed Service		27540—28000	Government	
10005—10100	Aeronautical Mobile	Transoceanic Flights	28000—29700	Amateur 10 Meters	
10100—10150	Amateur 30 Meters	CW/Data Only	29700—29800	Forestry Service	
10100—11175	Fixed Service		29800—29890	Fixed Service	
11175—11400	Aeronautical Mobile	Transoceanic Flights	29890—29910	Government	
11400—11650	Fixed Service		29910—30000	Fixed Service	
11650—12050	Shortwave Broadcast	25 Meters			

GADGETS GALORE

Ham radio operators tend to be inveterate gadgeteers, and the shortwave listening hobby brings a whole new set of interesting radios for you to consider.

Because ham HF transceivers include power hogging transmitters, there aren't many small HF ham rigs on the market. But the past decade has seen the introduction of ever-smaller and more conveniently sized shortwave receivers. One example is the Sony ICF-SW100 (see Fig 1-1). The SW100 is a full coverage (0.15-30 MHz) shortwave receiver featuring 50 memories and the much desired synchronous detector (more on this technology later). And the new generation of hand-held sized UHF/VHF scanners often include the shortwave spectrum (see the sidebar, "The DC-to-Daylight Radios Are Here! Almost"). If you've got the money, there are plenty of interesting places to put it.

Fig 1-1—The ultra-compact Sony ICF-SW100 receiver packs many handy features in a package small enough to fit in a coat pocket. (*photo courtesy Sony USA*)

Choosing a Receiver

Many current HF amateur transceivers feature a *general-coverage receiver* (usually 0.5-30 MHz). If you own one you're just a spin of the dial away from the joys of shortwave listening. If you're planning to purchase an amateur transceiver in the near future, make general-coverage receive capability a requirement. You won't regret it.

However, if you're just interested in browsing the shortwave bands, one of the new portables on the market may be for you. Widely considered the best of the low cost portables is the Sangean ATS-803a, which has been marketed by Radio Shack as the DX-440 in the past. More money buys more features and performance, such as the "DXer's delight": the Sony ICF-2010. This receiver is reknowned for its weak-signal capability. All of these radios are relatively large portables, approximately $12 \times 6.5 \times 2.5$ inches and weighing approximately 4 pounds each.

In the midsize range (approximately $7.5 \times 4.75 \times 1.5$ inches; 1.5 pounds), the Sony ICF-SW7600G (note the "g"), the Grundig Yacht Boy 400 and the Panasonic RGB-65 are all worthy contenders. Although it's not quite midsize, the Drake

SW-8 is another wide-range receiver with many useful features.

If you've browsed through ham radio ads, you may be surprised at the brands that are missing; Yaesu, ICOM and Kenwood are not major players in the battery powered portable shortwave receiver market. (ICOM does have a contender in the "dc-to-daylight" category, though.)

Be aware that many of the least-expensive shortwave radios can't demodulate single sideband, the mode of choice for HF ham radio and for utility/military transmissions. If these transmissions are of interest to you, be sure the radio you choose has SSB capability. All of the radios listed here can demodulate SSB.

Synchronous Detection

So what is this much-hyped "synchronous detection"? Putting it simply, a synchronous detector circuit can stabilize signals with fluctuating strengths and provide a clearly readable signal in the presence of severe adjacent channel interference. Since it is only used for AM reception, even a well-informed ham who doesn't listen to shortwave broadcasting probably won't know about it. (Perhaps that's an oxymoronic phrase… "a well informed ham who doesn't listen to shortwave broadcasting.") Even ancient hams who once transmitted AM probably won't be aware of synchronous detection. It didn't exist back then!

Amplitude modulation has three constituent parts: two identical sidebands and a carrier. If the shortwave receiver can't obtain adequate reception of both sidebands and the carrier, the signal quality can quickly deteriorate and become unreadable. The synchronous detection circuit automatically locks onto the transmitted carrier, mutes it and substitutes a carrier generated internally in the radio. Since this carrier is strong and stable, the effects of fluctuations on the received signal are reduced. Also, as a part of this process, it is now possible to receive a clear signal using only one of the sidebands. Therefore, the user can tune to the sideband opposite any station whose close-by transmission is splattering over on the station they want to hear (otherwise known as adjacent channel interference). This greatly improves readability in a crowded band. In addition, as synchronous detection substitutes its own carrier, any signal whose carrier is not precisely on the same frequency is substantially attenuated, without the loss in fidelity that a narrow filter would entail.

In actual use, the results can be dramatic. Signals that are unintelligible because of adjacent channel interference can

The Drake SW-8 is one of the leaders among larger shortwave receivers. It's designed for home and portable use. (*R.L. Drake Company*)

The DC-to-Daylight Receiver Is Here! Almost.

The ultimate in receiver coverage was always called "dc-to-daylight." In other words, it's coverage that reaches from the lowest possible radio frequencies all the way to light itself. The receivers that are out today do not hit that impossible goal, but many *do* cover a frequency range so wide as to boggle the mind. Amazing as it may seem, several models come in small hand-held size packages!

At the low end, these wide-range receivers bottom out at 100 to 500 kHz, which is about typical for a general-coverage shortwave receiver. But at the high end, they soar past the 30 MHz shortwave boundary deep into the VHF/UHF region. All of them go well past 1 GHz (1000 MHz), and the most advanced of them approach 2 GHz.

Actually, these receivers have sprung from an explosion in UHF/VHF scanner technology. Monitors of the multitudinous services in this spectrum have long sought portable radios that have an expansive number of memories and coverage for all of the popular voice modes. For years, scanners have encroached upon shortwave territory as low as 25 MHz. By the mid-90s, however, scanners began to expand into *full* shortwave spectrum coverage.

One of the first entries in this category was the AOR AR-1000, covering 500 kHz to 1300 MHz in 5-kHz steps with 1000 memories. Its small package (6.7 × 2.6 × 1.4 inches) included AM as well as wide- and narrow-band FM. The AR-1500 was essentially the same radio, but with a beat frequency oscillator (BFO) for reception of single sideband (SSB) and continuous wave (CW) signals.

Soon thereafter, ICOM introduced the R1, an even smaller radio (4 × 1.9 × 1.4 inches), which covered 100 kHz to 1300 MHz in steps as small as 0.5 kHz. However, it only demodulated AM, FM and WFM and only had 100 memories.

As early entries in this super high-tech arena, each of these radios had some deficiencies like poor sensitivity over the specified range, intermodulation problems and complicated programming. Although each of these have their ardent defenders, their well-publicized liabilities limited their appeal.

It wasn't long afterward before American readers of the British publication *Shortwave Magazine* noticed some ads for a small (6.1 × 2.6 × 1.5 inches) radio known as the Yupiteru MVT-7100. It covered 100 kHz to 1650 MHz in 50-Hz segments. Because it used true carrier injection to provide SSB reception, the full range of modes (AM/FM/WFM/LSB/USB) were programmable in its 1000 memories. A conversation about the radio commenced on one of the

You can have 100 kHz to 2 GHz coverage in a hand-held with the AOR 8000. (*photo courtesy Electronic Equipment Bank, Inc*)

national on-line services, and soon several enthusiasts had ordered 7100s from an accommodating British dealer.

In the wake of the rave reviews that followed, more of the Yupiteru 7100s were ordered from England and a phenomenon started. Here was a radio that was reasonably simple to use (considering all it would do), and that provided good sensitivity throughout its frequency range. A review on the shortwave show "Spectrum" actually said that it was as sensitive as the venerable Sony 2010, but as an owner of them both, I find that a bit excesive. However, it does provide shortwave reception commensurate with the respected Sangean ATS-803 (which, like it, doesn't have synchronous detection) in a much smaller package. And it's also a full featured VHF/UHF scanner!

The Yupi's position at the top of the heap was relatively short-lived. AOR introduced the AR-8000, which covers 100 kHz to 1.9 GHz in 50-Hz steps. It includes 1000 channels and true single sideband in a package as compact as the MVT-7100. Unlike the Yupi, it also features a four-line alphanumeric display and computer interface capability. Some users contend that while the 8000 has more features, the Yupi is more sensitive. It seems clear that both are worthy contenders in the "dc-to-daylight" category.

The AR-8000 is readily available to listeners in the United States, but its frequency coverage has been crippled. As a result of a law that went into effect in mid-1994, no scanning radio with coverage of the cellular phone frequencies (in the 800-900 MHz range) can be manufactured or imported into the United States (it is *not* illegal to own such a radio, however). AOR has released an FCC-certified version of the AR-8000 with the appropriate frequencies deleted, but Yupiteru has not chosen to do so. Some full coverage, AR-1000s, AR-1500s and R1s remain, but their future is cloudy.

Admittedly, receivers with a dc-to-daylight frequency range have been available to the government and military for years, but in desk-crushing sizes and weights—and at a cost approaching that of a Rolls-Royce. The ICOM R-9000's mid-90s street price of $4795 is barely equivalent to a Rolls-Royce hubcap, but it's still beyond the budgets of most hobbyists. For those who can afford it, the desktop-sized R-9000 covers 100 kHz to 2 GHz with 1000 memories, all modes and a built in spectrum analyzer. It offers a level of performance equal to or exceeding most of the equipment used by the FCC and other governmental monitors.

For hams without such deep bank accounts, the advent of affordable, hand-held, full-frequency receivers provides the ability to monitor everything from the local 2-meter repeater to the 80-meter sideband nets. And you get complete shortwave coverage to boot!

If money is no object, a high-end receiver such as the Watkins-Johnson HF-1000 may be for you—if you can cough up nearly $4000.

become easily and clearly readable. Even weak stations can be made much more readable when their fading carrier is replaced internally. Few hobbyists who've heard a radio with a well-designed synchronous detector will deny its substantial benefit for the reception of AM signals.

From "Death Before Digital" to "Digital Forever"

As a ham for over 25 years, I was long a believer that, "Real hams don't use digital readouts." When I finally got a radio with digital readout, I quickly changed my mind. For someone new to shortwave listening, a digital readout is a tremendous aid. The ability to accurately determine the frequencies of unidentified signals makes looking them up in frequency guides a breeze. Also, digital readout assists in "repeatability"—the ability to be sure that you can return to precisely the same frequency you were listening to last week.

With the vagaries of propagation, the signal might not be strong enough to listen to anyway, but at least you will be sure you are on the correct frequency. With the old, less precise analog-frequency readouts, when you couldn't hear the signal it was difficult to determine whether it was just bad propagation, or if you were just off frequency.

Getting to the frequency you want is easier with digital receivers, too. The calculator or telephone-type keypad on many modern receivers allows you to directly punch in the frequency you want. This makes it easy to jump from frequency to frequency to check propagation or programming.

But it can get easier still. Typical among modern digital shortwave receivers are *memories*, where frequencies and modes can be stored for quick access. Thus, your favorite BBC frequency—and other often-used frequencies—can be accessed by pressing a couple of buttons. This makes manual scanning of multiple frequencies (like the low-traffic utility frequencies) a breeze.

Yet another advantage of radios with digital readouts is that they usually use frequency synthesizers, which all but eliminate drift. ("Drift" is the tendency of a radio to change frequency as its internal components warm up.) Drift is all but nonexistent in modern digital frequency synthesized receivers, at least such drift as is discernible by ear. While a digital readout does not technically ensure that the radio uses frequency synthesis, the two are found together so commonly as to be virtually synonymous.

Oldies But Goodies?

Some well-meaning hams may advise you to get an older receiver at a hamfest to get your start in shortwave listening. They may speak with nostalgia of spinning the dials on a

---

Print Resources for the SWL

With the vast frequencies in the shortwave spectrum, the ever-changing international implications of shortwave radio and the constant changes in technology, you need as many information sources as you can get!. The shortwave shows for the hobbyist can be helpful and on-line computer services are available for those on the much touted Information Superhighway (see "On-line Computer Resources for SWLs). But what if you're a beginner with no radio or modem?

For these people and many others, multitudinous print resources for the shortwave listener come to the rescue. The two print mainstays of shortwave broadcast listening are the *World Radio TV Handbook* (BPI Communications, 1515 Broadway, New York, NY 10036) and *Passport to World Band Radio* (IBS North America, Box 300, Penn's Park, PA 18943). Both are published yearly and are available from ARRL Hq. *Passport to World Band Radio* is considered easier to navigate for the beginner, while the *World Radio and TV Handbook* is considered the definitive source for shortwave information.

Passport lists broadcasts by frequency as well as a 24-hour tour of selected transmissions to North America. The *World Radio and TV Handbook* provides a listing of transmissions by language and details of *everything* you could possibly want to know about each broadcaster. If you're new to shortwave, you should get a copy of each. Don't let your $300-$500 receiver go underutilized for lack of information!

Introductory books include *The Shortwave Listening Guidebook* by Harry Helms AA6FW (HighText Publications, Inc, PO Box 1489, Solana Beach, CA 92075) and *The Complete Shortwave Listener's Handbook* by Hank Bennett, Andrew Yoder, et al (TAB Books, Division of McGraw-Hill, Inc, Blue Ridge Summit, PA 17294). Helms' book (first published in 1990) is an excellent survey of all aspects of shortwave listening, and is highly recommended.

In addition to shortwave listening, there is, as you might guess, longwave listening. Below 515 kHz are broadcast stations, a low-powered unlicensed communications band from 160 to 190 kHz, beacons, fax and data transmissions. The Longwave Club of America, 45 Wildflower Road, Levittown, PA 10957 (telephone 215-945-0543) publishes the monthly newsletter *Lowdown*. They also compile a list of the aircraft and ship navigation beacons heard in this band. Contact Ken Styker, Editor, *Aero/Marine Beacon Guide*, 2856-G West Touhy Avenue, Chicago, IL 60645 for more information.

Monthly publications can help the SWL stay on top of the ever-changing shortwave world. Two standard monthlies are *Popular Communications* (76 North Broadway, Hicksville, NY 11801) and *Monitoring Times* (PO Box 98, Brasstown, NC 28902). "PopComm" is a slickly produced shortwave newsmagazine that can be found on many newsstands, while *Monitoring Times'* use of newsprint allows for slightly more current information in each issue. Both magazines cover the full range of shortwave listening interests, from international broadcasting to military to utilities to digital modes.

For those interested in specific information on utilities and the digital modes, two frequency guides stand out. *Ferrell's Confidential Frequency List* (Listening In, PO Box 123, Park Ridge, NJ 07656) carries a full listing by frequency, mode, call, location and mode from 1602 kHz to 25545 kHz. It also has a reverse listing sorted by call. It is available from ARRL Hq.

The *Klingenfuss Guide to Utility Stations* (Klingenfuss Publications, Hagenloher Str. 14, D-72070 Tuebingen, Germany) has its own frequency list along with detailed listings of RTTY and fax services, schedules and other useful information.

For those interested in the content of the shows on the international broadcasting stations, the *Guide To Shortwave Programs* by Kannon Shanmugan (Grove Enterprises, PO Box 98, Brasstown, NC 28902) is an invaluable resource. It lists the frequencies and programs that are on the air at any time of the day or year. Though many have tried to lay claim to the title, this is perhaps the closest approximation of a *TV Guide* for the shortwave bands. It also has a special section on shortwave shows for the hobbyist.

Hammarlund, Hallicrafters, National, and so on back in their early days in radio.

Although it sounds good at first, there are a couple of problems with this advice. Those older receivers don't have digital readouts or synchronous detection. They also don't have the stability borne of frequency synthesis, the evidence of which was the common practice in the old days of letting them warm up for an hour or so before doing any serious listening to get the drift stabilized. As has just been discussed, these features can make shortwave listening much more enjoyable.

Another situation negatively impacting the attractiveness of older receivers is that in recent years they've become the targets of collectors. This means that the good ones aren't cheap and the cheap ones aren't good (and often weren't very good even when they were new!).

Certainly, these old receivers have a special appeal that comes from a hernia-inducing big box full of tubes. They can also provide much of the heat your shack requires in the winter. I confess that a late 1950s Hammarlund HQ-180 (obtained in the mid-1970s) has a place of honor in my shack. But while these older receivers have excellent sensitivity and selectivity, you can usually achieve better performance and ease of use (not to mention a lower cost) with modern receivers.

If there is any exception to this rule, it's in the medium wave and long wave spectrum. Many of the newer shortwave receivers tend to suffer from poor performance at the AM broadcast band and below. The best of the oldies are great performers on these frequencies. If you intend to be especially active in medium or longwave monitoring, or if you must have that "real radio" look, an oldie can indeed be a goodie. Just don't expect it to be inexpensive.

Antennas

Since this is not a technical manual, we'll keep it simple. If you're using an amateur HF transceiver, an 80 or 40-meter dipole will work well for shortwave listening. If you use a portable shortwave receiver, adequate reception of international broadcasts is often possible using the attached whip antenna.

If you want maximum performance, put up a wire outside as high as possible and as long as possible. Practically, a wire between 25 and 75 feet long and 20 feet high (perhaps between two trees) will work well. If your portable receiver doesn't have an antenna jack, just use an alligator clip and connect the external antenna to the whip. Connecting an outside antenna to the whip will sometimes overload the receiver, so you may have to experiment to determine the optimum combination for the signals you want to hear.

While you may achieve acceptable results with the whip antenna, don't underestimate the value of a good antenna system. Many people like to spend the most of their money on the radio. It's certainly the element of the station that will draw "ohs and ahs" from the uninitiated. However, the more experienced operator knows that, while the antennas may not be as photogenic as radios, it's easier to get high-quality results with a good radio and a good antenna than with the best radio and a lousy antenna!

Living in a place where it is difficult to erect an outside antenna shouldn't prevent you from enjoying shortwave listening. An "invisible" outside antenna (constructed using a wire so thin as to be nearly invisible) can bring useful results. Invisible antennas are covered in many ham and shortwave antenna texts. Reading and experimenting will help you achieve the optimum performance for your location.

INTERNATIONAL SHORTWAVE BROADCASTERS

The international broadcasters operate the best known and easiest to find stations on the shortwave bands. They are often sponsored by governments, who are willing to spend substantial sums to spread their viewpoints around the world. Their booming signals, sometimes running in the half-megawatt range, provide music, news (sometimes tainted with propaganda), educational broadcasts and entertainment programming. Not only are their signals usually strong, their transmissions tend to be confined to identifiable frequency ranges.

Another reason that the international broadcasters may be easy to hear is that the transmitter may not be located where you think it is. Though you may be listening to Radio Netherlands, Deutsche Welle or Radio Taiwan, the transmitter might be located off the north coast of South America (the Netherlands Antilles is a popular location), or even from within North America via a rented transmitter site! Those listening for program content won't care where the transmitter is located, but DXers often seek transmissions that originate far from their locations.

Interval Signals

If you listen to an international shortwave broadcaster at the beginning of a transmission, or at the top of the hour, you may hear a repeating series of tones. These tones are known as "interval signals." They're designed to help you identify the station when they're transmitting in a language foreign to you, or when the signal is weak. The familiar tones which the National Broadcasting Company (NBC) still uses are a domestic example of an interval signal. In addition to interval signals, there is another type of musical identification: the musical signature.

Interval signals are short and often played repeatedly (sometimes to the point of irritation) for 2 to 5 minutes just before the top of each hour. The instrumentation of an interval signal is simple, usually just a piano, celesta, carillon or electronic keyboard. The musical signature is typically more fully orchestrated and is played just once per period. In most cases, interval signals are played hourly and musical signatures only at sign-on and sign-off, but there are numerous exceptions.

The tones or music are often related to the culture and history of the originating nation. The Voice of America uses "Yankee Doodle Dandy" and Radio Australia uses "Waltzing Maltilda," while the BBC uses a very regal-sounding tune called "Lilliburlero." Many of the European nations use classical music for their musical signatures. Not surprisingly,

ITU

Radio waves can not be kept within a country's borders, so it was evident very early in the development of radio that international treaties and agreements must be established to prevent chaos on the radio waves.

The ITU regulates the frequencies where the various services operate, but many countries are not bashful about moving their national broadcasting services outside the specified bands. That's why you need to scan outside the listed frequency ranges. You'll find many fascinating signals lurking there.

The shortwave broadcast bands are based upon agreements drawn up under the auspices of the International Telecommunication Union. The various frequency ranges are traditionally referred to using the wavelength (in meters) of one of the component frequencies. For example, 49 meters is the wavelength of 6122 kHz. Therefore, the 49-meter band extends from 5950 to 6200 kHz.

the religious stations often use hymns.

Hearing and identifying interval signals is another subset of the shortwave listening hobby. The level of interest is indicated by the fact that one pirate radio station uses the bogus call "WLIS"—"We Love Interval Signals."

World Time

If your operating has been confined to the VHF bands, you may not have much experience with what is known as *World Time*. Since shortwave signals reach around the world, listing program times can be a problem. Whose time do you use?

To eliminate this confusion, the time at the Royal Observatory in England at zero degree meridian (formerly at Greenwich) is used. It was called Greenwich Mean Time (GMT) for years, but now it's known as Coordinated Universal Time (UTC), World Time, Universal Time or Zulu (from the phonetic for the letter *z*, which the military uses to indicate UTC).

UTC was established by international agreement in 1972, and is governed by the International Bureau of Weights and Measures (BIPM) in Paris, France. It differs from local time by a specific number of hours. The number of hours depends on the number of time zones between your location and the location of the zero meridian. For time zones in the United States, UTC is 5 hours ahead of EST, 6 hours ahead of CST, 7 hours ahead of MST and 8 hours ahead of PST. That is, at 5AM EST, the UTC time is 10AM. When local time changes from Daylight Saving to standard time, or vice versa, UTC does not change. However, the difference between UTC and local time does change—by 1 hour.

UTC is expressed using the 24-hour clock system, sometimes called "military time." The hours are numbered beginning with 00 at midnight through 12 hours at noon to 23 hours and 59 minutes just before the next midnight. So, 7 PM is 1900 hours.

Another attribute that you must remember when using UTC is that (for the United States at least), since UTC reaches midnight before the American time zones, *the date changes earlier also*. Thus, since 7:00 PM EST (1900 EST) is midnight UTC, at 7:01 PM (1901) EST on December 31, it is 0001 UTC *on January 1st*. Many people keep 24-hour clocks set to UTC at their listening posts, which makes logging the correct time a breeze. But even experienced hobbyists sometimes forget to advance the date after 0000 UTC. Additional information on UTC is available in Chapter 3.

The Major Players

Before we delve into some of the major international broadcasters, a word about names is in order. Since we are dealing with foreign countries and foreign languages, these stations can be called by either their native name or the English variant of it. There are no definite rules, but some standards have arisen.

The "Voice of Germany" is almost always referred to as Deutsche Welle (and as with the classical composer Wagner, the 'w' is pronounced as a 'v'). The national stations of Cuba and Holland are referred to almost equally by their native and English names, Radio Habana Cuba/Radio Havana Cuba and Radio Nederland/Radio Netherlands, respectively. The French are very sensitive about maintaining the influence and purity of their language, so their national broadcaster is always called Radio France International and never Radio Francise or some other silly variant. (A note to our French readers. Although true, that was a joke!) HCJB's nickname "The Voice of the Andes" is occasionally called "La Voz De Los Andes. So if you see a name unfamiliar to you, be sure it's not just a native language rendition of a station whose name you already know!

BBC

If you can't hear the BBC on your shortwave receiver, the radio isn't working. The British Broadcasting Corporation's worldwide network of shortwave transmitters was established to serve the British Empire back in the days when the sun never set on it (it was known as the Empire Service). They continue to transmit powerhouse signals all across the shortwave spectrum. The BBC remains the standard by which all shortwave broadcasting operations are measured.

Long before the days of cable TV and CNN, the around-the-clock news reports of the BBC provided listeners with virtually constant access to late-breaking information. Even in today's news rich environment, those without cable TVs—or those wanting an alternative viewpoint—depend on the World Service (its name today) of the BBC.

Reliance on the BBC extends to more than news. If you want to know the proper pronunciation of a word, listen for it on the BBC. They have several people whose full-time job is to determine the correct English (perhaps as opposed to "American") pronunciation of any new word that comes up on their broadcasts!

Radio Moscow

If you listened to Radio Moscow prior to 1987, you *must* listen to a few broadcasts from Radio Moscow *today*, even if you aren't interested in shortwave listening. Borrow a portable shortwave receiver, or go to a friend's shack if you don't have shortwave capability. The change in the content of their newscasts and programming will flabbergast you.

The Voice of America facility in Greenville, North Carolina. (*photo courtesy of Voice of America*)

In the old days, the only time they weren't slamming the United States (and the western world in general) was when they were broadcasting fascinating reports on the wheat harvest in the Ukraine or another thrilling subject. They were also very tough on announcers whose tongues were loose. One announcer who slipped and called the Soviet action in Afghanistan "an invasion" was never heard from again. (Evidently, the official line was that they were in Afghanistan for humanitarian reasons, using Soviet battle tanks to plow fields and heavily armed Hind helicopters for crop spraying.)

But as the changes took place in Russia from 1987 to 1991, so did they at Radio Moscow. As was previously mentioned, monitoring Radio Moscow during the August 1991 coup and the insurrection of the legislature in September 1993 provided more accurate information than virtually any of the outside news sources.

Now Radio Moscow sounds very much like many western broadcasters. It's shocking to hear frank discussions of problems within Russia. Not to make light of their problems (every country has its share), but to go from such a controlled purveyor of the party-line to a relatively independent broadcaster has been an astounding transformation.

The downside of the change is that Russia and the Commonwealth of Independent States (CIS) are reducing the amount of programming on Radio Moscow. Some of the transmitter sites are being used by newly independent states for their independent broadcasts. Other transmitter sites are rented to anyone who will pay their price. At present, Radio Moscow is still a major presence on the shortwave bands, but substantially less so than in its heyday. What the future holds is very uncertain but, whatever happens, monitoring the evolution of Radio Moscow is a fascinating experience.

Note: At press time, an announcement was made that Radio Moscow International was changing its name to the Voice of Russia World Service. And thus the evolution continues!

Voice of America

The Voice of America (VOA) was established immediately after the end of World War II. After spending the war years listening to the domestic German propaganda broadcasts orchestrated by Joseph Goebbels, the US government was especially sensitive to the issue of broadcast content. As a result, the Voice of America was forbidden from creating broadcasts for consumption within the United States. The idea was to remove the temptation for the government to propagandize the general population.

However, in recent years this view has moderated and now VOA provides an excellent amount of programming information. They even offer access through computer bulletin-board systems (BBS). Though the VOA definitely broadcasts from the American point of view, it is generally respected as an authoritative source of information. VOA broadcasts were a critical lifeline to the outside world for those behind the "Iron Curtain" during the Cold War.

The end of the Cold War has been difficult for the VOA. Programming priority has shifted from Eastern Europe to emerging democracies of Africa and Latin America, but there is less urgency (and less money) supporting these broadcasts. The large VOA transmitter site in Bethany, Ohio, was decommissioned in late 1994 and the number of transmissions from the massive transmitter sites outside of Greenville, North Carolina, has dropped dramatically.

Should you be in the North Carolina area and have the time to drive to Greenville, a trip to one of the transmitter sites (they

The Voice of America site in Greenville, North Carolina, boasts massive transmitting antennas. (*photo courtesy of Voice of America*)

have two) will surprise and astound you. If you call ahead (tel 919-752-7115) they can set up a tour, but just driving by one of the sites will allow you to see an antenna field that no radio enthusiast will forget. Cut into the North Carolina pines is an open field approximately 3 miles long and 3 miles wide. The huge transmitter building sits about $1/2$ mile off the narrow secondary road, and off in the distance are groups of four 400-foot towers set in a semicircle with about a 1-mile radius spanning over 180° of arc. Each of these sets of four towers holds a massive curtain-array antenna. To change their signal pattern they don't rotate an antenna, they just switch to an antenna pointing in the needed direction. I have visited the site on several occasions and I never fail to be impressed.

The Voice of America headquarters in Washington, DC, also has an excellent tour, albeit one that is less impressive from a radio hardware point of view.

Selected frequencies (kHz): 5995, 6130, 7405, 9455, 11580, 11920, 13710, 15320, 15410, 15445, 17800.

Radio Netherlands

The *Happy Station*, home of the famous "Happy Station" program for well over half a century, is one of the most popular shortwave broadcasters in the world. Radio Netherlands International dates back to 1927 when Philips Radio Laboratories began broadcasting to the Dutch West Indies. In 1994, Radio Netherlands celebrated the 75th anniversary of Dutch radio, contending (in a friendly way) that the Dutch radio pioneer Steringa Idzerda invented radio . . . or radio broadcasting . . .or at least regularly scheduled radio broadcasts.

Even though not many encyclopedias recall Idzerda's contributions, Radio Netherlands had fun celebrating his work. Their goal was to prove that, "75 years on, there's a lot of life left in international radio." That's a sentiment sure to increase their popularity.

HCJB, Quito, Ecuador

Broadcasting from the equator at an elevation of more than 10,000 feet, the evangelical Christian "Voice of the Andes" dates back to the early 1930s. Compared to the sometimes strident style of Christian broadcasting common within the United States, HCJB is relatively understated. HCJB also has many programs popular with those who do not necessarily agree with the beliefs of their sponsors, World Radio Missionary Fellowship of Opa-Locka, Florida.

To combat some atmospheric problems caused by high-power broadcasting at such a high altitude, engineers at HCJB invented the cubical quad antenna in the late 1930s, an antenna that has become very popular with ham radio operators. The call letters HCJB adhere to the call sign allocations for Ecuador, but also stand for Heralding Christ Jesus' Blessings. HCJB sends a booming signal into the United States, and is typically an early catch for a new shortwave listener

Radio Habana Cuba

The Cold War lives on at Radio Habana Cuba, so don't tune in to hear praise of the western world in general or the United States in particular! Because of its proximity to the United States, it generally puts in a solid signal. Over the past couple of years, however, RHC's signal has suffered from distortion and weakness attributed to power brownouts and the lack of spare parts. Russia no longer pours money and resources into the island nation, which has lead to fuel and resource crises. The former Soviet Union is not infrequently criticized on the air for their abandoning the Communist philosophy (and with it, their substantial support for Cuba)!

This will be an interesting station to monitor to observe the changes that are sure to come in Cuba. And given its location, the implications for its northern neighbor (us!) give listeners in the United States substantial motivation to follow the trends.

Selected frequencies (kHz): 6000, 6010, 6060, 9510, 9550, 9820, 11760, 13700, 17760.

Deutsche Welle

The "Voice of Germany" was for many years the voice of West Germany. Since World War II it has emerged from the shadows of the propaganda broadcasts of Radio Berlin and is a solid source of news and analysis. Though its worldwide resources aren't as vast as those of the BBC, it is an emerging leader in international reporting.

Selected frequencies (kHz): 6985, 6145, 9640, 9700, 11810, 11890, 13610, 13770, 13790, 15205.

Radio France Internationale

With its renovated transmitters, monstrous ALLISS rotating antenna and increased English programming, RFI is becoming more available to the American audience. It retains strong African coverage from its colonial days and its clever programming treads a fine line between spontaneity and professionalism. For French language students, RFI offers an excellent source of practice direct from an unimpeachable source.

Selected frequencies (kHz): 6175, 9805, 11670, 11705, 15365, 17620, 17795.

Shortwave Shows for the Hobbyist

If you want to keep up with the fast changing world of short-wave radio, you need access to information resources that can quickly disseminate updated data. One excellent resource is on-line computer services and BBSs (see "On-line Computer Resources for SWLs"). However, these require the use of a computer and modem.

The shortwave broadcasters offer shows on a regular basis (usually weekly) that help the SWL stay on top of what's happening. And all the equipment the listener needs to access this

The huge ALLISS rotating antenna of Radio France Internationale—the largest rotating shortwave antenna system in the world! (*photo courtesy of Jacques d'Avignon*)

information is the same shortwave receiver they use to pursue the hobby.

The breadth of topics varies with each show, but among those covered are the latest happenings in the shortwave world, tips on improving the proficiency and enjoyment of shortwave listening, human interest stories relating to SWLing and other related communications and electronics topics. Information about satellite broadcasting and reception has been added to many of these shows.

In addition, some of the "mailbag shows" give listening tips, frequency changes and other information of interest to shortwave listeners.

Here's a selected list of the most popular shows...

Radio Sweden's "MediaScan," formerly known as "Sweden Calling DXers," is the world's oldest radio program about international broadcasting. Radio Sweden has presented this round-up of radio news, features and interviews on Tuesdays since 1948. It is currently broadcast on the first and third Tuesdays of the month.

"DXers Unlimited" is a program broadcast by Radio Habana Cuba. Host Arnie Coro, CO2KK, devotes a considerable amount of time on topics of interest to those new to shortwave radio. "DXers Unlimited" is totally nonpolitical, and probably the only nonpolitical program on RHC! "DXers Unlimited" offers new shows twice a week, on Sundays and Wednesdays.

"DX Party Line" from HCJB is for "dial-twiddlers" and those who "like to listen to far away countries just for the thrill of hearing a distant signal." Rich and Lisa McVicar host this popular show, which also has regular features for shortwave newcomers. Long a fixture on Saturdays, HCJB has recently shortened the Saturday show and added a mid-week (Wednesday) miniversion of "DX Party Line" called "The Latest Catch."

"Ham Radio Today" is also offered by HCJB. As its name suggests, "Ham Radio Today" focuses on issues of interest to hams. It's hosted by John Beck every Thursday.

"Musical Mailbag" on HCJB is hosted by the crew of Curt Cole, Gail Pfeffer, Ralph Kurtenbach and Giselle Russell. It airs each Saturday for a half hour of fun and frivolity. They read letters, play music, eat—and eat some more! Occasionally, while reading reception reports (between bites), they even have some shortwave listening information.

"Saludos Amigos." Now in its 10th year, HCJB's international friendship program is very successful at this goal. Part of the show has host Ken MacHarg reading letters, and in typical "mailbag" fashion, there are occasional gems for SWLs.

"Waveguide." Courtesy of the BBC, Waveguide covers technology from the English view, which means they are occasionally ahead of trends in the United States, and sometimes behind. What they always offer is a unique perspective that can shed new light on the changing world of technology.

"Spectrum" is a weekly show on WWCR covering the "spectrum" of electronics, ham radio, shortwave radio and related subjects. It's broadcast on Sunday mornings (Saturday evenings in US time zones.)

If it's on the air in this solar system, producer Jonathan Marks and the Radio Netherlands "Media Network" team are listening! Each Thursday (repeated on Fridays) this award-winning survey of communication developments draws from its network of more than 157 regular contributors around the globe, including Jim Cutler, Mike Bird, Arthur Cushen, Diana Janssen, Victor Goonetilleke and their "secret weapon," Andy Sennitt. Andy Sennitt just happens to be editor of the venerable (and Amsterdam based) *World Radio-TV Handbook*, the creation of which keeps him plugged into high quality information sources around the world. Media Network also covers the latest in the world of technology.

"World of Radio" is carried by various shortwave stations. Produced and hosted by Glenn Hauser, this show focuses rather narrowly on radio monitoring in shortwave, mediumwave and VHF.

There is little information in Voice of America's "Communications World" that isn't available to the American audience via numerous sources. Even so, news concerning changes at VOA is announced here and, with the state of flux at VOA nowadays, that makes "Communications World" a worthwhile show to monitor.

"Shortwave Merry Go Round" from Swiss Radio International is another old favorite among DX shows. Concentrating on shortwave radio from the listener's viewpoint, "Shortwave Merry Go Round" provides excellent information for beginners or experienced SWLs.

Listening and the Law

The first rule of radio monitoring has always been that no license is required to listen. There are virtually no rules regulating listening, so there was little to know, from a professional and legal standpoint.

The guiding rule of shortwave listening has always been the Communications Act of 1934 (Section 705), whose rule was simple: you could listen to anything. But unless the transmission was originated by a broadcaster, a hobbyist or a ship in distress, you were not to divulge the contents of the transmission nor use it for personal gain. (Don't get excited. The average hobbyist can listen for eons and not hear anything that can be readily used for personal gain). In point of fact, the spirit of the law was observed such that no one was prosecuted for telling their friends about an interesting signal they heard last night, as long as there was no negative effect to the parties of the transmission.

In 1986, due to pressure from lobbyists from the cellular telephone industry, Congress passed a law called the Electronic Communications Privacy Act (ECPA). This law made it illegal to listen to cellular telephone calls and several other types of communication that normally take place in the VHF/UHF spectrum. These signals are usually not of interest to SWLs. However, the ECPA also made it illegal to listen to remote broadcast transmissions and studio-to-transmitter links, communications that sometimes take place in the 26-30 MHz range.

There have been numerous reports of the reception of studio-to-transmitter communications, often hundreds of miles from the studio site. In one publicized example during the peak in the sunspot cycle, a Florida radio station was receiving (and soliciting!) reception reports from across the United States from its 100-watt shortwave studio-to-transmitter link. Interesting and innocent though this occurrence may be, it is *illegal* now.

To date there are no known instances of the casual listeners being prosecuted for "illegal reception." But just because the law has been ignored so far doesn't ensure that some technoramus (technological ignoramus) lawyer won't seek to make his reputation on such a case in the future. Don't allow yourself to become the guinea pig for such legal action. When in doubt, don't listen to a transmission if you think it falls under the umbrella of the ECPA. If you believe you may have illegally monitored a signal, keep the details to yourself!

MILITARY AND UTILITY STATIONS

Despite how the name sounds, utility monitoring doesn't involve radio traffic between crews from your local electric

---

Table 1-2
Mystic Star Frequencies

Radio frequencies used by high-ranking military and government officials when traveling on official business.

All frequencies shown in kHz

3032	6817	9023	11413	13585	20053	
3046	6803	9026	11441	13710	20154	
3076	6918	9043	11460	13823	20313	
3071	6927	9120	11466	13960	22723	
3116	6993	9158	11484	14715	23265	
3144	7316	9180	11488	14902	25578	
4721	7690	9270	11498	14913	26471	
4731	7735	9320	11545	15015		
4742	7765	9958	11596	15036		
4760	7813	9991	11615	15048		
5688	7858	10112	11627	15091		
5700	7997	10427	12324	15687		
5710	8040	10530	12317	16080		
5760	8060	10583	13201	16117		
5800	8162	10881	13204	16320		
5820	8170	11035	13214	16407		
6683	8967	11055	13215	17385		
6715	8992	11118	13241	17480		
6738	8993	11176	13247	17972		
6756	9007	11180	13412	17993		
6757	9014	11210	13440	18027		
6760	9017	11226	13455	18218		
6790	9018	11249	13457	19047		
6812	9020	11407	13485	20016		

and gas company. In the broad sense, utility monitoring is listening to any non-Amateur Radio transmissions not intended for public consumption.

This activity requires a more sophisticated level of equipment than shortwave broadcast monitoring. Because these transmissions are only intended for a select audience, the power level is lower than most broadcasters. This requires a better antenna, a more sensitive receiver—or both. In addition, most utility voice transmissions are in upper sideband (USB), a mode not included on low-end shortwave receivers. Radioteletype (RTTY) and the digital modes require outboard decoding equipment (usually including a computer) and a receiver with good selectivity and stability.

Military monitoring is a subset of utility monitoring. It encompasses the full spectrum of the utilities world: upper sideband, RTTY and digital modes as well as unique transmission modes. If you tried to get your hands on the equipment necessary to decode some of these transmissions, you'd probably end up on the FBI's Most Wanted List!

To those willing to go that extra step to acquire the equipment necessary for utility monitoring (the legal equipment!), the rewards are many. For example, the Mystic Star frequencies can yield very interesting listening as members of the Executive Branch and VIPs travel about the world. Although the planes that serve as Air Force One are equipped with sophisticated satellite communication gear, the shortwave equipment is still aboard and occasionally used for routine traffic, or to achieve a certain diplomatic effect.

Monitoring the Land and the Sea

Despite all of the high-tech equipment available to the US Air Force, the Global High Frequency System remains a backbone of communications. Long ago, during the height of the Cold War, when I first heard the transmission, "Skyking, Skyking, this is McDill, do not answer . . ." I was ready to head for the bomb shelter (except that my parents didn't have a bomb shelter). I knew enough to know that these transmissions concerned the (now defunct) Strategic Air Command and the American global nuclear forces, but not enough to know that the transmission occurred frequently.

Even with the major reorganization of the military air commands that took place in June 1992, the "Skyking..." transmissions can still be heard. However, frequency lists compiled prior to June 1992 cannot be considered accurate. Those listed in the accompanying "Voice Utility Sampler" have been compiled since the reorganization. Though many of these transmissions are cryptic, those transmissions that are "in the clear" can be fascinating.

The civilian air communications also provide interesting listening, and they are a little easier to understand. When flying over land, civilian aircraft use VHF frequencies, but on transoceanic and transpolar flights the shortwave frequencies come into use. Since English is the language of international aviation, translation is seldom a problem for the American listener.

The frequencies listed as "Air Traffic Control" on the "Utility Sampler" are used to transmit in-flight information such as weather queries, fuel consumption, updated estimated times of arrival and other related data. These frequencies are not in constant use; if you want to hear some activity on these bands choose a busy time (a time that you have reason to believe that air traffic will be heavy on the route you have chosen) and "park" on the associated frequency or do a little scanning.

For instance, you'll find that 5598, 5616, 11279, 13306 and 17946 kHz are listed for the North Atlantic. Remember, these transmissions are on upper sideband. Prime listening times are when the bulk of the jets are in the air, generally 7 PM to 12 AM EST (0000-0500 UTC) and 7 AM to 12 PM EST (1200-1700 UTC).

During this period, just sit on one of these frequencies (below 10 MHz in the evenings, above 10 MHz in the mornings), or use the memory function of your receiver to occasionally switch between them. Obviously, patience is required for this aspect of the hobby.

Some of the easiest catches in civilian shortwave aviation are the VOLMET or aviation weather transmissions. They are listed as "Aero WX" (for Aero Weather) on the accompanying "Utility Sampler." For example, New York Radio transmits on its assigned frequencies (3485 kHz, 6604 kHz, 10051 kHz, and so on) from the top of each hour to twenty minutes past each hour, and from thirty minutes past each hour to fifty minutes past each hour.

As with aircraft, both VHF and shortwave frequencies are used for maritime communications. While most aircraft transmissions use the back-and-forth simplex-type transmission found on Amateur Radio, quite a bit of the maritime communication takes place in full duplex.

Numbers Stations

If you tune around the utility shortwave frequencies, sooner or later you'll stumble across stations sending groups of numbers or letters in Spanish, English, Russian or German. These stations are known as *numbers stations*.

Ever since these stations were first noticed in the early 1960s, people have speculated about their purpose. Does the transmission contain encoded weather forecasts or shipping information? Could they be some sort of bizarre propagation beacon system? The possibilities are endless. Some people even thought the transmissions were part of a secret project communicating with UFOs!

Over time a consensus developed that these transmissions were messages from various intelligence agencies to their agents in the field. When radio direction-finding techniques located transmission sites in both (the former) East and West Germany, Nicaragua and Cuba, as well as on some military bases in the United States, this theory seemed largely confirmed. More recently, former spies have related how these transmissions were used to convey instructions. However, it continues to puzzle listeners that the level of numbers station activity has not decreased with the end of the Cold War, and that the stations on both sides of the former German border continue to operate.

Numbers stations are found as low as 2 MHz to as high as 26 MHz, using CW and voice transmissions. Though they can be heard at any time, activity tends to peak from about 0000 to 0800 UTC. Don't try to send a signal report, though. You'll never hear an identification!

PIRATES AND CLANDESTINE BROADCASTERS

Two unique types of shortwave stations are pirate and clandestine broadcasters. Though both generally operate illegally, they are quite different in their purposes. Of course, neither the American Radio Relay League nor I condone unlicensed radio broadcasting, but it's not illegal to listen to such stations!

K4ZAD'S Shortwave Voice Utility Sampler

(All frequencies are in kHz)

2182	Marine Emergency Calling Channel
2598	Canadian CG Marine Information Broadcasts
2670	USCG Marine Information Broadcasts
3413	Aero Weather—Shannon Ireland
3485	Aero WX—New York, NY and Gander, Newfoundland
4065	Inland River Towboats—WCM—Cincinnati
4125	Marine Ship Calling
4149	Marine Simplex Utility Channel 4B
4372	US Navy
4381	Great Lakes Ore Boats—WLC—Rogers City, MI (Ships on 4089)
4582	Civil Air Patrol—Emergency Channel
4722	RAF Aero Weather—Continuous
4725	US Air Force—Global High Frequency System
4742	RAF—Architect
5015	US Army Corps of Engineers—Net at 8:00 AM ET, M-F
5211	Federal Emergency Management Agency —Primary Night Channel
5505	Aero Weather—Shannon Ireland—Continuous
5598	Air Traffic Control—North Atlantic—NY, Gander, Shanwick
5616	Air Traffic Control—North Atlantic—Gander, Shanwick
5680	Search & Rescue Channel—Worldwide
5692	US Coast Guard—Chopper Ops.
5696	US Coast Guard—Air Ops.
5841	US Anti-Drug Agents
6215	Marine Ship Calling Ch./Utility Ch. (Ch. 606) —Shore on 6516
6230	Marine Simplex Utility Channel 6C
6510	River Towboats—WCM—Cincinnati (Skeds at 14:30-15:00 ET)
6577	Air Traffic Control—Caribbean—NY
6604	Aero WX—New York, NY and Gander, Newfoundland
6628	Air Traffic Control—North Atlantic—NY, Santa Maria
6676	Aero WX—Sydney, Singapore, Bangkok, Bombay
6679	Aero WX—Honolulu, Tokyo, Auckland, Hong Kong
6697	US Navy
6720	US Navy
6738	US Air Force—Global High Frequency System
6753	Canadian Military WX—Edmonton, Trenton, St. Johns
6812	USAF—A prime frequency for Air Force One
7527	US Anti-Drug Agents
7535	US Navy
7635	CAP—Nationwide Frequency (Command Net Weekdays at 1600 UTC)
8125	FAA—Eastern Net (Wednesdays at 10:45 AM ET)
8176	Sydney, Australia. Marine Radio —VIS—(early mornings)
8213	River Towboats—WCM—Cincinnati (Skeds at 13:00-14:00 ET)
8255	Marine Ship Calling (Ch. 821) —Shore Stations on 8779
8297	Marine Simplex Utility Channel 8B
8794	Great Lakes Ore Boats—WLC— Rogers City, MI—Ships on 8270
8825	Air Traffic Control—North Atlantic—NY, Gander, Shanwick
8828	Aero WX—Honolulu, Tokyo, Auckland, Hong Kong
8846	Air Traffic Control—Caribbean—NY
8864	Air Traffic Control—North Atlantic—Gander, Shanwick
8867	Air Traffic Control—Pacific—Honolulu, Auckland, Sydney, Nandi
8903	Air Traffic Control—Pacific & Africa
8906	Air Traffic Control—North Atlantic—NY, Santa Maria
8912	US Anti-Drug Agents
8957	Aero Weather—Shannon Ireland—Continuous
8967	US Air Force—Global High Frequency System
8980	US Coast Guard—Chopper Ops.
8984	US Coast Guard—Air Ops.

8993	US Air Force—Global High Frequency System
9023	Canadian Military & USAF NORAD
9032	RAF—Architect
10051	Aero WX—New York, NY and Gander, Newfoundland
10493	Federal Emergency Management Agency— Primary Day Channel
10780	USAF—NASA Support—Cape Radio— Primary Day Channel
11176	US Air Force—Global High Frequency System
11191	US Navy—Air Operations—Hershey at Key West, FL
11195	US Coast Guard—Air Ops.
11198	US Coast Guard—Chopper Ops.
11200	RAF Aero Weather—Continuous
11205	US Navy
11233	Canadian Military
11234	RAF—Architect
11255	US Navy
11267	US Navy
11279	Air Traffic Control—North Atlantic—NY, Gander, Shanwick
11282	Air Traffic Control—Pacific—San Francisco, Honolulu
11309	Air Traffic Control—North Atlantic—NY, Santa Maria
11384	Air Traffic Control—Pacific—Honolulu, Tokyo, Hong Kong
11387	Aero WX—Sydney, Singapore, Bangkok, Bombay
11396	Air Traffic Control—Caribbean—NY
11494	US Anti-Drug Agents
12290	Marine Ship Calling Ch. (Ch. 1221)—Shore Stations on 13137
12359	Marine Simplex Utility Channel 12C
13201	US Air Force—Global High Frequency System
13257	Canadian Military
13261	Air Traffic Control—Pacific—Honolulu, Auckland, Sydney, Nandi
13264	Aero Weather—Shannon Ireland—Continuous
13270	Aero WX—New York, NY and Gander, Newfoundland
13282	Aero WX—Honolulu, Tokyo, Auckland, Hong Kong
13288	Air Traffic Control—Pacific—San Francisco, Honolulu
13297	Air Traffic Control—Caribbean—NY
13300	Air Traffic Control—Pacific—Honolulu, Tokyo, Hong Kong
13306	Air Traffic Control—North Atlantic—NY, Gander, Shanwick
13312	Anti-Drug Agents and FAA Flight Tests and Commercial Flight Tests
13330	Air—Long Distance Operational Control —NY, Houston
13354	Air Traffic Control—Pacific—San Francisco, Honolulu
13457	FAA—Western Net (Wednesdays at 10:30 AM MT)
15015	US Air Force—Global High Frequency System
15867	US Anti-Drug Agents
16420	Marine Ship Calling Channel (Ch. 1621) —Shore Stations on 17302
16534	Marine Simplex Utility Channel 16C
17904	Air Traffic Control—Pacific—Honolulu, Tokyo, Hong Kong
17946	Air Traffic Control—North Atlantic—NY, Gander, Shanwick
17975	US Air Force—Global High Frequency System
18009	US Navy
22060	Marine Ship Calling Channel (Ch 2221) —Shore Stations on 22756
22171	Marine Simplex Utility Channel 22E
23287	US Navy

Frequencies courtesy of Tom McKee, K4ZAD, from his book, *The Other Shortwave* (Key Research, PO Box 846G, Cary, NC 27512).

Clandestine broadcasters are specifically political in nature. When a political group is out of power in a country, they will sometimes set up a shortwave station to broadcast to those sympathetic to their cause. These stations may be located within the target country, often illegally, or may be located in a nearby country. In this case, they may be operating legally, even though the host country may not officially acknowledge their existence (to avoid official diplomatic hostility with the target country).

Central America has traditionally been a hotbed of clandestine activity, with stations appearing and disappearing as the fortunes of their sponsoring groups change within the target countries. Africa and the Middle East have also had a sizable number of clandestine stations over the years, but since North America is far from their target audiences, they can be very hard to hear.

Most clandestines broadcast in the native language of the target country, so for Central American stations Spanish is the norm. However, some clandestine stations want to obtain money and support from within the United States. Such stations may use English, even when English is rarely spoken within the target country.

Clandestine stations may operate with regular schedules within the standard shortwave broadcast bands. Even when they are operating illegally within a country, they may have guerrilla forces protecting their station. These clandestines tend to be more brazen in their operations. Easy clandestine catches are antiCastro stations such as Radio Caiman on 9965 kHz and La Voz del CID on 9942 kHz.

Pirates are sometimes politically oriented, but typically they are people who broadcast simply for the joy of broadcasting. Their enthusiasm notwithstanding, the quality of pirate broadcasts range from highly professional programming to drivel that sounds like a drunk turned loose in a radio studio (and, indeed, they sometimes *are* inebriated!).

The most common programming formats of pirate stations are satire and comedy (sometimes unintentional). When political views are expressed, they tend to be extreme—at either end of the spectrum.

Pirate stations are most active in the evenings, on weekends and on holidays. They keep erratic schedules to help them avoid a discovery by the FCC. To prevent the FCC monitors from getting an accurate fix on their location, transmissions tend to be between 30 minutes and 1 hour in length. The common frequencies for pirate operations have been 7415 kHz and 7485 kHz, but recently legitimate broadcast stations have begun transmissions on these frequencies, causing the pirates to move. Remember, pirates often use old Amateur Radio transmitters from the 1950s running 100 to 200 W into temporary antennas. The major international broadcasters may run from 100,000 to 500,000 W into sophisticated antenna arrays. When the two clash on a frequency, there is little doubt whose signal will triumph.

Still, most pirate stations are operating between 7370 and 7490 kHz, so that's the best place to start hunting. Other fruitful places to tune are 15050 kHz on weekend afternoons and 1620 kHz evenings. Pirates usually use AM, but occasionally upper sideband broadcasts will be heard. Most weekends there will be between 2 and 20 North American pirates to be found (the most appear on holiday weekends), but since they don't have powerful signals, a good antenna and receiver are needed to hear them. Since pirates are already operating illegally, they aren't bound by FCC obscenity rules, so young children should be out of the room when you go hunting for pirates!

While pirates would contend that they operate in the spirit of "good fun," the FCC does not take such a casual attitude. Fines in excess of $10,000 are not uncommon and the revocation of any FCC licenses held (including Amateur Radio licenses) is virtually automatic. Whatever the appeal of running your own shortwave broadcast station may be, it's not worth your Amateur Radio career. It's okay to *listen* to pirates, but don't yield to the temptation to become one.

KEEPING TRACK

Shortwave Logs

One of the traditional rules of scientific experimentation has been, "If you don't record it, it didn't happen." The tradition of keeping a list of stations heard, the date and time, the frequency, conditions and transmission content in a "logbook" has been continued since the days when radio transmission and reception *were* scientific experiments.

For many years, the keeping of a logbook was the law. The Federal Communications Commission used to *mandate* that licensed ham radio operators keep a meticulous log of their transmissions and communications, but those regulations have long since been relaxed to the point where the keeping of a log is a request, not a requirement.

Despite the less restrictive regulations, the logbook still has many useful purposes in both ham radio and shortwave listening. The log is a permanent record of the reception achievements of the operator and the station. It is a reference resource for active frequencies, propagation and the comparison of different radios and antennas.

QSLs

So you've heard a rare station. How do you prove it? One way is to send a reception report to the station, including the date, time, frequency and information on the reception quality and programming content. (This is essentially the same information you record in a logbook.)

If they find your information to be accurate, they may send you a confirmation. The confirmation can take the form of a letter (especially with a utility type station), but the international broadcasters traditionally send picture postcards. These cards have become known as "QSL cards" (from the ham radio "Q" signal meaning, "acknowledging receipt.") Collecting QSLs is another subset of shortwave listening, along with collecting the sometimes exotic stamps that come with them.

The QSL card is a courtesy of the transmitting station. They aren't *obligated* to send them. To increase the chances of receiving one, and to defray the mailing costs of the broadcaster, many SWLs include an IRC (International Reply Coupon) to help defray the postage cost. An IRC can be exchanged for First-Class postage in most countries. They're available from most larger post offices.

The SINPO Code

The SINPO code is a way of quantifying reception conditions in a five-digit code, especially for use in reception reports to broadcasters. The ham radio equivalent is the RST system for Readability, Strength and Tone. The SINPO code covers Signal strength, Interference (from other stations), Noise (from atmospheric conditions), Propagation disturbance (or Fading, in the SINFO code),

WWV
Broadcast Format
Via Telephone (303) 499-7111
(Not a Toll-Free Number)

STATION ID

440 Hz 1-HOUR MARK

NIST RESERVED

STORM INFORMATION

NO AUDIO TONE

LOCATION
40°40'49.0"N; 105°02'27.0" W

STANDARD BROADCAST FREQUENCIES
AND RADIATED POWER
2.5 MHz — 2.5 kW   10 MHz — 10 kW
5 MHz — 10 kW   15 MHz — 10 kW
20 MHz — 2.5 kW

UT1 CORRECTIONS

FOR ADDITIONAL INFORMATION CONTACT
NIST RADIO STATION WWV
2000 EAST COUNTY RD. 58
FORT COLLINS, CO 80524
(303) 484-2372

GPS REPORTS

OMEGA REPORTS

GEOALERTS

STATION ID
30 MINUTES

SPECIAL ANNOUNCEMENT OR 500 Hz TONE
SILENT EXCEPT TICK
UTC VOICE ANNOUNCEMENT

600 Hz TONE
SILENT EXCEPT TICK
UTC VOICE ANNOUNCEMENT

□ BEGINNING OF EACH HOUR IS IDENTIFIED BY 0.8-SECOND-LONG, 1500-Hz TONE.

□ BEGINNING OF EACH MINUTE IS IDENTIFIED BY 0.8-SECOND-LONG, 1000-Hz TONE.

□ THE 29th AND 59th SECOND PULSES OF EACH MINUTE ARE OMITTED.

□ 440 Hz TONE IS OMITTED DURING FIRST HOUR OF EACH DAY.

WWV's hourly broadcast schedule (*courtesy of the National Institute of Standards and Technology*)

and Overall. The code is as follows:

(S)ignal

5   excellent
4   good
3   fair
2   poor
1   barely audible

(I)nterference

5   none
4   slight
3   moderate
2   severe
1   extreme

(N)oise

5   none
4   slight
3   moderate
2   severe
1   extreme

(P)ropagation

5   none
4   slight
3   moderate
2   severe
1   extreme

(O)verall

5   excellent
4   good
3   fair
2   poor
1   unusable

Table 3
Time and Frequency Stations

All stations operate 24 hours per day

Station	Location	Frequencies (MHz)
WWV	Fort Collins, Colorado	2.5, 5, 10, 15, 20
WWVH	Kekaha, Kauai, Hawaii	2.5, 5, 10, 15
JJY	Tokyo, Japan	2.5, 5, 8, 10, 15
CHU	Ottawa, Canada	3.330, 7.335, 14.67
VNG	Llandilo, Australia	2.5, 5, 15
ATA	New Delhi, India	5, 10, 15
BSF	Chung Li, Taiwan	5, 15
YVTO	Caracas, Venezuela	5
BPM	Xian, China	5, 10, 15
RID	Irkutsk, Siberia	5.004
RWM	Moscow, Russia	4.996

In recent years, many broadcasters have tried to steer listeners away from the SINPO code and toward the simpler SIO code. SIO deletes the extremes (1 and 5) and the noise and propagation categories, which were confusing to too many people to be useful. In sending reports to stations other than large international broadcasters, it is better to simply describe reception conditions in words.

WWV Time and Propagation Information

Want to know the *exact* time? The shortwave station WWV in Fort Collins, Colorado, is operated by the Time and Frequency Division of the National Institute of Standards and Technology and provides a precise time signal around the clock. Additional services include standard time intervals, standard frequencies, propagation forecasts, geophysical alerts, marine storm warnings, Global Positioning System (GPS) information, UT1 time corrections and BCD Time Codes.

Another station, WWVH, is located in Hawaii. Voice announcements are made from WWV and WWVH once every minute. Since both stations can be heard in some locations, a man's voice is used on WWV and a woman's voice is used on WWVH to reduce confusion. The WWVH announcement occurs first, at about 15 seconds before the minute. The WWV announcement follows at about 7.5 seconds before the minute. Though the announcements occur at different times, the tone markers are transmitted at the same time from both stations. However, they may not be received at exactly the same instant due to differences in the propagation delays from the two station sites.

The broadcasts from WWV and WWVH include a time code. The time code signal is 100 Hz removed from the main carrier and is called a *subcarrier*. The code pulses are sent out once per second. With a good signal and a high-quality receiver, you can hear the time code as low rumble in the audio. HF receivers that receive and decode this signal can automatically display the time of day.

WWV/WWVH also broadcasts propagation information at 18 minutes after each hour. Unfortunately, the information they provide isn't quite as simple as, "Propagation will be good today" or even, "Propagation will be fair below 15 MHz today." No, they give you *indices*. With a PhD in propagation and a year's worth of calculations, you can turn the indices into a meaningful prediction of radio conditions.

Perhaps it's not quite that bad, but understanding the indices does take some study. Both *The ARRL Handbook* and *The ARRL Antenna Book* offer details on how to interpret the propagation data provided by WWV. Should you decide to delve deeply into propagation theory and the indices, *The Shortwave Propagation Book* by George Jacobs, W3ASK, and Theodore J. Cohen, N4XX (CQ Publishing, 76 North Broadway, Hicksville, NY 11801) is an excellent text on the subject.

Because WWV and other time and frequency stations broadcast with substantial power around the clock on a variety of frequencies, they can provide you with information on propagation just by listening. Measuring the strength of the signal from Boulder, Colorado, on 5, 10 and 15 MHz gives an immediate, if imprecise, reading on how the ionosphere is doing.

The ionosphere is a fickle thing. A signal can be loud one moment and gone the next. Many shortwave listeners will tell you that there is one immutable law of propagation: the signal quality will deteriorate just before the station sends its identification! Actually, this is just another corollary of Murphy's Law. This corollary, well known to hams and engineers, states that the rarer a station is, the worse propagation will become as the station ID nears.

The ionosphere can be humbling at other ways. When propagation is at its worst, Radio Fredonia can put a megawatt into a 20-dB gain curtain array antenna and you can crank up your $20,000 ex-CIA receiver connected to the best receiving antenna and hear . . . *nothing*! When the signal is not there, it's just not there. All the radio horsepower in the world can't change that fact.

SHORTWAVE VS. SATELLITES

Some people have predicted the end of shortwave radio as

Even the Voice of America uses satellite technology to transfer signals between sites. (*photo courtesy Voice of America*)

satellite technology becomes more prevalent. Many countries have indeed cut back their shortwave operations and increased programming via satellite. Military and diplomatic operations have switched much of their communications to satellites. Is shortwave listening on its deathbed?

No. One major difference between shortwave transmissions and satellite transmissions is *infrastructure*. For satellite operations to work, a lot of complex equipment has to perform in concert. Satellite dishes have to be aimed precisely at both ends of the signal path. To make matters worse, the satellites function in one of the harshest environments imaginable. As the old saying goes, "The more complicated they make the plumbing, the easier it is to clog the pipes." That's why even the military, with their desire for secure communications, can still be heard frequently on the shortwave bands.

The need for infrastructure also requires approval from those who control the infrastructure. When Operation Desert Storm commenced, the Iraqi government immediately took control of the satellite uplinks. From that moment onward, they had exclusive control of all information coming out of the country.

Shortwave communication, on the other hand, is not so easy to control. Relatively simple equipment is all you need: a transmitter, a receiver and the appropriate antennas (along with a little help from the ionosphere). While it would be hard for someone in a controlled country to hide a satellite dish, a portable shortwave receiver and antenna can be easily concealed.

So for all of the convenience and quality that satellites provide, there will always be a need for shortwave operations. Your shortwave receivers won't lack for signals for a long time to come.

On-Line Computer Resources for SWLs

As one who started working with computers in high school in 1971 using a Teletype machine and an acoustic 110-baud modem to connect by phone with a time-sharing mainframe 30 miles away, I like to say I've been traveling the Information Superhighway since it was a dirt road. Dirt road or eight-lane freeway, there is no doubt that the depth and breadth of information available by computer is increasing at a geometric rate.

How else can you exchange detailed, exhaustive information with other hobbyists around the world, 24 hours a day, in a format you can quickly turn into printed "hard copy"? Where else can you be in instant communication with programmers of major international broadcast stations?

All this and more can be found among the electronic information resources available via computers and modems. Although well-equipped radio hobbyists almost always have microcomputers at their disposal, only now are many of them becoming aware of how they can use them to retrieve valuable information.

Long before all of the hype about the "Information Superhighway," an elaborate network of computer bulletin boards has been shuttling data through the world. This network is both formal and informal, publicly and privately financed. You'll find everything from professionally managed information resources to home-based computer systems running on shoestring budgets.

The integration of telecommunications capability and microcomputers has provided an important new way to transmit and exchange information. With just a microcomputer, an ordinary voice-type telephone line and modem, a computer bulletin board can be set up to facilitate interchange among target interest groups and/or the computer literate population in general.

There are also national commercial database services such as CompuServe and America Online. These systems are based on large centrally located mainframe computers that are hooked into data networks. Hundreds of communities can reach them via a local phone call. The Internet, a computer network originally developed for research, is now accessible to the public and offers connections around the world.

A growing number of computer bulletin boards (BBSs) are being specifically dedicated to shortwave and ham radio related topics. These BBSs provide such information as frequency and program schedules, reviews, modifications and information on new equipment, software and so on.

Even the "general interest" computer services can provide valuable resources for the radio hobbyist. Given the fact that radio hobbyists tend to be computer users also, most of the national on-line services quickly set up special areas just for hams and radio enthusiasts.

Computer information services allow for access to information in several ways. The most common methods of information dissemination are conference groups, electronic mail and files.

Conference Groups

No matter how knowledgeable you are about your special interests in radio, you'll occasionally be confronted with a seemingly unsolvable question. Electronic information services can provide you with a forum to get input on your questions from hobbyists from all over the country.

Most electronic information services have public message areas called "conferences" . . . also known as forums, newsgroups and SIGs (Special Interest Groups). Titles of some typical conference areas for the BBSs listed below are "Ham Radio," "Shortwave Radio," "Equipment and Antennas" and "Packet."

The user chooses the conference with the most appropriate title and types in ("posts") the question, usually addressed to ALL. Anyone accessing the conference area after that will see the message.

Through a concept known as *echoing*, many of the private BBSs relay messages across the country and around the world. Had a ham been involved in this process early in its development, they might have started the American Computer Relay League, given the similarity to the early days of radio. As it evolved, the most popular of these relay or echo systems came to be named after a dog, FIDOnet. The FIDO network consists of hundreds of local BBS computers relaying messages across the country by automatically calling each other during the early morning hours (when telephone rates are lowest).

Within days or hours, the original question will usually receive responses. The responses may produce more questions and will often yield an in-depth exploration of the topic by a number of knowledgeable responders. Of course, the responders are self-selecting, so there are no guarantees of their accuracy, but this same caveat applies to many seminars, books, magazine articles and the like.

Because of the delays inherent in relaying, messages on FIDOnet can take a couple of days to go cross-country, and a couple more days to receive replies from cross-country. Replies can take up to several days to reach you, depending upon your location and the location of your responders.

For this reason, Internet and the commercial services are a bit better for questions that need immediate responses. Of course, you have to pay for access to them (unless you work for an organization that provides Internet access). Also, on Internet or FIDOnet you occasionally may not get a response to a particularly obscure question. With the commercial services, most conference areas are overseen by *SysOps* (System Operators), knowledgeable people who assist their users. Even when they do not know the answer to your question, they'll usually attempt to connect you with someone who can.

Electronic Mail

The conferences are generally public posting areas (where everyone can read them), but if you have a private message you can send it somewhat confidentially to one person or several through e-mail (electronic mail). The "somewhat" modifier is used because, in addition to the intended recipient(s), the administrators of the service or BBS (and anyone they designate) can see the message.

Files

New computer hobbyists are often surprised by the amount of free programs available. In many computer or electronics magazines there are ads for software available for $3 to $9 per disk. Most of this software can be downloaded (transferred to your computer via the phone line) from computer bulletin boards (BBSs) for free.

The range of this "shareware" is as broad as the range of regularly distributed commercial software, and the quality and level of sophistication is often as good or better. Spreadsheet, communications, database and word-processing programs are all available, as well as a seemingly endless supply of games. The quality of this software varies widely, but you can't complain about the cost of acquiring it! Even the BBSs that emphasize radio and electronics have some general purpose programs available.

Although the number of programs dealing with radio and electronics is not as large as those on other topics, they are growing in number. Some of them can be very useful and some are worth exactly what you paid for them, but they all provide with some insight in the interface of computers and radios.

In many cases the author of the software gives the

(Continued on next page)

rights of usage freely, or only requests donations from satisfied users. Other authors consider shareware an alternative method of distributing and selling commercial software (allowing you to try the software instead of advertising). They may require a modest payment if you use the program regularly. Of course, given the circumstances, you're on the honor system.

Getting Started in Telecomputing

The basic tools needed for telecomputing are a computer with a serial interface, communications (terminal) software, a modem and a telephone line. A typical voice-type telephone line is all that is required. The expensive specialized "data-lines" you may have heard of are not needed for the type of computer communications covered here.

In almost any radio club you can find someone who is already connected to the computer services. They will be your best source for help in getting started. Although it's not as hard as some make it out to be, there are enough differences that having a knowledgeable person available for consultation is invaluable.

Once you get on-line, you'll never lack for reading material on your favorite hobby. In fact, you'll soon have to deal with the problem of *excess* information. That's a nice problem to have!

W3INK Ham Radio Landline BBS List

(Sorted by telephone area code. Subject to change without notice.)

Name	Telephone number	Comments
The DX Connection	201-692-3705	
BFWK BBS	201-941-3302	
Gallery's BBS (MAC)	202-298-6009	
Trumbull-Mini	203-261-6434	Fido 1:141/370
ARRL BBS	203-666-0578	
Teletalk	203-799-7454	Fido 1:141/328
Bill VE4UB	204-785-8518	
The Data Connection	205-601-0917	Fido 1:3625/465
The Bulletin Board BBS	205-758-5017	
Bears BBS	206-237-3472	
The AA6ED Packet Gateway	206-271-4657	
Bill & Moe's BBS	206-328-4522	Fido 1:343/159
The Precedent	206-355-1295	Fido 1:343/9
KA1SVC BBS	207-247-6225	
Cobra BBS	207-725-0899	
The Rebel BBS	208-997-3937	
WCPB	209-661-5355	Fido 1:10/45
Communication Specialties	212-645-8673	Fido 1:278/307
Kenwood Factory BBS	213-761-8284	
COM Port One	214-226-1181	Fido 1:124/7009
Datalink BBS AMSAT	214-394-7438	
Smoked Armadillos	214-669-9645	Fido 1:124/5118
Better Amateur Radio	216-237-8208	
The NorthEast Ohio Hamshack	216-686-0057	Fido 1:157/607
Cleveland Hamnet	216-942-7516	

Name	Telephone number	Comments
Radio Daze	219-256-2255	
3 WINKS BBS	301-216-0662	Fido 1:109/418
PC-HAM	301-593-9067	
WJ3P Ham Exchange BBS	301-831-5954	
WB3IKP BBS	302-798-8186	Fido 1:150/220.0
Mile High	303-431-1404	Fido 1:104/223
The Right Connections BBS	305-382-6687	Fido 1:135/63
Telcom Central	305-828-7909	Fido 1:135/23
FTW	307-328-1923	C-64
TP PACKET BBS(Twisted pair)	309-822-8781	
PC Haven	310-374-7929	Fido 1:102/137
GFRN BBS	310-541-2503	
On The Air	313-522-5349	
Air Studio	313-546-7045	Fido 1:120/216
The M.A.R.C. BBS	313-669-4279	
The Black Hole BBS	313-879-7387	Fido 1:120/36
The Shop BBS	314-449-2710	Fido 1:289/39
Dots-N-Dashes BBS	315-642-1220	Fido 1:2609/220
Ham-Net BBS	315-682-1824	Fido 1:260/375
N2JEU BBS	315-697-5123	Fido 1:260/304
Southside BBS	317-535-9097	Fido 1:231/30
Springhill Amateur Radio BBS	318-539-2398	
Tri-State Data Exchange	319-556-4536	Fido 1:283/610
The Roadrunner BBS	401-821-1457	
Aksarben ARC BBS	402-289-4658	
Alta Packet Info Net	403-464-5069	Fido 1:342/7
Thunder Bay Trading Post	404-516-1282	
WA4BRO'S Ham Board BBS	404-552-0868	
Retriever's Retreat	405-359-1540	Fido 1:147/2001
AVALON	405-636-4022	
N7GXP Hamshack	406-458-9379	
HAM BYTES	407-575-9680	
Pclogic	407-879-4823	
Atlanta Amateur Radio Club	404-393-3083	
Lockheed Arc BBS	404-949-0687	
Saratoga Clone	408-395-1402	
WBBS	409-447-4267	Fido 1:106/4267
The Amateur Radio BBS	410-661-2475	
NB3P BBS	410-750-6403	Fido 1:261/1453
Beaver Valley BBS	412-775-7536	
N3OBL Amateur Radio BBS	412-864-8183	Fido 1:129/238
Joe's Garage	414-453-5145	Fido 1:154/222
Milwaukee Heath Users Group	414-548-9866	
No-Name RBBS	415-481-0252	RBBS 8:914/101
BBS-JC	415-961-7250	
Nortown ARC BBS	416-223-2186	
The Black Hole BBS	419-228-7236	Fido 1:234/16
The Paragould Ham Net	501-239-2397	
Deckman's Exchange	502-267-7422	
MacHam	503-656-0510	
The Wireless BBS	503-692-7097	
Helpnet of Baton Rouge	504-273-3116	Fido 1:396/101
The Digital Cottage	504-897-6614	Fido 1:396/65
Salt Air	508-385-3427	
Ham Shack BBS	508-949-3590	
Think Tank II	509-244-6446	Fido 1:346/8

Name	Telephone number	Comments
Time & Eternity	510-886-5467	Fido 1:215/17
The Antenna Farm BBS	512-444-1052	
Hams on Health-Link	512-444-9908	Fido 1:382/5
The Armadillo BBS	512-837-2003	Fido 1:382/32
The Listening Post	513-474-3719	
VE2MMM Amateur Radio BBS	514-624-5651	Fido 1:167/230
The Ham Tech BBS	515-981-9116	
The Long Island Ham Radio BBS	516-368-4979	
Amateur Radio Selling Post	516-581-1896	
No Frills Plus Ham BBS	516-661-3643	
The Silicon Garden	516-736-6662	
N2BQF Amateur Connection	516-789-4902	Fido 1/2619:218
Datalink BBS	516-862-8764	Fido 1:107/241
Ham It Up	516-878-4906	
AA8ES Amateur Radio BBS	517-939-0169	Fido 1:239/1020
RADIO FREQS	518-782-0507	Fido 1:267/103
Kitchner-Waterloo AR BBS	519-578-9314	Fido 1:221/177
The Telegraph Station	602-844-9740	
The Pembroke Connection	603-485-8549	Fido 1:132/169
The Legal Beagle	603-883-4466	Fido 1:132/115
The Radio Wave BBS	609-764-0812	
The 5th Dimension BBS	610-827-7689	Fido 1:273/304
MadMax BBS	610-792-9667	Fido 1:2614/736
Ham>Link<	612-HAM-0000	Fido 1:282/100
NFB	612-726-1972	
The Lion's Den BBS	613-392-8294	Fido 1:249/303
Ham BBS	614-895-2553	
The Sounding Board	615-281-2104	
WA4JUB BBS	615-872-7511	
Pet Finders System	616-765-3372	
De Shop	617-233-0297	
DX ONLINE	617-592-8404	
Baystate BBS	617-598-6646	Fido 1:101/370
Tom's BBS	617-356-3538	Fido 1:101/470
JAFO! BBS	618-797-1675	Fido 1:2250/26
Radio Sport BBS	619-279-3921	
KB0DSW's Amateur Radio BBS	701-839-2163	Fido: 1:14/649
Ham/Shortwave BBS	702-368-0846	Fido 1:209/730
Frequency Forum	703-207-9622	Fido 1:109/239
The Technicians Exchange BBS	703-387-1780	
KC3OL BBS	703-689-7156	
AMRAD BBS	703-734-1387	
Das Spitzen Sparken Board	703-791-6198	
WB4YZA BBS	704-284-4854	
N7IJI	704-372-8925	
NC EMS BBS	704-637-6906	Fido 1:379/308
Radio Hobby Online	708-238-1901	Fido 1:115/747
Micro Overflow	708-355-6942	
Elk Grove Repeater	708-529-1586	
The Precision Board	708-980-9544	Fido 1:115/410
The Far Point Relay	713-463-8324	Fido 1:106/8324
The Politically Incorrect BBS	713-550-1070	Fido 1:106/1010
USS Pegasus—Houston node	713-777-0821	Fido 1:106/9636
InterFlex Systems Design	714-497-5860	
T.E.L. Net Systems #2	714-597-7858	Fido 1:207/107
F.O.G.	714-638-2298	Fido 1:103/100
The Byte Stops Here	715-341-9723	
The Shack Too	716-288-5848	Fido 1:260/232
Vector Board	716-544-1863	
NF2G Online	716-663-8478	Fido 1:260/218
HamCenter BBS	716-885-6604	
Vacuum Valley BBS	717-323-1645	
Tec-Board	717-561-8145	
KB7UV Roserver/PRMBS	718-956-7133	
The Mav Plus BBS	801-634-3655	Fido 1:15/10
We Serve Your Drives	802-453-6074	Fido 1:325/101
N1HUM's HAM BBS	802-479-2159	Fido 1:325\604
The HAM Connection	802-868-5274	
WV4B	803-223-5185	
KC4YJC	803-269-7899	
The Byte Bucket BBS	803-871-3076	
The Computer Forum BBS	804-471-3360	
HamdiNet BBS	804-496-3320	Fido 1:275/429
HamBones Fish House	804-665-7906	
The Listening Post	804-850-8586	Fido 1:271/263
CVACC	805-499-5699	Fido 1:102/1002
Simi Valley Am. Radio BBS	805-583-2282	
The Barn	809-781-9805	
The Ultimate Connection	810-232-1905	
KA9LQM's Ham Shack BBS	812 428-3352	
The Cat House	813-796-2486	Fido 1:3603/110
DataCOM Systems BBS	813-791-1454	
Pac-Comm, Inc. BBS	813-874-3078	
Solomon's Portico	813-954-5139	Fido 137/19
KC HamNet	816-459-9752	Fido 1:280/116
Revelstone HST	817-732-1767	Fido 1:130/402.0
N6MBR BBS	818-597-0641	Fido 1:102/1004
R&D BBS	819-772-2952	Fido 1:163/506
Database of Tenn. BBS	901-749-5256	
Marty's Place	903-753-0485	Fido 1:19/62
COPS-N-HAMS BBS	904-678-5998	
The Ham Shack	904-756-3273	
Ham Radio BBS	905-827-0704	Fido 1:259/304
Discovery Software BBS	907-248-8130	
The Micro Room	908-245-6614	Fido 1:107/919
The Attic BBS	908-396-0790	
Planet Shadowstar TBBS	908-494-3417	Fido 1:107/344
Microphone TBBS	908-494 3649	
Cape Fear BBS	910-270-3178	
Albany Amateur Radio Club	912-435-9466	Fido 1:3617/6
The Boarding House BBS	913-827-0744	
Coynet Amateur Radio BBS	914-485-3393	
Joe Brown's BBS	914-667-9385	Fido 1:2625/111
Hamnet BBS	915-653-9077	Fido 1:383/4
Galactic Crossroads BBS	916-334-5641	Fido 1:203/57
WA6RDH	916-678-1535	
QST	916-920-1288	Fido 1:203/730
DX-BBS	916-992-0923	
The Ham Radio Emporium	918-272-4327	Fido 1:170/801
Dark Oak BBS	919-380-7194	
Gadgets & Gizmos BBS	919-772-9745	
Raleigh ARS BBS	919-676-8469	

CHAPTER 2

The Amateur Radio Spectrum

PAUL RINALDO, W4RI
ARRL WASHINGTON OFFICE

The electromagnetic spectrum is a limited resource. Every kilohertz of the radio spectrum represents precious turf that is blood sport to those who lay claim to it. Fortunately, the spectrum is a non-depletable resource—one that if misused can be restored to normal as soon as the misuse stops. Every day, we have a fresh chance to use the spectrum intelligently.

Simply because the radio spectrum has been used in a certain way, changes are not impossible. Needs of the various radio services evolve with technological innovation and growth. There can be changes both in the frequency-band allocations made to the Amateur Service and how we use them. Also, to our benefit, signals transmitted from one spot on Earth propagate only to certain areas, so specific frequencies may be reused numerous times throughout the globe. Amateur Radio is richly endowed with a wide range of bands starting at 1.8 MHz and extending to above 300 GHz. Thus, we enjoy a veritable smorgasbord of bands with propagational "delicacies" of every type—direct wave, ground wave, sky wave, tropospheric scatter, meteor scatter—and we can supplement this with moon reflections and active artificial satellites.

Why do you need to know any of this? Well, you don't. But if you know the history of our frequency allocations, who the players are and what the process is, you may be able to help influence the process in the future.

INTERNATIONAL REGULATION OF THE SPECTRUM

Amateur Radio frequency band allocations don't just happen. Band allocation proposals must first crawl through a maze of national agencies and the International Telecommunication Union (ITU) with more adroitness than a computer-controlled mouse. Simultaneously, the proposals have to run the gauntlet of the competing interests of other spectrum users.

Treaties and Agreements

To bring some order to international relationships of all sorts, nations sign treaties and agreements. Otherwise (with respect to international communications) chaos, anarchy and bedlam would vie for supremacy over the radio spectrum. Pessimists think we already have some of that, but they haven't any idea how bad it could be *without* international treaties and agreements. The primary ones that affect the Amateur Service are:

• The International Telecommunication Convention, signed at Nairobi on November 6, 1982.

• The Radio Regulations annexed to the International Telecommunication Convention were signed at Geneva on December 6, 1979, and entered into force on January 1, 1982.

• Partial revisions of the Radio Regulations, relating to space and radio astronomy, were signed at Geneva on November 8, 1963, and July 17, 1971. They became effective on January 1, 1965, and January 1, 1973, respectively.

• The United States-Canada Agreement on Coordination and Use of Radio Frequencies above 30 MHz was agreed to in an exchange of notes at Ottawa on October 24, 1962. A revision to the Technical Annex to the Agreement, made in October 1964 at Washington, was put into effect in a note signed by the United States on June 16, 1965. Canada signed the note on June 24, 1965. It became effective on June 24, 1965. Another revision to this Agreement to add Arrangement E (between the Department of Communications of Canada and the National Telecommunications and Information Administration and the Federal Communications Commission of the United States on the use of the 406.1 to 430 MHz band in Canada-US Border Areas) was made by an exchange of notes signed by the United States on February 26, 1982, and Canada on April 7, 1982.

The International Telecommunication Union

The origins of the International Telecommunication Union (ITU) trace back to the invention of the telegraph in the 19th century. To establish an international telegraph network, it was necessary to reach agreement on uniform message handling and technical compatibility. Bordering European countries worked out some bilateral agreements. This eventually led to creation of the ITU at Paris in 1865 by the first International Telegraph Convention, which yielded agreement on basic telegraph regulations.

Starting with a conference in Berlin in 1906, attention focused on how to divide the radio spectrum by specific uses to

Glossary of Spectrum Management

Parenthetical abbreviations following definitions indicate the following sources:

(RR) ITU *Radio Regulations*
(NTIA) National Telecommunications and Information Administration *Manual of Regulations and Procedures for Federal Radio Frequency Management*

Access—Freedom to use, or protocol for using, the spectrum. (See Code, Frequency and Time Division Multiple Access.)

Adjacent Channel—The channel immediately above or below the reference channel.

AIRS—ARRL Interference Reporting System—collection of information regarding out-of-band transmissions in the amateur bands (formerly called Intruder Watch).

Alligator—An unbalanced repeater with a "big mouth" (high-power transmitter) and "small ears" (insensitive receiver).

Allocation (of a frequency band)—Entry in the Table of Frequency Allocations of a given frequency band for the purpose of its use by one or more (terrestrial or space) radiocommunication services or the radio astronomy service under specified conditions. This term shall also be applied to the frequency band concerned. (RR)

Note: Allocations are distributions of frequency bands to specific radio services. On the highest level, allocations are made by competent (legally qualified) World Administrative Radio Conferences. Certain specific uses for the spectrum are grouped together so that the probability of interference between major types of radio services is minimal. Domestic (US) allocations are made by the FCC and NTIA in coordination within the limits of the *Radio Regulations* and using broad principles... (Feller, *Planning an Electromagnetic Environment Model for Spectrum Management,* 1981, FCC)

Allotment (of a radio frequency or radio frequency channel)—Entry of a designated frequency channel in an agreed plan, adopted by a competent Conference, for use by one or more administrations for a (terrestrial or space) radiocommunication service in one or more identified countries or geographical areas and under specified conditions. (RR)

Note: Allotments are distributions of frequencies to areas or to countries. They provide for the orderly use of certain frequency bands. (Feller, ibid)

Amateur-Satellite Earth Station—An earth station in the Amateur-Satellite Service.

Amateur-Satellite Service—A radiocommunication service using space stations on earth satellites for the same purposes as those of the Amateur Service. (RR)

Amateur-Satellite Space Station—A space station in the Amateur-Satellite service.

Amateur Service—A radiocommunication service for the purpose of self-training, intercommunication and technical investigation carried out by amateurs, that is, by duly authorized persons interested in radio technique solely with a personal aim and without pecuniary interest. (RR)

Amateur Station—A station in the Amateur Service. (RR)

AMSL—Height above mean sea level.

Amplitude Compandored Single Sideband (ACSSB)—A single-sideband transmission system which uses a pilot to tell the receiver how much expansion is required at each instant, in order to recover the dynamic range of the original signal.

Assigned Frequency—The center of the Frequency Band assigned to a station.

Assigned Frequency Band—The frequency band within which the emission of a station is authorized; the width of the band equals the necessary bandwidth plus twice the absolute value of the frequency tolerance. Where space stations are concerned, the assigned frequency band includes twice the maximum Doppler shift that may occur in relation to any point of the Earth's surface. (RR)

Assignment (of a radio frequency or radio frequency channel)—Authorization given by an administration for a radio station to use a radio frequency or radio frequency channel under specified conditions. (RR)

Note: Assignments are frequency distributions to specific stations. When making assignments, administrations authorize a station to use one or more specific frequencies under specified conditions. (Feller, ibid) With rare exceptions, the FCC does not assign specific frequencies within the Amateur Service but permits amateur stations to operate on any frequency within allocated frequency bands.

Attenuation Ratio—The magnitude of the path loss.

Authorized Bandwidth—Authorized bandwidth is . . . the necessary bandwidth (bandwidth required for transmission and reception of intelligence) and does not include allowance for transmitter drift or Doppler shift. (NTIA)

Automatic Control—The use of devices and procedures for control of a station when it is transmitting so that compliance with the FCC Rules is achieved without the control operator being present at a control point. (FCC Rules, Section 97.3)

Automatic Power Control—Dynamic control of transmitter power level in order to limit interference to other stations and allow frequency reuse.

Automatic Transmitter Identification System (ATIS)—A device built into a transmitter that sends a unique identifier without action by the operator.

Availability—The characteristic denoting whether the resource is ready for immediate use, or the probability related to a period of time.

Balanced System—A system, usually a repeater, whose transmitting and receiving subsystems are designed for the same quality signal at the same range. (See *Alligator* and *Rabbit.*)

Band—(1) A range of frequencies. (2) One of the ranges of frequencies allocated to the Amateur Service.

Bandwidth—The width of a frequency band outside of which the mean power of the total emission is attenuated at least 26 dB below the mean power of the total emission, including allowances for transmitter drift or Doppler shift. (FCC Rules, Section 97.3)

Beacon, Engineering—A transmission or frequency used to convey technical information about the host system. In a satellite, it provides data on the health and welfare of the spacecraft.

Beacon (packet radio)—Periodic transmission of a frame giving station identification information—an early practice now deprecated as it adds to channel congestion.

Beacon, Propagation—A transmitting station that continuously or periodically emits a signal to permit listeners to determine whether a path is open and its quality.

Beam (antenna)—The major lobe of an antenna radiation pattern.

Beamwidth (antenna)—The width in degrees azimuth between the half-power points in the major lobe of the antenna radiation pattern.

Broadband—(See *Wideband*.)

Broadcasting—Transmissions intended for reception by the general public, either direct or relayed. (FCC Rules, Section 97.3)

Calling Frequency—A frequency used only for establishing contact, after which the stations change to another frequency to carry out their communication. Some band plans identify national and DX calling frequencies.

Carrier—A continuous radio frequency capable of being modulated by a baseband signal or subcarrier.

Carrier Power (of a radio transmitter)—The average power supplied to the antenna transmission line by a transmitter during one radio frequency cycle taken under the condition of no modulation. (RR)

Carrier-to-Noise Ratio—The ratio of the amplitude of the carrier vs the amplitude of noise, usually expressed in dB.

CCIR—International Radio Consultative Committee, a former agency of the International Telecommunication Union.

CCITT—International Telegraph and Telephone Consultative Committee, a former agency of the International Telecommunication Union.

Cell—A single service zone within a geographical grid. The network often has the structure of a honeycomb. This is the basis of the cellular telephone system which has base stations in the center of each cell to serve mobiles throughout the cell. Individual cells, if regular shapes, may be hexagonal, circular or elliptical. Cell shapes also may be modified according to terrain.

Cellular—A UHF radio system wherein groups of frequencies in a cell are reused in others in a controlled pattern and wherein a central computer is used to hand off control of a given mobile unit from cell to cell by means of landline trunks.

Centimetric Waves—Radio frequencies between 3 and 30 GHz. (SHF)

Channel—(1) A one- or two-way transmission path divided from any others in frequency or time. (2) A band of frequencies designated for a specific use.

Channel Loading—The total occupancy of a channel by all users affected by each other's operation.

Channel Spacing—The frequency interval between successive (usually adjacent) channels, measured from center to center or between other characteristic frequencies.

Channel, Voice (Speech) Grade—A frequency channel that can faithfully pass frequencies in the 300-3000 Hz audio range.

Characteristic Frequency—A frequency which can be easily identified and measured in a given emission. A carrier frequency may, for example, be designated as the characteristic frequency. (RR) (See also *Reference Frequency*.)

CISPR—International Special Committee on Radio Interference, an organ of the International Electrotechnical Commission (IEC).

Class of Emission—The set of characteristics of an emission, designated by standard symbols, e.g., type of modulation, modulating signal, type of information to be transmitted, and also if appropriate, any additional signal characteristics. (RR)

Closed Repeater—A repeater where use by nonmembers is discouraged.

Cochannel Interference—Interference on the same channel.

Code Division Multiple Access (CDMA)—A technique to permit multiple stations to access a given band by using different spread spectrum codes.

Coded Squelch—A system wherein radio receivers are equipped with devices which allow audio signals to appear at the receiver output only when a carrier modulated with a specific signal is received. (NTIA)

Coordination—The process of ascertaining from other users whether a proposed use of a radio frequency can occur without causing harmful interference.

Coordination Area—The area associated with an earth station outside of which a terrestrial station sharing the same frequency band neither causes nor is subject to interfering emissions greater than a permissible level. (RR)

Coordination Contour—The line enclosing the coordination area. (RR)

Coordination Distance—Distance on a given azimuth from an earth station beyond which a terrestrial station sharing the same frequency band neither causes nor is subject to interfering emissions greater than a permissible level. (RR)

Coordinated Station Operation—The repeater or auxiliary operation of an amateur station for which the transmitting and receiving frequencies have been implemented by the licensee in accordance with the recommendation of a frequency coordinator. (FCC Rules, Section 97.3)

Coordinated Universal Time (UTC)—Time scale, based on the second (SI), as defined and recommended by the CCIR and maintained by the International Time Bureau (BIH). For most practical purposes associated with the Radio Regulations, UTC is equivalent to mean solar time at the prime meridian (0° longitude), formerly expressed in GMT. (RR)

Critical Frequency—The frequency below which vertically directed radio waves are reflected back to earth.

Crossband—Two-way communication in which the frequencies lie in two distinctly separate frequency bands. (Compare In-band.)

Decametric—Radio frequencies between 3 and 30 MHz. (HF)

Decimetric Waves—Radio frequencies between 300 MHz and 3 GHz. (UHF)

Deviation (frequency)—The difference in frequency from the center of the channel to the upper and lower excursions of the carrier frequency.

Disaster Communication—Emergency communication relating to an unforeseen calamity.

Doppler shift—The change in apparent frequency resulting from relative motion between the source and observer.

Drift (frequency)—An undesired change in frequency, usually as a result of temperature change in frequency-determining components.

Duplex Operation—Operating method in which transmission is possible simultaneously in both directions of a telecommunication channel.

Duty Cycle—The ratio of the *on* time to the total time of an intermittent operation. (FM is said to have a 100% duty cycle, as its transmitter is at full power, when transmitting, regardless of the amount of modulation.)

Dynamic Frequency Sharing (or real-time frequency management)—Operation on a secondary basis where there is no possibility of a claim for interference-free communication made possible with frequency-agile transmitting and receiving equipment. (CCIR Report WARC-92, Section 5.4)

Dynamic Range—The ratio to the maximum signal level (with a specified amount of distortion) to the noise level (or minimum signal level), usually expressed in dB.

Earth Station—A station located either on the Earth's surface or within the major portion of the Earth's atmosphere and intended for communication:
—with one or more space stations; or
—with one or more stations of the same kind by means of one or more reflecting satellites or other objects in space. (RR) An amateur station located on, or within 50 km of, the Earth's surface intended for communications with space stations or with other Earth stations by means of one or more objects in space. (FCC Rules, Section 97.3)

Effective Isotropic Radiated Power (EIRP)—The product of the power supplied to the antenna and its gain relative to an isotropic antenna.

Equivalent Isotropically Radiated Power (EIRP or e.i.r.p.)—The product of the power supplied to the antenna and the antenna gain in a given direction relative to an isotropic antenna (absolute or isotropic gain). (RR)

Electromagnetic Compatibility (EMC)—The ability of electronic equipment to function in its environment without introducing additional disturbances into that environment detrimental to other electronic equipment.

Emergency Communication—Any Amateur Radio communication directly relating to the immediate safety or life of individuals or the immediate protection of property. (FCC Rules, Section 97.3)

Emission—Radiation produced, or the production of radiation, by a radio transmitting station. (RR)

European Council of Postal and Telecommunications Administrations (CEPT)—The European regional organization of post and telecommunications organizations.

Extremely High Frequencies (EHF)—Radio frequencies between 30 and 300 GHz. (Millimetric)

Facsimile—A form of telegraphy for the transmission of fixed images, with or without half-tones, with a view to their reproduction in a permanent form. In this definition the term telegraphy has the same general meaning as defined in the Convention. (RR)

Fading—Changes in signal strength generally caused by variations in the propagation medium.

Fading, Flat—Fading in which all frequencies within a channel fade together.

Fading, Selective—Fading in which frequencies within a channel fade somewhat independently.

Frequency Agility—The ability to shift the frequency of a transmitter and/or receiver rapidly to achieve a desired effect.

Frequency Coordinator—An entity, recognized in a local or regional area by amateur operators whose stations are eligible to be auxiliary or repeater stations, that recommends transmit/receive channels and associated operating and technical parameters for such stations in order to avoid or minimize potential interference. (FCC Rules, Section 97.3)

(Glossary continued on next page)

Frequency Diversity—A form of diversity reception using simultaneous transmission on two or more frequencies to take advantage of the tendency that the different frequencies fade independently.

Frequency Division Multiple Access (FDMA)—A technique to permit multiple stations to access a given band by breaking the band into channels.

Frequency, Input—(repeater usage) The center frequency to which the repeater receiver is tuned.

Frequency Offset—The difference (in hertz) between two frequencies, such as: (a) between one used for transmitting and one used for receiving, (b) between a beat-frequency and the signal of interest in a receiver, or (c) between a specific transmission frequency and the nominal channel frequency.

Frequence Optimum de Travail (FOT)—A radio frequency 15% below the maximum usable frequency that will provide communications 90% of the days. (Also called *Optimum Working Frequency, OWF*.)

Frequency, Output—(repeater usage) The center frequency to which the repeater transmitter is tuned.

Frequency Registration—The act of making a written record of frequency assignment or usage in a central registry or in a published form for distribution.

Frequency Reuse—The ability to use a frequency in a different geographical location without harmful interference.

Frequency Sharing—The common use of the same portion of the radio frequency spectrum by two or more users where a probability of interference exists. (NTIA)

Frequency-Shift Telegraphy—Telegraphy by frequency modulation in which the telegraph signal shifts the frequency of the carrier between predetermined values. (RR)

Frequency Tolerance—The maximum permissible departure by the center frequency of the frequency band occupied by an emission from the assigned frequency or, by the characteristic frequency of an emission from the reference frequency. The frequency tolerance is expressed in parts in 10^6 or in hertz. (RR)

Full Carrier Single-Sideband Emission—A single-sideband emission without suppression of the carrier. (RR)

Geostationary Satellite—A geosynchronous satellite whose circular and direct orbit lies in the plane of the Earth's equator and which thus remains fixed relative to the Earth; by extension, a satellite which remains approximately fixed relative to the Earth. (RR)

Geostationary-satellite Orbit (GSO)—The orbit in which a satellite must be placed to be a geostationary satellite. (RR)

Geosynchronous Satellite—An earth satellite whose period of revolution is equal to the period of rotation of the Earth about its axis. (RR)

Guard Band—A narrow band of frequencies surrounding a (usually weak-signal) channel (or band) that are not to be used in order to prevent interference.

Guard Channel (or Frequency)—A (calling) frequency designated to be monitored continuously or within scheduled times.

Harmful Interference—Interference which endangers the functioning of a radionavigation service or of other safety services or seriously degrades, obstructs, or repeatedly interrupts a radiocommunication service operating in accordance with these Regulations.

Hectometric Waves—Radio frequencies between 300 kHz and 3 MHz. (MF)

High Band—VHF FM jargon for the 2-meter Amateur Radio band.

Highest Possible Frequency (HPF)—A radio frequency 15% above the maximum usable frequency, considered to be usable 10% of the days.

High Frequency (HF)—Radio frequencies between 3 and 30 MHz. (Decametric)

Image Frequency—(superheterodyne receiver) An undesired signal frequency that differs from the desired frequency by 2 times the intermediate frequency (IF).

In-band—Two or more frequencies within the same audio- or radio-frequency band.

Independent Sideband—Modulation of a carrier in which each sideband carries different intelligence.

Interference—Unwanted signals. (See *Harmful Interference*)

International Amateur Radio Union (IARU)—The federation of national Amateur Radio societies worldwide founded in Paris in 1925. It has three regional organizations: Region 1) Africa, Europe, the Commonwealth of Independent States, Middle East (excluding Iran) and Mongolia; Region 2) North and South America, including Hawaii and the Johnston and international Secretariat is located in Newington, Connecticut.

InterAmerican Telecommunication (CITEL)—The regional telecommunications organization of the Americas, with headquarters in Washington, DC.

International Frequency Registration Board (IFRB)—Formerly, an ITU agency that recorded frequecny assignments made by different countries.

Intersymbol Interference (ISI)—Extraneous energy from the signal in one keying interval that interferes with another keying interval. (Data transmission)

Inverse-Square Law—(radio) The power varies with distance inversely as the square of that distance (a basic law of free-space propagation which says that as the distance is doubled, the signal will be one fourth or 6 dB weaker).

Jamming—Intentional radiation of a signal for the purpose of imparing the usefulness of radio reception.

Key Clicks—Undesired switching transients beyond the necessary bandwidth of a Morse code radio transmission caused by improperly shaped modulation envelopes.

Kilometric Waves—Radio frequencies between 30 and 300 kHz. (LF)

Left-Hand (or Anti-Clockwise) Polarized Wave—An elliptically or circularly polarized wave, in which the electric field vector, observed in the fixed plane, normal to the direction of propagation, whilst looking in the direction of propagation, rotates with time in a left-hand or anti-clockwise direction. (RR)

Low Band—VHF FM jargon for the 6-meter Amateur Radio band. HF jargon for the 160, 80 or 40-meter bands.

Low Earth Orbit (LEO)—A nongeostationary satellite orbit close to the Earth.

Lowest Usable (High) Frequency (LUF)—The minimum radio frequency that will provide satisfactory ionospheric communication between two points 90% of the days.

Low Frequencies (LF)—Radio frequencies between 30 and 300 kHz. (Kilometric)

Machine—VHF FM jargon for a repeater.

Maximum Usable Frequency (MUF)—The highest radio frequency that will provide satisfactory ionospheric communication between two points at a given time 50% of the days.

Mean Power (of a radio transmitter)—The average power supplied to the antenna transmission line by a transmitter during an interval of time sufficiently long compared with the lowest frequency encountered in the modulation taken under normal operating conditions. (RR)

Medium Frequencies (MF)—Radio frequencies between 300 kHz and 3 MHz. (Hectometric)

Meteor Burst Communications—Communications by the propagation of radio signals reflected by ionized meteor trails. (NTIA)

Meteor Scatter—Propagation via ionized meteor trails.

Metric Waves—Radio frequencies between 30 and 300 MHz. (VHF)

Microcell—A small cell covering a few city blocks.

Microwaves—Radio frequencies above 1 GHz.

Millimetric Waves—Radio frequencies between 30 and 300 GHz. (EHF)

Modulation—Variation of one signal with another, usually to convey intelligence.

Multipath—Propagation of a radio signal via two or more distinct paths, such that signals (at certain frequencies and times) combine out of phase and produce a distorted output in the receiver.

Myriametric Waves—Radio frequencies between 3 and 30 kHz. (VLF)

Narrow-band—(1) A band of frequencies which are smaller than a reference bandwidth. (2) A band of frequencies which lie within the passband of a receiver or measuring device.

National Radio Quiet Zone—The area in Maryland, Virginia and West Virginia bounded by 39° 15'N on the north, 78° 30'W on the east, 37° 30'N on the south and 80° 30'W on the west. (FCC Rules, Section 97.3)

Necessary Bandwidth—For a given class of emission, the width of the frequency band which is just sufficient to ensure the transmission of information at the rate and with the quality required under specified conditions. (RR)

Net Control Station—A radio station responsible for real-time direction of a radio net, including maintenance of circuit discipline necessary to expeditious handling of traffic.

Net Manager—An individual responsible for the mission and resource management of a radio net.

NTIA—National Telecommunications & Information Administration of the US Department of Commerce.

Occupied Bandwidth—The width of a frequency band such that, below the lower and above the upper frequency limits, the mean powers emitted are each equal to a specified percentage B/2 of the total mean power of a given emission. Unless otherwise specified by the CCIR for the appropriate class of emission, the value of B/2 should be taken as 0.5%. (RR)

Open Repeater—A repeater where transient operators are welcome.

Optimum Working Frequency (OWF)—(Same as *Frequence Optimum de Travail, FOT*)

Out-of-band Emission—Emission on a frequency or frequencies immediately outside the necessary bandwidth which results from the modulation process, but excluding spurious emission. (RR)

Paging—One-way signaling to portable receivers for alerting or transmission of brief messages.

Pair (frequency)—Associated input and output frequencies.

Peak Envelope Power (of a radio transmitter)—The average power supplied to the antenna transmission line by a transmitter during one radio frequency cycle at the crest of the modulation envelope taken under normal operating conditions. (RR)

Period (of a satellite)—The time elapsing between two consecutive passages of a satellite through a characteristic point on its orbit. (RR)

Pilot—A signal to control the characteristics of a transmission. (For example, a 3.1-kHz pilot is used for automatic frequency control reference and to indicate the amount of expansion needed upon reception of an amplitude compandored single-sideband signal.)

Polarization—The alignment of the electric lines of force of a wave with respect to earth. If the lines are perpendicular to earth, the polarization is said to be vertical; if parallel to earth, horizontal. Polarization may also be elliptical or circular (right-hand or left-hand).

Polarization Discrimination—Reduction in interference between two systems using antenna polarizations that are orthogonal to each other. Examples: one antenna vertical, the other horizontal; one antenna right-hand circular, the other left-hand circular.

Polarization Diversity—A form of diversity reception using separate vertically and horizontally polarized antennas to take advantage of the tendency of vertical and horizontal components to rotate and thus fade independently.

Power Flux-Density (p.f.d.)—Power density per unit cross-sectional area per unit of frequency. Usually expressed in dB(W/m²/Hz), although bandwidths other than 1 Hz, such as 4 kHz or 1 MHz, may be seen.

Propagation—The motion of a radio wave.

Protection Ratio—The minimum value of the wanted-to-unwanted signal ratio, usually expressed in decibels, at the receiver input determined under specified conditions such that a specified reception quality of the wanted signal is achieved at the receiver output. (RR)

Rabbit—An unbalanced repeater with "big ears" (sensitive receiver) and "small mouth" (low-power transmitter).

Radiation—The outward flow of energy from any source in the form of radio waves. (RR)

Radiocommunication—Telecommunication by means of radio waves. (RR)

Radiocommunication Service—A service . . . involving the transmission, emission and/or reception of radio waves for specific telecommunication purposes. In these regulations, unless otherwise stated, any radiocommunication service relates to terrestrial radiocommunication. (RR)

Radio Regulations Board (RRB)—A component of the ITU That REcords Frequency assignments made by different countries.

Raster—The pattern of spacing of channels in a specific group or band.

Reciprocity—(1) (radio propagation) The ability of radio signals to traverse the same path in both directions with the same results. (2) (mobile radio) Transmitting and receiving subsystems that have equal capability to work over the same path loss. (See *Balanced System*.)

Reduced Carrier Single-Sideband Emission—A single-sideband emission in which the degree of carrier suppression enables the carrier to be reconstituted and to be used for demodulation. (RR)

Reference Frequency—A frequency having a fixed and specific position with respect to the assigned frequency. The displacement of this frequency with respect to the assigned frequency has the same absolute value and sign that the displacement of the characteristic frequency has with respect to the center of the frequency band occupied by the emission. (RR) (See also *Characteristic Frequency*.)

Roaming—Moving between cells. (cellular radio)

Roving—Moving between grid squares. (amateur VHF-microwave)

Scanner—A radio receiver that can be programmed to sample different frequencies in a systematic sequence.

Selectivity—The ability of a radio receiver to discriminate against frequencies outside the desired passband.

Service Area—The area served by a radio station (typically a broadcast transmitter, base station or repeater).

Shift (frequency)—The difference between the frequency of the mark and space frequencies in frequency-shift-keyed modulation.

Sideband—The group of frequencies above or below the main carrier frequency which contain the product of modulation.

Signal-to-Interference (S/I) Ratio—The ratio of the amplitude of the desired vs interfering signals.

Silent Period—A designated time during which all stations must cease transmission and listen for distress traffic.

Simulcast—Simultaneous transmission on two or more frequencies or modulation modes. (Also known as quasi-synchronous transmission.)

Signal-to-Noise Ratio—The ratio of the amplitude of the signal vs noise, usually expressed in dB.

Single-Sideband Emission—An amplitude modulated emission with one sideband only. (RR)

Site Noise—Noise, external to the receiver, that limits reception at a location—a composite of co-channel interference, power-line noise, machine emissions, and spurious emissions from transmitters.

Space Station—A station located on an object which is beyond, intended to go beyond, or has been beyond, the major portion of the Earth's atmosphere. (RR) An amateur station located more than 50 km above the Earth's surface. (FCC Rules, Section 97.3)

Space Telecommand—The use of radiocommunication for the transmission of signals to a space station to initiate, modify or terminate functions of equipment on an associated space object, including the space station. (RR)

Space Telemetry—The use of telemetry for the transmission from a space station of results of measurements made in a spacecraft, including those relating to the functioning of the spacecraft. (RR)

Spectrum Efficiency—The ratio of the information delivered to the spectrum resources (bandwidth, time and physical space) used.

Spectrum Engineering—The science of effective use of the radio spectrum to achieve the desired results economically.

(Glossary continued on next page)

Spectrum Management—Efficient and judicious use of the radio spectrum through application of techniques such as spectrum engineering, frequency coordination, monitoring and measurement, computer data base, and analysis.

Spectrum Metric—A unit of measure of spectrum-resource use.

Spectrum Occupancy—A frequency is said to be occupied when the signal strength exceeds a given threshold. Spectrum Occupancy is normally expressed in terms of probability or as a percent.

Splatter—A form of adjacent-channel interference caused by overmodulation on peaks.

Spread Spectrum—A signal structuring technique that employs direct sequence, frequency hopping or a hybrid of these, which can be used for multiple access and/or multiple functions. This technique decreases the potential interference to other receivers while achieving privacy and increasing the immunity of spread spectrum receivers to noise and interference. Spread spectrum generally makes use of a sequential noise-like signal structure to spread the normally narrowband information signal over a relatively wide band of frequencies. The receiver correlates the signals to retrieve the original information signal. (NTIA)

Spurious Emission—Emission on a frequency or frequencies which are outside the necessary bandwidth and the level of which may be reduced without affecting the corresponding transmission of information. Spurious emissions include harmonic emissions, parasitic emissions, intermodulation products and frequency conversion products, but exclude out-of-band emissions. (RR)

Station—One or more transmitters or receivers or a combination of transmitters and receivers, including the accessory equipment, necessary at one location for carrying on a radiocommunication service, or the radio astronomy service. Each station shall be classified by the service in which it operates permanently or temporarily. (RR)

Subband—A part of a (frequency) band, usually having the same purpose or operator privileges.

Super High Frequencies (SHF)—Radio frequencies between 3 and 30 GHz. (Centimetric)

Suppressed Carrier Single-Sideband Emission—A single-sideband emission in which the carrier is virtually suppressed and not intended to be used for demodulation. (RR)

Telecommand—The use of telecommunication for the transmission of signals to initiate, modify or terminate functions of equipment at a distance. (RR)

Telecommunication—Any transmission, emission or reception of signs, signals, writing, images and sounds or intelligence of any nature by wire, radio, optical or other electromagnetic systems. (RR)

Telegraphy—A form of telecommunication which is concerned in any process providing transmission and reproduction at a distance of documentary matter, such as written or printed matter or fixed images, or the reproduction at a distance of any kind of information in such a form. For the purposes of the Radio Regulations, unless otherwise specified therein, telegraphy shall mean a form of telecommunication for the transmission of written matter by the use of a signal code. (RR)

Telemetry—The use of telecommunication for automatically indicating or recording measurements at a distance from the measuring instrument. (RR)

Telephony—A form of telecommunication set up for the transmission of speech or, in some cases, other sounds. (RR)

Television—A form of telecommunication for the transmission of transient images of fixed or moving objects. (RR)

Time-Bandwidth Product—The duration of a transmission multiplied by the bandwidth of the signal—a measure of how much bandwidth is used for how long.

Time Division Multiple Access (TDMA)—A technique to permit multiple stations to access a given band by nonoverlapping time sequenced bursts.

Transmitter Power—The peak envelope power (output) present at the antenna terminals (where the antenna feed line, or if no feed line is used, the antenna, would be connected) of the transmitter. The term "transmitter" includes any external radio frequency power amplifier which may be used. Peak envelope power is defined as the average power during one radio frequency cycle at the crest of the modulation envelope, taken under normal operating conditions. (FCC Rules, Section 97.3)

Trunking—A channel-utilization scheme whereby two stations automatically can select an unused channel out of a group.

Tuning, Tune-Up—Emission of a signal for the purpose of adjusting a radio transmitter.

Ultra High Frequencies (UHF)—Radio frequencies between 300 MHz and 3 GHz. (Decimetric)

Unwanted Emissions—Consist of spurious emissions and out-of-band emissions. (RR)

Very High Frequencies (VHF)—Radio frequencies between 30 and 300 MHz. (Metric)

Very Low Frequencies (VLF)—Radio frequencies between 3 and 30 kHz. (Myriametric)

Vestigial Sideband (VSB)—The sideband of an AM transmission that has been substantially suppressed and which rolls off gradually with distance from the carrier frequency.

World Administrative Radio Conference (WARC)—Former term for ITU radio conferences, of which WARC-92 was the last. (See WRC)

World Radiocommunication Conference (WRC)—New term for ITU radio conferences starting in 1993.

Wideband—(1) A band of frequencies which are larger than a reference bandwidth. (2) A band of frequencies which exceed the passband of a receiver or measuring device.

Working Frequency—A frequency designated for exchange of communications after contact is made on a calling frequency or guard channel.

minimize interference. In 1927, the ITU Washington Radio Conference of 27 maritime states was the beginning of international regulation of Amateur Radio. At that conference, the Table of Frequency Allocations was first devised, and six harmonically related bands from 1.715 to 60 MHz were allocated to amateurs. Some of the amateur bands were revised in this and subsequent conferences held in Cairo (1938), Atlantic City (1947) and Geneva (1959), in consideration of the needs of other radio services.

In 1932, a Telegraph and Telephone Conference and a Radiotelegraph Conference in Madrid merged into a single International Telecommunication Convention. The Radio Regulations resulting from this conference explicitly prohibited amateurs from handling international third-party messages unless specifically permitted by the two countries involved. US amateurs may exchange third-party messages with the countries listed in Chapter 5.

In Atlantic City in 1947, radical changes were made in the organization of the Union. The ITU became a specialized agency of the United Nations—the oldest such agency—and ITU headquarters was transferred from Berne to Geneva.

At Nice in 1989, the ITU established a High Level Committee (HLC) to review the structure and functioning of the Union. They produced a report in 1991 entitled "Tomorrow's ITU: The Challenges of Change," that said the ITU must reorganize itself and adopt new working methods to keep pace with technological change. A new ITU Constitution and new Convention were approved at the Addition Plenipotentiary Conference, Geneva, 1992. The new organization and procedures were fully implemented by 1994. The new structure of the ITU is shown in Fig 2-1.

Plenipotentiary Conferences

The ITU has Plenipotentiary Conferences every four years. "Plenipotentiary" is a conference that is fully empowered to do business. The conferences determine general policies, review the work of the Union, revise the Convention if necessary, elect the Members of the Union to serve on the ITU Council, and elect the Secretary-General, Deputy Secretary-General, the Directors of the Bureaus and members of the Radio Regulations Board.

The Three ITU Sectors

The Telecommunication Standardization Sector (ITU-T) studies technical, operating and tariff questions, and issues recommendations relating to the telephone and data networks. World Telecommunications Standardization Conferences (WTSCs) are held every four years. Familiar recommendations issued by ITU-T include the V-series on modems and X-series on data protocols. The work of the sector is done through its Study Groups and the Telecommunication Standardization Bureau (TSB).

The Telecommunication Development Sector (ITU-D) primarily assists developing countries in modernizing their telecommunications infrastructure. Its World Telecommunication Development Conferences (WTDCs) meet every four years. It has two Study Groups and a Telecommunication Development Bureau (BDT).

The Radiocommunication Sector (ITU-R) revises the international Radio Regulations, studies technical and operational questions, and issues recommendations relating to radio communication. The organization of the Radiocommunication Sector is shown in Fig 2-2. Radiocommunication Bureau (BR) Director Richard C. Kirby, WØLCT, was succeeded by Robert W. Jones, VE3CTM, in January 1995. The League and the International Amateur Radio Union monitor the work of the Radiocommunication Sector primarily through participation in Study Group 8, the Radiocommunication Advisory Group (RAG), Radiocommunication Assembly (RA), Conference Preparation Meetings (CPMs) and World Radiocommunication Conferences (WRCs)—now held every two years. ARRL contributes to the United States preparation for these ITU meetings and conferences.

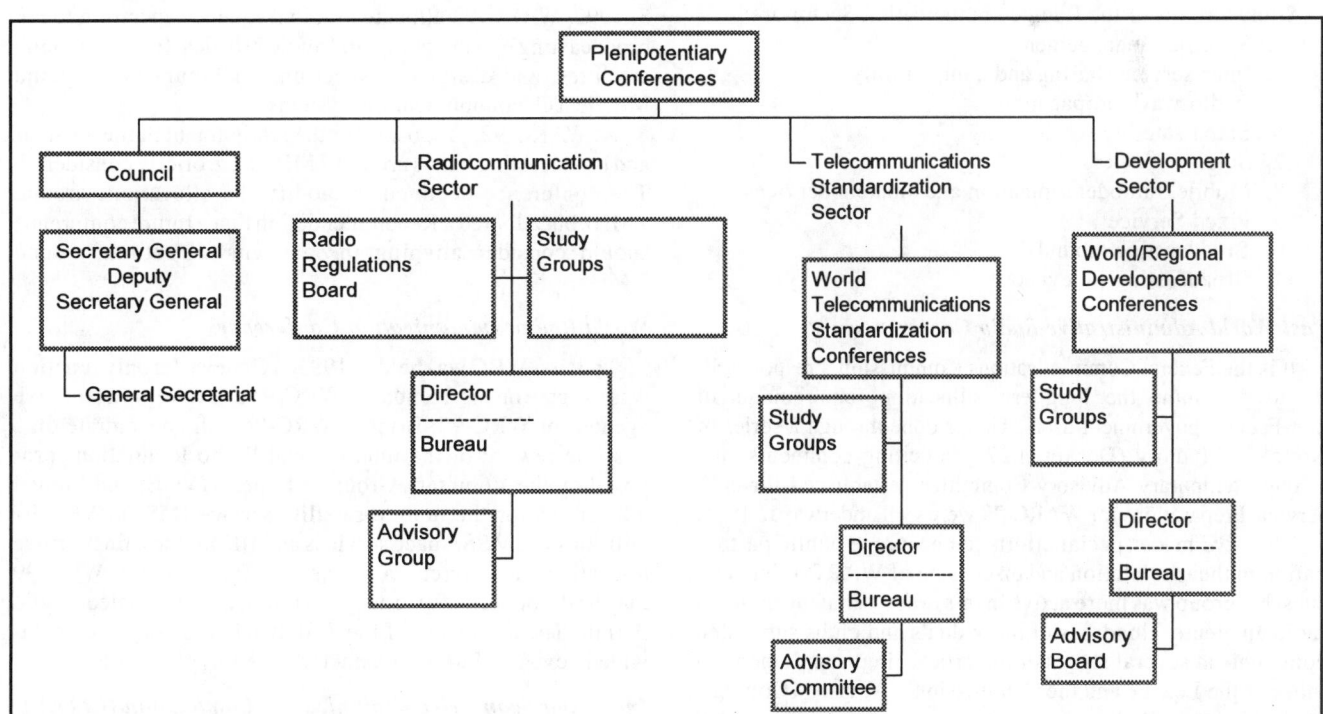

Fig 2-1—Organization chart of the International Telecommunication Union (ITU).

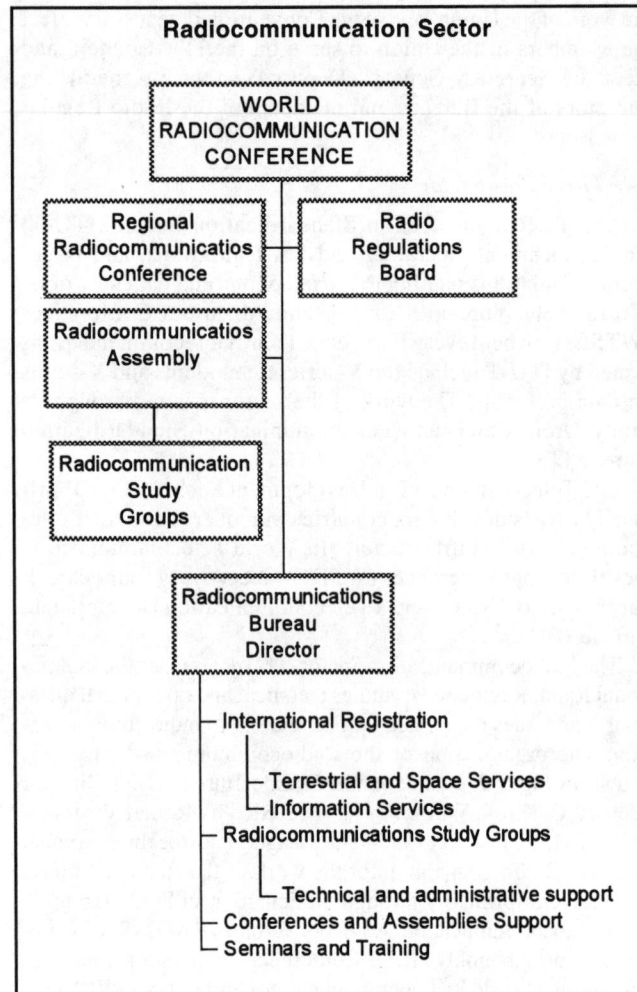

Radiocommunication Sector

- WORLD RADIOCOMMUNICATION CONFERENCE
  - Regional Radiocommunicatios Conference
  - Radio Regulations Board
  - Radiocommunicatios Assembly
  - Radiocommunication Study Groups
  - Radiocommunications Bureau Director
    - International Registration
      - Terrestrial and Space Services
      - Information Services
    - Radiocommunications Study Groups
      - Technical and administrative support
    - Conferences and Assemblies Support
    - Seminars and Training

Fig 2-2—Organization chart of the Radiocommunication Sector of the ITU.

Study Groups of the Radiocommunication Sector are:
1. Spectrum management
2. Inter-service sharing and compatibility
3. Radio-wave propagation
4. Fixed-satellite service
7. Science services
8. Mobile, radiodetermination and amateur services
9. Fixed Service
10. Broadcasting (sound)
11. Broadcasting (television)

Past World Administrative Radio Conferences

It is the Federal Communications Commission's responsibility to determine the long-term allocation requirements of non-Federal government users. This is done through a series of Notices of Inquiry (Docket 20271) soliciting comments, and through temporary Advisory Committees established for each service. Preparations for WARC-79 were well underway in 1974.

The FCC made special efforts to encourage public participation in the formulation and discussion of WARC-related issues. No group was more active in this process that the nation's radio amateurs. Hundreds of individuals and clubs submitted comments at several stages of the proceeding, supplementing those of the League and the Commission's Advisory Committee for Amateur Radio. The League's comments alone totaled nearly 200 pages.

The resultant 560-page Report and Order set forth the Commission's proposals. It was said to represent the combined thinking and agreement of the Commission and the various agencies of the Federal government. The document was forwarded to the Department of State with the recommendation that it be made the official US proposal for WARC-79.

IARU WARC-79 Objectives

Formal adoption of WARC objectives for Amateur Radio came at a series of meetings of the IARU Regions in 1975-76. In April 1976, IARU President Eaton had convened a meeting with representatives of the three IARU regional organizations and several national amateur Societies. One idea to emerge from this meeting was the need for a worldwide IARU working group to assist in guiding the preparatory work. This working group met several times over the following two years. One of their important products was a model position paper in English, French and Spanish that national Societies could use as a basis for their own formal submissions to their governments. Even more important were the efforts by IARU Member-Societies to get Amateur Radio representatives on their national delegations. This was successful in 17 countries.

The basic approach, basic strategies and organization used in preparing Amateur Radio for WARC-79 proved to be sound. When the Final Acts of the Conference were signed in December 1979, Amateur Radio emerged not only with its present bands intact, but with important new allocations at 10, 18 and 24 MHz as well! Further, there were relaxation of restrictions on some of the existing amateur bands, and additional segments were earmarked for amateur satellite work. WARC-79 was a tremendous team effort on the part of many, many persons who had the faith and the vision to work successfully toward a common goal. The degree of success is attributable almost entirely to the intensive preparation conducted by the IARU.

WARC-92

In Malaga-Torremolinos, Spain, February 3 through March 3, 1992, WARC-92 allocated radio frequency spectrum to HF broadcasting expansion, digital audio broadcasting and mobile satellites, and designated spectrum for Future Public Land Mobile Telecommunication Systems.

At WARC-92, proposals for the realignment of the amateur and broadcasting bands around 7 MHz were briefly considered. The conference declined to modify the allocations around 7 MHz but adopted a Recommendation that a future conference should consider aligning the amateur allocations around 7 MHz.

World Radiocommunication Conferences

The first WRC was held in 1993 in Geneva. Its only function was to agree on the agenda for WRC-95 and develop a tentative agenda for WRC-97. Briefly, WRC-95 will concentrate on a possible rewrite of the international Radio Regulations proposed by the Voluntary Group of Experts (VGE) and limited allocations for the mobile-satellite service (MSS). WRC-97 will consider MSS, space services and HF broadcasting service allocations, and agree on the agenda for WRC-99. WRC-99 may be the opportunity to adopt an International Amateur Radio Permit. The expansion of the 7-MHz band to 300 kHz worldwide is expected to be considered at WRC-95 or later.

Inter-American Telecommunication Commission (CITEL)

CITEL is the regional telecommunications organization for

the Americas. It is a permanent commission under the Organization of American States (OAS), with a secretariat in Washington, DC. The CITEL Assembly meets every four years, while its committees meet once or twice yearly. ARRL and IARU Region 2 are active participants in Permanent Consultative Committee no. 3 (PCC.III - Radiocommunications). Others are COM/CITEL (executive committee), PCC.I (telephone networks) and PCC.II (broadcasting).

TELECOMMUNICATIONS REGULATION IN THE UNITED STATES

The Communications Act of 1934, as amended, provides for the regulation of interstate and foreign commerce in communication by wire or radio. This Act is printed in Title 47 of the US Code, beginning with Section 151.

Federal Communications Commission

The FCC is responsible, under the Communications Act, for regulating all telecommunications except that of the Federal government. This includes the Amateur Radio Service and the Amateur-Satellite Service, which are regulated under Part 97 of the Commission's rules. (Part 97 is available in *The FCC Rule Book,* published by the ARRL.)

The new internal structure of the FCC in place in late 1994 is shown in Fig 2-3. The FCC components of most concern to Amateur Radio are:

- The five Commissioners
- Wireless Telecommunications Bureau
- Special Services Division
- Licensing Division, Gettysburg, PA
- International Bureau
- Office of Engineering and Technology
- Compliance and Information Bureau, Regional Offices, Field Offices and Monitoring System

Federal Government Telecommunications

The functions relating to assignment of frequencies to radio stations belonging to, and operated by, the United States Government were assigned to the Assistant Secretary of Commerce for Communications and Information (Administrator, National Telecommunications and Information Administration-NTIA) by Department of Commerce Organization Order 10-10 of May 9, 1978. See Figs 2-4 and 2-5 for NTIA organizational charts. Among other things, NTIA:

- coordinates telecommunications activities of the Executive Branch
- develops plans, policies and programs relating to international telecommunications issues
- coordinates preparations for US participation in international telecommunications conferences and negotiations
- develops, in cooperation with the FCC, a long-range plan for improved management of all electromagnetic spectrum resources, including jointly determining the National Table of Frequency Allocations.
- conducts telecommunications research and development.

Interdepartment Radio Advisory Committee

The Interdepartment Radio Advisory Committee (IRAC) traces back to a mutual agreement of Government departments on June 1, 1922. Its status is as defined on December 10, 1964, and continued per Executive Order 12046 of March 27, 1978. The NTIA appoints the officers of the IRAC and chairpersons

Fig 2-3—Organization chart of the Federal Communications Commission (FCC).

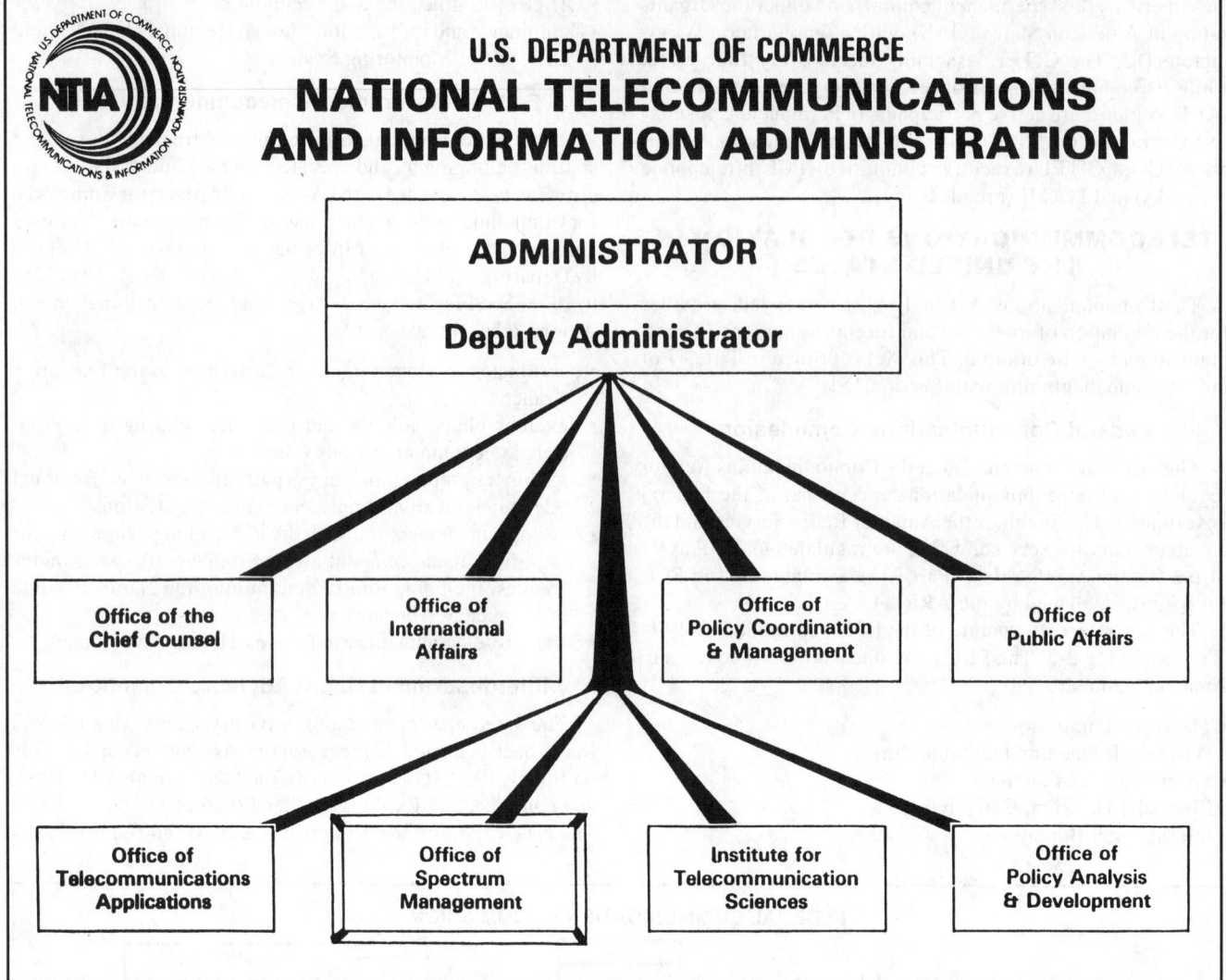

U.S. DEPARTMENT OF COMMERCE
NATIONAL TELECOMMUNICATIONS
AND INFORMATION ADMINISTRATION

Fig 2-4—Organization chart of the National Telecommunications and Information Administration (NTIA).

of its subcommittees. It is composed of representatives appointed by the following member departments and agencies: Agriculture, Air Force, Army, Coast Guard, Commerce, Energy, Federal Aviation Administration, Federal Emergency Management Agency, General Services Administration, Health and Human Services, Interior, Justice, National Aeronautics and Space Administration, National Science Foundation, Navy, State, Treasury, US Information Agency, US Postal Service and Veterans Administration.

The IRAC assists the Assistant Secretary of Commerce in assigning frequencies to US government radio stations and in developing and executing plans, procedures, and technical criteria. The FCC must coordinate with IRAC any actions that might affect the Federal government's use of the radio spectrum. It is reasonable to expect that FCC dockets involving such sharing will take a bit longer to consider than those solely under the FCC's wing.

Bibliography

Baldwin and Sumner, "The Geneva Story," *QST,* Feb 80, pp 52-61.

Feller, *Planning an Electromagnetic Environment Model for Spectrum Management* (Washington, DC: FCC, June 26, 1981).

ITU, *Radio Regulations* (Geneva, 1990), as amended by the Final Acts of WARC-92.

Kleinschmidt and Rinaldo, "WARC-92: What it Means to You," *QST,* Jun 91, p 16.

NTIA, *Manual of Regulations and Procedures for Federal Radio Frequency Management* (Washington, DC: Department of Commerce, May 1989).

Palm, *The FCC Rule Book*, ARRL, Newington, CT.

Rinaldo, "League WARC-92 Preparations Continue," *QST,* May 90, p 55.

Rinaldo, "WARC-92: Inside the United States Delegation," *QST,* May 92, p 28.

Sumner, "FCC WARC Proposals Finalized, "*QST,* Feb 79, p 55.

Sumner, "ITU Lays Technical Foundation for WARC-79," *QST,* Mar 79, p 56.

Sumner, "WARC-92 Draft Proposals Released," *QST*, Nov 90, p 42.

Sumner, "WARC Agenda Released," *QST*, Aug 90, p 9.

Sumner, "WARC-92 Finds Room for New Radio Services," *QST*, May 92, p 25.

AMATEUR FREQUENCY ALLOCATIONS AND BAND PLANS

There are bands of frequencies allocated to the Amateur

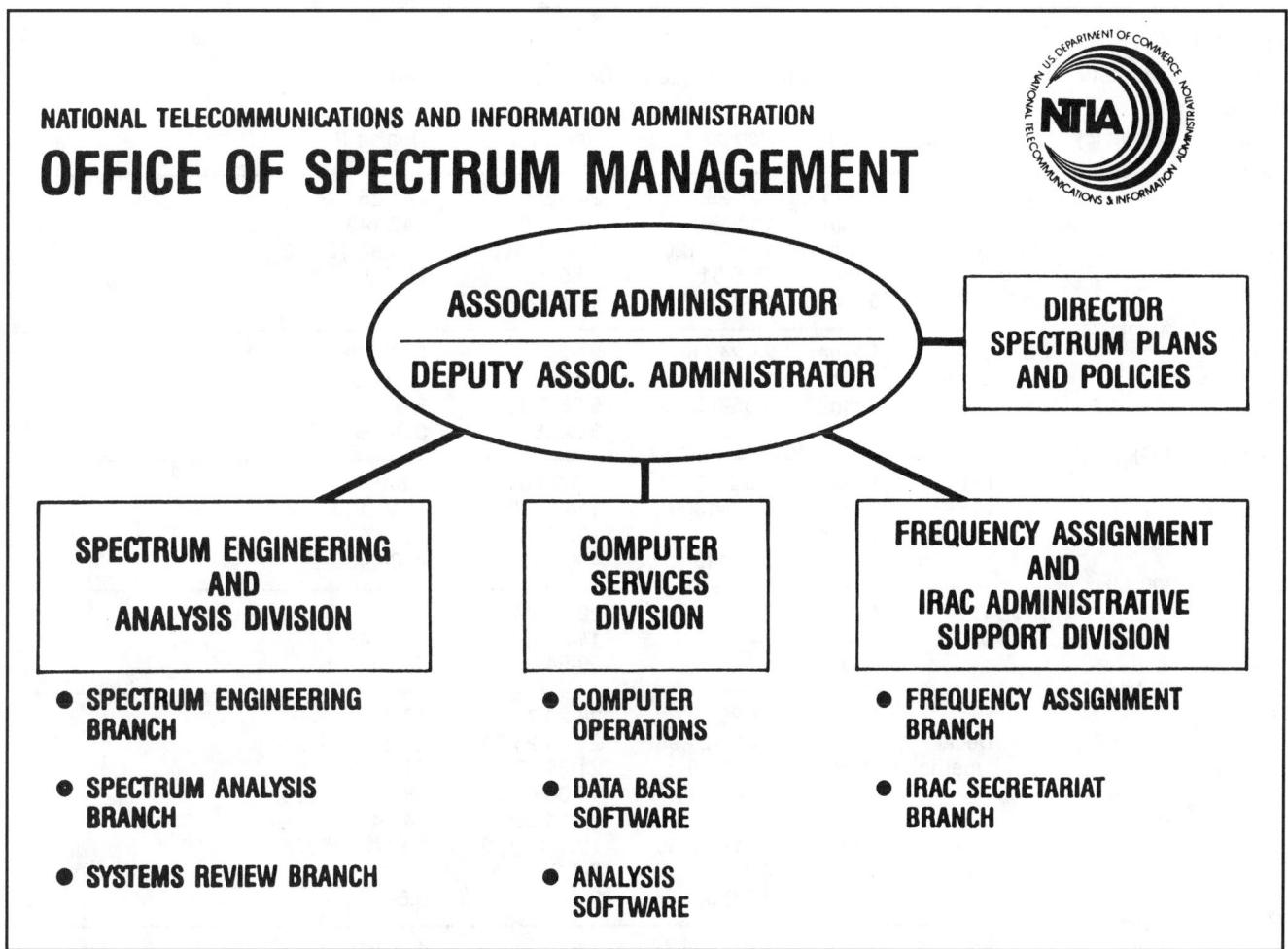

Fig 2-5—Organization chart of the NTIA Office of Spectrum Management.

Service extending from 1800 kHz to over 300 GHz. Table 2-1 will give you an overview of the radio spectrum, some nomenclature and the Amateur Service bands allocated in the ITU *Radio Regulations*.

Table 2-4 is a more-detailed, band-by-band picture of the radio spectrum. The part labeled "Allocations" is an extract of the ITU *Radio Regulations* Table of Frequency Allocations. All footnotes that pertain to the Amateur Service and other services sharing the same frequency bands are included. The part labeled "Amateur Radio Band Plans" includes the band plans formally adopted by the IARU Regions and the ARRL. In most of the high-frequency bands there are no ARRL band plans that were adopted by the ARRL Board of Directors.

There is a tendency for band plans to lag reality. This is due in part to the time needed to research, invite and digest comment from amateurs, arrive at a mix that will serve the diverse needs of the amateur community, and adopt a formal band plan. This is a process that can take a year or more on the national level and a similar period in the IARU. Nevertheless, new communications modes or popularity of existing ones can make a year-or-two-old band plan look obsolete. Such revolutionary change has taken place recently with the popularity of packet radio, particularly in the 20 and 2-meter bands. Changes of this magnitude cause the new users to scramble for frequencies and some of the existing mode users to draw their wagons in a circle. The national societies, their staffs and committees, and the IARU have the job of sorting out the contention for various frequencies and preparing new band plans. Fortunately, we

have not exhausted all possible ways of improving our management of the spectrum so that all Amateur Radio interests can be accommodated.

The 160-Meter Band

Those who have a fascination for 160-meter DX should take a close look at Chapter 5. It contains a country-by-country listing of the frequency allocations in this band. The basic problem with allocations in this band has been competition with the Radiolocation Service. New pressures are possible as a result of planned expansion of the medium-frequency broadcast band in the 1605-1705 kHz range.

The 80-Meter Band

While US amateurs enjoy the use of 3500-4000 kHz, not all countries allocate such a wide range of frequencies to the 80-meter band. There are fixed, mobile and broadcast operations, particularly in the upper part of the band.

The 40-Meter Band

The 40-meter band has a big problem: International broadcasting occupies the 7100-7300 kHz (41-meter) band in many parts of the world. During the daytime particularly when sunspots are high, broadcasting does not cause much interference to US amateurs. However, at night, especially when sunspot activity is low, the broadcast interference is heavy. Some countries allocate only the 7000-7100 kHz band to amateurs. Others, particularly in Region 2, allocate

Table 2-1

The Electromagnetic Spectrum with Amateur Service Frequency Bands by ITU Region

Wave-length	Frequency		Nomen-clature	Metric Band	Amateur Radio Bands by ITU Region		
					Region 1	Region 2	Region 3
1 mm	300 GHz						
		|	EHF	1 mm	241-250	241-250	241-250
		|	Milli-	2 mm	142-149	142-149	142-149
		M	metric	2.5 mm	119.98-120.02	119.98-120.02	119.98-120.02
		i		4 mm	75.5-81	75.5-81	75.5-81
		c		6 mm	47-47.2	47-47.2	47-47.2
1 cm	30 GHz	r					
		o	SHF	1.2 cm	24-24.25	24-24.25	24-24.25
		w	Centi-	3 cm	10-10.5	10-10.5	10-10.5
		a	metric	5 cm	5.65-5.85	5.65-5.925	5.65-5.85
		v		9 cm		3.3-3.5	3.3-3.5
10 cm	3 GHz	e					
		s	UHF	13 cm	2.3-2.45	2.3-2.45	2.3-2.45
		|	Deci-	23 cm	1240-1300	1240-1300	1240-1300
		|	metric	33 cm		902-928	
				70 cm	430-440	430-440	430-440
1	300 MHz						
			VHF	1.25 m		222-225	
			Metric	2 m	144-146	144-148	144-148
				6 m		50-54	50-54
10	30 MHz						
			HF	10 m	28-29.7	28-29.7	28-29.7
			Deca-	12 m	24.89-24.99	24.89-24.99	24.89-24.99
			metric	15 m	21-21.45	21-21.45	21-21.45
				17 m	18.068-18.168	18.068-18.168	18.068-18.168
				20 m	14-14.350	14-14.350	14-14.350
				30 m	10.1-10.150	10.1-10.150	10.1-10.150
				40 m	7-7.1	7-7.3	7-7.1
				80 m	3.5-3.8	3.5-4	3.5-3.9
100	3 MHz						
			MF	160 m	1.81-1.85	1.8-2	1.8-2
			Hecto-metric				
1000	300 kHz						
			LF				
			Kilo-metric				
10,000	30 kHz						
			VLF				
			Myria-metric				
100,000	3 kHz						

Note: This table should be used only for a general overview of where ITU Amateur Service and Amateur-Satellite Service frequencies fall within the radio spectrum. They do not necessarily agree with FCC allocations; for example, the 70-cm band is 420-450 MHz in the United States. Refer to subsequent tables for more specific information.

7100-7300 kHz as well, which at times is subject to interference. The result is that there is a great demand for frequencies in the 7000-7100 kHz segment. The effect is that there are two band plans overlaid on each other: ours is spread out over 7000-7300 kHz and a foreign one that compresses everything into 7000-7100 kHz.

At the IARU Region 2 conference in Curacao in September 1992, a new 40-meter band plan was adopted to align to the extent possible the digital segments despite the regional differences in allocations. See the Region 2 band plans later in Table 2-4.

The 30-Meter Band

The 30-meter band is an excellent band for CW and digital modes. The only problem is that US amateurs must not cause

harmful interference to the fixed operations outside the US. This restricts transmitter power output to 200 watts and is one reason for not having contests on this band.

The 20-Meter Band

The workhorse of DX is undoubtedly the 20-meter band. It offers excellent propagation to all parts of the world throughout the sunspot cycle and is virtually clean of interference from other services.

The 17-Meter Band

The 18068-18168 kHz band was awarded to amateurs on an exclusive basis, worldwide, at WARC-79. It was made available for amateur use in the US in January 1989.

The 15 and 12-Meter Bands

The 21000-21450 and 24890-24990 kHz bands are excellent for DX during the high part of the sunspot cycle. They also offer some openings throughout rest of the sunspot cycle.

The 10-Meter Band

This is an exclusive amateur band worldwide. Its popularity has risen sharply as a result of FCC PR Docket 86-161, better known as Novice Enhancement.

The VHF and Higher Bands

The 6-meter band is not universal, but the trend seems to be toward allocating it to amateurs as TV broadcasting vacates the 50-54 MHz band. Recent new 6-meter privileges for certain European amateurs adds to the worldwide interest in this band. It is also excellent for amateur exploitation of meteor-scatter communications using various modes including packet radio.

Two meters is heavily used throughout the world for CW, EME, SSB, FM and packet radio. Satellites occupy the 145.8-146 MHz segment.

Since August 1991, the 1.25-cm band has been reduced to 222-225 MHz, which is heavily used by repeaters. As it proved difficult to accommodate the former users of the 220-222 MHz band, the FCC has granted access to a shared band just below 220 MHz. Contact the ARRL for more information.

The 70-cm band is prime UHF spectrum. The 430-440 MHz band is virtually worldwide, whereas the 420-430 MHz and 440-450 MHz bands are not. Frequencies around 432 are used for weak-signal work, including EME, and the 435-438 MHz band is for amateur satellites. The 70-cm band is the lowest-frequency band that can be used for fast-scan television and spread spectrum emissions.

The 33-cm band (902-928 MHz) is widely shared with other services, including Location and Mounting Service (LMS), which is primary, and ISM (industrial, scientific and medical) equipment applications. A number of low-power devices including spread spectrum local area networks operate in this band under Part 15 of the FCC's Rules.

The 1240-1300 MHz band is used by amateurs for essentially all modes, including FM and packet. By regulation, the 1260-1270 MHz segment may be used only in the Earth-tospace direction when communicating with amateur satellites.

While the Amateur Service has a secondary allocation in the 2300-2450 MHz band in the international tables, in the United States, the allocation is 2390-2400 MHz and 2402-2417 MHz primary, and 2300-2310, 2400-2402 and 2417-2450 secondary. Most of the weak-signal work in the US takes place around 2304 MHz, while much of the satellite activity is in the 2400-2402 MHz segment.

The remaining microwave and millimeter bands are the territory of amateur experimenters. It is important that the Amateur Service and the Amateur-Satellite Service use these bands, and contribute to the state-of-the-art in order to retain them.

FREQUENCY COORDINATION

From the earliest days of Amateur Radio, hams simply listened on a frequency to see if it was in use, and if not, went ahead and transmitted on that frequency. Amateurs operating on HF still use this simple listen-before-transmit procedure of randomly *assigning* themselves a frequency for use over a short period of time. This simple procedure has served us well over the years, but is no longer the only "rule of the road."

HF CW and phone nets historically have had problems finding frequencies on which to conduct their regular operations. Some have come up with strong arguments supporting the need for nets to have priority over a two-station contact on the same frequency. The rationale is that nets have many stations that would be difficult to move to another frequency and that the needs of the many translate to more public good than just a few ragchewers. The other side of the argument is that the two stations have a right to conduct their communication. The principle of priority becomes a bit clearer when one set of users has emergency traffic and the other does not. Nevertheless, no one (including the FCC and the ARRL) has expressed willingness to say who gets a specific HF channel and who does not. This is rooted in the knowledge that stations and their needs come and go hour by hour, that ionospheric propagation changes hour by hour, and that most hams can bend a little in frequency and time to let both sets of users carry on.

Over time it became necessary to introduce a mild spectrum-management technique of frequency registration to let hams know where and when nets normally operate. This information is published annually in the ARRL *Net Directory*. This appearance of a listing in the *Net Directory* does not indicate an *assignment* by any authority. It is particularly useful when researching new frequencies for net or contest operations. It is of only limited use to individuals wishing to conduct random contacts, as the listen-before-transmit technique is more practical.

The Days Before Repeater Coordination

Two-meter FM repeaters became the rage in the 1960s. Just put your repeater on 34/94 (146.34 MHz input, 146.94 MHz output) and there was no problem... until someone else wanted to do the same thing nearby. "We got here first, so we are in the right; they are in the wrong. Why don't *they* find another frequency pair? What pair? That's *their* problem. Let's send two

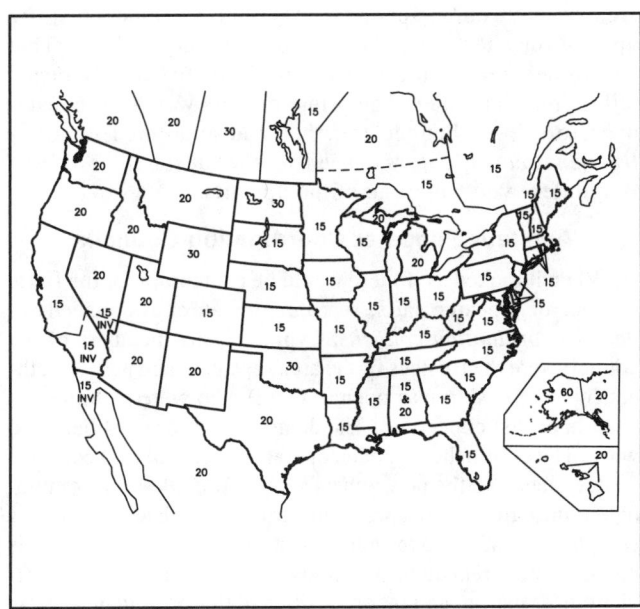

Fig 2-6—This map shows channel spacing in kHz. Please check with your Regional Frequency Coordinator for more information.

of our 250-pound guys over there and cut their coax!" No, that's not the way.

The early repeater contention problems were solved by having the latecomers find new frequencies. New transceivers typically were supplied with crystals for (146) 34, 52, 76 and 94. There were repeaters with inputs on 34 and outputs on 76 to avoid local simplex operation on 94. But, 76 became paired with a 16 input, and the 600-kHz offset was well on its way to becoming standard. Much of the early spadework in repeater band planning was done by the Texas VHF Society.

By early 1976, the repeater count in the US was more than 2000. The ARRL VHF Repeater Advisory Committee (VRAC) agreed on band plans that named repeater pairs in the 10-meter, 6-meter, 2-meter, 1.25-meter and 70-cm bands.

Many early tube-type amateur FM transceivers had seen service in the back of a police car or taxi. The land mobile services had been using 15-kHz deviation and had spaced their channels in a 30-kHz raster (30 kHz between channel center frequencies) but were in the process of converting to 5-kHz deviation and 15-kHz channel spacing. Amateurs followed suit by first setting up 30-kHz channel spacing with so-called *wideband* FM (15-kHz) then changing to 15-kHz channel spacing, which in that context was known as *narrow-band* FM. The new channels created in between the original 30-kHz spaced channels were called *splinter channels*. The change was accomplished in some older rigs by backing off on the transmitter deviation and somehow tightening up the receiver bandwidth. The problem was solved by new solid-state rigs from Japan, first with a few crystals and room for more, later with synthesizers.

Synthesized transceivers were able to tune to any frequency within the 2-meter band in 5-kHz steps. This opened up another possibility: It was now feasible to try some different channel spacings without causing people to buy several hundred dollars in new crystals. The result today is parts of the US use different spacings. A map showing 2-meter channel spacing throughout the US and Canada is shown in Fig 2-6.

This spreading out throughout the 2-meter and other bands to accommodate more repeaters had its limits, of course. It was clear that there would come a day when all the channels would be full, particularly in the metropolitan areas. What would happen then? Would latecomers just fire up new repeaters on top of established ones? Chaos! That was to be avoided at all costs, so amateurs actually began talking to each other. The prime movers were the repeater operators; they banned together and appointed *repeater frequency coordinators* to keep out the "upstarts." This was fine unless you were one of the upstarts.

Amateur Repeater Coordination Councils

When it looked as if there would be contention for the finite number of repeater pairs, repeater operators established *repeater coordination councils*. Typically, the members of the council were the trustees of the area repeaters and perhaps others (if membership was open to them) who were planning repeaters or just concerned. The democratic processes began to work. The established repeater operators were able to keep their frequencies, and the newcomers somehow could be accommodated until all the pairs were full. Most areas had a *frequency coordinator,* who had the administrative job of keeping records on all existing repeaters. The frequency coordinator kept sufficient information, and theoretically had the skills to determine whether a proposed repeater would interfere with existing repeaters. If, after polling established repeater operators, there appeared to be no unacceptable interference, the frequency

coordinator would *coordinate* the new repeater. This added a new term to the Amateur Radio lexicon: *coordinated repeater.*

What if a "noncoordinated" repeater owner is not satisfied with the decision made administratively by the coordinator? There are ways to solve this problem: The owner or trustee could become a member of the council—perhaps even volunteer for president. This way he or she could try to influence the frequency coordinator into favorable action. Never fire a frequency coordinator who is doing the job; where could you find another competent person to put in long hours for no pay?

But wait a minute! Is Repeater Coordination Council X legitimate? Don't they just represent the existing repeater owners, not prospective ones, not the repeater users, and not the other hams who might want to use the same frequencies for other purposes? Some have been challenged. Some have come and gone. A few frustrated individuals have started rival coordination entities. For a discussion of what happens when a frequency coordinator "doesn't," see the FM/RPT column in September 1986 *QST.*

Okay, "Let's ask the FCC," someone says. The FCC is not enthusiastic about this (and it is usually undesirable and not recommended from the amateur point of view). How would you like to be called in to settle a fight? After the FCC's in-boxes are empty and its phones stop ringing, you know what the bottom line will be: "Why don't you guys get together and work this thing out. *OUT."*

FCC PR Docket 8S-22

Coordinating one's choice of frequencies for a repeater has never been a licensing requirement. By early 1985 demand for frequencies had grown to the point where even an FCC predisposed to prefer "unregulation" decided something needed to be done. In February 1985, after many formal comments and replies the FCC issued PR Docket 85-22, which stated in part: ".... Most of the reported cases of amateur repeater-to-repeater interference appear to involve one or more non-coordinated repeaters. In the past two years we have had to resolve amateur repeater-to-repeater interference disputes in which at least one non-coordinated repeater was involved and in which the parties to the dispute could reach no amicable solution. We attribute the growing number of instances of amateur repeater-to-repeater interference and the need for increased intervention in these matters to the mounting pressure, which develops as the desirable repeater frequencies become fully assigned.

".... The commission should not recognize a single entity, such as a national frequency coordinator for amateur repeater operation. Such coordination activities should be performed by local, or regional, volunteer frequency coordinators with appropriate support to these coordinators to be provided by the League ...."

This FCC docket became the rationale behind the current Part 97 rules that affect amateur repeater coordination.

The League's Role in Coordination

While the ARRL successfully argued before the FCC against a national frequency coordinator, there remains a significant role for the League in the coordination process. For one thing, the ARRL has for years published the *Repeater Directory.* By 1994 it had over 20,000 entries including repeaters, packet radio and beacons.

The data base used in preparation of the *Directory* is accessible to frequency coordinators via a toll-free phone line. The design of the National Repeater Data Base (NRDB) was proposed by the Mid-Atlantic Repeater Council (TMARC), and was placed in service as designed in April of 1989—just in time for its "trial by fire"—the 1990 edition of the *Directory*!

After a year of use, it was determined that some changes and streamlining were necessary for the system. A new data base was designed by the Mid-America Coordination Council (MACC) in June of 1991. This "user-friendly" system is running smoothly and allows coordinators to actively use the data for coordination of repeaters near bordering states, thus lessening the chance for interference.

The fields used by the coordinators are as follows:

Band	29-52-144-222-440-ATV-902-1240-PKT-BEA
Output	146.94, etc
Input	146.34, etc
State	2-letter designators
Location	City/town
Call	Amateur call sign of repeater
Sponsor	Club or sponsoring organization
Last Update	Current date
Source	Coordinator or Council
Geographic Area	County
Notes	Open-Linked-Autopatch-etc

With the fields set up in this manner the coordinators have access to all pertinent coordination data, and the *Repeater Directory* can make use of the "general" information for publication annually.

PROPAGATION BEACONS

NCDXF Beacons

The worldwide beacon network on 14.100 MHz is probably the most useful group of beacons on any HF band. There are nine beacons transmitting in sequence every 10 minutes. Each beacon transmits for 1 minute, and the last minute in each 10-minute sequence is silent. The complete sequence is shown in Table 2-2.

Table 2-2

NCDXF Beacons and 10-minute Sequence
[Entire sequence repeats every 10 minutes]

Minute	Call	Location
00	4U1UN/B	United Nations, New York City
01	W6WX/B	Stanford University, CA
02	KH6O/B	Honolulu Community College, HI
03	JA2IGY	JARL, Mt Asama, Japan
04	4X6TU/B	Tel Aviv University, Israel
05	OH2B	Helsinki Technical University, Finland
06	CT3B	ARRM (Madeira Radio Society), Madeira Island
07	ZS6DN/B	Transvaal, South Africa
08	LU4AA/B	Radio Club Argentino, Buenos Aires
09	Silent	

[Note: Minutes 09-14 tentatively planned for a 15-minute sequence.]

The unique, and very interesting feature, of these beacons is that the transmitter power is varied from 100 watts to 1/10 of a watt in four steps during every one-minute transmission period. The transmission starts with QST DE 4u1N/B (in New York). Then four continuous tones are sent, for about nine seconds each, at 100, 10, 1 and 1/10 of a watt. Then the beacon signs off at full power (100 watts). If you listen for a while, you'll be pleasantly surprised at how effective 1/10 of a watt into a vertical can be.

Besides monitoring, the beacons also can be used for several do-it-yourself practical observations or projects. The beacons:

• provide an "in-band" time/frequency standard.
• are a means of comparing antennas or receivers by switching back and forth during the same nine-second dash. (Check DX versus short-range characteristics of different antennas. When a beacon signal is steady, rotate the beam antenna [from center to side] to check beamwidth.)
• enhance propagation investigation. (Record the lowest signal level copied from each beacon [including "no copy"] hour-to-hour and day-to-day.) How do your observations and interpretations of the beacon's reception compare with published propagation forecasts, or correlate with various A and K indices, WWV broadcasts, sunspot activity or just generally observed band activity? Is there any correlation at all? Can beacon signals be used to improve forecasts of conditions?

The beacon net was organized and financed by the Northern California DX Foundation (NCDXF). The current network and the underlying technical approach is described in *The NCDXF/IARO International Beacon Network*. Part 1 appeared in *QST*, October 1994, page 31 and Part 2 in November 1994, page 49. NCDXF asks all listeners making regular observations to send them to Al Lotze, W6RQ, who analyzes such reports and correlates them with solar and geomagnetic data. Send reports to W6RQ at 46 Cragmont Ave, San Francisco, CA 94116 USA.

28-MHz and Higher Beacons

The IARU is in the process of improving the International Beacon Project (IBP). The aim is to enhance the power and quality of those stations selected to receive IARU sponsorship and, with the cooperation of NCDXF, to extend the power-stepping time-shared networks to 15 and 10 meters. There will also be regional networks and continuous-duty stations on 10 meters to investigate the special properties of that band. Details of new stations and frequencies will be announced from time-to-time in *QST*. The old 10-meter beacon subband (28200-28300) is being phased out in favor of the new 28190-28225 kHz subband. See Table 2-3 for an all-source listing of amateur propagation beacons. Note: You are urged to avoid transmitting on beacon frequencies; interference reduces the value of the beacon service to other amateurs.

Table 2-3
Consolidated Beacon List

[This list is correct and complete to the best of our knowledge. Please send any updates to the attention of the Editor, ARRL Operating Manual.]

Freq	Call Sign	Location	Grid Square	Emission	Ant	Pol	ERP	Remarks
7.03	JA1IRL	Tokyo		A3J				
14.1	4U1UN/B	UN New York, NY	FN30	A1A	Omni		100	T+0 min
14.1	4X6TU/B	Tel Aviv		A1A	Omni		100	T+4 min
14.1	CT3B	Funchal		A1A	Omni		100	T+6 min
14.1	JA2IGY	Mt Asama		A1A	Omni		100	T+3 min
14.1	KH6O	Kane'ohe, HI	BL11cj	A1A	Omni		100	T+2
14.1	LU4AA/B	Buenos Aires		A1A	Omni		100	T+8 min
14.1	OH2B	Helsinki		A1A	Omni		100	T+5 min
14.1	W6WX/B	Stanford Univ, CA		A1A	Omni		100	T+1 min
14.1	ZS6DN/B	Transvaal		A1A	Omni		100	T+7 min
24.9205	KK6TG	St Helena	CM88qq	F1A	Omni	Horz	0.03	
28.05	PY2GOB	Sao Paulo			Omni	Vert	15	
28.175	VE3TEN	Ottawa, ON			Omni	Vert	10	Cont
28.191	5B4ZL	Cyprus			GP		10	
28.191	VE6YF	Edmonton, AB					10	
28.195	IY4M	Bologna	JN54qk	A1A	Omni	Vert	20	Robot
28.2	GB3SX	Crowborough	JO01bb	F1A	N/S	Horz	8	H+25&55
28.2	KF4MS	St Petersburg, FL			Omni	Vert	75	Cont
28.2	KG5YB	Tyler, TX	EM22	A1A	Vert	Vert	3	Cont
28.2	W6WX	California					100	
28.201	LU8ED	Argentina						
28.2015	EA7PS	Spain			GP		1	
28.202	KE5GY	Arlington, TX				Vert	5	
28.2025	ZS5VHF	Natal			Omni	Vert	5	
28.204	DLØIGI	West Germany			Dipl	Vert	100	
28.205	DLØIGI	Mt Predigtstuhl		F1A	Omni	Vert	100	Cont
28.206	KJ4X	Pickens, SC				Vert	2	
28.2075	KE4NL	Sarasota, FL		A1A		Vert	5	
28.2075	W8FKL	Venice, FL			Omni	Vert	10	Cont
28.208	WA1IOB	Marlboro, MA				Vert	75	
28.209	NX2O	Staten Island, NY	FN30	A1A	GP	Vert	10	Cont
28.21	3B8MS	Mauritius			Omni	Vert		Cont
28.21	K4KMZ	Elizabethtown, KY			Omni	Vert	20	Intrmt
28.21	KC4DPC	Wilmington, NC	FM14	A1A	Dpl	Horz	4	Cont
28.2125	EA6RCM	Palma De Mallorca			Yagi		4	
28.2125	ZD9GI	Gough Island		F1A	Omni	Vert		
28.215	GB3RAL	Slough	IO91rl	F1A	Omni	Vert	20	Cont
28.215	KA9SZX	Champaign, IL	EN50	A1A	Omni	Vert	1	Cont
28.215	LU4XI	Cape Horn						
28.216	GB3RAL	Slough Berkshire			GP		20	
28.2168	N6UGR	California					10	
28.2175	VE2TEN	Chicoutimi, PQ					4	
28.2175	WB9VMY/5	Oklahoma City, OK	EM15	A1A	Omni	Vert	2	Cont
28.218	W8UR	Mackinac Island, MI	EN75	A1A	Omni	Vert	0.5	Cont
28.219	PT8AA	Rio Branco			GP		5	
28.22	5B4CY	Zyyi		F1A	Omni	Vert	26	Cont
28.22	KB9DJA	Mooresville, IN		A1A	GP	Vert	10	Cont
28.22	LU4XS	Tierra Del Fuego			GP		2	
28.221	PY2GOB	Sao Paulo				Vert	15	
28.2215	K5PF	Apex, NC	FM05	A1A	Omni	Vert	16	1200Z-0300Z
28.222	W9UXO	Chicago, IL			Omni	Vert	10	Cont
28.2225	HG2BHA	Tapolca		F1A	Omni	Vert	10	Cont
28.2245	KA9UEX	West Bend, IN				Vert	2	
28.225	EA1EVE	Spain					1	
28.225	KW7Y	Everett, WA		A1A	Omni	Vert	4	Cont
28.225	PY2AMI	Sao Paulo			GP		5	
28.227	FX5TEN	Toulouse					8	
28.2275	EA6AU	Mallorca		A1A	Omni	Vert	10	Cont
28.229	EA6AU	Mallorca			GP		10	
28.23	N2ECB	Springfield, NJ			Yagi		25	
28.23	ZL2MHF	Mt Climie		F1A	Omni	Vert	50	Cont
28.231	N4LMZ	Mobile, AL			GP		2	
28.232	W7JPI	Sonoita, AZ			Yagi	Horz	5	Cont
28.2325	KD4EC	Jupiter, FL			Omni	Vert	7	Cont
28.235	VP9BA	Hamilton		F1A	Omni	Vert	10	Cont
28.237	NV6A	San Diego, CA		A1A	GP	Vert	0.5	Cont
28.237	YO2KHP	Timisoara					20	
28.2375	LA5TEN	Oslo		A1A	Omni	Vert	10	Cont
28.24	KB8JVH	Newark, OH		A1A	Yagi	Horz	5	
28.24	OA4CK	Lima	FH17mw	A1A			10	Cont
28.2405	5Z4ERR	Kenya						

Freq	Call Sign	Location	Grid Square	Emission	Ant	Pol	ERP	Remarks
28.2425	ZS1CTB	Capetown		F1A	Omni	Vert	20	
28.244	WA6APQ	Long Beach, CA		A1A	Vert	Vert	30	24 Hrs
28.244	WT8D	Hilton Head, SC					10	
28.2445	KB0DJJ	Omaha, NE			GP		10	
28.245	A92C	Bahrain		F1A	NW/SE	Horz		
28.246	N8KHE	Mackinaw City, MI	EN75	A1A	Omni	Vert	0.05	Cont
28.2475	EA2HB	San Sebastian			Omni	Vert	6	Intrmt
28.248	K1BZ	Belfast, ME			Omni	Vert	5	Cont
28.249	PI7BQC	Haarlem					2	
28.25	K0HTF	Des Moines, IA	EN31dx	A1A	Inv V		2.5	
28.25	N4MW	Memphis, TN					10	
28.25	WJ9Z	St Francis, WI	EN63	F1A	Omni	Vert	15	Cont
28.25	Z21ANB	Bulawayo		F1A	Omni	Vert	15	Cont
28.2505	4N3ZHK	Mt Kum				Vert	1	
28.252	WB4JHS	Durham, NC			Omni	Vert	7	Intrmt
28.252	WJ7X	Victoria, MN		A1A	GP	Vert	10	Cont
28.255	LU1UG	G'ral Pico	FF84dh		Omni	Vert	5	
28.255	WB4JHS	Durham, NC				Vert	7	
28.2575	DK0TEN	Arbeitsgen		F1A	Omni	Vert	40	Cont
28.259	WB9FVR	Pembroke Pines, FL			Dpl	Horz	1	
28.26	KA1NSV	Hyannis, MA	FN41	A1A	Dpl	Horz	80	
28.26	VK5WI	Adelaide		A1A	Omni	Vert	10	Cont
28.262	VK2RSY	Sydney NSW		A1A	Omni	Vert	25	Cont
28.263	N6PEQ	Tustin, CA		A1A	Dipl	Horz	2	Cont
28.264	VK6RWA	Perth, WA		A1A				Cont
28.265	N9JL	East Alton, IL	EM48xv	A1A	Yagi	Horz	1	Cont
28.266	VK6RTW	Albany, WA						Cont
28.268	KB4UPI	Birmingham, AL			¼wave	Vert	20	
28.2685	W9KFO	Eaton, IN			Omni	Vert	0.75	Intrmt
28.27	VK4RTL	Townsville						Cont
28.27	ZS6PW	Pretoria			Yagi		10	Cont
28.2725	9L1FTN	Freetown			Omni	Vert	10	Intrmt
28.275	AL7GQ	Denver, CO		A1A	Omni	Vert	1.5	Intrmt
28.275	ZS1LA	Stillbay			Yagi		10	
28.28	KB2BBW	Rutherford, NJ				Vert	5	
28.28	LU8EB	Argentina					5	
28.28	YV5AYV	Caracas	FK60ni	F1A	Yagi	Horz	10	
28.281	VE1MUF	Newfoundland						
28.282	OK0EG	Hradec Kralove			Dpl		10	
28.282	VE2HOT	Beaconsfield, PQ				Vert	5	
28.283	WD4LWG	Tampa, FL	EL87	F1A	Omni	Vert	2	
28.2835	EA7RCC	Cordova					10	
28.285	N2JNT	Troy, NY			GP		1	
28.285	VP8ADE	Antarctica			V Bm		8	Cont
28.286	KA1YE	Henrietta, NY		A1A	Omni	Vert	2	
28.286	KE2DI	Rochester, NY			Dpl		5	
28.286	KK4M	Las Vegas, NV	DM26		Omni	Vert	5	
28.286	N3GPP	Lancaster, PA					1	
28.2866	KK4M	Las Vegas, NV	DM26	A1A	Omni	Vert	5	
28.287	H44SI	Solomons			Omni		15	Cont
28.287	W8OMV	Ashville, NC			Omni	Vert	5	
28.288	W2NZH	Moorestown, NJ		A1A	Omni	Vert	3	Intrmt
28.29	VS6TEN	Mt Matilda		A1A	Omni	Vert	10	Cont
28.29	W3SV	Elverson, PA				Vert	10	
28.2925	LU2FFV	San Jorge			Omni	Vert	5	
28.294	WC8E	Deerpark, OH				Vert	10	
28.2945	KE0UL	Greeley, CO	DN70pj	A1A	Omni	Vert	5	
28.295	WB8UPN	Cincinnati, OH			Omni	Vert	10	
28.296	W3VD	Laurel, MD			Omni	Vert	1.5	
28.297	WA4DJS	Ft Lauderdale, FL			Omni	Horz	10	
28.299	N8CGY	Michigan			Dpl		5	
28.299	PY2AMI	Sao Paulo	GG67ig	A1A	Omni	Vert	10	Cont
28.3	KF4MS	St Petersburg, FL			GP		5	
28.3	ZS1LA	Still Bay		F1A	Yagi	Horz	20	Cont
28.303	PT7AAC	Fortaleza			GP		5	
28.315	ZS6DN	Irene			Omni	Vert	100	Cont
28.888	W6IRT	North Hollywood, CA		A1A	Omni	Vert	5	Intrmt
28.89	WD9GOE	Freeburg, IL						
28.992	DL0NF	Germany		A1A	E/W		1	
50.003	PY1RO	Rio de Janeiro		A1A				
50.005	H44HIR	Solomon Islands	RI00	A1A			25	
50.005	PY1AA	Brazil		A1A				
50.005	VE8R	NW Territory						
50.005	W0TH/KH6	Hawaii						
50.005	ZS1SIX	Cape Province			Omni		10	
50.005	ZS2SIX	Cape Province	KF25	A1A	Omni	Horz	10	

Freq	Call Sign	Location	Grid Square	Emission	Ant	Pol	ERP	Remarks
50.006	GB3RMK	Inverness		F1A	N/S	Horz	30	
50.007	TF3T	Iceland						
50.008	KØGUV	Park Rapids, MN	EN26	A1A				
50.009	ZS6PW	South Africa	KG44				15	
50.01	JA2IGY	Mile	PM84	A1A	GP		10	
50.01	ZS1STB	Still Bay	JF25	F1A	Yagi	Horz	40	
50.01	ZS6STB	Vereeniging						
50.015	P29BPL	Port Moresby	QI30	A1A	Dpl		30	
50.015	PJ2B	Bonaire						
50.015	SZ2DH	Athens	KM18		Yagi		10	>1300Z
50.02	GB3SIX	Anglesey, Wales	IO73tj	F1A	Yagi	Horz	25	
50.02	JA6YBR	Japan						
50.02	JA6ZIH	Kyushu	PM51	A1A	Sqlo		10	
50.02	JF6DEA	Japan	PM51					
50.02	SZ2DH	Athens						
50.02	WJ7X	Victoria, MN		A1A	GP	Vert	25	Cont
50.022	ZS6LW	Rosslyn	KG44		Yagi		65	
50.023	HH2PR	Haiti						
50.025	5Z4YV	Kenya						
50.025	6Y5RC	Jamaica		F1A	Yagi	Horz	40	
50.025	OH1SIX	Ikaalinen		A1A	Dpl	Horz	2	
50.025	ZS6LN	Pietersburg	KG46		Yagi		150	
50.025	ZS6SIX	Kempton Park						
50.027	JA7ZMA	Fukushima	QM07		Yagi		10	
50.028	JA7JMA	Japan	QM07		Yagi		10	
50.029	CTØWW	Joao Pesdueira	IN61	F1A	Dpl		40	
50.029	ZS6PW							
50.03	STØWW	Portugal	IN61	A1A	Dpl		40	
50.03	XE3VV	Yucatan	EL50ex	A1A	Omni	Vert	5	
50.03	ZS6PW	Pretoria		A1A	N/NNW			1000-2000Z
50.032	ZD8VHF	Ascension Is	LL22tb		Dpl		50	
50.033	LU8YYO	Argentina						
50.033	ZD8VHF	Ascension Is	II22	F1A	Vert	Vert	50	
50.035	EL2CA	Monrovia						
50.035	HC1JX	Quito						
50.035	ZB2VHF	Gibraltar	IM76he	A1A	Yagi	Horz	35	
50.038	FY7THF	Kourou	GJ34	F1A	GP		100	
50.04	CX8BE	Montivideo	GF15		Yagi		10	
50.04	WA6MHZ	San Diego, CA						
50.04	ZS6VHF	South Africa						
50.041	WA8KGG	Ohio						
50.045	DL3ZM/YV5	Caracas						
50.045	OX3VHF	Julianehaab	GP60qq	A1A	GP	Vert	20	
50.045	OX3VHF	Simiutaq	GP60	F1A	GP		20	
50.045	WB2MAI	New Jersey						
50.048	VE6ARC	Grand Prairie, AB	D075	A1A	Omni	Vert	50	
50.048	WA6IJZ	California		A1A		Vert	10	
50.05	GB3NHQ	London	IO91uq	F1A	Tstl	Horz	15	RSGB HQ
50.05	GB8NHQ	Potters Bar	EL09		Tstl		40	
50.05	KA4ZAY	Miami	EL95	A1A	Omni	Vert	10	
50.05	LU2DH	Argentina						
50.05	VE6NAB	Alberta						
50.05	VE8VHF	NW Territory						
50.05	WA1EXN	Maine						
50.05	ZS6DN	Pretoria	KG44					
50.05	ZS6LN	Petersburg					10	
50.053	VK3SIX							
50.055	WA9FEF	Chicago						
50.055	WB9VMY/5	Calumet, OK	EM15				8	
50.055	ZS6DN	Irene	KG44					
50.055	ZS6DN	Pretoria	KG44					
50.057	KH6HI	Honolulu, HI	BL01		Tstl		15	
50.057	TF3SIX	Iceland	HP94cc	FSK		Vert	50	
50.058	WA8IGY	Florida	EM79					
50.059	GB3RMK	Cornwall			Beam		15	
50.06	GB3RMK	Rosemarkie	IO77ud	F1A	Dpl	Horz	30	
50.06	K1NFE	Burlington, CT	FN31		Omni	Vert	25	
50.06	K4TQR	Birmingham, AL	EM63	A1A	Dip	Horz	1	Sun
50.06	KH6EGI	Hawaii						
50.06	KH6EQI	Hawaii						
50.06	KM4BB	Lower Oahu, HI						
50.06	PY2AA	Sao Paulo	GG66		GP	Vert	25	
50.06	WA8ONQ	Cincinnati, OH	EM79	A1A	Omni	Horz	2	
50.06	WA8ONQ	Middletown, OH	EM79		Tstl		2	Cont
50.06	ZS6DN	Pretoria			Yagi	Horz	100	
50.062	K1NFE	Burlington, CT	FN31		Tstl		15	Cont

Freq	Call Sign	Location	Grid Square	Emission	Ant	Pol	ERP	Remarks
50.062	N7BSN	Keyport, WA	CN87				10	
50.062	PY2AA	Sao Paulo	GG66		GP		40	
50.062	W3VD	Laurel, MD	FM19	A1A	Omni	Horz	10	
50.0628	KB6BKN	Novato, CA	CM88	A1A	Omni		6	1100-1700 UTC
50.063	GB3NGI	Northern Ireland	IO65	F1A	Dpl		18	
50.063	N4PZ	Sarasota, FL	EL87		Vert		0.4	Cont
50.064	N4PZ	Sarasota, FL	EL87	A1A	Omni	Vert	0.5	
50.064	N7DB	Boring, OR	CN85				30	
50.064	WA5UUD	New Orleans, LA						
50.064	WB4IJY	Florida						
50.065	KA0CDN	Aurora, CO	DM79		Halo		20	Cont
50.065	KA5FYI	Austin, TX	EM10	A1A	Yagi	Horz	10	
50.065	W0IJR/KA0CDN	Denver, CO	DM79	A1A	Omni	Horz	20	
50.065	W0MTK	Fruita, CO	DM59	A1A	Dpl	Horz	2	
50.065	W5VAS	Metairie, LA	EL19	A1A	Omni	Horz	1	
50.065	WB5ZRL	New Orleans, LA		A1A	Omni	Horz	2	
50.066	AL7C	Big Lake, AK	BP51	A1A	Quad	Horz		
50.066	GJ4HXJ	Jersey	IN89	F1A	Halo		10	
50.066	N6ANG	California						
50.066	WD7Z		DM24					
50.0664	KD4LP							
50.067	KD4LP	Morristown, TN	EM86		Omni		100	Cont
50.067	W7US	Arizona						
50.067	WA6IJZ	Oxnard, CA	DM04		Omni	Vert		
50.067	WB8IGY	Loveland, OH	EM79		Omni	Vert	2	
50.068	K6FV		CM87					
50.068	W4RFR	Fairview, TN	EM66		Halo	Horz	2	
50.069	W0BJ	North Platte, NE	DN91	A1A	Omni	Horz	5	
50.07	4U1ITU	Geneva						Prop
50.07	K0HTF	Des Moines, IA	EN31dx	A1A	Inv V		2	
50.07	KA4VEY	Harvest, AL	EM64		Omni	Vert	10	
50.07	KB4UPI	Birmingham, AL		A1A			1	
50.07	KH6HI	Central Oahu						
50.07	KK4M	Las Vegas, NV	DM26	A1A	Halo	Horz	5	
50.07	KS2T	Toms River, NJ	FM29vx	A1A	Omni	Horz	10	
50.07	N4MW	Memphis, TN					8	
50.07	VP9WB	Bermuda						
50.07	W0BJ	North Platte, NE	DN91				6	
50.07	W2CAP/1	Cape Cod, MA	FN41	A1A	Omni	Vert	5	
50.07	W4HHK	Collierville, TN	EM55			Horz	2	
50.07	WA2YTM	Rochester, NY	FN12		Omni	Horz	15	
50.07	WA7ECY	Troutdale, OR	CN35		Omni	Vert		
50.07	WB0CGH/5	Lewisville, TX	EM13	A1A	Omni	Horz	1.5	
50.07	WB4GJG	Danville, VA	FM06		Omni	Vert	1	
50.07	ZS3E	SW Africa						
50.071	W9FKO	Eaton, IN						
50.071	WA2YTM	Victor, NY	FN12	A1A	Omni	Horz	15	
50.072	KH6HI	Honolulu	BL01		Dip		80	Cont
50.072	KW2T	Palmyra, NY	FN13		Sql		0.25	Cont
50.072	VE1CCP	Prince Edward Island						
50.072	W9KFO	Eaton, IN						
50.072	WA2TYM	Canandaigua, NY	FN12		Tstl		15	Cont
50.073	VE1MQ	Fredericton, NB	FN65	A1A	Omni	Horz	10	
50.073	W7KMA	Arizona						
50.073	ZS4SA	South Africa	KG33		Yagi		20	Intrmt
50.074	KH6HI	Hawaii						
50.074	KH6HME	Honolulu, HI	BL 01		Tstl		15	
50.074	WB5DSH	Oklahoma City, OK	EM15		Halo		30	Intrmt
50.075	K7IHZ	Mesa, AZ	DM43		Sql		20	Intrmt
50.075	KP4EKG	Baymon, PR	FK68		Vert		10	Intrmt
50.075	LU1DMA	Mendoza	FF57		Quad		10	Cont
50.075	N5JM	New Orleans, LA	EL49	A1A	Omni	Vert	3	
50.075	PY2AMI	Sao Paulo	GG66		GP		5	
50.075	VS6SIX	Hong Kong	OL72	A1A	GP		30	
50.075	WB4OSN	Margate, FL	EL95	A1A	Omni			
50.075	WB4WTC	Martinsville, VA	FM06		Loop		10	Cont
50.075	ZS3E	SW Africa						
50.075	ZS5SIX	South Africa	KG50		Halo		10	
50.076	KL7GLK	Annapolis, MD	FM18	A1A	Omni	Vert	4	24-Hrs
50.077	N0LL	Smith Center, KS	EM09	A1A	Omni	Horz	30	
50.077	N5JM	New Orleans, LA	EL49		Vert		3	Intrmt
50.077	VE3DRL	Toronto, ON		A1A		Horz	30	
50.077	WB2CUS	Queens, NY	FN30		Yagi		1	Intrmt
50.077	WB2CUS/3	Mt Effort, PA	FN21		Sql		3	Intrmt
50.0775	W8UR	Mackinac Island, MI	EN75	A1A	Dip	Horz	2	Cont
50.078	PT7AAC	Fortaleza	HI06		GP		5	Cont

Freq	Call Sign	Location	Grid Square	Emission	Ant	Pol	ERP	Remarks
50.078	TI2NA	San Jose	EJ79		Dpl		20	
50.079	W6SKC/7	Nogales, AZ	DM41		Vert		50	Cont
50.08	HC8SIX	Galapagos	EI59		Vert	Vert	4	
50.08	KM4BB	Kona						
50.08	TI2NA	San Jose		A1A				
50.08	W1AW	Newington, CT	FN31	A1A	W	Horz	50	Cde Prac
50.08	WB4OOJ	Tampa, FL	EL87		Omni	Vert	10	
50.08	ZS5TR	Durban						
50.08	ZS5VHF	Durban		A1A	Omni	Horz	10	
50.082	HC8SIX	Galapagos	EI59		Vert		4	Cont
50.083	LU1DMA	Buenos Aires	GF06		Yagi		20	
50.085	9H1SIX	Valletta	JM75fv	F2A	Omni	Horz	25	
50.085	WA6JRA	California						
50.086	VE2STL	Val Belair, PQ	FN46		Omni	Vert	1.5	
50.086	VE2TH	Val Belair, PQ	FN46		Dip		3	
50.086	VP9MO	Plymoth	FK86		Yagi		10	
50.088	VE1ASJ	St Johns, NB						
50.088	VE1SIX	New Brunswick			GP		10	Intrmt
50.089	FC1MKY	La Cannet	JN33		Yagi		10	Cont
50.09	KJ6BZ	Johnson Is	AK56					
50.09	WA6JRA	California						Cont
50.091	9L1US	Sierra Leone	IJ38				100	
50.091	ZF2KZ	Grand Cayman	EK99		Yagi		30	Intrmt
50.092	W5GTP	New Iberia, LA	EM40		Yagi		30	
50.093	WA8FTA	Michigan						
50.095	K7IHZ	Arizona						
50.096	HD1QRC	Quito						
50.098	KG6JIH	Guam						
50.099	KG6DX	Barrigada	QK23		Yagi		20	Intrmt
50.099	TI2CF	San Jose	EJ79					
50.1	HC2FG	Guyaquil	FI07	A1A	J-Pole		8	
50.1	KG6DX	Guam	QK23	A1A	Yagi		40	
50.1	KH6EJQ	Hawaii						
50.1	PY5YD	Brazil						
50.1	VP5D	Turks & Caicos	FL31		Rng		6	Intrmt
50.103	N8AJD	Ohio						
50.104	K4EJQ	Tennessee						
50.105	ZS3E	Namibia	JG89					
50.109	JD1YAA	Minami						
50.11	A61XL	United Arb Em	LL74		Yagi		50	
50.11	A61XL	Untd Arab Emirates	LL74		Yagi		50	Intrmt
50.11	KG6DX	Barrigata	QK23		Yagi		20	Intrmt
50.11	KG6DX	Guam						
50.11	VE7SIX	Kaleden, BC		A1A				
50.11	ZL7C	Alaska						
50.11	ZS6LN	South Africa		A1A		Horz	100	Intrmt
50.112	JD1YAA	Japan						
50.112	W7KNP	Montana						
50.12	4S7EA	Sri Lanka	MJ97		Yagi		50	
50.12	ZS3E	SW Africa						
50.125	ZS3AK	SW Africa						
50.15	VE7RAN	Penticton, BC		A1A			100	
50.246	LU8MBL	Argentina						
50.321	ZS5SIX	South Africa	KG50		Halo		6	
50.44	K1NFE	Hartford, CT						
50.49	JG1ZGW	Tokyo	PM95	A1A	Dpl		10	
50.499	5B4CY	Cyprus	KM54	F1A	GP		15	
50.5	5B4CY	Cyprus	KM64pr	F1A	Omni	Vert	15	
50.555	WA9FEF	Chicago, IL		A1A				
50.904	ZS1STB	Stilbaai	KF05		Dpl		25	
50.945	ZS1SIX	Cape Province		F2A	Omni		8	
51.002	ZL1BPW	Auckland						
51.02	ZL1UHF	Nihotupu		F1A			25	
51.02	ZL1VHF	West Auckland	RF73		Dpl	Vert	25	
51.03	ZL2MHB	Fernhill		F2A			10	
51.199	VE7BEE	Penticton, BC	DN09		Vert		50	Cont
51.225	ZL2VHT	Inglewood		F1A			15	
52	K6FV	Woodside, CA	CM87	A1A				
52.013	P29BPL	Papua						
52.02	FK8KAB	New Caledonia						
52.033	P29BPL	Papua						
52.1	ZK2SIX	Niue						
52.15	VK0CK	Macquarie						
52.2	VK8VF	Darwin	PH57					
52.25	ZL2VHM	Pahiatua Track		F1A			8	
52.3	VK6RPH	Perth, WA						

Freq	Call Sign	Location	Grid Square	Emission	Ant	Pol	ERP	Remarks
52.3	VK6RTV	Perth, WA						
52.31	ZL3MHF	Christchurch		F2A			20	
52.32	VK6RTT	Carnarvon, WA					25	
52.32	VK6RTT	Wickham	OG89		J-Pole		25	
52.325	VK2RHV	Newcastle	QF57					
52.33	VK3RGG	Geelong Victoria		F1A	Omni	Horz	4	Exp
52.345	VK4ABP	Longreach	QG26		Vert	Vert	10	
52.35	VK6RTU	Kalgoorie	PF09					
52.37	VK7RST	Hobart	QE37					
52.41	VK0MA	Mawson	MC12					
52.42	VK2RSY	Sydney NSW				Horz	25	
52.425	VK2RGB	Gunnedah NSW						
52.435	VK3RMV	Hamelton	QF12					
52.44	VK4RTL	Townsville	QH30					
52.445	VR4RIK	Australia						
52.45	VK5VF	Mount Lofty	PF94		Tstl		20	
52.46	VK6RPH	Perth	OF78					
52.465	VK6RTW	Albany	OF84					
52.47	VK2RSY	Dural	QF56		Tstl		25	
52.47	VK7RNT	Launceston	QE38					
52.485	VK8RAS	Alice Sprngs	PG66					
52.49	ZL2SIX	Blenheim		F2A			10	
52.5	ZL2VHM	Palmerston						
52.51	ZL2MHF	Mt Climie		F1A			2	
70.03	GB3CTC	United Kingdom	IO71gy					
70.05	GB3BUX	United Kingdom	IO93be					
70.06	GB3ANG	United Kingdom	IO86mn					
70.112	5B4CY	Cyrpus	KM64pr					
70.12	ZB2VHF	Gibraltar	IM76he					
70.13	EI4RF	Ireland	IO63sn					
144.019	VK6RBS	Busselton						
144.047	W9IP	Northeast	FN24	A1A	Yagi		35	Aurora Mon
144.05	KA4ZAY	Miami, FL	EL95	A1A	Omni	Vert	20	
144.052	W3VD	Laurel, MD	FM19nd		Omni	Horz	15	
144.053	W0VB	Rochester, MN	EN34				100	
144.054	KC4IH	Marion, VA	EM96fv		Omni	Horz	3	
144.055	K0PP/7	Montana					40	
144.055	WB0QIY	Lincoln, NE					10	
144.055	WB9VMY/5	Calumet, OK	EM15				8	
144.056	KJ5Q	Oklahoma City, OK	EM15				13	
144.057	K1GRU	Rochester, NH	FN43	A1A	Yagi	Horz	5	1600-2400 UTC
144.057	K1GRU	Rochester, NH	FN43	A1A	Yagi		1	0700-1500 UTC
144.057	K1GRU	Rochester, NH	FN43	A1A	Yagi		5	1200-2400 GMT
144.058	W4RFR	Fairview, TN	EM66		Halo	Horz	0.25	
144.058	W4UDZ	Tennessee	EM65			Yagi	1	
144.064	W5VAJ	Louisiana						
144.139	5B4CY	Cyprus	KM64ht					
144.145	ZB2VHF	Gibraltar	IM76he					
144.17	KH6HME	Mauna Loa, HI	BK290		Yagi	Horz	10	
144.175	KP4EOR	Puerto Rico						
144.275	KD0DW	Longmont, CO	DN70 LF	A1A	Omni	Horz	180	Cont
144.277	KA0EWQ	Johnson, MN	EN15	A1A	Yagi	Horz	50	Cont SE
144.28	KG5YB	Tyler, TX	EM22	J2A	Vert	Vert	25	Cont
144.28	N4MW	Memphis, TN	EM55	A1A	Squal	Horz	3	
144.285	WD4LWG	Tampa, FL	EL87	F1A	Omni	Vert	100	
144.289	KL7GLK	Annapolis, MD	FM 18	A1A	Omni	Horz	2	Cont
144.29	K0HTF	Des Moines, IA	EN31dx	A1A	Inv V		1	
144.29	KK4M	Las Vegas, NV	DM26	F3	Omni	Vert	5	
144.29	N2PNO	Endwell, NY	FN22	A1A	Omni	Vert	2	1700-2100 UTC
144.293	W9IP	Canton, NY	FN24	A1A	Yagi		45	6-sec tone/ID
144.297	N4ZRW	Memphis, TN					4	
144.297	VE1MQ	Fredericton, NB	FN65	A1A	Yagi	Horz	5	N(win)SW(Sum)
144.4	VK4RTT	Mt Mowbullan						
144.41	VK1RCC	Canberra						
144.42	VK2RSY	Sydney NSW				Horz	25	
144.445	VK4RIK	Australia						
144.465	VK6RTW	Albany						
144.48	VK8VH	Darwin						
144.55	VK5RSE	Mt Gambier						
144.565	VK6RPB	Port Hedland						
144.8	OH8VHF	Finland	KP25tc					
144.8	VK5VF	Mt Lofty						
144.81	IS0A	Sardinia	JN40sx					
144.825	I0A	Italy	JN61es					
144.83	9V1VHF	Malta	JM75rv				1	
144.84	IT9G	Italy	JM68					

Freq	Call Sign	Location	Grid Square	Emission	Ant	Pol	ERP	Remarks
144.85	DLØUB	Germany	JO62ql					
144.855	LA5VHF	Norway	JP99kp					
144.86	LA1VHF	Norway	JO94gm					
144.865	HB9HB	Switzerland	JN37md					
144.867	EA1VHF	Spain	IN53ug					
144.87	LA2VHF	Norway	JP53eg					
144.875	SK2VHF	Sweden	JP94tf					
144.885	OY6VHF	Faroes	IP62na					
144.89	LA4VHF	Norway	JP20oq					
144.895	FXØTHF	France	JN08ml					
144.9	OH6VHF	Finland	KP02tg					
144.902	OX6VHF	Greenland	GP60		SE	Horz	10	
144.91	DLØPR	Germany	JO44jh					
144.915	GB3CTC	United Kingdom	IO70oj					
144.917	EA6VHF	Ibeza						
144.92	SK7VHF	Sweden	JO65sn					
144.925	GB3VHF	United Kingdom	JO01dh					
144.93	OZ7IGY	Denmark	JO55vo					
144.945	SP4VHG	Poland	JO71sw					
144.95	SK1VHF	Sweden	JO97cg					
144.95	VK2RCW	Sydney NSW					25	
144.96	SK4MPI	Sweden	JP70nj				25	
144.965	GB3LER	United Kingdom	IP90jd					
144.975	DLØSG	Germany	JN68eq					
144.975	GB3ANG	United Kingdom	IO86mn					
144.98	SP3VHC	Poland	JO94gm					
144.984	ON4VHF	Belgium	JO20fp					
144.985	Y41B	Germany	JO53rp					
145	VK6RPH	Perth, WA						
222.05	N4MW	Memphis, TN					1	
222.052	WB2IEY	Maryknoll, NY	FN31	A1A	Tnstl		15	
222.0525	KØHTF	Des Moines, IA	EN31dx	A1A	GP		2	
222.055	KK4M	Las Vegas, NV	DM26	F3	Omni	Vert	5	
222.055	WB4FQR	Woodbridge, VA	FM18	A1A	Tnstl	Horz	15	Cont
222.058	WB4NKM	Greenville, TN					10	
432.057	VK6RBS	Busselton						
432.07	NØUU	Abilene, KS	EM19		Whl		10	
432.071	W4RFR	Fairview, TN	EM66		Halos		10	
432.0715	K4MSK	Lookout Mtn, GA	EM85md	A1A	Omni	Horz	1	
432.0715	N3CX	Pennsburg, PA	FN20		Omni	Horz	1	
432.073	K5BYS	Lewisville, TX	EM12				0.01	
432.073	WA4PGI	Lynchburg, VA						
432.074	KH6HME	Mauna Loa, HI	BK29go		Yagi	Horz	10	
432.074	KK4M	Las Vegas, NV	DM26	F3	Omni	Vert	5	
432.075	WØUC	Eau Clair, WI	EN44hr		Yagi	Horz	4	
432.075	WA8ONQ	Ohio						
432.075	WBØQIY	Lincoln, NE	EN105		Omni	Horz	1	
432.077	NØII	Abilene, KS	EM19				7	
432.077	W4ODW	Niceville, FL	EM60		Omni	Horz	4	
432.078	W4HHK	Collierville, TN	EM55		Halos	Horz	5	
432.079	KB3NQ	Penndel, PA	FN20	A1A	Omni	Horz	12	
432.079	W4HHK	Tennessee						
432.095	VE1SMU	Nova Scotia						
432.16	VK6RPR	Nedlands						
432.3	W2UHI	Muskegon, MI	EN63vb	A1A	Omni	Horz	10	
432.34	KB3NQ	Penndel, PA	FN20	A1A	Omni	Horz	7	
432.385	WD4LWG	Tampa, FL	EL87	F1A	Omni	Vert	40	
432.395	N4MW	Memphis, TN					10	
432.41	VK1RBC	Canberra						
432.42	VK2RSY	Sydney NSW				Horz	15	
432.425	VK3RMB	Ballarat						
432.44	VK4RBB	Brisbane						
432.855	LA5UHF	Norway	JP99kp					
432.855	SK3UHF	Sweden	JP92qw					
432.86	LA1UHF	Norway	JO59jw					
432.865	OZ2UHF	Denmark	JO46jd					
432.87	FX4UHF	France	IN93eh					
432.87	LA2UHF	Norway	JP53eg					
432.88	LA3UHF	Norway	JO48bc					
432.885	OY6UHF	Faroes	IP62na					
432.89	LA4UHF	Norway	JO29pj					
432.895	OZ4UHF	Denmark	JO75je					
432.9	OH3UHF	Finland	KP11un					
432.91	GB3MLY	United Kingdom	IO93eo					
432.925	SK6UHF	Sweden	JO67bf					
432.93	OZ7IGY	Denmark	JO55vo					

Freq	Call Sign	Location	Grid Square	Emission	Ant	Pol	ERP	Remarks
432.935	OKØEA	Czech Republic	JO70ss					
432.96	SK4UHF	Sweden	JO79kh					
432.97	GB3CTC	United Kingdom	IO70oj					
432.975	SK5UHF	Sweden	JP80sa					
432.98	GB3ANG	United Kingdom	IO86mn					
432.983	OZ2ALS	Denmark	JO45vb					
432.984	HB9F	Switzerland	JN36xn					
902.071	K5BYS	Desert, TX	EM13		Omni	Horz	0.15	
902.125	K5BYS	Lewisville, TX	EM12		Omni	Horz	0.02	
902.125	KF5N	Ft Worth, TX	EM12		Omni	Horz	10	
902.28	N4MW	Memphis, TN		A1A				
903.08	K3SIW/9	Hanover Park, IL			E	Horz	5	
903.083	KK4M	Las Vegas, NV	DM26	F3	Omni	Vert	5	
903.083	N3XC	Pennsburg, PA	FN20		Omni	Horz	4	
903.09	K3SIW/9	Schaumburg, IL	EN52		CRef	Horz	30	Cont
903.837	WBØQIY	Lincoln, NE	EN10	A1A			0.1	
904	K4MSK	Mt Toxaway, NC	EM85					
1206.125	K5BYS	Lewisville, TX	EM12		Omni	Horz	0.02	
1294.52	JA2IGY	Ise City		F2				Susp
1294.52	JA4IGY	Hiroshima		F2				Cont
1294.52	JA7IGY	Fukushima		F2				Cont
1294.54	JAØIGY	Nagano City		F2				Cont
1294.54	JA6IGY	Kumamato City		F2				Cont
1294.56	JA9IGY	Kanazawa City		F2				Susp
1294.58	JA1IGY	Tokyo			F2			Cont
1294.58	JA8IGY	Hokkaido			F2			Cont
1296	KH6HME	Mauna	BK290		Yagi	Horz	20	
1296.025	WØPW	Boulder, CO	DM79		Yagi	Horz	4	
1296.055	W9YH	Urbana, IL	EN50		Yagi	Horz	10	Cont
1296.07	K3IVO	Ft Mead, MD	FM19pd		Omni	Horz	1	
1296.07	K3SIW/9	Schaumburg, IL	EN52		Yagi	Horz	20	Cont
1296.071	K5BYS	Lewisville, TX	EM12				0.01	
1296.085	W8MQW	East Lansing, MI	EN72	A1A	Omni	Horz	6	
1296.125	KF5N	Arlington, TX	EM12		Omni	Horz	2	
1296.125	KK4M	Las Vegas, NV	DM26	F3	Omni	Vert	5	
1296.1625	W7CNK	Okla City, OK	EM15					
1296.171	VK6RBS	Busselton						
1296.21	NØUU	Abilene, KS	EM19		Omni	Horz	0.1	
1296.2145	K4MSK	Lookout Mtn, GA	EM85md	A1A	Omni	Horz	2	
1296.2145	N3CX	Pennsburg, PA	FN20		Omni	Horz	2	
1296.222	KF5N	Ft Worth, TX	EM12		Slat		0.85	
1296.234	W4HHK	Collierville, TN	EM55				0.2	
1296.28	N4MW	Memphis, TN					10	
1296.3	W6KGS	California	DM13					
1296.41	VK1RBC	Canberra						
1296.42	VK2RSY	Sydney NSW			Omni	Horz	5	
1296.48	VK6RPR	Nedlands						
1296.81	GB3NWK	United Kingdom	JO01bi					
1296.83	GB3BPO	United Kingdom	JO02pb					
1296.854	DBØJO	Germany	JN36xn					
1296.87	GB3AND	United Kingdom	IO91gf					
1296.885	ON5UHF	Belgium	JO10un					
1296.89	GB3DUN	United Kingdom	IO91sv					
1296.91	GB3CLE	United Kingdom	IO82rl					
1296.92	DBØVC	United Kingdom	JO54bh					
1296.92	PAØQHN	Netherlands	JO22fh					
1296.925	SK6UHG	Sweden	JO57tq					
1296.93	GB3MLE	United Kingdom	IO91eo					
1296.93	OZ7IGY	Denmark	JO55vo					
1296.975	DBØJU	Germany	JO31bu					
1296.99	GB3EDN	United Kingdom	IO85jw		FH	Horz	10	Cont
2304.02	K3SIW/9	Schaumburg, IL	EN52		FH	Horz	10	Cont
2304.02	N3CX	Pennsburg, PA	FN20		SW	Horz	0.1	
2304.05	KB3NQ	Penndel, PA	FN20ND	A1A		Horz	4	
2304.064	K5BYS	Lewisville, TX	EM12		Omni	Horz	0.18	
2304.075	W9ZIH	Malta, IL	EN51		FH	Horz	5	Cont
2304.11	W6IFE	California	DM14kf		Omni	Horz	2.5	
2304.215	N4MW	Memphis, TN		A1A				Cont
2424.78	JA9IGY	Kanazawa City		F2				Susp
2424.82	JA2IGY	Ise City		F2				
2424.82	JA6IGY	Kumamoto City		F2				Cont
2424.82	JA7IGY	Fukushima		F2				Cont
2424.86	JA1IGY	Tokyo		F2				Cont
2424.86	JE5IGY	Hojo City		F2				Cont
3456.02	K3SIW/9	Schaumburg, IL	EN52		FH	Horz	0.5	Cont
3456.04	N4MW	Memphis, TN		A1A				

Freq	Call Sign	Location	Grid Square	Emission	Ant	Pol	ERP	Remarks
3456.572	K4MSK	Lookout Mtn, GA	EM85md	A1A	SW	Horz	0.001	
5756.9	JA3IGY	Kyoto		F2				
5759.95	K3SIW/9	Schaumburg, IL	EN52		FH	Horz	5	Cont
5759.95	W9ZIH	Mt Morris, IL	EN51		FH	Horz	10	Cont
5760.04	N4MW	Memphis, TN						Cont
5760.9295	K4MSK	Lookout Mtn, GA	EM85md	A1A	SW	Horz	8	
10000.265	WB6DNX	Santiago Peak, CA	DM13fr					
10250	N4IZW	La Mesa, CA	DM12ls		Omni	Vert	0.1	
10256	WA6EJO	Red Mountain, CA	DM04hi	F1A	Horn	Vert	5.6	Cont
10265	WB6DNX	Santiago Peak, CA	DM13FR					
10300	VK6RVF	Roleystone						
10368.02	KK6TG	St Helena	CM88qq	F1A		Horz	0.1	
10369.716	K4MSK	Lookout Mtn, GA	EM85md	A1A				
10369.716	K4MSK	Mt Toxaway, NC	EM85					

Table 2-4
Amateur Frequency Allocations and Band Plans

Unlike FCC rules, which must be obeyed if the operator is to avoid official sanctions (possibly including loss of license), band plans are voluntary guidelines developed through representative mechanisms. Regional band plans were first developed, and are periodically reviewed and updated, by the regional organizations of the International Amateur Radio Union (IARU). The IARU Administrative Council serves as a coordinator of regional policies, and seeks worldwide agreement whenever possible. However, because each region is autonomous, subtle differences in the regional band plans exist which cannot be easily explained in a simple chart. The charts of regional band plans shown here are intended to summarize the band plans of the three IARU regions, and should not be assumed to be definitive.

Emissions are to be confined to the band segments specified for them. See end of table for Emission Designators.

The regions differ in their treatment of data modes. In general, the term "DATA" used here includes Baudot, AMTOR, ASCII, packet radio, and similar modes. Where a segment of a band is specifically designated for packet radio, the intent generally is to indicate a preference that packet radio in that band should be limited to that segment.

SSTV generally includes FAX.

Region 3 has adopted the following policy with regard to its band plans:

(1) In all cases of conflict between a band plan and the national regulations of a country, the latter shall prevail.

(2) Nothing in these band plans shall be construed as prohibiting different national arrangements, provided that harmful interference is not caused to stations in the countries operating in accordance with the regional band plans.

(3) Notwithstanding item 2 above, member societies of Region 3 are strongly urged to use these regional band plans as a basis for their national band plans.

Allocations							
INTERNATIONAL			UNITED STATES				
Region 1 kHz	Region 2 kHz	Region 3 kHz	Band kHz	National Provisions	Government Allocation	Non-Government Allocation	Remarks
1800-1810 RADIOLOCATION 485 486 1810-1850 AMATEUR 490 491 492 493	1800-1850 AMATEUR	1800-2000 AMATEUR FIXED MOBILE except aeronautical mobile RADIONAVIGATION Radiolocation	1800-1900			AMATEUR	
1850-2000 FIXED MOBILE except aeronautical mobile 484 488 495	1850-2000 AMATEUR FIXED MOBILE except aeronautical mobile RADIOLOCATION RADIONAVIGATION 494	489	1900-2000	US290	RADIOLOCATION	RADIOLOCATION AMATEUR	

484—Some countries of Region 1 use radiodetermination systems in the bands 1606.5-1625 kHz, 1635-1800 kHz, 1850-2160 kHz, 2194-2300 kHz, 2502-2850 kHz and 3500-3800 kHz. The establishment and operation of such systems are subject to agreement obtained under the procedures set forth in Article 14. The radiated mean power of these stations shall not exceed 50 W

485—Additional allocation: in Angola, Bulgaria, Hungary, Mongolia, Nigeria, Poland, the Federal Republic of Germany, Chad, the Czech Republic and the CIS, the bands 1625-1635 kHz, 1800-1810 kHz and 2160-2170 kHz are also allocated to the fixed and land mobile services on a primary basis subject to agreement obtained under the procedure set forth in Article 14.

486—In Region 1, in the bands 1625-1635 kHz, 1800-1810 kHz and 2160-2170 kHz (except in the countries listed in No. 485 and those listed in No. 499 for the band 2160-2170 kHz), existing stations in the fixed and mobile, except aeronautical mobile, services (and stations of the aeronautical mobile (OR) service in the band 2160-2170 kHz) may continue to operate on a primary basis until satisfactory replacement assignments have been found and implemented in accordance with Resolution 38.

487—In Region 1, the establishment and operation of stations of the radiolocation service in the bands 1625-1635 kHz, 1800-1810 kHz and 2160-2170 kHz shall be subject to agreement obtained under the procedure set forth in Article 14 (see also No. 486). The radiated mean power of radiolocation stations shall not exceed 50 W. Pulse systems are prohibited.

488—In the Federal Republic of Germany, Denmark, Finland, Hungary, Ireland, Israel, Jordan, Malta, Norway, Poland, the United Kingdom, Sweden, the Czech Republic and the CIS administrations may allocate up to 200 kHz to their amateur service in the bands 1715-1800 kHz and 1850-2000 kHz. However, when allocating the bands within this range to their amateur service, administrations shall, after prior consultations with administrations of neighboring countries, take such steps as may be necessary to prevent harmful interference from their amateur service to the fixed and mobile services of other countries. The mean power of any amateur station shall not exceed 10 W.

489—In Region 3, the Loran system operates either on 1850 kHz or 1950 kHz, the bands occupied being 1825-1875 kHz and 1925-1975 kHz respectively. Other services to which the band 1800-2000 kHz is allocated may use any frequency therein on condition that no harmful interference is caused to the Loran system operating on 1850 kHz or 1950 kHz.

490—Alternative allocation: in the Federal Republic of Germany, Angola, Austria, Belgium, Bulgaria, Cameroon, the Congo, Denmark, Egypt, Spain, Ethiopia, France, Greece, Italy, Lebanon, Luxembourg, Malawi, the Netherlands, Portugal, Syria, Somalia, Tanzania, Tunisia, Turkey and the CIS, the band 1810-1830 kHz is allocated to the fixed and mobile, except aeronautical mobile, services on a primary basis.

491—Additional allocation: in Saudi Arabia, Iraq, Israel, Libya, Poland, Romania, Chad, the Czech Republic, Togo and Yugoslavia, the band 1810-1830 kHz is also allocated to the fixed and mobile, except aeronautical mobile, services on a primary basis.

492—In Region 1, the use of the band 1810-1850 kHz by the amateur service is subject to the condition that satisfactory replacement assignments have been found and implemented in accordance with Resolution 38, for frequencies to all existing stations of the fixed and mobile, except aeronautical mobile, services operating in this band (except for the stations of the countries listed in Nos. 490, 491 and 493). On completion of satisfactory transfer, the authorization to use the band 1810-1830 kHz by the amateur service in countries situated totally or partially north of 40° N shall be given only after consultation with the countries mentioned in Nos. 490 and 491 to define the necessary steps to be taken to prevent harmful interference between amateur stations and stations of other services operating in accordance with Nos. 490 and 491.

493—Alternative allocation: in Burundi and Lesotho, the band 1810-1850 kHz is allocated to the fixed and mobile, except aeronautical mobile, services on a primary basis.

494—Alternative allocation: in Argentina, Bolivia, Chile, Mexico, Paraguay, Peru, Uruguay and Venezuela, the band 1850-2000 kHz is allocated to the fixed, mobile, except aeronautical mobile, radiolocation and radionavigation services on a primary basis.

US290—In the band 1900-2000 kHz, amateur stations may continue to operate on a secondary basis to the Radiolocation Service, pending a decision as to their disposition through a future rule making proceeding in conjunction with implementation of the Standard Broadcasting Service in the 1625-1705 kHz band.

FCC Part 97 Privileges

License Class	Terrestrial location of the amateur radio station			Limitations*
	Region 1 kHz	Region 2 kHz	Region 3 kHz	
General Advanced Extra	1810-1850	1800-2000	1800-2000	

* See §§97.301-307 of the FCC rules for further details on authorized privileges for each license class, frequency sharing requirements and authorized modes.

Amateur Radio Band Plans

Region 1 kHz	Region 2 kHz	Region 3 kHz	ARRL* kHz
1810-1838 CW	1800-1830 CW/DATA	1800-1830 CW	1800-1840 CW/RTTY/NB
	1830-1840 CW/DATA DX CW Window	1830-1834 CW/RTTY/DX	
		1834-1840 CW	
1838-1840 CW/DATA			1830-1850 Intercontinental QSO's only
1840-1842 DATA/Phone/CW	1840-1850 Phone/CW DX Phone WIndow	1840-2000 Phone/CW	1840-2000 Phone SSTV WB CW
1842-2000 Phone/CW	1850-2000 Phone/CW		

*Adopted by the ARRL Board of Directors January 1986.

Allocations

INTERNATIONAL			UNITED STATES				
Region 1 kHz	Region 2 kHz	Region 3 kHz	Band kHz	National Provisions	Government Allocation	Non-Government Allocation	Remarks
3500-3800 AMATEUR 510 FIXED　MOBILE except　aeronautical　mobile	3500-3750 AMATEUR 510	3500-3900 AMATEUR 510 FIXED MOBILE	3500-4000	510		AMATEUR	
	509　511						
484	3750-4000 AMATEUR 510 FIXED MOBILE except　aeronoutical　mobile (R)						
3800-3900 FIXED AERONAUTICAL　MOBILE (OR)　LAND MOBILE							
3900-3950 AERONAUTICAL　MOBILE (OR)		3900-3950 AERONAUTICAL　MOBILE BROADCASTING					
513		3950-4000 FIXED BROADCASTING					
3950-4000 FIXED BROAD-CASTING							
	511　512　514　515	516					

484—Some countries of Region 1 use radiodetermination systems in the bands 1606.5-1625 kHz, 1635-1800 kHz, 1850-2160 kHz, 2194-2300 kHz, 2502-2850 kHz and 3500-3800 kHz. The establishment and operation of such systems are subject to agreement obtained under the procedures set forth in Article 14. The radiated mean power of these stations shall not exceed 50 W.

509—Additional allocation: in Honduras, Mexico, Peru and Venezuela, the band 3500-3750 kHz is also allocated to the fixed and mobile services on a primary basis.

510—For the use of the bands allocated to the amateur service at 3.5 MHz, 7.0 MHz, 10.1 MHz, 14.0 MHz, 18.068 MHz, 21.0 MHz, 24.89 MHz and 144 MHz in the event of natural disasters, see Resolution 640. (Resolution 640 may be found in The FCC Rule Book.)

511—Additional allocation: in Brazil, the band 3700-4000 kHz is also allocated to the radiolocation service on a primary basis.

512—Alternative allocation: in Argentina, Bolivia, Chile, Ecuador, Paraguay, Peru and Uruguay, the band 3750-4000 kHz is allocated to the fixed and mobile, except aeronautical mobile, services on a primary basis.

513—Alternative allocation: in Botswana, Lesotho, Malawi, Mozambique, Namibia, South Africa, Swaziland, Zambia and Zimbabwe, the band 3900-3950 kHz is allocated to the broadcasting service on a primary basis. The use of this band by the broadcasting service is subject to agreement obtained under the procedure set forth in Article 14 with neighboring countries having services operating in accordance with the Table.

514—Additional allocation: in Canada, the band 3950-4000 kHz is also allocated to the broadcasting service on a primary basis. The power of broadcasting stations operating in this band shall not exceed that necessary for a national service within the frontier of this country and shall not cause harmful interference to other services operating in accordance with the Table.

515—Additional allocation: in Greenland, the band 3950-4000 kHz is also allocated to the broadcasting service on a primary basis. The power of the broadcasting stations operating in this band shall not exceed that necessary for a national service and shall in no case exceed 5 kW.

516—In Region 3, the stations of those services to which the band 3995-4005 kHz is allocated may transmit standard frequency and time signals.

FCC Part 97 Privileges

License Class	Terrestrial location of the amateur radio station			
	Region 1 kHz	Region 2 kHz	Region 3 kHz	Limitations*
Novice**	3675-3725***	3675-3725***	3675-3725***	
General	3525-3750 —	3525-3750 3850-4000	3525-3750 3850-3900	
Advanced	3525-3750 3775-3800	3525-3750 3775-4000	3525-3750 3775-3900	
Extra	3500-3800	3500-4000	3500-3900	

*See §§97.301-307 of the FCC rules for further details on authorized privileges for each license class, frequency sharing requirements and authorized modes.

**Technicians having passed Element 1(A) (5 WPM code) are authorized to use Novice HF bands

***The Novice band was moved from 3700-3750 to 3675-3725 kHz March 16, 1991, per FCC PR Docket PR 90-100.

Amateur Radio Band Plans

Region 1 kHz	Region 2 kHz	Region 3† kHz	ARRL kHz
3500-3510 DX CW Contests	3500-3510 DX CW	3500-3510 DX CW	3500-3580 CW
3510-3560 CW Contests	3510-3525 CW	3510-3535 CW	
	3525-3580 CW (Phone permitted)	3535-3775 Phone/CW	
3560-3580 CW			
3580-3590 CW/DATA	3580-3620 DATA/CW (Phone permitted)		3580-3620 DATA/CW
3590-3600 CW/DATA/Packet			3590 DX/RTTY
3600-3620 DATA/Phone/CW Contests			
3620-3650 Phone/CW Contests	3620-3635 Packet/DATA/CW (Phone permitted)		3620-3635 Packet/DATA/CW
	3635-3750 CW (Phone permitted)		3635-3750 CW
3650-3700 Phone/CW			
3700-3730 Phone/CW Contests			
3730-3740 SSTV/Phone/CW Contests			
3740-3775 Phone/CW Contests	3750-3775 Phone (CW permitted)		3750-3790 Phone
3775-3800 DX Phone/CW	3775-3800 DX Phone (CW permitted)	3775-3800 DX/Phone/CW	3790-3800 Phone DX Window
	3800-3840 Phone (CW permitted)	3800-3900 Phone/CW	3800-4000 Phone
	3840-3850 SSTV/Phone (CW permitted)		3845 SSTV
	3850-4000 Phone (CW permitted)		

† In Region 3 countries where the total band available nationally is 100 kHz or less, phone operation may commence at 3525 kHz.

Allocations

INTERNATIONAL			UNITED STATES				
Region 1 *kHz*	*Region 2* *kHz*	*Region 3* *kHz*	*Band* *kHz*	*National* *Provisions*	*Government* *Allocation*	*Non-Government* *Allocation*	*Remarks*
7000-7100	AMATEUR 510 AMATEUR-SATELLITE 526 527		7000-7100	510		AMATEUR 510 AMATEUR-SATELLITE	
7100-7300 BROADCASTING	7100-7300 AMATEUR 510 528	7100-7300 BROADCASTING	7100-7300	510 528		AMATEUR	

510—For the use of the bands allocated to the amateur service at 3.5 MHz, 7.0 MHz, 10.1 MHz, 14.0 MHz, 18.068 MHz, 21.0 MHz, 24.89 MHz and 144 MHz in the event of natural disasters, see Resolution 640 and Resolution 641. Both may be found in *The FCC Rule Book*.

526—Additional allocation: in Angola, Iraq, Kenya, Rwanda, Somalia and Togo, the band 7000-7050 kHz is also allocated to the fixed service on a primary basis.

527—Alternative allocation: in Egypt, Ethiopia, Guinea, Libya, Madagascar, Malawi and Tanzania, the band 7000-7050 kHz is allocated to the fixed service on a primary basis.

528—The use of the band 7100-7300 kHz in Region 2 by the amateur service shall not impose constraints on the broadcasting service intended for use within Region 1 and Region 3.

FCC Part 97 Privileges

	Terrestrial location of the amateur radio station			
License *Class*	*Region 1* *kHz*	*Region 2* *kHz*	*Region 3* *kHz*	*Limitations**
Novice**	7050-7075	7100-7150	7050-7075	
General	7025-7100	7025-7150 7225-7300	7025-7100	
Advanced	7025-7100	7025-7300	7025-7100	
Extra	7000-7100	7000-7300	7000-7100	

*See §§97.301-307 of the FCC rules for further details on authorized privileges for each license class, frequency sharing requirements and authorized modes.

** Technicians having passed Element 1(A) (5 WPM code) are authorized to use Novice HF bands.

Amateur Radio Band Plans

Region 1 *kHz*	*Region 2* *kHz*	*Region 3* *kHz*	*ARRL* *kHz*
7000-7035 CW	7000-7035 CW	7000-7025 CW	7000-7080 CW
		7025-7030 DATA/CW	
		7030-7040 DATA/Phone/CW	
7035-7040 RTTY/SSTV/CW	7035-7040 DATA/CW		
7040-7045 RTTY/SSTV/CW Phone	7040-7050 Packet [Int'l]/ DATA/CW	7040-7100 Phone/CW	7040 DX RTTY
7045-7100 Phone/CW			
	7050-7080 Phone (CW permitted)		
	7080-7100 DX Phone (CW permitted)		7080-7100 RTTY/CW
	7100-7120 Packet/Phone (CW permitted)	7100-7300 Phone/CW (Secondary operation)	7100-7105 Packet/CW
			7105-7150 CW
	7120-7166 Phone (CW permitted)		7150-7300 Phone
	7166-7176 SSTV/Phone (CW permitted)		7171 SSTV
	7176-7300 Phone (CW permitted)		

Allocations

INTERNATIONAL			UNITED STATES				
Region 1 kHz	Region 2 kHz	Region 3 kHz	Band kHz	National Provisions	Government Allocation	Non-Government Allocation	Remarks
10100-10150	FIXED Amateur 510		10100-10150	US247 510		AMATEUR	

510—For the use of the bands allocated to the amateur service at 3.5 MHz, 7.0 MHz, 10.1 MHz, 14.0 MHz, 18.068 MHz, 21.0 MHz, 24.89 MHz and 144 MHz in the event of natural disasters, see Resulotion 640 at the end of the table.

US247—The band 10100-10150 kHz is allocated to the fixed service on a primary basis outside the United States and Possessions, Transmissions of stations in the amateur service shall not cause harmful interference to this fixed service use and stations in the amateur service shall make all necessary adjustments (including termination of transmission) if harmful interference is caused.

FCC Part 97 Privileges

License Class	Terrestrial location of the amateur radio station			
	Region 1 kHz	Region 2 kHz	Region 3 kHz	Limitations*
General Advanced Extra	10100-10150	10100-10150	10100-10150	

*See §§97.301-307 of the FCC rules for further details on authorized privileges for each license class, frequency sharing requirements and authorized modes.

Amateur Radio Band Plans

Region 1 kHz	Region 2 kHz	Region 3 kHz	ARRL kHz
10100-10140 CW	10100-10130 CW	10100-10140 CW	10100-10130 CW
10140-10150 DATA/CW	10130-10140 DATA/CW	10140-10150 DATA/CW	10130-10140 DATA/CW
	10140-10150 Packet/DATA/CW		10140-10150 Packet/DATA/CW

Allocations

INTERNATIONAL			UNITED STATES				
Region 1 kHz	Region 2 kHz	Region 3 kHz	Band kHz	National Provisions	Government Allocation	Non-Government Allocation	Remarks
14000-14250	AMATEUR 510 AMATEUR-SATELLITE		14000-14250	510		AMATEUR AMATEUR-SATELLITE	
14250-14350	AMATEUR 510 535		14250-14350	510		AMATEUR	

510—For the use of the bands allocated to the amateur service at 3.5 MHz, 7.0 MHz, 10.1 MHz, 14.0 MHz, 18.068 MHz, 21.0 MHz, 24.89 MHz and 144 MHz in the event of natural disasters, see Resulotion 640 at the end of the table.

535—Additional allocation: in Afghanistan, China, the Cote d'Ivoire, Iran and the CIS, the band 14250-14350 kHz is also allocated to the fixed service on a primary basis. Stations of the fixed service shall not use a radiated power exceeding 24 dBW.

FCC Part 97 Privileges

License Class	Terrestrial location of the amateur radio station			
	Region 1 kHz	Region 2 kHz	Region 3 kHz	Limitations*
General	14025-14150 14225-14350	14025-14150 14225-14350	14025-14150 14225-14350	
Advanced	14025-14150 14175-14350	14025-14150 14175-14350	14025-14150 14175-14350	
Extra	14000-14350	14000-14350	14000-14350	

*See §§97.301-307 of the FCC rules for further details on authorized privileges for each license class, frequency sharing requirements and authorized modes.

Amateur Radio Band Plans

Region 1 kHz	Region 2 kHz	Region 3 kHz	ARRL kHz
14000-14060 CW Contests	14000-14070 CW	14000-14070 CW	14000-14070 CW
14060-14070 CW			
14070-14089 DATA/CW	14070-14095 DATA/CW	14070-14099.5 DATA/CW	14070-14095 DATA/CW
14089-14099 Packet/DATA/CW	14095-14099.5 Packet/DATA/CW		14095-14099.5 Packet/DATA/CW
14099-14101 Beacons	14099.5-14100.5 Beacons	14099.5-14100.5 Beacons	14099.5-14100.5 Beacons
	14100.5-14112 Packet/DATA/ Phone	14100.5-14112 DATA/CW/Phone	14100.5-14112 Packet/CW
14101-14112 Packet/Phone/CW			
14112-14125 Phone/CW	14112-14150 Phone/CW permitted International traffic	14112-14225 Phone/CW	14112-14150 CW
14125-14225 Phone/CW Contests	14150-14225 Phone/CW permitted		14150-14350 Phone
14225-14235 Phone/SSTV/ CW Contests	14225-14235 SSTV/Phone/CW permitted	14225-14235 SSTV/CW/Phone	14230 SSTV
14235-14300 Phone/CW Contests	14235-14250 Phone/CW permitted	14235-14350 Phone/CW	
14300-14350 Phone/CW	14250-14340 Phone/CW permitted International traffic		
	14340-14350 Phone Emergency (CW permitted) International traffic		

Allocations							
INTERNATIONAL			**UNITED STATES**				
Region 1 kHz	*Region 2 kHz*	*Region 3 kHz*	*Band kHz*	*National Provisions*	*Government Allocation*	*Non-Government Allocation*	*Remarks*
18068-18168	AMATEUR 510 AMATEUR-SATELLITE 537 538		18068-18168	510		AMATEUR AMATEUR-SATELLITE	

510—For the use of the bands allocated to the amateur service at 3.5 MHz, 7.0 MHz, 10.1 MHz, 14.0 MHz, 18.068 MHz, 21.0 MHz, 24.89 MHz and 144 MHz in the event of natural disasters, see Resulotion 640 at the end of the table.

537—The band 18068-18168 kHz is allocated to the fixed service on a primary basis subject to the procedure described in Resolution 8. The use of this band by the amateur and amateur-satellite services shall be subject to the comple-tion of satisfactory transfer of all assignments to stations in the fixed service operating in this band and recorded in the Master Register, in accordance with the procedure described in Resolution 8.

538—Additional allocation: in the CIS, the band 18068-18168 kHz is also allo-cated to the fixed service on a primary basis for use within the boundary of the CIS, with a peak envelope power not exceeding 1 kW.

FCC Part 97 Privileges

License Class	Terrestrial location of the amateur radio station			
	Region 1 kHz	*Region 2 kHz*	*Region 3 kHz*	*Limitations**
General Advanced Extra	18068-18168	18068-18168	18068-18168	

*See §§97.301-307 of the FCC rules for further details on authorized privileges for each license class, frequency sharing requirements and authorized modes.

Amateur Radio Band Plans

Region 1 kHz	*Region 2 kHz*	*Region 3 kHz*	*ARRL kHz*
18068-18100 CW	18068-18100 CW	18068-18100 CW	18068-18100 CW
18100-18110 DATA/CW	18100-18105 DATA/CW	18100-18110 DATA/CW	18100-18105 DATA/CW
	18105-18110 Packet/DATA/CW		18105-18110 Packet/DATA/CW
18110-18168 Phone/CW	18110-18168 Phone/CW	18110-18168 Phone/CW	18110-18168 Phone/CW

Allocations

INTERNATIONAL			UNITED STATES				
Region 1 kHz	Region 2 kHz	Region 3 kHz	Band kHz	National Provisions	Government Allocation	Non-Government Allocation	Remarks
21000-21450	AMATEUR 510 AMATEUR-SATELLITE		21000-21450	510		AMATEUR AMATEUR-SATELLITE	

510—For the use of the bands allocated to the amateur service at 3.5 MHz, 7.0 MHz, 10.1 MHz, 14.0 MHz, 18.068 MHz, 21.0 MHz, 24.89 MHz and 144 MHz in the event of natural disasters, see Resulotion 640 at the end of the table.

FCC Part 97 Privileges

License Class	Terrestrial location of the amateur radio station			
	Region 1 kHz	Region 2 kHz	Region 3 kHz	Limitations*
Novice**	21100-21200	21100-21200	21100-21200	
General	21025-21200 21300-21450	21025-21200 21300-21450	21025-21200 21300-21450	
Advanced	21025-21200 21225-21450	21025-21200 21225-21450	21025-21200 21225-21450	
Extra	21000-21450	21000-21450	21000-21450	

* See §§97.301-307 of the FCC rules for further details on authorized privileges for each license class, frequency sharing requirements and authorized modes.

** Technicians having passed Element 1(A) (5 WPM code) are authorized to use Novice HF bands.

Amateur Radio Band Plans

Region 1 kHz	Region 2 kHz	Region 3 kHz	ARRL kHz
21000-21080 CW	21000-21070 CW	21000-21070 CW	21000-21070 CW
	21070-21090 DATA/CW	21070-21125 DATA/CW	21070-21090 DATA/CW
21080-21100 DATA/CW	21090-21125 Packet/DATA/CW		21090-21100 Packet/DATA/CW
21100-21120 Packet/DATA/CW			21100-21149.5 CW
21120-21149 CW	21125-21149.5 CW	21125-21149.5 CW	
21149-21151 Beacons	21149.5-21150.5 Beacons	21149.5-21150.5 Beacons	21149.5-21150.5 Beacons
21151-21335 Phone/CW	21150.5-21200 Phone/CW permitted International traffic	21150.5-21335 Phone/CW	21150.5-21200 CW
	21200-21300 Phone/CW permitted		21200-21450 Phone
	21300-21335 Phone/CW permitted International traffic		
21335-21345 SSTV/CW	21335-21345 SSTV/Phone (CW permitted)	21335-21345 SSTV/CW/Phone	21340 SSTV
21345-21450 Phone/CW	21345-21440 Phone (CW permitted)	21345-21450 Phone/CW	
	21440-21450 Phone Emergency (CW permitted)		

Allocations

INTERNATIONAL			UNITED STATES				
Region 1 kHz	Region 2 kHz	Region 3 kHz	Band kHz	National Provisions	Government Allocation	Non-Government Allocation	Remarks
24890-24990	AMATEUR 510 AMATEUR-SATELLITE 542 543		24890-24990	510		AMATEUR AMATEUR-SATELLITE	

510—For the use of the bands allocated to the amateur service at 3.5 MHz, 7.0 MHz, 10.1 MHz, 14.0 MHz, 18.068 MHz, 21.0 MHz, 24.89 MHz and 144 MHz in the event of natural disasters, see Resulotion 640 at the end of the table.

542—Additional allocation: in Kenya, the band 23600-24900 kHz is also allocated to the meteorological aids service (radiosondes) on a primary basis.

543—The band 24890-24990 kHz is allocated to the fixed and land mobile services on a primary basis subject to the procedure described in Resolution 8. The use of this band by the amateur and amateur-satellite services shall be subject to the completion of the satisfactory transfer of all assignments to fixed and land mobile stations operating in this band and recorded in the Master Register, in accordance with the procedure described in Resolution 8.

FCC Part 97 Privileges

License Class	Terrestrial location of the amateur radio station			
	Region 1 kHz	Region 2 kHz	Region 3 kHz	Limitations*
General Advanced Extra	24890-24990	24890-24990	24890-24990	

*See §§97.301-307 of the FCC rules for further details on authorized privileges for each license class, frequency sharing requirements and authorized modes.

Amateur Radio Band Plans

Region 1 kHz	Region 2 kHz	Region 3 kHz	ARRL kHz
24890-24920 CW	24890-24920 CW	24890-24920 CW	24890-24920 CW
24920-24930 DATA/CW	24920-24925 DATA/CW	24920-24930 DATA/CW	24920-24925 DATA/CW
	24925-24930 Packet/DATA/CW		24925-24930 Packet/DATA/CW
24930-24990 Phone/CW	24930-24990 Phone/CW	24930-24990 Phone/CW	24930-24990 Phone/CW

License Class	Terrestrial location of the amateur radio station			
	Region 1 kHz	Region 2 kHz	Region 3 kHz	Limitations*
Novice Technician	28100-28500	28100-28500	28100-28500	
General Advanced Extra	28000-29700	28000-29700	28000-29700	

* See §§97.301-307 of the FCC rules for further details on authorized privileges for each license class, frequency sharing requirements and authorized modes.

Amateur Radio Band Plans

Region 1 kHz	Region 2 kHz	Region 3 kHz	ARRL kHz
28000-28050 CW	28000-28070 CW	28000-28050 CW	28000-28070 CW
28050-28120 DATA/CW	28070-28120 DATA/CW	28050-28150 DATA/CW	28070-28120 DATA/CW
28120-28150 Packet/DATA/CW	28120-28189 Packet/DATA/CW		28120-28189 Packet/DATA/CW
28150-28190 CW		28150-28190 CW	
28190-28200 CW New Beacon Band	28189-28200 CW New Beacon Band	28190-28200 CW New Beacon Band	28189-28200 CW New Beacon Band
28200-28300 Phone/CW Old Beacon Band	28200-28300 CW Old Beacon Band	28200-28300 CW Old Beacon Band	28200-28300 CW Old Beacon Band
28300-28675 Phone/CW	28300-28670 Phone CW permitted	28300-28675 Phone/CW	28300-29300 Phone
28675-28685 SSTV/Phone/CW	28670-28690 SSTV/Phone CW permitted	28675-28685 SSTV/CW/Phone	28680 SSTV
28685-29200 Phone/CW	28690-29300 Phone CW permitted	28685-29300 Phone/CW	
29200-29300 Packet/Phone/CW			
29300-29550 Satellite down-links	29300-29510 Satellite	29300-29510 Satellite	29300-29510 Satellite
29550-29700 Phone/CW	29510-29700 FM Phone and repeaters	29510-29700 WB/FM/CW	29510-29590 Repeater inputs
			29600 FM simplex calling frequency
			29610-29700 Repeater outputs

Allocations							
INTERNATIONAL			UNITED STATES				
Region 1 MHz	Region 2 MHz	Region 3 MHz	Band MHz	National Provisions	Government Allocation	Non-Government Allocation	Remarks
47-68 BROADCASTING	50-54		50-54			AMEATEUR	
	AMATEUR						
553 554 555 559 561	556 557 558 560						

553—Additional allocation: in Hungary, Kenya, Mongolia, Czechoslovakia and the U.S.S.R., the bands 47-48.5 MHz and 56.5-58 MHz are also allocated to the fixed and land mobile services on a secondary basis.

554—Additional allocation: in Albania, the Federal Republic of Germany, Austria, Belgium, Bulgaria, Cote d'Ivoire, Denmark, Spain, Finland, France, Gabon, Greece, Ireland, Israel, Italy, Jordan, the Lebanon, Libya, Liechtenstein, Luxembourg, Madagascar, Mali, Malta, Morocco, Mauritania, Monaco, Nigeria, Norway, the Netherlands, Poland, the United Kingdom, Senegal, Sweden, Switzerland, Swaziland, Syria, Togo, Tunisia, Turkey and Yugoslavia, the band 47-68 MHz, and in Romania, the band 47-58 MHz, are also allocated to the land mobile service on a permitted basis. However, stations of the land mobile service in the countries mentioned in connection with each band referred to in this footnote shall not cause harmful interference to, or claim protection from, existing or planned broadcasting stations of countries other than those mentioned in connection with the band.

555—Additional allocation: in Angola, Cameroon, the Congo, Madagascar, Mozambique, Somalia, Sudan, Tanzania, Chad and Yemen, the band 47-68 MHz is also allocated to the fixed and mobile, except aeronautical mobile, services on a permitted basis.

556—Alternative allocation: in New Zealand, the band 50-51 MHz is allocated to the fixed, mobile and broadcasting services on a primary basis, the band 53-54 MHz is allocated to the fixed and mobile services on a primary basis.

557—Alternative allocation: in Afghanistan, Bangladesh, Brunei, India, Indonesia, Iran, Malaysia, Pakistan, Singapore and Thailand, the band 50-54 MHz is allocated to the fixed, mobile and broadcasting services on a primary basis.

558—Additional allocation: in Australia, China and the Democratic People's Republic of Korea, the band 50-54 MHz is also allocated to the broadcasting service on a primary basis.

559—Alternative allocation: in Botswana, Burundi, Lesotho, Malawi, Namibia, Rwanda, South Africa, Swaziland, Zaire, Zambia and Zimbabwe, the band 50-54 MHz is allocated to the amateur service on a primary basis.

560—Additional allocation: in New Zealand, the band 51-53 MHz is also allocated to the fixed and mobile services on a primary basis.

561—Additional allocation: in Botswana, Burundi, Lesotho, Malawi, Mali, Namibia, Rwanda, South Africa, Swaziland, Zaire, Zambia and Zimbabwe, the band 54-68 MHz is also allocated to the fixed and mobile, except aeronautical mobile, services on a primary basis.

FCC Part 97 Privileges

License Class	Terrestrial location of the amateur radio station			
	Region 1 MHz	Region 2 MHz	Region 3 MHz	Limitations*
Technician General Advanced Extra	—	50-54	50-54	

* See §§97.301-307 of the FCC rules for further details on authorized privileges for each license class, frequency sharing requirements and authorized modes.

Amateur Radio Band Plans

Region 1† MHz	Region 2 MHz	Region 3 MHz	ARRL MHz
50.00-50.10 CW		50.0-50.10 CW/Beacons	50.00-50.10 CW/Beacons
50.02-50.08 Beacons			50.06-50.08 Automatically Controlled beacons
50.09 Center of CW activity		50.10-54 CW/Phone/WB	50.10-50.30 SSB, CW
50.10-50.50 All narrow-band modes			50.10-50.125 DX Window
			50.125 SSB Calling
50.10-50.13 SSB/CW/DX			50.30-50.6 All Modes
50.11 DX calling			50.60-50.80 Nonvoice Communications
50.185 Center of activity for crossband working			50.62 Digital calling
50.20 SSB calling			50.80-51.00 Radio Control (20 kHz channels)
50.30 CW meteor scatter reference			51.00-51.10 Pacific DX Window
50.35 SSB meteor scatter reference			51.12-51.48 Repeater inputs (19 channels)
			51.12-51.18 Digital repeater inputs
50.50-51.00 All modes			51.50-51.60 Simplex (6 channels)
50.60 RTTY			51.62-51.98 Repeater outputs (19 channels)
50.62-50.75 Packet			51.62-51.68 Digital repeater outputs
51.00-52.00 All modes			52.0-52.48 Repeater inputs (except as noted; 23 channels)
51.11 VK/ZL calling			52.02, 52.04 FM simplex
			52.2 TEST PAIR (input)
51.41-51.59 FM			52.50-52.98 Repeater outputs (except as noted; 23 channels)
51.51 FM Calling			52.525 Primary FM simplex
			52.54 Secondary FM simplex
			52.7 TEST PAIR (output)
			53.0-53.48 Repeater inputs (except as noted; 19 channels)
			53.0 Remote base FM simplex
			53.02 Simplex
			53.1, 53.2, 53.3, 53.4 Radio remote control
			53.50-53.98 Repeater outputs (except as noted; 19 channels)
			53.5, 53.6, 53.7, 53.8 Radio remote control
			53.52, 53.9 Simplex

Allocations

INTERNATIONAL			UNITED STATES				
Region 1 MHz	Region 2 MHz	Region 3 MHz	Band MHz	National Provisions	Government Allocation	Non-Government Allocation	Remarks
144-146	AMATEUR AMATEUR-SATELLITE 605 606		144-146	510		AMATEUR AMATEUR-SATELLITE	
146-149.9 FIXED MOBILE except aeronautical mobile (R)	146-148 AMATEUR 607	146-148 AMATEUR FIXED MOBILE 607	146-148			AMATEUR	

510—For the use of the bands allocated to the amateur service at 3.5 MHz, 7.0 MHz, 10.1 MHz, 14.0 MHz, 18.068 MHz, 21.0 MHz, 24.89 MHz and 144 MHz in the event of natural disasters, see Resolution 640. (Resolution 640 may be found in *The FCC Rule Book*.)

605—Additional allocation: in Singapore, the band 144-145 MHz is also allocated to the fixed and mobile services on a primary basis. Such use is limited to systems in operation on or before 1 January 1980, which in any case shall cease by 31 December 1995.

606—Additional allocation: in China, the band 144-146 MHz is also allocated to the aeronautical mobile (OR) service on a secondary basis.

607—Alternative allocation: in Afghanistan, Bangladesh, Cuba, Guyana and India, the band 146-148 MHz is allocated to the fixed and mobile services on a primary basis.

FCC Part 97 Privileges

License Class	Terrestrial location of the amateur radio station			
	Region 1 MHz	Region 2 MHz	Region 3 MHz	Limitations*
Technician General Advanced Extra	144-146	144-148	144-148	

* See §§97.301-307 of the FCC rules for further details on authorized privileges for each license class, frequency sharing requirements and authorized modes.

Amateur Radio Band Plans

Region 1 MHz	Region 2 MHz	Region 3 MHz	ARRL* MHz
144.00-144.15 CW exclusive		144.00-144.025 EME	144.00-144.05 EME (CW)
144.15-144.50 CW/SSB			144.05-144.06 General CW and weak signals
144.30 SSB calling			144.06-144.10 General CW and weak signals
144.40 SSB meteor scatter reference frequency			144.10-144.20 EME and weak-signal SSB
144.50 SSTV calling			144.20 National calling
144.50-144.845 All modes			144.20-144.275 General SSB operation
144.60 RTTY calling			144.275-144.30 Beacons
144.625-144.675 Digital			144.30-144.50 New OSCAR subband
144.70 FAX calling			144.50-144.60 Linear translation inputs
144.75 ATV calling			144.60-144.90 FM repeater inputs
144.845-144.99 Beacons			144.90-145.10 Weak signal and FM simplex
144.99-145.20 FM repeater inputs			145.10-145.20 Linear translator outputs
145.20-145.60 FM simplex channels			145.10-145.50 FM repeater outputs
145.60-145.80 FM repeater outputs			145.50-145.80 Miscellaneous and experimental modes
145.80-146.00 Satellites		145.80-146.00 Satellite	145.80-146.00 OSCAR subband
			146.01-146.37 Repeater inputs
			146.40-146.58 Simplex**
			146.61-147.39 Repeater outputs
			147.42-147.57 Simplex
			147.60-147.99 Repeater inputs

Region 1:

Repeater (input/output):

				Simplex channels:	
145.00	R0	145.60	R0	145.25	S10
145.025	R1	145.625	R1	145.275	S11
145.05	R2	145.65	R2	145.30	S12
145.075	R3	145.675	R3	145.325	S13
145.10	R4	145.70	R4	145.35	S14
145.125	R5	145.725	R5	145.375	S15
145.15	R6	145.75	R6	145.40	S16
145.175	R7	145.775	R7	145.425	S17

Simplex channels:

145.45	S18
145.475	S19
145.50	S20 Most common simplex calling freq; once contact established move off frequency
145.525	S21
145.55	S22
145.575	S23

U.K. & European repeaters require a 1750 Hz (±25 Hz) tone burst of about 200 ms to open the repeater and in practice it is best to transmit this tone at the start of each transmission, especially in the U.K.

ARRL

*This band plan has been proposed by the ARRL VHF-UHF Advisory Committee.

**The frequency 146.40 MHz is used in some areas as a repeater input.

The following Packet Radio frequency recommendations were adopted by the ARRL Board of Directors in July 1987.

1) Automatic/unattended operations should be conducted on 145.01, 145.03, 145.05, 145.07 and 145.09 MHz.

    a) 145.01 should be reserved for inter-LAN use.

    b) Use of the remaining frequencies (above) should be determined by local user groups.

2) Additional frequencies within the 2-meter band may be designated for packet-radio use by local coordinators.

Packet Footnotes

Specific VHF/UHF channels recommended above may not be available in all areas of the US.

Prior to regular packet-radio use of any VHF/UHF channel, it is advisable to check with the local frequency coordinator.

The decision as to how the available channels are to be used should be based on coordination between local packet-radio users.

Repeater frequency pairs (input/output):

144.51/145.11	144.81/145.41	146.31/146.91
144.53/145.13	144.83/145.43	146.34/146.94
144.55/145.15	144.85/145.45	146.37/146.97
144.57/145.17	144.87/145.47	146.40 or 147.60/147.00*
144.59/145.19	144.89/145.49	146.43 or 147.63/147.03*
144.61/145.21	146.01/146.61	146.46 or 147.66/147.06*
144.63/145.23	146.04/146.64	147.69/147.09
144.65/145.25	146.07/146.67	147.72/147.12
144.67/145.27	146.10/146.70	147.75/147.15
144.69/145.29	146.13/146.73	147.78/147.18
144.71/145.31	146.16/146.76	147.81/147.21
144.73/145.33	146.19/146.79	147.84/147.24
144.75/145.35	146.22/146.82	147.87/147.27
144.77/145.37	146.25/146.85	147.90/147.30
144.79/145.39	146.28/146.88	147.93/147.33
		147.96/147.36
*Local Option		147.99/147.39

Some areas use 146.40-146.60 and 147.40-147.60 MHz for either simplex or repeater inputs and outputs. Frequency pairs in those areas are:

147.415/146.415	147.46/146.46	147.505/146.505
147.43/146.43	147.475/146.475	147.595/146.595
147.445/146.445	147.49/146.49	

Suggested 15-kHz splinter channels (input/output):

146.025/146.625	146.295/146.895	147.735/147.135
146.055/146.655	146.325/146.925	147.765/147.165
146.085/146.685	146.355/146.955	147.825/147.225
146.115/146.715	146.385/146.985	147.855/147.255
146.145/146.745	147.615/147.015	147.885/147.285
146.175/146.775	147.645/147.045	147.915/147.315
146.205/146.805	147.675/147.075	147.945/147.345
146.235/146.835	147.705/147.105	147.975/147.375
146.265/146.865		

Simplex frequencies:

*146.415	*146.475	146.535	146.595	147.465	147.525
*146.43	*146.49	146.55	147.42	147.48	147.54
*146.445	*146.505	146.565	147.435	147.495	147.555
*146.46	#146.52	146.58	147.45	147.51	147.57
					147.585

#National Simplex Frequency

*May also be a repeater (input/output). See repeater pairs listing.

Several states have chosen to realign the 146-148 MHz band, using 20 kHz spacing between channels. This choice was made to gain additional repeater pairs.

The transition from 30 to 20 kHz spacing is taking place on a case by case basis as the need for additional pairs occurs. Typically the repeater on an odd numbered pair will shift to 10 kHz, up or down, creating a new set on an even numbered channel. For example, the pair of 146.13/.73 would change to 146.12/72 or 146.14/.74 while the pairs of 146.10/.70 and 146.16/.76 would be left unchanged.

Allocations

INTERNATIONAL			UNITED STATES				
Region 1 MHz	Region 2 MHz	Region 3 MHz	Band MHz	National Provisions	Government Allocation	Non-Government Allocation	Remarks
223-230 BROADCASTING Fixed Mobile 622 628 629 631 632 633 634 635	220-225 AMATEUR FIXED MOBILE Radiolocation 627 225-235 FIXED MOBILE	223-230 FIXED MOBILE BROADCASTING AERONAUTICAL RADIONAVIGATION Radiolocation 636 637	216-220	627 G2 NG152	MARITIME MOBILE Aeronautical-Mobile Fixed Land Mobile Radiolocation	MARITIME MOBILE Aeronautical-Mobile Fixed Land Mobile	Amateur 219-220 MHz
			222-225	US243 627	Radiolocation	AMATEUR	

622—Different category of service: in the Federal Republic of Germany, Austria, Belgium, Denmark, Spain, Finland, France, Israel, Italy, Liechtenstein, Luxembourg, Monaco, Norway, the Netherlands, Portugal, the United Kingdom, Sweden, Switzerland, and Yemen, the band 223-230 MHz is allocated to the land mobile service on a permitted basis (see No. 425). However, the stations of the land mobile service shall not cause harmful interference to, nor claim protection from, broadcasting stations, existing or planned, in countries other than those listed in this footnote.

623—Additional allocation: in the Congo, Ethiopia, Gambia, Guinea, Kenya, Libya, Malawi, Mali, Uganda, Senagal, Sierra Leone, Somalia, Tanzania and Zimbabwe, the band 174-223 MHz is also allocated to the fixed and mobile services on a secondary basis.

625—Additional allocation: in Australia and Papua New Guinea, the bands 204-208 MHz and 222-223 MHz are also allocated to the aeronautical radionavigation service on a primary basis.

626—Additional allocation: in China, India and Thailand, the band 216-223 MHz is also allocated to the aeronautical radionavigation service on a primary basis and to the radiolocation service on a secondary basis.

627—In Region 2, the band 216-225 MHz is allocated to the radiolocation service on a primary basis until 1 January 1990. On and after 1 January 1990, no new stations in that service may be authorized. Stations authorized prior to 1 January 1990 may continue to operate on a secondary basis.

628—Additional allocation: in Somalia, the band 216-225 MHz is also allocated to the aeronautical radionavigation service on a primary basis, subject to not causing harmful interference to existing or planned broadcasting services in other countries.

629—Additional allocation: in Oman, the United Kingdom and Turkey, the band 216-235 MHz is also allocated to the radiolocation service on a secondary basis.

630—Additional allocation: in Japan, the band 222-223 MHz is also allocated to the aeronautical radionavigation service on a primary basis and to the radiolocation service on a secondary basis.

631—Different category of service: in Spain and Portugal, the band 223-230 MHz is allocated to the fixed service on a permitted basis (see No. 425). Stations of this service shall not cause harmful interference to, or claim protection from, broadcasting stations of other countries, whether existing or planned, that operate in accordance with the Table.

632—Additional allocation: in Saudi Arabia, Bahrain, the United Arab Emirates, Israel, Jordan, Oman, Qatar and Syria, the band 223-235 MHz is also allocated to the aeronautical radionavigation service on a permitted basis.

633—Additional allocation: in Spain and Portugal, the band 223-235 MHz is also allocated to the aeronautical radionavigation service on a permitted basis until 1 January 1990, subject to not causing harmful interference to existing or planned broadcasting stations in other countries.

634—Additional allocation: in Sweden, the band 223-235 MHz is also allocated to the aeronautical radionavigation service on a permitted basis until 1 January 1990, subject to agreement obtained under the procedure set forth in Article 14, and on condition that no harmful interference is caused to existing and planned broadcasting stations in other countries.

635—Alternative allocation: in Botswana, Lesotho, Namibia, South Africa, Swaziland and Zambia, the bands 223-238 MHz and 246-254 MHz are allocated to the broadcasting service on a primary basis subject to agreement obtained under the procedure set forth in Article 14.

636—Alternative allocation: in New Zealand, Western Samoa and the Niue and Cook Islands, the band 225-230 MHz is allocated to the fixed, mobile and aeronautical radionavigation services on a primary basis.

637—Additional allocation: in China, the band 225-235 MHz is also allocated to the radio astronomy service on a secondary basis.

US243—In the band 220-225 MHz, stations in the radiolocation service have priority until 1 January 1990.

G2—In the bands 216-225, 420-450 (except as provided by US217) 890-902, 928-942, 1300-1400, 2300-2450, 2700-2900, 5650-5925, and 9000-9200 MHz, the Government radiolocation is limited to the military services.

NG152—The band 219-220 MHz is also allocated to the amateur service on a secondary basis for stations participating, as forwarding stations, in point-to-point fixed digital message forwarding systems, including intercity packet backbone networks.

FCC Part 97 Privileges

License Class	Terrestrial location of the amateur radio station			
	Region 1 MHz	Region 2 MHz	Region 3 MHz	Limitations*
Novice	—	222-225	—	
Technician General Advanced Extra		219-220* 222-225		219-220 MHz: point-to-point digital message forwarding stations only. Stations must register with ARRL 30 days prior to activation.

*See §§97.301-307 of the FCC rules for further details on authorized privileges for each license class, frequency sharing requirements and authorized modes.

NOTE: Effective August 28, 1991 the FCC allocated 220 to 222 MHz on an exclusive basis to the Land Mobile Service and allocated 222 to 225 MHz exclusively to the Amateur Radio Service.

Amateur Radio Band Plans

Region 1 MHz	Region 2 MHz	Region 3 MHz	ARRL MHz
			219-220**** Digital Links
			222.0-222.15 Weak-signal modes
			222.0-222.025 EME
			222.05-222.06 Propagation beacons

			222.10
			  Calling frequency
			  (SSB & CW)

			222.10-222.15 Weak-signal SSB & CW
			222.15-222.25 General operation, CW or SSB, etc.
			222.25-223.38 Repeater inputs**
			223.40-223.52 Simplex***
			223.52-223.64 Digital, packet
			223.64-223.70 Links, control
			223.71-223.85 Simplex, packet, repeater outputs
			223.85-224.98 Repeater outputs**

**Repeater frequency pairs (input/output):

***Simplex frequencies:

222.26/223.86	222.50/224.10	222.74/224.34	222.96/224.56	223.18/224.78
222.28/223.88	222.52/224.12	222.76/224.36	222.98/224.58	223.20/224.80
222.30/223.90	222.54/224.14	222.78/224.38	223.00/224.60	223.22/224.82
222.32/223.92	222.56/224.16	222.80/224.40	223.02/224.62	223.24/224.84
222.34/223.94	222.58/224.18	222.82/224.42	223.04/224.64	223.26/224.86
222.36/223.96	222.60/224.20	222.84/224.44	223.06/224.66	223.28/224.88
222.38/223.98	222.62/224.22	222.86/222.46	223.08/224.68	223.30/224.90
222.40/224.00	222.64/224.24	222.88/224.48	223.10/224.70	223.32/224.92
222.42/224.02	222.66/224.26	222.90/224.50	223.12/224.72	223.34/224.94
222.44/224.04	222.68/224.28	222.92/224.52	223.14/224.74	223.36/224.96
222.46/224.06	222.70/224.30	222.94/224.54	223.16/224.76	223.38/224.98
222.48/224.08	222.72/224.32			

223.42	223.52	223.62	223.72	223.82
223.44	223.54	223.64	223.74	223.84
223.46	223.56	223.66	223.76	
223.48	223.58	223.68	223.78	
223.50‡	223.60	223.70	223.80	

‡National simplex frequency

****100-kHz channels for point-to point fixed digital message-forwarding systems with a maximum of 50 watts PEP. Notification of ARRL 30 days before operation is required, and coordination with Automated Maritime Telecommunication System (AMTS) Coast Stations may be required; see Section 97.303(e).

Channel A	219.050	Channel F	219.550
Channel B	219.150	Channel G	219.650
Channel C	219.250	Channel H	219.750
Channel D	219.350	Channel I	219.850
Channel E	219.450	Channel J	219.950

INTERNATIONAL			UNITED STATES				
Region 1 MHz	Region 2 MHz	Region 3 MHz	Band MHz	National Provisions	Government Allocation	Non-Government Allocation	Remarks
420-430			420-450	US7 US87 US217 US228 US230 664 668	RADIOLOCATION	Amateur	
	FIXED MOBILE Except aeronautical mobile Radiolocation 651 652 653						
430-440 AMATEUR RADIOLOCATION 653 654 655 656 657 658 659 661 662 663 664 665	430-440 RADIOLOCATION Amateur 653 658 659 660 663 664						
440-450	FIXED MOBILE except aeronautical mobile Radiolocation 651 652 653 666 667 668				G2 G8	NG135	

651—Different category of service: in Australia, the United States, India, Japan and the United Kingdom, the allocation of the bands 420-430 MHz and 440-450 MHz to the radio location service is on a primary basis (see No. 425).

652—Additional allocation: in Australia, the United States, Jamaica and the Philippines, the bands 420-430 MHz and 440-450 MHz are also allocated to the amateur service on a secondary basis.

653—Additional allocation: in China, India, the Federal Republic of Germany, the United Kingdom and the U.S.S.R., the band 420-460 MHz is also allocated to the aeronautical radionavigation service (radio altimeters) on a secondary basis.

654—Different category of service: in France, the allocation of the band 430-434 MHz to the amateur service is on a secondary basis (see No. 424).

655—Different category of service: in Denmark, Libya, Norway and Sweden, the allocation of the bands 430-432 MHz and 438-440 MHz to the radiolocation service is on a secondary basis (see No. 424).

656—Alternative allocation: in Denmark, Norway and Sweden, the bands 430-432 MHz and 438-440 MHz are allocated to the fixed and mobile, except aeronautical mobile, services on a primary basis.

657—Additional allocation: in Finland, Libya and Yugoslavia, the bands 430-432 MHz and 438-440 MHz are also allocated to the fixed and mobile, except aeronautical mobile, services on a primary basis.

658—Additional allocation: in Afghanistan, Algeria, Saudi Arabia, Bahrain, Bangladesh, Brunei, Burundi, Egypt, the United Arab Emirates, Ecuador, Ethiopia, Greece, Guinea, India, Indonesia, Iran, Iraq, Israel, Italy, Jordan, Kenya, Kuwait, Lebanon, Libya, Liechtenstein, Malaysia, Malta, Nigeria, Oman, Pakistan, the Philippines, Qatar, Syria, Singapore, Somalia, Switzerland, Tanzania, Thailand and Togo, the band 430-440 MHz is also allocated to the fixed service on a primary basis and the bands 430-435 MHz and 438-440 MHz are also allocated to the mobile, except aeronautical mobile, service on a primary basis.

659—Additional allocation: in Angola, Bulgaria, Cameroon, the Congo, Gabon, Hungary, Mali, Mongolia, Niger, Poland, the Federal Republic of Germany, Romania, Rwanda, Chad, Czechoslovakia and the U.S.S.R., the band 430-440 MHz is also allocated to the fixed service on a primary basis.

660—Different category of service: in Argentina, Colombia, Costa Rica, Cuba, Guyana, Honduras, Panama and Venezuela, the allocation of the band 430-440 MHz to the amateur service is on a primary basis (see No. 425).

661—In Region 1, except in the countries mentioned in No. 662, the band 433.05-434.79 MHz (centre frequency 433.92 MHz) is designated for industrial, scientific and medical (ISM) applications. The use of this frequency band for ISM applications shall be subject to special authorization by the administration concerned, in agreement with other administrations whose radiocommunication services might be affected. In applying this provision, administrations shall have due regard to the latest relevant CCIR Recommendations.

662—In the Federal Republic of Germany, Austria, Liechtenstein, Portugal, Switzerland and Yugoslavia, the band 433.05-434.79 MHz (centre frequency 433.92 MHz) is designated for industrial, scientific and medical (ISM) applications. Radiocommunication services of these countries operating within this band must accept harmful interference which may be caused by these applications. ISM equipment operating in this band is subject to the provisions of No. 1815.

663—Additional allocation: in Brazil, France and the French Overseas Departments in Region 2, and India, the band 433.75-434.25 MHz is also allocated to the space operation service (Earth-to-space) on a primary basis until 1 January 1990, subject to agreement obtained under the procedure set forth in Article 14. After 1 January 1990, the band 433.75-434.25 MHz will be allocated in the same countries to the same service on a secondary basis.

664—In the bands 435-438 MHz, 1260-1270 MHz, 2400-2450 MHz, 3400-3410 MHz (in Regions 2 and 3 only) and 5650-5670 MHz, the amateur-satellite service may operate subject to not causing harmful interference to other services operating in accordance with the Table (see No. 435). Administrations authorizing such use shall ensure that any harmful interference caused by emissions from a station in the amateur-satellite service is immediately eliminated in accordance with the provisions of No. 2741. The use of the bands 1260-1270 MHz and 5650-5670 MHz by the amateur-satellite service is limited to the Earth-to-space direction.

664A—In Mexico, the bands 430-435 MHz are also allocated on a primary basis to the land

mobile service, subject to agreement obtained under the procedure set forth in Article 14.

665—Additional allocation: in Austria, the band 438-440 MHz is also allocated to the fixed and mobile, except aeronautical mobile, services on a primary basis.

666—Additional allocation: in Canada, New Zealand and Papua New Guinea, the band 440-450 MHz is also allocated to the amateur service on a secondary basis.

667—Different category of service: in Canada, the allocation of the band 440-450 MHz to the radiolocation service is on a primary basis (see No. 425).

668—Subject to agreement obtained under the procedure set forth in Article 14, the band 449.75-450.25 MHz may be used for the space operation service (Earth-to-space) and the space research service (Earth-to-space).

US7—In the band 420-450 MHz and within the following areas, the peak envelope power output of a transmitter employed in the amateur service shall not exceed 50 watts, unless expressly authorized by the Commission after mutual agreement, on a case-by-case basis, between the Federal Communications Commission Engineer in Charge at the applicable District Office and the Military Area Frequency Coordinator at the applicable military base:

(a) Those portions of Texas and New Mexico bounded on the south by latitude 31° 45' North, on the east by 104° 00' West, on the north by latitude 34° 30' North, and on the west by longitude 107° 30' West;

(b) The entire State of Florida including the Key West area and the areas enclosed within a 200-mile radius of Patrick Air Force Base, Florida (latitude 28° 21' North, longitude 80° 43' West), and within a 200-mile radius of Eglin Air Force Base, Florida (latitude 30° 30' North, longitude 86° 30' West);

(c) The entire State of Arizona;

(d) Those portions of California and Nevada south of Latitude 37° 10' North, and the areas enclosed within a 200-mile radius of the Pacific Missile Test Center, Point Mugu, California (latitude 34° 09' North, longitude 119° 11' West).

(e) In the State of Massachusetts within a 160-kilometer (100 mile) radius around locations at Otis Air Force Base, Massachusetts (latitude 41° 45' North, longitude 70° 32' West).

(f) In the State of California within a 240-kilometer (150 mile) radius around locations at Beale Air Force Base, California (latitude 39° 08' North, longitude 121° 26' West).

(g) In the State of Alaska within a 160-kilometer (100 mile) radius of Clear, Alaska (latitude 64° 17' North, longitude 149° 10' West). (The Military Area Frequency Coordinator for this area is located at Elmendorf Air Force Base, Alaska.)

(h) In the State of North Dakota within a 160-kilometer (100 mile) radius of Concrete, North Dakota (latitude 48° 43' North, longitude 97° 54' West). (The Military Area Frequency Coordinator for this area can be contacted at: HQ SAC/SXOE, Offutt Air Force Base, Nebraska 68113.)

US87—The frequency 450 MHz, with maximum emission bandwidth of 500 kHz, may be used by Government and non-Government stations for space telecommand at specific locations, subject to such conditions as may be applied on a case-by-case basis.

US217—Pulse-ranging radiolocation systems may be authorized for Government and non-Government use in the 420-450 MHz band along the shorelines of Alaska and the contiguous 48 States. Spread spectrum radiolocation systems may be authorized in the 420-435 MHz portion of the band for operation within the contiguous 48 States and Alaska. Authorizations will be granted on a case-by-case basis; however, operations proposed to be located within the zones set forth in US228 should not expect to be accommodated. All stations operating in accordance with this provision will be secondary to stations operating in accordance with the Table of Frequency Allocations.

US228—Applicants of operation in the band 420 to 450 MHz under the provisions of US217 should not expect to be accommodated if their area of service is within the following geographic areas:

(a) Those portions of Texas and New Mexico bounded on the south by latitude 31° 45' North, on the east by 104° 00' West, on the north by latitude 34\0/30' North, and on the west by longitude 107° 30' West;

(b) The entire State of Florida including the Key West area and the areas enclosed within a 200-mile radius of Patrick Air Force Base, Florida (latitude 28° 21' North, longitude 80° 43' West), and within a 200-mile radius of Eglin Air Force Base, Florida (latitude 30° 30' North, longitude 86° 30' West);

(c) The entire State of Arizona;

(d) Those portions of California and Nevada south of Latitude 37° 10' North, and the areas enclosed within a 200-mile radius of the Pacific Missile Test Center, Point Mugu, California (latitude 34° 09' North, longitude 119° 11' West).

(e) In the State of Massachusetts within a 160-kilometer (100 mile) radius around locations at Otis Air Force Base, Massachusetts (latitude 41° 45' North, longitude 70° 32' West).

(f) In the State of California within a 240-kilometer (150 mile) radius around locations at Beale Air Force Base, California (latitude 39\0/08' North, longitude 121° 26' West).

(g) In the State of Alaska within a 160-kilometer (100 mile) radius of Clear, Alaska (latitude 64° 17' North, longitude 149° 10' West). (The Military Area Frequency Coordinator for this area is located at Elmendorf Air Force Base, Alaska.)

(h) In the State of North Dakota within a 160-kilometer (100 mile) radius of Concrete, North Dakota (latitude 48° 43' North, longitude 97° 54' West). (The Military Area Frequency Coordinator for this area can be contacted at: HQ SAC/SXOE, Offutt Air Force Base, Nebraska 68113.)

(i) In the States of Alabama, Florida, Georgia and South Carolina within a 200-kilometer (124 mile) radius of Warner Robins Air Force Base, Georgia (latitude 32° 38' North, longitude 83° 35' West).

(j) In the State of Texas within a 200-kilometer (124 mile) radius of Goodfellow Air Force Base, Texas (latitude 31° 25' North, longitude 100° 24' West).

G2—In the bands 216-225, 420-450 (except as provided by US217), 890-902, 928-942, 1300-1400, 2300-2450, 2700-2900, 5650-5925, and 9000-9200 MHz, the Government radiolocation is limited to the military services.

US230—Non-government land mobile service is allocated on a primary basis in the bands 422.1875-425.4875 and 427.1875-429.9875 MHz within 50 statute miles of Detroit, MI, and Cleveland, OH, and in the bands 423.8125-425.4875 and 428.8125-429.9875 MHz within 50 statute miles of Buffalo, NY.

G8—Low power government radio control operations are permitted in the band 420-450 MHz.

NG135—In the 420-430 MHz band the Amateur service is not allocated north of line A. (def. §2.1). All amateur radio stations shall operate north of line A in accordance with the Agreement between the United States and Canada.

FCC Part 97 Privileges

License Class	Terrestrial location of the amateur radio station			
	Region 1 MHz	Region 2 MHz	Region 3 MHz	Limitations*
Technician General Advanced Extra	430-440	420-450	420-450	

*See §§97.301-307 of the FCC rules for futher details on authorized privileges for each license class, frequency sharing requirement and authorized modes.

The following Packet Radio frequency recommendations were adopted by the ARRL Board of Directors in January 1988.

1) 100-kHz-bandwidth channels:

430.05	430.35	430.65
430.15	430.45	430.85
430.25	430.55	430.95

2) 25-kHz-bandwidth channels:

431.025	441.000	441.050
440.975	441.025	441.075

Amateur Radio Band Plans

Region 1† MHz	Region 2 MHz	Region 3 MHz	ARRL* MHz
			420.00-426.00 ATV repeater or simplex with 421.25-MHz video carrier, control links and experimental
		430.00-431.90 ATV simplex with 427.25-MHz video carrier	426.00-432.00
		431.90-432.24 EME	
432.00-432.15 CW only			432.00-432.07 EME
432.00-432.025 EME			432.07-432.10 Weak-signal CW
432.05 CW center of activity			432.10 Calling frequency
432.15-432.50 SSB/CW		432.24-435.00	432.10-432.30 Mixed mode and weak signal
432.20 SSB center of activity			432.30-432.40 Beacons
432.35 microwave talkback			
432.50 SSTV center of activity			432.40-433.00 Mixed mode and weak signals
432.50-432.80 All modes			433.00-435.00 Auxiliary/repeater links
432.60 RTTY			
432.70 FAX		435.00-438.00 Satellite	435.00-438.00 Satellite only (internationally)
432.80-433.00 Beacons		438.00-450.00	438.00-444.00 ATV repeater input with 439.25-MHz video carrier and repeater links
433.00-435.00			
433.30 RTTY			
433.50 FM calling			442.00-445 Repeater inputs and outputs (local option)
433.60 RTTY			445.00-447.00 Shared by auxiliary and control links, repeaters and simplex (local option)
435.00-438.00 Satellites			
438.00-440.00			446.00 National simplex frequency
			447.00-450.00 Repeater inputs and outputs (local option)

*This band is under review by the ARRL VHF-UHF Advisory Committee.

Allocations							
INTERNATIONAL			UNITED STATES				
Region 1 MHz	Region 2 MHz	Region 3 MHz	Band MHz	National Provisions	Government Allocation	Non-Government Allocation	Remarks
890-942 FIXED MOBILE except aeronautical mobile BROADCASTING 703 Radiolocation	902-928 FIXED Amateur Mobile except aeronautical mobile Radiolocation	890-942 FIXED MOBILE BROADCASTING Radiolocation	902-928	US215 US218 US267 US275 707	RADIOLOCATION	PRIVATE LAND MOBILE AMATEUR	(ISM 915 ± 13 MHz)
704	705 707 707A	706			G11 G59		

704—Additional allocation: in Bulgaria, Hungary, Mongolia, Poland, the Federal Republic of Germany, Romania, Czechoslovakia and the U.S.S.R., the band 862-960 MHz is also allocated to the aeronautical radionavigation service on a permitted basis until 1 January 1998. Up to this date, the aeronautical radionavigation service may use the band, subject to agreement obtained under the procedure set forth in Article 14. After this date, the aeronautical radionavigation service may continue to operate on a secondary basis.

705—Different category of service: in the United States, the allocation of the band 890-942 MHz to the radiolocation service is on a primary basis (see No. 425) and subject to agreement obtained under the procedure set forth in Article 14.

706—Different category of service: in Australia, the allocation of the band 890-942 MHz to the radiolocation service is on a primary basis (see No. 425).

707—In Region 2, the band 902-928 MHz (center frequency 915 MHz) is designated for industrial, scientific and medical (ISM) applications. Radiocommunication services operating within this band must accept harmful interference which may be caused by these applications. ISM equipment operating in this band is subject to the provisions of No. 1815.

707A—Different category of service: in Chile, the band 903-905 MHz is allocated to the mobile, except aeronautical mobile, service on a primary basis and is subject to agreement obtained under the procedure set forth in Article 14.

US215—Emissions from microwave ovens manufactured on and after January 1, 1980, for operation on the frequency 915 MHz must be confined within the band 902-928 MHz. Emissions from microwave ovens manufactured prior to January 1, 1980, for operation on the frequency 915 MHz must be confined within the band 902-940 MHz. Radiocommunications services operating within the band 928-940 MHz must accept any harmful interference that may be experienced from the operation of microwave ovens manufactured before January 1, 1980.

US218—The band segments 902-912 MHz and 918-928 MHz are available for Location and Monitoring Service (LMS) subject to not causing harmful interference to the operation of Government stations authorized in these bands. These systems must tolerate any interference from the operation of industrial, scientific, and medical (ISM) devices and the operation of Government stations authorized in these bands.

US267—In the band 902-928 MHz, amateur radio stations shall not operate within the States of Colorado and Wyoming, bounded by the area of: latitude 39° N to 42° N and longitude 103° W to 108° W.

US275—The band 902-928 MHz is allocated on a secondary basis to the amateur service subject to not causing harmful interference to the operations of Government stations authorized in this band or to the Location and Monitoring Service (LMS) systems. Stations in the amateur service must tolerate any interference from the operations of industrial, scientific and medical (ISM) devices, LMS systems, and the operations of Government stations authorized in this band.

G11—Government fixed and mobile radio services, including low power radio control operations, are permitted in the band 902-928 MHz on a secondary basis.

G59—In the bands 902-928 MHz, 3100-3300 MHz, 3500-3700 MHz, 5250-5350 MHz, 8500-9000 MHz, 9200-9300 MHz, 13.4-14.0 GHz, 15.7-17.7 GHz and 24.05-24.25 GHz, all Government non-military radiolocation shall be secondary to military radiolocation, except in the sub-band 15.7-16.2 GHz, airport surface detection equipment (ASDE) is permitted on a co-equal basis subject to coordination with the military departments.

FCC Part 97 Privileges				
	Terrestrial location of the amateur radio station			
License Class	Region 1 MHz	Region 2 MHz	Region 3 MHz	Limitations*
Technician General Advanced Extra	—	902-928	—	

*See §§97.301-307 of the FCC rules for further details on authorized privileges for each license class, frequency sharing requirements and authorized modes.

Amateur Radio Band Plans

Region 1 MHz	Region 2 MHz	Region 3 MHz	ARRL* MHz
			902.0-903.0 Weak signal communications
			902.1 Calling Frequency
			903.0-906.0 Digital
			906.0-909.0 FM Repeater outputs
			909.0-915.0 ATV
			915.0-918.0 Digital
			918.0-921.0 FM repeater inputs
			921.0-927.0 ATV
			927.0-928.0 FM simplex and links

*Adopted by the ARRL Board of Directors meeting in July 1989.

The following Packet Radio frequency recommendations were adopted by the ARRL Board of Directors in January 1988 as interim guidance.

Two 3-MHz-bandwidth channels are recommended for 1.5 Mbit/s links. They are 903-906 MHz and 915-918 MHz with 10.7-MHz spacing.

Packet Footnotes

Specific VHF/UHF channels recommended above may not be available in all areas of the US. These channels may interfere with existing weak-signal communications operating on or about 903.100 MHz.

Prior to regular packet-radio use of any VHF/UHF channel, it is advisable to check with the local frequency coordinator.

The decision as to how the available channels are to be used should be based on coordination between local packet-radio users.

Allocations

INTERNATIONAL			UNITED STATES				
Region 1 MHz	Region 2 MHz	Region 3 MHz	Band MHz	National Provisions	Government Allocation	Non-Government Allocation	Remarks
1240-1260	RADIOLOCATION RADIONAVIGATION-SATELLITE (Space-to-Earth) 710 Amateur 711 712 712A 713 714		1240-1300	664 713 714	RADIOLOCATION	Amateur	
1260-1300	RADIOLOCATION Amateur 664 711 712 712A 713 714				G56		

664— In the bands 435-438 MHz, 1260-1270 MHz, 2400-2450 MHz, 3400-3410 MHz (in Regions 2 and 3 only) and 5650-5670 MHz, the amateur-satellite service may operate subject to not causing harmful interference to other services operating in accordance with the Table (see No. 435). Administrations authorizing such use shall ensure that any harmful interference caused by emissions from a station in the amateur-satellite service is immediately eliminated in accordance with the provisions of No. 2741. The use of the bands 1260-1270 MHz and 5650-5670 MHz by the amateur-satellite service is limited to the Earth-to-space direction.

710—Use of the radionavigation-satellite service in the band 1215-1260 MHz shall be subject to the condition that no harmful interference is caused to the radionavigation service authorized under No. 712.

711—Additional allocation: in Afghanistan, Angola, Saudi Arabia, Bahrain, Bangladesh, Cameroon, China, the United Arab Emirates, Ethiopia, Guinea, Guyana, India, Indonesia, Iran, Iraq, Israel, Japan, Jordan, Kuwait, Lebanon, Libya, Malawi, Morocco, Mozambique, Nepal, Nigeria, Oman, Pakistan, the Philippines, Qatar, Syria, Somalia, Sudan, Sri Lanka, Chad, Thailand, Togo and Yemen, the band 1215-1300 MHz is also allocated to the fixed and mobile services on a primary basis.

712—Additional allocation: in Algeria, the Federal Republic of Germany, Austria, Bahrain, Belgium, Benin, Burundi, Cameroon, China, Denmark, the United Arab Emirates, France, Greece, India, Iran, Iraq, Kenya, Liechtenstein, Luxembourg, Mali, Mauritania, Norway, Oman, Pakistan, the Netherlands, Portugal, Qatar, Senegal, Somalia, Sudan, Sri Lanka, Sweden, Switzerland, Tanzania, Turkey and Yugoslavia, the band 1215-1300 MHz is also allocated to the radionavigation service on a primary basis.

712A—In Cuba, the band 1215-1300 MHz is also allocated to the radionavigation service on a primary basis subject to the agreement obtained under the procedure set forth in Article 14.

713—In the bands 1215-1300 MHz, 3100-3300 MHz, 5250-5350 MHz, 8550-8650 MHz, 9500-9800 MHz and 13.4-14.0 GHz, radiolocation stations installed on spacecraft may also be employed for the earth exploration-satellite and space research services on a secondary basis.

714—Additional allocation: in Canada and the United States, the bands 1240-1300 MHz and 1350-1370 MHz are also allocated to the aeronautical radionavigation service on a primary basis.

G56—Government radiolocation in the bands 1215-1300, 2900-3100, 5350-5650 and 9300-9500 MHz is primarily for the military services; however, limited secondary use is permitted by other Government agencies in support of experimentation and research programs. In addition, limited secondary use is permitted for survey operations in the band 2900-3100 MHz.

FCC Part 97 Privileges				
	Terrestrial location of the amateur radio station			
License Class	Region 1 MHz	Region 2 MHz	Region 3 MHz	Limitations*
Novice	1270-1295	1270-1295	1270-1295	
Technician General Advanced Extra	1240-1300	1240-1300	1240-1300	

*See §§97.301-307 of the FCC rules for further details on authorized privileges for each license class, frequency sharing requirements and authorized modes.

Amateur Radio Band Plans

Region 1 MHz	Region 2 MHz	Region 3 MHz	ARRL* MHz
			1240-1246 (2,3) ATV #1
			1246-1248 (2) NB FM point-to-point links and digital duplex with 1258-1260
			1248-1258 (2,8,9) Digital communications
			1252-1258 (2,3) ATV #2
			1258-1260 (2) NB FM point-to point links and digital, duplexed with 1246-1252
		1260-1270 Satellite	1260-1270 (5,6) Satellite uplinks ref WARC 1979 experimental, simplex ATV
		1270-1296	1270-1276 (2,4,9,10) Repeater inputs, FM and linear, paired with 1282-1288, 239 pairs, every 25 kHz, eg, 1270.025, 050, 075, etc.
			1271/1283 uncoordinated test pair
			1276-1282 (2,3) ATV #3
			1282-1288 (2,4,9,10) Repeater outputs paired with 1270-1276
			1288-1294 (6) WB experimental, simplex ATV
			1294-1295 (7) NB FM simplex, 25-kHz channels
			1294.5 National FM simplex calling
			1295-1297 NB weak-signal (no FM)
			1295.0-1295.8 SSTV, FAX, ACSSB experimental
			1295.8-1296.0 Reserved for EME, CW expansion
		1296-1297 EME	1296.0-1296.05 EME exclusive
			1296.07-1296.08 CW beacons
			1296.1 CW/SSB calling
			1296.4-1296.6 Crossband linear translator input
			1296.6-1296.8 Crossband linear translator output
			1296.8-1297.0 Experimental beacons (exclusive)
		1297-1300	1297-1300 Digital communications

*Adopted by the ARRL Board of Directors January 1985.

Footnotes:
1) Deleted
2) Coordinated assignments required.
3) ATV assignments should be made according to modulation type (for example), VSB-ATV, SSB-ATV, or combination. Coordination of multiple users of a single channel in a local area can be achieved through isolation by means of cross polarization and directional antennas. DSB ATV may be used, but only when local and regional activity levels permit. The excess bandwidths from such users are secondary to the assigned services.
4) Coordinate assignments with 100-kHz channels, beginning at the lower end of the segment until allocations are filled, then assign 50-kHz channels until allocations are filled before assigning 25-kHz channels.
5) Wide bandwidth experimental users are secondary to the satellite service and may be displaced upon the installation of any new satellites. Users are EIRP-limited to the noise floor of the satellites in service and may suffer interference from satellite uplinks.
6) Simplex services only; permanent users shall not be coordinated in this segment. High-altitude repeaters or other unattended fixed operations are not permitted.
7) Voice and non-voice operations.
8) Consult 47 C.F.R. 97.307 (FCC regulations) for allowable data rates and bandwidths.
9) Provide guard bands at the higher frequency end of segments, as required, to avoid interference to ATV.
10) 1274.0-1274.2 and 1286.0-1286.2 MHz are optionally reserved for contiguous Linear Translators supporting multiple narrow bandwidth users. These may also be duplexed with other non-coordinated band segments.

The following Packet Radio frequency recommendations were adopted by the ARRL Board of Directors in January 1988.
1) 2-MHz-bandwidth channels at:
1249.00   1251.00   1298.00

2) 100-kHz-bandwidth channels:
1299.05   1299.45   1299.75
1299.15   1299.55   1299.85
1299.25   1299.65   1299.95
1299.35

3) 25-kHz-bandwidth channels:
1294.025   1294.125
1294.050   1294.150
1294.075   1294.175
1294.100 National packet simplex calling

INTERNATIONAL			UNITED STATES				
Region 1 MHz	Region 2 MHz	Region 3 MHz	Band MHz	National Provisions	Government Allocation	Non-Government Allocation	Remarks
2300-2450 FIXED Amateur Mobile Radiolocation	2300-2450 FIXED MOBILE RADIOLOCATION Amateur		2300-2310	US253	RADIOLOCATION Fixed Mobile G2 US253	Amateur US253	
			2310-2360	US276 US327 US328	Radiolocation Mobile Fixed G2 G120 751B	BROADCASTING- SATELLITE Mobile 751B US328	Digital Audio Radio Services
			2360-2390	US276	RADIOLOCATION MOBILE Fixed G2 US276 G120	MOBILE US276	
			2390-2400	664 752	G122	AMATEUR	
			2400-2402	664 752	RADIOLOCATION G2	Amateur	
			2402-2417	664 752	G122	AMATEUR	
			2417-2450	664 752	RADIOLOCATION G2	Amateur	
664 751A 752	664 750B 751 751B 752						

664—In the bands 435-438 MHz, 1260-1270 MHz, 3400-3410 MHz (in Regions 2 and 3 only) and 5650-5670 MHz, the amateur-satellite service may operate subject to not causing harmful interference to other services operating in accordance with the Table (see No. 435). Administrations authorizing such use shall ensure that any harmful interference caused by emissions from a station in the amateur-satellite service is immediately eliminated in accordance with the provisions of No. 2741. The use of the bands 1260-1270 MHz and 5650-5670 MHz by the amateur-satellite service is limited to the Earth-to-space direction.

750B—Additional Allocation: in the United States of America and India, the band 2310-2360 MHz is also allocated to the broadcasting-satellite service (sound) and complementary terrestrial broadcasting service on a primary basis. Such use is limited to digital audio broadcasting and is subject to the provisions of Resolution 528.

751—In Australia, the United States and Papua New Guinea, the use of the band 2310-2390 MHz by the aeronautical mobile service for telemetry has priority over other uses by the mobile services.

751A—In France, the use of the band 2310-2360 MHz by the aeronautical mobile service for telemetry has priority over other uses by the mobile service.

751B—Space stations of the broadcasting-satellite service in the band 2310-2360 MHz operating in accordance with No. 750B that may affect services to which this band is allocated in other countries shall be coordinated and notified in accordance with Resolution 33. Complementary terrestrial broadcasting stations shall be subject to bilateral coordination with neighboring countries prior to their bringing into use.

752—The band 2400-2500 MHz (center frequency 2450 MHz) is designated for industrial, scientific and medical (ISM) applications. Radio services operating within this band must accept harmful interference which may be caused by these applications. ISM equipment operating in this band is subject to the provisions of No. 1815.

G2—In the bands 216-225, 420-450 (except as provided by US127), 890-902, 928-942, 1300-1400, 2300-2450, 2700-2900, 5650-5925, and 9000-9200 MHz the Government radiolocation is limited to the military services.

G120—Development of airborne primary radars in the band 2310-2390 MHz with peak transmitter power in excess of 250 watts for use in the United States is not permitted.

US327—The band 2310-2360 MHz is allocated to the broadcasting-satellite service (sound) and complementary terrestrial broadcasting service on a primary basis. Such use is limited to digital audio broadcasting and is subject to the provisions of Resolution 528.

US328—In the band 2310-2360 MHz, the mobile and radiolocation services are allocated on a primary basis until 1 January 1997 or until a broadcasting-satellite (sound) service has been brought into use in such a manner as to affect or be affected by the mobile and radiolocation sevices in those service areas, whichever is later. The broadcasting-satellite (sound) service during implementation should also take cognizance of the expendable and reusable launch vehicle frequencies 2312.5, 2332.5, and 2352.5 MHz, to minimize the impact on this mobile service use to the extent possible.

FCC Part 97 Privileges

| License Class | Terrestrial location of the amateur radio station | | | Limitations* |
	Region 1 MHz	Region 2 MHz	Region 3 MHz	
Technician General Advanced Extra	2300-2310 2390-2450	2300-2310 2390-2450	2300-2310 2390-2450	

*See §§97.301-307 of the FCC rules for further details on authorized privileges for each license class, frequency sharing requirements and authorized modes.

Amateur Radio Band Plans

Region 1 MHz	Region 2 MHz	Region 3 MHz	ARRL* MHz
			2300.0-2303.0 High-rate data (transmission rate ≥ 4800 bauds, duplex)
			2303.0-2303.5 Packet (transmission rate ≤ 2400 bauds, channel spacing = 25 kHz)
			2303.5-2303.8 TTY, packet (transmission rate ≤ 2400 bauds, bandwidth ≤ 2.5 kHz)
			2303.8-2303.9 Packet (bandwidth ≤ 2.5 kHz), TTY, CW, EME
			2303.9-2304.1 CW, EME
			2304.1-2034.2 CW, EME, SSB
			2304.2-2304.3 SSB, SSTV, fax, packet (bandwidth ≤ 2.5 kHz), AM, AMTOR
			2304.30-2304.32 Propagation beacon network
			2304.32-2304.40 General propagation beacons
			2304.400 Calling frequency
			2304.4-2304.5 SSB, SSTV, ACSSB, fax, packet (bandwidth ≤ 2.5 kHz), AM, AMTOR, experimental (bandwidth ≤ 2.5 kHz)
			2304.5-2304.7 Crossband linear translator input
			2304.7-2304.9 Crossband linear translator output
			2304.9-2305.0 Experimental beacons
			2305.0-2305.2 FM simplex (channel spacing = 25 kHz)
			2305.200 FM simplex calling frequency
			2305.2-2306.0 FM simplex (channel spacing = 25 kHz)
			2306.0-2309.0 FM repeaters (channel spacing = 25 kHz, input)
			2309.0-2310.0 Control and auxiliary links
			2390.0-2396.0 Fast-scan television
			2396.0-2399.0 High-rate data (transmission rate ≥ 4800 bauds, duplex)
			2399.0-2399.5 Packet (transmission rate ≤ 2400 bauds, channel spacing = 25 kHz)
			2399.5-2400.0 Control and auxiliary links
			2400.0-2403.0 Satellite
			2403.0-2408.0 Satellite, high-rate data (transmission rate ≥ 4800 bauds, duplex)
			2408.0-2410.0 Satellite
			2410.0-2413.0 FM repeaters (channel spacing = 25 kHz, output)
			2413.0-2418.0 High-rate data (transmission rate ≥ 4800 bauds, duplex)
			2418.0-2430.0 Fast-scan television
			2430.0-2433.0 Satellite
			2433.0-2438.0 Satellite, high-rate data (transmission rate ≥ 4800 bauds, duplex)
			2438.0-2450.0 Wide-band FM, fast-scan television, FM television, spread spectrum, experimental

*This band plan for 2300-2310 and 2390-2450 MHz was adopted by the Board of Directors at its January 1991 meeting. See Minute 56.

INTERNATIONAL			UNITED STATES				
Region 1 MHz	Region 2 MHz	Region 3 MHz	Band MHz	National Provisions	Government Allocation	Non-Government Allocation	Remarks
3300-3400 RADIOLOCATION	3300-3400 RADIOLOCATION Amateur Fixed Mobile	3300-3400 RADIOLOCATION Amateur	3300-3500	US108 664 778	RADIOLOCATION	Amateur Radiolocation	
778 779 780	778 780	778-779					
3400-3600 FIXED FIXED-SATELLITE (Space-to-Earth) Mobile Radiolocation	3400-3500 FIXED FIXED-SATELLITE (Space-to-Earth) Amateur Mobile Radiolocation 784						
781 782 785	664 783				G31		

664—In the bands 435-438 MHz, 1260-1270 MHz, 2400-2450 MHz, 3400-3410 MHz (in Regions 2 and 3 only) and 5650-5670 MHz, the amateur-satellite service may operate subject to not causing harmful interference to other services operating in accordance with the Table (see No. 435). Administrations authorizing such use shall ensure that any harmful interference caused by emissions from a station in the amateur-satellite service is immediately eliminated in accordance with the provisions of No. 2741. The use of the bands 1260-1270 MHz and 5650-5670 MHz by the amateur-satellite service is limited to the Earth-to-space direction.

778—In making assignments to stations of other services, administrations are urged to take all practicable steps to protect the spectral line observations of the radio astronomy service from harmful interference in the bands 3260-3267 MHz, 3332-3339 MHz, 3345.8-3352.5 MHz and 4825-4835 MHz. Emissions from space or airborne stations can be particularly serious sources of interference to the radio astronomy service (see Nos. 343 and 344 and Article 36).

779—Additional allocation: in Afghanistan, Saudi Arabia, Bahrain, Bangladesh, China, the Congo, the United Arab Emirates, India, Indonesia, Iran, Iraq, Israel, Japan, Kuwait, Lebanon, Libya, Malaysia, Oman, Pakistan, Qatar, Syria, Singapore, Sri Lanka and Thailand, the band 3300-3400 MHz is also allocated to the fixed and mobile services on a primary basis. The countries bordering the Mediterranean shall not claim protection for their fixed and mobile services from the radiolocation service.

780—Additional allocation: in Bulgaria, Cuba, Hungary, Mongolia, Poland, the Federal Republic of Germany, Romania, Czechoslovakia and the U.S.S.R., the band 3300-3400 MHz is also allocated to the radionavigation service on a primary basis.

781—Additional allocation: in the Federal Republic of Germany, Israel, Nigeria and the United Kingdom, the band 3400-3475 MHz is also allocated to the amateur service on a secondary basis.

782—Different category of service: in Austria, the allocation of the band 3400-3500 MHz to the radiolocation service is on a primary basis (see No. 425), subject to the agreement of the Administrations of the following countries: Hungary, Italy, the Federal Republic of Germany, Czechoslovakia and Yugoslavia. Such use is limited to ground-based stations. However, this Administration is urged to cease operations by 1985. After this date, this Administration shall take all practicable steps to protect the fixed-satellite service and coordination requirements shall not be imposed on the fixed-satellite service.

783—Different category of service: in Indonesia, Japan, Pakistan and Thailand, the allocation of the band 3400-3500 MHz to the mobile, except aeronautical mobile, service is on a primary basis (see No. 425).

784—In Regions 2 and 3, in the band 3400-3600 MHz, the radiolocation service is allocated on a primary basis. However, all administrations operating radiolocation systems in this band are urged to cease operations by 1985. Thereafter, administrations shall take all practicable steps to protect the fixed-satellite service and coordination requirements shall not be imposed on the fixed-satellite service.

785—In Denmark, Norway and the United Kingdom, the fixed, radiolocation and fixed-satellite services operate on a basis of equality of rights in the band 3400-3600 MHz. However, these Administrations operating radiolocation systems in this band are urged to cease operations by 1985. After this date, these Administrations shall take all practicable steps to protect the fixed-satellite service and coordination requirements shall not be imposed on the fixed-satellite service.

US108—Within the bands 3300-3500 MHz and 10000-10500 MHz, survey operations, using transmitters with a peak power not to exceed five watts into the antenna, may be authorized for Government and non-Government use on a secondary basis to other Government radiolocations operations.

G31—In the bands 3300-3500 MHz, the Government radiolocation is limited to the military services, except as provided by footnote US108.

FCC Part 97 Privileges				
	Terrestrial location of the amateur radio station			
License Class	Region 1 MHz	Region 2 MHz	Region 3 MHz	Limitations*
Technician General Advanced Extra	--	3300-3500	3300-3500	

Amateur Radio Band Plans			
Region 1† MHz	Region 2 MHz	Region 3 MHz	ARRL** MHz
			3456.3-3456.4 Propagation beacons

*See §§97.301-307 of the FCC rules for further details on authorized privileges for each license class, frequency sharing requirements and authorized modes.

**A beacon subband was adopted by the ARRL Board of Directors in July 1988.

INTERNATIONAL			UNITED STATES				
Region 1 MHz	Region 2 MHz	Region 3 MHz	Band MHz	National Provisions	Government Allocation	Non-Government Allocation	Remarks
5650-5725			5650-5850	664 806 808	RADIOLOCATION	Amateur	(ISM 5800 ±75 MHz)
	RADIOLOCATION Amateur Space Research (Deep space) 664 801 803 804 805						
5725-5850 FIXED-SATELLITE (Earth-to-space) RADIOLOCATION Amateur 801 803 805 806 807 808	5725-5850 RADIOLOCATION Amateur 803 805 806 808				G2		
5850-5925 FIXED FIXED-SATELLITE (Earth-to-space) MOBILE 806	5850-5925 FIXED FIXED-SATELLITE (Earth-to-space) MOBILE Amateur Radiolocation 806	5850-5925 FIXED FIXED-SATELLITE (Earth-to-space) MOBILE Radiolocation 806	5850-5925	US245 806	RADIOLOCATION G2	Amateur FIXED-SATELLITE (Earth-to-space)	

664—In the bands 435-438 MHz, 1260-1270 MHz, 2400-2450 MHz, 3400-3410 MHz (in Regions 2 and 3 only) and 5650-5670 MHz, the amateur-satellite service may operate subject to not causing harmful interference to other services operating in accordance with the Table (see No. 435). Administrations authorizing such use shall ensure that any harmful interference caused by emissions from a station in the amateur-satellite service is immediately eliminated in accordance with the provisions of No. 2741. The use of the bands 1260-1270 MHz and 5650-5670 MHz by the amateur-satellite service is limited to the Earth-to-space direction.

801—Additional allocation: in the United Kingdom, the band 5470-5850 MHz is also allocated to the land mobile service on a secondary basis. The power limits specified in Nos. 2502, 2505, 2506 and 2507 shall apply in the band 5725-5850 MHz.

802—Between 5600 MHz and 5650 MHz, ground-based radars used for meteorological purposes are authorized to operate on a basis of equality with stations of the maritime radionavigation service.

803—Additional allocation: in Afghanistan, Saudi Arabia, Bahrain, Bangladesh, Cameroon, the Central African Republic, China, the Congo, the Republic of Korea, Egypt, the United Arab Emirates, Gabon, Guinea, India, Indonesia, Iran, Iraq, Israel, Japan, Jordan, Kuwait, Lebanon, Libya, Madagascar, Malaysia, Malawi, Malta, Niger, Nigeria, Pakistan, the Philippines, Qatar, Syria, Singapore, Sri Lanka, Tanzania, Chad, Thailand and Yemen, the band 5650-5850 MHz is also allocated to the fixed and mobile services on a primary basis.

804—Different category of service: in Bulgaria, Cuba, Hungary, Mongolia, Poland, the Federal Republic of Germany, Czechoslovakia and the U.S.S.R., the allocation of the band 5670-5725 MHz to the space research service is on a primary basis (see No. 425).

805—Additional allocation: in Bulgaria, Cuba, Hungary, Mongolia, Poland, the Federal Republic of Germany, Czechoslovakia and the U.S.S.R., the band 5670-5850 MHz is also allocated to the fixed service on a primary basis.

806—The band 5725-5875 MHz (centre frequency 5800 MHz) is designated for industrial, scientific and medical (ISM) applications. Radiocommunication services operating within this band must accept harmful interference which may be caused by these applications. ISM equipment operating in this band is subject to the provisions of No. 1815.

807—Additional allocation: in the Federal Republic of Germany and in Cameroon, the band 5755-5850 MHz is also allocated to the fixed service on a primary basis.

808—The band 5830-5850 MHz is also allocated in the amateur-satellite service (space-to-Earth) on a secondary basis.

US245—The Fixed-Satellite Service is limited to International inter-Continental systems and subject to case-by-case electromagnetic compatibility analysis.

G2—In the bands 216-225, 420-450 (except as provided by US217), 890-902, 928-942, 1300-1400, 2300-2450, 2700-2900, 5650-5925, and 9000-9200 MHz, the Government radiolocation is limited to the military services.

FCC Part 97 Privileges				
	Terrestrial location of the amateur radio station			
License Class	Region 1 MHz	Region 2 MHz	Region 3 MHz	Limitations*
Technician General Advanced Extra	5650-5850	5650-5925	5650-5850	

Amateur Radio Band Plans			
Region 1† kHz	Region 2 kHz	Region 3 kHz	ARRL kHz
			5760.3-5760.4 Propagation beacons

*See §§97.301-307 of the FCC rules for further details on authorized privileges for each license class, frequency sharing requirements and authorized modes.

Allocations							
INTERNATIONAL			UNITED STATES				
Region 1 GHz	Region 2 GHz	Region 3 GHz	Band GHz	National Provisions	Government Allocation	Non-Government Allocation	Remarks
10-10.45 FIXED MOBILE RADIOLOCATION Amateur	10-10.45 RADIOLOCATION Amateur	10-10.45 FIXED MOBILE RADIOLOCATION Amateur	10-10.45	US58 US108 828	RADIOLOCATION	Amateur Radiolocation	
828	828 829	828			G32	NG42	
10.45-10.5	RADIOLOCATION Amateur Amateur-Satellite		10.45-10.5	US58 US108	RADIOLOCATION	RADIOLOCATION Amateur Amateur-Satellite	
	830				G32	NG42 NG134	

828—The band 9975-10025 MHz is also allocated to the meteorlogical-satellite service on a secondary basis for use by weather radars.

829—Additional allocation: in Costa Rica, Ecuador, Guatemala and Honduras, the band 10-10.45 GHz is also allocated to the fixed and mobile services on a primary basis.

830—Additional allocation: in the Federal Republic of Germany, Angola, China, Ecuador, Spain, Japan, Kenya, Morocco, Nigeria, Sweden, Tanzania and Thailand, the band 10.45-10.5 GHz is also allocated to the fixed and mobile services on a primary basis.

US58—In the band 10000-10500 MHz, pulsed emissions are prohibited, except for weather radars on board meteorological satellites in the band 10000-10025 MHz. The amateur service and the non-Government radiolocation service, which shall not cause harmful interference to the Government radiolocation service, are the only non-Government services permitted in this band. The non-Government radiolocation service is limited to survey operations as specified in footnote US108.

US108—Within the bands 3300-3500 MHz and 10000-10500 MHz, survey operations, using transmitters with a peak power not to exceed five watts into the antenna, may be authorized for Government and non-Government use on a secondary basis to other Government radiolocations operations.

G32—Except for weather radars on meteorological satellites in the band 9975-10025 MHz and for Government survey operations (see footnote US108). Government radiolocation in the band 10000-10500 MHz is limited to the military services.

NG42—Non-Government stations in the radiolocation service shall not cause harmful interference to the amateur service.

NG134—In the band 10.45 GHz non-Government stations in the radiolocation service shall not cause harmful interference to the amateur and amateur-satellite services.

FCC Part 97 Privileges				
	Terrestrial location of the amateur radio station			
License Class	Region 1 GHz	Region 2 GHz	Region 3 GHz	Limitations*
Technician General Advanced Extra	10.0-10.5	10.0-10.5	10.0-10.5	

Amateur Radio Band Plans			
Region 1† GHz	Region 2 GHz	Region 3 GHz	ARRL GHz
			10.00-10.5
			10.364* Calling Frequency

*See §§97.301-307 of the FCC rules for further details on authorized privileges for each license class, frequency sharing requirements and authorized modes.

*Adopted by ARRL Board of Directors January 1987.

Allocations

INTERNATIONAL			UNITED STATES				
Region 1 GHz	Region 2 GHz	Region 3 GHz	Band GHz	National Provisions	Government Allocation	Non-Government Allocation	Remarks
24-24.05	AMATEUR AMATEUR-SATELLITE 881		24-24.05	US211 881		AMATEUR AMATEUR-SATELLITE	
24.05-24.25	RADIOLOCATION Amatuer Earth Exploration-Satellite (Active) 881		24.05-24.25	US110 881	RADIOLOCATION Earth Exploration Satellite(Active) G59	Amateur Radiolocation Earth Exploration Satellite (Active)	(ISM 24.125 ± 125 MHz)

881— The band 24-24.25 GHz (centre frequency 24.125 GHz) is designated for industrial, scientific and medical (ISM) applications. Radiocommunication services operating within this band must accept harmful interference which may be caused by these applications. ISM equipment operating in this band is subject to the provisions of No. 1815..

US211—In the bands 1670-1690, 500-5250 MHz, and 10.7-11.7, 15.1365-15.35,15.4-15.7, 22.5-22.55, 24-24.05, 31.0-31.3, 31.8-32, 40.5-42.5, 84-86, 102-105, 116-126, 151-164, 176.5-182, 185-190, 231-235, 252-265 GHz, applicants for airborne or space station assignments are urged to take all practicable steps to protect radio astronomy observations in the adjacent bands from harmful interference; however, US74 applies.

G59—In the bands 902-928 MHz, 3100-3300 MHz, 3500-3700 MHz, 5250-5350 MHz, 8500-9000 MHz, 9200-9300 MHz, 13.4-14.0 GHz, 15.7-17.7 GHz and 24.05-24.25 GHz, all Government non-military radiolocation shall be secondary to military radiolocation, except in the sud-band 15.7-16.2 GHz airport surface detection equipment (ASDE) is permitted on a co-equal basis subject to coordination with the military departments.

FCC Part 97 Privileges

License Class	Terrestrial location of the amateur radio station			
	Region 1 GHz	Region 2 GHz	Region 3 GHz	Limitations*
Technician General Advanced Extra	24.00-24.25	24.00-24.25	24.00-24.25	

*See §§97.301-307 of the FCC rules for further details on authorized privileges for each license class, frequency sharing requirements and authorized modes.

Allocations

INTERNATIONAL			UNITED STATES				
Region 1 GHz	Region 2 GHz	Region 3 GHz	Band GHz	National Provisions	Government Allocation	Non-Government Allocation	Remarks
47-47.2	AMATEUR AMATEUR-SATELLITE		47-47.2			AMATEUR AMATEUR-SATELLITE	

FCC Part 97 Privileges

License Class	Terrestrial location of the amateur radio station			Limitations*
	Region 1 GHz	Region 2 GHz	Region 3 GHz	
Technician General Advanced Extra	47.0-47.2	47.0-47.2	47.0-47.2	

*See §§97.301-307 of the FCC rules for further details on authorized privileges for each license class, frequency sharing requirements and authorized modes.

FCC Part 97 Privileges

License Class	Terrestrial location of the amateur radio station			Limitations*
	Region 1 GHz	Region 2 GHz	Region 3 GHz	
Technician General Advanced Extra	75.5-81	75.5-81	75.5-81	

*See §§97.301-307 of the FCC rules for further details on authorized privileges for each license class, frequency sharing requirements and authorized modes.

Allocations

INTERNATIONAL			UNITED STATES				
Region 1 GHz	Region 2 GHz	Region 3 GHz	Band GHz	National Provisions	Government Allocation	Non-Government Allocation	Remarks
75.5-76	AMATEUR AMATEUR-SATELLITE		75.5-76	912		AMATEUR AMATEUR-SATELLITE	
76-81	RADIOLOCATION Amateur Amateur-Satellite 912		76-81		RADIOLOCATION	RADIOLOCATION Amateur Amateur-Satellite	

912—In the band 78-79 GHz radars located on space stations may be operated on a primary basis in the Earth exploration-satellite service and in the space research service.

Allocations

INTERNATIONAL			UNITED STATES				
Region 1 GHz	Region 2 GHz	Region 3 GHz	Band GHz	National Provisions	Government Allocation	Non-Government Allocation	Remarks
116-126	EARTH EXPLORATION SATELLITE (Passive) FIXED INTER-SATELLITE MOBILE 909 SPACE RESEARCH (Passive) 722 915 916		116-126	US211 US263 722 909 915 916	FIXED INTER-SATELLITE MOBILE EARTH EXPLORATION SATELLITE (Passive) SPACE REA SPACE RESEARCH (Passive)	FIXED MOBILE INTER-SATELLITE EARTH EXPLORATION- SATELLITE (Passive) SPACE REA SPACE RESEARCH (Passive)	(ISM 122.5 GHz ± 500 MHz)

722— In the bands 1400-1727 MHz, 101-120 GHz and 197-220 GHz, passive research is being conducted by some countries in a program for the search for international emissions of the extra-terrestrial origin.

909—In the bands 54.25-58.2 GHz, 59-64 Ghz, 116-134 GHz, 170-182 GHz and 185-190 GHz, stations in the aeronautical mobile service may be operated subject to not causing harmful interference to the inter-satellite service (see No. 435).

915—The band 119.98-120.02 GHz is also allocated to the amateur service on a secondary basis.

916—The band 122-123 GHz (centre frequency 122.5 GHz) is designated for industrial, scientific, and medical (ISM) applications. The use of this frequency band for ISM applications shall be subject to special authorization by the administration concerned in agreement with other administrations whose radiocommunications services might be affected. In applying this provision administrations shall have due regard to the latest relevant CCIR Recommendations.

US211—In the bands 1670-1690, 500-5250 MHz, and 10.7-11.7, 15.1365-15.35, 15.4-15.7, 22.5-22.55, 24-24.05, 31-31.03, 31.8-32, 40.5-42.5, 84-86, 102-105, 116-126, 151-164, 176.5-182, 185-190, 231-235, 252-265 GHz, applicants for airborne or space station assignments are urged to take all practicable steps to protect radio astronomy observations in the adjacent bands from harmful interference; however, US74 applies.

US263—In the frequency bands 21.2-21.4, 22.21-22.5, 36-37, 50.2-50.4, 54.25-58.2, 116-126, 150-151, 174.5-176.5, 200-202 and 235-238 GHz, the Space Research and the Earth Exploration-Satellite Services shall not receive protection from the Fixed and Mobile Services operating in accordance with the Table of Frequency Allocations.

FCC Part 97 Privileges				
	Terrestrial location of the amateur radio station			
License Class	Region 1 GHz	Region 2 GHz	Region 3 GHz	Limitations*
Technician General Advanced Extra	119.98-120.02	119.98-120.02	119.98-120.02	

*See §§97.301-307 of the FCC rules for further details on authorized privileges for each license class, frequency sharing requirements and authorized modes.

Allocations

INTERNATIONAL			UNITED STATES				
Region 1 GHz	Region 2 GHz	Region 3 GHz	Band GHz	National Provisions	Government Allocation	Non-Government Allocation	Remarks
142-144	AMATEUR AMATEUR-SATELLITE		142-144	918		AMATEUR AMATEUR-SATELLITE	
144-149	RADIOLOCATION Amateur Amateur-Satellite 918		144-149		RADIOLOCATION	RADIOLOCATION Amateur Amateur-Satellite	

918— The bands 140.69-140.98 GHz, 144.68-144.98 GHz, 145.45-147.75 GHz and 146.82-147.12 GHz are also allocated to the radio astronomy services on a primary basis for spectral line observations. In making assignments to stations of other services to which the bands are allocated, administrations are urged to take all practicable steps to protect the radio astronomy service from harmful interference. Emissions from space or airborne stations can be particularly serious sources of interference to the radio astronomy service (see Nos. 343 and 344 and Article 36).

FCC Part 97 Privileges

License Class	Terrestrial location of the amateur radio station			
	Region 1 GHz	Region 2 GHz	Region 3 GHz	Limitations*
Technician General Advanced Extra	142-149	142-149	142-149	

*See §§97.301-307 of the FCC rules for further details on authorized privileges for each license class, frequency sharing requirements and authorized modes.

Allocations

INTERNATIONAL			UNITED STATES				
Region 1 GHz	Region 2 GHz	Region 3 GHz	Band GHz	National Provisions	Government Allocation	Non-Government Allocation	Remarks
241-248			241-248	922	RADIOLOCATION	RADIOLOCATION Amateur Amateur-Satellite	(ISM 245 GHz ± 1 GHz)
	RADIOLOCATION Amateur Amateur-Satellite 922						
248-250			248-250			AMATEUR AMATEUR-SATELLITE	
	AMATEUR AMATEUR-SATELLITE						

922— The band 244-246 GHz (centre frequency 245 GHz) is designated for industrial, scientific, and medical (ISM) applications. The use of this frequency band for ISM applications shall be subject to special authorization by the administration concerned in agreement with other administrations whose radiocommunication services might be affected. In applying this provision administrations shall have due regard to the latest relevant CCIR Recommendations.

FCC Part 97 Privileges

License Class	Terrestrial location of the amateur radio station			
	Region 1 GHz	Region 2 GHz	Region 3 GHz	Limitations*
Technician General Advanced Extra	241-250	241-250	241-250	

*See §§97.301-307 of the FCC rules for further details on authorized privileges for each license class, frequency sharing requirements and authorized modes.

ITU Radio Regulations RR 346

Where, in adjacent Regions or sub-Regions, a band of frequencies is allocated to different services of the same category. . . the basic principle is the equality of right to operate. Accordingly, the stations of each service in one Region or sub-Region must operate so as not to cause harmful interference to services in the other Regions or sub-Regions.

ITU Resolution No. 640

Relating to the International Use of Radiocommunications, in the Event of Natural Disasters, in Frequency Bands Allocated to the Amateur Service

The World Administrative Radio Conference, Geneva, 1979.

considering

a) that in the event of natural disaster normal communication systems are frequently overloaded, damaged or completed disrupted;

b) that rapid establishment of communication is essential to facilitate worldwide relief actions;

c) that the amateur bands are not bound by international plans or notification procedures, and are therefore well adapted for short-term use in emergency cases.

d) that international disaster communications would be facilitated by temporary use of certain frequency bands allocated to the amateur service;

e) that under those circumstances the stations of the amateur service, because of their widespread distribution and their demonstrated capacity in such cases, can assist in meeting essential communication needs;

f) the existence of national and regional amateur emergency networks using frequencies throughout the bands allocated to the amateur service.

g) that, in the event of a natural disaster, direct communication between amateur stations and other stations might enable vital communications to be carried out until normal communications are restored;

recognizing

that the rights and responsibilities for communications in the event of a natural disaster rest with the administration involved;

resolves

1. that the bands allocated to the amateur service which are specified in No. 510 may be used by administrations to meet the needs of international disaster communications;

2. that such use of these bands shall be only for communications in relation to relief operations in connection with natural disasters;

3. that the use of specified bands allocated to the amateur service by non-amateur stations for disaster communications shall be limited to the duration of the emergency and to the specific geographical areas as defined by the responsible authority of the affected country;

4. the disaster communications shall take place within the disaster area and between the disaster area and the permanent headquarters of the organization providing relief;

5. that such communications shall be carried out only with the consent of the administration of the country in which the disaster has occurred;

6. that relief communications provided from outside the country in which the disaster has occurred shall not replace existing national or international amateur emergency networks.

7. that close cooperation is desirable between amateur stations and the stations of other radio services which may find it necessary to use amateur frequencies in disaster communications.

8. that such international relief communications shall avoid, as far as practicable, interference to the amateur service networks;

invites administrations

1. to provide for the needs of international disaster communications;

2 to provide for the needs of emergency communications within their national regulations.

Emission Designators

Emissions are classified by the following basic characteristics:

(1) First symbol—type of modulation of the main carrier

(1.1) Emission of an unmodulated carrier — N

(1.2) Emission in which the main carrier is amplitude modulated (including cases where subcarriers are angle modulated)

    (1.2.1) Double sideband — A

    (1.2.2) Single sideband, full carrier — H

    (1.2.3) Single sideband, reduced or variable-level carrier — R

    (1.2.4) Single sideband, suppressed carrier — J

    (1.2.5) Independent sidebands — B

    (1.2.6) Vestigial sideband — C

(1.3) Emission in which the main carrier is angle modulated

    (1.3.1) Frequency modulation — F

    (1.3.2) Phase modulation — G

(1.4) Emission in which the main carrier is amplitude and angle modulated either simultaneously or in a pre-established sequence. — D

(1.5) Emission of pulses[1]

    (1.5.1) Sequence of unmodulated pulses — P

    (1.5.2) A sequence of pulses

      (1.5.2.1) modulated in amplitude — K

      (1.5.2.2) modulated in width/duration — L

      (1.5.2.3) modulated in position/phase — M

      (1.5.2.4) in which the carrier is angle modulated during the period of the pulse — Q

      (1.5.2.5) which is the combination of the foregoing or is produced by other means — V

(1.6) Cases not covered above, in which an emission consists of the main carrier modulated, either simultaneously or in a preestablished sequence, in a combination of two or more of the following modes: amplitude, angle, pulse — W

(1.7) Cases not otherwise covered — X

(2) Second symbol—nature of signal(s) modulating the main carrier

(2.1) No modulating signal — 0

(2.2) A single channel containing quantized or digital information without the use of a modulating subcarrier[2] — 1

(2.3) A single channel containing quantized or digital information with the use of a modulating subcarrier[2] — 2

(2.4) A single channel containing analog information — 3

(2.5) Two or more channels containing quantized or digital information — 7

(2.6) Two or more channels containing analog information — 8

(2.7) Composite system with one or more channels containing quantized or digital information, together with one or more channels containing analog information — 9

(2.8) Cases not otherwise covered — X

(3) Third symbol—type of information to be transmitted[3]

(3.1) No information transmitted — N

(3.2) Telegraphy—for aural reception — A

(3.3) Telegraphy—for automatic reception — B

(3.4) Facsimile — C

(3.5) Data transmission, telemetry, telecommand — D

(3.6) Telephony (including sound broadcasting) — E

(3.7) Television (video) — F

(3.8) Combination of the above — W

(3.9) Cases not otherwise covered — X

The following two optional characteristics may be added for a more complete description of an emission:

(4) Fourth symbol—detail of signal(s)

(4.1) Two-condition code with elements of differing numbers and/or durations

(4.2) Two-condition code with elements of the same number and duration without error correction

(4.3) Two-condition code with elements of the same number and duration with error correction

(4.4) Four-condition code in which each condition represents a signal element (of one or more bits)

(4.5) Multi-condition code in which each condition represents a signal element (of one or more bits)

(4.6) Multi-condition code in which each condition or combination of conditions represents a character

(4.7) Sound of broadcasting quality (monophonic)

(4.8) Sound of broadcasting quality (stereophonic or quadraphonic)

(4.9) Sound of commercial quality (excluding categories K and L below)

(4.10) Sound of commercial quality with the use of frequency inversion or band-splitting

(4.11) Sound of commercial quality with separate frequency-modulated signals to control the level of demodulated signal

(4.12) Monochrome

(4.13) Color

(4.14) Combination of the above

(4.15) Cases not otherwise covered

(5) Fifth symbol—nature of multiplexing

(5.1) None

(5.2) Code-division multiplex[4]

(5.3) Frequency-division multiplex

(5.4) Time-division multiplex

(5.5) Combination of frequency-division multiplex and time-division multiplex

(5.6) Other types of multiplexing

Notes

[1]Emissions where the main carrier is directly modulated by a signal that has been coded into quantized form (for example, pulse code modulation) should be designated under (1.2) or (1.3).

[2]This excludes time-division multiplex.

[3]In this context the word "information" does not include information of a constant, unvarying nature such as is provided by standard-frequency emissions, continuous-wave and pulse radars, and so forth.

[4]This includes bandwidth-expansion techniques.

Emission Types

Emissions may also be referred to by the following emission types:

CW
First Symbol: A, C, H, J or R
Second Symbol: 1
Third Symbol: A or B
Also J2A and J2B.

MCW
First Symbol: A, C, D, F, G or R
Second Symbol: 2
Third Symbol: A or B

Phone
First Symbol: A, C, D, F, G, H, J or R
Second Symbol: 1, 2 or 3
Third Symbol: E
Also first symbol B and second symbol 7, 8 or 9.

Image
First Symbol: A, C, D, F, G, H, J or R
Second Symbol: 1, 2 or 3
Third Symbol: C or F
Also first symbol B; second symbol 7, 8 or 9; and third symbol W.

RTTY/Data
First Symbol: A, C, D, F, G, H, J or R
Second Symbol: 1 (2, 7 or 9 above 51 MHz)
Third Symbol: B (RTTY) or D (Data) (W above 51 MHz)
J2B is permitted anywhere F1B is permitted; J2D is permitted anywhere F1D is permitted.

Pulse
First Symbol: P, K, L, M, Q, V or W
Second Symbol: 0, 1, 2, 3, 7, 8, 9 or X
Third Symbol: A, B, C, D, E, F, N, W or X

SS (Spread Spectrum)
First Symbol: A, C, D, F, G, H, J or R
Second Symbol: X
Third Symbol: X

Test (Emissions containing no information)
First Symbol: Any
Second Symbol: Any
Third Symbol: N

Basic Operating

BILL JENNINGS, K1WJ
PAUL DANZER, N1II
and
THE ARRL STAFF

There are many faces to Amateur Radio. Tire of one and there is another awaiting you. This constant stream of new ideas, new challenges and new skills is one reason why so many people become hams and remain in the hobby for the rest of their lives.

There is one basic idea linking the various opportunities in ham radio—that of communication. Talking, sending pictures and data, building and testing transmitters and receivers all assume sooner or later you are going to operate these pieces of equipment in the various modes.

This manual covers all the most popular modes of ham radio. The basics of operating a ham station are discussed in Chapter 3. If you are a new ham or one just rejoining the hobby after a long absence, this material should help you decide what to operate, where to operate and how to operate.

Think of your participation in Amateur Radio as a never-ending journey—there's always something new to explore, always something new to do.

LET'S GET STARTED

Perhaps you already have your license or you anticipate passing the exam shortly. Now it's time to plan to get on-the-air. This can seem to be an overwhelming task, and your head is probably spinning with all kinds of questions ranging from "What kind of antenna should I use?" to "How do I make my first contact?"

Every newcomer needs help. Even experienced "Old Timers" once had those seemingly embarrassing questions when they began. After all, everyone has to start somewhere.

Who Are Hams?

Because of the diverse activities found within ham radio, almost everyone can find something to pique his or her interests. The 14-year-old high school student can keep in contact with friends across town. A 70-year-old retired engineer can chase those last elusive countries to put him on the DXCC (DX Century Club) Honor Roll. A Midwestern homemaker can show off her skills at contesting by racking up consistent QSO rates of 60 contacts per hour in the ARRL November Sweepstakes. You never know *who* you might run into.

Amateur Radio is a Service!

If you have read the Public Service column in *QST*, you've no doubt heard of the exploits of hams from all walks of life who've selflessly donated their time by providing emergency communication. Many, many more hams provide this service than are given recognition by the media, but that's all part of being a ham—the intrinsic reward is the satisfaction of doing a job well.

You may or may not be called at some time to provide this service to your community. But by being prepared, honing your on-the-air operating skills to their sharpest, maximizing your equipment to obtain the best from it, and being prepared to pitch in should the need arise, you will be ready. In so doing, you will derive untold hours of satisfaction from the exciting hobby we know as Amateur Radio.

Experience? Get On the Air

Actual on-the-air operating experience is the best teacher. In this section, we will try to give you enough of the basics to be able to make it through your first QSO. If nothing else, read the following material for additional reference notations. Learn by doing and don't be afraid to ask questions of someone who might be able to help. After all, we're in this hobby together,

Timothy Totten, KJ4VH, enjoys the annual contest known as the "Sweepstakes." No, he's not operating CW and voice at the same time. The boom microphone just comes along with the headset.

In Pursuit of . . . DX

"I'll never forget the thrill of my first DX contact. It was on CW in the Novice portion of 15 meters. I heard a G3 (a station from England) calling CQ, and no one answered right away so I decided to give it a try. Success! With a lump in my throat and sweat on my palms, I managed to complete my first DX QSO and become hopelessly hooked. DX is great!"

Most hams can tell a similar story. DX is Amateur Radio shorthand for long distance; furthermore, DX is universally understood by hams to be a station in a foreign country. Chasing DX has steadily become one of the most popular activities in our hobby, and you—if you operate on the HF bands—can get in on the fun!

Within your Novice/Technician Plus frequency allocations, you can work DX almost as easily as an Extra Class ham. You don't need a super kilowatt station and a huge antenna installation. The beauty of DXing is that your operating skills can overcome deficiencies in your station equipment. Pick the right frequency band at the right time-of-day (or night) and the DX stations will be there—ready to talk to you.

Keep a few tips in mind. If you hear a lone DX station calling CQ, work it. If there is a pileup (a lot of stations trying to work the same station), don't jump right in. Listen for five or six contacts so you can figure out how the DX station is handling the pileup. Remember to keep your calls short, and don't call the DX station after he has gone back to someone else. Try not to send the DX station's call more than once, if at all. It's better to sign just your call. If the DX station is working "tailenders" (stations who call immediately at the end of the DX station's last QSO), try it. If the station is not working tailenders, don't disrupt the proceedings by trying it.

When the DX station is working only stations in a call area other than your own, perhaps conditions are not yet quite right for you. Listen, and follow the DX station's instructions to the letter. Be courteous and fair to all the stations on frequency; using good operating procedures is every bit as important as logging that DX.

Check the "How's DX" column in *QST*. It contains tips on propagation and news of interest to DXers. And don't forget to listen to the weekly W1AW DX bulletin for the latest in DX operations. Good DX!

---

and assistance is only as far away as the closest ham. How do you develop good operating habits? This *ARRL Operating Manual* is an excellent place to start. An entire chapter is devoted to each of the major Amateur Radio activities, everything from working DX to space communications—each chapter written by a ham with considerable experience in that area of hamdom. Experience, even through the words of others, is a powerful teacher. Notice the table of contents; you'll be amazed at the diversity and amount of good solid reference material you have at your fingertips.

Amateur Radio Publications

The American Radio Relay League (ARRL), publisher of this book, and the only nonprofit organization to serve the over 600,000 Amateur Radio operators (hams) in the United States, produces an entire line of reference manuals and specialty publications. It also publishes the world-renowned monthly journal, *QST*. Just about anything you may need to know—whether it be from the technical or operating sides of the hobby—can be found in an ARRL publication or obtained directly from the HQ staff. The League's basic beginner's publication is *Now You're Talking!* You only need write to ARRL

HQ, 225 Main St, Newington, CT 06111, for complete information on all of the League's services.

QST Magazine

The ARRL journal, *QST*, published since 1914, is an excellent source of technical, operating, regulatory and general Amateur Radio information. Since it is published each month, *QST* is a timely source of information all hams can use, and a significant benefit of League membership. Write to the ARRL at the address given above for current membership rates.

On-the-Air Bulletins

Up-to-the-minute information on everything of immediate interest to amateurs—from Federal Communications Commission policymaking decisions to propagation predictions, to news of DXpeditions—is transmitted in W1AW bulletins. W1AW is the Amateur Radio station maintained at ARRL HQ in Newington, Connecticut. Bulletins are transmitted at regularly scheduled intervals on CW (Morse code), phone (voice) and several digital (teletype) modes on various frequencies. In addition there are Morse-code practice broadcasts. A schedule of W1AW transmissions appears regularly in *QST* and may also be obtained from ARRL HQ (please include a self-addressed, stamped envelope).

Local Amateurs/Clubs

Don't overlook the greatest source of information of all—the experience of your fellow hams. There is nothing hams like to do better than share the "vast" wealth of experience they have in the hobby. Being human, most hams like to brag a little about their on-the-air exploits. A general question on a particular area of operation to a ham who has experience in that area is likely to bring you all kinds of data. A dedicated DXer will talk for hours on techniques for working a rare DX station. Similarly, a traffic handler will be only too happy to give you hints on efficient net procedures.

For the new ham (or potential ham), the problem may not be in asking the proper questions, but in finding another amateur of whom to ask the questions. That's where Amateur Radio clubs play an important role.

As a potential ham or a new ham, you might not know of an amateur in your immediate area. A good choice would be to try to find the time and meeting place for a nearby Amateur Radio club. What better source of information for the not-yet-licensed and the newly licensed than a whole club full of experienced hams? Simply write to the Educational Activities Department at ARRL HQ for referral to an ARRL-affiliated Amateur Radio club near you. The League is more than happy to supply this information. If you are not yet licensed, your local Amateur Radio club is the place to find Morse code and electronics theory courses that will prepare you for the FCC exams.

The ARRL Educational Activities Department offers a wide range of information for prospective hams. They have a toll-free telephone number: 800-32NEWHAM.

Listen Listen Listen...!

"Experience is the best teacher." The most efficient way to learn to do something is to actually pitch right in and give it a try. Since there are rules and an established way of doing things in the Amateur Radio Service, you'll learn most effectively by just plain, old-fashioned listening. Use a receiver (yours or one borrowed from a fellow amateur) and your ears. Tune the amateur frequency bands, and just listen, listen, listen. Copy as many QSOs as you can. Learn how the operators there conduct

Your First Contact

Are your palms sweaty, hands shaking and is there a queasy herd of butterflies (wearing spikes, perhaps) performing maneuvers in your stomach? Chances are you're facing your first Amateur Radio QSO. If so, take heart! Although few may admit it, the vast majority of hams felt the same way before firing up the rig for their first contact.

Although nervousness is natural, there are some preparations that can make things go a bit more smoothly. Practicing QSOs with a friend, or perhaps with several members of the local Amateur Radio Club is a good way to ease the jitters.

Another thing some people find helpful is to write down information that you will use during the QSO in advance. One nervous Novice made notes of her own QSO data on index cards. One card contained her QTH and another her name (no kidding!); but after a half dozen contacts, she gradually forgot to use the cards. So will you.

A good security blanket to have is an experienced operator in your radio shack during your first few QSOs. You'll find that after your initial nervousness wears off you are able to do just fine by yourself. Even so, you'll have honored a ham friend (perhaps your teacher or your "Elmer") by allowing him the privilege of sharing your initial QSO. Indeed you may also want your Elmer to help you write up an equipment tune-up checklist so you can put a properly adjusted signal on the air every time, even when Elmer isn't around.

themselves, see what works for other operators and what doesn't, and incorporate those good points in your own operating habits. In other words, "copy the best and forget the rest."

Especially when going onto a new band or trying a new mode, take the time to listen to how those already established operators conduct themselves. Try to recognize good operating habits and incorporate them into your own style. You'll be surprised how much operator savvy you can pick up just by listening. A little common sense goes a long way in helping decide what is and what isn't good operating practice.

On-the-Air Experience for Newcomers

When it is time for you to take the plunge, fire up the rig and make your first QSO, go for it! Nothing can compare with actual on-the-air experience. Remember, everyone on the air today had to make their first QSO at some time or other. They had the same tentativeness as you might have now, but they

Jo Ann Simpson, KA3WPD, was a new operator when this picture was taken. All hams are a little nervous during their first few QSOs.

made it through fine. So will you.

It's the nature of the ham to be friendly, especially towards other members of this wonderful fraternity we all joined when we passed that examination. So trust the operator at the other end of your first QSO to understand your feelings and be as helpful as possible. After all, that operator was in your shoes once, too!

LET'S BUILD A STATION

The task of selecting and buying a ham station can seem overwhelming. You may be new to ham radio. Or perhaps you're planning to get back on the air after an absence of a few years. There are so many different possible station configurations it is hard to choose the very "best" station for you. In addition all hams face constraints such as having a limited budget to spend for ham gear or having restrictions on the size and location of antennas.

The first step in selecting your new station is to make up a list. The list simply answers a few questions. What sort of operation do you want—HF or VHF? How much room—how big can your shack or operating position be? Do you have room for long antennas, such as HF dipoles or high antennas such as HF verticals and VHF/UHF arrays? How much do you plan to spend, including rig, furniture, coax and wire?

Some of these questions may not apply to your situation, and you may need to list the answers to other questions not discussed here. The result of making this list is to give you an idea of what you want as well as your limitations.

The next step is to do some research. You have many sources of information on ham gear:

1) *Radio club members.* Ask members of your local radio club about their personal preferences in gear and antennas. Be prepared for a great volume of input. Every amateur has an opinion on the "best" equipment and antennas. Years of experimentation usually go into finding just the right station equipment to meet a particular amateur's needs. Listen and take note of each ham's choices and reasons for selecting a particular kind of gear; there's a lot of experience, time, effort and money behind each of those choices.

2) *Hands-on experience.* Try to use as many different pieces of gear as possible before you decide. Ask a nearby ham friend or one or more of your fellow radio club members if you can use their station. Try your club station. Note what you like and what features you don't care for in each of the stations you tried.

3) *Advertisements. QST is* chock-full of ads for all the newest up-to-date equipment as well as some premium used gear. Read the ads, and don't be afraid to contact the manufacturers of the gear for further information. Compare specifications and prices to get the best deal. If there is a dealer close to you pay him a visit.

Buying Your Station Equipment

After you've made your final choice (it's not really final, as most hams will trade station equipment several times during their ham careers), consider the sources where you might get the best deal, whether it be a new or used transceiver. Remember to include shipping and handling in the cost, and inquire about warrantee service. Several possible sources of equipment are:

1) *Local amateurs.* Many hams will have spare used gear and may be willing to part with this gear at a reasonable price. Be sure you know what a particular rig is going for on the open market before settling on a final price. If you are new to the

hobby and you buy a rig from a local club member, you may be able to talk him or her into "Elmering" you (helping you) with the rig's installation and operation.

2) **Hamfests/flea markets.** Many radio clubs run conventions called hamfests. Usually one of the big attractions of these events is the flea market or equipment sales. Much used gear, usually in passable shape, can be found at reasonable rates. Local distributors and manufacturers of new ham gear sometimes show up at these events to sell equipment at fair prices.

3) **Local electronics dealers.** If you are lucky enough to live near an electronics distributor who handles a line or two of ham gear, so much the better. The dealer can usually answer any questions you may have, will usually have a demonstration unit and will be only too happy to assist you to buy your new gear.

4) **Mail order.** There are many mail order ham equipment distributors to choose from. Some deal in new equipment, some deal in used equipment and some deal in both. Check the ads in the ham publications, such as those found in *QST*, for the equipment you want. The ads are also an excellent place to find out the going price of a particular piece of gear. Since prices sometimes change very fast many dealers list an 800 number to call for a price. Don't be shy—call their number.

A detailed treatment of used equipment and getting on the air quickly and effectively is in the ARRL's *Now You're Talking!*. The ARRL *Radio Buyer's Sourcebook* contains dozens of in-depth Product Reviews from *QST* magazine. Other *Sourcebook* articles will help you understand how the tests were conducted and what the results mean. If you're looking for a rig you should have a copy of the *Sourcebook*.

Take your time in deciding which gear to get, and even consider home brewing your gear. There's a lot of satisfaction to be had in telling the operator on the other end of your QSO (contact) "the rig here is home brew." *The ARRL Handbook for Radio Amateurs* has extensive coverage of construction and basic-to-advanced electronics theory information.

QRP vs QRO

Many hams find the low initial cost of low power (*QRP*) equipment on HF very attractive. In the ham radio community, there are both QRO (power output up to the full legal limit) and QRP (5 watts output) enthusiasts. Most hams run around 100 watts—basically the power level of most typical transceivers. There are times when it is necessary to run the full legal power limit to establish and maintain solid communications or compete effectively in a DX contest on certain bands. Most often the 100-watt level or less is more than enough to provide excellent contacts.

On VHF the situation is slightly different. Unless you are trying to work several hundred miles in an opening or operate a weak signal mode, a few watts (5 to 10) are generally enough power when coupled with a good antenna.

VHF/UHF Gear

If you're interested in voice and maybe packet radio, all you need is an FM transceiver. FM transceivers are available for hand-held and mobile use; the mobile rig can be used at home if you have a 12-V battery or suitable power supply. Fig 3-1 illustrates the choices available for a VHF or UHF station and how portable, mobile and home (fixed) rigs and antennas can be used.

1) **Hand-held transceivers.** Hand-held transceivers put out from under a watt to 5 watts or more. If the hand-held you're considering is capable of high-power operation, you'll want to be able to switch to low power when possible, to conserve the battery. Most hand-helds offer a **HIGH/LOW** power switch. Good used hand-held transceivers sell from $150 and up. New transceivers start at $250. See *The ARRL Radio Buyer's Sourcebook* for more information. Remember: with higher power levels you will want to use a remote antenna. Radiation of a lot of power in front of your eyes or near your head is not a good idea. This important topic is discussed in the Safety chapter of *The ARRL Handbook for Radio Amateurs,* 1995 and later editions.

2) **Mobile transceivers** have power outputs ranging from 10 to 50 watts. In populated areas with many repeaters, any more than 10 watts is probably unnecessary. Mobile transceivers are available used from $150 and up. New transceivers start at about $250 for a 10-watt unit. A mobile transceiver can be used indoors, too, if you have an adequate power supply. Selection of a mobile antenna is just as important as selection of the rig. Most hams are very disappointed when they try to use a hand-held transceiver in a car with the rubber ducky antenna attached to the rig.

3) **Fixed rigs.** As shown in Fig 3-1, a mobile rig connected to a power supply or 12-V automotive battery makes a good fixed station. It can also be disconnected and moved into the car for mobile operation. Powering the rig from a battery has the added advantage of allowing emergency operation when the local power lines go down in a storm.

4) **VHF packet radio.** Most VHF packet-radio operation takes place on the 2-meter and 440-MHz bands. A mobile transceiver makes a good packet-radio station rig, too. You'll need a computer or terminal, and a terminal node controller (TNC). Used TNCs are available for $70 or so, but new ones are only slightly more expensive. Newer units may offer "mailbox" features, where other hams can leave messages for you when you aren't home.

HF Equipment

1) **Separate transmitter and receiver.** Before single-sideband caught on, most amateur stations featured a separate transmitter and receiver. In many cases, the transmitter was homemade and the receiver was a commercial model. Most of this equipment is over 30 years old now, and maintenance problems are common. There wasn't the emphasis on compact, lightweight equipment we see now, and the receiver alone was usually larger than most modern transceivers. Some hams who

Sherman Harrison, KV4F, had to make quite a few decisions before he came up with this collection of equipment.

Fig 3-1—There are many choices of equipment for use with repeaters. A hand-held transceiver is the most versatile but it is usually low power and may require a number of accessories to use from a car or from home.

were licensed many years ago like to use this older gear for nostalgic reasons. It often appears on used equipment shelves and at hamfests. Our advice: Leave this equipment to people who enjoy working with older equipment or buy one of the receivers as a spare or for SWLing.

2) *Used separate* transmitter/receiver *combinations*. At one time all the major equipment manufacturers sold separate receivers and transmitters to "transceive" together. Examples are the Heath SB401/SB303, Drake T-4/R-4, Kenwood T-599/R-599 and Collins 32S-3/75S-3. Each of these rigs was a classic in its time, but all are outdated now. They lack many refinements common on newer transceivers, including the ability to work on

the WARC bands (30, 17 and 12 meters). All the transmitters mentioned, and most of the receivers, used vacuum tubes.

Years and high temperatures can really take its toll on electronic equipment. Tubes are expensive and harder to find. Capacitors tend to degenerate and finding replacements rated at the operating voltages of tubes can be a real chore. If you can find a good deal and know the equipment is in good working order (don't take the seller's word unless you know him or her well!), you may be able to pick up a pair for under $200. Don't be surprised at some of the prices quoted. Some rigs, such as the Collins S-line, command high prices from collectors. If you're looking for a first rig, the cost of a used rig such as the

A Hobby with No Barriers

One of the outstanding characteristics of hamming is that anyone can enjoy it. Amateur Radio has a special appeal to persons with visual or physical disabilities: it provides a means of people-to-people contact on a basis of absolute equality. Ham radio has always been a natural way of making new friends, renewing old acquaintances and visiting—via radio—people and places all over the world. It is perfect for those with limited mobility. There are hundreds of individuals with disabilities active in Amateur Radio. They have shown that with the proper training, patience and determination, almost any physical disability can be overcome.

The HANDI-Hams

A nationwide group specializes in helping persons with disabilities become a part of the hobby—the HANDI-Hams. Bruce Humphrys, KØHR, a former director of the HANDI-Hams, described the program in *QST*: "Courage Center is a comprehensive rehabilitation facility for persons with physical, speech, hearing and visual handicaps, based in Golden Valley, Minnesota. Aside from the HANDI-Ham System, Courage Center offers a tremendously wide array of services for handicapped people . . . the HANDI-Ham system is one of the Center's services—serving more than 7000 people all over the world. The system provides three direct (and many indirect) services: (1) educational material, fraternity and close personal attention; (2) Amateur Radio equipment on loan; and (3) specially designed devices for ease of station control. The system relies on a trained cadre of radio amateurs to help provide these services. We not only help new students to get their first license, but also help handicapped licensed hams to upgrade."

Every year, the HANDI-Hams hold radio camps for students with disabilities who are studying for new or upgraded ham tickets. Amateur exams are conducted after each session.

Contact the Courage HANDI-Hams, 3915 Golden Valley Rd, Golden Valley, MN 55422, Telephone 612-520-4515, for further information.

Getting the Ticket

All candidates for an amateur license must prove their ability to send and receive the International Morse code, and must pass a test on radio regulations and theory.

It is possible to exempt individuals with severe handicaps or disabilities from the 13 or 20-wpm code tests, or to use special accommodation procedures for these tests.

Section 97.509(h) describes these procedures: "The administering Volunteer Examiners must accommodate an examinee whose physical disabilities require a special examination procedure. The administering VEs may require a physician's certification or statement indicating the nature of the disability before determining which, if any, special procedures must be used." These special accommodation procedures, at the discretion of the Volunteer Examiner Team, could mean administering the examination at a place convenient or comfortable to the examinee (even at his or her bedside).

If the disability is so severe that the examinee is unable to pass the examination even with the special accommodation procedures, then a Morse code exemption for the 13 or 20-wpm code element may be obtained. For the purpose of obtaining an exemption for the higher speed code examination, the FCC defines a "severe handicap" as a disability that extends more than 365 days after the certification. (The 5-wpm code element cannot be waived.) The FCC states that an applicant for a 13 or 20-wpm Morse code exemption must submit to the VE Team an FCC Form 610 signed by a physician. The form (available from ARRL HQ) must be dated November 1993 or later.

ARRL Involvement

ARRL Headquarters helps enable persons with disabilities to prepare for FCC amateur exams. It serves as an information center and clearinghouse for this purpose, and maintains a database containing lists of operating aids and clubs, and other pertinent information. In addition, *QST* publishes technical and general-interest articles of special interest to persons with disabilities as well as a comprehensive booklet, *Source Book for the Disabled.* Contact ARRL HQ for information on obtaining a copy in printed or electronic format.

Membership in the League is available at a reduced rate without *QST* to persons who are blind. This gives already licensed blind members an opportunity to take an active role in ARRL activities, vote in ARRL Director and SM elections, run for League offices and otherwise share in the privileges of membership. Unlicensed blind persons, too, may show their support for the work of the ARRL through an associate membership. For more information, write the Program for the Disabled Coordinator at ARRL HQ.

QST is available in digital form directly from the Library of Congress. Contact your local library for details.

S-line is such that you're better off considering a more modern, used transceiver.

3) *Transceivers.* HF transceivers have been common for over 25 years. Like all older gear, older transceivers are likely to have maintenance problems. Mobile use subjects a radio to a great deal of vibration and wide ranges of temperature. A rig showing signs of having been used for mobiling may not be a good choice for your first station.

The modern age of transceivers can be divided in half. Transceivers manufactured in the '70s were partly solid state. The transmitters usually had three tubes: two in the power amplifier and another serving as the driver stage. Some transceivers from this era had built-in ac power supplies. Because they are newer, there may be fewer maintenance problems. However, mechanical parts, such as those used in the tuning assembly, may be impossible to obtain. Make sure everything works properly before you buy. Transceivers from this era probably won't operate on the 30, 17 and 12-meter bands.

The second era of modern transceivers feature solid-state construction. Tuning mechanisms have fewer mechanical parts

to wear and all-band operation is common. Many feature general-coverage receivers that continuously tune from below 100 kHz to 30 MHz. With the general-coverage receiver you can listen to time- and frequency-standard stations like WWV and WWVH and enjoy the variety of short-wave broadcasting (see Chapter 1).

New transceivers don't usually offer many more features than the better-equipped used gear mentioned in the previous paragraph. Of course, new equipment comes with a factory warranty, and you know no one opened the case to make modifications.

4) *Homemade equipment.* Beginners usually lack the skill and experience to design or build their first stations, but if you have a background in electronics, don't pass up the chance to build ("home brew") your own. Many good circuits appear in *The ARRL Handbook for Radio Amateurs.*

Accessories

A few accessories found in almost all stations, are needed either to help set up and test the equipment or to operate it.

Plans for a station should include some of these accessories.

SWR Indicator

This device is handy for testing an antenna and feed line when the antenna is first erected, and later to make sure the antenna is still in good shape. If a Transmatch is used with a multiband antenna system, such as a Windom or G5RV, an SWR indicator is essential for ensuring the Transmatch is adjusted for a reasonable SWR. Many modern HF transceivers include an SWR indicator. External SWR meters can cost as little as $30. If you plan to operate VHF and HF you may have to buy one for HF and a second for VHF/UHF.

Transmatch (Antenna Tuner/Antenna Coupler)

There are so many Transmatches available on the market that you might think you can't get by without one. Manufactured Transmatches cost between $75 and $1500. The circuits are pretty simple, though, and many first-time home-brew projects are Transmatches. *(The ARRL Handbook for Radio Amateurs* contains plans for building Transmatches.) By carefully shopping at hamfest flea markets you can often assemble your own for less than $50.

Keys, Keyers and Paddles

If you're interested in CW, you have to have some means of sending the code. You should start with a straight key until you feel you have the proper rhythm. Then you may wish to buy a keyer and paddles, or use your computer to send code. Some modern rigs have keyers built in, so all you need is a paddle. Paddles may be "standard" or "iambic." The iambic type requires less hand motion but usually a longer learning period.

Keyers cost from $50 to $250, and paddles cost about the same. Code-transmitting programs are available for the common computers; see the ads in *QST.*

Computers

Computers have become a very common part of the ham shack. Some are used "on-the-air"; that is, to send and receive Morse code, fax, slow-scan TV or digital modes. Others are used for logging and record keeping. Hams use various types of computers, ranging from old *VIC-20s* to *UNIX*-based work stations. The most popular type—having the most software available—is the PC- or MS DOS-based unit.

When considering a computer for the shack remember inside the computer is an oscillator—the "clock"—and it has the capability of generating signals right in the middle of some ham bands. Other computers are very sensitive to the presence of RF such as generated by your transmitter a foot or so away. For help with this topic look at the ARRL book *Radio Frequency Interference—How to Find It and Fix It.* It contains an entire chapter on computer problems in the shack.

Test Meter

An inexpensive multimeter capable of measuring voltage, current and resistance is very helpful around the shack. High accuracy is not needed for most problems. A $15 to $30 unit will pay for itself the first time you need to check the integrity of a coax connector you've just installed.

Station Setup

How you set up your station is determined by how much space you have, and how much equipment you have to squeeze into it. The table should be about 30 inches high and 30 inches deep. An old desk makes a good operating table. Build shelves for your equipment. Radios stacked on top of one another can't "breathe" and may overheat. It's also easier to change cables and move equipment around when you use shelves.

Place your key far enough from the edge of the table so your entire arm is supported, to prevent fatigue. A microphone can be mounted on a stand placed on the table or on an extension that reaches in from the back or side. The best way to test the arrangement of your rig is to sit in the operating chair and operate the rig's controls. If the knobs are too low try placing spacers or blocks under the front feet of the rig. Too high? Try placing the spacers under the rear legs.

Check that you can see the frequency dial. While it may be nice to place a keyboard right in front of your position, it may cause considerable strain on your back, shoulder and arms if you have to reach across the keyboard and to operate the rig. In this case consider a "T" or "U" shaped operating position. If this is your first shack remember it won't be the last time you will rearrange the equipment.

Be sure to have an effective earth ground routed to your station. Use as many grounds as you can locate, bonding them together electrically and bringing the connection plate to the operating position. If you live on a higher floor and a ground is not available, use a cold-water pipe if you can.

At least one manufacturer makes an "artificial ground," which is really a Transmatch for ground radials. While this device probably won't let you have a stronger signal, it may remove RF from the equipment cases.

Connections from the various ground points to the shack should be as short as possible. Since RF currents flow mainly on the surfaces of conductors, ground conductors such as shield braid from RG-8 coaxial cable or wide strips of flashing copper are best.

Safety should be a prime consideration. Many hams will not plug in their equipment for the first time until all the cases have been tied to ground. Use a master switch and make sure other people in the house know where it is and how to use it. Consider running your shack through a *GFI* (ground fault interrupt) outlet. Unfortunately not all ham gear, especially older units, will operate with these devices. However,

The top shelf of KD0GT's operating position is filled with various accessories. Martin Eichenberger operates from this neat setup.

German station DK3KD has all the cables and ground connections behind the equipment.

by detecting unbalanced line currents and shutting off the voltage when unbalance occurs, they could save your life.

Antennas

There are so many varieties of amateur antennas you could write books about them. We did: *The ARRL Antenna Book* and *W1FB's Antenna Notebook.* Whatever antennas you select, install them safely. Don't endanger your life or someone else's for your hobby!

Antennas are important. The best (and biggest) transmitter in the world will not do any good if the signal is not radiated into the air. A good rule of thumb to follow is "always erect as much antenna as possible." The better your antenna array, the better will be your radiated signal. A good antenna system will make up for inadequacies or shortcomings in station equipment. A less-sensitive receiver "hears better" with a good antenna system and a "bigger" antenna system will make a QRP (low power) station sound a lot louder at the receiving end.

If you are thinking of a tower talk to a few local hams before starting construction (or applying for a building permit). Local rules and ordinances may have a large impact on your plans.

Apply common sense to your antennas! Many hams act as though the world will end if they put up an antenna and measure an SWR greater than 1.5:1. For most purposes an SWR of 5:1 or even 10:1 is perfectly acceptable at HF with good quality feed lines 100 feet or less in length.

First VHF/UHF Antennas

One of the nice things about VHF and UHF operation is that often the simplest antennas, if mounted high enough, will do an excellent job. Ground planes, "J-poles" and simple beams can either be purchased at a reasonable cost or often constructed in a few minutes. If you want to test a 2-element quad for 144-MHz, just take the design from the *ARRL Antenna Book,* build it from scrap wood and heavy gauge copper or aluminum wire, and run a few tests. The unit you built may not stay up for a long time in bad weather but it will be fine for determining if this is the sort of antenna you want to put up "permanently."

A simple ground-plane antenna is shown in Fig 3-2. It can be mounted by taping the feed line and bottom connector to a pole so that the antenna extends over the top of the pole. It can also be suspended by a cord by lengthening the vertical element and bending the extra length into a loop. One end of the cord is fastened to the loop and the other end run over a tree branch.

If you are willing to make a temporary antenna—which may

last only a few months or a year—you can substitute ordinary wire from a wire clothes hanger. It may rust but as long as the connections are tight it will operate well.

Simple HF Antenna

The most popular "first" antenna is the half-wavelength dipole. It consists of a half wavelength of antenna wire with a feed line and an insulator at its center. See Fig 3-3.

For a new ham's activities, a resonant dipole system is a good selection. It is very easy to erect and has a low *SWR* (standing wave ratio—a measure of how well an antenna is tuned to the transmitted frequency) for its desired cut-to-length operating range. A single-band dipole fed with low-loss feed line can also be used on other bands. In fact with a Transmatch, a balun, and a random length center fed dipole you can actually operate on any HF band. The trick is to use a low-loss feed line and provide a match with a tuning network or Transmatch.

Where to Put the Antenna?

Think about your antenna location next. Remember: Your antenna should be as high and as far away from surrounding trees and structures as possible. Never put an antenna around or even near power lines! The dipole will require one support at each of its ends (perhaps trees, poles or even house or garage eaves), so survey your potential antenna site with this in mind. If you find space is so limited that you can't put up a straight-line dipole, don't give up. You can bend the ends of the dipole and still make plenty of contacts.

Some hams have been known to become very frustrated when they cannot put up a 135-foot long dipole that looks like

Frequency (MHz)	A (Inches)	B (Inches)
146	19–5/16	18–11/16
225	12–5/8	12
445	6–3/8	5–3/4

Fig 3-2—A simple ground-plane antenna for VHF. The elements are made from 3/32 or 1/16-in brass welding rod or #10 or #12 copper wire.

it came out of an engineering design manual. Most dipoles are very close to the ground and still work well. Many hams successfully use a dipole suspended from the middle, called an "inverted V." The performance of this antenna is so satisfactory they are not even aware of the large losses this configuration may have as compared to the ideal dipole.

You can put up an antenna under almost any circumstances, but you may need to use your imagination. Look at *The ARRL Antenna Book, Now You're Talking!, W1FB's Antenna Notebook* and *QST* for many workable antenna ideas.

Antenna Parts

If this is the first time you have tried to put up a dipole by yourself the following parts list will give you some guidance.

1) *Antenna wire.* #12 or #14 hard-drawn Copperweld (copper-clad steel) is preferred, so the antenna won't stretch. It will be strong enough to support itself as well as the weight of the feed line connected at its center. Always buy plenty of wire. It never goes to waste!

2) *Insulators.* You need one center and two end antenna insulators for a simple dipole.

3) *Clamp.* Large enough to fit over two widths of your coaxial cable.

4) *Coax.* Feed line made of a center conductor surrounded by an insulating dielectric. This in turn is surrounded by a braid called the shield. You need RG-58/U (or /AU or /BU or polyfoam) or RG-8U. Look for coax with heavy braid shield such as Belden, Saxton or Times Wire and Cable. Stay away from cheap cable from unknown manufacturers, or too-good-to-be-true deals on "surplus cable."

A good alternative is balanced "open-wire line." This may be actually two parallel pieces of wire connected and spaced with plastic rods or enclosed in a plastic jacket, similar to TV 300-Ω wire but with pieces of the center plastic removed. If you are sure you want to build a single band dipole stick to the coax, otherwise consider the Transmatch, balun and open-wire configuration.

5) *Connector.* Connects the feed line (coax) to your rig. This will probably be a PL-259, standard on most rigs. If your radio needs another kind, check your radio's instruction manual for installation information. You also need connectors for the coax lines between your Transmatch, SWR-meter and your rig.

6) *Electrical tape and coax sealant.* This is needed to cover the antenna ends and joints to make them waterproof. Otherwise water can get into coax and "short" it, ruining your antenna system.

7) *Rope.* You need enough to tie the ends of the antenna to a supporting structure. Rope may be used; however, it does degenerate with time and weather. You can use conductive guy wire. Conventional wisdom says you should make sure the length of the guy wire is not resonant in a ham band. In practice, it's doubtful you'll see any difference in antenna performance with any length of guy wire.

8) *SWR meter.* This is essential for your station and especially for antenna adjustment. SWR meters are readily available and inexpensive, making them easier to buy than build.

Gather all the parts you'll need for your chosen antenna. Almost everything is available from your local electronics store or from suppliers advertising in *QST.* When this is done, the fun—actually putting together your antenna—can begin.

Assembly is quite simple. Your dipole consists of two lengths of wire, each approximately ¼ wavelength long at your chosen operating frequency. These two wires are connected in the center, at an insulator, to the feed line. In our antenna, the feed line is coaxial cable, and it brings the signals to and from your radio. Calculate the length of the half-wave dipole by using this formula: antenna length in feet = 468/frequency in MHz. (The Novice/Technician Plus-band information below has approximate lengths already calculated.) Now measure the antenna wire, keeping it as straight as possible.

Antenna Lengths in Feet

(cut to center of Novice/Technician Plus band)

	½ wavelength	¼ wavelength
80 m	126'6"	63'3"
40 m	65'8"	32'10"
15 m	22'2"	11'1"
10 m	16'7"	8'3"

Remember to add about 1 foot to each end of the dipole for tuning adjustment.

Putting It Together

Carefully assemble your antenna, paying special attention to waterproofing the coax. Don't solder the antenna ends until later, when retuning is completed. Just twist them for now. Route the coax to your station, remembering to keep it as unobtrusive as possible. Cut the coax to a length that will leave some excess for strain relief so your rig won't be pulled around during strong winds! Install the connector(s) according to the diagrams. Now connect your SWR bridge between the feed line (coax) and the transmitter.

One trick used by old-timers is the addition of a 10,000-Ω resistor, soldered directly across the center insulator of the dipole. An ohmmeter connected from the one side to the feed line to the other at the ground end should measure this value of 10,000 Ω. If it measures an open circuit it means the feed line is disconnected or broken. A short circuit means the feed line or the connector is shorted. The resistor has no effect on the antenna or SWR. Left in place it acts as a handy check on the antenna and feed line.

There are several ways to see if you built the antenna correctly. One very direct way is to buy or borrow an antenna noise bridge. Details of these handy devices are in the *ARRL Handbook.* The bridge will tell you the resonant frequency of the antenna. If the frequency is too low you should shorten the length of the antenna (cut equal amounts from both ends). Resonant frequency too high? The antenna must be lengthened.

The other common method is the use of an SWR meter. If it shows an SWR of 2:1 or less at your desired operating frequency, your antenna system is "tuned," and you can go ahead and operate. If your SWR is greater than 2:1, you must "retune" your antenna to obtain a lower SWR. Disconnect the transmitter, and try shortening the antenna a few inches on each end. Keep notes! Reconnect the transmitter, and check the SWR. If the SWR gets lower, continue shortening the ends until you get the lowest possible SWR reading. Be careful not to shorten it too much! As the last step, when you are sure the length of the dipole is correct, solder the wire wrapped around the end insulators.

Problems and Cures

An SWR reading of greater than 4:1 for a simple single band resonant dipole probably means something is wrong beyond merely length misadjustments. Check to see if your coax is open or shorted. Make sure your antenna isn't touching anything and all your connections are sound.

When all systems are "go," get on the air and operate. As

Build Your Own Antenna

To connect the coax to the center insulator, cut a few inches of the outer covering off the coax. Next, separate the copper braid from the inner conductor and insulation. After you've done that, twist the strands of copper together to form a single wire. Remove about half the insulation covering the inner conductor and bend it away from the twisted strands of copper braid. Loop the cable over the insulator as shown and solder the braid to one half the antenna and the inner conductor to the other. Be sure to tape all connections securely for waterproofing; the braid can soak up water like a sponge or wick, making the coax useless after a while. B and C show how the wire is connected to various types of insulators at the ends, and D shows the connection of the feed line at the center.

Examples of simple but effective wire antennas. A horizontal dipole is shown at A. The legs can be drooped to form an "inverted V" as shown at B. A sloping dipole (sloper) is illustrated at C. The feed line should come away from the sloper at a 90° angle for best results. If the supporting mast is metal, the antenna will have some directivity in the direction of the slope.

Using a PL-259 Connector

If you are using RG-58 or RG-59 with a PL-259 connector, you will need to use an adapter, as shown here. This material courtesy of Amphenol Electronic Components, RF Division, Bunker Ramo Corp.

1) Cut end of cable even. Remove vinyl jacket ¾"—don't nick braid. Slide coupling ring and adapter on cable.

2) Fan braid slightly and fold back over cable.

3) Position adapter to dimension shown. Press braid down over body of adapter and trim to 3/8". Bare 5/8" of center conductor. Tin exposed center conductor.

Fig 3-3—Building your own HF dipole antenna is a popular project.

(D)

$$L(\text{FEET}) \approx \frac{468}{f(\text{MHz})}$$

TREE, MAST, TOWER ETC

50- OR 72-Ω COAX

SUPPORT POLE

6 TO 7 FEET MINIMUM

TO STATION

SLOPING DIPOLE

(C)

4) Screw the plug assembly on adapter. Solder braid to shell through solder holes. Solder conductor to contact sleeve.

5) Screw coupling ring on plug assembly.

you settle into that first QSO with your new antenna, enjoy those feelings of pride, accomplishment and fun that will naturally follow. After all, that's what Amateur Radio is all about.

OPERATING—WHAT, WHERE AND HOW

We're finally getting down to the real nitty-gritty. You have read a little about the basics of ham radio and have some ideas about selecting a rig and station equipment and erecting a decent antenna system.

Now it's time to learn a little about how to go about playing the ham radio game. For everyone to be able to communicate with everyone else effectively, reasonable operating procedures are needed.

A brief explanation follows, describing the major modes of ham radio communication and a few of the procedures and conventions in use on the air today. Other chapters of this manual discuss in great detail the operating procedures for many specialized modes of communication. Here we are going to look at voice and CW (Morse code) operating.

Spend as much time as possible listening to other operators on the bands. Find out what works and what doesn't. Copy the good habits of the good communicators and incorporate those habits into your own operating style. Good operating habits will make you an effective communicator for as long as you pound that CW key or speak into that microphone.

Where To Operate

One of the first questions you may have is where to operate? The ham bands are defined by FCC rules and these rules limit where you can operate. Fig 3-4 lists the frequencies each class of licensee can use for various modes. Within these rules hams have agreed to limit their operation to certain frequencies for certain modes. Table 3-1 shows these agreements.

If you're a new ham and hold a Technician or higher class license, you will probably do some operating on 2 meters, the most popular ham band. Although packet radio is also very common, the most popular mode is 2-meter FM using repeaters. Chapter 11 of this manual goes into great detail on this type of operation. The sidebar *Picking a Band* gives the characteristics of the various ham bands.

CW Operating

CW (continuous wave, Morse code) is the universal language, and a common bond of hamdom. Many hams take pride in their code proficiency. It takes practice to master the art of sending good code on a hand key, bug (semi-automatic key) or electronic keyer. You have to practice to get that smooth rhythm, practice to get that smooth spacing between words and characters, and practice to learn the sound of whole words and phrases, rather than just individual letters.

CW is an expedient mode of communications. CW transmitters are simpler devices than their phone counterparts, and a CW signal can usually get through very heavy QRM (interference on the band) much more effectively than a phone signal. Therefore, shortcuts and abbreviations are used during a CW QSO. Many of the abbreviations hams use have developed within the ham fraternity, while some are borrowed from the old-time telegraph operators. Q signals are among the most useful of these abbreviations. A list of the most popular Q signals is in Table 3-2.

You don't have to sit down and memorize this list. Copy the list and keep the copy on your operating table. You can also request Form FSD-218 from the ARRL. It contains a handy list of Q signals and abbreviations printed on cardboard for mount-

Table 3-1

The "Considerate Operator's Frequency Guide"

The following frequencies are generally recognized for certain modes or activities (all frequencies are in MHz).

Nothing in the rules recognizes a net's, group's or any individual's special privilege to any specific frequency. Section 97.101(b) of the Rules states that "Each station licensee and each control operator must cooperate in selecting transmitting channels and in making the most effective use of the amateur service frequencies. No frequency will be assigned for the exclusive use of any station." No one "owns" a frequency.

It's good practice—and plain old common sense—for any operator, regardless of mode, to check to see if the frequency is in use prior to engaging operation. If you are there first, other operators should make an effort to protect you from interference to the extent possible given that 100% interference-free operation is an unrealistic expectation in today's congested bands.

1.800-1.830	CW, data and other narrowband modes
1.810	QRP CW calling frequency
1.830-1.840	CW, data and other narrowband modes, intercontinental QSOs only
1.840-1.850	CW; SSB, SSTV and other wideband modes, intercontinental QSOs only
1.850-2.000	CW; phone, SSTV and other wideband modes
1.910	QRP SSB calling frequency
3.560	QRP CW calling frequency
3.590	RTTY DX
3.580-3.620	Data
3.620-3.635	Automatically controlled data stations
3.710	QRP Novice/Technician CW calling frequency
3.790-3.800	DX window
3.845	SSTV
3.885	AM calling frequency
3.985	QRP SSB calling frequency
7.040	RTTY DX
	QRP CW calling frequency
7.080-7.100	Data
7.100-7.105	Automatically controlled data stations
7.110	QRP Novice/Technician CW calling frequency
7.171	SSTV
7.285	QRP SSB calling frequency
7.290	AM calling frequency

10.106	QRP CW calling frequency
10.130-10.140	Data
10.140-10.150	Automatically controlled data stations
14.060	QRP CW calling frequency
14.070-14.095	Data
14.095-14.0995	Automatically controlled data stations
14.100	NCDXF/IARU beacons
14.1005-14.112	Automatically controlled data stations
14.230	SSTV
14.285	QRP SSB calling frequency
14.286	AM calling frequency
18.100-18.105	Data
18.105-18.110	Automatically controlled data stations
21.060	QRP CW calling frequency
21.070-21.100	Data
21.090-21.100	Automatically controlled data stations
21.340	SSTV
21.385	QRP SSB calling frequency
24.920-24.925	Data
24.925-24.930	Automatically controlled data stations
28.060	QRP CW calling frequency
28.070-28.120	Data
28.120-28.189	Automatically controlled data stations
28.190-28.225	Beacons
28.385	QRP SSB calling frequency
28.680	SSTV
29.000-29.200	AM
29.300-29.510	Satellite downlinks
29.520-29.580	Repeater inputs
29.600	FM simplex
29.620-29.680	Repeater outputs

Note

ARRL band plans for frequencies above 28.300 MHz are shown in *The ARRL Repeater Directory* and *The FCC Rule Book*. For detailed packet frequencies, see *QST*, September 1987, page 54, and March 1988, page 51.

Table 3-2

Q Signals

These Q signals are the ones used most often on the air. (Q abbreviations take the form of questions only when they are sent followed by a question mark.)

QRG Will you tell me my exact frequency (or that of ___)? Your exact frequency (or that of ___) is ___ kHz.

QRL Are you busy? I am busy (or I am busy with ___). Please do not interfere.

QRM Is my transmission being interfered with? Your transmission is being interfered with ___ (1. Nil; 2. Slightly; 3. Moderately; 4. Severely; 5. Extremely.)

QRN Are you troubled by static? I am troubled by static ___. (1-5 as under QRM.)

QRO Shall I increase power? Increase power.

QRP Shall I decrease power? Decrease power.

QRQ Shall I send faster? Send faster (___ WPM).

QRS Shall I send more slowly? Send more slowly (___ WPM).

QRT Shall I stop sending? Stop sending.

QRU Have you anything for me? I have nothing for you.

QRV Are you ready? I am ready.

QRX When will you call me again? I will call you again at ___ hours (on ___ kHz).

QRZ Who is calling me? You are being called by ___(on ___kHz).

QSB Are my signals fading? Your signals are fading.

QSK Can you hear me between your signals and if so can I break in on your transmission? I can hear you between signals; break in on my transmission.

QSL Can you acknowledge receipt (of a message or transmission)? I am acknowledging receipt.

QSN Did you hear me (or ___) on ___kHz? I did hear you (or ___) on ___kHz.

QSO Can you communicate with ___direct or by relay? I can communicate with ___ direct (or relay through ___).

QSP Will you relay to ___? I will relay to ___.

QST General call preceding a message addressed to all amateurs and ARRL members. This is in effect "CQ ARRL."

QSX Will you listen to ___on ___kHz? I am listening to _____ on ___kHz.

QSY Shall I change to transmission on another frequency? Change to transmission on another frequency (or on ___ kHz).

QTB Do you agree with my counting of words? I do not agree with your counting of words. I will repeat the first letter or digit of each word or group.

QTC How many messages have you to send? I have ___ messages for you (or for ___).

QTH What is your location? My location is ___.

QTR What is the correct time? The time is ___.

160 METERS

E,A,G

1800 — 1900 — 2000 kHz

Amateur stations operating at 1900–2000 kHz must not cause harmful interference to the radiolocation service and are afforded no protection from radiolocation operations.

80 METERS

3525 · 3675 · 3725 · 3775 · 3850

N,T+ / G / A / E

3500 — 3750 — 4000 kHz

5167.5 kHz (SSB only): Alaska emergency use only.

40 METERS

7025 · 7100 · 7150 · 7225

N,T+ / G * / A * / E *

7000 — 7150 — 7300 kHz

* Phone operation is allowed on 7075–7100 kHz in Puerto Rico, US Virgin islands and areas of the Caribbean south of 20 degrees north latitude; and in Hawaii and areas near ITU Region 3, including Alaska.

30 METERS

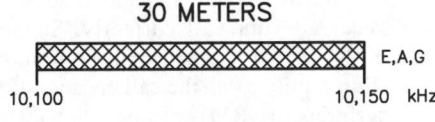

E,A,G

10,100 — 10,150 kHz

Maximum power on 30 meters is 200 watts PEP output. Amateurs must avoid interference to the fixed service outside the US.

20 METERS

14,025 · 14,150 · 14,175 · 14,225

G / A / E

14,000 — 14,150 — 14,350 kHz

17 METERS

18,068 · 18,110 · 18,168 kHz

E,A,G

15 METERS

21,025 · 21,100 · 21,200 · 21,225 · 21,300

N,T+ / G / A / E

21,000 — 21,200 — 21,450 kHz

12 METERS

24,890 · 24,930 · 24,990 kHz

E,A,G

10 METERS

28,100 · 28,500

N,T+ / E,A,G

28,000 · 28,300 — 29,700 kHz

Novices and Technicians are limited to 200 watts PEP output on 10 meters.

6 METERS

50.1

E,A,G,T+,T

50.0 — 54.0 MHz

2 METERS

144.1

E,A,G,T+,T

144.0 — 148.0 MHz

1.25 METERS

E,A,G,T+,T,N

222.0 — 225.0 MHz

Novices are limited to 25 watts PEP output from 222 to 225 MHz.

70 CENTIMETERS **

E,A,G,T+,T

420.0 — 450.0 MHz

33 CENTIMETERS **

E,A,G,T+,T

902.0 — 928.0 MHz

23 CENTIMETERS **

1270 · 1295

N

E,A,G,T+,T

1240 — 1300 MHz

Novices are limited to 5 watts PEP output from 1270 to 1295 MHz.

US AMATEUR BANDS

December 20, 1994

US AMATEUR POWER LIMITS

At all times, transmitter power should be kept down to that necessary to carry out the desired communications. Power is rated in watts PEP output. Unless otherwise stated, the maximum power output is 1500 W. Power for all license classes is limited to 200 W in the 10,100–10,150 kHz band and in all Novice subbands below 28,100 kHz. Novices and Technicians are restricted to 200 W in the 28,100–28,500 kHz subbands. In addition, Novices are restricted to 25 W in the 222–225 MHz band and 5 W in the 1270–1295 MHz subband.

Operators with Technician class licenses and above may operate on all bands above 50 MHz. For more detailed information see The FCC Rule Book.

KEY

- ▓ = CW, RTTY and data
- ▒ = CW, RTTY, data, MCW, test, phone and image
- █ = CW, phone and image
- ⧄ = CW and SSB
- ⧅ = CW, RTTY, data, phone, and image
- ▢ = CW only

E = EXTRA CLASS
A = ADVANCED
G = GENERAL
T+ = TECHNICIAN PLUS
T = TECHNICIAN
N = NOVICE

** Geographical and power restrictions apply to these bands. See The FCC Rule Book for more information about your area.

Above 23 Centimeters:

All licensees except Novices are authorized all modes on the following frequencies:
2300–2310 MHz
2390–2450 MHz
3300–3500 MHz
5650–5925 MHz
10.0–10.5 GHz
24.0–24.25 GHz
47.0–47.2 GHz
75.5–81.0 GHz
119.98–120.02 GHz
142–149 GHz
241–250 GHz
All above 300 GHz

For band plans and sharing arrangements, see The ARRL Operating Manual or The FCC Rule Book.

Fig 3-4—Frequency Allocation Chart.

Table 3-3
Some Abbreviations for CW Work

Although abbreviations help to cut down unnecessary transmission, make it a rule not to abbreviate unnecessarily when working an operator of unknown experience.

AA	All after	OC	Old chap
AB	All before	OM	Old man
ABT	About	OP-OPR	Operator
ADR	Address	OT	Old timer; old top
AGN	Again	PBL	Preamble
ANT	Antenna	PSE	Please
BCI	Broadcast interference	PWR	Power
BCL	Broadcast listener	PX	Press
BK	Break; break me; break in	R	Received as transmitted; are
BN	All between; been	RCD	Received
BUG	Semi-automatic key	RCVR (RX)	Receiver
B4	Before	REF	Refer to; referring to; reference
C	Yes	RFI	Radio frequency interference
CFM	Confirm; I confirm	RIG	Station equipment
CK	Check	RPT	Repeat; I repeat; report
CL	I am closing my station; call	RTTY	Radioteletype
CLD-CLG	Called; calling	RX	Receiver
CQ	Calling any station	SASE	Self-addressed, stamped envelope
CUD	Could	SED	Said
CUL	See you later	SIG	Signature; signal
CW	Continuous wave	SINE	Operator's personal initials or
	(i.e., radiotelegraph)		nickname
DLD-DLVD	Delivered	SKED	Schedule
DR	Dear	SRI	Sorry
DX	Distance, foreign countries	SSB	Single sideband
ES	And, &	SVC	Service; prefix to service message
FB	Fine business, excellent	T	Zero
FM	Frequency modulation	TFC	Traffic
GA	Go ahead (or resume sending)	TMW	Tomorrow
GB	Good-by	TNX-TKS	Thanks
GBA	Give better address	TT	That
GE	Good evening	TU	Thank you
GG	Going	TVI	Television interference
GM	Good morning	TX	Transmitter
GN	Good night	TXT	Text
GND	Ground	UR-URS	Your; you're; yours
GUD	Good	VFO	Variable-frequency oscillator
HI	The telegraphic laugh; high	VY	Very
HR	Here, hear	WA	Word after
HV	Have	WB	Word before
HW	How	WD-WDS	Word; words
LID	A poor operator	WKD-WKG	Worked; working
MA, MILS	Milliamperes	WL	Well; will
MSG	Message; prefix to radiogram	WUD	Would
N	No	WX	Weather
NCS	Net control station	XCVR	Transceiver
ND	Nothing doing	XMTR (TX)	Transmitter
NIL	Nothing; I have nothing for you	XTAL	Crystal
NM	No more	XYL (YF)	Wife
NR	Number	YL	Young lady
NW	Now; I resume transmission	73	Best regards
OB	Old boy	88	Love and kisses

ing at your operating position.

After using some of the Q signals and abbreviations a few times you will quickly learn the most popular of them without needing any reference. With time, you'll find that as your CW proficiency rises, you will be able to communicate almost as quickly on CW as you can on the voice modes. For more information on CW operating, see *Morse Code: The Essential Language,* published by ARRL. A list of common abbreviations is in Table 3-3.

It may not seem that way to you now, but your CW sending and receiving speed will rise very quickly with on-the-air practice. For your first few QSOs, carefully choose to answer the calls from stations that are sending at a speed you can copy (perhaps another first timer on the band?). Courtesy on the ham bands dictates that an operator will slow his or her code speed to accommodate another operator. Don't be afraid to call some-

one who is sending just a bit faster than you can copy comfortably. That operator will generally slow down to meet your CW speed. Helping each other is the name of the game in ham radio.

To increase your speed, you may continue to copy the code practice sessions from W1AW, the ARRL HQ station. It might be a good idea to spend some time sending in step (on a code-practice oscillator—not on the air, of course) with W1AW transmissions; it may help develop your sending ability.

Correct CW Procedures

The best way to establish a contact, especially at first, is to listen until you hear someone calling CQ. CQ means, "I wish to contact any amateur station." Avoid the common operating pitfall of calling CQ endlessly; it clutters up the air and drives off potential new friends. The typical CQ would go like this: CQ CQ CQ DE K5RC K5RC K5RC K. The letter K is an invitation for any station to go ahead. If there is no answer, pause for 10 or 20 seconds and repeat the call.

If you hear a CQ, wait until the ham finishes transmitting (by ending with the letter K), then call him, thus: K5RC K5RC DE K3YL K3YL $\overline{AR}$ ($\overline{AR}$ is equivalent to "over"). In answer to your call, the called station will begin the reply by sending K3YL DE K5RC R....

That R ("roger") means that he has received your call correctly. Suppose K5RC heard someone calling him, but didn't quite catch the call because of interference (QRM) or static (QRN). Then he might come back with QRZ? DE K5RC K (Who is calling me?).

The QSO

During the contact, it is necessary to identify your station only once every 10 minutes. Keep the contact on a friendly and cordial level, remembering that the conversation is not private and many others, including non-amateurs, may be listening. It may be helpful at the beginning to have a fully written-out script typical of the first couple of exchanges in front of you. A typical first transmission might sound like this: K3YL DE K5RC R TNX CALL. RST 599 599 QTH HOUSTON TX NAME TOM. HW? K3YL DE K5RC $\overline{KN}$. This is the basic exchange that begins most QSOs. Once these basics are exchanged the conversation can turn in almost any direction. Many people talk about their jobs, other hobbies, families, travel experiences, and so on.

Both on CW and phone, it is possible to be informal, friendly and conversational; this is what makes the Amateur Radio QSO enjoyable. During the contact, when you want the other station to take a turn, the recommended signal is $\overline{KN}$, meaning that you want only the contacted station to come back to you. If you don't mind someone else signing in, just K ("go") is sufficient.

Table 3-4
ARRL Procedural Signals

Situation	CW	Voice
check for a clear frequency	QRL?	Is the frequency in use?
seek contact with any station	CQ	CQ
after call to a specific named station or to indicate the end of a message	$\overline{AR}$	over, end of message
invite any station to transmit	K	go
invite a specific named station to transmit	$\overline{KN}$	go only
invite receiving station to transmit	BK	back to you
all received correctly	R	received
please stand by	$\overline{AS}$	wait, stand by
end of contact (sent before call sign)	$\overline{SK}$	clear
going off the air	CL	closing station

Additional RTTY prosigns

SK QRZ—Ending contact, but listening on frequency.
SK KN—Ending contact, but listening for one last transmission from the other station.
SK SZ—Signing off and listening on the frequency for any other calls.

Ending the QSO

When you decide to end the contact or the other ham expresses the desire to end it, don't keep talking. Briefly express your thanks for the contact: TNX QSO or TNX CHAT—and then sign out: 73 $\overline{SK}$ WA1WTB DE K5KG. If you are leaving the air, add CL to the end, right after your call sign.

These ending signals, which indicate to the casual listener the "status of the contact," establish Amateur Radio as a cordial and fraternal hobby. At the same time they foster orderliness and denote organization. These signals have no legal standing; FCC regulations say little about our internal procedures.

Table 3-4 contains a brief list of these signals. You can also request Form FSD-220 from the ARRL. It contains this list plus other handy information for the operating position.

Phone Operating Procedures

These phone or voice operating procedures apply to operation on the HF bands as well as SSB on VHF/UHF. Procedures used on repeaters are different since the operation there is channelized—that is, anyone listening to the repeater will hear you as soon as you begin to transmit. Therefore there is no need to call CQ. Each repeater may use a slightly different procedure. There is a complete discussion of repeater operations in Chapter 11 of this book.

Learning phone procedures for HF is exactly like learning CW operating procedure. Listen to what others are doing, and incorporate their good habits into your own operating style.

Use common sense in your day-to-day phone QSOs, too:

1) Listen before transmitting. Ask if the frequency is in use before making a call on any particular frequency.

2) Give your call sign as needed, using the approved ITU (International Telecommunication Union) Phonetics. These phonetics are given in a table later in this chapter..

3) Make sure your signal is "clean." Do not keep your microphone gain turned up too high. If you have a speech processor use it only when you are sure it is properly adjusted. Don't take a chance of transmitting spurious signals.

4) Keep your transmissions as short as possible to give as many operators as possible a chance to use the frequency spectrum.

5) Give honest signal reports. Chapter 17 has a list of what the various RST reports mean.

Whatever band, mode or type of operating some are undertaking, there are three fundamental things to remember. The first is that courtesy costs very little and is very often amply rewarded by bringing out the best in others. The second is that the aim of each radio contact should be 100% effective communication. The good operator is never satisfied with anything less. The third is that the "private" conversation with another station is actually in public. Keep in mind that many amateurs are uncomfortable discussing so-called controversial subjects over the air. Also, never give any information on the air that might be of assistance to the criminally inclined. As an example, never state when you are going to be out of town!

Correct phone operation is more challenging than it first may appear, even though it does not require the use of code or special abbreviations and prosigns. This may be because operators may have acquired some imperfect habits in their pronunciation, intonation and phraseology even before entering Amateur Radio! To these might be added a whole new set of clichés and mannerisms derived from listening to below-par operators.

Using the proper procedure is very important. Voice operators say what they want to have understood, while CW operators have to spell it out or abbreviate. Since the speed of transmission on phone is generally between 150 and 200 words per minute, the matter of readability and understandability is critical to good communication. The good voice operator uses operating habits that are beyond reproach.

It is important to speak clearly and not too quickly. This is particularly important when talking to a DX station who does not fully understand our language.

Avoid using CW abbreviations (including "HI") and Q signals on phone, although QRZ (for "who is calling?") has become accepted. Otherwise, plain language should be used. Keep jargon to a minimum. Some hams use "we" instead of "I," "handle" instead of "name" and "Roger" instead of "that's correct." These expressions are not necessary and do not contribute to better operating. No doubt you will hear many more.

Albert Payson, KB7IPU, is second from the right in the photo. He is showing off his shack to friends (left-to-right) Ryan, Garrett, and Matt.

Procedure

There are two ways to initiate a voice contact: *call* (a general call to any station) or *answer a CQ*. If activity on a band seems low and you have a reasonable signal, a CQ call may be worthwhile.

Before calling CQ, it is important to find a frequency that appears unoccupied by any other station. This may not be easy, particularly in crowded band conditions. Listen carefully—perhaps a weak DX station is on frequency.

Always listen before transmitting. Make sure the frequency isn't being used *before* you come barging in. If, after a reasonable time, the frequency seems clear, ask if the frequency is in use, followed by your call: "Is the frequency in use? This is N2EEC." If, as far as you can determine, no one responds, you are ready to make your call.

As in CW operation, CQ calls should be kept short. Long calls are unnecessary. If no one answers you can always call again. If you do transmit a long call you may interfere with stations that were already on the frequency but whom you didn't hear in the initial check. In addition, any stations intending to reply to the call may become impatient and move to another frequency. If two or three calls produce no answer, there may be interference on the frequency. It's also possible that the band isn't open.

An example of a CQ call would be:

"CQ CQ Calling CQ. This is N2EEC, November-Two-Echo-Echo-Charlie, November-Two-Echo-Echo-Charlie, calling CQ and standing by."

When replying to a CQ both call signs should be given clearly. Use the ITU phonetic alphabet (see table later in this chapter) to make sure the other station gets your call correct. Phonetics are necessary when calling into a DX pileup and initially in most HF contacts but not usually used when calling into an FM repeater.

When calling a specific station, it is good practice to keep calls short and to say the call sign of the station called once or twice only, followed by your call repeated twice. VOX (voice operated switch) operation is helpful because, if properly adjusted, it enables you to listen between words, so that you know what is happening on the frequency. "N2EEC N2EEC, this is W2GD, Whiskey-Two-Golf-Delta, Over."

Once contact has been established, it is no longer necessary to use the phonetic alphabet or sign the other station's call. According to FCC regulations, you need only sign your call every 10 minutes, or at the conclusion of the contact. (The exception is handling international third-party traffic; you must sign both calls in this instance.) A normal two-way conversation can thus be enjoyed, without the need for continual identification. The words "Over" or "Go Ahead" are used at the end of a transmission to show you are ready for a reply from the other station.

Signal reports on phone are two-digit numbers using the RS portion of the RST system (no tone report is required). The maximum signal report would be "59"; that is, readability 5, strength 9. On FM repeaters, RS reports are not appropriate—when a signal has fully captured the repeater, this is called "full quieting."

Conducting the Contact

Aside from signal strength, it is customary (as in a typical CW QSO described elsewhere in this chapter) to exchange name, location and a brief description of the rig and antenna. Keep in mind that many hams will not know the characteristics of your *Loundenboomer 27A3* and your *SignalSquirter 4*. Therefore, you may be better off just saying that your rig runs 150 watts input and the antenna is a trap dipole. Once these routine details are out of the way, you can discuss virtually *anything* the two of you find appropriate and interesting.

DX can be worked on any HF band as well as occasionally on 6 meters. However as good conditions return to 10 meters, worldwide communication on a daily basis will again be commonplace on this band. Ten meters is an outstanding DX band when conditions are right. A particular advantage of 10 for DX work is that effective beam-type antennas tend to be small and light, making for relatively easy installation.

Keep in mind that while many overseas amateurs have an exceptional command of English they may not be familiar with many of our colloquialisms. Because of the language differences, some DX stations are more comfortable with the "barebones" type contact, and you should be sensitive to their pref-

Bill Baker is usually WA9L from the US. In this photo he is operating V31DX from Belize in a DX contest. Bill is using a lap-top computer to log his contacts.

erences. In unsettled conditions, it may be necessary to keep the whole contact short. The good operator takes these factors into account when expanding on a basic contact.

When the time comes to end the contact, end it. Thank the other operator (once) for the pleasure of the contact and say good-bye: "73, N9ABC THIS IS N2EEC, CLEAR." This is all that is required. Unless the other amateur is a good friend, there is no need to start sending best wishes to everyone in the household including the family dog! Nor is this the time to start digging up extra comments on the contact that will require a "final final" from the other station (there may be other stations waiting to call in). Please understand that during a band opening on 10 meters or on VHF, it is crucial to keep contacts brief so as many stations as possible can work the DX coming through.

Additional Recommendations

Listen with care. It is very natural to answer the loudest station that calls, but sometimes you will have to dig deep into the noise and interference to hear the other station. Not all amateurs can run a kilowatt.

Use VOX or push-to-talk (PTT). If you use VOX, don't defeat its purpose by saying "aaah" to keep the relay closed. If you use PTT, let go of the mike button every so often to make sure you are not "doubling" with the other station. Don't filibuster.

Talk at a constant level. Don't "ride" the mike gain. Try to keep the same distance from the microphone. It is a good idea to keep the mike gain down to eliminate room noise. Follow the manufacturer's instructions for use of the microphone. Some require close talking, while some need to be turned at an angle to the speaker's mouth. Speech processing (sometimes built in to contemporary transceivers or available as an outboard accessory) is often a mixed blessing. It can help you cut through the interference and static, but if too much is used, the audio quality suffers greatly. Tests should be made to determine the maximum level that can be used effectively, and this should be noted or marked on the control. Be ready to turn it down or off if it is not really required during a contact.

The speed of voice transmission (with perfect accuracy) depends almost entirely on the skill of the two operators concerned. Use a rate of speech that allows perfect understanding as well as permitting the receiving operator to record the information.

VHF/UHF SSB

"Weak-signal" VHF/UHF operating practices are not very different than those common on HF. Pileups can occur in good conditions, and because these openings can be short-lived, it is good procedure to limit the contact to exchange of signal reports or grid squares and location details. Other details can be left to the QSL card. As mentioned above, in this way more stations get a chance to work the DX station.

When conditions are good *call CQ sparingly,* if at all, unless you have an extremely potent signal. The old adage that "if you can't hear 'em, you can't work 'em" is just as true at VHF as HF. All too often, stations can be heard calling CQ DX on the same frequency as distant stations that they cannot hear.

One feature of VHF operating that's rarely found on the HF bands is the use of *calling frequencies.* The idea is that each frequency provides a "meeting place" for operators using the same mode. Once contact has been set up, a change to another frequency (the working frequency) is arranged so others can use the calling frequency. Table 3-5 contains the current VHF and UHF calling frequencies. See Chapter 12 for additional details.

Repeater Customs

A repeater is a device that receives a signal on one frequency and simultaneously transmits (repeats) it on another frequency. In a sense, it is a robot transmitter. Often located on top of a tall building or high mountain, VHF/UHF repeaters greatly extend the operating coverage of amateurs using mobile and hand-held transceivers.

To use a repeater, you must have a transceiver with the capability of transmitting on the receiver's *input frequency* (the frequency that the repeater listens on) and receiving on the repeater's *output frequency* (the frequency the repeater transmits on). This capability can be acquired by dialing the correct frequency and selecting the proper offset (frequency difference between input and output).

When you have the frequency capability (and most rigs to-

Table 3-5
VHF/UHF/EFH Calling Frequencies

Band (MHz)	Calling Frequency
50	50.125 SSB 50.620 digital (packet) 52.525 National FM simplex frequency
144	144.010 EME 144.100, 144.110 CW 144.200 SSB 146.520 National FM simplex frequency
222	222.100 CW/SSB 223.500 National FM simplex frequency
432	432.010 EME 432.100 CW/SSB 446.000 National FM simplex frequency
902	902.100 CW,SSB 903.1 Alternate CW, SSB 906.500 National FM simplex frequency
1296	1294.500 National FM simplex frequency 1296.100 CW/SSB
2304	2304.1
10000	10364

VHF/UHF Activity Nights

Some areas do not have enough VHF/UHF activity to support contacts at all times. This schedule is intended to help VHF/UHF operators make contact. This is only a starting point; check with others in your area to see if local hams have a different schedule.

Band (MHz)	Day	Local Time
50	Sunday	6 PM
144	Monday	7 PM
222	Tuesday	8 PM
432	Wednesday	9 PM
902	Friday	9 PM
1296	Thursday	10 PM

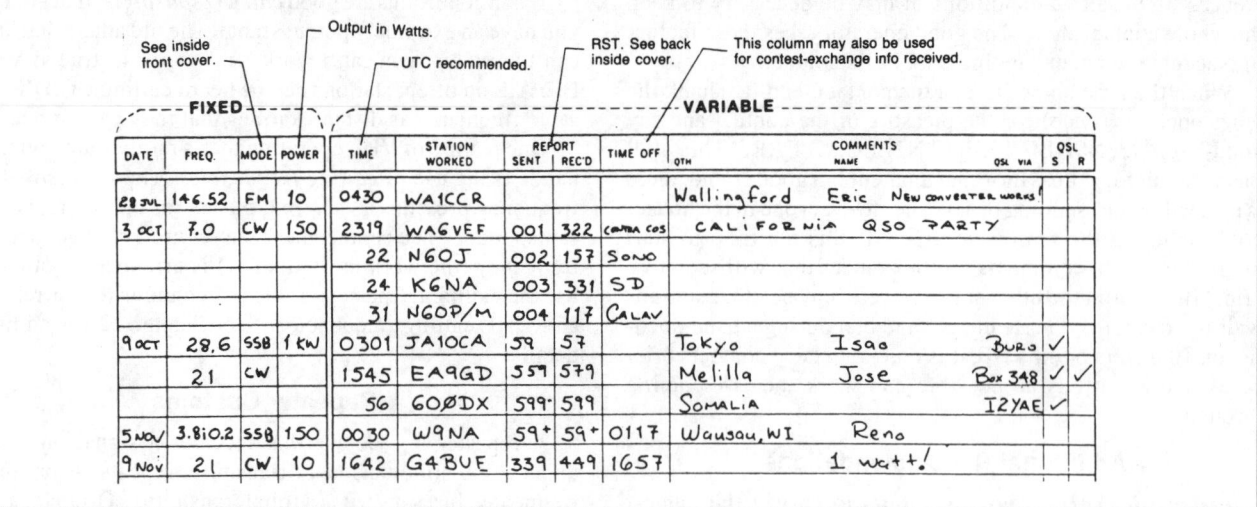

See inside front cover. — Output in Watts. — UTC recommended. — RST. See back inside cover. — This column may also be used for contest-exchange info received.

--- FIXED --- --- VARIABLE ---

DATE	FREQ.	MODE	POWER	TIME	STATION WORKED	REPORT SENT	REPORT REC'D	TIME OFF	QTH	COMMENTS NAME	QSL VIA	QSL S	QSL R
28 JUL	146.52	FM	10	0430	WA1CCR				Wallingford	Eric NEW CONVERTER WORKS!			
3 OCT	7.0	CW	150	2319	WA6VEF	001	322	CONTRA COS	CALIFORNIA QSO PARTY				
				22	N6OJ	002	157	SONO					
				24	K6NA	003	331	SD					
				31	N6OP/M	004	117	CALAV					
9 OCT	28.6	SSB	1 KW	0301	JA1OCA	59	57		Tokyo	Isao	BURO	✓	
	21	CW		1545	EA9GD	559	579		Melilla	Jose	Box 348	✓	✓
				56	6OØDX	599	599		Somalia		I2YAE		✓
5 NOV	3.810.2	SSB	150	0030	W9NA	59+	59+	0117	Wausau, WI	Reno			
9 NOV	21	CW	10	1642	G4BUE	339	449	1657		1 watt!			

Fig 3-5—The ARRL Log Book is adaptable to all types of operating.

UTC Explained

Ever hear of Greenwich Mean Time? How about Coordinated Universal Time? Do you know if it is light or dark at 0400 hours? If you answered no to any of these questions, you'd better read on!

Keeping track of time can be pretty confusing when you are talking to other hams around the world. Europe, for example, is anywhere from 4 to 11 hours ahead of us here in North America. Over the years, the time at Greenwich, England, has been universally recognized as *the* standard time in all international affairs, including ham radio. (We measure longitude on the surface of the earth in degrees east or west of the Prime Meridian, which runs approximately through Greenwich, and which is halfway around the world from the International Date Line.) This means that wherever you are, you and the station you contact will be able to reference a common date and time easily. Mass confusion would occur if everyone used their own local time.

Coordinated Universal Time (abbreviated UTC) is the name for what used to be called Greenwich Mean Time.

Twenty-four-hour time lets you avoid the equally confusing question about AM and PM. If you hear someone say he made a contact at 0400 hours UTC, you will know immediately that this was 4 hours past midnight, UTC, since the new day always starts just after midnight. Likewise, a contact made at 1500 hours UTC was 15 hours past midnight, or 3 PM (15 − 12 = 3).

Maybe you have begun to figure it out: Each day starts at midnight, 0000 hours. Noon is 1200 hours, and the afternoon hours merely go on from there. You can think of it as adding 12 hours to the normal PM time—3 PM is 1500 hours, 9:30 PM is 2130 hours, and so on. However you learn it, be sure to use the time everyone else does—UTC. See chart below.

The photo shows a specially made clock, with an hour hand that goes around only once every day, instead of twice a day like a normal clock. Clocks with a digital readout that show time in a 24-hour format are quite popular as a station accessory.

UTC	EDT/AST	CDT/EST	MDT/CST	PDT/MST	PST
0000*	2000	1900	1800	1700	1600
0100	2100	2000	1900	1800	1700
0200	2200	2100	2000	1900	1800
0300	2300	2200	2100	2000	1900
0400	0000 *	2300	2200	2100	2000
0500	0100	0000 *	2300	2200	2100
0600	0200	0100	0000 *	2300	2200
0700	0300	0200	0100	0000 *	2300
0800	0400	0300	0200	0100	0000 *
0900	0500	0400	0300	0200	0100
1000	0600	0500	0400	0300	0200
1100	0700	0600	0500	0400	0300
1200	0800	0700	0600	0500	0400
1300	0900	0800	0700	0600	0500
1400	1000	0900	0800	0700	0600
1500	1100	1000	0900	0800	0700
1600	1200	1100	1000	0900	0800
1700	1300	1200	1100	1000	0900
1800	1400	1300	1200	1100	1000
1900	1500	1400	1300	1200	1100
2000	1600	1500	1400	1300	1200
2100	1700	1600	1500	1400	1300
2200	1800	1700	1600	1500	1400
2300	1900	1800	1700	1600	1500
2400	2000	1900	1800	1700	1600

Time changes one hour with each change of 15° in longitude. The five time zones in the U.S. proper and Canada roughly follow these lines.

*0000 and 2400 are interchangeable. 2400 is associated with the date of the day ending, 0000 with the day just starting.

Fig 3-6—Time is recorded both in log books and on QSL cards. UTC is always used.

day are fully synthesized), all you need do is key the microphone button and you turn on (access) the repeater. Some repeaters have limited access, requiring the transmission of a subaudible tone, a series of tones or bursts to gain access. However, most repeaters are open. Most repeaters briefly transmit a carrier after a user has stopped transmitting to inform the user that he is actually accessing a repeater.

After acquiring the ability to access a repeater, you should become acquainted with the operating practices of this unique mode. See Chapter 11 for specifics on operator practices.

Specialized Communications

Specialized communications, including satellite communications and the digital modes (including radioteletype, AMTOR and packet radio), all have distinct procedures. This manual thoroughly covers these modes in other chapters, while additional information can be found in *The ARRL Handbook for Radio Amateurs.*

Logging and QSLing

Although the FCC does not require logging except for certain specialized occurrences, you can still benefit by keeping an accurate log (see Fig 3-5). *The ARRL Logbook,* on sale at your local radio bookstore or directly from ARRL HQ, provides a good method for maintaining your log data at your fingertips.

The log entry should include:

1) The call sign of the station worked.

2) The date and time of the QSO. Always use UTC (sometimes also called GMT or Zulu time) when entering the date and time. Use UTC whenever you need a time or date in your hamming activities. The use of UTC helps all hams avoid confusion through conversion to local time. Fig 3-6 explains the UTC system.

3) The frequency or frequency band on which the QSO took place.

4) The emission mode of communication.

5) Signal reports sent and received.

6) Any miscellaneous data, such as the other operator's name or QTH that you care to record.

The FCC has promulgated a rather elaborate system of emission designators (see Chapter 2 on spectrum management). For logging simplicity, the following common abbreviations are often used:

Abbreviation	Explanation
CW	telegraphy on pure continuous wave
MCW	tone-modulated telegraphy
SSB	single-sideband suppressed carrier
AM,DSB	double-sideband with full, reduced or suppressed carrier
FAX	facsimile
FM	frequency- or phase-modulated telephony
RTTY	radioteletype
AMTOR	time diversity radioteletype
P	pulse
TV or SSTV	television or slow-scan television

Computer Logging

If you are into personal computing, there are many log-keeping programs that will take your input, format your log file and even give you a printed output. In addition, computer programs for contest applications are discussed in Chapter 7.

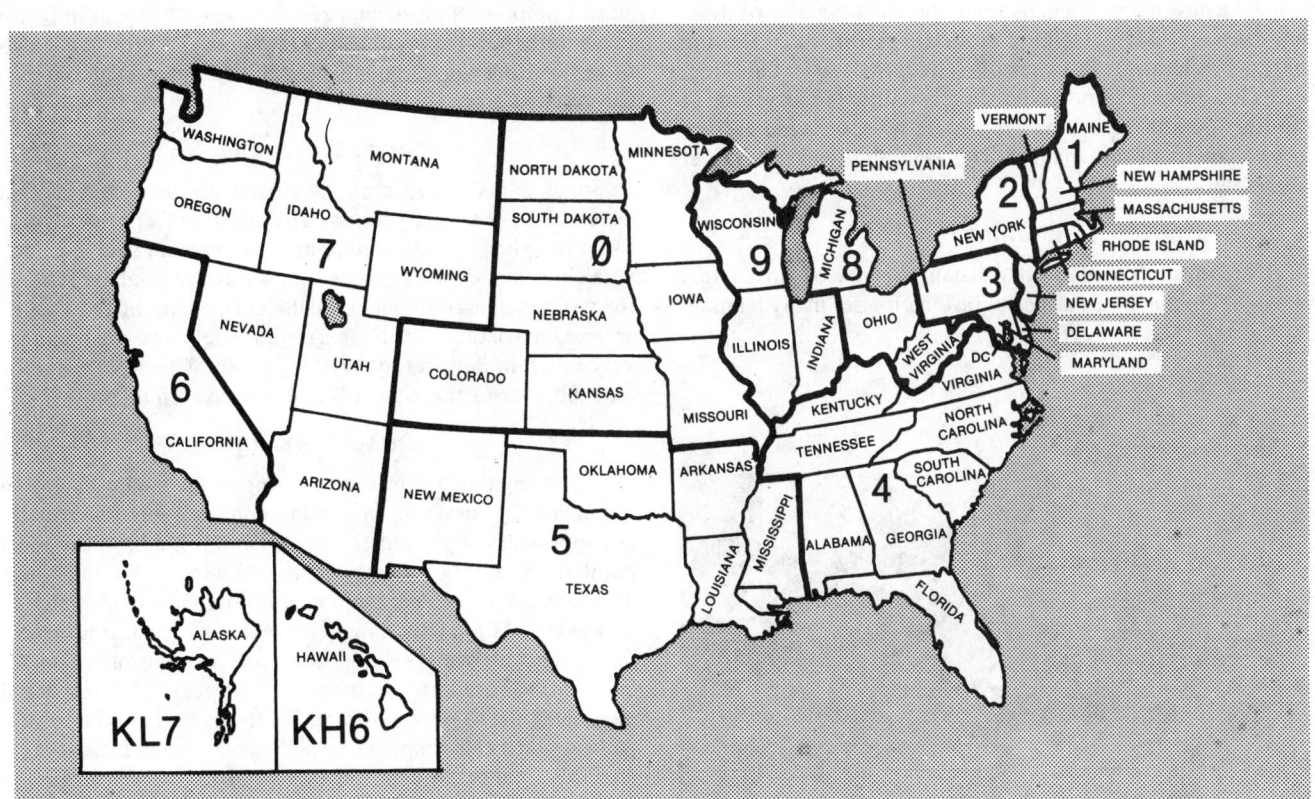

Fig 3-7—The 10 US Call Districts established by the FCC. At present, the number in each ham call tells you the district the ham was in when the call was issued.

A well-kept log will help you preserve your fondest ham radio memories for years. It will also serve as a bookkeeping system should you embark upon a quest for ham radio awards, or a complete collection of QSL cards.

QSLing

The QSL card is the final courtesy of a QSO. It confirms specific details about your two-way contact with another ham. Whether you want the other station's QSL as a memento of an enjoyable QSO, or for an operating award, it's wise to have your own QSL cards and know how to fill them out. That way, when you send your card to the other station, it will result in the desired outcome (his card sent to you).

Your QSL

Your QSL card makes a statement about you. It may also hang in ham radio shacks all over the world. So you will want to choose carefully the style of QSL that you have printed.

There are many QSL vendors listed in the Ham Ad section of *QST* each month. A nominal fee will bring you many samples from which to choose or you may design and/or print your own style. The choice is up to you. See the accompanying sidebar for more information on QSLs.

That just about finishes our ever so slight foray into operating procedures. Your knowledge will grow quickly with your on-the-air experience. Enjoy the learning process. Ham radio is such a diverse activity that the learning process never stops—there is always something new to learn.

ON-THE-AIR ACTIVITIES

One of the advantages of Amateur Radio is the number of activities open to us. Often it's hard to decide which to do first. Many hams will spend a few months or even years on one of them—become proficient—and then jump to another to have a new experience. Among the most popular of these activities are DXing (talking to distant stations), contesting and awards hunting. Ragchewing (chatting), traffic handling, QRP (low power operation) and packet radio are just some of the other popular activities. Most of these are covered in complete chapters of this manual. This section gives you a brief overview.

DXing

DX (ham shorthand for long distance) is one of those terms in ham radio that holds a special fascination for many hams. It can be correctly defined in different ways by different people. In the beginning, to most amateurs DX is the lure of seeing how far away you can establish a QSO—the greater the distance, the better. DX is a personal achievement, bettering some previous "best distance worked," involving a set of self-imposed rules.

DX can also be competing on a larger scale, trying to break the pileups on a rare DX (foreign) station and aiming for one of the DX-oriented awards. DXing can be a full-time goal for some hams and a just-for-fun challenge for others. No matter if you turn into a serious DXer or just have a little fun, DXing is one of the more fascinating aspects of Amateur Radio. See Chapter 5 for more particulars on DXing.

Contesting

Contesting is to Amateur Radio what the Olympic Games are to worldwide amateur athletic competition: a showcase to display talent and learned skills, as well as a stimulus for further achievement through competition. Increased operating skills and greater efficiency may be the end result of Amateur Radio contesting but fun is the most common experience.

The contest operator is also likely to have one of the better signals on the band—not necessarily the most elaborate station equipment, but a signal enhanced by constant experimentation and improvement. Contest operation encourages station and operator efficiency.

Nearly every contest has competitors vying to see which one can work the most stations (depending on the rules of the particular contest) in a given time frame. In some contests, the top-scoring stations have consistently worked 100 or more stations per hour for the entire 48-hour contest period.

The ARRL contest program is so diverse that it holds some appeal for almost every operator—the beginning contester and the old hand, the newest Technician and oldest Extra-Class veteran, "Top Band [160 meters] Buff" and microwave enthusiast. Complete contest entry rules and results appear in *QST* a few months before the contest. See Chapter 7 for further details on the exciting world of contesting. There is at least one contest for every interest.

Awards Hunting

Some hams spend most of their time in a contest that never ends—earning awards. The ARRL offers the awards listed in Table 3-6. Most other national amateur societies, private clubs and contest groups sponsor awards and certificates for various operating accomplishments. Many of the awards are very handsome certificates or intricately designed plaques very much in demand by awards chasers. Chapter 8 lists the most popular of these awards and how to earn them.

Rag Chewing

"Chewing the rag" refers to getting on the air and spending minutes (or hours!) in interesting conversation on virtually any and every topic imaginable. Without a doubt, the most popular operating activity is ragchewing. The rag chew may be something as simple as a brief chat on a 2-meter or 440-MHz FM repeater as you drive across town. It also may be a group of friends who have been meeting on 15-meter SSB every Saturday afternoon for 20 years. The essential element is the same—hams talking to each other on any subject that interests them. The ARRL awards a certificate that declares membership in the *Rag Chewer's Club* earned by carrying on a conversation of half an hour or longer. Most hams will tell you that it is not difficult to chat back and forth for this period or longer!

This station is used by Jack Reed, WA7LNW, for both contests and working DX from the heart of Arizona.

The Final Courtesy

QSL cards are a tradition in ham radio. Exchanging QSLs is fun, and they can serve as needed confirmations for many operating awards. You'll probably want your own QSL cards, so look in *QST* for companies that sell them, or you may even want to make your own. Your QSL should be attractive, yet straightforward. All necessary QSO information on one side of the card will make answering a QSL a relatively simple matter.

A good QSL card should be the standard 3.5 × 5.5-inch size (standard post card size) and should contain the following information.

1) Your call. If you were portable or mobile during the contact, this should be indicated on the card.

2) The geographical location of your station. Again, portables/mobiles should indicate where they were during the contact.

3) The call of the station you worked. This isn't as simple as it sounds. Errors are very common here. Make sure it is clear if the call contains the numeral *1 (one)*, capital letter *I ("eye")* or lower case letter *l ("el".)*

4) Date and time of the contact. Use UTC for both and be sure you convert the time properly. It is best to write out the date *in words* to avoid ambiguity. Use May 10, 1995 or 10 May 1995, rather than 5/10/95. Most DX stations will use 8-2-95 to mean February 8, 1995. The day is written before the month in many parts of the world.

5) Frequency. The band in wavelength (meters) or approximate frequency in kHz or MHz is required.

6) Mode of operation. Use accepted abbreviations, but be specific. CW, SSB, RTTY and AM are clear and acceptable. FCC emission designations are not always understood by DX stations.

7) Signal report.

8) Leave no doubt the QSL is confirming a two-way contact by using language such as "confirming two-way QSO with," or "2 X" or "2-Way" before the other station's call. Other items, such as your rig, antenna and so on, are optional.

9) If you make any errors filling out the QSL, destroy the card and start over. Do not make corrections or mark-overs on the card, as such cards are not acceptable for awards purposes.

Now comes the problem of how to get your QSLs to the DX station. Sending them directly can be expensive, so, many amateurs use the ARRL's Overseas QSL Service. This is an ongoing service for ARRL members to send DX QSL cards to foreign countries at a minimum of cost and effort.

To receive QSL cards from DX (overseas) stations, the ARRL sponsors incoming QSL Bureaus, provided free for all amateurs throughout the United States and Canada. Each call area has its own bureau staffed totally by volunteers. To expedite the handling of your QSL cards (both incoming and outgoing), be sure to follow the bureau's requirements at all times. See Chapter 5 for further details on the ARRL Overseas QSL Service and the ARRL QSL Bureau System.

Table 3-6
ARRL Operating Awards

Award	Qualification
Friendship Award	Contact 26 stations with calls ending A through Z.
Rag Chewer's Club	A single contact ½ hour or longer
Worked All States (WAS)	QSLs from all 50 US states
Worked All Continents (WAC)	QSLs from all six continents
DX Century Club (DXCC)	QSLs from at least 100 foreign countries
VHF/UHF Century Club (VUCC)	QSLs from many grid squares
Extra Class Certificate	Hold an Amateur Extra Class license
A-1 Operator Club	Recommendation by two A-1 operators
Code Proficiency	One minute of perfect copy from W1AW qualifying run
Old Timers Club	Held an Amateur Radio license at least 20 years prior
ARRL Membership	ARRL membership for 25, 40, 50, 60 or 70 years

able public resource that it is. Traffic handling is covered in detail in Chapter 15.

Nets are regular gatherings of hams who share a mutual interest and use the net (short for "network") to further that interest. *Nets* most commonly meet to pass traffic or participate in one of the many other ham activities from awards chasing and DXing to just plain old ragchewing. County- and State-hunting nets are very popular since they provide a frequency to work that "49th" and "50th" state for the WAS (Worked All States) award. DX nets provide the same opportunity for those interested in working DX stations without the competition of contests or random calling.

There are several nets dedicated to Novice or slow speed CW operation that can help a newcomer sharpen operating skills. The *ARRL Net Directory*, a compilation of public service nets, is available from ARRL HQ for $2.

QRP Operation

QRP is the Q signal for low power. Sent with a question mark QRP? it means "Shall I lower my power?" This Q signal is also used to describe low power operation—some hams like the challenge of using lower power to talk to other hams.

Packet Radio and Digital Modes

BC (Before Computers) hams used RTTY (teletype) as their only digital mode. As computers became more popular the number of digital modes expanded. The most common and popular is packet radio.

Packet radio requires either a full computer or a terminal (display and keyboard) at each end in addition to a packet radio controller and the radios themselves. Chapter 10 discusses this mode. If you are interested in packet, the ARRL publication *Your Packet Companion*, by Steve Ford, WB8IMY, is an invaluable guide for the newcomer.

Traffic Handling and Nets

Traffic handling involves passing messages to others via the amateur bands. Hams handle *third-party traffic* (messages for nonhams) in both routine situations and in times of disaster. Public-service communications make Amateur Radio the valu-

Picking a Band

160 and 80 Meters

Eighty meters, and its single-sideband neighbor, 75 meters, are favorites for ragchewing. I frequently check out the 80-meter Novice/Technician CW subband. There I find both newcomers as well as old-timers trying to work the rust out of their fists! Around 3600 kHz you'll find the digital modes, RTTY, AMTOR and packet. The QRP frequency is 3560 kHz. If you hear a weak signal calling CQ near 3560, crank down your power and give a call! Another favorite frequency is 3579.5 kHz. If you live in the eastern half of North America, listen for W1AW on 3581 (CW), 3625 (RTTY/AMTOR/packet) or 3999 kHz (SSB). W1AW runs 1000 watts to a modest antenna—an inverted V at 60 feet. If you can copy W1AW, you can probably work the East Coast, even with low power. An outside antenna is a definite plus on these bands, however.

Even if you can't chase DX, you'll find plenty to do on either band. Ionospheric absorption is greatest during the day, thus local contacts are common. At night, contacts over 200 miles away are more frequent, even with a poor antenna. Summer lightning storms make for noisy conditions in the summer while winter is much quieter. You may also be troubled by electrical noise here. A horizontally polarized antenna, especially one as far from the building as possible, will pick up less electrical noise.

"Top Band," as 160 meters is often called, is similar to 80 meters. QSOs here tend to be a bit more relaxed with less QRM. DX is frequent at the bottom of the band. Don't let the length of a half wave dipole for 160 keep you off the band; a 25 or 50-foot "long wire" can give you surprisingly good results. One favorite trick is to connect together the center conductor and shield of the coax feed line of a 40 or 80-meter dipole and load the resulting antenna as a "T."

40 and 30 Meters

I must confess to being biased in favor of these bands, especially 40 meters. If I could only have a receiver that covered one band, it would be 40. Running 10 watts from my East Coast apartment (indoor antenna) I can work European hams, ragchew up and down the coast, check into Saturday morning QRP nets, and listen to foreign broadcast stations besides. Yes, 40 is a little crowded. Look at the bright side: you won't be lonely. I think it's possible to work someone on 40 any time of the day or night.

From a modest station, daytime operation on 40 is easier than at night. The "skip" (distance you can work) during the day is shorter, but that helps reduce interference. Most other countries have SSB privileges down to 7050 kHz. During the day, they, and the broadcasters, won't bother you much. Outside the US, QRPers hang around 7030 kHz; in the US we use 7040 kHz. RTTY/ AMTOR/packet operators work around 7060 to 7100 kHz. Don't miss the Novice/Technician CW subband from 7100 to 7150 kHz. Even when the foreign broadcasters are booming in, you can usually find someone to work there. Forty is a good band for daytime mobile SSB operation, too. You'll find plenty of activity, and propagation condi-

tions tend to be stable enough to allow you to ragchew as you roll along.

The 30-meter band has propagation similar to 40 meters. Skip distances tend to be a little longer on 30 meters, and it's not so crowded. At present, stations in the US are limited to 200-watts output on this band. DX stations seem to like the low end of the band, from 10100 to 10115 kHz. Ragchewers often congregate above 10115. We share 30 meters with other services, so be sure you don't interfere with them. SSB isn't allowed on 30, but you can use CW and RTTY/AMTOR/packet. There is no agreed-upon QRP frequency on 30 meters.

20 Meters

As much as I like 40 and 30 meters, I have many fond memories of 20 meters as well. When I upgraded my license to General in 1963, I made a bee line to 20 meters. To this day, I can't stay away for long. A 20-meter dipole is only 33 feet long, and that doesn't have to be in a straight line. Many US hams have worked their first European or Australian contacts with a dipole and 100 watts.

While many hams think 20 meters is the workhorse band, it isn't always reliable. At the bottom of the sunspot cycle, 20 meters may be usable for a few hours a day. During years of higher solar activity though, 20 is often open all night.

There's plenty of room on the band. CW ragchewers hang out from 14025 to 14070 kHz, where you start hearing RTTY/AMTOR/packet stations. The international QRP frequency is 14060 kHz. The sideband part of the band is sometimes pretty busy and then it may be difficult to make a contact with low power or a modest antenna. Look above 14250 for ragchewers. Impromptu discussion groups sometimes spring up on 20 SSB, and make for interesting listening, even if you don't participate. If you like photographs, look around 14235 kHz for slow-scan TV. You'll need some extra equipment (as discussed in Chapter 16 of this book) to see the pictures.

17, 15 and 12 Meters

Except during years of high solar activity, you'll do most of your operating during daylight hours. Propagation is usually better during the winter months.

Seventeen and 12 meters are relatively new amateur bands, so they aren't as crowded as 15. Fifteen, though, is not nearly as crowded as 20. On 15, the QRP calling frequency is 21060 kHz. Don't overlook the Novice/ Technician CW subband from 21100 to 21150 kHz. No special frequencies are used for QRP operation on 17 and 12. SSB operation is much easier on 17 and 12, because of lower activity. Low activity doesn't mean no activity; when those bands are open, you'll find plenty of stations to work. You only need one at a time, after all. RTTY/ AMTOR/packet operation is found from 21070 to 21100 kHz, and around 18100 and 24920 kHz.

Some Russian RS-series satellites operate on 15 and 10 meters. Because of their low orbits, it isn't possible to work more than about 3000 miles on the RS satellites.

However, their low orbits make it possible to use these satellites with low power and simple antennas.

Practical indoor, outdoor, mobile or portable antennas for these bands are simple to build and install. It's even possible to make indoor beam antennas for the range of 18 to 25 MHz.

10 Meters

The 10-meter band stretches from 28000 to 29700 kHz. During years of high solar activity, 10 to 25-watt transceivers will fetch plenty of contacts. When the sun is quiet, there are still occasional openings of a few thousand miles. Ten meters also benefits from sporadic-E propagation. You'll find most sporadic-E openings in the summer, but they can happen anytime. Sporadic-E openings happen suddenly and end just as quickly. You won't be able to ragchew very long, but you'll be amazed at how many stations you can work.

SSB activity is heaviest in the Novice/Technician subband from 28300 to 28500 kHz. The Novice/Technician CW subband is a good place to look for activity, as is the QRP calling frequency at 28060 kHz. You can operate 1200-baud packet radio on 10 meters, whereas we're limited to 300 baud on the lower bands. RTTY/AMTOR/packet operation takes place from 28070 to 28100 kHz.

Higher in the band, above 29000 kHz, you'll find amateur FM stations and repeaters, and the amateur satellite subband.

Operating On 50 MHz and Above

The VHF/UHF/microwave bands offer many advantages to the low-power operator. The biggest plus is the relatively smaller antennas used. A good-sized 2-meter beam will easily fit in a closet when not in use. Portable and mobile operation on these bands is also easy and fun.

6 Meters

Six meters is perhaps the most interesting amateur band. When solar activity is high, worldwide QSOs are common. When solar activity is low, however, the opportunities for long-distance communication decrease. Sporadic-E propagation, which I mentioned earlier, is the most reliable DX mode during periods of low solar activity.

With small antennas, like 3-element beams, it's possible to work 1000 miles on sporadic E. Three-element 6-meter beams don't fit well inside houses or apartments, but you might be able to put one in an attic or crawl space. Even if you can only use a dipole, you'll be able to work locals, and snag some more-distant stations when the band opens.

Just about any mode found on the HF bands is used on 6 meters. CW and SSB operation take place on the lower part of the band. Higher up you'll find FM simplex and repeater stations. Another mode you'll sometimes find on 6 meters is radio control (RC) of model planes, helicopters, boats and cars.

2 Meters

Simply stated, 2 meters is the most popular ham band in North America. From just about any point in the US, you can probably work someone on 2 meters, 24 hours a day. Most hams know about 2-meter FM and packet radio operation, but CW and SSB are used here too. There's even an amateur satellite subband on 2 meters.

CW and SSB operation is done mostly with horizontally polarized antennas. FM and packet operators use vertical polarization, while satellites can be worked with either. A popular 2-meter antenna called a halo is perfect for indoor or mobile use on CW or SSB. The omnidirectional halo has no gain, but you'll be able to work locals, and up to 100 miles during band openings.

FM and packet usually require only a simple vertical antenna. The *ARRL Repeater Directory* will tell you what repeaters are available in your area. This book lists all repeaters reported to ARRL for the bands from 50 to 10,000 MHz. It also lists packet radio nodes and digipeaters. Nodes and digipeaters are amateur stations that allow you to send packet messages over long distances. Unlike FM repeaters, which have separate input and output frequencies, packet radio repeater stations use the same frequency for both.

Packet radio and VHF were made for each other. Packet has become so popular that ARRL publishes several books on the subject including *Your Gateway to Packet Radio* and *Your Packet Companion*.

The 222 and 430-MHz Bands

Every mode used on 2 meters is found on 222 except satellite communication. The 430-MHz or 70-cm band is becoming more and more popular, too. Multiband hand-held and mobile transceivers are available at prices only slightly higher than single-band rigs. If you think you'd like to try these bands in addition to 2 meters, look into a multiband rig.

One mode you'll find on 70 cm that isn't allowed on the lower frequencies is fast scan amateur television (ATV). Assuming you already have a broadcast TV set, all you need is a receive converter, transmitter, antenna and camera. Inexpensive cameras designed for home-video use are fine for ATV. ATV repeaters may be found in larger metropolitan areas. They're listed in the *ARRL Repeater Directory*.

23 Cm and Up

As you go higher in frequency, the relative size of antennas gets smaller. This fact allows you to use very high-gain antennas that aren't very big. Commercial equipment is available for the bands through 10,000 MHz (10 GHz). You'll also find easy-to-build kits (the tuned circuits are etched onto the circuit boards).

Because antennas are so small, it's possible to have 20 to 30-dB gain antennas that fit in your car's trunk. In comparison, a big 20-meter beam might offer only 10-dB gain. Operating from the field with battery-powered equipment is very popular, especially during VHF/UHF/microwave contests. Thanks to high-gain antennas, contacts over several hundred miles are possible with equipment running 1 or 2 watts.—*Jim Kearman, KR1S, from Low Profile Amateur Radio, published by the ARRL*

Less is More...Or Enjoying QRP

Are you a newly licensed radio amateur trying to get on the air with a minimum dollar investment? Are you a certified old-timer with 5BWAS, 5BWAC and 5BDXCC under your belt and have found hamming has lost some of its excitement? Are you somewhere in between Novice and veteran, and want to introduce a new spice to your ham palate? If you fit into any of these categories, read on.

Operating QRP (low power, defined by ARRL as 10 watts input or 5 watts output) is a popular modus operandi of thousands of hams. The thrill of communicating at low power levels is perhaps best described as being similar to the excitement experienced during your first QSO. Interestingly enough, the level of enjoyment proportionately goes up as your power goes down.

Many amateurs believe QRP is a relatively new avenue of hamming that started in the '70s with the introduction of the Ten-Tec PM series and the Heathkit HW-7 transceivers. In reality, QRP dates back at least 60 years before the birth of these rigs. In the early days of Amateur Radio, particularly during the Depression years in the '30s, amateurs didn't have a lot of money to invest in their hobby. "Make do with what you have" and professional scrounging were the order of the day to acquire the necessary parts to build up a station. This, more often than not, limited the station transmitter to one of the QRP variety.

A look at page 45 of May 1933 *QST* reveals an article entitled "Low Power Records," listing success stories hams of the day were enjoying using QRP. Actual QSOs not just hello-good-bye contacts, at 30-mW input, were being enjoyed by these pioneer QRPers. Low-power operating is, therefore, nothing new and might well be one of the few remaining common bonds with a simpler time.

The first thing anyone contemplating a jump into the QRP sport should do is cast off the notion you must run high power (QRO). A more enlightened attitude is needed. Consider your QRP operating as an adventure, a challenge, a unique and very personal voyage on the airwaves—riding a leaf instead of a supersonic jet. It's a gentle form of communication; think "heart and soul," not "blood and guts." Shoot down your DX prey with a pea-shooter rather than a double-barrel shotgun. A positive frame of mind will set the stage for an enjoyable time with QRP. Here are some important ground rules:

1) Listen, listen, listen.
2) Call other stations, don't call CQ.
3) Expect less-than-optimum signal reports.
4) Be persistent and patient.
5) Know when to quit.

Listen to the bands and try to figure out what the prevailing propagation is. Is the skip short or long? Who's working who? Is there much interference, static or fading (QRM, QRN or QSB respectively) present? A quick analysis of the band conditions should be the first thing you do when sitting down for a session of QRP operating. Listening will help you decide what band to operate. A thorough study of the radio-wave propagation information in this and other ARRL publications will help you understand what you hear. Always listen, listen, listen!

Call CQ as a last resort! Most hams prefer to answer a strong signal, which you probably will not have. You will be much better off answering a CQ. Try answering someone like this: WJ1Z DE KA1CV/QRP or WJ1Z DE KR1S/2W K. This tells WJ1Z why you aren't doing a meltdown on his headphones!

Don't be discouraged if you receive signal reports like RST 249 or RS 33. With less than 5 watts output, you can't expect to be overloading the receiver front ends out there in DX-land. Have faith, for you will get more than

your fair share of very respectable reports. The ultimate ego gratification of a 599 or 59 will be yours if you keep at it!

If at first you don't succeed, then try again. And again. This QRP stuff is a game of persistence, so don't give up if you don't get an answer to your call on the first try. Don't be discouraged. Make up your mind—instant success just isn't part of the plan. That's what makes it so much fun.

If band conditions are inadequate, if your dipole has become the prey of a falling tree limb, if your neighbor has just started up a noise-generating appliance, if the dog is barking, the baby is crying and your spouse is out shopping, give it up! The last thing a QRPer needs is such distractions. Go tend the baby and see what the hound is upset about. Know when to quit. Sometimes you will have days like this, so know when to take a break from hamming. Try again tomorrow (or at least wait until your spouse returns from the store).

There is a relatively small selection to choose from if you plan to purchase a QRP-only transceiver. Only a limited amount of commercially made (or kit) equipment specifically designed for QRP operating is available. However, *QST* Ham Ads or flea markets are usually fruitful hunting grounds for good "preowned" radios at very reasonable prices.

Making its long awaited return to the amateur equipment scene is the Ten-Tec Argonaut II: a completely updated version of the popular 505, 509 and 515 Argonauts of the 1970s. With the addition of 160, 30, 17 and 12 meters, as well as 1990s' technology, the Argonaut II is a "big little rig" for the QRPer.

The earlier Argonauts still command respectable prices on the used market and are considered by some to be the rig of rigs for QRP. All featured full CW and SSB coverage on the "traditional" 80 to 10-meter bands. In addition, don't overlook Ten-Tec's old PM series. They were basic barebones units and are excellent for getting started.

The Heathkit HW-7, -8 and -9 transceivers were CW-only units covering the lower portions of various HF bands. Direct-conversion receivers were the weak links in the HW-7s and -8s. Heath recognized this deficiency and came out with an improved superhet receiver design for the HW-9.

Some of the Japanese multimode HF transceivers can be purchased with a low-power amplifier (Japan has a 10-watt amateur license class). These are full-feature transceivers, with all the advantages of modern technology, *sans* the full-power amplifier.

If you already have the typical amateur transceiver running 100 to 200 watts, never fear! All you have to do is reduce the output power by cranking down the drive or *RF output* control. (An accurate wattmeter also comes in handy.) With some rigs, minor tinkering with ALC circuitry can give you smooth level control well below 1 watt output.

Many amateurs have as much fun building QRP transmitters as they do operating them! Let's face it—nobody has the resources to build a deluxe, professional-quality transceiver. A two- or three-stage CW transmitter, on the other hand, is certainly within the technical ability of many of us. See the ARRL publication *QRP Classics* if you have the urge to roll your own!

A very useful tool for the QRP station is the wattmeter. A commercially built unit can be found for a reasonable price. A basic single meter unit with switchable forward and reverse power is a good way to start. In time you may want to add another meter to eliminate the need to switch back and forth between forward and reverse power. Save the switch and use it to change the power range of the meter. This way you can have one range for a 5-watt full scale, and the other a 1-watt scale. You can calibrate this wattmeter with a VTVM (vacuum-tube voltmeter), a simple home brew dummy load and an RF probe. Not only will you be saving some hard-earned bucks, but you will be gaining experience in designing, building, modifying and calibrating test equipment.

With QRP, your antenna is going to be much more important and instrumental in your success than if you run QRO. Running 100 watts into a random wire will net you plenty of solid contacts. But when you reduce power into the same wire, your signal effectiveness will decrease, too. As a result, the old axiom of putting up the biggest antenna you can muster, as high and as in the clear as possible, means more to the QRPer than someone running 100 or 1000 watts. The important thing is to optimize your antenna to your own personal circumstances. Many operators have reported amazing results using less-than-optimum antenna systems, but this is not to say you should be lax in your antenna installation. By running QRP, you are already reducing your effective radiated power (ERP); no need for a further (unintentional) reduction by cutting corners on your antenna system.

For 160 to 30 meters, dipoles generally will work well. Height is always nice, so do the best you can. Loops, end-fed wires and verticals are also used on these bands. The popular HF DX bands, 20 to 10 meters, deserve some serious thought as to rotatable gain antennas—Yagis or quads. Although this train of thought usually leads to a considerable outlay of cash, you will benefit in several ways. A 1-watt signal to a 10-dB gain Yagi will give you an effective radiated output of about 10-watt! That's just like having an amplifier that needs no power to run. A directional antenna is a reciprocal device as well. It is effective on both received signals and the transmitted signals. Listening to Europeans is so much more fun when you don't have to hear them along with signals from other unwanted directions.

The VHF and UHF bands are ripe for QRP operating and some serious antenna-design work. Amateurs with even very limited space can put up a high-gain Yagi for 6 or 2 meters or above. Many of the headaches of the HF bands are virtually nonexistent on VHF/UHF. The creative QRPer should be able to have a lot of success on these bands.

If you can't swing a big Yagi, either physically or financially, just get a decent wire antenna in the air. They all work, just some better than others. Again, the important thing is to put up the best system you possibly can.

Now that you're hooked, what to do about it? General operating and ragchewing are always fun, but do not limit yourself to just one kind of operating. Study the rest of this *Operating Manual,* and you should have enough info to attack several fronts with your mighty QRP signal! Here are a few suggestions.

Chase DX. Don't plan on busting a 250 station pileup, but very respectable DX can be worked with an extremely modest station and antenna. If you are cagey enough and have a few tricks up your sleeve (see Chapter 5 on DX operating techniques), then DX will indeed be yours.

Get in a contest. During the course of a contest weekend, activity on the bands is very high. What better time to try out a new transmitter or antenna? Go ahead and jump in—the water's fine. How do you compete against the super contest stations? You don't. You can compete with other QRPers or, better yet, compete against yourself. Set a personal goal—100 contacts, 30 states or countries or whatever—and keep a record of your results. Try to beat your previous record during the next contest. Once you get involved with QRP contesting, though, watch out—it's addicting!

Chase VHF/UHF grid squares. First, pack up your equipment and drive up to a mountain or hilltop location.

On VHF, if the band is open, your single watt will sound like 100 watts! With a couple of watts output to a gain antenna at a high elevation, you should be able to bend a few S-meter needles.

Go portable. Put the rig in your knapsack take a hike up a hill or mountain and set up your rig. Throw a wire over a tree limb and load 'er up. Working QRP from a scenic spot is hard to beat. Battery powered and away from all man-made electrical noise, you'll be able to hear what your receiver really can do. You can operate Field Day—not once a year, but every time you go for a hike!

Try HF digital modes, SSB or SSTV. Low-power operating does not mean you have to operate CW exclusively. If you like PacTOR, then you'll have it using QRP. The same goes for SSB, SSTV or other models. Further, operating SSB is an excellent way to spread the QRP word. More people seem to be "reading the mail" in the voice portions of the bands, so what better place to be a representative for the QRP movement?

You might be wondering how low can you go. Conduct minimum-power-level experiments and log your results. Get on the air with a fellow QRPer and, while monitoring your output power, see just how low in power you can go while still maintaining communications. You will be amazed at the possibilities. If the band conditions are right, there will be no difference between 1 watt and 5 watts; you may even be able to communicate at 10- or 20-mW. That's less power than some flashlights run. This type of threshold experimentation can be one of the most ego gratifying experiences in your Amateur Radio career.

On the bands, CW QRP activity centers around 1810, 3560, 7040 (7030 for QRP DX), 10, 106, 14,060, 21,060 and 28,060 kHz. Phone operation is around 3985, 7285, 14,285, 21,385 and 28,885 kHz. The WARC bands, 10/18/24 MHz, are also hotbeds of QRP activity. Novice and Technician Plus licensees should check 3710, 7110, 21,110 and 28,110 kHz.

QRP operating can result in earning some awards with a special QRP endorsement. There are many hams who have WAS and WAC under their belt using no more than 2 watts. DXCC has even been achieved by a few hardy QRP souls.

Several clubs cater to the QRPer. Most sponsor contests and awards. A number of clubs also publish newsletters. Joining one (or all!) of these clubs can give the interested amateur a much more rounded view of the QRP game.

The *QRP ARCI* (Amateur Radio Club International) offers membership for US amateurs at $12 per year and publishes the newsletter *QRP Quarterly.* Contact:
Michael Bryce
2225 Mayflower NW
Massillon, OH 44647

The *Michigan QRP Club* dues are $7 for US members. The club prints an informative newsletter, *The Five Watter.* Contact:
Membership Chairperson
564 Georgia Street
Marysville, MI 48040

When the bands are dead, the *real* QRPer either hits the workbench or sits down with a good book. Check out *W1FB's QRP Notebook* by Doug DeMaw, and *QRP Classics.* Both are available from your local radio bookstore or ARRL HQ.

So what's holding you back? Join in on the fun and thrill of communicating with low power. Amaze your fellow amateurs and outdo the "big gun" across town with your operating accomplishments. Be a rebel; Quit Running Power!—Jeff Bauer, WA1MBK

AM: The Good Old Days are Now

Tim Walker, N2GIG, is quite proud of the waxed front panels of his 250-watt AM rig. Clyde, his 2nd operator and trusted assistant, can't seem to take his eyes off the 6-foot rack containing the recycled transmitter from commercial broadcast station WSK. The call letters WADP in the photograph were assigned to the station's remote pick-up transmitter.

Tim and many others enjoy restoring old *amplitude modulated* (AM) equipment for use on the ham bands. While it is not the sort of transmitter you would want to use during a DX contest, there are many AM fans who operate—some call it *broadcast*—on a regular basis in round tables with similarly minded friends. Transmissions tend to be a bit longer than SSB conversations, and the audio quality on many of the older rigs is often warmer and more realistic than a 300 to 2400-Hz system.

Favorite operating frequencies include the upper portions of the 160-meter band, 3.885, 7.290, 14.286, 21.390 and the range between 29.000 and 29.200 MHz. Occasionally you will hear an OT (*Old-Timer*) using a rig he saved from the late 1930s or 1940s. Other times the station operator is a *young squirt* who just enjoys using vacuum tube equipment which both heat and light the operating shack.

The table contains the names and addresses of several groups and publications that feature AM radio. Periodic meetings and flea markets make some of the older equipment available to anyone interested. One caution, however—if you manage to get your hands on a vintage receiver such as a Hammarlund SP-600, National NC-300, Collins 75A-1,2,3 or 4 or even a military surplus R-390 series unit—you might have to reinforce the operating table. The good news, however, is that compact, modern transceivers often include AM.

Clyde, with his back to the camera, is admiring both his master, Tim Walker, N2GIG, and Tim's classic AM transmitter.

Publications:	Groups
Electric Radio	*AM International*
PO Box 57	9 Dean Ave
Hesperus, CO 81326	Bow, NH 03304
monthly publication	active AM group
AM Press/Exchange	*AM Radio Network*
2116 Old Dover Road	Box 73
Woodlawn, TN 37191	West Friendship, MD 21794
bimonthly publication	informal associations

ARRL On-The-Air Services

Up-to-the-minute information on everything of immediate interest to amateurs—from Federal Communications Commission policy-making decisions to propagation predictions, to news of DXpeditions—is transmitted on a timely basis in W1AW bulletins. W1AW is the Amateur Radio station maintained at ARRL HQ in Newington, Connecticut. W1AW bulletins are transmitted at regularly scheduled intervals on CW (Morse code), phone (voice) and RTTY (teletype) on various frequencies, as is a comprehensive Morse-code practice agenda. A detailed schedule of W1AW transmissions is in Chapter 17 and also appears regularly, with changes if any, in the ARRL monthly magazine *QST*. The schedule may also be obtained on request from ARRL HQ (please include a self-addressed, stamped envelope).

RULES AND REGULATIONS

The government of the United States grants us, as radio amateurs, certain privileges. In order to retain these privileges, we must conform to the rules and regulations in Part 97 of the Federal Communications Commission (FCC) rules.

This chapter will summarize only the most obvious rules as they pertain to day-to-day operation. This is by no means a complete discussion of all the FCC rules and regulations. For more complete information on the FCC rules as they pertain to Amateur Radio, along with the complete text of Part 97, see *The FCC Rule Book,* published by ARRL. (Contact ARRL HQ for ordering information.) See also Spectrum Management, Chapter 2 of this book.

Your License

You must have a license granted by the FCC to operate an Amateur Radio station in the United States, in any of its territories or possessions, or from any vessel or aircraft registered in the United States. The FCC sets no minimum or maximum age requirement for obtaining a license, nor does it require that an applicant be a US citizen.

An Amateur Radio license incorporates two kinds of authorization. For the operator, the license grants operator privileges. The license also authorizes an Amateur Radio station—operation of transmitting equipment.

Table 3-7
Amateur Operator Licenses†

Class	Code Test	Written Examination	Privileges
Novice	5 WPM (Element 1A)	Novice theory and regulations (Element 2)	Telegraphy on 3675-3725, 7100-7150 and 31,100-21,200 kHz with 200 watts PEP output maximum; telegraphy, RTTY and data on 28,000-28,300 kHz and telegraphy and SSB voice on 28,300-28,500 kHz with 200 W PEP max; all amateur modes authorized on 222-225 MHz, 25 W PEP max; all amateur modes authorized on 1270-1295 MHz, 5 W PEP max.
Technician		Novice theory and regulations; Technician-level theory and regulations. (Elements 2 and 3A)*	All amateur privileges above 50.0 MHz.
Technician Plus	5 WPM (Element 1A)	Novice theory and regulations; Technician-level theory and regulations. (Elements 2 and 3A)*	All amateur privileges above 50 MHz. Telegraphy on 3675-3725, 7100-7150 and 21,100-21,200 kHz with 200 watts PEP output maximum; telegraphy, RTTY and data on 28,100-28,300 kHz and telegraphy and SSB on 28,300-28,500 kHz with 200 watts PEP max.
General	13 WPM (Element 1B)	Novice theory and regulations; Technician and General theory and regulations. (Elements 2, 3A and 3B)	All amateur privileges except those reserved for Advanced and Amateur Extra class; see Fig 3-4.
Advanced	13 WPM (Element 1B)	All lower exam elements, plus Advanced theory. (Elements 2, 3A, 3B and 4A)	All amateur privileges except those reserved for Amateur Extra class; see Fig 3-4.
Amateur Extra	20 WPM (Element 1C)	All lower exam elements, plus Extra-class theory (Elements 2, 3A, 3B, 4A and 4B)	All amateur privileges; see Fig 3-4.

†A licensed radio amateur will be required to pass only those elements that are not included in the examination for the amateur license currently held.

*If you hold a valid Technician class license issued before March 21, 1987, you also have credit for Element 3B. You must be able to prove your Technician license was issued before March 21, 1987 to claim this credit.

Alien Amateur Radio operators, if they're licensed by a country with which the United States has signed a reciprocal-operating agreement and are citizens of that country, may apply to the FCC for permission to operate without having to pass the FCC amateur examinations.

The United States has six classes of operator licenses: Novice, Technician, Technician Plus, General, Advanced and Extra. There are two entry-level classes of license. The Novice license requires a minimum of technical knowledge and Morse code proficiency. The Novice license authorizes operation on six of the amateur frequency bands, with a variety of modes. The Technician license requires that applicants pass the Novice and Technician written tests, but a code examination is not required. The Technician license authorizes operation on all frequencies above 30 MHz. Upon passing the 5 wpm code test, a Technician becomes a Technician Plus, which includes all Novice HF privileges. Successive classes of license require more technical knowledge and/or Morse code proficiency. But each license rewards the licensee with more privileges, including expanded frequency allocations and more modes of operation. See Table 3-7.

License Renewal/Modification

Once you have a license, you can keep it for your lifetime, but it must be renewed every 10 years.

1) To renew your license, complete FCC Form 610 (available from ARRL HQ; SASE, please) being careful not to miss any applicable boxes or blanks. Type or print neatly in pen. As of early 1995, only November 1993 and later editions of FCC Form 610 may be used.

2) Remember to attach a photocopy (or the original) of your license to the Form 610.

3) Mail your application to FCC, 1270 Fairfield Road, Gettysburg, PA 17325-7245. There is no fee for renewing an amateur license.

4) Application to renew should be made no more than 90 days before expiration.

5) It's a good idea to retain a copy of your application as proof of filing before the expiration date of your license. If your renewal application arrives at the FCC's Gettysburg office before the expiration date, you may continue to operate until the new license arrives. Otherwise you may not operate until you receive a new license.

6) If your license has already expired, it is still possible to renew. The FCC allows a 2-year grace period after the expiration date before *both* the license and the call sign expire.

How to Apply for an Amateur Radio License

If you are applying for any class of amateur license, you must appear before a team of three accredited Volunteer Examiners. If you want the latest list of test sessions near you, send an SASE to the Volunteer Examiner Department at ARRL HQ. The tests are typically given by local Amateur Radio Clubs and are assembled from a standard list of questions issued by the FCC. These questions and answers appear in the ARRL *License Manuals* series. If you are physically unable to travel to an examination point, you may make special arrangements for taking your exam. Special examination procedures are available for persons with severe disabilities. See Chapter 3.

Complete details on the Volunteer Examining program are available from ARRL HQ. Also, see *The FCC Rule Book* published by the ARRL.—*KJ4KB*

7) If you are simply modifying your license (change of address, for example), you must also fill out Form 610. If you have any questions, contact the Regulatory Information Branch, ARRL HQ.

Information about the status of your application can be obtained by calling the FCC Consumer Assistance line, 717-337-1212 or 1-800-322-1117.

If your license is lost, mutilated or destroyed, you must apply to the FCC for a duplicate. Send a letter to the FCC giving the circumstances and details under which the license was lost, mutilated or destroyed. After receiving your letter, the FCC will issue you a duplicate license bearing the same expiration date as the original license. If, after applying, you find the original license, either the original or the duplicate must be returned to the FCC.

US AMATEUR CALL SIGNS

The International Telecommunication Union (ITU) radio regulations outline the basic principles used in forming amateur call signs. According to these regulations, an amateur call sign must consist of one or two letters (sometimes the first or second may be a number) as a prefix, followed by a number and then a suffix of not more than three letters. Refer to Chapter 17 for a complete listing of international call-sign prefixes.

Every US Amateur Radio station call sign is a combination of a 1 or 2-letter prefix, a number and a 1, 2 or 3-letter suffix. The first letter of every US Amateur Radio call sign is always an A, K, N or W. For example, in the call sign W1AW, the W is the prefix, 1 is the number, and AW is the suffix.

For Amateur Radio stations located within the continental United States, the number in the call sign designates the geographic area in which the station was *originally* licensed. The call-sign districts for the continental US are illustrated in Fig 3-7. The prefix/ numeral designators for US Amateur Radio districts outside the continental US are shown in Table 3-8.

The FCC used to require the number (or in the case of US stations located outside the continental US, the prefix/number) to correspond to the call-sign district in which the station was located. In 1978, however, the FCC relaxed the rules and now allows hams to retain their present call sign even if the station location is moved permanently to another call-sign district. It's routine to hear station call signs on the air that do not correspond to the call-sign district in which the station is located. It is no longer necessary to identify a station as "portable," "mobile," "fixed" or "temporary."

Some amateurs choose to make their locations clear by including an indicator at the end of their call signs. For example, the voice identification, "This is W9KDR, permanent W1" or "fixed W1 " means that the station, though originally licensed in the ninth call district, is now licensed to an address in the first call district. When identifying on CW, W9KDR/1 could be used to clarify the station's location.

On-The-Air Identification Requirements

The amateur rules specifically prohibit unidentified transmissions, but what is required for proper station identification? An amateur station must be identified at the end of a transmission or series of transmissions. You must also identify at intervals not to exceed 10 minutes during a single transmission or a series of transmissions of more than 10 minutes' duration. The identification must be given using the International Morse code or, when identifying on phone, in the English language. At the end of an exchange of third-party communications (that is, on behalf of anyone other than the

Table 3-8
FCC-Allocated Prefixes for Areas Outside the Continental US

Prefix	Location
AH1, KH1, NH1, WH1	Baker, Howland Is
AH2, KH2, NH2, WH2	Guam
AH3, KH3, NH3, WH3	Johnson I
AH4, KH4, NH4, WH4	Midway I
AH5K, KH5K, NH5K, WH5K	Kingman Reef
AH5, KH5, NH5, WH5 (except K suffix)	Palmyra, Jarvis Is
AH6, KH6, NH6, WH6	Hawaii
AH7, KH7, NH7, WH7	Kure I
AH8, KH8, NH8, WH8	American Somoas
AH9, KH9, NH9, WH9	Wake I
AHØ, KHØ, NHØ, WHØ	Northern Mariana Is
AL7, KL7, NL7, WL7	Alaska
KP1, NP1, WP1	Navassa I
KP2, NP2, WP2	Virgin Is
KP4, NP4, WP4	Puerto Rico
KP5, NP5, WP5	Desecheo

control operator) with a station located in a foreign country, you must also give the call sign of the station with which third-party traffic was exchanged.

Temporary Designator

Temporary designators must be used after a call sign under some circumstances. These designators are used for "instant upgrading," an FCC rule that permits holders of an amateur license who successfully pass an examination for a higher class of license to use the new privileges immediately after leaving the test session. The accredited Volunteer Examiners issue the amateur a certificate of successful completion, which authorizes the new privileges for one year or until the issuance of the new, upgraded, permanent license (whichever comes first). The

Note to Canadian Readers

For purely practical reasons, this chapter was written to reflect the regulations established by the Federal Communications Commission for US amateurs. Canadian amateurs are regulated by Industry Canada (IC). Though the privileges of Canadian and US amateurs are generally similar, there are some differences.

There is one operator certificate available in Canada (the Radio Amateur Operator's Certificate) and it is available in four versions. The Basic Qualification is the entry level license and requires that the applicant pass a test composed of 100 questions. The Morse Code 5 wpm Qualification requires that the applicant pass the code test with fewer than 5 errors and the Morse Code 12 wpm has the same requirements, but at a higher speed. The Advanced Qualification is the most difficult to obtain and it consists of 50 questions. The Basic Qualification is mandatory and code is not required. The other three levels of qualification are voluntary. There is no minimum age requirement.

More information on licensing in Canada can be obtained from Radio Amateurs of Canada, Inc (RAC), PO Box 356, Kingston, ON K7L 4W2, Canada, tel 613-634-4184 or -7118. RAC publishes the *Canadian Amateur Study Guide* as well as the *Basic Question Bank*.

Guest Operating

An FCC rules interpretation allows the person *in physical control* of an Amateur Radio station to use his or her own call sign when guest operating at another station. Of course, the guest operator is bound by the frequency privileges of his or her own operator's license, no matter what class of license the station licensee may hold. For example, Joan, KB6MOZ, a Novice, is visiting Stan, N6MP, an Extra Class licensee. Joan may use her own call sign at N6MP's station, but she must stay within the Novice subbands.

When a ham is operating from a club station, the club call is usually used. Again, the operator may never exceed the privileges of his or her own operator's license. The club station trustee and/or the club members may decide to allow individual amateurs to use their own call signs at the club station, but it is optional. In cases where it is desirable to retain the identity of the club station (W1AW at ARRL HQ, for example), the club may require amateurs to use the club call sign at all times.

In the rare instance where the guest operator at a club station holds a higher class of license than the club station trustee, the guest operator must use the club call sign and his/her own call sign. For example, Billy, KR1R, who holds an Amateur Extra Class license visits the Norfolk Novice Radio Club station, KA4CVX. Because he wants to operate the club station outside the Novice subbands, Billy would sign KA4CVX/KR1R on CW and "KA4CVX, KR1R controlling" on phone. (Of course, this situation would prevail only if the club requires that their club call sign be used at all times.)

temporary amateur permit also sets forth a two-letter indicator for the upgraded class of license. For example, "temporary AA" means the operator has upgraded to Advanced.

When operating under the authority of a temporary amateur permit with privileges authorized by the permit, but which exceed the privileges of the licensee's permanent station license, the station must be identified in the following manner:

1) On phone, by the transmission of the station call sign, followed by a word such as "temporary" followed by the special indicator shown on the certificate, appropriate to the newly earned class of license.

2) On CW and digital modes, by the transmission of the station call sign, followed by the slant mark, followed by the special indicator shown on the temporary amateur permit, for example, KB5NOW/AA.

Additional Identification Requirements

To meet the identification requirements of the amateur rules, the call sign must be transmitted on each frequency being used. If identification is made by an automatic device used only for identification by CW, the code speed must not exceed 20 words per minute. While the FCC does not require use of a specific list of words to aid in sending one's call sign on phone, it does encourage the use of a nationally or internationally recognized standard phonetic alphabet as an aid for correct phone identification. The ITU phonetic alphabet appears in Chapter 17.

Permitted Communications

The US amateur rules clearly set forth the kinds of stations with which amateurs may communicate. These stations are:

1) other amateur stations;
2) stations in other FCC-licensed services during emergencies, and with RACES stations.

3) *any* station that is authorized by the FCC to communicate with amateurs.

Additionally, certain "one-way" transmissions are allowed, including those for tuning up your rig, calling CQ, while in remote control operation, during emergency communications, for code practice and bulletins, and for retransmission of space-shuttle communications.

Every amateur should operate with one eye on the basis and purpose of the Amateur Radio Service, as set forth in Section 97.1 of the amateur rules. Adhere to these fundamental principles and you can't go wrong!

§97.1 Basis and Purpose

The rules and regulations in this part are designed to provide an amateur radio service having a fundamental purpose as expressed in the following principles:

(a) Recognition and enhancement of the value of the amateur service to the public as a voluntary noncommercial communication service, particularly with respect to providing emergency communications.

(b) Continuation and extension of the amateur's proven ability to contribute to the advancement of the radio art.

(c) Encouragement and improvement of the amateur radio service through rules which provide for advancing skills in both the communication and technical phases of the art.

(d) Expansion of the existing reservoir within the amateur radio service of trained operators, technicians, and electronics experts.

(e) Continuation and extension of the amateur's unique ability to enhance international goodwill.

Third-Party Communications

Third-party traffic is defined in the US amateur rules as, "a message from the control operator (first party) of an amateur station to another amateur station control operator (second party) on behalf of another person (third party)." In other words, sending or receiving any communication on behalf of anyone other than yourself or the control operator of the other station. If you allow anyone to say "hello" into the microphone at your station, that is third-party traffic. Sending or receiving formal radiogram messages is third-party traffic (see Chapter 15). Autopatching and phone patching (interconnecting your radio to the telephone system) are also third-party traffic. You, the control operator of the station, are the first party; the control operator at the other station is the second party; and anyone else participating in the two-way communication from either station is a third party.

Certain kinds of third-party traffic are strictly prohibited. All amateurs should know that they must *not* handle the following kinds of third-party traffic:

1) international third-party traffic, except with countries that allow it or except where the third party is a licensed amateur eligible to be a control operator of the station;

2) third-party traffic involving material compensation, direct or indirect, paid or promised to any party;

The FCC rules were changed in 1993 giving amateurs greater flexibility in their public service and personal communications. Amateurs should look at Section 97.113 when trying to decide if a particular use of the ham bands is allowed:

1) Is is expressly prohibited in the rules?
2) Is there compensation—is anyone being paid?
3) Does the control operator or his employer have a financial interest in the activity using the ham bands?
4) Are communications transmitted on a regular basis which

could be reasonably furnished through other radio services?

Third-party traffic may be handled by US amateur stations with US possessions and territories where amateur stations are also licensed by the FCC. This message handling is subject to the same criteria as domestic traffic in the states. There must be no pecuniary (financial) interest, direct or indirect, to any party. Note: Only amateur stations identified by authorized call signs having a one- or two-letter prefix beginning with AA-AL, K, W or N are licensed by the US Government. Third-party traffic may be handled freely with these stations.

Third-party traffic is prohibited with amateur stations in the rest of the world (see Chapter 5 for a list of countries that do permit third-party traffic handling). This is not through reluctance on the part of the US Government, but because of prohibitions on the part of the other governments. The amateur at the other end is commonly forbidden to handle any third-party traffic. US amateurs must abide by this restriction and must have no participation in the handling of third-party traffic in such cases.

Active participation by a third party: One of the best ways to get a nonham interested in Amateur Radio is to allow him or her to speak into a microphone and experience first-hand the excitement of amateur communication. While the US amateur rules allow active participation by third parties, the control operator must make certain that the communications abide by all the restrictions pertaining to third-party traffic. For example, the third party may not talk about business and may not communicate with any country that has not signed a third-party agreement with the US. Also, the control operator may not leave the controls of the station while the third party is participating in amateur communications. The amateur rules allow third-party participation only if a control operator is present at the control point—and if he or she is continuously monitoring and supervising the third party's participation.

Third-party participation is not the same as designating another licensed amateur to be control operator of a station. Control operators designated by a station licensee may operate the station to the extent authorized by the control operator's class of license.

In the 1980s, the FCC modified the rules to close a loophole that previously allowed amateurs who have had their licenses suspended or revoked to participate in Amateur Radio communications as third parties.

Logging

Although there is no FCC requirement that amateurs keep logs of routine operating activities, amateurs are encouraged to keep voluntary logs, and FCC Engineers-In-Charge may still mandate log-keeping in specific cases. Details of nonroutine operations, such as repeater data when certain power limits are exceeded, auxiliary and remote control operations, and names and call signs of guest control operators must still be kept as part of the official station records.

Authorized Power

Generally, an Amateur Radio transmitter may be operated with a PEP *(peak envelope power)* output of up to 1500 watts. However, there are circumstances under which less power must be used. These are:

1) Under all circumstances, an amateur is required to use the minimum amount of transmitter power necessary to carry out the desired communications.

2) Novices are limited to a maximum of 5 watts PEP output on the 1270-MHz band, 25 watts PEP on the 222-MHz band, and 200 watts PEP on the 80, 40, 15 and 10-meter bands. The 200-watt limitation also applies to any licensed radio amateur who operates in the 80, 40 and 15-meter Novice subbands and to any amateur operating on the 30-meter band. The rules limit the maximum transmitter output power in the Amateur Radio Service to 1500-watts PEP. Higher-class licensees may use 1500-watts PEP in the Novice sections of 28 MHz, 222 MHz and 1270 MHz.

3) Certain power limitations apply at VHF and UHF. See §97.313 of the FCC rules.

Amateur Operation in a Foreign Country

Operation may not be conducted from within the jurisdiction of a foreign government unless that foreign government has granted permission for such operation. Some countries have allowed US amateurs to operate while visiting; however, not all countries permit foreigners this privilege. Every country has the right to determine who may and who may not operate an Amateur Radio station within its jurisdiction. There are severe penalties in some countries for operating transmitting equipment without the proper authority. See Chapter 6 for details on overseas operating.

Canada-US reciprocal operating: The most liberal agreement between two countries allowing reciprocal operating for their radio amateurs is between Canada and the US. No formal application or permit is required for US amateurs to operate amateur equipment in Canada, nor is any formal paperwork required for Canadian amateurs to operate in the United States. It's automatic. US radio amateurs may operate in Canada with the mode and frequency privileges authorized them in the United States. Canadian radio amateurs may operate in the US with the mode and frequency privileges authorized to them in Canada, but not exceeding the privileges of a US Extra Class operator. In all cases, however, visitors must stay within the band and mode restrictions of the host country.

Other reciprocal operating agreements: Except for Canada, US amateurs must obtain a written permit to be allowed to operate in any foreign country. The administrations of some countries have signed reciprocal operating agreements with the US. Such agreements facilitate the application procedures for getting permission to operate. These agreements, because they are reciprocal, also allow foreign amateurs the opportunity to receive permission to operate while in the US. If you want to operate a station in a foreign country, ARRL HQ can help you obtain forms and information. Send an SASE to ARRL HQ's Regulatory Information Branch for details. Include one unit of postage for every country requested. Many countries require a US licensee to hold a General or higher-class license to qualify for a permit to operate. You will be required to pay a licensing fee in order to obtain a permit, which must be obtained before equipment is taken into a country. Plan as far ahead as possible. See Chapter 6 for a current list of the countries that have signed reciprocal operating agreements with the US.

Radio Frequency Interference

The FCC authorizes radio amateurs to use up to 1500 watts PEP output and many different bands of the radio spectrum. However, these privileges do not come without some effort on the part of the license applicant. A prerequisite for getting on the air is passing an examination that tests the applicant's technical knowledge. One area of special importance is knowing about radio frequency interference (RFI), its causes and its cures. The FCC expects radio amateurs to identify and solve most RFI problems, but RFI is not always a simple matter.

Much has been written about RFI. For example, the material in the ARRL book *Radio Frequency Interference: How to Find It and Fix It* examines this subject from both a legal and technical standpoint. Also, *The ARRL Handbook for Radio Amateurs* devotes an entire chapter to solving radio frequency interference problems. If you are the recipient of interference complaints, these two books will be of interest to you.

A growing number of amateurs are unjustly accused of causing RFI. This occurs when an amateur transmitter is operating properly, and the responsibility for the interference falls with the electronic device experiencing the interference. For example, some television sets receive signals they are not supposed to because the manufacturer decided that reducing the per-unit cost was more important than incorporating adequate shielding in the circuit design. In these cases, the transmitter is operating with a "clean" signal; it is the television set that must be brought up to today's engineering standards. This usually means that the only effective cure must be performed by adding a filter at the television set. Trying to make your neighbor understand that it is his equipment that is at fault is usually no easy task, however.

Public Law 97-259 allows the FCC to require standards for unwanted-signal-rejection filtering in electronic home-entertainment devices. Also, some manufacturers have made some progress in dealing with this problem by providing free assistance to owners of their products who experience RFI. However, the onus for preventing the interference continues to fall on amateurs.

Today, the interference problem is of huge proportions. This can be attributed to the rapid growth of the consumer electronics industry and the failure of manufacturers to include proper shielding or filtering in the home-entertainment equipment.

If you are accused of causing RFI:

1) Check your log. Were you operating at that time? (A complete log, although no longer required by the FCC, is very useful in interference situations.)

2) Check with your own nonamateur equipment. If you are not interfering with your own TV set, chances are the problem lies with your neighbor's receiver and not with your transmitter.

3) Solicit the cooperation of your neighbor in testing to determine the exact cause of the interference.

4) Check with your local radio club for a TVI committee or other assistance or contact your ARRL Section Manager for a referral to an ARRL Technical Coordinator or Technical Specialist in your area.

5) Request RFI assistance from the manufacturer of the home-entertainment device.

6) Read the ARRL *Radio Frequency Interference* book and the Interference chapter in *The ARRL Handbook for Radio Amateurs*.

7) Be prompt, courteous and helpful; Amateur Radio's reputation is at stake, as well as your own.

8) For further assistance, write ARRL HQ. Include as much detail as you can about the symptoms and circumstances, but avoid emotional commentary.

On the Horizon

A basic appreciation of the rules will provide amateurs with increased levels of enjoyment. The rules have evolved over a long time and will continue to evolve. The deregulation of the Amateur Radio Service has had a desirable impact. The FCC proposes additional rules changes periodically; check *QST* for news of late-breaking events.

The rules are dynamic because Amateur Radio is dynamic—constantly changing to meet and create new communications technologies to better serve the public and society.

CHAPTER 4

Antenna Orientation

It is probably no news to most people nowadays that true direction from one place to another is not what it appears to be on the old Mercator school map. On such a map, if one starts "west" from Wichita, Kansas (the approximate center of the continental US) he winds up in the neighborhood of Beijing, People's Republic of China. Actually, as a minute's experiment with a strip of paper on a small globe will show, a signal starting due west from Wichita never hits China at all but rather passes near Perth, in Western Australia.

"The shortest distance between two points is a straight line" is true only on a flat surface. The determination of the shortest path between two points on the surface of a sphere is a bit more complicated. Imagine a plane that intersects two points on the surface and the center of the sphere. The intersection of the plane and the sphere describes a circle on the surface of the sphere that is defined as a great circle. The shortest distance between the points follows the path of the great circle. The direction or bearing from your location to another point on the Earth is the direction of a great circle as it passes through your location on its way to the other point.

If, therefore, we want to determine the direction of some distant point from our own location, the ordinary Mercator projection alone is utterly useless. True bearing, however, may be found in several ways: by using a special type of world map that does show true direction from a specific location to other parts of the world, by working directly from a globe or by using mathematics.

DETERMINING TRUE NORTH

Determining the direction of distant points is of little use to amateurs erecting a directive array unless they can put up the array itself in the desired direction. This, in turn, demands a knowledge of the direction of true north (as against magnetic north), since all our directions from a globe or map are worked in terms of true north.

A number of ways may be available to amateurs for determining true north from their location. Frequently, the streets of a city or town are laid out, quite accurately, in north-south and east-west directions. A visit to the office of your city or county engineer will enable you to determine whether or not this is the case for the street in front of or parallel to your own lot. Or from such a visit it is often possible to locate some landmark, such as a factory chimney or

GERALD (JERRY) HALL, K1TD
181 BRIMFIELD RD
WETHERSFIELD, CT 06109-3309

and

CHUCK HUTCHINSON, K8CH
ARRL HEADQUARTERS

church spire, that lies true north with respect to your house. If you cannot get true north by such means, three other methods are available: compass, pole star and sun.

By Compass

Get as large a compass as you can; it is difficult, though not impossible, to get satisfactory results with the "pocket" type. In any event, the compass must have not more than 2 degrees per division.

It must be remembered that the compass points to magnetic north, not true north. The amount by which magnetic north differs from true north in a particular location is known as variation. Your city engineer's office or the flight office at a nearby airport can tell you the magnetic variation for your locality. The information is also available from US Geological Survey topographic maps for your locality. These may be available in the orienteering equipment section of your local sporting goods store, or may be on file at your local library. When correcting your "compass north," do so *opposite* to the direction of the variation. For instance, if the variation for your locality is 12° west (meaning that the compass points 12° west of north), then true north is found by counting 12° east of north as shown on the compass.

When taking the bearing, make sure that the compass is located well away from ironwork, fencing, pipes, etc. Place the instrument on a wooden tripod or support of some sort, at a convenient height as near eye level as possible. Make yourself a sighting stick from a flat stick about 2 feet long with a nail driven upright in each end (for use as "sights") and then, after the needle of the compass has settled down, carefully lay this stick across the face of the compass—with the necessary allowance for variation—to line it up on true north. *Be sure you apply the variation correctly.*

This same sighting-stick and compass rig can also be used in laying out directions for supporting poles for antennas in other directions—provided, of course, that the compass dial is graduated in degrees.

By the Pole Star

Many amateurs in the Northern Hemisphere use the pole star, Polaris, in determining the direction of true north. An advantage is that the pole star is never more than 0.8° from true north, so that in practice no corrections are necessary. Disadvantages are that some people have difficulty identifying the pole star, and that because of its comparatively high angle above the horizon at high northerly latitudes, it is not always easy to "sight" on it accurately. Polaris is not visible in the Southern Hemisphere. In any event, if visible, it is a handy check on the direction secured by other means.

By the Sun

With some slight preparation, the sun can be used easily for determination of true north. One of the most satisfactory methods is described below. The method is based on the fact that exactly at noon, local time, the sun bears due south (in the northern latitudes), so at that time the shadow of a vertical stick or rod will bear north. The resulting shadow direction, incidentally, is *true* north.

Two corrections to your Standard Time must be made to determine the exact moment of local true noon.

The first is longitude correction. Standard Time is time at some particular meridian of longitude: EST is based on the 75th meridian, CST on the 90th meridian, MST on the 105th meridian and PST on the 120th meridian. From an atlas or perhaps Table 4-3, determine the difference between your longitude and the longitude of your time meridian. Getting this to the nearest 1/4° of longitude is close enough. Example: Newington, Connecticut, which runs on 75th meridian time (EST) is at 72.75° longitude, or a difference of 2.25°. Now for each 1/4° of longitude, figure one minute of time; thus 2.25° is equivalent to nine minutes of time (there are 60 "angle" minutes to a degree, so that each degree of longitude equals four minutes of time). *Subtract* this correction from noon if you are *east* of your time meridian; *add* if you are *west*.

To the resulting time, apply a further correction for the date from Table 4-1. The resulting time is the time, by Standard Time, when it will be true noon at your location. Put up your vertical stick (use a plumb bob to make sure it is actually vertical), check your watch with Standard Time, and, at the time indicated from your calculations, mark the position of the shadow. That is true north.

In the case of Newington, if we wanted correct time for true noon on October 20: First, subtracting the longitude correction—because we are east of the time meridian—we get 11:51 AM; then, applying the further correction of –15 minutes from Table 4-1, we get 11:36 AM EST (12:36 PM EDST) as the time of true noon at Newington on October 20.

AZIMUTHAL MAPS

While the Mercator projection does not show true directions, it is possible to make up a map that will show true bearings for all parts of the world from any single point. Three such maps are reproduced in this chapter. Fig 4-1 shows directions from Washington, DC, Fig 4-2 gives directions from San Francisco and Fig 4-3 (a simplified version of the ARRL Amateur

Fig 4-1—Azimuthal map centered on Washington, DC.

Fig 4-2—Azimuthal map centered on San Francisco, California.

Table 4-1
Time Correction for Various Dates of the Year

To get time of true noon, apply to clock time as indicated by the sign.

Date		Min	Date		Min
Jan	1	+ 4	Jul	10	+ 5
	10	+ 8		20	+ 6
	20	+11		30	+ 6
	30	+13			
Feb	10	+14	Aug	10	+ 5
	20	+14		20	+ 3
	28	+13		30	+ 1
Mar	10	+10	Sep	10	– 3
	20	+ 8		20	– 7
	30	+ 4		30	–10
Apr	10	+ 1	Oct	10	–13
	20	– 1		20	–15
	30	– 3		30	–16
May	10	– 4	Nov	10	–16
	20	– 4		20	–14
	30	– 3		30	–11
Jun	10	– 1	Dec	10	– 7
	20	+ 1		20	– 2
	30	+ 4		30	+ 3

Fig 4-3—Azimuthal map centered on Wichita, Kansas. Copyright by Rand McNally & Co, Chicago. Reproduction License No. 394.

Radio map of the world) gives directions from the approximate center of the United States—Wichita, Kansas.

For anyone living in the immediate vicinity (within 150 miles) of any of these three reference points, the directions as taken from maps will have a high degree of accuracy. However, one or the other of the three maps will suffice for any location in the United States for all except the most accurate work; simply choose the map whose reference point is nearest you. Greatest errors will arise when your location is to one side or the other of a line between the reference point and the destination point; if your location is near or on the resulting line, there will be little or no error.

By tracing the directional pattern of the antenna system on a sheet of tissue paper, then placing the paper over the azimuthal map with the origin of the pattern at one's location, the "coverage" of the antenna will be readily evident. This is a particularly useful technique when a multi-lobed antenna, such as any of the long single-wire systems, is to be laid out so the main lobes cover as many desirable directions as possible. Often a set of such patterns will be of considerable assistance in determining what length antenna to put up, as well as the direction in which it should run.

The current edition of the ARRL Amateur Radio Map of the World, entirely different in concept and design from any other radio amateurs' map, contains a wealth of information especially useful to amateurs. A special type of azimuthal projection made by Rand-McNally to ARRL specifications, it gives great-circle bearings from the geographical center of the continental United States, as well as the great-circle distance measurement in miles and kilometers, within an accuracy of 2%. The map shows principal cities of the world, local time zones, WAC divisions, index of DXCC countries and amateur prefixes throughout the world. The map is large enough to be easily readable from the operating position, 31 × 41 inches, and is printed in six colors on heavy paper. The map is available from ARRL HQ; write for details.

The *Radio Amateur's Callbook* includes great-circle maps and tables, and another Callbook publication, *The Radio Amateur's World Atlas,* features a polar projection world map, maps of the continents and world amateur prefixes. The maps are in color.

Bill Johnston, N5KR, offers computer-calculated and computer-drawn great-circle maps; an extensive selection of these fine maps for various areas of the world appears in Chapter 17. An 11- × 14-inch map can be custom made for your location. Write to PO Box 370, White Sands, NM 88003.

WORKING FROM A GLOBE

Bearings for beam-heading purposes may be determined easily from an ordinary globe with nothing more complicated than a small school protractor of the type available in any school-supply or stationery store. For best results, however, the globe should be at least 8 inches in diameter. A thin strip of paper may be used for a straightedge to determine the great-circle path between your location and any other location on the Earth's surface. The bearing from your location may be determined with the aid of a protractor. For convenience, a paper-scale circle calibrated in degrees of bearing may be made and affixed over the point representing your location on the globe. The 0° mark of this scale should point toward the North Pole.

A SIMPLIFIED DIRECTION FINDER

A simplified direction finder may be made by removing a globe from its brazen meridian (semi-circular support) and remounting it in the manner shown in Fig 4-4. Drill a hole that will accept the support at your location on the globe, and another hole directly opposite the first. This second hole will have the same latitude as yours but will be on the other side of the equator (north latitude vs south latitude). Its longitude will be opposite in direction from yours from the Greenwich or 0° meridian, east vs west, and will be equal to 180° minus your

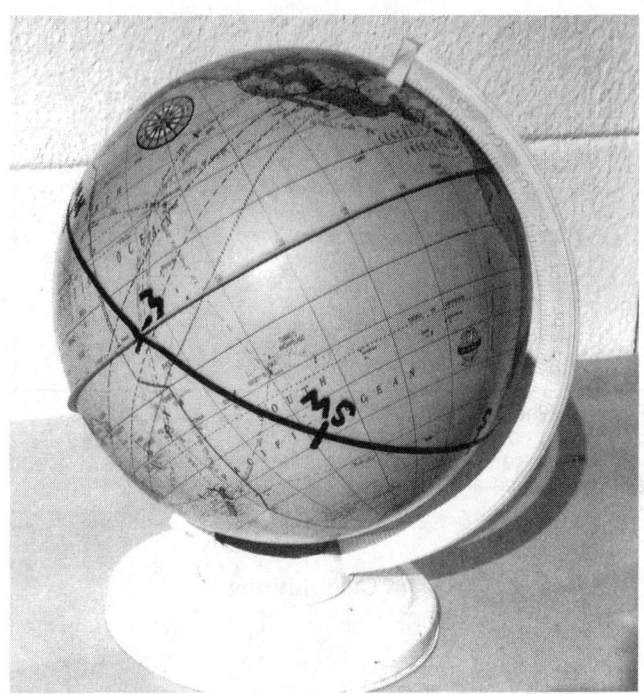

Fig 4-4—A simple direction finder made by modifying a globe. Bearing and distance to other locations from yours may be determined quickly after modification, no calculations being required.

longitude. For example, if your location is 42° N lat, 72° W long, the point opposite yours on the globe is 42° S lat, 180 –72 or 108° E long.

Once the holes are drilled, remount the globe with your location in the position formerly occupied by the North Pole. By rotating the globe until the distant point of interest lies beneath the brazen meridian, this support may be used to indicate the great-circle path. A new "equator" calibrated in a manner to indicate the bearing may be added with India ink, as shown in Fig 4-4, or a small protractor-like scale may be added at the top of the globe over your location. A distance scale can be affixed to the brazen meridian so that both the bearing and distance to other locations may be readily determined (12,500 miles or 40,000 km to the semicircle).

DIRECTION AND DISTANCE BY TRIGONOMETRY

The methods to be described will give the bearing and distance as accurately as one cares to compute them. All that is required is a table of latitude and longitude information, such as Table 4-3, and a calculator or computer with trigonometric functions. The latitude and longitude for any other location can be taken from a map of the area in question.

Fig 4-5 will help you to visualize the nature of the situation. That sketch represents the path between points situated relatively such as Wichita, Kansas, USA (at point A), and Perth, Western Australia (at point B). In using these equations, northerly latitudes are taken as negative. Also, westerly longitudes are taken as positive, and easterly longitudes are taken as negative. *In all calculations*, the appropriate signs are to be retained. *All additions and subtractions throughout the procedure are to be made algebraically*. Thus, if a negative-value number is subtracted from a positive-value number, the resultant will be positive, and it will be the sum of the two absolute values, and so on.

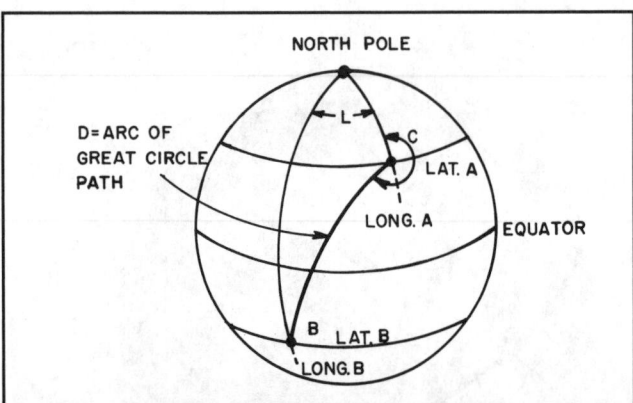

Fig 4-5—The various terms used in the equations for determining bearing and distance. North latitudes and west longitudes are taken as positive, while south latitudes and east longitudes are taken as negative.

The Calculations

The two equations we'll be using for these calculations are:

$$\cos D = \sin A \sin B + \cos A \cos B \cos L \qquad \text{(Eq 1)}$$

$$\cos C = \frac{\sin B - \sin A \cos D}{\cos A \sin D} \qquad \text{(Eq 2)}$$

where
A = *your* latitude in degrees
B = latitude of the other location in degrees
L = *your* longitude minus that of the other location (algebraic difference)
D = distance along path in degrees of arc
C = true bearing from north if the value for sin L is positive. If sin L is negative, true bearing is 360 – C.

The term *cos* is an abbreviation for cosine, and the term *sin* is an abbreviation for sine. A knowledge of the meanings of these terms isn't necessary for their use here.

The actual calculating procedure uses, first, Eq 1 to determine the angular value for D, in degrees. From this value the path-length distance may be determined in miles or kilometers. Next, Eq 2 is used to determine the bearing angle.

Using the Wichita-to-Perth example mentioned earlier, refer to Fig 4-5 to see how the equations are used. From Table 4-3, it can be seen that the location of Wichita is 37.7° north latitude, 97.3° west longitude. Similarly, Perth is located at 31.9° south latitude, 115.8° east longitude. Your location is in Wichita. Values for use in the equations are as follows:

A = lat A = + 37.7°
B = lat B = –31.9°
L = long A – long B
= 97.3° – (–115.8°) = 213.1°

Solving Eq 1, cos D = sin 37.7° × sin (–31.9°) + cos 37.7° × cos (–31.9°) × cos 213.1°. D = 152.4°. Each degree along the path equals 60.0359 nautical miles. Therefore, 152.4° of arc is equivalent to 60.0359 × 152.4 = 9,147 nautical miles. To convert to statute miles, multiply degrees by 69.0826. If the distance is desired in kilometers, multiply degrees by 111.1775. This means that the Wichita-to-Perth distance is 10,525 miles or 16,939 kilometers.

Solving Eq 2

$$\cos C = \frac{\sin (-31.9°) - \sin 37.7° \cos 152.4°}{\cos 37.7° \sin 152.4°}$$

C = 87.9°. Because the sin of L (213.1°) is negative, however, the correct value for C is 360° - 87.9° = 272.1°. Thus, the true bearing from Wichita to Perth is 272.1° and the distance is 10,525 statute miles. If the bearing from Perth were desired, it would be necessary only to work through Eq 2, interchanging latitude values for A and B. Because of the way L is defined (now –213.1°), sin L will be positive, and it will not be necessary to subtract from 360° to get the true bearing at Perth, which is 68.6°.

These equations give information for the great-circle bearing and distance for the shortest path. For long-path work, the bearing will be 180° away from the answers obtained.

The equations described above may be used for any two points on the Earth's surface—both locations in the Northern Hemisphere, both locations in the Southern Hemisphere, either or both on the equator, and so on. The equations themselves are exact, not being based on any approximations. However, there are some cases where practical limitations exist in the accuracy of the results obtained from Eq 2, in relation to the number of significant figures used during calculations. (Round-off errors in calculators and computers during computations will effectively reduce the number of significant figures in the resulting answers.) These cases are where both locations are near or at exact opposite points on the Earth (antipodes), where the

Table 4-2

BASIC Language Program for Determining Bearings and Distances

```
100 REM * * *  BEARINGS.BAS  * * *
110 REM  By Jerry Hall, K1TD, ARRL HQ, 1991
120 DEFDBL A-E, L-M, O, Y: DEFSTR T-W
130 K = 111.1775: M = 57.2957795131#: N = 60.0359: S = 69.08261
140 T = "": V = "###.#": W = "##,###"
150 GOTO 380
160 REM * * * *  Subroutine to parse latitude/longitude string  * * * *
170    DIR$ = RIGHT$(U,1): X = VAL(U)
180    IF DIR$ = "S" OR DIR$ = "s" OR DIR$ = "E" OR DIR$ = "e" THEN X = -X
190 RETURN
200 REM * * * *  Math package subroutine  * * * *
210    L = (L1 - L2)/M
220    E = SIN(A) * SIN(B) + COS(A) * COS(B) * COS(L)
230    D = -ATN(E/SQR(1 - E * E)) + 1.57079
240    C = (SIN(B) - SIN(A) * E)/(COS(A) * SIN(D))
250    IF C >= 1 THEN C = 0: GOTO 300
260    IF C <= -1 THEN C = 180: GOTO 300
270    C = -ATN(C/SQR(1 - C * C)) + 1.57079
280    C = C * M
290    IF SIN(L) < 0 THEN C = 360 - C
300 RETURN
310 REM * * * *  Subroutine to display values  * * * *
320    PRINT USING V; AZ1;
330    PRINT SPC(6) USING V; AZ2;
340    PRINT SPC(5) USING W; S * DIST * M;: PRINT " statute mi."
350    PRINT TAB(35) USING W; K * DIST * M;: PRINT " kilometers"
360    PRINT TAB(35) USING W; N * DIST * M;: PRINT " naut. mi."
370 RETURN
380 REM * * * MAIN PROGRAM * * *
390 CLS
400 PRINT "PROGRAM TO CALCULATE GREAT CIRCLE DISTANCES AND BEARINGS"
410 PRINT
420 PRINT "Enter latitudes and longitudes in degrees and decimals, followed"
430 PRINT "by a letter designating direction (N, S, E, W), such as '41.72N'"
440 PRINT
450 IF T = "Y" OR T = "y" THEN 500
460 INPUT "Your latitude (degrees/decimal and direction)"; U
470 GOSUB 160: YOURLAT = X / M
480 INPUT "Your longitude (degrees/decimal and direction)"; U
490 GOSUB 160: YOURLONG = X
500 INPUT "Other latitude (degrees/decimal and direction)"; U
510 GOSUB 160: OTHERLAT = X / M
520 INPUT "Other longitude (degrees/decimal and direction)"; U
530 GOSUB 160: OTHERLONG = X
540 A = YOURLAT: B = OTHERLAT: L1 = YOURLONG: L2 = OTHERLONG
550 GOSUB 200: C1 = C: D1 = D: D2 = 360/M - D
560 A = OTHERLAT: B = YOURLAT: L1 = OTHERLONG: L2 = YOURLONG
570 GOSUB 200: C2 = C
580 PRINT
590 PRINT SPC(13); "Your        Other"
600 PRINT SPC(12); "bearing,    bearing"
610 PRINT SPC(12); "degrees     degrees     Distance"
620 PRINT SPC(12); "-------     -------     --------"
630 AZ1 = C1: AZ2 = C2
640 DIST = D1: PRINT "Short path:   ";: GOSUB 310
650 IF C1 > 180 THEN AZ1 = C1 - 180 ELSE AZ1 = C1 + 180
660 IF C2 > 180 THEN AZ2 = C2 - 180 ELSE AZ2 = C2 + 180
670 DIST = D2: PRINT "Long path:    ";: GOSUB 310
680 PRINT
690 INPUT "Do another calculation from same location (Y/N)"; T
700 IF T = "Y" OR T = "y" THEN 410
710 END
```

locations are close together, or where your location is at or near one of the poles. (At the poles, all directions are either south or north, anyway.) More specifically, these situations exist when lat A is near +90°, or where D is near 0° or 180°.

Table 4-2 is a BASIC language program for calculating bearings and distances by computer. The program is written for an IBM-PC/XT/AT or compatible computer using BASIC or GWBASIC. The program calculates the great-circle bearings and distances for both ends of the path, for both the short path and the long path. There are numerous versions of BASIC, and it may be necessary for you to modify portions of the program for use with your computer system. Table 4-3 shows latitude and longitude for various US and Canadian cities as well as other areas of the world. The data is arranged alphanumerically by call prefixes; DX information generally follows the ARRL DXCC list. Table 4-3 also includes bearings from the eastern, central and western regions of the United States to the DX location specified.

Acknowledgments

Data for Table 4-3 was compiled from various sources, including *The World Almanac and Book of Facts 1990*, Newspaper Enterprise Associates, Inc, New York. Special thanks to Jerry Cross, N4NO.

BIBLIOGRAPHY

Source material and more extended discussions of topics covered in this chapter can be found in the references given below.

Davis, "A Simplified Direction Finder," Hints and Kinks, May 1972 *QST*.

Hall, "Bearing and Distance Calculations by Sleight of Hand," August 1973 *QST*.

Klopf, "A Bearing and Distance Calculator," March 1971 *QST*.

Norton, *Norton's Star Atlas and Reference Handbook*, Gail and Inglis, London, England; also published in the US by Sky Publishing Corp, Cambridge, Massachusetts.

Table 4-3
Latitude and Longitude of Various US/Canadian Cities and DX Locations with Bearing from East, Central, and Western USA

Pref	State/Province/Country/City	Lat	Long	fm E USA	fm C USA	fm W USA
VE1	New Brunswick, Saint John	45.3 N	66.1 W	58.0	62.6	64.8
	Nova Scotia, Halifax	44.6 N	63.6 W	63.8	64.5	65.4
	Prince Edward Island, Charlottetown	46.2 N	63.1 W	57.7	61.0	62.7
VE2	Quebec, Montreal	45.5 N	73.6 W	38.4	59.7	65.7
	Quebec City	46.8 N	71.2 W	40.3	57.3	63.0
VE3	Ontario, London	43.0 N	81.3 W	342.3	63.2	71.9
	Ottawa	45.4 N	75.7 W	28.9	58.8	66.1
	Sudbury	46.5 N	81.0 W	353.9	50.4	63.9
	Toronto	43.7 N	79.4 W	6.9	62.0	70.1
VE4	Manitoba, Winnipeg	49.9 N	97.1 W	315.1	2.8	47.0
VE5	Saskatchewan, Regina	50.5 N	104.6 W	309.6	341.5	33.5
	Saskatoon	52.1 N	106.7 W	312.4	339.5	24.8
VE6	Alberta, Calgary	51.0 N	114.1 W	306.4	324.0	6.2
	Edmonton	53.5 N	113.5 W	312.1	330.5	6.3
VE7	British Columbia, Prince George	53.9 N	122.8 W	310.4	321.2	344.0
	Prince Rupert	54.3 N	130.3 W	310.4	317.2	330.9
	Vancouver	49.3 N	123.1 W	301.7	310.2	333.9
VE8	Northwest Territories, Yellowknife	62.5 N	114.4 W	329.0	343.0	1.9
	Resolute	74.7 N	95.0 W	353.2	1.3	9.3
VY1	Yukon, Whitehorse	60.7 N	135.1 W	320.6	326.6	336.6
VO1	Newfoundland, St. John's	47.6 N	52.7 W	59.7	58.8	58.4
VO2	Labrador, Goose Bay	53.3 N	60.4 W	38.5	47.0	51.2
W1	Connecticut, Hartford	41.8 N	72.7 W	69.6	70.9	72.4
	Maine, Bangor	44.8 N	68.8 W	56.2	63.3	66.2
	Portland	43.7 N	70.3 W	59.7	65.9	68.4
	Massachusetts, Boston	42.4 N	71.1 W	67.4	69.3	70.9
	New Hampshire, Concord	43.2 N	71.5 W	60.5	67.0	69.6
	Rhode Island, Providence	41.8 N	71.4 W	71.7	71.0	72.1
	Vermont, Montpelier	44.3 N	72.6 W	49.5	63.6	67.8
W2	New Jersey, Atlantic City	39.4 N	74.4 W	96.1	78.3	77.4
	New York, Albany	42.7 N	73.8 W	57.9	68.0	71.0
	Buffalo	42.9 N	78.9 W	15.6	65.3	71.7
	New York City	40.8 N	74.0 W	78.1	73.9	74.7
	Syracuse	43.1 N	76.2 W	41.3	65.9	70.8
W3	Delaware, Wilmington	39.7 N	75.5 W	93.5	77.4	77.2
	District of Columbia, Washington	38.9 N	77.0 W	114.4	80.3	79.3
	Maryland, Baltimore	39.3 N	76.6 W	103.9	78.9	78.4
	Pennsylvania, Harrisburg	40.3 N	76.9 W	81.8	75.4	76.6
	Philadelphia	39.9 N	75.2 W	90.0	76.7	76.7
	Pittsburgh	40.4 N	80.0 W	15.4	74.6	77.3
	Scranton	41.4 N	75.7 W	65.4	71.8	74.0
W4	Alabama, Montgomery	32.4 N	86.3 W	215.7	116.9	98.3
	Florida, Jacksonville	30.3 N	81.7 W	188.7	114.9	98.4
	Miami	25.8 N	80.2 W	180.9	123.9	104.5
	Pensacola	30.4 N	87.2 W	213.7	127.2	103.3
	Georgia, Atlanta	33.8 N	84.4 W	210.9	106.8	93.8
	Savannah	32.1 N	81.1 W	186.8	108.1	94.7
	Kentucky, Lexington	38.0 N	84.5 W	241.7	85.8	84.5
	Louisville	38.2 N	85.8 W	250.1	85.0	84.6
	North Carolina, Charlotte	35.2 N	80.8 W	187.9	96.2	88.5
	Raleigh	35.8 N	78.6 W	164.8	92.1	86.1
	Wilmington	34.2 N	77.9 W	163.2	97.1	88.6
	South Carolina, Columbia	34.0 N	81.0 W	188.0	101.1	91.0
	Tennessee, Knoxville	36.0 N	83.9 W	218.8	95.8	88.7
	Memphis	35.1 N	90.1 W	241.7	112.2	95.2
	Nashville	36.2 N	86.8 W	236.8	98.0	90.1
	Virginia, Norfolk	36.9 N	76.3 W	135.7	87.0	82.8
	Richmond	37.5 N	77.4 W	140.1	85.4	82.2
W5	Arkansas, Little Rock	34.7 N	92.3 W	245.4	124.0	98.2
	Louisiana, New Orleans	29.9 N	90.1 W	222.4	138.7	107.5
	Shreveport	32.5 N	93.7 W	240.0	146.2	105.7
	Mississippi, Jackson	32.3 N	90.2 W	230.0	129.5	102.2
	New Mexico, Albuquerque	35.1 N	106.7 W	265.4	250.1	120.7
	Oklahoma, Oklahoma City	35.5 N	97.5 W	257.5	170.5	101.3
	Texas, Abilene	32.5 N	99.7 W	250.8	194.7	114.7
	Amarillo	35.2 N	101.8 W	261.4	228.6	108.6
	Dallas	32.8 N	96.8 W	247.2	168.9	109.0
	El Paso	31.8 N	106.5 W	257.3	230.9	133.9
	San Antonio	29.4 N	98.5 W	240.7	183.0	121.1
W6	California, Los Angeles	34.1 N	118.2 W	271.3	262.7	197.3
	San Francisco	37.8 N	122.4 W	280.1	277.0	248.2
W7	Arizona, Flagstaff	35.2 N	111.7 W	269.3	259.9	143.3
	Phoenix	33.5 N	112.1 W	266.0	252.8	153.1
	Idaho, Boise	43.6 N	116.2 W	289.6	297.9	357.2
	Pocatello	42.9 N	112.5 W	287.5	298.5	41.0
	Montana, Billings	45.8 N	108.5 W	295.0	318.4	40.9
	Butte	46.0 N	112.5 W	295.1	311.3	21.9
	Great Falls	47.5 N	111.3 W	298.9	318.6	22.8
	Nevada, Las Vegas	36.2 N	115.1 W	273.5	267.7	169.0
	Reno	39.5 N	119.8 W	282.1	281.9	261.5
	Oregon, Portland	45.5 N	122.7 W	294.4	300.1	320.4
	Utah, Salt Lake City	40.8 N	111.9 W	282.3	289.0	74.3
	Washington, Seattle	47.6 N	122.3 W	298.4	306.3	331.2
	Spokane	47.7 N	117.4 W	298.6	310.7	353.0
	Wyoming, Cheyenne	41.1 N	104.8 W	281.4	302.7	79.0
	Sheridan	44.8 N	107.0 W	292.5	318.1	51.2
W8	Michigan, Detroit	42.3 N	83.0 W	316.5	64.7	73.8
	Grand Rapids	43.0 N	85.7 W	307.0	58.0	72.5
	Sault Ste. Marie	46.5 N	84.4 W	335.2	45.4	63.6
	Traverse City	44.8 N	85.6 W	321.1	49.8	67.8
	Ohio, Cincinnati	39.1 N	84.5 W	256.9	79.9	81.9
	Cleveland	41.5 N	81.7 W	319.8	69.3	75.4
	Columbus	40.0 N	83.0 W	271.0	75.6	79.2
	West Virginia, Charleston	38.4 N	81.6 W	218.4	83.1	82.3
W9	Illinois, Chicago	41.9 N	87.6 W	290.8	60.7	75.6
	Indiana, Indianapolis	39.8 N	86.2 W	269.6	75.2	80.8
	Wisconsin, Green Bay	44.5 N	88.0 W	309.9	45.9	68.5
	Milwaukee	43.0 N	87.9 W	299.5	53.7	72.6
W0	Colorado, Denver	39.7 N	105.0 W	277.2	289.5	88.5
	Grand Junction	39.1 N	108.6 W	276.9	280.8	96.6
	Iowa, Des Moines	41.6 N	93.6 W	283.2	41.8	77.3
	Kansas, Pratt	37.7 N	98.7 W	267.0	242.1	94.2
	Wichita	37.7 N	97.3 W	265.8	117.9	93.0
	Minnesota, Duluth	46.8 N	92.1 W	311.8	24.4	60.6
	Minneapolis	45.0 N	93.3 W	301.4	25.2	65.8
	Missouri, Columbia	39.0 N	92.3 W	267.9	75.6	85.5
	Kansas City	39.1 N	94.6 W	270.1	66.5	86.2
	St. Louis	38.6 N	90.2 W	263.2	82.0	85.7
	Nebraska, North Platte	41.1 N	100.8 W	280.7	326.0	79.6
	Omaha	41.3 N	95.9 W	281.3	25.5	78.6
	North Dakota, Fargo	46.9 N	96.8 W	305.1	5.3	57.2
	South Dakota, Rapid City	44.1 N	103.2 W	291.0	328.9	62.5
1A0	S.M.O.M	41.9 N	12.4 E	54.4	45.5	35.8
1S	Spratly Is.	8.8 N	111.9 E	344.6	322.8	306.5
3A	Monaco	43.7 N	7.4 E	54.9	46.4	37.6
3B6	Agalega	10.4 S	56.6 E	64.6	46.2	14.5
3B7	St. Brandon	16.3 S	59.8 E	67.5	48.1	10.0
3B8	Mauritius	20.3 S	57.5 E	74.2	57.1	17.7
3B9	Rodriguez Is.	19.7 S	63.4 E	67.9	46.6	1.7
3C	Equatorial Guinea, Bata	1.8 N	9.8 E	88.7	77.4	63.7
	Malabo	3.8 N	8.8 E	87.9	76.5	63.0
3C0	Annobon Is.	1.5 S	5.6 E	94.0	82.7	69.6
3D2	Fiji Is., Suva	18.1 S	178.4 E	263.0	251.8	240.5
3D2	Conway Reef	21.4 S	174.4 E	262.5	251.5	240.5
3D2	Rotuma Is.	12.3 S	177.7 E	268.2	256.8	245.1
3DA	Swaziland, Mbabane	26.3 S	31.1 E	99.0	90.1	73.5
3V	Tunisia, Tunis	36.8 N	10.2 E	60.1	50.5	40.3
3W, XV	Vietnam, Ho Chi Minh City (Saigon)	10.8 N	106.7 E	351.5	329.5	312.4
	Hanoi	21.0 N	105.8 E	353.8	335.0	319.2
3X	Guinea, Conakry	9.5 N	13.7 W	98.1	85.9	74.8
3Y	Bouvet	54.5 S	3.4 E	139.1	135.1	131.0
3Y	Peter I Is.	68.8 S	90.6 W	184.0	177.2	170.5
4J, 4K	Azerbaijan, Baku	40.4 N	49.9 E	35.8	24.0	10.8
4L	Georgia, Tbilisi	41.7 N	44.8 E	38.0	26.9	14.3
4S	Sri Lanka, Colombo	7.0 N	79.9 E	26.2	2.9	339.0
4U	ITU Geneva	46.2 N	6.2 E	52.8	44.9	36.6
4U	United Nations Hq.	40.8 N	74.0 W	78.1	73.9	74.7
4X, 4Z	Israel, Jerusalem	31.8 N	35.2 E	50.4	38.7	24.9
5A	Libya, Tripoli	32.5 N	12.5 E	62.7	52.5	41.4
	Benghazi	32.1 N	20.0 E	59.0	48.4	36.4
5B	Cyprus, Nicosia	35.2 N	33.4 E	49.1	37.9	25.0
5H	Tanzania, Dar es Salaam	7.0 S	39.5 E	75.7	62.1	40.2

Pref	State/Province/Country/City	Lat	Long	fm E USA	fm C USA	fm W USA
5N	Nigeria, Lagos	6.5 N	3.4 E	89.2	77.8	65.2
5R	Madagascar, Antananarivo	18.9 S	47.5 E	80.8	67.2	38.7
5T	Mauritania, Nouakchott	18.1 N	16.0 W	92.0	80.2	69.8
5U	Niger, Niamey	13.5 N	2.0 E	84.6	73.3	61.2
5V	Togo, Lome	5.8 N	1.2 E	91.2	79.8	67.3
5W	Western Samoa, Apia	13.5 S	171.8 W	260.7	249.4	236.6
5X	Uganda, Kampala	0.3 N	32.5 E	74.9	62.0	43.4
5Z	Kenya, Nairobi	1.3 S	376.8 E	86.6	75.2	60.3
6W	Senegal, Dakar	14.7 N	17.5 W	96.2	83.9	73.3
6Y	Jamaica, Kingston	18.0 N	76.8 W	171.9	131.3	111.5
7O	Yemen, Aden	12.8 N	45.0 E	56.5	41.9	22.6
	Sanaa	15.4 N	44.2 E	55.3	41.1	22.5
7P	Lesotho, Maseru	29.3 S	27.5 E	103.9	95.9	81.7
7Q	Malawi, Lilongwe	14.0 S	33.8 E	85.7	74.0	54.1
	Blantyre	15.8 S	35.0 E	86.5	74.9	54.5
7X	Algeria, Algiers	36.7 N	3.0 E	63.6	54.3	44.7
8P	Barbados, Bridgetown	13.1 N	59.6 W	140.6	115.7	102.0
8Q	Maldive Is.	4.4 N	73.4 E	35.2	12.6	346.7
8R	Guyana, Georgetown	6.8 N	58.2 W	143.7	120.6	106.5
9A	Croatia, Zagreb	45.8 N	16.0 E	49.3	40.7	31.4
9G	Ghana, Accra	5.5 N	0.2 W	92.3	80.9	68.6
9H	Malta	36.0 N	14.4 E	58.7	48.9	38.0
9J	Zambia, Lusaka	15.4 S	28.3 E	90.6	79.6	61.9
9K	Kuwait	29.5 N	47.8 E	43.7	30.5	14.9
9L	Sierra Leone, Freetown	8.5 N	13.2 W	98.6	86.4	75.2
9M2	West Malaysia, Kuala Lumpur	3.2 N	101.6 E	357.7	331.7	312.1
9M6,8	East Malaysia, Sabah, Sandakan(9M6)	5.8 N	118.1 E	335.7	314.4	299.3
	Sarawak, Kuching (9M8)	1.6 N	110.3 E	344.7	320.0	302.8
9N	Nepal, Kathmandu	27.7 N	85.3 E	13.9	356.8	340.1
9Q	Zaire, Kinshasa	4.3 S	15.3 E	89.9	78.7	64.0
	Kisangani	0.5 N	25.2 E	79.7	67.7	51.0
	Lubumbashi	11.7 S	27.5 E	87.9	76.6	59.0
9U	Burundi, Bujumbura	3.3 S	29.3 E	79.9	67.6	49.6
9V	Singapore	1.3 N	103.8 E	354.3	327.8	308.6
9X	Rwanda, Kigali	2.0 S	30.1 E	78.3	65.9	47.7
9Y	Trinidad & Tobago , Port of Spain	10.5 N	61.3 W	145.5	120.2	105.7
A2	Botswana, Gaborone	24.8 S	25.9 E	100.5	91.4	76.2
A3	Tonga, Nukualofa	21.1 S	175.2 W	256.6	245.7	234.0
A4	Oman, Masqat	23.6 N	58.6 E	39.0	23.6	5.5
A5	Bhutan, Thimbu	27.3 N	89.4 E	10.2	352.8	336.3
A6	United Arab Emirates, Abu Dhabi	24.5 N	54.2 E	41.9	27.3	9.8
A7	Qatar, Ad-Dawhah	25.3 N	51.5 E	43.5	29.3	12.3
A9	Bahrein, Al-Manamah	26.2 N	50.6 E	43.7	29.7	13.0
AP	Pakistan, Karachi	24.9 N	67.1 E	31.2	15.0	356.9
AP	Islamabad	33.7 N	73.2 E	22.6	7.7	352.1
BV	Taiwan, Taipei	25.1 N	121.5 E	339.0	323.0	309.9
BY	Peoples Rep. of China, Beijing	40.0 N	116.4 E	347.4	334.2	322.6
	Harbin	45.8 N	126.7 E	341.7	330.6	320.6
	Shanghai	31.2 N	121.5 E	341.0	326.3	313.8
	Fuzhou	26.1 N	119.3 E	341.4	325.3	312.0
	Xian	34.3 N	108.9 E	352.4	337.4	324.2
	Chongqing	29.8 N	106.5 E	354.0	337.7	323.5
	Chengdu	30.7 N	104.1 E	356.3	340.1	325.8
	Lhasa	29.7 N	91.2 E	8.1	351.4	335.6
	Urumqi	43.8 N	87.6 E	9.0	355.9	343.2
	Kashi	39.5 N	76.0 E	18.5	4.7	350.6
C2	Nauru	0.5 S	166.9 E	284.9	272.8	261.2
C3	Andorra	42.5 N	1.5 E	58.5	50.1	41.6
C5	The Gambia, Banjul	13.5 N	16.7 W	96.7	84.4	73.7
C6	Bahamas, Nassau	25.1 N	77.4 W	170.9	120.5	103.0
C9	Mozambique, Maputo	26.0 S	32.6 E	97.8	88.8	71.4
	Mozambique	15.1 S	40.7 E	82.0	69.3	45.9
CE	Chile, Santiago	33.5 S	70.8 W	172.0	156.9	143.5
CE0	Easter Island	27.1 S	109.4 W	207.3	191.1	173.6
CE0	Juan Fernandez	33.6 S	78.8 W	179.0	163.4	149.4
CE0	San Felix	26.3 S	80.1 W	180.1	162.5	146.9
CM,CO	Cuba, Havana	23.1 N	82.4 W	187.6	133.7	110.7
CN	Morocco, Casablanca	33.6 N	7.5 W	71.7	62.1	53.1
CP	Bolivia, La Paz	16.5 S	68.4 W	166.8	147.3	131.8
CT	Portugal, Lisbon	38.7 N	9.2 W	66.9	58.3	50.1
CT3	Madeira Islands, Funchal	32.6 N	16.9 W	77.4	67.5	59.1
CU	Azores , Ponta Delgada	37.7 N	25.7 W	75.0	66.1	59.2
CX	Uruguay, Montevideo	34.9 S	56.2 W	160.2	146.5	134.8
CY0	Sable Is.	43.8 N	60.0 W	69.1	66.3	65.7
CY9	St. Paul Is.	47.2 N	60.1 W	56.8	59.3	60.6
D2	Angola, Luanda	8.8 S	13.2 E	94.7	83.8	69.6
D4	Cape Verde, Praia	14.9 N	23.5 W	100.3	87.3	76.9
D6	Comoros, Moroni	11.8 S	43.7 E	76.7	62.8	38.1
DA-DL,Y2-Y9	Fed. Rep. of Germany, Bonn	50.7 N	7.0 E	47.9	40.7	33.1
	Berlin	52.5 N	13.4 E	43.9	36.6	28.8
DU	Philippines, Manilla	14.6 N	121.0 E	335.9	317.4	303.2
E3	Eritrea, Asmara	15.3 N	38.9 E	59.3	45.8	28.2
EA	Spain, Madrid	40.4 N	3.7 W	62.8	54.3	45.9
EA6	Balearic Is., Palma	39.5 N	2.6 E	61.0	52.2	43.1
EA8	Canary Is., Las Palmas	28.4 N	14.3 W	80.7	70.2	60.9
EA9	Ceuta & Melilla, Ceuta	35.9 N	5.3 W	68.2	59.0	50.1
	Melilla	35.3 N	3.0 W	67.8	58.5	49.2
EI	Ireland, Dublin	53.3 N	6.3 W	48.6	42.9	37.1
EK	Armenia, Yerevan	40.3 N	44.5 E	39.1	27.8	14.9
EL	Liberia, Monrovia	6.3 N	10.8 W	98.7	86.7	75.3
EP	Iran, Tehran	35.8 N	51.8 E	37.2	24.6	10.1
ER	Moldova, Kishinev	47.0 N	28.8 E	42.6	33.4	23.2
ES	Estonia, Tallinn	59.4 N	24.8 E	33.5	26.8	19.4
ET	Ethiopia, Addis Ababa	9.0 N	38.7 E	63.9	50.1	31.3
EU,EV,EW	Belarus, Minsk	53.9 N	27.6 E	37.4	29.5	20.7
EX	Kyrgyzstan, Frunze	42.9 N	74.6 E	18.4	5.5	352.2
EY	Tajikistan, Samarkand	39.7 N	66.8 E	25.0	11.9	357.8
	Dushanbe	39.1 N	68.8 E	23.8	10.4	356.2
EZ	Turkmenistan, Ashkhabad	38.0 N	58.4 E	31.6	18.7	4.5
F	France, Paris	48.8 N	2.3 E	51.5	44.2	36.7
FG	Guadeloupe	16.0 N	61.7 W	141.2	114.5	100.8
FJ, FS	St. Martin	18.1 N	63.1 W	141.4	113.4	99.8
FH	Mayotte	13.0 S	45.3 E	76.6	62.5	36.6
FK	New Caledonia, Noumia	22.3 S	166.5 E	266.4	255.1	245.1
FM	Martinique	14.6 N	61.0 W	141.4	115.5	101.7
FO	Clipperton Is.	10.3 N	109.2 W	229.1	202.9	166.7
FO	Fr. Polynesia, Tahiti	17.6 S	159.5 W	249.9	238.3	224.1
	Rurutu, (Austral Is.)	22.5 S	151.3 E	276.2	263.4	254.0
	Hiva Oa, (Marquesas Is.)	9.9 S	139.0 W	241.5	227.5	208.3
FP	St. Pierre & Miquelon, St. Pierre	46.7 N	56.0 W	61.0	60.5	60.4
FR/G	Glorioso	11.5 S	47.3 E	73.6	58.7	32.0
FR/J,E	Juan de Nova	17.0 S	42.8 E	82.3	69.5	44.7
	Europa	22.3 S	40.4 E	89.3	78.3	55.5
FR	Reunion	21.1 S	55.6 E	76.8	60.9	23.2
FR/T	Tromelin	15.9 S	54.4 E	72.0	55.3	21.9
FT8W	Crozet	46.0 S	52.0 E	116.0	119.5	128.4
FT8X	Kerguelen	49.3 S	69.2 E	123.5	144.9	199.9
FT8Y	Antarctica, Dumont D'Urville	66.6 S	140.0 E	206.7	209.6	211.0
FT8Z	Amsterdam & St. Paul Is., Amsterdam	37.7 S	77.6 E	89.7	86.4	277.9
FW	Wallis & Futuna Is., Wallis	13.3 S	176.3 W	263.7	252.4	240.1
FY	Fr. Guiana, Cayenne	4.9 N	52.3 W	137.3	116.9	103.7
G	England, London	51.5 N	0.1 W	49.2	42.6	35.9
GD	Isle of Man	54.3 N	4.5 W	46.9	41.3	35.6
GI	Northern Ireland, Belfast	54.6 N	5.9 W	46.8	41.5	35.9
GJ	Jersey	49.3 N	2.2 W	52.3	45.5	38.5
GM	Scotland, Glasgow	55.8 N	4.3 W	45.0	39.8	34.2
	Aberdeen	57.2 N	2.1 W	42.9	37.7	32.2
GU	Guernsey	49.5 N	2.7 W	52.2	45.5	38.6
GW	Wales, Cardiff	51.5 N	3.2 W	50.0	43.7	37.3
H4	Solomon Islands, Honiara	9.4 S	160.0 E	282.7	269.9	259.0
HA	Hungary, Budapest	47.5 N	19.1 E	46.5	38.0	28.7
HB	Switzerland, Bern	47.0 N	7.5 E	51.5	43.7	35.3
HB0	Liechtenstein	47.2 N	9.6 E	50.6	42.6	34.1
HC	Ecuador, Quito	0.2 S	78.0 W	176.9	149.5	129.6
HC8	Galapagos Is.	0.5 S	90.5 W	195.9	168.1	143.7
HH	Haiti, Port-Au-Prince	18.5 N	72.3 W	160.6	123.9	106.8
HI	Dominican Republic, Santo Domingo	18.5 N	70.0 W	155.3	120.8	104.8
HK	Colombia, Bogota	4.6 N	74.1 W	169.9	141.0	122.0
HK0	Malpelo Is.	4.0 N	81.1 W	181.9	151.4	129.6
HK0	San Adreas	12.5 N	81.7 W	183.6	146.0	122.7
HL	Korea, Seoul	37.5 N	127.0 E	338.6	325.8	314.6
HP	Panama, Panama	9.0 N	79.5 W	179.0	145.3	123.6
HR	Honduras, Tegucigalpa	14.1 N	87.2 W	195.7	155.3	127.6
HS	Thailand, Bangkok	13.8 N	100.5 E	359.4	337.8	319.8
HV	Vatican City	41.9 N	12.5 E	54.4	45.4	35.7
HZ, 7Z	Saudi Arabia, Dharan	26.3 N	50.0 E	44.1	30.2	13.6
	Mecca	21.5 N	39.8 E	54.4	41.2	24.8
	Riyadh	24.6 N	46.7 E	47.5	33.8	17.1
I	Italy, Rome	41.9 N	12.5 E	54.4	45.4	35.7
	Trieste	45.7 N	13.8 E	50.3	41.9	32.7

Pref	State/Province/Country/City	Lat	Long	fm E USA	fm C USA	fm W USA
	Sicily	37.5 N	14.0 E	57.6	48.0	37.4
IS	Sardinia, Cagliari	39.2 N	9.1 E	58.4	49.2	39.4
J2	Djibouti, Djibouti	11.6 N	43.2 E	58.7	44.3	25.1
J3	Grenada	12.0 N	61.8 W	145.1	119.1	104.7
J5	Guinea-Bissau, Bissau	11.9 N	15.6 W	97.3	85.1	74.2
J6	St. Lucia	13.9 N	61.0 W	142.1	116.2	102.3
J7	Dominica	15.4 N	61.3 W	141.1	114.8	101.1
J8	St. Vincent	13.3 N	61.3 W	143.1	117.2	103.1
JA-JS	Japan, Tokyo	35.7 N	139.8 E	328.5	316.6	306.1
	Nagasaki	32.8 N	129.9 E	334.6	321.1	309.6
	Sapporo	43.1 N	141.4 E	331.1	320.7	311.3
JD1	Minami Torishima	24.3 N	154.0 E	311.8	299.9	289.1
JD1	Ogasawara, Kazan Is.	27.5 N	141.0 E	323.4	310.3	299.1
JT	Mongolia, Ulan Bator	47.9 N	106.9 E	355.4	343.6	332.7
JW	Svalbard, Spitsbergen	78.8 N	16.0 E	14.2	12.2	9.8
JX	Jan Mayen	71.0 N	8.3 W	25.1	23.6	21.5
JY	Jordan, Amman	32.0 N	35.9 E	49.8	38.1	24.2
KC4	Antarctica, Byrd Station	80.0 S	120.0 W	187.6	184.2	180.8
	McMurdo Sound	77.7 S	166.7 E	195.7	195.6	194.9
	Palmer Station	64.8 S	64.0 W	173.0	165.6	158.7
KC6	Belau, Yap	9.5 N	138.2 E	315.7	299.7	287.5
	Koror	7.3 N	134.5 E	317.9	300.9	288.4
KG4	Guantanamo Bay	19.9 N	75.2 W	167.0	126.1	107.8
KH0	Mariana Is., Saipan	15.2 N	145.8 E	312.5	298.4	286.9
KH1	Baker, Howland Is.	0.5 N	176.0 W	274.2	262.9	250.0
KH2	Guam, Agana	13.5 N	144.8 E	312.2	297.9	286.2
KH3	Johnston Is.	17.0 N	168.5 W	282.3	272.2	258.8
KH4	Midway Is.	28.2 N	177.4 W	296.5	287.5	276.7
KH5	Palmyra Is.	5.9 N	162.1 W	269.5	258.2	243.0
KH5K	Kingman Reef	7.5 N	162.8 W	271.2	260.1	245.0
KH6	Hawaii, Hilo	19.7 N	155.1 W	276.4	266.4	250.4
	Honolulu	21.3 N	157.9 W	279.5	269.9	254.9
KH7	Kure Is.	28.4 N	178.4 W	297.2	288.2	277.5
KH8	American Samoa, Pago Pago	14.3 S	170.8 W	259.4	248.2	235.3
KH9	Wake Is.	19.3 N	166.6 E	299.7	288.5	277.5
KL7	Alaska, Adak	51.8 N	176.6 W	316.5	311.9	307.1
	Anchorage	61.2 N	150.0 W	321.3	323.2	327.0
	Fairbanks	64.8 N	147.9 W	326.2	329.1	334.0
	Juneau	58.3 N	134.4 W	316.8	322.7	333.4
	Nome	64.5 N	165.4 W	327.3	326.9	327.4
KP1	Navassa Is.	18.4 N	75.0 W	167.3	127.9	109.3
KP2	Virgin Islands, Charlotte Amalie	18.3 N	64.9 W	144.6	115.0	100.9
KP4	Puerto Rico, San Juan	18.5 N	66.2 W	147.0	116.2	101.7
KP5	Desecheo Is.	18.3 N	67.5 W	150.0	118.0	103.0
LA-LJ	Norway, Oslo	60.0 N	10.7 E	36.8	31.2	25.1
LU	Argentina, Buenos Aires	34.6 S	58.4 W	161.9	147.9	136.0
LX	Luxembourg	49.6 N	6.2 E	49.3	42.0	34.3
LY	Lithuania, Vilna	54.5 N	25.5 E	37.7	30.0	21.5
LZ	Bulgaria, Sofia	42.7 N	23.3 E	48.7	39.2	28.6
OA	Peru, Lima	12.1 S	77.1 W	176.4	154.4	136.6
OD	Lebanon, Beirut	33.9 N	35.5 E	48.8	37.3	23.8
OE	Austria, Vienna	48.2 N	16.3 E	47.0	38.8	29.8
OH	Finland, Helsinki	60.2 N	25.0 E	32.7	26.1	18.9
OH0	Aland Is.	60.2 N	20.0 E	34.2	27.9	21.1
OJ0	Market Reef	60.3 N	19.0 E	34.4	28.2	21.5
OK,OL	Czech Rep., Prague	50.1 N	14.4 E	45.9	38.1	29.7
OM	Slovak, Rep., Bratislava	48.0 N	17.0 E	46.9	38.6	29.6
ON	Belgium, Brussels	50.9 N	4.4 E	48.5	41.6	34.2
OX	Greenland, Godthaab	64.2 N	51.7 W	25.0	31.0	34.6
	Thule	76.6 N	68.8 W	4.3	10.0	14.8
OY	Faroe Islands, Torshavn	62.0 N	6.8 W	37.3	33.8	29.8
OZ	Denmark, Copenhagen	55.7 N	12.6 E	40.9	34.3	27.1
P2	Papua New Guinea, Madang	5.2 S	145.6 E	298.0	282.7	271.4
	Port Moresby	9.4 S	147.1 E	293.0	278.1	267.1
P4	Aruba, Oranjestad	12.1 N	69.0 W	157.8	127.3	110.5
PA-PI	Netherlands, Amsterdam	52.4 N	4.9 E	46.7	40.0	32.9
PJ2,4,9	Netherlands Antilles, Willemstad	12.6 N	70.1 W	159.6	128.2	111.0
PJ5-8	St. Maarten and Saba, St. Maarten	17.7 N	63.2 W	142.1	114.0	100.3
PY	Brazil, Brasilia	15.8 S	47.9 W	145.1	128.8	116.2
	Rio De Janeiro	23.0 S	43.2 W	144.5	130.2	118.3
	Natal	6.0 S	35.2 W	127.3	112.3	100.5
	Manaus	3.1 S	60.2 W	152.3	130.8	116.0
	Porto Alegre	30.1 S	51.2 W	154.4	140.3	128.4
PY0	Fernando De Noronha	3.9 S	32.4 W	123.3	108.7	97.1
PY0	St. Peter & St. Paul Rocks	1.0 N	29.4 W	117.1	102.8	91.4
PY0	Trindade & Martin Vaz Is., Trindade	20.5 S	29.3 W	131.9	119.1	107.9
PZ	Suriname, Paramaribo	5.8 N	55.2 W	140.3	118.7	105.1
R1FJ_	Franz Josef Land	80.0 N	53.0 E	8.7	5.5	2.2
R1MV_	Malyj Vysotskij Is.	60.6 N	28.6 E	31.2	24.5	17.1
S0	Western Sahara, Smara	26.4 N	11.4 W	81.1	70.4	60.7
S2	Bangladesh, Dacca	23.7 N	90.4 E	9.8	351.3	334.0
S5	Slovenia, Ljubljana	46.0 N	14.5 E	49.8	41.3	32.1
S7	Seychelles, Victoria	4.6 S	55.5 E	60.5	42.5	14.4
S9	Sao Tome	0.3 N	6.7 E	91.9	80.6	67.3
SM	Sweden, Stockholm	59.3 N	18.1 E	35.6	29.3	22.5
SP	Poland, Krakow	50.0 N	20.0 E	43.9	35.7	26.8
	Warsaw	52.2 N	21.0 E	41.5	33.6	25.1
ST	Sudan, Khartoum	15.6 N	32.5 E	63.5	50.8	34.4
ST0	Southern Sudan, Juba	5.0 N	31.6 E	71.9	59.1	41.3
SU	Egypt, Cairo	30.0 N	31.4 E	54.0	42.4	28.7
SV	Greece, Athens	38.0 N	23.7 E	52.3	42.2	30.7
SV/A	Mount Athos	40.2 N	24.3 E	50.2	40.4	29.2
SV5	Dodecanese, Rhodes	36.4 N	28.2 E	51.1	40.5	28.3
SV9	Crete	35.4 N	25.2 E	53.5	43.0	30.9
T2	Tuvalu, Funafuti	8.7 S	178.6 E	270.6	259.1	247.1
T30	West Kiribati, Bonriki	1.4 N	173.2 E	282.1	270.4	258.5
T31	Central Kiribati, Kanton	2.8 S	171.7 W	268.9	257.6	244.2
T32	East Kiribati, Christmas Is.	1.9 N	157.4 W	263.3	251.5	235.3
T33	Banaba Is.	0.5 S	169.4 E	283.2	271.2	259.6
T5	Somalia, Mogadishu	2.1 N	45.4 E	63.8	48.7	26.6
T7	San Marino	43.9 N	12.3 E	52.6	44.0	34.6
T9	Bosnia-Herzegovina, Sarajevo	43.9 N	18.4 E	50.0	41.0	31.0
TA	Turkey, Ankara	39.9 N	32.9 E	46.0	35.4	23.4
	Istanbul	41.2 N	29.0 E	47.1	37.0	25.6
TF	Iceland, Reykjavik	64.1 N	22.0 W	34.4	33.4	31.6
TG	Guatemala, Guatemala City	14.6 N	90.5 W	202.9	162.1	131.5
TI	Costa Rica, San Jose	9.9 N	84.0 W	187.8	152.3	127.8
TI9	Cocos Is.	5.6 N	87.0 W	192.2	160.1	135.1
TJ	Cameroon, Yaounde	3.9 N	11.5 E	86.0	74.6	60.8
TK	Corsica	42.0 N	9.0 E	55.8	47.1	37.7
TL	Central African Republic, Bangui	4.4 N	18.6 E	81.1	69.4	54.4
TN	Congo, Brazzaville	4.3 S	15.3 E	89.9	78.7	64.0
TR	Gabon, Libreville	0.4 N	9.5 E	90.0	78.7	65.1
TT	Chad, N'Djamena	12.1 N	15.0 E	77.5	66.0	52.2
TU	Côte d'Ivoire, Abidjan	5.3 N	4.0 W	95.0	83.4	71.4
TY	Benin, Porto Novo	6.5 N	2.6 E	89.7	78.3	65.8
TZ	Mali, Bamako	12.7 N	8.0 W	91.6	79.9	68.7
UA	Russia, European, Leningrad (UA1)	59.9 N	30.3 E	31.3	24.3	16.6
	Archangel (UA1)	64.6 N	40.5 E	24.0	17.4	10.3
	Murmansk (UA1)	69.0 N	33.1 E	22.3	17.1	11.4
	Moscow (UA3)	55.8 N	37.6 E	31.9	23.6	14.6
	Kuybyshev (UA4)	53.2 N	50.1 E	28.0	18.5	8.3
	Rostov (UA6)	47.5 N	39.5 E	37.1	27.2	16.3
UA2	Kaliningrad	55.0 N	20.5 E	39.1	31.8	23.8
UA9,0	Russia, Asiatic, Novosibirsk (UA9)	55.0 N	82.9 E	9.8	359.5	349.2
	Perm (UA9)	58.0 N	56.3 E	22.2	13.4	4.1
	Omsk (UA9)	55.0 N	73.4 E	15.0	4.9	354.6
	Norilsk (UA0)	69.3 N	88.1 E	4.4	357.7	351.1
	Irkutsk (UA0)	52.3 N	104.3 E	357.4	346.6	336.4
	Vladivostok (UA0)	43.2 N	131.9 E	337.3	326.1	316.1
	Petropavlovsk (UA0)	53.0 N	158.7 E	327.7	320.7	314.0
	Khabarovsk (UA0)	48.5 N	135.1 E	337.6	327.6	318.6
	Krasnoyarsk (UA0)	56.0 N	92.8 E	4.0	354.0	344.2
	Yakutsk (UA0)	62.0 N	129.7 E	346.1	338.7	331.9
	Wrangel Island (UA0)	71.0 N	179.5 W	337.1	335.8	335.2
	Kyzyl (UA0Y)	51.7 N	94.5 E	3.4	352.3	341.1
UJ-UM	Uzbekistan, Bukhoro	39.8 N	64.4 E	26.6	13.7	359.5
	Tashkent	41.2 N	69.3 E	22.7	9.7	356.0
UN-UQ	Kazakhstan, Alma-Ata	43.3 N	76.9 E	16.6	3.8	350.6
UR -UZ,EM-EO	Ukraine, Kiev	50.4 N	30.5 E	39.1	30.4	20.7
V2	Antigua & Barbuda, St. Johns	17.1 N	61.8 W	140.2	113.3	99.8
V3	Belize, Belmopan	17.3 N	88.8 W	201.1	156.2	126.1
V4	St. Kitts & Nevis	17.3 N	62.6 W	141.4	113.9	100.2
V5	Namibia, Windhoek	22.6 S	17.1 E	103.6	94.2	80.6
V6	Micronesia, Ponape	6.9 N	158.3 E	296.8	283.9	272.6
V7	Marshall Islands, Kwajalein	9.1 N	167.3 E	291.9	280.1	268.5
V8	Brunei, Bandar Seri Begawan	4.9 N	114.9 E	339.5	317.1	301.3
VK	Australia, Canberra (VK1)	35.3 S	149.1 E	260.9	250.9	243.9
	Sydney (VK2)	33.9 S	151.2 E	261.8	251.6	244.2
	Melbourne (VK3)	37.8 S	145.0 E	258.8	249.6	243.4

Pref	State/Province/Country/City	Lat	Long	fm E USA	fm C USA	fm W USA	Pref	State/Province/Country/City	Lat	Long	fm E USA	fm C USA	fm W USA
	Brisbane (VK4)	27.5 S	153.0 E	269.1	257.5	248.8	YB-YD	Indonesia, Jakarta	6.2 S	106.8 E	348.0	318.5	299.8
	Adelaide (VK5)	34.9 S	138.6 E	267.1	255.8	249.3		Medan	3.6 N	98.7 E	1.9	335.8	315.4
	Perth (VK6)	31.9 S	115.8 E	297.5	272.2	264.2		Pontianak	0.0	109.3 E	345.7	320.0	302.5
	Hobart, Tasmania (VK7)	42.9 S	147.3 E	249.4	242.4	237.3		Jayapura	2.6 S	140.7 E	304.8	288.4	276.6
	Darwin (VK8)	12.5 S	130.9 E	306.6	287.0	275.1	YI	Iraq, Baghdad	33.0 N	44.5 E	43.7	31.3	16.8
VK0	Heard Is.	53.0 S	73.4 E	134.6	161.1	203.1	YJ	Vanuatu, Port Vila	17.5 S	168.3 E	269.6	258.0	247.4
VK0	Macquarie Is.	54.7 S	158.8 E	228.8	225.0	221.3	YK	Syria, Damascus	33.5 N	36.3 E	48.5	36.9	23.4
VK9C	Cocos-Keeling Is.	12.2 S	96.8 E	6.7	329.0	304.7	YL	Latvia, Riga	57.0 N	24.1 E	36.0	28.8	21.0
VK9L	Lord Howe Is.	31.6 S	159.1 E	260.7	250.4	242.1	YN	Nicaragua, Managua	12.0 N	86.0 W	192.4	154.4	128.2
VK9M	Mellish Reef	17.6 S	155.8 E	278.1	265.3	255.2	YO	Romania, Bucharest	44.4 N	26.1 E	46.0	36.6	26.0
VK9N	Norfolk Is.	29.0 S	168.0 E	258.8	248.5	239.1	YS	El Salvador, San Salvador	13.7 N	89.2 W	199.7	159.8	130.6
VK9W	Willis Is.	16.3 S	149.5 E	284.3	270.3	260.1	YU	Yugoslavia, Belgrade	44.9 N	20.5 E	48.2	39.2	29.2
VK9X	Christmas Is.	10.5 S	105.7 E	348.7	316.1	296.9	YV	Venezuela, Caracas	10.5 N	67.0 W	155.1	126.5	110.3
VP2E	Anguilla	18.3 N	63.0 W	141.0	113.0	99.5	YV0	Aves Is.	15.7 N	63.7 W	145.1	117.0	102.7
VP2M	Montserrat	16.8 N	62.2 W	141.2	114.1	100.4	Z2	Zimbabwe, Harare	17.8 S	31.0 E	91.0	80.2	61.7
VP2V	British Virgin Is., Tortola	18.4 N	64.6 W	144.0	114.6	100.6	Z3	Macedonia, (ex Yugoslav), Skopje	42.0 N	21.4 E	50.2	40.7	30.2
VP5	Turks & Caicos Islands, Grand Turk	21.4 N	71.2 W	155.5	118.1	102.5	ZA	Albania, Tirane	41.3 N	19.8 E	51.6	42.1	31.6
VP8	Falkland Islands, Stanley	51.7 S	57.9 W	166.5	156.3	147.0	ZB2	Gibraltar	36.1 N	5.4 W	68.1	58.9	50.0
VP8	So. Georgia Is.	54.3 S	36.8 W	156.0	147.8	140.4	ZC4	British Cyprus	34.6 N	33.0 E	49.7	38.5	25.5
VP8	So. Orkney Is.	60.6 S	45.5 W	163.3	155.9	149.1	ZD7	St. Helena	16.0 S	5.9 W	112.4	101.5	89.9
VP8	So. Sandwich Islands, Saunders Is.	57.8 S	26.7 W	153.4	146.8	140.8	ZD8	Ascension Is.	8.0 S	14.4 W	112.3	100.2	88.7
VP8	So. Shetland Is., King George Is.	62.0 S	58.3 W	169.7	161.9	154.7	ZD9	Tristan da Cunha	37.1 S	12.3 W	131.7	122.8	113.7
VP9	Bermuda	32.3 N	64.7 W	117.2	91.7	84.0	ZF	Cayman Is.	19.5 N	81.2 W	183.2	137.0	114.2
VQ9	Chagos, Diego Garcia	7.3 S	72.4 E	44.5	18.2	344.8	ZK1	No. Cook Is., Manihiki	10.4 S	161.0 W	256.3	244.5	229.9
VR2,VS6	Hong Kong	22.3 N	114.3 E	345.2	327.6	313.3	ZK1	So. Cook Is., Rarotonga	21.2 S	159.8 W	247.3	235.9	222.3
VR6	Pitcairn Is.	25.1 S	130.1 W	224.9	210.8	193.9	ZK2	Niue	19.0 S	168.9 W	254.6	243.4	230.8
VU	India, Bombay	19.0 N	72.8 E	28.7	10.3	350.3	ZK3	Tokelaus, Atafu	8.4 S	172.7 W	265.3	253.9	240.9
	Calcutta	22.6 N	88.4 E	12.0	353.2	335.4	ZL	New Zealand, Auckland (ZL1)	36.9 S	174.8 E	247.2	238.1	229.3
	New Delhi	28.6 N	77.2 E	21.0	6.8	347.7		Wellington (ZL2)	41.3 S	174.8 E	242.4	234.1	226.0
	Bangalore	13.0 N	77.6 E	26.3	5.5	343.7		Christchurch (ZL3)	43.5 S	172.6 E	240.7	233.0	225.5
VU	Andaman Islands, Port Blair	11.7 N	92.8 E	8.9	346.3	326.3		Dunedin (ZL4)	45.9 S	170.5 E	238.6	231.4	224.6
VU	Laccadive Is.	10.0 N	73.0 E	32.7	11.8	348.5	ZL5	Antarctica, Scott Base	77.9 S	166.4 E	195.4	195.4	194.7
XE	Mexico, Mexico City (XE1)	19.4 N	99.1 W	224.1	183.3	139.9	ZL7	Chatham Is.	44.0 S	176.5 W	236.1	228.0	219.5
	Chihuahua (XE2)	28.7 N	106.0 W	250.1	218.0	140.9	ZL8	Kermadec Is.	29.3 S	177.9 W	251.0	240.8	230.3
	Merida (XE3)	21.0 N	89.7 W	206.4	154.8	122.5	ZL9	Auckland & Campbell Is., Auckland	50.7 S	166.5 E	233.5	227.8	222.2
XF4	Revilla Gigedo	19.0 N	111.5 W	241.4	215.5	168.2		Campbell Is.	52.5 S	169.1 E	230.5	225.1	219.6
XT	Burkina Faso, Ouagadougou	12.4 N	1.6 W	87.7	76.3	64.5	ZP	Paraguay, Asuncion	25.3 S	57.7 W	158.4	142.5	129.4
XU	Cambodia, Phnom Penh	11.7 N	104.8 E	354.0	332.1	314.6	ZS	South Africa, Cape Town (ZS1)	33.9 S	18.4 E	113.0	105.9	95.2
XW	Laos, Viangchan	18.0 N	102.6 E	357.1	337.1	320.3		Port Elizabeth (ZS2)	34.0 S	25.7 E	109.6	102.9	91.1
XX9	Macao	22.2 N	113.6 E	345.9	328.2	313.8		Bloemfontein (ZS4)	29.2 S	26.1 E	104.6	96.6	82.7
XZ	Myanmar, Yangon	16.8 N	96.0 E	4.6	343.8	325.7		Durban (ZS5)	29.9 S	30.9 E	102.7	94.9	79.8
YA	Afghanistan, Kandahar	31.0 N	65.8 E	29.5	14.7	358.4		Johannesburg (ZS6)	26.2 S	28.1 E	100.6	91.8	76.2
	Kabul	34.4 N	69.2 E	25.4	11.0	355.6	ZS8	Marion Is.	46.8 S	37.8 E	120.3	119.7	118.4

CHAPTER 5

DXing

BOB LOCHER, W9KNI
7850 DRY CREEK RD
GEYSERVILLE, CA 95441

Almost every radio amateur at one time or another becomes interested in DX, surely the oldest activity of Amateur Radio. After all, Guglielmo Marconi, the first amateur, went on the first DXpedition—to W1 land! And there hardly breathes an operator who is not interested in working a station farther away than his normal range, even if only through the local repeater.

But for many amateurs, pushing a high-frequency (HF) signal as far as it will go, and working for contacts with stations in countries never before logged, is the primary interest. DXing is one of the most exciting and satisfying aspects of Amateur Radio, and one that virtually any amateur with a working HF station can pursue.

Is your station capable of working DX? If your station can work both coasts on a consistent basis on 40, 20 or 15 meters, you are capable of working DX, especially on CW. You do not need a B-I-G linear amplifier, nor do you need a Yagi on a tower. Of course, such things help, and can at times help a lot, but one of the really nice things about DXing is that it is based on the art of the possible.

To be more specific, if you are running one of the typical 100-watt transceivers, and a decent dipole, inverted V, vertical or better antenna on the bands from 40-10 meters, you can work DX. And, with such equipment, you can be assured that you will be able to work 95% of the stations you can hear. Much DX has been worked with similar antennas and 5 watts or less. Of course, if you hear a really rare one buried in a howling pileup, your chances of snagging a QSO are slim, but if you trip across the same station a day later calling CQ in solitary splendor, your chances of nailing him are very good.

Before going further, let's agree on a definition of what DX really is, at least for purposes of this chapter. Although the classic definition of DX is "distance," today it generally means amateur stations in foreign countries. Typically, this suggests countries thousands of miles away, but not always! For example, for nearly half the amateurs in the United States, the French-owned St Pierre and Miquelon Islands off the coast of Newfoundland are closer than Seattle, yet these islands, using the prefix FP, are considered DX by most amateurs, while the W7 in Seattle is not. And again, Clipperton Island, with the prefix FO, lying off the coast of Mexico, is considered to be a very rare DX country, even though it is closer to San Francisco than New York City is. So, for most HF work, any station in a foreign country is considered to be DX.

Certainly, a JT1 in Ulan Bator, Mongolia, is going to raise a good deal more excitement on the bands than an HI8 in the Dominican Republic, but both are considered DX. Again, working a rare country excites some amateurs, while other amateurs are equally pleased with a friendly chat with a G3 in London. But, it has been noticed more than once that the amateur claiming to welcome equally a contact with England or with Mongolia already has cards from both England and Mongolia hanging on the wall!

Which raises the question of why people try to work DX. For some, it is a simple matter of satisfaction. They find a ragchew with a station in another land interesting and rewarding in its own right, and have no interest beyond that. For others, a contact with a country never before worked provides a thrill unmatched by any other QSO. Likely, such an amateur will avidly pursue the QSL card as a trophy for the wall and for

Bob Eshleman, W4DR, is a dentist, teacher . . . and one of the most successful DXers in the United States. (*photo by James Cain, K1TN*)

the premier award of DXing, the ARRL DX Century Club (DXCC).

And for you? If you don't know the answer, be glad. If the thought of a contact with a foreign land excites your fancy, you are in for a lot of fun seeking the answer.

TUNING THE BANDS

Let's slip the headphones on and have a look around the bands, and see if a worthwhile bit of DX to snag can be located. It's a late Friday afternoon in April. With the implicit blessing of my boss, I head home early. I grab a cup of coffee, settle the headphones over my ears, and turn the gear on. The antenna rotator shows that the tribander is pointed at 45°, the nominal short path bearing for Europe from the Midwest. I move the transceiver to 20 meters, switch to the CW filter, set the AGC (automatic gain control) to "fast," the RF gain on full, move the dial to the bottom of the CW band, and start tuning slowly.

The first few kilohertz are fairly empty of activity, but as I move a bit higher up the band, the activity seems to build. There are several loud stations coming in, and a brief listen to each confirms my expectation that they are Americans and Canadians. One is calling CQ, another is working the West Coast. Nothing interesting. I keep tuning higher.

There—there's a station calling CQ. It's an OH6 from Finland. Hmmm. I listen carefully to his signal. He's an S8. Nice and steady with a clean note.

Whenever you are checking out a band, especially 20 meters, the quality of signal from a northern European, such as an SM, OH, LA, or a UA1, will tell you a lot about the band in a hurry. If the signal is a bit fluttery, with a trace of auroral buzz to it, there are at least some signs of disturbance. It is therefore unlikely that there will be a transpolar opening, depending on how strong the auroral effect is on the North European signal. On the other hand, if the signals from the high latitude Europeans are strong and clean, with any luck there will be an excellent transpolar opening into such areas as Central Asia, India, Pakistan, Sri Lanka and so on, at the right time for propagation to those areas. From late spring through early fall, this will be an hour or so after sunrise and an hour or so after sunset.

I keep moving. There's a weak signal signing with a W4. Okay, it's a 4X4, Israel. Another one I don't need. There's another CQ. Right, it's an EI3, Ireland. I don't need that either.

I come to a W2 calling CQ DX, and listen to see what he raises by way of answer. His signal is strong and has a definite bit of backscatter trace to it, a raspy, weaker overlay, spread a few hertz above and below his signal. The upper layers of the ionosphere reflect a portion of his signal back toward me. The uneven reflections yield a Doppler shift "hash" mixed with cleaner bounces of his signal, causing the rasp. (The effect is more frequently heard on the signals of stations using Yagis or quads pointed to the same area that the receiving station antenna is aimed at. And, generally, the more the backscatter from nearby stations [for which, read "North American"], the better conditions are likely to be.)

There, the W2 signs. I listen. Yes, there are several stations calling him. I hear a DK3, Germany. And there—there's an ON8 in Belgium. Ah, the W2 starts coming back to the DK3. Not surprising; the DK3 is dead on the W2's frequency, and started his call almost immediately after the W2 signed. And his speed is virtually identical to the W2's. On the other hand, the ON8 started his call several seconds later. He's two or three hundred hertz above the W2's frequency, and is calling at a slower speed.

I smile to myself; the act of analyzing why the DK3 got the

QSO instead of the ON8 is a good way to learn how to be effective at calling other stations, DX or otherwise. But the lessons learned are no good unless I am smart enough to practice the lessons for myself.

I don't need a QSO with a DK3. Today I am after more exotic game. I tune on.

Hmmm. I come across six or eight stations calling someone. I listen; this might be interesting. Ah, there—someone just below the callers' frequency comes back to a WB8, someone with a good, clean signal, with no backscatter or flutter. He just sends the WB8's call, and 5NN K, the 5NN being CW shorthand for 599. I hear the WB8 coming back to him, via backscatter, R TU 5NN 5NN 73 BK. The other station responds, R TU QRZ? and that's all he sends. All the other stations start calling again.

I continue listening carefully and with interest, but I make no effort to call. I want to see who the DX station is before I commit myself to the mini-pileup. The DX station puts away a W3, then a VE3. Then, as he signs with the VE3, he sends

QSL VIA I2CRG I2CRG NW QRZ UP 1 UP 1 DE 5L2EQ K

A 5L2? What in the heck is that? I glance quickly at my *DXCC Countries List* (which you'll find in Chapter 17) and don't find it; 5L is not a standard amateur call sign, apparently. I check the back of the log book for the ITU call sign allocations table (which also appears in Chapter 17), and look up 5L there. Yes, there it is; 5LA-5MZ is assigned to Liberia, in West Africa. (5L also is listed in the prefix cross-reference section of the *Countries List*.) Darn. I need a lot of African countries, especially the "Terrible Ts," but Liberia I don't need. I have an EL2 in the log and on the wall. Well, ELs aren't particularly rare, but the unusual prefix does help explain the mini-pileup. No doubt some of the stations calling want the special prefix for one of the prefix hunter's awards, while others calling just like the idea of a fast QSO with an out-of-the-ordinary station. But I'm after bigger game. I tune on.

I come across UA2FAM from Kaliningrad signing clear with a W9. I check my list; I have worked several UA2s but have no QSL as yet to show for it. I check my QSL list; UA2FAM is not one of the stations I have worked, so I better give him a call.

The W9 is finishing up the QSO. He is not very strong, coming in on weak groundwave, but I hear him well enough. I set my keyer speed to match his. Next I throw the VOX switch to OFF on my transceiver, check to ensure that my RIT (receiver incremental tuning) control is in the OFF position, as it should always be when tuning, and touch my paddle to send a string of dits that are heard only in my monitor and don't go out over the air.

I move the transceiver dial until the W9's signal is identical in tone to my monitor, then flip on the RIT. Carefully I retune the signal by adjusting the RIT control until he is back where he was when I first heard him. My CW signal is now zero beat with the W9, so I know I will transmit on the same frequency, where, after all, I know the UA2 is listening. I turn the VOX back on.

The W9 finishes, and the UA2 returns with the traditional "dit dit." I wait a brief moment to see that the UA2 does not send a CQ or a QRZ. I hear nothing, so I give him a call, UA2FAM DE W9KNI W9KNI AR. I listen; hah, I got him! I log the details of the QSO; the UA2, like most Russian amateurs, keeps the contact short, and after the promise of an exchange of QSLs, I am soon tuning the band again. Though the UA2 wasn't a new country for me, if he is the first UA2 that

I receive a QSL from, it will be as good as the first QSO with a new one.

I look out the window; darkness is gathering. It will be dinnertime shortly, but the band seems very good, and I don't want to miss the opportunity it offers. I arrange with the family for a quick dinner a bit later, and return to the rig with a fresh cup of coffee.

I glance at my frequency, rather high in the CW portion of the band. I quickly tune up higher a few more kilohertz then, having found nothing of interest, I return to the bottom edge again, and start tuning higher.

The very bottom of the band is as quiet as before, but about 3 kilohertz in from the edge, things start to pick up. There's a UN7 calling CQ. UN7s are relatively common on 20 meters, but they are a harbinger of rarer DX. They come through on the same path as the rarer EXs, EYs and EZs, as well as a path close to the path for India and the islands of the Indian Ocean.

There's a CQ—not real strong, but enough for copy. I dig for the call, but have trouble getting it at first. I adjust the frequency a bit and he gets clearer. Yes, it's EX3FZ. That's a new one for me! Quickly, I start lining him up. Oops; my RIT is still on from the last QSO. But it must be close to the transmit frequency; I trimmed only a couple hundred hertz for that UA2. (While looking for DX, never tune a transceiver with the RIT on unless you have it set at a specific tone for a specific reason.)

I turn off the RIT; the tone of the EX3 shifts, but he's still inside my crystal filter's passband, so I get away with my mistake. I switch off the VOX and hit the paddle, and adjust the frequency of the transceiver until the EX3 is at the same tone as the monitor. There, that's set. Good thing he's making that CQ a long call. I switch the VOX and the RIT back on, but before I retune the RIT, he finishes his call. No matter, I'm ready for him. I call, EX3FZ DE W9KNI W9KNI W9KNI $\overline{AR}$, adjusting my keyer speed to match the speed of the EX3 as I start my transmission. I finish my call, and listen. Darn! He's already coming back to someone else—a WA5.

I copy all the data that he is sending the WA5. That way, if I have trouble with QRM when I get him, I'll still have his name and QTH. Yes, he is copying the WA5 RST 579, his QTH is Frunze, and his name is Victor. He turns it back to the WA5. I wince, as a wave of RF sends my S meter slamming into the peg. Guess we must have short skip. And if the WA5 is anywhere near as strong for the EX3 as he is in my receiver, it's no wonder he beat me out. I temporarily switch my RIT out again, and find the WA5 within a hundred hertz of my monitor note; that means I was close to being on the same frequency anyhow. The WA5 is giving the EX3 a 579 report. I can only say if he's hearing the EX3 a real 579, he's doing a lot better than I am. The EX3 is really 559 at best here.

The WA5 is describing his station. Yes, a kilowatt and a five-element Yagi at 30 meters. Okay, that's 100 feet, give or take, against my little three-element tribander at 45 feet, and his kW against my 100 watts. No wonder I got beat out. I decide not to move my transmit frequency; I was close enough to the WA5's frequency that there's no reason to move. But there, they are signing now; maybe it's my turn this time. I call the EX3 again, then listen.

As the transceiver goes back to receive, I hear another station for a moment, a KA4. Guess he must have also been calling the EX3. There's the EX3 coming back.

Phooey. Beaten again.

This time the EX3 is coming back to a K8. Should I wait around again, or should I go tuning again, looking for another new one? Maybe I would find another EX3, as far as that goes.

But if I do, I'll probably lose any chance of snagging Victor. I decide to ride it out a bit longer. Besides, I think his signal is starting to come up a bit; the path is getting better. I glance at my antenna rotator. Darn, it's still at 45°. I check the beam heading on my beam chart. Okay, I should be at 13°. I turn the antenna more north, and the EX3 seems to pick up another half S-unit, bringing him to a solid 569 now. I decide to stay after him.

The EX3 signs clear with the K8, and I wait for the EX3's "dit dit." Two stations start calling the EX3 while I wait, and they are both on almost exactly the same frequency as the K8 was—and where my transmit frequency is. If I call there, all I'm going to do is get QRMed by those other stations. Quickly, blindly, I move my frequency up about 300 Hz. I know that I am moving away from where the EX3 was listening, but with my not-overly-strong signal, there's no way I'm likely to blow those other callers away. So a fast move seems to be my best bet. I make my call a little longer as well, hoping to give the EX3 more time to tune off those other stations and find me.

I listen again, moving the RIT back down to the EX3's frequency. Yes, there are still two stations calling him, right on his frequency. I can make out that the EX3 is in there. He must have come back to someone, but there's no way I'm going to copy him with those fellows still calling him dead on his frequency. Suddenly, one of them stops transmitting, and then, almost immediately, so does the other station. . . 569 $\overline{BT}$ QTH HR FRUNZE FRUNZE ES NAME IS VICTOR VICTOR QSL SURE VIA BURO HW COPY W9KNI DE EX3FZ $\overline{KN}$. All right—I got him!

R EX3FZ $\overline{DE}$ W9KNI R FB VICTOR ES TNX QSO ES FIRST EX3 $\overline{BT}$ UR RST 569 569 HR QTH NR CHICAGO NR CHICAGO ES NAME IS BOB BOB PSE QSL ES TNX NEW COUNTRY DR VICTOR HW CPY EX3FZ DE W9KNI KN.

R W9KNI DE EX3FZ R R FB BOB ES TNX REPORT FM NR CHICAGO ES GLAD BE UR FIRST EX3 $\overline{BT}$ QSL SURE QRU NW 73 ES GOOD DX DSW $\overline{SK}$ W9KNI DE EX3FZ TU.

I start breathing again; a real "goody" in the log. I look over at the UTC (Coordinated Universal Time) clock, and get him noted in. Let's see here, the clock shows 0039 Zulu (ie, UTC). I must have started my QSO with Victor at about 0034, and I enter that time into the log. But 0034 is after midnight Zulu

Worked Ascension Island lately? Then you've probably worked one of these guys, who've provided all the activity during recent times. Gathered for their annual meeting at the VC Club are (l-r) Andy Chadwick, ZD8VJ/G4ZVJ; Jim Neiger, ZD8Z/N6TJ; and Mike Wadsworth, ZD8M/G3UOF. (*photo by Mike Wadsworth, G3UOF*)

time; that means I have to put tomorrow's date in the log for the date of the QSO.

(A fellow who does a lot of QSL managing for rare DX stations told me once that more stations are not found in a DX log because of either using local time instead of UTC or because they got the date wrong than for any other reason. Although he checks for wrong dates and thus salvages a lot of contacts for stations, he thinks a lot of managers don't bother or just don't have the time to check. Ever since he told me that, I've been extra careful to see that the date is rendered correctly in my log book, being sure that I don't accidentally enter the log late when it's already tomorrow in UTC.)

Well, that's a good one. I listen on my frequency. There are four or five stations calling the EX3 now. There, he comes back to the KA4 that was calling him on the same frequency as whoever the other station was. The EX3 gives him a 589. That's two S-units better than my report was! But I got the EX3 first, because I moved away from the QRM. Maybe I learned something here. I start tuning up the band again. Conditions seem pretty good tonight; maybe I'll get lucky again.

20 METERS: THE QUEEN OF THE DX BANDS

There is a lot of DX to be worked on any of the HF bands. Twenty-meter capability is not necessary to achieve a high country total, but it helps! No matter at what point we are in the sunspot cycle, there is DX to be worked on 20 at some hour every day, with only an occasional lapse of a day or so when massive solar storms take over. Twenty meters is the DX crossroads of the Amateur Radio world. At the high point of the solar cycle, 20 is often open around the world around the clock—the DXer's ultimate dream. And, at the bottom of the cycle, especially in the summer months, 20 is often the only band where everyone has a real chance of working DX. If there is going to be a path open to a remote DX spot that is to be workable during difficult conditions, 20 meters will almost always be the band that opens the path.

All of this, of course, means there are a lot of serious DXers on 20, many with very large antenna arrays and using maximum power levels. But even the station with modest capabilities can work a great deal of DX on 20 meters. There assuredly are people on the DXCC Honor Roll who got there using 100 watts or less with simple antennas, such as dipoles or verticals. You can, too, especially on CW.

The best times of the year on 20 are the spring and fall, during which time the sunrise and sunset times are the best for long-haul DX, but with openings to all parts of the world at various times. In summer, 20 is poor in the four to six hours around local noon time, and often open the rest of the time. In winter, 20 is excellent from sunrise to sunset, and often some hours into the evening. And, always watch for long-path openings around sunrise and sunset! (See the sunrise/sunset tables in Chapter 17.)

Amateurs interested in learning how to become effective DXers should spend a major part of their time on 20 meters (assuming that your license class permits you to operate on the band). Although all of the HF bands, including 160 meters, offer at least some DX opportunities, the reliability of 20 meters and the quality of the competition make it the ideal DX learning ground.

Back to the Bands

Well, that EX3 was a good one! The thrill of a new country always makes me feel great. I run upstairs from my basement to tell the family of my triumph, grab another cup of coffee and

head back to the shack. I'm going hunting again! We have a good band opening and I'm going to keep after it.

I slowly wind the transceiver knob higher up the band looking for the weak, watery signals that spell DX. There's one, not really so weak, giving out a report. I listen; when he says QTH ALMA ATA, I tune on. He's a UN7, and one that I have hanging on the wall. The funny thing is that Alma Ata is all of 10 miles from the border of EX3, but it counts as a UN7. I keep moving.

I come across a loud signal, FB KHANRI QSL SURE VIA UR MGR ES TNX NEW COUNTRY. Needless to say, thanking the other station for the new country makes me all ears! A new country for whoever the loud station is might make it a new country for me, too. He continues. 73 ES GOOD DX SK S21AC DE WB5FZH. Wow! S2, Bangladesh, a very rare country indeed. I listen carefully, as I look over my gear. Yes, the RIT is off, the keyer is at the right speed, and so on. And there the S21 is, a solid S6, coming back to the WB5, FB BILL TNX QSO ES 73 NW QRZ? DE S21AC AR. By the time he finishes sending QRZ, I'm ready for him. I hope no one else has noticed him.

I pause for a brief moment after he signs—and my mouth drops open. The frequency suddenly sounds as if it were the focal point of a volcano. My poor S meter is buried against the pin. There must be hundreds of stations calling the S21 all at once, all obviously desperate for the QSO. I shake my head sadly, knowing that there's no hope for me in this one. My barefoot transceiver and low little tribander just aren't about to beat out that mob. I don't even hang around to see who snags the S21, knowing that there's no way it is going to be me. After all, DXing is supposed to be the art of the possible. A happier thought occurs to me. With all of the interest in the S21, maybe there's some other good bit of DX lurking around somewhere—without a pileup.

The pileup is so strong that I have to move up more than 5 kHz to get clear of the pounding QRM. I start tuning again, listening carefully to each station. It seems that I was right about the S21 pileup attracting all the attention on the band; there are very few W stations around. Now, if I could only find a good one to catch without the competition! There's a weak CQ, one with a bit of warble to it. I dig him out of the mud, but it's only a UA9. Of course, "only" a UA9 means I have UA9 confirmed. I keep on looking.

The sound of RTTY tells me that I have tuned up past 14,070 kHz and into the digital area. I have a very quick look on up to 14,100 to see that there are no stray CW signals that should be inspected, then move the rig back to the bottom of the band. Almost immediately, I come across a fairly strong signal, but it's clearly of DX origin, as indicated by its watery sound (often marking a station on the other side of the North Pole). OK DR SHANTI QSL SURE VIA BURO 73 ES DSW SK 4S7WP DE UK0QAC K. Pay dirt! The UK0 is working a 4S7, Sri Lanka. But can I hear the 4S7?

I listen intently. Yes, there he is—and good copy, too. He's a clean 559 signal standing out against a relatively quiet band. As he signs clear with the Siberian, I quickly line up on him. I swing the antenna a touch to dead north, make sure my RIT is off, switch the VOX off, tune the receiver while keying the monitor so the monitor note is the same as the 4S7's (which is the same as the UK0's), switch the RIT and VOX back on and wait. Yes, there's the "dit dit" from the UK0. I listen on my frequency for a moment; no loud signals calling, but another watery signal starts calling the 4S7, probably another Siberian.

I start my call. 4S7WP DE W9KNI W9KNI W9KNI AR. I

make my call longer than normal, figuring that the other station calling is probably louder in Sri Lanka than I am. If all goes well, the curiosity of the 4S7 will make him want to see who the weaker station calling is. I listen. There is a pause, then I hear that most glorious of musical symphonies — my call coming back from a new country! My long call must have worked. W9KNI DE 4S7WP GM OM ES TNX QSO BT UR RST . . .

And so it goes. This DXing style is how the amateur with less than an outstanding signal gets a respectable DXCC country-total: by tuning, listening carefully to every station and being ready to pounce quickly whenever a new one is found. And, by recognizing a lost cause when he sees one—knowing when the game is over, and that it's time to move on to the next signal.

There is one other important lesson. The DXer quickly learns to tell if a signal is from a DX station or a local. But, if the station heard is not DX, the DXer must always be aware that the station heard might be *working* rare DX. On a good DX band, most QSOs are between two stations that are DX to each other. So, if you hear a VE or W in QSO on 20 meters and the band is open for DX, chances are good that that station is working DX, maybe even rare DX.

There are some telltale signs to watch for when listening to QSOs. When you find a loud station promising to QSL direct or through a manager, or asking for QSL instructions, be alert! If you hear a loud signal with a fist or voice that sounds nervous or excited, take note. There is a reason, and you want to find out what it is. Chances are good that someone is in QSO with a new one, and possibly one that would be new for you.

Tuning Techniques

You've noticed references to turning off the RIT and VOX as part of the routine of zeroing the other station's frequency. For the CW operator and would-be DXer, there is no skill more important than learning how to put your transmitter or transceiver exactly on another station's frequency. The frustration and downfall of many a potential DXer comes from a failure to understand the proper technique. DXers using split transmitter-and-receiver combinations can ignore this section, as can those with rigs having specific CW zero-beating switches. If you are fortunate enough to own such a rig, be sure to study the manual and affirm that you truly understand how to use it.

If you own one of the common American or Japanese transceivers that lack a specific zero-beating function, the following techniques will enable you to match, or zero beat, the frequency of a station. First, see that the RIT, and if you have it, the XIT (transmitter incremental tuning) is turned off. Remember, you should never engage in general tuning with the RIT on when you may need to zero beat a station. The RIT is to be used only after you are zeroed, not before!

If your transceiver has passband tuning, be sure the passband tuning is at the center of its range or not far off, so you're certain that you are not listening to the wrong side of the BFO (beat frequency oscillator) from where the transmitter offset is placed. Never tune with the passband on the wrong side, unless you fancy a career as an SWL. But be warned—DXCC is not offered to SWLs!

Next, turn off the VOX. On most rigs, you can now use your keyer to send a string of dits that will be heard in the CW monitor while the receiver continues to operate. The audio frequency (AF) of the monitor matches the offset of the CW note from your transceiver's transmitter in almost any transceiver built today. Adjust the transceiver frequency until the audio tone of the station you wish to zero beat is identical in tone to

the transceiver's CW monitor note. You are now at zero beat with the other station, and when you transmit, you are exactly on that station's frequency.

For your own good, please be sure you understand this material. If you are listening on the wrong side of the BFO, or using the RIT and failing to turn it off, you could be calling DX as much as 2 kHz off, a fatal error when the DX station is listening with a 500-Hz filter. (Additional information on zero-beating techniques can be found in *The Complete DXer*, referenced at the end of this chapter.)

The Morning Shift

I brush the sleep out of my eyes as I settle into the operating chair, shivering slightly in my heavy robe. The glare of the rising sun reflects off the snow and into my basement window. I turn the antenna to the southwest, set the transceiver at the bottom edge of 20-meter SSB and begin tuning up the band. It's obvious that the band is in decent shape; the low background rumble in my receiver indicates a high activity level on the band. Immediately, I come across a station speaking with a definite accent, almost surely Australian. He signs over to a WB3. Yes, it's a VK6 in Perth, Western Australia. He's up late. I keep tuning.

There's a strong signal with a pure midwest American accent. He's talking about the weather in a very calm and unexcited voice. I go on. I trip over a small pileup of stations. Must be some DX there; yes, I hear a station coming back. Wonder where he is? Oh, right, he's a P29 in Papua New Guinea. Not very strong, either. But then I remember that my antenna is pointed southwest for the long path into the Middle East, rather than a bit north of west as would be correct for a P29. Since I already have a P29, I keep on moving.

There's a station passing out a report, and the operator has a different accent. Ah, it's a 3B8 on Mauritius in the Indian Ocean. I have one confirmed from a QSO on 40-meter CW (one of my prouder moments on that band), but I do not have it on SSB. I check the bearing on my great-circle map and glance at

The Chinese Radio Sports Association conducts its first DXpedition from Scarborough Reef in the South China Sea.

the antenna rotator. I move the antenna a hair to the north, so that my bearing is a little south of west, giving me the correct bearing for a long-path shot at 3B8.

He signs with his last QSO, and I call, "W9KNI, W9KNI, OVER." I listen. There are several other stations calling him. I can still hear him, although I cannot yet copy him because of the QRM (interference). The frequency clears. "5 and 8, 5 and 8, KB9IY go ahead, please."

I listen; yes, there's KB9IY. He lives not too far from me, and is very loud. He runs a linear, and I'm not disappointed at getting beaten out. He is dead on the frequency of the 3B8, so I don't need to move. He gives the 3B8 a report, and signs clear all in one transmission. That's good; apparently the 3B8 is interested in working a lot of stations, so I have a better chance of getting him. He clears with KB9IY, and calls QRZ.

I call him. "Whiskey Nine Kilowatt, Norway, India. W9KNI, Over," and listen again. He's already back to someone. Yes, another W8. He turns it over to the W8, who is weak, and coming through on backscatter. I am able to copy him well enough to realize that he is nearly a kilohertz lower than the 3B8's frequency, and I confirm it by tuning him in with the RIT control temporarily engaged. Quickly, I note the frequency displayed, kill the RIT and move my dial so that my frequency is the same. Just as I do, the W8 turns it back to the 3B8. I move the RIT back to copy the 3B8 properly. He signs clear with the W8, and I call—transmitting on the W8's frequency.

DX Operating Procedures and Guidelines

For many years DX operating procedures have consisted of split-frequency operation and transceive-type operation. Recently, several other methods have emerged which, if not used properly, can cause frustration, unhappiness and undue interference. These operations are known as lists, DX nets and roulette.

Split-Frequency Operation — Best Procedure of All

Split-frequency operation is, by far, the best method to use. It results in less confusion, interference and frustration than other procedures. In split-frequency operation, the DX station transmits on one frequency and receives on another. This method gives the highest QSO rate, allowing more amateurs to contact the DX station. Anyone considering going on a DXpedition should plan to take proper equipment along so that split-frequency operation can be used. Any DX operator planning to purchase a transmitter-transceiver should purchase one that has split-frequency capability or has the provisions for use of an external VFO.

It is recommended that a DX station operating phone transmit outside the US phone subbands, when possible, and "listen up" (listen higher in frequency) for US stations and "listen down" (listen lower in frequency) for other non-US stations. Examples: A phone DX station transmits on 14.190 MHz, listens for US stations from 14.200 to 14.230 MHz and listens for others from 14.180 to 14.160 MHz. A CW DX station transmits on 14.020 MHz and listens up from 14.025 to 14.035 MHz.

No more bandwidth should be specified than the absolute minimum that permits rapid separation of the calling stations. Not everyone on the band is interested in calling the DX station, and the interests of others should be respected.

Transceive Operation—
Requires Experience and Control

In this method of DX operation, the DX station transmits and receives on the same frequency. This takes less band width than split-frequency operation, but more so than operating split. Its success depends on operator control and experience. There are three important points that lead to a successful transceive operation: (1) Experience and maintaining proper control; (2) A good to excellent signal; and (3) Knowing when and how to divide a pileup.

Handling a pileup is an art, and you must maintain firm control of the situation and know what to do to regain control should it be lost. There are some DX operators who can handle a pileup no matter how big it gets. But, there are times when it is advantageous to divide the pileup in order to reduce it to manageable limits. There are many good methods now in use to divide a pileup. The most common method is by call districts.

Example: The DX station requests calls from the first US call district. After working a specific number (usually three to 10), the DX station requests calls from the second US call district, moving on to the other areas until he or she is finished with W0s. Then he requests non-US stations, working a specific number of these.

Whatever sequence is used, it can be repeated as long as the DX station desires. However, once a sequence is started, it should be continued until completed. This gives everyone a fair chance to have an opportunity to contact the DX station.

If the DX station's time is limited, the number of QSOs per call area can be reduced to, for example, three QSOs per district instead of 10. Whichever way the DX station decides to divide a pileup, he must be very firm and not talk to stations out of turn, even though some will call out of turn. To do so encourages others to follow, leading to on-the-air bickering, and, in short order, chaos. To prevent stations from calling out of turn, the DX station should announce often that he or she is dividing the pileup and the method being used to do so.

Another important factor in a successful transceive operation is that the DX station should have a good signal, so that calling stations will know who the DX station has acknowledged. Weak DX stations tend to be overcome by their own pileups and no one calling can be sure who has been acknowledged, if anyone. When this happens, QSO rates go down and confusion takes over. Then, the best approach for the DX station is silence until the pileup reduces itself to a low level or goes away. The DX operator regains control by announcing that he or she is going to divide the pileup and the method that he is going to use. Again, whichever method the DX operator uses, he must stick to it and give everyone a fair chance to contact him.

DX Networks—More Orderly, But Slower

DX nets have been set up for many purposes. Some are for certain areas, club members, and so on. Some are for working DX only. They range from fully "open" nets for all to those controlled as to who can check in and when. For the most part, the DX operators who participate in these DX nets are permanent residents. They check in regularly and, with a reasonable effort, almost anyone can obtain a contact with them. If the DX station desires to participate in these DX nets, he or she has the right to do so.

Any net operated for the sole purpose of working DX should be operated in such a manner that all non-DX check-ins are given a fair chance to work the DX. Many of the guidelines used in transceiver operation can be applied to DX net operations. Rare or semi-rare DX stations with limited time should refrain from using DX nets because of the resulting low QSO rate. DX nets are helpful to those DX operators who do not have the proper equipment to operate split frequency and/or the experience to operate transceive only.

I dump my call in, "Whiskey Nine Kilowatt, Norway, India, W9KNI, Over." The smoke clears as he is giving a report—and my chest swells with pride when he turns it over to me. A new one on phone! I log in the details and run up the stairs to grab another cup of coffee. A few minutes later, I'm tuning again.

As I move up the band, I hear a good deal of DX, but nothing really juicy. I was hoping to nail one of the Middle East stations that I need, such as a 9K2 or an HZ1. All I hear are 4X4s, a loud UA6 and several other East Europeans. Obviously, the band is open to the Middle East, but there are no players for me to chase. I finish tuning up past 14,300 and move back down to the bottom of the phone portion of the band.

I find a French station calling a W6, apparently on schedule,

and about where I found the VK6 before. Just above him, I run into a pileup, with all stations on a single frequency. The QRM clears rapidly, and I hear the DX stations come back to a W4 with a report.

The DX station has a big signal, running well over S9, which would help explain the size of the pileup. The W4 responds with thanks, but without signing the DX station's call. The DX station comes back saying, "Roger, 73. QRZ 4s from HZ1AB."

The frequency explodes with stations calling the HZ1. This is one that I need. If he is going by call areas, he's sure to be on the 9th call area sooner or later. For a new country like that, I'm going to sit and wait—and see if I can learn anything about this fellow's operating habits in the meantime.

Fig 5-1—An example of split-frequency operation. We are trying to work TL0BQ, Central African Republic, who is transmitting on 14.195 MHz. TL0BQ says he is listening for stations to call him from 14.210 to 14.240 MHz. We listen a bit and find: W0UA calling at 14.212, N4NW calling at 14.228 and N9MM calling at 14.237. We write these calls and frequencies down. Then, when TL0BQ works on of the stations that we are using for a benchmark, say N9MM at 14.237, we can tune our transmitter quickly to this frequency and have a good shot at working TL0BQ

Lists—Limited Use Only

List-style DX operations are, in some cases, the only way to work a DX station. However, it should be understood that lists should be used only when other DX operating procedures are not possible. Here are some guidelines that can make list operations run smoothly, be fair to almost everyone, and reduce interference to all.

Conditions under which lists might be used: (1) DX operator cannot speak English or understand it well or at all; (2) DX operator is inexperienced; or (3) DX station has a very weak signal because of poor antenna or location, low power or poor propagation.

Take a list for a DX station only if he asks or desires to do so. Do not pressure him. If it is determined that he is inexperienced in DX operating procedures, one could suggest a list operation, but leave the decision up to the DX station.

Take lists only for real-time operation. Old lists taken on a previous day or on different bands should not be used. Many on these lists do not show up, leading to frustration to those on frequency. It wastes time. Pre-arranged lists should not be used. Lists that are not accessible to all who can hear the DX station should be avoided.

Only stations designated by the DX station to take lists should do so. When you take a list for a DX station, propagation differences should be taken into account. If necessary, the MC (master of ceremonies or list-taker) should direct one or two alternates in different areas to pick up the stations whom the MC cannot copy because of propagation. This gives all a fair chance to get on the list.

The MC should have excellent copy on the DX station, so that control can be maintained at all times. Lists should conform to the DX station's desires as to content—number per call area, length and so on. For example, take a short

list, about 20 stations or so, run them, then come back and pick up a second group of 20. Repeat this procedure for as long as the DX station is willing to do so.

Never relay signal reports and call-sign corrections. If it is really a QSO, the DX station should be able to copy the call and report of the station he is working without the help of the MC.

The MC should be clear and concise in giving instructions as to what he and/or the DX station desires. The instructions should be repeated often to reduce confusion and to inform those who have just arrived on the frequency.

When working a DX station in a list operation, conform to the DX station's desires as to length of QSO. If the DX station is giving only signal reports, do likewise and do not give your name, QTH or a weather report. If the DX station gives reports and his name, be courteous and give him the report and your name; but in any case, conduct the QSO clearly and quickly, so others can also have a chance to work him. Think of others.

Remember, list operations should be used only when required, not as a replacement for other unaided DX operating procedures.

Roulette—Should *Not* Be Used

At times, a DX station may appear on a predetermined frequency and announce that he will listen for callers within a certain band segment and will answer callers on their own frequency. This shotgun technique is referred to as *roulette*. This method is not at all recommended for DX operating and should be highly discouraged. It causes confusion, encourages bootleggers, has a very poor QSO rate, results in questionable QSOs and produces a lot of unnecessary QRM.—*John Kanode, N4MM*

As he keeps working 4s, I begin to realize that he is the loud station with the American Midwest accent that I heard earlier! I blew it. The pileup I heard earlier was probably far smaller then. I would have had a much better chance of getting him into the log.

Well, you live and learn. At least I hope I do. I should have never passed over that station without seeing who and where he was. As I listen, he describes his station; stacked rhombics at 150 feet and a full-bore linear. No wonder he's so loud. Finally, he goes on to the W5s, and a new pileup jumps into being on the frequency. I listen as he takes six W5s, and moves on to the 6s. I cringe mentally when he calls for the W6s, expecting the frequency to be obliterated. But, strangely, there are no W6s calling. The band must not be open out there yet. Finally, he works one lone W6 operating portable 4, and then goes on to the W7s.

After only a few W7s, he moves on to the W8s, and the situation returns to normal—seemingly hundreds of eager DXers, all trying to score. I begin to get nervous about my chances. With the fellow working people as long as I have been listening to him, and with the size of the pileup calling in, by now there's sure to be a load of W9s in wait. And I don't have the signal to beat out hordes of fellow W9s.

After six W8s, our turn comes; my fears prove to be well founded. I pause for a moment before calling, and see my S meter move over to the pin. I call frantically, shouting into the microphone like everyone else, and, like for most of the W9s calling, to no avail. After six QSOs, he's on to the suffering WØs. I can't complain; I had my chance. But, I mentally kick myself. I know that I had a much better chance earlier, and I failed to take advantage of it.

Propagation-Determining Tools

Great-circle bearings and long-path propagation are two of the more interesting technical elements of DXing. Most amateurs were taught about great-circle routes in school. Strings were pulled across globes to demonstrate that the shortest route between two points on the Earth is rarely a single compass bearing. Instead, it's a constantly changing bearing going across a globe. Remember? Our signals travel through the ionosphere in the same manner, seeking the shortest route, which is a great-circle bearing.

Special maps utilizing an *azimuthal equidistant projection* possess the unique property of showing great-circle bearings and paths as a straight line. Also, they show distances accurately from the center of the map to all points. (A set of fine azimuthal maps produced by Bill Johnston, N5KR, appears in Chapter 17.) A Great Circle World Map designed especially for DXers is offered by the ARRL. This map is constructed around the geographical center of the United States and offers accurate beam headings worldwide for amateurs in the contiguous states and the southern parts of Canada. It includes country prefixes and zone information, and should be prominently visible on the wall of any DXer's station.

To use these maps for determining the short-path bearings to any point on the globe, simply draw a real or imaginary line from the center of the map to the point where you wish to work. Extend the line to the edge of the map and read the bearing from that point. That is the correct direction to point your antenna.

An alternative approach is to make or purchase a list of computerized beam headings from your location. Computer programs for this are in several ARRL publications, or customized lists are available from several suppliers advertising in *QST*. The *Radio Amateur Callbook* includes lists of bearings

for a number of major American cities.

Long-path propagation, mentioned above, is one of the more fascinating aspects of DXing. Of course, when we transmit, our signals travel in a straight-line great-circle path. What happens when our signals travel to the point on the opposite side of the globe, 12,000 + miles away? Do they stop dead, or fall off the edge? Of course not; they keep on going in the same direction exactly as they were, sometimes a great deal farther, still strong enough for communication. And, often, when they travel in such a fashion, the propagation to that farther point via short-path propagation is nil. Indeed, thanks to seasonal and diurnal variations, the long-path propagation to some points is often more reliable at certain seasons than short-path propagation!

For example, from the American Midwest in the winter months, propagation on 20 meters to the Middle East and Southern Asia is best in the hours after sunrise, with the station antenna pointed southwest, on the reverse bearing for the Middle East, and south for Southern Asia. In summer months, Southern Africa is often best worked in the morning hours with the antenna aimed straight west. Although long-path propagation is most consistently usable on 20 meters, especially for stations with relatively modest antennas, the phenomenon can be used to advantage on all bands from 160 to 6 meters. The bearing for long-path propagation is 180° opposite that of the short-path bearing.

Sometimes, both the long and the short paths are open, especially in the spring and fall seasons. If you are listening to a DX station on the short path and notice an echo on his signal, try turning the antenna to the long path. If long-path propagation is superior, signals can show dramatic improvement. Your opportunity to work him is equally enhanced!

There are two other propagation anomalies that occur reliably enough such that DXers should be aware and take advantage of them. They are *grayline* propagation enhancement and *crooked-path* propagation. Grayline enhancement takes advantage of the fact that propagation conditions are best in twilight conditions, thanks to solar absorption dropping greatly in twilight while upper atmosphere ionization has not yet weakened enough to cause a reduction in the MUF (maximum useable frequency).

At any given moment, there is a circle or band of twilight

You don't need four hands to work a DX contest, but it helps! It looks like Carlos Rodriguez, SMØKCO, is well equipped for the job. (*photo by Henryk Kotowski, SMØJHF*)

girdling the globe somewhere. All stations in this twilight band are capable of communicating more easily with other stations in the twilight band—especially if they can point a directional gain antenna down the band. Stations in the twilight band will often find communication possible over paths as long as 8000 to 10,000 miles on three or even four bands at the same time. (A comprehensive table of sunrise-sunset times for each DXCC country appears in Chapter 17.) Very useful tools demonstrating grayline times anywhere on the globe are also available from Xantec Inc, PO Box 834, Madison Square Station, New York, NY 10159, and MFJ Enterprises, PO Box 494, Mississippi State, MS 39762.

Crooked-path propagation is a unique phenomenon that also can be useful to DXers, especially those with directive antennas. Sometimes, signals will be observed coming in from directions that are unrelated to short-path or long-path bearings. Almost always, such signals are weak and watery, but they are workable. Often, crooked-path propagation is combined with grayline enhancement. For example, in December and January, at sunrise in the American Midwest, it is sunset in India and Sri Lanka. Aiming a 40 or 20-meter Yagi southwest, pointing down the grayline belt, will often turn up workable signals from that part of the world, even though the great-circle bearings are far removed from the actual antenna bearing. Indeed, the southwest crooked-path opening on 40 meters in winter can be one of the most exciting DX activities of a season.

Again, crooked-path openings can be found on other occasions. Another example can occur during the late fall and early spring contest seasons. Stations in North America, anxious to add multipliers to their score, will look for Europeans on 10 meters when the only band opening is to Southern Africa. By pointing directional antennas at the bearing to Southern Africa, the DXer will sometimes hear faint Europeans who can be worked to advantage, while turning the antenna to Europe results in the loss of reception. Note that this opening is usually very weak, and one that requires better-than-normal stations at both ends. Crooked-path openings are generally more useful on the lower frequencies.

International Third-Party Traffic— Proceed with Caution

Occasionally, DX stations may ask you to handle a third-party message to a friend or relative in the States. This is all right as long as the US has an official Third-Party Traffic Agreement with that particular country, and the traffic is noncommercial and of a personal, unimportant nature. During an emergency, our State Department will often work out a special temporary agreement with the country involved, but in normal times, never handle traffic without first making sure it is legally permitted.

As of January 1996, US Amateurs May Handle Third-Party Traffic With:

C5	The Gambia	V2	Antigua and Barbuda
CE	Chile		
CO	Cuba	V3	Belize
CP	Bolivia	V4	St Christopher and Nevis
CX	Uruguay		
D6	Comoros	V6	Micronesia
DU	Philippines	V7	Marshall Islands
EL	Liberia	VE	Canada
GB*	Great Britain	VK	Australia
HC	Ecuador	VR6**	Pitcairn Island
HH	Haiti	XE	Mexico
HI	Dominican Republic	YN	Nicaragua
HK	Colombia	YS	El Salvador
HP	Panama	YV	Venezuela
HR	Honduras	ZP	Paraguay
J3	Grenada	3DA	Swaziland
J6	St Lucia	4U1ITU	ITU, Geneva
J7	Dominica	4U1VIC	VIC, Vienna
J8	St Vincent	4X	Israel
JY	Jordan	6Y	Jamaica
LU	Argentina	8R	Guyana
OA	Peru	9G	Ghana
PY	Brazil	9L	Sierra Leone
T9	Bosnia-Herzegovina	9Y	Trinidad and Tobago
TG	Guatemala		
TI	Costa Rica		

* Third-party traffic permitted between US amateurs and special-events stations in the United Kingdom having the prefix GB only. (GB3 stations are not included in this agreement.)

** Since 1970, there has been an informal agreement between the United Kingdom and the US, permitting Pitcairn and US amateurs to exchange messages concerning medical emergencies, urgent need for equipment or supplies, and private or personal matters of island residents.

Please note that Region 2 of the International Amateur Radio Union (IARU) has recommended that international traffic on the 20 and 15-meter bands be conducted on the following frequencies:

14.100-14.150 MHz
14.250-14.350 MHz
21.150-21.200 MHz
21.300-21.450 MHz

The IARU is the alliance of Amateur Radio Societies from around the world; Region 2 comprises Member-Societies in North, South and Central America, and the Caribbean.

Note: At the end of an exchange of third-party traffic with a station located in a foreign country, an FCC-licensed amateur must transmit the call sign of the foreign station as well as his own call.

Checking 15 Meters

It's 10 AM on a fine, rainy April morning. It's fine because I can sit in the shack most of the day chasing DX instead of feeling guilty about undone yard work. I have the antenna on Europe, and I'm looking across 15 meters, hoping for a pleasant surprise—one of the elusive Middle East stations that I need. I have already checked out the phone band, and now I'm down at the other end to have a look on CW.

The band is good; there were plenty of Europeans, with northern Africans mixed, but nothing that I need. Ten meters is essentially dead, with only a few Central and South Americans coming through. Again, there's nothing that I need. I start working my way up from the bottom of the CW band.

One of the nice things about 15 and 10 meters is that there usually is not much short skip, so that the great majority of stations heard, especially the stronger ones, are DX. Of course, that makes things tougher on the "tune for the pileup" crowd, but for DXers willing to listen carefully to the weak signals, it can be a real advantage. As I move through the band, I unearth loads of DX signals, mostly Europeans. But again, nothing that I need

At 21,055 my luck changes. I hear a station, about 569, running antenna tests with another station. The station doing the tests carries the funny little noises propagation can add to a signal that has come a long way. They're subtle shifts of frequency caused by slight Doppler effects from different paths and reflection areas propagating the signal. On top of that, he has a distinctive chirp on the first letter of each word; probably a power-supply regulation problem.

After several repetitions of NW ANT 1 ANT 1 NW ANT 2

1996 100 Most-Wanted DXCC Countries

This list is compiled annually by the ARRL DXCC department based on the number of stations claiming credit for working various countries.

Rank	DXCC Prefix	Country	Rank	DXCC Prefix	Country
1	P5	North Korea	51	YA	Afghanistan
2	E3	Eritrea	52	XU	Kampuchea (Cambodia)
3	VKØh	Heard Island	53	YI	Iraq
4	7O	Yemen	54	3C	Equatorial Guinea
5	A5	Bhutan	55	VK9w	Willis Island
6	Z3	Macedonia	56	TY	Benin
7	5A	Libya	57	VK9X	Christmas Island
8	VU/a	Andaman & Nicobar Island	58	VK9C	Cocos-Keeling Island
9	T9	Bosnia-Herzegovina	59	ZK3	Tokelau Islands
10	FR/T	Tromelin island	60	CEØZ	Juan Fernandez Island
11	OK	Czech Republic	61	3B9	Rodriguez Island
12	VKØm	Macquarie Island	62	A6	United Arab Emirates
13	SV/A	Mount Athos	63	JX	Jan Mayen
14	ZL8	Kermadec Island	64	ET	Ethiopia
15	OM	Slovak Republic	65	9N	Nepal
16	FR/G	Glorioso Island	66	TI9	Cocos Island
17	VU/l	Laccadive Islands	67	1S	Spratly Islands
18	3V8	Tunisia	68	YK	Syria
19	ZS8MI	Prince Edward & Marion Islands	69	CEØX	San Felix Island
20	ZL9	Auckland & Campbell Island	70	5X	Uganda
21	3B7	Agalega & St Brandon Islands	71	8Q	Maldive Islands
22	3Y	Bouvet Island	72	T5	Somalia
23	FR/J	Juan De Nova, Europa	73	3D2r	Rotuma
24	FT5W	Crozet Island	74	KH4	Midway Island
25	FT5Z	Amsterdam & St Paul Island	75	ST	Sudan
26	TN	Congo	76	TL	Cameroon
27	VP8sg	South Georgia Island	77	JD1	Minami Torishima
28	TT	Chad	78	ZD9	Tristan Da Cunha & Gough Is
29	3CØ	Pagalu Island (Annobon)	79	ZC4	UK Bases On Cyprus
30	FT5X	Kerguelen Island	80	KH5	Palmyra, Jarvis Islands
31	PYØs	St Peter and Paul Rocks	81	D6	Comoros
32	9U	Burundi	82	4K	Armenia
33	XW	Laos	83	XV	Vietnam
34	XZ	Burma (Myanmar)	84	CYØ	Sable Island
35	PYØt	Trindade & Martin Vaz Islands	85	S9	Sao Tome & Principe
36	HKØm	Malpelo Island	86	9X	Rwanda
37	9A	Croatia	87	XT	Burkina Faso (Upper Volta)
38	STØ	Southern Sudan	88	R1M	M-V Island
39	S5	Slovenia	89	ZK1	North Cook Islands (Manihiki)
40	T31	Cent. Kiribati (Brit. Phoenix)	90	3A	Monaco
41	KH5K	Kingman Reef	91	A7	Qatar
42	3D2c	Conway Reef	92	KH9	Wake Island
43	5R8	Madagascar	93	5V	Togo
44	S2	Bangladesh	94	3X	Guinea
45	VP8so	South Orkney Islands	95	C2	Nauru
46	EP	Iran	96	YVØ	Aves Island
47	SØ	Western Sahara (Rio De Oro)	97	VP8	South Shetland Islands
48	1AØKM	Sov Military Order Of Malta	98	JD1	Ogasawara
49	VP8ss	South Sandwich Islands	99	T8	Belau (W. Caroline Is.)
50	T33	Banaba (Ocean Island)	100	FOØc	Clipperton Island

ANT 2, he signs, WA4RNE DE 5R8AP KN. Oh boy, a 5R8 on Madagascar! An all-time new one!

I listen carefully for the WA4. Yes, there he is, about 559, almost dead zero with the 5R8. The WA4 is telling the 5RX that the two antennas appear about the same, which was the same report that I would have sent, given the chance. And that's a chance I hope to get. I check my transmit frequency, getting zero beat with the WA4, and can only sit and wait for the end of the QSO. The 5R8 and the WA4 seem to be friends, chatting back and forth easily, and I begin to suspect that they have a regular schedule. I make notes on my little blackboard of the time, frequency and date; if they are meeting a schedule, next week I'll be meeting it as well, unless I get lucky today.

After what seems like an interminable wait (actually about 10 minutes), they sign clear. I listen for a moment before I start my call to see if I have much competition on the frequency. I discover there can't be more than 15 or 20 stations calling that I can hear, with doubtless more that I can't. Not knowing what else to do, I call, but I have a nasty feeling about how effective my call is going to be. I listen; there, yes, he's coming back to someone, yes, an F3 in France. Well, that's not really surprising; I expect a 5R8 is pretty good DX almost anywhere. He is sending to the F3 in French. He now turns it back to the French station. I look, and almost immediately find the F3, about 500 Hz above the 5R8's frequency.

That's interesting; apparently the 5R8 is willing to tune off his frequency. He listened up a half kHz. I can hear the F3 pretty well; he's about the same strength as the 5R8. That means that likely everyone in the States at least can hear the F3 as well as I do, and I have no doubt that the serious DXers in the pileup will be zero-beating the F3 station's frequency. And I'm virtually positive that at least several of the stations calling will be significantly louder in Madagascar than I am. In short, for me to call on the F station's frequency is probably a waste of time.

Hmmm. The 5R8 tuned higher to avoid the bulk of the pileup. He has several choices, assuming that he intends to stick on the frequency and deal with the pileup. He can tune higher yet to work the next station, he can try to fish someone out of the pile on the French station's frequency or he can tune lower. If all goes well, he will avoid working any more stations on his frequency, or the QRM there will become unbearable. I decide that my best bet is to call him higher, perhaps another 500 Hz above the French station's frequency.

I move my transmit frequency, and wait for the next call. The 5R8 signs with the F3, and at least 40 of us call. I give him a 1×3 (5R8AL DE W9KNI W9KNI W9KNI AR) and listen. Yes, he's coming back to someone, but the call is covered up by those still calling him on his frequency. He gives a report, his name and QSL manager, and turns it over to a KA1. I let my breath out, as probably do the 40 other operators in unison. I look for the KA1. Yes, there he is. My transmit frequency is almost exactly the same as that of the successful KA1. I was on the right track. I guessed right, but so did the KA1. When it came down to a horse race, I got beat.

What to do now, for my next call? That one is easy; chances are the 5R8 will again tune a bit higher. But now that pattern is more obvious to a lot of the operators calling him in the pileup, and I have no doubt that a number of them will be calling the 5R8 another half kHz above the KA1. So if I want to have my signal stand out in solitary splendor, I need to try something a bit different.

I could call 1 kHz above the KA1. That would be enough to move me above the new pileup group. Or, I could go back down to where the F station made the previous contact, figuring the 5R8 to tune down from the pileup. This is more likely if he doesn't have an external VFO and has to rely on his RIT to tune through the pileup. But I obviously don't know what he's using for equipment. I decide to try calling 1 kHz higher.

He finishes with the KA1. I pause a moment and listen on my transmit frequency before I start my call. It appears to be quiet. Maybe this time it's my turn. I give him a 1×3 call again, and listen. And again, he's back to someone, and again buried by stations still blindly calling him on his own frequency. That distinctive chirp on the opening character of a word makes his signal stand out from the rest, and that helps find him. The smoke clears, and the whole lot of us wait to see who made it through. Darn, it's a WB2.

I look for the WB2's frequency. There he is, down where the F8 station worked him two QSOs ago. That suggests that he is probably going to tune a fairly narrow range above his own frequency, probably using RIT, unless he also tunes below his own frequency. If I'm right, and he is using an RIT control, that's his only other option. But he hasn't yet worked anyone below himself. I decide to try calling him above his frequency again, about a half kHz above the WB2, and I move my transmit frequency. There, he's clearing with the WB2; he's stepping up the tempo a bit.

I give him another 1×3 call; obviously he knows his own call, and there is no doubt about who the pileup on and above his frequency is for. I listen. Once again he comes back, but to whom? We all wait.

And . . . I wait in vain. A K3 puts him away that time. I find the K3's frequency. Darn, almost dead zero with me. I guessed right again, but couldn't pull the trigger. Let's set up a half kHz

above the K3; that's about the frequency where he worked the KA1, and the highest frequency that he worked anyone on. I begin to be distinctly unhappy that all he has worked so far has been East Coast stations; that does not bode well. It suggests that the East Coast has a definite propagation edge on the Midwest. But there, the contact is briefer. Good, he's going faster; he's clearing with the K3. I call again.

Okay, he's back. The pileup sweats out the wait to see who got him; hello, it's a WBØ. Where is he? I look all around the area above the 5R8, but can't find the WBØ. Darn! I look below the 5R8, and there, nearly 1 kHz below the 5R8's frequency is the WBØ, "fat, dumb and happy" with a new one. I grunt as I move my transmitter a half kHz below the WBØ. Sure hope that I'm the only operator to find the WBØ. If I am, my chances ought to be pretty good now. I give a 1×3 call.

There, he's back. The gang on his frequency have finally learned enough to try something different; I can hear him spelling out a W5's call. Phooey! Okay, where's the W5? I look below again; he and I were real close in frequency. I move down 1 kHz. Wait a minute. What's the 5R8 saying now? He's going QRT. Shucks. Yup, there he goes with a CL.

I give him one last long call, hoping against hope that he will decide that he needs a W9 in the log worse than he needs to go to sleep. But even as I call, I know it's futile. And I'm right. I sigh and go over to the little blackboard to add some notes on the 5R8. That was the first time that I've ever heard him. Although I didn't get him, I learned a lot. I know what his signal sounds like and what his pileup operating habits are. Next time I'll be ready. It looks like he may have a sked with that WA4 next Sunday—and I'll be there, ready and waiting. My experience of how he deals with a pileup will give me a big edge over people who haven't a clue. There's hope!

Except in years near the bottom of the sunspot cycle, 15 meters is one of the best amateur DX bands available. Every band has its own "flavor," and in some respects, 15 is somewhat bland. It lacks the reliability of 20 meters and the seductive appeal of 10 meters. But it is a solid performer with plenty of room and few commercial intruders. Antennas can be smaller and still be effective. And short skip is rarer, indicating that most strong stations on the band are DX.

In Eastern Europe, club stations like YO3KWT in Bucharest, Romania, are active in contests. Lix, YO3NP, mans the controls.

Fifteen rarely opens before sunrise in the best of times, and is usually disappointing in mid-summer, except in the early evening hours. In the better parts of the sunspot cycle, it is a terrific DX band. And, it is a great band for Novice and Technician Plus licensees to get their first taste of CW DX. Some have even attained DXCC on 15 meters, and no doubt more will.

Working a New One

I'm just getting ready for bed on a mid-August evening and decide to have a last look at the *DX PacketCluster* screen before I retire. I look at the last entry, and my heart skips a beat. "C97/OH3QT 14195.0 Listening Up." I glance at the time the report was logged—seven minutes ago. And the originating call is a WB8. I groan. That means the whole country is already reading the entry by now. But a C9! Mozambique! That would be an all-time new one. And that should be a good path this time of year.

I hit the master switch and the linear switch with one hand while I scrape my operating chair out from under the desk with the other. As the linear warms, I snuggle the headset comfortably over my ears, and swing the antenna straight east for the C9. If that's not the exact bearing, it certainly is close enough. I move the transceiver down to 14,195. Yes, there he is, nice and loud.

"Roger, QSL via OH3QT. QRZ? Listening 14210 to 225, from OH3QT, portable C-Ninety-Seven."

Okay, he wants to do it that way. I kick the external VFO in, and move it up to 14,210 to have a look. Wow! I'm not the only operator to read the cluster screen before going to bed. I run smack into a wall of QRM. Most of it is not too loud; long skip must be in. I can't really copy any of the stations calling, and there sure seems to be a lot of them. I glance over at the linear. Yes, the ready light is on, and it is tuned for 20-meter SSB. I'm ready. I move the VFO experimentally to 14,220. I flip the VFO control switch back to the transceiver VFO, and listen back on the C9's frequency. He's calling QRZ ("Who's calling me") again.

I simply sign my call, "Whiskey Nine Kilowatt, Norway, Italy, W9KNI, over." There, he's back to another W9. Yes, it's W9DWQ, a member of my DX club. When I hear the OH3 say the W9-part of the call sign, I'm thrilled for a tiny moment, thinking that I had him. But it was W9DWQ. Anyhow, 'DWQ will be strong enough that I can easily find his transmit frequency.

I flip the switch to the external VFO, and look for 'DWQ. Yes, there he is on 221. I leave my VFO on him, and switch back to the OH3/C97. He clears with 'DWQ and I call again. He's back, to a WA5. He turns it over to the WA5; I flip to the external VFO, starting on my own frequency. Nope, not on my frequency anyhow. I tune a bit higher up, and find the WA5 about 2 kHz above my own frequency. Okay, the OH3 moved up 2 after he worked W9DWQ, so I need to move 2 up from the WA5's frequency. Hmmm, that puts me right on 225, the OH3's upper calling limit.

The OH3 signs with the WA5, and I call. Nope, a KB8 got him. Hope I can find the KB8 on backscatter. I flip back to the external VFO. There he is, right on my frequency. So I guessed right, but got beaten. Now what? If the OH follows his own instructions, he'll stay between .210 and .225. But he is moving about 2 kHz per QSO. So the question is, will he stop at .225 like he says, or ignore that and keep tuning higher out of his prescribed tuning area? Perhaps he'll tune down in 2-kHz increments, more or less, from 225. Or he may simply go back to 210 and start over again, working his way up the band.

I decide that (a) he probably means what he says, and

DX, Pileups and Common Sense

A couple of recent DXpeditions have generated more than the usual amount of mail. The common thread in the letters is that we have a real problem, and something needs to be done about it. The problem, as defined by most of the correspondents, is that individual DXpeditions are taking up altogether too much of the bands as they spread out calling stations so as to make it easier to pick out call signs from the pile. The problem usually is associated with split-frequency SSB operation, particularly on 20 meters, but the CW band has seen its abuses as well.

The reason this is a problem is that not everyone on the band shares the single-minded goal of contacting that particular DX station at that particular moment. Nets, ragchews and a host of other worthwhile activities are going on at the same time. These operators are understandably miffed when they are set upon by a horde of people shouting their call signs ad infinitum, sometimes without listening on their transmitting frequency. It's not deliberate or malicious interference, but it's certainly thoughtless. Tempers flare, there's retaliation, things escalate and we have another incident that wreaks havoc on Amateur Radio's image and does nothing good for our collective blood pressure.

The individual in the best position to do something about the problem is the DXpedition operator. He or she specifies the range of listening frequencies and the stations (divided by call sign numerically, alphabetically or geographically) whose calls will be answered, and otherwise manages the pileup. A talented operator can keep hundreds, even thousands, of stations in line while making contacts at the rate of four or more per minute, without taking up more than a few kilohertz of the band. It takes a rare combination of good nature, fairness and operating skill, but it can be done. The result is a well-ordered pileup that is a joy to behold, fun to be in and doesn't get in anyone else's way.

Is it too much to hope for that operators on rare DXpeditions will all measure up to that ideal? Probably, since the people with the time and money for such ventures aren't automatically the best operators (though there are a number who do mighty well). For the rest of us who may find ourselves on the other end of a massive pileup, though, a little common sense will go a long way. Nearly everyone in a pileup will follow instructions and not get out of line if things are being handled fairly and if contacts are being made at a reasonable rate. Nearly everyone not in a pileup will take pains to avoid causing interference as long as it's not "polluting" too much of the band.

How much is too much? Reasonable people will differ, but 10 kHz strikes us as being sufficient for nearly any situation, and over the years we've been on both ends a few times. Clearly, there's no justification for "listening 200 to 350," sending CW stations into RTTY territory, or being so nonspecific with instructions that the whole band gets clobbered. An expedition to a rare spot is a significant event in Amateur Radio, but everything else that goes on in our bands can't be expected to stop for it.

Those of us who stay at home aren't completely helpless, either. Expeditions to rare spots generally require financial backing, and the talent and temperament of operators being sent certainly deserves more than a passing thought when sponsorship is considered. DX organizations ought to be especially sensitive to this.

Will things get better? We certainly hope so. If not, the non-DX part of the amateur community is going to press the League, if not the FCC, for measures that will protect its interests. A little self-restraint by DXers will go a long way toward avoiding restraints designed by others.—*David Sumner, K1ZZ*

(b) that the QRM is mostly at the top end of his tuning range as stations discover that is where he has been making QSOs, so (c) I decide he will move back to 210 and start moving up again. So I set up on 210 and try there. I call, then listen. He's back to another W9. It's WD9AHJ; he's in my club, too, and lives nearby. Yes, I find him at 223. Darned! Okay, I guessed wrong, but Ron didn't. Good for him anyhow. Clearly the fellow is tuning down from the top now. I'll try 221.

I set up on 221 and make my call. Phooey! He's back to a WD4. He turns it over to the WD4. Well, I was within a few hundred hertz of him, so I was about right. I'll go to 219. I leave the external VFO in for a moment and can tell when the OH3 goes QRZ. The roar in my headphones is as perfectly orches-trated as if Georg Solti was directing the Chicago Symphony on frequency. I listen for an instant and hear several stations right on my frequency. I slide down half a kHz; it sounds clearer, and I call. Bingo!

"W9KNI, you're 5-by-9. W9KNI go ahead."

"Roger, from W9KNI. You're 5-and-9, thanks for the new one, over."

"W9KNI, Roger, QSL. QRZ from Ocean-Henry-Three-Quebec-Tango portable-Charlie-Ninety-Seven, listening 210 to 225."

I sit in my chair, stupidly grinning from ear to ear. I turn on the 2-meter rig and go to the DX repeater frequency, just in time to hear W9DWQ congratulating me.

SSB or CW?

During periods of above-average propagation condi-tions, Novices and Technicians will find more DX on 10-meter phone than on either of their 15 or 10-meter CW subbands. But, as go the sunspots, so goes 10 meters. When solar activity is low, the 15-meter Novice/Technician CW subband is more productive. Novices and Technicians should not overlook their CW subbands when the sun is active either, because some choice DX occasionally shows up there. (Beginners in other countries have to study the code, too.) And, because activity in the Novice 15 and 10-meter CW subbands is lower than in the 10-meter phone band, there is less competition and a greater chance for a QSO. Further, if you are really interested in DXing, you need a higher class of license and the ability to sling out the dits and dahs.

This opinion is not just the raving of an old-time CW nut! I could fill three books with the gory details of DXpeditions and other rare ones I have called for hours on SSB, when I just couldn't break the pileup. When the DX finally got on CW, I often got through on the first call. SSB receivers use wider bandwidths by necessity and all the other strong signals calling simply overload the automatic gain control (AGC) circuit. CW allows the use of much narrower filters. Also, a good CW operator develops the ability to separate signals in his brain. This ability is no mystery, because with CW we're dealing with single tones, not a mish-mash of voices.

If you have a General license or higher, you can operate on the other HF bands not available to Novices and Technicians. There, you should also concentrate on CW, especially if you don't have a big signal. True, some of the rarer DX stations sometimes operate in the Extra Class subbands, but a majority of those operators listen higher in the band, where General and Advanced class amateurs can transmit. If you get hooked on DXing, the first time you hear a really rare station working in the Extra Class subband, you'll be surprised how fast you can get your code speed up to snuff! Meanwhile, you can readily work more than 100 countries without going into the Extra Class subband.

What about SSB? True enough, there is plenty of DX on SSB. Almost every country I hear people fighting over on SSB, however, I have easily worked on CW (and usually on more than one band). Enough said. CW is a different means of communication and no one masters it overnight. Most of us old-timers couldn't wait to get on phone (in the days when Novices were limited to CW). Many of us found our way back to CW again, though, simply because we like it better. No matter how wild the pileup may be on CW, it just doesn't sound as bad as even a low-key pileup on SSB.

Even SSB-only operators have an incentive to get their code speed up, if only to pass the Extra Class exam. There is always something happening in the Extra Class phone subbands and you won't want to miss it.

Why is there so much DX on CW? Most American radio amateurs can afford a rig that works both modes, but this isn't true in many foreign countries. In many parts of the world, amateurs can't buy commercial equipment at any price and obtaining crystal filters and other parts neces-sary to build SSB gear is difficult, if not impossible.

You won't work these folks on any other mode. CW is harder at first; you know how to speak with your voice—now you must learn to speak with your key. I am not suggesting you give up SSB. Quite the opposite: I am hoping you won't give up CW. Being flexible is the mark of a true-blue DXer.

Getting By on CW

Let's say you decide to take the plunge and you're listening on a CW subband. Suddenly, you hear a bunch of stations sending their calls for a few seconds, spread out over several kilohertz. Then, just as suddenly, all is quiet. You tune lower in frequency and hear a weaker signal sending "5NN" ("N" is an abbreviation for "9"), but it sounds more like a machine-gun burst. There's your DX station. Who is it? Stick around, you'll hear the call sign soon enough.

Now he's calling another station, but you can only copy part of the call sign because he's going so fast. If it were your call sign he was sending, would you recognize it? If not, no sense adding to the QRM by calling him. Find another station. But for now, let's say you can follow what he's sending. There he goes again, he just sent "TU." That's his way of acknowledging the other station's report. He's listening up the band for another call now—yes, he just let loose with another burst. Wow, you really have to concentrate!

Why not listen above him, and see if you can find the station he's working. There's someone else, sending at the same machine-gun speed, sending his call at the end of his report. Okay, back to the DX station again. A few contacts later, you've got his routine figured out and you can copy most of what he's sending.

Assuming your license allows you to transmit on this frequency, why not give him a shout? You know about how long he listens before he picks out a call. And you know whether he tunes up, down, or randomly after each contact Here's a hint: If you send your call a little slower than he's sending, he might slow down when he calls you. Sometimes a slowly sent call sign stands out in a pileup of fast callers, too.

Okay, you say, here goes. "KR1S." Nope, someone else got him. Let's see, the station he called was a little lower in frequency than me. Guess I'll stand pat for one more call. Here we go again. "KR1S." Yikes! He just called me! Gosh, "5NN de KR1S TU." He sent "TU." I guess he heard me. Not exactly long winded, is he! Oh, yeah, what was his call sign? I figured it out a few minutes ago and wrote it down . . . here it is. Okay, put it in the log and don't forget the UTC time and date. Hey! I just worked a new one on CW! Gee, that wasn't too bad. Where'd I put that W1AW code-practice schedule, anyway?— *From* The DXCC Companion, *by Jim Kearman, KR1S*

"Roger. Thanks, Ed. Excuse me, I better get this into the log book before I forget what time it is. Got to thank the packet system for that one; I was about to go to bed."

I neglected to write down the QSO time for a good one once. The next day, I had to talk to the other fellows in the local "wolf pack" who had worked the station, so that I could guesstimate my log book entry. I was lucky; the guess was close enough that the DX station found me in his log. The QSO time on the QSL I got back was eight minutes off from the time in my log. I wasn't complaining at all! I was lucky as it was. But now I make sure that I *always* write the QSO time/date down immediately.

Packet radio has had a great effect on DXing. While a lot of DXers decry the effect it has had on the sport of DXing, virtually no DXer can afford to be without it. The packet screen is constantly updating with the latest spottings of rare DX, giving the most idle amateur a quick feeling for the pulse of the band. Also, DX *PacketClusters* offer other useful information including QSL routes, addresses, a data base of DX spots from the last week or more, messages, DX news and more. (See Chapter 10 for further information on packet operating and commands.)

But while packet is mesmerizing, particularly for the new DXer, it is an imperfect tool. The system is often overloaded with DX many DXers don't want to know about, such as relatively common stations. This information slows the system. Often the information delivered is many minutes old. By the time you get to a called in frequency, you find a huge pileup of other packet users who got the information ahead of you. Few

The ARRL Outgoing QSL Service

One of the greatest bargains of League membership is being able to use the ARRL Outgoing QSL Service to conveniently send your DX QSL cards overseas to foreign QSL bureaus. Your ticket for using this service is your *QST* address label and just $3 per pound. You can't even get a deal like that at your local warehouse supermarket! And the potential savings over the substantial cost of individual QSLing is equal to many times the price of your annual dues. Your cards are sorted promptly by the Outgoing Bureau staff, and cards are on their way overseas usually within a week of arrival at ARRL HQ. More than two million cards are handled by the Service each year!

QSL cards are shipped to QSL bureaus throughout the world, which are typically maintained by the national Amateur Radio Society of each country. While no cards are sent to individuals or individual QSL managers, keep in mind that what you might lose in speed is more than made up in the convenience and savings of not having to address and mail QSL cards separately. (In the case of DXpeditions and/or active DX stations that use US QSL managers, a better approach is to QSL directly to the QSL manager. The various DX newsletters and other publications, are good sources of up-to-date QSL manager information.)

As postage costs become increasingly prohibitive, don't go broke before you're even halfway toward making DXCC. There's a better and cheaper way—"QSL VIA BURO" through the ARRL Outgoing QSL Service!

How To Use the ARRL Outgoing QSL Service

1) Presort your DX QSLs alphabetically by call-sign prefix (AP, C6, CE, DL, F, G, JA, LU, PY, 5N, 9Y and so on). Addresses are not required.

2) Enclose the address label from your current copy of *QST,* or a copy of the label. This shows that you are a current ARRL member.

3) Enclose payment of $3 per pound (or less) of cards—approximately 155 cards weigh 1 pound. Note: A package of ten (10) cards or less costs only $1. Please pay by check (or money order) and write your call sign on the check. Do not send cash.

4) Include only the cards, address label and check in the package. Wrap the package securely and address it to the ARRL Outgoing QSL Service, 225 Main St, Newington CT 06111.

5) Family members may also use the service by enclosing their QSLs with those of the primary member. Include the appropriate fee with each individual's cards and indicate "family membership" on the primary member's *QST* address label or copy of the label.

6) Blind members who do not receive *QST* need only include the appropriate fee along with a note indicating the cards are from a blind member.

7) ARRL affiliated-club stations may use the service when submitting club QSLs by indicating the club name. Club secretaries should check affiliation papers to ensure that affiliation is current. In addition to club station QSLs through this service, affiliated clubs may also "pool" their members' individual QSL cards to save even more. Each club member using this service must also be a League member. Cards should be sorted "en masse" by prefix and a *QST* label or label copy enclosed for each ARRL member sending cards.

Recommended QSL-Card Dimensions

The efficient operation of the worldwide system of QSL bureaus requires that cards be easy to handle and sort. Cards of unusual dimensions, either much larger or much smaller than normal, slow the work of the bureaus. Most of which is done by unpaid volunteers. A review of the cards received by the ARRL Service indicates that most fall in the following range: height = $2^3/4$ to $4^1/4$ inches (70 to 110 mm), width = $4^3/4$ to $6^1/4$ inches (120 to 160 mm). Cards in this range can be easily sorted, stacked and packaged. Cards outside this range create problems; in particular, the larger cards often cannot be handled without folding or otherwise damaging them. In the interest of efficient operation of the worldwide QSL bureau system, it is recommended that cards entering the system be limited to the range of dimensions given. Note: IARU Region 2 has suggested the following dimensions as optimum—height = $3^1/2$ inches (90 mm), width = $5^1/2$ inches (140 mm).

Countries Not Served by the Outgoing QSL Service

Approximately 260 DXCC countries are served by the ARRL Outgoing QSL Service; as detailed in the *ARRL*

A typical day's work at the ARRL Outgoing QSL Service at Headquarters.

things are more frustrating to a DXer than to unearth a rare one and watch in exasperation as the call sign pops up on the 'cluster. Within minutes a horde of hungry stations from all over the country are jostling each other, creating an instant pileup that is growing by the minute. Without fail, the other stations have better propagation, trashing the lone wolf's prospects.

That said, packet is a marvelous tool. But the DXer must remember that packet is a tool to help the DXer, not the end product. By all means, use packet to keep in touch with what others are finding on the bands. And you should contribute by calling in the significant DX you find. Even so, never fail to do your own tuning and listening too, unless you like huge pile-ups. The easiest DX to work is the station calling CQ, and packet will virtually never hand you a good one that way!

QSLing

As the old saying goes, "The job ain't done 'til the paper work is finished." For the competitive DXer, the paperwork consists of getting the QSL in hand from the rare DX station. And, as any experienced DXer will tell you, getting the QSL can be more difficult than getting the QSO was in the first place. But, it is a part of the game, one that the successful DXer must be good at.

Fortunately, getting confirmations from three quarters or more of the new ones that you work is easy. The DX station may have a stateside QSL manager, or may handle his own cards in a prompt and orderly fashion. And the DXer who cares enough to go after the card in a proper manner will soon be enjoying the thrill of finding a new one in his mailbox.

DXCC Countries List. This includes nearly every active country. As noted previously, cards are forwarded from the ARRL Outgoing Service to a counterpart bureau in each of these countries. In some cases, there is no bureau in a particular country and cards, therefore, cannot be forwarded (although cards can, of course, be forwarded to their QSL Managers in served countries). For this reason, the ARRL Outgoing Service cannot forward cards to the following countries:

A5	Bhutan
A6	United Arab Emirates
D2	Angola
EP	Iran
HZ	Saudi Arabia
J5	Guinea-Bissau
KC4	US bases in Antarctica
KC6	Belau
KH0	Mariana Islands
KH1	Baker and Howland Is
KH4	Midway I
KH5	Palmyra and Jarvis Is
KH7	Kure I
KH8	American Samoa
KH9	Wake I
KP1	Navassa I
KP5	Desecheo I
P5	North Korea
S2	Bangladesh
S7	Seychelles
T2	Tuvalu
T3	Kiribati
T5	Somalia
TJ	Cameroon
TL	Central African Rep
TN	Congo
TT	Chad
TY	Benin
V6 (KC6)	Micronesia
VP2M	Montserrat
XU	Kampuchea
XW	Laos
XX9	Macao
XZ (1Z)	Myanmar (Burma)
YA	Afghanistan
ZD7	St Helena
ZD9	Tristan da Cunha
ZK3	Tokelau
3C	Equatorial Guinea
3C0	Pagalu Is
3V	Tunisia
3W, XV	Vietnam
3X	Guinea
5A	Libya
5H	Tanzania
5R	Madagascar
5T	Mauritania
5U	Niger
7O, 4W	Yemen
7Q	Malawi
8Q	Maldives
9G	Ghana
9N	Nepal
9Q	Zaire
9U	Burundi
9X	Rwanda

QSLs from some recent DXpeditions.

Note: The ARRL Outgoing QSL Service CANNOT forward stamps, IRCs or "green stamps" (cash) to foreign QSL bureaus. It is also not used to exchange QSL cards within the 48 contiguous states.

The ARRL Incoming QSL Bureau System

Purpose

Within the US and Canada, the ARRL DX QSL Bureau System is made up of numerous call area bureaus that act as central clearing houses for QSLs arriving from foreign countries. These "incoming" bureaus are staffed by volunteers. The service is free and ARRL membership is not required.

How It Works

Most countries have "outgoing" QSL bureaus that operate in much the same manner as the ARRL Outgoing QSL Service. The member sends his cards to his outgoing bureau where they are packaged and shipped to the appropriate countries.

A majority of the DX QSLs are shipped directly to the individual incoming bureaus where volunteers sort the incoming QSLs by the first letter of the call-sign suffix. One individual may be assigned the responsibility of handling one or more letters of the alphabet.

Claiming Your QSLs

Send a 5×7½ or 6×9-inch self-addressed, stamped envelope (SASE) to the bureau serving your call-sign district. Print your call sign neatly in the upper-left corner of the envelope. A preferred way to send envelopes is to affix a First-Class stamp and clip extra postage to the envelope. Then, if you receive more than 1 oz of cards, they can be sent in the single package.

Some incoming bureaus sell envelopes or postage credits in addition to the normal SASE handling. They provide the proper envelope and postage upon the prepayment of a certain fee. The exact arrangements can be obtained by sending your inquiry with an SASE to your area bureau. A list of bureaus appears below.

Helpful Hints

Good cooperation between the DXer and the bureau is important to ensure a smooth flow of cards. Remember that the people who work in the area bureaus are volunteers. They are providing you with a valuable service. With that thought in mind, please pay close attention to the following DOs and DON'Ts.

DOs

Do keep self-addressed 5×7½ or 6×9-inch envelopes on file at your bureau, with your call in the upper left corner, and affix at least one unit of First-Class postage.

Do send the bureau enough postage to cover envelopes on file and enough to take care of possible postage-rate increases.

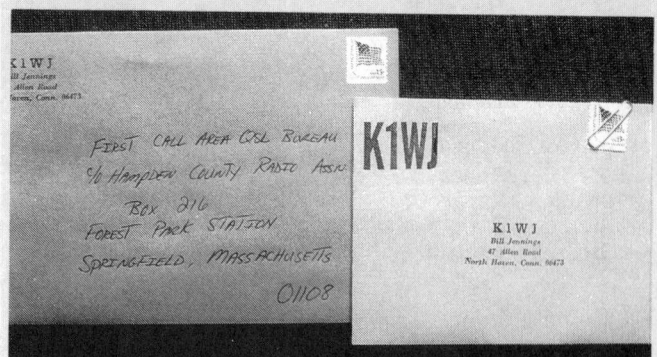

Remember to keep a supply of envelopes on file with your incoming QSL bureau (see text for address). Affix one unit of First-Class postage to your self-addressed, stamped envelope, put your call sign in the upper-left corner of the envelope and clip additional postage to the envelope if you are expecting many cards.

Do respond quickly to any bureau request for envelopes, stamps or money. Unclaimed card backlogs are the bureau's biggest problem.

Do notify the bureau of your new call as you upgrade. Send envelopes with new call in addition to envelopes with old call.

Do include an SASE with any information request to the bureau.

Do notify the bureau in writing if you don't want your cards.

DON'Ts

Don't send domestic US-to-US cards to your call-area bureau.

Don't expect DX cards to arrive for several months after the QSO. Overseas delivery is very slow. Many cards coming from overseas bureaus are over a year old.

Don't send your outgoing DX cards to your call-area bureau.

Don't send envelopes to your "portable" bureau. For example, AA2Z/1 sends envelopes to the W2 bureau, not the W1 bureau.

Don't send SASEs to the ARRL Outgoing QSL Service.

ARRL Incoming DX QSL Bureau Addresses

First Call Area: All calls*
W1QSL Bureau
Yankee Clipper Contest Club
PO Box 80216
Springfield, MA 01138-0216

Second Call Area: All calls*
ARRL 2nd District QSL Bureau
NJDXA
PO Box 599
Morris Plains, NJ 07950

Third Call Area: All calls
Pennsylvania DX Association
PO Box 100
York Haven, PA 17370-0100

Fourth Call Area: All single-letter prefixes (K4, N4, W4)
Mecklenburg Amateur Radio Club
PO Box DX
Charlotte, NC 28220

Fourth Call Area: All two-letter prefixes (AA4, KB4, NC4, WD4, etc)
Sterling Park Amateur Radio Club
Call Box 599
Sterling, VA 20167

Fifth Call Area: All calls*
ARRL Fifth (5th) District DX QSL Bureau
PO Box 50625
Midland, TX 79710

Sixth Call Area: All calls*
ARRL Sixth (6th) District DX QSL Bureau
PO Box 1460
Sun Valley, CA 91352

Seventh Call Area: All calls*
Willamette Valley DX Club, Inc
PO Box 555
Portland, OR 97207

Eighth Call Area: All calls
8th Area QSL Bureau
PO Box 182165
Columbus, OH 43218-2165

Ninth Call Area: All calls*

Northern Illinois DX Assn
Box 519
Elmhurst, IL 60126

Zero Call Area: All calls*

WØQSL Bureau
PO Box 4798
Overland Park, KS 66204

Puerto Rico: All calls*

KP4 QSL Bureau
PO Box 1061
San Juan, PR 00902

US Virgin Islands: All calls

Virgin Islands ARC
GPO Box 11360
Charlotte, Amalie
Virgin Islands 00801

Hawaiian Islands: All calls*

Wayne Jones, NH6GJ
PO Box 788
Wahiawa, HI 96786

Alaska: All calls*

Alaska QSL Bureau
4304 Garfield St
Anchorage, AK 99503

Guam

MARC
Box 445
Agana, GU 96910

SWL

Mike Witkowski, WDX9JFT
4206 Nebel St
Stevens Point, WI 54481

QSL Cards for Canada may be sent to:

RAC Incoming QSL Bureau
Box 51
St John, NB E2L 3X1

QSL cards may also be sent to the individual bureaus:

VE1, VEØ, VY2* James Wade, VE1DH
   PO Box 141
   Petitcodiac, NB E0A 2H0
VE2 A. G. Daemen, VE2IJ
   2960 Douglas Ave
   Montreal, PQ H3R 2E3
VE3 Garry Hammond, VE3XN
   5 McLaren Ave
   Listowel, ON N4W 3K1
VE4 Adam Romanchuk, VE4SN
   26 Morrison St
   Winnipeg, MB R2V 3B4
VE5* Bj. Madsen, VE5FX
   739 Washington Dr
   Weyburn, SK S4H 3C7
VE6* Norm Waltho, VE6VW
   PO Box 1890
   Morinville, AB TOG 1PO
VE7* Dennis Livesey, VE7DK
   8309 112 St
   Delta, BC V4C 4W7
VE8* Rolf Ziemann, VE8RZ
   2 Taylor Rd
   Yellowknife, NWT X1A 2K9
VY1 W. L. Champagne, VY1AY
   PO Box 4597
   Whitehorse, YU Y1A 2R8
VO1, VO2 Roland Peddle, VO1BD
   PO Box 6
   St John's, NF A1C 5H5

*These bureaus sell envelopes or postage credits. Send an SASE to the bureau for further information.

There are at least two routes you can use in pursuit of the QSL. The most inexpensive is, of course, the ARRL Outgoing Bureau, which can be surprisingly effective (see the accompanying sidebar). Of course, the return trip does take a while, with total round-trip times commonly one or two years. It is the route that many DXers use for the bulk of their cards, except for the all-time new ones or DX stations with US QSL managers.

Almost every DXer is in the DXCC chase, but many DXers also pursue other DX awards which require possession of valid QSL cards as proof of contact. (For details on DXCC and many other operating awards, see Chapter 8.) A DXer interested in even a small number of different operating awards will often find the expense of a direct airmail QSL—with provision for return air postage—to be more than he or she is comfortably prepared to pay. In such situations, QSLing via the League's Outgoing Bureau is the best and most cost-effective way to go.

But, for the DXer intent on gaining the DXCC Honor Roll as fast as humanly possible, the direct airmail QSL with return envelope and provision for return postage is considered appropriate (unless the DX station has a QSL manager). For this, the DXer will need a set of current *Callbooks*. The *Callbooks* include all the reference information that the DXer needs for QSLing, as well as a good deal of useful reference information. If the DX station asks for QSL direct or via CBA (*Callbook* address), the address information is almost always available in the *Callbook*.

It is unreasonable to expect a rare DX station in a Third World country to respond to every direct airmailed QSL and pay the postage out of his own pocket. A few stations do just that, but if you expect to get a card mailed directly from a foreign station, you are going to have to expect to make provision for return postage. To accomplish this, the DXer has three choices:

❑ He can affix unused postage stamps from the foreign country on his return envelope in an amount appropriate for an airmail reply. Such stamps can often be purchased from a stamp dealer.
❑ The second choice is to enclose an International Reply Coupon, which can be purchased at any post office. Under recently changed international postal rules, an IRC can be exchanged at the local post office for enough local postage to pay for the return of a QSL card via air mail. The present selling price of an IRC is 95 cents US.
❑ The third choice is the famous "green stamp," a euphemism for a dollar bill. Almost everywhere, a dollar exchanged into the local currency is about enough to pay for an airmail response. But there are problems with the use of the dollar bill. In some countries, possession of US currency is illegal, largely because of currency restrictions, but illegal nonetheless. Sending a dollar to DX stations in such countries can reportedly get the DX operator in serious trouble. Never send a "green stamp" unless you are absolutely sure that it will not cause problems for the DX station. If you are not sure, either ask or don't send the "stamp."

Also, theft from the mails is a growing problem in some countries. Once a postal employee with larceny in his heart discovers that envelopes addressed to "Amateur Radio DX1DX" often have money in them, you can imagine how many of the envelopes with QSLs actually get through. Indeed, in some countries the chance of getting a QSL through to a station in any way, except as a post card or via registered mail at $6 plus per toss, can become remote. The DXer should feel pleased, however, that he is helping to stimulate the local economy! Although IRCs can also disappear, they are often more successful, since they are harder to convert into the local currency and are easier to conceal in an envelope. Be sure that your envelope going overseas with either IRCs or "green stamps" is opaque in bright sunshine.

A special problem area as this edition is being prepared is getting QSLs from operators in the various countries of the former Soviet Union. As these countries strive to reshape their economies and their governments, mail service has suffered a great deal. It can be very difficult indeed to get mail to the station at the other end, let alone getting a response. QSL bureaus are in some cases nonexistent. And they suffer the same problems sending and receiving mail that individual amateurs face. The DXer intent on getting that QSL needs to be flexible in these situations.

There is no easy solution available as this is being written.

The best thing to do is ask the DX station. He or she surely knows what might work to get a card to them. Some stations have QSL managers in places where reliable mail service is simpler. Sometimes a friendly amateur in one of these countries can help you get a card from another—their mail is often able to get through where envelopes with western stamps are opened, emptied and tossed. So, developing a friendship with such a station can pay dividends above and beyond the fraternal rewards.

The easiest and usually most reliable way to get a QSL is from a DX station's manager, where such an arrangement is available. A QSL manager is another amateur, almost invariably in another, more common country, who receives the DX station's logs, either by mail or over the air, and who handles all QSLs for that DX station. If you work a station with a manager, you send the manager a QSL for the DX station, as well

So You Want To Be a QSL Manager!

There have been many fine articles directed at new and old-time DXers on the proper ways to use the services of QSL managers—and service is exactly what a QSL manager provides.

Let's determine which DX station will be interested in your efforts. First, there are the well-equipped stations who have the time to work quite a few stateside and other DX stations, but simply do not have the time to spend filling out QSLs. Second, there are fine DX operators who make many QSLs, but simply have no desire to send or receive any, even though they have the time and financial means to do so. Third, many DX operators find that QSLing is a financial burden. We often do not realize the financial sacrifice so many of our DX friends have had to make just to get a transceiver and a good antenna.

What personal qualifications are necessary to make a good QSL manager? A good manager will be a DXer. This will give him or her the insight into, and understanding of, the feelings of those he'll be trying to help. A DXer will also know what is happening on the bands and will probably subscribe to one or more of the DX newsletters. Time is also very important. If you are pressed for time by your occupation or family, it is best that you do not consider this added demand on your time. And, remember, time spent on QSL manager duties takes that much time away from your own operating activities.

It takes months and years to become well-known and

to have your call published in the several QSL manager directories. Don't start unless you are prepared to stay with it for a long time.

Being a QSL manager will cost you money if you intend to do a good job. Regardless of the SASEs received and an occasional dollar bill from some thoughtful DXer, you will find that you will have to lay out some money. You will always have the cost of mailing cards you have received to the DX station. There is also the cost of printing the QSLs for the DX station who just can't afford the expense. In some cases, you will have to send postage money to the DX station so he can send you the logs. Expenses are probably the main reason why many QSL managers "throw in the towel" after just a few months or a year on the job. This can leave hundreds of DXers waiting for QSL cards that they will never receive, and can also give managers in general a bad name. If spending a few extra dollars will be a financial burden to you, don't consider becoming a QSL manager.

A good manager must be able to deal with detailed paperwork and be very organized. Now, if you are still interested in becoming a QSL manager, how do you go about finding a DX station who is interested in using your services? You must "sell yourself" and your abilities to the DX station. After all, you will be his representative to the DX community, and he has every right to be able to feel that he can trust you and depend on you.

 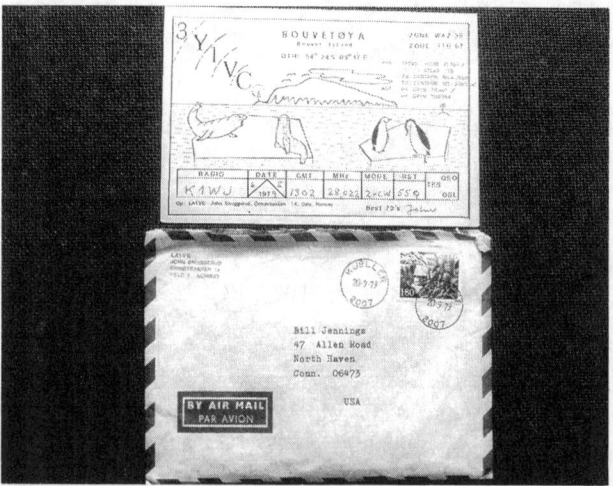

The left photo shows what was sent to the QSL manager of 3Y1VC (Bouvet I). Note the self-addressed envelope and the IRCs for return postage. On the right: 3Y1VC's QSL was received promptly.

as a self-addressed return envelope with provision for return postage. The manager's address is usually listed in the *Callbook*.

When sending a QSL to a DX station's QSL manager, there are a few tips that will make that volunteer's job easier. First, don't even bother to send a QSL (to a QSL manager or to any DX station) unless you show the QSO time and date in UTC. Be sure that the self-addressed stamped envelope (SASE) that you supply is large enough to accommodate a normal-size QSL—unless you like folded QSLs. And be sure that the return envelope has adequate postage.

Another technique that will help the manager is to write the call of the station that the QSL is for, as well as the date, time and band of the QSO on the back of the return envelope. That way, the manager can arrange the cards for easiest processing while he waits for the logs to come in from the DX station.

Which brings up another point. If you are slow in getting your card from the manager, don't harass him for the card. The QSL manager is at the mercy of the mails, of the DX station and a number of other factors over which he has no control. He is a volunteer, doing a tough job. Treat him kindly. After a wait of some months, or if you have not gotten the card while everyone else in town has been showing off theirs for weeks, mail a polite note to the manager explaining the circumstances, and enclose an SASE for his reply. And don't forget the QSO information.

Not every DX station with a manager announces that fact on the air. Anytime you are trying to get a QSL from a new one, you should check to see if the DX station has a manager. The *GOLIST QSL Manager List* offers the most complete information about current DX stations having QSL managers available, and is published bimonthly in *The DX Magazine*, (see

Type up a description of yourself with a friendly offer to be a QSL manager, stating why you wish to be a manager and why you would make a good job of the required duties for as long as required. When you QSL a DX station who does not have a manager, and for whom you'd like to assume those duties, mail your typed description along with a picture of yourself and your QSL to him. If you have made a good presentation of yourself, in time you will have made some good contacts as well as having made some good DX friends.

Let's say you got the manager's job. Now you are ready to start. Requests are coming in for QSLs for "your" DX station. The DX station is telling his contacts to QSL to your call and, finally, the log sheets of QSLs that he has sent to you arrive. No two managers organize their work the same way, but the first thing they all have to do is open the mail. See who the card is for, note the date and call on the back of the envelope and file all requests by station (if you manage for more than one station) and date.

You will need some supplies. Stock up on envelopes and postage stamps. Remember that all envelopes must meet the postal code for size. From the local stationery store, get a glue stick, adhesive tape, a roll of gummed paper tape, scissors, a moistener to wet the glue on the envelopes, and a rubber stamp with your name, call and address.

Most DXers will make it easy for you to send them their QSLs. They will have their call sign and all QSO information printed on the same side of their QSL. They will also have the correct date and time in UTC. And, most importantly, they will enclose an SASE with their QSL request. The DXers will also keep in mind that it may take several months for you to get the logs from the DX station for verification.

Overseas stations should send you a self-addressed envelope with two IRCs for airmail return or one IRC for seamail return. Some will only send you an IRC or two without the self-addressed envelope. For these, take your scissors and cut out the return address on their envelope and, using your glue stick, attach that name and return address to one of your stock of envelopes. Never rewrite the names and addresses unless you absolutely have to. About 25% of the overseas stations will QSL you via the ARRL QSL Bureau for your call area. Be sure to inform your district QSL Bureau of the calls of the stations you manage. Keep plenty of proper-size envelopes and postage at your bureau. Return overseas cards received via the bureau through the ARRL outgoing QSL Bureau. This will save you the expense of sending direct to the overseas bureaus. Be sure to answer all the SWL (shortwave listener) QSLs received. These SWL cards mean a lot to prospective amateurs in many countries.

About 10% of the stateside stations will fail to send you an SASE or even a postage stamp. For these, you must pay the postage yourself or else return the DX QSL via the bureau and hope the requesting station has an envelope at his or her district QSL bureau. Otherwise, he'll never get that QSL.

Carefully check the requests for cards with your DX station's logs. Very often the time and or date on the QSL is off considerably, so make every effort to find the QSO in the logs. If you can't find the QSO, you can do either of two things: send the card back to the station making the request with the notation "not in log, please check time and date," or you can send the card direct to the DX station for him to decide whether to QSL. The best thing to do is to return the card to the station making the request. This is the most disappointing thing that you as manager will have to do. Operators requesting QSLs from your DX station really believe they have made the contact, but in the QRM of the pileup, it is easy to be mistaken. This happens to all DXers at some time or another. Don't just disregard the request; let the DXer know that he is not in the log.

Fill out all QSLs. Be sure that the date, frequency and mode are correct. Use a ballpoint or felt-tipped pen. Sign your name or initials and your call sign, then stamp the back of the QSL with your rubber stamp to add authenticity.

Some DX stations that you manage for will tell you they do not want any cards received for them. It's best to keep these cards, so store them away in a dry place in case the DX station changes his mind at some future time. For those DX stations who want their incoming cards sent to them, wrap the cards in heavy wrapping paper (grocery bags work fine) seal securely with gummed wrapping tape in packets that weigh less than two pounds (approximately 310 QSLs). Send them via seamail with the required customs declaration stating that the contents are "used radio postcards" with no monetary value. This is the most economical way to send the cards to your DX station. Airmail costs over twice as much.

Keep up with your QSLing. Send out the QSLs every day if you have the logs for confirmation. If you get far behind, the job will seem more like drudgery than a pleasant chore. If necessary, enlist the aid of your XYL/OM or an amateur friend or two to help out. Remember, some anxious DXer is waiting for the mail every day.

In spite of all the work, it will never cease to be rewarding when you tune in on your DX friends and hear them say to QSL via your call sign, or most of all, when you open an envelope and find a note from an appreciative DXer who simply says "thanks, OM."

Now, get with it, do a good job, and good luck!—*Don Brickley, W7OK* address listing at end). The list is also available through many DX *PacketCluster* systems, and through telephone BBSs.

The printed list is especially valuable as an operating tool to the DXer interested in a score. The smart operators will put a premium on working stations in countries that have QSL managers, knowing that their chances of securing a QSL are greatly enhanced. Write for complete information to: *GOLIST*, PO Box 2306, Paducah, KY 42002-2306. Be sure to include a self-addressed, stamped business-sized envelope.

Eavesdropping on 40

I settle into the comfortable old captain's chair in front of the rig, put on the headphones and move the antenna selector switch to my new southwest 40-meter sloper. It sure would be nice to have one of those "shorty 40" Yagis, but the conventional wisdom of the local DX gang seems to be that a set of good slopers is an excellent DX antenna system. With a bit of help from my son, we were able to get the slopers up yesterday, before the winter storms.

I glance at my clock. It's just after 6 AM local, 1200 UTC. I warm one hand on my coffee cup while the other hand starts tuning the receiver. I'm out to see if my new southwest sloper is any good for the grayline DX I've been hearing about on 40. I was on 40 last night, but results from testing the north sloper were inconclusive. But then I am not so sure that the band was very good either. And, to get the feed line for one of the slopers, I had to take the old inverted V down. I have no A/B test that I can run; only a set of three slopers.

I set the receiver down at 7 MHz and begin tuning slowly up the band. I'm sure glad that I got my Extra Class license last fall. It has opened up a whole new world of DX opportunities for me, especially on the CW bands. The Extra has done more for my DXCC country total than buying a linear, and as much for my DXing as putting up a tribander.

There's a K1 calling someone. He's about an hour further into daylight than I am, so I'll get some idea of how good a band opening is by hearing what he's working. My chance may come a bit later. I listen; yes, there's someone coming back. It's a JA7, weak and watery sounding. I tune on.

I hear a weak CQ, just barely above the noise. I listen carefully. Okay, it's a UA9. I switch the north sloper in for a moment to see if he gets any stronger, but he almost disappears. That's interesting. I thought slopers were supposed to have cardioid patterns, which would certainly include a UA9 on a north-facing sloper—unless he isn't coming through on the short path. Yes, I guess he could be coming through on a crooked path, since it's sunset somewhere in Siberia. Interesting.

I tune on; I don't need a UA9, but the experience intrigues me. There are three or four stations calling someone. Ah, he's coming back. Good grief! It's VS6DO! I've heard him on 40 before, but never better than Q2 to Q3, and there he is an honest 559. Maybe there is something to these slopers! I decide to try to put him in the log. I have a QSL from VS6 on 20 meters, but nailing one on 40 would be real satisfaction for me. I find the station that he's working, a W4, and set my transmit frequency just above the W4. The VS6 is working people quickly, passing out reports and QSL information, a K7 manager. If I can nail him, it means an easy QSL to boot!

The W4 signs, the VS6 goes QRZ, and I give a 2× call. I listen. He's back to a WA8. He turns it to the WA8. I listen; yes, there he is, down about a half kHz. I move my transmit frequency down a bit, closer to the WA8 but not on top of him. The VS6 clears with him and calls QRZ again. I call him and wait

to see what's happening. Hah! I got him! I note the details in the log, grinning from ear to ear. Guess that new sloper is okay!

I start moving up the band again. I come across a VK with good signal strength clearing with a WB5. I find a mini pileup on an FK8, but I already have one on 40. Then I strike paydirt—a weak signal calling CQ. I listen carefully, and dig out his call sign. It's VU2TEC! I've never heard a VU2 on 40 with my old inverted V, and here's one calling CQ, RST 449. Guess that new sloper is doing the job. I set up on his frequency, and when he ends his call, I'm ready. I pause a moment, and, hearing nothing, call him with a 1×3. I listen, anxiously. There, he's coming back, but to a UKØ. Darn. Well, not really a surprise. A UKØ is certainly much closer than a W9, and much stronger.

I wait out the QSO with the UKØ. When the VU2 turns it over, I try to hear the UKØ so that I can be sure that I transmit on the UKØ's frequency after he has finished the QSO. After all, that is where the VU2 certainly is listening. But I can find no trace of the UKØ. I keep careful watch. There, the VU2 is back. He's promising to QSL and is starting to sign clear. I'm nervous. This is real DX.

The VU signs clear with the UKØ, and I don't know what to do. Is the VU copying the closing transmission of the UKØ? If so, will he give a simple "dit dit" at the end of the QSO, or will he call QRZ, or will he call CQ again? I wait, unsure.

I decide to give him a short call: TEC DE W9KNI W9KNI $\overline{AR}$. I wait a moment. R 73 DR IVAN $\overline{SK}$ NW QRZ W9? DE VU2TEC $\overline{KN}$. All right! I call him nervously, giving a 1×3, and turning it back to him. There it is, beautiful music, R W9KNI DE VU2TEC . . .

Forty meters is one of the best DX bands available, an important complement to 20 meters. It offers consistent DX openings, especially when 20 meters is dead. Although the primary DX activity on 40 is on CW, there is a fair amount of DX worked on SSB as well. But, because the bulk of the serious DX activity takes place in the first 25 kHz of 40 meters, an Extra Class license is essential for the serious DXer.

Not so long ago, any form of Yagi on 40 meters was rare, but now many DXers have two-element Yagis, making the pileups more competitive. Those with such antennas are especially enthusiastic about the band, but there are still a lot of DXers using various forms of wire antennas on 40 to good advantage. Often, DX signals are weak on 40, and at other times display good signal strength, but against a background of high noise levels. And, many of the DXers on 40 are very proficient at the game. But these factors combine to make 40 a superb DX band where the working of a new one is always a special thrill.

CONTEST!

As I ease into the operating chair, I'm a touch anxious; it's 2350 Zulu, 10 minutes until the start of the CW weekend of the ARRL DX Contest. I've decided to start out on 20 meters. A quick listen confirms that the band is open. I glance at the antenna rotator; yes, I'm pointed into the northwest, the great-circle direct path for Southeast Asia.

My strategy is clear. I'm only interested in new countries. I look at the clock. It's 2354. Time to run for a fresh cup of coffee. I listen across a few kilohertz near the bottom of the band. There are carriers everywhere as stations tune their linears for the last watt of power. Then, with about 30 seconds left before the start, there suddenly seem to be dozens of stations calling CQ TEST. Fascinated, I turn the selectivity to the 8 kHz position, and hear about 12 stations calling CQ TEST all at once. As I listen, I watch the clock. Magically, as the clock goes to 0000, the CQ calls are ended, and there is a brief mo-

Working the pileups from ZD9SXW.

ment of silence. Then, stations lucky enough to have had an answer to their CQs start acknowledging with their contest exchanges, while others start calling CQ again.

A few minutes at the bottom of the band shows me that the big gun contest stations are all intent on staking out their turf in the bottom 15 or 20 kHz, but as I tune higher, things start to loosen up. The QRM seems deafening at first, but I find that I can copy plenty of weaker stations between the big guns. I slowly work up the band, examining every signal, trying to determine if it comes from a country that I need.

I hear JAs by the score, Siberians by the bunch. I trip across an HL (Korea) passing out reports to an eager following, and a little higher a DU (Philippines) station is proving to be most popular. But I don't need any of them. Then, I come across a tremendous pileup. I listen carefully, wondering what manner of DX could attract such a throng. The pile seems to be several kilohertz across and I cannot find the target. I tune back down a moment, trying to find the lower side of the pileup, and in doing so find AH2G (Guam) passing out a report. His signal is terrific, well over S9. Could he be the target? I open up the selectivity as he signs with a station. Wow! The whole pileup calls him.

I am mesmerized for a few minutes. I listen to a couple rounds, then start calling him myself. After about 10 minutes with no joy, I come back to my senses, and laugh. After all, I don't need it. I look at the AH2 card on the wall for reassurance. Just because he's loud, and apparently a superb operator, is not enough reason for me to depart from my game plan. I start tuning again.

After passing two more JAs, I hit big, in the form of V85AK. Brunei is a new one, if I can nail him. I set up as he calls his CQ, hoping that no one else will find him. He signs, and I drop my call in. He's back to a W5. The W5 is just about on the same frequency as I am, so I hold my ground and do not move. There, he finishes, and the V85 calls a QRZ. I wince, as I anticipate a pileup forming that will dwarf that of the AH2. I pause for a moment before I begin my call, waiting to hear just how loud the pileup is.

There is so much silence that I almost blow my chance. I call, and listen incredulously as he comes back, W9KNI R TU 579150 579150 W9KNI K̄N̄.

I barely manage to respond, R V85AK DE W9KNI R TU NEW COUNTRY 559ILL 559ILL QSL 73 DE W9KNI K̄N̄.

He comes right back, R W9KNI FB QSL OK BOX 73

BANDAR S̄K̄ QRZ DE V85AK K.

Wow, just like that! I log him in, and listen. There is no one calling him at all now. There he goes again, calling CQ. Okay, now he has a taker, a WA6. I glance back down the band for a moment. Yes, that AH2 has a bigger pileup than ever now, while the relatively rare V85 goes unchallenged, begging for QSOs. I guess it's the AH2's big signal. But you'll get no complaint from me!

Contests for the DXer

Contests offer the DXer an unparalleled opportunity to snag new countries and to practice pileup techniques. (But remember—if you are "practicing" in a pileup for a station whose country you don't need, you just may be missing one a few kilohertz away that you do need.)

The best contest for country hunting is the CQ World Wide Contest, with the phone section held the fourth weekend of October, and CW the fourth weekend of November, which is usually, but not always, the weekend after the US Thanksgiving holiday.

The ARRL DX Contest is a close second in importance, with the CW weekend the third full weekend of February, and the phone weekend the first full weekend of March. This contest is of particular interest to American and Canadian DXers, because DX stations can work only Ws and VEs. Unlike the CQ contest, DX stations are not allowed to work each other for points.

Another popular event on the contest calendar is the All Asia Contest, with phone and CW dates announced in the major magazines. This contest is for Asian stations "against the world," meaning that non-Asians can work only Asian stations, and Asians can work only non-Asian stations. It is a fine contest—unless you already have all the Asian countries worked and confirmed. Of course, there are a number of other contests that will be of great interest, such as the CQ WPX Contest and the IARU HF World Championship, especially to the DXer with the smaller country total.

Countries with enough amateurs to organize and promote a DX contest are usually countries that are not considered rare. Still, many contests, such as the WAE (Worked All Europe), offer an occasional pleasant surprise (many amateurs still need ZA) and, in any case, offer good operating practice and the chance of new band-countries. Details and dates on such contests are published regularly in *QST*.

Contests, including the minor ones, are especially fine opportunities for the QRP operator to make a mark. An alert and aggressive operator in a contest will regularly find even rare stations calling CQ, with the chance of a one-on-one shot at a QSO. If you can't get a QSO under such circumstances, something is usually wrong.

Persons seriously interested in contesting should refer to Chapter 7 for further information. Even if you're not particularly interested in competing in the contest from a scoring standpoint, you should nevertheless be aware of the opportunity to snag new countries and get experience under fire. Rules for all major contests are published in the Contest Corral column of *QST*.

Concluding Notes

The scope of this book obviously does not permit an exhaustive study of DXing. It is designed more as a basic introduction that points you in the proper direction. On that basis, here— briefly— are a few important notes.

First, you'll note that I emphasized the 40, 20 and 15-meter

160 Meters (1.8-2.0 MHz)

Devotees of DXing on 160 meters, aka Top Band, tend to be viewed by others as gluttons for punishment. Why else would anyone stay up to all hours of the night (and early morning) just to get an earful of static? Those who see it that way should probably move back to the safer regions of 20 meters. But for those who really are up to the challenge, 160 meters is irresistible.

Top Band was at one time an exclusive club. It became known as the "gentleman's band" because some of the shenanigans often heard in 20-meter pileups just did not happen on 1.8 MHz. That "gentleman's band" concept continues today, but the level of competition has increased manifold with the proliferation of transceivers with 160-meter capability. Participants are also discovering that acres of real estate aren't a necessity to compete. The two most common antennas found are a high inverted V, for those with towers of 80 feet and higher, and the inverted L, the favorite of the "tribander at 50 feet" crowd, who can locate the antenna next to the tower and extend the horizontal section to a support such as a tree.

DXing operating habits on 160 have also undergone some dramatic change with the burgeoning population and worldwide frequency allocations in a state of flux. To keep pace, the ARRL Board of Directors adopted a modified band plan in 1986. This plan recommends that CW, RTTY and other narrow-band modes be used from 1800 to 1840 kHz, while CW, SSB, SSTV and other wideband modes should be used from 1840 to 2000 kHz. Because the 1830 to 1850 segment is the most common international allocation in the band, amateurs are encouraged to use it only for intercontinental QSOs.

Transceive operation, both CW and phone, is commonplace between 1830 and 1850 kHz. The segment 1850 to 1855 kHz, is often used by Eastern European countries that are restricted to that portion of the band. General phone operation is by convention used above 1855.

The other band segment of significance is 1907.5 to 1912.5 kHz, the window for Japanese stations. JA stations will announce where they are listening, usually on the low end. VK and ZL stations will be found on CW at our sunrise, near the bottom of the band, just up from 1800 kHz, best during the Northern Hemisphere's winter season. Russian 160-meter allocations are 1830 to 1930 kHz on a secondary basis. 1830 to 1860 is CW only, 1860 to 1900 is CW and LSB, and 1900 to 1930 CW, LSB and AM.

160 meters is strictly a nighttime DX band. Knowledge and use of grayline sunset and sunrise information is probably of more use on this band than any other. The equinoxes can be especially productive. DXing on Top Band offers a greater challenge than any other HF band. There is no better reward than to have another sleepless night produce a "new one." Where it once took years to accumulate 100 countries for Top Band DXCC, DXers are now doing it in a single operating season—meaning fall through spring.

80/75 Meters (3.5-4.0 MHz)

It's called 80 meters in the CW portion of the band and 75 meters in the phone part of the band. We'll refer to it simply as "80."

80 (like 160) is a nighttime band with best DX times being an hour before sunset and an hour after sunrise. Those hours between sunset and sunrise are not a vast wasteland, either. More-than-adequate possibilities to work some medium- and long-haul DX exist during the hours of darkness. Again as on 160, long-haul DX paths between two given areas are enhanced when sunset at one end of the DX path corresponds with sunrise in the area at the other end. This is called shadow edge or grayline DXing. See the sunrise/sunset tables in Chapter 17. It's hard to imagine that the daytime short-hop ragchewing and net operation that seems to be the mainstay of 80 meters gives way to a world of some "heavy-duty" DXing by night. Several 80-meter DXers have worked from 200 to well over 300 countries. If they can do it, so can you, with a little time and effort. On 80, an Advanced or Extra Class license can be the difference between success and failure in your DXing efforts, although a respectable country total can be amassed with one of the other license classes. (To gauge propagation conditions to Eastern Europe, check the Russian beacon continuously sending the letter U in Morse code on 3636.3 kHz.)

Eighty-meter WAC (Worked All Continents) and racking up 100 countries for your Single-Band and Five-Band DXCC are well within the grasp of any operator with even a modest station/antenna setup and a little bit of operating savvy. Antennas for 80 can range from simple dipoles and trapped verticals to elaborate delta loops and quad configurations. The rig should have split-frequency capabilities if you are interested in working 75-meter phone DX. To work some of the countries that can't transmit in the US phone band, an external receiver or external VFO would definitely work to your advantage.

The low end of the US phone band (3.750 to 3.800 MHz) is a hotbed of DX activity. Extras can operate from 3.750 and up, while the Advanced portion starts at 3.775 MHz. The segment 3.790 to 3.800 MHz is considered a "DX Window." Note that many DX stations will operate split frequency, transmitting below 3.750 and listening above 3.800 MHz for callers. So the knowledgeable 80-meter DXer should listen well below 3.800 for a DX station who may wish to work the US using split-frequency operation. (Check also for SSB DX calling CQ around 3.640, listening in the American band.) On CW, the majority of DX stations work in the 3.50- to 3.525-MHz range. An accompanying table shows the 80-meter phone frequency allocations from around the world.

For those interested in serious DXing on 80 meters there is a definitive work on the subject: *Antennas and Techniques for Low-Band DXing* by John Devoldere, ON4UN, published by the ARRL. The book is probably the most complete work on the topic, written by an operator with 300 countries worked on 80!

40 Meters (7.0-7.3 MHz)

Forty, like the other low bands, 160 and 80 meters, is bands. These major bands are useful DX bands for new DXers almost anywhere in North America, a situation perhaps not as true of 80 or 160 meters. Which is not to say that DX cannot be worked on those bands, but for the newcomer operating with 150 watts from the American Midwest, for example, it is a more challenging proposition. Chasing DX on 160 meters especially requires that good operators follow established practices that are unique to the band. See accompanying sidebar on these "low" bands.

Ten meters is a very popular band when the sunspot cycle favors it. It's a band where even low-powered stations with simple antennas can work the world. (Novice and Technician Plus licensees have certainly found this out via their 10-meter SSB privileges.) But DXers who concentrate on 10 meters are people who'll need to find another hobby every few years! The band is a very seductive one when it is wide open and very dead when closed.

basically a nighttime and early-morning DX band. Again, grayline-DXing methods prove most fruitful.

On 40, though, the DXer faces a new enemy. Propagation and band characteristics are not the only obstacles to overcome. Forty meters is (like 160 and 80) a shared band; shortwave broadcast stations have assigned frequencies on 40 also. Thus, as the band conditions begin to change and become more favorable for working DX, it seems as if every broadcast station in existence is trying to claim a piece of 40 meters as its own. The bc stations use several kilohertz of bandwidth and legally run tens or hundreds of kilowatts of power. Strangely enough, it's possible to tell when 40 is open to certain areas by the strength of a received bc signal.

At 7 MHz, rotatable/steerable antenna arrays start to become more manageable in size, and some of the simpler wire and vertical antennas that work well on 80 meters do a creditable job on 40. By running near or at full legal transmitter power to an antenna array, one can put out quite a respectable signal on 40 meters. The beam antennas make it possible to explore different signal paths utilized on the higher bands, such as the long-path signal. Long path is a phenomenon in which a signal will peak if you steer your antenna so that it is facing approximately 180° away from the azimuthal heading (beam heading, or short-path beam heading) as determined from a great-circle projection centered on your QTH. For example, if you usually point your antenna at 120° to work a station in Africa, a long-path opening would cause the same African station's signal to peak up when you rotated your antenna to a bearing of about 300°. It's worth checking for longpath signals, especially when signals seem to be down early in the morning or early evening.

As they do on most of the other bands, the CW DXers like to stay in the lower 25 kHz of the 40-meter band. But since there is such strong competition for space on 7 MHz DX stations can regularly be found on CW much higher in the band—as high as 7.050 MHz or so.

Working DX on 40 phone will quickly bring home the need to have the ability to work split frequency. The US phone band extends from 7.150 to 7.300 MHz, and South and Central Americans as well as Africans can be found and worked here transceive (on their own frequency, not split). European and some Asian stations must be worked split frequency on 40. The Europeans can be found transmitting between 7.040 and 7.100 MHz, listening for US amateurs in the US phone segment.

When the band is quiet and not crowded, it is not necessary to have a "killer" signal to work DX easily. Many QRPers (operators who run a maximum of 10 watts input or 5 watts output) are quite successful in achieving respectable country totals on this band. QRPers on 40, as on most of the other bands, tend to be CW operators and gather at or near 7.040 MHz. Forty meters, for all its quirks, is quite an interesting band and has a lot to offer for a serious or casual DXer.

Thanks to the ARRL and other IARU societies, the 1979 WARC brought us three exciting new bands at 10, 18 and 24 MHz. These bands are significant to the DXer. Among other things, they offer a quiet sanctuary from weekend contests.

Each band is located—in terms of frequency—approximately midway between two of the existing amateur bands and, as might be expected, their propagation characteristics are intermediate between those of their neighbors. As solar activity

(L-R) Two active Guadeloupe hams are FG5BT and FG5BG (*FS5UQ photo*)

declines, there is an increasing tendency for the MUF on northerly paths to fall below 10 MHz during darkness.

Apart from this, absorption is the main factor affecting DXing, which will generally be best when much of the path is in darkness or, if not, is reasonably close to twilight. DX openings therefore follow the same pattern as those on 40 meters, but because absorption is lower, they start earlier and end later than on 40, and peak signals will be stronger. As to 18 and 24 MHz, while there are definite differences with different levels of solar and geomagnetic activity, the propagation characteristics of these two bands show marked similarities to those of their immediate neighbors. The 24-MHz band remains open for DX on days even when 28 MHz is not useful, while the 18-MHz band is good for daytime DX.

Another aspect of DXing not referenced above are the DX newsletters. Three are published in the US at present, each with its own viewpoint and following, as well as a monthly magazine. They are:

DX Bulletin, PO Box 50, Fulton, CA 95439-0050 (weekly)
QRZ DX, PO Box 16522, Asheville, NC 28816-6522
Long Island DX Bulletin, PO Box 50, Fulton, CA 95439-0050 (biweekly)
The DX Magazine, PO Box 50, Fulton, CA 95439-0050 (monthly)

All of these bulletins will be happy to quote you subscription rates and send you sample copies upon receipt of your SASE with postage for three ounces. A subscription to a DX newsletter is an excellent investment for the operator seriously interested in working new countries.

Worldwide Top-Band Frequency Allocations

This list shows international frequency allocations on 160 meters. Information was garnered from the International Amateur Radio Union (IARU) Annual Reports submitted by IARU Member-Societies, as compiled by ARRL International Programs Manager Nao Akiyama, NX1L/JH1VRQ. Some other countries, with or without IARU representation, may give their amateurs access to 160 meters. Since this information reflects administering government regulations, these frequencies may apply to several DXCC countries within that jurisdiction. For example, the listing for G (United Kingdom) applies to G, GD, GI, GJ, GM, GU and GW.

Country	Privileges (kHz)	Notes
A2	1810-2000	32 W PEP Max
A4	1800-2000	CW,10 W max
A9	1810-1850	
AP	1800-2000	
BV	1800-2000	
C2	1800-4000	
C3	1810-4875	
C6	1800-2000	
CE	1800-1850	
CP	1800-1850	
CT	1830-1850	CW & RTTY 60 W max
DL	1815-1835	SSB: 1832-1835
	1850-1890	
DU	1800-2000	SSB: 1900-2000
EA	1830-1850	
EI	1800-2000	
F	1830-1850	
FR	1810-1850	
FG,FJ,FM,FS,FY	1800-2000	
FO	1800-2000	
G	1810-2000	10 W max
H4	1800-2000	
HA	1830-1850	CW, 10 W max
HB	1810-1850	
HBØ	1810-1850	
HI	1800-2000	
HK	1800-2000	
HL	1800-1825	
HP	1800-2000	
HR	1800-2000	
I	1830-1850	
J2	1810-1850	
J3	1800-2000	
JA	1907.5-1912.5	CW only
LA	1820-1850	CW only, 15 W max
LU	1800-1810	CW
	1810-1850	CW & Ph
LX	1810-1850	
OA	1800-2000	
OE	1810-1950	100 W max SSB: 1840-1850
OH	1820-1845	10 W max
	1915-1955	10 W max
OK	1750-1950	10 W output max SSB: l820-1950
ON	1830-1850	10 W max
OY	1830-1850	CW, 10 W max
OZ	1830-1850	CW, 10 W max
P2	1800-1866	
	1874-2000	
PA	1825-1835	
PJ	1800-2000	
PY	1800-1850	
PZ	1800-2000	
SM	1830-1845	CW: 10 W input max
SP	1810-1980	10 W input max (except 1830-1850)
SU	1815-1890	75 W dc input max
T7	1810-1850	
TA	1810-1850	30 W max ERP
TG	1800-2000	
TI	1800-1850	
TR	1810-1850	
U	1830-1930	Low power ph: 1860-1930
V2	1800-2000	
V3	1800-2000	
V8	1800-2000	
VE	1800-2000	

Another subject covered only in the most general terms has been propagation. The committed DXer will learn through experience the propagation patterns of his favorite bands. Even so, a study of the physical phenomena that make DX propagation possible will enhance both the DXer's interest and knowledge of when to look for that A51 or another needed country. The chapter on antenna orientation in this book provides a good introduction. Also see *The ARRL Handbook* for a

VK	1800-1866	Ph: 1825-1866	
	1874-1875		
VP2M	1800-2000	CW & SSB only	
VP9	1800-1825		
	1875-1900		
VS6	1800-2000	CW only	
VU	1820-1860		
W,K,N,AA-AK	1800-2000	See ARRL-recommended band plan	
XE	1800-2000		
Y2-9	1810-1950	10 W max ph: 1900-1950	
YB	1800-2000	SSB: 1840-2000	
YK	1830-1850		
YJ	1800-2000		
YN	1800-2000		
YS	1800-2000		
YU	1810-1850	SSB: 1830-1850	
ZB2	1800-2000		
ZF	1800-2000	75 W max	
ZL	1803-1857		
	1863-1950		
ZP	1800-2000		
ZS	1810-1850		
3A	1830-1850		
3B	1810-2000	10 W max	
3D2	1800-1850		
4S	1800-2000		
4X	1810-2000		
5B	1800-2000	Ph: 1900-2000	
5N	1800-2000	10 W max	
5W	1800-2000	Ph: 1850-2000	
5Z	1830-1850		
6W	1810-1850		
7P	1800-2000	10 W max	
9H	1810-2000	10 W max	
9J	1810-1850	10 W dc input max	
9K	1810-1850		
9L	1800-2000	10 W max	
9M	1800-2000		
9V	1800-2000	10 W input max	
9Y	1800-2000		

80-Meter Phone Allocations Throughout the World

Area	Frequency (MHz)
Africa	3.600-3.800
Asia (most)	3.600-3.900
Australia	3.535-3.700, 3.794-3.800
Canada	3.725-4.000
Europe (except Russia)	3.600-3.800
Japan	3.525-3.575, 3.791-3.805
Pacific	3.600-3.900
Russia	3.600-3.650
South America	3.500-3.800
USA and possessions	3.750-4.000

IARU Region 1 has officially recommended that 3.790-3.800 MHz be used only for intercontinental contacts.
Note: There are several exceptions, including license-class restrictions, to this table.

published by CQ Communications, and the *Radio Amateur's Guide to the Ionosphere*, by Leo McNamara, published by Krieger Publishing.

Of course, a truly comprehensive discussion of such a complex matter as DXing is impossible in the space allocated to one chapter, and there are significant subjects that have not been covered at all, such as buying gear for maximum DX benefit, antennas, site selection and CQ calls. [See Chapter 8 for a complete survey of international awards.] Those amateurs getting started in DXing should obtain *The DXCC Companion*, by Jim Kearman, KR1S, published by ARRL. And no DXer aspiring to success should be without *The Complete DX'er* by Bob Locher, W9KNI, published by Idiom Press and available from the ARRL.

Good hunting!

more detailed treatment. Two good books on the subject for the DXer are *The Shortwave Propagation Handbook*, second edition, by George Jacobs, W3ASK, and Theodore Cohen, N4XX,

CHAPTER 6

Overseas DXing/DXpeditions

CARL HENSON, WB4ZNH
7765 HALL RD
FAIRBURN, GA 30213

"Y ou want to go on a what?" asks your spouse. "D-X-pedition," you enunciate very clearly. "What in the world is that?" comes the not unexpected response.

"Well, honey, you know all those DX stations all over the world that I've been chasing the last couple of years…"

"Uh huh."

"Well, I would like to try being chased myself for a change."

She grins and replies, "Come now, you're too old for that!"

"Seriously, this year instead of taking the kids to DisneyWorld, I want to go on a DXpedition!"

Now she knows you are serious. The grin is gone and in its place is a hurt abandoned look. "You want to leave me and the children while you go off and get hurt playing Indiana Jones?"

My goodness, you think, this woman can sure make you feel guilty. "No, darlin'," you soothe, "I want to leave the kids at your mother's and I want us to spend a week on some beautiful tropical island."

"But what about your DXpedition?"

"That will be my DXpedition!"

"It doesn't sound so bad; you'll just be talking on your radio at night like you do now?"

"Well," not wanting to limit your activity, "maybe a *little* more than that; after all, I will be on vacation!"

Now that the stage has been set, let's plan a DXpedition: where to go, when to go, what to take, what to expect, and some obligations. If everything else goes as it should, you will then be free to enjoy being the much sought-after DX. While these ideas are written primarily for amateurs in the US, they could apply anywhere with little difficulty. This chapter will attempt to follow the same chronological flow as would be done in actually planning and executing a DXpedition.

There are five general categories: (1) planning and preparation, (2) travel and arrival, (3) documentation and assembly of the station, (4) DX operating, and (5) postDXpedition requirements.

There are some things about operating that deserve a little thought before going further. At a certain level of competency, all ham radio operators know how to operate, and all DXers know how to DX. There is no denying this. One purpose of this manual, particularly this chapter and the previous chapter on domestic DXing, is to help you become a better operator. Whether you actually go on a DXpedition or not, the philosophies expressed here are aimed at improving the quality of operating.

At this point, it is appropriate to define some terms. For the purpose of this chapter, a DX station, shortened to "the DX" or just "DX," is an Amateur Radio station, on land, in an ARRL DXCC country. DX can be both a noun and a verb. For example: DXers DX when they pursue the elusive DX. DXing is the activity of trying to contact DX stations and it is pursued by DXers. "To DX" is to engage in DXing. You should realize that the DX (the hunted) can also be and often is a DXer (the hunter). Certainly there are quite proper domestic uses of the term "DX" that won't be addressed in this chapter.

Let us begin with the most logical point, choosing your destination.

PLANNING AND PREPARATION

Where to Go

Most DXers know which countries are the rarest and consequently the ones from which an Amateur Radio operation would generate a really large pileup. Many DXers never consider going on a DXpedition because they can't be the one to put on an operation from one of those very rare DXCC countries. This is a sad mistake! While it is certainly normal to pursue lofty goals, you should note that, with only minor exceptions, there are few differences in pileups other than their size. Because of the many facets of DXing, it is quite possible to conduct a truly exciting DXpedition from a country that already has an active ham population! This is because many resident hams prefer rag chewing to pileups, and will welcome your efforts to take the pressure off of them!

Many relatively easy-to-contact DXCC countries are quite rare on certain modes (eg, CW or RTTY) or bands, such as 160, 80 or 40 meters. Likewise, by choosing operating times and frequencies carefully, some nice operating can be had from what might be considered common countries, which are actually rare in other geographical areas. Some examples are the

Caribbean area (rare to Japanese amateurs), Europe (difficult to contact from the West Coast of the US) and the Pacific Islands (which offer some challenge to be worked from the East Coast). From time to time, the DX newsletters will poll their subscribers and list, in descending order, the most needed or the rarest DXCC countries. Numerous DX clubs maintain similar lists of the countries needed by their members, usually in conjunction with a telephone, 2-meter FM or PacketCluster DX alert system. Both of these can be good sources of information to help you decide which location to choose for your DXpedition. Just remember that if all of your information comes from one geographical location, then the results will be distorted, favoring that area.

Another person to help you choose a DX location is your travel agent. In the present competitive status of airline fares, a wide variety of distant locations are within reach of some rather modest budgets. The travel agent charges you nothing, but receives a commission on the sale of your ticket. This is an excellent source of information and savings that you should not ignore. If your trip is in conjunction with business, a little research is still in order. You might be able to change from the most direct routing and include a stopover in a semirare DXCC country and not increase the cost of your air travel at all. In addition, your business destination might be located only a short distance from a relatively good DX operating location. The business traveler/DXer arriving in Hong Kong will find some DXing fun there, but might experience a "major league" DXpedition in nearby Macao (XX9), just a short and inexpensive trip by hydrofoil, jetfoil or ferry.

When to Go

If your sojourn is employment related, then you might not have a choice of times. But if you do have a choice, then you will probably want to choose a time during the Northern Hemisphere's winter. With the Earth closer to the sun during the winter months, the F layer of the ionosphere (the one more likely to bring DX) will be more favorably charged. In the summer, those long days allow the D layer (commonly blamed for absorbing signals) to build up. The third, and possibly the most significant, reason for planning a DXpedition for the wintertime is to get away from snow and cold weather!

Another appropriate time for a DXpedition is during a major contest. You don't have to be a serious competitor to enjoy yourself in a contest. But even so, an examination of the published results will make it clear that it is quite easy to enter a category that will ensure the DXpedition a nice operating award. The results of a previous contest can also be used to help select the destination, but they can also be very misleading since there is no certainty that the same countries will be activated from year to year (check the DX newsletters to find out who is going where).

Local weather conditions at the chosen DX location can either be devastating or richly rewarding to the operation. Some geographical regions have no winter or summer; instead they have only a wet season and a dry season. The rainy season can ruin QSO totals on lower frequency bands such as 160 and 80 meters, not to mention multiplying the risk of contracting malaria in susceptible areas many times over!

If this is to be a specialty-band-related trip, it is important to remember that the lower the frequency, the more critical the selection of the time for the operation. It would be prudent to examine such DX aids as the sunrise/sunset tables that appear in this book. This would help to ensure that the goals for the operation are realistic, and that every con-

sideration has been taken to optimize the results.

Learning What to Expect

Assuming that a tentative choice of one (or more) of the DXCC countries has been made as the destination, it is time now to research and investigate so as to not receive any surprises which might ruin your adventure. If the choice is a place where tourism blossoms, there will be relatively few pitfalls, and these should be very minor. If you receive a tourist visa from the embassy of the destination country late in the year, and it is numbered in single, double or triple digits, then it is safe to assume that tourism there is rare (usually for a good reason) and that some difficulty will be encountered.

Licensing is essential, of course, but it might occur at the beginning of the planning stage or after arrival at the DX location. It is easy to see the possibilities of failure if licensing is attempted only after arrival.

Some countries can be activated only in this most precarious way, because it is simply impossible with certain countries to reach the correct person via telephone or to receive an answer to a letter. In nations such as these, it is very important to reach the person who actually has the authority to grant the permission. A lesser official cannot say "yes," but if he says "no," then it will be very difficult to get a reversal of that decision.

ARRL HQ maintains a great deal of information regarding licensing in other countries*. This should adequately cover the countries where most tourists go and, surprisingly, it can be of significant benefit in the more exotic places. For example, the difference in receiving permission to operate from a former European colony might very well hinge on having a license from the former colonial power. On the other hand, depending upon the relationship between the two countries, that same license might be a ticket to sure failure. A little history study at your local library might make a significant difference. It will also increase your enjoyment of the trip!

Begin your research anywhere convenient, but do not accept either a glowing or a gloomy report at face value without evaluating the source and possible motives for the report. The US State Department will give a conservative outlook, since some of their duties encompass trying to rescue US citizens in distress. Representatives of the host nation will never volunteer anything that might be construed as derogatory about their own nation. Your evaluation should also weigh quite heavily on your own international experience (or lack thereof). Do not choose a so-called developing country as a destination unless you are a seasoned traveler. (A possible exception to this would be when a very dependable, healthy and resourceful person is going to meet your flight and assist you in-country.)

Other amateurs who have operated from the particular country are an important source of information. They understand what you want to do (it's unlikely that anyone else will understand or care). If the ill-fated Spratly Islands DXpedition had contacted the previous successful expedition (ISIDX) before them, IS might be much lower on the list of DXCC countries needed and, far more important, that awful tragedy might never have occurred. (When the ISIDX group approached one of the few islands of the Spratly Group large enough to land on, they discovered that it was occupied. Not only was it occupied, but the inhabitants were not at all sociable. Gunfire has never been a sign of open friendship! Luckily, no one was hit, and several days later the DXpedition located an uninhabited tiny patch of

*For a copy please send an SASE. Include one unit of postage for each country of interest.

Operating Overseas—A Brief Tutorial

Do you need a permit to operate in a country with which the United States holds a reciprocal-operating agreement?

Yes, you do, except in the case of Canada, where your FCC license is automatically valid. Generally, the existence of such an agreement relieves you only from taking an examination for a license and minimizes the paperwork necessary for a permit. On the other hand, the absence of such an agreement does not necessarily mean you cannot be licensed.

Where do you apply for a permit, and how?

Send a business-size SASE to the ARRL Regulatory Information Branch at HQ and indicate which country you're planning to visit. Include one unit of postage for each country of interest. You'll be sent instructions on how to apply and (if available) an application form. If sufficient up-to-date information isn't on hand, you will at least receive the address of the licensing administration or the national Amateur Radio Society (usually our International Amateur Radio Union—IARU—Sister Society) of the country from which information can be obtained.

If you do not have a specific application form, write a letter requesting a permit. This should include the purpose of your trip, dates and places of your stay, passport information and the equipment you intend to use. You should also attach a copy of your FCC license and, in some cases, a letter attesting to your character signed by your local chief of police (or equivalent).

If a certified copy of your license is required, a notary public should be able to provide you with one if he can compare a photocopy with your original (and is given the phone number of the FCC Consumer Assistance Line, 717-337-1212 or 1-800-322-1117 to obtain confirmation of the validity of the license). If you're a League member and the foreign administration concerned honors this, you may request ARRL HQ to write the appropriate administration attesting to your membership and license. Be sure to provide HQ with a photocopy of your license and an SASE.

How is an application fee paid for?

Some administrations charge for permits. It is best to send a bank draft drawn on a bank that does business in the country concerned. Or draw an international postal money order and send a copy of the receipt, unless otherwise instructed. Personal checks are rarely accepted in foreign countries.

Is it possible to apply on short notice, less than one month in advance of my projected trip?

You may be taking a chance. But if you do, and if you enclose a fee (or if you're sure there is no charge), you should furnish your address in the host country where you would like your permit, if granted, to be mailed. If you're not sure whether a fee is charged, you should promise to pay any charge when you pick up the permit upon arrival in that country.

I've applied well in advance, but I haven't received a permit. What to do?

Write or call the administration. If you prefer calling but do not have the telephone number, call their embassy or consulate in the United States and you will be furnished the number.

I don't have a permit, but want to take a transceiver into a foreign country.

For most countries, you cannot take in such equipment without a license or permit. If you've already applied for the permit and expect to pick it up on arrival in the country, show the customs officer a copy of your application. If you intend to apply for a permit later or are just carrying the equipment before or after operating in some other country, show your FCC license (and permit of the other country). This will help you verify that you are carrying communications equipment lawfully. It is quite likely, however, that your gear will be held at customs until you obtain a permit or are leaving the country. Do not try to take in any equipment secretly.

I'd like to operate "on the way" (ship or plane). How can I do this?

ARRL HQ can furnish you information on how to conduct international maritime-mobile operation. If your ship isn't one of US registry, let ARRL know the nationality. You'll need a reciprocal-operating permit before you seek the captain's permission. Amateur operation aboard a commercial passenger aircraft in the US must comply with §97.11 of the FCC rules.

Is ARRL's information accurate and up-to-date?

In the case of countries that normally license foreign amateurs, ARRL contacts the administration (or IARU Sister Societies) from time to time to confirm the information on file. Various DX news sources are followed closely to ascertain the current licensing situation. It isn't possible to be 100% up-to-date, but HQ does try.

How do I find answers to other questions?

ARRL's Regulatory Information Branch keeps track of any changes regarding reciprocal-operating and/or third-party-traffic agreements. Additionally, you can obtain a list of amateur frequencies and subbands outside of the United States. For any information request, please enclose an SASE and ask either specific questions or for a copy of the general information package.

There are cautions to keep in mind, of course. Inquiries about regulatory matters (privileges, etc) should be addressed to the administration concerned. Queries about purely amateur matters (repeaters, local clubs, etc) should go to the national Amateur Radio Society. Subjects of general matters (importing a general-coverage receiver, the limit of territorial waters, etc) should go to the embassy or consulate here in the United States.

While signing an FCC call /MM, /AM or /VE (Canada), you must observe the FCC-authorized portions that are part of the amateur frequencies of the region concerned. You must never use the 50-MHz band while cruising the Mediterranean or work SSB on 14.110 MHz, etc. But, this does not apply when you are authorized by a foreign administration (except Canada) to sign a portable FCC call, such as G/W1AW. You are then bound only by the rules and regulations of the host country.

Further specific information regarding travel, etc, within certain countries (particularly information generally not available from a travel agency or from the diplomatic mission of the country concerned) may be obtained from the US State Department. Travel advisories, with the primary purpose of alerting the public to adverse conditions in specific countries, can be obtained from the Citizens Emergency Center, Room 4811 NS, Department of State, Washington, DC 20520 (telephone 202-647-5225).

Have a good trip!—*Nao Akiyama, NX1L/JH1VRQ, International Programs Manager, ARRL*

Note to Canadian readers: For information on how to obtain an amateur license or operating permit in most foreign countries, contact RAC, Box 356, Kingston, ON K7L 4W2 (tel 613-545-9100).

sand and coral large enough to install some antennas and still have room from which to operate. Several years later another group went back to the same original island, apparently without benefit of the knowledge gained by the ISIDX crew. They were greeted by more accurate gunfire. The boat was hit, and the gasoline tanks exploded. Three persons were killed, and the survivors drifted for days in the South China Sea before being rescued by a passing merchant vessel.) Some good sources for this information would be any QSL route listing, the Callbook, The W6GO/K6HHD QSL Managers List and several other Amateur Radio magazines and newsletters (see Chapter 5 for a discussion of DX publications).

Here's to Your Health

Quite possibly the most important research to be undertaken for a DXpedition is medical. This advice is not intended to frighten you, but to impress upon you the need to be careful. If you are cautious, there is little chance of harm. However, disregard prudent health precautions and you are assured of discomfort and possibly far worse. There are two major sources of information: Your family doctor and the Centers for Disease Control and Prevention (CDCP) in Atlanta, Georgia. The CDCP is part of the Federal government and offers much good information about conditions in other parts of the world. Call the CDCP for information about your destination, even if thousands of tourists go there annually. Phone calls are inexpensive, and your health is everything! Some suggestions of subjects to inquire about include cholera, yellow fever, malaria and chloroquin-resistant malaria. Your local health department can help out with some vaccinations but the really serious ones will have to be given by your doctor. Don't put this off until the last minute. Cholera and yellow fever shots are quite unpleasant, and traveling with a sore arm or hip can do little to improve your spirits. Furthermore, cholera is a two-shot series that requires one to four weeks between vaccinations.

Here are some good guidelines that, while they might not apply everywhere, won't hurt you anywhere.

1) Do not swim, bathe or walk barefooted in or near fresh water. Parasites can cause debilitating disease.

2) Do not drink water that has not been brought to a rolling boil for at least 30 minutes or treated with water-purification tablets. Remember: Ice cubes are water.

3) Do not eat food that may have been in contact with untreated water or handled by anyone without good personal hygiene habits.

4) Do not get too romantic. Romance may be good for the soul, but exotic QTHs have some very exotic levels of disease!

5) Do not stand under mango trees. Parasitic larvae can drop onto your hair and scalp.

6) Do take some form of anti-malaria medicine, and do use insect repellent, especially if in an area where malaria and falaria (river blindness) might be prevalent.

7) Do take a first-aid kit. Talk to your family doctor and the Centers for Disease Control about its contents. A common side effect of the primary anti-malarial used is diarrhea. Water purification tablets can kill the good little creatures inside of you, resulting in diarrhea. Deaths from cholera result from dehydration caused by vomiting and diarrhea. (The most common travelers' ailment is diarrhea. Your doctor can prescribe an effective medicine to combat it.) Drink lots of (safe) water as well as control diarrhea to prevent dehydration. Sunscreen is another valuable asset to include in the first-aid kit; sunburn is an insidious and serious problem. Some additional suggestions for the first-aid kit include: Cough drops (especially for phone operators), lip balm, aspirin, an ear syringe, adhesive bandage material, antibiotic ointment and disinfectant.

Your Equipment

The selection and preparation of equipment can make the difference between success and failure, at least as far as your own evaluation of dollars spent as compared to QSOs. Almost all DXpeditions go through some stage of air travel. The maximum allowable checked luggage for flights originating from the US is three bags: Two checked and one carry-on. The total dimensions (length + width + height) of the checked luggage cannot exceed 109 inches. One single bag cannot exceed 62 inches in total dimensions. The maximum allowable weight is 70 pounds per bag. Your carry-on baggage cannot exceed 45 pounds, and it must be of a size to fit underneath the airplane seat. Working within these parameters, make the most intelligent choice of equipment for your DXpedition possible. (Note: The non US baggage weight limitation is 44 lb, so if your flight originates outside of the US and has a destination outside of the US, it will be necessary to figure within these more restrictive confines.) Always check with your airline for any last minute changes.

A DXpeditioner will have many equipment requirements for any operation. Think of a DXpedition as going out on Field Day, with the single exception that after you leave home it will not be possible to acquire anything else—either from home or from a store. Of course, radio, antennas, documents and personal effects are very broad and general classifications, but everything needed could be lumped together under one of these. Let us take them one at a time.

The Radio

Selection of the rig to be used is quite important, especially if there is any doubt about the stability of the current source to be used. The power supply should have a tapped transformer in the not-unlikely event that the voltage source will be somewhat different than the standard choices of 120/240 volts offered by some manufacturers. (From personal experience, using a Yaesu FT-101B in The Gambia in 1978: The power source was an old, old 208-volt generator, and there was a significant voltage drop along the line. My VOM read 190 volts! The Yaesu's power output on 160 meters, which was marginal to begin with, was about 20 watts. By careful selection of the voltage taps, however, normal operation was restored.)

No one in their right mind would trust an unprotected radio to the cargo hold of even the finest, most conscientious air carrier. The radio then automatically becomes carry-on luggage. Very seldom does carry-on luggage get weighed, so it is more of a weight saving to carry the heavier items and pack the lighter ones. Indeed, it is more prudent to carry a piece of equipment that has a built-in power supply than to carry the radio and pack the supply and VFO. Many of the Amateur Radio manufacturers are now building solid-state equipment which is compact enough to carry both radio and power supply in a large briefcase. A built-in second VFO or frequency memory circuit goes a long way toward conserving both space and weight. Vince Thompson, K5VT, and Erik Sjolund, SMØAGD, both fine DXpedition operators, have chosen this solid-state type of equipment (others prefer the older hybrid transceivers). Vince made his choice because he prefers to keep his radio out of sight so as to not invite questions from customs officials, while Erik likes the memory features. Erik discovered several years ago that when operating split frequency, it was easier to work the pileup when everyone was on the same frequency or spread out on several specific frequencies.

If the DXpedition will use more than one radio, these radios should be checked, while operating in close proximity to each other, for interference problems. ARRL Field Day is an excellent testing ground for this aspect of equipment performance. When this problem exists, it still might be

Modification of the Hy-Gain TH3 Jr

To modify a TH3 Jr to fit into a 38-inch cardboard box, it is necessary to make a total of 10 cuts and splices. There are six 6-foot main element halves, two 6-foot boom halves and two trap assemblies, which are longer than 3 feet. A drill, an inexpensive tubing cutter, a hacksaw, a tape measure, an old pocket knife and a smooth file are all the tools required. Several feet of different size tubing is needed for making the splices.

Tubing and pipe are different. Tubing is measured on the outside diameter (OD) and pipe on the inside diameter (ID). Tubing dimensions are expressed in nominal wall thickness in thousandths of an inch, and pipe dimensions by its "schedule." The result is that tubing ID is determined by tubing OD minus its wall thickness times 2. Commercial antennas are constructed of tubing, so to modify one to fit the dimension of your box, you will need aluminum tubing only. You will cut any piece which is too long (greater than 36 inches) into a desired length. Then during assembly at your destination, splice the two pieces together with a short aluminum sleeve and clamp it, returning that antenna part to its original dimension.

A good choice of tubing is 6061T6. The "6061" refers to the alloy and the "T6" refers to the tensile strength. This alloy is very common and inexpensive, and T6 is an excellent strength for antennas. The wall thickness for inner splices is not critical, but it should not be really thin. If some outside splices are used (the TH3 Jr uses both inside and outside splices with the same size tubing), the wall thickness can be critical. Never use 0.058-in. or larger for outside splices. The tubing increases in size steps of 1/16 of an inch, or 0.0625-in., so if you want to splice a ¾-inch tube on the inside, select a piece of 5/8-inch tubing with about 0.035-in. wall.

It will be necessary to sand down the splice (a long and difficult task) or have it "turned down" in a machine shop when splicing thick-walled tubing. This is because there is a "wall" on each side of a diameter, so the ID of ¾-inch (0.058-in.) wall is calculated by 0.750-in. (¾-inch) minus 0.116-in. (2 × 0.058-in). You might note that 1/8 of an inch is 0.125-in., so it appears that 0.058-in. wall thickness will give a side clearance of 0.009-in. Not really; it gives 0.0045-in. on each side, and, since wall thicknesses are described as nominal, it might be less. Aluminum is soft, and this whole operation is not very precise anyway. The result of using 0.058-in. wall thickness is a probability that two pieces of aluminum will get stuck together. Once this happens, the chances of getting them apart are somewhere between slim and none. Most commercial antennas are made from 0.035-in. wall and should be no trouble to modify. The exception to this is the boom of the Hy-Gain TH3 Jr. It is thick-walled; and the two splice pieces must be reduced in OD.

Once the antenna and splicing sleeves have been cut, each piece must be "deburred." The small blade of an old pocket knife will do fine for the ID, but a smooth file will be required for the OD of the cut. Rough edges should be completely eliminated since they can cut you or cause the same sticking together of antenna parts. Slots will be made in the ends of antenna material (except the boom) to be spliced if it is an inside splice or in the splice material if it

is an outside splice. This will enable you to clamp the slotted aluminum tightly onto the other piece. Cut twice with a hacksaw very close to each other (1/16 inch) and pull the little piece between the cuts outward (not inward), creating a small slot. Pulling the piece outward will make it easier to deburr. Deburring the slots is more difficult, but do it thoroughly. (Alternately, it is just as good to drill screw holes through the antenna part and the splice, and attach them with screws; but then each splice end becomes "matched," and it is necessary to do something to identify each antenna piece and its matching splice. Drilled splices can need deburring more so than slotted splices, if the drill size was too small for the screw size you use for the job.)

It will be preferable to splice the boom sections using screws, since the thickness of the material makes it difficult to compress. Simply insert the splice half way into one of the boom pieces to be joined, and drill a hole through both pieces. Install a sheetmetal screw in the hole. Next, slide the other piece of boom over the splice material, and drill a hole through that section of boom and the splice. Lock these together with another sheetmetal screw.

Aluminum tubing can be purchased several ways. Occasionally, ads in the ham magazines will feature a company selling antenna aluminum to the public. Otherwise, look in the Yellow Pages of the telephone book under "Metals." If you choose slotted splices, the local auto parts or hardware store will be able to supply you with hose clamps. If you choose a Hy-Gain antenna, replace the supplied clamps with hose clamps.

Using the same techniques, a short mast can easily be constructed. Stores that sell tubing usually have a minimum, so there can be an added incentive to building yourself a 12-foot or so mast. This gives you something to attach to another structure and costs little in space and weight.

It is advisable to splice mast sections by screws. This will ensure that the mast sections don't slide together under compression, causing a binding problem on disassembly and future use. Some of those spare clamps can readily attach your mast to a railing or flag pole too large for your antenna boom-to-mast clamp.

possible to work around it by using the radio which suffers the most interference on the lower frequency or sometimes on a band which has no harmonic relationship to the band where the interfering radio is operating. Since the intermediate frequency (IF) mixer might be involved, the solution might be as simple as finding the right band combinations by trial and error. Ferrite beads on both dc power cables, the microphone cable and the coax might help.

"Bells and whistles" might be nice to have on your home station radio, but an excess of internal options can complicate your life at a DX location which has a poor earth (or ground)

connection for the electrical power source. Most of the rest of the world uses 220 volts ac which can transfer power without the third wire as is required in the US. It is common for technicians in developing nations to quit wiring as soon as power is established and completely ignore that very important third wire.

The Antennas

While it is not the purpose of this chapter to endorse any particular product, some (either by accident or design) are more ideally suited for DXpeditions. Therefore, the first antennas discussed will be triband beams. Two different models were tried simply because of their small size and their modest acquisition cost at a local hamfest. These antennas are the Mosley TA-33 Jr and the Hy-Gain TH3 Jr. No electrical comparison was made, and the assumption is that they are equivalent antennas. The way both of them come from the factory is not very acceptable for a DXpedition, especially one on a budget, though.

The longest pieces of both antennas are 6 feet or 72 inches. The container for this antenna should stay under a total dimension of 62 inches to avoid airline excess baggage charges. Already the length is 10 inches too long, without consideration for width and height. A box 38 inches long, 12 inches wide and 12 inches high would meet this 62-inch dimension requirement, and this would require only one splice in each section of antenna longer than 3 feet or 36 inches. This allows for 1-inch-thick ends of the box. (A box this size is large enough to hold other luggage or equipment as well as the antenna.) The Mosley antenna requires several more splices than the Hy-Gain antenna, which means more work at the operating site; it also needs more different sizes of splicing material. The detailed instructions included in this chapter (see accompanying sidebar) are for modifying the Hy-Gain TH3 Jr. The two Mosley TA-33 Jr antennas modified in this manner have found their way to the Caribbean Sea and Indian Ocean areas and are serving Amateur Radio well there.

You can see in the picture that Hy-Gain used a total of eight 6-foot-long pieces of aluminum in the construction of their antenna. This is probably to take advantage of the fact that aluminum tubing is sold in 12-foot lengths. To modify any antenna with splices requires a few simple hand tools and only a moderate amount of knowledge of tubing. For a box in which to put that antenna, you might try a commercial box-manufacturing firm. Look in the Yellow Pages of the phone book under "Boxes."

Any antenna constructed of aluminum tubing can be modified easily to fit standard luggage dimensions. The most common variety, aside from Yagis, is the vertical. Probably the easiest to adapt and smallest would be the trap vertical. These antennas certainly have their place, but it is probably not on a DXpedition. Other types of verticals might perform well with a suitable ground plane, but unless installing dipoles and a beam is not possible, a trap vertical is not recommended.

Simplest and easiest to build, hardest to see and a moderately good performer is the trusty dipole. Refer to *The ARRL Antenna Book* for a wide variety of inexpensive wire antennas. A ZL-Special would make an excellent rotatable antenna for a DXpedition, especially if bamboo flourishes at your destination. An exceptionally good antenna is a three-element beam constructed of wire; though probably not rotatable, it will perform almost as well as one made of aluminum tubing. An inverted L antenna should work nicely on lower frequency bands, but necessitates your taking along an impedance-matching device. There are several small antenna tuners on the market, but this represents still another package to take along.

Any wire antennas taken on a DXpedition should be made of insulated wire. Insulation will not adversely affect their performance, but you might get your antenna much closer to high-voltage electrical wires than you would under saner circumstances. Additionally, uninsulated antenna wire could hardly double for anything else except guy wire, where insulated wire could have several alternate uses. Plan to take some extra wire also; you might need to construct another antenna or add another guy wire.

Miscellaneous Supplies

Plenty of coaxial cable (figure on about 100 feet per antenna) is recommended for several reasons. It might be necessary to locate the antennas quite some distance from the actual station. If more than one radio is operating at the same time and there is an interference problem, it might help to roll an RF choke in the coax at several places. If you have some clip-on ferrite chokes (or if you've assembled your coax with ferrite beads beforehand), these would serve the same purpose. Place the chokes as necessary to break up the feed-line into nonresonant lengths. Try placing one choke at the antenna, one at the radio and a couple spaced at nonresonant distances from the radio. (Antenna currents along the outside of the feed line can cause an interference problem, if only by having the effect of bringing the RF field closer to the other transceiver.) The last and possibly the most important reason is the possibility that you might meet a local amateur who has great need for such a simple commodity!

Miscellaneous supplies and equipment should include tools to assemble everything you are taking with you, connectors for joining two coaxial fittings, extension cord(s), a multiple electrical outlet, a foreign voltage converter (with adapter plugs), wire cutters, a pocket knife, tools for opening up your radio, your radio's owner's manual, a 220-volt soldering iron (120-volt will do if you absolutely can't find a 220-volt iron and are very careful to unplug it just as soon as it gets hot enough to melt solder), solder, nylon cord (to support wire antennas and to turn your triband beam), strapping tape, several hose clamps (a variety of sizes, along with the strapping tape can be used to secure antennas, lights and numerous other things), a Boy Scout compass and a map (a great circle centered on your destination might be hard to find so a Polar projection will do nicely. To find approximate directions, just orient it with your location directly underneath the North Pole).

Documents

The necessary and nice-to-have papers are relatively few, but as usual there are some simple items which might make things go much smoother for you.

Your passports are especially important. Yes, *passports*—if your journey includes travel into countries which are at odds with each other. The US State Department will issue more than one passport if you need it. Reserve your short-term passport for entry into countries such as the Republic of South Africa (RSA). (Never present a passport with an RSA stamp or visa at almost any other African nation's border or embassy.)

Try to have all of the visas for all of the countries you might visit before you go. There are numerous visa services in both Washington, DC, and New York City, which can assist you and expedite the handling of your papers. These companies can "walk through" a passport application in days instead of the weeks it normally takes. They do charge a fee for each service they offer. Visa service companies normally have travel agents as their clients, but they will take individuals, if you are willing to pay up front.

Extra passport pictures can make the difference if you wish to alter your travel plans in mid-stream or if it was not possible to get all of the necessary visas before beginning your travel. For example, suppose you planned a trip to Nigeria (5N) and you entered the country, but then left Nigeria for another country. If you wanted to return to Nigeria, you would need an additional Nigerian visa since they don't (at this writing) give multiple-entry visas. Also, never pack your passport in your suitcase or leave it in your hotel room.

Your amateur licenses and several photocopies of them should be with your important papers. If all goes well, you will have obtained operating permission before beginning a DXpedition. Take several photocopies of whatever you have. When customs inspectors see your equipment, they will also want to see some papers. They will go away with your papers, and since you just got there, you might not be able to find or identify the person to whom you gave them. You won't worry as much if you have more copies. (Photocopy machines are a rarity in parts of the world generally associated with DXpeditions.)

Your International Certificate of Vaccination (which you should obtain from your local health department), if lost, might mean that you would have to be reinoculated. It is theorized that dirty medical needles in Africa are responsible for the spread of disease in that continent! You do not want to have to be reinoculated!

Personal Effects

Your personal effects include all of your clothing, which should be appropriate to the climate where you are heading. The desert requires long sleeves, not short, to protect your arms from sunburn in the daytime and keep you warm at night (deserts get very cold). The only price you pay is a little sweat. Cotton is preferable to polyester for coolness and comfort. Revealing or skimpy clothing for women is not recommended for travel in Islamic countries (a woman in a conservative one-piece bathing suit is considered to be nude). Always take very comfortable and broken-in walking shoes.

Immersion water heaters can be of great service to you. Don't expect to boil water for 30 minutes with them, but it is quite nice to make instant coffee or instant soup which you should also take along. If no 220-volt models are to be found, it is possible to put two 120-volt heaters in series with no ill effects. (This is a simple trick using the adapter cord described in the accompanying sidebar.)

In addition to instant drinks and soups, a small amount of snacks such as granola bars might be useful in case mealtime coincides with a band opening that is too good to miss. Food might just not be affordable because of artificially controlled exchange rates, or, as in some countries, there may simply not be enough food. If a person can go without any food for a couple of months and survive, then you should have no major difficulty living on instant soup, powdered milk and granola bars for a couple of weeks.

A small battery-operated or key-wound alarm clock will help you keep your log accurately, help you meet schedules and/or avoid sleeping through band openings, and keep you from missing your plane when it comes time to go home! If your clock operates off of ac voltage, make sure that it has its own crystal time base. (Clocks which use the line voltage's 60-cycle frequency are slightly better than useless if the power source is an overloaded 50-cycle generator.) A wristwatch alarm might not be loud enough to wake a tired and weary DXer.

Here is a list of other items which ought to be considered for taking along as well: toilet tissue, laundry powder, a universal bathtub stopper (flat round rubber), eating utensils, candles, matches (or disposable lighter), flashlight and spare batteries (or disposable flashlights), water-purification tablets and a couple of collapsible water containers. Many of the things you might need will be found where camping supplies are sold in your hometown. Of course, the rarer your destination, the more important the list becomes.

Plenty of logbooks or logsheets are a must! You might want to design your own logs, similar to a contest log, containing the minimum amount of information for the maximum number of contacts. Several cheap black (not blue) ink ballpoint pens will ensure that your logs will photocopy accurately, but using sharpened pencils is often preferred since this permits erasures and corrections while you're operating (make sure you obtain an inexpensive, battery-powered pencil sharpener to go with the pencils). And by photocopying your logs when you return, you and your QSL manager can each have a set.

TRAVEL AND ARRIVAL

As previously mentioned, a travel agent will help you immensely and will not charge for the services. The travel agent will be paid by the airlines and other service companies that receive your business. Some knowledge is recommended on your part to ensure that the agent provides the service you desire.

As discussed earlier, the 70 pounds maximum per bag is for flights to and from the US. This higher limitation is required by the US Government, while, outside the US, the limitation is set by the International Airline Transport Authority (IATA). Membership by airlines in IATA is voluntary, and it should be realized that excess-baggage charges are a significant portion of international airline revenues.

For these reasons, it might be impossible to avoid excess baggage charges on your return trip from the DXpedition. Even checking the luggage straight through to the US might not be enough, unless only one carrier is used. Sometimes that doesn't work either. Plan on taking enough money with you to cover the possibility of excess-baggage charges, just in case.

A woman's purse would never be considered as "carry-on" or "hand" baggage. A giant purse will hold many breakable things such as a VSWR bridge, VOM, spare tubes, microphone—and her normal purse. A man's toiletry kit would be in the same category. What a perfect place to carry all of those heavy tools that would otherwise skyrocket those excess-baggage charges! It may be cumbersome and uncomfortable, but it will keep more cash in your pocket!

The fewer airlines and fewer connections involved in any trip, the less likely are travel-connected problems. It is also a good idea to pack in a balanced way, so that, if only one piece of luggage arrives at your destination, you will have at least your minimum personal requirements and enough equipment

Countries with which the US shares Reciprocal Operating Agreements

A2	Botswana	HR	Honduras	PA	Netherlands	XE	Mexico
C6	Bahamas	HS	Thailand	PJ	Netherlands Antilles	YB	Indonesia
CE	Chile	I	Italy	PY	Brazil	YN	Nicaragua
CP	Bolivia	J3	Grenada	PZ	Suriname	YU	Yugoslavia
CT	Portugal	J6	St Lucia	S7	Seychelles	YV	Venezuela
CX	Uruguay	J7	Commonwealth of	SM	Sweden	YS	El Salvador
DL	Germany		Dominica	SV	Greece	ZL	New Zealand
DU	Philippines	J8	St Vincent	T2	Tuvalu	ZP	Paraguay
EA	Spain	JA	Japan	T3	Kiribati	ZS	South Africa
EI	Ireland	JY	Jordan	TF	Iceland	3A	Monaco
EL	Liberia	LA	Norway	TG	Guatemala	3D	Fiji
F	France*	LU	Argentina	T1	Costa Rica	4X	Israel
G	United Kingdom**	LX	Luxembourg	V2	Antigua & Barbuda	5B	Cyprus
H4	Solomon Islands	OA	Peru	V3	Belize	6Y	Jamaica
HB	Switzerland	OE	Austria	V6	Micronesia	8P	Barbados
HC	Ecuador	OH	Finland	V7	Marshall Is	8R	Guyana
HH	Haiti	ON	Belgium	VE	Canada	9K	Kuwait
HI	Dominican Republic	OZ	Denmark	VK	Australia	9L	Sierra Leone
HK	Colombia	P2	Papua New Guinea	VU	India	9Y	Trinidad &
HP	Panama						Tobago

Notes: An automatic reciprocal agreement exists between the US and Canada, so there is no need to apply for a permit. Simply sign your US call followed by a slantbar and the Canadian letter/number identifier.
*Including French Guiana, French Polynesia, Guadeloupe, Amsterdam Island, Saint-Paul Island, Crozet Island, Kerguelen Island, Martinique, New Caledonia, Reunion, St. Pierre and Miquelon, and Wallis and Futuna Islands.
**Including Bermuda, British Virgin Islands, Cayman Islands, Channel Islands (including Guernsey and Jersey), and Falkland Islands (including South Georgia Islands and South Sandwich Islands), Gibraltar, Isle of Man, Montserrat, St. Helena (including Ascension Island, Gough Island and Tristan Da Cunha Island), Northern Ireland, and the Turks and Caicos Islands.

to put one station on the air.

As risky as it might seem, it might be advisable to take at least part of your money in cash. There are places in the world where traveler's checks and credit cards are not honored. If you take a significant amount of cash, purchase a money belt or similar pickpocket-proof way of carrying the bulk of your money. Of course, keep out enough for taxis, tips and other incidental expenses. Several international airports in the Western world have bad reputations for thieves, and some third-world airports can be downright dangerous for the unwary.

Strapping tape wrapped tightly around your suitcase might keep thieves from looking inside. Protecting your person might require a little more imagination. One large infamous airport has been the scene of such a regularly occurring ruse that it is worthwhile to mention here. You are standing in the terminal and an official-looking person walks up and demands to see some identification. Dutifully you show your passport and other papers. The person examines them carefully, returns them and leaves. (Or the person reads your name from a luggage tag.) A few minutes later another person arrives and asks if you are Mr, Mrs or Ms (insert your name) and, when given an affirmative answer, says that he has been sent to pick you up. If this person is legitimate, then he will have something from the party who was supposed to meet you. If he can't produce something to satisfy you, then don't go with him; he is probably a thief, who obtained your name from the "official."

Don't misunderstand: International travel is not dangerous in and of itself. There are some hot spots around the world to be avoided, and many places where close attention should be paid to activities around you. It should also be noted that it is highly illegal to take photographs at the airports of nearly all of the so-called developing nations. Many of these countries are overly sensitive about their security and about possible spies. Carrying a radio makes you doubly suspicious.

Upon arrival at your destination, you and your equipment will have to go through customs and immigration. If operating permission has not already been secured, be prepared to leave your radio equipment "in bond" at the airport or some customs warehouse. This is usually quite safe, as long as you get a receipt for whatever is left there. Some countries, such as the Bahama Islands, may require you to post a sizeable cash bond even though you have operating permission. This is to ensure that you are not importing merchandise to sell and are just trying to avoid import taxation. You won't be helping any local amateurs at all if you get caught illegally importing Amateur Radio equipment into their country!

Some countries, especially the poorer developing nations, have a problem with the authorities soliciting bribes. The residents of these countries should be entitled to good government, whether or not they get it. Don't contribute to the problem by paying bribes!

Many hotels and private residences in the Caribbean actively solicit hams to stay at their facilities for DXpedition vacations. One hotel in Freeport, Bahamas, has a ham radio suite on the top floor with specially drilled holes to accommodate coax from the rooftop. Some hotels will also gladly provide postcards for QSL cards. Almost all hotels will try their best to help you, as long as other guests are not disturbed and your antennas don't present a safety hazard.

(text continues on page 6-13)

Electric Current Abroad

[Information courtesy of the US Department of Commerce.]

The characteristics of electric current—type (alternating or direct), number of phases, frequency (hertz) and voltage—found in major foreign cities are listed in this report. In addition, the stability of the frequency and the number of wires to a commercial or residential installation are given where available.

The current characteristics and other data furnished relate to domestic and commercial service only. It does not include special commercial installations involving relatively high voltage requirements, nor does it refer to any industrial installations.

Persons who are planning to use or ship appliances abroad should acquaint themselves with the characteristics of the electric supply available in the area in which the appliance is to be used. It may be less expensive to buy a new appliance than to purchase the auxiliary equipment needed to make presently owned appliances usable. A transformer may be obtained locally to correct the voltage if required; however, if the operation of an appliance will require exact timing or speed and if the frequency of the foreign electricity supply is other than that for which a United States appliance has been designed, it is advisable that an appliance designed especially for the foreign frequency to be used. Auxiliary equipment to change frequency is too bulky and expensive for the average residential or commercial installation.

Some foreign hotels have a special circuit, providing approximately 120 volts, for the convenience of guests using electric shavers and other low-wattage appliances made in the US. Such circuits have markers at the convenience outlets indicating the voltage.

Among the nominal voltages indicated in this report, the lower voltages shown are used primarily for lighting and smaller appliances, while the higher voltages are used primarily for air conditioners, heating and other large appliances.

Readers are reminded that the list of characteristics presented here were compiled over a period of months from a large number of sources. There is, consequently, some possibility of errors and omissions for which the Department of Commerce cannot assume responsibility.

Readers are further reminded that the information presented here should not be taken as final in the case of industrial or highly specialized commercial installations. The Department of Commerce regrets that it is impossible for it to maintain complete data on every foreign industrial installation. It is recommended that for special equipment required for commercial use or any heavy equipment required for industrial use, the current characteristics for the area of installation be obtained from the end user.

[Editor's Note: There are often different *sizes* in use within these plug types, and the traveler might even find a mixture of these types being used in developing countries. Former French colonies tend to use French-style plugs. Voltage adapters that cover almost anything are usually on sale at major European airports.]

Key to Terms Used

Type of Current—ac indicates alternating current and dc, direct current.

Frequency—Shown in number of hertz (cycles per second). Even if voltages are similar, a 60-hertz US clock or tape recorder will not function properly on 50-hertz current.

Number of phases—1 and 3 are the conventional phases which may be available.

Nominal voltages—Direct current nominal voltages, for example, are: 110/220 and 120/240. The lower voltage is always one-half of the higher voltage. On a direct current installation, the lower voltage requires 2 wires while the higher voltage requires 3 wires.

Alternating current nominal voltage: Alternating current is normally distributed either through 3-phase wye ("star") or delta ("triangle"), 4-wire secondary distribution systems. In the wye or star distribution system the nominal voltage examples are: 120/208, 127/220, 220/380 and 230/400. The

TYPE A
FLAT BLADE ATTACHMENT PLUG

TYPE B
FLAT BLADES WITH ROUND GROUNDING PIN

TYPE C
ROUND PIN ATTACHMENT PLUG

TYPE D
ROUND PINS WITH GROUND

TYPE E
ROUND PIN PLUG AND RECEPTACLE WITH MALE GROUNDING PIN

TYPE F
"SCHUKO" PLUG AND RECEPTACLE WITH SIDE GROUNDING CONTACTS

TYPE G
RECTANGULAR BLADE PLUG

TYPE H
OBLIQUE FLAT BLADES

TYPE I
OBLIQUE FLAT BLADES WITH GROUND

TYPE J
OBLIQUE FLAT BLADES WITH GROUND

higher voltage is 1.732 (the square root of 3) times the lower voltage. In a delta or triangle system, 110/220 and 230/460 are examples of nominal voltages. The higher voltage is always double the lower voltage.

The higher voltage is obtained by using 2- or 3-phase wires and the neutral wire while the lower voltage is the voltage between the neutral wire and one phase wire. The higher voltage may be single or 3 phase while the lower voltage is always single phase and used primarily for lighting and for small appliances.

In this report, the term nominal voltage is used to denote the reported voltage in use in the majority of residential and commercial establishments in the city or area named.

Type of attachment plug in use—Attachment plugs used throughout the world come in various forms, dimensions and configurations too numerous to describe in this report. This report does, however, attempt to point out the basic and most commonly used types of plugs by country. Adapters may be purchased to change from the American type to other types. The availability of these adapters is shown for each country immediately after the key to the type of plug used.

Number of wires to the consumer—The number of wires which may be used by the consumer is shown. Normally, a single phase, 220/380 volt system or 127/220 system will have two wires if only the lower voltage is available (one phase wire and the neutral). It will have three wires if both the higher and lower voltages are available (two phase wires and the neutral) and where three phase motors will be used, four wires will be available for the higher voltage (the three phases wires and the neutral wire).

Frequency stability—A "Yes" in the column indicates that the frequency stability is good and that service interruptions are rare.

Types of Electric Plugs by Country

Country	Plug type shown	Adapters readily available
Afghanistan	D	Yes
Algeria	C & D	No
American Samoa	A,B,F & I	Yes
Angola	C	—
Antigua	A,B & G	Yes
Argentina	C & G	—
Australia	I	No
Austria	F	No
Azores	C & D	Yes
Bahamas	A & B	Yes
Bahrain	D & G	Yes
Bangladesh	C & D	No
Barbados	A & B	—
Belgium	C & E	No
Belize	A & B	—
Benin	D	—
Bermuda	A,B,G & I	No
Bolivia	A,B,C & F	—
Botswana	D & G	Yes
Brazil	A,B,C,D,E,F, G,H,I & J	Yes
Bulgaria	C & F	No
Burma [Myanmar]	D & G	—
Burundi	C,E & F	No
Cambodia [Kampuchea]	No information	
Cameroon	C & E	—
Canada	A & B	—
Canary Islands	C & E	Yes
Cape Verde, Rep of	C & F	No
Cayman Islands	A & B	—
Central African Rep	C	—
Chad	D,E, & F	No
Channel Islands	C & G	Yes
Chile	C	Yes
China, People's Rep of	C & J	No
Colombia	A & C	Yes
Congo	C	—
Costa Rica	A,D,I & J	Yes
Cyprus	G	Yes
Czechoslovakia	E	No
Denmark	C	No

Country	Plug type shown	Adapters readily available
Djibouti	C & E	No
Dominica	G	—
Dominican Rep	A & J	No
Ecuador	A & C	—
Egypt	C	No
El Salvador	A,B,G,I & J	No
England	C & G	Yes
Equatorial Guinea	C	No
Ethiopia	C,D & F	—
Faeroe Islands	C	No
Fiji	I	Yes
Finland	C & F	No
France	C,E,F, & G	Yes
French Guiana	C,D & E	No
Gabon	C & E	—
Gambia, The	G	Yes
Germany	F & C	—
Ghana	C,D & G	No
Gibraltar	C & G	Yes
Greece	C,D & F	No
Greenland	C	No
Grenada	C,D & G	Yes
Guadeloupe	C,D & E	Yes
Guam	A & B	—
Guatemala	A & I	—
Guinea	C & E	No
Guinea-Bissau	C,D & F	No
Guyana	A,B,C,D & G	Yes
Haiti	A,B,I & J	—
Honduras	A & B	—
Hong Kong	D	No
Hungary	F	—
Iceland	C & F	Yes
India	C & D	—
Indonesia	C,E, & F	Yes
Iran	No Information	—
Iraq	C,D & G	No
Ireland	F & G	—
Isle of Man	C & G	Yes
Israel	C,D & J	Yes

Electric Current Abroad—Characteristics

This is a simplified version (with footnotes omitted) of the detailed listing appearing in *Electric Current Abroad*, published by the US Department of Commerce. For further particulars (all such information is subject to change at any moment), obtain the latest edition of this booklet from the Commerce Department. The column heading key can be found on page 6-13.

[1]	[2]	[3]	[4]	[5]	[6]
Afghanistan					
Kabul	ac 50	1,3	220/380	2,4	Yes
Algeria					
Algiers	ac 50	1,3	127/220 220/380 120/240	2,4	Yes
American Samoa					
	ac 60	1,3	240/480	2,3,4	Yes
Angola	ac 50	1,3	220/380	2,4	Yes
Antigua	ac 60	1,3	230/400	2,3,4	Yes
Argentina					
Buenos Aires	ac 50	1,3	220/380	2,4	Yes
	dc		220/440	2,3	Yes
Australia					
Canberra	ac 50	1,3	240/415	2,3,4	Yes

[1]	[2]	[3]	[4]	[5]	[6]
Austria					
Vienna	ac 50	1,3	220/380	2,4	Yes
Azores	ac 50	1,3	220/380	2,3,4	Yes
Bahamas					
Nassau	ac 60	1,3	120/240 120/208	2,3,4	Yes
Bahrain	ac 50	1,3	230/400	2,3,4	Yes
Bangladesh					
Dacca	ac 50	1,3	220/380	3,4	No
Barbados					
Bridgetown	ac 50	1,3	115/230 115/200	3,4	Yes
Belgium					
Brussels	ac 50	1,3	127/220 220/380	2,3,4	Yes
Belize	ac 60	1,3	110/220	2,3,4	Yes
Benin	ac 50	1,3	220/380	2,4	Yes
Bermuda	ac 60	1,3	120/240 120/208	2,3,4	Yes

Country	Plug type shown	Adapters readily available
Italy	C & F	Yes
Ivory Coast	C	—
Jamaica	A & B	—
Japan	A	—
Jerusalem	C,D & J	Yes
Jordan	C,F & G	Yes
Kenya	D & G	—
Korea	A & F	Yes
Kuwait	C,D & G	Yes
Laos	A & C	Yes
Lebanon	C	—
Lesotho	C	No
Liberia	A & G	No
Libya	D	—
Luxembourg	C & F	Yes
Macao	C & D	—
Madagascar	C & E	No
Madeira	C & D	Yes
Majorca Island	C & E	Yes
Malawi	G	—
Malaysia	G	Yes
Maldives	D	—
Mali	C & E	—
Malta	G	No
Martinique	C,D & E	Yes
Mauritania	C	—
Mauritius	C,D & G	—
Mexico	A	—
Monaco	C,D,E & F	Yes
Montserrat	A,B & G	Yes
Morocco	C,D,E & F	—
Mozambique	C & F	—
Nepal	D	Yes
Netherlands	C & F	No
Netherlands Antilles	A,B,C,D & F	Yes
New Caledonia	No Information	
New Zealand	I	Yes
Nicaragua	A	—
Niger	C	No
Nigeria	D & G	—
Northern Ireland	C & G	Yes
Norway	C & F	No
Okinawa	A	—
Oman	D & G	Yes
Pakistan	C & D	No
Panama	A & B	—
Papua New Guinea	H & I	No
Paraguay	C	No
Peru	A	—
Philippines	A,B,C,D & J	Yes
Poland	C,E & F	No
Portugal	C & D	—
Puerto Rico	A & B	—
Qatar	D & G	Yes
Romania	C & F	—
Rwanda	C & J	No
St Kitts-Nevis	D & G	Yes
St Lucia	G	—
St Vincent	G	—
Saudi Arabia	A,B,C,D,E,F,G, H,I & J	No
Scotland	C & G	Yes
Senegal	C & F	No
Seychelles	D & G	No
Sierra Leone	D & G	—
Singapore	C & G	Yes
Somalia	C & D	Yes
South Africa, and Southwest Africa/Namibia	D	—
Spain	C & E	Yes
Sri Lanka	D	No
Sudan	C,F & G	No
Suriname	C & F	Yes
Swaziland	D	—
Sweden	C & F	No
Switzerland	C & F	No
Syria	C	Yes
Tahiti	A	—
Taiwan	A,B,H,I & J	Yes
Tanzania	D & G	—
Thailand	A & C	No
Togo	C & E	—
Tonga	D & I	Yes
Trinidad & Tobago	A,B,D,G & I	Yes
Tunisia	C	—
Turkey	C,E & F	—
Uganda	G	—
USSR	C & I	No
United Arab Emirates	D & G	Yes
Upper Volta [Burkina Faso]	C	—
Uruguay	C & I	Yes
Venezuela	I & J	—
Virgin Islands (American)	A & B	—
Wales	C & G	Yes
Western Samoa	No Information	
Yemen	A, C & D	—
Yugoslavia	C & F	—
Zaire	E	Yes
Zambia	G	—
Zimbabwe	D & G	No

[1]	[2]	[3]	[4]	[5]	[6]
Bolivia La Paz	ac 50	1,3	115/230 220/380	2,3	Yes
Botswana Gaborone	ac 50	1,3	220/380	2,4	Yes
Brazil Brazilia	ac 60	1,3	220/380	2,3,4	Yes
Bulgaria Sofia	ac 50	1,3	220/380	2,4	No
Burkina Faso Ouagadougou	ac 50	1,3	220/380	2,4	No
Burma [Myanmar] Rangoon	ac 50	1,3	230/400	2,4	No
Burundi	ac 50	1,3	220/380	2,4	No
Cambodia [Kampuchea] Phnom-Penh	ac 50	1,3	220/380	2,3,4	No
Cameroon Yaounde	ac 50	1,3	127/220 220/380	2,4	Yes
Canary Islands	ac 50	1,3	127/220 220/380	2,3,4	Yes
Cape Verde	ac 50	1,3	220/380	2,3,4	No
Cayman Islands	ac 60	1,3	120/240	2,3	Yes
Central African Republic Bangui	ac 50	1,3	220/380	2,4	Yes
Chad	ac 50	1,3	220/380	2,4	No
Channel Islands	ac 50	1,3	240/415	2,4	Yes
Chile Santiago	ac 50	1,3	220/380	2,3,4	Yes
China, People's Rep of Beijing	ac 50	1,3	220/380	2,4	No
Colombia Bogota	ac 60	1,3	150/240	2,3,4	Yes
	ac 60	1,3	110/220	2,3,4	Yes
Congo, Rep of Brazzaville	ac 50	1,3	220/380	2,4	Yes
Costa Rica San Jose	ac 60	1,3	120/240	2,3,4	Yes

Overseas DXing/DXpeditions   6-11

[1]	[2]	[3]	[4]	[5]	[6]
Cyprus	ac 50	1,3	240/415	2,4	Yes
Czechoslovakia					
Prague	ac 50	1,3	220/380	2,3,4	Yes
Denmark					
Copenhagen	ac 50	1,3	220/380	2,3,4	Yes
Djibouti	ac 50	1,3	220/380	2,4	Yes
Dominica	ac 50	1,3	230/400	2,4	N.A.
Dominican Republic					
Santo Domingo	ac 60	1,3	110/220	2,3	Yes
Ecuador					
Quito	ac 60	1,3	120/208	2,3,4	Yes
			127/220		
Egypt					
Cairo	ac 50	1,3	220/380	2,3,4	No
El Salvador					
San Salvador	ac 60	1,3	115/230	2,3	Yes
Equatorial Guinea	ac 50	1	220	2	No
Ethiopia					
Addis Ababa	ac 50	1,3	220/380	2,4	Yes
Faeroe Islands	ac 50	1,3	220/380	2,3,4	Yes
Fiji	ac 50	1,3	240/415	2,3,4	Yes
Finland					
Helsinki	ac 50	1,3	220/380	2,4	Yes
France					
Paris	ac 50	1,3	115/230	2,4	Yes
French Guiana					
Cayenne	ac 50	1,3	220/380	2,3,4	No
Gabon					
Libreville	ac 50	1,3	220/380	2,4	Yes
Gambia, The					
Banjul	ac 50	1,3	220/380	2,4	No
Germany					
Berlin	ac 50	1,3	220/380	2,4	Yes
Ghana					
Accra	ac 50	1,3	220/440	2,3,4	No
Gibraltar	ac 50	1,3	240/415	2,4	Yes
Greece					
Athens	ac 50	1,3	220/380	2,4	Yes
Greenland					
Godthaab	ac 50	1,3	220/380	2,3,4	Yes
Grenada	ac 50	1,3	230/400	2,4	N.A.
Guadeloupe	ac 50	1,3	220/380	2,3,4	Yes
Guam	ac 60	1,3	110/220	3,4	Yes
			120/208		
Guatemala					
Guatemala City	ac 60	1,3	120/240	2,3,4	Yes
Guinea					
Conakry	ac 50	1,3	220/380	2,3,4	No
Guinea-Bissau	ac 50	1,3	220/380	2,3,4	No
Guyana					
Georgetown	ac 50	1,3	110/220	2,3,4	Yes
Haiti					
Port-au-Prince	ac 60	1,3	110/220	2,3,4	No
Honduras					
Tegucigalpa	ac 60	1,3	110/220	2,3	No
Hong Kong	ac 50	1,3	200/346	2,3,4	Yes
Hungary					
Budapest	ac 50	1,3	220/380	2,4	Yes
Iceland					
Reykjavik	ac 50	1,3	220/380	2,3,4	Yes
India					
New Delhi	ac 50	1,3	230/400	2,4	Yes
			230/415		
Indonesia					
Jakarta	ac 50	1,3	220/380	2,4	Yes
Iran					
Tehran	ac 50	1,3	220/380	2,3,4	Yes
Iraq					
Baghdad	ac 50	1,3	220/380	2,4	Yes
Ireland (Eire)					
Dublin	ac 50	1,3	220/380	2,4	Yes
Isle of Man					
Douglas	ac 50	1,3	240/415	2,4	Yes
Israel					
Tel Aviv	ac 50	1,3	230/400	2,4	Yes
Italy					
Rome	ac 50	1,3	127/220	2,4	Yes
			220/380		
Ivory Coast					
Abidjan	ac 50	1,3	220/380	2,4	Yes
Jamaica					
Kingston	ac 50	1,3	110/220	2,3,4	Yes

[1]	[2]	[3]	[4]	[5]	[6]
Japan					
Tokyo	ac 50	1,3	100/200	2,3	Yes
Jordan					
Amman	ac 50	1,3	220/380	2,3,4	Yes
Kenya					
Nairobi	ac 50	1,3	240/415	2,4	Yes
Korea	ac 60	1,2	220/380	2,3,4	Yes
Kuwait	ac 50	1,3	240/415	2,4	Yes
Laos					
Vientiane	ac 50	1,3	220/380	2,4	Yes
Lebanon					
Beirut	ac 50	1,3	110/190	2,4	No
Lesotho					
Maseru	ac 50	1,3	220/380	2,4	Yes
Liberia					
Monrovia	ac 60	1,3	120/240	2,3,4	No
			120/208		
Libya					
Tripoli	ac 50	1,3	127/220	2,4	Yes
Luxembourg	ac 50	1,3	120/208	3,4,5	Yes
			220/380		
Macao	ac 50	1,3	220/380	2,3	Yes
Madagascar					
Tananarive	ac 50	1,3	127/220	2,3,4	Yes
			220/380		
Madeira					
Funchal	ac 50	1,3	220/380	2,3,4	Yes
	dc		220/440	2,3	Yes
Majorca Island					
Palma de					
Majorca	ac 50	1,3	127/220	2,3,4	Yes
			220/380		
Malawi					
Blantyre	ac 50	1,3	230/400	3,4	Yes
Malaysia					
Kuala Lumpur	ac 50	1,3	240/415	2,4	Yes
Maldives	ac 50	1,3	230/400	2,4	Yes
Mali, Rep of					
Bamako	ac 50	1,3	220/380	2,4	No
Malta					
Valletta	ac 50	1,3	240/415	2,4	Yes
Mexico					
Mexico, DF	ac 60	1,3	127/220	2,3,4	No
Monaco					
Monte-Carlo	ac 50	1,3	127/220	2,4	Yes
			220/380		
Montserrat	ac 60	1,3	230/400	2,4	NA
Morocco					
Rabat	ac 50	1,3	127/220	2,4	Yess
Mozambique	ac 50	1,3	220/380	2,3,4	Yes
Nepal					
Kathmandu	ac 50	1,3	220/440	2,4	No
Netherlands					
The Hague	ac 50	1,3	220/380	2,4	Yes
Netherlands Antilles					
Curacao	ac 50	1,3	223/380	2,3,4	Yes
New Caledonia					
Noumea	ac 50	1,3	220/380	2,3,4	Yes
New Zealand					
Auckland	ac 50	1,3	230/400	2,3,4	Yes
Nicaragua					
Managua	ac 60	1,3	120/240	2,3,4	Yes
Niger					
Niamey	ac 50	1,3	220/380	2,3,4	No
Nigeria					
Lagos	ac 50	1,3	230/415	2,4	Yes
Norway					
Oslo	ac 50	1,3	230	2,3	Yes
Okinawa	ac 60	1	120/240	2,3	Yes
			100/200		
Oman					
Muscat	ac 50	1,3	240/415	2,4	Yes
Pakistan					
Rawalpindi	ac 50	1,3	230/400	2,3,4	No
Panama					
Panama City	ac 60	1,3	115/230	2,3,4	Yes
			126/208		
Papua New Guinea	ac 50	1,3	240/415	2,4	Yes
Paraguay					
Asuncion	ac 50	1,3	220/380	2,4	Yes

[1]	[2]	[3]	[4]	[5]	[6]
Peru					
Lima	ac 60	1,3	220	2,3	Yes
Philippines					
Manila	ac 60	1,3	115/230 110/220	2,3,4	Yes
Poland					
Warsaw	ac 50	1,3	220/380	2,4	No
Portugal					
Lisbon	ac 50	1,3	220/380	2,3,4	Yes
Puerto Rico					
San Juan	ac 60	1,3	120/240	2,3,4	Yes
Qatar					
Doha	ac 50	1,3	240/415	2,4	No
Romania					
Bucharest	ac 50	1,3	220/380	2,4	Yes
Rwanda	ac 50	1,3	220/380	2,4	Yes
St Kitts and Nevis	ac 60	1,3	230/400	2,4	Yes
St Lucia	ac 50	1,3	240/416	2,4	NA
St Vincent	ac 50	1,3	230/400	2,4	NA
Saudi Arabia					
Jidda	ac 60	1,3	127/220	2,4	No
Senegal					
Dakar	ac 50	1,3	127/220	2,3,4	No
Seychelles	ac 50	1,3	240	2,3	Yes
Sierra Leone					
Freetown	ac 50	1,3	230/400	2,4	No
Singapore	ac 50	1,3	230/400	2,4	Yes
Somalia					
Mogadishu	ac 50	1,3	220/380	2,4	No
South Africa and Southwest Africa/Namibia					
Cape Town		1,3	220/380	2,4	Yes
Spain					
Madrid	ac 50	1,3	127/220 220/380	2,3,4	Yes
Sri Lanka					
Colombo	ac 50	1,3	230/400	2,4	Yes
Sudan					
Khartoum	ac 50	1,3	240/415	2,4	Yes
Suriname	ac 60	1,3	115/230	2,3,4	Yes
Swaziland	ac 50	1,3	230/400	2,4	Yes
Sweden					
Stockholm	ac 50	1,3	220/380	2,3,4	Yes
Switzerland					
Bern	ac 50	1,3	220/380	2,3,4	Yes
Syria					
Damascus	ac 50	1,3	220/380	2,3	No
Tahiti					
Papeete	ac 60	1,3	127/220	2,3,4	No
Taiwan					
Taipei	ac 60	1,3	110/220	2,3,4	Yes
Tanzania					
Dar es Salaam	ac 50	1,3	230/400	2,3,4	Yes

[1]	[2]	[3]	[4]	[5]	[6]
Thailand					
Bangkok	ac 50	1,3	220/380	2,3,4	Yes
Togo					
Lome	ac 50	1,3	127/220 220/380	2,4	Yes
Tonga	ac 50	1,3	240/415	2,3,4	NA
Trinidad and Tobago	ac 60	1,3	115/230 230/400	2,3,4	Yes
Tunisia					
Tunis	ac 50	1,3	127/220 220/380	2,4	Yes
Turkey					
Ankara	ac 50	1,3	220/380	2,3,4	Yes
Uganda					
Kampala	ac 50	1,3	240/415	3,4	Yes
USSR					
Moscow	ac 50	1,3	220/380	NA	No
United Arab Emirates					
Abu Dhabi	ac 50	1,3	240/415	2,3,4	No
United Kingdom					
London	ac 50	1	240/480	2,3	Yes
		3	240/415	4	Yes
Wales	ac 50	1,3	240/415	2,4	Yes
Northern Ireland	ac 50	1,3	220/380 230/400	2,4	Yes
Uruguay					
Montevideo	ac 50	1,3	220	2,3	Yes
Venezuela					
Caracas	ac 60	1,3	120/240	2,3,4	Yes
Viet Nam					
Saigon-Cholon	ac 50	1,3	120/208 220/380	2,4	No
Virgin Islands (US)	ac 60	1,3	120/240	2,3,4	Yes
Western Samoa	ac 50	1,3	230/400	2,3,4	Yes
Yemen	ac 50	1,3	220/380	2,4	No
Yugoslavia					
Belgrade	ac 50	1,3	220/380	2,4	Yes
Zaire, Rep of					
Kinshasa	ac 50	1,3	220/380	2,3,4	No
Zambia	ac 50	1,3	220/380	2,4	Yes
Zimbabwe					
Harare	ac 50	1,3	220/380	2,3,4	Yes

KEY
[1] country/city
[2] type and frequency of current
[3] number of phases
[4] nominal voltage
[5] number of wires
[6] frequency stability—stable enough for electric clocks

(continued from page 6-8)

Documentation and Station Assembly

If you received operating permission before your arrival, you should already have several extra copies. Many hotels will be far less reluctant to allow an Amateur Radio antenna installation on their roof if they are given a copy of the permission which they can present to any inquiring government security personnel. If the permission is to be granted after arrival, explain to the hotel what your plans are, emphasizing that nothing will be put up before permission is in hand. Most places find this quite acceptable. After receiving the permission, try to find a way to duplicate it. If it is not possible to photocopy the permission, then never let it leave your sight. Do not leave it in the hotel room. Chances are that it would be quite safe but, if not, then you might never get it replaced. Hotels, banks and your embassy are the three most likely places to find a photocopy machine. The US Embassy is generally not in the business of photocopying

for tourists. However, if there is no other machine around, then the Embassy probably would not mind making a copy or two for you, since they would be interested in being apprised of your whereabouts.

Documentation for the ARRL DXCC Desk is required only if the DXpedition is to a country or place where there has previously been permission denied or which has been inactive for so long as to indicate that Amateur Radio is not permitted there. Some places are rare because of their inaccessibility rather than the difficulty in obtaining permission to operate. Acceptable documentation in these instances might include an original photograph showing members of the DXpedition at some clearly identifiable landmark that proves the DXpedition was really there. Receipts for passage or passport stamps are also good "documentation" for this purpose. See Chapter 8 for details on DXCC accreditation.

If operating permission is granted at the location and photocopies are obtained, it might not be a bad idea to just mail a copy right then to the DXCC Desk at ARRL HQ in

Newington. That way the ARRL has a copy of the permission and a canceled stamp showing the date and place from which it was mailed. The HQ staff isn't concerned about documentation for operations from countries that normally permit Amateur Radio operations. You can have a great time operating from Jamaica, for example, and not have to worry about providing documentation to the ARRL DXCC Desk.

Another piece of documentation which could be of significance is a currency exchange transaction record if it is required in the particular country. Some nations protect their currency from the free trade market with a fixed rate of exchange. The purpose of this is to control their economy by preventing too much of their nation's wealth from leaving.

Because of this control, there will be people who have more of the local currency than they need and who want to exchange it for a currency which can be used in other countries. They will be willing to exchange it at a more favorable rate than the bank. Therefore, the country has rules requiring you to keep your copy of the transaction record until you leave the country. You can be made to account for all the money you brought into the country. You are supposed to have all the foreign exchange currency (dollars, pounds, yen, etc) which you declared upon entry minus that which shows up on the transaction record as having been exchanged. Some countries have very severe penalties for even first offenders. You certainly won't do yourself any good, or ham radio any good for that matter, if you get caught dealing in the black market merely to save a few dollars!

With permission in hand and the nod from the hotel, you must now assemble the station and get it on the air. Assuming there is no one else on the DXpedition or that you know locally to assist, it could be quite difficult to put up the antennas. One thing about developing countries is that human labor is usually plentiful and inexpensive. The hotel will probably be happy to provide you with young men for climbing trees and otherwise installing antennas. For this service, you will rightfully be expected to pay. Try to find out from the hotel manager or some other person familiar with the local economy an appropriate amount to pay. It is important that you pay a fair, even generous amount, but it could be even more important that you not create jealousies by paying too much.

OPERATING TECHNIQUES

Like it or not, the actions of the sought-after DX stations (DXpeditioners and residents alike) have a great and profound influence on the operating techniques of DXers all over the world. If you go on a DXpedition, you have a responsibility to make every effort to be a positive influence on the sport of DXing.

DXers will generally do whatever is necessary to work DX. A "good" DXer will draw the line at illegal or unethical operating practices. Some poor operating techniques get used so much that a gray area exists in the minds of many DXers as to what is or isn't ethical (good sportsmanship and fair play). It is easy to understand how this came about. Isn't it true that the object of any competition is to come away with the prize? DX operators should never "reward" poor operating since it reinforces undesirable behavior.

If a station became active from the mythical DXCC country of Outer Baldonia, and the operator of that station could be active only for one hour a day, there would be frenzy upon the bands. It would be impossible for that operator to work all the DXers who want to contact that country. How then will the competition be conducted? Within the minor limitations of the DXCC rules, the decision will be made by the DX station. Whatever DXers hear other people doing that works is what they will do. Therefore, when you come on the air from whatever DXCC country you choose for your DXpedition, the responsibility of good sportsmanship and fair play rests squarely in your hands. Possibly more important than the operating during that one operation, people will learn (or reinforce earlier learning) about how to work other DX stations. If all this chapter is able to do is start DX operators thinking about how their operating affects DXing, then it will be worthwhile.

The next few paragraphs are some thoughts and reasoning about preferences in operating practices. There are several methods of conducting a DX operation which might seem to be inconsistent with other ways. In fact, the different skill levels of the individual operators (or other circumstances) make what would be totally inappropriate for one ideal for the other. Because of this, there can be few hard and fast rules.

Finally, now after all of the research, licensing, planning, preparation, traveling and hassles, you have the first antenna in the air and it is time for the fun to begin! You rub your hands together with anticipation as you hurry back to your hotel room.

You flip on the switch, and your receiver leaps to life. In seconds, you are ready to operate. After making the initial contacts, a number of interested DXers have found you. Soon there are quite a few people calling you. You have a pileup. Now a decision must be made as to how to handle it.

Assuming that you have two VFOs or a frequency memory circuit that can function as a VFO, you simply say, "Okay everyone, I will transmit here and listen up 5 to 10 kHz." (On CW, you might send QSX UP 5.) Almost all DXers would prefer to listen on one frequency to the station they are working while actually transmitting on another. This makes it easier to hear the station being worked. Here are some suggestions regarding split-frequency operation:

1) Always try to operate within the band limitations of the DXers you are working. This way, if the DXer makes a mistake and calls on your frequency, he or she will not have transmitted "out of band." This automatically eliminates the interference from the "helpful" operator who feels a need to warn the out-of-band offender. It might also prevent that DXer from getting into trouble with his or her own licensing authority.

2) Try to keep your transmit frequency as close as possible to the frequency or frequencies to which you are listening. This will make it easier for people who are spinning the dial between frequencies trying to work you without a separate VFO. Of course, stay far enough away so that your pileup doesn't interfere with your own transmissions.

3) It may even be a good idea to listen on only one frequency. Erik Sjolund, SM0AGD, discovered that on SSB the overlapping interference caused by stations above and below the station he was trying to hear was much worse than having all of the stations on one frequency or several separated specific frequencies. By having all the stations piled up in one spot, it is relatively easy to take the loudest stations first. The only difficulty results when there are several stations of equal or nearly equal signal strength.

4) The amount of total spectrum used should be reasonable. Never use more than you actually need. It would be hard to imagine a situation where more than 20 kHz would be needed. As an alternative to spreading the pileup out more,

it might be worthwhile to reduce your own output power. This might reduce the number of people hearing your signal and thus reduce the size of the pileup. This certainly ought to be tried if the radio has only one VFO. Remember that non-DXers have a right to use the band as well. They probably won't realize what you did for them, but they will surely realize it if you don't do it for them!

5) It might become necessary to bring the pileup back to your frequency for a while in the event that adjacent activity starts to move in, keeping the pileup from hearing you.

6) The pileup can also be divided several ways. This is difficult or perhaps impossible to do and maintain good sportsmanship and fair play. The most common method is to go by call districts when working the US. But no two call districts have the same number of amateurs. If the same number of contacts are made from each district, then the more populous 4th and 6th districts are "cheated," and it is easy to see how some people in these areas, or other areas for that matter, could rationalize poor operating habits to compensate for being "cheated."

Since the FCC chose to have us keep our call signs when we move into another district, a station in this situation gets two chances to call: Once when the district is called that matches the number in the call sign, and once again when the district is called where that stateside DXer actually lives. In the latter instance, he or she usually identifies as "portable." If the DXer is ethical and only calls with the district where he or she lives, then there is a good possibility, paradoxically enough, that he will be ignored for calling out of turn! A poor reward for exhibiting good sportsmanship and fair play!

Another way to divide the pileup is by the last letter of the call sign. But is it proper to ask a station to identify by transmitting only part of its identification? Probably not. It may also be technically illegal, depending upon the identification rules of the country (or countries) concerned.

7) Try to keep your instructions to the pileup down to a minimum. The more the DX station talks, the fewer stations he or she is working.

8) Try to end each QSO the same way. Some operators prefer TU or simply K on CW. On phone, the call sign and QRZ? are commonly used. The important part is making it clear when it is time to call.

9) Remember that many, many more people will hear you than will hear the station you are working. If a DXer operates poorly or unethically and it makes you angry, try not to chastise that person on the air. For one thing, it may have been an honest misunderstanding; for another, you will seem like the "heavy" to many people listening to the one-sided conversation. It usually won't help control the pileup, wastes your time and robs you of enjoyment.

10) People will ask you to do things for them. They will ask to do things for you. Some will be genuinely helpful, others will have less commendable motives. You will be asked to listen on other bands, make schedules for friends and friends of friends, work different modes, check into a particular net or operate from a list. Do those things which coincide with your goals for your DXpedition. Do not do those which are at odds with your goals or which you feel do not promote good sportsmanship and fair play. Always be polite but firm. Be assured, it will be difficult!

11) Make thoughtful decisions about the ground rules you lay down for the pileup, then modify them only if absolutely necessary. If the DX station calls for 7s and then answers a W2, the other operators in the pileup have no reasonable

expectation that stations will be answered in proper turn in the future. In that case, they feel that they must now call no matter which number is in their call sign. Suddenly, a no-holds-barred situation is in effect.

These suggestions apply to transceive operating as well. There are some additional complications with the single-VFO situation which are accentuated by not being able use separate transmit and receive frequencies.

Typical Operating Situations

The Tail-ender

If done properly, this technique (on the part of the DXers in the pileup) is extremely helpful to the DX station. Some years ago, proper "tail-ending" was the rule and not the exception. Unfortunately, in recent years improper tail-ending has worked so well for some people that now many employ this very poor operating technique. The result is that it disrupts and slows the entire operation.

Let us examine a QSO in order to see the effect. The DX station is Igor, L3ØA, from the Republic of Outer Baldonia. The pileup calls and Igor answers, "K1XA, 5 by 9!"

"QSL, Igor! You are also 5 by 9. L3ØA, this is K1XA."

"WB4ZNH."

"WB4ZNH, 5 by 9!" Igor was able to identify the station and give a signal report and get one back without interruption. He was also saved the time and trouble of asking QRZ.

Now let's tail-end wrong. The pile-up calls and Igor answers, "K1XA, 5 by 9!"

"WB4ZNH!"

"...Igor! You are..."

"WB4 Zulu November Hotel!" "...XA."

"Please WB4ZNH, stand by! K1XA, again my report please." Igor was able to identify the station and give a signal report, but WB4ZNH called and prevented Igor from receiving his report from K1XA. Then Igor made a severe tactical error. He identified the offending station, which means that WB4ZNH has "broken" the pileup and will most likely be the next station worked. But what is the cost? The extra time spent in this exercise reduced the number of total DXers worked by 33%! Where Igor contacted two, he could have worked three. Worse yet, all of the amateurs in that pileup now realize that if they call while another station is talking, they stand a better chance of getting through and getting that precious QSL card from Outer Baldonia. They will also attempt to use this disruptive tactic in the future in other DX pileups.

If Igor had a separate VFO, he could merely turn the dial and start over, but even then some people will find the station he is working before the QSO is completed.

Igor had several alternatives. Possibly the best approach would have been to ignore WB4ZNH and have just asked K1XA to repeat his signal report. In any case, WB4ZNH should not be rewarded with a QSO until his operating practices conform to the standard of good sportsmanship and fair play. That is easy to say but difficult to do without the aid of the second VFO. Some DX stations dealing with this problem attempt to punish the offender by keeping a "blacklist" of stations which will not receive a QSL card, ever. This is excessive punishment that will more likely damage the reputation of the DX station more than that of the original offender. Other DX stations simply tell the offender to come back tomorrow for they will not receive a QSO that day. This is a lot more reasonable than the blacklist, but uses up a lot of time.

Inadequate Operators

Fortunately when you consider our numbers, there really aren't many; it just seems like there are, sometimes. No matter what your personal opinion of a particular amateur might be, never fail to work any DXer if that person is operating correctly. It can be difficult to bring yourself to have a QSO with someone for whom you have little regard, but you will feel better in the long run. You will be setting the example, so make it a good one.

Intentional Interference

It is embarrassing to admit that this negative aspect of DXing exists, but there are ways to minimize its effects. For your own sense of respect for the hobby and for DXing, it is very important to remember for every amateur calling when they should be listening and every amateur actively intentionally causing trouble, there are thousands (yes, thousands) of good operators quietly enduring the same agony as you.

The best way to deal with intentional interference is don't deal with it. Ignore it. Above all, don't quit operating. That might be the goal of the person interfering, and he must not be allowed to control any aspect of Amateur Radio. Whatever the ratio of troublemakers to good people, there seems to be an equal spread throughout society.

This is another argument for not splitting the pileup: If the call-sign number is the determining factor of the pileup division, then 9/10ths of the troublemakers (if they are also DXers) are free to engage in mischief. If they are calling in the pileup, then chances are they are not causing trouble.

[It is interesting to note that, on numerous occasions when DXers stopped answering my calls, someone in the pileup would say "Interference on your frequency," and I would be totally unable to hear the signal causing the problem. It would seem to substantiate the previous conclusion.]

Talking to Your Friends

It will be necessary or desirable sometimes to have a QSO much longer than the "QSL QRZ?" type. DX operators are hams, too. When it comes time for the schedule with your QSL manager back home, or whatever the reason, and you are taking a break from DXing, every attempt should be made to inform the DXers there on frequency as to what is going to transpire. Remember, some of these folks have possibly been waiting hours and even days to talk with you. When you're at the DX end, you are most certainly entitled to conduct the operation as you please, but that does not relieve the obligation to be polite and thoughtful.

It is always important to give respect if one is to expect to receive respect. Tell the pileup that you are leaving only when it is time to go. Any earlier than that creates a certain frantic frenzy which deteriorates operating efficiency. Thoughtfulness dictates that they be told when and where to expect your return to the game. If you see that it won't be possible to begin operating then, at least try to go there at the prescribed time and announce a new time. If the non-DXing QSO occurs on the pileup frequency, then be careful to acknowledge only those who are a party to that conversation. Otherwise, try to avoid lengthy discussions on the pileup frequency. It is much better to have a prearranged frequency for this purpose.

Multiple Operators and Duplicate Contacts

If your goals include contacting the maximum number of

DXers possible, then some attention must be paid to "insurance" contacts. To avoid getting their QSL card back marked "Sri, not in log," most DXers will try to make more than one contact; if any mistake occurred on one QSO, they will have the "insurance" of another contact. Most of these DXers will be considerate of others and make the additional contact only on another mode or band. Depending on the situation, this seems to be quite ethical operating. These same operators will have no compunction whatsoever about working another operator of the same DXpedition if another call sign is involved. To passively combat this wasteful duplication of contacts, try to work out an operating schedule where no two operators work the same bands or band/modes. In almost all cases, it is not practical to keep track of duplicate contacts during a DXpedition. The only opportunity to influence future operating is during the QSLing stage of the operation, and you should make your intentions clear.

POST-DXPEDITION CHORES

Finally, as always, all good things must come to an end. It is important that, when you leave this DX location, the government or licensing authority will have no reason at all to be reluctant about permitting another such operation. It is always good to express your gratitude. While you're there, the rules laid down by the licensing authority should have been adhered to without exception.

If the DXpedition operators are good amateurs, then there will be some new friends left behind. This is always kind of sad; but remember, this might be the spark to create another ham radio operator, or DXer, if the new friend is already an amateur!

If there are local amateurs, be sensitive to their needs, and, if possible, leave behind for their use any equipment or hardware which might be difficult for them to obtain.

A QSL manager is a good idea for all but the very smallest of DXpeditions. Select a QSL manager who is capable of handling the expected influx of QSL cards. It is important that the QSL manager have a positive outlook with similar ideals as your own. The manager should be as accessible to you as possible.

While the logs and your memory are still fresh, go over them and make any notes that will aid the QSL manager. In the heat of the pileup, your penmanship can deteriorate considerably. This is a good time to add clarity to the logs.

If all goes well, plenty of photographs will have been taken. Start to work on a slide program about your DXpedition. There are many DXers who will never have the opportunity to do what you have done, and the closest they will ever come to it is through your pictures. (A cassette narration is also nice for those beyond your traveling range. A CW sidetone makes an excellent signal to advance to the next slide.)

QSL cards should be printed as soon as possible. Most people don't stop to think that, in many cases, the DXpedition won't even know the call sign to be used until a day or two before the actual operating begins. Also, it is not until after the fact, when your QSO totals are in, that a guess of the quantity of QSL cards to be ordered can be made. Surprisingly enough, unless the DXpedition is to a fairly rare spot, QSL requests will usually only equal about 50% of the QSO total.

Not only is it poor form to request money on the air, but the DXCC rules require ethical behavior when it comes to QSLing procedures. Donations should never be a condition for getting a QSL. DXpeditions can be very expensive, so they should be undertaken only with the expectation that the operators themselves will bear all of the expense. Some DXers are under the mistaken impression that donations sent with the QSL cards are a nice source of income. It is okay if they think that, but if you are planning a DXpedition, it is important to realize that if you are lucky, "green stamps" might cover the cost of QSL cards and postage, but nothing more.

The cards should be sorted as they arrive (another good reason to have a manager, since the operators will still be on DXpedition when the first cards arrive). In the first category should be placed the cards which include return postage. The second contains the cards without postage and cards from the QSL bureau. The third and final category is the duplicate-contact people.

Some sort of "first-in, first-out" schedule should be established. The QSL cards for the operators with excessive duplicate contacts should go out last, and there should be included a brief and matter-of-fact note explaining the delay. While in the DXpedition country, you can buy some low-denomination stamps as souvenirs to be included with the cards for those people who were especially helpful during the operation or assisted in some other way.

Some operations choose to have their QSL manager in a country with a lower population, since stateside or European amateurs are more likely to send a donation where international postage rates apply rather than domestic ones. Typically, these operations then send the cards to some domestic point for distribution. While this is not really unethical, it is considered by many to be poor form.

Conclusion

DX operating and DXpeditions can be very satisfying and rewarding activities. These should not be taken lightly, since many people will look at you in the spotlight and form very

Be ready for the result. Roger Western, G3SXW, operated as ZD9SXW on Tristan da Cunha. Over 4000 QSL cards ended up on his kitchen table!

distinct opinions about Amateur Radio, DXing and you personally. With just a little effort, though, all three can benefit.

Contests

CLARKE GREENE, K1JX
92 CYNTHIA LN, B2
MIDDLETOWN, CT 06457

The word "contest" brings to mind a number of images. Extravaganzas sponsored by various magazine publishers, where "you may have already won $100,000" may be most familiar. If you've ever had any dealings with the legal system or the legal profession, the term "contest your claim" may hold special significance (particularly depending on the outcome of that contest). Many sporting events are zealously described as "titanic contest(s) between close rivals."

To many radio amateurs, this same word describes an event encompassing none of these images, but at the same time, all of them. An Amateur Radio contest is an operating event, held over a predefined time period where the goal is...to enjoy yourself. Of course everyone has their own definition of enjoyment; that diversity of definition only adds to the appeal.

The first Amateur Radio contests were operating periods set aside specifically to attempt transoceanic communication between North America and Europe, just after World War I. Through this coordination effort, many of the participants successfully made some of the very first intercontinental contacts. Ultimately, this activity evolved into the ARRL International DX Contest. Although the technology and results have changed "somewhat," the object is still the same: Contact as many stations in as many areas of the globe as possible during the contest period.

Two other early operating activities were conceived as intense exercises to train radio amateurs in traffic handling and emergency communications. These are now known as the ARRL Sweepstakes and Field Day. The only real difference between the two is that Field Day emphasizes portable operation as might be required at a disaster site.

These examples demonstrate the two primary motivations for establishing radio contests: advancing the state of the art in radio amateur communications and advancing the expertise level of the Amateur Radio operator. Today, these same two goals provide the inspiration for sponsoring organizations to continue support of Amateur Radio contests. Many of the hams who take part in contests use them as opportunities to work a few new states or collect a few new countries. Others test the effectiveness of a new antenna or other piece of gear. Some work at increasing code speed. There are even some who take all of this very seriously and are out to *win*. In the final analysis, each participant gains expertise, learns a little more about propagation, and becomes a better operator, more equipped to perform effectively in the case of an emergency or to explore some new area of the Amateur Radio art, all while having a good time.

PHILOSOPHY OF EFFECTIVE CONTESTING

For many hams, participation in contests is very casual. If they happen to be on the air and hear a flurry of activity that can only be the result of a contest or big DXpedition, they'll figure out what's going on and pass out some contacts to the hams who are seriously going at it. Indeed, if it were not for these casual operators who are willing to get on the air and work a few contesters, most contests would be extremely slow and boring after the first few hours (see accompanying sidebar "Contesting for Non-Contesters"). For casual contesters, who may not have the time or inclination to compete (at least on that particular weekend), little forethought or planning is required. Getting on the air and knowing the contest exchange (or at least be willing to have someone tell you the exchange) is adequate preparation.

By contrast, serious contesting, at any level, requires some advance preparation. Preparation for a contest entails a number of considerations, and truly is an interdisciplinary subject. Unlike many forms of competition, the success of a radio amateur contest effort is a synergistic product of the station, the operator and a dose of luck as well. An inefficient operator at the controls of a giant station at a great location may not fare as well as a superior operator at a mediocre station. An experienced, competent operator at an outstanding station might completely dominate his or her class, or may have an electrical outage three hours into the contest and have to give up. One thing is certain, though; an inexperienced operator at a poorly prepared station not only should expect dismal results, but shouldn't expect to have fun, either.

Bill Clark, N7PXF, works with Mark Bienz, KB7TMS, at the Keno ARC Field Day site.

For Novices, Techs and New Hams

One Saturday evening you turn on the rig, and find the band jammed packed with signals. What's this, you say? Some national emergency? A convention of crazies? You may be close on the second guess—it's probably a contest.

What is it about contests that attracts people to them? For many it's a chance to work stations in rare places, whether it be DX locations, states or VHF grid squares. For others, it's the competition—participation in a sport where it's you against the other guy, whether it's across the street, the state, or the world. How well can you, your mind, and your equipment hold up against someone else? It's the mental equivalent to chess, yet can require the physical endurance of a marathon. Others use contests as a chance to try out new equipment, experiment with antennas or participate in a group effort with their radio club. On VHF, you may work stations on bands in places that would never see any activity apart from the contest weekends.

If you ask a die-hard contester, they'll probably boil it all down to one reason, though—excitement. Whether you run with the big boys, or just get on for an hour or two without even keeping track of points, contesting can be exciting. If you've never been bitten by the bug, perhaps we can strip away some of the mystique about this segment of the hobby for you.

What is a contest? Simply put, a ham radio contest is an event where hams try to contact as many other hams as possible in a given time under certain conditions. These rules can restrict the bands you can work (such as VHF/UHF QSO Parties); the places you can work (in the ARRL DX Contest, DX stations work US and Canadian stations *only*, and vice versa) or how many times you can work them (such as the November Sweepstakes, where you can only work a station once.) There about as many different contests as you can imagine, ranging from "major" events, such as ARRL DX, Field Day, November SS, and *CQ WW*, to smaller events such as state QSO parties. Each one has its own charms and personality.

How do you start? If you're like most folks, you've probably ran across a contest and wanted to make a contact but did not know what to do. First, find out which contest it is! On any given weekend there can be several small contests. The Contest Corral column in *QST* gives a good list of upcoming events and a thumbnail sketch of the rules. **Table 7-1** is a list of the major contests and when they are held.

The rules will have the next thing you're looking for—the *exchange*. Simply put, this is what the stations will be sending to one another. These can range from serial numbers, (in other words, a count of how many contacts you've made so far), to your state, county, or name. Some contest exchanges include your grid square (a letter-number combination used to determine the 1°×2° area of the world you live in—used in VHF/UHF Contests) or your zone in some DX contests. The VHF Operating chapter of this manual contains a grid square map you can use to find your own grid square or the location—and therefore the beam heading—of other stations. The DX zone map is in the reference section of this manual.

Once you think you have a good idea of what information is included in the exchange, it's time to make some contacts. There are two techniques folks use to do this: *Hunt and Pounce*, and *Running*.

Hunt and Pounce is just like its name—you hunt up and down the band looking for stations to work who are calling CQ. Then you pounce on them (call them quickly.) There's not much to it, on the face of it. Almost everyone starts at the bottom of the band, and works their way up to the top. If you're new at this, you will find you need to tune slowly. It takes a while to get used to hearing many signals close together. If you hear a station calling, just zero-beat it and send your call sign. Just once is sufficient—more than once and you may miss who he comes back to, and will probably QRM the QSO. When he comes back to you, don't expect a rag-chew! In contesting, the name of the game is speed, and they'll be looking to work as many stations as they can.

The *run* is where you stay in one spot and call CQ. This can have the advantage of working large amounts of stations in a short amount of time. Be careful though! Handling a pileup is not without its dangers and troubles, and it takes an experienced hand to be able to do it right. If you're just starting out, you may want to take some time and listen to how some other stations on the air do it, and perhaps pick up some helpful hints. At the same time, though, don't be afraid to call CQ! Many times a band may seem quiet, when in actuality everyone else on the band is listening.

What are the best contests to start with? Many people get their start during Field Day. Others try the relatively low-intensity stuto QSO parties to get their feet wet. The 3 major VHF contests (the January VHF Sweepstakes, and the June and September VHF QSO Parties) are perfect opportunities for you to check things out. In the more populated areas of the country, you can make lots of QSOs with just a hand held!

Don't hesitate to jump into the most popular of the HF contests. There are usually a few spots on one band or another offering you a chance to participate without being buried in the fray. The November Sweepstakes is a great chance to start as it doesn't take much of a station to be able to make contacts. You can even get a nice pin for making 100 contacts, and there's slow-speed segments set aside for ops who can't roll along at 35 wpm on CW.

If you've never been to Field Day, you're missing some of the best contest and camaraderie ham radio has to offer! When sunspots are at their peak, the 10-Meter Contest in December brings stations from all over the country and the world on the air for the weekend, and hams trying to make DXCC or WAS can find their totals taking a big jump.

Last, but not least, are the smaller contests and QSO parties. These range from ones sponsored on a state level (such as the California or Pennsylvania QSO Parties), to ones run by a country's ham society (such as the Worked All Europe, sponsored by DARC in Germany, or the All-Asian DX Contest, sponsored by JARL). These events offer a chance to make a large number of contacts in an afternoon, and may provide activity from places you wouldn't hear on the air normally.

Whatever contest strikes your fancy, remember these two maxims—get on the air, and have fun! Once you try it, you may find yourself a contest addict in no time. See you in the pileups!—NF1J

Setting Goals

The very first step when entering competition is to set a goal for yourself. As mentioned earlier, a goal might be to improve your code speed by 10 words per minute or to work the last two states you need for the Worked All States award. These would certainly be very worthwhile achievements, and a contest is an excellent arena in which to attain them.

If you're more oriented toward the competitive aspects of contests, then you should establish a competitive goal for yourself. The goal should be reasonably attainable, or you'll just get discouraged when you don't achieve it. Don't set the goal too low or you won't progress, or, more importantly, draw any satisfaction out of achieving it.

The best types of goals are relative goals; that is, how well you can do compared to a particular station in your area who might have similar capabilities to your own. Absolute goals,

Table 7-1

Major Contests

Month	Contest	Scope	Exchange	For more Information
Jan	ARRL VHF Sweepstakes	Primarily W/VE	Grid-square locator	Dec QST
Jan	CQ Worldwide 160-Meter Contest (CW)	International	W/VE: signal report and state/province; DX: signal report and country	Dec CQ; Contest Corral, Jan QST
Jan	ARRL RTTY Roundup	International	W/VE: Signal report and state/province; DX: signal report and serial number	Dec QST
Feb	ARRL Novice Roundup	Novices/Techs work others	Signal report and ARRL section	Jan QST
Feb	ARRL International DX Contest (CW)	W/VE stns work DX stns	W/VE: signal report and state/province; DX: signal report and power	Dec QST
Feb	CQ Worldwide 160-Meter Contest (phone)	International	See above	See above
Mar	ARRL International DX Contest (phone)	International	See above	Dec QST
Mar	Spring RTTY Contest	International	UTC, signal report and consecutive QSO serial number	Contest Corral, May QST
Mar	CQ WPX Contest (phone)	International	Signal report and consecutive QSO serial number	Jan CQ; Contest Corral, Feb QST
May	Russian CQ-M Contest (phone and CW)	International	Signal report and consecutive QSO serial number	Contest Corral, Apr QST
May	CQ WPX Contest (CW)	International	See above	See above
Jun	ARRL June VHF QSO Party	International	See above	May QST
Jun	All Asian DX Contest (phone)	Asian stns work	Signal report and age	Contest Corral Jun QST
Jun	ARRL Field Day	Primarily W/VE	Transmitter "class" and ARRL	May QST
Jul	IARU HF World Championship (phone and CW)	International	Signal report and ITU zone	April QST
Jul	CQ VHF WPX Contest	International	Consecutive QSO serial number and call sign	Feb CQ; Contest Corral Jul QST
Aug	Worked All Europe (CW)	EU stns work others	Signal report and consecutive QSO serial number	Contest Corral, May QST
Sep	Worked All Europe (phone)	See above	See above	See above
Sep	ARRL September VHF QSO Party	International	Grid-square locator	Aug QST
Oct	CQ Worldwide DX Contest (phone)	International	Signal report and CQ zone	Sep CQ; Contest Corral, Oct QST
Nov	ARRL Sweepstakes (CW)	W/VE	Consecutive QSO serial number, power level designator, call sign last 2 digits of the year you were licensed, ARRL section.	Oct QST
Nov	ARRL Sweepstakes (phone)	W/VE		Oct QST
Nov	CQ Worldwide DX Contest (CW)	International	See above	See above
Dec	ARRL 160-Meter Contest (CW)	International	W/VE: signal report and ARRL Section; DX: signal report	Nov QST
Dec	ARRL 10-Meter Contest (phone and CW)	International	W/VE: signal report and state/province; DX: signal report and consecutive QSO serial number	Nov QST

K5MA used this antenna farm for a multi-op station. The two towers hold several monoband antennas and VHF arrays.

such as making 250 contacts on 28 MHz during a DX contest, may prove to be totally unreasonable; the band might not even open during periods of low sunspot activity or might be so good that stations using a converted CB set and a mobile whip antenna are capable of making 500 contacts. By setting your sights on an absolute goal, you probably are preventing yourself from doing the best that you can.

Another reason for setting a relative goal as opposed to an absolute goal is the wide disparity between different geographical areas. For example, during a DX contest, because of the vast differences in propagation to population centers around the world, the only real relationship between a station competing from the East Coast of the United States and one from the West Coast during a DX contest is the coincidence that both stations happen to be active during the same time period.

The immense Amateur Radio population in Japan is readily accessible to the average western United States station, so typically the West Coast results show a higher number of contacts than do East Coast results. The East Coast stations, who may only have a short period of useful propagation to Japan, usually have good propagation to Europe. While the amateur population in Europe is less than that of Japan, the population is distributed among a large number of countries, each counting as separate multipliers.

Since the final score is the product of the number of contacts and the number of multipliers, the bottom-line score (with a lot of multipliers and a fair number of contacts) of an East Coast station may be similar to the bottom-line score of a West Coast station (with a lot of contacts, but relatively few multipliers). However, the scores are not comparable, since the conditions are so different. Are football championships comparable to baseball championships? The obvious answer is no; each is an achievement unto itself. By setting your contest goals relative to a nearby station, you eliminate the difficulty of competing with something you have no control over—propagation.

The ideal goal is to compete with a station similar in capability to yours, operated by someone who has similar experience and talent to your own. The absolute optimum goal would be to compete with someone you know personally; after the contest is over you can then compare notes, learn from each other, and spur each other on to new levels of achievement in the next contest.

Compete in a class that is commensurate with your own abilities as well as your station's. For example, if you live on a suburban lot and at present only have a three-element 15-meter monoband beam antenna, it is pointless to enter the all-band category in a DX contest. Similarly, if you don't own (or wish to own) a high-power linear amplifier for the HF bands, compete in a low-power category.

The contest sponsors make it easy for you to be judged only against your "peers." Not only are there multiple classes (such as monoband 15 meters, or transmitter power below 150 watts) that you can enter, but results are geographically processed. You only compete against other stations in your own geographical area (ARRL section, state, or call area, depending on the contest). The contest sponsors give awards based on these classes and geographical distributions.

Finally, be realistic. Don't try to set the contest world afire in your first attempt. Most of the top competitors not only share your drive and desire, but also have vast experience coupled with that desire. After all, there is a reason they got to the top. Don't forget, as you yourself advance, that the competition isn't standing still either. In the event you don't find yourself challenged enough, set higher goals or choose a new station to compete with.

If you find that you consistently win all the contests in every category you enter by a wide margin, please contact the editor of this book for possible employment in writing the contest chapter of the next edition!

The Fundamentals

Once you have set your goal, the next step is to decide exactly what tools you need to attain it. First, the fundamentals.

What kind of contest are you going to enter? Most contests fall into four somewhat overlapping categories:

Domestic—contact as many stations as possible during the contest period in the United States and Canada. The ARRL Sweepstakes (SS) is an example of a domestic contest.

DX—contact as many stations during the contest period outside your own country. The ARRL International DX Contest and the *CQ* Magazine World Wide DX Contest are examples of DX contests.

VHF—contact as many stations as possible on the amateur bands above 50 MHz during the contest period. The ARRL VHF QSO Parties and the *CQ* Magazine VHF WPX Contest are examples of VHF contests.

Specialty—contact as many stations as possible during the contest period. Usually, all contacts must be made with operators who either are a member of a particular group or are using a particular transmission medium. The ARRL 10-Meter Contest and the *CQ* Magazine 160-Meter Contest are examples of contests in which operation is restricted to a particular transmission medium (in these two cases, a single amateur band). The RTTY Roundup requires all contacts to be made using radioteletype communications.

Once you have analyzed the format of the contest (and have read and understood the rules), you are in a position to decide how to maximize points for that contest. For example, in the ARRL Sweepstakes, a maximum of 75 multipliers can be worked, and each station may be worked only once, regardless of band. Clearly the emphasis is on making contacts in the SS. Where are these contacts most likely to come from?

An examination of the *Radio Amateur Callbook* for the United States gives a good clue. The licensed amateur population per state is presented in tabular form. As might be expected, the amateur population is densest around the major population centers like the Northeast corridor (Washington, DC through Boston) and Southern California. From this information, you might conclude that a percentage of your contacts could come from these population centers. The *QST* summary of the 1990 SS results confirms this, and adds the information that there are some hotbeds of Sweepstakes activity not apparent from the amateur census figures. For example, the Northern California Contest Club over the past decade has consistently taken part in the SS as a major club activity. As a result, the number of Sweepstakes participants from Northern California is proportionally much higher than almost any other area of the United States and Canada.

Now that you have an idea where to geographically find the stations you will be attempting to work in bulk, the next subject to examine is how to go about reaching them. Here, a knowledge of propagation is essential. If you aren't already reasonably well versed in the fundamentals of propagation, you might want to stop now and consult Chapter 4, which provides an excellent introduction to the subject. If you want more detailed information, see *The ARRL Antenna Book* or *The ARRL Handbook for Radio Amateurs* (this suggestion applies regardless of whether you are interested in DX or domestic contests, or contesting on the HF bands or on VHF and above).

From your knowledge of propagation, you know that the higher frequency bands—typically 20 through 10 meters— are generally the best bands for long-distance contacts (over 1000 miles), particularly during the day. From a location in Connecticut (that is, if you're from Connecticut), this means that the higher frequency bands will be good for working the Midwest and the West Coast during the sunlight hours. Since your demographic study indicates that the largest percentage of stations available for working during the SS are in the Northeast, 20 through 10 meters aren't going to be as productive (particularly for long periods of time) as 80 and 40 meters where the optimum working range is shorter. This suggests that a station in Connecticut probably would want to emphasize 80 and 40 meters as much as possible to maximize the contact total. Because of ionospheric absorption during daylight hours, 80 meters is pretty much restricted to nighttime use; 40 meters becomes the daytime band of choice by default. At night, both 40 and 80 meters have good propagation and should be covered. The remainder of the time, particularly during the day, should be spent on the higher bands. The ideal mix (in theory, anyway) is to spend an equivalent amount of time on each band in proportion to the active amateur population workable on that band from your location (eg, if 38% of the potential contacts workable from your station are from the West

Coast, you should spend 38% of your time on bands with propagation to the West Coast).

Of course, this analysis assumes that your station is equally effective on all bands and that there are no factors other than propagation involved. Unfortunately, neither is usually the case. Particularly on phone, the 40-meter band is flooded with foreign broadcast stations after European sunset. Very few amateur stations are loud enough to be heard through these broadcast signals, and even fewer have the ability to effectively hear stations responding through the raucous QRM (interference). This makes 40 meters a tough row to hoe. Fortunately (at least from the East), the skip zone on 40 meters stretches out dramatically by the time the broadcasters become unbearably loud. Late at night, stations in the Northeast are no longer reliably workable, and the exodus to 80 meters begins.

On 80 meters, the QRM is ferocious, but at least none of the stations there run megawatts, play music or broadcast a carrier. On 80 meters late at night, the goal is to work as many East Coast stations as possible, because during the day that feat will be more difficult.

In summary then, based on the analysis, if you live in the Northeast, maximize your effectiveness on 80 and 40 meters so that you can penetrate into the area with the greatest contact potential—the Northeast.

A station on the West Coast has a different perspective. For him (or her), the prime bands are 20 through 10 meters where

The 1992 EME contest felt the presence of K1FO with this set of 24 Yagis on 432 MHz.

propagation is good to the East Coast. This is a blessing in a number of ways. First, hot action during the day and little or no action late at night means that the West Coast operator can probably sleep when most people normally sleep— at night. Second, the higher frequency bands are generally quieter and certainly don't have the interference problems that 40 meters does. Third, antennas for the higher frequency bands are physically smaller than they are for 80 and 40 meters; the average ham in an urban or suburban area has a better chance to install an effective antenna on 20/15/10 meters. Since DXing is one of the more popular amateur activities, and DX is generally easier to work on the higher bands, the average ham probably has his or her most effective antenna system on some combination of these three bands. Therefore, the average West Coast ham's most effective bands are the bands with the greatest point potential. By the same reasoning, there are more active hams with good stations in the Northeast on the higher bands ripe for the plucking by West Coast stations. Fourth, there is less chance for TVI and other interference problems during the day than at night, so more "casual contesters" get on the air. The astute West Coast station, who may have made a similar study to the one you made, makes use of this advantage to get a higher score.

Are the odds heavily weighted in favor of the West Coast (never mind the South or the Midwest, which we didn't even examine)? Yes and no. From the *QST* SS summary, it's obvious that the East Coast typically fares poorly compared to the competition at points west. The odds become a little more favorable to the east when sunspot conditions deteriorate and the higher bands aren't open for as many hours coast to coast. But, so what? Earlier it was decided that it's impossible to compete with everybody everywhere, and it is preferable to set your goal to compete with someone in our own area. This is just another case in point.

An analysis of this type can be (and should be) made for each contest in which you plan to participate. If you're really serious in your own way, your station design (especially your antenna system), your operating "game plan," and even your choice of contests will be defined by an analysis of the fundamentals of the contest. In engineering and scientific terms, this is known as "defining the problem."

Now that the first stage is complete, and you know what you want to do, you can try to figure out how to go about doing it.

Chiharu, JA3DLE, is a regular participant in RTTY contests from Japan.

STATION PREPARATION

If you are serious about competing in contests, chances are that your station and antenna system already are (or will soon be) oriented toward producing higher contest scores. Unfortunately, very few of us have the luxury of being able to install the ideal contest antenna system. The key to building the most effective contest station within your own means is to analyze your objectives (see the previous section of this chapter), and take advantage of whatever resources you might have at hand to achieve these objectives.

The prime purpose in optimizing your station is to maximize your station's effectiveness into those areas where there is the greatest point potential. This may sound too simple to be accurate, but the concept really does apply, no matter what rung of the contest career-ladder you may be standing on. Station effectiveness is merely a relative term for describing your ability to hear and be heard (in that order of importance).

Antennas

The subject of engineering and installing antenna systems is far too complex to cover in a book like this one; the best we can do here is examine what characteristics might be desirable for a contest-oriented antenna farm. Antenna design is a very popular Amateur Radio topic, and antenna-related articles appear almost every month in the various amateur magazines. For a solid basis in the fundamentals of this subject, *The ARRL Antenna Book, The ARRL Handbook for Radio Amateurs, Yagi Antenna Design*, by the late Jim Lawson, W2PV, and the more theoretically oriented *Antennas* and *Electromagnetics*, both by John D. Kraus (W8JK), are texts well worth studying. They all deserve a place on your bookshelf.

Simply stated, the idea in designing an antenna system for successful contesting is to obtain the greatest antenna gain possible over the geographical area where there is the greatest point potential. Determining the means to this end requires knowledge of propagation and how a signal might travel from your location to the target location. From the bottom of 160 meters up through at least 6 meters, signals propagate beyond the line of sight by means of high-altitude refraction of the transmitted electromagnetic wave front. Above 6 meters, most long distance propagation is the result of relatively low-altitude refraction. The causes for the refraction vary from band to band and between the various propagation modes. For a more detailed description of the processes involved and their effects, see the propagation references mentioned earlier.

Once the wave propagation path from you to the target area is understood, the best antenna system can be chosen by identifying a system whose characteristics produce a radiation pattern that matches the propagation path. The closer the transmitted wave follows the propagation path, the greater the signal strength that arrives at the target site. The reciprocal holds true as well for reception; the greater the energy captured from the propagation path, the greater the received signal strength. Like a lot of other subjects discussed so far, this one sounds very simple. In fact, given a large enough budget (for both antenna hardware and real estate), it is possible to build an ideal antenna for any given propagation path.

Unfortunately, propagation paths change hourly, daily, monthly and yearly. Some of the changes are subtle; some are manifest. The real challenge in building the optimum antenna system is to be able to accommodate these changes as they occur, rather than having to string new wires or plant new towers as the band changes. As the ionosphere goes through its changes, the geometric path that will support propagation changes. If the propagation path is no longer optimum for your antenna system, signals will fade (the old familiar QSB), perhaps until the "match" between your antenna system's pattern and the propagation path is so bad that the band will effectively close between you and that area. This effect explains why a station only a few miles from you might be able to carry on a conversation with another station that to you is inaudible. Perhaps your neighbor's antenna system has a different vertical radiation pattern than your system does, and the distant station's signals are arriving at a radiation angle at which your antenna system has a severe null. This might be because your neighbor's antenna is at a different height above ground, or because the ground beneath his or her antenna falls away at a different angle from your own, effectively modifying the vertical radiation pattern.

With the ionosphere performing its undulation act, no one "ideal" antenna is ever really optimum. The situation is similar to trying to shoot a moving target from a rocking boat; you know approximately the direction in which to aim, but not much more. In fact, if you are considering the situation up through 6 meters, where almost all antenna systems are at best only a few wavelengths above ground, an antenna's vertical radiation pattern is a series of peaks and nulls, as shown in the various antenna texts. This effect is caused by wave interference, both constructive and destructive, between the portion of the antenna's radiation lobes that are directed up away from the ground and the portion of the antenna's radiation lobes that reflect from the ground's surface. When one of the peaks in the antenna system's vertical radiation pattern matches the propagation path to a particular point, then signals will propagate well, with minimum loss. When the propagation path falls into one of the antenna system's nulls, where the antenna system response is very low, the transmission path loss is high, perhaps to the point that the band will appear "closed." Usually, neither of these extremes is the case, and the transmission path loss (and resultant signal levels) falls somewhere between the two.

Your goal is to be as loud as you can into your target area (to attract attention and get people to hear you) for as long as you can (to have exposure to the maximum number of stations). Thus, your antenna-system effectiveness is determined not by how loud signals are at their peak (when the propagation path and an antenna lobe match well), but rather by signal-strength consistency over a period of time. An antenna system that produces signal peaks of S9 + 30 dB for 20 percent of the time and signal fades that drop into the noise the rest of the time isn't nearly as effective as a system that produces consistent S9 signals all the time. It's obvious that you can make contacts only as long as the band is open; long band openings produce a lot more contacts than loud band openings. An effective antenna system might not make you any louder, but it should let you take advantage of any band opening, no matter how poor the propagation.

There are a number of ways to accommodate the changing ionosphere and maximize band-opening time. The general strategy is to use an antenna system that employs some sort of radiation-pattern steering mechanism to match the antenna pattern to the propagation path. Almost everyone who uses a "beam" antenna of some type has either an electronic or mechanical rotation facility to turn the beam in azimuth to favor the desired transmission path. However, to match an antenna system's radiation pattern with the propagation path, control of the vertical radiation pattern is needed. The most common way to achieve this is to employ antennas at different heights. Since the ground reflection patterns, for even identical antennas, change with height above ground, a variety of antennas at different heights produce a wide selection of vertical radiation patterns. The appropriate antenna for a given propagation path can then be chosen experimentally.

Another vertical-pattern steering technique is to employ a phased multiple-antenna system (such as "stacked" dipoles or beams), and electrically vary the phase relationship between the antenna elements. The actual shape of the radiation pattern then can be changed as needed. While this approach is very effective, the space required and the mechanical enormity of stacked antennas forces most station owners to employ the simpler arrangement of switched multiple antennas.

Keeping Your Ears Clean

So far, all the discussion has been aimed toward being as loud as you can into your chosen target areas, for as long as possible. Fortunately (thanks to Mother Nature), an antenna system that provides the best transmit performance also provides the best receive performance, as far as signal strength is concerned. Unfortunately, the reality of the amateur bands is that the signals you want to hear occupy the same RF spectrum as noise and other interfering signals. In fact, you are often limited in receive capability not by low signal levels, but rather by interference and noise. As far as your receiver is concerned, it doesn't know the difference between the signal you want to hear and the other junk on the same frequency. Once noise or interference makes it as far as your receiver's input terminals, there is often no way to separate the signal you want to hear from everything else on frequency (more on this later). The only panacea for interference ills is to prevent the interference from reaching your receiver in the first place. The most effective way to do this is through your antenna system.

The basic physical process that allows your antenna system to produce gain is a very simple one: Energy from one direction in an antenna's pattern is redirected into another direction. The total energy in the antenna pattern is fixed; it is merely redis-

John, SV1AHV, operates from SV1SV, the official HQ station of the RAAG. *(SV1BDO photo)*

Nancy Bluett, KD4DMZ, is another new contest operator. She proudly shows the photographer her full 10-meter SSB log sheet.

tributed. A gain is developed in one direction, "negative gain" is developed in another direction. This process is not only useful in making you louder toward one direction, but it is also the means to reduce undesired signals and noise from another direction. Most every beam antenna has a null directly off the "back" that is between 10 and 30 decibels below the strength of the main front lobe. Response off the "side" of the antenna is even less, as compared to the main front lobe. The rejection of signals off the main "beam" lobe is one of the more compelling reasons for using a directional antenna. Not only are you louder in the preferred direction, but you cause less interference and are interfered with less in other directions. The cliche "you can't work 'em if you can't hear 'em" really is true. The use of a directional antenna is a very effective step in beating the implications of the cliche.

Particularly on the lower frequencies, where antennas are very large to begin with, it is often very difficult, if not impossible, to install beam antennas (imagine a three-element 160-meter beam at a height of a half wavelength!). Reduced size antennas work reasonably well, but as an antenna's size is reduced, so is its radiation resistance. The antenna's losses increase as the radiation resistance drops. In most practical situations, the theoretical gain offered by a dramatically "shrunken" antenna is more than offset by the increased losses; as far as transmitting is concerned, the simpler dipole is at least as effective as the more complicated directional array. In addition, the simpler antenna probably can be installed at a greater height where it also might be more effective.

Even in a situation where a large directional antenna is out of the question, it is still possible to effectively employ a directional antenna to cut down received interference. The solution is quite simple; use the directional antenna for receiving only. As pointed out earlier, signal strength is usually not the limit-

ing factor in hearing well; what really matters is the signal-to-noise/interference ratio. There are a number of directional antenna designs that are not very efficient and offer "negative gain" compared to a dipole. While these particular antennas would be poor transmitting antennas, they can be very effective for receiving. For example, signals received on a directional loop antenna might be 20 decibels weaker than the same signals as received on a dipole. However, by orienting the loop properly, an interfering signal might be reduced in strength by 40 decibels. The overall improvement in signal-to-noise/interference ratio is then 20 decibels, a substantial margin. In addition, many of these directional antennas also are capable of vertical radiation pattern-steering, giving further rejection of undesired signals. Some designs that fall into this category are the Beverage antenna, the Adcock antenna and the tuned loop.

Many of the same principles apply for both HF work and VHF work. On the VHF bands, where the wavelengths are physically short, even modest-size antenna systems develop prodigious gain compared to their HF counterparts. The effective radiated power (ERP) transmitted by most VHF contest stations is usually several times greater than even the largest HF amateur contest stations. That, coupled with the higher gain needed for best sensitivity in a VHF receiver, provides the breeding ground for severe receiver-overload conditions. Aside from improving receiver dynamic range, the most effective way to reduce receiver overloading is simply to reduce received signal strength (selectively, of course). An antenna system with a "clean" pattern (that is, few lobes aside from the desired main lobe) can offer effective rejection of an overpowering signal—when the antenna isn't pointed in the direction of the offending station!

Station Electronics

Today's Amateur Radio market offers a wide variety of electronic marvels for HF, VHF, UHF, CW, SSB, ASCII, RTTY, FM, SSTV, ATV and every other transmission mode and frequency range imaginable. The rigs scan, autotune and operate break-in. They offer passband tuning, notch filters, digital readout, speech processing, computer control, RIT, XIT, VOX and audio filters either built-in or as options. The combinations and possibilities are endless and can be very confusing even to the knowledgeable and experienced radio veteran. Furthermore, how any of these features find utility in contest work is a very confusing issue. Well, don't expect to find any definitive answers here. Most of the answers are a matter of your own personal taste and your pocketbook (although high priced doesn't necessarily mean better). There are a few areas that you should especially examine, however, if you are purchasing a radio specifically for contest use.

Unlike even a decade ago, virtually the only form of transmission equipment commercially available is the transceiver. In the past, transmitter and receiver "separates" were the standard bill of fare in the amateur station. When transceivers first became available, they were generally viewed as compromise units, primarily intended for portable or mobile use, mostly on phone. Many didn't have the capability for optional narrow CW receive filters. Few offered any facility for adjusting the receive frequency independently of the transmit frequency; in fact, on many designs, CW operation was included only as a poorly conceived afterthought. Most of the early HF transceivers were made for operation on a few of the amateur bands, perhaps only one. Among contesters, transceivers were a concession to portability or finances.

Earlier, the cliche "you can't work 'em if you can't hear

'em" was mentioned. The wisdom in this is pretty obvious. For most Amateur Radio work, the need to hear what is on frequency right now doesn't usually exist. In most activities, imminence isn't a priority. For example, if a DX station happens to fade for a few minutes, normally even a ravenous DXer can stand by for a couple of minutes until the band peaks up again. Since a contest is a timed event, a contester can't afford to spend a lot of time trying to dig through the QRM to hear the other station's information. The use of a high-performance receiver can go a long way toward helping the contester dig the weak ones out of the noise and QRM.

The most basic function of a receiver is to allow the operator to hear what is being transmitted. Again, this is one of those ultra-fundamental statements that is so obvious that little is made of it; the real underlying meaning does have particular significance here. The concept of hearing implies not only detecting the presence of sound, but also being able to interpret the sound and understand what was sent. This is because detecting a signal is only half the battle; you have to understand what is being sent for the transmission to be of value. A more descriptive version of our cliche might be "you can't work 'em if you can't hear 'em *and* understand 'em." Keep this in mind when evaluating a receiver.

One of the most audible indications of contest activity is just that, contest activity. The bands become very crowded, and signal levels are ferociously high. The combination of high signal-levels and lots of them creates a lot of QRM. The same combination causes receiver overload and exacerbates otherwise relatively subtle problems like reciprocal mixing. Clearly, there is enough interference to contend with on the bands without generating a dose of interference within your own station. Strive to get the very best receiver that you possibly can.

Many features such as multiple internal VFOs, receiver passband tuning and frequency "memory," just to name a few, are primarily the beneficiary of one major change in the commercial design of Amateur Radio equipment, the phase-locked loop (PLL). All frequency-control functions in modern transceivers are controlled by an internal PLL system. One of the main advantages of the PLL scheme is the use of a single frequency-controlling element, which allows the inclusion of these advanced features. Another big advantage is the possible manufacturing economies offered by digital technology within the PLL. That's the good news. The bad news is that PLLs have inherent deficiencies that may have a direct bearing on the performance of a transceiver for contest use.

One of the negative side-effects of using a PLL is the generation of "phase noise sidebands." These sidebands cause a receiver phenomenon known as reciprocal mixing. This might be best described as a rushing noise detectable at frequencies away from a strong received signal. On CW, the reciprocal mixing products have a whooshing sound that follows the keying of a loud keyed signal on the band. Of course, these reciprocal mixing products don't really exist on the amateur bands; they are generated within the transceiver itself (actually, a transmitter controlled by a PLL system also transmits phase noise sidebands). The extent to which phase noise affects receiver performance is a function of receiver design and component quality, just as other receiver performance characteristics are a function of design and quality. How much is too much? That depends on what you are willing to accept. Most manufacturers don't give meaningful specifications on the phase noise performance of their products. One of the best sources of this data is the Product Review column in *QST*; be sure to consider this very important parameter when choosing a new rig.

There are a number of other parameters that require study when evaluating the performance of a particular rig. Receiver intermodulation distortion performance, receiver sensitivity and receiver selectivity performance are all characteristics of importance to contest operators. Unfortunately, manufacturer's specifications aren't all made to the same test standards, and can't be conveniently compared. Again, perhaps the best source of information is the Product Review column in *QST*; all the equipment is measured to the same set of standards using the same test equipment. Other specifications of interest to the contest operator are received audio distortion, AGC (automatic gain control) threshold, AGC range, AGC audio output vs RF input level performance, and frequency readout accuracy. Most of these characteristics are not specified anywhere, and many can be subjective; if you can, try out a piece of equipment yourself before you decide to purchase it.

The same cliche can be modified to describe the philosophy you should embrace with regard to your transmitting equipment; "you can't work 'em if you can't be understood." This particularly applies to speech processing on phone. Forget for a moment the effect of an improperly adjusted speech processor on your transmitted bandwidth; think of how you sound to the station at the other end. It is currently in vogue to use speech processing—lots of it—in an attempt to sound loud. Unfortunately, most of this sophisticated processing does raise the average level of a transmitted signal, but primarily by adding distortion products (which does make the wattmeter hang higher). It can be scientifically demonstrated that, except under very weak signal conditions when the signal-to-white noise ratio approaches unity, speech processing adds nothing to the readability of an SSB signal. One of the main reasons amateurs are deluded into using a lot of processing is the on-the-air reports they receive; they hear how "loud" they sound. Indeed, if high blower-noise levels and lots of distortion constitutes loudness, then cranked-up speech processors make you sound loud.

In fairness, speech processors do have their place. Very few of us have ideal radio voices, complete with an optimum distribution of audio energy. This, coupled with less than optimum microphones, can make you sound somewhat anemic on phone. A speech processor used with discretion can help equalize the energy distribution of your voice as transmitted. Overzealousness with the processing equipment does make the wattmeter needle stand straight up, but it actually detracts from

Tim, K3LR, was one of three operators in the ARRL International DX Contest (CW). The others were Scott, WR3G and Tim, W3YQ.

Computers and Contesting—The Wave of the Present

Computers and contesting have come a long way in a short time. Programs that came out 10 years ago are already antiques when compared to the software packages on the market now. What do these programs do? How do they work? Which one is right for you? Let's try to answer some of these questions, which can confound even the experienced contester.

What do they do?

The first program function is to check whether you have worked a particular station. In the old days of paper logging, everyone's biggest fear was working too many duplicate calls. This could even mean disqualification from the contest. All the current programs on the market have some form of tracking to let you know what stations you have worked on what bands. Most even have what is called a "check partial." This function allows you to enter part of a call sign. The computer then checks to see if you have worked anyone on the current band with a call sign similar to your entry.

By identifying duplicates you can save a lot of time calling someone in a pileup whom you have already worked. It also can prove handy when *passing multipliers*—knowing instantly on what bands you've worked a station. This gives you the ability to ask the station to work you on other bands open at that time.
Thus the dreaded "How could I miss a 6Y5 on 15 the entire contest?" syndrome can be prevented.

Another handy function is automatic score keeping. Some contests have extremely complicated rules and point schemes. It can be a real head-scratcher trying to figure out what's what. Nowadays you can just look at a window on your screen and automatically know exactly how you're doing.

The biggest advantage of these programs is the elimination of paperwork. No longer do you have to recopy messy log sheets or puzzle over those strange forms. Contesting software will automatically generate just what you need, whether it be log sheets, dupe sheets, or even the files needed for electronic submission. This allows you to eliminate all the paperwork altogether! (See the box elsewhere on electronic submissions for ARRL contests.)

All of these packages give you additional features. Most show you how fast you're working stations by providing

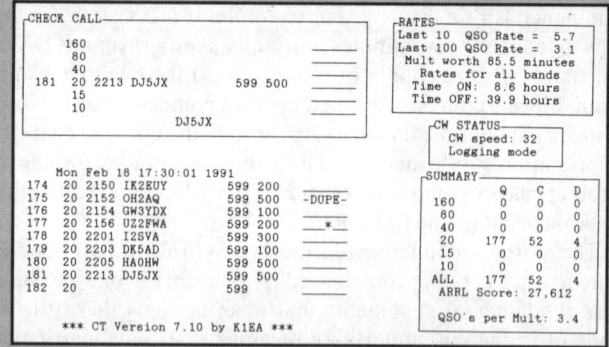

The *CT* software system configured for typical contest operation.

some kind of rate function—either a box, a chart or both. Virtually all of them have a QSL function to print QSL labels on command at the end of the contest! If you're a statistics lover, contesting software packages can generate enough figures to keep you happy until the next contest rolls around.

There are three other features of on-line logging software that should not be overlooked. The most popular is the ability to send and receive messages to and from your local *PacketCluster* via your TNC and 2-meter radio. If you're running in a multioperator or one of the various single operator-assisted classes, this will allow you to see and share spots at a moment's notice.

The second handy function is the ability to control your radio from your computer. Given the right connections and equipment, you can instantly change bands, work spots and monitor pileups from the keyboard. No more working someone and logging them on the wrong band.

The third (and most widely used) feature is the ability to send CW with your computer. The software acts as a memory keyer, and automatically sends all the information with just the touch of a key. The computer can also be configured to automatically tell a station if it's a dupe. For phone operators, two recent innovations have been the *Digital Voice Processor* or DVP from K1EA and the *Voice*

your ability to be heard and understood at the other end. Particularly in a DX contest, when there is interference galore, and the operator at the other end may not understand English too well, you should try to make yourself as understandable as you possibly can.

As far as amplifiers are concerned, your choice is determined by personal preference and budget. Most HF contests have categories reserved specifically for stations running low power. Many operators choose to run less than the legal limit (either "low power," which is usually defined as approximately 150 watts output, or QRP, which is usually defined as either 10 watts input or 5 watts output) to add an additional challenge to their contesting. The subtleties of operating using a few watts vs a thousand appeals to a large segment of the contest community.

On the VHF bands, where signals are often weak, lowpower operation is truly a handicap. That doesn't mean you can't have fun operating a VHF contest using low power (particularly from a mountaintop). But if making a high score is a priority for you, you'd better consider the reality of highpower operation. Many of the propagation modes at VHF and above have for-

mally and collectively been given the title "weak signal modes." To take advantage of these modes on a consistent basis, you need high power in addition to large antennas. For example, on 6 meters, one of the more useful propagation modes for communicating beyond 300 miles is generically known as "scatter." Scatter signals peak around dawn when the Earth collects its morning breakfast of meteors. It is quite possible for even a very modest 6-meter station (ie, 10 watts to a small beam) to contact some of the larger 6-meter contest stations through clever operating. The system-gain capabilities at the large station tend to overcome the lower system gain found at the smaller station. Since there is a relatively modest number of "large" 6-meter stations, this does tend to limit the number of stations workable by a small station.

Station Accessories and Productivity Boosters

During a contest, your main activity should be making contacts, not performing mental gymnastics. There are a number of station accessories that can relieve you from having to *think* during a contest (if you don't understand why you don't want to think during a contest, just wait; you'll quickly learn from

Blaster software from LTA. The DVP is a plug-in board for your computer. It functions as a voice keyer and recorder, allowing you to send and receive audio from your keyboard. The LTA software fills the same purpose by using your computer's sound card.

How do they work?

Virtually all the programs used by contesters today were written for PCs. Some packages have different hardware requirements. It is always a good idea to check with the software supplier before plunking down your hard-earned money on a program that requires more computer than you own. You almost certainly need a hard drive. The size of these packages and all the different contests they cover take up a lot of disk space. The files they generate during a contest require even more!

Once you have installed your program, the rest is very simple. The first screen is the start-up screen where you enter all the pertinent information such as contest, call sign, class and address. At this point you are all set up and ready to contest. Most of your functions (such as changing a band or sending a CW message) are controlled by your computer's function keys.

You log stations by moving from field to field and typing in the information as you go. Hit <enter>, and you are on to the next contact. On CW (if you've set up to key the rig), typing the call and hitting the <insert> key will automatically send the station you're working the full exchange. All programs have some form of on-line help, and there are keyboard overlays for some of the most popular programs. It is a good idea to try the program for the first time by running your own mini-contest (off the air) before the day of the big contest.

Which one is right for me?

This depends on your preferences. Some folks are happy with software they have used for years, that just covers the few contests they always operate. Others want the latest and greatest offering. Any package you buy should allow you to at least work the major (ARRL DX, CQ WW, SS, WPX, Field Day and VHF) contests; most also cover smaller events like state QSO parties. You should make certain your software allows you to send CW from your keyboard, and it will talk to your TNC. (Not all do!) You want to have a package with a "post contest" function. This allows you to go in and correct mistakes you may notice after the heat of the battle has faded. Any good software program should also allow you to

Amanda seems to be content with the 160-meter contest operating of her mistress Stephanie Koles, KD2RZ.

submit your log electronically by generating the proper files.

What do the "Big Gun" contesters use? There are three software packages that are the most popular with contesters. *CT*, written by Ken Wolff, K1EA, has grown to be extremely popular. Contact K1EA Software, 5 Mount Royal Ave, Marlborough MA 01752-1935 for details. LTA, PO Box 77, New Bedford, PA 16140 is the source of the second package called *NA*. The third program gaining favor is the *TR Logging Software*, written by Larry "Tree" Tyree, N6TR. It's available from George Fremin III, WB5VZL, 913 Ramona St, Austin, TX 78704. Any one of these should serve you quite well. You also want to check the advertisers in *QST* for other packages that might serve your purpose.—*Warren Stankiewicz, NF1J, Assistant Contest Manager*

experience!).

Before worrying about these brain-conserving gadgets, however, you should give the rest of your body some consideration. First on your priority list should be a comfortable chair; well, not too comfortable. You'll be sitting for long periods, and if you value your physical well-being, the investment in a good chair is small compared to the benefits. A chair designed for use by a secretary, who also sits for hours at a work station, is an excellent choice for an operating-position chair. Since your major activities during a contest are tuning your receiver and writing in the log, it's obvious that you'll be spending a lot of time in an erect sitting position. For best comfort over long periods, the chair you select should offer good back support (primarily in the lumbar region) and good leg support when you are sitting upright. Also, the chair shouldn't be too confining.

It may sound like this subject is being overly stressed, but there are few things more debilitating to a contest effort than being physically uncomfortable.

Since you will be looking at your station clock every time you make a log entry, a digital clock set to UTC should be placed in a prominent position where you won't have to crane your neck to view it. The digital readout aspect is important, since you won't have to interpret the clock reading as you would with a normal clock face. Although telling time may seem second nature to you normally, most people don't actually discern the exact time when reading their watch (as you have to when making a log entry). Rather, they view the clock face in a relative manner, relating their daily schedule to the position of the hands on the clock face (this phenomenon may explain the current trend away from digital watches). In a contest, you don't necessarily want to know the time relative to the world; you want only a bookkeeping reference for your log. In fact, it's quite common to lose your sense of time during a contest; time relates only to band propagation, schedules, strategic planning and (thank goodness) the end of the contest.

As a matter of practice, *always* use headphones for receiving. Aside from sparing the rest of the household from the contest din, headphones allow you to concentrate better than would a speaker. The frequency response of headphones is usually superior to the normal speaker placed in the typical acoustic environment found in an amateur station, and noise

More Operators Means More Fun

In contests, there seem to be two types. First there is the lone wolf who struggles through the weekend alone, fighting single handedly against QRM, equipment failures and the demands of sleep and family. Next there is the group operator, who joins with his friends for a multioperator effort. If you've never tried a multiop, as they are known, or even thought about it, then perhaps this will open your eyes to an entirely different way of contesting.

A good example of a multiop effort is a club during Field Day. It's more than just a bunch of people trying to make contacts together—it's an *atmosphere*, a camaraderie that comes with a team coming together trying to accomplish a specific purpose. There are two basic types of multioperator efforts—*single transmitter*, where you use only one transmitter at a time and *multitransmitter*, where you use more than one at a time. There are numerous rules and restrictions for each contest.

How hard (or easy) is multiop? It depends on what you're setting out to do. In contesting there are advantages and disadvantages to everything. How you approach it, and how you make the best use of the resources at hand, spell out the difference between a good time and the weekend of unmemorable frustration. One thing is certain; with a group even the biggest mistakes become the basis for stories to be told over and over!

First, let's cover the good part of operating multiop. Top on anyone's list is *sleep*. With more than one operator, you can usually try to schedule things so everyone gets a break when they need one, and no one gets so bored or tired they begin to make mistakes. Having more operators allows you to work stations on other bands more easily, or listen to multipliers. In multi-two or multi-multi, you can have signals on as many bands as you have rigs and antennas. You can also use a technique known as the *pass*: asking a station to move from one band to allow you to make certain you work that multiplier on every band possible. The advent of *PacketCluster* has enabled groups to find (and give out information on) DX stations to an extent unheard of in the old days of 2-meter FM local nets.

Best of all, it offers a great opportunity for you or other members of your club to cooperate in a contest. If you're new to the hobby, or aren't able to put up much of a station, a good multiop effort can get you some experience to pick up some important contest skills, or the chance to operate better equipment than you normally use. Besides, it's a lot of fun! If you've have a good station, but can't get interested in working through an entire contest (48 hours sometimes seems to get longer and longer) then perhaps you ought to consider opening up your shack to your local group. After all, any old-timer will tell you it takes time, skill and patience to be a good contester—this is your golden opportunity to show someone the *right* way to do things! We're sure there are plenty of hams in your community who would jump at the chance to operate from a station with a nice antenna farm, or with one of the latest rigs or amps.

How hard can it be? Well, at times, very. The challenges of running or operating a multioperator station can be formidable, and it takes a lot of time and preparation to make things flow smoothly. Here's some of the common pitfalls:

Logging: Almost everyone uses computers these days. When it comes to multioperating, though, nightmares crop up unexpectedly. Does everyone know how to use the software? If not, are there handy "cheat sheets" nearby the operating positions? Have you checked to make sure the software works on all the computers in the shack? More than one effort has been stymied when the PC someone brought wouldn't run the contest software on hand. (This usually happens when the clock shows 2354Z—6 minutes to contest start time.) Some computers generate noise in your receivers; others will lock up the first time you key-down the amplifier. If you're linking computers, or connecting them to radios, packet TNCs or keyers, the cabling between different ports and connectors can be a complicated nightmare! (Hint: make certain you have *plenty* of adapters of all shapes and sizes!)

Creature comforts. Do you have a place for people to sleep or relax, if need be? Operators in the shack should be left alone to operate—have a separate room for people to rest, eat, sleep or tell war stories. Make certain to stock enough food and beverages to last the weekend, or your operators may desert you! Of course, the need for well-developed toilet

---

emanating from outside the station is sealed out. Many experienced operators claim that the use of headphones offers at least 10-dB signal-to-noise improvement over the use of a speaker. On phone, the use of headphones also eliminates most, if not all, of the problems associated with VOX operation.

The important feature to look for in headphones is comfort. Many modern stereo headphones are quite comfortable and can be worn for hours without causing pain or fatigue. The sound quality of the better units far surpasses any speaker that you might consider using. One note: Most amateurs seem content to use inexpensive ($10) headphones for contesting. These cheap units don't sound very good, aren't very durable, and certainly aren't too comfortable. It seems foolish to invest hundreds, if not thousands, of dollars in electronic equipment (not even counting the cost of an antenna system) and then spend very little on inadequate headphones. A station is only as effective as its weakest link.

For CW operation, a memory keyer is a sound investment. Many operators find that their physical coordination tends to go awry late in the contest, and their sending skills trail off. Since you want the operator at the other end to understand you clearly, the memory keyer can provide an edge to the tired contester. Another big advantage of a memory keyer is that it frees you from sending redundant information. Most contest

exchanges can be preprogrammed into the keyer. When it's time for you to send your information, you merely have to press the button and you're able to sit back and catch up on the paperwork of dupe sheets, multiplier sheets and logs. Life is less frantic (and less tiring) when you use a memory keyer.

Along the same lines, choose a paddle that you find comfortable, and somehow prevent it from wandering around the operating table as you send. Taping a piece of sandpaper, rough side up, where you want the paddle to sit is a good measure to prevent paddle creep.

For phone operation, use a microphone that works effectively in combination with your own voice and your transmitter. Close matching of microphone characteristics with your voice can provide far more "punch" in your audio than any speech processor. If you can, mount the microphone on a boom so that the microphone base doesn't take up valuable writing space. Be sure to route the microphone cable in such a way as to avoid the writing surface. Most operators prefer to use the VOX (voice-operated switch) feature on their rigs, rather than PTT (push-to-talk), since VOX permits both hands to be free for logging and so on.

Perhaps the ultimate solution is the use of a headset mounted boom microphone. The advantages to this arrangement are numerous. The microphone is always optimally placed relative facilities has wiped out more than one operation. Operators at the famed top multi-multi station N2RM may be content to use an outdoor or porta-potty in February; your operators may not be such dedicated contesters.

How well lit is your shack? Are the operating positions comfortable? Is the ambient noise level good? Remember, in a phone contest, this room could have six people calling CQ at the top of their lungs.

The biggest hurdles to overcome are technical. Multiple stations means multiple interference, and it takes quite a bit of work to overcome some of these problems. To operate on more than one band, you need to be able to *hear* on more than one band. Harmonics from your 40-meter transmitter can trash 15; 80 can wipe out 20 or 10. It's not easy! The solutions are too complex to cover in any detail here, but let's cover some basics.

To start off, ground everything. Even if you don't think it needs to be. Use hefty grounds, too—a strap made of tinned braid an inch wide will work a lot better than #22 AWG wire. Ground everything in the station together, and to the main ground for your shack. Second is the physical placement of your antennas—two different antennas on one tower can lead to various problems if they're not done right. Most of the biggest multitransmitter stations have at least two towers for their farm, with adequate spacing between any two antennas for any given two bands in order to reduce the interference.

Inside the station there is a vital accessory you should not overlook. The well-equipped multiop uses a device known as a *passive receive filer*. These go in your receiver's line to eliminate unwanted signals from other bands.

Most widely known are the famed W3LPL filters, used at Frank Donovan's Maryland megastation, based on an LC design. Other stations use ones they're developed themselves, or one of the many commercial versions now appearing on the open market. These improvements can go a long way in turning a weekend of potential nightmares into a lean, mean, contesting machine.

Whatever your aim—whether to be the king of the mountain, tops in your club, or to train some new operators, enjoy yourselves! Many hands can make light work—and for a bigger score!—*Warren Stankiewicz, NF1J*

The new guys take over. Brad Parker (in the rear) and Patrick Conroy were still waiting for their call letters to come in the mail when they operated N1II at Field Day for the Greater Norwalk Amateur Radio Club. A few weeks later the mailman told Brad he was KB1AKA and Patrick he was N1TFA. Peter Braun, N1RRA, took the photograph.

to your mouth, but never obstructs your view of the log materials or the rig. Since the microphone is in a fixed position relative to your head, you are free to move about and stay comfortable. Finally, the use of a headset-type microphone eliminates the possibility of bashing yourself in the mouth with the microphone when you're not paying close attention (don't laugh—it happens).

A very good accessory to use with a microphone is an equalizer. Within limits, an equalizer can compensate for an imperfect match between the energy distribution in your voice and the frequency response of the microphone.

Some operators prefer to use a foot switch to control transmit/receive switching (a sort of "STT"—step to talk). Footswitch operation is very fast, and of course doesn't tie up the use of your hands. The foot switch can be wired in parallel with the PTT connections within the transmitter microphone connector. Very often this arrangement allows control on CW as well as on phone.

Since a large part of a contest effort is devoted to logging and other paperwork, having an electric pencil sharpener nearby can be quite handy. Similarly, a pad of scrap paper finds numerous uses over a contest weekend.

There are lots of other station accessories that you might find useful. One item that should not be considered an optional accessory is good station lighting. Good lighting not only keeps your eyes healthy, but it also reduces fatigue. Fluorescent lighting, when used alone, causes many people to get headaches when they concentrate. High-level incandescent lighting is very tiring. A good compromise is a combination of fluorescent and incandescent lighting. Many operators find that local lighting at the operating position, with the rest of the room held at reduced lighting levels, allows them to concentrate more on operating. Additionally, the use of a dimmer to control light levels can reduce fatigue tremendously (try to avoid the solid-state variety; they tend to generate noise in the RF spectrum).

PREPARING FOR THE CONTEST

In addition to preparing your station for a contest effort, you must also prepare yourself for the main event. Practicing contest operating is quite a challenge; the only realistic simulation for a contest (until recently, anyway—see accompanying sidebar) *is* a contest. There is never the QRM level or amount of activity during non-contest periods that there is during a contest. Most of the skills you are interested in developing really can be honed only under fire.

To maintain their already substantial operating skills, many successful contesters put in at least token efforts in the less-active contests that seem to pop up every weekend. They gain

Contest Entries and the Electronic Age

Since 1989, the ARRL Contest Branch has accepted entries electronically. Since there are so many software packages out there, and different methods of transmission—from diskette to BBS to e-mail—it can be confusing, or at least a little intimidating, to figure out how to do it.

To submit a log electronically for an ARRL contest, you need two things. First is a log file. It must have the name CALLSIGN.LOG (such as K1ZZ.LOG). The second thing needed is a summary sheet. This can be the standard piece of paper if you are mailing in a disk. If you're sending it to HIRAM (the ARRL BBS) or by e-mail, you need a file with the same summary information.

Your .LOG file must be an MS-DOS file and follow a certain format. It has to be in the "Standard File Format for Submission of Contest Log Data," as shown in the accompanying sidebar. It can't be a binary (.BIN) or "all" (.ALL) file; nor should you name another file CALLSIGN.LOG. Fortunately, almost all of the major contesting software packages (see the sidebar "Computers and Contesting—the Wave of the Present" for details) will automatically generate the proper files for you—check your software manual.

Once you have the summary and the log file, three options are open for you. You can mail the disk in with the summary sheet; you can upload them as separate files to the ARRL BBS (203-665-0090, 14.4-8-N-1), or you can send them by e-mail to the Contest Branch Internet address: contest@arrl.org. (If you use the Internet, you can send your log and summary files as one file; on the ARRL BBS you can also ZIP them together.) Any of these methods should work just fine.

But wait! What if something happens to my disk?—the ARRL Contest Branch is not a government bureaucracy— if we hit a snag with a file, we won't disqualify you for it. What if my file gets lost? All e-mail submissions are acknowledged with a short message. If you don't get this message in about 2 weeks or so, try again! Many people who mail entries in to ARRL have also found a self-addressed, stamped postcard is a good way to keep tabs on their log. As soon as your entry is processed, we mail the card back to you, attenuating any worries by several tens of dBs.

Several other contest sponsors (*CQ*, WAE, NAQP and others) are now accepting entries electronically. Most sponsors are still in the infant stages in this area, so we strongly suggest you check with them on the formats and files they require for submission.

Where's the payoff in all this? Time. You no longer have to work your way through reams of paper. This saves you time and effort. It also reduces the massive job the Contest Branch has to do every year in processing the 16 contests it sponsors. These contests generate roughly 16,000 pieces of mail in a year—a lot of paper for two and a half people to sort through. The more time (and money) we can save, the more you, the members, save as well.

And from all of us, thank *you* for helping!—*NF1J*

valuable practice by participating, their call becomes more familiar to the rest of the contest fraternity, and they have fun. These "minor" contests are the best training ground for the "major" contests. Additionally, don't forget that just because you consider the East Podunk QSO Party to be a minor event doesn't mean that there aren't some operators out there who treat the East Podunk QSO Party like the Olympics. Get on and give them a contact; they might return the favor some day.

There are some exercises that you can perform to sharpen your contest skills during non-contest periods. One of the greatest skills a contest operator can develop is the ability to copy weak signals through heavy interference. Not many amateurs spend their time looking for QRM, but you might actually benefit from doing just that. If you hear a big pileup somewhere, spend some time listening. Try to pick calls out. Listen to see who gets through the pileup first, and how he or she does it. The benefits here are threefold. First, you get practice at pulling signals through QRM. Second, you learn about some of your potential contest competition and their effectiveness in breaking through pileups. Third, you get to examine firsthand the various techniques people use to get through. Try to analyze and remember the *successful* approaches for your own use later. Take notes if you have to.

Once you reach the point where you have no difficulty copying all the stations calling in a pileup, try adding your own QRM—but not by getting on the air and adding to the chaos! Increase your receiver filter bandwidth a step wider than you normally use. For example, don't listen to a CW pileup using your narrowest code filter. Listen on the SSB filter. If you're listening to an SSB pileup, switch over to the AM filter. Using a wider filter forces you, the operator, to separate the signals in your mind, instead of letting the receiver provide the interference rejection. After enough practice, you'll find that your brain is a very effective filter. When contest time comes along and you go back to the narrower filters, you'll marvel at the selectivity. You may find that your *cerebral* filter scheme works so well that, particularly on CW, you don't need as narrow an electronic filter as previously. If you can attain this level of proficiency, you gain an added advantage over most of the competition. Since your receive filter is wider, "chasing" contacts is much easier because you can effectively tune the band faster than someone with a narrow filter.

Strategy plays an important role in successful contest operating. The decision to be on a particular band at a particular time can be crucial to the outcome of a competition, particularly when you make the right move and the competition doesn't. Your strategic decisions should be made on the basis of hard facts and educated choices (in reality, a lot of luck will be involved, too). To have a full complement of facts available during the contest, you have to do yet more research before the contest. There are two areas where you can gather information prior to the contest that will be of value during the contest. The first is propagation.

During the weeks prior to the contest, study the bands. At what times do they open to where? From what direction? For how long? Keep records of your observations. The most significant times to observe are the few days just prior to the contest, and 28 days before the contest period (the sun rotates once every 28 days; propagation trends tend to follow the same pattern). Once you have completed your study, plot your results in graphical form. Use the resulting presentation in preparing a tentative operating plan for the contest. Of course, like the weather, you truly can't predict band openings with a high degree of reliability, so be prepared to modify your strategy according to the propagation you actually encounter. Be flexible.

Speaking of the weather, you should also follow the weather forecasts for the days immediately preceding the contest. Most enhanced VHF propagation conditions are the result of slowly moving weather systems; weather-map study can help you predict potential favorable directions in a VHF contest. HF contest enthusiasts should also follow the weather; passing storm fronts cause QRN (static), particularly on the lower HF bands. If you expect a front to pass through your area during the contest, plan to emphasize the higher frequency bands during that period to avoid the QRN. Plan your lower frequency operation around the QRN. Some television stations now show foreign weather conditions as well as local forecasts; for DX contest operating, use the foreign weather information to plan your lower frequency operation around QRN at the DX end, so the DX stations can hear you.

The second area where you can gather useful contest information before the contest is in the study of activity patterns. Having a knowledge of certain stations' operating habits can truly be a blessing with regard to finding rare contest multipliers. Is there a particular gathering frequency for groups of stations in a particular geographical area (such as for severe weather reporting in tropical areas)? Does a station located in a rare multiplier habitually get on the air just before his or her bedtime every evening, perhaps way up in frequency on the band? Is a particular station willing to change bands with you for a new multiplier? Is someone going to a rare area specifically for the contest? All this information can be put to good use if you document it carefully and put the data in a form that you can understand well into the contest.

Sometime before the contest, prepare your logging materials for contest use. Obviously you can't fill in contact data in advance, but you can put in some constant information like your own call sign, the contest starting date, a consecutive number for each log page, the dupe sheet frequency band information and the like. Not only will this step save time during the contest, it will also lead to less confusion during the contest, allowing you to concentrate on making contacts.

It's also a good idea before the contest to have your station all set up for the event. Adjust the VOX controls on the transmitter, set your clock, calibrate your receiver, replace the burned-out light bulb over the operating position, program the memory keyer, note the optimum band markings for quick tuneup on each band, and any other maintenance tasks that might need doing. You certainly don't want to perform maintenance during the contest; you have enough to do just in operating and fixing the inevitable visits from Murphy.

Be sure to read and thoroughly understand the latest contest rules for the contest in question. Perhaps the rules have changed since last year or your recollection of the existing rules is rather fuzzy. In either case, being clear on the rules might prevent you from making a fatal mistake that might make your hard work during the contest weekend for naught. Finally, prepare yourself mentally for the contest. As most sports psychologists will tell you, your mental attitude often is the determining factor in how well you perform in a competitive activity. Convince yourself that you can do well by putting in the effort. Picture in your mind what you're going to do in each situation you might encounter during the contest. Run through it over and over until it becomes second nature. You have to determine for yourself how to get motivated, since everyone is different. One thing is for sure, though; if you aren't prepared to do your best, you won't.

OPERATING THE CONTEST

The big moment finally arrives. Now what? Regardless of whether you are operating a VHF contest, an HF DX contest, the Novice Roundup, or whatever, there is one fundamental guideline that you should follow in a contest: Do what gains you the most points. "Another obvious suggestion," you're probably thinking. It may be obvious, but if you truly follow it, you'll end up with the highest score possible for your station on that weekend.

For example, early in the contest you come across a rare

multiplier. He is just casually working a few stations at a rate of about 12 an hour. He is on an "oddball" frequency, away from most of the contest activity. What should you do? If you decide that the only way you'll work the multiplier during the whole weekend is to stick it out and spend the time, you'll be missing the highest activity point in the contest. The time might be well spent. Of course, you could also spend 45 minutes and never work him; think of the contacts you lost and will never recover.

Decisions like this need to be made almost constantly during a contest. Your decision should be based on considering that simple fundamental. If you feel that propagation might be favorable right then, that your station is loud enough, and that the addition of another multiplier is more beneficial to the final score than running up your contact total at that point, then by all means try to work the multiplier. Your gamble may or may not work. You might decide that you can more easily work a number of stations in the same time that it might take to work the multiplier, more than enough contacts to compensate for the "lost" multiplier. Your best choice then probably is to forget about the multiplier; he might be easier to work later. Decisions like these are the very essence of contesting, and only

OH2MM seized the opportunity to use these antennas and operate at EA8EA in the 1993 ARRL DX Competition.

The Contest Ain't Over Til the Paperwork Gets Done

If you're new to contesting, or just mildly interested, you may be a bit hesitant to plunk down some money for a fancy program that you spend more time scratching your head at than actually using. Never fear—the art of logging by paper still exists, and it's fairly simple to learn. And no, you won't be a dinosaur, either!

Every contest sponsor will supply forms by mail. If you can't get them, use standard log sheets, and copy all the information needed. Just make sure your sheets have the band, date, time, call, the exchanges sent and received, point value, and some mark denoting if a contact is a multiplier. The ARRL DX Contest log sheet (**Fig 7-1**) is a typical log form. These log sheets, a summary sheet (which

we'll cover later) and a cross-check sheet is all you'll need.

Start off with a good pen. Some folks prefer a good collection of sharp pencils, but it's easy for pencil writing to smear and be hard to read. Remember, you want your log to be as legible as possible—not only for your benefit, but for those who will have to look at it later. A scribbled log can attract unwanted attention! If you're using a pen and make a mistake, just cross it out and start on another line—no problem.

Most important, however, are the *cross-check* or *dupe sheets*. These are the key to your success or failure in logging. Unless you're one of those with a truly phenomenal memory, the need to tell at a glance who you worked on

Fig 7-1—A typical written log page. Notice the band, time, and multiplier entries.

Fig 7-2—Dupe sheets are needed to avoid working a station twice and violating contest rules. This one is for a DX contest.

Fig 7-3—The form for stateside contests differs from the DX contest form but the intent is the same.

ARRL INTERNATIONAL DX CONTEST

CW Third FULL weekend in February—Phone First FULL weekend in March

CALL USED **WB8ISZ**   CW ☒

            PHONE ☐   ARRL SECTION (p. 8 QST) **OH**

Call of operator if different from call used

Separate logs and summaries must be submitted for each mode of operation.

		Valid QSOs	DXCC Countries
☐ Novice/Technician			
Single operator	160	12	11
☒ All band	80	107	51
☐ Single band (circle one) 160 80 40 20 15 10	40	279	60
☐ QRP (5W or less output)	20	561	81
☐ Assisted	15	375	69
Multioperator	10	65	37
☐ Single transmitter			
☐ Two transmitter	Total	1399	309
☐ Unlimited			

Multiops list ALL operators:

Please enclose logs, dupe sheets, photos and soapbox comments with your entry and mail within 30 days after the phone contest to: ARRL Contest Branch, 225 Main Street, Newington, CT 06111.

$\underline{1399}$ (Valid QSOs) × 3 = $\underline{4197}$ (QSO Points) × $\underline{309}$ (DXCC Countries) =

$\underline{1,296,873}$ Claimed Score.

POWER OUTPUT—CHECK ANY USED: ☐ 5 watts or less   ☐ 6-150   ☒ Over 150

YES ☒   AFFILIATED CLUB PARTICIPATION   NO ☐

ARRL AFFILIATED CLUB NAME **The Wabash Gang**

(Note: Club secretary must submit list of eligibles in order for club aggregate to be listed. 3 or more entries must be submitted.)

"I have observed all competition rules as well as all regulations established for Amateur Radio in my country. My report is correct and true to the best of my knowledge. I agree to be bound by the decisions of the ARRL Awards Committee."

Date **28 Feb. 1991**   Signature **David Patterson**   Call **WB8ISZ**

NAME: **David Patterson**   CALL: **WB8ISZ**

ADDRESS: **3834 Willow Creek Dr.**

Dayton, OH 45415

1. Check log for duplicate QSOs.
2. Copy ALL QSO info carefully. A penalty is assessed for incorrectly copied QSO info.
3. Observe mailing deadline.
4. Dupe/check sheet must accompany all entries of 500 or more QSOs.

MCS-53 (590)   Printed in U.S.A.

Fig 7-4—The standard summary sheet lets the contest judges see at a glance who you are and what you did.

what band, or whether you have worked this particular station before, is vital. Work the same station twice, and it will cost you. Work too many stations twice, and you'll find yourself disqualified, a term that strikes fear into the strongest of contesters everywhere. In fact, if you're submitting a paper log most contest sponsors *require* you to send in cross-check sheets. **Fig 7-2** is a typical DX contest dupe sheet and **Fig 7-3** a stateside contest dupe sheet.

What should you do? First, work the station. As you complete the contact, fill in all the information on the log sheet, taking special care to mark the band, time, exact call sign, and the information you received. Immediately enter the call into the right spot on your dupe sheet. Failure to do so now can lead to problems later! As the contest progresses, you should be able to quickly look and see if you've worked HZ1AB on 40 before, or if the station that just called you is a duplicate contact or not.

When the contest is over, check your log! Make certain every contact in your log shows up somewhere on the dupe sheet—it's a good way to make sure you didn't work anybody twice. Look it over carefully to see if you made any mistakes. Then total up your QSOs, points, and multipliers, and fill out your summary sheet. Your summary sheet (**Fig 7-4**) is a form containing the name of the contest you're entering, call sign and operating class (single operator, low power, single band, whatever.) It includes the count of QSOs, points and multipliers you're claiming for each band or mode. Finally it contains the arithmetic that you used to determine your final score, your name and address and a disclaimer. The disclaimer is a signed statement that you obeyed the rules of the contest, and agree to be bound by any decisions of the contest sponsor. Stuff it all into an envelope, add enough stamps, and mail it off! Now, all you have to do is sit back and wait for the results to appear.

you can decide based on your knowledge of propagation, your station's capabilities, contest strategy and your own capabilities. Indeed, preparation, experience and raw talent determine your ability to make the right decision when needed.

There are some operating hints that might make your strategic decisions easier to make. First, learn the relative point value of a new multiplier compared to a new QSO. Simply dividing the number of QSOs made at that point in the contest by the number of multipliers worked at that point in the contest gives a ratio that represents the number of contacts a multiplier is worth at that point. You can do this during the contest with a calculator or slide rule, but arithmetic exercises when you're already fatigued likely will be very time consuming, and may not yield accurate results. Instead, a prepared chart, showing the number of QSOs along one axis, the number of multipliers along the other axis and the QSO/multiplier ratio at the intersections, is much easier and more reliable to use during a contest (in fact, add that to your list of items to prepare before the contest).

At each point in a contest, the QSO/multiplier ratio changes depending on the contest format and your own performance. For example, during a DX contest, where there is a relatively high number of multipliers available, the QSO/multiplier ratio might be around three. That means each new multiplier worked at that point is equivalent to making three (nonmultiplier) QSOs, as far as your point total is concerned. In the November Sweepstakes, in which there is a finite number of multipliers available (77), the QSO/multiplier ratio might be over 10. In

that circumstance, it would require making 10 QSOs to equal the point value of one new multiplier.

By applying your knowledge of propagation, your intimate familiarity with your station's capabilities, and your best estimate of how the propagation and activity is evolving in the contest, you can use the QSO/multiplier ratio information to decide whether to look for contacts or to try jumping into a pileup for a new multiplier. During high activity and good propagation periods, making contacts is probably a better tack, since you might never be able to recover the band opening time later on. Spending 10 minutes in a pileup when the band is hot and you have the potential to make a lot of contacts is a poor choice compared to spending 10 minutes in a pileup when the activity or propagation has slackened, and you only might be able to make two contacts during that same 10 minutes.

The key principle is to utilize your band-opening time wisely, as you have little or no control over propagation and have to work what you can when you can. This strongly implies that any unusually good propagation should be taken advantage of immediately, since it may never return. At VHF especially, where band openings tend to be short-lived (anywhere from a second to a couple of hours), enhanced propagation is a sure sign to stop whatever you're doing and work what you can during the opening. You may not get another chance. The value of band-opening time is inversely proportional to its duration.

As an example, consider 6-meter scatter propagation during a VHF contest morning. A meteor burst might make a 1000-

mile-distant station audible (and potentially workable) for 20 seconds. Unless you make the contact during that time period, you're out of luck. You should call quickly and succinctly and exchange contest reports the same way. If you make yourself understood the first time, there is no need for repeats that you might not have propagation for. There is no time to give the other station's call (which he or she presumably already knows) five times, your call eight times and the instruction to go ahead six times. By the third "go ahead," the burst is over, the contact is lost, and the operator at the other end is frustrated over the failure to complete the contact successfully. The next time there is a meteor burst, that other operator might tell you to "go away," and then try to work someone quicker on the draw than you were.

HF contesters are just as guilty of bad (or no) timing as VHF contesters. How many times have you heard the din of a pileup go on for two or three minutes after the station being called has already come back to someone? Who gained there? The moral: Don't repeat unless asked to.

There are two general approaches to making contacts—calling CQ (and "running" people/getting responses) and hunting. Each has its own advantages and disadvantages, as you might expect. Calling CQ effectively requires getting the attention of other stations and motivating them to call you. Getting attention at the other end is most easily accomplished by wielding a big signal. Big signals always seem to get answered (as do signals emanating from rare multipliers). Of course, you may not be fortunate enough to possess a big signal, so calling CQ may not be as productive for you. Here again, knowledge of your station's capabilities is required to make that decision.

Motivating respondents to call is usually easy early in the contest while you still haven't drained the pool of active contesters. The contesters are already motivated to call; they want the QSO. The strain begins *after* you have worked the serious competitors, and the casual operator is all that remains. Here you have to be prepared to answer questions about the contest, and perhaps tell the operator at the other end what information you need for a valid contest QSO.

There is one operating strategy to keep in mind here: Don't scare anyone off. Try to be as polite as you can, and give the information at a reasonable pace. Sending an exchange at 50 words per minute or speaking as fast as you can may be fine when you are in contact with another contest devotee, but is totally lost on most everyone else. Not only will they not be able to acknowledge your transmitted information, they (as well as anyone else listening) might become discouraged or even incensed, disappearing without giving you your QSO.

There seemingly are as many hunting techniques as there are contest operators. While everyone has a favorite approach, there is one generalization about the successful versions; they are all organized. The operator starts at one end of the band and methodically scans to the other end, while trying to contact all the stations not already worked in between. Maintenance of an updated duplicate check sheet (dupe sheet) is a necessity for hunting. It is not only embarrassing but a waste of time to call

someone already worked. Some operators maintain a "map" of the band that lists the call signs of the stations holding forth on the various frequencies. A map such as this allows the hunting operator to skip over frequencies that he or she knows are occupied by someone already in the log. This can save considerable time, especially on bands that are clogged with big pileups; spending 10 minutes to work a station that turns out to be a duplicate contact doesn't add much excitement to the contest.

Successful contest operating can be summarized very easily: Always spend your time where you can produce the most points and treat that time as if it is precious—because it is! Always strive to make your log entries an accurate and complete representation of what was sent to you. Accurate communication is the very basis for the contest; inaccurate communication can be the basis for disqualification.

AFTER THE CONTEST

Afterwards, every participant faces the same task, regardless of his or her performance in the contest—preparing the log for submission. The proficient and ambitious contester adds a second task—analyzing the results. The first task shouldn't be taken lightly. Many contests have been won or lost as a result of accurate or inaccurate log checking before submission. Log preparation is a lot like preparing income tax returns. The accuracy of your submission has a lot to do with the outcome. While you may be able to get away with "creative logging" for a year (or contest) or two, eventually you will be revealed. From that point on, your return (or log) will be carefully scrutinized on a routine basis. What is probably worse about contest logs than income-tax returns is that once you are disqualified, you will never have any credibility with your fellow contesters. That in itself might be a more severe punishment than a few measly years in prison for defrauding the government!

A proper log submission consists of a photocopy of the original log sheets, with invalid and duplicate contacts so marked. Don't erase these contacts, just mark them so that the log checkers can identify and verify them by cross-checking with other logs. It is virtually imperative to "re-dupe" a log before submission. Any duplicate check-sheet made during the contest is bound to be inaccurate and should only be considered an operating aid. In these modern times, computers are increasingly being used to eliminate duplicate contacts from contest logs prior to submission (be sure to double check for keyboarding errors, however).

A contest log might be considered the history of a "lost" weekend. While there is no explicit description whatsoever of the events of the weekend, a clear record is available for the experienced to interpret. Your best source of information to use while preparing for the next contest is your record of the last contest. What mistakes did you make? What bands were the most or least productive and why? What equipment worked well? What equipment failed? Responding to these questions should keep you busy until the next contest. And you can be sure that the next contest will be more productive and satisfying than the last.

CHAPTER 8

Operating Awards

STEVE FORD, WB8IMY
ARRL HEADQUARTERS

A wards hunting is a significant part of the life-support system of Amateur Radio operating. It's a major motivating force of so many of the contacts that occur on the bands day after day. It takes skillful operating to qualify, and the reward of having a beautiful certificate or plaque on your ham-shack wall commemorating your achievement is very gratifying. (On the other hand, you don't necessarily have to seek them actively; just pull out your shoebox of QSLs on a cold, winter afternoon, and see what gems you already have on hand.) Aside from expanding your Amateur Radio-related knowledge, it's also a fascinating way to learn about the geography, history or political structure of another country, or perhaps even your own.

This chapter provides detailed information on awards sponsored by ARRL and other major US organizations. Following that is perhaps the most comprehensive collection of international awards information ever gathered under one cover, obtained by the editor through the good offices of International Amateur Radio Union (IARU) Societies, the sister associations of the League. In so doing, the awards data base has been expanded like never before.

There are some basic considerations to keep in mind when applying for awards. Always carefully read the rules, so that your application complies fully. Use the standard award application if possible, and make sure your application is neat and legible, and indicates clearly what you are applying for. Official rules and application materials are available directly from the organization sponsoring the particular award; always include an SASE (self-addressed, stamped envelope) or, in the case of international awards, a self-addressed envelope with IRCs (International Reply Coupons, available from your local Post Office) when making such requests. Sufficient return postage should also be included when directing awards-related correspondence to Awards Managers, many (if not most) of whom are volunteers. So above all, be patient!

As to QSL cards, if they are required to be included with your application, send them the safest possible way and always include sufficient return postage for their return the same way. (A return-postage chart for ARRL awards appears in Chapter 17 of this book). It is vital that you check your cards carefully before mailing them; make sure each card contains your call and other substantiating information (band, mode, and so on). Above all, don't send cards that are altered or have mark-overs, even if such modifications are made by the amateur filling out the card. Altered cards, even if such alterations are made in "good faith," are not accept-able on this no-fault basis. If you are unsure about a particular card, don't submit it. Secure a replacement.

None of the above is meant to diminish your enthusiasm for awards hunting. Just the opposite, since this chapter has been painstakingly put together to make awards hunting even more enjoyable. These are just helpful hints to make things even more fun for all concerned. Chasing awards is a robust facet of hamming that makes each and every QSO a key element in your present or future Amateur Radio success.

ARRL AWARDS

To make Amateur Radio QSOing more enjoyable and to add challenge, the League sponsors awards for operating achievement, some of which are among the most popular awards in ham radio. Please note that with the exception of RCC and Code Proficiency, applicants in the US and possessions, Canada and Puerto Rico must be League members to participate.

Rag Chewers Club (RCC)

RCC is often the first operating award a new ham aims for. This is designed to encourage friendly contacts of more sub-

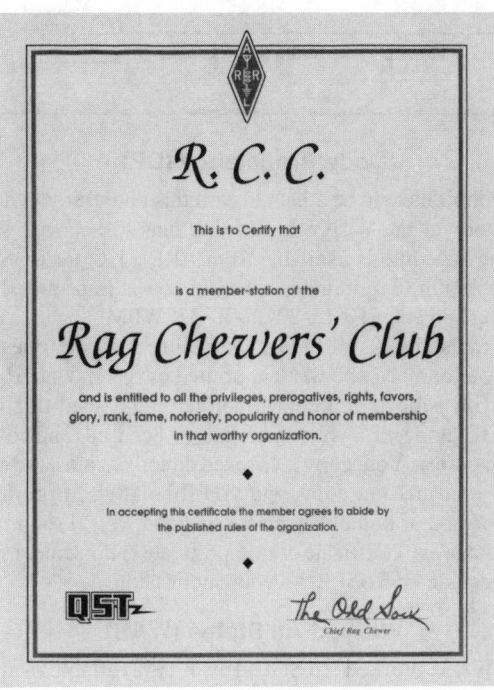

stance than the hello-goodbye type QSO. RCC has just one requirement: "Chew the rag" over the air for at least one solid half hour. If you want to obtain the RCC certificate, report the QSO to ARRL HQ, with a business-size SASE, and you'll soon be issued the distinctive blue award. If you want to nominate someone else for membership, send the nomination to him or her, not to ARRL HQ. If this person wants the award, he or she can send the nomination to HQ. This way, no one gets an unwanted certificate, and confirmed ragchewers can still nominate those they think are qualified. RCC is a free ARRL service to all amateur licensees.

Friendship Award

The purpose of this award is to encourage friendly, conversational contacts between radio amateurs. The Friendship Award is available to any ARRL member who submits log copies that show two-way communication with 26 stations whose call signs end with each of the 26 letters of the alphabet. (For example: W4RA, KØORB, W3ABC, K1ZZ.) QSL cards are not required.

Logs must indicate the contact date, call sign, name, location *and some other fact about the person you contacted.* (For example: age, other interests, occupation and so on.) Contacts can be made on any frequency or mode. All contacts must be made after November 1, 1993. For a Friendship Award application, send a self-addressed envelope with one unit of First-Class postage to: ARRL, 225 Main St, Newington, CT 06111.

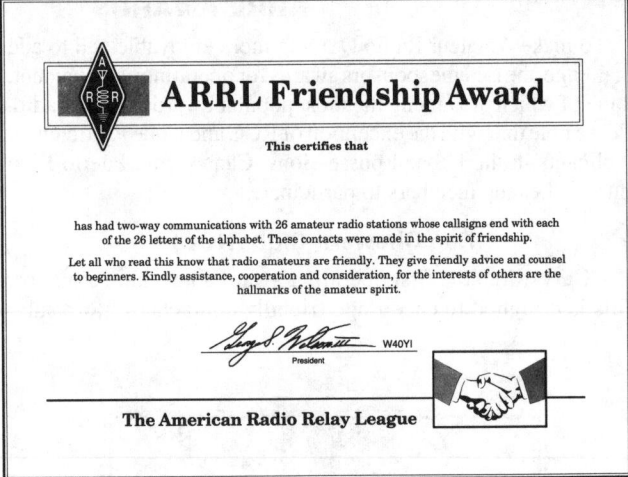

Code Proficiency (CP)

You don't have to be a ham to earn this one. But you do have to copy one of the W1AW qualifying runs. (A current W1AW operating schedule is available from ARRL HQ for an SASE.) Twice a month, five minutes worth of text is transmitted at the following speeds: 10-15-20-25-30-35 WPM. Twice yearly, W1AW transmits 40-WPM runs. To qualify, just copy one minute solid. Your copy can be written, printed or typed. Underline the minute you believe you copied perfectly and send this text to ARRL HQ along with your name, call (if licensed) and complete mailing address. Your copy is checked directly against the official W1AW transmission copy, and you'll be advised promptly if you've passed or failed. If the news is good, you'll soon receive either your initial certificate or an appropriate endorsement sticker. Please include an SASE with your submission.

Worked All States (WAS)

The WAS (Worked All States) award is available to all ama-

teurs worldwide who submit proof with written confirmation of having contacted each of the 50 states of the United States of America. The WAS awards program includes 10 different and separately numbered awards as listed below. In addition, endorsement stickers are available as listed below.

Two-way communications must be established on amateur bands with each state. Specialty awards and endorsements must be two-way (2×) on that band and/or mode. There is no minimum signal report required. Any or all amateur bands may be used for general WAS. The District of Columbia may be counted for Maryland.

Contacts must all be made from the same location or from locations no two of which are more than 50 miles apart, which is affirmed by signature of the applicant on the application. Club-station applicants, please include clearly the club name and call sign of the club station (or trustee).

Contacts may be made over any period of years. Contacts

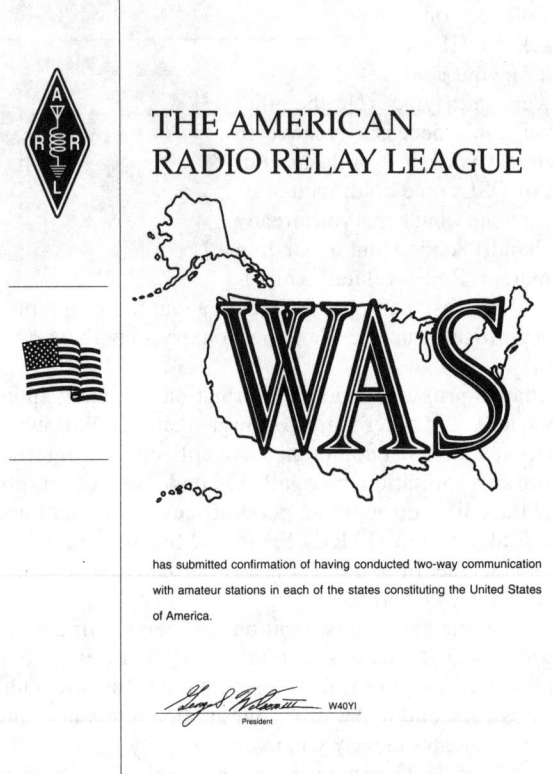

must be confirmed in writing, preferably in the form of QSL cards. Written confirmations must be submitted (no photocopies). Confirmations must show your call and indicate that two-way communication was established. Applicants for specialty awards or endorsements must submit confirmations that clearly confirm two-way contact on the specialty mode/band. Contacts made with Alaska must be dated January 3, 1959 or later, and with Hawaii dated August 21, 1959 or after.

Specialty awards (numbered separately) are available for OSCAR satellites, SSTV, RTTY, 432 MHz, 222 MHz, 144 MHz, 50 MHz, and 160 meters. In addition, the "75-Meter 2-letter Extra Class" award is available to any Extra Class amateur who has worked all states consisting of 1 × 2, 2 × 1 or 2 × 2 call signs on the Extra Class portion of 75 meters. Endorsement stickers for the basic mixed mode/band award and any of the specialty awards are available for SSB, CW, EME, Novice, QRP, packet and any single band. The Novice endorsement is available for the applicant who has worked all states as a Novice licensee. QRP is defined as 10-watts input (or 5-watts output) as used by the applicant only and is affirmed by signature of the applicant on the application.

Contacts made through "repeater" devices or any other power relay method cannot be used for WAS confirmation. A separate WAS is available for OSCAR contacts. All stations contacted must be "land stations." Contact with ships (anchored or otherwise) and aircraft cannot be counted.

A W/VE applicant must be an ARRL member to participate in the WAS program. DX stations are exempt from this requirement.

HQ reserves the right to "spot call" for inspection of cards (at ARRL expense) of applications verified by an HF Awards Manager. The purpose of this is not to question the integrity of any individual, but rather to ensure the overall integrity of the program. More difficult-to-be-attained specialty awards (222-MHz WAS, for example) are more likely to be so called. Failure of the applicant to respond to such a spot check will result in nonissuance of the WAS certificate.

Disqualification: False statements on the WAS application or submission of forged or altered cards may result in disqualification. ARRL does not attempt to determine who has altered a submitted card; therefore do not submit any marked-over cards. The decision of the ARRL Awards Committee in such cases is final.

Application Procedure (please follow carefully): Confirmations (QSLs) and application form (MCS-217) may be submitted to an approved ARRL Special Service Club HF Awards Manager. ARRL Special Service Clubs appoint HF Awards Managers whose names/addresses are on file at HQ. If you do not know of an HF Awards Manager in your local area, call a club officer to see if one has been appointed or contact HQ. If you can have your application so verified locally, you need not submit your cards to HQ. Otherwise, send your application and cards to HQ, as indicated on the application form (reproduced in Chapter 17).

Be sure that when cards are presented for verification (either locally or to HQ) they are sorted alphabetically by state, as listed on the back of application form MCS-217.

All QSL cards sent to HQ must be accompanied by sufficient postage for their safe return (postage guidelines appear in Chapter 17).

Five-Band WAS (5BWAS)

This award is designed to foster more uniform activity throughout the bands, encourage the development of better antennas and generally offer a challenge to both newcomers and veterans. The basic WAS rules apply, including cards being checked in the field by Awards Managers; in addition, 5BWAS carries a start date of January 1, 1970. Unlike WAS, 5BWAS is a one-time-only award; no band/mode endorsements are available. Contacts made on 10/18/24 MHz are not valid for 5BWAS. In addition to the 5BWAS certificate, a 5BWAS plaque is available at an additional charge.

VHF/UHF Century Club (VUCC)

The VHF/UHF Century Club Award is awarded for contact with a minimum number of "Maidenhead" 2° × 1° grid-square locators per band as indicated below. Grid squares are designated by a combination of two letters and two numbers. More information on grid squares can be found in January 1983 *QST*,

pp 49-51 (reprint available from HQ on request), *The ARRL World Grid Locator Atlas* and the *ARRL Grid Locator for North America* (see latest *QST* for prices and ordering information). The VUCC certificate and endorsements are available to League members in Canada, the US and possessions, and Puerto Rico, and to other amateurs worldwide. Only those contacts dated January 1, 1983 and later are creditable for VUCC purposes. (A VUCC lapel pin is available for a nominal charge of $2 US.)

The minimum number of squares needed to qualify initially for each individual band award is as follows: 50 MHz—100; 144 MHz—100; Satellite—100; 222 MHz—50; 432 MHz—50; 902 MHz—25; 1296 MHz—25; 2.3 GHz—10; 3.4 GHz—5; 5.7 GHz—5; 10 GHz—5; 24 GHz—5; 47 GHz—5; 75 GHz—5; 119 GHz—5; 142 GHz—5; 241 GHz—5; laser—5. Certificates for 222 and 432 MHz are designated as Half Century, 902 and 1296 MHz as Quarter Century, and those above as SHF Awards. Individual band awards are endorsable in the following in-

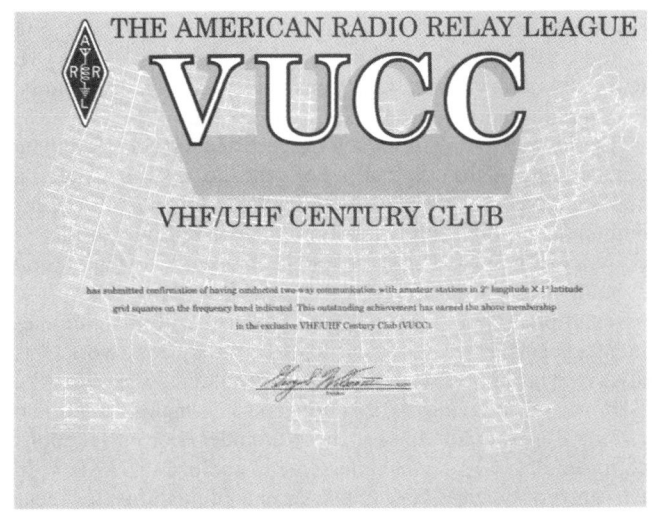

crements: 50 and 144 MHz—25 credits; 222 and 432 MHz— 10 credits; 902 MHz and above—5 credits. There are no specialty endorsements such as "CW only," etc.

Separate bands are considered as separate awards. No cross-band contacts permitted—except for the Satellite award. No contacts through active "repeater" or satellite devices, or any other power relay methods, are permitted. Contacts with aeronautical mobiles (in the air) *do not* count. Contacts with stations shipboard (located on a body of water) *do* count.

Stations who claim to operate from more than one grid square simultaneously (such as from the intersection of four grid squares) must be physically present in more than one square to give multiple square credit with a single contact. This requires the operator to know precisely where the intersection lines are located and placing the station exactly on the boundary to meet this test. To achieve this precision work requires either current markers permanently in place or the precision work of a professional surveyor. Operators of such stations should be prepared to provide some evidence of meeting this test if called upon to do so. Multiple QSL cards are not required.

For VUCC awards on 50 through 1296 MHz, all contacts must be made from a location or locations within the same grid square or locations in different grid squares no more than 50 miles apart. For SHF awards, contacts must be made from a single location, defined as within a 300-meter-diameter circle.

Application Procedure (please follow carefully): Confirmations (QSLs) and application forms (MCS-259 and MCS-260) must be submitted to an approved VHF Awards Manager for certification. ARRL Special Service Clubs appoint VHF Managers whose names are on file at HQ. If you do not know of an Awards Manager in your area, HQ will give you the name of the closest Manager. Do not send cards to HQ.

For the convenience of the Awards Manager in checking cards, applicants may indicate in pencil (pencil *only*) the grid square locator on the address side of cards that do not clearly indicate the grid locator. The applicant affirms that he/she has accurately determined the proper locator from the address information given on the card by signing the affirmation statement on the application.

Cards must be sorted alphabetically by field and numerically from 00 to 99 within that field.

Where it is necessary to mail cards for certification, postage equal to the amount needed to send them must be included for return of cards along with a separate self-addressed mailing label. In addition, US and Canadian applicants will enclose $1 US for any initial application to cover postage and packaging of the certificate. For endorsements, enclose only an SASE with two units of First-Class postage. When mailing cards, registered or certified mail is recommended.

Enclosed with the initial VUCC certificate from HQ will be a photocopy of the original list of grid squares for which the applicant has received credit (MCS-259). When applying for endorsement, the applicant will indicate in red on that photocopy those new grid squares for which credit is sought, and submit cards for certification to an Awards Manager. A new, updated photocopy listing will be returned with the endorsement sticker. Thus, a current list of grid squares worked is always in the hands of the VUCC award holder, available to the VHF Manager during certification, and a permanent historical record always maintained at HQ. Reminder: For initial application enclose $1, and for endorsements enclose an SASE (with two units of postage). For endorsement applications, it is necessary to submit only those MCS-259s that indicate new grid squares worked since the previous submission (indicated in red).

Disqualification: Altered/forged confirmations or fraudulent applications submitted may result in disqualification of the applicant from VUCC participation by action of the ARRL Awards Committee. The applicant affirms he/she has abided by all the rules of membership in the VUCC and agrees to be bound by the decisions of the ARRL Awards Committee.

Decisions of the ARRL Awards Committee regarding interpretation of the rules here printed or later amended shall be final.

Operating Ethics: Fair play and good sportsmanship in operating are required of all VUCC members.

Worked All Continents (WAC)

In recognition of international two-way Amateur Radio communication, the International Amateur Radio Union (IARU) issues Worked All Continents certificates to Amateur Radio Stations of the world. Qualification for the WAC award is based on examination by the International Secretariat, or a Member-Society, of the IARU, that the applicant has received QSL cards from other amateur stations in each of the six continental areas of the world (see the ARRL DXCC Countries List in Chapter 17 for a complete listing of continents). All contacts must be made from the same country or separate territory within the same continental area of the world. All

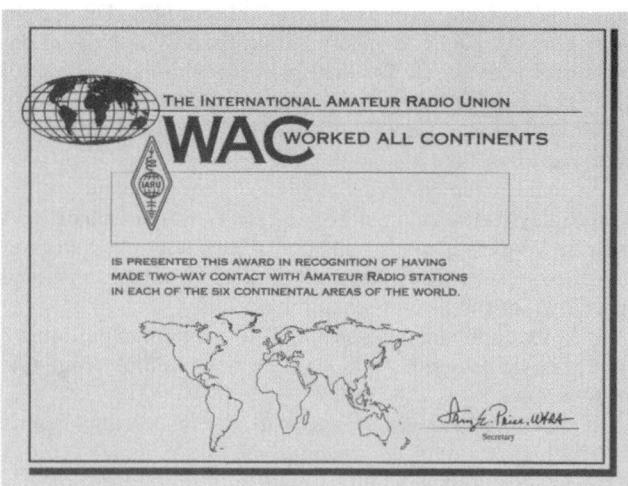

QSL cards (no photocopies) must show the mode and/or band for any endorsement applied for.

WAC certificates are issued as follows:
- Basic Certificate (mixed mode)
- CW Certificate
- Phone Certificate
- SSTV Certificate
- RTTY Certificate
- FAX Certificate
- Satellite Certificate
- 5-Band Certificate

WAC endorsement stickers are issued as follows:
- 6-Band Sticker
- QRP Sticker
- 1.8-MHz Sticker
- 3.5-MHz Sticker
- 50-MHz Sticker
- 144-MHz Sticker
- 430-MHz Sticker
- Any higher band

Contacts made on 10/18/24 MHz or via satellites are void for the 5-band certificate and 6-band sticker. All contacts for the QRP sticker must be made on or after January 1, 1985, while running a maximum power of 5-watts output or 10-watts input.

For amateurs in the United States or countries without IARU representation, applications should be sent to the IARU International Secretariat, PO Box AAA, Newington, CT 06111, USA. After verification, the cards will be returned, and the award sent soon afterward. There is no application fee; however, sufficient return postage for the cards, in the form of a self-addressed, stamped envelope or IRCs, is required. US amateurs must have current ARRL membership. [A sample application form and a return-postage chart appear in Chapter 17.] All other applicants must be members of their national Amateur Radio Society affiliated with IARU and must apply through the Society only. Note: The DXCC Countries List in Chapter 17 includes a continent designation for each DXCC country.

DX Century Club (DXCC)

DXCC is the premier operation award in Amateur Radio. The initial DXCC certificate (a nominal fee of $2 US is charged for the DXCC lapel pin) is available to League members in Canada, the US and possessions, and Puerto Rico, and all amateurs in the rest of the world. There are 12 separate DXCC awards available:

Mixed (general type)—contacts must be made using any mode since November 14, 1945.

Phone—contacts must be made using radiotelephone since November 14, 1945. Confirmations for cross-mode contacts for this award must be dated before October 1, 1981. Confirmations need not indicate two-way (2×) to be credited.

CW—contacts must be made using CW since January 1, 1975. Confirmations for cross-mode contacts for this award must be dated before October 1, 1981. Confirmations need not indicate two-way (2×) to be credited.

RTTY—contacts must be made using radioteletype since November 15, 1945. Confirmations for cross-mode contacts for this award must be dated before October 1, 1981. Confirmations need not indicate two-way (2×) to be credited.

160 meter, 80 meter, 40 meter, 10 meter, 6 meter, 2 meter—contacts must be made since November 15, 1945.

Satellite—contacts must be made using satellites since March 1, 1965 (non-endorsable).

DXCC Fee Schedule

Effective January 1, 1994, all amateurs applying for their very first DXCC Award will be charged a one-time Registration Fee of $10. This same fee applies to both ARRL members and foreign non-members, and both will receive one DXCC certificate and a DXCC pin. Applicants must provide funds for postage charges for QSL return.

(a) A $5 shipping and handling fee will be charged for each additional DXCC certificate issued, whether new or replacement. A DXCC pin will be included with each certificate.

(b) Endorsements and new applications may be presented at ARRL HQ, and at certain ARRL conventions. When presented in this manner, such applications shall be limited to 110 cards maximum, and a $2 handling charge will apply.

(c) Each ARRL member will be allowed one submission in each calendar year at no cost (except as in [b] above, or return postage). This annual submission may include any number of QSL cards for any number of DXCC Awards, and may be a combination of new and endorsement applications. Fees as in (a) above will apply for additional new DXCC Awards.

(d) Foreign non-members will be allowed the same annual submission as ARRL members, however, they will be charged a $10 DXCC Award fee, in addition to return postage charges. Fees in (a) and (b) may also apply.

(e) DXCC participants who wish to submit more than once

per year will be charged a DXCC fee for each additional submission made during the remainder of the calendar year. These fees are dependent upon membership status: ARRL members: $10 foreign non-members: $20. Additionally, return postage must be provided by applicant, and charges from (a) and (b) above may be applied.

Confirmations (QSL cards) must be submitted directly to ARRL HQ for all countries claimed. Confirmations for a total of 100 or more countries must be included with first application. Contacts made on all amateur bands are valid for DXCC. **[Contacts made on 10/18/24 MHz are not valid for 5-Band DXCC.]**

The ARRL DXCC Countries List criteria (see below) will be used in determining what constitutes a "country."

Confirmations must be accompanied by a list of claimed DXCC countries and stations to aid in checking and for future reference (the required DXCC application materials are available from ARRL HQ for an SASE).

Endorsement stickers for affixing to certificates or pins will be awarded as additional credits are granted. For the Mixed, Phone, CW, RTTY and 10-Meter DXCC, these stickers are in exact multiples of 25, ie, 125, 150, etc, between 100 and 250 DXCC countries; in multiples of 10 between 250 and 300, and in multiples of 5 above 300 DXCC countries. For 160-Meter, 80-Meter, 40-Meter, 6-Meter, 2-Meter and Satellite DXCC, the stickers are in exact multiples of 10 starting at 100 and multiples of 5 above 200.

All contacts must be made with amateur stations working in the authorized amateur bands or with other stations licensed to work amateurs. Contacts made through "repeater" devices or any other power relay method (aside from Satellite DXCC) are invalid for DXCC credit.

In countries where amateurs are licensed in the normal manner, credit may be claimed only for stations using regular government-assigned call signs or portable call signs where reciprocal agreements exist or the host government has so authorized portable operation. No credit may be claimed for contacts with stations in any country that has temporarily or permanently closed down Amateur Radio operations by special government edict where amateur licenses were formerly issued in the normal manner. Some countries, in spite of such prohibitions, issue authorizations that are acceptable.

All stations must be "land stations." Contacts with ships and boats, anchored or under way, and airborne aircraft, cannot be counted.

All stations must be contacted from the same DXCC country.

All contacts must be made by the same station licensee. However, contact may have been made under different call signs in the same country if the licensee for all was the same.

Any altered, forged or otherwise invalid confirmations submitted by an applicant for DXCC credit may result in disqualification of the applicant. Any holder of a DXCC award submitting altered, forged or otherwise invalid confirmations may forfeit the right to continued DXCC membership. The ARRL Awards Committee shall rule in these matters and may also determine the eligibility of any DXCC applicant who was ever barred from DXCC to reapply, and the conditions for such application.

(a) Fair play and good sportsmanship in operating are required of all DXCC members. In the event of specific objections relative to continued poor operating ethics, an individual may be disqualified from DXCC by action of the ARRL Awards Committee.

(b) Credit for contacts with individuals who have displayed continued poor operating ethics may be disallowed by action of the ARRL Awards Committee.

For (a) and (b) above, "operating" includes confirmation procedures and/or documentation submitted for DXCC accreditation.

Each DXCC applicant must stipulate that he/she has observed all DXCC rules as well as all pertinent governmental regulations established for Amateur Radio in the country or countries concerned, and agrees to be bound by the decisions of the ARRL Awards Committee. Decisions of the ARRL Awards Committee regarding interpretations of the DXCC rules (either currently in effect or later amended) shall be final.

All new DXCC applications must contain sufficient postage in the form of US currency, check or money order. For DXCC endorsements, sufficient funds for return postage is available on request from ARRL HQ (and is reproduced in Chapter 17). ARRL membership is not required of foreign applicants.

Official DXCC application forms are required. These may be obtained from the DXCC Desk at ARRL HQ. Please include a business-size SASE. **The complete DXCC rules appear in Chapter 17.**

5BDXCC

A Five-Band DXCC Award has been established to encourage more uniform DX activity throughout the amateur bands, encourage the development of more versatile antenna systems and equipment, provide a challenge for DXers, and enhance amateur-band occupancy. The basic DXCC rules apply, although the starting data for valid QSOs is January 1, 1969.

The 5BDXCC certificate is issued after the applicant submits a minimum of 500 QSLs representing two-way contact with 100 different DXCC countries on each of five Amateur Radio bands. Phone and CW segments of the band do not count as separate bands for this award. Confirmations made on any legal mode are acceptable, but no cross-mode or cross-band contacts are acceptable. Contacts made on 10/18/24 MHz are not valid for 5BDXCC. All QSLs must be checked by the ARRL HQ DXCC Desk. 5BDXCC is endorsable for additional bands: 160 meters, 17 meters, 12 meters, 6 meters, 2 meters. In addition to the 5BDXCC certificate, a 5BDXCC plaque is available at an extra charge of $25.

Countries List Criteria

The ARRL DXCC Countries List is the result of progressive changes in DXing since 1945. The full list will not necessarily conform completely with the current criteria since some of the listings were recognized from pre-WW II or were accredited from earlier versions of the criteria. While the general policy has remained the same, specific mileages and additional points have been added to the criteria over the years. The specific mileages in Point 2(a) and Point 3, mentioned in the following criteria, have been used in considerations made April 1960 and after. The specific mileage in Point 2(b) has been used in considerations made April 1963 and after.

When an area in question meets *at least one* of the following three points, it is eligible as a separate country listing for the DXCC Countries List. These criteria address considerations by virtue of Government [Point 1] or geographical separation [Points 2 and 3], while Point 4 addresses ineligible areas. All distances are given in statute miles.

Point 1, GOVERNMENT

An independent country or nation-state having *sovereignty*

(that is, a body politic or society united together, occupying a definite territory and having a definite population, politically organized and controlled under one exclusive regime, and engaging in foreign relations—including the capacity to carry out obligations of international law and applicable international agreements) constitutes a separate DXCC country by reason of **Government**. They *may* be indicated by membership in the United Nations (UN). However, some nations that possess the attributes of sovereignty are *not* members of the UN, although these nations may have been *recognized* by a number of UN-member nations. Recognition is the formal act of one nation committing itself to treat an entity as a sovereign state. There are some entities that have been admitted to the UN that lack the requisite attributes of sovereignty and, as a result, are *not* recognized by a number of UN-member nations.

Other entities which are not totally independent may also be considered for separate DXCC country status by reason of Government. Included are Territories, Protectorates, Dependencies, Associated States, and so on. Such an entity may delegate to another country or international organization a measure of its authority (such as the conduct of its foreign relations in whole or in part, or other functions such as customs, communications or diplomatic protection) *without* such an entity is individually considered, based on all the available facts in the particular case. In making a reasonable determination as to whether a sufficient degree of sovereignty exists for DXCC purposes, the following characteristics (list not necessarily all-inclusive) are taken into consideration:

(a) Membership in specialized agencies of the UN, such as the International Telecommunication Union (ITU).

(b) Authorized use of ITU-assigned call sign prefixes.

(c) Diplomatic relations (entering into international agreements and/or supporting embassies and consulates), and maintaining a standing army.

(d) Regulation of foreign trade and commerce, customs, immigration and licensing (including landing and operating permits), and the issuance of currency and stamps.

An entity that qualifies under Point 1, but consists of two or more separated land areas, will be considered a single DXCC country (since none of these areas alone retains an independent capacity to carry out the obligations of sovereignty), *unless* the areas can qualify under Points 2 or 3.

Point 2, SEPARATION BY WATER

An island or a group of islands which is part of a DXCC country established by reason of **Government**, Point 1, is considered as a separate DXCC country under the following conditions.

(a) The island or islands are situated off shore, geographically separated by a minimum of 225 miles of open water from a continent, another island or group of islands that make up *any part* of the "parent" DXCC country.

For any *additional* island or islands to qualify as an additional separate DXCC country or countries, such must qualify under Point 2(b).

(b) This point applies to the "second" island or island grouping geographically separated from the "first" DXCC country created under Point 2(a). For the second island or island grouping to qualify, at least a 500-mile separation of open water from the first is required, as well as meeting the 225-mile requirement of (a) from the "parent." For any subsequent island(s) to qualify, the 500-mile separation would again have to be met. This precludes, for example, using the 225-mile measurement *for each* of several islands from the parent country to make several DXCC countries.

Point 3, SEPARATION BY ANOTHER DXCC COUNTRY

(a) Contiguous land mass: Where a country, such as that covered by Point 1, is totally separated by an intervening DXCC country into two areas which are at least 75 miles apart, *two* DXCC countries result. This straight line measurement is made at the closest point, and may include inland lakes and seas (that are part of the country) in the measurement. International waters may be included in the separation but do not contribute to the 75-mile minimum requirement.

(b) Islands: Where two islands, of the government under Point 1, are totally separated by an intervening DXCC country (also under Point 1), *each* island counts as a separate DXCC country. No minimum distance is required. The test for total separation means that a straight line cannot be drawn from any point on one island to any point on the other island without passing through another DXCC country. This intervening country may be part of either island, another island, or part of a continent.

Point 4, INELIGIBLE AREAS

(a) Any area which is unclaimed or unowned by any recognized government does not count as a separate DXCC country.

(b) Any area which is classified as a Demilitarized Zone, Neutral Zone or Buffer Zone does not count as a separate DXCC country.

(c) The following do not count as a separate DXCC country from the host country: Embassies, consulates and extraterritorial legal entities of any nature, including, but not limited to, monuments, offices of the United Nations agencies or related organizations, other intergovernmental organizations or diplomatic missions.

A-1 Operator Club (A-1 OP)

Membership in this elite group attests to superior competence and performance in the many facets of Amateur Radio operation: CW, phone, procedures, copying ability, judgment and courtesy. You must be recommended for the certification independently by two amateurs who already are A-1 Ops. This honor is unsolicited; it is earned through the continuous observance of the very highest operating standards.

Old-Timers' Club (OTC)

In recognition of amateurs who have held an amateur license 20-or-more years (lapses permitted), a suitable award is available—OTC. If you qualify as an "old-timer," you'll find the necessary paperwork pretty easy. Drop a note to HQ (with an SASE) with the date of your first amateur license and your call then and now. HQ will verify the information, and if you're eligible, you'll soon receive your OTC certificate by return mail.

Contests

The League sponsors exciting on-the-air contests (see Chapter 7) for developing operating skills from 160 meters to 10 GHz, and participation certificates are available for all levels of activity, from casual participation to "all-out war"! Contests are also the "fast-track" way to make those needed QSOs for ARRL awards.

Plaques

• Those who qualify for either 5BWAS and/or 5BDXCC are eligible for a handsome individually engraved 9 × 12-inch walnut plaque. Further information, including required fee, is included in the 5BWAS or 5BDXCC application materials.

• ARRL International DX Contest Awards Program. Beautiful plaques can be won for specific achievements in the ARRL DX Contest. Details appear in *QST*.

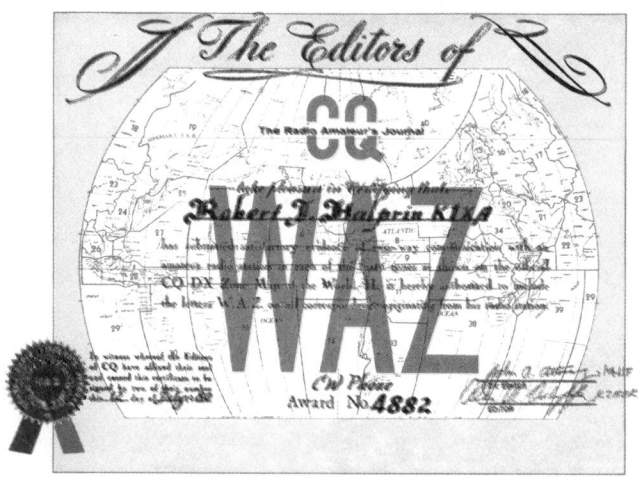

RULES FOR *CQ* MAGAZINE'S DX AWARDS
Worked all Zones (WAZ)

The CQ WAZ Award will be issued to any licensed amateur station presenting proof of contact with the 40 zones of the world. This proof shall consist of proper QSL cards, which may be checked by any of the authorized CQ checkpoints or sent directly to the WAZ Award Manager. Many of the major DX clubs in the United States and Canada and most national Amateur Radio societies abroad are authorized CQ check points. If in doubt, consult the WAZ Award Manager or the *CQ* Magazine DX editor. Any legal type of emission may be used, providing communication was established after November 15, 1945.

The official CQ WAZ Zone Map and the printed zone list will be used to determine the zone in which a station is located. [A DXCC Countries List, which includes CQ Zones, appears in Chapter 17 of this book.] Confirmation must be accompanied by a list of claimed zones, using CQ form 1479, showing the call letters of the station contacted within each zone. The list should also clearly show the applicant's name, call letters and complete mailing address. The applicant should indicate the type of award for which he or she is applying, such as All SSB, All CW, Mixed, All RTTY. In remote locations and in foreign countries, a handwritten list may be submitted and will be accepted for processing, provided the above information is shown.

All contacts must be made with licensed, land-based, amateur stations operating in authorized amateur bands, 160-10 meters. All contacts submitted by the applicant must be made from within the same country. It is recommended that each QSL clearly show the station's zone number. When the applicant submits cards for multiple call signs, evidence should be provided to show that he or she also held those call letters. Any altered or forged confirmations will result in permanent disqualification of the applicant.

A processing fee ($4 for subscribers—a recent *CQ* mailing label must be included; $10 for nonsubscribers) and a self-addressed envelope with (sufficient postage or IRCs to return the QSL cards by the class of mail desired and indicated) must accompany each application. IRCs equal in redemption value to the processing fee are acceptable. Checks can be made out to the WAZ Award Manager.

In addition to the conventional certificate for all bands and modes, specially endorsed and numbered certificates are available for phone (including AM), SSB and CW operation. The phone certificate requires that all contacts be two-way phone, the SSB certificate requires that all contacts be two-way SSB and the CW certificate requires that all contacts be two-way CW.

If at the time of the original application, a note is made pertaining to the possibility of a subsequent application for an endorsement or special certificate, only the missing confirmations required for that endorsement need be submitted with the later application, providing a copy of the original authorization signed by the WAZ Manager is enclosed.

Decisions of the CQ DX Awards Advisory Committee on any matter pertaining to the administration of this award will be final.

All applications should be sent to the WAZ Award Manager after the QSL cards have been checked by an authorized CQ checkpoint. Zone maps, printed rules and application forms are available from the WAZ Award Manager. Send a business-size (4 × 9-inch), self-addressed envelope with two units of First-Class postage, or a self-addressed envelope and 3 IRCs.

Single Band WAZ

Effective January 1, 1973, special WAZ Awards will be issued to licensed amateur stations presenting proof of contact with the 40 zones of the world on 80, 40, 20, 15 and 10 meters. Contacts for a Single Band WAZ award must have been made after 0000 hours UTC January 1, 1973. Single-band certificates will be awarded for both two-way phone, including SSB and two-way CW.

5 Band WAZ

Effective January 1, 1979, the CQ DX Department, in cooperation with the CQ DX Awards Advisory Committee, announced a most challenging DX award—5 Band WAZ. Applicants who succeed in presenting proof of contact with the 40 zones of the world on the five HF bands—80, 40, 20, 15 and 10 meters (for a total of 200)—will receive a special certificate in recognition of this achievement.

These rules are in effect as of July 1, 1979, and supersede all other rules, Five Band WAZ will be offered for any combination of CW, SSB, phone or RTTY contacts, mixed mode only. Separate awards will not be offered for the different modes. Contacts must have been made after 0000 UTC January 1, 1979. Proof of contact shall consist of proper QSL cards checked only by the WAZ Award Manager. The first plateau will be a total of 150 zones on a combination of the five bands. Applicants should use a separate sheet for each frequency band, using CQ Form 1479.

A regular WAZ or Single Band WAZ is a prerequisite for a 5 Band WAZ certificate. All applications should show the applicant's WAZ number. After the 150 zone certificate is earned, each 10 additional zones requires the submission of QSL cards and a $1 fee. The final objective is 200 zones for a complete 5 Band WAZ. The applicant has a choice of paying a fee for a plaque and/or applying for an endorsement sticker commemorating this achievement.

All applications should be sent to the WAZ Award Manager. The 5 Band Award is governed by the same rules as the regular WAZ Award and uses the same zone boundaries.

WARC Bands WAZ

Effective January 1, 1991, single band WAZ Awards were issued to amateurs presenting proof of contact with the 40 zones of the world on any *one* of the WARC bands: 30, 17 or 12 meters. (Each band constitutes a separate award and may be applied for separately.) This award is available for Mixed

Mode, SSB, RTTY or CW. Contacts for each WARC WAZ Award must have been made after each station involved in the contact had permission from its licensing authority to operate on the band and mode.

RTTY WAZ

Special WAZ Awards are issued to Amateur Radio stations presenting proof of contact with the 40 world zones using RTTY. For the mixed band award, QSL cards must show a date of November 15, 1945 or later. The RTTY WAZ is also available with a single band endorsement for 80, 40, 20, 15 or 10 meters. QSL cards submitted for single band endorsements must show a date of January 1, 1973 or later.

WNZ

WNZ stands for "Worked Novice Zones" and is available *only* to holders of a US Novice or Technician license. Proof of contact with at least 25 of the 40 CQ zones as described by the WAZ rules is required. All contacts must be made using the 80, 40, 15 and 10-meter Novice bands. In addition, all contacts must be made while holding a Novice or Technician license, although the application may be submitted at a later date. Contacts must be made prior to receiving authorization to operate with higher class privileges. The WNZ is available as a Mixed Mode, SSB or CW award. It may also be endorsed for a single band. The WNZ award may be used to fulfill part of the application requirement for the WAZ Award when all 40 zones are confirmed.

The basic award can be obtained by submitting QSL cards for 25 zones. The processing fee is $5 for all applicants. All QSL cards must show a date of January 1, 1952 or later. Use CQ Form 1479 to apply for this award.

160 Meter WAZ

The WAZ Award for 160 meters requires that the applicant submit directly to the WAZ Manager QSL cards from at least 30 zones. All QSL cards must be dated January 1, 1975 or later and a $5 fee must accompany all applications. The 160 WAZ is a mixed mode award only. The basic 160 WAZ Award may be secured by submitting QSLs from 30 zones. Stickers for 35, 36, 37, 38, 39 and 40 zones can be obtained from the WAZ Manager upon submission of the QSL cards and $2 for each sticker.

Satellite WAZ

The Satellite WAZ Award is issued to Amateur Radio stations submitting proof of contact with all 40 CQ zones through any Amateur Radio satellite. The award is available for mixed mode only. QSL cards must show a date of January 1, 1989 or later.

The Prefix Award Program

WPX

The CQ WPX Award recognizes the accomplishments of confirmed QSOs with the many prefixes used by amateurs throughout the world. Separate distinctively marked certificates are available for two-way SSB, CW and mixed modes, as well as the VPX Award for shortwave listeners and the WPNX Award for Novices.

All applications for WPX certificates (and endorsements) must be submitted on the official application form, *CQ* 1051A. This form can be obtained by sending a self-addressed stamped, business-size (4 × 9-inch) envelope to the WPX Award Manager, Norm Koch, K6ZDL, PO Box 593, Clovis, NM 88101. All QSOs must be made from the same country. All call letters

must be in strict alphabetical order and the entire call letters must be shown. All entries must be clearly legible. Certificates are issued for the following modes and number of prefixes. Cross-mode QSOs are not valid for the CW or 2¥ SSB certificates. Mixed (CW/phone only): 400 prefixes confirmed. CW: 300 prefixes confirmed. Separate applications are required for each mode. Cards need not be sent, but they must be in the possession of the applicant. Any and all cards may be requested by the WPX Award Manager or by the CQ DX Committee. The application fee for each certificate is $4 for subscribers (subscribers must include a recent *CQ* mailing label) and $10 for nonsubscribers, or the equivalent in IRCs. All applications and endorsements should be sent to the WPX Award Manager.

Prefix endorsements are issued for each 50 additional prefixes submitted. Band endorsements are available for working the following numbers of prefixes on the various bands: 1.8 MHZ—50, 3.5 MHz—175, 7 MHz—250, 14 MHz—300, 21 MHz—300, 28 MHz—300. Continental endorsements are given for working the following numbers of prefixes in the respective continents: North America—160, South America—95, Europe—160, Africa—90, Asia—75, Oceania—60. Endorsement applications must be submitted on CQ Form 1051A. Use separate applications for each mode and be sure to specify the mode of your endorsement application. For prefix endorsements, list only additional call letters confirmed since the last endorsement application. A self-addressed, stamped envelope or proper IRCs for surface or airmail return is required, and $1 or 5 IRCs for each endorsement sticker.

Prefixes: The two or three letter/numeral combinations which form the first part of any amateur call will be considered the prefix. Any difference in the numbering, lettering, or order of same shall constitute a separate prefix. The following would be considered different: W2, WA2, WB2, WN2, WV2, K2 and KN2. Any prefix will be considered legitimate if its use was licensed or permitted by the governing authority in that country after November 15, 1945. A suffix would designate portable operation in another country or call area and would count only if it is the normal prefix used in that area. For example, K4IIF/KP4 would count as KP4. However, KP4XX/7 would not count as KP7, since this is not a normal prefix. Suffixes such as /M, /MM, /AM, /A and /P are not counted as prefixes. An exception to this rule is granted for portable operation within the issued call area. Thus, contact with a special prefix such as WS2JRA/2 counts for WS2; however, WS2JRA/3 would count for W3. All calls without numbers will be assigned an arbitrary Ø plus the first two letters to constitute a prefix. For example, RAEM counts as RAØ, AIR as AIØ, UPOL as UPØ. All portable suffixes that contain no numerals will be assigned an arbitrary Ø. For example, W4BPD/LX counts as LXØ, and WA6QGW/PX counts as PXØ.

WPNX

The WPNX Award can be earned by US Novices who work 100 different prefixes prior to upgrading. The application may be submitted after receiving the higher license, providing the actual contacts were made as a Novice. Prefixes worked for the WPNX Award may be used later for credit toward the WPX Award. The rules for the WPNX Award are the same as for the WPX Award except that only 100 prefixes must be confirmed. Applications are sent to the WPX Award Manager.

WPX Honor Roll

The WPX Honor Roll recognizes those operators and stations that maintain a high standing in confirmed, current prefixes. The rules therefore, reflect the belief that Honor Roll membership should be accessible to all active radio amateurs and not to be unduly advantageous to the "old-timers." With the exceptions listed below, all general rules for WPX apply toward Honor Roll credit. A minimum of 600 prefixes is required to be eligible for the Honor Roll.

Only current prefixes may be counted toward WPX Honor Roll standings, those prefixes to be listed and updated annually in *CQ* or to be available from the WPX Award Manager. Special-issue prefixes (such as OF, OS 4A) will be considered current for as long as they are assigned to a particular country and deducted as credit for Honor Roll standings after cessation of their use or assignment. Honor Roll applicants must submit their list of current prefixes (entire call required) separate from their regular WPX applications. Use regular Form 1051 and indicate "Honor Roll" at the top of the form. Forms may be obtained by sending a business-size SASE or one IRC (foreign stations send extra postage or IRC if airmail desired) to the WPX Award Manager. A separate application must be made for each mode. Endorsements for the Honor Roll may be made for 10 or more prefixes. An SASE or IRC should be included. For prefixes by countries, see Chapter 17.

WPX Award of Excellence

This is the ultimate award for the prefix DXer. The requirements are 1000 prefixes mixed mode, 600 prefixes SSB, 600 prefixes CW, all six continental endorsements, and the 5 band endorsements 80-10 meters. A special 160-meter endorsement bar is also available. The WPX Plaque fee is $60, and the 160-meter bar is $5.25.

The CQ DX Awards Program

The CQ CW DX Award and CQ SSB DX Award are issued to any amateur station submitting proof of contact with 100 or more countries on CW or SSB. The CQ DX RTTY Award is issued to any amateur station submitting proof of contact with 100 countries on RTTY. Applications should be submitted on the official CQ DX Award application (Form 1067B). All QSOs must be 2 × SSB, 2 × CW or 2 × RTTY. Cross-mode or one-way QSOs are not valid for the CQ DX Awards. QSLs must be listed in alphabetical order by prefix, and all QSOs must be dated after November 15, 1945. QSL cards must be verified by one of the authorized checkpoints for the CQ DX Awards or must be included with the application. If the cards are sent directly to the CQ DX Awards Manager, Billy Williams, N4UF, PO Box 9673, Jacksonville, FL 32208, postage for return by First-Class mail must be included. If certified or registered mail return is desired, sufficient postage should be included. Country endorsements for 150, 200, 250, 275, 300, 310 and 320 countries will be issued.

Any altered or forged confirmations will result in permanent disqualification of the applicant. Fair play and good sportsmanship in operating are required for all amateurs working toward CQ DX Awards. Continued use of poor ethics will result in disqualification of the applicant. A fee of $4 for subscribers (subscribers must include a recent *CQ* mailing label with their application) or $10 for nonsubscribers, or the equivalent in IRCs, is required for each award to defray the cost of the certificate and handling. An SASE or 1 IRC is required for each endorsement.

The ARRL DXCC Countries List (see Chapter 17) constitutes the basis for the CQ DX Award country status. Deleted countries will not be valid for the CQ DX Awards. Once a

country has lost its status as a current country, it will automatically be deleted from our records. All contacts must be with licensed land-based amateur stations working in authorized amateur bands. Contacts with ships and aircraft cannot be counted. Decisions of the CQ DX Advisory Committee on any matter pertaining to the administration of these awards shall be final.

To promote multiband usage and special operating skills, special endorsements are available for a fee of $1 each:

- A 28-MHz endorsement for 100 or more countries confirmed on the 28-MHz band.
- A 3.5/7-MHz endorsement for 100 or more countries confirmed using any combination of the 3.5 and 7-MHz bands.
- A 1.8-MHz endorsement for 50 or more countries confirmed on the 1.8-MHz band.
- A QRP endorsement for 50 or more countries confirmed using 5-watts input or less.
- A Mobile endorsement for 50 or more countries confirmed while operating mobile. The call-area requirement is waived for this endorsement.
- An SSTV endorsement for 50 or more countries confirmed using two-way slow scan TV.
- An OSCAR endorsement for 50 countries confirmed via amateur satellite.

(After the basic award is issued, only a listing of confirmed QSOs is required for these seven special endorsements. However, specific QSLs may be requested by Award Manager N4UF.)

The CQ DX Honor Roll will list all stations with a total of 275 countries or more. Separate Honor Rolls will be maintained for SSB and CW. To remain on the Honor Roll, a station's country total must be updated annually.

USA-CA Rules and Program

The United States of America Counties Award, also sponsored by *CQ*, is issued for confirmed two-way radio contacts with specified numbers of US counties under rules and conditions below. [Note: A complete list of US counties appears in Chapter 17.]

The USA-CA is issued in seven (7) different classes, each a separate achievement as endorsed on the basic certificate by use of special seals for higher class. Also, special endorsements will be made for all one band or mode operations subject to the rules.

Class	Countries Required	States Required
USA-500	500	Any
USA-1000	1000	25
USA-1500	1500	45
USA-2000	2000	50
USA-2500	2500	50
USA-3000	3000	50

USA 3076-CA for ALL counties and Special Honors Plaque [$40]

USA-CA is available to all licensed amateurs everywhere in the world and is issued to them as individuals for all county contacts made, regardless of calls held, operating QTHs or dates. All contacts must be confirmed by QSL, and such QSLs must be in one's possession for identification by certification officials. Any QSL card found to be altered in any way disqualifies the applicant. QSOs via repeaters, satellites, and/or endorsement requested. A completed USA-CA Record Book constitutes medium of basic application and becomes the property of *CQ* for record purposes. On subsequent applications for either higher classes or for special endorsements, the applicant may use additional Record Books to list required data or may make up own alphabetical list conforming to requirements. Record Books are to be obtained directly from *CQ*, 76 North Broadway, Hicksville, NY 11801, for $1.25 each. It is recommended that two be obtained, one for application use and one for personal file copy.

Make Record Book entries necessary for county identity and enter other log data necessary to satisfy any special endorsements (band-mode) requested. Have the certification form provided signed by two licensed amateurs (General class or higher) or an official of a national-level radio organization or affiliated club verifying the QSL cards for all contacts as

moonbounce and phone patches are not valid for USA-CA. So-called "team" contacts, wherein one person acknowledges a signal report and another returns a signal report, while both amateur call signs are logged, are not valid for USA-CA. Acceptable contact can be made with only one station at a time.

The National Zip Code & Directory of Post Offices will be helpful in some cases in determining identity of counties of contacts as ascertained by name of nearest municipality. Publication No. 65, Stock no. 039-000-00264-7, is available at your local Post Office or from the Superintendent of Documents, US Government Printing Office, Washington, DC 20402, for $12, but will be shipped only to US or Canada.

Unless otherwise indicated on QSL cards, the QTH printed on cards will determine county identity. For mobile and portable operations, the postmark shall identify the county unless information stated on QSL cards makes other positive identity. In the case of cities, parks or reservations not within counties proper, applicants may claim any one of adjoining counties for credit (once).

The USA-CA program will be administered by a *CQ* staff member acting as USA-CA Custodian, and all applications and related correspondence should be sent directly to the custodian at his or her QTH. Decisions of the Custodian in administering these rules and their interpretation, including future amendments, are final.

The scope of USA-CA makes it mandatory that special Record Books be used for application. For this purpose, *CQ* has provided a 64-page 4¼- × 11-inch Record Book which contains application and certification forms and which provides record-log space meeting the conditions of any class award

listed have been seen. The USA-CA Custodian reserves the right to request any specific cards to satisfy any doubt whatever. In such cases, the applicant should send sufficient postage for return of cards by registered mail. Send the original completed Record Book (not a copy) and certification forms and handling fee. Fee for nonsubscribers to *CQ* is $10 US or 40 IRCs; for subscribers, the fee is $4 or 12 IRCs. (Subscribers, please include recent *CQ* mailing label.) Send applications to USA-CA Custodian, Norm Van Raay, WA3RTY, Star Route 40, Pleasant Mount, PA 18453.

For later applications for higher-class seals, send Record Book or self-prepared list per rules and $1.25 or 6 IRCs handling charge. For application for later special endorsements (band/mode) where certificates must be returned for endorsement, send certificate and $1.50 or 8 IRCs for handling charges.

Note: At the time any USA-CA award certificate is being processed, there are no charges other than the basic fee, regardless of number of endorsements or seals; likewise, one may skip lower classes of USA-CA and get higher classes without losing any lower awards credits or paying any fee for them. Also note: IRCs are not accepted from US stations.

[The Mobile Emergency and County Hunters Net meets on 14,336 kHz SSB every day and on 3866 kHz evenings during the winter. The CW County Hunters Net meets on 14,066.5 MHz daily.]

GERATOL NET

The ARRL offers many endorsements to the popular Worked All States (WAS) award. One of them is the "75-Meter 2-letter Extra Class" Award, available to any Extra Class amateur who has worked all states consisting of 1×2, 2×1 or 2×2 call signs on the Extra Class portion of 75 meters.

DIRECTORS AWARD
GERATOL NET

Date_____   Award No._____

By these presents be it known that on this date

SAMPLE

has submitted satisfactory evidence of prevailing
against the adverse reception conditions in the
75 Meter Phone Band and Conquering the Uncertainties
of the Ionospheric Propagation of Electromagnetic Waves
by exchanging Geratol numbers with 100 Geratol net members.

Be it therefore resolved that

SAMPLE

be awarded this Directors Award
and may persue the endorsements thereto.

Awards Administrator

During the winter of 1977, a separate award—the "Unbelievable Operating Achievement Award" (otherwise known as the GERATOL Net award)—was established by N4BA, K5BG, W7RQ and W0GX. On March 10, 1978, an informal net went into operation to provide a meeting place and to assist others with the completion of both the ARRL WAS and the new award developed by N4BA and company.

Around this time, someone suggested that since most of those working toward the unique 2-letter WAS award were senior citizens, the name "Geritol Net" might be appropriate!

That name stuck; the spelling has been changed to the acronym "GERATOL" (Greetings Extra Radio Amateurs—Tired of Operating Lately?). In many ways, this describes the essence of the GERATOL Net. More than one "GERATOLLER" will tell you that the fellowship and challenge has revitalized their amateur operations and allowed a welcome escape from the mundane "Roger 59" DX world.

The GERATOL Net is quite active during the good 75-meter propagation months, from October through April. As many as 48 states have checked into the net during the course of an evening, including Alaska and Hawaii. Using three net-control shifts, it convenes around 0100 UTC each evening, and routinely runs until after 0700. Meeting around 3767 kHz, the net still pursues the original purpose developed by its founders. No dues or membership fees are required. DX stations are welcome to check in and work any W or VE, but W/VEs are not allowed to call DX stations (for obvious reasons).

To qualify for the GERATOL Award, you first must qualify for the ARRL 75-Meter 2-Letter WAS Award. Once you have secured the ARRL WAS Award, you are eligible to apply for your GERATOL Award. Include your ARRL WAS certificate serial number and date, your name as you want it to appear on the GERATOL certificate, and $3 to defray printing and mailing costs, and send to Bruce Winkelman, AA5CO, 3014 W 68th Pl, Tulsa, OK 74132.

Once you have secured the basic GERATOL certificate, 12 interesting and challenging endorsements are available to add to it. The honor system is used, so no QSL cards are required. A brief description of each follows:

1) WAS with 1×2 call signs, K prefixes only.

2) 48 states with 1×2 call signs, N prefixes only.

3) 48 states with 1×2 call signs, W prefixes only.

4) WAS with 1×2 call signs, with the suffix of each containing each operator's initials.

5) WAS with 1×2 call signs, with the suffixes containing the postal abbreviations of the states.

6) WAS with 1×2 call signs, with the suffixes containing double letters twice through the alphabet.

7) WAS with 2×1 call signs.

8) WAS with other stations that have already qualified for the ARRL 75-Meter 2-Letter WAS award.

9) WAS with 2×1 call signs, with the suffixes containing each letter twice through the alphabet.

10) WAS with Extra Class 1×3 call signs.

11) WAS with 2×1 call signs, A prefixes only.

12) WAS with 2×1 call signs, K prefixes only.

13) WAS with YLs only. Contact must be made after 0500 UTC, February 13, 1990.

14) WAS with 2×1 call signs, N prefixes only.

Endorsements 15 through 25 require that contacts be made after 0500 UTC, February 13, 1990.

15) WAS with 2×1 call signs, W prefixes only.

16) 48 contiguous states with double letter prefixes. (example: AA4GN, NN8P)

17) WAS with 2×3 call signs.

18) WAS with state capitols (1×2, 2×1 and 2×2 Extra Class call signs).

19) WAS with QCWA members. (Collect QCWA numbers)

20) WAS with 1×2 call signs (2×2 Extra Class in HI and AK) and exchange GERATOL numbers. Operators must have a GERATOL number to collect numbers.

21) WAS with 2×1 call signs and exchange GERATOL numbers. Operators must have a GERATOL number to collect numbers.

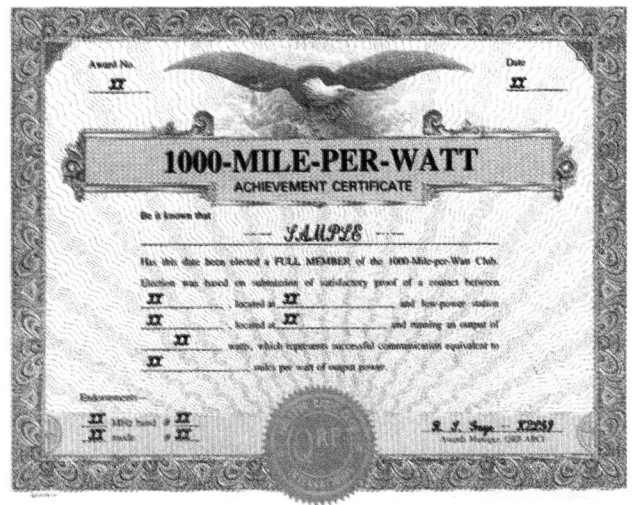

22) WAS five times over with 1 × 2, 2 × 1 and 2 × 2 Extra Class format call signs. This endorsement requires 250 total contacts, 5 in each state.

23) Century DX: Work all 50 states at least once with no more than 3 contacts in any state (100 contacts required). Collect number and last letter on 2 × 1 call signs, last two letters on 1 × 2 and 2 × 2 call signs that represent a DX country prefix. (See current ARRL DXCC Country List.) For example, K9JA, the JA = Japan; WC5T, the 5T = Mauritania.

24) WAS with 2 × 2 call signs, K prefixes only.

25) 48 contiguous states mobile with any call-sign format.

For further information on endorsements, and on any GERATOL Net details, send your request with a business-size SASE to Richard Beran, WØYTZ, 300 Valley View Dr, Ord, NE 68862.

The "Directors Award" is available for working 100 GERATOL members and exchanging 100 GERATOL numbers. Additional recognition is available for each group of 100 contacts verified above the initial 100 contact requirement. The operator must have a GERATOL number to exchange numbers.

The GERATOL Net also is involved in a parallel operation, the Canadian 2/80 Award. Requirements for the award consist of working the following Canadian provinces and territories:

VO1	Newfoundland	VE3	Ontario
VO2	Labrador	VE4	Manitoba
VE1	New Brunswick	VE5	Saskatchewan
VE1	Nova Scotia	VE6	Alberta
VE1	Prince Edward Island	VE7	British Columbia
VE2	Quebec	VE8	Northwest Territories

Contacts may be made anywhere on 75/80 meters on any mode. All QSL cards must show a contact date of 0000 UTC November 1, 1972 or later. Stations worked must have 2-letter call-sign suffixes. Once you have all 12 QSL cards, submit them to Norm Smith, VE3EJQ, 2/80 Contest Coordinator, 1849 Cora Dr, RR #3, Cavan, ON L0A 1C0, Canada. Include $3 US for handling.

So take a shot of GERATOL and join the group on 75 meters to experience a new operating challenge.

QRP AMATEUR RADIO CLUB INTERNATIONAL (ARCI)

The objective of the QRP ARCI, Inc awards program is to demonstrate that power is no substitute for skill, encouraging full enjoyment of Amateur Radio while running the minimum power necessary to complete a QSO and reduce QRM on crowded amateur bands. Requirements for the following awards are set forth below:

QRP-25. Issued to any amateur for working 25 members of the QRP ARCI while those members were running a power output of 5 watts or less. Endorsements are offered for 50, 100 and every 100 thereafter. To apply, send full log info with the list of the member-number in order of the member-number.

WAC-QRP. Issued to any amateur for confirmed contacts with stations in all six continents. A power output of 5 watts or less must have been used by the applicant for all QSOs. To apply, send full log info with photocopies of QSLs or a confirmed list.

WAS-QRP. Issued to any amateur for confirmed QSOs with each of the 50 states, while running a power output of 5 watts or less. Basic award issued for 20 states, endorsement seals for 30, 40 and 50. To apply, send full log data with photocopies of QSLs or a confirmed list.

1000-Mile-Per Watt (km/W). Issued to any amateur transmitting from or receiving the transmissions of a low-power station such that the great circle bearing distance between the two, divided by the power output of the low-power station, equals or exceeds 1000-miles-per-watt. Additional certificates can be earned with different modes/bands. To apply, send full log info and photocopies of QSLs or certified list including RST and power used on both sides, band, mode and exact (as nearly as possible) QTH on both sides.

QRP-Net (QNI-25). Issued to those members completing 25 check-ins into any individual QRP ARCI net. Subsequent 25 QNIs in another net will earn an endorsement seal. Net managers send a list of those qualifying for the QNI certificate to the Net Manager at the end of each month. Awards are issued FREE to those qualified by the Awards Chairman.

Notes:

1) A two-way QRP seal will be issued for all the above awards, except QRP-25, if log data indicates the power output on both sides of all QSOs was 5 watts or less.

2) Other endorsement seals available—one band, one mode, natural power, Novice. Please specify endorsements desired.

3) Certified list (GCR). QRP ARCI will accept as satisfactory proof of confirmed QSOs and that the QSLs are on hand as claimed by the applicant if the list is signed by (a) a radio club official, or (b) two Amateur Radio operators of General class or higher, or (c) a notary public, or (d) a CPA (Certified Public Accountant). If you must send QSLs, be sure to include postage for their return.

4) QRP ARCI member-numbers are not published. The Awards Manager will accept as satisfactory proof for any of the club awards a QSO with a club member giving his membership number and power output in the log data. If the QRP number and power level are not given, a QSL is required for confirmation.

5) Fee structure

   (a) original certificate and seal—$2 or 10 IRCs.

   (b) subsequent endorsement seals—$1 or five IRCs.

   (c) return postage for QSLs or other application material to be returned.

6) Send applications and the required fee to QRP ARCI Awards Chairman, Bob Gaye, K2LGJ, 25 Hampton Parkway, Buffalo, NY 14217-1217.

TEN-TEN

Amateurs operating on 10 meters are often bewildered by

requests for "10-10 numbers." 10-10 numbers are assigned by the 10-10 International Net, Inc. A number is available to any amateur who works ten 10-10 members and submits the log data to the appropriate 10-10 Call Area Manager. Once the log data has been received and approved, the applicant will be issued his own 10-10 number, which can be used to exchange with others on 10 meters to obtain a "Bar" for each 100 numbers collected. 10-10 numbers are issued for life, and well over 70,000 10-10 numbers have been issued internationally.

The purpose of 10-10 is to promote interest and activity on the 10-meter band. 10-10 holds a net meeting each day, except Sunday, on 28.800 MHz at 1800Z. (An alternate net is held each Monday at 1800 UTC on 28.380 MHz.) All amateurs, 10-10 members or not, are invited to check in on the daily net. In addition to holding both CW and SSB contests twice each year, 10-10 has more than 220 Chapters which hold net meetings at least once each week. Most 10-10 Chapters issued certificates for working members of that Chapter. If chasing certificates

and awards is your "thing," obtaining a 10-10 number is highly recommended. For information on how to join 10-10, including an application form and the latest list of 10-10 Chapters, meeting times and frequencies, send $1 and two First-Class stamps and an address label to Mike Elliott, KF7ZQ, 9832 W Gurdon Court, Boise, ID 83704.

Besides the collection of "Bars" for each 100 10-10 members worked, 10-10 offers some additional awards for contacts on 10 meters. 10-10 Worked All States (10-10 WAS) is awarded in recognition of confirmed contacts with members in all 50 states on 10 meters. QSL card confirmations must indicate the 10-10 numbers of stations worked. For information and application, send an SASE to Susan Brackeen, KA1CAD, Rte 6, Box 151, Boonville, MS 38829. The 10-10 Countries Award is issued for confirmed contacts with 10-10 members in 25 differ-

ent countries on the 10-meter band. Endorsement may also be earned for additional countries worked. For information on this award and an application form, send an SASE to Alan Sherman, K1AS, RR 4, Box 422, Danielson, CT 06239.

WORLD ITU ZONE AWARD

This award will be issued to amateurs who work the prescribed number of ITU (International Telecommunication Union) zones of the world. There are 90 ITU zones. QSL cards are suggested, but not required. If QSLs are not submitted, a certified list of contacts (verified by a local club official or two licensed amateurs) shall be submitted.

The ITU Award is issued in the following classes:

Class D—40 zones
Class C—50 zones
Class B—65 zones
Class A—75 zones
Class AA—90 zones

The official ITU zone list [see Chapter 17 for an ITU zone list] and map will be the criteria for determining zones.

Fees are $2 US postpaid for the basic award, including endorsements requested at the same time, and 50 cents US postpaid for each future endorsement.

Applications should be sent to John Lee, K6YK, 3654 Three Oaks Rd, Stockton, CA 95205.

INTERNATIONAL AWARDS

The following awards information was obtained from International Amateur Radio Union (IARU) Societies around the world, who kindly provided the rules of the Amateur Radio awards they sponsor. [Not all of the IARU Societies are, of course, fortunate enough to have the resources necessary for awards sponsorship.] In addition to the HF awards described below, many IARU Societies also sponsor localized and/or VHF-only awards, or VHF-endorsed versions of their HF awards. Further, shortwave listeners (SWLs) are eligible for many, if not most, of these awards as well. Since awards rules have been necessarily summarized for publication, further details can be secured by writing directly to the Awards Manager(s) of the particular Society or Societies concerned. When addressing correspondence to an Awards Manager, it is a good idea to enclose a self-addressed envelope and at least 2 IRCs (International Reply Coupons).

As a general principle, QSLs are not required to be included with the application. Ordinarily, a list of the appropriate QSLs with a signed certification by an official of a national-level Amateur Radio Society or two officers of a local Amateur Radio Club stating that the valid QSLs are in possession of the applicant will be honored by the IARU Society sponsoring the award. This list should show the call sign, date, time, frequency, mode and signal report for each contact. This is often referred to as a "GCR" (General Certification Rules) list. Awards Managers may reserve the right to inspect QSL cards on rare occasions, and any decisions made by the Awards Managers as to the validity of applications are final.

When applying for a particular award, type or neatly print your name, call sign and mailing address. State precisely what award and/or endorsement you are applying for, and enclose the appropriate application forms and the required fee (usually in IRCs) to defray postage and stationery costs. You should also state that you have complied with the rules of the particular award as well as the regulations established for Amateur Radio in your country.

ARGENTINA (LU)

Radio Club Argentino (RCA) awards will be issued to any

recognized radio amateur holding an official license. Each request must be accompanied by a log-extract or list of the appropriate contacts for the award. The list must be accompanied by the corresponding QSL cards or by a statement from an officer of the national Amateur Radio Society stating that the necessary cards have been checked. All QSOs must be made November 20, 1945 or later, except where otherwise noted. Contacts with mobile stations will not be accepted (except for the CEMA and CEMARA awards). The fee for each certificate is 10 IRCs (or $5 US). The fee for endorsements is 4 IRCs ($2 US). Awards applications should be sent to: Radio Club Argentino, Awards Manager Sr. Ricardo Schroder, LU8AEJ, Casilla de Correo 97, (1000) Buenos Aires, Argentina.

101 (101 Paises)

Awarded to all amateurs for QSOs with 101 different DXCC countries. Endorsements for 121, 141, 161, etc. Different certificates issued for CW, Phone and Mixed modes—one band or mixed bands. Five bands—"laureated." One LU contact required.

TPA (Todos Paises America)

Awarded to amateurs for QSOs made on any band or mode with 21 American Republics and Canada (22 contacts): CE, CO, CP, CX, HC, HH, HI, HK, HP, HR, LU, OA, PY, TG, TI, VE, W, XE, YN, YS, YV, ZP.

CCC (Cinco Continentes Comunicados)

Awarded to amateurs for QSOs with America, Africa, Asia, Australia and Europe (North American stations must contact South American stations for America; South Americans must contact North Americans). All contacts must be made on two bands, once on each band. "Laureated"—making the required contacts on three or more bands. Different certificates are available for CW and phone.

CAA (Certificado Antartico Argentino)

Awarded to amateurs for working at least one LU Antarctic base (LU-Z). LU stations must work three bases. Endorsement stickers are issued for working additional bases, and separate certificates are issued for CW and phone. All bands.

CEMA (Certificado Moviles Argentinas)

Awarded to amateurs for working 25 LU mobile stations, any band or mode.

CEMARA (Certificado Moviles Armada Argentina)

The Servicio de Radioaficionados de la Armada (SARA) and the Radio Club Argentino jointly issue this award to amateurs who work 25 Maritime Mobile stations, 10 of which must be LUs (including five from the Argentine Navy). The 15 other Maritime Mobiles stations can be from any country. Starting date: January 1, 1960. All bands. Separate awards issued for CW, AM or SSB. No mixed modes.

CA (Certificado Argentino)

Awarded to non-LU amateurs who work 100 LU stations. Any mode or band.

RA (Republica Argentina)

Awarded to all amateurs working 18 LU stations to form the words Republica Argentina with the first letter of each call-sign suffix. Example: LU1RA, LU6EAM, LU2PB, LU5US, LU9BBA, LU3LA, LU2IO, LU9CAM, LU4AA, and so on. All modes/bands. Starting date: January 1, 1965.

TRA (Toda Republica Argentina)

Awarded to amateurs working all 25 Argentine districts:

LU1AA-LU9CZZ	Federal Capital
LU1DA-LU9EZZ	Buenos Aires
LU1FA-LU9FZZ	Santa Fe
LU1GA-LU9GOZ	Chaco
LU1GP-LU9GZZ	Formosa
LU1HA-LU9HZZ	Cordoba
LU1IA-LU9IZZ	Misiones
LU1JA-LU9JZZ	Entre Hios
LU1KA-LU9KZZ	Tucuman
LU1LA-LU9LZZ	Corrientes
LU1MA-LU9MZZ	Mendoza
LU1NA-LU9NZZ	Santiago del Estero
LU1OA-LU9OZZ	Salta
LU1PA-LU9PZZ	San Juan
LU1QA-LU9QZZ	San Luis
LU1RA-LU9RZZ	Catamarca
LU1SA-LU9SZZ	La Rioja
LU1TA-LU9TZZ	Jujuy
LU1UA-LU9UZZ	La Pampa
LU1VA-LU9VZZ	Rio Negro
LU1WA-LU9WZZ	Chubut
LU1XA-LU9XOZ	Santa Cruz
LU1XP-LU9XAA	Tiera del Fuego
LU1YA-LU9YZZ	Neuquen
LU1ZA-LU9ZZZ	Antarctic Bases

LU 10 DL (LU 10 doble letras)

Awarded to amateurs working 10 stations (at least one must be an LU) with double letter prefixes from 1 to 0 without repeating the pair of letters (such as PY1AA, CP2FF, LU5BB). Any band/mode. Starting date: January 1, 1965.

RCA (Radio Club Argentino)

Awarded to amateurs working stations (at least one must be an LU) to form the words Radio Club Argentino using only the first two letters after the call-sign suffix. Examples: CT1RA, OK3DI, PY2OC, W8LU, LU3BAC. Any band/mode. Starting date: January 1, 1965.

AUSTRALIA (VK)

The *Worked All VK Call Areas (WAVKCA) Award* is offered by the Wireless Institute of Australia (WIA) for working and confirming contacts with the various call areas of the Commonwealth of Australia. All VK amateurs and all amateurs who are members of their IARU-affiliated National Society are eligible for the award. Twenty-two (22) QSOs are required (73 for VK amateurs) as follows:

Territory	Call Area	Required QSOs/QSLs
Australian Antarctica		
Heard Island	VKØ	1
Macquarie Island		
Australian Capital Territory	VK1	1
State of New South Wales	VK2	3
State of Victoria	VK3	3
State of Queensland		
Thursday Island	VK4	3
State of South Australia	VK5	3
State of Western Australia	VK6	3
State of Tasmania		
King Island	VK7	3

Flinders Island
Northern Territory   VK8   1
[All VK9 Territories]   VK9   1

Contacts must have been made from January 1, 1946 on.

Any band/mode, but crossband contacts not allowed. Maritime and aeronautical mobile contacts do not count, but land-mobile and -portable contacts are valid provided the QSL card shows the station location at time of contact. Send a certified log extract (in lieu of the actual QSLs), verified by an official of the applicant's IARU Society or two licensed amateurs, application and 8 IRCs to: The Federal Awards Manager, Wireless Institute of Australia, John Keller, VK3DP, PO Box 175, South Caulfield, Victoria 3161, Australia.

In addition, approximately 60 awards are issued by local clubs in Australia. Contact the Federal Awards Manager for further information.

AUSTRIA (OE)

Except where otherwise noted, applications for Austrian awards should be sent to the Austrian IARU Society (OVSV) as follows: OVSV HQ, International Awards Manager, Theresiengasse 11, A-1180 Vienna, Austria.

WAOE (Worked All Austria)

North American amateurs must work eight of the nine OE call areas on any band (different requirements apply to European applicants). Awards are issued as mixed and SSB. A separate 160-meter award, WAOE-160, is also available, for working four OE call areas on 160. Send verified list of QSLs and 10 IRCs.

OE 100

This certificate is issued for working at least 100 different OE stations. Contacts must be made on any mode and band that OE stations can legally operate. All contacts must be made from the same country and must be made after April 1, 1954. Endorsement stickers are available for 200, 300…different OE contacts. Send certified list and 10 IRCs (2 IRCs for each endorsement).

WLOE (Worked District Locators in Austria)

Each OE call district has its own three-digit "District Locator" (ADL). For the basic award, 30 different ADLs must be worked in at least six different OE call districts (OE1-OE9). ADLs in the block 001-099 count as a separate call district. Contacts from January 1, 1968 are valid.

The WLOE award is available for mixed HF and CW HF.

Endorsement stickers are issued for each additional 10 ADLs and a special 160-meter sticker for working 10 ADLs on 160. QSLs are not required for contacts made during the WAOE 160-Meter International Contest.

To apply, send a certified list of QSLs and 10 IRCs (4 IRCs for each endorsement sticker). Note: Only OVSV members have an ADL number and only those stations count for the award.

WPXZ 15 (Worked Prefix Zone 15 Awards)

This award is offered for working stations in three classes:
Class 1—15 countries and 50 prefixes
Class 2—12 countries and 40 prefixes
Class 3—8 countries and 30 prefixes

Contacts must be made with the following countries (by prefix): HA, HV, I, IS, IT, OH, OK, T7, TK, UA2, UR, UP, UQ, YU, ZA, 1A0, 4U1VIC, 9H. Starting date: January 1, 1958. Send verified list of QSLs and 10 IRCs to OVSV—LV OE1, Eisvogelgasse 4, A-1060 Vienna, Austria. Endorsement stickers—2 IRCs.

WIEN Diplom (Vienna Award)

This award requires contacting Vienna districts as follows: Class 1—all 23 districts, Class 2—15 districts. Vienna stations (OE1) should be asked to clearly mark their district on their QSL card. All bands/modes. Send certified list and 10 IRCs to OVSV—LV OE1, Eisvogelgasse 4, A-1060 Vienna, Austria.

WDRA (Worked Danube River)

This award is issued for working countries adjoining the Danube River. To qualify, work 15 DL stations, seven OE stations (one of which must be an OE1), five HA stations, three YO stations, three YU stations, two OK stations, and one each from LZ and UO5. Send certified list and 10 IRCs to OVSV—LV OE1, Eisvogelgasse 4, A-1060 Vienna, Austria.

ACA (Austrian Large Cities Award)

For this award, you qualify for one point for each Austrian large city worked. The award is issued in three classes:

Class 1—100 points
Class 2—70 points
Class 3—40 points

Send certified list and 10 IRCs to OVSV—LV OE1, Eisvogelgasse 4, A-1060 Vienna, Austria.

AMRS (Austrian Military Radio Section)

This award is issued for working members of the Austrian Military Radio Section, a subdivision of OVSV, for a total of 20 points. Each AMRS club station counts as one point, AMRS club stations count two points, and stations in United Nations duty count three points. To apply, send certified list to Werner Hafner, OE8HFL, Hausergasse 30/1/7/16, A-9500 Villach, Austria.

A variety of localized awards are also available. Contact OVSV HQ for details.

BELGIUM (ON)

The Union Belge des Amateurs-Emetteurs (UBA) sponsors the *Worked All Belgian Provinces (WABP)* award for amateurs who have worked and confirmed contacts with each of the nine ON provinces on two bands. To apply, send a certified list of QSLs and 10 IRCs (or $3 US) to: ON5KL, UBA-HF Awards

Manager, Van Campenhout Mat, PO Box 400, 8400 Oostende-1, Belgium.

ON Provinces:

AN—Antwerp	LG—Liege	OV—East Flanders
BT—Brabant	LM—Linbourg	NR—Nanur
HT—Hainaut	LU—Luxembourg	WV—West Flanders

BERMUDA (VP9)

Applications, lists of contacts, QSLs and other required materials for Bermuda awards should be sent to: The Awards Manager, Radio Society of Bermuda, Box HM 275, Hamilton 5, Bermuda.

Bermuda 100 Club

This award is a mahogany plaque with the Bermuda Coat of Arms embossed thereon in full color. An engraved plate for affixing to the plaque is provided for each additional 50 stations worked up to 200. It is suitably imprinted with the name and call sign of the recipient. The rules are as follows: 100 different VP9 call signs must be worked on any band or mode from the same QTH (note: A reissued VP9 call sign held by two different operators counts as two call signs if both operators have been worked. However, only one contact credit is allowed for working the same station mobile, portable, etc); there is no time limit, and the award is free. A letter of application, together with a list of contacts and the relevant QSL cards, is required.

Worked All Bermuda Award

This award is a print of an antique map of Bermuda, signed by the Governor. The award is issued to licensed amateurs who submit proof (valid QSL cards) of contacts with all nine Bermuda parishes: Hamilton, Devonshire, Paget, Pembroke, Sandys, Smith's, Southampton, St George's and Warwick. The following rules apply: Applications must be accompanied by valid QSL cards from each parish, only one of which can be from a mobile or other non-fixed station; there is no time limit for the award and it is free; VP9 stations are not eligible; all contacts must be made from the same QTH or within a 25-mile radius thereof. A letter of application listing stations contacted and all nine QSL cards required.

BRAZIL (PY)

To apply for these awards sponsored by the Liga de Amadores Brasileiros de Radio Emissao (LABRE), the relevant log information must be checked by the Awards Manager of the applicant's National Society or by two licensed amateurs. All contacts must be with land stations. Send the application and 10 IRCs for each award to: Awards Manager, LABRE HQ, PO Box 07-0004, 70359-Brasilia, DF, Brazil.

WAB (Worked All Brazil)

This award is available for working and confirming con-

tacts with Brazilian stations in all 26 Brazilian states and the capital city, Brasilia. A special ribbon (to attach to the certificate) is available for working and confirming the Ocean Islands.

WAO (Worked All Oceans)

This award is available for working and confirming contacts with nine Brazilian Geographic Regions, as follows:

1st Region—PY1-PP1
2nd Region—PY2-PP2-PT2-PQ2
3rd Region—PY3
4th Region—PY4
5th Region—PY5-PP5
6th Region—PY6-PP6
7th Region—PY7,PP7,PR7,PS7,PT7
8th Region—PY8,PP8,PR8,PS8,PT8,PQ8,PV8,PW8
9th Region—PY9,PT9,PYØ and 21 countries of the Atlantic Ocean

WAA (Worked All America)

This award is available for working and confirming contacts with 45 countries in the American Geographic Area; one contact must be with Brazil.

DBDX (Brazilian DX Award)

This award is available for working and confirming contacts with a minimum of 20 different DXCC countries. One contact must be Brazil. Endorsement stickers are available for

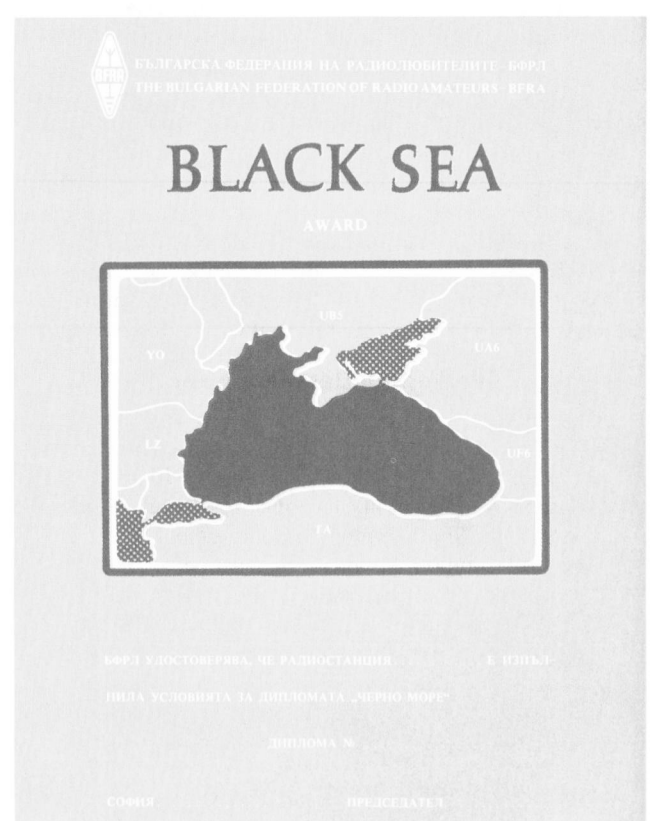

additional countries in groups of 10. All contacts must be made on 160, 80 or 40 meters. Four kinds of certificates are issued—phone/CW, phone only, CW only, and RTTY.

BULGARIA (LZ)

The Bulgarian Federation of Radio Amateurs (BFRA) offers four certificates for two-way contacts on CW, SSB/AM or mixed modes. Applications should include a list of claimed QSOs verified by two licensed radio amateurs or local radio club authorities indicating stations worked, date, time, band and mode, together with 10 IRCs. Send all applications to: Bulgarian Federation of Radio Amateurs, Awards Manager, PO Box 830, 1000 Sofia, Bulgaria.

5-Band LZ Award

To qualify, work one LZ1/LZ3 and one LZ2/LZ4/LZ6 stations on each band, 80-10 meters. Contacts must be made after January 1, 1979.

W 100 LZ Award

To qualify, work 100 QSOs with different LZ stations during one calendar year. Contacts must be made after January 1, 1979.

Black Sea Award

To qualify, work a total of 60 different amateur stations in countries bordering the Black Sea. QSOs with LZ, TA, YO, UA6, UB5, UX, UT and UU are mandatory. All contacts must be made after January 1, 1979.

SOFIA Award

To qualify, 100 total points must be earned by working amateur stations in Sofia, the capital of Bulgaria. Points are allocated as follows: 80 meters—15 points, 40 meters—five points, 20 meters—one point, 15 meters—two points, 10 meters—three points (for Europeans, two points are earned for working Sofia stations on 80, 40, 15 and 10 meters, and one point for 20-meter contacts). All contacts must be made after January 1, 1979. Stations may be worked only once per band, regardless of mode. Send station list and 10 IRCs to the Awards Manager.

CHILE (CE)

WACE (Worked All Chile)

This award is issued by the Radio Club de Chile to all amateurs who can submit satisfactory evidence of having established two-way contact with CE stations in each of the 10 zones in which the Chilean territory is divided. The application must contain a list of the stations worked, including date, signal report and mode. QSL cards are not necessary if the list is verified and signed by a recognized radio club or national Amateur Radio organization. All contacts must be made from the same "country." Eight IRCs required. Send the application to: Awards Manager, Radio Club De Chile, Casilla 13630, Santiago, Chile.

Diploma Republica de Chile (Republic of Chile Award)

To qualify, applicants must make contact with 16 different CE stations from any Chilean zone so as to form the phrase "Republica de Chile" with the last letter of the suffix of their call signs. All contacts are valid starting January 1, 1986 on any band and mode. Applicants may send QSL cards or preferably a list of QSOs duly certified by an IARU Member-Society. QSLs or QSO list must be arranged in order to spell Republica de Chile. Fee: 8 IRCs. Send applications to the Radio Club de Chile Awards Manager at the address above.

COLOMBIA (HK)

Unless otherwise noted, please send all applications to: Liga Colombiana de Radioaficionados (LCRA), PO Box 584, Bogota, Colombia.

CHK Award

Contact and confirm different HK stations as follows: North and South American stations must contact 50 HK stations, elsewhere 25 different HK stations. Send log data and 8 IRCs.

ZHK Award

Contact and confirm the 10 Colombian call areas as follows:

LIGA COLOMBIANA DE RADIOAFICIONADOS

North and South American stations must work 9 call areas, elsewhere 8 call areas. Log data required.

HK5 Diploma

American stations must contact 12 HK5 stations, others must contact 8 HK5s. Contacts are valid after January 1, 1957.

City of Cali Foundation Diploma

This diploma is awarded for contacts made on July 25 (Sectional Five Anniversary) with HK5VD, 5J5VD, 5K5VD and five other HK5 stations. Applications to LCRA, Sectional Cali, Apartado 6149, Cali, Colombia.

100 HK3 Diploma

One hundred contacts with HK3 stations are required.

COSTA RICA (TI)

All applications should be sent to: Awards Manager, Radio Club de Costa Rica (RCCR), PO Box 999, Heredia, Costa Rica.

Diploma FRACAP

The principal reasons for the creation of this diploma are to increase the international awareness of FRACAP (the Federation of Radio Amateurs of Central America and Panama) and to raise the interest and knowledge of the geographical location of its member-countries—Costa Rica, El Salvador, Guatemala, Honduras, Nicaragua and Panama. The requirements are two-way confirmed contacts (any band/mode) with each of the six FRACAP countries. Each contact must be made with an amateur who is a member of a FRACAP-affiliated radio club. Contacts are valid

after August 16, 1985. The application must be accompanied by either the QSL cards or a list verified by the radio club of which the applicant is a member. Ten IRCs (or $5 US) required.

TTI Diploma

To qualify, two-way QSOs (any mode) must be made with seven of the eight Costa Rican call areas: TI2—San Jose, TI3—Cartago, TI4—Heredia, TI5—Alajuela, TI6—Limon, TI7—Guanacaste, TI8—Puntarenas, TI9—Isla del Coco. A QSO with the official RCCR club station, TI0RC, can be used to replace one call area. Verified list of QSLs and 10 IRCs (or $4 US) required.

CUBA (CO)

To apply for awards sponsored by the Federacion de Radioaficionados de Cuba (FRC), send a certified list of QSLs and 10 IRCs (or $2 US) for each award. All bands/modes are acceptable. Address applications to: FRC, Awards Department, PO Box 1, Habana 1, Cuba.

America Award

To qualify, the applicant must work 45 countries and islands in America for Class III; 50 countries for Class II; and 51+ for Class I. A contact with Cuba is mandatory; KG4 contacts are not acceptable.

Cuba Award

To qualify, the applicant must work all eight Cuban call sign districts, CM1/CO1-CM8/CO8. Cuban Amateur Radio Club stations (three-letter suffixes) can be substituted for up to three missing districts. Contacts are valid after January 1, 1959.

Caribbean Award

To qualify, the applicant must work 20 or more of the 32 countries in the Caribbean, including CO, HK, HP, HR, TI, TG, V3, XE, YN and YV. CO contacts are mandatory; KG4 contacts are not acceptable. Contacts are valid after January 1, 1959.

Cuba DX Group

To qualify, the applicant must work four official members of the Cuba DX Group for a total of four points (one point per member). Aspiring members of the DX Group count a $^1/_2$ point. Contacts are valid after September 1, 1980.

CYPRUS (ZC4)

The *Cyprus Award* Certificate, sponsored by the Cyprus Amateur Radio Society, is awarded to any licensed amateur outside Cyprus who makes a specified number of two-way contacts with licensed amateurs on the island of Cyprus. To reduce as far as possible any advantage accruing to stations by reason of their geographical location and to encourage activity on the less-frequently used bands, the certificate will be awarded on a points basis determined by CQ Zone location and bands used as shown below:

Zone	1.8	3.5	7	14	21	28
	Points Scored Per Contact					
20	8	2	1	1	2	4
1, 2, 3, 6, 7, 10, 12, 19, 24, 25, 26, 27, 28, 29, 30, 31 and 32	16	8	4	2	4	8
All Other Zones	8	4	2	1	2	4

The total number of points required to win the award depends on the bands used. If all contacts are made on only one band, 32 points are required. If the contacts are made on any two bands, 24 points are required. If the contacts are made on any three bands, 16 points are required, and any four bands, 12 points are required. Any mode may be used. Contacts must be made after April 1, 1973. Contacts with any one Cyprus station can only count once per band.

To claim the award, copies of log entries should be submitted under the following heading: DATE/TIME UTC, STATION WORKED, FREQUENCY BAND, SIGNAL REPORTS IN AND OUT. Each log sheet should be headed with the applicant's call sign, zone number and complete mailing address, preferably typed or printed in block capitals. Either the appropriate QSLs or a statement from the applicant's National Society certifying that the QSL cards have been produced to them is required. In countries without a National Society, a similar statement signed by two other amateurs will suffice. Log sheets and 10 IRCs should be sent to: Awards Manager, Cyprus Amateur Radio Society, PO Box 1267, Limassol, Cyprus.

CZECH REPUBLIC (OK)

The following awards are issued by the Czech Radio Club. The fee for all awards is 10 IRCs. For endorsement stickers, send a confirmed list and 4 IRCs, and indicate the number and issue date of the basic award. QSL cards may be sent with the application; cards need not be sent when the national-level Amateur Radio Society or club has confirmed possession of the listed QSLs by the applicant. [Cards for the P75P must contain locations (QTH) of listed stations.] Applications should be sent to: Czech Radio Club, Awards Manager, PO Box 69, 113 27 Praha 1, Czech Republic.

S6S (Six Continents)

To qualify, work and confirm contacts with at least one amateur station located in each of the six continents (as defined by IARU) since January 1, 1950; all CW, all phone, all RTTY, or all SSTV. Endorsement stickers for the basic certificate are available for single bands (80, 40, 20, 15, 10).

P75P (Worked 75 Zones)

To qualify, work and confirm contacts with at least one

amateur fixed station location in each of the ITU zones since January 1, 1960, in three classes: 1st Class—70 zones, 2nd Class—60 zones, 3rd class—50 zones.

100 OK

To qualify, work and confirm contacts with at least 100 different OK stations since January 1, 1954. Endorsement stickers are available for every additional 100 stations confirmed up to 500.

DENMARK (OZ)

The following awards are sponsored by Eksperimenterende Danske Radioamatorer (EDR):

Cross Country Award

This award is issued for all CW or all phone. European amateurs must make 50 points, all others 40 points. The call sign is used as the basis for the award. Each call prefix, OZ1 to OZ9, and OX3, must be contacted. Three contacts with each call prefix permitted on each band (two contacts permitted for European stations), with the exception of OX3 (nine contacts permitted). Each contact counts one point, with the exception of 70 cm, where each contact counts two points. Starting date: April 1, 1970. Send a list certified by the Awards Manager of the National Society and 5 IRCs (or $2 US) to: EDR Awards Manager, Tage Eilman, OZ1WL, PO Box 213, 5100 Odense C, Denmark.

Greenland Award

This award is issued in three categories—CW, phone or mixed, in three classes:

Class 1—five different locations and 15 different stations.
Class 2—four different locations and 10 different stations.
Class 3—three different locations and five different stations.

Only two-way CW or phone QSOs with OX3 stations made after January 1, 1978 count. One contact per band with the same station permitted. All amateur bands (3.5 to 1296 MHz) are valid. Crossband QSOs do not count. Minimum report 33 (phone), 338 (CW). Contacts with portable or mobile stations are not valid. A detailed list of contacts (certified by the Awards Manager of a National Society) and 5 IRCs (or $2 US) should be sent to OZ1WL (see address above).

Fairytale Award

Radio amateurs in Odense, the native town of Hans Christian Andersen, the fairytale writer, issue the Fairytale Award. Amateurs outside of Scandinavia are eligible for the award after making nine two-way CW QSOs with nine different OZ

stations; each OZ call area (1-9) must be represented, a minimum of three must be from Odense (separate rules apply to those in LA/OZ/SM). A contact with OZ3FYN, the EDR Odense Division club station, may be used to replace one call area. Only two-way CW QSOs since December 6, 1967, the centenary of Hans Christian Andersen's being appointed an honorary citizen of Odense, count for the award. All amateur bands (3.5 to 1296 MHz) are valid; crossband QSOs do not count, nor do mobile/portable contacts. Minimum signal report is 338. To apply, send a certified list of QSOs and 6 IRCs to: E. Hansen, OZ7XG, 14 Sophus Bauditz Vej, DK 5210, Odense, Denmark.

FINLAND (OH)

General rules for OH awards, sponsored by the Finnish IARU Society, Suomen Radioamatoorilitto (SRAL), are as follows:

All contacts must be made with Finnish fixed stations after June 6, 1947 (after February 1, 1967 for OHA 500). Do not send QSLs; send a list of QSLs checked by two amateurs or the Awards Manager of the national Amateur Radio Society. This list must include the call sign, date, time, band, mode and report for each QSO/QSL. Eight IRCs (or $2 US) are required for each award. For further information and application forms, contact Jukka Kovanen, OH3GZ, SRAL Awards Manager, Varuskunta 47 as 11, SF-11310, Riihimaki 31, Finland.

Finnmaid—Work three OH YLs.
OHA—Work 15 OH stations in at least five call areas.
OHA 100—Work 100 OH stations in all 10 call areas using at least two bands.
OHA 300—Work at least 300 OH stations.
OHA 500—Work at least 500 OH stations.
OHA 600—Work at least 600 OH stations.

FRANCE (F)

The following material is translated from *Les Diplomes Francais* by F6AXP, published by Reseau des Emetteurs Francais (REF).

General application requirements:

Each application must be dated and contain a signed statement that all awards rules have been complied with. The application must be addressed to the appropriate Awards Manager.

The application must include a detailed list of QSOs, indicating call sign, date, frequency and mode for each QSO.

The applicant must provide evidence in the form of QSLs or a certified list of QSOs/QSLs verified by an official of the applicant's National Society. A self-addressed envelope must be provided when sending QSLs. When applying for an endorsement, the number of the basic certificate and mode must be supplied.

DDFM (French Metropolitan Departments Diploma)

This award is issued in these classes:

- DDFM phone
- DDFM CW
- DDFM mobile or special mention

For applicants outside of France, it is necessary to work and confirm 40 different departments on the same band. Endorsement stickers are available in multiples of ten; an "excellence" endorsement sticker is issued for working all 96 departments on one band. QSLs are required with all applications and must clearly indicate the department of the station contacted. Contacts are valid from June 30, 1957 (January 1, 1982, for the new WARC bands). For the basic award, 12 IRCs are required; 7 IRCs are required for each endorsement. For further details, see REF Form C3-01-7 (2/82). Awards Manager: Christian Coupas, F6EDW, PO Box 83, F. 63307, Thiers-Cedex, France. A 5-band award *(5BDDFM)* is also available for working 300 departments using a combination of five bands (at least 10 contacts must be made per band). A personalized and numbered plaque is issued to each 5BDDFM recipient; 68 IRCs are required for both the return of the cards and the plaque. Contact F6EDW for further details.

French Departments		Province
01	Ain	Rhone-Alpes
02	Aisne	Picardie
03	Allier	Auvergne
04	Alpes Haute Provence	Provence-Cote D'Azur
05	Alpes (Hautes)	Provence-Cote d'Azur
06	Alpes-Maritimes	Provence-Cote d'Azur
07	Ardeche	Rhone-Alpes
08	Ardennes	Champagne
09	Ariege	Midi-Pyrenees
10	Aube	Champagne
11	Aude	Laguedoc-Roussillon
12	Aveyron	Midi-Pyrenees
13	Bouches-du-Rhone	Provence-Cote d'Azur
14	Calvados	Basse Normandie
15	Cantal	Auvergne
16	Charente	Poitou-Charente
17	Charente-Maritime	Poitou-Charente
18	Cher	Centre
19	Correze	Limousin
2A	Corse Sud	Corse
2B	Corse (Haute)	Corse
21	Cote d'Or	Bourgogne
22	Cotes du Nord	Bretagne
23	Creuse	Limousin
24	Dordogne	Aquitaine
25	Doubs	Franche-Comte
26	Drome	Rhone-Alpes
27	Eure	Haute Normandie
28	Eure-et-Loir	Centre
29	Finistere	Bretagne
30	Gard	Languedoc-Roussillon
31	Haute-Garonne	Midi-Pyrenees
32	Gers	Midi-Pyrenees
33	Gironde	Aquitaine
34	Herault	Languedoc-Roussillon
35	Ille-et-Vilaine	Bretagne
36	Indre	Centre
37	Indre-et-Loire	Centre
38	Isere	Rhone-Alpes
39	Jura	Franche-Comte
40	Landes	Aquitaine
41	Loir-et-Cher	Centre
42	Loire	Rhone-Alpes
43	Haute-Loire	Auvergne
44	Loire-Atlantique	Pays de Loire
45	Loiret	Centre
46	Lot	Midi-Pyrenees
47	Lot-et-Garonne	Aquitaine
48	Lozere	Languedoc-Roussillon
49	Maine-et-Loire	Pays de la Loire
50	Manche	Basse-Normandie
51	Marne	Champagne
52	Marne (Haute)	Champagne
53	Mayenne	Pays de Loire
54	Meurthe-et-Moselle	Lorraine
55	Meuse	Lorraine
56	Morbihan	Bretagne
57	Moselle	Lorraine
58	Nievre	Bourgogne
59	Nord	Nord
60	Oise	Picardie
61	Orne	Basse-Normandie
62	Pas-de-Calais	Nord
63	Puy-de-Dome	Auvergne
64	Pyrenees-Atlantique	Aquitaine
65	Pyrenees (Hautes)	Midi-Pyrenees
66	Pyrenees-Orientales	Languedoc-Roussillon
67	Bas-Rhin	Alsace
68	Haut-Rhin	Alsace

69	Rhone	Rhone-Alpes
70	Saone (Haute)	Franche-Comte
71	Saone-et-Loire	Bourgogne
72	Sarthe	Pays de Loire
73	Savoie	Rhone-Alpes
74	Haute-Savoie	Rhone-Alpes
75	Ville de Paris	Ile-de-France
76	Seine-Maritime	Haute-Normandie
77	Seine-et-Marne	Ile-de-France
78	Yvelines	Ile-de-France
79	Deux-Sevres	Poitou-Charentes
80	Somme	Picardie
81	Tarn	Midi-Pyrenees
82	Tarn-et-Garonne	Midi-Pyrenees
83	Var	Provence-Cote d'Azur
84	Vaucluse	Provence-Cote de'Azur
85	Vendee	Pays de Loire
86	Vienne	Poitou-Charente
87	Haute-Vienne	Limousin
88	Vosges	Lorraine
89	Yonne	Bourgogne
90	Territoire de Belfort	Franche-Comte
91	Essone	Ile-de-France
92	Hauts-de-Seine	Ile-de-France
93	Seine-Saint-Denis	Ile-de-France
94	Val-de-Marie	Ile-de-France
95	Val-d'Oise	Ile-de-France

[Note: Many attractive certificates are also available for contacts with individual departments. For particulars, see Chapter II of *Les Diplomes Francais*; for information on how to obtain this booklet, contact REF, PO Box 2129, F37021, Tours-Cedex, France.]

DPF (French Provinces Diploma)

This award, for working a required number of French provinces, is issued in these classes:

- DPF CW
- DPF phone
- DPF "Special Mention" for any single band or mode.

For the DPF HF, the applicant must work and confirm 25 provinces. Contacts are valid from January 1, 1951 (WARC-band contacts valid from January 1, 1982). QSLs must be submitted with the application, and the cards must clearly indicate the provinces of the stations contacted. For further details, see REF Form C3-01-5 (2/82). To cover the postage for returning the QSLs and the certificate, 12 IRCs are required. Awards Manager: F6EDW (see address above).

5BDPF (5-Band Province Diploma)

To qualify, work and confirm contacts with stations in 22 different provinces on five HF bands. For further details, see REF Form C3-01-8 (2/82). A personalized and numbered plaque is sent to each successful applicant. For return of QSLs and shipping for the plaque, 68 IRCs are required. Awards Manager: F6EDW (see address above).

French Provinces	Includes These Departments (By Number)
1—Alsace	67, 68
2—Aquitaine	24, 33, 40, 47, 54
3—Auvergne	3, 15, 43, 63
4—Basse-Normandie	14, 50, 61
5—Bourgogne	21, 58, 71, 89
6—Bretagne	22, 29, 35, 56
7—Centre	18, 28, 36, 37, 41, 45
8—Champagne	8, 10, 51, 52
9—Corse	2A, 2B
10—Franche-Comte	25, 39, 70, 90
11—Haute-Normandie	27, 76
12—Languedoc-Roussillon	11, 30, 34, 48, 66
13—Limousin	19, 23, 87
14—Lorraine	54, 55, 57, 88
15—Midi-Pyrenees	9, 12, 31, 32, 46, 65, 81, 82
16—Nord	59, 62
17—Pays-de-Loire	44, 49, 53, 72, 85
18—Picardie	2, 60, 80
19—Poitou-Charentes	16, 17, 79, 86
20—Provence Cote-d'Azur	4, 5, 6, 13, 83, 84
21—Ile-de-France	75, 77, 78, 91, 92, 93, 94, 95
22—Rhone-Alpes	1, 7, 26, 38, 42, 69, 73, 74

DUF (French-speaking "Universe" Diploma)

This award, for working stations in the French-speaking world, is available in four classes:

DUF1—for confirmed contacts with five countries in three continents.

DUF2—for confirmed contacts with eight countries in four continents.

DUF3—for confirmed contacts with 10 countries in five continents.

DUF4—for confirmed contacts with 20 countries in six continents.

QSL cards must accompany the application and must clearly indicate the country and continent. Contacts are valid from April 1, 1946, unless otherwise indicated. For more details, see REF Form C3-01-9 (2/82). Awards Manager: Edmond Dubois, F9IL, 33 Les Tamaris, F34340, Marseillan, France.

5BDUF

A personalized and numbered plaque is awarded to those qualifying for 5BDUF, which requires confirmed contacts with 30 countries in six continents on bands higher than 7 MHz and confirmed contacts with 15 countries in at least five continents on 7 MHz and each band below. Contacts are valid from January 1, 1981, and QSLs that clearly indicate the country and continent must accompany the application; 68 IRCs, to cover cost of returning the QSLs and sending the plaque, must also be included. For further details, see REF Form C3-01-10 (2/82). Awards Manager: F9IL (see address above).

DUF Countries [simplified list]:

Europe—C3, DA, F, TK, 3A.

Africa—CN, D6 (after 5 July 1975), FH (after 5 July 1975), FR/R, FR/E (after 1 August 1968), FR/G (after 25 June 1960), FR/J (after 25 June 1960), FR/T (after 25 June 1960), J2, TJ (after 1 January 1960), TL (after 13 August 1960), TN (after 15 June 1960), TR (after 17 August 1960), TT (after 11 August 1960), TU (after 7 August 1960), TY (after 1 August 1960), TZ (after 20 June 1960), XT (after 5 August 1960), 3V, 3X (after 1 October 1958), 5R (after 14 October 1958), 5T (after 20 June 1960), 5U (after 3 August 1960), 5V (after 27 April 1960), 6W (after 20 June 1960), 7X (after 1 July 1962).

Asia—XU, XV, XW.

North America—FG, FM, FS, FO, FP.

South America—FY.

Oceania—FK, FO, FW, YJ.

Austral Continent—FT8W, FT8X, FT8Y, FT8Z.

DOTOM (French Overseas Departments and Territories Diploma)

This award is issued for working the following overseas departments (FG, FS, FM, FP, FR) and the following overseas territories (FH, FK, FO, FW, 3Y). All contacts are valid from January 1, 1982, and QSLs and 12 IRCs must accompany the application. For further details, see REF Form C3-01-13 (2/82). Awards Manager: Max Pomel, F6AXP, PO Box 73, 63370 Lempdes, France.

DTA (Southern French Lands Diploma)

To qualify for this award, three confirmed contacts are required with FT8W, FTX, FT8Y and FT8Z. The *DTA Excellence* diploma is issued for four confirmed contacts with each country. QSLs and 10 IRCs must accompany the application. For further details, see REF Form C3-01-14 (2/82). Awards Manager: F6AXP (see address above).

DTC (CW Class C Diploma)

This award is issued for making 1000 CW QSOs. All bands and all countries are valid. The following endorsements are available:

DTC 1—2000 additional CW QSOs.
DTC 2—4000 additional CW QSOs (for a total of 5000).
DTC 3—9000 additional CW QSOs (for a total of 10,000).
DTC 4—14,000 additional CW QSOs (for a total of 15,000).
DTC REF HONOR—20,000 CW QSOs.

To apply, send certified list of QSOs (no QSLs) and 12 IRCs (36 IRCs for DTC HONOR). Awards Manager: Thierry Mazel, F6CUK, Puy Chany, F-63360, St Beauzire, France.

F-CW-500

This award is issued for working 500 different French metropolitan stations on CW. Endorsements are available in 100-QSO increments. Applications must include a certified list of QSOs in alphanumerical order by call (with complete QSO data) and 10 IRCs. Awards Manager: F6CUK (see address above).

YL De France Diploma

This award is issued in three classes:

Class 1—work five French YLs, plus one YL contact on three different continents.
Class 2—work 100 YLs, including five French YLs, on three different continents.
Class 3—work 500 YLs, including five French YLs, on six continents.

QSL cards and 12 IRCs (to defray costs of returning the QSLs and sending the certificate) must be included in each application. Awards Manager: Gilda Le Gall, F6FMO, Quilvidic, Mellac, F29130 Quimperlé, France.

GABON (TR)

Association Gabonaise des Radio-Amateurs (AGRA) issues three different certificates as follows:

DDG1—for confirmed contacts with eight different TR stations; any mode, any band.
DDG2—for confirmed contacts with 12 different TR stations on at least three different HF bands, any mode.
DDG Special—for one confirmed TR contact on each of five different HF bands; two of the five bands must be 160, 80 and/or 40 meters.

Minimum signal report: 33(9).

To apply, send a certified list of QSLs signed by two amateurs or an official of the applicant's National Society and 10 IRCs to: AGRA Diplome Manager, PO Box 1826, Libreville, Gabon.

GERMANY (DL)

DLD Award

The DLD Award is an official award of the German Amateur Radio Club (DARC). It is available to all licensed radio amateurs and shortwave listeners (SWL). The names of new award holders will be published in the DARC magazine *CQ-DL*.

All members of DARC, its associate club VFDB, and club stations of both organizations are issued a *District Location Code* (DOK). To qualify for DLD, applicants must submit QSL cards from licensed radio amateurs showing a certain number of DOKs worked or—for SWLs—heard.

DLD Award Classes and Modes

• DLD is issued separately for each amateur band.
• DLD is issued in different classes on each band as follows: DLD 100, DLD 200, DLD 300, DLD 400, DLD 500 (with lapel badge), DLD 600, DLD 700, DLD 800, DLD 900 and DLD 1000 (with engraved badge of honor).
• For SWLs, the awards are known as DLD-SWL 100, DLD-SWL 200 and so on up to DLD-SWL 1000.
• All DLD Awards may be issued for mixed modes, or may be endorsed for single mode operation, providing that this is supported by QSL cards.

Conditions of Issue

• All modes permitted by the applicant's license may be used.
• The DLD Award will be issued initially when the applicant submits evidence of confirmed contacts with 100 different DOKs on a single band. For each additional 100 DOKs on the same band, the applicant may apply for the next class of DLD. Applicants may skip awards if they wish. That is, it is not necessary to apply for a separate award for each 100 DOKs worked.
• A DOK will only count if the station worked or heard is located in the Federal Republic of Germany at the time of the contact. Stations have only one DOK each and must only give out the DOK under which they are registered with DARC QSL Bureau. Special Event DOKs will be published in *CQ-DL*.
• There will be a charge for the issue of the DLD Awards. Detail of charges will be published from time to time in *CQ-DL* and payment should be included with the application, or sent to the DARC bank by bank transfer.

Deutscher Amateur-Radio-Club e.V.
DLD-Diplome, Postfach 11 55
D-3507 Baunatal 1
Federal Republic of Germany

Postal Giro Account 2571 55-302 Hannover
Sort Code (BLZ) 260 100 30

Applications for DLD

All valid DOKs are listed in the official DOK List which is obtainable from DARC publications and is used as the application form. It is recommended that you use a separate list for each band. You may order the DOK List by sending an address label and $7 US or 5 IRCs per list to DARC at the address mentioned above.

A computer generated list will be accepted providing it uses the same format as the application form issued by DARC. All applications must be verified by the applicant's local club or by an official Award Manager before forwarding the DARC-HQ. The DLD Award will be issued once the relevant fee is received.

DIG Award Program

All DIG awards and trophies may be applied for by licensed radio amateurs and SWLs around the world. The awards are available either without endorsement for contacts on all bands and modes (mixed), or for contacts on CW only or VHF only. Applicants must be in possession of the QSL cards, but need not send them. It is sufficient to sent a GCR list (General Certification Rule) signed by two witnesses. Please enclose a return address label.

DIG Award Manager Addresses:

DF8BQ, Dieter Weckmann, Alte Reihe 28, 27313 Dorverden, Germany

DH1PAL, Werner Theis, Tilsiter Str 16, 53879 Euskirchen, Germany

DJ8OT, Eberhard Warnecke, PO Box 10 12 44, 42512 Velbert, Germany

DJ8VC, Alfons Niehoff, Ernst-Hase-Weg 6, 48282 Emsdetten, Germany

DK7ZT, Bernd Muller, Weitershauser Str 11, 35041 Marburg, Germany

DL1YCA, Dieter Petring, Bruderstr 52, 32584 Lohne, Germany

DL6YBY, Uwe Lusmoller, PO Box 150, 45713 Hallern, Germany

DL8JS, Walter Hymmen, PO Box 1925, 32219 Bunde, Germany

DL9HC, Wolfgang Landgraf, Weidenstr 18, 68526 Ladenburg, Germany

DL9XW, Hans-Peter Gunther, PO Box 1406, 48504 Nordhorn, Germany

DIG Membership List

The DIG membership list will be published during the first quarter of each year and may be obtained from the DIG secretary, DJ8OT. A return mail label and 4 IRCs for surface mail (or 8 IRCs for air mail) are required for all countries.

DIG Diplom 77

The DIG Diplom 77 is a DARC-recognized award and may be applied for by all stations. All contacts after January 1, 1977 count. Applicants need to have worked (heard) 77 DIG members in at least 7 different countries, but only 49 (7 × 7 = 49) DIG members in the applicant's own country.

Mixed, CW and VHF awards are available. No QSLs required. Send GCR list, return mail label and $7 US or 10 IRCs to DARC, Postfach 1155, D-3507, Baunatal 1, Germany.

European Prefixes Award (EUPXA)

The EUPXA is a DARC-recognized award and may be applied for by all stations who can prove contacts with 100 European stations with different prefixes. Prefixes are defined according to WPX rules. Endorsements are available for each additional 50 prefixes.

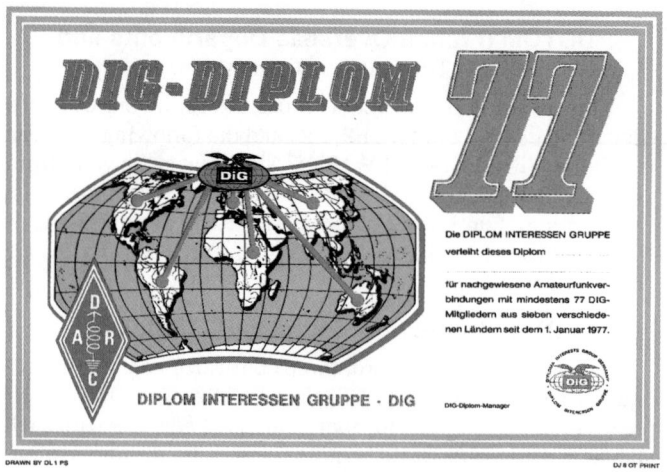

Mixed, CW and VHF awards are available. No QSLs required. Send a GCR list, return mailing label and $7 US or 10 IRCs (endorsements only require a label and 1 IRC) to the Award Manager: DJ8VC

Familia Award

The Familia Award is a DARC-recognized award and may be applied for by all stations. The applicant must obtain 100 points after January 1, 1980. All contacts with at least two licensed amateurs of the same family count 1 point for each contact. Mixed, CW and VHF awards are available. No QSLs required. Send a GCR list, return mailing label and $7 US or 10 IRCs to DARC, Postfach 1155, D-3507, Baunatal 1, Germany.

Germany Award

The Germany Award is a DARC-recognized award which commemorates the unification of West and East Germany into the Federal Republic of Germany.

This award may be applied for by all stations. The award requires proof of radio contacts with at least five different Amateur Radio stations in each German federal state, using at least two Amateur Radio bands, after 3 October 1990. No band or mode restrictions.

The 16 German federal states are: Baden-Wurttemberg, Bayern, Berlin, Brandenburg, Bremen, Hamburg, Hessen, Mecklenburg-Vorpommern, Niedersachsen, Nordrhein-Westfalen, Rheinland-Pfalz, Saarland, Sachsen, Sachsen-Anhalt, Schleswig-Holstein and Thuringen.

Mixed, CW and VHF awards are available. No QSLs required. Send a GCR list, return mailing label and $7 US or 10 IRCs to the Award Manager: DL1YCA

International Airport Award (IAPA)

The IAPA is a DARC-recognized award and may be applied for by all stations who can prove contacts with Amateur Radio amateur stations in 50 different towns that have airports with

scheduled international services. Only one contact may be made from the applicant's country. Contacts must be with fixed stations and must have taken place after January 1, 1973. All six continents must be worked.

Mixed, CW and VHF awards are available. No QSLs required. Send a GCR list, return mailing label and $7 US or 10 IRCs to the Award Manager: DL8JS.

Two Modes Award (TMA)

The TMA is a DARC-recognized award and may be applied for by all stations who have made CW contacts with 50 different countries (including Germany and six continents) and phone contacts with the *same* 50 countries.

Mixed, CW and VHF awards are available. No QSLs required. Send a GCR list, return mailing label and $7 US or 10 IRCs to DARC, Postfach 1155, D-3507, Baunatal 1, Germany.

Worked DIG Members (W-DIG-M)

The W-DIG-M may be applied for in one of 3 classes. You must be able to prove contacts with the required number of DIG members as shown below.

	DIG Members		
	Class 3	Class 2	Class 1
European stations must work	50	75	100
All other stations must work	15	30	50

Each member counts only once regardless of call sign used. Expeditions calls do not count separately. Endorsements are available for each additional 100 members worked. Endorsements are upgrades to Class 1 only.

Mixed, CW and VHF awards are available. No QSLs required. Send a GCR list, return mailing label and $7 US or 10 IRCs to the Award Manager: DH1PAL

Worked DX Stations (WDXS)

The WDXS is a DARC-recognized award and may be applied for by all stations who can prove contacts with Amateur Radio stations after January 1, 1964 as follows:

Class 4: European stations work 200 DX stations including 20 stations on 80/40 meters. All other stations work 200 European stations including ten stations on 80/40 meters

Class 3: European stations work 500 DX stations including 50 stations on 80/40 meters. All other stations work 500 European stations including 25 stations on 80/40 meters

Class 2: European stations work 1000 DX stations including 50 stations on 80/40 meters. All other stations work 1000 European stations including 50 stations on 80/40 meters

Class 1: European stations work 2000 DX stations including 100 stations on 40 meters and 20 contacts on 80 meters. All other stations work 2000 European stations including 100 stations on 40 meters and 20 contacts on 80 meters

Mixed, CW and VHF awards are available. No QSLs required. Send a GCR list, return mailing label and $7 US or 10 IRCs to the Award Manager: DL8JS

Worked German Large Cities (WGLC)

The WGLC may be applied for by all stations who have prove contacts with radio amateurs in large German towns (more than 100,000 inhabitants). The award is available in three classes:

	Class 3	Class 2	Class 1
European stations	20 cities	40 cities	60 cities
VHF stations	20 cities	30 cities	40 cities
All other stations	10 cities	20 cities	30 cities

Each town can be listed in the application only once. All contacts after January 1, 1962 count for this award. Mixed, CW and VHF awards are available. No QSLs required. Send a GCR list, return mailing label and $7 US or 10 IRCs to the Award Manager: DK7ZT.

The following large German towns count for WGLC:

Aachen, Augsburg, Bergisch-Gladbach, Berlin, Bielefeld, Bochum, Bonn, Bottrop, Braunschweig, Bremen, Bremerhaven, Chemnitz, Cottbus, Darmstadt, Dessau, Dortmund, Dresden, Dusseldorf, Duisburg, Erfurt, Erlangen, Essen, Frankfurt/Main, Freiburg, Furth, Gelsenkirchen, Gera, Gottingen, Hagen, Halle/Saale, Hamburg, Hamm, Hannover, Heidelberg, Heilbronn, Herne, Hildesheim Ingolstadt, Jena, Kaiserslautern, Karlsruhe, Kassel, Kiel, Koblenz, Koln, Krefeld, Leipzig, Leverkusen, Ludwigshafen, Lubeck, Magdeburg, Mainz, Mannheim, Monchengladbach, Mulheim/Ruhr, Munchen, Munster, Neuss, Nurnberg, Oberhausen, Offenbach, Oldenburg, Osnabruck, Paderborn, Pforzheim, Potsdam, Recklinghausen, Regensburg, Remscheid, Rheydt, Rostock, Saarbrucken, Salzgitter, Schwerin, Siegen, Solingen, Stuttgart, Trier, Ulm, Wanne-Eickel, Wiesbaden, Wilhelmshaven, Witten, Wolfsburg, Wurzburg, Wuppertal, Zwickau.

DIG TROPHIES

The top honors of the DIG are the DIG Trophy, the DIG CW Plaque, the DIG VHF Plaque and the DIG Trophy 1000. All applications for these awards must be sent to the Award Manager: DL9XW

All QSL cards from licensed DIG members count one point. Contacts with holders of a DIG Trophy and/or DIG plaques (specially marked in the DIG membership list) count two points. Contacts with DIG club stations and DIG special event stations (CR3DIG, DA0DIG, DF0DIG, DK0DIG, DL0DIG, DL0SDE, DL0XYL, DL0YL, HB9DIG, LX9DIG, OE1XDC, OK5DIG, OL5DIG, Pl4DIG, 4X4DIG, 4X25DIG) count three points. Stations can be mobile, fixed or portable. QSL cards must show the member's DIG number and/or the DIG logo.

You do not need to submit QSLs. Simply send a GCR list, return mailing label and $27 US or 20 IRCs for each award. There is no charge for the DIG Trophy 1000. The DIG Trophy and DIG Plaques will be sent after the call sign has been engraved. The DIG Trophy 1000 will be presented at a DIG meeting only.

DIG Trophy

The DIG Trophy will be awarded to an individual only once. To earn the trophy you must be able to prove that you are in possession of at least four different DIG awards as well as having obtained 500 points by collecting QSL cards from DIG members. The DIG Trophy is a solid brass wall plaque weighing 400 grams.

DIG CW Plaque

The DIG CW Plaque will be awarded to an individual only once. To earn the plaque you must be able to prove that you are in possession of at least three different DIG awards with a CW endorsement, as well as having obtained 250 points by collecting QSL cards from DIG members. All contacts have to be made in CW only by both operators. The DIG CW Plaque is red-and-gold solid brass weighing 400 grams.

DIG VHF Plaque

The DIG VHF Plaque may be awarded to an individual only once. You must possess at least three different DIG awards with VHF endorsements. In addition, you must obtain 250 points by collecting QSL cards from DIG members. All contacts must be made on the VHF/UHF bands only by both operators. The DIG VHF Plaque is blue-and-gold solid brass weighing 400 grams.

DIG Trophy 1000

This is the most prestigious DIG award available. To earn the DIG Trophy 1000 you must be able to prove that you are in possession of *all* DIG awards in the highest classifications. In addition, you must show proof of contact with at least 1000 different DIG members. There is no charge for this exclusive trophy.

DARC DX Awards

The official awards of the German Amateur Radio Club (DARC) are processed by the department for DX and HF Contesting. They may be claimed by all licensed amateurs and SWLs.

All contacts must be made from the same country. All cards must be in the applicant's possession. Any submission of forged or altered cards can lead to the disqualification from the awards program. Awards of any class may be endorsed as "YL-Award," "Via Satellite" or "QRP" if the applicant can submit the necessary credits. Awards for club stations will be issued under the club's name, not that of the applicant.

Applicants from outside Germany must submit a GCR list (a list of cards verified by the applicant's local club or by an official Awards Manager). If in doubt, the manager may require the submission of some or all the cards for verification. Applications should be sent to the respective Award Manager. New holders of the awards will be published in the official journal of the DARC, *CQ-DL*.

The service charge for any one of the awards is $22.50 US. For an application for one or more stickers, $7.50 US. For any of the plaques, if sent by mail, $60 US ($37.50 if claimed personally at the HAM Radio Convention in Friedrichshafen). Rules, the country list and application forms can be obtained from the managers by sending a return mailing label plus return postage (dollars or IRCs).

It is best to use the official application forms issued by the DARC, but applications containing all the relevant data will likewise be accepted. A complete set of application forms is available for $6 US or 5 IRCs from Eberhard Warnecke, DJ8OT, Postfach 10 12 44. Velbert, D-42512, Germany.

EU-DX Award

The basic idea of this award is a proportional combination of European and DX contacts in **one calendar year**. It is issued in the following classes: 2×CW. 2×SSB and mixed modes.

A minimum of 50 points is required for the EU-DX Award per year; 20 points must be obtained by contacts with European countries and 30 points by contacts with countries outside Europe. Each country counts one point, two points on 80 and 160 meters. All bands can be used. European countries are defined in the WAE country list; non-European countries in the valid DXCC list.

Bonus points can be obtained for any additional block of four European and six non-European countries within the same calendar year. Each year's score may be added to obtain the EU-DX 500 badge. Stickers will be awarded for any block of 50 points, regardless of the annual scores.

Example:

20 EU points in 1994 + 30 DX the same year = 50 points for that year
28 EU points in 1993 + 42 DX the same year = 70 points for that year
32 EU points in 1992 + 48 DX the same year = 80 points for that year
Total: 200 points (four stickers)

EU-DX 1000 Trophy (Plaque)

The trophy is awarded for a total of 1000 points, no matter how many years you need to reach the total. EU-DX Manager: Dietmar Kasper, DL3DXX, Birkenweg 3, Pirna-Jessen, D-01796, Germany.

EUROPE Award

This certificate is awarded for working or hearing amateurs in **European countries** for a total score of at least 100 points. The idea is to work or hear as many European countries as possible on different bands in different calendar years. There are no band or mode restrictions.

Each confirmed European country counts one point per amateur band. Only the current and the five previous years are valid. Contacts older than five years are not valid. The total score must reflect the sum of the annual score for the year of application, and the five preceding years.

EUROPE Award Honor Roll

Each certificate holder with an actual score of at least 300 points becomes eligible for the Honor Roll, which is published regularly in *CQ-DL*. Cards can be submitted every year, before December 1, to be considered for the Honor Roll.

EUROPE 300 Trophy (Plaque)

Holders of the Europe Award may obtain the Europe 300 Trophy. Applicants must achieve 300 country points counting each country on each band only once.

Examples:
50 countries on 6 bands = 300 points
10 countries on 9 bands + 20 countries on 6 bands + 15 countries on 5 bands + 5 countries on 3 bands = 300 points. Europe Award Manager: Franz Berndt, DL9GFB, Heinrich-Heine-Str. 1. Bad Doberan, D-18209.

WORKED All Europe (WAE)

The oldest and most renowned of all DARC certificates is awarded for contacts with amateur stations in European coun-

tries and European islands. Each confirmed country counts one point per band, with a maximum of five claimed bands per country. DX stations may count two points for any contact on 160 or 80 meters. The award is issued in two divisions: two-way telegraphy (CW) and two-way phone (SSB/AM/FM).

Classes:
WAE III: at least 40 counties and 100 points
WAE II: at least 50 countries and 150 points
WAE I: at least 55 countries and 175 points
Holders of the WAE I get a special WAE badge.

WAE Top Trophy (Plaque)

Awarded to stations who have worked at least 70 countries and earned 300 points

Unlike WAE Classes I through III, deleted countries may not be counted for WAE Top plaque. WAE Award Manager: Michael Adaszewski, DL5ARX, Postfach, Suhl 3, D-98509, Germany

WAE COUNTRY LIST

deleted countries:

1	9S4	Saar	from 08.11.1947	to 31.12.1956
2	I1	Trieste		to 31.03.1957
3	UN	Karelo-Finnish Rep.		to 30.06.1960
4	DL	Germany		to 16.09.1973
5	Y2	GDR	from 17.09.1973	to 02.10.1990
6	UAl N	Karelia	from 01.07.1960	to 31.12.1991
7	OK	CSSR		to 31.12.1992

valid countries:

1	C3	Andorra
2	CT	Portugal
3	CU	Azores
4	DL	Germany
5	EA	Spain
6	EA6	Balearic Islands
7	EI	Ireland
8	ER	Moldova
9	ES	Estonia
10	EU	White Russia
11	F	France
12	G	England
13	GD	Isle of Man
14	GI	Northern Ireland
15	GJ	Jersey
16	GM	Scotland
17	GM	Shetland Islands
18	GU	Guernsey&Dependencies
19	GW	Wales
20	HA	Hungary
21	HB	Switzerland
22	HBØ	Liechtenstein
23	HV	Vatican
24	I	Italy
25	IS	Sardinia
26	IT	Sicily
27	1W	Bear Island
28	JW	Svalbard
29	JX	Jan Mayen
30	LA	Norway
31	LX	Luxembourg
32	LY	Lithuania
33	LZ	Bulgaria
34	OE	Austria
35	OH	Finland
36	OHØ	Aland Islands
37	OJØ	Market Reef
38	OK	Czech Republic after 01.01.1993
39	OM	Slovak Republic after 01.01.1993
40	ON	Belgium
41	OY	Faroe Islands
42	OZ	Denmark
43	PA	Netherlands
44	R1/FJL	Franz Josef Land
45	Rl/MVI	Malyj Vysotskij Island
46	RA	Russia (European part)
47	RA2	Kaliningradsk
48	S5	Slovenia after 15.01.1992
49	SM	Sweden
50	SP	Poland
51	SV	Greece
52	SV5	Rhodes
53	SV9	Crete
54	SY	Mt. Athos
55	T7	San Marino
56	T9	Bosnia-Herzegovina after 07.04.1992
57	TA1	Turkey (European part)
58	TF	Iceland
59	TK	Corsica
60	UR	Ukraine
61	YL	Latvia
62	YO	Romania
63	YU	Yugoslavia
64	Z3	Macedonia
65	ZA	Albania
66	ZB	Gibraltar
67	1A	S.M.O.M.
68	3A	Monaco
69	4U/ITU	UN Geneva
70	4U/VIC	UN Vienna
71	9A	Croatia after 15.01 1992
72	9H	Malta

GIBRALTAR (ZB2)

The *ZB2 Award* is issued to amateurs who work at least five different ZB2 stations on any band at any time. QSLs are not necessary; certified log entries are acceptable. Send log and $5 US to defray costs. The *ZB2BU Award* is issued to amateurs who work Gibraltar Amateur Radio Society club station ZB2BU on three different bands (one contact per band is sufficient). Otherwise, same rules apply to the ZB2 Award. Send applications to: Awards Manager, Gibraltar Amateur Radio Society, PO Box 292, Hargraves Parade, Gibraltar.

GREECE (SV)

The Athenean Award

Issued for contacts with 25 Amateur Radio stations in the Athens area. The diploma is issued in any modes and in three classes:

1st Class—160 and 80 meters
2nd Class—40 and 30 meters
3rd Class—all other bands

Endorsement stickers are available for each 25 new contacts.

You do not need to send your QSLs. Applications must be accompanied by a certified list of QSLs (checked and signed by two amateurs), together with 10 IRCs (4 IRCs for each endorsement sticker), or $5 US. Send your application to: RAAG, Award Manager, PO Box 3564, Athens GR-102 10, Greece.

Greek Islands Award

The award is available for working and confirming 10 contacts with at least three different groups of Greek islands as follows:

Criti
Dodekanisos
Ionian
Evia
Kiklades
Sporades
Lesvos
Chios
Thasos-Samothraki
Ikaria-Limnos

Awards are issued for two-way SSB, two-way CW, mixed mode, single mode, mixed band or single band.

You do not need to send your QSLs. Applications must be accompanied by a certified list of QSLs (checked and signed by two amateurs), together with 10 IRCs (4 IRCs for each endorsement sticker), or $5 US. Send your application to: RAAG, Award Manager, PO Box 3564, Athens GR-102 10, Greece.

RAAG Award

The award is available for working and confirming contacts with at least seven different Greek stations from the nine SV call areas. You do not need to send your QSLs. Applications must be accompanied by a certified list of QSLs (checked and signed by two amateurs), together with 10 IRCs (4 IRCs for each endorsement sticker), or $5 US. Send your application to: RAAG, Award Manager, PO Box 3564, Athens GR-102 10, Greece.

GUATEMALA (TG)

The Club de Radioaficionados de Guatemala sponsors the *WATG Award* for working all TG zones (consisting of TG4-TGØ). Send either the QSL cards or photocopies to the Awards Manager, along with 10 IRCs (or $10 US) to Awards Manager,

Club de Radioaficionados de Guatemala (CRAG), PO Box 115, Guatemala City, Guatemala.

HONG KONG (VS6)

To apply for awards sponsored by the Hong Kong Amateur Radio Transmitting Society (HARTS), send application and

certified log extracts (no QSL cards required) to: Awards Manager, HARTS, GPO Box 541, Hong Kong.

Catch 22 Award

To qualify, the applicant must submit verified evidence of two-way contacts with amateur stations located in the 22nd Parallel of Latitude North. The award is issued in three classes:

Class 3—contacts with at least 15 countries.
Class 2—contacts with at least 20 countries.
Class 1—contacts with at least 25 countries.

A VS6 contact is mandatory. Contacts must be made after January 1, 1980. Band/mode endorsement stickers are available. Fee: $7 US; endorsements $1 US.

22nd Parallel countries: A4, A6, BV, BY, C6, CN, CO, HZ, KH6, S2, ST, SU, TT, TZ, VS6, VU, XE, XW, XX9, XZ, 3W, 5A, 5T, 5U, 7X.

Nine Dragons Award

To qualify, work one country in Zone 18, 19 and 24-30, for a total of nine zones. The Zone 24 contact must be with a VS6. Contacts are valid after January 1, 1979. Fee: $3 US or 25 IRCs.

Firecracker Award

To qualify, work six different VS6 stations. Contacts are valid after January 1, 1964. Fee: $2 US or 10 IRCs.

HUNGARY (HA)

For awards sponsored by the Hungarian Radioamateur Society (MRASZ), all amateur bands and modes may be used, except repeaters. Each station may be contacted only once on any band and any mode. Contacts may be made from any location within the same DXCC country. Applications must include a log showing the name, call sign and QTH of applicant, and the following information: station worked, UTC, band, mode, report. With the exception of Szeged Festival, Dunaferr and HCS awards, each application must be accompanied by a statement from the applicant's National Society, or from any two amateurs, that the valid QSL cards listed are in the possession of the applicant. Applications must be sent to the individual Awards Managers as indicated below. General correspondence may be sent to: Hungarian Radioamateur Society, Awards Committee, PO Box 22, Tiszakecske H-6061, Hungary.

BD (Balaton Diploma)

This award is issued by the Radioclub Siofok for QSOs made on or after January 1, 1967. The applicant must make 15 points, including at least one contact with a Siofok club member (Europeans must make 30 points, including at least two club members). Each club member counts five points. Fee: 10 IRCs.

Siofok club members (counting five points) are as follows: HA/HG 3s KGJ KHL GI GJ GQ HE HL HQ HZ IG IK IQ IS NG; HA4XW; HA6NP; HA8UA.

The following stations, residing in the Lake Balaton area, count three points: HA/HG 1s KXX XA XX ZY; 2s KRQ RC RQ SH Y YRC; 3s KHB KHO GG GO HK HO HU.

The following stations, in Zala, Veszprem and Somogy county, count one point: HA/HG 1s KRA-KRZ, KXA-KXZ, KZA-KZZ, RA-RZ, XA-XZ, ZA-ZZ, DRA-DZZ; 2s KPA-KTZ, PA-TZ, ENA-EZZ; 3s KGA-KIZ, GA-IZ, FLA-FSZ.

Awards Manager: Jozsef Turjanyi, HA3GJ, PO Box 78, Siofok H-8601, Hungary.

BPA (Budapest Award)

This award is issued by the Radioamateur Society of Budapest for QSOs with HA5/HG5 stations made on or after January 1, 1959. Requirements:

• European applicants must work 75 different HA/HG5 stations
• Applicants outside Europe must work 25 different HA5/HG5 stations.

There is no restriction as to the use of repeaters, satellites or moon reflections over 500 km. Fee: 10 IRCs.

Awards Manager: Verebes Janosne, HA5YR, PO Box 64, Budapest H-1475, Hungary.

DD (Dunakanyar Diploma)

This award is issued by the Radioamateur Society of Pest County for QSOs with HA7/HG7 stations made on or after January 1, 1970. The applicant must confirm five HA7/HG7 contacts (Europeans must confirm 20). Fee: 6 IRCs.

Awards Manager: HA7PL, PRASZ, PO Box 36, Budapest H-1387, Hungary.

HCS (Hungarian Castle Series)

This award, granted for QSOs made on or after January 1, 1968, is issued directly by the MRASZ in three classes:

Class I, Gold—for all QSLs, number 1-36.
Class II, Silver—for QSLs numbers 1-24 or 12-36.
Class III, Bronze—for QSLs numbers 1-12 or 13-24 or 25-36.

QSL numbers by district:
HA1/HG1 7-22-25-31.
HA2/HG2 6-8-12-15-21-23-30-32-35.
HA3/HG3 3-14-23-30-32-33-35.
HA4/HG4 17-23-30-32-35.
HA5/HG5 1-13-36.
HA6-HG6 4-10-11-34.
HA7/HG7 2-5-19.
HA8/HG8 16-20-24.
HA9/HG9 18-27-28-29.
HA0/HG0 9-26-29.

Applicants must include the QSL number "slice," cut from each QSL card, along with the application and the appropriate number of IRCs (5 for bronze, 8 for silver, 10 for gold).

Awards Manager: Janos Retkes, HA8UB, PO Box 22, Tiszakecske H-6061, Hungary.

Pannonia Award

This award is issued by the Radioamateur Society of Gyor-Sopron County for contacts made on or after January 1, 1966. The applicant must submit proof of two QSOs each with HA/HG1, HA/HG2, HA/HG3, and HA/HG4, for a total of eight contacts, on more than one band. Fee: 10 IRCs.

Awards Manager: Radioclub HA1KSA, PO Box 79, Gyor H-9001, Hungary.

Savaria Award

This award is issued by the Radioamateur Society of VAS County for contacts with HA1/HG1 stations made on or after January 1, 1976. The applicant must confirm 10 different HA1/HG1 QSOs (Europeans must confirm 20). Fee: 10 IRCs. Awards Manager: Savaria Radioclub, Puskas T.u.7, Szombathely H-9700, Hungary.

Szeged Festival Award

This award is issued yearly by the Radioamateur Society of

Csongrad County for QSOs made between July 1 and August 31. The applicant must make five points (Europeans, 10 points). Fee: 5 IRCs.

Stations in Szeged count two points: HA/HG8 CA CB CD CH CP CT CV CX CZ DC DE DF DP DQ DR DT DZ EK EL KCC KCK KDA.

Stations in Csongrad County (HA/HG8 CA-FZ, KCA-KFZ, LSA-LZZ) count one point.

Awards Manager: Imre Kelemen, HA8CH, PO Box 673, Szeged H-6701, Hungary.

VTA (Videoton Award)

This award is issued by the Videoton Radioclub for QSOs with HA4/HG4 stations made on or after January 1, 1969. Contacts must be confirmed by special QSL cards representing broadcast (BC) and TV receivers and computers respectively (there are three groups of special cards, 3-4-3).

Classes:
Gold—all ten cards [fee: 5 IRCs]
Silver—complete set of any two groups. [fee: 3 IRCs]
Bronze—one complete set of any group [fee: 2 IRCs]

Awards Manager: Halmi Belane, HA4XP, Berkes Lpt 40, Szekesfehervar H-8000, Hungary.

HRD (Hungarian Rummy Diploma)

The Rummy awards are issued by the Radio Amateur Society of Somogy County. Contacts are valid on or after September 1, 1972 [for HRD-108 and HCD, contacts are valid on or after April 4, 1980]. Fee: 5 IRCs.

The *basic HRD award* is issued in three degrees:
Bronze degree—"hand rummy," collecting 14 cards in accordance with the rules of the game.
Silver degree—full collection of one of the four series, plus one joker of the same color.
Gold degree—full pack, 54 cards.

The *HRD-108* award requires two packs of QSL cards from 108 different stations. For the *Hungarian Canasta Diploma (HCD)*, three canastas, 21 cards, have to be confirmed from different call signs in accordance with the rules of the game. The canasta contains seven cards of the same figure; two of them can be equivalent (eg, seven cards of figure 5, seven cards of figure 8 and seven cards of kings). Not more than three cards can be substituted by the four jokers and the eight "little jokers" in one canasta.

Allocation of the HFD cards:

HA/HG Call Areas	Spades/Hearts R/a	Diamonds/Clubs R/b
1	A	J
2	2	J
3	3	J
4	4	J
5	5	Q
6	6	Q
7	7	K
8	8	K
9	9	K
0	10	K
???	red and black Joker = Y =	

Awards Manager: Dr Janos Mihalyfy, HA3GA, PO Box 173, Kaposvar H-7401, Hungary.

WHD (Worked Hungarian Districts)

This award is issued directly by MRASZ for QSOs made with HA/HG stations on or after January 1, 1958. The applicant must confirm two QSOs with at least five different HA call districts on two bands, for a total of 10 contacts (Europeans must confirm two QSOs with at least eight different districts). Fee: 5 IRCs.

Awards Manager: HA8UB, PO Box 22, Tiszakecske H-6061, Hungary.

HADXC (HUNGARIAN DX CHAPTER) AWARDS

The general rules for HADXC awards are essentially the same as for MRASZ awards. The fee for all HADXC awards is 12 IRCs. Send all correspondence to: HADXC Awards Manager, HA8UB (see address above).

WHADXCA (Worked Hungarian DX Chapter Award)

Applicants must work HADXC members on or after January 1, 1986, for a total of 15 points (each club member counts one point, and club station HG3DXC counts two points). Europeans must make 20 points.

WAHA-HF (Worked All Hungarian Award, HF)

This award, for contacts made on or after January 1, 1980, is issued in three classes:

Class III—250 points required
Class II—500 points required
Class I—1000 points required

Contacts with club station HG3DXC counts 10 points, HADXC members count five points, and all other Hungarian stations count one point.

HADXC Members: HG1S, HA1TJ, HG1W, HA1YA, HG1Z, HA2RD, HG3DXC, HA3GJ, HA3NU, HA4XH, HA4XT, HA4XX, HA4YD, HG5A, HA5AM, HA5DW, HA5FM, HA5HO, HA5MY, HG6N, HA6NF, HG6V, HG7B, HA8ET, HG8U, HA8UB, HG9R, HA9RU, HAØLZ, HAØMM.

ICELAND (TF)

The *Icelandic Radio Amateurs Award (IRAA)* is available to amateurs outside of Iceland. Only contacts with amateurs who are Icelandic citizens operating from Icelandic territory are valid. There is no date limit for QSOs. QSL cards or certified photocopies of the QSL cards must be submitted together with a complete list of the log entries for the contacts. Mixed-mode and crossband contacts are not valid, except for amateur satellites. Each TF station may be contacted only once per band per mode. [TF Novices run 5 watts input and are permitted to operate on three bands, 3500-3600 kHz, 7000-7040 kHz and 21100-21150 kHz, and are identified by a three-letter suffix ending in N.] Total points needed to qualify for the award depend upon the applicant's ITU zone:

I. ITU Zones 5, 9, 18-20, 27-29	= 98 points
II. ITU Zones 1-4, 6-8, 21-26, 30, 31, 36, 37	= 48 points
III. ITU Zones 10-13, 32-35, 38-40, 46-48	= 28 points
IV. ITU Zones 14-16, 41-45, 49-75	= 18 points

Points by band and mode are calculated according to the following chart:

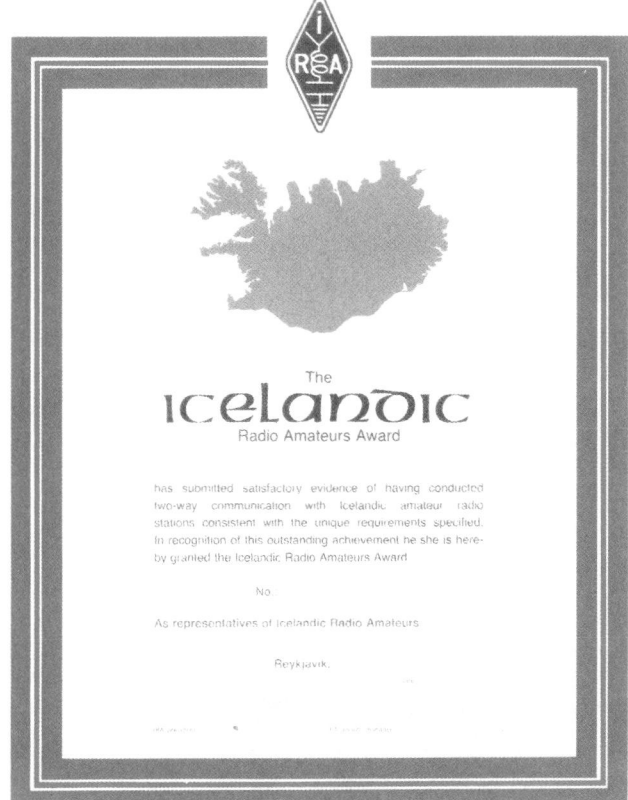

Norway (LA), Finland (OH), Aland Islands (OHØ), Marked Reef (OJØ), Greenland (OX), Faroe Islands (OY), Denmark (OZ), Sweden (SM) and Iceland (TF). One confirmed contact with Iceland is mandatory. There are no band limitations, but all QSOs must use the same mode. The award is issued in three classes:

Class A—One confirmed contact with each of the countries listed above.
Class B—One confirmed contact with eight of the countries listed above as well as two others in CQ Zone 40.
Class C—One confirmed contact with five of the countries listed above as well as one other in CQ Zone 40.

Contacts must be between fixed stations only.

A list verified by two radio amateurs or local club officials must be submitted. The list must show: stations worked, date, year, time (UTC), band, report and mode. The fee is 8 IRCs or $5 US.

Send applications to: Brynjolfur Jonsson, TF5BW, PO Box 121, IS-602 Akureyri, Iceland.

Zone 40

This award is based on confirmed contacts with the following DXCC countries located in CQ Zone 40: Svalbard (JW), Jan Mayen (JX), Franz Josef Land (UA1), Greenland (OX), and Iceland (TF). There are no band limitations, but all QSOs must use the same mode.

To qualify for the award, you must be able to confirm at least one contact with a station located in each of the countries listed above. (European stations must work three Icelandic stations.) Contacts must be between fixed stations only.

A list verified by two radio amateurs or local club officials must be submitted. The list must show: stations worked, date, year, time (UTC), band, report and mode. The fee is 15 IRCs or $10 US.

Send applications to: Brynjolfur Jonsson, TF5BW, PO Box 121, IS-602 Akureyri, Iceland.

INDIA (VU)

The WRI (Worked All India) award is issued by the Amateur Radio Society of India. It is awarded for contacts on any HF band 160 to 10 meters with stations in mainland India, the Laccadives, and the Andaman and Nicobar islands.

Original QSL cards or certified logs must be submitted to the awards manager along with a letter stating the award desired.

Division of Points by Bands and Modes

Band	Novice	CW	RTTY	SSTV	SSB	Via Satellite
1.8		10	8	8	6	
3.5	32	8	6	6	4	
7	24	6	5	5	3	
10		5	4	0	0	
14		3	2	2	1	8
18		4	3	3	2	
21	16	5	4	4	3	
24		6	5	5	4	
28		7	6	6	5	

Send application materials and 14 IRCs to: Islenzkir Radioamatorar (IRA), Awards Manager, PO Box 121, IS-602 Akureyri, Iceland.

Worked All Nordic Countries

This award is based on confirmed contacts with the following Nordic DXCC countries: Svalbard (JW), Jan Mayen (JX),

Contacts with the same Indian stations on different bands will count for points, but different modes in the same band do *not* count. QSOs must take place between two fixed stations. Contacts with portable, maritime or aeronautical mobile stations are not eligible. A minimum of 100 points are required to earn the award.

Award Points

Band	VU2, VU3	VU4, VU7	Special prefix stations
160	3	5	4
80	2	4	3
40	2	4	3
20—10	1	3	2

The award fee is 5 IRCs. Send all applications to:

Vice President and Awards Manager, Amateur Radio Society of India, 40 Ghalib Apts, Parwana Rd, Pitampura, Delhi 110 034, India.

INDONESIA (YB)

The general rules for awards sponsored by ORARI (Organisasi Amatir Radio Indonesia) are as follows: Awards are issued for two-way SSB, two-way CW, mixed mode or single mode, mixed or single band. To be valid, contacts must be made on or after July 9, 1968. Only contacts with land stations are valid. Applications must be accompanied by a list of QSL cards in alphabetical order indicating station worked, dates, bands and modes, and a statement from the applicant's National Society, club station or from any two amateurs, stipulating that the correct QSL cards of the contact listed are in the possession of the applicant. Include $8 (US) or 16 IRCs with each application. Send applications to: ORARI National QSL & Awards Bureau, PO Box 96, Jakarta 10002, Indonesia.

JA (Jakarta Award)

To qualify, work a total of 20 stations, including at least one club station in Jakarta (Ø call area), the capital of Indonesia. [YBØ club stations: YBØZAA, YBØZAB, YBØZAD, YBØZAE, YBØZAF, YBØZBA, YBØZBB, YBØZCA, YBØZCB, YBØZCD, YBØZCE, YBØZDB, YBØZDC, YBØZDD, YBØZDE, YBØZDG, YBØZEA, YBØZEE, BYØZZ.]

WAIA (Worked All Indonesia Award)

To qualify, work two stations in each of the Indonesian call areas (1-Ø, a total of 20 contacts.

WTEA (Worked the Equator Award)

To qualify, confirmed contacts are required with the DXCC countries along the equator: C2, HC, HC8, HK, KH1/KB6, PP-PY, PYØ (St Peter & Paul), S9, T3Ø, T31, T32, T5, TN, TR,

YB5, YB7, YB8, 5X, 5Z, 8Q, 9Q [YB5, YB7, YB8 contacts are mandatory]. The WTEA is issued in three classes:

Class I—for confirmed contacts with 15 countries
Class II—for confirmed contacts with 12 countries
Class III—for confirmed contacts with 8 countries

IRELAND (EI)

The Irish Radio Transmitters Society (IRTS) has just introduced its first award, *WEIC*. The WEIC Award is available to licensed amateurs who have worked stations located in at least 20 of the 26 different counties of Ireland (EI/EJ). Contacts from January 1, 1982 are valid. Applications for the WEIC

Award must be accompanied by 10 IRCs, a QSO list and a statement from the applicant's National Society Awards Manager that correctly filled-in QSL cards are in the possession of the applicant. If this is not possible, the applicant must submit all QSLs concerned.

Counties of Ireland: Carlow, Cavan, Clare, Cork, Donegal, Dublin, Galway, Kerry, Kildare, Kilkenny, Laois, Leitrim, Limerick, Longford, Louth, Mayo, Meath, Monaghan, Offaly, Roscommon, Sligo, Tipperary, Waterford, Westmeath, Wexford, Wicklow.

Send applications to: Irish Radio Transmitters Society, PO Box 462, Dublin 9, Ireland.

ISRAEL (4X)

The following awards are sponsored by the Israel Amateur Radio Club (IARC). Applications must be accompanied by a certified list of QSLs (checked and signed by two amateurs) together with 7 IRCs and should be sent to: Mark Stern, 4Z4KX, Israel Awards Manager, Israel Amateur Radio Club, PO Box 17600, 61176 Tel-Aviv, Israel.

The 4 × 4 = 16 Award

This award can be obtained by sending a certified list of at least 16 QSLs that confirm contacts with 16 different Israeli stations. No restrictions as to modes or bands, but contacts must be made on at least four different bands.

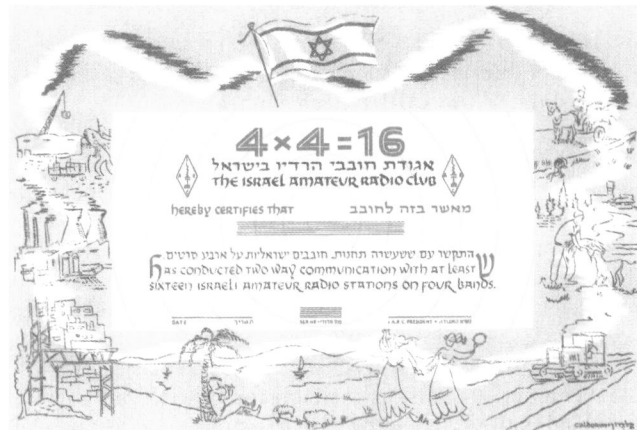

The Israel Award

This award can be obtained by making 25 points, as follows: Each QSO with an Israeli station above 10 MHz counts one point, and each contact below 10 MHz counts two points. Each station may be worked only once per band. No restrictions as to modes or bands, but contacts must be made after January 1, 1983.

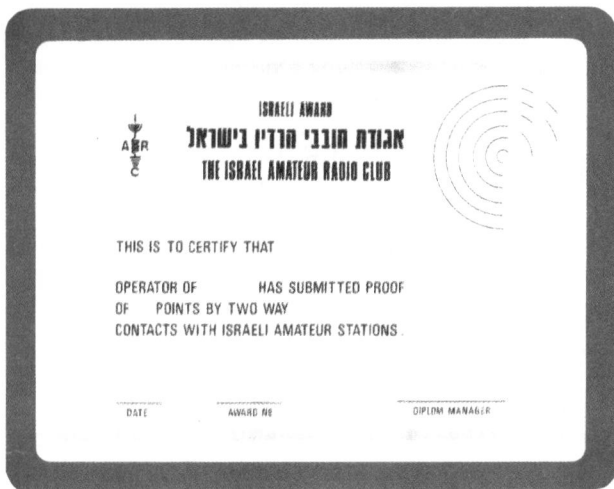

The Jerusalem Award

Send a certified list of four contacts with stations located in Jerusalem. Each certificate will be signed by the mayor of Jerusalem. Applications and 7 IRCs should be sent to Dr Milton Gordon, 4X6AA, Box 4079, 91040 Jerusalem, Israel.

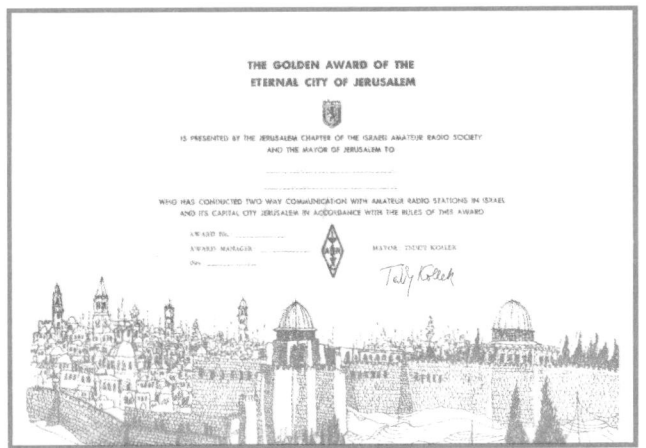

The Tel-Aviv Award

This award is issued in commemoration of the anniversary of the city of Tel-Aviv. To obtain this award, 15 points are required, as follows: Stations in Tel-Aviv count one point, stations located in Jaffa count two points, special-events station 4X75TA counts five points. No restrictions as to bands or modes, but each station may be worked only once on each band, and all contacts must be made after January 1, 1984. Each certificate will be signed by the mayor of Tel-Aviv. Applica-

tions with certified log extract and $3 US should be sent to Shlomo Mussali, 4X6LM, PO Box 8225, 61081 Tel Aviv-Jaffa, Israel.

The Kiryat Ono Award

This award is available to any amateur who works at least four stations located in Kiryat-Ono. Valid contacts must take place after March 1, 1990 on any HF band or mode. Send certified list and $3 US or 7 IRCs to: Joseph Obstfeld, 4X6KJ, PO Box 873, Kiryat-Ono 55000, Israel.

The Rishon Le Zion Award

This award can be obtained by working at least seven stations located in Rishon Le Zion. Valid contacts must take place after January 1, 1990 on any HF band or mode. Send certified list with $3 US or 7 IRCs to: Ehud Levin, 4X1EL, PO Box 466, Rishon Le Zion, 75103, Israel.

Nazareth Illit "Shalom" Award

The award features landmarks of the City of Nazareth and is signed by the mayor of the city. To earn the award, you must collect at least 35 contact points. 4Z4SZ, located in Nazareth, is an obligatory contact and counts for 10 points. Stations 4X4CD, 4Z4KX, 4X1AT, 4X6XJ, 4Z4RJ and 4Z4RM count for 5 points each. All other Israeli stations count as point each. No band or mode limitations. Send your log with $3 US or 5 IRCs to: 4Z4SZ, PO Box 1144, Nazareth Illit 17000, Israel.

The Holyland Award

This attractive award is printed on gold-anodized aluminum, silk-screened in two colors showing a panorama of Jerusalem. Each award is personally printed for the winner. Complete rules and qualifications are available from M. Webman, 4X4JU, PO Box 8181, Petah Tiqwa 49651, Israel.

ITALY (I)

To apply for the following awards sponsored by Associazione Radioamatori Italiani (ARI), send all applications to: ARI Awards Manager, Via Scarlatti 31, 20124 Milano, Italy.

WAIP (Worked All Italian Provinces)

The WAIP is awarded for confirmed contacts with fixed stations in at least 60 Italian provinces (75 provinces for Italian amateurs). Contacts must be made since January 1, 1949, and the same station may be worked more than once if located in different provinces. Endorsements are available for CW, phone, mixed and RTTY. Minimum signal report 33(8). A log extract and 10 IRCs (or $3 US) must be included with the application; QSLs are not necessary if the log extract is certified by the Awards Manager of the applicant's National Society or two amateurs.

Italian provinces are as follows:

Agrigento, Alessandria, Ancona, Aosta, Arezzo, Ascoli P, Asti, Avellino, Bari, Belluno, Benevento, Bergamo, Bologna, Bolzano, Brescia, Brindisi, Cagliari, Caltanissetta, Campobasso, Caserta, Catania, Catanzaro, Chieti, Como, Cosenza, Cremona, Cuneo, Enna, Ferrara, Firenze, Foggia, Forli, Frosinone, Genova, Gorizia, Grosseto, Imperia, Isernia, L'Aquila, La Spezia, Latina, Lecce, Livorno, Lucca, Macerata, Mantova, Massa, Matera, Messina, Milano, Modena, Napoli, Novara, Nuoro, Oristano, Padova, Palermo, Parma, Pavia, Perugia, Pesaro, Pescara, Piacenza, Pisa, Pistoia, Pordenone, Potenza, Ragusa, Ravenna, Reggio C, Reggio E, Rieti, Roma, Rovigo, Salerno, Sassari, Savona, Siena, Siracusa, Sondrio, Taranto, Teramo, Terni, Torino, Trapani, Trento, Treviso, Trieste, Udine, Varese, Venezia, Vercelli, Verona, Vicenza, Viterbo.

CDM (Certificato del Mediterraneo)

The CDM is awarded for confirmed contacts with fixed stations in at least 22 Mediterranean countries excluding Italy, plus at least 30 amateur stations located on the Italian mainland. Contacts must be made since June 1, 1952, and stations may be worked only once. Endorsements are available for phone, CW, mixed and RTTY. Minimum signal report —33(8). A log extract and 10 IRCs (or $3 US) must be included with application; QSLs are not necessary if the log extract is certified by either the Awards Manager of the applicant's National Society or two amateurs.

CDM countries: EA, EA6, EA9, CN, F, ISØ, IT, I/Trieste

(before December 31, 1957), OD5, SU, SV, SV5, SV9, TA, TK, YK, YU, ZA, ZB, ZC4/5B, 3A, 3V, 4X, 5A, 7X, 9H.

IIA (Italian Islands Award)

Applicants must work and confirm contacts with Italian islands in the following archipelagos: IA1—Liguri Group, IA5—Tuscan Arch, IBØ—Ponziano Arch, IC8—Napoli Arch, ID9—Eolie Arch, IE9—Ustica, IF9—Egadi Arch, IG9—Pelagie Arch, IH9—Pantelleria, IJ7—Cheradi Arch, IL7—Tremiti Arch, IMØ—Maddalena Arch, ISØ—Sardinia, IT9—Sicily. Contacts must be made starting January 1, 1970, on SSB, CW, RTTY or mixed modes.

To qualify, applicants must make 10 points (Italian applicants must make 40 points, all other Europeans 20 points). In general, each island counts one point, but some count two-or-more points. Further details can be obtained from the ARI Awards Manager for a self-addressed envelope and one IRC. Point values usually appear on the QSL cards. Note: Each island counts one-or-more points even if in the same archipelago, and islands can be worked more than once if a different mode is used. *IIA Honor Roll* requires 60 points. *Five-Band IIA* requires 10 contacts with 10 different Italian islands on each of five HF bands. QSLs are normally not required, but must be available on request. Application and log extract must be accompanied by 30 IRCs (or $8 US).

JAPAN (JA)

General information for awards issued by the Japan Amateur Radio League (JARL): Each application must be accompanied by a list of QSL cards, stating the call signs of each station worked, date, band and mode. Each list must be accompanied by a statement from the applicant's National Society or from any two amateurs that the valid QSL cards are in the possession of the applicant. If such a statement is not available, the applicant must submit the actual QSL cards. Eight IRCs are required for each award (2 additional IRCs for airmail delivery). If QSL cards are submitted, sufficient funds for their return must also be included. A list of cities and guns, along with the Asian Countries list, is available for 3 IRCs. Only contacts with land stations (including mobile stations on a river or lake) are valid. Contacts with KA stations are not acceptable. For all JARL awards, contacts are acceptable if made July 29, 1952, or later, with the following exceptions:

- Satellite endorsement—December 15, 1972
- SSTV endorsement—April 10, 1973
- RTTY endorsement—August 8, 1968
- Okinawa Prefecture contacts—May 15, 1972

With the exception of the ADXA award, all JARL awards featured in this section offer endorsements for 160 meters, 80 meters, 40 meters, 30 meters, 20 meters, 15 meters, 10 meters, 6 meters, 2 meters, 70 cm and 23 cm. Mode endorsements are available for CW, AM, SSB, FM, SSTV, RTTY, ATV and fax. Satellite and QRP (1 watt or less) endorsements are available as well.

Send all applications to: Japan Amateur Radio League, Awards Section, 1-14-2 Sugamo, Toshima, Tokyo 170, Japan.

AJD (All Japan Districts)

Work and confirm amateur stations in each of the Japanese call areas (1-Ø).

WAJA (Worked All Japanese Prefectures Award)

Work and confirm amateur stations in each of the 47 prefectures (administrative districts) of Japan listed immediately below. The list of QSLs should be arranged in order of WAJA reference number in lieu of prefecture name. Prefectures:

Call Area 1—Tokyo, Kanagawa, Chiba, Saitama, Ibaraki, Tochigi, Gumma, Yamanashi
Call Area 2—Shizuoka, Gifu, Aichi, Mie
Call Area 3—Kyoto, Shiga, Nara, Osaka, Wakayama, Hyogo
Call Area 4—Okayama, Shimane, Yamaguchi, Tottori, Hiroshima
Call Area 5—Kagawa, Tokushima, Ehime, Kochi
Call Area 6—Fukuoka, Saga, Nagasaki, Kumamoto, Oita, Miyazaki, Kogoshima, Okinawa
Call Area 7—Aomori, Iwate, Akita, Yamagata, Miyagi, Fukushima
Call Area 8—Hokkaido
Call Area 9—Toyama, Fukui, Ishikawa
Call Area Ø—Niigata, Nagano

JCC (Japan Century Cities)

Work and confirm amateur stations in each of at least 100 different Japanese cities. JCC-200, 300, 400, 500 and 600 will be issued as separate awards. The list of QSLs should be arranged in order of JCC reference number in lieu of city name.

JCG (Japan Century Guns)

Same rules, including endorsements, as JCC (above), with guns replacing cities. [Note: Japan has, as administrative districts, 47 prefectures, which are divided into cities, towns and villages. The gun, not being an administrative district, is a regional congregation of towns and villages.]

ADXA (Asian DX Award)

Work and confirm amateur stations in at least 30 Asian countries, including Japan, in accordance with the ARRL DXCC List and the Asian Countries List. The QSL list should be arranged in order by country per the Asian Countries List. Endorsements: 160 meters, 80 meters, SSTV, RTTY, fax.

WACA (Worked All Cities Award)

Work and confirm amateur stations in each of all the cities of Japan that are in existence on the day when the final contact claimed for the award is made. The QSL list should be arranged in order of JCC reference number. A satellite endorsement is available.

WAGA (Worked All Guns Award)

Same rules as WACA (above), with guns replacing cities.

JORDAN (JY)

To apply for these Royal Jordanian Radio Amateur Society (RJRAS) awards, applications must be accompanied by photocopies of the QSL cards (or certification) and 10 IRCs. Send applications to: Awards Manager, PO Box 1055, Amman, Jordan.

The Silver Award

This award, issued by King Hussein, JY1, requires proof of working six different JY prefixes. Photocopies of QSL cards (or certification) and 10 IRCs required.

The Coral Award

This award, issued by King Hussein, JY1, is for amateurs who visit Jordan and make a QSO from Aqaba.

The Arabian Knights Award

This award, issued by the Arab Radio Amateur League (ARL) and presented by JY1, is for working 10 Arab countries, one of which must be either JY1 or JY2, since January 1, 1971. [The Arab countries are A4, A6, A7, A9, CN, HZ, J2, JY, OD, ST, SU, T5, YI, YK, 3V, 4W, 5A, 5T, 7O, 7X, 9K.]

KENYA (5Z)

The Radio Society of Kenya (RSK) issues the *Kenyan Award* for contacts made after December 31, 1977. Ten points are necessary to qualify, as follows: Each contact with a 5Z station who must be a member of RSK = two points, and a contact with club station 5Z4RS = five points. All bands/modes. To apply, send a log photocopy, witnessed and signed by a responsible official of the local radio society/club or the licensing authority, and 15 IRCs (or $8 US). A self-addressed, adhesive label must be enclosed with the letter of application, which should be addressed to RSK and marked "Kenyan Award" at the top left-hand corner of the envelope. Send application to: The Radio Society of Kenya, PO Box 45681, Nairobi, Kenya.

KOREA (HL)

General rules for Korean Amateur Radio League (KARL) Awards:

Eight IRCs are required for each award, 4 IRCs for each endorsement. If QSL cards are submitted, IRCs for return postage must be provided. Separate endorsements for specific bands/modes, and QRP, are available. Contacts with HL stations (except HL9) are acceptable on or after February 3, 1959. Contacts with HL9s are not acceptable. All contacts must be made from the same call area. Mail application to: Korean Amateur Radio League, CPO Box 162, Seoul 100, Korea.

HLA (HL Award)

The HLA is issued for confirmed contacts with HL stations (except HL9s) in the following classes.

Class K—5 QSLs required
Class O—10 QSLs required
Class R—20 QSLs required
Class E—30 QSLs required
Class A—50 QSLs required

Endorsement stickers are available in multiples of 50 upon submission of the QSL cards.

AKA (All Korea Award)

The AKA Award is issued for confirmed contacts with HL stations in each of the seven Korean call areas (1,2,3,4,5,8,Ø).

KDN (Korean District Number Award)

The KDN Award is issued for confirmed contacts with HL stations located in each of the 50 different cities, guns or gus in Korea. The award will be issued in multiples of 50 (KDN 50, 100, 150, etc) upon submission of the QSL cards accompanied by a list prepared in order of KDN reference numbers.

APA (All Province Award)

The APA is issued for confirmed contacts with HL stations located in each of the different special cities and provinces in Korea. The area codes for each city and/or province are as follows:

Area Code	Province and/or City
1	City of Seoul
2	Inchon City, Kyonggi-do, Kangwon-do
3	Chungchongnam-do, Chungchongbuk-do
4	Chollanam-do, Chollabuk-do, Cheju-do
5	Pusan City, Taegu City, Kyongsangnam-do, Kyongsangbuk-do

KUWAIT (9K)

The *Kuwait Award* is issued by the Kuwait Amateur Radio Society (KARS) for working 10 different 9K stations. There is no date limitation, any band/mode may be used, and all contacts must be made from the same location or within a radius of 100 km of the original location. Send 5 IRCs and certified list of contacts to: Awards Manager, Kuwait Amateur Radio Society, PO Box 5240, Safat, Kuwait, State of Kuwait.

Kuwait National and Liberation Day Award

On February 25 and 26, the state of Kuwait celebrates its National and Liberation days. To mark this event, the Kuwait

Six Counties Award

Work and confirm two-way contacts with at least one amateur station in six counties in Liberia on at least two different bands since April 1, 1964.

West African Countries Award

Work and confirm two-way contacts with at least one amateur station in eight West African countries and at least five amateur stations in Liberia on at least two different bands since January 1, 1962.

Work ECOWAS Countries Award

Work and confirm two-way contacts with at least one amateur station in each of the 16 member-countries of the Economic Order of West African States (ECOWAS), including Liberia, on at least three different bands since May 28, 1975. [ECOWAS countries: C5, EL, D4, J5, TU, TY, TZ, XT, 3C, 5N, 5T, 5U, 5V, 6W, 9G, 9L.]

Amateur Radio Society offers the National and Liberation Day contest and award.

The contest begins at 0000 UTC February 25 and ends at 0000 UTC on the last day of February. To qualify for the award, you must:

1. Make two contacts with Kuwait Amateur Radio Society station 9K2RA/NLD and one contact with any other Kuwait station signing /NLD.

or

2. Make three contacts with three different Kuwait stations.

Submit a certified copy of the logbook along with 5 IRCs or $3 US to: The Award Manager, Kuwait Amateur Radio Society, PO Box 5240, Safat 1305, Kuwait.

LIBERIA (EL)

For awards sponsored by the Liberia Radio Amateur Association (LRAA), 20 IRCs (or $5 US) is required. QSL cards need not be submitted; instead, satisfactory evidence is a certified list of claimed contacts, signed by an officer of the applicant's National Society, or two amateurs of at least General class. This list must include each station worked, date, time, band, mode and signal report. Send all applications to: Awards Manager, Liberia Radio Amateur Association, PO Box 987, Monrovia, Liberia.

Worked All Liberia Award

Work and confirm two-way contacts with at least one amateur station in each of Liberia's nine counties on at least three amateur bands since April 1, 1964.

LUXEMBOURG (LX)

Applications for awards issued by the Reseau Luxembourgeois des Amateurs d'Ondes Courtes (RL) should be sent to: Awards Manager, A. Engelhardt, LX2EA, RL, PO Box 1352, L-1013 Luxembourg. Stations worked must be listed on the application in alphanumeric order.

LX Award

To commemorate 50 years of Amateur Radio activity in the Grand Duchy of Luxembourg, the LX Award is issued for contacts with LX amateurs since January 1, 1951. Applicants outside of Europe must make 20 points (contacts with LX stations on 160/80/40/30 meters count two points, and one point on 20/17/15/12/10 meters). Europeans must make 30 points (at least 20% on 160/80/40 meters).

LX station can be worked once on each band in different modes. Fifteen points (10 for Europeans) can be counted for working the same LX station on five bands. No mode restrictions. No QSLs required; send a detailed list showing the calls of the LX stations worked, and date, time, band and mode, countersigned by two licensed amateurs or the Awards Manager of the National Society. Ten IRCs (or $8 US) is required for each award.

European Community Award

This is the official RL diploma to commemorate the 25th anniversary of the European Community. Each contact with a station with one of the European Community member-countries, made on or after the date of the country's entry into the European community (see list below), counts one point. Non-European applicants must make 50 points; each member-country must be worked at least once, and at least three LX stations must be worked (Europeans need 100 points, must work each member-country at least once, five LX stations, and no more than 10 contacts are allowed with stations in the applicant's own country). No band/mode restrictions.

Each station may be counted only once. No more than 20% of the QSO points may be obtained by contacts with the same member-country. A contact with special station LXØRL may replace a missing contact with any of the member-countries. Contacts made via active earthbound reflectors or repeaters are not valid. To apply, send a list of QSLs (confirmed by two licensed amateurs, by one club official or by a notary) and 10 IRCs (or $8 US).

Date of entry into the European Community by prefix:
March 25, 1957—DL, F, I, LX, ON, PA
January 1, 1973—EI, G, OZ
January 1, 1981—SV
January 1, 1986—CT, EA

MALAYSIA (9M)

The Malaysian Amateur Radio Transmitters' Society (MARTS) issued the *Worked All Malaysia Award* to amateurs who work and confirm 10 9M2 contacts, and one each with 9M6 and 9M8. The award and endorsements are issued for two-way SSB, two-way RTTY, two-way CW, mixed or single band, mixed or single mode. All contacts must be made starting August 31, 1957. Send a certified list of QSLs and 10 IRCs ($5 US) to: MARTS Awards Manager, Eshee Razak, 9M2FK, PO Box 13, Penang, Malaysia.

MALTA (9H)

To apply for diplomas issued by the Malta Amateur Radio

League (MARL), send a certified list of QSL cards and 20 IRCs (or $5 US) to: Malta Amateur Radio League (MARL), PO Box 575, Valletta, Malta.

[Note: For European applicants, send 15 IRCs or $4 US]

Dip-MED Award

Work and confirm a minimum of 15 of the 26 Mediterranean countries (see list below). A 9H QSO is mandatory.

[Mediterranean countries, by prefix: CN, EA, EA6, EA9, F, I, IS, IT, OD, SU, SV, SV5, SV9, TA, TK, TZ, YK, ZA, ZB, 3A, 3V, 4X, 5A, 5B, 7X.]

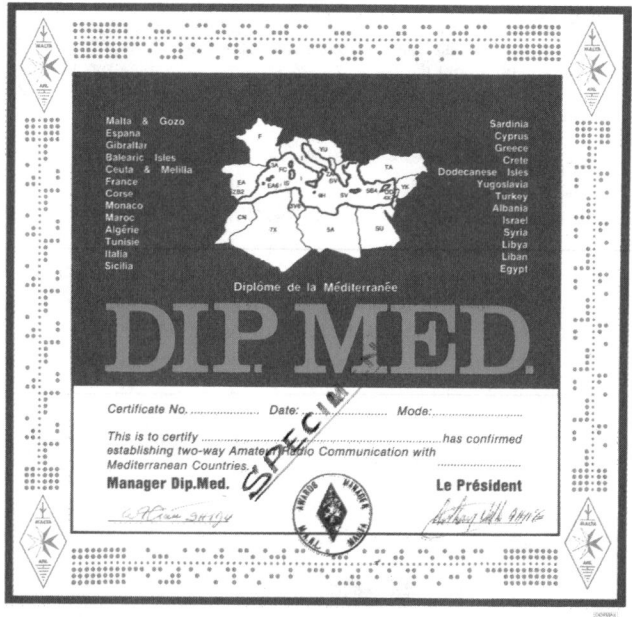

9H Diploma

To qualify for the 9H Diploma, applicants outside Europe need 5 points (Europeans need 10 points). Each QSO with a 9H1 station counts one point, each QSO with a 9H4 counts two points. A QSO with club station 9H1MRL counts two points.

MONACO (3A)

The Association des Radio-Amateurs de Monaco offers the following awards:

PMA (Principality of Monaco)

The *Principality of Monaco* award is available in one class, for having worked or heard three resident stations of Monaco since January 1, 1980. Contacts via repeater are not valid. All bands and all modes may be utilized. To apply for the award, send QSLs or logbook photocopy along with $8 US (20 IRCs) to: Claude Passet, 3A2LF, 7 rue de la Turbie, MC 98000, Monaco.

3A-CW Award

The *3A-CW* award is issued by the 3A-CW group. It is available to radio amateurs and SWLs.

All contacts or SWL reports with stations in Monaco must be on CW **ONLY**. Each QSO or SWL report counts as 1 point. "Joker" stations count 2 points each when worked for the first time only. Joker stations are 3A0CW, 3A2ARM, 3A200SM, 3A7A, 3A8A, 3A9A and 3A7JO. All stations may be worked only once per band and all bands may be utilized with no minimum RST. Reports must be dated after January 1, 1981.

The award is available in 3 classes: Basic, First Class and Excellence. Award requirements are as follows:

European applicants— Basic: 4 points
First Class: 8 points
Excellence: 12 points
All other applicants— Basic: 2 points
First Class: 4 points
Excellence: 6 points

The application fee is $8 US or 20 IRCs (add $3 US or 8 IRCs for registered air mail). For award upgrades, the application fee is $2 US or 5 IRCs. Submit QSLs to: Claude Passet, 3A2LF, 7 rue de la Turbie, MC 98000, Monaco.

(WASEC) Worked All Small European Countries Award

The *Worked All Small European Countries* award is offered by the Association de Radio-Amateurs de Monaco for working (or hearing) small European countries. The WASEC award is available in two classes:

Class I—All eight small European countries
Class II—Three small European countries and Monaco

Small European countries include: C31, HB0, LX, 9H, IA0, 3A, T7 and HV. All bands and modes can be utilized, but contacts via repeaters are not valid.

To apply for the award, send QSLs or logbook photocopy along with $8 US (20 IRCs) to: Claude Passet, 3A2LF, 7 rue de la Turbie, MC 98000, Monaco.

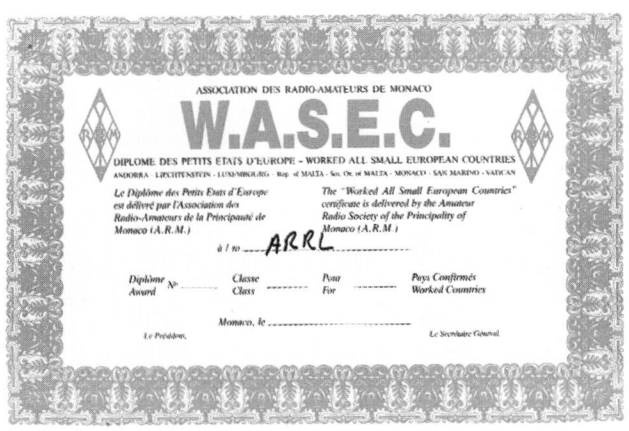

NETHERLANDS (PA)

Applications for awards issued by VERON (Vereniging voor Experimenteel Radio Onderzoek in Nederland) should be sent to: Traffic Bureau, VERON, S. Wybenga, PA3DKE, Pr. Bernhardl.60 8501 JG Joure, Netherlands.

PACC

Work and confirm contacts with 100 different PA/PI stations. Endorsement stickers are available for 200-900 contacts. Contacts must be made June 1, 1945 and after. Certified list of QSLs and 8 IRCs required.

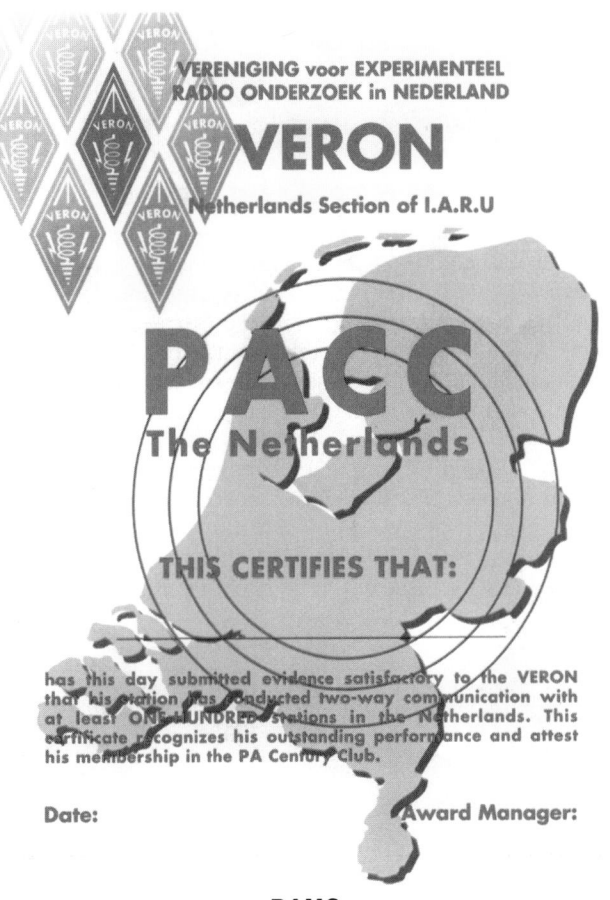

PAMC

Same rules as above, except 1000 different contacts required.

QSL Region Certificate

Work and confirm each of the 50 Dutch QSL Regions on the

HF bands. Contacts September 1, 1980 and after are valid, and the QSL Region has to be mentioned on the QSL card. Certified list and 8 IRCs required.

NETHERLANDS ANTILLES (PJ)

Vereniging voor Experimenteel Radio Onderzoek in Nederlandse Antillen (VERONA) sponsors the *Curacao Certificate*, issued to any amateur who works and confirms contacts with at least three licensed amateurs on the island of Curacao. All contacts afeter January 1, 1967 are valid. Allbands/modes, signal reports must show at least readability 3. The applicants can make the necessary contacts from any location in the same call area. Send certified list of QSOs (certified by two officers of a local Amateur Radio club or by three licensed amateurs in the absence of a club) and $3 US to: VERONA, PO Box 3383, Curacao, Netherlands Antilles.

NEW ZEALAND (ZL)

To apply for these many fine awards sponsored by the New Zealand Association of Radio Transmitters (NZART), QSL cards are not required as NZART stresses the Honor System. It is sufficient merely to certify that the QSL was legitimately made. $2 US is required for each award (except WAP and 5X5 which are $3); 2 IRCs and a self-addressed envelope required for each separate endorsement. Appropriate application forms may be obtained from the Awards Manager for 2 IRCs. Please address all applications and correspondence to: NZART Awards Manager, PO Box 108, Gisborne 3815, New Zealand.

ENZART

This prestige award, to honor NZART, is issued in three classes:

Basic Award—200 points (endorsements available for 250, 300, and 350 points).
Honour Award—400 points (with endorsements for each additional 50 points).
Honour Plaque—500 points.

It is essential to clearly determine that the operator contacted is a financial member of NZART and on the voting roll of the branch concerned. Points are allocated as follows:

Phone QSO, 1 point; CW QSO, 2 points; club station, 2 points; YL operator, 1 point; mobile QSO, 1 point; AMSAT QSO, 1 point; Branch President, 1 point. All contacts must be made November 1, 1976 or later. Contacts claimed with a Branch Station must be from a club-room or other permanent headquarters, or alternatively when the station is being operated in a contest, from an exhibition, or other activity of a similar kind. (Contacts as a result of call-sign "swapping" at a member's QTH are not eligible.) Contacts with mobiles are valid only when the mobile station is within 30 miles of his Branch's club-rooms or his own QTH. The same station may be claimed in several point categories, but each must be for a separate contact, and the contacts must be not less than 24 hours apart. Contacts must be "phone to phone" or "CW to CW," and contacts made on awards nets are not valid.

ZL applicants must be members of NZART, and all applicants must certify that each contact made fulfills award requirements. No QSL cards need be submitted; applications must be made on the special NZART Check-list (available for 10 cents and a large SASE). Each certificate requires $1 US; endorsements are free, but an SASE is required.

WAZL (Worked All New Zealand)

To qualify for the WAZL award, work 35 branches of NZART (ZLs must work 45). Special endorsements are available if WAZL is completed within a 12-month period. Note: Mobiles operating out of their own Branch area must use the Branch number of the area in which they are operating.

NZART Branches

01	Ashburton
02	Auckland
03	Western Suburbs
04	Cambridge
05	Christchurch
06	Dannevirke
07	
08	East Southland
09	Egmont
10	Franklin
11	Gisborne
12	Hamilton
13	Hastings-Havelock Nth
14	Hawera
15	Hawke's Bay Central
16	Horowhenua
17	Huntly
18	Hutt Valley
19	Inglewood
20	Manawatu
21	Manakau
22	Marlborough
23	Marton
24	Motueka

25	Napier
26	Nelson
27	New Plymouth
28	Northland
29	North Shore
30	Otago
31	Pahiatua
32	Rahotu Coastal
33	Rotorua
34	South Cantebury
35	South Otago
36	South Westland
37	Southland
38	Taumarunui
39	Tauranga
40	Te Awamuta
41	Thames Valley
42	Titahi Bay
43	Waihi
44	Matamata Radio Club
45	Waimarino
46	Wairarapa
47	Waitara
48	Wanganui
49	Westland
50	Wellington
51	Eastern Bay of Plenty
52	Northern Hawkes Bay
53	Te Puke
54	Patea
55	Waitomo
56	Christchurch West
57	Tokoroa
58	Helensville
59	Mangakino
60	Taupo
61	Central Otago
62	Reefton-Buller
63	Upper Hutt
64	North Otago
65	Papakura
66	Auckland VHF
67	Kawerau
68	North Canterbury
69	Kapiti
70	Fielding
71	Rodney
72	Opotiki
73	North Wairoa
74	Wellington VHF
75	Queenstown
76	Kaikoura
77	Te Aroha & Districts
78	Far North
79	Howick & Districts
80	Hibiscus Coast
81	Waikato VHF Group
82	Southern Wairarapa

NZA (New Zealand Award)

The NZA is available to all radio amateurs other than ZLs. Required contacts: 35 contacts with ZL1, plus 35 contacts with ZL2, plus 20 contacts with ZL3, plus 10 contacts with ZL4, plus one contact with a ZL "territory" (either NZ Antarctica,

Chatham, Kermadec or Campbell). This makes a minimum of 101 required contacts. Twenty ordinary ZL contacts can be substituted for the territory contact, if desired. All contacts must be dated from December 8, 1945.

NZC (New Zealand Counties)

The basic NZC award requires contacts with 20 different New Zealand counties. Endorsements for 40, 60, 80 and 100 counties are available, with a special certificate for all 112 counties. A counties map and checking sheet are available from NZART. Initial certificate with any or all endorsements to 100 costs $1 US. Separate endorsements require 1 IRC plus self-addressed envelope. The special checking sheet must be used and are available from NZART for 1 IRC. The "NZC 224," to recognize the outstanding achievement of "double 112" (different stations contacted in the relevant counties for each NZC 112), is available to those who already hold two NZC 112 awards. Cost of the plaque is $12.

5×5 (Five by Five)

This premier award has been instituted to recognize the increasing interest in five-band DX operation. The initial award requires that the same station be contacted on five bands; first endorsement after another five has been contacted (a total of 10), with the 20 endorsement requiring another 10, and so on to 100. Certified list with full QSO data and $2 (for initial application and for each endorsement) required. Contacts must date from 1945.

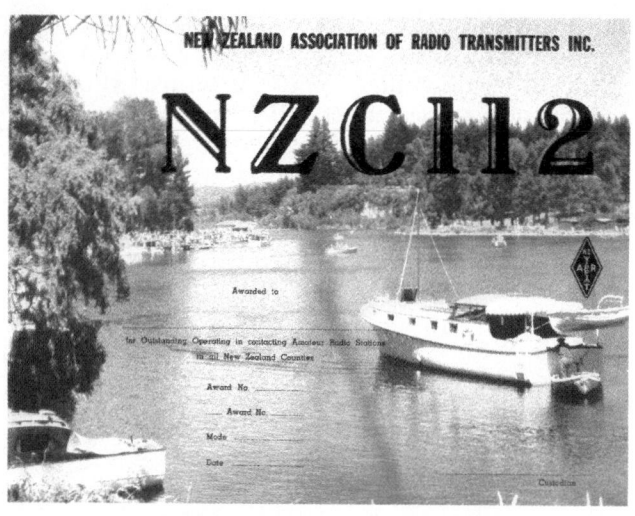

(NZLA (NZ Lakeside Award)

To qualify for the basic award, contacts must be made with stations operating on the shores of 10 freshwater lakes in New Zealand. Endorsements for each additional lake up to 40; Honour Award for 50 lakes, with further endorsements for each additional 10. Stations worked must be located within one kilometer of the water or in a town on the lake shore. Complete rules are with the check sheets (available from ZL2GX for an IRC and a self-addressed envelope) which must be used when applying. Contacts must be dated May 1, 1976 or later.

WAP (Worked All Pacific)

Contacts required with 30 different Oceania countries (as indicated on the ARRL DXCC List). Fee $1.50 US.

Five Band WAP Plaque

To qualify, contact any 30 Oceania countries, each on five different bands, for a total of 150 contacts. $8 US required.

ZLA Award

To qualify, contacts are required with Auckland City/ZL1, Wellington City/ZL2, Christchurch/ZL3, Dunedin City/ZL4, Antarctica/ZL5, Chatham Island/ZL7, Kermadec Island/ZL8 and Campbell Island/ZL9.

MIA (Minor Islands Award)

To qualify, five contacts with minor islands off the New

Zealand coast are required (eg, Kapiti Island, Portland Island, Great Barrier Island, Little Barrier Island, Waiheke Island, D'Urville Island, Rangitoto Island, Kawau Island, Motutapu Island). The so-called islands in the Auckland area that are connected to the mainland by a causeway are not eligible.

ZL Awards

ZL1—the basic award requires contacts with 125 ZL1 stations in the post-war era. Endorsements are available for 175 and 250.

ZL2—the basic award requires contacts with 100 different ZL2 stations in the post-war era. Endorsements available for 150 and 200.

ZL3—the basic award requires contacts with 50 different ZL3 stations in the post-war era. Endorsements available for 75 and 100.

ZL4—the basic award requires contacts with 25 different ZL4 stations in the post-war era. Endorsements available for 35 and 50.

[Certificates can be endorsed for any band/mode.]

Tiki Award

To qualify, work five different ZLs, each on five different bands, in the post-war era.

CJC (Captain James Cook Award)

To perpetuate the memory of this world-famous navigator and seaman, the CJC award is issued in three classes:
• Basic "Sailor Class"—requires contacts with G in Yorkshire, FO in Fr Polynesia, ZL2, VK2, KH6.
• "Officer Class"—requires all "Sailor Class" contacts plus ZL1, ZL3, ZL4, VK3, VK4, P2 and any Antarctic station.
• "Command Class"—requires all "Officer Class" contacts plus any five of the following, VE2, VO, YJ, FK, CE8, KL7, A3.

Antipodes Award

This award recognizes all-around achievement as follows:
1) *Technical*
    (a) The applicant's station (receiver, transmitter, antenna) must be home built, completely of amateur design and construction, which the applicant has personally developed and constructed.
    (b) Commercially constructed equipment or kits from any source are not valid.
2) *Operating*
The applicant must make three QSOs with stations in the

general "Antipodes continental area" (the opposite side of the Earth) from the applicant.

[Note: The following awards are not officially sponsored by NZART but are supported by NZART branches.]

IARU Region 3 Operating Award

Contacts made after April 5, 1982, are valid; a certified list in lieu of QSL cards required. The basic award requires seven countries, the Silver Star endorsement requires 15 countries, and the Gold Star endorsement requires 20 countries. Any mode/band endorsement is available. Send $1 US ($2 airmail) and application to ZL2GX. Eligible countries: A3, DU, H4, HL, HS, JA, KH1/2/8/9/Ø, P2, VK, VS6, VU, YB, ZL, 3D2, 4S, 5W, 9M, 9V. Applications to NZART Awards Manager ZL2GX (see address above).

Regional Awards

Auckland Branch Certificate

For overseas stations only. To qualify, work eight members of the Auckland Branch (02) since January 1, 1957. Branch station ZL1AA counts as two members. No QSLs required. (Note: not all Auckland stations are members of the Auckland Branch.) Send certified list and $1 US to Awards Manager, Auckland Branch, Box 18-003, Glen Innes, Auckland, New Zealand.

Worked New Zealand Novices

For the Basic Award, QSO 10 ZL Novices, including one each from ZL1, ZL2, ZL3, ZL4. Endorsements for 25 (two QSOs from each call area), 50 (five QSOs from each call area), 100 (ten QSOs from each call area). All contacts must be made on or after October 1, 1979. Certificates may be endorsed all CW or phone. No QSLs required; send certified list with relevant QSO data. [Note: The ZL Novice band is 3525-3575 kHz.] Send applications to Awards Manager, Auckland Branch (see address above). One dollar for each certificate required no charge for endorsements.

Christchurch Award

Work five stations in Christchurch. Award fee: $1 US. Applications to Awards Manager, Box 1733, Christchurch, New Zealand.

Bay of Plenty Award

Work five stations in the Bay of Plenty area (Opotiki, Rotorua, Tauranga, Whakatane). Send certified list of QSOs and 8 IRCs to ZL1BJP, 36 Western Rd, Otumoetai, Tauranga, New Zealand.

Horowhenua Award

Work stations in the Branch 15 area (Levin, Otaki, Waikanae, Shannon) with bonus points in certain categories. Involved rules; contact E. Tombs, ZL2IG, Ihakara, RD 1, Levin, New Zealand for details. Please include self-addressed envelope and 1 IRC.

Kapiti Award

Work stations in the Kapiti Branch (69) as follows: ITU Region 1 applicants need five contacts, Region 2 applicants seven contacts, and Region 3 applicants nine contacts [except for VK/ZL, which requires 15 contacts on 160 through 10 meters]. Contacts with club station ZL2KB count double (two contacts). Net and repeater contacts invalid. One contact with any one club member permitted. Special endorsements for single band, all phone, all CW. Contacts must be made January 1, 1976 or later. Certified list and 3 IRCs (or 50 cents US) to Box 1608, Paraparaumu Beach, New Zealand.

Manawatu Award

To qualify, work five stations who are members of the Manawatu Branch of NZART. Certified list and 5 IRCs to ZL2AFT, 431 Albert St, Palmerston, New Zealand.

Papakura Radio Club "21" Award

To qualify, work 20 members of the Papakura Radio Club, plus ZL1VK. Contacts must date from January 1, 1977. Band/mode endorsements. Apply to Ms C. Johnson, ZL1AJL, 63 Red Hill, Papakura, New Zealand.

Gisborne Award

To qualify, work two stations in Gisborne (Branch 11). Apply to ZL2GX (see address above).

ENDEAVOUR GISBORNE AWARD

No .... Presented by Gisborne Branch NZART Date ....
No ... to ...
Mode .... for radio contacts with Gisborne City Custodian
"The First City of the Sun"

NIGERIA (5N)

Unless otherwise noted, contacts for these awards sponsored by the Nigerian Amateur Radio Society (NARS) must be made on or after January 1, 1980. To apply, send a certified log (that must be verified by the applicant's National Society) and 10 IRCs (or $4 US) to: Awards Manager, NARS, PO Box 2873, Lagos, Nigeria.

WNS (Worked All Nigeria States Award)

To qualify, the applicant must work one amateur station from each of the 19 states in the Federal Republic of Nigeria:

5NØ—Lagos State and Federal Territory Abuja
5N1—Ogun State and Oyo State
5N2—Kwara State and Niger State
5N3—Ondo State and Bendel State
5N4—Anambra State and Rivers State
5N5—Imo State and Cross River State
5N6—Benue State and Plateau State
5N7—Bauchi State and Gongola State
5N8—Bornu State and Kano State
5N9—Kaduna State and Sokoto State

5NDX (Outstanding Nigeria DX Award)

This award is issued in three classes:
First Class—work 100 5N stations.
Second Class—work 50 5N stations.
Third Class—work 20 5N stations.

NARS President's station (5NØAAJ) counts as five stations, XYL stations count as two stations and club stations count as two stations.

NOA (Nigerian October Award)

The month of October has special significance in Nigerian history, and in commemoration, 5N amateurs are authorized to use a special prefix every year in the month of October (eg, 5N20 for 1980, 5N21 for 1981, 5N27 for 1987). To qualify for the award, work at least five different 5N stations using the special prefix in the month of October of that particular year. This award is effective October 1, 1980.

WNZ (Worked All Nigeria Zones Award)

To qualify, work one station in each of the 10 5N call areas. Usual rules apply.

NORWAY (LA)

The following awards are sponsored by the Norsk Radio Ralae Liga [Norwegian Radio Relay League], NRRL:

WALA (Worked All Norway)

Contacts with LA and LB stations made after January 1, 1950, are valid for this award. Applicants must work and confirm contacts with each of the 19 Norwegian counties ("flykers") on any band (Scandinavian applicants must mak two contacts, on separate bands, with each county). All contacts must be made from the same QTH, within a radius of 100 km. Contacts via repeaters or satellites are not valid. Contacts may be made on all legal modes and may be endorsed as appropriate. Cross-band contacts are not valid. Contacts with Arctic stations (JW or JX) count and may be substituted for W, X and/or Y counties. Applications should include a list of stations worked and QSL cards or a certified list of QSLs (verified by the applicant's National Society) that includes date and time in UTC, call sign, signal report and QTH of station worked, along with other relevant information in the case of endorsements. Send application and 10 IRCs (or $5 US) to either the Norwegian Radio Relay League, PO Box 21 Refstad, N-0513 Oslo, Norway, or NRRL Awards Manager Oystein Skaaden, LA7OEA, Steinsviksvn, 19A, N-5353, Straume, Norway.

Norwegian Counties:

A	Oslo	R	Hordaland
B	Ostfold	S	Sogn ug Fjordane
C	Akershus	T	More og Romsdal
D	Hedmark	U	Sor-Trondelag

E	Oppland	V	Nord-Trondelag
F	Buskerud	W	Nordland
Z	Vestfold	X	Troms
H	Telemark	Y	Finnmark
I	Aust-Agder	JW	Svalbard/Bear Island
K	Vest-Agder	JX	Jan Mayen
L	Rogaland		

WNC (Worked Norwegian Cities)

The WNC certificate is issued in three classes:

Class 3—work 5 cities (Europeans, 10 cities)
Class 2—work 10 cities (Europeans, 20 cities)
Class 1—work 15 cities (Europeans, 30 cities)

No date, band or mode limitations, but QSOs with LF, LH and LJ stations are not valid. Send a certified list verified by two amateurs and 10 IRCs (or $5 US) to Larvik Society of NRRL, Awards Manager, PO Box 59, N-3251 Larvik, Norway.

Valid Norwegian Cities:

Arendal	Mandal
Bergen	Molde
Bodo	Mosjoen
Drammen	Moss
Egersund	Mo i Rana
Fredrikstad	Namsos
Gjovik	Narvik
Grimstad	Notodden
Hammerfest	Oslo
Halden	Porsgrunn
Hamar	Sarpsborg
Harstad	Sandnes
Haugesund	Sandefjord
Holmestrand	Stavanger
Horten	Skien
Kongsberg	Steinkjer
Kongsvinger	Trondheim
Kristiansand S	Tonsberg
Kristiansand N	Tromso
Kragero	Vardo
Larvik	Alesund
Lillehammer	

W 100 LA (Worked One Hundred LA Amateurs)

To qualify, applicants must work 100 different LA/LB stations. Contacts made after January 1, 1984 are valid. The awards issued for CW, phone or mixed modes. All valid amateur bands may be used (10, 18 and 24 MHz will not be available before January 1, 1989). LF, LH and LJ stations do not count. A list showing full details of the contacts should be certified by the Awards Manager of the applicant's National Society. The application must contain call sign, date, band, RST and modes, and must be accompanied by 10 IRCs (or $5 US). Send applications to Awards Manager, Stavangergruppen av NRRL, PO Box 354, 4001 Stavanger, Norway.

PANAMA (HP)

The *Panama International Award* is sponsored by the Panamanian Amateur Radio League (LPRA). The award is available to all licensed radio amateurs according to the following rules:

1) The award will be issued to applicants who can prove contact with at least one Panamanian amateur in each of the nine Panamanian call areas (1-9) on any band or mode.

2) Contacts with official LPRA club stations (regardless of location) may be substituted for as many as three call area contacts.

3) All contacts must have taken place after January 1, 1978.

4) Applicants must send the list of stations worked to the LPRA-HP Bureau, PO Box 175, Panama City 9A, Panama. If

the applicant is not an LPRA member, QSL cards must also be submitted (they will be returned).

5) The fee for the *Panama International Award* is $3 US in the form of an international money order or equivalent International Reply Coupons. The money order must be payable to "Liga Panamena de Radioaficionados Nacional."

PAPUA NEW GUINEA (P2)

The Papua New Guinea Amateur Radio Society (PNGARS) sponsors *The Bird of Paradise Award* to encourage contacts with amateurs in Papua New Guinea, to advance the art of radio communication, and to help foster international friendship and goodwill. Applicants outside of Oceania must contact at least five amateurs in P2, with at least three of them located in different provinces. Amateurs in Oceania (ITU Zones 27-32) must contact seven stations in P2, at least five located in different provinces. The National Capital District (Port Moresby) is a separate province. The PNGARS official station, P29PNG, which is only activated on special occasions, may be substituted for any province. Contacts must be made after September 16, 1975 (the date of independence). Any band/mode; special

endorsements for specific bands and modes are available if specifically requested. QSL cards are not required; send certified log entries and $5 US to offset printing and postage costs.

The PNG provinces are as follows:

National Capital District	Chimbu
East New Britain	Gulf
West New Britain	Manus
Southern Highlands	Madan
Eastern Highlands	Northern
Western Highlands	Western
New Ireland	East Sepik
Moroba	Milne Bay
Central	North Solomons
Enge	West Sepik

Send all applications to: The Awards Committee, PNGARS, PO Box 204, Port Moresby, Papua New Guinea.

PARAGUAY (ZP)

To qualify for awards sponsored by the Radio Club Paraguayo, all contacts must be made after May 15, 1952. Send a certified list of QSLs and 5 IRCs to: Radio Club Paraguayo, Awards Manager, PO Box 512, Asuncion, Paraguay.

All Mediterranean Counties Awards

This award is issued for confirmed contacts with Mediterranean countries as follows:

Class A—41 countries
Class B—30 countries
Class C—20 countries

A ZP contact is mandatory for each class.

Countries: A2, A5, AC3, C31, CP, HA, HB, HBØ, HV, JT, LX, M1, OE, OK, TL, TT, TZ, UC2, UD6, UG6, UH8, UI8, UJ8, UL7, UM8, UO5, XT, XW8, YA, ZE, ZP, 3D6, 4U1ITU, 5U7, 5X5, 7P8, 7Q7, 9J2, 9N1, 9U5, 9X5.

Tropics of Cancer and Capricorn Award

This award is issued for confirmed contacts with countries touched by the Tropic of Cancer and Capricorn as follows:

Class A—28 countries
Class B—20 countries
Class C—12 countries

A ZP contact is mandatory for each class.

Countries: Tropic of Cancer—A4, A6, BV, BY, C6, EA9 (Sahara), KH6, S2, S3, SU, TZ, VU, XE, XZ, 5A, 5T5, 5U7,

7X, 7Z. Tropic of Capricorn—A2, CE, C9, LU, PY, VK, ZP, ZS, ZS3, 5R8.

All Zone 11 Prefixes Award

This award is issued for confirmed contacts with stations in CQ Zone 11 as follows:

Class A—30 different prefixes
Class B—19 different prefixes
Class C—12 different prefixes

[Prefixes consist of ZP1-ZP9, PY1-PYØ, and the special prefixes issued for the CQ WPX Contest.]

Diploma Sud-America

This award is issued for confirmed contacts with countries located in ITU Zones 12, 13, 14, 15, 16 and 73 as follows:

Class A—33 countries and six ITU zones
Class B—25 countries and six ITU zones
Class C—18 countries and five ITU zones

A ZP contact is mandatory for each class.
Countries:
Zone 12—CP1, CP8, CP9, FY, HC, HC8, HK, HKØ/Malpelo, OA, PZ, YV, 8R
Zone 13—PY6, PY7, PY8, PYØ/Fernando, PYØ/St Peter & Paul
Zone 14—CE1, CE2, CE3, CE4, CE5, CEØX/San Felix, CEØZ/Juan Fernandez, CP2, CP3, CP4, CP5, CP6, CP7, CX, LU-A/U/Y, ZP
Zone 15—PY1, PY2, PY3, PY4, PY5, PY9, PYØ/Trindade
Zone 16—CE6, CE7, CE8, VP8/Falklands, LU-V-W/X
Zone 73—KC4USP/Palmer, LU-Z, CE9AA/AM, VP8/S Georgia, VP8/S Orkney, VP8/S Sandwich, VP8/S Shetland

DP (Diploma Paraguay)

This award is issued for confirmed contacts with five different ZP stations (South Americans must contact 15 ZP stations).

WAZP (Worked All ZP)

This award is issued for confirmed contacts with one station in each of the nine ZP call areas.

CRCP (Certificado Radio Club Paraguayo)

This award is issued for confirmed contacts with 15 different ZP stations (South Americans must contact 50 different stations).

ZP 100, ZP 150, and ZP 200 Awards

These awards are issued for confirmed contacts with 100, 150 and 200 different ZP stations.

ZP3 Award

This award is issued for confirmed contacts with two different ZP3s (CE, CX, LU and PY must work five different ZP3s).

DDP (Diploma Departamentos del Paraguay)

This award is issued for confirmed contacts with the capital of Paraguay and the different departments in which Paraguay is divided, as follows:

Class A—20 contacts
Class B—16 contacts
Class C—12 contacts

Paraguay Departments, by prefix:

ZP1—Boqueron, Chaco, Nueva Asuncion
ZP2—Alto Paraguay, Pte Hayes
ZP3—Amambay, Concepcion
ZP4—Canendiyu, San Pedro
ZP5—Asuncion (nation's capital)
ZP6—Central, Cordilleras, Paraguari
ZP7—Caaguazu, Caazapa, Guaira
ZP8—Misiones, Neembucu
ZP9—Alto Parana, Itapua

PHILIPPINES (DU)

The Philippine Amateur Radio Association (PARA) sponsors the UN-DU Award in commemoration of the formation of the United Nations in 1945, of which the Philippines is a charter member.

For US applicants, the procedure is as follows:

The applicant must work and confirm 100 or more QSOs with amateurs in member-nations of the UN, including the applicant's own country. All QSOs must be dated on or after the date of the particular country's admission to the UN.

Send K6EDV the original QSL cards, photocopies of the QSL cards and an alphabetical list to be certified by K6EDV, Pete Peterson, 4830 Nebraska Lane, Fair Oaks, CA 95628. [Application forms, including the UN admission date for each country, are available from K6EDV for an SASE.] All material sent to K6EDV must include sufficient return postage; all submitted QSL cards shall be returned by K6EDV after they have been checked and the application has been certified. When returned, the applicant then sends the alphabetical list and the photocopies of the QSL cards to the Philippine Amateur Radio Association, PO Box 4083, Manila, Philippines. American amateurs can send the original QSLs to PARA if desired; all other amateurs must send original QSLs directly to PARA.

The fee for the UN-DU Award is $16 US, which also covers return air mail postage. This amount must be by certified check or postal money order, made out to PARA, and sent directly to PARA with the application.

The initial application is for 100 countries; endorsements can be submitted to K6EDV in groups of five. The original QSLs and photocopies are required when applying for endorsements.

The UN-DU Award is granted for mixed modes and for single modes (SSB, CW, RTTY, SSTV, satellite).

POLAND (SP)

To apply for awards issued by the Polski Zwiazek Krotkofalowcow, PZK, please include 10 IRCs for each award to cover the cost of each certificate and postage. All contacts, mixed or single band, mixed or single mode, are valid (exception—satellite and/or repeater contacts are invalid). To apply, send a certified list of QSLs (in lieu of QSL cards) and the IRCs to: PZK Awards Manager, PO Box 320, PL 00-950 Warszawa 1, Poland.

The POLSKA Award

This award is available in three classes:

Class 1—for confirmed contacts with all provinces (49) in Poland.
Class 2—for confirmed contacts with 35 provinces.
Class 3—for confirmed contacts with 20 provinces.

All contacts are valid from June 1, 1975, and must be confirmed by QSL cards. The applicant's QSL list must be in alphabetical order by province abbreviation (see below). When applying for a higher class, please supply the number of your diploma of the next lower class and a list of your additional contacts.

POLSKI ZWIĄZEK KRÓTKOFALOWCÓW

Abbreviations, and prefixes, of the 49 Polish provinces:

BB/SP9	JG/SP6	LG/SP6	PR/SP8	TA/SP9
BK/SP4	KA/SP9	LO/SP4	PT/SP7	TG/SP7
BP/SP8	KI/SP7	LU/SP8	RA/SP7	TO/SP2
BY/SP2	KL/SP3	NS/SP9	RZ/SP8	WA/SP5
CH/SP8	KN/SP3	OL/SP4	SE/SP5	WB/SP6
CI/SP5	KO/SP1	OP/SP6	SI/SP7	WL/SP2
CZ/SP9	KR/SP9	OS/SP5	SK/SP7	WR/SP6
EL/SP2	KS/SP8	PI/SP3	SL/SP1	ZA/SP8
GD/SP2	LD/SP7	PL/SP5	SU/SP4	ZG/SP3
GO/SP3	LE/SP3	PO/SP3	SZ/SP1	

The AC 15 Z Award

This award is issued for working and confirming contacts with at least 23 countries/call areas located in *CQ* Zone 15: HA, HV, I, IS, IT, OE (two call areas), OH (three call areas), OHØ, OJØ, OK, SP (four call areas), T7, TK, UA2, UP, UQ, UR, YU (three call areas), ZA, 9H.

Contacts with four call areas in Poland are mandatory. All contacts are valid from January 1, 1985. The list of QSLs must be in alphabetical order by country.

The W 21 M Award

This award is issued for working and confirming contacts with at least 16 countries situated on the meridian 21 east: A2, D2, JW, LA, OH, OHØ, OK, SP, SV, TL, TT, UA2, UP, UQ, UR, YO, YU, ZS, ZS3, 5A, 9Q. Contacts with Poland are mandatory. All contacts are valid from January 1, 1985. The list of QSLs must be in alphabetical order by country.

SP DX Award

The SP DX Club issues a certificate attesting to honorary SP DX Club membership. To qualify, work 10 (15 for European applicants) regular SPDX members after October 1, 1959. Send certified list and 10 IRCs to: SP DX Club of PZK, Awards Manager SP9PT, PO Box 131, 44-201 Rybnik, Poland.

ROMANIA (YO)

Awards issued by the Federatia Romana de Radioamatorism (Romanian Radioamateur Federation), FRR, are issued for CW, AM, SSB, RTTY or mixed modes; awards are also issued for single bands (80-10 meters) and for any combination of bands. Contacts are valid after August 23, 1949. [Note: Special requirements apply to European applicants; contact FRR for details.] Applications must include a certified list of QSLs (no cards) and 7 IRCs sent to Romanian Radioamateur Federation, PO Box 22-50, R-71100 Bucharest, Romania.

Romania Award

This award is issued for working at least 30 different YO stations, each in a different YO county, Bucharest (capital city) included. All eight YO districts (YO2-YO9) must also be represented.

County		District
AB	Alba	YO5
AR	Arad	YO2
AG	Arges	YO7
BC	Bacau	YO8
BH	Bihor	YO5
BN	Bistrita-Nasaud	YO5
BT	Botosani	YO8
BR	Braila	YO4
BV	Brasov	YO6
BZ	Buzau	YO9
CL	Calarasi	YO9
CS	Caras-Severin	YO2
CJ	Cluj	YO5
CT	Constanta	YO4
CV	Covasna	YO6
DB	Dimbovita	YO9
DJ	Dolj	YO7
GL	Galati	YO4
GJ	Gorj	YO7
GR	Giurgiu	YO9
HD	Hunedoara	YO2
HR	Harghita	YO6
IL	Ialomita	YO9
IS	Iasi	YO8
NM	Maramures	YO5
MH	Mehedinti	YO5
MS	Mures	YO6
NT	Neamt	YO8
OT	Olt	YO7
PH	Prahova	YO9
SJ	Salaj	YO5
SM	Satu Mare	YO5
SB	Sibiu	YO6
SV	Suceava	YO8
TR	Teleorman	YO9
TM	Timis	YO2
TL	Tulcea	YO4
VS	Vaslui	YO8
VL	Vilcea	YO7
VR	Vrancea	YO4

Bucuresti Award

This award is issued for working 10 different YO3 stations. The Jubiliar Bucuresti award is issued for working five different YO3 stations annually, between December 1 and December 31. During this time there are various activities celebrating the national day of Romania. All YO stations located in the city of Bucharest use the YO3 prefix. Application must be mailed by March 31 of the next year.

FEDERATIA ROMÂNĂ DE RADIOAMATORISM

DIPLOMA BUCUREȘTI

Se conferă stației _____
operator _____
pentru legăturile ce au fost efectuate cu stații de radioamator din Municipiul București, Capitala României.

PREȘEDINTE, _____   Data _____
                        Nr. _____

YO-AD (YO—All Districts)

This award is issued in three classes for working all YO districts:

Class I—work all eight YO districts, three QSOs/district
Class II—work six YO districts, two QSOs/district
Class III—work three YO districts, one QSO/district

YO-AM (YO—Alma Mater)

This award is issued for working YO stations located in the university centers of Romania. The following cities are university centers: Bucharest, Cluj, Timisoara, Brasov, Lasi, Constanta, Oradea, Baia Mare, Ploiesti, Sibiu, Bacau, Pitesti, Galati, Suceava, Petrosani, Hunedoara, Tirgu Mures, Craiova, Resita. North American applicants must make 10 points; contacts on 80 meters = 8 points, 40 meters = 3 points, 20 meters = 1 point, 15 meters = 2 points, 10 meters = 4 points. A station can be worked on more than one band.

YO-BZ (YO—Balkans Zone of Peace)

This award is issued for working stations in LZ, SV, SV9, SV/A, TA, YO, YU and ZA in three classes:

Class I—work at least four countries, including 10 call districts and three YO call districts.
Class II—work at least three countries, including eight call districts and three YO call districts.
Class III—work at least two countries, including six call districts and three YO call districts.

YO-DC (YO—Double Call)

This award is issued for working 26 "double-letter" call

sign suffixes using all letters from AA to ZZ respectively. This includes three-letter suffixes, such as AAT, BBZ, BCC. Applicants outside of Europe must work at least two YO stations with double calls.

YO-CM (YO—Chess Mate)

This award is issued for working all 64 stations as follows: On the first horizontal row of a chessboard, fill in the call signs of eight different YO stations you have worked. On each corresponding vertical row, fill in the call signs of seven other DX stations (YO okay) whose suffixes will include at least one similar letter, and always the same, from the suffix of the first listed YO call sign. For example, the first list YO station is YO3JP; on the following seven vertical squares, you might list YO3AAJ, JA1BJA, DL1CJ, YO3JW, I1SJ, SP5JJ, W8ACJ (the letter J being the "key" of this series). If the first YO call signs listed on the first horizontal row belong to YO DX Club members, you will be eligible for the special *YO—CM Master award*. In this case, you may use the same YO DX Club member call sign several times if you have worked it on several bands.

YO-DR (YO—Danube River)

This award is issued for working different stations located in countries along the River Danube (DL, HA, LZ, OE, OK, YO, YU, UA) on two bands. Two QSOs are required with each country except YO (three QSOs required). At least three QSOs must be with stations located in cities on the Danube River.

YO-DX-C (YO—DX Club Members)

This award is issued for working two YO DX Club members or honorary members. A membership list is available on request for 1 IRC. Working YODXC members during the annual YO DX Contest on the first weekend in August can qualify you as an honorary member.

YO-LC (YO—Large Cities)

This award is issued for working YO stations located in the large cities in three classes:

Class I—work 20 cities
Class II—work 10 cities
Class III—work 5 cities

YO large cities:
YO2—Arad, Deva, Hunedoara, Lugoj, Petrosani, Resita, Timisoara
YO3—Bucharest
YO4—Barila, Constanta, Focsani, Galati, Tecuci, Tulcea
YO5—Alba Iulia, Baia Mare, Bistrita, Cluj, Dej, Oradea, Satu Mare, Sighetul Marmatiei, Turda, Zalau
YO6—Brasaov, Medias, Miercurea Ciuc, Odorheiul Secuiesc, Sfintu Gheorghe, Sibu, Sighsoara, Tirgu Mures
YO7—Craiova, Pitesti, Rimnicu Vilcea, Slatina, Tirgu Jiu, Drobeta-Turnu Severin
YO8—Bacau, Birlad, Botosani, Lasi, Oras, Gheorghe Gheorghiu-Dej, Piatra Neamt, Roman, Suceava
YO9—Alexandria, Buzau, Calarasi, Giurgiu, Ploiesti, Slobozia, Tirgoviste, Turnu Magurele

YO-NC (YO—Namesake Calls)

This award is issued for working five different stations having the same one- or two-letter suffix as the applicant. For instance, KN5A may apply for the award if he worked LA1A, BV2A, HP1A, OZ3A, EL2A; N1JW may apply if he worked ON4JW, UA0JW, YO3YW, DL6JW, VK9JW. Stations with three-letter suffixes need to work only three different stations with the same suffix.

YO-5 On 5 (YO—5 Continents on 5 Bands)

This award is issued for working one station in each continent (other than the continent of the applicant) on five bands. One continent may be replaced by a YO station.

Single-Band Awards:

YO-10 × 10—work 10 different YO stations on 10 meters
YO-15 × 15—work 15 different YO stations on 15 meters
YO-20 × 20—work 20 different YO stations on 20 meters
YO-40 × 40—work 40 different YO stations on 40 meters
YO-80 × 80—work 80 different YO stations on 80 meters

YO-100

Work 100 different YO stations, regardless of band. This award is available in further multiples of 100 (200, 300...).

YO-20 Z (YO—Zone 20)

This award is issued in three classes for working Zone 20 countries: LZ, JY, OD, SV, SV5, SV9, SV/A, YO, 4X, 5B4, ZC4, YK, TA.

Class I—work six countries
Class II—work four countries
Class III—work two countries
For all classes, a YO contact is mandatory.

YO-25 M (YO—25 Meridian)

This award is issued in three classes for working the follow-ing countries situated on 25 east meridian: Antarctica, A2, ES, LA, LY, LZ, OH, ST, SU, SV, STØ, SV9, TL, UA, UB, UC, V5, YO, ZS, 5A, 9J, 9Q.

Class I—work 12 countries
Class II—work eight countries
Class III—work five countries
For all classes, a YO contact is mandatory.

YO-45 P (YO—45 Parallel)

This award is issued in three classes for working the follow-ing countries situated on 45 north parallel: BY, F, I, JA, JT, UA, UAØ, UB5, UI, UL, VE, W, YO, YU.

Class I—work eight countries
Class II—work six countries
Class III—work five countries
For all classes, a YO contact is mandatory

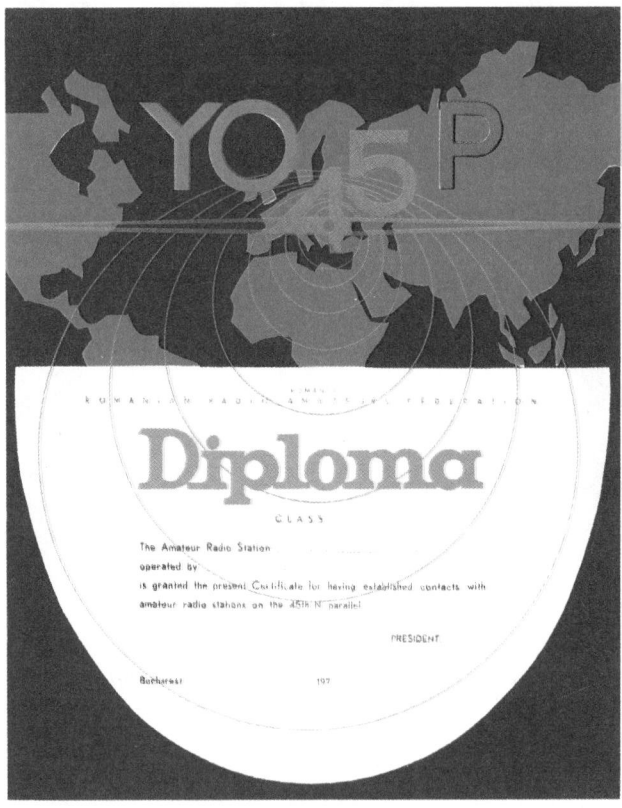

SENEGAL (6W)

The *Senegalese Award* is issued by the Association des

Radio-Amateurs du Senegal (ARAS) to amateurs who contact five different 6W stations. The request for the award must be forwarded with either a certified list of QSL cards or photocopies of the QSL cards and 10 IRCs to: Awards Manager, 6W1JN, ARAS, PO Box 971, Dakar, Senegal.

SOUTH AFRICA (ZS)

To qualify for awards sponsored by the South African Radio League (SARL), the applicant must submit a list of QSLs endorsed by the Awards Manager of his National Society and 10 IRCs. Send applications to: Awards Manager, South African Radio League, PO Box 807, Houghton 2041, South Africa.

AAA (All Africa Award)

This award is issued to amateurs working and confirming two-way contact with each of the ZS call areas, as well as one contact with V51, A2, 7P, 3DA, plus one contact each from 25 different areas of the remaining groups of country prefixes (this list is available from the SARL Awards Manager for a self-addressed envelope and 1 IRC). All contacts must be made after November 1945, and must be land stations only (islands do not count).

WAZS (Worked All ZS Call Areas)

This award is issued to amateurs working and confirming two-way contacts with 100 South African amateur stations as follows: ZS1/ZS7/ZS8/ZS9—15 contacts, ZS2—10 contacts, ZS6—50 contacts. Contacts may be made on CW, phone or mixed modes; endorsements are available for all CW or all phone. All contacts must be made since January 1, 1958.

SPAIN (EA)

To apply for awards sponsored by the Union de

Radioaficionados Espanoles (URE), applications must be accompanied by a log sheet certified by the Awards Manager of an IARU Society and 5 IRCs unless otherwise noted. Send all applications to: URE, PO Box 220, 28080 Madrid, Spain.

Espana Award

To qualify, the applicant must meet the following requirements:

HF (CW and Phone)—work a total of 80 different EA stations in 30 provinces. At least three different HF bands must be utilized. The 80 station total must include 10 stations contacted in each of the following call districts: 1, 2, 3, 4, 5, 7 and 8. Five stations must also be contacted in call districts 6 and 9 each. Stations may be worked only once.

VHF/UHF/SHF (CW and phone)—work a total of 30 EA stations in 12 provinces and four call districts. The total number of contacts per province must not exceed five. Contacts with the same station on different VHF/UHF/SHF bands are permitted.

Repeater and satellite contacts are not valid, nor are contacts with or from mobiles. Contacts must be made after January 1, 1952. Application must be accompanied by log sheet containing QSO details arranged in call district order and alphabetical order.

TPEA Award

This award is issued for working and confirming all 52 Spanish provinces. All HF bands; no mobile contacts. Repeat and satellite contacts are not acceptable. Contacts must be made after March 1, 1979. The special *5-Band TPEA Plaque* may be obtained by those amateurs contacting all 52 Spanish provinces on five bands (80-10 meters). Applications must be accompanied by the QSL cards and 25 IRCs in addition to the log sheet.

Provinces of Spain:

EA1—Asturias (O), Avila (AV), Burgos (BU), Cantabria (S), La Coruna (C), Leon (LE), Lugo (LU), Orense (OR), Palencia (P), Pontevedra (PO), La Rioja (LO), Salamanca (SA), Segovia (SG), Soria (SO), Valladolid (VA), Zamora (ZA)

EA2—Alva (VI), Guipuzcoa (SS), Huesca (HU), Navarra (NA), Teruel (TE), Vizcaya (BI), Zaragoza (Z)

EA3—Barcelona (B), Gerona (GE), Lerida (L), Tarragona (T)

EA4—Bardajoz (BA), Caceres (CC), Ciudad Real (CR), Cuenca (CU), Guadalajara (GU), Madrid (M), Toledo (TO)

EA5—Albacete (AB), Alicante (A), Castellon (CS), Murcia (MU), Valencia (V)

EA6—Baleares (PM)

EA7—Almeria (AL), Cadiz (CA), Cordoba (CO), Granada (GR), Huelva (H), Jaen (J), Malaga (MA), Sevilla (SE)

EA8—Las Palmas (GC), Tenerife (TF)

EA9—Ceuta (CE), Melilla (ML)

CIA Award

This award is issued in two categories:

Gold—work 20 Iberoamerican countries, plus Portugal and Spain, for a total of 22 required QSOs.

Silver—work 15 Iberoamerican countries, plus Portugal and Spain, for a total of 17 required QSOs.

[Iberoamerican countries: CE, CO, CP, CX, HC, HI, HK, HP, HR, HT, KP4, LU, OA, PY, TG, TI, YS, YV, XE, ZP.]

100-EA-CW Award

This award is issued for making 100 points, as follows:

• Applicants in CQ Zones 14, 15, 16, 30 and 33 must make 100 CW contacts with EA stations (each QSO = one point).

• Applicants in CQ Zones 4, 5, 8, 9, 11, 17, 18, 21, 22, 34, 35, 36 and 37 must make 50 CW contacts with EA stations (each QSO = two points).

• Applicants in all other CQ Zones must make 25 CW QSOs (each QSO = four points).

Each applicant must make contacts in at least seven of the EA call districts, and the contacts must be made on at least three bands. The same station may be worked on different bands if there are three days between contacts (except during the EA DX CW Contest). Contacts are valid from January 1, 1966. A silver medal is awarded for making 500 points, and a gold medal is awarded for making 1000 points.

EA DX100 Award

To qualify for this award, the applicant must work at least 100 countries; contacts with EA, EA6, EA8 and EA9 are mandatory. Separate awards are issued for phone and CW; all HF bands can be used. Contacts are valid from April 1, 1949. All DXCC and WAE (see below) countries are valid; deleted countries are not valid. Endorsements are issued in 25 or 50 country increments; endorsements are issued in single increments once 300 accredited countries have been attained. QSL cards must accompany the log sheet (which must be filled out in alphabetical order).

TD-EA-CW Award

The Villarreal section of URE sponsors this award, which is available for working and confirming the nine EA call districts on CW (9 contacts). Contacts are valid from January 1, 1976, and can be made on any authorized amateur HF band. Contacts made through repeaters, satellites and the like are not valid; contacts made from or with mobiles are not valid. Contacts must be made from the same DXCC country with the same call sign. Endorsements: *5B-TD-EA-CW* (work all nine call districts on five bands, 45 contacts); *160-TD-EA-CW* (work all nine call districts on 160 meters). Send log and $3 (or the equivalent in IRCs) to Delegacion Local de URE, La Mura 67, Villarreal (Castellon), Spain.

SRI LANKA (4S)

The Radio Society of Sri Lanka (RSSL) issues the *Worked 4S7*

Award for working 10 different 4S7 stations. Further details can be obtained from RSSL, PO Box 907, Colombo, Sri Lanka.

SURINAME (PZ)

The Vereniging van Radio Amateurs in Suriname, VRAS, issues one operating award for making two-way contacts with three PZ stations. Certified log data and 10 IRCs (or $5 US) for defrayment of postage and printing of the certificate should be sent to VRAS, PO Box 1153, Paramaribo, Suriname.

SWEDEN (SM)

The Sveriges Sandareamatorer (SSA) sponsors the following operating awards:

WASA (Worked All Sweden)

The WASA award will be issued to licensed radio amateurs for verified contacts with Swedish counties (laens) and call sign districts made after January 1, 1988. Swedish applicants shall be members of SSA and overseas applicants must be members of their own country's IARU affiliated society (ARRL in the United States). All contacts must be made from the same QTH and/or within a radius of 150 km from that QTH. Each individual contact must be made with the same band and mode, but the same station may be contacted on several different bands. All contacts must be made with land-based stations. Earth-based repeater contacts are not valid.

Separate awards will be issued for HF, 144 MHz, 432 MHz, 1296 MHz and satellites. For HF, 1.8, 3.5, 7, 10, 14, 18, 21, 24 and 28 MHz are counted as separate bands.

Within every group, separate awards can also be issued for different classes. Stickers can be gained for CW, Phone and RTTY as well. All contacts must be verified with QSL cards or equivalent, on which there is sufficient information to accurately determine the laen/call sign district worked. Applications must consist of QSL cards and the list of laens/call sign districts in alphabetical/numeric order. Overseas applicants may have their QSLs checked by IARU awards managers in their own countries, if such a person is available. The fee for each award is $5 US or 10 IRCs.

Applications should be addressed to: WASA, Diploma Manager, SSA, Ostmarksgatan 43, S-123 42, Farsta, Sweden. For additional information, the official WASA record book is available directly from the SSA at a cost of $4 US.

WASA Award Requirements:

HF—European applicants   **All others**

		All others
Class 3	All laens on 2 different bands	All call districts (0-7)
Class 2	All laens on 3 different bands	All laens
Class 1	All laens on 4 different bands	All laens on 2 different bands

144 MHz—(All applicants)

Class 2	All call sign districts
Class 1	All laens
Shield	Five different stations in each laen

432 MHz—(All applicants)

Class 2	All call sign districts
Class 1	All laens
Shield	Three different stations in each laen

1296 MHz—(All applicants)

Class 1	All call sign districts
Shield	All laens

Satellite

Class 1	All call sign districts
Class 2	All laens
Shield	All laens with two modes each

SLA (Swedish Locator Award)

The SLA is issued by the SSA to licensed radio amateurs for verified contacts made with the various locator squares in Sweden as defined by the Maidenhead system. Qualified contacts must have been made after January 1, 1988. The award is also issued to SWLs on an equivalent basis.

Swedish applicants must be SSA members and all other applicants must be members of their own country's IARU affiliated society (ARRL in the United States). All amateur bands may be utilized, but contacts through earth-based repeaters are not permitted.

SLA Award Requirements:

Basic award—25 squares
Sticker—35 squares
Sticker—45 squares
Sticker—55 squares
Sticker—60 squares
Sticker—all squares

Endorsements can be obtained for individual bands and modes.

The fee for the basic award is $5 US or 10 IRCs and $1 US or 2 IRCs for each sticker application. More information is available in the official SLA record book. The record book can be ordered directly from SSA at a cost of $3 US.

SLA award applications and record book orders should be sent to: SLA Award Manager, SSA, Ostmarksgatan 43, S-123 42 Farsta, Sweden.

Field Award

The Swedish Amateur Radio Society will issue the Field Award to licensed radio amateurs and short wave listeners for verified contacts with **fields** as defined by the locator system adopted after January 1, 1985 (Maidenhead locator). Contacts on or later than this date are valid for the award.

The field Award is issued in four classes:

BRONZE	(basic award)	100 fields verified
SILVER	(sticker)	200 fields verified
GOLD	(sticker)	300 fields verified
PLATINUM	(sticker)	All 324 fields verified

All Amateur Radio bands and modes are permitted. Endorsements will not be issued. All contacts must be made with terrestrial stations.

Contacts must be verified by QSL cards or their equivalent,

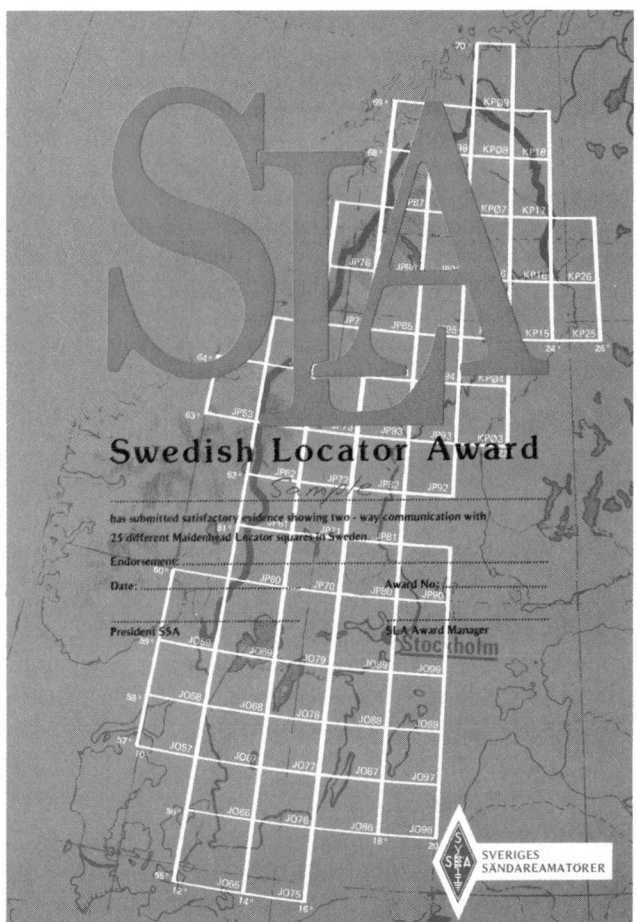

HASA (Heard All Sweden Award)

HASA will be issued by SSA to all shortwave listeners (SWLs) for verified reports of stations in Swedish laens and call districts for contacts made after January 1, 1988. The award is issued in the classes and groups corresponding to the rules for the Worked All Sweden Award (WASA). However, no shields will be issued.

SWITZERLAND (HB)

The *Helvetia Award* is issued by the USKA (Union of Swiss Short Wave Amateurs) for working and confirming contacts with all 26 Swiss cantons since January 1, 1979. The award is issued in four categories: mixed phone and CW, CW only, RTTY, and SSTV. Cross-mode and cross-band contacts are not valid.

The cantons, preceded by their abbreviation, are as follows:

AG—Aargau
AI—Appenzell Inner Rhoden
AR—Appenzell Outer Rhoden
BE—Berne
BL—Basle-Country
BS—Basle-City
FR—Fribourg
GE—Geneva
GL—Glaris
GR—Grisons
JU—Jura
LU—Lucerne
NE—Neuchatel
NW—Nidwalden
OW—Obwalden
SG—St Gall
SH—Schaffhausen
SO—Solothurn
SZ—Schwyz
TG—Thurgau
TI—Ticino
UR—Uri
VD—Vaud
VS—Valais
ZG—Zug
ZH—Zurich

on which the field or position is clearly stated with such accuracy that the field can be determined. The term "position" refers to latitude and longitude, or to a place name.

If there is any uncertainty about a field, the SAA may demand further information before approving the contact. If the uncertainty remains, the contact will not be approved.

QSL cards must be sent along with a GCR list containing the information from each QSL card, which is required for approval. The GCR list must be verified by the applicant's National Awards Manager or other official in the applicant's IARU society (ARRL in the United States). The fee is $4 US or 10 IRCs. An informative Field Award record book is available from SSA for $4 US.

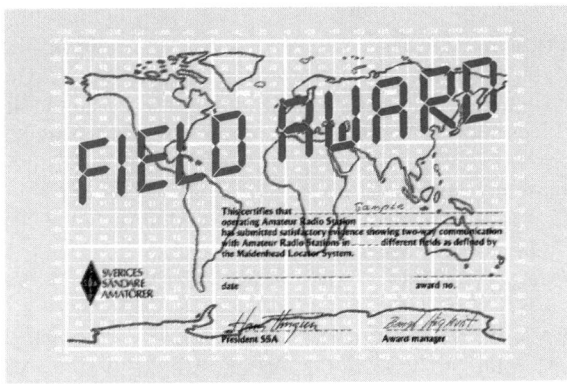

Applications and record book orders may be sent to: Field Award Manager, SSA, Ostmarksgatan 43, S-123 42 Farsta, Sweden.

QSLs should indicate the canton and they should be sent, along with a list of the QSLs (with complete QSO data), to: Kurt Bindschedler, HB9MX, Strahleggweg 28, 8400 Winterthur, Switzerland.

The award is free, but sufficient IRCs should be included to cover return postage for QSLs.

THAILAND (HS)

The Radio Amateur Society of Thailand (RAST) sponsors the *Siam Award*. To qualify, contact 10 different HS stations, any mode. To apply, send certified list and $5 US (for postage) to: Awards Manager, Radio Amateur Society of Thailand, GPO Box 2008, Bangkok 10501, Thailand.

UNITED KINGDOM (G)

The Radio Society of Great Britain (RSGB) sponsors awards in the HF and VHF categories, administered by separate Awards Managers. The following general rules apply to HF awards:

Applicants from the UK, Channel Islands and Isle of Man must be RSGB members and, as proof of membership, should provide a recent address label from the RSGB journal, *Radio Communication*. All other applicants need not be members of RSGB, but should include a label if they are members. Each application from within the UK must be accompanied by QSL cards. All others must include QSLs in the case of a special plaque; otherwise, a statement from the applicant's National Society, or a statement by two officers of a local society affiliated with the National Society, that the necessary cards have been checked will be accepted. Each application from a non-

member of RSGB must include 12 IRCs or $4 US per certificate. Applicants submitting QSL cards must also include sufficient payment to cover their return.

For RSGB HF Awards, all contacts must be made by the holder of the call sign on bands below 30 MHz. Contacts may be made from any location in the same geographical call area or, if no such area exists, then from the same country. Except where otherwise indicated, credit will be given for confirmed contacts made on or after November 15, 1945. Contacts with land-mobile stations will be accepted provided that the exact location of each station at the time of contact is clearly stated on the evidence submitted. By decision of the RSGB's HF Committee, credit will not be given for contacts made on the 10, 18 and 24-MHz bands. (This decision will be reviewed when the bands become freely available to amateurs worldwide and restricted power limits are removed.) Credit will be given for contacts made entirely on one mode or a combination of modes. Certificate endorsements for single modes and/or single bands are available. No credit for cross-band or cross-mode contacts.

Send the appropriate fee, QSL cards (if necessary) and application to: RSGB HF Awards Manager, Steve Emlyn-Jones, GW4BKG, PO Box 20, Bridgend, Mid Glamorgan CF31 4YL, Wales, United Kingdom.

Additional information on HF awards as well as a 5-Band Commonwealth checklist can be obtained by sending a self-addressed envelope to: Radio Society of Great Britain (RSGB), Lambda House, Cranborne Road, Potters Bar, Herts, EN6 3JW, England.

CCC (Commonwealth Century Club)

To apply, work and confirm contacts with amateurs in at least 100 British Commonwealth call areas on the current list [see below] since January 1, 1984. An engraved plaque is available to all recipients upon payment of a contributory charge. Applicants working all of the Commonwealth call areas on the current list will be eligible for a suitably engraved plaque (charge to be determined).

5BCC (5-Band Commonwealth Century Club)

This award, available in five classes, is available to any amateur who has worked the requisite number of stations in the call area listed since November 15, 1945 using all five bands (80-10 meters). Each station should be located in a different call area per band. The five classes are as follows:

5BCCC Supreme—500 stations
Class 1—450 stations
Class 2—450 stations, minimum 50 per band
Class 3—300 stations, minimum 40 per band
Class 4—200 stations, minimum 30 per band

Certificates are issued for each class. In addition, those qualifying for Class 1 are eligible for an engraved plaque. Those qualifying for 5BCC also receive an engraved plaque (additional fees apply). Applicants are required to pay the standard fees for any lower classes of certificates not already held.

Commonwealth Call Areas:

A2, A3, C2, C5, C6, G, GD, GI, GJ, GM, GU, GW, H4, J3, J6, J7, J8, P2, S2, S7, T2, T3Ø, T31, T32, V2, V3, V4, V8, VE1/Mar, VE1/Sable, VE1/St Paul, VE2, VE3, VE4, VE5, VE6, VE7, VE8, VK1, VK2, VK3, VK4, VK5, VK6, VK7, VK8, VK9L, VK9N, VK9X, VK9Y, VK9, VK9Z, VKØ/H, VKØ/M VKØ/A, VO1, VO2, VP2E, VP2M, VP2V, VP5, VP8/Falklands, VP8/S Georgia, VP8/S Orkney, VP8/S Sandwich, VP8/S Shetland, VP9, VQ9, VR6, VS6, VY1, VU, VU7/A, VU7/L, YJ, Z2, ZB, ZC4, ZD7, ZD8, ZD9, ZF, ZK1/N, ZK1/S, ZK2, ZK3, ZL1, ZL2, ZL3, ZL4, ZL7, ZL8,

ZL9, 3B6/3B7, 3B8, 3B9, 3D2, 3D6, 4S, 5B, 5H, 5N, 5W, 5X, 5Z, 6Y, 7P, 7Q, 8P, 8Q, 8R, 9G, 9H, 9J, 9L, 9M2, 9M6/9M8, 9V, 9Y. [Contacts with 5B valid only after March 12, 1961; contacts with 8Q valid only after July 8, 1982.]

WITUZ (Worked ITU Zones)

To qualify, work and confirm contacts with land-based amateur stations located in at least 70 of the 75 ITU zones. QSOs must be made January 1, 1984 or later to be valid. Those qualifying for WITUZ are eligible for an engraved plaque. For those working all 75 ITU zones, a suitably engraved plaque is also awarded. Additional fees apply for plaques.

5BWITUZ (5 Band Worked ITU Zones)

This award, available in five classes, is available to any amateur who has confirmed contacts since November 15, 1945 with the requisite number of land-based amateur stations located in the 75 ITU zones, using all bands (80-10 meters). Each station should be located in a different ITU zone per band. The five classes are as follows:

Supreme—350 stations
Class 1—325 stations
Class 2—300 stations, minimum 50 per band
Class 3—250 stations, minimum 40 per band
Class 4—200 stations, minimum 30 per band

Certificates are issued for each class. Class 1 winners and Supreme winners are eligible for engraved plaques. Additional fees apply. Applicants are required to pay the standard fees for any lower classes of certificates not already held. [Note: Minami Torishima (JD) lies outside the 75 ITU zones; a confirmed contact will be accepted as credit for one missing zone for WITUZ and as one missing zone per band for 5BWITUZ.]

Islands on the Air (IOTA)

The IOTA award program was created by Geoff Watts, a leading British SWL, in late 1964. In March 1985, it was taken over by the RSGB at his request. The program is well established and increasingly popular, currently consisting of 17 separate awards. Any amateur may qualify by working and confirming (after November 15, 1945) the required number of amateur stations located on islands worldwide. Many of the islands are DXCC countries in their own right; others are not, but by meeting particular eligibility criteria, they also count for credit. IOTA is an evolving program with new islands being activated all the time.

The following awards are available:

• IOTA 100 Islands Of The World
• IOTA 200 Islands Of The World
• IOTA 300 Islands Of The World
• IOTA 400 Islands Of The World
• IOTA 500 Islands Of The World
• IOTA 600 Islands Of The World
• IOTA Arctic Islands (AI)
• IOTA British Isles (BI)
• IOTA West Indies (WI)
• IOTA Africa (AF)
• IOTA Antarctica (AN)
• IOTA Asia (AS)
• IOTA Europe (EU)
• IOTA North America (NA)
• IOTA Oceania (OC)
• IOTA South America (SA)

• IOTA World Diploma (WW)

It is recommended that those wishing to participate in this award program send for the 54 page *IOTA Directory*. The *Directory* lists all islands available for IOTA credit and provides full information on all awards. It is available for $12 US or 18 IRCs. Mail all requests to: Roger Balister, G3KMA, IOTA Director, La Quinta, Mimbridge, Chobham, Woking, Surrey GU24 8AR, England. (If payment is made via check, the check must be issued by a United Kingdom bank. Non-UK checks will not be accepted.)

IARU Region 1 Award

This award is available is three classes for confirmed contacts with stations located in the requisite number of countries whose National Societies are members of IARU Region 1. The three classes for contacts are as follows:

Class 1—all IARU member-countries on list
Class 2—45 member-countries
Class 3—30 member-countries

IARU Region 1 member countries: A2, A4, A9, C3, C5, CN, CT, DL, EA, EI, EL, F, G, HA, HB, HB0, I, J2, JY, LA, LX, LZ, OD, OE, OH, OK, ON, OY, OZ, PA, SM, SP, SU, SV, SZ, T7, TA, TF, TR, TU, UA, YK, YO, YU, Z2, ZB, ZS, 3A, 3B8, 3D6, 4X, 5B, 5N, TZ, 6W, 7P, 7X, 9G, 9H, 9J, 9K, 9L.

A special version of this award is available in the same three classes for confirmed contacts on 10 meters since July 1, 1983.

28-MHz Counties Award

This award is available for confirmed 28-MHz contacts since April 1, 1983 with amateur stations located in 40 counties/regions of the UK, Channel Islands and the Isle of Man. Stickers are available for confirming 60 and all 177 counties/regions.

[Information on the RSGB's extensive VHF awards program can be obtained from VHF Awards Manager Jan L. Cornes, G4OUT; please include 2 IRCs and a self-addressed envelope. Address all correspondence to: RSGB VHF Awards Manager, 6 Haywood Heights, Little Haywood, Stafford ST18 0UR, England.]

VANUATU (YJ)

The Vanuatu Amateur Radio Society (VARS) issues the *Vanuatu Award* for working at least six YJ8 stations who are members of VARS. Contacts made from July 30, 1980, Vanuatu

Independence Day, are acceptable. Contacts may be made by any mode. Endorsements for single band, single mode, or additional stations worked are available. Two contacts with any one YJ8 station will be accepted provided that the contacts are made on different days, bands or modes. Send a certified log extract and 10 IRCs (or $2 US) to: Awards Manager, VARS, PO Box 665, Port Vila, Vanuatu.

HF Digital Communications

STEVE FORD, WB8IMY
ARRL HEADQUARTERS

If you've spent time exploring the Amateur Radio HF bands, you've probably heard some mysterious sounds— especially the odd noises that you'll find just above the CW portions of the bands. No doubt you've listened to the warbling, musical signals of Baudot radioteletype, otherwise known as *RTTY* (pronounced "ritty"). If you twisted the dial a bit further, you may have heard a chorus of electronic crickets. These are the chirping dialogs of *AMTOR* (*AM*ateur *T*eleprinting *O*ver *R*adio) stations. Perhaps you also picked up the sounds of *PacTOR*, *G-TOR* or *CLOVER*. And what were those raspy, high-pitched bursts? Those are the unmistakable signatures of HF *packet*.

These are the primary HF digital modes. They're called digital modes because the communication involves an exchange of data between one station and another. In the case of RTTY, for example, letters typed on a keyboard are translated into data by a computer or data terminal. Another device, usually a *multimode communications processor (MCP)*, accepts this data and converts it to whatever encoded audio tones are required. The tones are sent to the transmitter and away they go! At the receiving end the same process occurs in reverse: The tones are translated back into data and displayed as text on a computer or terminal screen.

AN HF DIGITAL HISTORY LESSON

In the sense that we commonly interpret the word "digital" today, RTTY has the distinction of being the granddaddy of HF digital communications. RTTY dates back to World War II when the military began connecting mechanical teletype machines to HF radios. The idea was to send printed communications into areas where telephone lines didn't exist.

At first they tried simple on/off keying to send the text, but that didn't work very well. The receiving equipment couldn't always tell the difference between a signal and a burst of noise. After some further experimentation, the designers switched to *frequency shift keying (FSK)*. This approach used two specific tones to indicate the on/off (MARK/SPACE) signals. FSK was a success and RTTY as we know it today was born.

Hams adopted RTTY after the war ended and by the early '50s it was a well-established mode. Initially it found a home on VHF, but later became more popular on the HF bands.

(RTTY is seldom heard on the VHF bands today.)

For several decades, hams relied on surplus teletype machines for their RTTY stations. These mechanical monsters were slow, noisy and often dirty (they had a nasty tendency to drip oil on the floor!). Operators had to read the text on paper as it was printed. The keyboard was a bit unusual, mainly due to the nature of the Baudot code.

With Baudot, all letters are capitalized (upper case). All numbers from 0 through 9 are available along with some limited punctuation. To send numbers or punctuation a special

Multimode communications processors (MCPs) such as these offer RTTY, AMTOR, packet, PacTOR, CW and many other modes in a single device.

Fig 9-1—The old teleprinters featured a keyboard layout similar to the one shown here. Notice the FIGS key in the lower left corner. You had to press this key before sending punctuation or numbers. Pressing the LTRS key returned you to the "letters" mode. Although these vintage units have all but disappeared, the need to shift from FIGS to LTRS and back again remains. This is now handled automatically by software.

FIGS character must be sent *first*. To return to alphabetical letters, a LTRS character must be sent. On the original teleprinters you had to press the FIGS key whenever you wanted to send numbers or punctuation. To return to alphabetical letters, you had to press the LTRS key (see Fig 9-1). You can imagine how difficult it must have been to master the keyboards of those old machines!

Some ancient teletype machines remain, but most RTTY enthusiasts rely on computerized systems. The Baudot code is still the same, however, and the FIGS/LTRS shift is still required. No need to worry about your typing skills. The FIGS/LTRS shift is handled automatically by most systems. All you have to do is type and your software and hardware will take care of everything else. Rather than reading the text on flowing sheets of paper, you'll see it on your monitor screen.

RTTY has fulfilled its promise of transmitting the written word throughout the world. When band conditions are good and signals are optimal, RTTY is efficient and accurate. But what happens when conditions are less than ideal? That's where RTTY shows its weak side. Interference from other transmitters, fading and electrical noise cause errors in RTTY communications. Mild interference will cause a few letters to be deleted here and there. Severe interference, however, can turn the entire text to gibberish!

Commercial maritime communication systems relied on RTTY for decades, but the inability to detect errors was a persistent problem. It could even have life-threatening consequences. What if a maritime weather service tried to alert ships of an approaching storm? The RTTY-transmitted warning had to be repeated over and over to give the ships a decent chance of copying the entire message.

THE ADVENT OF AMTOR

The pressure to create a more reliable teletype system lead to the development of *TOR* (Teleprinting Over Radio), commonly known today as *SITOR* (Simplex Teleprinting Over Radio). Instead of sending the text in one long transmission, the TOR method sends only a few characters at a time. The receiving station checks for errors using a bit-ratio-checking scheme. If all characters are received error-free, the receiving station sends an acknowledgment or ACK signal and the next few characters are transmitted. If an error is detected, a *nonacknowledgment* or *NAK* signal is sent. This tells the transmitting station to repeat the characters. The result is digital communication *without* errors—a major improvement over previous RTTY systems. The rapid error-checking dialog—known as *Mode A* or *ARQ*—creates the distinctive chirping sounds associated with SITOR communications.

In the early 1980s, the Federal Communications Commission approved SITOR techniques for Amateur Radio use. Peter Martinez, G3PLX, adapted SITOR coding and developed AMTOR. AMTOR was tailor-made for the personal-computer era and it quickly became popular. (The advent of PCs also made *ASCII* RTTY possible using a *complete* character set including upper and lower case letters. However, its use remains limited today because of its lack of error detection and correction.)

Packet, PacTOR, CLOVER and G-TOR

Packet also appeared on the scene in the early 1980s, although it was (and still is) most popular on VHF and UHF. Twenty meters is the most popular band for HF packet, but you'll encounter this mode on other bands, too. Every MCP on the market today has HF packet capability. So do a number of packet-only TNCs. The widespread availability of equipment is the engine that's driven much of the packet activity on the HF bands.

Three more HF digital modes debuted in the early 1990s: PacTOR, G-TOR and CLOVER. PacTOR is a fusion of AMTOR and packet. It features the capabilities of packet (upper/lower case characters, binary transmission and so on) with an ACK/NAK system similar to AMTOR. G-TOR and CLOVER use complicated modulation and coding schemes to overcome the problems associated with HF digital communications. All three modes boast relatively efficient operation (CLOVER is the most efficient of the group).

THE NATURE OF FSK

If you want to become a competent HF digital operator, it helps to understand frequency-shift keying, or FSK. Let's start with data. I'm sure you've heard that the fundamental language of all computers is binary *machine code*. In a binary-number system, you're only dealing with 0s and 1s. This is a natural situation for a computer since it's comprised of a multitude of solid-state *logic* switches that can only be *on* or *off* ("high" or "low"). So, an "on" condition represents a binary 1 while an "off" condition represents a binary 0.

If you use wires to connect two computers, the on/off voltage states are communicated from one machine to another easily. But let's make the situation more complicated and move the computers several hundred miles apart. Now what are you going to do? Radio seems like a natural choice, but you can't send high/low voltage states over the air. . .or can you?

What if you translated the changing voltages to changing *tones*? You could use 2,125 Hz to represent a binary 1 and 2,295 Hz to represent a binary 0. Feed those tones to the audio input of an SSB transceiver operating on lower sideband (LSB), for example, and they'll be transmitted as signals at specific

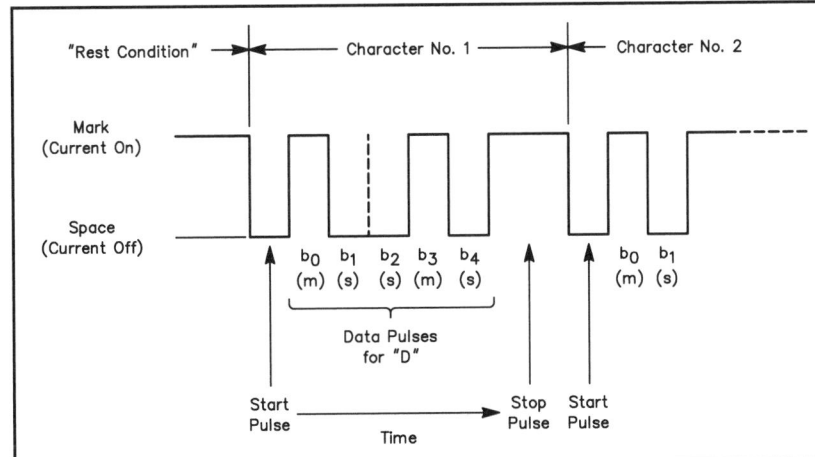

Fig 9-2—This is a diagram of RTTY MARK and SPACE signals as the letter "D" is sent. A start pulse begins the character, followed by the 5 bits that define it (b_0 through b_4). By looking at the bits, you can see that "D" is MARK-SPACE-SPACE-MARK-SPACE. A stop pulse signals the end of the character and another start pulse begins a new character. These MARK and SPACE pulses appear as shifting audio tones at the receiver. They are fed to a multimode communications processor (MCP) to be decoded back into data and displayed as text.

points below the *suppressed carrier frequency*. The 2125-Hz tone will create a signal 2125 Hz below the suppressed carrier. The 2295-Hz tone will create a signal 2295 Hz below the suppressed carrier. Subtract the frequency of the high signal from the low signal and you get 170 Hz. In other words, the signal *shifts* 170 Hz to represent a 1 or 0. Shifting voltages have become shifting signal frequencies! As we discussed previously, the signal that represents a binary 1 is called the MARK. The signal that represents a binary 0 is called the SPACE (see Fig 9-2).

At the receiving end of the path, you'll need to convert the signals back into binary high/low voltage states. FSK demodulators are designed with audio filters to detect the MARK and SPACE tones and produce corresponding data pulses. Feed those data pulses to a computer running *terminal* software and—trumpet fanfare please!—text appears on the screen.

Most HF digital modes use frequency shift keying to pass information from one station to another. A 170-Hz shift between the MARK and SPACE tones is the Amateur Radio standard for RTTY and AMTOR. A few hams use an 850-Hz shift, an artifact from the early VHF RTTY days. Commercial SITOR stations often use a 425-Hz shift. Most MCPs available today offer all three shifts. However, many MCPs fudge the 170-Hz shift to accommodate HF packet as well as PacTOR (see the sidebar "Shifty MCPs").

Shifty MCPs

Most multimode communications processors (MCPs) use a 200-Hz shift for RTTY operation (mark = 2110 Hz, space = 2310 Hz). These tones are used for HF packet radio as well. They are compatible with standard 170-Hz shift tones (2125/2295 Hz) since the center frequency (2210 Hz) is the same. In weak-signal conditions, using a 200-Hz shift places you at a disadvantage. Both you and the other station (possibly using 170-Hz shift) must resort to *straddle tuning* to achieve true transceive operation. This is tricky and often results in less than optimum performance.

It's possible to obtain optimum RTTY performance by retuning the MCP receive filters and transmitter tones to match the 2125/2295 Hz standard. While this technique offers a noticeable improvement, you should *not* attempt the adjustment unless you are thoroughly familiar with your MCP and possess all the necessary test equipment. (Most MCP manuals *do not* describe this procedure.) A botched realignment will probably void your MCP warranty!

FSK or AFSK?

In our previous example, the shifting tones were supplied to the transmitter via an audio input jack. Technically speaking, this is audio FSK or AFSK. There is another method known as *direct* FSK. What's the difference? If you're using direct FSK, the data pulses from your MCP are *not* converted to MARK and SPACE tones. Instead, the MCP supplies these high and low logic levels to the FSK input of an SSB transmitter where they're used to shift the frequency of the master oscillator up and down. The result at the receiver is the same: MARK and SPACE tones.

So which is better—direct FSK or AFSK? Some hams believe that FSK is a "purer" method of transmitting because it minimizes distortion and harmonics. Other hams argue that if the transmitter isn't overmodulated, AFSK transmissions are just as pure. You'll find large numbers of amateurs using one method or the other, although the majority seem to favor the AFSK approach because many transceivers do not feature an FSK input. The important thing to remember is that whether you use AFSK or direct FSK, the result at the receiving station is the *same*.

BAUDOT RTTY

Before you can enjoy a RTTY conversation, you must be capable of tuning the signals properly. Every MCP features some sort of tuning indicator, depending on the type of equipment you're using. In days gone by, RTTY operators would attach oscilloscopes to their terminal units and tune the signals until they saw the classic "crossed bananas" display (see Fig 9-3). As technology advanced, many terminal units in-

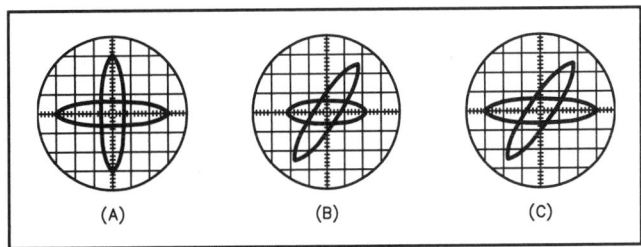

Fig 9-3—Oscilloscope-type tuning indicators produce patterns like these. Pattern A is the classic "crossed bananas," showing that the RTTY signal is tuned properly. At B the receiver is slightly off frequency, while C indicates that the transmitting station is using a shift that differs from the MCP setting.

cluded tiny built-in oscilloscopes that performed the same function.

MCPs often use LED indicators. Some units feature an indicator comprised of several LEDs arranged in a horizontal bar. When the LEDs at opposite ends of the bar flash in sync with the RTTY signal, you know you have it tuned properly.

What Is *That Signal?*

Most RTTY operators use lower sideband transmissions with a 170-Hz frequency shift between the MARK and SPACE signals. The commonly used data rate is 60 wpm, often expressed as 45 baud. But what if you stumble upon two operators who aren't conforming to "conventional" practices? Your indicator says you're tuned in properly, but nothing coherent prints on your screen!

It looks like you'll need to do a little detective work. Check the following:

❏ Is the signal "upside down"? The RF frequency of the MARK signal is usually higher than the RF frequency of the SPACE signal, but there is no law that dictates this standard. With most MCPs, all it takes is a push of a button or keyboard key to invert the normal MARK/SPACE frequency relationship.

❏ Are the operators really using 170-Hz shift at 45 baud? For example, many RTTY operators prefer to run at 75 baud (100 wpm) when exchanging lengthy files. To complicate matters, an operator may also decide to use an 850-Hz shift.

It may be of some comfort to know that these situations are uncommon. You may encounter RTTY at 75 baud or higher, but most operators stick to 45 baud. The use of inverted MARK/SPACE signals and odd frequency shifts is relatively rare.

Listen to the Action

Let's say you've successfully tuned in a RTTY signal. Unless you happen to be monitoring during a contest, a typical RTTY conversation may look like this:

I FINALLY TOOK DOWN MY OLD VERTICAL AND REPLACED IT WITH A DIPOLE CUT FOR 40 METERS. YOU KNOW WHAT I FOUND OUT? I CAN FORCE FEED THAT DIPOLE WITH MY ANTENNA TUNER AND USE IT ON 20, 15 AND 10 METERS. N9GOR DE WA1MBK K

WA1MBK DE N9GOR....REALLY? THAT SOUNDS PRETTY INCREDIBLE. I HAD NO IDEA YOU COULD DO THAT. I BET THE SWR IS VERY HIGH ON THE OTHER BANDS SINCE THE DIPOLE WOULD NOT BE RESONANT ON THOSE FREQUENCIES. WHAT ABOUT THE RF YOU LOSE IN YOUR FEED LINE? WA1MBK DE N9GOR K

N9GOR DE WA1MBK...FEED LINE LOSS IS NOT A PROBLEM EVEN WITH A HIGH SWR. AS LONG AS YOU USE LOW LOSS COAX OR OPEN WIRE FEED LINE, THE LOSS IS SO SMALL IT DOES NOT MATTER. A GOOD ANTENNA TUNER IS THE KEY...N9GOR DE WA1MBK K

And so it goes, back and forth at a leisurely pace. When an operator is not sending data, you'll hear a continuous tone or a rhythmic *bee-bee-bee-bee* signal. The rhythmic signal is known as the *diddle*. The MCP is simply switching back and forth between MARK and SPACE signals while awaiting more data from the operator.

The call signs at the beginning and end of each transmission are optional, as long as you identify at least once every 10 minutes. Still, some habits are hard to break and you'll find that RTTY operators often open and close their transmissions with an exchange of call signs. This is made easier by the fact that many modern MCPs feature automatic call sign exchange

The well-equipped HF digital station of Francis Fallon, N2FF.

capability. You simply enter the call sign of the other station *once*. After that, you can generate the entire <his call sign>-DE-<your call sign> sequence by pressing a single key.

Did you notice the K used to signify the end of each transmission? It's a RTTY custom to use CW prosigns in conversations. Depending on the operator, he or she may send a K (over to you), AR K (end of message, over to you) or KN (over to you *only*). When the conversation is over, it's common to use SK to signal the end of the contact. RTTY operators also adopt the CW custom of abbreviating words. In the sample conversation shown above, "frequencies" may be sent as "freqs."

It's Your Turn

What if you're eavesdropping on a QSO and you see that it's about to end?

N9GOR DE WB8IMY...I HAVE TO RUN, WAYNE. DINNER WILL BE READY IN AN HOUR AND I STILL HAVE NOT FINISHED MOWING THE LAWN. HAVE A GOOD WEEKEND AND I HOPE TO CHAT WITH YOU AGAIN....N9GOR DE WB8IMY SK

WB8IMY DE N9GOR...NO PROBLEM. GLAD I HAD THE CHANCE TO DISCUSS ANTENNAS WITH YOU. I LEARN SOMETHING EVERY DAY. SEE YOU LATER...WB8IMY DE N9GOR SK

Looks like N9GOR is still available. Why not give him a call? Since you were able to copy the transmissions, chances are good that your MCP is already set for the proper data rate, shift and MARK/SPACE frequency relationship. (A quick check never hurts, though!) Switch your MCP to the transmit mode and start typing...

N9GOR N9GOR DE N6ATQ N6ATQ N6ATQ

N9GOR N9GOR DE N6ATQ N6ATQ N6ATQ K K

[switch back to receive]

Make sure to repeat your call sign several times. After all, N9GOR knows his own call, but he doesn't know *yours*. Send your transmission in several short lines rather than one long line.

If the other operator is able to copy your signal, you may see a response like this:

N6ATQ N6ATQ DE N9GOR N9GOR...THANKS FOR THE CALL. NAME HERE IS WAYNE WAYNE AND I AM LOCATED IN MILWAUKEE MILWAUKEE WISCONSIN WISCONSIN. YOUR RST RST IS 579 579...BACK TO YOU...N6ATQ DE N9GOR K K

Notice how N9GOR sends the important information *twice*. He doesn't know how well you are receiving his signal, so he wants to make sure that you won't miss his name, city, state and so on. This is always a good technique to use when you aren't sure of the signal path between your station and another.

Wayne has sent your signal report (RST) and it's 579. The first digit from the left is your readability (R), the second is relative strength (S) and the third is tonal quality (T). In this case, a 579 RST means that he is receiving you very well, your signal strength is moderate and your RTTY tones are good. A perfect RST would be a 599, but a 579 is fine. You can be reasonably certain that he is receiving everything you're sending. Go ahead and tell him who you are and where you are. Don't forget to give him a signal report, too.

[switch to transmit]

N9GOR DE N6ATQ....HELLO WAYNE. NAME HERE IS CRAIG CRAIG AND I AM IN ESCONDIDO ESCONDIDO CALIFORNIA CALIFORNIA. YOUR RST RST IS 599 599. THIS IS MY FIRST RTTY CONTACT. I AM USING A MODEL PK232 MCP AND A KENWOOD TS-820S TRANSCEIVER. ANTENNA IS A DIPOLE UP 30 FEET. SO HOW COPY? N9GOR DE N6ATQ K

[switch to receive]

These are the preliminaries of most RTTY contacts. Some hams are very proud of their station equipment and will give you a brief rundown of their entire setup. In the earlier days of RTTY, messages such as these were created in advance and stored on reels of paper tape. When fed to a teleprinter, the holes punched in the paper tape were translated into MARK and SPACE signals for transmission.

If you own a computer, you can create and store "canned" messages of your own and save them on diskette or magnetic tape. Depending on the type of software you're using, a single keystroke will send the entire message automatically! Many RTTY operators store and send their station descriptions in this manner, although the process is still known by its old namesake: the *brag tape*. Consult your software manual to learn how to create your own brag tapes and other stored messages.

Is There Anybody Out There?

If you can't find someone to talk to, consider calling CQ. You never know what you'll turn up...

[switch transmitter on]

CQ CQ CQ CQ CQ CQ CQ DE WB8IMY WB8IMY WB8IMY

CQ CQ CQ CQ CQ CQ CQ DE WB8IMY WB8IMY WB8IMY

CQ CQ CQ CQ CQ CQ CQ DE WB8IMY WB8IMY WB8IMY K K

[switch to receive]

A CQ should be long enough to attract attention, but short enough to avoid boring the other station. Repeat your call sign often so the operator on the other end has a decent chance of getting it right. You may have to send your CQ more than once before you're noticed. Storing your CQ on disk makes it easy to send it again without retyping.

If you don't receive an answer, don't lose hope. Just move to another frequency or band. Before calling CQ, it's good practice to ask if the frequency is in use. Perhaps you've just tuned onto a frequency and it *seems* to be unoccupied. Don't let appearances fool you! You often hear only one side of a con-

Bad RTTY Habits

As you scan the RTTY subbands, you'll find some operators sending long streams of RYs at the beginning of their transmissions.

RYRYRYRYRYRYRYRYRYRYRYRYRYRYRYRYRYRYRY CQ CQ CQ CQ CQ CQ CQ CQ DE WB8XYZ WB8XYZ WB8XYZ K K

This is another artifact from the early days of RTTY when it was necessary to make sure that mechanical teleprinters were ready to copy a transmission. In the modern era of computers and data terminals, this is unnecessary. Many operators find these long RY streams highly irritating since they do nothing but waste time and band space. Try to avoid the RY habit. If you have something to say, such as calling CQ, go ahead and say it. You don't need to send a meaningless string of letters in advance.

You may also find operators who make frequent use of the date/time function included in their MCPs. Many units have this feature but, thankfully, most operators don't use it except in contest or public service situations. What does it do? It inserts the time and date (local or UTC) as part of the transmitted text. Usually it appears at the end of a transmission, but may pop up at the beginning. You may also see it at the end of stored messages.

Unless you're in desperate need of a clock, do you really want to know what time it is? Probably not. Now turn the tables. Do you think the other station cares to be informed of the date and time? I think you get the idea! Time and date "stamping" is another waste of RF in most cases. Veteran RTTY operators consider it a nuisance, although most will be too polite to tell you!

versation—and that side may be listening at the moment!

Checking the frequency is easy. For example, I'd send:

QRL? QRL? DE WB8IMY WB8IMY K K

If the frequency is occupied, someone will let me know right away. On the other hand, if no one replies, I can assume it's safe for me to go ahead and call CQ.

Working the DX Pileups

Sooner or later you're bound to encounter the fascinating phenomenon known as the *DX pileup*. You'll know when you've found a pileup because it will sound like pure pandemonium.

Pileups are the result of a desirable DX station coming on the air. What makes a DX station worthy of a pileup? If the operator is in a country that is not heard on the air often, that country is considered "rare DX." Any transmission from that part of the globe is a major event and it quickly attracts hordes of hams eager to make a contact.

The first CQs from a rare DX station snag the few hams who are lucky enough to be near the frequency at the time. More join the fray as they discover what's going on. DX *PacketClusters* also sound the alarm on the VHF frequencies and bring operators by the dozens. Within minutes you have a huge number of stations chasing the same goal: the hapless DX operator. As soon as he finishes a contact, everyone starts calling at once! You can imagine how this would sound in your receiver.

When you hear a pileup in progress, the first thing to do is monitor the exchanges. Determine the DX station's call sign and see if you can copy his signal. When the FR5ZU/G group began calling for contacts from Reunion Island, here is how it looked at my station.

DE XE1/JA1QXY..XE1/JA1QXY..XE1/JA1QXY..BK BK

XE1/JA1QXY DE FR5ZU/G...UR 599 599 BK TO U...KN

QSL UR 559-559 TKS QSO DE XE1/JA1QXY

XE1/JA1QXY sends his call sign several times and is heard by FR5ZU. Signal reports are exchanged quickly and the contact is over in a matter of seconds! Notice the heavy use of Q-signals and abbreviated words to speed the process. The DX station is fair game once again and two operators slug it out for the prize…

DE W2JGR DE W2JGR DE W2JGR DE W2JGR K

FR5ZU/G DE NJØM NJØM NJØM PSE K

NJØM uses the traditional approach while W2JGR tries repeating his call sign preceded by DE ("from"). His tactic pays off and he wins.

W2JGR DE FR5ZU/G...GOOD MORNING...UR 569 569...BK

FR5ZU/G DE W2JGR...TNX...UR 579 579 NAME JULES... QSL??? BK

FR5ZU passes along a signal report and shoots it right back to W2JGR. W2JGR gives his report and sends his name (Jules) as well. At the end of his transmission, he asks if the DX station copied everything ("QSL?").

QSL ES 73...FR5ZU/G QRZ KK

FR5ZU sends a quick "QSL" to mean, "Yes, I got it all" along with his best wishes. He immediately sends QRZ to signal that he is ready for another contact. That's NJØM's cue to try again!

DE NJØM NJØM NJØM PSE KK

What you've seen here is less than five minutes of a DX pileup that lasted over an hour! Pity the DX operator at the other end of this melee. He has to do his best to sort out readable call signs among the rampaging signals. Sometimes the interference is so severe, he sees nothing but a wild jumble of letters on his screen.

The best you can do is be patient and keep calling. The rules of DX courtesy say that you shouldn't transmit if a contact has already been established. If you keep transmitting, you may get the attention of the DX station in a way you'd never expect—he'll refuse to answer you for the remainder of the operation! Also, if the DX operator tries to control the mayhem through techniques such as working stations by call sign areas (1s, 2s, 3s and so on), don't buck the system. All you'll manage to do is anger the person you're trying to contact!

It's a "Split Decision"

What if you discover a rare DX station, but you don't hear a pileup? No matter how long you listen, you only seem to copy one side of the conversation—his! This is the telltale sign of a DX station that is *working split*. In other words, he is listening on one frequency and transmitting on another.

For some DX stations, working split is the only way to manage a pileup. This is especially true when a pileup gets too large and begins to disintegrate into chaos. Without the ability to work split frequencies, the DX station may be buried under a torrent of competing signals. Even if he manages to sort out a call sign, making contact is difficult because of interference from other stations who are continuing to call.

A good DX operator will always make it clear that he is working split and will indicate where he's listening for replies. You may see something like this:

CQ CQ DE 5U7M 5U7M UP 10

or…

QRZ DE 5U7M LISTENING 21.085 TO 21.090

The DX operator, 5U7M, is telling everyone that he is listening for calls 10 kHz above this frequency ("UP 10"), or that he is listening between 21.085 and 21.090 MHz specifically.

To contact a DX station using a split-frequency scheme, you'll need a transceiver that can transmit on a frequency other than the receive frequency. Many modern transceivers feature *dual VFOs* for this purpose. Other rigs have the capability to use a *remote VFO* (a second VFO in a separate enclosure).

If you have remote or dual-VFO capability, leave your receiver tuned to the DX station and move your transmit frequency to the desired spot. Check your equipment carefully and make sure you know exactly where you are transmitting and receiving. Whatever you do, don't transmit on the DX station's frequency.

AMTOR

AMTOR is a close cousin to RTTY. What sets it apart is its ability to detect errors. Unlike RTTY, AMTOR provides HF digital communications with a consistent level of accuracy.

Two types of AMTOR are used most often in Amateur Radio: *FEC* (or *Mode B*) and *ARQ* (or *Mode A*). When you hear those odd *chirp-chirp* signals, you're listening to ARQ AMTOR. FEC, on the other hand, sounds like very fast RTTY. Before you make your first AMTOR contact, let's take a moment to talk about both modes.

It Looks so Nice, I Sent it Twice

FEC stands for *Forward Error Correction*. In AMTOR applications it is a synchronized mode where every character is sent twice. The receiving station takes full responsibility for sorting out the errors. This means that an FEC transmission is strictly a one-way affair. The receiving station cannot ask for a repeat if it detects an error.

Most AMTOR operators use FEC to send CQ in the hope of establishing an ARQ connection. Although it's not a common practice, you can also use FEC to carry on complete conversations. Like RTTY, *any* SSB transceiver is capable of operating in the FEC mode. Just be sure to watch your output power! FEC is a 100% duty-cycle mode. You may need to decrease your output by as much as 50% to avoid damaging your equipment.

Shaking Hands with ARQ

AMTOR ARQ is a totally different mode compared to FEC. ARQ is a generic data term that means *Automatic Repeat reQuest*. In the ARQ mode, the *information sending station* (*ISS*) sends its text in bits and pieces. It transmits a group of three characters in a single burst and then switches to the receive mode. The *information-receiving station* (*IRS*) checks the characters for the 4:3 bit ratio and then transmits a control character. The control character means "Acknowledged. Send the next three," (*ACK*) or, "Not acknowledged. Repeat the last three" (*NAK*). If the ISS doesn't receive a reply (due to fading signals or interference), it repeats the characters anyway. Each station gets its turn to be the IRS or ISS (see Fig 9-4).

It's fair to say that when you're using AMTOR ARQ there are really two conversations taking place. First, there is the conversation between you and the other amateur. Secondly, there is the conversation—often referred to as *handshaking*—going on between your hardware and the hardware at the other station.

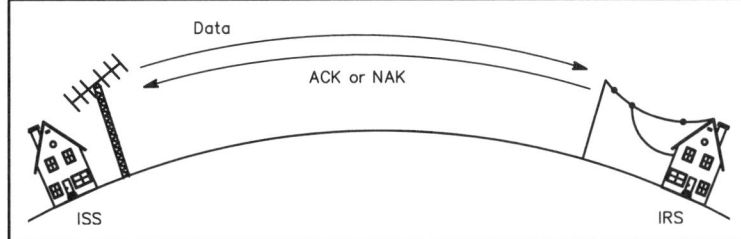

Fig 9-4—AMTOR, CLOVER, PacTOR and G-TOR use a system of acknowledgments (ACKs) and nonacknowledgments (NAKs) to send information. The information sending station (ISS) sends a group of data bits to the information receiving station (IRS). If the data is received without errors, an ACK is sent and the next group is transmitted. A NAK means that errors occurred and the data must be sent again.

Eavesdropping on AMTOR

Most MCPs include some sort of *listen* mode. The listen mode allows you to tune to an ARQ contact in progress and print the text *without* being part of the link. You need a bit of patience when using the listen mode since it doesn't include the error-correcting feature. You'll see plenty of errors and repeated characters, but at least you can follow the discussion.

Listen mode also works differently, depending on the MCP you're using. The listen mode in some controllers automatically senses and switches to ARQ or FEC. Other units monitor only ARQ transmissions in the listen mode and must be manually switched to the standby mode to monitor FEC signals.

Switch your transceiver to LSB and your MCP to the listen or *monitor* mode. Now hunt for a chirping ARQ signal. I suggest that you look for a strong signal between 14.070 and 14.080 MHz. (Twenty meters isn't the only band with AMTOR activity, but it's one of the most popular.) When you've tuned the signal correctly—remember to watch your indicator—you should see characters within 15 to 20 seconds. If not, try another signal. On some controllers, you may have to reset the listen mode to restart the synchronizing process. With practice, you should be able to copy an ARQ signal with ease. The listen mode may drop out of sync at times, especially when one station ends its transmission and switches from ISS to IRS. This is normal since the listen mode can only synchronize to one station at a time.

Now, find an FEC signal. At first, FEC may be hard to differentiate from RTTY. An FEC transmission includes special synchronizing characters, but they're often sent only once per line of text. It can take 10 seconds or more to receive these synchronizing characters—longer if you miss them or get a noise burst when they're sent. Some MCPs send extra FEC synchronizing signals and these signals will be easier to receive. Practice tuning FEC before you attempt your first AMTOR contact.

Calling CQ

Switch your MCP to the FEC mode. (Some MCPs omit this step. They permit you to send FEC while in the ARQ mode.) An AMTOR CQ should include both your call sign *and* your SELCAL code (see the sidebar, "What's Your SELCAL?").

[switch your transmitter on]

[send a blank line]

CQ CQ CQ DE WB8IMY WB8IMY WB8IMY (WIMY WIMY WIMY)

CQ CQ CQ DE WB8IMY WB8IMY WB8IMY (WIMY WIMY WIMY)

CQ CQ CQ DE WB8IMY WB8IMY WB8IMY (WIMY WIMY WIMY) K

[switch back to receive]

The last step is very important! Your AMTOR controller must return to *standby* to be ready to receive an ARQ call. Most AMTOR controllers have two different ways to end an FEC transmission: one command returns the controller to the standby mode and another returns to the FEC mode. Be sure to check your manual.

Also remember to send your FEC CQ with lots of short lines terminated with carriage returns. Why? It allows receiving MCPs to quickly synchronize with your signal. The faster they synchronize, the sooner they'll copy your message.

If someone wants to chat with you, they'll call you in ARQ mode using the SELCAL code you sent (WIMY in our example). If you don't get an answer, try again. Keep your calls short—don't get fancy or long-winded.

Let's assume that someone is answering your call. You hear the familiar chirp-chirp-chirp as they send your SELCAL over and over. In the meantime, your MCP is decoding the signals. Does the received SELCAL match yours? Yes! At this point you may hear a tone or chime followed by a message telling you that you're linked to the other station.

The AMTOR dance has begun, but who calls the tune? Well, a station that answers (in ARQ) the FEC call of another station is the *master*. It sets the timing parameters and the other station—the *slave*—must synchronize to its signals. In this case, the station that called you is the master and you're the slave. Throughout the conversation, the master and slave roles remain fixed.

Anatomy of an ARQ Conversation

Once an ARQ link begins, the exchange of ACKs and NAKs goes on *continuously*—even if no one is sending information. If you and the other operator decided to leave your keyboards and grab a snack, your stations would chirp mindlessly back and forth to each other.

When the link is first established, you're the IRS. Just wait patiently for the other operator to send his greeting...

WB8IMY DE KU7G...

HELLO! MY NAME IS BOB AND I LIVE IN WASHINGTON STATE IN THE SHADOW OF MOUNT ST HELENS. YOUR RST IS 589. BACK TO YOU....+?

Notice how Bob leads off with his call sign and yours. This is important since all you've exchanged so far are you SELCALs. He sends a brief greeting and ends it with **+?**. This strange-looking character is the *over* command. It allows the stations to trade places from ISS to IRS. Bob is turning the link over to you so you can send text.

Over to You

There is a very important point to remember about AMTOR ARQ: *each station must turn the link over to the other station at the end of every transmission.* (This has nothing to do with the master/slave timing relationship.) The normal procedure is to type **+?** to switch the link. With many MCPs, the **+?** is sent by tapping a single key. This automatically sends a control code that says, "Let's turn the link around. You're the ISS now." When the link switches, you'll hear a distinct change in the chirping rhythm.

Recognizing that there are times when the IRS operator (receiving station) would like to immediately break in and make a comment, most MCPs include a *forced over* command. A forced over causes an immediate link reversal, even if the ISS operator is still typing or has text in his transmit buffer. The exact command used to cause a forced over varies between units. Use the forced over sparingly; it's rarely needed, but very handy at times.

The Conversation Continues

So Bob has turned the link over to you. Now you're the ISS and he's the IRS. Why not send a short greeting and a signal report, too?

HELLO, BOB. YOU ARE SOUNDING FINE HERE IN CONNECTICUT. I LIVE IN A TOWN CALLED WALLINGFORD AND MY NAME IS STEVE. YOUR RST IS 599.

There is no need to repeat the information as you might do

during a RTTY exchange. With the ARQ ACK/NAK system, the other station either receives your text or doesn't! With rare exception, the text is never garbled on the receiving end. If there is trouble on the frequency (due to fading or interference), the flow of incoming or outgoing text will slow down or stop altogether.

It sounds like Bob lives in an interesting place. Maybe you should ask him about it.

WERE YOU LIVING NEAR MOUNT ST HELENS WHEN IT ERUPTED BACK IN 1979? KU7G DE WB8IMY +?

You remembered to turn the link over. Now Bob can respond.

YES, I WAS JUST MOVING INTO MY NEW HOME WHEN THE MOUNTAIN BLEW ITS TOP. THERE WERE ASHES EVERYWHERE! I COULDNT DRIVE TO MY NEW JOB BECAUSE OF ALL THE ASH IN THE ENGINE. I STILL HAVE SOME OF IT IN JARS IN MY BASEMENT! +?

Now you have the start of a fascinating conversation! As with RTTY, you can type your comments while you're receiving his. When the link turns over, your system will begin sending the pretyped text automatically. All good things must come to an end, however...

WELL, STEVE, I HAVE TO GET UP VERY EARLY TOMORROW AND DRIVE ALL THE WAY TO BOISE, IDAHO. I THINK I SHOULD GET TO BED SOON OR I WILL NEVER HEAR THE ALARM CLOCK. THANKS FOR THE GREAT CONVERSATION. I REALLY ENJOYED IT. GO AHEAD AND MAKE YOUR FINAL COMMENTS AND THEN YOU CAN DOWN THE LINK. 73...WB8IMY DE KU7G SK +?

Bob says you can "down the link." What does that mean? It simply means that he is asking you to send the *end* command that will terminate the ARQ link between your stations.

GOODNIGHT, BOB. HAVE A GOOD TRIP. HOPE TO LINK UP WITH YOU AGAIN ONE OF THESE DAYS. 73...KU7G DE WB8IMY SK

Most MCPs feature a single keystroke that sends the end command (some label it "disconnect").

Answering an AMTOR CQ

What if you're prowling through the RTTY/AMTOR subbands and you see someone calling CQ? The first order of business is to load the *other station's* SELCAL code. Be careful not to change your own in the process! Obviously, there are two SELCAL codes involved: your SELCAL and the SELCAL of the other station. To help keep this straight, MCPs label your SELCAL as MYCALL, MYA, MYSEL or LOCAL CALL (LC).

Most MCPs will prompt you to enter the SELCAL of the other station:

SELCAL? _____

Some units include a call directory that lists several calls in a menu format. Still other controllers label the SELCAL to be sent as the remote call (RC), or sometimes *HISCALL*. Consult your manual to determine the correct label and the proper procedure to enter the other station's SELCAL.

Once you've entered the SELCAL, your station will begin chirping it over and over. If the link isn't established within a certain period of time, the transmissions stop. With luck, the other station will hear your signals and start the link. You're the calling station, so you're the master now. All the timing parameters for this conversation will be dictated by *your* station. Since you started the conversation, you're also the ISS at the moment. Send your opening remarks and start having fun!

APLink

APLink is a bulletin board system (BBS) program designed by Vic Poor, W5SMM, to provide message store-and-forward capability for AMTOR users. APLink allows stored messages to be shared between two computer ports—one port for an HF AMTOR controller and another for a VHF packet TNC. Messages can be read or stored by an AMTOR station on HF, or by a packet station on VHF. As a result, APLink provides a connection between AMTOR operators and the VHF packet network.

The APLink Scanning BBS

A scanning APLink BBS uses the programmable memories and scanning options of newer HF transceivers. When the BBS is not in use, its receiver continuously scans the programmed frequencies. A BBS station often scans four or five bands and two or three different frequencies within each band. A complete scan can take 15 to 30 seconds.

The scan pauses for 2 to 5 seconds on each frequency. If the BBS station hears its own SELCAL being sent by a potential user, scanning ceases and an ARQ link is established. If the frequency is busy, or the BBS SELCAL is not heard, the BBS resumes scanning.

This frequency-scanning technique allows one BBS to serve stations at varying distances—regardless of propagation conditions. If you can't connect to an APLink BBS on 20 meters, for example, try again on another band. Eventually you'll find a band that offers a good path between you and the APLink BBS.

If you want to connect to a scanning APLink BBS, you must: (1) know the exact frequencies the BBS is scanning, and (2) be able to accurately set your transmitter to those frequencies. Scanning BBS stations have files that list the exact frequencies they monitor. When you first connect to an APLink BBS, download its frequency list so you'll know where to find it the next time.

Before attempting to use an APLink station, I suggest you listen to several of the listed frequencies and see if you can hear the BBS communicating with other stations. This will tell you whether the BBS is busy (calling it will do no good if it is), and the quality of the BBS station's signal at your QTH. When you finally call the BBS, choose a frequency where its signal is strong and stable.

Most APLink operating commands will be familiar to packet users. As is the case with many packet BBS systems these days, APLink stations have disabled their command list response. This means that you may only see **GA +?** when you initially establish the link ("GA" means "Go Ahead").

Sending a private message via APLink is very similar to packet, but AT is used in place of @. For example, on an APLink BBS you'd enter: SP WAØQKL AT WØAMN rather than SP WAØQKL @ WØAMN.

HF PACKET

Noise and interference are deadly to packet, and the HF bands have both in abundance. All it takes is the corruption of *one* bit of data and an entire packet data frame is ruined. The receiving station won't accept it, which means the transmitting station must try again. When you listen to packet signals, the first thing you notice is that some of the bursts are relatively long—up to 3 seconds or more. That's an eternity when you're sending data. It's also more than enough time for a pop of static or a blip of interference to wreck everything.

The HF bands are also prone to fading. Even a slight fade is enough to cause data loss. By combining the effects of noise, interference and fading, you set the stage for frustration. When conditions deteriorate, throughput suffers. The act of sending a simple greeting can take several minutes. (Oops! Lost a couple of bits! We'll have to send that frame again. And again. And again…)

This is why so many hams ultimately turn to other digital modes such as AMTOR, G-TOR, PacTOR and CLOVER when they need to pass information efficiently on HF. This is not to say that packet doesn't have its place on the HF bands. When signal conditions are good, it performs quite well. There are still many VHF and UHF bulletin boards throughout the world that depend on HF packet to move mail over great distances. And there are great opportunities to enjoy fascinating conversations on HF packet. The key to success is patience!

An HF Packet Tour

Before you enjoy your first HF packet conversation, spend some time prowling through the bands. HF packet can be a little confusing at first glance

Set up your TNC or MCP so that you can monitor all packet data—even packets not intended for you. In some units the MCOM command is used (MCOM ON). In others it is part of the MONITOR command. Check your TNC or MCP manual.

When you find a packet signal, tune *very carefully*. If your TNC or MCP has a bargraph type tuning indicator, you're ahead of the game. You can use it to get your radio on the correct frequency with little trouble. Packet signals can be tuned without an indicator, but it is quite difficult. With practice, however, you'll learn to do it by ear. That is, you'll know you're close when you hear the packet signal at a certain pitch.

Don't expect to see information on your screen right away. You may have to wait for the next transmission (after you've stopped tuning) before data is displayed. And if conditions are so-so, you may have to wait even longer.

Mail Forwarding

Of all the activity you'll see on HF packet, *mail forwarding* is the most common. The global packet system is made up of VHF and UHF networks that serve local areas. The bulletin boards in these networks pass mail between each other on VHF and UHF frequencies when they can, but sometimes the routes require leaps of hundreds or thousands of miles. Unless the bulletin board operator has the ability to relay mail via amateur satellite, the next best choice is HF packet.

The act of relaying packet mail from one point to another is known as *forwarding*. Messages and bulletins travel on VHF or UHF until they reach stations with HF packet forwarding capability. A station must have more than an HF transceiver and a TNC to be a forwarder. An HF mail forwarder keeps appointments with other stations on specific frequencies. Elaborate software checks each incoming message from the local network and "decides" which HF station should receive it. The messages are stored until they can be relayed via HF packet to the appropriate station.

Let's say that you've stumbled upon some forwarding in progress. It may look like this:

WØXK>N4HOG:

ö_îywl*°_ÛòÊ$£¶ØæHÅ´G^.2%T´JÜ·_y7™o,ÖÓ

WØXK>N4HOG:

ç
‰»_áY∂lÚÎî˚_x"°xö®;=»8À‡¿Í¥c_CaueJó &__ÓÛúÿ_OQçy@}/
YÔÀÂ_J›3—sé ¿ –Ÿ‹miV=

WØXK>N4HOG:

Ïø,pßÁ%-%ŸAÜ¥k6M—ß2ŒF˜‰•_A'Rq+>3åQ±9M_òa&
1¿£:ûµ_…ûÔø ·k#@ ,Ù~RM$8ËÌ-†Xµ

No, there's nothing wrong with your computer! These stations are using a type of data compression to make it easier to send large amounts of data. The result on your screen, however, looks like this. If you see stations passing traffic without compression, this is more typical of what you'll see:

XF3R>XE1M-15:

Another ham (80 years old!) told me about a mode called PACTOR.

XF3R>XE1M-15:

Another ham (80 years old!) told me about a mode called PACTOR.

XF3R>XE1M-15:

Another ham (80 years old!) told me about a mode called PACTOR.

XF3R>XE1M-15:

I never heard of this before!

XF3R>XE1M-15:

Has anybody further info?

XF3R>XE1M-15:

Thanx in advance.

XF3R>XE1M-15:

73...Joss.

Notice that parts of this message were repeated several times. I bet you can guess why! Noise, interference or fading must have caused errors on the receiving end. A few repeats were required to get the message through.

Bulletin Boards and Mailboxes

HF packet gives you the opportunity to connect with bulletin board systems (PBBSs) throughout the world. By connecting with a bulletin board, you can view messages and bulletins concerning every topic imaginable! Of course, you can do the same thing on VHF packet, but DX bulletin boards offer a perspective you won't find elsewhere. What are the big issues among hams in Argentina, for example? Connect to an Argentinean PBBS and find out!

PBBS activity is easy to spot. Most hams request a list of the latest bulletins when they connect, so look for any station that seems to be sending such a list. Here is an actual example copied on 20 meters:

VE4GQ>WB4TDB:

8965	BF	1967 ALL@MAN VE4SYG 07-Jun MRS BYLAWS
8964	B$	1587 SALE@ALLCAN VO2APL 04-Jun LOTS OF THINGS FOR SALE
8961	B$	1012 SALE@ALLCAN VO2APL
8960	B$	414WANTED@CANADA VE4YE 07-Jun
8956	B$	2986 RSGB@CANADA VE6EQ 05-Jun RSGB MAIN News 5th June

| 8955 | B$ | 2603 RAC@CANADA VE3LVO 05-Jun $$$ STONEWALLING? $$$ |

Some PBBSs are able to serve more than one station at a time, so you may see several messages going to one station, then another.

As you comb through the packet signals you may also run into data that *looks* like it's coming from a PBBS, but it isn't. For example:

[KPC2-5.00-HM$]

16000 BYTES AVAILABLE

Thanks for checking in. Please leave a message and I'll get back to you. 73, Bill/W1KKF.

ENTER COMMAND: B,J,K,L,R,S, or Help >

This is the typical sign-on message you'll receive from a packet *mailbox*. These mailbox functions are common features on TNCs and MCPs. Depending on the model you've purchased, you may have one at your fingertips right now!

Although they look similar, there are major differences between PBBSs and mailboxes. Notice, for example, that this packet mailbox has only 16,000 bytes of data storage available. Packet mailboxes store all of their incoming and outgoing messages in solid-state RAM (random access memory). The typical RAM capacity of a packet mailbox is only about 32 kbytes. True PBBSs, on the other hand, usually store data on *hard disks* with several hundred *megabytes* of capacity.

You'll also notice that the *command line* (ENTER COMMAND: B,J,K,L,R,S, or Help >) is much shorter than those you'll see on most packet bulletin boards. That's because packet mailboxes have very limited capabilities. This one will allow you only to send mail (S), read mail (R), list messages (L), kill messages (K), see a list of stations heard recently (J) and disconnect (B).

Using a PBBS

Connecting to a PBBS is the same as connecting to any other packet station. Monitor the packet frequencies and watch for bulletin board activity. When you find one, make your connect request and have fun! Let's pretend we've discovered the W1NRG-4 PBBS.

CONNECT W1NRG-4

If the other station hears you, you'll be rewarded with:

*** CONNECTED TO W1NRG-4

Very good! You've made the connection. Now the PBBS will send its sign-on information to you.

Welcome to W1NRG's MSYS PBBS in Wallingford, CT Enter command: A,B,C,D,G,H,I,J,K,L,M,N,P,R,S,T,U,V,W,X,?,* >

That's a big list of commands and we don't have the space in this book to discuss each one. Would you like to see what bulletins are available on the PBBS? The *LB* (List Bulletins) command will do the trick. You can read a message by sending the letter *R* followed by the message number.

Live Conversations

Live conversations are common on HF packet, perhaps more common than on VHF and UHF packet. You'll recognize them right away because they're just lines of text flying back and forth from one station to another. For example:

WB8IMY>N6ATQ:
So did you get your new vertical up last weekend? >>>

N6ATQ>WB8IMY:
No, it rained like crazy. There's no way I was going to lay radial wires in the mud! >>>

Note the use of >>> at the end of each line. Many packeteers use this symbol to let the other stations know that it's their turn to comment. You'll also see BTU (back to you) and K used in the same manner.

Calling CQ

Before you attempt an HF packet connection, check your receive audio level. Adjust the level control until the RCV or DCD indicator on your TNC or MCP glows when a packet is received (tune in some packet signals to test this). If the indicator blinks in response to noise on the frequency, you have the audio level set too high.

The next step is to send a *CQ* (general call) to flush other hams out of the woodwork. You can send your CQ as a single, *unconnected* packet, but you must be in the converse mode to do so. (If you're in the command mode, enter CONV at the cmd: prompt.) When you're not connected to another station, everything you enter while in the converse mode is sent as an unconnected packet. Any station that receives the packet without errors will be able to read it. Since it's sent without an established connection, an ACK from another station isn't required or expected.

Sending a CQ this way is easy. Just type a brief message and hit your ENTER or RETURN key. For example:

CQ, CQ from NØMZR in Palmyra, Missouri. K

The problem with this approach is that the CQ is sent only once. What if a burst of interference prevents someone from receiving it? What if you sent your CQ just seconds before a lonely packeteer arrived on the frequency? To send CQ more than once, you'd have to type the entire line over and over.

Fortunately, there's an easier way to send multiple CQs. Your TNC or MCP is equipped with a BEACON function. When the beacon is activated, your TNC will transmit unconnected packets repeatedly until you turn it off.

The content of your beacon packet is determined by your BTEXT parameter. BTEXT stands for "beacon text" and it contains the actual text of your CQ message. Simply switch to the command mode and enter something like this:

BTEXT CQ CQ from N6ATQ in Escondido, California

When you're ready to start, enter the BEACON command along with the beacon *rate*.

BEACON EVERY *N*

N usually equals a certain interval of time. In some TNCs and MCPs, for example, 1 is equal to 10 seconds. So, BEACON EVERY 1 will cause the TNC to transmit an unconnected packet containing your BTEXT message every 10 seconds. BEACON EVERY 3, on the other hand, would cause the packet to be transmitted every 30 seconds (10 × 3). Timing can differ from one unit to another, so check your manual carefully.

What beacon rate should you select? If the frequency is busy, keep your beacon rate low (once per minute or longer). If the frequency is unoccupied, you can beacon at a much faster rate. The key is knowing when to stop. Sending beacons on a quiet frequency should net a response within a few minutes. If it doesn't, turn off the beacon (BEACON EVERY 0) and try another frequency. If someone answers your call, you may need to switch to the command mode and turn off the beacon.

PACTOR

Every MCP on the market today offers PacTOR as an operating mode. As a result, PacTOR has become one of the most popular forms of amateur HF digital communication.

Like packet, PacTOR supports the complete ASCII character set. This means that you can send upper- and lower-case letters as well as binary files (computer software and other information). You cannot do this with RTTY or AMTOR.

PacTOR sends error-free information by using a handshaking system. When the data is received intact, the receiving station sends an ACK signal (for *acknowledgment*). If the data contains errors, a NAK is sent (for *nonacknowledgment*). In simple terms, ACK means, "I've received the last group of characters okay. Send the next group." NAK means, "There are errors in the last group of characters, send them again." This back-and-forth data conversation sounds like crickets chirping. In the case of PacTOR, the long chirp is the data and the short chirp is the ACK or NAK. AMTOR and PacTOR sound similar when you hear them, but PacTOR is the mode with the extended chirps.

Memory ARQ

With AMTOR and packet, a group of characters must be repeated over and over if that's what it takes to deliver the information error-free. This results in slow communications, especially when conditions are poor.

PacTOR handles the challenge of sending error-free data in a different and interesting way. As with AMTOR, each character block is sent and acknowledged if it's received intact. If signal fading or interference destroys some of the data, a NAK is sent and the block is repeated—just like AMTOR. The big difference, however, involves *memory*.

When a PacTOR controller receives a mangled character block, it analyzes the parts and temporarily memorizes whatever information appears to be error-free. If the block is shot full of holes on the next transmission as well, the controller quickly compares the new data fragments with what it has memorized. It fills the gaps as much as possible and then, if necessary, asks for another repeat. Eventually, the controller gathers enough fragments to construct the entire block. This is a tremendous improvement over AMTOR and packet where the character block is often repeated many times before it finally makes it through unscathed. PacTOR's *memory ARQ* feature dramatically reduces the need to make repeat transmissions of damaged data. This translates into much higher throughput.

With memory ARQ, and the ability to automatically change data rates, PacTOR allows you to enjoy conversations under conditions that would easily disrupt AMTOR or packet. Any ham, even those with very modest stations, can use PacTOR to communicate effectively on the HF bands. Amplifiers, beams and other expensive toys are definitely *not* required to enter the world of PacTOR.

Calling CQ on PacTOR

If you read the AMTOR section in this chapter, you'll recall that you send CQ on AMTOR using the *FEC* mode. The same is true for PacTOR, although PacTOR FEC is unique to this mode. An AMTOR station cannot read a PacTOR FEC transmission.

Tune through the digital subbands until you hear what sounds like a fast RTTY signal. As you lock-in the signal, your MCP will indicate that it is trying to synchronize with the FEC transmission. Within a short time you should begin to see text on your screen.

You can always identify an AMTOR CQ because the operator must list his or her SELCAL in the text.

CQ CQ CQ CQ CQ DE WB8IMY WB8IMY (WIMY) (WIMY) K K

That's easy enough, but what about this?

CQ CQ CQ CQ CQ DE WB8IMY WB8IMY WB8IMY K K

Is this a PacTOR or G-TOR CQ? There's no way to tell. Both modes use complete call signs rather than SELCALs. To alleviate the confusion, many HF digital operators are adding their *mode* to their CQ message. This is a good practice to follow. Put your MCP in the *PacTOR standby* mode and call CQ.

[Switch your MCP to FEC transmit]

[Send a blank line]

CQ PacTOR CQ PacTOR CQ PacTOR DE WB8IMY WB8IMY WB8IMY

CQ PacTOR CQ PacTOR CQ PacTOR DE WB8IMY WB8IMY WB8IMY

CQ PacTOR CQ PacTOR CQ PacTOR DE WB8IMY WB8IMY WB8IMY

[Switch your MCP back to receive]

Any station that's monitoring your signal will know exactly which mode you're using—and will reply accordingly. When you send CQ, use several short lines of text rather than a few long lines. This helps stations synchronize to your signal more easily.

Once you make contact, the conversation flows along just like AMTOR. The station that answered your call is the *ISS* (information sending station) and you're the *IRS* (information receiving station). When he sends his "hello," he'll probably use the *over* command to flip the link. (Consult your MCP manual and/or software documentation concerning the over command.) Now you're the ISS and he is the IRS. Introduce yourself and ask a question about where he lives, or what he does for a living. Use the over command to flip the link again. A conversation is underway!

Bill Price, WA4MCZ, is a picture of concentration as he operates the ARRL RTTY Roundup contest.

Answering a CQ

Answering a CQ in PacTOR is straightforward. Depending on the software you're using, it may be as simple as entering:

CALL N1BKE

or,

CONNECT N1BKE

at the cmd: prompt. Some types of software streamline the process even further. There may be pop-up boxes where you simply enter a call sign.

PacTOR has a clever feature that comes in handy when you're trying to connect to a distant station. If the station is more than about 5,000 miles away, use the exclamation mark (!) in your connect request. This forces the MCP to lengthen the time it waits to receive an ACK signal from the station, allowing more time for your signal to reach him and his ACK to return. For example:

CONNECT !HS1CHG

Working a PacTOR BBS

Although live conversations are the rule with PacTOR, there are lots of BBSs as well. You'll find that they're almost identical to packet systems. When you connect to a PacTOR BBS, you'll see something like this:

Welcome to WB8SVN's PacTOR BBS in Placentia, CA. Enter command: A,B,C,D,G,H,I,J,K,L,M,N,P,R,S,T,U,V,W,X,?,* >

After you enter a command you *do not* need to send the over code. The BBS will flip the link automatically after it receives the command from you.

You can use PacTOR BBSs to exchange messages with other PacTOR operators. And with PacTOR's ability to handle binary data, you can even download small programs from the BBS and run them on your computer!

Many AMTOR APLinks and BBSs now respond to PacTOR calls as well. If you spot an AMTOR operator using an APLink, give it a try after he's finished. There's a good chance that it will answer your PacTOR call.

Finally, most PacTOR-capable MCPs have mailbox functions. You can use your PacTOR mailbox as a convenient way to pick up messages from friends when you're not at the keyboard.

CLOVER

CLOVER is an advanced HF digital communication system which Ray Petit, W7GHM, developed in a joint venture with HAL Communications of Urbana, Illinois. CLOVER uses a four-tone modulation scheme. Depending on signal conditions, any of *ten* modulation formats can be selected manually or automatically. Six of the modulation systems employ phase-shift modulation (PSM); two use amplitude-shift modulation (ASM); and two use frequency-shift modulation (FSM). Each tone is phase- and/or amplitude-modulated as a separate, narrow-bandwidth data channel. As you might guess, the resulting CLOVER signal is very complex!

For example, when the tone pulses are modulated using quadrature phase-shift modulation (QPSM), the differential phase of each tone shifts in 90° increments. Two bits of data are carried by each tone for a total of eight bits in each 32-ms frame. The resulting block data rate is about 250 bits per second. CLOVER is capable of even higher data rates when using 16-phase, four-amplitude modulation (16P4A). In this format, CLOVER perks along at 750 bit/s.

The complex, higher-speed modulation systems are used when conditions are favorable. When the going gets rough, CLOVER automatically brings several slower (but more robust) modes into play.

Even with these ingenious adaptive modulation systems, errors are bound to occur. That's where CLOVER's Reed-Solomon coding fills the gaps. Reed-Solomon coding is used in all CLOVER modes. Errors are detected at the receiving station by comparing check bytes that are inserted in each block of transmitted text. When operating in the ARQ mode, CLOVER's damaged data can often be reconstructed without the need to request repeat transmissions! This is a major departure from the techniques used by packet, AMTOR and PacTOR. Of course, CLOVER can't always repair data; repeat transmissions—which CLOVER handles automatically—are sometimes required to get everything right.

With the combination of adaptive modulation systems and Reed-Solomon coding, CLOVER boasts remarkable performance—even under the worst HF conditions. The only Amateur Radio digital mode with the potential to approach CLOVER's performance is PacTOR-2, and possibly G-TOR (under certain conditions).

CLOVER Handshaking

As you may recall, PacTOR and AMTOR both use an *over* command to switch the link so that one station can send while the other receives. CLOVER links must be switched as well, but the switching takes place *without* using *over* commands.

When two CLOVER stations make contact, they can send limited amounts of data to each other (up to 30 characters in each block) in what is known as the *chat mode*. If the amount of data waiting for transmission at one station exceeds 30 characters, CLOVER automatically switches to the *block data mode*. The transmitted blocks immediately become larger and are sent much faster. The other station, however, remains in the chat mode. Because of precise frame timing, all of this takes place without the need for either operator to change settings, or send *over* commands. The CLOVER controllers at both stations "know" when to switch from transmit to receive and vice versa. And what if both stations have large amounts of data to send at the same time? Then they *both* switch to the data block mode. This high degree of efficiency is transparent to you, the operator. All you have to do is type your comments or select the

file you want to send—CLOVER takes care of everything else!

CLOVER features an FEC mode similar to that used by AMTOR, PacTOR and G-TOR. You use the CLOVER FEC to call CQ, or to send transmissions that can be received by several stations at once. (In the CLOVER ARQ mode, only two stations can communicate at a time.) CLOVER shares another characteristic with AMTOR: the use of SELCALs. When attempting to contact another CLOVER station, you must send its SELCAL first.

What Do I Need to Run CLOVER?

The requirements for a CLOVER station differ substantially from those of other HF digital modes. They are:

❑ An SSB transceiver. The transceiver must be very stable (less than 30 Hz drift per hour). It should also include a frequency display with 10-Hz resolution. The audio output from the CLOVER controller is fed to the audio input of the transceiver (CLOVER uses AFSK, not FSK). Receive audio is supplied to the controller from the external speaker jack or other source.

❑ An IBM-PC computer or compatible. The computer must be at least a 286-level machine.

❑ A PCI-4000 CLOVER controller board. All CLOVER controllers are available exclusively from HAL Communications. The HAL PCI-4000 CLOVER controller is installed *inside* the computer using any available expansion slot (see Fig 9-5). The board uses a dual-microprocessor design and digital signal processing to achieve signal modulation and demodulation.

❑ HAL PC-CLOVER software. This is supplied by HAL Communications and is included with every PCI-4000 controller. It is *not* a terminal program. The PC-CLOVER software is the instruction set of the PCI-4000 itself. It's loaded into the PCI-4000's memory each time you decide to operate. This approach makes it easy to update the PCI-4000 in the future. You simply buy a new diskette or download the software from a BBS.

CLOVER Conversations

When it comes to on-the-air operating, CLOVER is different from any of the modes we've discussed so far. To call CQ, for example, you switch to the MODE menu, highlight CQ and press ENTER. The PCI-4000 sends a CW identification fol-

Fig 9-5—This is the HAL Communications PCI-4000 CLOVER controller. It fits conveniently inside your IBM-PC or compatible computer. The only external wiring are audio cables to and from your SSB transceiver.

lowed by a raucous stream of data. Unlike other digital modes, you do not see "CQ CQ CQ" flowing across your screen. In fact, you see nothing at all.

The PCI-4000 sends CQ in the form of data signals that appear as CQ "flags" to other CLOVER stations. When another CLOVER operator tunes in your signal, all he sees is a statement on his screen announcing that you are calling CQ. At that point he can ignore you or press a single key to establish a CLOVER connection.

Once the conversation has started, it is similar to HF packet. Because there are no *over* commands, you need to let the other station know when you've completed a statement. For example:

Hello! My name is Steve

and I live in Wallingford, Connecticut. I am new to

CLOVER. What do you think of it? >>>

Without >>>, BTU, K or a similar symbol at the end of my statement, the other operator might inadvertently jump in after the end of the first line. This can be very confusing for everyone!

While you're watching the conversation, it's easy to get distracted by the receive/transmit status table in the upper right corner of the screen. The table displays the modulation format in use at the moment, the signal-to-noise ratio, tuning error, phase dispersion, error-correction capacity and transmitter output power (as a percentage of full output. The table is split into horizontal rows labeled "MY" and "HIS." Not only do you see your own parameters changing, you see the changes taking place *at the other station!* (CLOVER accomplishes this feat by periodically swapping station data.) Who is enjoying the best receive conditions? Which station is doing the greatest amount of error correcting at the moment? Just look at the table.

Adaptive Operating

The PCI-4000's adaptive behavior is fascinating. As signal conditions fluctuate during a conversation, the PCI-4000 raises and lowers the audio level to the input of your transceiver. This causes a corresponding change in output power.

For example, when my CLOVER station learns that you are receiving me extremely well (based on the signal-to-noise and error-correction data supplied by your station) it may automatically drop my power to a mere 6% of full output. If conditions deteriorate, my output may jump to 50% or even 100%. In other words, CLOVER uses only as much power as it needs to maintain the link at any given time.

G-TOR

On New Year's Day 1994, WØXI and WK5M of the Kantronics Corporation transmitted a 9718-byte file from their station in Kansas to WA4EGT in California on 20 meters. Despite mediocre band conditions, the file was transferred in just 5 minutes and 20 seconds. They used a digital mode never heard before on the HF bands: *G-TOR*. Immediately thereafter, the file was transmitted again, this time using PacTOR. It took 20 minutes and 15 seconds—nearly four times slower. Throughout the month of January these tests were repeated with over 1,000,000 bytes transferred error-free. The average character/second rate for G-TOR was 23.7; for PacTOR 8.64. More testing and tweaking followed until March 1, 1994, when G-TOR made its official debut.

Equipment Requirements

As this book went to press, G-TOR was available in only one multimode controller: the Kantronics *KAM Plus,* specifically those units sold after March 1, 1994. Owners of earlier KAM Plus MCPs can upgrade to G-TOR by simply purchasing and installing a version 7.0P EPROM chip. Hams who own older KAMs can also add G-TOR, but they must purchase the version 7.0E enhancement board. If you already own a KAM with an enhancement board, all you need is a 7.0E EPROM. Kantronics has also offered to license G-TOR to other MCP manufacturers and several have shown interest. In 1995, MFJ Enterprises announced that they would add G-TOR to their popular 1278 MCP.

If you have an SSB transceiver that will switch from transmit to receive in less than 100 ms (most will), you should be able to use it for G-TOR with little difficulty. You can operate G-TOR in direct FSK or AFSK. Most G-TOR operators are using AFSK, however.

Who Put the "G" in G-TOR?

G-TOR is an acronym for *Golay-coded Teleprinting Over Radio.* Golay coding is the error-correction system created by M.J.E. Golay and used by the *Voyager* spacecraft. Sending billions of bytes of data across the Solar System required a scheme to ensure that the information could be recovered despite errors caused by interference, noise, and so on.

To create G-TOR, Kantronics combined the Golay coding system with full-frame data interleaving, on-demand Huffman compression, run-length encoding, a variable data rate capability (100 to 300 bit/s) and 16-bit CRC error detection. G-TOR system timing is liberal enough to permit long-distance communication.

G-TOR in Action

G-TOR is surprisingly easy to operate. If you're familiar with AMTOR or PacTOR operating, G-TOR is essentially the same. When you're in the G-TOR mode, you use AMTOR FEC to call CQ. This allows your transmission to be copied by as many stations as possible. If someone wants to talk to you, they respond in G-TOR using your full call sign.

When you call CQ, let the stations know that you're fishing for G-TOR contacts. Remember that AMTOR stations may be tuning in. Here's an example of a typical G-TOR CQ:

CQ CQ CQ CQ CQ—G-TOR

CQ CQ CQ CQ CQ—G-TOR

CQ CQ CQ CQ CQ—G-TOR

DE WB8IMY WB8IMY WB8IMY

Standing by for G-TOR calls. K K

Once the G-TOR link is established, the station that called you is the ISS, or *information sending station.* It's up to him to speak first. While he is transmitting, you're the *information receiving station,* or IRS. You flip these IRS/ISS roles back and forth by sending an "over" code. Like PacTOR, G-TOR can transfer ASCII or binary information

If you copy someone calling CQ G-TOR, you simply switch to the G-TOR standby mode. Establishing a connection is as easy as entering the word G-TOR followed by the call sign of the station you wish to contact. For example:

G-TOR WB8IMY

You can monitor the G-TOR data rate by glancing at the front panel of the KAM Plus. When the STA LED is off, the rate is 100 bit/s. A flashing STA indicates 200 bit/s and a steady STA means that you're perking along at 300 bit/s. The data rates change automatically based on the quality of the link. Links always begin at 100 bit/s. If the number of correctly received frames exceeds a preset value, the receiving station will request a speed increase to 200 or 300 bit/s. If the link deteriorates, the data rate will automatically ratchet downward. You can also set the data rate manually.

Monitoring G-TOR

The KAM Plus (and all other MCPs) include a *listen* mode for monitoring AMTOR and PacTOR conversations. You simply tune in a signal, switch to the listen mode and follow the chat.

The Kantronics KAM Plus does *not* feature a G-TOR listen mode. The task of decoding a G-TOR transmission while performing other functions (such as receiving packet on the KAM's VHF port) is too much for the microprocessor.

Kantronics provides a solution with their *G-MON* software. By using *G-MON*, you can eavesdrop on G-TOR links with *any* KAM that has WEFAX capability. You can also use *G-MON* with the WEFAX function of other Kantronics TNCs such as the KPC-2 and KPC-3 to monitor G-TOR activity. *G-MON* is provided with new KAM Plus units as well as their upgraded *HostMaster II+* software. You can also purchase *G-MON* separately from Kantronics for a nominal fee.

G-MON is essentially a terminal program. By placing the KAM Plus in the *GSCAN* mode, you can monitor data frames being exchanged between two linked G-TOR stations. Your computer is doing quite a bit of processing under *G-MON*, so you must run the software on a reasonably fast machine—preferably a 286 or faster PC. Using *G-MON*, you can eavesdrop on G-TOR conversations. There is no error correction under *G-MON*, so you'll receive a number of mutilated frames. Even so, you'll be able to follow the conversations and identify the stations most of the time.

G-TOR signals sound similar to PacTOR when you hear them on the air. The difference is the G-TOR data bursts are somewhat longer—nearly two seconds in length. With a little practice you'll be able to sort out the signals with ease.

Packet Radio

I n a very short time, packet has become one of the most popular operating modes in Amateur Radio. Since packet experiments began, in 1978, loads of packet equipment have been sold and, today, most active hams have some kind of packet installation in their ham shacks.

A big reason for packet's popularity was its timing. It came on the scene at the beginning of the home-computer revolution. Hams have a natural curiosity when it comes to new electronic gadgets, so a lot of hams built and bought computers, many intending to use them for ham radio applications.

A substantial amount of ham radio software was written for home computers. Some of the software even performed communications (CW and RTTY, for example). However, there was something lacking in the communications software. It wasn't that the software couldn't do CW and RTTY, but that CW and RTTY did not use the full potential of the computer as a communications tool. The computer hardly worked up a sweat when performing these tasks. It had an appetite for something faster.

For a number of years, a mode of computer communication called packet *switching* was being used commercially to transfer data between computers at high speeds. Although packet switching usually operated over wire lines of some sort, there

STAN HORZEPA, WA1LOU
1 GLEN AVE
WOLCOTT, CT 06716-1442

had been some success using packet switching over radio waves. It was only a matter of time before someone got the idea to try it over Amateur Radio as well. The rest is history.

Besides speed, packet offers other attractive advantages.

❏ It operates without error. Whatever you send is received perfectly.

❏ It uses the radio spectrum efficiently. Multiple communications may be conducted by multiple stations on the same frequency at the same time.

❏ It provides time-shifting communication. By storing messages on packet bulletin boards (PBBS) or mailboxes, stations can communicate with other stations who are not on the air.

❏ It operates over networks when necessary. Any station can access packet networks and greatly expand their communication capabilities.

❏ With all of these features, hams now are using packet for numerous diverse applications including traffic handling, satellite QSOs, DXing, emergency communications and more.

There seems to be a niche for every ham in the packet mode. See if you can find yours.

THE PACKET STATION

There are three basic parts of a packet station: the terminal, the radio equipment and the terminal node controller (TNC). Put these parts together and you are on the air.

Terminal

A terminal interfaces the user to the TNC. By means of the terminal's keyboard, the user enters commands that control the

Bonnie King, KD4BNV, keeps in touch with friends through packet. (*AC4HF photo*)

Fig 10-1—The components and interconnections of a typical packet station.

TNC. Once communications are established, the user enters information on that same keyboard for transmission to the other station. Via its display, the terminal allows the user to read the TNC's responses to commands and to read the information that's being sent by the other station.

Computers Acting Like Terminals

In some cases, real terminals are used in packet stations. However, in the typical packet station, computers *emulating* terminals are used. Terminal emulation is achieved by running *terminal emulation software* on the computer. A variety of terminal software is available, from programs used to access mainframe computers over wire links to software that provides access to bulletin board systems (BBSs) via telephone lines. There is even software available that is specifically designed for packet applications. (For simplicity's sake, the remainder of this chapter will use the terms *DTE* and *terminal* when referring to a computer emulating a terminal, as well as a real terminal.)

If you have software for your computer that allows it to use

A Software Sampler

Today, nearly all computers may be programmed to function as DTEs. The software that does this transformation may turn a computer into a bare bones, dumb terminal or a very sophisticated data-communications device. The drawback with such software is that it is primarily designed to operate with telephone modems, not packet modems (TNCs). As a result, they include telephone data communications features that are useless to the packet user and lack features that the packet user would find useful.

There is data communications software that is optimized for packet operation. This software was written with packet in mind, so the user does not have to be compromised by telephone modem software any longer. (Refer to the accompanying figures to see the difference between terminal emulation software that is and is not optimized for packet communications.) Beyond packet terminal emulation software, some computers have been programmed to emulate TNCs, too.

These emulators are available as commercial packages (you pay for it before you use it), shareware (you use it before you pay for it and only if you continue to use it), and freeware (you use it for free). A sampling of this software for our more popular computers follows.

Apple II

Apple Packet (APR) is an Apple II (II/II+/IIe/IIc/IIgs) packet terminal program for TNC 2s that may be had by sending a blank, 5.25- or 3.5-inch formatted disk and a postage-paid, self-addressed disk-mailer to Larry East, W1HUE, 1355 S Rimline Dr, Idaho Falls, ID 83401.

Apple Macintosh

MacRATT is a full-featured packet terminal program that is optimized for the AEA PK-232 and PK-88 TNCs, however, it works well with other TNCs, too. It is available from AEA, PO Box C-2160, Lynnwood, WA 98036.

Savant is a packet program that features a multiple window, multiple connection interface. It works with any TNC in the KISS mode and it will work without a TNC if you use an external modem and SoftKiss, a modem software driver. Savant, SoftKiss and an external modem called PacketMac modem are available from Sigma Design Associates, PO Box 49085, Colorado Springs, CO 80949-9085.

Commodore C64/C128

TNC64, for TNC 1s and 2s, is available from the Texas Packet Society (TPRS), PO Box 50238, Denton, TX 76206-0238.

DIGICOM 64 is a TNC emulation program that was written by the Digicom BAYCOM Team in Germany. It and its accompanying files are available from the HamNet data library of CompuServe.

IBM PC

PaKet is a full-featured packet terminal program that runs under DOS, OS/2 and UNIX-DOS/Merge. North American hams can obtain PaKet for $25 from Jim Flannery, WB0NZW, 8098 S Carr Ct, Littleton, CO 80123-5304.

Often, the packet operator is stuck using software designed for telephone data communication. Here, an Apple Macintosh computer is running White Knight 11.10, displays a window interspersed with received data and transmitted data. The software includes unnecessary telephone-specific commands like dial, redial and hang-up.

The interface between a packet operator and a TNC is friendlier when communications software designed for packet is used. Here, an Apple Macintosh computer running MacRATT 2.0 displays one window (the "Send data..." window) containing transmitted data, a second window (the "Monitor" window) containing all of the data monitored on the channel and a third window (the "Connection A..." window) containing received data. The pull-down menus across the top of the screen access TNC commands.

BAYCOM is a TNC emulation program that was written by Johannes Kneip, DG3RBU, and Florian Radlherr, DL8MBT. It is available from the HamNet data library of CompuServe.

a telephone modem, it can probably be used for packet, too. After all, a TNC is essentially an "intelligent" modem that communicates over the airwaves rather than over telephone lines. Such software will usually serve you well in the packet mode.

Software that is specifically designed for packet terminal emulation offers features that are optimized for packet communications. For example, if you communicate with more than one station simultaneously (a multiconnect situation), some packet terminal programs allow you to open separate communication windows for each multiple connection. No telephone line software package offers that feature!

If you become dissatisfied using telephone line modem software in the packet mode, consider getting a program specifically designed for packet. The sidebar titled "A Software Sampler" lists examples of available software.

The Real Thing

Computers emulating terminals are not the only kind of DTEs you will find in a packet station. Some hams use real terminals! Surplus DTEs are inexpensive and plentiful (go to any ham/computer flea market and see what I mean). A preowned DTE can cost less than a preowned computer, and if you use a real DTE for packet you don't have to tie up your computer doing terminal emulation.

The disadvantage of using a real DTE is that it lacks features that computer terminal emulation software provides. (Some "intelligent" DTEs do include some of the features of terminal emulation software, but these DTEs are much more expensive than their "dumb" brethren.) The solution is to use a computer that runs more than one program simultaneously (this is called multitasking). Then you can run a terminal emulator for packet while your computer is doing other more important chores such as duping your Field Day log or running PBBS software!

Facing the Interface Issue

Whatever DTE you use, it must be capable of interfacing with your TNC. Almost all TNCs use the EIA/TIA-232-E ("EIA-232" for short) interface with its ubiquitous 25-pin connector to provide a serial connection to a DTE, so your DTE should have an EIA-232 interface as well.

Most real DTEs have an EIA-232 interface, so connecting them to a TNC is simple. On the other hand, some computers don't necessarily have EIA-232 interfaces. Some computers have no interface at all and EIA-232 may be an option that's available through the addition of a PC board. (If you have an IBM PC or compatible, you can add a TNC-on-an-expansion-card to your PC instead of the EIA-232 interface and avoid the serial-interface middleman altogether.)

There are computers that use other interfaces. For example, TIA/EIA-422-B ("EIA-422" for short) is found on some computers like the Macintosh. EIA-422 is close enough to EIA-232 that it can be made to work by properly wiring the interfaces. (Cables that interface a Macintosh computer to a telephone line modem will interface a Mac to a TNC, too.)

Some older computers had a TTL interface, which required signal conversion before they could be interfaced to EIA-232. Since some TTL-interfaced computers were very popular (for example, the Commodore C64), a number of older TNC models offered the TTL interface as an alternative to the EIA interface. If your computer uses TTL, find a pre-owned TNC that is TTL compatible.

Although EIA-232 supports 25 signals, any TNC you use will need only eight of them (the signals on pins 2-8 and pin 20) and many TNCs can get by using even fewer signals. Check the TNC's manual and see what you can do to economize your DTE-to-TNC cabling.

The Radio

There can be a lot of equipment at the RF end of a packet station. Some of it is of little concern to us. For example, as long as the antennas and feed line are capable of putting a signal on the desired packet frequency, it satisfies our requirements. Other RF hardware needs closer inspection, however.

Our primary concern is the radio equipment's receive-to-transmit and transmit-to-receive turnaround time. A TNC can switch between the transmit and receive modes very quickly. So quickly, in fact, that it must wait for the RF equipment to switch before it can continue to communicate.

According to a study conducted by Tucson Amateur Packet (TAPR), most amateur FM radios have receive-to-transmit and transmit-to-receive turnaround times between 150 and 400 milliseconds (ms). This dramatically reduces the amount of data that can be sent and increases the chance that two or more stations will interfere with one another. Such delays slow down what is intended to be a fast mode of communication.

The physical switching of an antenna, internally in a transceiver or externally with a separate transmitter and receiver, affects the turnaround time. The older the transceiver, the more likely that switching is performed mechanically by a relay. If a separate transmitter and receiver are used with one antenna, a mechanical relay is probably performing the switching function as well. In addition, if an external power amplifier and/or receive preamplifier is used, more mechanical switching may be involved.

With newer equipment, the switching is often accomplished electronically. This speeds up the process, but the improvement may be compromised by the frequency synthesizer circuitry that is also found in the newer equipment. After switching between the transmit and receive modes, synthesizers require some time to lock on frequency before they are ready. New equipment is being designed with synthesizers that can lock more quickly. (Older RF equipment does not use fre-

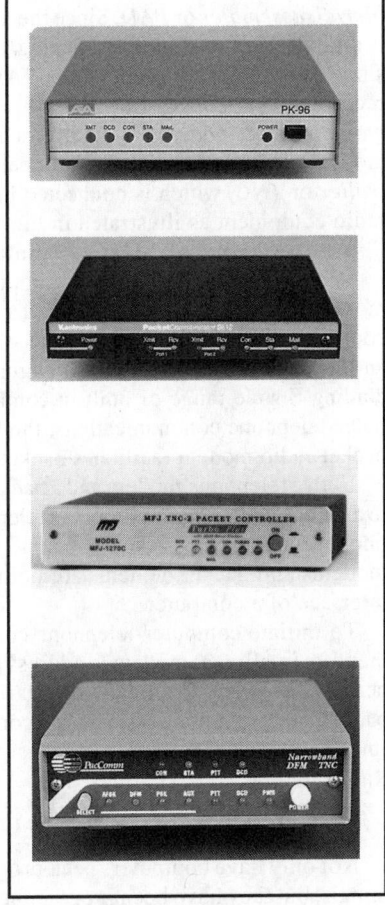

A potpourri of TNCs. From top to bottom: AEA PK-96, Kantronics KPC-9612, MFJ-1270C and the PacComm TNC/NB96. All four TNCs are capable of 1200 and 9600 bit/s operation.

quency synthesis; therefore, it does not suffer from this delay.) If you're hunting for a new transceiver for packet applications, keep this feature in mind.

Another problem cited by TAPR is that the modem-to-radio interface of most radios used for packet depends on audio response filters and audio levels *intended for microphones and speakers*. More often than not, this leads to incorrect deviation of the transmitted signal, noise and hum on the audio and so on. Splatter filters and deviation limiters distort frequency response and further reduce the performance of the packet-radio system. You are stuck in this environment unless you want to modify the radio. Performing surgery on your typical modern VHF/UHF FM voice transceiver may be difficult because of the use of LSIs, surface mounting and miniaturization.

Instead of using a typical amateur transceiver for packet, there are alternatives on the market today that solve many of the RF equipment problems. Check the pages of *QST* for news, reviews and advertisements for these radios that are designed with packet in mind.

The TNC

Packet is the communication mode in which the output of a DTE is assembled into bundles or "packets" according to a set of rules or a *protocol*. Each packet is transmitted to a remote radio station where it's checked for its integrity, then disassembled and its information fed to a terminal for you to read.

Packets are assembled and disassembled by a packet *assembler/disassembler* or *PAD*. Since the input and output of a PAD is digital, there is a digital-to-analog/analog-to-digital conversion between output/input of the PAD and radio transmitter/receiver. This conversion is performed by a *modulator-demodulator* or *modem*. Typically, a PAD and its associated modem are packaged into one unit called a *terminal node controller* or *TNC*, which is connected between the DTE and the radio equipment as illustrated in Fig 10-1.

You can think of a TNC as a very intelligent modem. Whereas a telephone modem permits a computer to communicate by means of the telephone network, a packet modem permits a computer to communicate by means of a radio network. Just as an intelligent telephone modem augments its functions by including a wide range of built-in commands to facilitate computer-telephone communications, the built-in intelligence of a packet-radio modem facilitates packet-radio communications.

Both a telephone modem and a packet modem are connected to a communications medium: the telephone modem to the telephone line and the packet "modem" to the radio. Telephone modems and packet modems are both connected to the serial interface of a computer.

To initiate computer-telephone communication, you command an intelligent modem to address (dial the telephone number of) the other computer. Similarly, to initiate computer packet-radio communications, you command a TNC to address (make a connection with the call sign of) another Amateur Radio station.

Computers Acting Like TNCs

Not only have computers been programmed to emulate terminals, but they have been programmed to emulate TNCs, too. Once upon a time, Bob Richardson, W4UCH, programmed the Radio Shack TRS-80 Model I computer to perform the functions of a TNC in software. All you had to do is connect the appropriate modem between the TRS-80's serial port and your radio and you were on the air.

The TRS-80 Model I is a computer museum piece today, but Bob's idea is still viable and has been implemented in Commodore, IBM and Macintosh computers. As in W4UCH's implementation, once you have the software, simply connect a modem to your computer's serial port and radio equipment and run the software.

See the sidebar "A Software Sampler" for a list of available TNC-emulation software. Modems in kit and assembled form are available from a number of sources for approximately $50. Typically, when you purchase the modem, it is bundled with terminal-emulation software.

Packet Protocols and Commands

The manner in which packet communication is conducted is called a *protocol*. The protocol consists of a standard set of rules and procedures that are programmed into each TNC so that they all will communicate in a compatible manner. In packet today, the most widely used protocol is called "AX.25." However, in the past, AX.25 had a competitor known as the "VADCG protocol."

After the Canadian Department of Communications authorized amateur packet in 1978, the VADCG protocol and a companion TNC were developed by Doug Lockhart, VE7APU, and the organization he founded, Vancouver Amateur Digital Communications Group (VADCG). Two years later, the FCC authorized amateur packet in the United States and, after a year of experimentation, the US packet pioneers agreed on a new protocol: AX.25.

The original TNC that was designed by VADCG supported only VADCG protocol (the only protocol that existed at the time). These TNCs were commonly known as "VADCG" TNCs. When AX.25 was implemented for the first time (by TAPR in their TNC 1) the VADCG protocol was also included. (VADCG TNCs outnumbered the new TAPR TNCs at that time.) The TNC 1 also had a new design and command set to support AX.25.

In 1985, TAPR introduced an AX.25-only TNC, the *TNC 2*, which used an expanded TNC 1 command set (memory limitations precluded implementation of VADCG protocol). The TNC 2 was quickly cloned by various manufacturers. After cloning, these manufacturers added new features to the clones, which resulted in "TNC 2 compatibles." Some TNC 2 compatibles not only had new bells and whistles for packet, but also permitted operation in other modes.

It was during the time that these clones and compatibles became available that the popularity of packet took off! Tens of thousands of these units were sold worldwide and the TNC 2 command set became predominant in packet. As a result, this chapter uses TNC 2 commands throughout its description of packet-radio operation.

To configure and control a TNC, you enter commands into the keyboard of a DTE. Commands may be entered only when the TNC's command prompt (cmd:) is displayed by your DTE. When this prompt is displayed, type the desired command and follow it with a carriage return (<CR>). For example, to command the TNC to disconnect, you type "DISCONNE" and a carriage return at the command prompt. In this chapter, entering commands will be represented as:

cmd: DISCONNE <CR>

To save keystrokes, many TNC commands may be used in an abbreviated version. For example, instead of typing DISCONNE for the disconnect command, you can simply type the letter D. In this chapter, each command will be printed partially in uppercase characters and partially in lowercase characters. The uppercase and lowercase characters together represent the full name of the command, whereas the uppercase characters alone represent the abbreviated version of the command. For example, the disconnect command will be represented as:

Disconne

where Disconne is the full name of the command and D is the abbreviated version.

Tailoring the TNC to Your Station

Before you transmit your first packet, you must configure the TNC to your needs. These include the requirements of your station equipment, specifically your DTE and your radio, and the requirements of the radio spectrum you will use, either HF or VHF/UHF.

DTE Parameters

Your TNC must be compatible with your DTE's operating parameters.

Data Rate—Set the data rate of your TNC's serial port to the bit/s rate of your DTE (your DTE is connected to the TNC's serial port). On the TNC 2, the data rate (300, 1200, 2400, 4800 or 9600 bit/s) is set by means of a rear panel DIP switch (positions 1 through 5) while the TNC is powered off. (All other parameters are set while the TNC is powered on.)

Note that many TNC 2 compatibles set their data rate using "autobaud" (the TNC automatically adapts to the data rate of the DTE). Typically, this is accomplished by entering carriage returns at the DTE keyboard after powering the TNC. After receiving some carriage returns, the TNC calculates the correct data rate, sets itself to that rate and sends its sign-on preamble to the DTE at the correct data rate.

Echo—The echo function causes the TNC to print on your DTE's display each character that you type into your DTE's keyboard. This TNC function should be turned on or off depending on whether your DTE does or does not provide this function. (Echo *on* is the TNC default.)

If two characters are displayed for each one that you type into the keyboard (you type HELLO and HHEELLLLOO is displayed), then both your DTE and TNC are echoing, so turn off the TNC's echo by typing:

cmd: Echo OFF <CR>

If nothing is displayed when you type, then the echo function is turned off in both your DTE and TNC, so turn on the TNC's echo function by typing:

cmd: Echo ON <CR>

Character Length—Set the TNC to the character length used by your DTE (7 bits per character is the default value.)

To select 7 bits per character, type:

cmd: AWlen 7 <CR>

To select 8 bits per character, type:

cmd: AWlen 8 <CR>

Parity—Set the TNC to the parity used by your DTE. (Even parity is the TNC default.)

To select even parity, type:

cmd: PARity 3 <CR>

To select odd parity, type:

cmd: PARity 1 <CR>

To select no parity, type:

cmd: PARity 0 <CR> or

cmd: PARity 2 <CR>

Screen Length—Set the TNC to the maximum number of characters that is displayed on each line of your DTE (80 characters per line is the TNC default), by typing:

cmd: SCREENLN n <CR>

where n is number 0 to 255, representing the number of characters per line displayed by your DTE.

Line Feeds—The automatic line-feed function causes the TNC to insert a line feed after each carriage return that is received in incoming packets. This TNC function should be turned on or off depending on whether your DTE does or does not provide this function. (Automatic line feed *on* is the TNC default.)

If a blank line follows each line of received text, then both your DTE and TNC are adding line feeds. Turn off the TNC's automatic line feed function by typing:

cmd: AUtolf OFF <CR>

If each line of received text is printed over the previously received line of text, then neither your DTE or TNC are adding line feeds, so turn on the TNC's automatic line feed function by typing:

cmd: AUtolf ON <CR>

Lowercase and Uppercase Characters—This TNC function normally sends both lowercase and uppercase characters to your DTE. If your DTE does not accept lowercase characters, this TNC function may be set to automatically translate all lowercase characters to uppercase characters.

If your DTE accepts both lowercase and uppercase characters, type:

cmd: LCok ON <CR>

Control Characters

Control characters, that is, characters that are typed while simultaneously pressing your DTE's control <CTRL> key, perform a number of useful functions. Some of the TNC default selections for these characters are the same as those that are commonly used by other DTEs. If your DTE is unable to generate a particular character or if you simply wish to select a different character to perform the same function, the following commands allow you to change the default character to another ASCII character of your choosing. (Note that the following list also includes control characters <CR>, <DELETE> and <I>, which are not used in conjunction with the <CTRL> key.)

Default Character	Command	Function
<CR>	SEndpac	Send previous keyboard input as a packet
<CTRL-C>	COMmand	Enter command mode (from converse mode)
<CTRL-Q>	STArt	Restart sending to DTE (from TNC)
<CTRL-Q>	XON	Restart sending to TNC (from DTE)
<CTRL-R>	REDisplay	Redisplay typed line before <CR>
<CTRL-S>	STOp	Stop sending to DTE (from TNC)
<CTRL-S>	XOFF	Stop sending to TNC (from DTE)
<CTRL-V>	PASs	Insert special character
<CTRL-X>	CANline	Cancel typed line before <CR>
<CTRL-Y>	CANPac	Cancel typed packet before <CR>
<DELETE>	DELete	Delete typed character
<I>	STReamsw	Change streams

If your DTE accepts only uppercase characters, type:

cmd: LCok OFF <CR>

Those are the most important parameters that you need to set to configure your TNC for compatibility with your DTE. By setting these parameters correctly, your DTE should now display text from your TNC in a clear, legible manner. If not, you should recheck these parameters and also check the physical connection between your TNC and DTE.

(In addition to the previously described DTE parameter commands, there are other DTE-related commands that allow you to select ASCII characters that perform various functions: input editing, flow control and so on. Refer to the sidebar titled "Control Characters" for more information.

Radio Parameters

Use TNC commands to customize it to the needs of your radio equipment and the radio spectrum that you are using.

Data Rate—Set the radio-port data rate of the TNC to the data rate you will use (300 bit/s for HF, 1200 or 9600 bit/s for VHF/UHF). On the TNC 2, the radio-port data rate is set by means of a rear-panel DIP switch positions 6 through 9 while the TNC is powered off. (All other parameters are set while the TNC is powered on.) With some TNC 2 compatibles, you set the radio-port data rate by means of software, so, refer to the manual that accompanies your TNC if it does not use a rear-panel DIP switch for setting the radio-port data rate.

By default, the TNC data rate is 1200 bit/s because this is the data rate that is used on the most popular packet-radio subband: 2-meter FM. The FCC allows 1200-bit/s operation anywhere above 28 MHz, so you may operate on 10 meters or 23 centimeters without adjusting this timing parameter. If you wish to operate below 28 MHz, however, the maximum legal data rate is 300 bit/s; this means that an adjustment of the radio-port data rate is necessary.

Turnaround Time—As discussed earlier, when a transceiver switches from receive to transmit, there is a short delay after the switching begins and before intelligence can be sent over the air. Similarly, when a transceiver switches from transmit to receive, there is a short delay after the switching begins and before intelligence can be received. Although this delay is only measured in milliseconds (ms), it is critical in packet because a TNC is capable of enabling a transmitter and sending intelligence almost instantaneously. Therefore, some delay must be programmed into the TNC to be compatible with the receive-to-transmit turnaround time of your transceiver. This delay may be varied from 0 to 1200 ms by typing:

cmd: TX delay n <CR>

where n is number 0 to 120 representing a delay in 10-ms increments [30 (300 ms) is the TNC default]. For example, to set the delay to 500 ms, set n to 50 (500 / 10 ms = 50 ms).

Your transceiver's turnaround time may be specified in your transceiver's manual. Typically, it is not specified, so you will have to experiment to find a delay that is suitable. In general, if your transceiver uses solid-state switching, you may set the delay relatively low. If your transceiver uses a mechanical relay for switching, you should set the delay higher. Also, synthesized transceivers require a greater delay than crystal-controlled transceivers. If you are using an external amplifier, you must take its switching delay into account, too.

Maximum Number of Unacknowledged Packets and Packet Length—The MAXframe and Paclen parameters are critical TNC parameters that you should adjust depending on the operating conditions (propagation and channel activity).

MAXframe selects the maximum number of outstanding unacknowledged packets the TNC will allow at any time. In other words, if MAXframe is set to 4, the TNC may send as many as four packets without receiving acknowledgments for any of them. Once the MAXframe limit is reached, however, the TNC will not send a new packet until one of the outstanding packets is acknowledged.

Paclen selects the maximum number of bytes of data in each packet. The TNC will never send a packet longer than the selected Paclen value. As data is entered from the DTE to the TNC, the TNC counts each byte of data. When the Paclen value is attained, the TNC makes up a packet containing the data, sends it over the air, and begins counting the number of bytes of data for the next packet. The TNC will send packets shorter than the selected Paclen value when it is specifically commanded to do so (whenever the SEndpac control character <CR> is entered).

The default values for the MAXframe and Paclen parameters (4 outstanding packets each 128 bytes long) are selected for good VHF operating conditions. When you are operating on HF or when VHF operating conditions are less than optimal (there is a high level of channel activity), the MAXframe and Paclen values should be reduced.

At 300 bit/s, it takes approximately four times as long to send the same packet as it does at 1200 bit/s. If you reduce the packet length by half, it still takes approximately twice as long to send a packet at 300 bit/s than at 1200 bit/s. This means that your packet is on the air approximately twice as long and the chances of encountering interference are twice as great! Add the fact that HF conditions are usually not optimal for packet operation, and you may end up with a lot of long unacknowledged packets hanging out there in the air. The best thing to do is to shorten the Paclen parameter and set the MAXframe parameter to 1 to force the TNC to send one short packet, wait for an acknowledgment, and then deal with the next short packet.

To change the value of the MAXframe parameter, type:

cmd: MAXframe n <CR>

where n is a number from 1 to 7 representing the maximum number of outstanding unacknowledged packets.

To change the value of the Paclen parameter, type:

cmd: Paclen n <CR>

where n is a number from 0 to 255 representing the maximum number of bytes of data in each packet. Note that 0 actually represents 256 bytes.

Number of Packet Retries—When a TNC sends a packet, it waits a preset time for an acknowledgment that its packet was received without error. If the time limit is reached, the TNC again tries to obtain an acknowledgment from the receiving TNC by resending the packet. When the maximum number of retries is reached, the sending TNC enters the *disconnected* state. The FRack parameter sets the amount of time between retries, and the REtry parameter controls the number of allowable retransmission attempts.

The REtry and FRack parameters should be adjusted upward or downward, depending on the operating conditions. If conditions are good (good propagation and a low level of activity on the channel), the REtry and FRack parameters may be adjusted downward. If a packet cannot get through after one or two attempts under good conditions, there is probably an insurmountable problem with the link. For example, the intended receiving station may have gone off the air, so you might as well abandon the effort immediately.

If conditions are marginal (marginal propagation and/or a medium level of activity on the channel), it may only take a little longer to get the packet through to the intended receiving station, so the REtry and FRack parameters may be adjusted upward.

If conditions are poor (poor propagation and/or a high level of activity on the channel), wait until conditions improve. You already have two strikes against you: it is very difficult to get the packet through to the intended receiving station because of poor propagation, and your packets are likely to collide with other packets because of the crowded channel.

By default, the REtry parameter value is set to 10 retries, but that may be changed by typing:

cmd: REtry n <CR>

where n is a number from 0 to 15. Zero represents an infinite number of retries and 1 to 15 represents the maximum number of retries that will be attempted before the TNC stops repeating a packet. Note that the REtry parameter does not include the initial transmission of the packet; it only represents the number of retries after the initial packet transmission.

By default, the FRack parameter value is set to 3 seconds, but this may be changed by typing:

cmd: FRack n <CR>

where n is a number from 1 to 15, representing the number of seconds that the TNC will wait for a packet acknowledgment before it again tries to obtain an acknowledgment. Note that the TNC automatically adjusts this value higher depending on the number of digipeaters used in the selected path of the packet according to the formula:

$$FRack \times (2 \times dr + 1) = adjusted\ FRack$$

where dr is the number of digipeaters in the selected path. For example, if FRack is set to 4 seconds and 2 digipeaters are in the selected path, the adjusted FRack value is 20 seconds (4 sec × ((2 × 2 digipeaters) + 1) = 20 sec).

Digipeater Timing—When one station's packet collides with another station's packet, each must take steps to rectify the problem. When one station's packet collides with a digipeater's packet, the digipeater is not responsible for requesting a retransmission; the station that originated the packet must keep track of the outstanding packets. In cases where several digipeaters are used in a link, a packet may be lost by any one of them.

Digipeated packets are at a distinct disadvantage. To help counteract this, a timing parameter is included in the TNC to give digipeated packets a break. When a TNC is digipeating packets, it sends its packets as soon as the channel is clear of activity. If a TNC is originating its own packets, however, it waits a selected time period (set by the DWait command parameter value) after the channel is clear before it sends its packets. As a result, when a channel is clear, the digipeater always transmits first. This reduces the chance of digipeated packets colliding with non-digipeated TNC packets. The default setting of the DWait parameter is 16, which represents 160 ms. To change this timing parameter, type:

cmd: DWait n <CR>

where n is a number from 0 to 250 representing a delay in 10 ms increments. For example, to select a delay of 120 ms, set n to

12 (120 ms / 10 ms = 12). Note that DWait should be similarly applied in network node environments (in most areas, network nodes have replaced digipeaters).

To work effectively, all of the TNCs on a channel used by one or more digipeaters should use the same DWait value. Unless you know that the packet operators in your area are using a DWait value other than the default, you should leave your TNC set at the default value.

Prioritized Acknowledgment—TNCs that are compatible with TAPR TNC 2 software release 1.1.7 or later are capable of using the channel-sharing protocol called "prioritized acknowledgment." As its name implies, this protocol gives priority to packet acknowledgments (ACKs) on a channel.

An ACK indicates that a packet has been received correctly. By giving an ACK priority, the station that sent the packet will receive the ACK more quickly and, as a result, will be less likely to resend the packet. (On channels where ACKs are not given priority, ACKs may be delayed long enough to cause the waiting station to give up and resend the packet, thus wasting precious time on the channel.)

Configuring a TNC for prioritized acknowledgment involves setting the following parameters as indicated.

For 1200-bit/s VHF FM:

ACkprior	ON
ACKTime	14
DEAdtime	33
DWait	33
FRack	8
MAXframe	1 to 7
RESptime	0
SLots	3

For 300-bit/s HF:

ACkprior	ON
ACKTime	52
DEAdtime	8
DWait	8
FRack	16
MAXframe	1
Paclen	32 to 128
RESptime	0
SLots	3

These settings are recommended by TAPR as a starting point. Once you are on the air, you may have to adjust these settings as conditions warrant. Also, check your TNC's manual in case it recommends different settings.

Station Identification—Perhaps the most important TNC command is the one that inserts your call sign in the TNC. To program your TNC with your call sign, type:

cmd: MYcall WA1LOU <CR>

where WA1LOU is the call sign of your station.

An optional *secondary station identification* (SSID), number 1 through 15, may be appended to your call sign by typing a hyphen and the desired number immediately after the call sign in the Mycall command (for example, WA1LOU-4). SSIDs are used to differentiate digital repeaters (digipeaters), network nodes, PBBSs, and other secondary packet station operations from individual packet stations that use the same call sign. For example, the author's individual packet station is identified as WA1LOU (or WA1LOU-0 because 0 is assumed to be the SSID unless specified otherwise), whereas, his PBBS may be identified as WA1LOU-4 (to differentiate it from WA1LOU-0).

ON-THE-AIR

Once your TNC is tailored to your station and your mode of operation, you can send your first packet. The following paragraphs describe how to initiate and conduct packet communications from the keyboard of your DTE.

Connecting and Disconnecting

To contact another station you must make a connection with that station. To do so, type:

cmd: Connect WØRLI <CR>

where WØRLI is the call sign of the station to contact.

If WØRLI's packet station is on the air and receives your connect request, your stations will exchange packets to set up a connection between stations. When the connection is completed, your DTE will display:

*** CONNECTED to WØRLI

and your TNC automatically switches to the Converse Mode.

Now, everything you type into the DTE keyboard is "packeted" and sent to the other station. A packet is sent whenever you enter a carriage return <CR> (the TNC default for the Sendpac character), or whenever the byte length of your keyboard input equals the number of bytes selected with the Paclen command (the default is 128 bytes).

When you are finished conversing with the other station, return to the Command Mode by typing <CTRL-C> (which is the TNC default for the Command character). When the command prompt (cmd:) is displayed, type:

cmd: Disconne <CR>

and your station will exchange packets with the other station to break the connection. When the connection is broken, your DTE displays:

*** DISCONNECTED

If, in mid-contact, you wish to enter a command and then continue with the contact, enter the Command Mode by typing

<CTRL-C>. When the command prompt (cmd:) is displayed, enter the desired command. When you are ready to return to your contact, exit the Command Mode and type:

cmd: CONVers <CR>

and you are back in the Converse Mode. Anything you type now is packeted and sent to the other station.

If, for some reason, the other station does not respond to your initial connect request, your TNC will resend the request. It will continue to do so until the number of additional attempts equals the number selected with the Retry command (the default is 10 attempts).

When the number of additional attempts exceeds the Retry command selection, your TNC stops sending connect requests and your DTE displays:

*** retry count exceeded

*** DISCONNECTED

(Retry not only sets the maximum number of times that a connect request is resent, it also sets the number of times any packet is resent without acknowledgment.)

Monitoring Mode

A rule of thumb in Amateur Radio says that you should listen before you transmit. In support of this rule, your TNC has an array of commands for the monitoring mode.

The Monitor command causes the TNC to display on your DTE the contents of each packet received by your station while in the Command Mode. The display will include the call sign of the station originating each packet and the call sign of the station that is the intended recipient of each packet. A monitored packet will be displayed as:

WS1O>KU7G: DATA

where WS1O is the call sign of the station originating the packet, KU7G is the call sign of intended destination of the packet and DATA is the contents of the packet. (The Monitor function is on by default.) To enable (or disable) this function, type:

cmd: Monitor ON (or OFF) <CR>

The Mrpt command makes the monitoring function more revealing by displaying the call sign(s) of the station(s) that repeat a monitored packet, in addition to the call signs of the originating and destination stations. With Mrpt on, a monitored packet is displayed as:

WS1O>KU7G,KA6M-9*,WA8DED-5:DATA

where WS1O is the station originating the packet, KU7G is the intended destination of the packet, KA6M-9 and WA8DED-5 are the stations repeating the packet, and DATA is the packet's contents. The asterisk indicates that KA6M-9 is actually transmitting the packet you are receiving. (The Mrpt function is on by default.) To enable (or disable) this function, type:

cmd: MRpt ON (or OFF) <CR>

Other monitoring commands that may be useful are:

MAll (default: on) allows you to monitor both connected and unconnected packets;

MCOM (default: off) allows you to monitor connect requests, disconnect requests, connect/disconnect acknowledgments and non-connect/disconnect acknowledgments, as well as packets containing data;

219—220 MHz: For Network Relays *Only*

In 1995 the FCC granted amateur access to 219 to 220 MHz on a secondary basis. The ruling limits usage to digital transmissions *only*—specifically to stations that act as message relays within networks. Intercity packet backbone networks are typical examples. Maximum signal bandwidth is 100 kHz and transmitter output power is limited to 50 W PEP.

The FCC also stipulates that you must give written notification of your intent to operate a network relay in this segment *30 days before you actually place it on the air.* You must send your notification to the ARRL, 225 Main St, Newington, CT 06111-1494. We'll add your information to a public database maintained at League Headquarters.

Because of our secondary status in this segment, hams must not cause interference to the operation of Automated Marine Telecommunications Systems (AMTS), television broadcasting on channels 11 and 13, Interactive Video and Data Service systems, Land Mobile Service systems or any other service that has a primary allocation on these frequencies. As secondary users, we're not protected from interference *from* these services.

Despite these limitations, 219 to 220 MHz is a valuable segment for packet networking. It relieves crowding on the 222-MHz band and provides room for the expansion of high-speed networks. Let's use it wisely!—*WB8IMY*

MCon (default: on) display packets from stations on frequency, while connected to another station;

MHeard lists the last 18 stations received on frequency by typing:

cmd: MHeard <CR>

MHClear clears the list of received stations that are logged for Mheard command recall by typing:

cmd: MHClear <CR>

MStamp (default: off) stamps the date and time of each monitored packet. To use this function, the date and time must be entered into the TNC using the *Daytime* command, by typing:

cmd: DAytime yymmddhhmm <CR>

where yy is the last two digits of the year (00-99), mm is two digits representing the month (00-12), dd is the day of the month (00-31), hh is the hour (00-23) and mm is the minute of the hour (00-59). (Note that your TNC has a 24-hour, rather than a 12-hour clock.) For example, to enter March 8, 1951, 1:30 PM, type:

cmd: 5103081330 <CR>

where 51 is the last two digits of 1951, 03 represents the third month (March), 08 represents the eighth day of the month, 13 represents the thirteenth hour of the day and 30 represents the thirtieth minute of the hour.

The *Budlist* and *Lcalls* commands may be used together to limit your monitoring. Use the Lcalls command to list a maximum of eight stations that you do or do not wish to monitor by typing:

cmd: LCAlls aaaaaa,bbbbbb...,hhhhhh <CR>

where aaaaaa, bbbbbb..., hhhhhh are the call signs of stations you do or do not wish to monitor.

Next, use the Budlist command to limit your monitoring according to the Lcalls list of stations. Your DTE will only display packets originating from stations in the Lcalls list if you type:

cmd: BUDlist ON <CR>

Your DTE will display packets of all stations except the stations in the Lcalls list (the TNC default). if you type:

cmd: BUDlist OFF <CR>

Most of your monitoring needs will be met with the wide range of TNC monitoring commands at your disposal.

VHF/UHF Operating

Today, most amateur packet activity occurs at VHF, on 2 meters to be specific. The most commonly used data rate is 1200 bit/s with frequency modulated AFSK mark and space tones of 1200 and 2200 Hz, respectively. This is referred to as the "Bell 202" telephone modem standard.

The majority of TNCs are optimized for VHF/UHF FM operation, so getting on the air is a simple matter of turning on your packet equipment and tuning to your favorite packet frequency.

Most activity hovers on or near 145.010 MHz including digipeater, network node, PBBS and direct connect contacts. (Refer to the sidebar entitled "Packet Frequencies" for information concerning other packet hot spots.) In some areas, an effort has been undertaken to establish band plans for packet. If there is a packet band plan in effect in your area, it is recom-

mended that you comply with it. If there is no established band plan in your area, the following guidelines should be observed:

If you are conducting a direct connect contact (sans digipeater or network node), move your contact to an unused frequency. It is very inefficient to try to exchange packets on a frequency where other stations, especially digipeaters and network nodes, are also exchanging packets. The competition slows down your station's ability to exchange packets and, in turn, you are also slowing down all of the other stations. You should use a frequency occupied by a digipeater or network node only if you are using that repeater or node.

VHF/UHF packet operation is similar to VHF/UHF FM voice operation. Under normal propagation conditions, you may communicate only with stations that are in line of sight of your station. By using a repeater or node, the line-of-sight limitation is expanded. For example, if a repeater is within your line of sight, you may communicate with other stations that are also within line of sight of the same repeater.

Packet expands upon this by permitting the simultaneous use of eight digipeaters and an infinite number of nodes. Packets are relayed from one station to another station, even though the destination may be at the end of a long (and distant!) chain of nodes or digipeaters.

Besides digipeaters and nodes, some packet operations are conducted using voice repeaters. Such operations usually take place in the absence of voice activity and with the blessing of the sponsor of the voice repeater. It is imperative that you monitor the repeater before initiating packet-radio operation. One way to wear out your welcome quickly is to send a few packets in the middle of a voice conversation! To take full advantage of the voice repeater, you should set your TNC's *Axdelay* and *Axhang* parameters correctly and, when entering the Connect command, do not include Via and the call sign of the voice repeater.

Packet Networking

The packet network is a complex evolutionary system of packet stations that have been organized to transfer packets between local or distant points. Depending on the locations of those points, the packets being transferred between them may travel through a simple network consisting of one station operating on one frequency. They may also travel through a variety of networks consisting of tens of stations operating on HF, VHF and UHF.

In the past, if you wanted to connect to another packet station, you had to know the exact path through the network to make the connection. If one or more stations were in that path, you had to know (and list) each station's call sign and use them when you invoked the Connect command.

Can you imagine what would happen if you had to know the name of each intermediary post office for every letter you mailed? If you accidentally omitted a single name, your letter would not be delivered!

Luckily, the post office does not require that you be familiar with its vast network to mail a letter. Similarly, in many parts of the packet network, you do not have to be familiar with the network to make a connection with another station.

The PBBS automatic mail-forwarding system is part of the packet network. Sending a message via the PBBS automatic mail-forwarding system is similar to sending a letter through the postal system. In either case, you do not have to know who will be handling your mail before it is delivered. All you need to know is the identity of the intended recipient, and the address where the intended recipient picks up his or her mail.

For example, to send mail via the postal system, you address the envelope with the intended recipient's identification (the recipient's name) followed by the address where the recipient picks up his mail. To send mail via the PBBS automatic mail-forwarding system, you address the mail with the intended recipient's Amateur Radio identification (the recipient's call sign) followed by the at-sign (@) and the address where the recipient picks up packet mail. Packet operators usually receive their mail at their local PBBS, so the address for a packet message is simply the call sign of the destination PBBS. Once you have properly addressed your mail, the system (either postal or PBBS) does the rest, automatically forwarding your mail to its destination.

Digipeaters Are History!

Only a few years ago, packet networking was completely dependent on digipeaters. Digipeating is a function built into every AX.25-compatible TNC that permits the TNC to receive, temporarily store and retransmit (repeat) the packet transmissions of other stations. A digipeater repeats only transmissions that are specifically addressed for routing through that digipeater, as opposed to a typical voice repeater that retransmits everything it hears.

In light of the other packet networking options that are available today, digipeating is a very basic form of networking. Only eight digipeater stations can be used between any two points that are attempting to transfer packets. In order to use the digipeaters, the call sign of each must be known and specified when invoking the initial Connect command. Even though digipeating is rudimentary in comparison to what is available today, it still serves a purpose when an intermediary station or two is needed in a pinch to complete a connection.

When you are unable to make a connection with another station, the problem may be caused by terrain or propagation that prevents your signal from being received by the other station. The digipeater func-tion provides the ability to circumvent this problem. You only need to determine which on-the-air packet stations can send and receive signals between your station and the station you are trying to reach.

Once the existence of an intermediary station is known, type:

cmd: Connect W4RI Via NK6K-5

where W4RI is the call sign of the station to connect and NK6K-5 is the call sign of the station that will act as the digipeater.

When NK6K-5 receives your connect request, it automatically enters the digipeater mode and stores your request in memory until the frequency is quiet. It then retransmits your request to W4RI on the same frequency. If W4RI's packet station is on the air and receives your connect request, your station and his will exchange packets through NK6K-5 to set up a connection between stations. Once the connection is established, your DTE displays:

*** CONNECTED to W4RI VIA NK6K-5

Your stations continue to use the facilities of the digipeater until the connection is broken.

Digital and voice repeaters both repeat, but the similarity ends there. A digipeater receives and transmits on the same frequency (whereas a voice repeater receives and transmits on different frequencies). As a result, a digipeater operates half-duplex; that is, it does not receive and transmit at the same time (as compared to a voice repeater, which transmits whatever it receives simultaneously).

If one station in the digipeater mode is insufficient to establish a connection, as many as eight stations can be called up on for digipeater operation in your connect request. Additional digipeaters are appended to the Connect command separated by commas. (For example, by typing "Connect W4RI Via NK6K-5,W3IWI-1" after the command prompt, your TNC will send the W4RI connect request to NK6K-5 which relays it to W3IWI-1. Then, W3IWI-1 relays it to W4RI.)

It is recommended that you not use more than one or two digipeaters at any one time because digipeater throughput decreases dramatically as the number of digipeaters increases. Each time you use a digipeater, you are competing with other stations using the same digipeater. Each competing station has the potential to generate a packet that may collide with your packet!

Any packet station can act as a digipeater. This occurs automatically without any intervention by the operator of the station being used as a digipeater. You do not need his permission, only his cooperation because he can disable his station's digipeater function by means of the Digipeat command. (In the spirit of Amateur Radio, most packet operators leave the digipeater function enabled and only disable it under special circumstances.)

Similar to VHF/UHF voice repeaters, some stations are set up as dedicated digipeaters. They are usually set up in good radio locations by packet clubs. Aside from location, the other advantage of a dedicated digipeater is that it is always there (barring a calamity). Stations do not have to depend on the whims of other packet stations, which may or may not be on the air when their digipeater functions are needed most.

NET/ROM and TheNet

A network *node* is a relay station that allows you to connect to stations that are not connectable directly (a function similar to that of a digipeater). A node differs from the digipeater function because it is more "intelligent."

Say you want to connect to station X, who is five digipeater hops away (digipeaters A, B, C, D and E). To connect to X, you enter the command "Connect X via A, B, C, D, E." If A, B, C, D and E were network nodes instead of digipeaters, however,

---

Packet Frequencies

On HF, most packet activity straddles the traditional RTTY subbands using 300-bit/s AFSK on LSB. The list of RTTY subbands in the United States follows.

1800-1830 kHz	14070-14099.5 kHz
3605-3645 kHz	18100-18110 kHz
7080-7100 kHz	21070-21100 kHz
10140-10150 kHz	28070-28150 kHz

Hotbeds of HF packet activity are 3606, 3630, 3642, 7093, 7097, 10145, 14101-14105, 21099-21105 and 28099-28105 kHz.

On 6 meters, 50.62 through 51.78 MHz is the packet subband with 51.70 MHz serving as the 6-meter packet calling frequency.

On 2 meters, 145.010 MHz (FM) is "the" packet frequency. Because of 145.010's popularity, packet activity has spread upward and downward to 144.910, 144.930, 144.950, 144.970, 144.990, 145.030, 145.050, 145.070, and 145.090 MHz as well as to unused simplex voice channels and on voice and RTTY repeaters.

On 222 MHz, 223.52 through 223.64 MHz is the packet subband. Another segment exists from 219 to 220 MHz, but this is restricted to network links only.

On 440 MHz, 430.05-431.025 and 440.975-441.075 MHz are the packet subbands. Above 450 MHz, there are packet subbands at 903-906, 915-918, 1248-1252, 1296 and 1297-1300 MHz.

---

you would first connect with node A, then you would command node A to connect with node E. Once connected to node E, you would command node E to connect with station X.

If you use digipeaters, you have to know the path between you and station X (via A, B, C, D and E). If you use nodes, the nodes already "know" the path to other nodes. All you have to know is which node is local to station X. If a path changes (a node disappears or a new one appears), the node is aware of the change because nodes exchange information with each other regularly.

Another advantage of nodes is that their throughput is better than digipeaters. With digipeaters, only the last station in the circuit acknowledges received packets. With nodes, each node acknowledges each packet it receives. If a packet is lost in a chain of digipeaters, your TNC is not aware of that fact until it fails to receive an acknowledgment from the station at the opposite end of the circuit. If a packet is lost in a node sequence, the node that fails to receive an acknowledgment will retransmit the lost packet until the next node in the circuit acknowledges it. As you might guess, the recovery of lost packets is lightning fast.

Packet nodes were the brainchild of Ron Raikes, WA8DED, who released NET/ROM to the packet community in 1987. NET/ROM became very popular because it was inexpensive and easy to install (just replace one of your TNC's EPROMs with a NET/ROM EPROM). Since then, NET/ROM has begotten several children (some legitimate, some not) that now inhabit every nook and cranny of the packet world. *TheNet* is NET/ROM's best known offspring.

Many nodes use a mnemonic identifier instead of their FCC-given call signs. Typically, the identifier is a three- to six-character acronym that identifies the node's location. The reason for using identifiers is to simplify what the user has to remember. For example, instead of having to remember that XK1NG is the node located on Kong Island, all you have to remember is the node's mnemonic identifier, KONG.

To use a node, you first connect to your local node by entering a Connect command using your local node's call sign or mnemonic identifier. For example, to connect with the Kong Island node, type either:

cmd: Connect XK1NG <CR>

or

cmd: Connect KONG <CR>

After you receive the connected message from your local node, you may connect to another station that is also local to your node, or you may connect to another node.

To connect to another station, enter the Connect command followed by the call sign of the other station. To connect to another node, enter the Connect command followed by the call sign or mnemonic identifier of the other node. After you connect to the other node, use the Connect command again to connect to a station that is local to the other node. When you disconnect from another station, the node(s) disconnect automatically.

The Connect command used to connect to a node and the Connect command used to ask a node to connect to another station are different. When you connect to a node, you are using a TNC command. When you ask a node to connect to another station, you are using a node command. In order for a node to receive a node command, it must pass through your TNC as data (your TNC must not interpret a node command as if it were a TNC command). To ensure that this does not occur, do not send a node command *until you connect to a node* (after you receive the node connection message). If you send the node command too soon, the TNC will try to interpret it and, when it can't, will send you an error message. Note that after you connect to your local node, you will not receive another command prompt until you disconnect from the node, so don't wait for a command prompt before sending commands to a node.

Nodes will respond to other commands besides Connect. Here is a quick summary of other node commands.

CQ lets you call CQ through a local or distant node. When you send CQ through a node, you may send along a maximum of 77 characters of text to indicate who is calling CQ. (For example, "CQ—This is Bruce in downtown Gotham.")

Ident lists the node's mnemonic identifier.

Nodes lists the other known nodes by their mnemonic identifier, call sign and SSID. Nodes followed by the mnemonic identifier or call sign and SSID of a known node (for example, NODES K1WJ-1) lists routing information about that node.

Parms lists the settings of the node's parameters.

Routes lists instructions for routing to other nodes.

Users lists the call signs of who is using the node.

KA-Node

KA-Node is the Kantronics implementation of a node-to-node acknowledgment protocol. It is available in every Kantronics TNC (except the original KPC-1), and the Kantronics All Mode (KAM) controller. As explained in the description of NET/ROM and TheNet above, node-to-node acknowledgment provides improved throughput over the standard AX.25 end-to-end acknowledgment. Besides node-to-node acknowledgment, KA-Node allows you to gateway from one port to another when you are connected to a dual-port KA-Node (available in some Kantronics models).

Whereas KA-Node, NET/ROM and TheNet are similar in that they all offer node-to-node acknowledgment, they are dissimilar in other ways. The most important difference is that KA-Node does not perform automatic routing as does NET/ROM and TheNet. The user must command the KA-Node as to the desired path to another node or station. In this way, KA-Node is more like an AX.25 digipeater than a NET/ROM or TheNet node. Although a KA-Node does not achieve the same functionality of NET/ROM and TheNet, it is still an improvement over digipeating and, as a result, it is a popular packet tool.

ROSE

The *RATS Open System Environment* (ROSE) was developed over a number of years by The Radio Amateur Telecommunications Society (RATS) of New Jersey. The current elements of ROSE are:

ROSE X.25 Packet Switch—This is the X.25 Packet Networking solution for the TAPR TNC 2 and its clones. The ROSE Switch is based upon international standard protocols as defined by CCITT and ISO. It was written by Tom Moulton, W2VY.

ROSErver/PRMBS, the Packet MailBox System—ROSErver/PRMBS is a WØRLI-compatible PBBS with an advanced user interface including an Internet-compatible mailer. PRMBS provides the user and SysOp with many powerful features.

ROSErver/OCS On-line Callbook Server—OCS provides either a text-based or an efficient graphic user interface depending on the terminal capabilities of the user. Developed by Keith Sproul, WU2Z, and Mark Sproul, KB2ICI, this advanced packet server operates on a Macintosh computer.

IP Address Coordinators

The following list denotes the IP address coordinators for Amateur Radio throughout the United States. To obtain an IP address, contact the coordinator for your locale. (An IP address is necessary for each station you wish to operate using TCP/IP packet.)

State	Coordinator
AK	John Stannard, KL7JL
AL	Richard Elling, KB4HB
AR	Richard Duncan, WD5B
AZ	David Dodell, WB7TPY
CA, Antelope Valley/Kern County	Dana Myers, KK6JQ
CA, Los Angeles—San Fernando Valley	Jeff Angus, WA6FWI
CA, Orange County	Terry Neal, AA6TN
CA, Sacramento	Bob Meyer, K6RTV
CA, Santa Barbara/Ventura	Don Jacob, WB5EKU
CA, San Bernardino and Riverside	Geoffrey Joy, KE6QH
CA, San Diego	Brian Kantor, WB6CYT
CA, Silicon Valley—San Francisco	Douglas Thom, N6OYU
CO (northeast)	Fred Schneider, K0YUM
CO (southeast)	Bdale Garbee, N3EUA
CO (western)	Bob Ludtke, K9MWM
CT	Bill Lyman, N1NWP
DC	Don Bennett, K4NGC
DE	John DeGood, NU3E
FL	Bruce La Pointe, WD4HIM
GA	Doug Drye, KD4NC
HI and Pacific islands	John Shalamskas, KJ9U
IA	Ron Breitwisch, KC0OX
ID	Steven King, KD7RO
IL (central and southern)	Chuck Henderson, WB9UUS
IL (northern)	Ken Stritzel, WA9AEK
IN	Jacques Kubley, KA9FJS
KS	Dale Puckett, K0HYD
KY	Tyler Barnett, N4TY
LA	James Dugal, N5KNX
MA (center and eastern)	Johnathan Vail, N1DXG
MA (western)	Bob Wilson, KA1XN
MD	Howard Leadmon, WB3FFV
ME	Carl Ingerson, N1DXM
MI (upper peninsula)	Pat Davis, KD9UU
MI (lower peninsula)	Jeff King, WB8WKA
MN	Gary Sharp, WD0HEB
MN (Minneapolis only)	Andy Warner, N0REN
MO	Stan Wilson, AK0B
MS	John Martin, KB5GGO
MT	Steven Elwood, N7GXP
NC (eastern)	Mark Bitterlich, WA3JPY
NC (western)	Charles Layno, WB4WOR
ND	Steven Elwood, N7GXP
NE	Mike Nickolaus, NF0N
NH	Gary Grebus, K8LT
NJ (northern)	Dave Trulli, NN2Z
NJ (southern)	Bob Applegate, WA2ZZX
NM	J Gary Bender, WS5N
NY (eastern)	Bob Bellini, N2IGU
NY (western)	Paul Gerwitz, WA2WPI
New York City and Long Island	Bob Foxworth, K2EUH
NV (southern)	Earl Petersen, KF7TI
NV (northern)	Bill Healy, N8KHN
OH	Gary Sanders, N8EMR
OK	Joe Buswell, K5JB
OR	Ron Henderson, WA7TAS
OR (northwest and Vancouver, WA)	Tom Kloos, WS7S
PA (eastern)	Doug Crompton, WA3DSP
PA (western)	Bob Hoffman, N3CVL
PR	Karl Wagner, KP4QG
RI	Charles Greene, W1CG
SC	Mike Abbott, N4QXV
SD	Steven Elwood, N7GXP
TN	Mark J. Bailey, N4XHX
TX (northern)	Don Adkins, KD5QN
TX(southern)	Kurt Freiberger, WB5BBW
TX (western)	Rod Huckabay, KA5EJX
UT	Matt Simmons, KG7MH
VA	Jim DeArras, WA4ONG
VA (Charlottesville)	Jon Gefaell, KD4CQY
VI	Bernie McDonnell, NP2W
VT	Ralph Stetson, KD1R
WA (eastern)	Steven King, KD7RO
WA (western)	Dennis Goodwin, KB7DZ
WI	Pat Davis, KD9UU
WV	Rich Clemens, KB8AOB
WY	Reid Fletcher, WB7CJO

ROSE/BBC Bulletin Broadcast Controller—BBC provides a mechanism for simultaneous reception of packet bulletins by several stations with complete data integrity and error correction. ROSE/BBC protocol was written by Gordon Beattie, N2DSY, and the software has been implemented for UNIX by Marsh Gosnell, AD2H.

ROSE/STS Station Traffic System—A complete NTS message management system, especially suited for packet, STS runs on MS-DOS and UNIX computers and is written by Frank Warren, KB4CYC.

Networks made up of ROSE X.25 Packet Switches provide amateurs with a vehicle through which they can easily establish packet connections with other stations. Connections through a ROSE Network can be thought of as a "reliable data pipe."

Using ROSE X.25 Packet Networks is similar to using the telephone network. Phones are easy to use. All one has to do is find a phone (the entry point into the network), dial the desired telephone number (the "network address" of the exit point from the network) and ask, by name, for the desired party (the person's "User ID"). A call from New York to California may go through Chicago, St. Louis or Atlanta, but all that matters to the caller is that it goes through automatically and reliably.

With ROSE Networks, users must know their entry point into the network (their local ROSE Switch), the exit point to reach the desired station (the ROSE address of the Switch used by that station) and the call sign of the desired station ("User ID"). Like the telephone network, the ROSE Network determines the path taken by the call.

ROSE Addresses follow the international standards used by commercial X.25 data networks. These addresses are based upon the telephone numbering plan used in each country. In North America, this is the telephone Area Code and Exchange. Since Area Codes and Exchanges are assigned geographically, ROSE addresses indicate a geographical area, too.

ROSE Networks are easy to use, as you will see in the following example.

In the ROSE Network, there is the KB2ICI-7 ROSErver/OCS On-line Callbook Server at ROSE Address 609426 (central New Jersey) and there is a ROSE Switch in New York City, KB7UV-3. A user of the KB7UV-3 ROSE Switch can connect to the On-line Callbook Server by sending the following command to a TNC:

C KB2ICI-7 VIA KB7UV-3,609426

where KB7UV-3 is the call sign of the entry ROSE Switch and 609426 is the desired ROSE address. The user's TNC will first display the "connected to" message from the TNC, followed by a network message that acknowledges the call request:

*** CONNECTED to KB2ICI-7 VIA KB7UV-3,609426
   Call being Setup

An additional message will be sent by the network once the connection has been established:

Call Complete to KB2ICI-7 @ 3100609426

This message includes the call sign of the station connected to (KB2ICI-7) and the complete ROSE address of the switch that made the final link (3100609426). This "complete" address also includes country identification information ("3100" is the Amateur Radio Data Network Identification Code [DNIC] for the US).

At this point, there is a connection between the two stations through the network. Either station can end the connection by simply sending the Disconnect command to their TNC. The network will advise the other station of this by sending the message:

*** Call Clearing *** 0000 3100609426 Remote Station cleared connection

ROSE X.25 Networks also inform users of other conditions by means of similar messages (other station busy, and so on). Each of these messages, as indicated in the example above, includes a ROSE address that indicates the point in the network where the condition described in the message has occurred.

Help is always available to users of ROSE Networks. Each ROSE Switch has informative text stored in memory that may be obtained by connecting directly to the switch, then sending a carriage return or waiting 30 seconds. The switch will send you the text, which usually includes information on the network and basic instructions on switch operation.

ROSE Networks can also have directories of Services (PBBSs, ROSE Switches, etc.) and other information available to users (if implemented by the network managers). This is done through the ROSE INFO application and special ROSE addresses.

The RATS ROSE Network uses special addresses made up of the Area Code and either "555" or "411" for directories. The 555 directories are for services and 411 directories are for users. If a user of the N2DSY-3 ROSE Switch wants information on services available in Southern New Jersey (Area Code 609), the user sends the following command to a TNC:

C INFO VIA N2DSY-3, 609555

This is automatically routed to the switch containing the directory information. Once the call was completed, the directory text is sent to the user.

Optionally, ROSE Networks can have other useful applications. The most popular are HEARD and USERS. HEARD provides a list of stations heard by a given switch and USERS provides a list of switch activity. Both are used in the same way, that is, you initiate a connection to the desired application at the ROSE address of interest and, if the application is loaded at that switch, the requested information will be sent.

TCP/IP

TCP/IP for amateur packet is based on a networking scheme known as the Defense Advanced Research Project Agency's (DARPA) Transmission Control Protocol (TCP) and Internet Protocol (IP). Despite what the name "TCP/IP" implies, TCP and IP are only two parts of a collection or *suite* of protocols that comprise the complete DARPA protocol used for landline data communications networking between mainframe computers worldwide.

Phil Karn, KA9Q, wrote the amateur packet implementation of TCP/IP (for DOS-based computers) and called it *NET*. Later, he rewrote and refined the software and called it *NOS*. Phil graciously allowed his handiwork to be used freely for Amateur Radio (non-commercial) purposes.

Like the TNC-emulation software mentioned earlier, TCP/IP software emulates many of the functions of a TNC. As a result, your computer is no longer limited to the functions programmed into the TNC's read-only memory, and it may be programmed to do much more.

One of the main functions of the TCP/IP software is to allow each TCP/IP station to function as an intelligent network node. These nodes are "intelligent" because they automatically route packets to their intended destination without any operator intervention or direction. TCP/IP frees the operator of the task of figuring out which string of digipeaters, or local and remote network nodes, or alphanumeric addresses must be used to get packets delivered to their intended destinations. The operator simply commands the software to communicate with station X and the network does the work, automatically determining the route to get to the other station.

TCP/IP software improves data throughput on busy channels by using "intelligent backoff" techniques, rather than the pseudorandom method used by plain AX.25. TCP/IP senses when a frequency gets busy and waits longer between sending

Packet Organizations

The following Amateur Radio organizations are devoted entirely to amateur packet or are devoted supporters of the mode. They are an excellent source for additional packet information. The name of the organization's periodical, if any, follows the organization's address.

Amateur Radio Research and Development Corp (AMRAD), PO Drawer 6148, McLean, VA 22106-6148, *AMRAD Newsletter*

American Radio Relay League (ARRL), 225 Main St, Newington, CT 06111, *QST* and *QEX*

Chicago Amateur Packet Association (CAPRA), PO Box 8251, Rolling Meadows, IL 60008, *The CAPRA Beacon*

Northern California Packet Association (NCPA), PO Box 61716, Sunnyvale, CA 94088-1761, *NCPA Download*

Northwest Amateur Packet Association (NAPRA), PO Box 70405, Bellevue, WA 98007, *Dedicated Link*

Radio Amateur Telecommunications Society (RATS), PO Box 93, Park Ridge, NJ 07656-0093

Southern California Digital Communications Council (SCDCC), PO Box 2744, Huntington Beach, CA 92647, *I-Frame*

Texas Packet Society (TPRS), PO Box 50238, Denton, TX 76206-0238, *TPRS Quarterly Report*

Tucson Amateur Packet (TAPR), 8987-309 E Tanque Verde Rd, #337, Tucson, AZ 85749-9399, *Packet Status Register*

HamNet is the Amateur Radio group on CompuServe (5000 Arlington Centre Blvd, Columbus, OH 43220, telephone 800-848-8199) that is accessible with a modem over the telephone line. *HamNet* has a very active section devoted to packet with a roster of members that is a who's who of the packet-radio community.

packets. As the channel's activity dies down, your station automatically starts sending more rapidly. This reduces collisions and retries and gets everyone's packets moving smoother and more efficiently.

Besides intelligent networking, the TCP/IP software has a packet-radio terminal function that allows you to communicate keyboard-to-keyboard with another packet operator. A file-transfer function, FTP, allows you to send and receive (upload and download) ASCII (text) and binary (executables, graphics, etc) files to and from any other TCP/IP station. The TCP/IP software includes a built-in mailbox that automatically sends and delivers mail between TCP/IP stations and "regular" PBBSs. What's more, TCP/IP software is multitasking on any machine. All of the aforementioned functions, and more, can be used simultaneously.

Here is what you need to get started with TCP/IP:

❏ An IP address, which is a unique number assigned to the computer used at your packet station for communications over the TCP/IP network. To get an IP address, contact your local IP address coordinator. Refer to the sidebar titled "IP Address Coordinators" to locate your local coordinator.

❏ Since the TCP/IP software emulates many of the functions of a TNC, you may have thought that it would allow you to get on packet without it. Well, you still need a TNC and it must be one that supports the *KISS* mode. The good news is that virtually all current TNCs support KISS and many of the older ones that don't support KISS can be made "KISSable" by updating the ROM containing the TNC software. KISS is the acronym for Keep It Simple, Stupid. When you invoke the KISS mode, it makes your TNC less intelligent. (If the software programmed in your TNC's ROM is compatible with TAPR TNC 2 software release 1.1.6 or later, then it is KISSable.)

❏ TCP/IP software. There are many versions of the software and there are many sources. Some of the sources are TAPR (8987-309 E Tanque Verde Rd, #337, Tucson, AZ 85749-9399), ham-oriented landline BBSs, such as N8EMR's (telephone 614-895-2553) or KA1SVW's ChowdaNet (telephone 401-331-0907) and the Amateur Radio sections of commercial computer services such as CompuServe, GEnie and America Online.

Once you obtain these three items, you set up and use the software as described by its accompanying documentation. (A very useful book that details the setup and operation of NOS is *NOSintro* by Ian Wade, G3NRW. The book is available from the ARRL.)

After you have everything set, let's see if it works!

Before you run your TCP/IP program, you must place your TNC in the KISS mode. To do so, at the TNC's normal command prompt, type:

cmd: KISS ON <CR>

After your TNC is in the KISS mode, power off the TNC and then power it on. As your TNC initializes itself, its STA and CON front panel LEDs will blink three times to indicate that it is in the KISS mode. Some TNCs only require you to enter *RESTART* or *RESET*. You don't need to turn them on and off. Check your TNC manual to be sure.

Now, run your TCP/IP program and, after the system prompt appears (net>), type:

FINGER (insert your call sign) @ (insert your call sign)

The system will read and display the file that you created in your \FINGER subdirectory. (The FINGER file is a short description of you and your TCP/IP station.) Now it's time to

perform a test to see if it will work on the air. ("FINGERing" yourself is an operation that occurs within the confines of your computer; nothing is sent over the air.)

Try to get a complete DOMAIN.TXT or HOSTS.NET file from another station—K1ZZ in this example. (DOMAIN.TXT or HOSTS.NET files are look-up tables of the TCP/IP stations that, at a minimum, list each station's call sign and IP address.) You can achieve this by using the File Transfer Protocol (or FTP). At the system prompt, type:

ftp k1zz

If all goes well, your computer indicates that a session has been "Established" and that "k1zz.ampr.org FTP" is "ready" for you to log on. After entering "anonymous," K1ZZ asks that you "Enter PASS command," so you enter pass <your call sign> in response. ("Pass" is short for password.)

The "Logged in" message should appear shortly indicating that you are in K1ZZ's *public* subdirectory. Send "dir" to get a list of the contents (a "directory") of the public subdirectory. If the directory contains a DOMAIN.TXT or HOSTS.NET file, you can get a copy by entering "get domain.txt" or "get hosts.net."

After entering the get command, your computer indicates that a data connection for retrieval of domain.txt or hosts.net is opening, then displays nothing while the file is actually being transferred. The file transfer can go on for quite a while. The keying of your transmitter is the only indication that anything is happening. When the transfer is completed, your computer displays "Get complete" followed by the number of bytes transferred and the "File sent ok" message. At the system prompt, enter close to end the FTP session and log off K1ZZ.

There are misconceptions about TCP/IP that have caused some potential TCP/IP users to steer clear of it. Let me assure you that the following *aren't* true:

• You must keep your radio equipment and computer on 24 hours a day.

For the viability of the TCP/IP network (not to mention your local electric company), around-the-clock operation is preferable, but it's not a necessity as far as receiving mail is concerned. If your TCP/IP station is on the air all the time, it is always ready to receive traffic heading its way. If it is not active around the clock, however, other TCP/IP stations can intercept and hold your station's traffic for automatic delivery to you when your station is on.

• You must use a DOS-based computer.

Although most computers using TCP/IP are DOS-based, versions of TCP/IP for other computers do exist. I use an Apple Macintosh, and there are versions for Commodore Amiga and Atari ST computers.

• You must use a fast computer with a lot of RAM and hard disk storage.

Relatively slow computers, such as IBM PCs and Macintosh Pluses are used successfully in the amateur TCP/IP world. Fast clocks and oodles of memory and disk storage are nice, but you can TCP/IP without them.

• TCP/IP cuts you off from the rest of the packet world.

Many of the bulletins distributed in the AX.25 packet world also get distributed in the TCP/IP world. Moreover, TCP/IP bulletin distribution is better because, instead of wading through all the bulletins on your local PBBS trying to find something of interest to read, TCP/IP automatically routes only bulletins of interest to your TCP/IP mailbox. For instance, I'm on the mailing list for Macintosh computer-related bulletins and messages and, as a result, all Mac-related bulletins and messages addressed to the Macintosh mailing list are automatically delivered to my mailbox. You can be on as many mailing lists as you desire and your mailbox will runneth over.

You aren't cut off from AX.25 mail forwarding, because many AX.25-to-TCP/IP gateway stations and PBBSs will forward your mail between the networks.

My station (IP address 44.88.4.8) has been TCP/IPing 24 hours a day for years with few problems. The S-meter panel light on my 2-meter rig is the only casualty I can recall. I admit to pulling the plug when thunderstorms pass through downtown Wolcott, or when I'm away for extended periods of time (like during my treks to Dayton). Other than that, my station is on all the time.

TexNet

The Texas Packet Society (TPRS) has developed a packet network called *TexNet*. It's composed of dual-port network-control processors (NCP) that provide AX.25-compatible, 1200-bit/s user-access on 2 meters and node-to-node linking at 9600 bit/s on 70-cm. This high-speed network is transparent to users, but results in a higher user data throughput than can be achieved over a 1200-bit/s network.

TexNet firmware can support 256 nodes within one network. Each node can provide several services including access to the TexNet network, a bulletin board, digipeaters, weather information and a conference bridge that provides a connected-mode round table service.

To use TexNet, you first connect with a local network node. You accomplish this by using the Connect command followed by the call sign of the local node and the appropriate SSID. Different SSIDs provide different services as delineated in the following list.

SSID	Service
0	Digipeating
2 and 3	Conference bridging
4	Network access
5	Local node terminal interfacing, also known as Local Node Console
6	Packet Message System or PMS (TexNet's PBBS service)
8 and 9	Cross-band digipeating

If you wish to use TexNet to communicate with another station, you access the network by using the Connect command followed by the call sign of the local node and the SSID of 4 (for example, Connect W5ABC-4). After you connect with a local node, you will receive the Network Cmd prompt. At the prompt, you can ask the network to set up communications with the other station by invoking the Connect command (with your TNC in the Converse Mode) followed by the call sign of the other station and the alias of that station's local node (Connect WA1LOU @ WOLCOTT). Local node alias' are usually based on city names. Your local node commences communications with the WOLCOTT node at 9600 bit/s to set up a connection with WA1LOU. At this speed, it will not be long before communication is established (assuming that WA1LOU is currently active on the WOLCOTT node's 2-meter frequency).

In addition to high-speed network communications, TexNet provides users with a variety of other useful services. You access these services by invoking the Connect command (again with your TNC in the Converse Mode) followed by the call sign of the node and the appropriate SSID. The following services are accessible via TexNet.

Conference Bridging—Each TexNet node supports two independent local conference bridges (SSIDs 2 and 3). These bridges allow three or more stations at a local node to communicate with each other in roundtable fashion. (They do not provide conferencing between stations located at different nodes in the network.)

Packet Message System (PMS)—Typically, each TexNet network is supported by one PMS. One PMS per network eliminates message forwarding, which is one of the prime culprits that bogs down throughput in an AX.25 multi-PBBS network. Because there is only one PMS in the network, you can also access it from any node on the network by simply invoking the Message command at the Network Cmd prompt.

Local Node Console—This function allows you to access the node "directly," that is, by means of the node's controller: the TexNet Network Control Processor (NCP) board.

Digipeating—When all else fails, TexNet provides the old standby: standard AX.25 digipeating. Normally, you should avoid this function in favor of the network and conference bridging functions.

Cross-band Digipeating—If other channels are supported by the local node, this function allows you to access them.

The following summarizes the commands that allow you to use TexNet from its Network Cmd prompt.

Bye—disconnects you from a TexNet node.

Locations Served—lists all the TexNet nodes on the network that you can access.

Connect W5ABC @ Location—connects you to station W5ABC at the TexNet node whose alias is Location.

Connect % W5ABC @ Location—makes a connection across TexNet, where Location is the alias of the far end node. To access a secondary port, append a comma and the port number to Location (for example, Connect NØCCW @ ALAMO, 2). The percent sign (%) is used for automated stations that require a disconnect upon failure. (This command permits users to connect to anything that is not part of the network. Typically a node will permit connections with as many as two digipeaters on either side of a connection. When a user requests a connection, the call sign of the local node is used with "*** LINKED TO" followed by the call sign of the far end station.)

Connect CQ @ Location—sends CQ at the remote TexNet node whose alias is Location.

Statistics @ Location—lists information about activity over a 24-hour period at the TexNet node whose alias is Location.

Statistics Yesterday @ Location—lists statistics from the previous day for the TexNet node whose alias is Location.

Message—connects you to the PMS.

Message @ Location—connects you to a PMS system at the TexNet node whose alias is Location rather than to the network's default PMS.

Weather—connects you to the PMS designated as the weather server.

Route @ Location—lists the routing table of the TexNet node whose alias is Location.

VHF/UHF Operating Procedures

The packet network consists of an assortment of systems. TexNet has haciendas and the ROSE blooms in some garden states. TCP/IP has its defenders and plain vanilla digipeaters still perk along, while NET/ROM and its brood are everywhere. Meanwhile, a variety of bulletin boards continue to push megabytes of mail through this pipeline of dissimilar plumbing. It may seem hard to believe that it all works!

But, it could work better, especially those parts of the network that are burdened by user applications operating at slower data rates. For example, user-to-user and user-to-PBBS communications operate at 1200 bit/s when the user is typing on a keyboard at a typical rate of 25 words per minute, or approximately 23 bits/s. If everyone conformed to a few rules for operation on such a user-intensive network, the network could be utilized more efficiently.

Say No to Nodes

The first rule states that you shouldn't use the network unless it is absolutely necessary. If fewer people used the network, there would be less demand for its use, less contention between its users, and, as a result, it would operate more efficiently.

If you are connected to another station and you do not need a digipeater or network node to maintain the connection, move off the network frequency. Similarly, if you are a user who is connected to a PBBS and you do not need a digipeater or node to maintain that connection, then do not use a digipeater or node to conduct your communication. Freeing up a digipeater or node will promote increased network efficiency, especially if you can maintain the connection using reduced output power.

The converse of the first rule is the second rule, which states that a user should utilize the network when it provides the only means of maintaining a reliable connection with a "local" or "remote" user or a "local" PBBS.

Users and PBBSs are considered to be local if they can all access the same digipeater or node directly without using any intervening digipeater(s) or node(s). On the other hand, users or PBBSs are considered remote if their nearest digipeater or node is beyond the normal operating range of your station, or your local PBBS. In other words, to reach the remote station (or vice versa), intervening distant digipeaters or nodes would have to be employed to make the connection.

If you must use a digipeater or node to maintain a connection with a local user or PBBS, then, by all means, use the network facilities to maintain the connection. Similarly, it is permissible to use the network facilities to maintain a connection with a remote user. However, using the network to communicate with a remote PBBS leads us to the third rule: Do not attempt communication with a remote PBBS. To enforce this rule, many PBBSs are already configured to reject access by remote users; that is, the PBBS will not accept connect requests from users beyond a SysOp-selectable maximum number of intervening digipeaters or nodes. For example, if the maximum number is set to two, users who require three or more digipeaters or nodes to connect to that PBBS will have their connect requests rejected.

The purpose of the third rule is simply a matter of supply and demand. While accessing PBBSs is a very popular pursuit among packet operators, the supply of PBBSs is low relative to the demand. The number of PBBSs that can exist in any one locality is limited by the number of frequencies that are available and the amount of co-channel interference. (Co-channel interference results in a decrease in the network's overall efficiency.)

To solve this supply and demand problem, preferential treatment is given to local users of the PBBS especially when the PBBS is located in a well-populated area. On the other hand, if a PBBS is located in the "boonies," then there is no great supply and demand problem and remote users need not be excluded. In such a case, a remote PBBS is considered to be a "local" PBBS for a remote user.

You can send traffic to a remote PBBS without making a connection to the remote system itself. All you have to do is leave a message on your local PBBS addressed to the station at (or "@")

the remote PBBS. The network will automatically forward the message to the remote PBBS as quickly as possible (see "Packet BBS Operating Procedures" for more information).

Incidentally, don't worry about missing something of interest that may be posted on a remote PBBS. General interest bulletins are usually sent to all of the PBBSs in the particular geographic area of interest, whether that area is a county, state, region, nation or planet.

HF Operating Procedures

HF packet is very different than VHF/UHF packet. A data rate of 300 bit/s is used rather than 1200 bit/s and the AFSK tones are not frequency modulated; LSB is used instead (1600 and 1800 Hz are commonly used for mark and space, respectively.)

The majority of TNCs are optimized for VHF/UHF FM operation, so getting on HF requires more effort than getting on VHF/UHF. In some TNCs, there are two modems, so switching between VHF/UHF and HF is painless. Other TNCs are not so versatile. The modems in these TNCs must be modified and recalibrated for HF operation or, TNC permitting, an external modem must be used that is compatible with HF packet operation.

While some TNC parameters may be set to whatever suits your particular VHF/UHF operation, the setting of these same parameters is much more critical for HF operation. HF packet is affected by interference, both atmospheric and man-made, and by the fickleness of propagation. Low signal-to-noise ratios also makes HF packet more difficult. Luckily, certain TNC parameters may be set to compensate for these conditions.

To increase throughput and decrease congestion on the frequency, your packets should be short. In addition, no more than one unacknowledged packet should be outstanding at any time. Use the Paclen and MAXframe commands, as described earlier in this chapter, to shorten the length of your packets and minimize the number of unacknowledged packets.

HF packet is adversely affected by spurious noise, which is considered by the TNC as a valid signal. This causes the TNC timing parameters to be reset each time such a "signal" is detected, resulting in a packet transmission delay for no legitimate reason. To avoid this, simply disable all of the timing parameters by setting DWait and TXdelay to zero, as described earlier in this chapter.

Changing the frequency of a VHF/UHF receiver is simply a matter of twisting the dial, but HF packet requires a steady hand at the controls. Tune your receiver very slowly (in 10 Hz increments, if possible) until your DTE begins displaying packets. When tuning, do not change frequency until a whole packet is received. If you shift frequency in mid-packet, that packet will not be received properly and will not be displayed on your DTE even if you were on the correct frequency before or after the frequency shift. Some TNCs have special indicators that make tuning easier.

The same rule of thumb that applies to VHF/UHF packet operation also applies to HF operation. Move to an unused frequency if there is other packet activity on the frequency you are presently using. Often, one frequency may be used as a calling frequency where stations transmit packets to attract the attention of other stations. Once a connection is established, the stations often move to another frequency where there is less activity. This clears the calling frequency and increases their own stations' throughput.

Space Communications

Packet is a space-age mode of communication and, befitting such a mode, you can find packet in space as well as on land.

WØRLI Mailbox Command Set

The following commands are available with version 10.11 of the WØRLI Mailbox public-domain software.

General commands:

B	Log off PBBS.
Jx	Display call signs of stations recently heard or connected on TNC port x.
N x	Enter your name (x) in system (12 characters maximum).
NE	Toggle between short and extended command menu.
NH x	Enter the call sign (x) of the PBBS where you normally send and receive mail.
NQ x	Enter your location (x).
NZ n	Enter your ZIP Code (n).
P x	Display information concerning station whose call sign is x.
S	Display PBBS status.
T	Ring bell at the SYSOP's DTE for one minute.

Information commands:

? *	Display description of all PBBS commands.
?	Display summary of all PBBS commands.
? x	Display summary of command x.
H *	Display description of all PBBS commands.
H	Display summary of all PBBS commands.
H x	Display description of command x.
I	Display information about PBBS.
I x	Display information about station whose call sign is x.
IL	Display list of local users of the PBBS.
IZ n	List users at ZIP Code n.
V	Display PBBS software version.

Message commands:

K n	Kill message numbered n.
KM	Kill all messages addressed to you that you have read.
KT n	Kill NTS traffic numbered n.
L	List all messages entered since you last logged on PBBS.
L n	List message numbered n and messages numbered higher than n.
L< x	List messages from station whose call sign is x.
L> x	List messages addressed to station whose call sign is x.
L@ x	List messages addressed for forwarding to PBBS whose call sign is x.
L n1 n2	List messages numbered n1 through n2.
LA n	List the first n messages stored on PBBS.
LB	List all bulletin messages.
LF	List all messages that have been forwarded.
LL n	List the last n messages stored on PBBS.
LM	List all messages addressed to you.
LT	List all NTS traffic.
R n	Read message numbered n.

RH n	Read message numbered n with full message header displayed.
RM	Read all messages addressed to you that you have not read.
S x @ y	Send a message to station whose call sign is x at PBBS whose call sign is y.
S x	Send message to station whose call sign is x at this PBBS.
SB x	Send a bulletin message to x at this PBBS.
SB x @ y	Send a bulletin message to x at PBBS whose call sign is y.
SP x @ y	Send a private message to station whose call sign is x at PBBS whose call sign is y.
SP x	Send a private message to station whose call sign is x at this PBBS.
SR	Send a message in response to a message you have just read.
ST x @ y	Send an NTS message to station whose call sign is x at PBBS whose call sign is y.
ST x	Send an NTS message to station whose call sign is x at this PBBS.

File transfer commands:

Dx y	From directory named x, download file named y.
U x	Upload file named x.
W	List what directories are available.
Wx	List what files are available in directory named x.
Wx y	List files in directory named x whose file name matches y.

Port commands:

C x y	Via port x, send connect request to station whose call sign is y.
C x	Send data via port x.
CM x y	Send message numbered x to station whose call sign is y.
CM x y @ z	Send message numbered x to station whose call sign is y at PBBS whose call sign is z.
Mx	Monitor port x.

Roundtable commands:

RT		Initiate roundtable function.
<ESC>	D x	Allows roundtable control station to disconnect station from roundtable whose call sign is x.
<ESC>	H	Obtain assistance.
<ESC>	P	Display ports available to roundtable.
<ESC>	N x	Enter your name (x).
<ESC>	Q x	Enter your location (x).
<ESC>	U	Display list of stations in roundtable.

Packet in space has taken place aboard the American space shuttles, the Russian space station and Amateur Radio satellites.

Space Shuttles and Stations

Shuttle Amateur Radio EXperiments (SAREX) have been conducted by astronauts holding Amateur Radio licenses on a number of American space shuttle missions. Typical shuttle operations feature unique robot software that allow the maximum number of stations to contact the shuttle via packet. This software recognizes a terrestrial connect request, sends a sequential QSO number to the station, summarily disconnects the station and then logs the contact. Beacons listing successful connections are transmitted intermittently.

Russian cosmonauts often conduct live operator-to-operator packet contacts using an Amateur Radio station aboard the space station *Mir*. When a cosmonaut is not on hand to operate, the station is switched into the PBBS mode to receive messages and news from Earth.

Four MicroSATs and two UoSATs mounted on the launch platform of the Ariane rocket that propelled them into space. *(photo courtesy of Joe Kasser)*

Musa Manarov, UV3AM, was active on packet from the Soviet *Mir* space station (U2MIR) on 145.55 MHz. He is shown here preparing for a mission at Moscow club station UK3R. *(UW3AX photo via WA2LQQ)*

modems in order to use some of these satellites. Radio equipment for various VHF/UHF bands must be available, too. With a properly equipped station, using a satellite's store-and-forward mode is similar to using a land-based PBBS.

Refer to Chapter 13 ("Satellites") for more information.

APPLICATIONS

Now that we have this high-speed, error-free mode of communications up and running, what can we do with it? The answer is "plenty." The following paragraphs describe some of the more popular pursuits that have been applied to the packet mode.

Packet BBS Operating Procedures

In 1982 there was a warehouse in Texas full of surplus Xerox 820-I computers. The "820" was a single board computer using the Zilog Z80 central processing unit (CPU). It ran under the CP/M operating system and featured 64-kBytes of RAM, one parallel port, two serial ports, a disk controller, and an 80 × 24 video display format. As surplus, the computer could be purchased for as little as $50!

Hank Oredson, WØRLI, became the father of PBBS when he decided to put this bargain computer to good use by writing software that permitted it to function as a BBS; not a telephone BBS, but a packet BBS. In no time, WØRLI PBBSs began appearing throughout packet country, not solely because the 820 was inexpensive, but also because Hank's software worked well.

PBBSs became so popular that other hams wrote PBBS software for their favorite computers. Jeff Jacobsen, WA7MBL, wrote the first PBBS software for the most popular computer in Amateur Radio, the IBM PC, and when WØRLI's code began outgrowing the Xerox 820-I, Hank rewrote his software for the IBM PC as well. Today, the IBM PC has a wide variety of PBBS software available for it.

So, what does a PBBS do? Its mailbox function allows you to post mail on the system for later retrieval by the addressee. You can also retrieve mail that's addressed to you. The mailbox does not limit your message posting to stations that frequent the local PBBS. If you know a station that checks into a *distant* PBBS, you can post a message to that station and it will be routed automatically to its destination.

Besides storing individual messages, files of interest to the general packet population may be stored as well. Here you will find packet network maps, programs, ARRL bulletins, newsletters and so on.

To use a PBBS, you must find one. On HF, a number of PBBSs are active in the packet subbands at various times of the day (refer to the sidebar "Packet Frequencies"). On VHF/UHF, there is probably at least one active PBBS on each packet channel.

The accompanying sidebar "WØRLI Mailbox Command Set" describes the commands of WØRLI's fabled software, and the PBBS procedures described below are based on this command set as well. If your local PBBS is not a WØRLI Mailbox, the sidebar and following description will still be applicable because most PBBSs use similar, if not identical,

No special packet equipment is required to contact the American shuttle packet stations or the *Mir* space station. Both use standard TNCs operating at 1200 bit/s. Shuttle operations usually occur on split 2-meter frequencies and a simple Connect command with the appropriate call sign on the right frequency is all that is needed to initiate a connection with the manned space vehicle. The frequencies are published prior to launch. The *Mir* space station can be found on 145.55 MHz simplex.

Amateur Satellites

After years of packet experimentation in space via manmade satellites, the groundwork has been laid for a packet network that relies on TNCs in orbit around the Earth. There are many packet satellites in operation (with more to come) and some of these are being configured to be integral parts of the worldwide packet network. The typical operating mode of a packet satellite is the digital store-and-forward mode (a ground station uploads data to the satellite and the satellite stores the data until it can be relayed to another ground station at a later time).

The satellites use a variety of modulation schemes and data rates with uplinks and downlinks on different frequency bands. As a result, standard TNCs must be outfitted with different commands for their basic operations.

Logging On

Once you locate a PBBS, you must "log on" in order to use it. Logging onto a PBBS is as simple as initiating a contact with any other packet station. Just use the Connect command. For example:

cmd: Connect K8KA-4 <CR>

where K8KA-4 is the call sign and SSID of the PBBS. After you are connected to the PBBS, the PBBS preamble is displayed on your DTE followed by a request for commands from you.

Logging Off

Whenever you are finished using a PBBS, you log off the system by using the B (for "bye") command. When the PBBS receives the B command, it logs you off the system and disconnects.

Reading the Mail

In order to read messages that are posted on the PBBS, you must know what messages are available. If a message stored on the PBBS is addressed to you, you are informed of that fact when you log onto the system. By using the RM (for "read mine") command, you can read whatever mail is waiting for you. When the PBBS receives the RM command, it retrieves each message and sends it to your DTE. After you have read your messages, use the KM (for "kill mine") command to delete all of the messages you have read. (Note that you can only "kill" messages originated by you or addressed to you.)

To read other messages that are on the PBBS, use the L (for "list") command to obtain a list of all of the messages that have been stored on the PBBS since the last time you logged onto the system. The PBBS lists each message by its message number. If you wish to read a particular message, use the R (for "read") command by typing:

R n <CR>

where n is the number of the message to be read.

Sending Mail

To send a message by means of a PBBS, use the SP (for "send personal") command. When the PBBS receives the command, it asks you for the call sign of the addressee. To post the message for retrieval on the local PBBS, simply type the call sign of the station to receive the message followed by <CR>. If you want to have your message forwarded to a station at another PBBS, type the call sign of the station to receive the message, followed by the at-sign (@), the call sign of the destination PBBS and <CR>. For example:

K9NG @ KV7D <CR>

where K9NG is the call sign of the station that is to receive the message and KV7D is the call sign of the destination PBBS where the message will be sent for retrieval). To save a step, you can address the message at the same time you invoke the SP command (for example, SP K9NG @ KV7D).

To assist the mail-forwarding network, you can append geographic information to the message's address. For example, to give the message addressed to K9NG @ KV7D a little push, you can append .AZ.US.NA to KV7D, which results in an address of K9NG @ KV7D.AZ.US.NA. (AZ, US and NA are the abbreviations for the state, country and continent, respectively, in which KV7D is located.) This is known as *hierarchical addressing*.

Next, the PBBS asks you for the title of the message. Type a short one- or two-word title that represents the contents of the message followed by <CR>. Next, type the actual contents of the message. After you have typed the last line of the message, type a <CTRL-Z> and <CR> to indicate the end of the message. When the PBBS receives the <CTRL-Z>, it stores the message for later retrieval or forwarding. To learn more about sending mail and bulletins via the packet network, pick up a copy of *Your Packet Companion* by Steve Ford, WB8IMY, at your favorite dealer. This book is also available from ARRL Headquarters directly.

PBBS Procedures

Those are the basic commands for using a PBBS. A few basic operating rules need to be mentioned also.

If the PBBS does not respond to a command immediately, be patient and do not resend it. One of the unique features of packet is that whatever you send is received perfectly at the other end. To achieve this result sometimes takes a number of attempts, especially if there is a lot of activity on frequency. If you send the same command twice, the PBBS will eventually receive it twice and will respond twice! If the response is a long one, your repeated command will waste valuable time.

Do not perform lengthy operations during prime operating hours. Use good judgment before deciding to perform a time-consuming task. To save time, you can send more than one command at a time by preceding the <CR> that follows each command with the pass character. <CTRL-V> is the default pass character which, in this case, prevents each <CR> from causing the commands to be sent individually. For example, to send the RM, KM and L commands at one time, you would type:

RM <CTRL-V> <CR> KM <CTRL-V> <CR> L <CR>

The last <CR> is not preceded by a <CTRL-V> because you want the last <CR> to force the transmission of this complete packet. The <CTRL-V>s preceding the other <CR>s prevent those <CR>s from forcing transmission of each command individually. Without the <CTRL-V>s, RM would be sent in one packet, KM would be sent as a second packet and L would be sent as a third packet.

Public Service Communications

Packet has become integral in serving the public by providing communications during emergencies and disasters. With portable computers, TNCs and radio equipment, packet stations have been rapidly set up at disaster sites to get all-important messages on the air and to their destinations quickly and accurately. Accuracy is very important in emergency communications. When a packet station sends NEED A SURGEON, you can be sure that the station receiving the message will not have a fish delivered to the originating station!

Since each packet station is a self-contained digipeater, packet networks can be set up quickly in times of emergencies. The value of packet for emergency communications has been recognized by various public service agencies and some have allocated funds towards the purchase of packet equipment to be available to hams in times of emergencies.

Hams have worked hard to establish a means of efficiently handling traffic through the packet network. As more and more packet stations appear on the air each day, the packet network expands. Each new station has the potential of being an outlet for emergency traffic. The mail-forwarding function of the PBBS network is a key part of the packet traffic system. This system is constantly being refined to handle greater volumes of emergency traffic whenever the need arises.

DX Packet Cluster Command Set

The following commands are available with Pavillion Software's DX PacketCluster software:

ANNOUNCE	Make an announcement.
A x	Send message x to all stations connected to the local node.
A/F x	Send message x to all stations connected to the cluster.
A/x y	Send message y to stations connected to node x.
A/x y	Send message y to stations on distribution list x.
BYE	Disconnect from cluster.
B	Disconnect from cluster.
CONFERENCE	Enter the conference mode on the local node.
CONFER	Enter the conference mode on the local node. Send <CTRL-Z> or /EXIT to terminate conference mode.
CONFER/ F	Enter the conference mode on the cluster. Send <CTRL-Z> or /EXIT to terminate conference mode.
DELETE	Delete a message.
DE	Delete last message you read.
DE n	Delete message numbered n.
DIRECTORY	List active messages on local node.
DIR/ALL	List all active messages on local node.
DIR/BULLETIN	List active messages addressed to "all."
DIR/n	List the n most recent active messages.
DIR/NEW	List active messages added since you last invoked the DIR command.
DIR/OWN	List active messages addressed from or to you.
DX	Announce DX station.
DX x y z	Announce DX station whose call sign is x on frequency y followed by comment z, e.g., DX SP1N 14.205 up 2.
DX/a x y z	Announce DX station whose call sign is x on frequency y followed by comment z with credit given to station whose call sign is a, e.g., DX/K1CC SP1N 14.205 up 2.
FINDFILE	Find file.
FI x	Ask the node to find file named x.
HELP or ?	Display a summary of all commands.
HELP x	Display help for command x.
READ	Read message.
R	Read oldest message not read by you.
R n	Read message numbered n.
R/x y	Read file named y stored in file area named x.
REPLY	Reply to the last message read by you.
REP	Reply to the last message read by you.
REP/D	Reply to and delete the last message read by you.
SEND	Send a message.
S/P	Send a private message.
S/NOP	Send a public message.
SET	Set user-specific parameters.
SE/A	Indicate that your DTE is ANSI-compatible.
SE/A/ALT	Indicate that your DTE is reverse video ANSI-compatible.
SE/H	Indicate that you are in your radio shack.
SE/L a b c d e f	Set your station's latitude as a degrees b minutes c north or south and longitude d degrees e minutes f east or west, e.g., SE/L 41 33 N 73 0 W.
SE/N x	Set your name as x.
SE/NEED x	Store in database that you need country(s) whose prefix(s) is x on CW and SSB, e.g., SE/NEED XX9.
SE/NEED/BAND = (x) y	Store in database that on frequency band(s) x, you need country(s) whose prefix(s) is y, e.g., SE/NEED/BAND = (10) YA.
SE/NEED/x y	Store in database that in mode x (where x equals CW, SSB or RTTY), you need country(s) whose prefix(s) is y, e.g., SE/NEED/RTTY YA.
SE/NEED/x/BAND = (y) z	Store in database that in mode x (where x equals CW, SSB or RTTY) on frequency band(s) y, you need country(s) whose prefix(s) is z, e.g., SE/NEED/RTTY/BAND = (10) ZS9.
SE/NOA	Indicate that your DTE is not ANSI-compatible.
SE/NOH	Indicate that you are not in your shack.
SE/Q x	Set your QTH as location x.
SHOW	Display requested information.
SH/A	Display names of files in archive file area.
SH/B	Display names of files in bulletin file area.
SH/C	Display physical configuration of cluster.
SH/C x	Display station connected to node whose call sign is x.
SH/CL	Display names of nodes in clusters, number of local users, number of total users and highest number of connected stations.
SH/COM	Display available Show commands.
SH/DX	Display the last five DX announcements.
SH/DX x	Display the last five DX announcements for frequency band x.
SH/DX/n	Display the last n DX announcements.
SH/DX/n x	Display the last n DX announcements for frequency band x.
SH/FI	Display names of files in general files area.
SH/FO	Display mail-forwarding database.
SH/H x	Display heading and distance to country whose prefix is x.
SH/I	Display status of inactivity function and inactivity timer value.
SH/LOC	Display your station's longitude and latitude.
SH/LOC x	Display the longitude and latitude of station whose call sign is x.
SH/LOG	Display last five entries in cluster's log.
SH/LOG n	Display last n entries in cluster's log.
SH/M x	Display MUF for country whose prefix is x.
SH/NE x	Display needed countries for station whose call sign is x.
SH/NE x	Display stations needing country whose prefix is x.
SH/NE/x	Display needed countries for mode x where x equals CW, SSB or RTTY.
SH/NO	Display system notice.
SH/P x	Display prefix(s) starting with letter(s) x.
SH/S x	Display sunrise and sunset times for country whose prefix is x.
SH/U	Display call signs of stations connected to the cluster.
SH/V	Display version of the cluster software.
SH/W	Display last five WWV propagation announcements.
TALK	Talk to another station.
T x	Talk to station whose call sign is x. Send <CTRL-Z> to terminate talk mode.
T x y	Send one-line message y to station whose call sign is x.
TYPE	Display a file.
TY/x y	Display file named y stored in file area named x.
TY/x/n y	Display n lines of file named y stored in file area named x.
UPDATE	Update a custom database.
UPDATE/x	Update the database named x.
UPDATE/x/APPEND	Add text to your entry in the database named x.
UPLOAD	Upload a file.
UP x	Upload a file named x.
UP/B x	Upload a bulletin named x.
UP/F x	Upload a file named x.
WWV	Announce and log WWV propagation information
W SF = xxx, A = yy, K = zz,a	Announce and log WWV propagation information where xxx is the solar flux, yy is the A-index, zz is the K-index and a is the forecast.

DX *PacketCluster*

The DX community has a tendency to share information about the latest and greatest DX. That is the reason why there are so many successful Amateur Radio newsletters devoted to spreading the word about DX. The problem with newsletters is that their news is not always current. DXpedition plans may change at the last minute or a new country may pop on the air without warning.

DX spotting helps alleviate this problem. When someone spots new DX, he or she shares that fact by making an announcement on the local DX spotting frequency. During the 1970s and 1980s, the DX spotting frequency was typically a 2-meter FM voice repeater. Packet and DX *PacketCluster* software have changed all that forever and DX spotting has become one of the most popular applications of packet communication.

A DX *PacketCluster* consists of specialized software running on a computer attached to a TNC and a transceiver. The software allows multiple stations to connect to the *PacketCluster* station or "node." It also allows each *PacketCluster* node to connect to other nodes to form a *PacketCluster* network. When one station connects to a node and makes an announcement, all stations connected to that node and all the nodes in the network receive the same announcement.

For example, let's say that K3NA announces that 9H3JR is operating on 7005.0 kHz. If K3NA connects to a *PacketCluster* node that is part of your local *PacketCluster* network, you will be able to copy his announcement even if you cannot copy K3NA directly.

If you wanted to talk with K3NA, the *PacketCluster* permits you to do that, too. What if you wanted to know K3NA's location? You can use the *PacketCluster* to ask K3NA exactly where he is located. Besides conversing with any station on the *PacketCluster* network, users may conduct conferences (discussions involving more than two people).

Another function of the *PacketCluster* is its bulletin board system that operates in a fashion similar to a standard PBBS. Like other PBBSs, the *PacketCluster* system tells you when you have new mail. To read mail, or any message or bulletin, you use the Read command. To post a message or bulletin of your own, you use the Send command. To obtain a list of the messages and bulletins that are on the board, you use the Directory command instead of the PBBS List command, and to remove a message or bulletin from the board, you use the Delete command instead of the PBBS Kill command. The other major difference between the *PacketCluster* and WØRLI-compatible systems is that the *PacketCluster* BBS does not support mail forwarding. (The sidebar "DX PacketCluster Command Set" describes all of the *PacketCluster* commands.)

Perhaps, the most powerful command in the *PacketCluster* command set is the Show command (SH for short). Invoke the Show DX (SH/DX) command and the *PacketCluster* will list the last five DX announcements including the DX station's operating frequency, call sign, the time and date of the announcement, other pertinent information concerning the DX station (long path, LOUD, weak and so on). and the call sign of the station that made the announcement. If you wish to limit the Show DX command to one band, you may do so. For example, "SH/DX 10" causes the *PacketCluster* to recount the last five DX announcements for 10 meters only. You also may limit the Show DX command to one country. For example, "SH/DX SP*" will cause the *PacketCluster* to list the last five announcements for Polish stations only.

When V51NAM is at the bottom of a pileup on 20 meters, you can use the *PacketCluster* to get some ammunition before you try to work him. Assuming you have already entered your longitude and latitude in the *PacketCluster*'s database (using the SET/L command), invoke the Show Heading command for Namibia (SH/HEADING V51). The *PacketCluster* will perform some calculations and send results that will look something like this:

    V51 Namibia: 107 degs
    Q dist: 7238 mi, 11649 km
    Reciprocal heading: 309 degs

Now you can use the Show Sun command (SH/SUN V51) and the *PacketCluster* will respond with:

V51 Namibia Sunrise: 0455Z Sunset: 1703Z

The Show Sun command is also handy for finding out your local sunrise and sunset. Simply invoke the command without appending a prefix (SH/SUN) and the *PacketCluster* will send:

<your call> QTH Q Sunrise: 1055Z Sunset: 2303Z

Finally, try using the Show MUF command (SH/MUF V51) and the *PacketCluster* will respond:

Namibia propagation: MUF: 22.7 MHz LUF: 2.3 MHz

If you work V51NAM, you could invoke the Show QSL (SH/QSL) command to obtain QSL information for that station. After invoking the command, the *PacketCluster* would search its database for the QSL route for V51NAM.

Other variations of the Show command include the Show WWV command to obtain WWV propagation information and the Show Users command to obtain a list of all the other stations connected to the *PacketCluster* node or network. The Show command also may access any database installed on your local *PacketCluster*. Such databases may include the contest or DX club membership rosters, contest information, DX news, FCC rules and regulations, IRC data, QSL bureau addresses and so on.

As you can see, the *PacketCluster* is a powerful tool. Although similar tools existed in the past, only packet could make it as powerful as it is today. Faster packets will only make it more powerful in the future.

Automatic Packet Reporting System

Long-time packeteer Bob Bruninga, WB4APR, developed the Automatic Packet Reporting System (APRS), which allows packet radio to track real-time events. It deviates markedly from message- and text-transfer applications and concentrates instead on the *graphic* display of station and object locations and movements. By running APRS software, you see the locations of other APRS stations on a computer-generated map.

APRS is the result of Bob's experience using packet for real-time communications and support for public-service events. Packet has great potential, but so far, has been used mostly for passing large volumes of message traffic from point-to-point or into the national traffic-distribution system. It has been difficult to apply packet to real-time events where information has a short life span. This is because several steps are involved in preparing and passing messages, including decisions about routing and connectivity.

Bob developed APRS based on his observation that operators at most events, emergencies, exercises, weather nets and general communications spend more time concerned with where things (people, stations and so on) are—and where they are going—than any other category. Regardless of whether APRS is the best solution, it demonstrates the potential for tactical communications via packet. Furthermore, it attempts to define a standard format using unconnected information (UI) frames for object positioning and reporting, which could be universally applicable in Amateur Radio. (UI frames can be thought of as packet "broadcasts." The information is sent to no one in particular and can be received by anyone.)

APRS avoids the complexity and limitations of trying to maintain a connected network. It permits any number of stations to participate and exchange data, just as voice users do on a phone net. Any station that has information to contribute simply transmits it and all stations receive it and log it.

Although APRS' mapping capability was developed to display the movement of hand-held global-positioning satellite (GPS) navigation devices interfaced to TNCs, most features evolved from earlier efforts to support real-time packet communications at special events. Any person in the network, upon determining where an object is located, can move his cursor and mark the object on his map screen. This action is then transmitted to all screens in the network, so everyone gains, at a glance, the combined knowledge of all network participants. Furthermore, the map screen retains this information for future reference. This means that moving objects can be dead-reckoned to their current locations with one keystroke—based on their previous positions—and this can be accomplished without using a single GPS device.

The availability of GPS receiver *cards* is icing on the cake. With a GPS card, TNC and hand-held transceiver stuffed in a cigar box, almost any object can be tracked by packet stations running APRS software. For example, such cigar-box technology was installed in a football helmet for the Army-Navy football game run. You can also place these boxes on bicycles for a marathon event, and, of course, in automobiles. The GPS receiver picks up precise position information from military satellites. This data is fed to the TNC which, in turn, incorporates the data into a UI frame. The radio then transmits the data to all receiving stations within range.

Bob's system is an excellent tool for triangulating the location of a hidden transmitter. A jammer command displays the intersection of bearing lines from a number of reporting stations. To use APRS in this manner, each station having a bearing on a jammer enters it into his APRS system. His station then reports its location and its bearing relative to the jammer. All stations running APRS can simply hit the J key to display the intersection of these bearing lines. Furthermore, if a direction-finding vehicle has a GPS or LORAN-C device on board, it can be tracked and directed to the location of the jammer.

APRS may also be a solution to the effective use of orbiting packet-radio digipeaters, such as on space shuttles, *Mir*, AO-21 and other satellites. The problem with space digipeaters is saturation on the uplink channel, which makes the use of a normal connected protocol impractical. For a connected contact, five successive and successful packet transmissions are required. Not only does APRS reduce this to one packet, it also capitalizes on its most fascinating aspects—its ability to display the locations of stations on a map. If all stations simply insert their latitude and longitude as the first 19 characters of their packet beacon texts, everyone monitoring the satellite with APRS software will see the location of every successful uplink. Because satellites are rapidly moving objects, the locations of successful uplink stations will move progressively along the ground track.

Bob's software is powerful and can perform a variety of tasks, yet it is simple to use. As a result, it has the potential to become an important application in amateur packet.

The APRS software is distributed as shareware and may be copied for any amateur application. Registered copies for IBM PCs are available for $19 from Bob Bruninga, WB4APR, 115 Old Farm Ct, Glen Burnie, MD 21060. A Macintosh version of APRS, MacAPRS, was developed by Keith Sproul, WU2Z, and Mark Sproul, KB2ICI.

Other Applications

Packet has begotten a host of other useful Amateur Radio applications. Database applications are particularly popular. One well-known database application is called *White Pages*. White Pages provided a solution to a problem that faced packet operators at the onset of PBBS mail forwarding. That is, how do you know who uses which PBBS? Now, by sending a short message to a White Pages database, you can find out which PBBS your fellow packet operator calls home.

Another popular database application came about with the advent of call sign directories on CD-ROMs. It wasn't long before packet call sign directories came on the air.

With enough patience and perseverance, almost any program that can run on a computer can be made to operate with packet. In this regard, the ultimate packet application may be NM1D's DOSgate, which allows you to run software over the air on a remote IBM-class computer.

THE FUTURE

As it exists today, packet is only beginning to realize its full potential. When higher data rates, improved networks, better radio equipment and more flexible protocols are in place, more complex and fascinating applications will be accessible.

Glossary of Packet Radio Terms

<CR>—abbreviation for carriage return

AFSK—abbreviation for audio-frequency-shift keying

Amateur Radio Research and Development Corporation (AMRAD)—a Maryland-based Amateur Radio organization that was instrumental in early United States packet radio developments including the adoption of the AX.25 protocol

amateur X.25 (AX.25)—the Link layer packet radio protocol based on the CCITT X.25 packet-switching protocol

American National Standard Code for Information Interchange (ASCII)—a seven-bit digital code used in computer and radioteleprinter applications

AMRAD—abbreviation for Amateur Radio Research and Development Corporation

AMSAT—abbreviation for The Radio Amateur Satellite Corporation

ASCII—abbreviation for American National Standard Code for Information Interchange

asynchronous—a data transmission timing technique that adds extra bits of information to indicate the beginning and end of each transmitted character

audio-frequency-shift keying (AFSK)—a method of transmitting digital information by switching between two audio tones fed into the transmitter microphone input

autobaud—the ability of a communications device to automatically adapt to whatever data rate is used by the terminal connected to it

automatic line feed—a DTE or DCE (TNC) function that causes a line feed control character to be sent whenever a carriage return control character is sent

AX.25—the Link layer packet radio protocol based on the CCITT X.25 packet-switching protocol

baud—a unit of signaling speed equal to one pulse (event or symbol) per second in a single-channel transmission

baud rate—the speed at which information is transferred, usually expressed in bauds or bits per second; also called bit rate or data rate

BBS—abbreviation for bulletin-board system

Bdale's Mailer (BM)—an application in the *KA9Q Internet Protocol Package* written by Bdale Garbee, N3EUA, that serves as a front-end to the Simple Mail Transfer Protocol (SMTP)

beacon—a TNC function that permits a station to automatically send unconnected packets at regular intervals

Bell 103—the designation for the telephone company standard for modems that transfer data at 300 bauds using 200-Hz frequency-shift-keyed tones at 1170 and 2125 Hz; commonly used for HF packet applications

Bell 202—the designation for the telephone company standard for modems that transfer data at 1200 bauds using frequency-shift-keyed tones at 1200 and 2200 Hz; commonly used for VHF packet applications

binary—the base two number system that uses the numerals 0 and 1

bit—binary digit, a signal that is either on/one or off/zero; bits are combined to represent alphanumeric and control characters for data communications

bit rate—the speed at which information is transferred, usually expressed in bauds or bits per second; also called baud rate and data rate

bit/s—abbreviation for bits per second

bits per character—the number of bits that are combined to represent alphanumeric and control characters for data communications

bits per second (bit/s)—a measure of the speed at which information is transferred

BM—abbreviation for *Bdale's Mailer*

buffer—a portion of computer memory that is set aside to temporarily store data

bulletin-board system (BBS)—a computer system where messages and files can be stored for other users

byte—a group of bits, usually eight in number

calling frequency—an operating frequency that is used only for establishing communications; once communications have been established on the calling frequency, the operators switch to a less-used frequency

carriage return (<CR>)—a DTE key or a control character that is used to indicate the end of a line of typed information; it causes the DTE display to begin printing at the left-hand margin

CCITT—abbreviation for International Telegraph and Telephone Consultative Committee

clone—a device that duplicates another device

collision—a condition that occurs when two or more packet radio transmissions occur at the same time; when a collision occurs, neither packet reaches its destination

command mode—the TNC operating mode where the TNC is waiting for command input from the user

connect—to establish a communications link (a connection) between two packet radio stations

connection protocol—a Network layer protocol that sets up and maintains a clearly defined path for the transfer of packets between the source and destination during a single data communications session; also called virtual circuit protocol

connectionless protocol—a Network layer protocol that transfers each packet independently along the best available route; also called datagram protocol

data circuit-terminating equipment, data communications equipment (DCE)—the device that provides communications between a DTE and radio equipment or telephone lines

data communications software—a computer program that causes a computer to function as a DTE for the purpose of transferring data over a communications medium

data rate—the speed at which information is transferred, usually expressed in bauds or bits per second; also called baud rate and bit rate

data terminal equipment (DTE)—a device that is used to interface between a human and a computer to allow the human to exchange information with the computer

datagram protocol—a Network layer protocol that transfers each packet independently along the best available route; also called connectionless protocol

DCE—abbreviation for data circuit-terminating equipment or data communications equipment

default—the state of a TNC parameter after the TNC is initially turned on or reset

demodulation—the process of retrieving information from a modulated carrier

destination—the intended recipient of a packet frame

digipeater—digital repeater, a device that receives, temporarily stores and then retransmits (repeats) packet radio transmissions that are specifically addressed for routing through the digipeater

download—to receive files from a PBBS or other packet radio station

DTE—abbreviation for data terminal equipment

dumb terminal—a simple DTE that provides only basic input and output functions

echo—a DTE and DCE (TNC) function that prints each character typed at the DTE keyboard on the display

EIA—abbreviation for Electronics Industries Association

EIA-232-D—the current EIA standard for DTE-to-DCE (TNC) interfacing that specifies the interface signals and their electrical characteristics; it replaces EIA RS-232-C

Electronic Industries Association (EIA)—an organization composed of representatives of the United States electronics industry; the EIA is involved in formulating data communication standards

emulate—to mimic; often used to describe software that attempts to perform the function of hardware

end-to-end acknowledgment—the networking protocol in which only the destination station or node informs the originating station or node that it has received a packet correctly

ENTER—a key on a computer keyboard that causes the computer to accept the information previously typed at its keyboard

enter—to use a key (e.g., the ENTER key) on a computer keyboard to cause the computer to accept the information previously typed at its keyboard

EPROM—abbreviation for erasable programmable ROM

erasable programmable ROM—read-only memory whose contents can be deleted (by ultraviolet light) and replaced

escape, escape code, escape characters—a control character or sequence of alphanumeric characters that are typed at a DTE keyboard to cause the DCE (TNC) to exit the current operating mode

File Transport Protocol (FTP)—a protocol in the KA9Q Internet Protocol Package that allows the user to transfer files to or from the computer at another node

flow control—the stopping and restarting of the transfer of characters between the DTE and a DCE (TNC)

frequency-shift keying (FSK)—a method of transmitting digital information by switching an RF carrier between two separate frequencies

FSK—abbreviation for frequency-shift keying

FTP—abbreviation for File Transport Protocol

gateway—a device or software function that allows packet radio stations on different operating frequencies to communicate with each other

half-duplex—a communications mode where you alternately transmit and receive

hardware flow control—flow control that is controlled by DCE- (TNC) and DTE-originated signals on the serial interface between the two devices

International Telegraph and Telephone Consultative Committee (CCITT)—an International Tele-communication Union agency involved in formulating international data communication standards

Internet Protocol (IP)—a Level 3 or Network layer protocol included in the KA9Q Internet Protocol Package that tracks network node addresses, routes outgoing packets and recognizes received packets

IP—abbreviation for Internet Protocol

IP address—a 32-bit binary number that is assigned to each computer at each packet radio station in the TCP/IP network as that computer's identification for the routing of packets by the Internet Protocol

KA-Node—Kantronics' implementation of a node-to-node acknowledgment network protocol

KISS—an acronym for "Keep It Simple, Stupid," a Link layer non-protocol for serial input and output that supports Serial Line Interface Protocol (SLIP), written by Mike Chepponis, K3MC

layer—one level of the seven level hierarchy of the OSI-RM

line feed—a control signal that causes a DTE display to begin printing on the next line

Link layer—Level 2 of OSI-RM that arranges data bits into frames and provides for the errorless transfer of the frames over a communications link

log off—to inform a PBBS that you are finished using the system

log on—to inform a PBBS that you wish to begin using the system

mail-forwarding—a PBBS function that allows its users to send mail to the users of other PBBSs

mailbox—a packet radio message receiving and sending system at an individual station

mark—one of the two elements in the binary start-stop code; it often represents the on-state or 1

message header—the non-data portion of a packet radio message that contains the message number, type, status, intended recipient and other routing information

mnemonic identifier—an alternative to a network node's call sign consisting of three-to-six letters that usually identify the node's location

modem—modulator-demodulator; an electronic device that permits digital equipment to use analog communications media for data communications

modulation—the process of varying a carrier to represent information

multiple connections—the ability to establish and maintain connections with more than one station simultaneously

NET.EXE—a TCP/IP implementation for the IBM PC written by Phil Karn, KA9Q

NET/ROM—a Network and Transport layer implementation for the TNC 2 written by Ron Raikes, WA8DED; it features node-to-node acknowledgment

Network layer—Level 3 of OSI-RM that routes frames through a network of links

network—a system of interconnected packet radio stations designed for the efficient transfer of packets over long distances

node—a junction point within a network

node-to-node acknowledgment—the networking protocol in which each node in the path of a packet informs the previous node in the path that it has received a packet correctly

Open Systems Interconnection Reference Model (OSI-RM)—a model formulated by the International Standards Organization that permits different computer systems to communicate with each other as long as the communications protocols used by the computer systems adhere to the model

OSI-RM—abbreviation for Open Systems Interconnection Reference Model

packet assembler/disassembler (PAD)—a device that accepts data from a DTE and formats it into packet frames for transmission via a communications medium; the PAD also accepts packet frames received via a communications medium, extracts data from the packet frame and transfers the data to a DTE

packet radio bulletin-board system (PBBS)—a BBS that is accessed via packet radio

PAD—abbreviation for packet assembler/disassembler

parallel port—an interconnection that transfers bit-encoded information character-by-character or byte-by-byte in parallel

parity—a method of checking the accuracy of a received character by adding an extra bit, the parity bit, to each character

path—the route between two connected packet radio stations consisting of digipeaters, nodes and other packet stations

PBBS—abbreviation for packet radio bulletin-board system

phase-shift keying (PSK)—a method of transmitting digital information by varying the phase of a carrier between two values

Physical layer—the lowest level (Level 1) of OSI-RM; the Physical layer involves the mechanical and electrical transfer of data bits from one device to another

port—a circuit that allows a device to communicate with external devices

prioritized acknowledgment— a variation of the AX.25 protocol that gives priority to packet acknowledgments in order to improve channel usage

protocol—a set of recognized procedures

PSK—abbreviation for phase-shift keying

The Radio Amateur Satellite Corporation (AMSAT)—an Amateur Radio organization that was instrumental in the early United States packet radio developments; also responsible for United States amateur satellite developments

The Radio Amateur Telecommunications Society (RATS)—an Amateur Radio organization that was responsible for the development of the RATS Open System Environment (*ROSE*) X.25 Packet Switch protocol

RAM—abbreviation for random-access memory

random-access memory (RAM)—a data storage device that can be written to and read from

RATS—abbreviation for The Radio Amateur Tele-communications Society

RATS Open Systems Environment (ROSE)—a networking protocol based on the virtual circuit or connection protocol; it uses node-to-node acknowledgment and network node addressing based on the CCITT X.121 numbering scheme

read-only memory (ROM)—a data storage device that can only be read

ROM—abbreviation for read-only memory

ROSE—abbreviation for RATS Open Systems Environment

roundtable—conversations between more than two stations

RS-232-C—the former Electronics Industries Association standard for DTE-to-DCE (TNC) interfacing that specifies the interface signals and their electrical characteristics; it has been replaced by EIA-232-D

RTTY—abbreviation for radioteletype

secondary station identifier (SSID)—a number that follows a packet radio station's call sign to differentiate between two or more packet radio stations operating under the same call sign

Serial Line Interface Protocol (SLIP)—a Level 2 or Link layer protocol included in the *KA9Q Internet Protocol Package* that provides a simple serial data transfer between the Network layer protocols and the EIA-232-D interface of the Physical layer

serial—the transfer of bit-encoded information bit-by-bit

serial port—an interconnection that transfers bit-encoded information bit-by-bit (serially); the TNC connection for a terminal or computer

Simple Mail Transfer Protocol (SMTP)—a protocol included in the *KA9Q Internet Protocol Package* that provides an automatic mail-forwarding function

simplex—a communications mode where you transmit and receive on the same frequency

SLIP—abbreviation for Serial Line Interface Protocol

SMTP—abbreviation for Simple Mail Transfer Protocol

source—the station that originates a packet frame

space—one of the two elements in the binary start-stop code; it often represents the off-state or 0

SSID—abbreviation for secondary station identifier

store-and-forward—the process of receiving and holding data from one radio station for the purposes of relaying that data to another station at a later time

stream—one connection between two stations in a multiple connection application

streamswitch—a character that indicates a change in the stream being addressed in multiple connection packet radio applications

synchronous—a means of data transmission timing that uses a modem's internal clock to synchronize data

SYSOP—abbreviation for system operator

system operator (SYSOP)—an individual who runs and maintains a bulletin-board system

TAPR—abbreviation for Tucson Amateur Packet Radio Corporation

TCP—abbreviation for Transmission Control Protocol

TCP/IP—abbreviation for Transmission Control Protocol/Internet Protocol; the Defense Advanced Research Projects Agency protocols that were proposed as possible Network and Transport layer amateur packet radio protocols by Phil Karn, KA9Q, in his *Internet Protocol Package*

Telnet—a terminal-emulation protocol included in the *KA9Q Internet Protocol Package* that provides communications with the live operator of another node

terminal—short for data terminal equipment or a computer emulating data terminal equipment

terminal emulation software—a computer program that causes a computer to function as a DTE for the purpose of transferring data over a communications medium

terminal node controller (TNC)—an Amateur Radio packet radio assembler/disassembler; may or may not include a modem

TexNet—a packet radio network developed by the Texas Packet Radio Society; composed of dual-port network control processors that provide 2-meter AX.25 compatible user access and 9600-bit/s node-to-node linking on 70-cm

TNC—abbreviation for terminal-node controller

TNC 1—the first TAPR TNC that was made available to the general public; it was based on the 6809 microprocessor

TNC 2—the second TAPR TNC that was made available to the general public; based on a Z80 microprocessor; its design was the most popular in amateur packet radio history

Transmission Control Protocol (TCP)—a Level 4 or Transport layer protocol included in the *KA9Q Internet Protocol Package* that assures data integrity between the points of origination and destination

Transport layer—Level 4 of OSI-RM that maintains a connection that is transparent to the source and destination

TTL interface—a digital interconnection based on transistor-transistor logic (TTL)

Tucson Amateur Packet Radio Corporation (TAPR)—the Arizona-based Amateur Radio organization that was instrumental in packet radio protocol and hardware developments in the United States

turnaround time—the time required to switch between the receive and transmit modes

unconnected packets—packets transmitted from a source station with no specific destination station being addressed; used for beacons, CQs, and round-table communications

upload—to send files to a PBBS or other packet radio station

VADCG—abbreviation for Vancouver Amateur Digital Communications Group

VADCG protocol—the first amateur packet radio protocol developed by Doug Lockhart, VE7APU, and VADCG; also known as the V-1 or Vancouver protocol

VADCG TNC—the first amateur packet radio PAD; developed by Doug Lockhart, VE7APU, and VADCG

Vancouver Amateur Digital Communications Group (VADCG)—a Canada-based Amateur Radio organization that was responsible for the first popular packet radio protocol and hardware developments in Amateur Radio

Vancouver protocol—the first amateur packet radio protocol developed by Doug Lockhart, VE7APU, and VADCG; also known as the V-1 or VADCG protocol

virtual circuit—the appearance of a direct connection between the source and destination of a packet

virtual circuit protocol—a Network layer protocol that sets up and maintains a clearly defined path for the transfer of packets between a source and destination during a single data communications session; also called connection protocol

WØRLI Mailbox—public domain PBBS software written by Hank Oredson, WØRLI

X.25—a packet switching protocol formulated by the CCITT

Xerox 820-1—a computer that used a Z80 microprocessor and ran the CP/M operating system; it was a popular computer for packet radio applications, including the first PBBSs (using software written by Hank Oredson, WØRLI)

Xoff—transmitter off, a flow control character used in ASCII file transfers; Xoff commands the transmitter to stop sending data

Xon—transmitter on, a flow control character used in ASCII file transfers; Xon commands the transmitter to continue sending data

CHAPTER 11

FM and Repeaters

BRIAN BATTLES, WS1O
ARRL HEADQUARTERS

Pizza and beer, Abbott and Costello, FM and repeaters: Some things make natural combinations. In Amateur Radio, the combination of frequency modulation (FM) and HF/UHF repeaters has made it in a big way. FM and repeaters have been the most popular communication modes in ham radio since the early 1970s and are in the forefront of the hobby today. Nearly every amateur has used an FM repeater at some point.

The reason for the popularity of FM and repeaters is their versatility. They offer something for everyone. HF DXers swap information with fellow DXers, traffic handlers pass traffic between the local and section levels, and packet and radioteletype enthusiasts communicate digitally using repeaters dedicated to their specific modes. If you just like to talk, strike up a conversation on any repeater. Adding to the versatility of repeater operation is the ability to communicate while walking, driving or simply relaxing in the comfort of your home. You can even boost your range on HF through the use of a 10-meter FM repeater.

What's a Repeater?

The term "repeater" can be confusing to newcomers. A standard repeater is simply a relay station. It consists of a separate "input" receiver and an "output" transmitter connected to each other and tuned to two separate frequencies within the same band. When the receiver picks up a signal on the input frequency, it simultaneously retransmits the same signal on the output frequency. In this way a repeater forms a link between two stations that may not be able to communicate with each other directly.

A repeater system may include connections to receiver/transmitter combinations on other bands. For example, a 2-meter repeater linked to the 70-cm band may receive on 147.69 MHz and transmit on 147.09, while it also receives on 449.625 and transmits on 444.625. If a signal is present on 147.69 or on 449.625, it's retransmitted on both 147.09 and 449.625. These are called *crosslinked* repeaters, and they may include coverage for several bands. Such systems may operate on all available bands at all times, or have remote-control features that allow the licensee or selected users to control which crossband links are on or off at a particular time.

There are more complex repeaters that form integrated wide-coverage systems. These consist of machines miles apart that are connected by two-way VHF or UHF links. Such

A New Way to Start

One of the most exciting developments in the history of Amateur Radio came about in December 1990 when the FCC announced it had created the codeless Technician class license. The ruling took effect on February 14, 1991.

For the first time since ham radio licenses were instituted, US amateurs could get on the air without passing a Morse code test. A candidate who passes written exam Elements 2 and 3A receives a Tech ticket and may operate with all amateur privileges above 30 MHz. Having an alternative to the traditional Novice class of entry-level license appeals to people who are not as interested in CW, HF and DX activities and prefer to explore VHF, UHF and microwave communications.

In most parts of the US, it's increasingly difficult to locate space on the VHF and UHF bands to establish new repeaters. Even so, the existing 144, 222, 440 and 1270 MHz repeaters are not too busy to allow more

amateurs to use them. Although 2-meter packet has saturated many regions, more hams are discovering and exploiting the vast potential of this mode on higher frequencies.

The codeless Technician license is a great opportunity for FM and repeater enthusiasts. Not only has the door been opened for technically minded hams to experiment on VHF, UHF and microwaves, thousands of people who are initially only interested in local communications are also able to gain access to FM repeaters. This includes friends and family members of hams who enjoy being able to get in on the excitement and rewards of Amateur Radio public service, emergency communications support, traffic handling and keeping in touch with their ham friends. In addition, repeater-oriented clubs can welcome greater numbers of new members to join their activities, enhance projects and support repeaters.

The Ins and Outs of Repeater Operating

Repeater users should know how repeaters work. Users who are not familiar with the workings of a repeater may unintentionally misuse the equipment and interfere with other users.

A repeater is like any other Amateur Radio station. It has a receiver, a transmitter and an antenna. The difference is that a repeater's receiver and transmitter are tuned to different frequencies and the output of the receiver is connected (through a carrier-operated relay or *COR*) to the input of the transmitter; thus, anything the receiver hears is simultaneously retransmitted, or *repeated*, by the transmitter.

Your transceiver's transmitter and receiver are also tuned to different frequencies. Your transmitter is tuned to the repeater's receive or *input* frequency, so whatever you transmit is received by the repeater. Meanwhile, your receiver is tuned to the repeater's transmit or *output* frequency, so whatever the repeater transmits (repeats) will be received by your transceiver.

To tune an FM transceiver, you turn its frequency control to the transmit/output frequency of the desired repeater. Your transmitter's frequency automatically follows the tuning of your receiver and is adjusted to the repeater's receive/input frequency. (The frequency display on your transceiver will indicate its *receive* frequency.) Usually, you must also select a switch position that chooses between the "simplex" and "duplex" mode. In the simplex mode, your transmitter and receiver are tuned to the same frequency. In the duplex mode, your transmitter and receiver are tuned to different frequencies, as required for repeater operation.

The frequency separation between the transmitter and receiver differs with each repeater band. On 144 MHz, the standard separation between the transmitter and receiver frequencies is 600 kHz; on 222 MHz, the separation is 1.6 MHz; and on 450 MHz, the separation is 5 MHz. A 2-meter repeater with an input frequency of 146.34 MHz has an output of 146.94 MHz (146.34 MHz + 600 kHz = 146.94 MHz [146.34 + 0.6 = 146.94]). Hams refer to a repeater by its output frequency. For example, the repeater mentioned above would be called "the 94 machine."

Transmitter-receiver frequency separation is usually built into a modern transceiver's memory chips. You *do* have to know whether the output frequency is higher or lower than the input frequency, however. In some segments of the 2-meter band, the input frequency is 600 kHz lower than the output frequency (as in our 146.34/146.94 repeater example above), while in other parts of the band, the input frequency is 600 kHz higher than the output

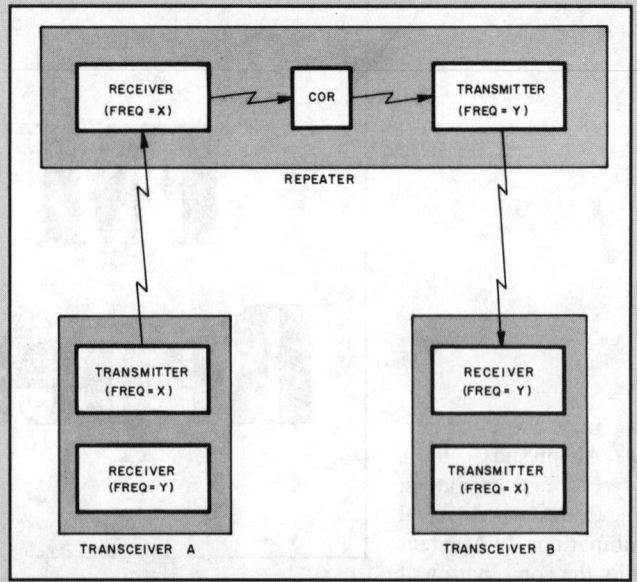

Fig A—Block diagrams of a typical FM repeater and FM transceivers show the relationship of frequency X, the repeater's input frequency, and frequency Y, the repeater's output frequency. Transceiver A transmits on frequency X and is received by the repeater's receiver, which is also tuned to frequency X. The repeater's transmitter retransmits (repeats) the received signal on frequency Y and is received by transceiver B's receiver, which is also tuned to frequency Y.

frequency. For example, a repeater with an output frequency of 147.09 MHz has an input frequency of 147.69 MHz.

Usually, when you select the duplex mode on your transceiver, you also have to select whether your transmitter frequency will be higher or lower than your receiver frequency. The simplex/duplex switch designated ± performs this function; the + position sets the transmitter frequency higher than the receiver frequency, while the − position sets the transmitter lower than the receiver frequency. The position between the + and − often selects the simplex mode.

---

systems allow operators to use a local repeater to make contacts with hams in distant cities or states. For example, the Evergreen Intertie includes more than 23 repeater stations throughout California, Oregon, Washington, Idaho, Montana, British Columbia and Alberta. An amateur on Mt Shasta can chat with a ham in Edmonton—even though both hams are using low-power hand-held transceivers! Integrated wide-coverage repeater systems allow users to turn links on and off as needed and activate telephone autopatching and other features by using the dual-tone multifrequency (DTMF) keypads on their radios.

Licensing

Except for the Novice VHF/UHF subbands, you need at least a Technician class license (or higher) to operate in the VHF and UHF Amateur Radio spectrum (above 30 MHz). A General class license (or higher) is required to operate 10-meter FM

and repeaters. Technician class hams may also use VHF/UHF repeaters that transmit on frequencies not normally permitted for use by Technicians. For example, a 450-MHz repeater may have an output on 29.640 MHz. Novices have considerable VHF/UHF privileges: All authorized modes from 222-225 MHz (25 watts maximum) and 1270-1295 MHz (5 watts). Novices can operate through all existing repeaters in those two subbands. For details, see Chapter 2.

OPERATING

FM repeaters provide the means to communicate efficiently. Before making your first transmission in the world of FM and repeater communications, however, you should be aware of some basic operating techniques.

Finding a Repeater

To use a repeater, you must know one exists. There are

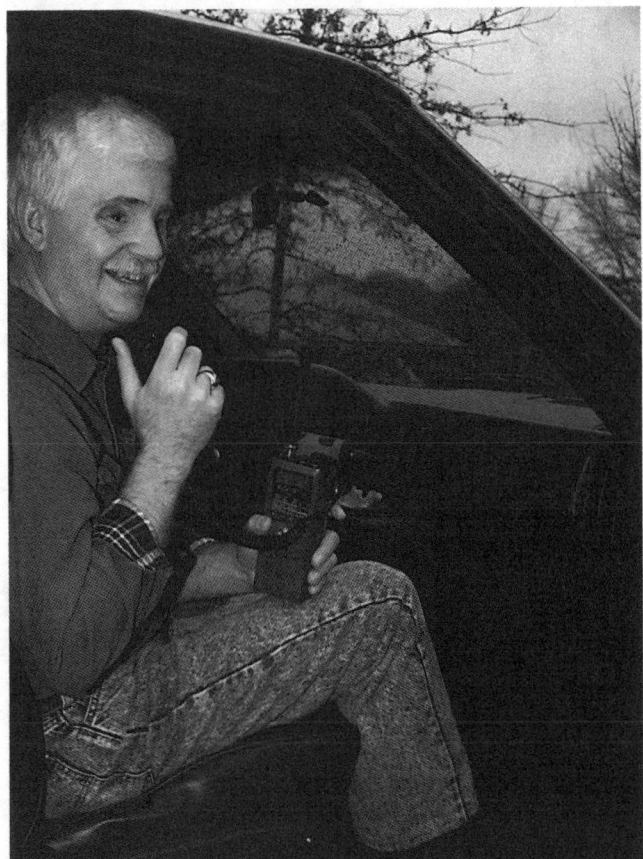

John Kirkendol, NØKJT, lost his sight in an accident. He enjoys 2-meter FM mobile from the passenger seat of a friend's car.

FM and Repeater-Speak

Here are some definitions of terms used in the world of Amateur Radio FM and repeaters:

access code: one or more numbers and/or symbols that are keyed into the repeater with a telephone tone pad to activate a repeater function, such as an autopatch.

autopatch: a device that interfaces a repeater to the telephone system to permit repeater users to make telephone calls. Often just called a "patch."

break: the word used to interrupt a conversation on a repeater only to indicate that there is an emergency.

carrier-operated relay (COR): a device that causes the repeater to transmit in response to a received signal.

channel: the pair of frequencies (input and output) used by a repeater.

closed repeater: a repeater whose access is limited to a select group. (see *open repeater*)

control operator: the Amateur Radio operator who is designated to "control" the operation of the repeater, as required by FCC regulations.

courtesy beeper: an audible indication that a repeater user may go ahead and transmit.

coverage: the geographic area within which the repeater provides communications.

digipeater, digital repeater: a packet radio repeater.

duplex: a mode of communication in which you transmit on one frequency and receive on another frequency.

frequency coordinator: an individual or group responsible for assigning channels to new repeaters without interference to existing repeaters.

full duplex: a mode of communication in which you transmit and receive simultaneously.

full quieting: a received signal that contains no noise.

half duplex: a mode of communication in which you transmit at one time and receive at another time.

hand-held: a portable transceiver small enough to fit in the palm of your hand, clipped to your belt or even in a shirt pocket.

input frequency: the frequency of the repeater's receiver (and your transceiver's transmitter).

key up: to turn on a repeater by transmitting on its input frequency.

machine: a repeater system (slang).

magnetic mount, mag-mount: an antenna with a magnetic base that permits quick installation and removal from a motor vehicle or other metal surface.

NiCd: a nickel-cadmium battery that may be recharged many times; often used to power portable transceivers. Pronounced "NYE-cad."

open repeater: a repeater whose access is not limited.

output frequency: the frequency of the repeater's transmitter (and your transceiver's receiver).

over: a word used to indicate the end of a voice transmission.

radio direction finding (RDF): the art and science of locating a hidden transmitter.

Repeater Directory: an annual ARRL publication that lists repeaters in the US, Canada and other areas.

separation, split: the difference (in kHz) between a repeater's transmitter and receiver frequencies. Repeaters that use unusual separations, such as 1 MHz on 2 meters, are sometimes said to have "odd splits."

simplex: a mode of communication in which you transmit and receive on the same frequency.

time-out: to cause the repeater or a repeater function to turn off because you have transmitted for too long.

timer: a device that measures the length of each transmission and causes the repeater or a repeater function to turn off after a transmission has exceeded a certain length.

tone pad: an array of 12 or 16 numbered keys that generate the standard telephone dual-tone multifrequency (DTMF) dialing signals. Resembles a standard telephone keypad.

various ways to find a repeater. Local hams can provide information about repeater activity or you can consult a repeater listing. Various clubs, sometimes those associated with the local frequency coordinator, publish statewide and regional directories of repeaters. Each spring, the ARRL publishes *The ARRL Repeater Directory*, a comprehensive listing of repeaters throughout the US, Canada and other parts of the world. Besides identifying local repeater activity, the *Directory* is handy for finding repeaters during vacations and business trips. Once you find a repeater to use, listen and familiarize yourself with its operating procedures.

Your First Transmission

If the repeater is quiet, pick up your microphone, press the switch, and transmit your call sign as "N6ATQ listening" or "N6ATQ monitoring" to attract someone's attention. After you stop transmitting, the repeater sends an unmodulated carrier for a couple of seconds to let you know it's working. If anyone is interested in talking to you, they'll call after your initial transmission. Some repeaters have specific rules for making yourself heard, but usually your call sign is all you need.

Don't call CQ to initiate a conversation on a repeater. It takes a lot longer to complete a long CQ than to simply transmit your call sign. (In some areas, a solitary "CQ" is permissible.) Efficient communication is the goal. You're not trying to attract the attention of someone who is casually tuning his receiver across the band. Except for scanner operation, there isn't much tuning through the repeater bands.

Whether you're at a public event, emergency situation, hamfest or shopping at the mall, a hand-held transceiver keeps you in touch.

If you want to join a conversation already in progress, transmit your call sign during a break between transmissions. The station that transmits next should acknowledge you. Don't use the word "break" to join a conversation (unless it's the operating practice in your area). "Break" usually suggests an emergency and indicates that all stations should stand by for the station with emergency traffic.

If you want to call another station and the repeater is inactive, simply call the other station. (For example, "WB8SVN, this is WB8OFR.") If the repeater is active, but the conversation in progress sounds as though it's about to end, be patient and wait until it's over before calling another station. If the conversation sounds like it's going to continue for a while, transmit your call sign between transmissions. After one of the other hams acknowledges you, politely ask to make a quick call. Usually, the other stations will acquiesce. Make your call short. If your friend responds to your call, ask her to move to a simplex frequency or another repeater, or to stand by until the present conversation is over. Thank the other users for letting you interrupt them to place your call.

Acknowledging Stations

If you're in the midst of a conversation and a station transmits its call sign between transmissions, the next station in queue to transmit should acknowledge that station and permit the newcomer to make a call or join the conversation. It's discourteous not to acknowledge him and it's impolite to acknowledge him but not let him speak. You never know; the calling station may need to use the repeater immediately. He may have an emergency on his hands, so let him make a transmission promptly.

The Pause That Refreshes

A brief pause before you begin each transmission allows other stations to participate in the conversation. Don't key your microphone as soon as someone else releases his. If your exchanges are too quick, you'll block other stations from getting in.

The "courtesy beepers" on some repeaters compel users to leave spaces between transmissions. The beep sounds a second or two after each transmission to permit new stations to transmit their call signs in the intervening time period. The conversation may continue only after the beep sounds. If a station is too quick and begins transmitting before the beep, the repeater may respond to the violation by shutting down!

Brevity

Keep each transmission as short as possible. Short transmissions permit more people to use the repeater. All repeaters promote this practice by having timers that "time-out," temporarily shutting down the repeater whenever the length of a transmission exceeds the preset time limit. With this in the back of their minds, most users keep their transmissions brief.

Learn the length of the repeater's timer and stay well within its limits. The length may vary with each repeater; some are as short as 15 seconds and others are as long as three minutes. Some repeaters vary their timer length depending on the amount of traffic on frequency; the more traffic, the shorter the timer. The other purpose of a repeater timer is to prevent extraneous signals from holding the repeater on the air continuously, potentially causing damage to the repeater's transmitter.

Because of the nature of FM radio, if more than one signal is on the same frequency at one time, it creates a muffled buzz or an unnerving squawk. If two hams try to talk on a repeater at once, the resulting noise is known as a "double." If you're in a roundtable conversation, it is easy to lose track of which station is next in line to talk. There's one simple solution to eradicate this problem forever: *Always pass off to another ham by name or call sign.* Saying, "What do you think, Jennifer?" or "Go ahead, WS1O" eliminates confusion and avoids doubling. Try to hand off to whoever is next in the queue, although picking out anyone in the roundtable is better than just tossing the repeater up for grabs and inviting chaos.

The key to professional-sounding FM repeater operation is to be brisk and to the point, and to leave plenty of room for others. Keep it moving. Don't drone; dart in and out. Don't hem and haw or be reluctant to "yield the floor." Your turn will come again in a moment. Turn it over, pause for others, get things rolling. Snappy, clearheaded exchanges sound sharp and are more enjoyable for your QSO partners.

Identification

You must give your call sign at the end of each transmission or series of transmissions and at least every 10 minutes during the course of a contact. You don't have to transmit the call sign of any other station, including the one you're contacting. (Exception: You must transmit the other station's call sign when passing third-party traffic to a foreign country.)

It's illegal to transmit without identification. Aside from breaking FCC rules, it's poor operating practice to key your microphone to turn on a repeater without identifying your station. This is called "kerchunking" the repeater. If you don't want to have a conversation, but simply want to check whether your radio works or if you're able to access a particular repeater, just say "KB4ZKD testing." This way you accomplish what you want to do legally.

Go Simplex

Simplex is a fancy-sounding word for a direct contact on a single frequency. After you've made a contact on a repeater, move the conversation to a simplex frequency, if possible.

The function of a repeater is to provide communications between stations not able to communicate directly because of terrain or equipment limitations; see Fig 11-1. If stations are able to communicate without a repeater, they shouldn't use a repeater. Always use simplex whenever possible so that the repeater will be available for stations that need its facilities.

Simplex communications offer a degree of privacy impossible to achieve on a repeater. There's also no timer to worry about or courtesy beep to wait for. When selecting a frequency, make sure it's designated for FM simplex operation. Each band is subdivided for specific modes of operation, such as satellite communications and weak-signal CW and SSB. If you select a simplex frequency indiscriminately, you may interfere with stations operating in other modes (and you may not be aware of it).

VHF and UHF FM are also acceptable modes for contesting in sweepstakes, sprints, Field Day and other events. The only stipulation is that contacts must be made without repeaters and generally conducted off the national simplex or calling frequencies. Watch for announcements in *QST* and bring that

Fig 11-1—The mountain blocks direct (simplex) communication between the two mobile stations. When mobile A transmits on 146.34 MHz the repeater receives this signal and retransmits it on 146.94 MHz. Mobile B's receiver is tuned to 146.94 to capture this signal from the repeater. When it's mobile B's turn to talk it also transmits on 146.34 MHz and mobile A listens for the retransmitted signal on 146.94 MHz.

The Great Frequency Hunt—Putting a New Repeater on the Air

Amateurs contemplating putting a repeater on the air may wonder how to obtain the frequency pairs for their repeaters. *Frequency coordination* is the recommended method. Across the US and Canada, volunteer individuals and groups have taken on the task of recommending frequencies for proposed repeaters within their "jurisdiction." These frequency coordinators keep extensive records on repeater operation in their area and adjacent areas. With this information, they're usually able to recommend the best pair of frequencies for proposed repeaters.

In cases of repeater-to-repeater interference, the two repeater stations are equally and fully responsible for resolving the interference unless one repeater is coordinated and the other is not. In that case, the noncoordinated repeater has primary responsibility for resolving the interference (according to the FCC).

Because of the amount of work involved in finding the best frequency pair, there is a time lag between the request for a frequency pair and receipt of the coordinated channel. If you plan to request a pair of frequencies, do it early. Consult the latest edition of *The ARRL Repeater Directory* for the listing of frequency coordinators.

The ARRL Board of Directors created the Spectrum Management Committee (SMC) in 1993. It has the following charter:

1) Restrict its activities to frequencies above 30 MHz and shall not engage in bandplanning although it may be tasked with examination and comment on such problems.

2) Examine and consider concerns of amateur operators with regard to mode, digital protocol, frequency utilization, impending technical improvement, applicable governmental regulations and activities, IARU actions and other matters referred to it by the Board of Directors.

3) When band-planning needs are identified, the committee will recommend to the Board creation of an Ad-Hoc committee to pursue and solve the problem.

Band Plans

Band plan refers to an agreement among concerned VHF and UHF operators about how each Amateur Radio band should be arranged. The goal of a band plan is to reduce interference between the modes sharing each band. Besides FM repeater and simplex activity, CW, SSB, AM, satellite, amateur television (ATV) and radio control operations also use these bands. (For example, a powerful FM signal at 144.08 MHz could spoil someone else's long-distance CW contact.) The VHF and UHF bands offer a wide variety of amateur activities, so hams have agreed to set aside space for each type.

The most heavily used FM activity is on 2 meters, so a description of the 2-meter band plan will be useful. The band is divided according to the following proposal by the ARRL VHF-UHF Advisory Committee:

Frequency (MHz)	Operation
144.00-144.05	EME (moonbounce) CW
144.05-144.10	General CW and weak signals
144.10-144.20	EME and weak-signal SSB
144.20	National calling frequency
144.20-144.275	General SSB operation
144.275-144.30	Propagation beacons
144.30-144.50	OSCAR subband
144.50-144.60	Linear translator inputs
144.60-144.90	FM repeater inputs
144.90-145.10*	Weak signal and FM simplex
145.10-145.20	Linear translator outputs
145.20-145.50	FM repeater outputs
145.50-145.80	Miscellaneous and experimental modes
145.80-146.00	OSCAR subband
146.01 -146.40	Repeater inputs
146.415-146.595	Simplex
146.61 -147.39	Repeater outputs
147.42-147.57	Simplex
147.60-147.99	Repeater inputs

*The following packet radio frequency recommendations were adopted by the ARRL Board of Directors in July 1987:

1) Automatic/unattended operations should be conducted on 145.01, 145.03, 145.05, 145.07 and 145.09 MHz.

    a) 145.01 should be reserved for interLAN use.

    b) Use of the remaining frequencies (above) should be determined by local user groups.

2) Additional frequencies in the 2-meter band may be designated for packet radio use by local coordinators.

Specific channels recommended above may not be applicable in all areas of the US. For example, amateurs in many areas conduct TCP/IP operations on 144.91 MHz and many *PacketCluster* nodes congregate on 144.95 MHz.

Prior to establishing regular packet radio use on any VHF/UHF channel, it's advisable to check with the local frequency coordinator. The decision on how the available channels are used should be based on coordination between local packet users.

Note: Please contact your regional frequency coordinator for information on channel availability.

Repeater Frequency Pairs (input/output):

144.51/145.11	144.83/145.43	146.37/146.97
144.53/145.13	144.85/145.45	146.40 or 147.60/147.00*
144.55/145.15	144.87/145.47	146.43 or 147.63/147.03*
144.57/145.17	144.89/145.49	146.46 or 147.66/147.06*
144.59/145.19	146.01/146.61	147.69/147.09
144.61/145.21	146.04/146.64	147.72/147.12
144.63/145.23	146.07/146.67	147.75/147.15
144.65/145.25	146.10/146.70	147.78/147.18
144.67/145.27	146.13/146.73	147.81/147.21
144.69/145.29	146.16/146.76	147.84/147.24
144.71/145.31	146.19/146.79	147.87/147.27
144.73/145.33	146.22/146.82	147.90/147.30
144.75/145.35	146.25/146.85	147.93/147.33
144.77/145.37	146.28/146.88	147.96/147.36
144.79/145.39	146.31/146.91	147.99/147.39
144.81/145.41	146.34/146.94	

*local option

Some areas use 146.40-146.60 and 147.40-147.60 MHz for simplex or for repeater inputs and outputs. Frequency pairs in those areas are:

147.415/146.415	147.475/146.475
147.43/146.43	147.49/146.49
147.445/146.445	147.505/146.505
147.46/146.46	147.595/146.595

Suggested 15-kHz Splinter Channels (input/output):

146.025/146.625	146.265/146.865	147.735/147.135
146.055/146.655	146.295/146.895	147.765/147.165
146.085/146.685	146.325/146.925	147.825/147.225
146.115/146.715	146.355/146.955	147.855/147.255
146.145/146.745	146.385/146.985	147.885/147.285
146.175/146.775	147.615/147.015	147.915/147.315
146.205/146.805	147.645/147.045	147.945/147.345
146.235/146.835	147.675/147.075	147.975/147.375
	147.705/147.105	

Simplex Frequencies:

*146.415	146.535	147.48
*146.43	146.55	147.495
*146.445	146.565	147.51
*146.46	146.58	147.525
*146.475	146.595	147.54
*146.49	147.42	147.555
*146.505	147.435	147.57
**146.52	147.45	147.585
	147.465	

*May also be a repeater (input/output). See repeater pairs listing.

**National simplex frequency

Several states have chosen to realign the 146-148 MHz subband, using 20-kHz spacing between channels. This choice was made to gain additional repeater pairs.

The transition from 30- to 20-kHz spacing is taking place on a case-by-case basis as the need for additional pairs occurs. Typically, the repeater on an odd-numbered pair will shift 10 kHz up or down, creating a new set on an even-numbered channel. For example, the pair of 146.13/.73 MHz would change to 146.12/.72 or 146.14/.74, while the pairs of 146.10/.70 and 146.16/.76 would be left unchanged.

2- or 6-meter FM rig out to a mountaintop and call "CQ contest" sometime!

Fixed Stations and Prime Time

Repeaters are intended to enhance mobile and portable communications. During mobile operating prime time, fixed stations should yield to mobile stations. (Some repeaters have an explicit policy to this effect.) But when you're operating as a fixed station, don't abandon the repeater completely; monitor mobile activity. Your help may be needed in an emergency.

No Free Lunch

It takes time, money, knowledge and energy to operate a reliable repeater system. How often do you stop to think of what goes into the machine you conveniently key up any time, 24 hours a day, 365 days a year? Nobody should feel compelled to join any group, and you can use thousands of repeaters across the US without joining any clubs. On the other hand, if you frequent a system—or just want to contribute to the cost of its upkeep so it can be counted on in an emergency—support your local repeater.

Many repeaters are run by clubs whose members pool their resources to operate the repeater. Some members provide the equipment that makes up the repeater, other members provide funds to pay the electric, phone, maintenance and site-rental bills, and others provide manpower at the repeater site whenever their help is needed.

Limited Access

Most Amateur Radio repeaters are open to all users. There are no restrictions on the use of the repeater's facilities. Limited-access repeaters do exist, however. Although such operations go against the spirit of our hobby, "closed" repeaters are legal according to FCC regulations.

There are "open" repeaters that require the use of special codes or subaudible tones to gain access. The reason for limiting the access to some "open" repeaters is to prevent interference. In cases where extraneous transmissions often activate the repeater, limiting access is the only way to resolve the problem.

How is access to these repeaters controlled? Most often, via a technique called *continuous tone-controlled squelch system (CTCSS)*. Many hams refer to CTCSS as *PL*, Motorola's trademark term that stands for Private Line, used on commercial radio gear, with a series of alphanumeric names that designate each of the tones. There are 42 standard CTCSS tone frequencies, as developed by the Electronic Industries Association (EIA). When a transmitter is configured for CTCSS, it sends a subaudible tone along with the transmitted voice or other signals. The frequency of the CTCSS tone is below the lowest audio frequency other stations will pass to their speakers, but it is sensed by a suitably equipped repeater. The repeater is programmed to respond only to carriers that send the proper tone. This effectively locks out signals that don't carry the correct CTCSS tone. Most modern VHF and UHF transceivers include the necessary circuitry to generate CTCSS tones, so if you know the one you need, you can simply program it on your rig.

There's a complete table of CTCSS frequencies in *The ARRL Repeater Directory*, and many repeaters listed in that book publish their CTCSS tones.

Minimum Power

The VHF and UHF repeater bands are paradise for QRP fans. Make it a habit to run your transceiver on a low-power setting. There's usually no need to pump out heavy watts on VHF or UHF FM if you're within a reasonable range of a repeater or other station operating simplex.

Inspect your station regularly for loose connections, broken wires, antenna problems, intermittent grounds and other potential weak spots. Mobile installations are most prone to wear and damage. A well-designed antenna, quality feed line and properly installed connectors will improve your transmitted and received signals. Never use a high-power external amplifier for a local contact. If a few watts won't bring up the 2-meter repeater five miles away, 160 watts probably won't cut it, either.

When you're on a repeater frequency, use the minimum power necessary to maintain communications to avoid the possibility of accessing distant repeaters on the same frequency. Using minimum power is not only a courtesy to the distant repeaters, but an FCC requirement.

Autopatch

An autopatch is a device that allows hams to make telephone calls through a repeater. In most repeater autopatch systems, the user simply has to generate the standard telephone company tones to access and dial through the system. This is usually done by interfacing a telephone tone pad with the user's transceiver. Dual-tone multifrequency (DTMF) tone keypads are often mounted on the front of hand-held portable transceivers or on the front of fixed or mobile transceiver microphones.

There are a number of important rules that govern autopatch operation. See the *ARRL Autopatch Guidelines* sidebar in this chapter for a thorough treatment of the subject.

If you have a legitimate reason to use the autopatch, how do you use it? First, you must access (turn on) the autopatch, usually by pressing a designated key or combination of keys on the tone pad. Check to make sure the frequency is clear before you begin and be sure to identify your station. The # (pound) and * (star) signs are common access keys. When you hear a dial tone, you've successfully accessed the autopatch. Now, punch in the telephone number you wish to call.

When you establish a call, tell the person you're calling what you're doing so that they'll know they're going over the ham bands. If you just say, "Hi, I'm calling you from the car," they might think you have a mobile cellular telephone. If the person isn't familiar with ham radio, you may want to explain that the conversation is not private and may be overheard. They should also be told that they have to wait their turn to talk. Use the word *over* to inform the person at the other end of the line that you've finished talking and that he or she can speak. Many autopatches have timers that terminate the connection after a certain period of time, so keep your telephone conversation short and end the autopatch as quickly as possible.

The procedure for turning off the autopatch is similar to the procedure for accessing it. A key or combination of keys must be punched to return the repeater to normal operation. Don't forget to identify when you finish. Autopatch users should consult the repeater group that sponsors the autopatch for specific information about access codes and timer specifics.

Traffic

Traffic handling is a natural for repeater operation. Where else can you find as many local outlets for the traffic coming down from the Section-level nets? Everyone who has a trans-

ARRL Autopatch Guidelines

Autopatch operation involves using a repeater as an interface to a local telephone exchange. Hams operating mobile or portable stations are able to use the autopatch to access the telephone system and place a call. Hams use autopatches to report traffic accidents, fires and other emergencies. There's no way to calculate the value of the lives and property saved by the intelligent use of autopatch facilities in emergencies. The public interest has been well served by amateurs with interconnect capabilities. As with any privilege, this one can be abused and the penalty for abuse could be the loss of the privilege for all amateurs. The suggested guidelines here are based on conventions that have been in use for years on a local or regional basis throughout the country. The ideas they represent have widespread support in the amateur community. Amateurs are urged to observe these standards carefully so our traditional freedom from government regulation may be preserved as much as possible.

1) Although it's not the intent of the FCC Rules to let Amateur Radio operation be used to conduct an individual's or an organization's commercial affairs, autopatching involving business affairs may be conducted on Amateur Radio. (The FCC has stated that it considers nonprofit and noncommercial organizations "businesses.") On the other hand, amateurs are strictly prohibited from accepting any form of payment for operating their ham transmitters, they may not use Amateur Radio to conduct any form of business in which they have a financial interest and they may not use Amateur Radio in a way that economically benefits their employers.

Amateurs should generally avoid using Amateur Radio for any purpose that may be perceived as abuse of the privilege. The point of allowing hams to involve themselves in "business" communication is to make it more convenient and to remove obstacles from ham operations in support of public service activities. Before this rule was revised in 1993, it was often technically illegal for amateurs to participate in many charitable and community service events because the FCC regarded any organization, commercial or noncommercial, as a business with respect to the rules, and prohibited hams from making any communications in any way facilitate the business affairs of any party. That meant that operating a talk-in station for a local nonprofit radio club's hamfest constituted a violation!

So now it's legal to use ham frequencies, including autopatch facilities, to communicate in such a way as to facilitate a business transaction. The distinction is essentially whether the amateur operator or his employer has a financial stake in the communication. This means that a ham may use a patch to call someone about a club event or activity, to make a dentist appointment, to order a pizza or to see if a load of dry cleaning is ready to be picked up. In such situations, the ham isn't in it for the money. However, no one may use the ham bands to dispatch taxicabs or delivery vans, to send paid messages, to place a sales call to a customer, or to cover news stories for the local media (except in emergencies if no other means of communication is available). If the ham is paid for or will profit from the communication, it may not be conducted on an amateur frequency. That's why there are telephones and commercial business radio services available.

Use care in calling a business telephone via an amateur autopatch. Calls may be legally made to one's office to receive or to leave personal messages, although using Amateur Radio to avoid the cost of public telephones, commercial cellular telephones or two-way business-band radio isn't considered appropriate to the purpose of the amateur service. Calls made in the interests of highway safety, such as for the removal of injured persons from the scene of an accident or for the removal of a disabled vehicle from a hazardous location, are clearly permitted.

A final word on business communications: Just because the FCC says that a ham can place a call involving business matters on a repeater or autopatch doesn't mean that a repeater licensee or control operator must allow you to do so! If he or she prefers to restrict all such contacts, he or she has the right to terminate your access to the system. A club, for instance, may decide that it would rather not have members order commercial goods over the repeater autopatch and may vote to forbid members from doing so. The radio station's licensee and control operator are responsible for what goes over the air and have the right to refuse anyone access to the station for any reason.

2) All interconnections must be made in accordance with telephone company rules and fee schedules (tariffs). If you have trouble obtaining information about them from telephone company representatives, the tariffs are available for public inspection at your telephone company office. Although some local telephone companies consider Amateur Radio organizations to be commercial entities and subject to business telephone rates, many repeater organizations, as noncommercial volunteer public service groups, have successfully arranged for telephone lines at repeater sites to be charged at the lower residential rate.

3) Autopatches should not be made solely to avoid telephone toll charges. Autopatches should never be made when normal telephone service could be just as easily used. The primary purpose of an autopatch is to provide vital, convenient access to authorities during emergencies. Operators should exercise care, judgment and restraint in placing routine calls.

4) Third parties (nonhams) should not be put on the air until the responsible control operator has explained to them the nature of Amateur Radio. Control of the station must never be relinquished to an unlicensed person. Permitting a person you don't know well to conduct a patch in a language you don't understand amounts to relinquishing control because you don't know whether what they are discussing is permitted by FCC rules.

5) Autopatches must be terminated immediately in the event of any illegality or impropriety.

6) Station identification must be strictly observed.

7) Phone patches should be kept as brief as possible, as a courtesy to other amateurs; the amateur bands are intended primarily for communication among radio amateurs, not to permit hams to communicate with nonhams who can only be reached by telephone.

8) If you have any doubt as to the legality or advisability of a patch, don't make it. Compliance with these guidelines will help ensure that amateur autopatch privileges will continue to be available in the future, which helps the Amateur Radio service contribute to the public interest.

ceiver capable of repeater operation is a potential traffic handler. As a result, local traffic nets have flourished on repeaters.

The procedures for handling traffic on a repeater are about the same as handling traffic anywhere else (consult Chapter 15 for full details). However, there are two things unique to repeater traffic handling: time-out timers and only occasional use of phonetics.

Repeaters have timers that shut down the repeater if a transmission is too long. Therefore, when you relay a message by repeater, release your push-to-talk (PTT) switch during natural breaks in the message to reset the timer. If you read the message in one long breath without resetting the timer, the repeater may shut down in the middle of your message. Then you'll hold up the net—and earn a red face—while you repeat the message.

Under optimal signal conditions, the audio quality of the FM mode of communications is excellent. It lacks the noise, static and interference common to other modes. Therefore, the use of phonetics and repetition is not necessary when relaying a message over a repeater. The only time phonetics are necessary is to spell out an unfamiliar word or words with similar-sounding letters. (An exception: The word "emergency" is always spelled out when it appears in the preamble of a formal radiogram.) On the rare occasion that a receiving station misses something, she can ask you to fill in the missing information.

The efficiency provided by repeaters has made repeater traffic nets popular. There are probably one or more active repeater traffic nets in your area (consult the latest edition of *The ARRL Net Directory* to find a local repeater traffic net). If there's no repeater traffic net in your vicinity, why not fill the need by starting one yourself? Contact your ARRL Section Manager (SM) or Section Traffic Manager (STM) for details. *QST* lists your local officials; check page 8 or Section News in the current issue.

RTTY

Digital operation is not as common as voice, but it's popular on VHF and UHF FM repeaters. With FM, there's minimal noise and interference, making it possible to send and receive error-free RTTY messages. Above 50 MHz, any digital code is permissible, whereas HF RTTY operators are limited to three digital codes: Baudot/Murray, ASCII and AMTOR. (Note: Although any digital code is permissible above 50 MHz, station identification at the end of each communication and every 10 minutes during any communication must be made using voice [where permitted], CW, Baudot/Murray, ASCII or AMTOR.)

The operating procedures for RTTY repeater operation are similar to HF RTTY procedures (consult Chapter 9 for details). One difference, though, is the length of transmissions. RTTY operators must keep their transmissions short enough so they don't time-out the repeater. On many RTTY repeaters, circuits are configured to disable the timer when a RTTY signal is detected.

Some repeaters used for RTTY are also used for voice communications. Under these circumstances, RTTY operators must take care not to disrupt voice communications by indiscriminately transmitting a RTTY message.

The operating procedures for radioteletype repeater operation are simple, so the transition from HF to VHF/UHF with RTTY should be easy.

Packet Radio

Packet radio on VHF and UHF FM is a digital communications mode that has become tremendously popular in the past

HF Repeaters—Your DX Connection!

Imagine having a QSO with a station in Ecuador while you're on your way to work. Instead of a large HF rig under your dashboard, there is only a small transceiver that might easily be mistaken for a 2-meter FM unit at first glance. You're only running about 10 or 20 watts, yet the Ecuadorian station is giving you a "59" report. How is this possible?

Well, all the FM repeater action isn't confined to the VHF and UHF bands. Narrowband voice FM is permitted from 28.3-29.7 MHz and wider-band FM voice is allowed from 29.0- 29.7 MHz. The upper end of the 29.0-29.7 MHz segment is populated with a number of FM repeaters. When propagation conditions are good, these repeaters can provide worldwide coverage!

It can be confusing to use a 10-meter FM repeater during good conditions. Your input signal may activate distant repeaters as easily as those close by. You may also hear multiple QSOs as several different machines battle for the repeaters' (and your receiver's) front end! Sometimes band openings are as much a burden as they are a blessing. In 1980, the ARRL Board of Directors adopted a set of CTCSS access tones to be used voluntarily to provide a uniform nationwide system for 10-meter repeaters (see the table in *The ARRL Repeater Directory*).

In 1987, the FCC adopted the Novice Enhancement plan that gave Novices and Technicians their first taste of voice privileges on HF. Using SSB on 28.3-28.5 MHz, Novice and Technician Plus licensees have worked hundreds of DX countries from before dawn to past local sunset.

Even at the nadir of sunspot-cycle-related propagation, band conditions will make frequent dramatic, unpredictable shifts that produce remarkable openings. How will we know 10 is open if nobody's there to listen—or transmit? That's where 10-meter FM repeaters linked to 2 meters, 222 MHz and 440 MHz are an asset. When 10 pops open to Europe, Asia, Africa or Australia, you'll hear it first if you're monitoring your local HF-to-VHF repeater.

Where are They?

Most 10-meter FM repeaters operate between 29.520-29.680, with the FM simplex calling frequency at 29.600 MHz. (Avoid using 29.3-29.510 MHz for terrestrial contacts—the ARRL recommends this segment be reserved exclusively for amateur satellite downlinks. Because these downlinks use CW or SSB, a powerful local FM signal would easily wipe out a weak-signal QSO, so respect this portion.)

There are four standard repeater frequency pairs, as follows (input/output MHz):

    29.520/29.620
    29.540/29.640
    29.560/29.660
    29.580/29.680

Not all 10-meter repeater activity is confined to hams with General or higher-class licenses. Technicians can use these machines via VHF or UHF links. (Novices can access 10-meter repeaters from 222-225 or 1270-1295 MHz links.)

few years. Hams all over the world are using terminal node controllers (TNCs) and their personal computers to create networks through which they can exchange messages and program files, handle traffic and enjoy direct keyboard-to-keyboard QSOs. Just as there are dedicated voice repeaters, there are dedicated packet *nodes* that perform the same func-

An HT at 10,000 feet comes in handy for Dan Cui, N1CJD. He uses it as he pilots a "Paramotor," which is a powered paraglider. Not all hams will want to repeat Dan's operation.

Direction Finding

Radio direction finding (RDF), also known as foxhunting, bunny hunting and hidden-transmitter hunting, has become a popular radiosport in VHF and UHF FM operation. Someone hides a transmitter in the woods, a park or other inconspicuous place, and the troops, with direction-finding equipment in hand, go out to try to be the first to find the hidden transmitter. Many hamfests and club outings include a hidden-transmitter hunt as part of their agenda, and the competition is vigorous.

RDF also has a serious side. Repeater operators use these skills to track down repeater jammers and unlicensed intruders. It's also used to track down stolen transceivers that suddenly pop up on the air operated by people who appear unfamiliar with Amateur Radio operating procedures.

Locating a hidden transmitter is an art and, like any other art, practice makes perfect. So, if you have the opportunity to participate in a foxhunt, take advantage of it to hone your direction-finding skills for the real thing.

Further reading:

J. D. Moell and T. N. Curlee, *Transmitter Hunting* (Blue Ridge Summit, PA: Tab Books, 1987).
R. Schetgen, Ed., *The 1995 ARRL Handbook* (Newington, CT: ARRL, 1994), Chapters 2 and 23.

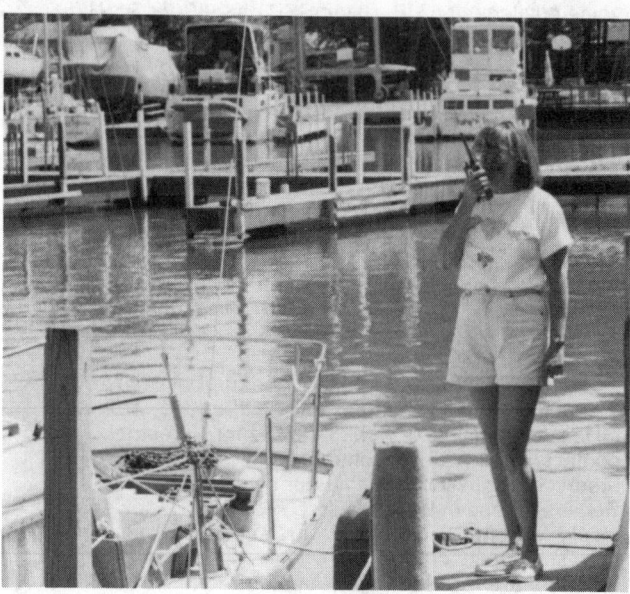

The family that talks together cruises together. Here Harriet Raynor, KB8JVY, has a dockside QSO with her spouse George, KB8JVZ.

tion as their voice counterparts. That is, they act as intermediaries between stations that otherwise could not communicate. Packet nodes use the same frequency for transmission and reception, yet more than one—and usually several—data exchanges can be conducted on the same frequency (using one repeater). How's that for efficient repeater communications?

The operating procedures for packet are different from other forms of Amateur Radio communications because you command your TNC to do various tasks. The commands differ slightly, depending on which type of TNC you use (and there are a lot of commands available to do a lot of things besides communications). They all work roughly the same way, however.

To call another packet radio station, you enter a one or two-letter command followed by the call sign of the station you wish to contact; for example, C WR1B. Once you have entered the command and call sign, your TNC will attempt to contact the other station automatically.

To use a network node to contact another station, you must first connect to the node (for example, C KA1JES-1 where KA1JES-1 is the call sign of the node). Once you are connected to the node, you need only enter C WR1B and the node will complete the connection and relay the data between you and WR1B. As with voice FM and repeater operation, it's advisable in packet radio operation to use a node only if necessary.

You can call CQ on packet radio by entering a CQ message in your TNC and commanding the TNC to transmit the message over the air. Don't send a CQ message through several nodes, though. Trying to maintain a QSO through more than two nodes is a difficult proposition at best. When the frequency is very active, it is often impossible.

If your TNC establishes contact with another station (or if someone else's TNC establishes contact with yours), your terminal will display a message to that effect (***CONNECTED TO WS1O). Once you make contact, what you type at your keyboard is transmitted to the other station after you enter each control character or carriage return (depending on your TNC). During prime-time operating hours, it's good procedure to keep your messages short because you're sharing the channel with other stations. Short transmissions are more likely to make it through to the other station than long transmissions. (Short transmissions are less likely to collide with ones emanating from another conversation on the same frequency; a collision causes your TNC to repeat the transmission.)

To turn the conversation over to the other station, end your transmission with >>, K, KN, ? or something else that lets the other operator know that you're done. To end a conversation, command your TNC to do the dirty work (the simple command D for disconnect, does the job). When that task is completed, the TNC will indicate it by displaying ***DISCONNECTED on your screen.

Another form of packet radio communications is transmission control protocol/internet protocol (TCP/IP). This is based on a more sophisticated computer networking model and permits some fancier functions for the user. To use TCP/IP, configure your TNC to turn control over to your computer by issuing a command to put the TNC in *KISS* mode (KISS stands for Keep It Simple, Stupid). The necessary software, called *NET* or *NOS*, is freely available from telephone and packet bulletin boards or from other sources by mail. The TCP/IP world interfaces with "regular" packet through gateway setups in many areas. For more information, see Chapter 10.

Satellite Gateway

A satellite gateway is a specially equipped station that links your local FM repeater to an Amateur Radio satellite. This makes it easier to try out a ham satellite without having to set up your own new equipment. It's your "gateway" to OSCAR operation because it allows you to use the simplest ham radio station, a hand-held transceiver, to work stations around the world.

Operating through a gateway is easy. The Operations Controller (OC) is in charge of what goes into the repeater and keeps things in order. A "list" operation works well to ensure minimum confusion by those trying to use the gateway through the repeater. The OC instructs users to keep each contact within a maximum length, to speak clearly, and preferably, to say "over" to avoid doubling. Good discipline in gateway operation is essential. The signals from the gateway are heard over a large portion of the Earth, so efforts to minimize confusion are desirable.

It's permissible to call CQ through a repeater to work someone by satellite. More stations will respond to "CQ from WS1O via gateway WB2NOM" than "QRZ OSCAR?" Beyond these considerations, operating through a satellite gateway is straightforward. Common sense dictates the rest.

FM—The Fun Mode

Monitor your local repeater and listen to how these procedures (and variations thereof) work in actual practice; you may

Rick Applegate, N7WKS, combined the use of a 440 MHz H-T with a paddle mounted satellite antenna on a canoe trip. He worked in the FM mode through OSCARs 13 and 21. A separate 2-meter rig was used for the downlink. A different paddle was used for the canoe.

discover that the local repeater community has its own special operating quirks. A little reading (of this chapter) and a little listening (to the local repeater) will prepare you to communicate efficiently and have fun with FM and repeaters.

VHF/UHF Operating

MICHAEL OWEN, W9IP
STAR ROUTE BOX 60
CANTON, NY 13617

The radio spectrum between 30 and 3,000 MHz is one of the greatest resources available to the radio amateur. The VHF and UHF amateur bands are a haven for ragchewers and experimenters alike; new modes of emission, new antennas and state-of-the-art equipment are all developed in this territory. Plenty of commercial equipment is available for the more popular bands, and "rolling your own" is very popular as well. Propagation conditions may change rapidly and seemingly unpredictably, but the keen observer can take advantage of subtle clues to make the most of the bands. Most North American hams are already well acquainted with 2-meter or 440-MHz FM. For many, channelized repeater operation is their first exposure to VHF or UHF. However, FM is only part of the story! This chapter will discuss the "other end" of the VHF/UHF bands, the low end, where almost all weak-signal SSB and CW activity takes place. It will discuss the bands from 50 MHz up through microwaves, but will emphasize the lower VHF frequency bands because the vast majority of amateur activity occurs there

HOW ARE THE BANDS ORGANIZED?

One of the keys to using this immense resource properly is knowing how the bands are organized. For example, if you are unaware of the calling frequencies on most of the VHF bands, you might spend weeks tuning around before you find someone to talk to! But, knowing the best frequencies and times, you will have little trouble making plenty of contacts, working DX, and enjoying the "world above 50 MHz." See Fig 12-1 for suggested band use in the VHF/UHF range.

Currently, activity on VHF/UHF is concentrated in the two lower VHF bands, 6 and 2 meters (50 and 144 MHz). The number of active stations on these bands is about equal. Above the 2-meter band, there are considerably more active stations on 432 MHz than any other. Because it is not available worldwide, 222 MHz is not inhabited as much as it could be. The 902-928 MHz band is gaining in popularity and 1296 MHz has quite a few occupants near major cities. For US amateurs, the 2300-MHz band is split into two parts (2300-2310 and 2390-2450 MHz) and many European amateurs have access only to the frequencies above 2320 MHz. Higher frequencies, above 3300 MHz, are the territory of the true experimenters because little commercial equipment is available for those bands. See accompanying sidebar on the microwave bands.

The VHF/UHF bands are roomy. That is, there is lots of frequency spectrum available for ragchewing, experimenting or working DX. Although some of the more popular frequencies occasionally get very crowded, most VHF/UHF allocations are available all the time. This means that whatever you like to do—chat with your friends across town, test amateur television or bounce signals off the moon—there is usually plenty of spectrum available. VHF/UHF is a great resource!

The key to enjoyable use of this resource is to know how everyone else is using it, and to follow their lead. Basically, this means to listen first. Pay attention to the segments of the band already in use, and follow the operating practices that experienced operators are using. This way, you won't interfere with ongoing use of the band by others, and you'll "fit in" right away.

All of the bands between 50 and 1296 MHz have widely accepted calling frequencies (see Table 12-1). These are the frequencies most operators monitor most of the time, where you just park the receiver when you're doing something else around the shack. If someone wants to call you, or someone calls CQ, he or she will use the calling frequency and you will already be "on frequency" to hear them call.

The most important thing to remember about the calling frequencies is that they are not for ragchewing. After all, if a dozen other stations want to have a place to monitor for calls, it's really impolite to carry on a long-winded conversation on that frequency, isn't it?

In most areas of the country, everyone uses the calling frequency to establish a contact, and then the two stations move up or down a few tens of kHz to chat. This way, everyone can share the calling frequency without having to listen to each other's QSOs. You can easily tell if the band is open by monitoring the call signs of the stations making contact on the calling frequency—you sure don't need to hear their whole QSO!

On 6 meters, a *DX window* has been established in order to reduce interference to DX stations. This window, which extends from 50.100-50.125 MHz, is intended for DX QSOs only. The DX calling frequency is 50.110 MHz. If you make a DX contact and expect to ragchew, you should move up a few kHz to clear the DX calling frequency. US and Canadian 6-meter operators should use the domestic calling frequency of 50.125 MHz for non-DX work. Once again, when contact is established, you should move off the calling frequency as quickly as possible.

Fig 12-1—Suggested US VHF/UHF band usage (in MHz). On 222 MHz and above, activity is usually centered around the calling frequency. On the 50-MHz band, a DX window exists between 50.100 and 50.125 MHz. US and Canadian operators should leave this window clear except when working or calling DX. Domestic and local SSB QSOs should take place above 50.125 MHz.

ACTIVITY NIGHTS

Although you can scare up a QSO on 50 or 144 MHz almost any evening (especially during the summer), in some areas of the country there is not always enough activity to make it easy to find someone to talk with. Therefore, informal "activity nights" have been established in many parts of the country so you will know when to expect some activity. Each band has its own night.

There is a lot of variation in activity nights from place to place, so don't trust the following list to be entirely accurate. Check with someone in your area to find out about local activity nights.

Activity nights are particularly important for 222 MHz and above. On these bands, there is often little activity during most "normal" nights. If you have just finished a new transverter or antenna for one of these bands, you will have a much better chance to try them out during the band's weekly activity night. That doesn't mean there is no activity on other nights, especially if the band is open. It may just take longer to get someone's attention during other times

Local VHF/UHF nets often meet during activity nights. Two

Microwaves

The region above 3000 MHz is known as the microwave spectrum. Although European amateurs have made tremendous use of this rich resource, few Americans have done so. Little commercial equipment is available and entirely different construction techniques are required if amateurs are to build their own equipment. Nevertheless, the microwave bands are receiving a lot of attention from the experimentally minded among us. We can expect a considerable amount of progress as more amateurs try out the microwave bands.

10 GHz is the most popular microwave band, and VHF enthusiasts—such as Cheryl, KA1IXI—find 10-GHz mountain-topping particularly effective. (NA1L photo)

The most popular microwave band in the USA is the 10-GHz (3 cm) band. Integrated QRP transmitter/receiver/horn antenna assemblies using Gunn diodes are available commercially. They provide a reliable range of several kilometers using wideband FM. Using narrowband techniques (SSB/CW) and high-gain dish antennas, hams have made QSOs of over 1000 miles. These are particularly remarkable since the power levels involved were less than 10 watts. Other microwave bands are less popular and have little activity at the present time. However, as designs for SSB/CW gear become more widely available, activity is certain to spread.

Most microwave activity is by prearranged schedules (at least in the US). This is partly because activity is so sparse and partly because antenna beamwidths are so narrow. With a 5° beamwidth, you would have to call CQ 72 times to cover a complete circle!

national organizations, SMIRK (Six Meter International Radio Klub) and SWOT (Sidewinders on Two), run nets in many parts of the country. These nets provide a meeting place for active users of the 50- or 144-MHz bands. In addition, the

Common Activity Nights

Band (MHz)	Day	Local Time
50	Sunday	6:00 PM
144	Monday	7:00 PM
222	Tuesday	8:00 PM
432	Wednesday	9:00 PM
902	Friday	9:00 PM
1296	Thursday	10:00 PM

Table 12-1
North American Calling Frequencies

Band (MHz)	Calling Frequency
50	50.110 DX calling
	50.125 US, local
144	144.010 EME calling
	144.100, 144.110 CW
	144.200 SSB calling
222	222.100 CW/SSB
432	432.010 EME calling
	432.100 CW/SSB
902	903.1
1296	1296.100 CW/SSB
2304	2304.1
10,000	10,368

Southeastern VHF Society runs a net on 432 MHz. For those whose location is far away from the net control's location, the nets may provide a means of telling if your station is operating up to snuff, or if propagation is enhanced. Furthermore, you can sometimes catch a rare state or grid square checking into the net. For information on the meeting times and frequencies of the nets which are run by SMIRK and SWOT, ask other local occupants of the bands in your area or write to these organizations directly: SMIRK c/o Ray Clark, K5ZMS, 7158 Stone Fence Dr, San Antonio, TX 78227; SWOT c/o Harry Arsenault, K1PLR, 48 Crane Rd, North Stamford, CT 06902.

WHERE AM I?

One of the first things you will notice when you tune the low end of any VHF band is that most QSOs include an exchange of "grid squares." What are grid squares? Well, first of all, grid squares aren't squares. They're more like rectangles, and they're just a way of dividing up the surface of the Earth. Grid squares are a shorthand means of describing your general location anywhere on the Earth. (For example, instead of trying to tell a distant station that I'm in Canton, New York, I tell them I'm in "grid square FN24." It sounds strange, but FN24 is a lot easier to locate on a map than my small town!)

Grid squares were developed by an international group at a conference in Maidenhead, England, so they are sometimes called "Maidenhead" grid squares.

Grid squares are based on latitude and longitude. The world is first divided into 324 very large areas called "fields," which cover 10∞ of latitude by 20° of longitude. Each field is divided into 100 "squares" from which grid squares get their name. Grid squares are 1° × 2° in size. When the full grid square designator is used, even smaller "sub-square" divisions are made, down to areas only a few kilometers wide. For most of us, knowing about "squares" is sufficient.

Grid squares are coded with a 2-letter/2-number/2-letter code (such as FN24KP). Most people just use the first four characters (such as FN24). This handy designator uniquely identifies the grid square; no two have the same identifying code. This is the biggest advantage of the Maidenhead system over the older QRA-locator system used for decades in Europe. The European system is not applicable worldwide because the same QRA designator might be used for more than one place.

There are several ways to find out your own grid square identifier. You can start by consulting Tables 12-2 and 12-3. By following the instructions shown in the tables, you will be able to locate your own grid square identifier. The hardest part is finding your QTH on a good map that has latitude and longitude on it; the rest is easy. Most high-quality road maps have this on the margins. Or, you could go a step further and purchase the topographic map of your immediate area from the US Geological Survey. For maps, write to US Geological Survey, Map Distribution, Federal Center, Building 41, Denver, CO 80225. Once you have your latitude and longitude, the rest is a snap!

The ARRL publishes a map (see Fig 12-2) of the continental United States and most populated areas of Canada, on which grid squares are marked. This map is available from ARRL for $1. It really helps to keep a copy of the map handy to help locate stations as you work them. If you are keeping track of grid squares for the purpose of VUCC (the VHF/UHF Century Club award—see below), you can color in each square as you work it. Some operators use a light color when they work the grid for the first time, then color it more darkly when they receive a QSL. Others use color or pattern schemes to indicate different propagation modes. [Note: The ARRL also publishes a *World Grid Locator Atlas*, available for $5.]

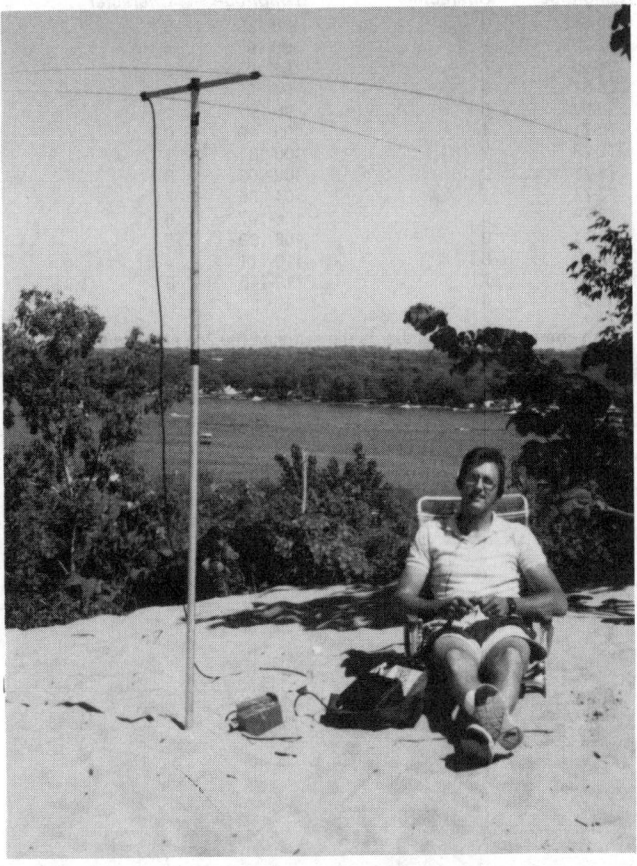

Pete, N8YEL, enjoys portable VHF operating along the shores of Lake Michigan.

Table 12-2
How to Determine Your Grid Location†

†For those geographical areas not encompassed here, a complete explanation appears in the April 1982, issue of *The Lunar Letter*, entitled "Worldwide QTH Locator System Proposed by Region 1," by Lance Collister, WA1JXN.

1st and 2nd characters: Read directly from the map.

3rd character: Take the number of whole degrees west longitude, and consult the following chart.

Degrees West Longitude	Third Character	Degrees West Longitude	Third Character	Degrees West Longitude	Third Character
60-61	9	88- 89	5	114-115	2
62-63	8	90- 91	4	116-117	1
64-65	7	92- 93	3	118-119	0
66-67	6	94- 95	2	120-121	9
68-69	5	96- 97	1	122-123	8
70-71	4	98- 99	0	124-125	7
72-73	3	100-101	9	126-127	6
74-75	2	102-103	8	128-129	5
76-77	1	104-105	7	130-131	4
78-79	0	106-107	6	132-133	3
80-81	9	108-109	5	134-135	2
82-83	8	110-111	4	136-137	1
84-85	7	112-113	3	138-139	0
86-87	6				

4th character: This number is the same as the *2nd single digit* of your latitude. For example, if your latitude is 41° N, the 4th character is 1; for 29° N, it's 9, etc.

This four-character (2-letter, 2-number) designator indicates your 2° × 1° square for VUCC award purposes.

Table 12-3
More Precise Locator

To indicate location more precisely, the addition of 5th and 6th characters will define the *sub-square*, measuring about 4 × 3 miles. Longitude-latitude coordinates on maps, such as U.S. Department of the Interior Surveys, can be extrapolated to the nearest tenth of a minute, necessary for this level of locator precision. *This is not necessary in the VUCC awards program.*

5th character: If your number of degrees longitude is an *odd* number, see Fig. A. If your number of degrees longitude is an *even* number, see Fig. B.

Odd Longitude° (Fig. A)

Minutes W. Longitude	5th Character
0- 5	L
5-10	K
10-15	J
15-20	I
20-25	H
25-30	G
30-35	F
35-40	E
40-45	D
45-50	C
50-55	B
55-60	A

Even Longitude° (Fig. B)

Minutes W. Longitude	5th Character
0- 5	X
5-10	W
10-15	V
15-20	U
20-25	T
25-30	S
30-35	R
35-40	Q
40-45	P
45-50	O
50-55	N
55-60	M

6th character: Take the number of *minutes of latitude* (following the number of degrees) and consult the following chart.

Minutes N. Latitude	6th Character
0- 2.5	A
2.5- 5.0	B
5.0- 7.5	C
7.5-10.0	D
10.0-12.5	E
12.5-15.0	F
15.0-17.5	G
17.5-20.0	H
20.0-22.5	I
22.5-25.0	J
25.0-27.5	K
27.5-30.0	L
30.0-32.5	M
32.5-35.0	N
35.0-37.5	O
37.5-40.0	P
40.0-42.5	Q
42.5-45.0	R
45.0-47.5	S
47.5-50.0	T
50.0-52.5	U
52.5-55.0	V
55.0-57.5	W
57.5-60.0	X

Fig 12-2—ARRL Grid Locator Map

WA6YBT's elegant station layout: a 2-meter multimode trans-ceiver, antenna tuner and HF rig, 2-meter and 70-cm transverter with a solid-state 70-cm amplifier on top of it, antenna rotator controls and a wattmeter. *(WA6YBT photo)*

If you own a computer, a program known as *GRIDLOC* will allow you to determine your grid square if you know your lati-tude and longitude. The program also works in reverse, provid-ing latitude and longitude for any grid square identifier. *GRIDLOC* is just one of 26 handy programs that are available on the *UHF/Microwave Experimenter's Software* diskette (3 1/2-inch, for IBM-PCs and compatibles). The diskette is avail-able from the ARRL for $10.

PROPAGATION

Normal Conditions

If you are new to the world above 50 MHz, you might won-der what sort of range is considered "normal." To a large ex-tent, your range on VHF is determined by your location and the quality of your station. After all, you can't expect the same performance from a 10-watt rig and a small antenna on the roof as you might from a kilowatt and stacked beams at 100 feet.

For the sake of discussion, consider a more-or-less "typi-cal" station. On 2-meter SSB, a hypothetical typical station would probably consist of a low-powered rig, perhaps a mul-timode rig (SSB/CW/FM), followed by a 100-watt amplifier. The antenna of our typical station might be a single long Yagi at around 50 feet, fed with low-loss coax.

Using SSB or CW, how far could this station cover on an average night? Location plays a big role, but it's probably safe to estimate that you could talk to a similarly equipped station about 200 miles away almost 100% of the time. Naturally, higher-power stations with tall antennas and low-noise receive preamps will have a greater range than this, up to a practical maximum of about 350-400 miles in the Midwest (less in the hilly West and East).

On 222 MHz, a similar station might expect to cover about the same distance, and somewhat less (150 miles) on 432 MHz. This assumes normal propagation conditions and a reasonably unobstructed horizon. This range is a lot greater than you would get for noise-free communication on FM, and it represents the sort of capability the typical station should seek. Increase the height of the antenna to 80 feet and the range might increase to 250 miles, probably more, depending on your location. That's not bad for reliable communication!

Band Openings and DX

The main thrill of the VHF and UHF bands for most of us is the occasional "band opening," when signals from far away are received as if they are next door. DX of well over 1000 miles on 6 meters is commonplace during the summer, and occurs at least once or twice a year on 144, 222 and 432 MHz.

DX propagation on the VHF/UHF bands is strongly influenced by the seasons. Summer and fall are definitely the most active times in the spectrum above 50 MHz, although band openings occur at other times as well.

The following is a review of the different types of "band openings," what bands they affect and how they may be predicted. Later on, how to operate these modes will be discussed. Remember that there is a lot of variation, and that no two band openings are alike. This uncertainty is part of what makes VHF/UHF interesting! Here is a list of the main types:

1) *Tropospheric—or simply "tropo "—openings.* Tropo is the most common form of DX-producing propagation on the bands above 144 MHz. It comes in several forms, depending on local and regional weather patterns. This is because it is caused by the weather. Tropo may cover only a few hundred miles, or it may include huge areas of the country at once.

a) *"Radiation Inversion"* (mostly summer). This common type of tropo is caused by the Earth cooling off in the evening. The air just above the Earth's surface also cools, while the air a few hundred meters above remains warm. This creates an inversion that refracts VHF/UHF radio signals. If there is little or no wind, you will notice a gradual improvement in the strength of signals out to a range of 100-200 miles (less in hilly areas) as evening passes into night. This form of tropo mostly affects the bands above 144 MHz, and is seldom noticeable at 50 MHz.

b) *Broad, regional tropo openings* (late summer and early fall). These are the DXer's dream! In one of these openings, stations as far away as 1200 miles (maybe more!) are brought into range on VHF/UHF. The broad, regional type of tropo opening is caused by stagnation of a large, slow-moving high-pressure system. The stations that benefit the most are often on the south or western sides (the so-called "back side") of the system. This sort of sluggish weather system may often be forecast just by looking at a weather map.

c) *Wave-cyclone tropo* (spring). These openings don't usually last long. They are brought about by an advancing cold front that interacts with the warm sector ahead of it. The resulting contrast in air temperatures may cause thunderstorms, but sometimes don't. If conditions are just right, a band opening may result. These openings usually involve stations in the warm sector ahead of the cold front. You may feel that there's a pipeline between you and a DX station 1000 miles or more away. You may not be able to contact anybody else!

There are several other types of tropo openings, but space doesn't allow more than a brief discussion here. Coastal breezes sometimes cause long, narrow tropo openings along the East and Gulf Coasts. Rarely, cold fronts may cause brief openings as they slide under warmer air. US West Coast VHFers are always on the alert for the California-to-Hawaii duct that permits 2500-mile DX. For a very complete discussion of all the major forms of weather-related VHF propagation, see "The Weather Which Brings VHF DX" by Emil Pocock, W3EP, May 1983 *QST.*

2) *The Scatter modes.* Long-distance communication on the VHF and lower UHF bands is possible using the ionosphere. Some modes are within the reach of modest stations, whereas others require very large antennas and high power. They all have one thing in common: They take advantage of the scattering of radio waves whenever there is an irregularity in the uppermost atmosphere or ionosphere. Very briefly, the main types are:

W9IP/2 in Western New York: 8 × 19-element RIW 432-MHz EME array fed with open-wire line.

a) *Meteor scatter.* When meteors enter the Earth's atmosphere, they ionize a small trail through the E layer. This ionization doesn't last long, often only a second. However, before it dissipates, the ionization can scatter, or sometimes reflect, VHF radio waves. This sort of propagation may be identified by the fact that it's very brief, but signals may be rather strong. It is quite common on 6 meters, less so on 2 meters, and rare indeed on 222 and 432 MHz. No successful QSOs have been completed at higher frequencies. Operating techniques for this mode are discussed later.

b) *Tropo scatter.* This takes place in the troposphere, and is the result of wave scattering from bundles of turbulent air. Signals are always weak, with a large amount of fading. Depending on frequency, tropo scatter may be assisted by scatter in the D and E layers of the ionosphere. Contacts out to about 1000 miles are possible, but difficult; definitely the territory of well-equipped stations.

3) *Sporadic E (often called "E_s " for "E sporadic").* This type of propagation is the most spectacular DX-producer on the 50-MHz band, where it may occur almost every day during late June and July and early August. A short E_s season also occurs during December-January. Sporadic E is most common in mid-morning and again around sunset, during the summer months, but it can occur any time, any date. E_s occurs once or twice a year on the 2-meter band in most areas.

E_s results from small patches of ionization in the E layer of the ionosphere. These patches move about, intensify or disappear altogether. The fast-moving patches of ionization appear to follow a general SE-NW track as they fly across the mid-latitude Northern Hemisphere.

Reflections from E_s patches make coast-to-coast DX possible on 6 meters. Single-hop distances of 1200-1400 miles are most common, but multi-hop openings are common on 6 meters. Multi-hop has also been observed on 2 meters. It is exceedingly rare, but it can provide the most spectacular 144-MHz DX available without moonbounce. QSOs above 2 meters are possible, but they are exceedingly rare.

E_s signals are usually quite strong, but they may fade away altogether within a short time. On 6 meters, some days are seemingly filled with E skip. Strong signals from all parts of the country can pour in for hours and hours, or sometimes only a small area is covered. However, it is also common to go through "dry periods" when there may be no E_s for a week or more, even in the middle of the summer. That's one reason they call it "sporadic" E!

Sporadic E openings usually last only about half an hour on 2 meters, but some have lasted much longer. They are much more rare on 2 meters than on 6. Probably fewer than one in 20 strong E_s openings on 6 meters ever materialize on 2 meters.

4) *Aurora ("Au")*. The aurora borealis, or "northern lights," is a beautiful spectacle which is seen occasionally by those who live in Canada, the northern part of the USA and Europe. Similar "southern lights" are sometimes visible in the southernmost parts of South America and Africa. The aurora is caused by the Earth intercepting a massive number of charged particles thrown from the Sun during a solar "storm." These particles are funnelled into the polar regions of the Earth by its magnetic field. As the charged particles interact with the upper atmosphere, the air glows, which we see as the aurora. And, important to VHFers, these particles also provide an irregular, moving curtain of ionization which can propagate signals for many hundreds of miles.

Like sporadic E, aurora is much more common on 6 meters than on 2 meters. Nevertheless, 2-meter Au is far more common than 2-meter sporadic E, at least above 40° N latitude. Au is also possible on 222 and 432 MHz, and many tremendous DX contacts have been made on these bands. Current record distances are over 1000 miles on 144, 222 and 432 MHz.

Aurora can be predicted by listening to radio station WWV at the National Institute of Science and Technology. At 18 minutes past each hour, WWV transmits a summary of the condition of the Earth's geomagnetic field. If the K index is 4 or above, you should watch for Au. However, as many VHFers have learned, a high K index is no guarantee of an aurora. Similarly, K indices of only 3 have occasionally produced spectacular radio auroras at middle latitudes. When in doubt, point the antenna north and listen!

Auroral DX signals are highly distorted. CW is the most practical mode, although SSB is sometimes used. Stations point their antennas generally northwards and listen for the telltale hissing note that is characteristic of auroral signals. More about operating Au is presented later.

5) *Transequatorial propagation (TE)*. This strange mode is responsible for almost unbelievable DX on 50 through 222 MHz. It is a strictly north-south path, crossing the equator. Stations which are equally spaced north and south of the geomagnetic equator have communicated over 4800 miles on 2 meters. Apparently some form of ionospheric ducting is responsible, although there is little agreement on the exact mechanism. TE is most common during and after periods of geomagnetic unrest. Signals have a weak, fluttery quality. This form of propagation is quite worthy of continued experimentation by amateurs.

6) *EME, or Earth-Moon-Earth, often called "moonbounce."* This is the ultimate VHF/UHF DX medium! Moonbouncers use the Moon as a passive reflector for their signals, and QSO distance is limited only by the diameter of the Earth. Any two stations who can simultaneously see the Moon may be able to work each other via EME. QSOs between the USA and Europe or Japan are commonplace on VHF and UHF by using this mode. That's DX!

Previously the territory of only the biggest and most serious VHFers, moonbounce has now become more widely popular. Thanks to the efforts of pioneer moonbouncers such as Bob Sutherland, W6PO, and Al Katz, K2UYH, hundreds of stations are active, mainly on 144 and 432 MHz. This huge increase in activity is largely the result of the "snowball" effect: More stations are EME-capable, so there's more incentive for others to "get on the Moon." Also, several individuals have assembled gigantic antenna arrays, which make up for the inadequacy of smaller antennas. The result is that even modestly equipped VHF stations (150 watts and one or two Yagis) are capable of making a few moonbounce contacts. Activity is constantly increasing. There is even an EME contest in which moonbouncers compete on an international scale.

Moonbounce requires larger antennas than most terrestrial VHF/UHF work. In addition, you must have a high-power transmitting amplifier and a low-noise receiving preamplifier if you wish to be able to work more than the biggest guns. A "typical" EME station on 144 MHz probably consists of four long-boom Yagi antennas on an azimuth-elevation mount (for pointing at the Moon), a kilowatt amplifier and a GaAsFET preamplifier. On 432 MHz, the "average" antenna is eight long Yagis. You can make contacts with a smaller antenna, but they will be with only larger stations on the other end. Yagi antennas aren't the only type available; collinears and quagis are also widely used. Several VHFers have also built large parabolic dish antennas.

HOW DO I OPERATE ON VHF/UHF?

Normal Conditions

The most important rule to follow, on VHF/UHF like all other amateur bands, is to listen first. Even on the relatively uncrowded VHF bands, interference is common near the calling frequencies. The first thing to do when you switch on the radio is to tune around, listening for activity. Of course, the calling frequencies are the best place to start listening. If you listen for a few minutes, you'll probably hear someone make a call, even if the band isn't open. If you don't hear anyone, then it's time to make some noise yourself!

Stirring up activity on the lower VHF bands is usually just a matter of pointing the antenna and calling CQ. Because most VHF beams are rather narrow, you might have to call CQ in several directions before you find someone. Several short CQs

WA8NLC's 17-foot dish for 1296-MHz EME.

We thought they only did this at airports! Jack, N3DQZ, deices his VHF and UHF antennas after a nasty winter storm.

are always more productive than one long-winded CQ. But don't make CQs too short; you have to give the other station time to turn the antenna toward you.

Give your rotator lots of exercise; don't point the antenna at the same place all the time. You never know if a new station or some DX might be available at some odd beam heading. VHFers in out-of-the-way locations, far from major cities, monitor the bands in the hope of hearing you.

Band Openings

How about DX? What is the best way to work DX when the band is open? There's no simple answer. Each main type of band opening or propagation mode requires its own techniques. This is natural, because the strength and duration of openings vary considerably. For example, you wouldn't expect to operate the same way during a 10-second meteor-burst QSO as during a three-day tropo opening.

The following is a review of the different main types of propagation and descriptions of the ways that most VHFers take advantage of them.

Tropo

Tropospheric propagation—usually called "tropo"—is usually most noticeable on the 144- through 1296-MHz bands. It often lasts at least several hours, and sometimes several days, so there's no big hurry to make contacts. You have time to listen carefully and determine how the opening is developing, how it is drifting and who is active. For stations on the West Coast, the openings usually extend up the coast and inland a couple of hundred miles or so. Therefore, the biggest question is how far north or south, and how far east, the opening will extend. In the middle part of the country and the East Coast, openings sometimes extend for well over a thousand miles, and in several directions.

[Recalling the ARRL VHF QSO Party in September 1979, in which the Midwest, South and East Coast experienced one of the greatest tropo openings in VHF history, my multioperator contest log shows literally dozens of QSOs with stations more than 750 miles away, and several over 1000 miles. The VHF bands were open over a huge area; within one hour, for example, stations in southern Texas and Vermont were worked from my QTH in Illinois on 2 meters. The 432-MHz band was almost as good.]

East of the Rockies, big tropo openings occur at least once

a year, although they aren't often as massive or long-lived as the September '79 opening. Unfortunately, amateurs in the mountainous West seldom experience tropo of this scale. In California, most openings are either along the coast or between the mainland and Hawaii. The latter, which accounts for the world's record tropo distances (greater than 2500 miles!) on VHF and several UHF bands, is rare.

To take best advantage of tropo openings, you need to know how they develop with time. Here are some tips: First, remember that most weather systems drift to the east with time. Large high-pressure systems tend to drift east and spread out as they mature. The long-haul DX may be available just before the opening terminates in your area. In other words, it may appear that the opening gets better and better and then all of a sudden you're out of it. If you're in the middle of the opening, sometimes stations will be "talking over your head." Be patient. Your time will come as the opening drifts east. Second, remember that most tropo-type openings improve as the evening progresses. If the opening is good and the weather is stable at sunset, then start the coffee pot, because you'll be DXing all night! In fact, the best tropo DX seems to come just around sunrise. But don't forget that if nobody is on the air, it may seem like the opening has ended. In the small hours of the morning, you might have trouble finding anyone on the air.

Tropo/Ionospheric Scatter

Tropo scatter usually requires prearranged schedules, a lot of patience and good equipment at each end of the path. It is often assisted by D-layer scattering, airplanes or even meteor scatter. Some form of tropo scatter is usable on all the VHF and UHF bands. The best means to take advantage of tropo scatter is to use medium-speed CW and follow a sequence in transmit-receive between the two stations; 30-second sequences seem to work best. One station transmits for the first 30 seconds of each minute, and the other station listens, then they trade-off. If you're lucky, a passing airplane or some other anomaly will briefly enhance signals enough to make the QSO.

Meteor Scatter

Meteor scatter is very widely used on 50 and 144 MHz, and it has been used on 222 and 432 as well. Operation with this exciting mode of DX comes under two main headings: prearranged schedules and random contacts. Either SSB or CW may be used, although SSB is more popular in North America. European meteor-scatter enthusiasts use very high speed CW (50-100 WPM) which is recorded on tape.

Most meteor-scatter work is done during major meteor showers, and many stations "run skeds" (prearranged schedules) with others in states or grid squares that they need. In a "sked," 15-second transmit-receive sequences are the norm for North America (Europeans use longer sequences). One station, almost always the westernmost, will "take the first and third." This means that he/she transmits during the first and third 15-second interval of each minute, and the other station transmits during the other two segments. This is a very simple procedure that ensures that only one station is transmitting when a meteor falls. See accompanying sidebar.

A specific frequency, far removed from local activity centers, is chosen when the sked is set up (on the HF bands or on the telephone). It is important that both stations have accurate frequency readout and synchronized clocks, but with today's technology this is not the big problem that it once was. Schedules normally run for $1/2$ hour or 1 hour. On 222 and 432 MHz, however, two-hour skeds are not uncommon because many fewer meteor trails are available at those frequencies. Meteor-scatter QSOs on 222 and 432 are well earned!

The best way to get the feel for the meteor-scatter QSO format is to listen to a couple of skeds between experienced operators. Then, look in *QST* or ask around for the name/call sign of veteran meteor-scatter operators in the 800-1000 mile range from you (this is the easiest distance for meteor scatter). Call them on the telephone or catch them on one of the national VHF nets and arrange a sked. After you cut your teeth on "easy" skeds, you'll be ready for more difficult DX.

A lot of stations make plenty of meteor-scatter QSOs without the help of skeds. Especially during major meteor showers, VHFers congregate near the calling frequency of each band. There they wait for meteor bursts like hunters waiting for ducks. Energetic operators call CQ while everyone else listens; when the big "Blue Whizzer" meteors blast in, the band comes alive with dozens of quick QSOs. For a brief time, normally five seconds to perhaps 30 seconds, 2 meters may sound like 20 meters! Then the band is quiet again. . . until the next meteor burst!

The quality of shower-related meteor-scatter DX depends on three factors. These three factors are well known or can be predicted. The most important is the "radiant effect." This refers to where the shower's radiant is located in the sky. The radiant is the spot in the sky from which the meteors appear to fall. If the radiant is below the horizon, or too high in the sky, you will hear very few meteors. The most productive spot for the radiant is at an elevation of about 45° and an azimuth of 90° from the path you're trying to work. The second important factor is the velocity of the meteors. Slow meteors cannot ionize sufficiently to propagate 144- or 222-MHz signals, no matter how many meteors there are. For 144 MHz, meteors slower than 50 km/s are usually inadequate (see accompanying sidebar detailing major meteor showers). Third, the shower will have a peak in the number of meteors that the Earth intercepts. However, because the peak of many meteor showers is more than a day in length, the exact time of the peak is not as important as most people think. Two interesting references are "Improving Meteor Scatter Communications" by Joe Reisert, W1JR, in June 1984 *Ham Radio*, and "VHF Meteor Scatter: an Astronomical Perspective," by M. R. Owen, W9IP, in June 1986 *QST*.

It takes a lot of persistence and a good station to be successful with "random" meteor scatter. This is mainly because you must overcome tremendous QRM in addition to the fluctuations of meteor propagation. At least 100 watts is necessary for much success in meteor scatter, and a full kilowatt will help a lot. One or two Yagis, stacked vertically, is a good antenna system.

In populated areas, it can be difficult to hear incoming meteor-scatter DX if many local stations are calling CQ. Therefore, many areas observe 15-second sequencing for random meteor-scatter QSOs, just as for skeds. Those who want to call CQ do so at the same time so everyone can listen for responses between transmissions. Sometimes a bit of peer pressure is necessary to keep everyone together, but it pays off in more QSOs for all. The same QSO format is used for scheduled and "random" meteor-scatter QSOs.

Sporadic E

Sporadic E (E_s) propagation is very common on 6 meters during the summer and very early fall, and rare at other times. It is always rare on 2 meters, but the greatest chance of it happening is during July and early August. On 6 meters, the band

Meteor-Scatter Procedure

In a meteor-scatter QSO, neither station can hear the other except when a meteor trail exists to scatter or reflect their signals. The two stations take turns transmitting so that they can be sure of hearing the other if a meteor happens to fall. They agree beforehand on the sequence of transmission. One station agrees to transmit the 1st and 3rd 15 seconds of each minute, and the other station takes the 2nd and 4th. It is standard procedure for the westernmost station to transmit during the 1st and 3rd.

It's important to have a format for transmissions so you know what the other station has heard. This format is used by most US stations:

Transmitting	Means you have copied:
Call signs	nothing, or only partial calls
Calls plus signal report (or grid square or state)	full calls—both sets
ROGER plus signal report (or grid square or state)	full calls, plus signal report (or grid square or state)
ROGER	ROGER from other station

Remember, for a valid QSO to take place, you must exchange full call signs, some piece of information, and acknowledgment. Too many meteor QSOs have not been completed for lack of ROGERs. Don't quit too soon; be sure the other station has received your acknowledgment. Often, stations will add "73" when they want to indicate that they have heard the other station's ROGER.

Until a few years ago, it was universal practice to give a signal report which indicated the length of the meteor burst. "S1" meant that you were just hearing pings, "S2" meant 1-5 second bursts, and so on. Unfortunately, virtually everyone was sending "S2," so there was no mystery at all, and no significant information being exchanged.

Grid squares have become popular as the "piece of information" in meteor-scatter QSOs. More and more stations are sending their grid-square designator instead of "S2." This is especially true on "random" meteor-scatter QSOs, where you might not know in advance where the station is located. Other stations prefer to give their state instead. For an excellent summary of meteor-scatter procedure, see "Meteor Scatter Communications" by Clarke Greene, K1JX, in January 1986 *QST* (pp 14-17).

may open for hours on end, constantly shifting, disappearing briefly, then reappearing. Signals are often very strong, and 10-watt stations can work 1000 miles with ease. On 2 meters, the openings are equally strong but often very short; a half-hour sporadic E opening on 2 meters is rare but wonderful.

Predicting these openings in advance is very difficult, if not impossible. A very weak correlation may exist between sporadic E and large thunderstorms, but this might be because big storms and E skip are both more common during the summer. The best way to alert yourself to the possibility of sporadic E is to monitor lower frequencies, because this type of propagation usually rises in frequency. An opening may begin as "short skip" on 10 meters, then rise to include 6 meters. Then, distant FM broadcast and TV stations will become audible as the maximum frequency of the E cloud reflections goes higher. Then, if you're very lucky indeed, the E cloud will create an opening on 2 meters. If you are monitoring 6 meters, you will notice that the skip distance gets shorter and shorter as the E_s MUF (maximum usable frequency) goes higher. If you hear loud 6-meter signals from 400 or fewer miles away, watch 2 meters for some action.

You'll need to be quick on your toes to take advantage of sporadic E. You might not even notice an E_s band opening in

Meteor Showers

Every day, the Earth is bombarded by billions of tiny grains of interplanetary debris, called meteors. They create short-lived trails of E-layer ionization which can be used as reflectors for VHF radio waves. On a normal morning, careful listeners can hear about 3-5 meteor pings (short bursts of meteor-reflected signal) per hour on 2 meters.

At several times during the year, the Earth passes through huge clouds of concentrated meteoric debris, and VHFers enjoy a "meteor shower." During meteor showers, 2-meter operators may hear 50 or more pings and bursts per hour. Here are some data on the major meteor showers of the year. Other showers also occur, but they are very minor.

Major Meteor Showers

Shower	Date range	Peak date	Time above quarter max	Approximate visual rate	Speed km/s	Best Path	Time (local)
Quadrantids	Jan 1-6	Jan 3/4	14 hours	40-150	41	NE-SW;SE-NW	13-15;05-07
Eta Aquarids	Apr 21-May 12	May 4/5	3 days	10-40	65	NE-SW;E-W;SE-NW	05-07;06-09;09-11
Arietids	May 29-Jun 19	Jun 7	?	60	37	N-S	06-07 & 13-14
Perseids	Jul 23-Aug 20	Aug 12	4.6 days	50-100	59	NE-SW;SE-NW	09-11;01-03
Orionids	Oct 2-Nov 7	Oct 22	2 days	10-70	66	NE-SW;N-S;NW-SE	01-03;01-02 & 07-09; 07-08
Geminids	Dec 4-16	Dec 13/14	2.6 days	50-80	34	N-S	22-24 & 05-07

Meteor-Shower Peaks

These calculated predictions of meteor-shower peak times/dates are based on the latest astronomical data. Nevertheless, meteor showers are notoriously fickle about meeting their predicted peaks. Therefore, don't put too much faith in these estimates. The time at which the shower's radiant is at a 45° elevation, perpendicular to the path you want to work, is much more important than the actual peak of the shower. Best times and paths are listed above. These values are calculated for Kansas City and are valid for the entire USA because the time indicated is local standard time. One hour before and after the given time ranges should generally be good as well.

Meteor Shower Peak Times (UTC): 1995-1998

Note: The first line indicates month-day (eg, 1-3 is January 3) and the second line (eg, 2330) indicates UTC.

Shower	1995	1996	1997	1998	Shower	1995	1996	1997	1998
Quadrantids	1-4	1-4	1-3	1-3	Perseids	8-13	8-12	8-12	8-12
	0100	0700	1300	1900		0030	0700	1300	1900
Eta Aquarids	5-6	5-5	5-6	5-6	Orionids	10-22	10-21	10-22	10-22
	1530	2130	0400	1000		1230	1900	0100	0700
Arietids	6-7	6-7	6-7	6-7	Geminids	12-14	12-13	12-13	12-14
	2100	0300	0900	1500		0700	1300	1930	0130

progress because signals are so strong ("DX is never that loud!"). Quick, brief calls and short exchanges are needed if you are to make the most of the opening. Don't spend a half hour calling a long-winded CQ. The opening will be gone by the time you're through. Once the E_s opening is in progress, most stations don't waste a lot of time hunting around. Just work everything you can hear, and quickly. Almost all stations will be DX, so you can't lose!

Don't worry if you aren't running a lot of power and a huge antenna; sporadic E is the greatest equalizer in VHF. The size of your station hardly matters at all. It's mostly location that counts! You may be having the greatest opening of all time, and your friend a few miles away might hear little or nothing. That's the way E_s is. Next week, your buddy may be the lucky one and you will be left out, so make the most of it!

Aurora

Aurora (Au) is a propagation mode which favors stations at high latitudes. It is a wonderful blessing for those who must suffer through long, cold winters because other forms of propagation are rare during the winter. Aurora can come at almost any time of the year. New England stations get Au on 2 meters about five to 10 times a year, whereas stations in Tennessee get it once a year if they're lucky. Central and Southern California almost never see aurora.

The MUF of the aurora seems to rise quickly, so don't wait for the lower-frequency VHF bands to get exhausted before moving up in frequency. Check the higher bands right away.

You'll notice Au by its characteristic hiss. Signals are distorted by reflection and scattering off the rapidly moving curtain of ionization. They sound like they are being transmitted by a leaking high-pressure steam vent rather than radio. SSB voice signals are so badly distorted that often you cannot understand them unless the speaker talks very slowly and clearly. The amount of distortion increases with frequency. Most 50-MHz Au contacts are made on SSB, where distortion is the least. On 144, 222 and 432 MHz, CW is the only really useful means of communicating via AU.

If you suspect Au, tune to the CW calling frequency (144.100) or the SSB calling frequency (144.200) and listen with the antenna to the north. Maybe you'll hear some signals. Try swinging the antenna as much as 45° either side of due north to peak signals. In general, the longest-distance DX stations peak the farthest away from due north. Also, it is possible to work stations far south of you by using the aurora; in that case your antenna is often pointed north.

High power isn't necessary for aurora, but it helps. Ten-watt stations have made Au QSOs but it takes a lot of perseverance. If you have low power, try calling only the loudest stations you hear. The weak ones probably won't be able to hear you. Increasing your power to 100 watts will greatly improve your chances of making Au QSOs. As with most short-lived DX openings, it pays to keep transmissions brief.

Aurora openings may last only a few minutes or they may last many hours, and the opening may return the next night, too. If WWV indicates a geomagnetic storm, begin listening on 2 meters in the late afternoon. Many spectacular Au openings begin before sunset and continue all evening. If you get the feeling that the Au has faded away, don't give up too soon. Aurora has a habit of "dying" and then returning several times, often around midnight. And remember: If you experience a terrific Au opening, look for an encore performance about 27-28 days later, because of the rotation of the sun.

Propagation Indicators

Active VHFers keep a careful eye on various propagation indicators to tell if the VHF bands will be open. The kind of indicator you monitor is related to the kind of propagation you are expecting. For example, during the summer it pays to watch closely for sporadic E because openings may be very brief. If your area only gets one or two per year on 2 meters, you sure don't want to miss them!

Many forms of VHF/UHF propagation develop first at low frequency and then move upward to include the higher bands. Aurora is a good example. Usually it is heard first on 10 meters, then 6 meters, then 2 meters. Depending on your location and the intensity of the aurora, the time delay between hearing it on 6 meters and its appearance on 2 meters may be only a few minutes, to as much as an hour, later. Still, it shows up first at low frequency. The same is true of sporadic E; it will be noticed first on 10 meters, then 6, then 2.

Tropospheric propagation, particularly tropo ducting, acts in just the reverse manner. Inversions and ducts form at higher frequencies first. As the inversion layer grows in thickness, it refracts lower and lower frequencies. However, most of us don't notice this because amateur activity drops off with higher frequencies, and atmospheric losses increase. So, it may be true that the band opens high in frequency first, but it may appear to open first on 2 meters because that is where much of the country's VHF activity is. Six-meter tropo openings are

Charles Suckling, G3WDG, (left) operating 10-GHz EME with Kent Britain, WA5VJB, using Kent's 3-meter dish.

very rare because few inversion layers ever develop sufficient thickness to enhance such long-wavelength signals.

How can you monitor for band openings? The best way is to take advantage of the "beacons" which are inadvertently provided by commercial TV and FM broadcast stations. Television Channels 2 through 6 (54-88 MHz) are great for catching sporadic E. As the E opening develops, you will see one (or more) stations appear on each channel. First, you may see Channel 2 get cluttered with stations 1000 miles or more away, then the higher channels may follow. If you see strong DX stations on Channel 6, better get the 2-meter rig warmed up. Channel 7 is at 175 MHz, so you rarely see any sporadic E there or on any higher channels. If you do, however, it means that a major 2-meter opening is in progress!

The gap between TV Channel 6 and 7 is occupied partly by the FM broadcast band (88-108 MHz). Monitoring that spectrum will give you a similar "feel" for propagation conditions.

Several amateurs have built converters to monitor TV video carrier frequencies. A system like this can be used to keep track of meteor showers. It's a way of checking to tell if meteors are plentiful or not, even if there appears to be little activity on the amateur VHF bands. It also can alert you to aurora and sporadic E. A variety of systems for monitoring propagation have been discussed by Joe Reisert, W1JR, in the June 1984 issue of *Ham Radio*.

EME: Earth-Moon-Earth

EME, or moonbounce, is available any time the Moon is in a favorable position. Fortunately, the Moon's position may be easily calculated in advance, so you always know when this form of DX will be ready. It's not actually that simple, of course, because the Earth's geomagnetic field can play havoc with EME signals as they are leaving the Earth's atmosphere and as they return. Not only can absorption (path loss) vary, but the polarization of the radio waves can rotate, causing abnormally high path losses at some times. Still, most EME activity is predictable.

You will find many more signals "off the Moon" during times when the Moon is nearest the Earth (perigee), when it is overhead at relatively high northern latitudes ("positive declination"), and when the Moon is nearly at Full phase. The one weekend per month which has the best combination of these three factors is informally designated the "activity" or "skeds" weekend, and most EMEers will be on the air then. This is particularly true for 432 EME; 144-MHz EME is active during the week and on non-skeds weekends as well. See accompanying sidebar on EME operating practices.

Hilltopping and Portable Operation

One of the nice things about the VHF/UHF bands is that antennas are relatively small, and station equipment can be packed up and easily transported. Portable operation, commonly called "hilltopping" or "mountaintopping," is a favorite activity for many amateurs. This is especially true during VHF and UHF contests, where a station can be very popular by being located in a rare grid square. If you are on a hilltop or mountaintop as well, you will have a very competitive signal. See accompanying sidebar for further information.

Hilltopping is fun and exciting because hills elevate your antenna far above surrounding terrain and therefore your VHF/UHF range is greatly extended. If you live in a low-lying area such as a valley, a drive up to the top of a nearby hill or mountain will have the same effect as buying a new tower and antenna, a high-power amplifier and a preamplifier, all in one!

Moonbounce Operating Practices

After traveling 400,000 km, bouncing off a poorly reflective Moon, and returning 400,000 km, EME signals are quite weak. A large antenna, high transmitter power, low-noise preamplifier and very careful listening are all essential for EME. Nevertheless, hundreds of amateurs have made EME contacts, and their numbers are growing.

The most popular band for EME is 144 MHz, followed by 432 MHz. Other bands with regular EME activity are 1296 and 2304/2320 MHz.

Moonbounce QSO procedure is different on 144 and 432 MHz. On 144 MHz, schedule transmissions are 2 minutes long, whereas on 432 they are 2½ minutes long. In addition, the meaning of signal reports is different on the two bands. This difference in procedure is somewhat confusing to newcomers. Most EMEers operate on only one band, so they grow accustomed to the procedure on that band pretty quickly.

Signal Report	Meaning	
	144 MHz	*432 MHz*
T	Signal just detectable	Portions of calls copyable
M	Portions of calls copyable	Complete calls copied
O	Both calls fully copied	Good signal, easily copied
R	Both calls, and "O" signal report copied	Calls and report copied

What does this difference in reporting mean? Well, the biggest difference is that "M" reports aren't good enough for a valid QSO on 2 meters but they are good enough on 432. So long as everyone understands the system, then there is no confusion. On both bands, if signals are really good, then normal RST reports are exchanged.

The majority of EME QSOs are made without any prearranged schedules. Almost always, there is no rigid transmitting time-slot sequencing. Just as in CW QSOs on HF, you transmit when the other station turns it over to you. This is particularly true during EME contests, where time-slot transmissions slow down the exchange of information. Why take 10-15 minutes when two or three will do?

On the other hand, many QSOs on EME are made with the assistance of "skeds." This is especially the case for newcomers and rare DX stations.

On 144 MHz, each station transmits for 2 minutes, then listens as the other station transmits. Which 2-minute sequence you transmit in is agreed to in advance. The common terminology is "even" and "odd." The even station transmits the 2nd, 4th, 6th, 8th...2-minute segment of the hour, while the odd station transmits the others (1st, 3rd, 5th...). Therefore, 0030-0032 is an "even" slot, even if the sked begins on the half hour. In most cases, the westernmost station takes the odd sequence. An operating chart, similar to many which have been published, will help you keep track:

Transmitting Sequence

Odd (eastern)	Even (western)
00-02	02-04
04-06	06-08
08-10	10-12
12-14	14-16
16-18	18-20
20-22	22-24
24-26	26-28
28-30	30-32
32-34	34-36
36-38	38-40
40-42	42-44
44-46	46-48
48-50	50-52
52-54	54-56
56-58	58-00

During the schedule, at the point when you've copied portions of both calls, the last 30 seconds of your 2-minute sequence is reserved for signal reports; otherwise, call sets are transmitted for the full 2 minutes.

On 432 MHz, sequences are longer. Each transmitting slot is 2½ minutes. You either transmit "first" or "second." Naturally, "first" means that you transmit for the first 2½ minutes of each 5 minutes.

144-MHz Procedure—2-Min Sequence

Period	1½ minutes	30 seconds
1	Calls (W6XXX DE W1XXX)	
2	W1XXX DE W6XXX	T T T T
3	W6XXX DE W1XXX	O O O O
4	RO RO RO RO	DE W1XXX K
5	R R R R R	DE W6XXX K
6	QRZ? EME	DE W1XXX K

432-MHz Procedure—2½-Min Sequence

Period	2 minutes	30 seconds
1	VE7BBG DE K2UYH	
2	K2UYH DE VE7BBG	
3	VE7BBG DE K2UYH	T T T
4	K2UYH DE VE7BBG	M M M
5	RM RM RM RM	DE K2UYH K
6	R R R R R	DE VE7BBG SK

On 222 MHz, some stations use the 144-MHz procedure and some use the 432-MHz procedure. Which one is used is determined in advance. On 1296 and above, the 432 procedure is always used.

EME operation generally takes place in the lowest parts of the VHF bands: 144.000-144.070; 222.000-222.025; 432.000-432.070; 1296.000-1296.050. Terrestrial QSOs are strongly discouraged in these portions of the bands.

The popularity of hilltopping has grown as more stations acquire solid-state rigs and amplifiers. The "box and brick" (compact multimode VHF rig and solid-state 100-200 watt amplifier) combination is ideally suited for mobile and portable operation. You need no other power source than a car battery, and even with a simple antenna your signals will be outstanding.

Many VHF/UHFers drive to the top of hills or mountains and set up their station. A hilltop park, rest area or farmer's field are equally good sites, so long as they are clear of trees and obstructions. You should watch out for high-power FM or TV broadcasters who may also be taking advantage of the hill's good location; their powerful signals may cause inter-modula-tion problems in your receiver.

Antennas, on a couple of 10-foot mast sections, may be turned by hand as the operator sits in the passenger seat of the car. A few hours of operating like this can be wonderfully enjoyable and can net you a lot of good VHF/UHF DX. In fact, some VHF enthusiasts have very modest home stations but rather elaborate hilltopping stations. When they notice that band conditions are improving, they hop in the car and "head for the hills." There, they have a really excellent site and can make many more QSOs.

In some places, there are no roads to the tops of hills or mountains where you might wish to operate. In this case, it is a simple (but sometimes strenuous) affair to hike to the top, carrying the car battery, rig and antenna. Many hilltoppers have had great fun by setting up on the top of a firetower on a hilltop, relying on a battery for power.

Some VHF/UHFers, especially contesters, like to take the entire station, high power amplifiers and all, to hilltops for extended operation. They may stay there for several days, camping out and DXing. Probably the most outstanding example of this kind of operation is put on regularly by the group at W2SZ/1. Dozens of operators and helpers assemble this multiband station, often with moonbounce capability, and operate major VHF and UHF contests from Mount Greylock in Western Massachusetts. Their winning scores in virtually every VHF contest they have entered testifies to their skill and the effectiveness of hilltopping!

Contests

The greatest amount of activity on the VHF/UHF bands occurs during contests. VHF/UHF contests are scheduled for some of the best propagation dates during the year. Not only are propagation conditions generally good, but activity is always very high. Many stations "come out of the woodwork" just for the contest, and many individuals and groups go hilltopping to rare states or grid squares.

A VHF/UHF contest is a challenging but friendly battle between you, your station, other contesters, propagation and Murphy's Law. Your score is determined by a combination of skill and luck. It is not always the biggest or loudest station that scores well. The ability to listen, switch bands quickly and to take advantage of rapidly changing propagation conditions is more important than brute strength.

There are quite a few VHF/UHF contests. Some are for all the VHF and UHF bands, while others are for one band only. Some run for entire weekends and others for only a few hours. Despite their differences, all contests share a basic similarity. In all North American contests, your score is determined by the number of contacts (or more precisely, the number of QSO points) you make, multiplied by some "multiplier," which may be ARRL DXCC countries, grid squares, call-sign prefixes or some other factor. In most of the current major contests, you keep track of QSOs and multipliers by band. In other words, you can work the same station on each band for separate QSO and multiplier credit.

Listed below are the major contests of the VHF/UHF realm in North America. Other contests are popular in Europe and Asia, but these are not listed here because information about them is not usually available to most of us. All the listed contests are open to all licensed amateurs, regardless of their affiliation with any club or organization. Detailed rules for each contest are published in *QST*, *CQ* and the newsletters of the sponsoring organizations. For more information about contesting, see Chapter 7.

1) *ARRL VHF Sweepstakes*: This one-weekend contest occurs in January. It is favored by clubs because it permits club members to pool their individual scores for the club's total. In addition, individuals and multioperator groups compete. Scores are determined by total QSO points per band multiplied by the number of grid squares worked per band.

2) *ARRL Spring Sprints*: These contests, as their names imply, are short: Only 4 hours. There is a separate Sprint for each of the major VHF and UHF bands. Each one takes place on a different evening, a little more than a week apart. The dates vary, but are usually in the spring. The evenings for the contests are chosen to coincide with each band's activity night. So, the 2-meter Sprint takes place on a Monday, the 222-MHz Sprint on

CT1WW is one of the most widely heard 6-meter European stations in the US. Tiago Frederico's shack holds antennas for several VHF bands and is also the location of the CT0WW 6-meter beacon. At 800 meters above sea level, Tiago has a clear shot in all directions.

Tuesday the next week, the 432-MHz Sprint on Wednesday a week later than that, 902 MHz on the Friday after that, and the 1296-MHz Sprint on the following Thursday. The 6-meter Sprint usually is held on a Saturday at the end of the sequence, to place it closer to the sporadic E season. Your score is the number of QSO points multiplied by the number of grid squares worked in the Sprint's 4-hour time period. Sprint scores are reported in *NCJ*, the *National Contest Journal* published by the ARRL.

3) *ARRL June VHF QSO Party*: This contest is the highlight of the contest season for most VHFers. Scheduled for the second full weekend of June, the "June Contest" sees the most activity of any North American VHF contest. Conditions on all the VHF bands are usually good, with 6 meters leading the way. This contest covers all the VHF and UHF bands, and QSOs and multipliers are accumulated for each band. Your score is determined by multiplying QSO points per band by grid squares per band. Singleband and multiband awards are given to high-scoring individuals in each ARRL section. In addition, multioperator groups compete against each other (and the competition can be fierce!).

4) *CQ Magazine Worldwide VHF WPX Contest*: This contest is based upon call sign prefixes as multipliers and is international in scope. Your score is the number of different prefixes worked per band times the number of QSO points per band. All VHF bands and UHF bands through 1296 MHz are permitted. A multitude of awards are available, based on power level, number of operators and number of bands employed.

5) *ARRL UHF Contest*: As its name suggests, this contest is restricted to the UHF bands (plus 222 MHz). It takes place over a full weekend in August. All UHF bands are permitted, and grid squares are the multiplier. Less equipment is involved for this contest than for contests which cover all VHF/UHF bands, so many groups go hilltopping for the UHF Contest.

6) *ARRL September VHF QSO Party*: The rules for this contest are identical to those for the June VHF QSO Party. This contest is also very popular, and many multioperator groups travel to rare states and grid squares for it. By the second weekend of September, most 6-meter sporadic E has disappeared, but tropo conditions are often extremely good. Therefore, the bands above 144 MHz are the scene of tremendous activity during the September contest. Some of the best tropo openings of recent decades have taken place during this contest, and they are made even better by the high level of activity.

7) *ARRL EME Contest*: This contest is devoted to

VHF Mountaintopping for the '90s

Recently, VHF has seen a resurgence of an activity as old as VHF itself—mountaintopping. Few are blessed with a home operating location that facilitates a total command of the frequency. Consequently, ardent VHFers construct bigger and bigger arrays and amplifiers for the home station in order to produce that booming signal. Some, however, utilize the "great equalizer" to compete on an equal or superior footing with the big home stations—namely, a mountaintop location. Here, perched high above all the home stations, a small portable rig with a single Yagi antenna only a few feet off the ground suddenly sounds like a kilowatt feeding a killer antenna installed at home. Simple equipment performs amazingly well from a mountaintop QTH on VHF.

A mountaintop expedition can vary from a spur-of-the-moment Sunday afternoon picnic to a full-fledged weekend contest. Quick trips can also be conducted during band openings. Since a contest optimizes the opportunity to work a lot of stations on VHF, this sidebar is mostly a "how-to" for weekend contest operation conducted by one or two people. But this can be scaled down to a mountaintop stay of shorter duration.

The Times Are a Changin'

Old-timers will remember the drudgery of lugging "boat anchors" up rocky crevices. Dragging equipment and generators weighing hundreds of pounds up steep mountainsides was no picnic. Those who suffered sprained backs soon gave it up. The advent of solid-state equipment, however, has made mountaintopping a far less strenuous activity. With a greater selection of such compact commercially available equipment, VHF has become a hotbed of activity. Even some of the highly competitive HF types have found new worlds to conquer above 50 MHz. A key factor in this revival has been the introduction of a worldwide grid locator system—now much in vogue of VHF. The use of grid squares in the major VHF contests has tickled the innermost secret desire of every radio amateur—to be on the receiving end of a DX pileup. Now instead of going on safari to some distant DX land, you can head for the mountains—some nearby mountaintop located in a rare grid square.

Choosing a Site

Choosing a mountaintop site involves considering how far you want to travel to get there, accessibility to the top of the mountain and its all important grid-square location. Ideally, your mountain is only a short driving distance away, towers into the cirrusphere, has a six-lane interstate to the top and rises within a grid square that has never been on the air before!

Obviously, some of these considerations may have to be compromised. Your first step to finding VHF heaven involves extensive study of a road atlas. How far do you want to travel? Where are the mountains? How high are they? Can you drive to the top? Draw in the grid line boundaries so you can tell which square it is in. Ask some active VHFers which are the difficult squares to work. When you start zeroing in on a potential site, you may want to get a topographic survey map of the area to determine access roads and direction of "drop-off" from the summit.

I've never operated from a mountaintop without first scoping it out in person. Access is most important. Thus far, I've operated from sites which I have reached by car, ferry, gondola, 4-wheel drive, motor home and hiking. Unless you are going with mini-radios and gel-cells, you want to get there without backpacking it. A passable road to the top is ideal. When checking out a potential site, bring a compass and 2-meter FM hand-held. A call on 146.52-MHz simplex should tell you how good the location is. Are you blocked in any direction? Is it already "RF-city" with commercial installations—a potential source of interference? Will you be able to clear any trees with a lightweight mast? Then the prime requisite: Is there a picnic table permanently at the site? If not, plan on bringing an operating table and chair—which adds considerable bulk and weight to transport.

Once you've selected an operating site, be sure you have secured the necessary permission to use the site. This may simply require verbal permission from some authority or the owner, or, it could involve a lengthy exchange of correspondence with a state environmental agency of forests and parks and the signing of a liability release. But be sure you have permission. The last thing you want is the local sheriff shining a flashlight in your eyes at 3 AM, rousting you out of a fantastic tropo opening on 2 meters. You'll find rangers on fire watch most helpful in pointing out how to obtain necessary permission.

Power Source

Unless you are awfully lucky to find a location that will permit you to just "plug in," make plans for providing your own power. With a single-band operation from a car (with antenna mast mounted just outside the car window), the car battery will probably suffice. Run a set of heavy-duty jumper cables directly to the car battery. Even a solid-state "brick" amplifier can be run off the car battery without ill effects. Just in case, park the car facing downhill!

For a more serious effort utilizing several VHF and UHF bands, a small generator is recommended. If the word generator conjures up an image of an ugly engine block from a 1947 LaSalle, then tune in to the modern world. Small, even attractive, generators in the 500-1000 W category, that look more like American Tourister luggage, are now available. Mine is a 650-watt beauty that weighs in at 43 pounds, and runs for four hours on a half gallon of petrol. And quiet? You can hear the wings of a Monarch butterfly flutter at 50 paces. A not-filled 5-gallon gerry-can provides more than enough flammable juice for a contest weekend.

Equipment

A mountaintop location effectively places your antenna atop a natural tower of hundreds or perhaps thousands of feet. With this height advantage, compact, lightweight, low-powered radios that can be boosted up to the 50- to 100-W range with solid-state amplifiers will perform nicely. Low-powered portable transceivers are manufactured for just this purpose. The popular 10- and 25-watt multimode rigs are also quite adequate. Many discontinued models can be obtained at a substantial savings through the Ham Ads section of *QST*. Use of transverters operating with mobile-type HF radios should also be considered.

If you don't have any sizable trees to get over, you can use simple mast sections that fit together. I use 5-foot sections available at the popular shack of radios. They are easily transportable. Important too is the method of antenna rotation. If you can install the antenna mast right next to the operating position, do it. It will save all the hassle of installing motorized rotators. Nothing beats the "Armstrong" method for speed and simplicity. I use a cross-piece of aluminum tubing mounted with U-bolts to the mast at arm level. See Fig 12-3. This provides instantaneous antenna-peaking capability—a necessity on V/UHF. While home stations are twirling their antennas in every direction trying to peak a weak signal, I've already worked him!

Installing antennas for several bands on the same mast is recommended. They should be oriented in the same direction. Many contacts on UHF are the result of moving stations over from other bands. For example, in a contest if you move a multiplier to 432 MHz after first working on 2 meters, and both antennas are on the same mast, you will first want to peak the signal on 144 MHz. Then, when you move to 432 MHz where the antennas are probably a bit more sharp and propagation perhaps marginal, both antennas will be pointing at each other for maximum signal. This can make the difference in whether the contact is made.

Fig 12-3—A closeup view of the "Armstrong" rotator, used for quick peaking of signals.

If your mountaintop operation involves staying overnight, additional attention must be paid to having the proper survival equipment. The most luxurious way to go is a van or RV. Otherwise, a tent will be required. For the rugged outdoors type, this can be as appealing as the radio part. I find that cooking steaks over a campfire with a canopy of stars overhead (while a programmable keyer is calling CQ) is half the fun. But keep this aspect of the operation also

as simple as possible. Champagne and caviar can be held for another time. I've also found out the hard way that one can expect heavy winds on mountaintops. Large tents blow down easily in such weather.

Further on the subject of weather, just because you're topping it in July or August, don't expect it will always be T-shirt and shorts weather. No matter what the season, expect to need a heavy jacket after dark. I always bring a heavy flannel shirt and ski jacket for night, and shorts in the daytime. And bring lightweight raingear, just in case. And depending on the habitat, don't be surprised to be introduced to a critter or two, especially after dark!

Getting Started

Okay, you've read this far and are beginning to say to yourself: "Self, I think I'd like to try that." But there is a little voice of caution in you that says: "Don't go bonkers until you've sampled a little first." Good advice!

Start out by setting up on an easily accessible mountain for an afternoon during a contest period on a single band. For a first effort, I recommend 2 meters. With so many 10-W multimode rigs out there in radioland, 2 meters is your "bread and butter" band. Using a multi-element Yagi a few feet off the ground of a strategically located mountain or hill can whet your appetite. I first got hooked by operating from the side of a highway on Hogback Mountain, Vermont, with a 3-W portable 2-meter radio to a 30-W brick and 11-element Yagi. I was astounded by the results, with contacts hundreds of miles away. This launched my interest in acquiring more equipment for portable mountaintop use, each operation adding a new band or better antenna. The basic formula of keeping it lightweight and simple has prevailed, however.

Now what's holding you back from operating from Mount Everest?—*John F. Lindholm, W1XX*

A Checklist of Typical Items Needed for a Weekend Portable Mountaintop Operation

Radio Equipment
- [ ] Transceivers for each band
- [ ] Solid-state amplifiers
- [ ] Antennas
- [ ] Coax (Belden 9913 or equiv)
- [ ] Keyer
- [ ] Paddle
- [ ] Antenna masts
- [ ] Earphones
- [ ] Coax connector cables
- [ ] Coax adapters
- [ ] dc cables with plugs
- [ ] Power supply (ies)
- [ ] Fuses
- [ ] Antenna rotator crosspiece
- [ ] Microphones
- [ ] Key line with plug
- [ ] Multi-ac plug outlet
- [ ] SWR meter
- [ ] DVM
- [ ] Multi-dc plug box
- [ ] Clip leads

Tools
- [ ] Wrenches
- [ ] Pliers
- [ ] Screwdrivers
- [ ] Hammer

Power Source
- [ ] Generator
- [ ] Power cable
- [ ] Gasoline
- [ ] Oil
- [ ] Jumper cables (if on car battery)
- [ ] Gas funnel

Personal
- [ ] Toothbrush/toothpaste
- [ ] Soap
- [ ] Towel
- [ ] Suntan lotion
- [ ] Change of clothes
- [ ] Rain gear
- [ ] Hat
- [ ] Warm jacket
- [ ] Alarm clock
- [ ] Bug spray
- [ ] Insect repellant
- [ ] Electric shaver
- [ ] Toilet paper

Camping Equipment
- [ ] Matches
- [ ] Tent stakes (enough for tent and masts)
- [ ] Tarpaulin canopy
- [ ] Tent
- [ ] Ground cloth
- [ ] Extra rain cover
- [ ] Cot
- [ ] Lantern
- [ ] Flashlight
- [ ] Pot and pan set
- [ ] Pot holder
- [ ] Spatula
- [ ] Can opener
- [ ] Water bottle (5 gal)
- [ ] Rope
- [ ] Charcoal briquettes
- [ ] Table and chair (if needed)
- [ ] Sleeping bag
- [ ] Food and drink
- [ ] Knife and fork set
- [ ] Paper plates
- [ ] Paper towels
- [ ] Paper cups
- [ ] Cooler with ice
- [ ] Cook stove
- [ ] Stove fuel
- [ ] Funnel for fuel
- [ ] Fluorescent-type battery lantern
- [ ] Old newspaper
- [ ] Coffee cup
- [ ] Aluminum foil
- [ ] Trash bag

Miscellaneous
- [ ] Compass
- [ ] 24-hour clock
- [ ] Black plastic tape
- [ ] Masking tape
- [ ] Logs
- [ ] Grid square maps
- [ ] Pencils (many)
- [ ] Clipboard
- [ ] Highlighter

One of the many multiop contest efforts by N2SB, this during the VHF Sweepstakes.

moonbounce. It takes place over two full weekends which are spaced almost a month apart, usually in the fall. The date of the contest is different each year because of the variable phase of the Moon. The dates are usually chosen by active moonbouncers to coincide with the best combination of high lunar declination, perigee and the full phase of the Moon. This contest is international in scope. Your score is the number of QSO points made via EME per band, multiplied by the number of US call districts and ARRL countries per band. Hundreds of EMEers participate in this challenging test of moonbounce capability.

Contests are lots of fun, whether you're actively competing or not. You don't have to be a full-time competitor to participate or to enjoy yourself! In fact, most of the participants in contests aren't really "in" the contest. Lots of operators get on the air to pass out points, to have fun for a while, and to listen for rare DX. Others try their hardest for the entire contest, keep track of their score and send their logs in for awards. Either way, the contest is a fun challenge.

If you don't plan on being a serious competitor, then you just need to know what the "exchange" is. The exchange is the minimum information that must be passed between each station to validly count the QSO in the contest. In all of the terrestrial VHF contests sponsored by the ARRL, you need to pass only your grid square designator to the other station, and receive theirs, plus acknowledgment, to count the QSO. Other contests require some different information, such as serial numbers or signal reports, so it pays to check the rules to make sure.

A serious contest effort requires dedication and effort, as well as a station that can withstand a real workout. Contests are a challenge to operators and equipment alike. A good contest score is the result of hard work, a good station and favorable propagation. A good score is something to be proud of, especially if there is lots of stiff competition. And with the popularity of VHF and UHF contests these days, competition is always stiff!

If you are a competitor in the contest, you will probably need to keep a "dupe sheet" which helps prevent duplicate QSOs. A dupe sheet is a large piece of paper on which you record each call sign that you work in a sort of a matrix so you can check it quickly. If the other station's call sign is already in the dupe sheet, working him/her again won't count for contest credit. Don't rely on your memory. After 24 hours of contesting, most of us have trouble remembering our own names!

Also, if you make more than 200 QSOs in an ARRL VHF contest, you must submit your dupe sheet along with your logs. Dupe sheets, as well as log sheets, are available from ARRL HQ if you send a large self-addressed, stamped envelope (SASE) with your request.

Many operators don't feel that their stations are competitive on all the VHF and UHF bands. Not to worry; in most contests, it is possible to compete on only one band if you want to. This has the advantage of concentrating your efforts where your station is the strongest, allowing you to devote full time to just one band. You don't need to be high powered to compete as a single-band entry. Location makes a lot of difference, and hilltopping single-banders have had tremendous success, particularly if they go to a rare grid square.

Other operators like to try for all-band competition. In this case, it's a real advantage to be able to hop from one band to another. You can quickly check 6 meters for activity while also tuning 2 meters. Or, if you work a rare grid on one band, you can take advantage of the opportunity by asking the other station to switch bands right then. Some contesters work one station on all possible bands within two minutes by band hopping! This is a speedy way to increase your grids and QSOs.

Multioperator stations, in which more than one person performs operating tasks such as logging, dupe checking or making contacts, are very popular in VHF/UHF contests. Multiop contest entries come in two varieties: *Limited* and *Unlimited*. Limited multiop entries can compete on as many bands as they want during the contest, but submit scores for their best four bands. This category is a good environment for small groups of contesters who have the equipment necessary to operate on only a few UHF or microwave bands. On the other hand, Unlimited multiop stations compete on *all* VHF, UHF and microwave bands. Another class of competition is the *Rover Class*. A Rover is a station that operates from more than one grid square during the contest. Usually, Rovers operate from a vehicle that takes advantage of hilltops in rare grid squares, moving from one to the other throughout the contest period. This class is becoming very popular.

In ARRL contests, all multiop stations are also multiband. For multiop stations, the ability to operate several bands simultaneously is an obvious advantage. Being able to pass messages from one operating position to another is also important. It allows one band-station to alert another to changing propagation conditions. For example, the 6-meter position may notice the beginning of an aurora several minutes before it will appear on 2 meters. If the 6-meter operator can alert the other operators, valuable time can be saved.

Similarly, it is advantageous to be able to pass "referrals" from one station to another in a multiop contest effort. A referral is an on-the-spot schedule to call another station at a particular time and frequency. It's the same as a quick band change for single op stations. For example, if you are working the multioperator contest station W8VP on 2 meters, you might request an immediate sked on 222. If the 2-meter operator at W8VP can pass a quick message to the 222 station, they'll be looking for you, and you will probably make the QSO promptly.

During VHF/UHF contests, the above-listed times for activity nights are often observed for special activity, particularly on 222 MHz and up. In other words, look for increased activity on 222 MHz at 8 PM, 432 MHz at 9 PM, and 1296 MHz at 10 PM local during major contests. In some areas, this scheme also applies to morning (AM) as well. Bigger stations sometimes turn their antennas in accordance with the minute hand of the clock, looking north at the beginning of the hour, east at 15 minutes past, and so forth.

SOURCES OF INFORMATION

Many VHF/UHF operators like to keep abreast of the latest happenings on the bands such as new DX records, band openings or new designs for equipment. There are many excellent sources for current information about VHF/UHF. In addition, they provide a way to share ideas and ask questions, and for newcomers to become familiar with operation above 50 MHz.

The main sources are nets, newsletters and published columns. Each has its own use and appeal; active VHFers usually seek out at least one of them.

Nets

Several nets meet regularly on the HF and VHF bands so that VHF/UHFers can chat with each other. These are listed below. It's a good idea to listen first, before checking in the first time, so you'll know the format of the net. Some nets like to get urgent news, scheduling information and other "hot" topics out of the way early, and save questions and discussion until later. Others are more free-form. You can learn quite a bit just by tuning in to these nets for a few weeks. Regular participants in the nets are often very knowledgeable and experienced. The technical discussions which sometimes take place can be very informative.

1) *Central States VHF Net.* Meets each Sunday at 0230 UTC, 3.818 MHz. Open to all interested stations. A terrific source of information, as well as skeds for meteor scatter and EME. Operates informally during major meteor showers.

2) *144 and 432 MHz EME Nets.* These meet each Saturday and Sunday at 1700 and 1600 UTC, respectively, on 14.345 MHz. These are international nets which serve those interested in moonbounce. Each net meeting is usually occupied by a review of recent EME conditions, news from participants and setting up schedules for future EME work. The 432 EME net is one hour long, whereas the 2-meter net often continues for several hours.

3) *Sidewinders On Two (SWOT) Net.* This net meets on 2-meter SSB, at regular intervals which vary by region. It is a good place to find active 2-meter operators. For information about the SWOT net in your area, consult the SWOT newsletter (c/o Howard Hallman, WD5DJT, 3230 Springfield, Lancaster, TX 75134).

4) *East Coast 432 Net.* This net is called each Wednesday at

9-11 PM local on 432.090 MHz. It is particularly popular in the Southeast, although they have had checkins from as many as 30 states in the past few years.

5) *Local Nets.* Lots of these nets, both formal and informal, are scattered across the country. It's impossible to tabulate their meeting times and frequencies. The best thing to do is to get on the air and ask other VHF operators about nets in your area.

Awards

Several awards have helped to spur activity on the VHF/UHF bands. The most popular are WAS, WAC and VUCC.

It was long thought that WAS, Worked All States, was an impossible dream for VHFers. After all, it was "common knowledge" that no VHFer could work more than a thousand miles or so. To think of working coast-to-coast was silly, and to consider Alaska and Hawaii was crazy. To work all 50 states was just impossible. Fortunately, not everyone listens to "common knowledge," and WAS has been attained by several hundred amateurs on the 50, 144, 222 and 432-MHz bands.

WAS requires that you work each state in the United States and confirm the QSOs with QSL cards. The quest for this award has probably been responsible for much of the technical advancement of VHF/UHFers during the past three decades.

This impressive station belongs to Dan Gautschi, HB9CRQ. With a 2-meter DXCC award on his wall, Dan is a member of a select group!

Jim Mead, WB2BYW, (left) and Tom Williams, WA1MBA, explore the frontiers of microwave communication. In 1994 they made the first amateur contact on 120 GHz over a distance of 1.15 km.

Moonbounce activity has benefited most directly because WAS on any band above 144 MHz requires EME capability. (It should be noted, however, that a handful of diehards have worked 48 states on 2 meters *without* moonbounce!)

The 6-meter band has been host to more than a hundred WAS-achievers. During the peak of the most recent sunspot cycle, F_2 openings made transcontinental contacts common for many months. Amateurs in the continental US worked Alaska, Hawaii and tons of DX during that time. As of late 1991, there is little hope of widespread F_2 for the next few years as we gradually head into the sunspot minimum. Nevertheless, multi-hop E_s openings occur each summer to provide the slim chance of WAS on 6 meters.

Using portable moonbounce stations, several enterprising groups have mounted EME-DXpeditions to rare states. This has allowed quite a few hard-working VHF/UHFers to complete WAS, even when there was no resident EME activity in some states. At last count, over 100 stations had received WAS on 2 meters. At least one station, W1JR, has achieved WAS on 50, 144 and 432 MHz.

WAC, Worked All Continents, has also been achieved on 6 meters with the assistance of F_2 propagation. The chances of sporadic E of sufficient distance for catching all continents are almost nil, so we'll have to wait a while for more WAC awards on this band.

WAC is available to EMEers on 144, 432 and perhaps 1296 MHz. Stations from all continents have been active on these bands in recent years, although 1296 is not fully represented. In fact, several top-scoring stations in the annual EME contests have made QSOs on all continents during a single weekend. The rarest continent is probably South America, where only a small handful of EMEers are active. WAC is not possible on 222 or 902 MHz because these bands are not authorized for amateur use outside of ITU Region 2 (North and South America).

ARRL breathed new life into the VHF and UHF bands in January 1983 when the League began sponsoring the VUCC (VHF/UHF Century Club) awards. These awards, based upon grid squares, rejuvenated the VHF and UHF bands tremendously. Somewhat later, the required exchange in all terrestrial ARRL-sponsored VHF and UHF contests was changed to include grid squares, and activity in those contests increased as well. You can do a lot of grid-hunting during a good contest!

The VUCC award is based on working and confirming a certain number of grid squares on the VHF/UHF bands. For 50 and 144 MHz, the number is 100. On 222 and 432 MHz, the number is 50, and on 902 and 1296 MHz the minimum number is 25. 2.3-GHz operators need to work 10 grids, and five grids are required on the higher microwave bands. This is quite a challenge, and qualifying for the VUCC award is a real accomplishment!

VUCC endorsement stickers are available for those who work specified numbers of additional grids above the minimum required for the basic award. Several stations have exceeded 300 grids on 6 meters and 200 on 2 meters, and a few have worked over 100 grids on 432. Impressive!

Another award should be mentioned although only a handful of VHFers can hope to attain it: *DXCC*. The DX Century Club award, based on working 100 ARRL DX countries, has been a popular goal among low-band (HF) DXers for decades. Many people have considered it preposterous to even think that DXCC could ever be achieved by VHFers. However, DXCC has been conquered by a number of 50-MHz stations. Excellent F_2 propagation during 1989-1990 and the avalanche of 6-meter operating permits in Europe made many of these DXCC awards

N4GJV, North Carolina, put up this 70-cm EME array, consisting of 16 13-element Yagis each with a wood boom.

possible. Even more impressive has been the achievement of a few hams who have earned DXCC on the 144-MHz band. Moonbounce, plus a lot of patience and hard work, did the trick for W5UN. For specific details concerning the ARRL awards program, see Chapter 8.

Problems

With the ever-increasing sharing of the VHF and UHF spectrum by commercial, industrial and private radio services, a certain amount of interference with amateur operation is al-

JA0JCJ is rightfully proud of this antenna system, 4 Yagis on 144 MHz and 4 Yagis on 432 MHz.

most inevitable. Amateurs are well acquainted with interference, and so we normally solve interference problems by ourselves.

Several main types of interference are common. The first is our old friend, television interference (TVI). Fundamental overload, particularly of TV Channels 2 and 3 from 6-meter transmitters, is still common. Some Channel 12 and 13 viewing is bothered by 222-MHz transmissions. Fortunately, as modern TV manufacturers have slowly improved the quality of their sets, the amount of TVI is beginning to decline.

A more common form of TVI is called CATVI, Cable Television Interference. It results from many cable systems distributing their signals, inside shielded cables, on frequencies which are allocated to amateurs. As long as the cable remains a "closed system," in which all of their signal stays inside the cable and our signals stay out, then everything is usually okay. However, it is a sad fact that many cable systems are deteriorating because of age, and have begun to leak. When a cable system leaks, your perfectly clean and proper VHF signal can get into the cable and cause enormous amounts of mischief. By the same token, the cable company's signals can leak out and interfere with legitimate amateur reception.

RFI—radio frequency interference—is a common complaint of owners of unshielded or poorly designed electronic entertainment equipment. Amateur transmissions, especially high power, may be picked up and rectified, causing very annoying problems. RFI may include stereos, video-cassette recorders and telephones. The symptoms of RFI usually include muffled noises which coincide with keying or SSB voice peaks, or partial to total disruption of VCR pictures.

This chapter cannot cover all the complex methods which are used to track down and correct TVI/RFI problems. However, a few general principles may help in beginning your search for a solution:

1) With ordinary TVI, be sure your transmitter is clean, all coax connectors are tightened, and a good dc and RF ground is provided in your shack—before you look elsewhere for the cause of the problem. Then, find out if TVI affects all televisions or just one. If it's just one, then the problem is probably in the set and not your station.

2) With CATVI, remember that it is the cable company's responsibility to keep its system "closed" to the limits of FCC rules. Unfortunately, the FCC's limits are loose enough that in some cases there will be interference-causing leakage from the system which is still within FCC limits. In that case, there is no easy solution to the problem. However, you may be pleasantly surprised by the cooperative attitude of some cable TV operators. (Cable company technicians often have worked overtime trying to solve CATVI complaints, but not everyone is so fortunate.)

3) In dealing with RFI, the main goal is to keep your RF out of the entertainment system. This is often solved rather simply by bypassing the speaker, microphone and power leads with disc ceramic capacitors. In other cases, particularly some telephones, you must also employ RF chokes. For further information on RFI and CATVI, consult *Radio Frequency Interference: How to Find It and Fix It,* published by ARRL.

Several other types of interference plague VHF/UHF amateurs. These are the receive-only kinds of interference, which just affect your receiving capability. One form has already been mentioned: CATV leakage. For example, cable channel "E" is distributed on a frequency in the middle of the amateur 2-meter band. If the cable system leaks, you may experience very disruptive interference. Reducing leakage or perhaps eliminating the use of channel "E" may be satisfactory. This problem has vexed many stalwart VHFers already, and no end is in sight.

Scanner "birdies" may be a problem in your area. All scanners are superheterodyne-type receivers which generate local oscillator signals, just as your receiving system does. Unfortunately, many scanners have inadequate shielding between the local oscillator and the scanner's antenna. The result is radiation of the oscillator's signal each time its channel is scanned—a very annoying chirp, swoosh or buzz sound every second or so. If the scanner's local oscillator frequency happens to fall near a frequency you're listening to, you'll hear the scanner instead. Amateurs can easily pick up 10-15 scanner birdies within 5 kHz of the 2-meter calling frequency (for example, a New York State Police frequency is about 10.7 MHz above 144.200!). Very little can be done about this problem aside from installing tuned traps in everyone's scanners—an unattractive prospect to most scanner owners.

Many amateurs who operate 432 MHz have lived with radar interference for years. Radar interference is identified by a very rapid burst of noise which sounds vaguely like ignition noise, repeated on a regular basis. Although some radars are being phased out (no pun intended), many remain. Amateurs are secondary users of the 420-450 MHz band, so we must accept this interference. The only solution may be directional antennas which may null out the interference, not a very satisfying alternative in some cases. 432-MHz EMEers sometimes hear radar interference off the Moon!

NEW FRONTIERS

The VHF/UHF spectrum offers amateurs a real opportunity to contribute to their own knowledge as well as the advancement of science in general. We as amateurs are capable of sev-

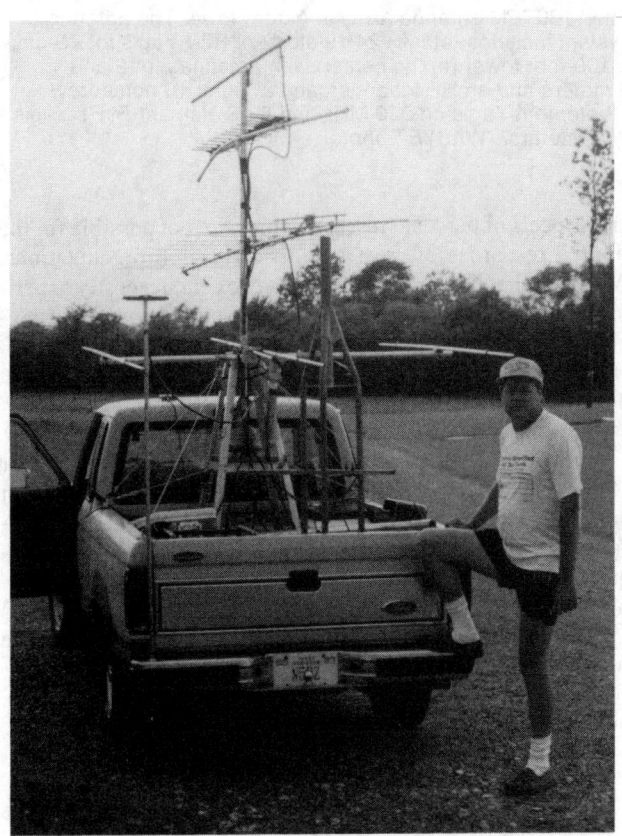

Ralph, N5RZ, poses beside his "rover machine."

The WBØTEM antenna array in Akron, Iowa. The antenna system includes (at left) 24 19-element RIW Yagis for 70-cm, a 100-foot tower (in the center) with antennas for 2 and 6 meters and an HF tribander, and at right, 16 homebrew 13-element Yagis on 220 MHz and 8 16-element F9FT Yagis on 2 meters. *(WA6YBT photo)*

N3BFL, with W3PM logging, operated 2-meter SSB during Field Day. *(WB3AMY photo)*

eral aspects of personal research which are not possible for the limited resources of most scientific research organizations. Although these organizations are able to conduct costly experiments, they are limited in the time available, geographic coverage and, sometimes, by the "it's impossible" syndrome. Hams, on the air 24 hours a day, scattered across the globe, don't know that some aspects of propagation are "impossible," so they occasionally happen for us.

For example, it is commonly "known" that VHF meteor scatter is limited to frequencies below about 150 MHz, yet hams have made dozens of contacts at 222 and 432 MHz. A spirit of curiosity and a willingness to learn is all that is required to turn VHF/UHF operating into an interesting scientific investigation.

What are some of the topics amateurs can contribute to? The following list includes just a few of the possibilities:

1) *Aurora on 902 or 1296.* Many aurora contacts have been made on 432 MHz; is Au "impossible" on 902? 1296?

2) *Unusual forms of F-layer ionospheric propagation*—FAI, TE and what else? Little is known about their characteristics, how they relate to overall geomagnetic activity and frequency. Amateurs can discover a lot here.

3) *Multiple-hop sporadic E on 50 MHz.* Single-hop propagation on this band is a daily occurrence during the summer. In recent years, multiple-hop openings have not been uncommon. Now that many European countries are beginning to allow 6-meter amateur operation, we can find out how many hops are actually possible!

4) *Meteor scatter.* Astronomers are very interested in the orbits of comets and their swarms of debris which give rise to meteor showers. However, it is often difficult for astronomers to observe meteors (because of clouds, moonlight and so on). Amateurs can make very substantial contributions to the study of meteors by keeping accurate records of meteor-scatter contacts.

5) *Polarization of E-layer signals.* Do VHF signals rotate in polarity when reflected by sporadic-E or meteor-trail ionization? Are vertical antennas better than horizontal antennas for E-layer signals?

6) *How effective is diversity reception* in different types of VHF/UHF reception? Stations with two or more antennas could investigate.

7) *New modes of data transfer.* Packet, AMTOR and the like—what are some of the limits to their use? Who knows how many amateurs may attempt to send packet information via EME.

In addition to this list, one should consider the exciting challenges available to amateurs in ITU Region 2 on the 902-MHz band. Its propagation characteristics are a compromise between those of 432 MHz and 1296 MHz. High power levels are more easily obtained at 902 than 1296 because several vacuum tubes are usable at 902 but not much higher. Also, antennas are still very small, so this is an ideal band for experimentation. The field is wide open for those who want to design equipment and antennas for this band.

Chapter 13

Satellites

JON BLOOM, KE3Z

ARRL HEADQUARTERS

Whe n asked about satellites, many amateurs will respond by saying, "Satellites are really interesting, but they're too complicated for me." Those who have experienced the fascination of amateur satellite operation know differently! Like any specialized activity, satellite operation has its unique requirements, and like many facets of Amateur Radio, going whole-hog with an elaborate station can be costly. In truth, you don't need to invest a lot of time or money to get involved in amateur satellite operation—there are a number of reasonably painless ways to get started.

SATELLITES: ORBITING REPEATERS

Most amateurs are familiar with repeater stations that retransmit signals to provide wider coverage. This is essentially the function of an amateur satellite as well. Of course, while a repeater antenna may be as much as a few thousand meters above the surrounding terrain, the satellite is hundreds or thousands of kilometers above the surface of the Earth. The area of the Earth that the satellite's signals can reach is therefore much larger than the coverage area of even the best Earth-bound repeaters. It is this characteristic of satellites that makes them attractive for communications. Most amateur satellites act either as analog repeaters, retransmitting signals exactly as they are received, or as packet store-and-forward systems that receive whole messages from ground stations for later relay. We'll concentrate on the analog satellites first.

Analog Satellites

Analog satellites contain *linear transponders*. These are devices that retransmit a band of frequencies, usually 50 to 100 kHz wide. Since the linear transponder retransmits the entire band, a number of signals may be retransmitted simultaneously. For example, if four SSB signals (each separated by 20 kHz) were transmitted to the satellite, the satellite would retransmit all four signals—still separated by 20 kHz each. Just like a terrestrial repeater, the retransmission takes place on a frequency that is different from the one on which the signals were originally received.

In the case of amateur satellites the difference between the transmit and receive frequencies is similar to what you might encounter on a cross-band terrestrial repeater. In other words, retransmission occurs on a different *band* from the original signal. For example, a transmission received by the satellite on 2 meters might be retransmitted on 10 meters. This cross-band operation allows the use of simple filters in the satellite to keep its transmitter from interfering with its receiver. Cross-band operation also has the happy effect of allowing ground stations to use simple filters in the same way. Because it's relatively easy to do, most satellite stations operate *full duplex*, meaning they can receive while transmitting. (The phrase *satellite stations* means amateur stations that use the satellite to relay their signals.) In most instances the built-in receiver filtering is sufficient to allow full-duplex operation.

Getting Started with the RS Satellites

Many amateurs were introduced to satellite operating through the use of the

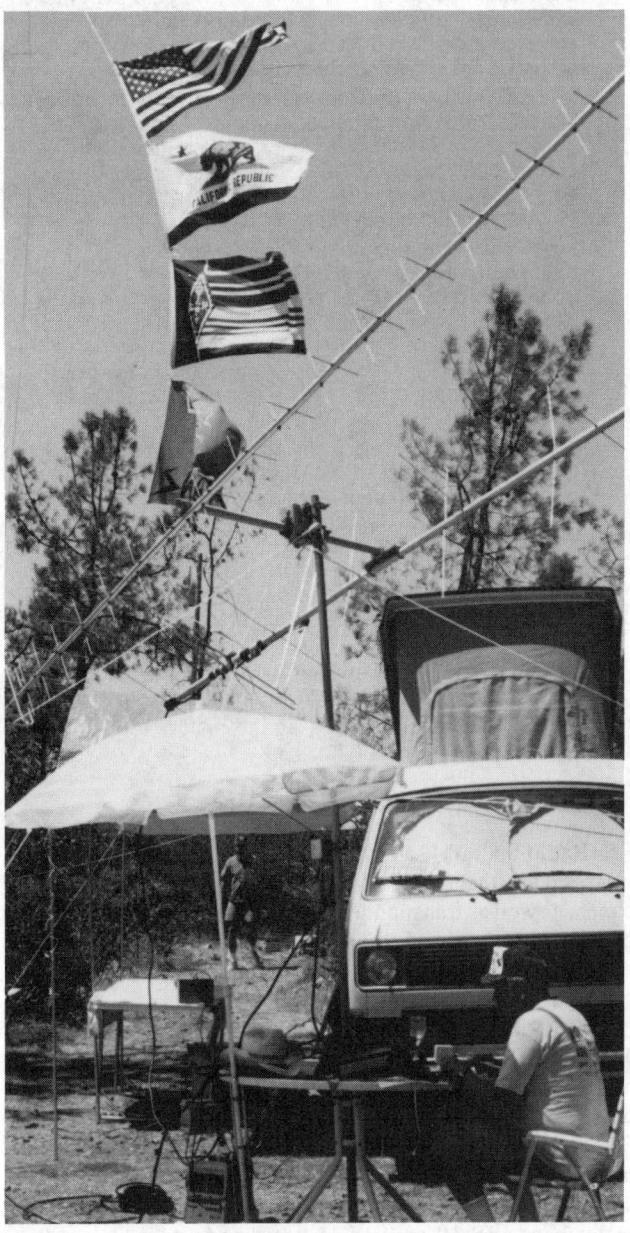

Gordy Fuller, WB6OVH, works OSCAR 13 during Field Day. *(WB6OVH photo)*

Amateur Satellites: A History

OSCAR 1, the first of the "Phase 1" satellites, was launched on December 12, 1961. It sent information concerning its internal temperature back to Earth on the 2-meter band. The now traditional CW identification HI was varied in code speed as the temperature changed. The 0.10-watt transmitter onboard discharged its batteries after only three weeks, but the mission was so successful that construction of a second satellite was begun.

OSCAR 2 was launched on June 2, 1962. Virtually identical to OSCAR 1, Amateur Radio's second venture into space lasted 18 days.

OSCAR 3, launched March 9, 1965, was the world's first free access communications satellite. During its two-week life, over 100 pioneering amateurs in 16 countries communicated through the 1-watt linear transponder on 2 meters.

OSCAR 4 was launched on December 21, 1965. More advanced in design than OSCAR 3, this satellite had a 2-meter-to-70-cm linear transponder with an output of 3 watts. Unfortunately, a launch-vehicle defect placed OSCAR 4 into a high elliptical orbit, which prevented widespread amateur use. A handful of hams did communicate through it, however, including the first US to USSR satellite contact of any kind.

OSCAR 5 was built by students at Melbourne University, in Australia. A new amateur organization, AMSAT, prepared the satellite for launch and coordinated ground activities. OSCAR 5 transmitted telemetry about its operating parameters on both 2 meters and 10 meters. Its batteries lasted for over a month.

OSCAR 6 was the first of the Phase 2 satellites. Launched on October 15, 1972, it carried a 1-watt-output 2-meter-to-10-meter Mode A linear transponder. Not only was the satellite magnetically stabilized, but the internal battery package was continuously recharged by solar cells. Tens of thousands of contacts were made during the nearly five-year lifespan of this satellite.

OSCAR 7 was built by hams from West Germany, Canada, Australia and the United States. Two elaborate linear transponders each ran 2-watts output; one was a 2-meter-to-10-meter Mode A unit like the one on OSCAR 6, the other a 70-cm-to-2-meter Mode B unit. Internal circuitry, as well as ground commands, controlled this sophisticated satellite's functions. OSCAR 7 was launched on November 15, 1974.

OSCAR 8, a cooperative effort of United States, Japanese, German and Canadian amateurs, was launched on March 5, 1978. The two transponders on board were a 2-meter-to-10-meter Mode A unit as used before, and a 2-meter-to-70-cm Mode J unit. Both had an output of 2 watts.

Radio Sputniks 1 and 2 were launched from the Soviet Union on October 26, 1978. Each satellite carried a 2-meter-to-10-meter transponder with extremely high sensitivity. Like their OSCAR counterparts, the RS satellites transmitted telemetry data relating to their well-being, and were commandable from the ground.

OSCAR Phase 3-A, the first of a planned series of satellites that would finally realize the dream of reliable high-altitude long-distance satellite communications over extended periods, was launched on May 23, 1980, a date that unfortunately came to be known as "Black Friday." Phase 3-A piggybacked aboard the European Space Agency Ariane LO2 rocket. A few minutes after liftoff from the ESA launch facility in Kourou, French Guiana, the Ariane rocket failed, dumping Phase 3-A and the dream that rode with it into the Atlantic.

OSCAR 9, built by a group of radio amateurs and educators at the University of Surrey in England, went aloft in October 1981 as part of a secondary payload aboard a NASA Delta rocket. This was a scientific/educational low-orbit satellite containing many experiments but no amateur transponders. These included HF beacons at 7.050, 14.002, 21.002 and 29.510 MHz for propagation studies, general and engineering data beacons on 2 meters and 70 cm, and two additional beacons on 13 cm and 3 cm. An Earth-imaging camera, magnetometer, synthesized voice telemetry capability and onboard computer rounded out the experimental hardware. UoSAT-OSCAR 9 became a "Silent Key" on October 13, 1989

Radio Sputniks 3-8 were launched simultaneously aboard a single vehicle in December 1981. Several of these RS satellites carried 2-meter-to-10-meter transponders that permitted communications over distances greater than 5000 miles. Also aboard two of them was a unique device nicknamed "Robot" that could automatically handle a CW QSO with terrestrial stations.

Iskra 2, was launched manually by two Soviet cosmonauts from the *Salyut 7* space station in May 1982. With the call sign RK02, Iskra 2 sported a telemetry beacon on 10 meters and a 15-meter-to-10-meter HF transponder. Iskra 2 was destroyed on reentering the atmosphere a few weeks after launch.

Iskra 3, launched in November 1982 from *Salyut 7*, was even shorter lived than its predecessor.

OSCAR 10, the second Phase 3 satellite, was launched on June 16, 1983, aboard an ESA Ariane rocket, and was successfully placed in its initial elliptical orbit.

RS (Radio Sputnik) satellites. These satellites contain analog transponders that receive and retransmit signals within the HF bands, as well as transponders operating at VHF. The presence of HF-only transponders means that many amateurs already have the equipment they need to operate through an RS satellite.

The RS satellites are built and launched in the Soviet Union. There have been a number of RS satellites in orbit. At the time of this writing, satellites RS-10/11, RS-12/13 and RS-15 are operating (see sidebar: *Amateur Satellites: A History*). RS-10/11, RS-15 and RS-12/13 are of primary interest to the beginner. RS-12 carries "Mode K" transponders, which receive signals on the 15-meter band and retransmit on the 10-meter band. RS-10 and RS-15 use "mode A" transponders, which receive on 2 meters and retransmit on 10 meters. (The use of "mode" terminology is one of the most confusing aspects of amateur satellite operation to the newcomer. Don't let it scare you away. The sidebar *Satellite Modes Demystified* should clear up any confusion.)

A Matter of Perspective: When you transmit a signal to a satellite, the satellite retransmits it, just like a repeater does. If you can receive while you transmit (highly recommended), you'll hear your own signal coming back from the satellite a fraction of a second after you transmit. Even dyed-in-the-wool satellite operators get a thrill from hearing their own signal being repeated by the satellite as it comes into view!

This brings up an important point. Unlike commercial television and telephone relay satellites, an amateur satellite is not always immediately accessible. Commercial satellites are *geostationary*, which means that they appear to be motionless

OSCAR 10 has proved to be a communications resource for radio amateurs throughout the world. OSCAR 10's computer (IHU) began showing the accumulated effects of near-Earth space particle radiation in mid-1986. Faulty operation of the IHU means that the satellite has become uncommandable and uncontrollable.

OSCAR 11, another scientific/educational low-orbit satellite like OSCAR 9, was built at the University of Surrey in England and launched on March 1, 1984. UoSAT-OSCAR 11 has also demonstrated the feasibility of store-and-forward packet digital communications and is fully operational as this is written.

OSCAR 12 or *Fuji-OSCAR 12* (FO-12) was the first Japanese OSCAR, a joint effort between the Japan Amateur Radio League (JARL) and the Japan AMSAT (JAMSAT) organization. It carried a Mode J transponder that served two functions: linear SSB service (Mode JA) and an advanced packet store-and-forward global message service with a 1.5-megabyte RAM storage bank called Mode JD (for Mode J digital).

Radio Sputnik 10/11 carries a dual designation because it consists of two electronics packages integrated into a single spacecraft. It was launched on June 23, 1987 and, like the Iskra satellites, carries 15-meter-to-10-meter transponders. These work in conjunction with a Mode A transponder to provide a variety of HF/VHF operating modes.

OSCAR 13, the third Phase 3 satellite and the second operational one, enjoyed a flawless launch and deployment on June 15, 1988. Its subsequent "burn" to final orbit also was flawless. OSCAR 13 has since provided worldwide communications using Modes B, J, L (23 cm to 70 cm) and S (70 cm to 13 cm). An experimental packet transponder called RUDAK, contributed by AMSAT-DL, failed after launch.

OSCAR 14 and 15, built by the University of Surrey were launched on January 22, 1990. OSCAR 14 carries a 2-meter-to-70-cm store-and-forward packet-radio transponder called the Packet Communications Experiment (PCE) along with scientific experiments. OSCAR 15, which carried only experiments and a solid-state TV camera failed catastrophically shortly after launch.

OSCARs 16 through 19 are the Microsats. Designed by a cooperative effort of amateurs in the Western Hemisphere, these tiny satellites (approximately 9-inch cubes) were launched, along with OSCARs 14 and 15, aboard a single Ariane rocket. At the time, this single launch more than doubled the number of operational amateur satellites! PACSAT-OSCAR 16, sponsored by

AMSAT-NA, carries a 2-meter-to-70-cm store-and-forward packet transponder, as does Webersat-OSCAR 18 and Lusat-OSCAR 19. Webersat, sponsored by Weber State University, carries a small TV camera for Earth imaging. LUSAT, sponsored by AMSAT-LU (Argentina), includes a 70-cm CW beacon transmitter. DOVE-OSCAR-17 was sponsored by AMSAT-Brazil. It consists of a 2-meter transmitter that can transmit packet or digitized voice messages.

OSCAR-20, launched on February 7, 1990 is almost a clone of OSCAR 12. It has a somewhat improved power subsystem for more consistent operation.

RS-12/13, like RS-10/11, carries multiple HF/VHF transponders. It was launched on February 5, 1991.

OSCAR 21, also known as RS-14, was the first Soviet amateur satellite to carry an OSCAR designation. It was deactivated in November 1994.

UoSAT-OSCAR 22 was designed by the University of Surrey in the United Kingdom. It provided packet store and forward capability at 9600 bit/s as well as a high resolution camera.

KITSAT-OSCAR 23 was the first Korean amateur satellite to carry an OSCAR designation. Launched in 1992, it offered packet capability at 9600 bit/s. KITSAT also provided views from *two* onboard cameras.

ARSENE, launched May 12, 1993, failed several weeks later. It was intended to function as a packet digipeater.

KITSAT-OSCAR 25 is a virtual clone of OSCAR 23. Launched in September 1993, it also has imaging cameras as well as 9600 bit/s packet capability.

ITAMSAT-OSCAR 26 was launched in September 1993 along with OSCAR 25. It was the first Italian amateur satellite. It operates primarily as a 1200 bit/s store-and-forward packet system.

AMRAD-OSCAR 27 was also a launch companion to OSCARs 25 and 26. It is an experimental module that operates as part of *Eyesat*, a commercial satellite. AMRAD-OSCAR 27 is intended to be used as a test bed for new Amateur Radio technology. When this book went to press, it was functioning as an FM repeater.

PoSAT-OSCAR 28 is the first Portuguese amateur satellite. It was launched along with OSCARs 25, 26 and 27. Like OSCAR 27, it is part of a commercial satellite. The satellite is functioning in its commercial mode, but amateur activity has been minimal. At the time of this writing, its future as a ham satellite is in question.

RS-15 was launched in December 1994. The satellite carries a Mode A transponder.

from our perspective. On the other hand, amateur satellites, including the RS series, utilize *low-Earth orbits* (LEOs). The LEO orbit describes a circle, with the satellites traveling about 1000 km above the surface at a speed that causes the satellite to complete the circle about every hour and a half. As the satellite orbits, the Earth rotates on its axis, bringing different portions of the Earth "in view" of the satellite at different times of the day.

From our perspective a LEO satellite rises above the horizon, travels across the sky in an arc and then sets again. It may do so six to eight times a day. For "passes" in which the satellite goes nearly overhead, this rise and set cycle takes 15 or 20 minutes. On some orbits the satellite's path is such that it rises only a short distance above the horizon, much like the winter sun near the Arctic Circle. As you might expect, the time the

satellite is in view is much shorter. The total amount of time that any particular LEO satellite is available for use at a given location is perhaps an hour or so each day. It doesn't seem like a long time, but it is more than enough to provide outstanding operating enjoyment on a regular basis.

Working the Satellite: When using the RS-12 mode K transponders you don't need to know precisely where the satellite is positioned in the sky. After all, you aren't likely to be using narrow beamwidth antennas. (Often heard on RS-12/13: "Antenna here is a dipole.") Mainly, you want to know when the satellite will be in view. Of course, 15 and 10-meter signals are subject to ionospheric bending, so it pays to listen for the satellite before and after the predicted visibility period.

Once you've determined when the satellite is due to rise above the horizon at your location, listen for the satellite's

Satellite Modes Demystified

Since there are a number of amateur satellites operating, and since they operate on a variety of bands, some way of distinguishing the various combinations of bands became necessary. The first amateur transponder in space was the one carried by OSCAR 6. It received uplink signals at 2 meters and retransmitted them on the downlink at 10 meters. When OSCAR 7 was launched it included a transponder with a 70-cm uplink and a 2-meter downlink. To distinguish between the two, the 2-meter-to-10-meter transponder operation was called "Mode A" and the 70-cm-to-2-meter operation "Mode B."

The launch of OSCAR 8 complicated matters by using a transponder with a 2-meter uplink and a 70-cm downlink, just the reverse of the Mode-B transponder of OSCAR 7. This new transponder was built by Japanese amateurs, and in their honor the new setup was dubbed "Mode J."

OSCAR 10 presented yet another possibility with its 23-cm-to-70-cm transponder. The 23-cm band is in that portion of the spectrum called "L band" by microwave engineers, so this uplink/downlink setup acquired the name "Mode L." A similar situation occurred when OSCAR 13 deployed a 70-cm-to-13-cm transponder. Since the 13-cm band is in the S-band part of the microwave spectrum, this mode is called—you guessed it—"Mode S."

The digital satellites, the first of which was the Japanese OSCAR 12, threw a monkey wrench into this (relatively) simple naming scheme. For the first time, amateurs had to contend with transponders that were different in kind rather than just frequency. FO-12 (and its successor, FO-20) carried both an analog transponder and a digital packet-radio transponder. Both transponders received uplink signals on 2 meters and transmitted on 70 cm, conforming to the Mode J band selection. To indicate the band selection while distinguishing between the analog and digital transponders, the analog operation is called Mode JA and the digital operation Mode JD.

Finally, the RS-12 satellite uses a unique uplink/downlink band pair: 15 meters to 10 meters (Mode K).

Some of the satellites can operate two transponders in conjunction. For example, signals in two separate uplink bands are combined and transmitted on a single downlink. When this is done, the operation is designated by a two-letter mode that encompasses the two transponders. For example, OSCAR 13 can combine its Mode B and Mode S transponders into a single downlink band. This operation is known as Mode BS.

Now, isn't that simple?

It's Very Easy to Work RS-10!
(Reprinted from *OSCAR Satellite Report*)

Think you need much to work RS-10? Wrong! I called CQ this morning using a Kenwood TH-26AT hand-held and its 4-inch "rubber duck" antenna, sending CW with the mike button. Back came VE3DJ in Ancaster, Ontario, totally without prearrangement, and we completed a QSO giving me a 569 report! Power output at this end was about 2 watts for the QSO, but I subsequently heard my downlink Q5 with half a watt output. I used my station receiver for the [10-meter] downlink with a long-wire antenna. The rubber duck was indoors!

This morning I called CQ again on RS-10 with my TH-26AT's push-to-talk button and was answered by W1SJM in New Hampshire, who gave me a 449 report. No rubber duck this time, though. RS-10 was only 17 degrees high to the west, so I needed a really large antenna: an AEA half-wave whip!—*by Ray Soifer, W2RS*

---

telemetry beacon (see Table 13-1). This signal is transmitted constantly by the satellite and carries information about the state of the satellite's systems, such as its battery voltage, solar-panel currents, temperatures and so on. You should hear it just as the satellite rises above the horizon. As soon as you can hear the beacon, start tuning across the downlink passband.

On an active day you should pick up several signals. They will sound like normal amateur voice and CW contacts. Nothing unusual about them at all, except… it sounds like the signals are slowly drifting downward in frequency! That's the effect of Doppler shift. It's not too serious on the 10-meter downlink, but it can be a challenge when the downlink is at 70 cm because the shift is proportional to the transmitted frequency. (More about the Doppler factor later.) Now tune your transmitter's frequency to the satellite's uplink passband. Send a series of dits while you tune the transmitter frequency (yes, the *transmitter* frequency). With any luck you'll soon hear your own signal coming back down from the satellite. A tentative CQ might bring a call from another station that also has the satellite in view. Give it a try. Did you hear an answer?

Congratulations! You've just become a satellite operator.

Movin' On Up—Mode A: While Mode K is fun and exciting, there is much to be said for VHF operation. The noise levels at both the satellite and ground station tend to be lower, allowing better reception. If you have an all-mode (CW and SSB) 2-meter transceiver and an HF transceiver, you're set to operate Mode A on RS-10 or RS-15. (While FM signals will pass through the satellites' transponders, use of FM is strongly discouraged. Because FM transmissions use full power at all times, they require a disproportionate amount of the satellite's available downlink power. A satellite can handle many more simultaneous SSB signals than FM signals, and even one FM signal can degrade the usability of the satellite. But there *is* a way you can use your FM transmitter: as a CW transmitter. See the sidebar: *It's Very Easy to Work RS-10!*)

If you don't have a 2-meter all-mode receiver but you do have a general-coverage HF receiver, you might consider trying a method described by K2UBC in *The Satellite Experimenter's Handbook*, 2nd Edition. He describes how to "pick off" a signal from your FM receiver's IF and connect it to your HF receiver to get CW and SSB reception.

The only significant difference when using a 2-meter uplink or downlink is that the Doppler shift becomes more pronounced. Doppler shift is the same effect you hear from a high-speed passing truck: the truck's sound is higher than normal as it approaches, lower as it retreats from you. The same thing happens to radio signals traveling to and from the satellite. See the sidebar: *Doppler Shift*.

Exotica: Phase 3 Satellites

The limitations of LEO satellites, especially their brief periods of availability, are overcome by a class of satellites called "Phase 3." The name comes from the various phases in the development of amateur satellites. The earliest ones, during Phase 1, contained beacon and telemetry transmitters, but not transponders. These early satellites were all in circular, low-Earth orbits—as were the Phase 2 satellites, which carried communications transponders.

Phase 3 satellites are not in low-Earth orbits. Rather, their orbits describe an ellipse. A Phase 3 satellite swings within a few hundred kilometers of the Earth's surface at one end of the ellipse (the *perigee*) and streaks out to 30,000 km or so at the other end (the *apogee*). The physics of an orbiting body dictates that the satellite spends much more of its time near apogee than perigee. Therefore, the Phase 3 satellites spend most of their

Table 13-1
Analog Transponder Frequencies

RS Satellites

	RS-10	RS-11	RS-12	RS-13	RS-15
Mode A					
Uplink	145.860-145.900	145.910-145.950	145.910-145.950	145.960-146.000	145.858-145.898
Downlink	29.360-29.400	29.410-29.450	29.410-29.450	29.460-29.500	29.354-29.394
Beacons	29.357/29.403	29.407/29.453	29.408/29.454	29.458/29.504	29.353/29.399
Mode A Robot					
Uplink	145.820	145.830	145.830	145.840	
Downlink	29.357/29.403	29.407/29.453	29.454	29.504	
Mode K					
Uplink	21.160-21.200	21.210-21.250	21.210-21.250	21.260-21.300	
Downlink	29.360-29.400	29.410-29.450	29.410-29.450	29.460-29.500	
Beacons	29.357/29.403	29.403/29.453	29.408/29.454	29.458/29.504	
Mode K Robot					
Uplink	21.120	21.130	21.129	21.138	
Downlink	29.357/29.403	29.403/29.453	29.454	29.504	
Mode T					
Uplink	21.160-21.200	21.210-21.250	21.210-21.250	21.260-21.300	
Downlink	145.860-145.900	145.910-145.950	145.910-145.950	145.960-146.000	
Beacons	145.857/145.903	145.907/145.953	145.912/145.958	145.862/145.908	
Mode T Robot					
Uplink	21.120	21.130	21.129	21.138	
Downlink	145.857/145.903	145.907/145.953	145.958	145.908	

Phase 3 Satellites

Satellite	Mode	Uplink (MHz)	Downlink (MHz)
AO-10	B	435.030-435.180	145.825-145.975
	Beacon		145.810
AO-13	B	435.420-435.570	145.825-145.975
	S	435.601-435.637	2400.711-2400.747
	Beacons		145.812, 145.985
			435.651
			2400.325

Other Satellites

Satellite	Mode	Uplink (MHz)	Downlink (MHz)
FO-20	J(A)	145.900-146.000	435.900-435.800
	Beacon		435.795

time at very high altitudes. From a typical point in the Northern Hemisphere, a particular Phase 3 satellite is available for more than 10 hours per day. This is a remarkable improvement over the LEO satellites! And because the Phase 3 satellite is so much higher, it is visible from a greater fraction of the Earth's surface, too. The result is a vast improvement in the communications capability of the satellite.

There is a downside, however. The greater distance to the Phase 3 satellite means that more transmitted power is needed to access it, and a weaker signal is received from the satellite at the ground station. (This problem is alleviated somewhat by the use of gain antennas on the satellite.) The signal levels are such that operation using the first generation of Phase 3 satellites is feasible only with ground-station antennas that exhibit significant gain (10 dBi or more). Such directional antennas must be pointed directly at the satellite. The satellite's position in the sky changes over time, however, so the antenna's position must change as well. In fact, the satellite changes both its bearing from the ground station (azimuth) and its height

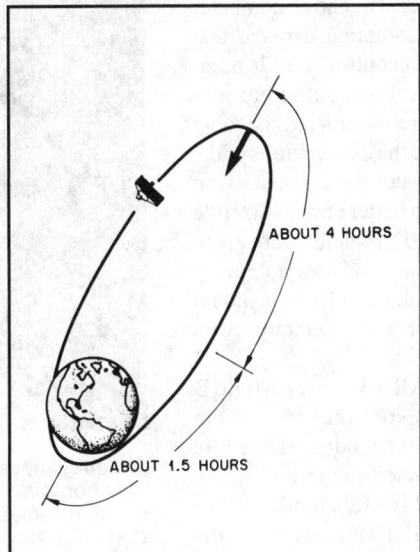

ABOUT 4 HOURS

ABOUT 1.5 HOURS

Fig 13-1—The orbit of AMSAT OSCAR 13 provides hours of coverage on each orbit to stations in the Northern Hemisphere.

Doppler Shift

Doppler shifting of signals is caused by the relative motion between you and the satellite. As the satellite moves toward you, the frequency of the downlink signals will increase by a small amount as the velocity of the satellite adds to the velocity of the transmitted signal. As the satellite passes overhead and starts to move away from you, there will be a rapid drop in frequency of a few kilohertz, much the same way as the tone of a car horn or a train whistle drops as the vehicle moves past the observer. Martin Davidoff, K2UBC, provides a very complete and understandable discussion of Doppler shift in Chapter 13 of *The Satellite Experimenter's Handbook.* This brief outline on Doppler shifts provides highlights of the effect.

The Doppler effect is different for stations located at different distances from the satellite because the relative velocity of the satellite with respect to the observer is dependent on the observer's distance from the satellite. The result of all this is that signals passing through the satellite transponder shift slowly around the calculated downlink frequency. Locating your downlink signal is more than a simple computation, since tuning is needed to compensate for the Doppler shift.

Doppler shift through a transponder becomes the sum of the Doppler shifts of both the uplink and downlink signals. In the case of an inverting type transponder (as in OSCAR 10 Mode B), a Doppler-shifted increase in the uplink frequency causes a corresponding decrease in downlink frequency, so the resultant Doppler shift is the *difference* of the Doppler shifts, rather than the *sum.* The shifts tend to cancel.

While Doppler shifts can be observed on OSCARs 10 and 13, it's not nearly so apparent as it was for OSCARs 7 and 8. The time required to observe a given change in Doppler shifts is much longer for Phase 3 satellites than for the low-altitude satellites, and is therefore much less of a problem. For instance, the total OSCAR 10 Doppler shift is approximately 4.5 kHz over the period of time that normal operations are conducted, approximately four hours on each side of apogee. This rate of change amounts to approximately 9 Hz per minute. For OSCAR 8, operators experienced a total shift of up to 12 kHz over a period of 16 minutes, a rate of approximately 750 Hz per minute!

It must be pointed out that the complete whole-orbit Doppler-shift curve is not a straight line. However, over the portion of the orbit that encompasses the majority of operations, a straight-line approximation will suffice quite well. These ranges cover approximately MA = 35 to MA = 220, and the total shift averages to be 4529 Hz, or 8.96 Hz/minute.

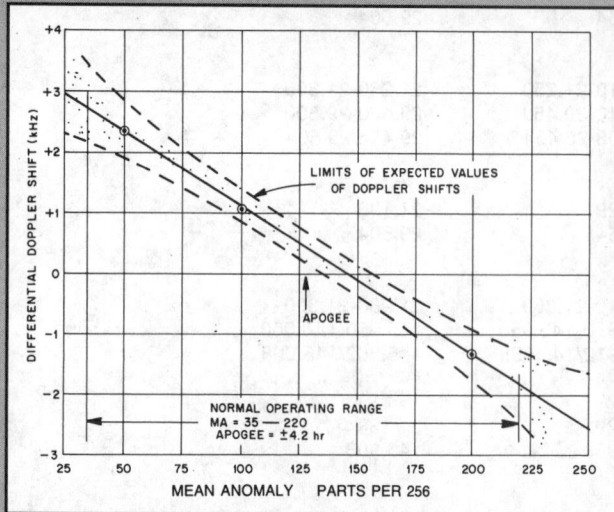

OSCAR 10 Mode B differential Doppler shift, normal for MA operating range.

above the horizon (elevation). Most ground stations that access Phase 3 satellites use antenna systems that can be rotated in both the azimuth and elevation planes.

At the time of this writing, two Phase 3 satellites have been successfully launched: OSCAR 10 and OSCAR 13. OSCAR 10 is only intermittently available; its control computer suffered accumulated radiation damage that rendered the satellite uncontrollable. It occasionally operates when it gets sufficient sunlight. OSCAR 13, on the other hand, is going strong. It has lived up to the expectations of the designers who built it and the amateurs who use it. And another even better Phase 3 satellite is in the works: Phase 3-D. (Satellites are given an OSCAR designation only after achieving orbit.) Scheduled for launch in 1996, it should provide excellent coverage for many years thereafter.

OSCAR 13: OSCAR 13 carries Mode B and S transponders. Note that the 70-cm band is common to these modes. Mode B is the most commonly used mode and many stations are equipped for this mode exclusively. Mode B ground stations need di-

The first image from cameras aboard KITSAT-OSCAR 23. This shot was taken only hours after it achieved orbit. The long, narrow objects in the image are the satellite's antennas. Similar images are also available from cameras on OSCARs 22 and 25. *(Courtesy Harold Price, NK6K)*

rectional 2-meter and 70-cm antennas (circularly polarized satellite antennas are ideal), a 2-meter CW/SSB receiver, a mast-mounted preamp and a 435-MHz CW/SSB transmitter. Some means of rotating the antennas is highly desirable, but because of the slow relative motion of the satellite you need to move the antenna only every 15 minutes or so. Mode B is popular because OSCAR 13 operates in that mode most of the time. The other modes depend on highly directional antennas onboard the satellite. These antennas are aimed at the Earth only when the satellite is near apogee. It is at this point in its orbit that the more exotic modes take over. But as the satellite departs from apogee, it switches to Mode B using its less directional antennas. (The satellite's available power doesn't allow it to operate all of its transponders continuously.)

Setting up a ground station for Phase 3 operation isn't all that difficult, but it's probably best to start at a less complex level. The RS satellites are a fine choice. If you're a packet enthusiast, you may want to start with the digital satellites. *The Satellite Experimenter's Handbook*, 2nd Edition, contains a wealth of information on setting up your station. If you want to start with Phase 3 satellites, you should get a copy of *A Beginner's Guide to OSCAR 13* from AMSAT.[1]

What About Phase 4?

It has long been the dream of amateur satellite designers to launch a geostationary amateur satellite. Fixed, high-gain antennas such as parabolic dishes could be pointed at the satellite, greatly reducing the complexity of the ground station. Unfortunately, the costs of producing and maintaining such satellites puts them out of the reach of the amateur community—for now.

You may wonder who sponsors these amateur satellites. The answer is that amateurs themselves provide most of the sponsorship of satellites built in the US and Europe. The organization that has performed most of this work in the last 20 years or so is AMSAT, the Radio Amateur Satellite Corporation, a nonprofit amateur group. If you are interested in satellite operation, join AMSAT. Aside from helping to sponsor future satellites, you'll receive *The AMSAT Journal* and discounts on AMSAT software and other membership benefits. For more information send a self-addressed, stamped envelope to AMSAT, PO Box 27, Washington, DC 20044.

PACSATs: Digital Wonders

Technology marches on, and anyone who has observed Amateur Radio in the past decade will recognize that the most fundamental change has been the advent of packet radio. Even in packet's infancy, visionaries saw the potential marriage of packet and satellite technology. The combination of the two has resulted in the PACSAT: a satellite carrying a packet radio transponder and a computer. PACSATs operate in a fundamentally different way from satellites with analog transponders.

A ground station transmits a message (digitally, of course) to the satellite. The satellite stores the entire message in its onboard memory, which typically can hold several million characters of message text. Later, when the satellite is over the ground station for which the message is intended, it transmits the message to that station. This kind of *store-and-forward* operation provides true worldwide communications using low-Earth-orbit satellites. Because PACSATs can hold a lot of data, and because they are optimized for transmitting data rather than voice or CW, they provide an unsurpassed bulletin trans-

mission system.

The software in the on-board computer works in conjunction with personal computers at ground stations to deliver error-free copies of bulletins from around the world. As the unique capabilities of PACSATs become better understood, they are being integrated into terrestrial packet radio networks. For the first time, hundreds of thousands of amateurs can have their messages relayed via amateur satellites.

In general, packet satellite operation is similar to terrestrial packet. The same AX.25 protocol is used to communicate with the satellite, but the radio equipment is somewhat different. The combination of weaker signals and Doppler shifting burdens a satellite-transmitted packet with more critical modem requirements. To solve this problem, phase-shift keying (PSK) or direct FSK modulation techniques are used. For AO-16, WO-18, LO-19 and IO-26, a PSK modem is used in conjunction with a 2-meter FM transmitter and a 70-cm SSB receiver. OSCARs 22, 23 and 25 require 9600 bit/s FSK modems with 2-meter and 70-cm FM rigs. The modems typically control the frequency of the receiver via the UP/DOWN pins on the microphone connector. Although this automatic tuning isn't strictly required, one quickly tires of adjusting the frequency to compensate for the large Doppler shift of the downlink signal! Most modern receivers that feature computer control capability can be used to directly compensate for Doppler shift.

To date, most of the store-and-forward PACSATs use a common scheme for communicating with ground stations. This scheme requires special-purpose software at the ground station. (Software packages are available from AMSAT.) For those interested in creating their own software to access these satellites, the specifics have been published in the *Proceedings of the 9th Computer Networking Conference*, available from the ARRL.

FINDING A SATELLITE

In order to use satellites, you'll need to have some idea when each one is "visible" at your location. Unfortunately, the time at which a particular satellite is above your horizon varies from day to day. So does its location in the sky from your point of view. Therefore, you need some means of determining where and when the satellite is going to appear. If you are using steerable directional antennas, you also need to know the satellite's exact position in the sky.

There are two common techniques used for determining a satellite's position: the graphical method and the use of computer-based calculations. The graphical method relies on precomputed depictions of the satellite's *ground track*, the path along the ground. The ground track is plotted on a map that contains a *range circle*. This circle is centered at your location and shows the satellite's elevation above the horizon at the point where its ground track falls within the circle. Together, the ground track and range circle can be used to give a reasonably accurate indication of when the satellite will be available to you. But to use the graphical method, some reference point along the ground track must be known. That is, you must know the satellite's location above the Earth at some particular point in time.

For LEO satellites we use the *equator crossing*, or EQX point. Each time a satellite circles the Earth, its ground track crosses the equator—once on its way from south to north (the *northbound* EQX) and again when it travels from north to south (the *southbound* EQX). The first northbound equatorial crossing of each day (UTC time) is the *reference orbit*. If we know the time and the longitude at which this crossing occurs, we can

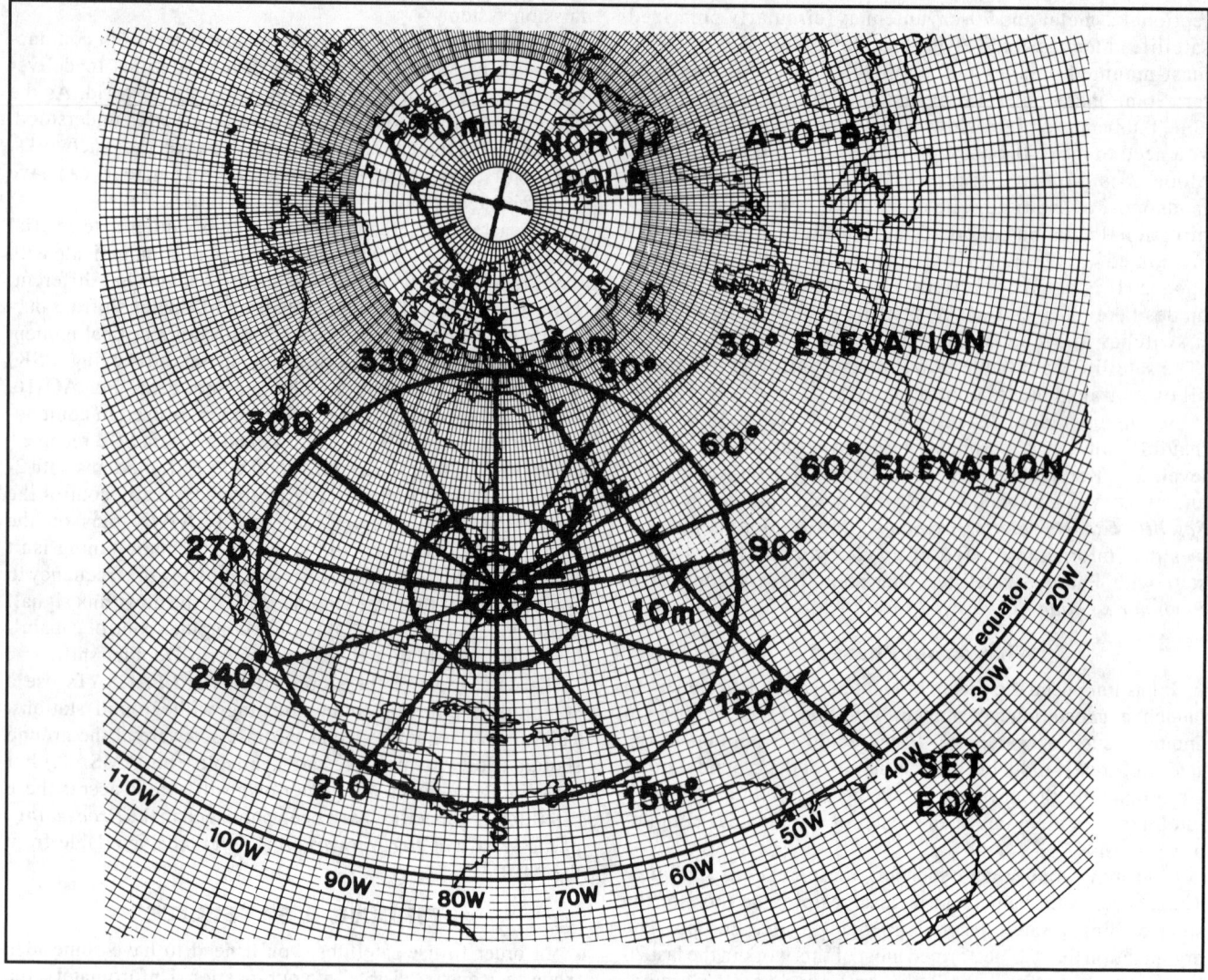

Fig 13-2—A graphical tracking aid can determine when the satellite can be used from your location. *The Satellite Experimenter's Handbook*, 2nd Edition, provides the maps, ground tracks and range circles needed to construct such an aid.

use the ground track and range circle to determine the location of the satellite for that orbit and for a number of orbits thereafter. After a dozen orbits or so, the graphical technique tends to lose accuracy and we need to know a new reference point. Practically speaking, we can use each day's reference orbit to track the satellite on that day only.

Computer-based calculation of the satellite position is the most common technique used by amateurs today. The calculations can be performed by most popular home computers using one of a variety of programs. Commercial programs that calculate satellite positions are available from AMSAT at a modest cost. For computers that have graphic display capability, there are programs that will display the satellite orbit on a map of the Earth.

The techniques used to mathematically calculate satellite position are well outside the scope of this book but are thoroughly explained in *The Satellite Experimenter's Handbook*. For our purposes, it is sufficient to know that in order to perform these calculations, the computer programs require data, called *Keplerian elements*, that mathematically describe the orbit of the satellite.

Since a satellite's orbit is affected by the inconstant gravity of the Earth, moon and sun, and by atmospheric drag in the case of the lower satellites, it tends to gradually change. This means that a new set of Keplerian elements are required from time to time. AMSAT issues Keplerian elements for a variety of satellites on a weekly basis. These are distributed by packet and landline bulletin-board systems, and in the AMSAT newsletter. They are also transmitted by W1AW using RTTY modes.

SETTING UP A SATELLITE STATION

A basic station for use with an analog transponder is shown at the beginning of this chapter. If you want to try AO-10 or AO-13, you'll probably need to add receiving preamplifiers and appropriate switching circuitry. (The preamps must be located on the tower near the antennas to minimize transmission line losses.) Since CW and SSB are the required modes, you may want to start on either Mode K or Mode A. Both modes require a 10-meter receiving system at the ground station. While some ground stations use specialized 10-meter satellite antennas such as a dipole above a reflecting screen or a "turnstile" design, these aren't really necessary to get started. An existing dipole, vertical or small Yagi antenna will work just fine.

Modern amateur receivers have sufficient sensitivity, selec-

09/18/91 16:03:23.81 UTC   < 74. ao-16   >

```
                Azimuth  Elevation  Range (km)  Doppler  Offp      Next Set
W1AW            21.999°   24.489°   1600.747    +8736              UR 00:10:45

Lat: 52.190°  Lon: -65.593°  Alt:  799.506  Phs:161.8  Mod:
123.3 km  ESE  of  Labrador City, Canada
```

Fig 13-3—AMSAT's InstantTrack is an example of one of the modern computer programs that track satellites and can control the antenna system.

tivity and stability for 10-meter satellite operation, but be careful of older equipment or inexpensive shortwave receivers. They may be lacking in one or more of these characteristics. Almost any 15-meter CW/SSB transmitter can be used for Mode K and, again, existing antennas should prove adequate. For Mode A you can still use existing antennas, but best performance will result from using a circularly polarized antenna such as a turnstile. (See the sidebar: *Circular Polarization*.) For LEO satellites you need only a few watts of 2-meter transmitter power, so most CW/SSB transmitters should do the job "barefoot."

Mode B requires 2-meter reception and 70-cm transmission capability. There are numerous ways of setting up Mode-B stations, ranging from the use of transverters along with HF transmitters and receivers to the purchase of multiband satellite transceivers. But these are the easiest of the VHF-and-above modes for the beginner. Equipment for the 2-meter and 70-cm bands is easier to find, cheaper to buy and easier to build than equipment for higher frequency bands.

One important factor to consider at VHF and above is transmission line loss. Losses tend to increase with frequency. A 100-foot run of coax that works perfectly well at 10 meters, for example, may be all but useless at 70 cm. Setting up a satellite station for Mode B or the higher-frequency modes requires a bit of engineering. It isn't difficult, but you have to consider the antennas, transmission line(s) and radios as a whole. See *The Satellite Experimenter's Handbook* for complete details.

To operate via the digital satellites you will also need a packet terminal node controller (TNC). Several of the satellites use a 1200-baud PSK downlink and 1200-baud Manchester FSK uplink scheme. At the ground station, this system requires a 70-cm SSB receiver and a 2-meter FM transmitter. A specialized modem must be connected between the TNC and the radio. 1200-baud PSK/FSK modems are available from the following sources:

Tucson Amateur Packet Radio
8987-309 E Tanque Verde Rd No. 337
Tucson, AZ 85749-9399
tel 817-383-0000

Pac-Comm, Inc
4413 N Hesperides St
Tampa, FL 33614-7618
tel 813-874-2980

These same sources can supply modems for use with the 9600-baud packet satellites.

A few of the digital satellites are actually intended for monitoring purposes only. For example, DOVE (or DO-17) simply transmits. It is capable of transmitting digitized voice messages or packet telemetry. DOVE's primary mission is education. Its target audience is schoolchildren and others interested in learning about satellites and Amateur Radio. DOVE also transmits telemetry and short bulletin messages uploaded by the command stations. These transmissions use 1200-baud *AFSK* packet—the same kind of transmission used

Circular Polarization

As VHF-and-above operators know, antenna polarization is critical to effective communications between stations. This presents a problem when using a satellite because the satellite changes its orientation with respect to the ground station as the satellite moves. Using horizontal or vertical polarization would result in deep fades when the satellite's antenna was cross-polarized to that of the ground station.

Fortunately, there's an answer: circular polarization (CP). A CP signal rotates in polarization once per cycle. That is, the polarization is constantly changing. This kind of polarization is immune to orientation changes of the satellite (assuming the satellite antenna is pointed directly at the ground station while the satellite spins). Circular polarization can be "right hand" (RHCP) or "left hand" (LHCP). Amateur satellites that use CP antennas typically use RHCP. While you can receive CP signals on a linearly polarized antenna, you lose about half the signal as compared to using a CP antenna.

Although RHCP is the norm, you may experience a change to LHCP signals when the spacecraft antenna is pointed away from you; often the antenna sidelobes have odd polarizations. Being able to switch between RHCP and LHCP is a useful antenna feature, and many satellite antennas have this option.

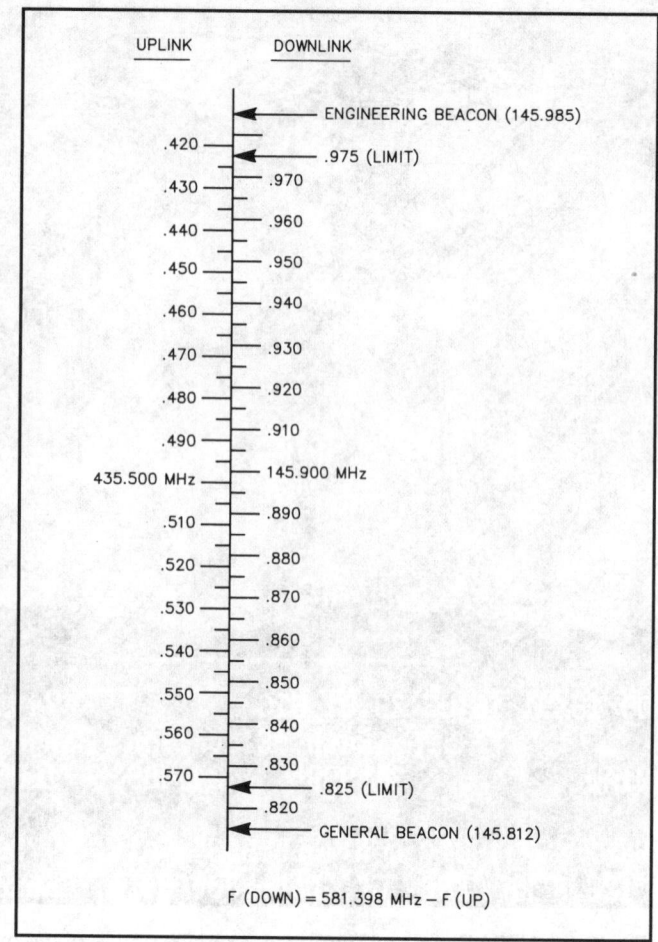

Fig 13-4—OSCAR 13 Mode B Frequency Translation Chart.

by terrestrial VHF packet stations. The strong signals of DOVE, coupled with the compatibility of its packet transmissions, means that literally tens of thousands of stations are already equipped to receive DOVE. If you operate 2-meter packet, your station is DOVE-capable right now! AMSAT has software that will help decode the telemetry from DOVE. Monitoring DOVE is a great way to get your feet wet in digital satellite operation.

Other transmit-only satellites include UO-11 and WO-18. UO-11 transmits on 145.825 MHz sending mostly telemetry, although it also can send digitized speech. While UO-11 does contain a digital store-and-forward system, its use is restricted because of the relatively small amount (128 kBytes) of on-board memory. UO-11 also includes several scientific/educational experiments.

WEBERSAT (WO-18) transmits using the 1200-baud PSK standard. It contains an on-board camera for Earth imaging. Again, software for decoding and viewing WEBERSAT images is available from AMSAT.

SATELLITE OPERATING

Linear Transponders

Operating via an analog transponder is much like operating on the HF bands. Then again, it's also similar to repeater operation since the transponder does essentially the same job as a repeater. But satellite operation is unique in many ways. The most dramatic difference is that in satellite operation you can hear the satellite's transponder repeating your own signal. While this is marvelous in itself, it leads to some different operating techniques. This is particularly true when you couple this characteristic with the need to compensate the operating frequency for the effects of Doppler shift.

To compensate for Doppler, the two stations in a QSO must occasionally adjust frequency. But there is yet another problem: Since the amount of frequency shift you experience is dependent on the satellite's speed *relative to your location*, and since the satellite has different relative speeds at different stations, two QSOs that were originally separated in frequency can "walk" into one another as the relative speeds change.

To minimize this problem, satellite operators use a technique that requires the least amount of retuning. One of the two stations in the QSO is chosen as a reference. This station holds the *lower* of its two frequencies (transmit or receive) constant. Since this station can hear its own signals (full duplex), it adjusts the *higher* frequency to match. That is, if the lower frequency is the receive (as in Mode B, for example), that remains constant while the transmit frequency is "swooped" onto the receive frequency. The transmit frequency is subsequently readjusted to keep the downlink signal stable. If the *transmit* frequency is the lower one (as in Mode J), it is held constant while the receive frequency is adjusted to match. Because different stations experience different Doppler shifts, only one of the stations can use this procedure. The other station adjusts its receive frequency to receive the first station, then adjusts its transmit frequency to match.

The technique described above works best when the reference station experiences the least amount of Doppler drift. Of the two stations, the one furthest removed from the ground track experiences the least drift. If you are using a tracking method that tells you the maximum elevation of the pass, you can compare that to the maximum elevation of the other station. The station with the lesser number is further from ground track of the satellite. It's worth noting that the Phase 3 satellites don't suffer from severe Doppler effects except when near perigee. We generally use the same procedure, though, since it still minimizes what Doppler effects are present.

Finding the Downlink Signal

Since a satellite's linear transponders aren't channelized,

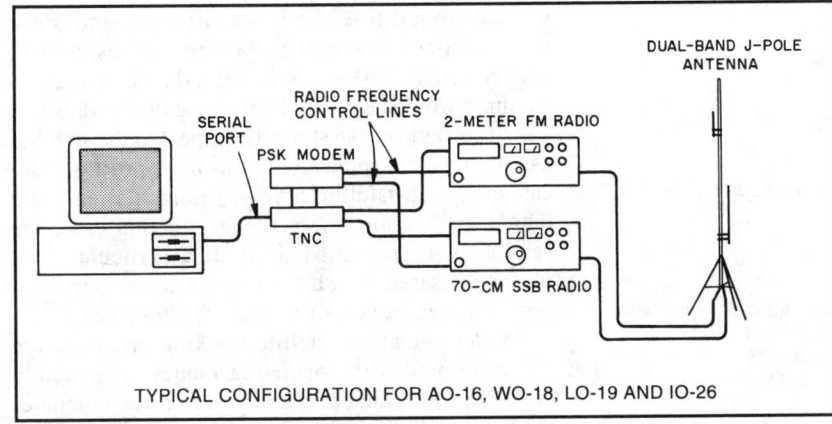

Fig 13-5—A typical Microsat ground station. For reception of DO-17 (DOVE) telemetry, however, only a 2-meter-FM transceiver and a standard 1200-baud TNC are required.

you'll need to be able to figure out where your signal will appear in the downlink passband when you're transmitting on a particular uplink frequency. While Doppler shift complicates this question, you can easily determine the frequency within a few kilohertz. This is usually sufficient since you can hear the downlink signal yourself and tune it exactly.

The linear transponder acts like a mixer. That is, it adds the received frequency to, or subtracts it from, a local oscillator. If you know the frequency of the transponder's oscillator, determining the downlink frequency for any given uplink frequency is a snap. For example, AO-13's Mode B oscillator operates at 581.398 MHz. The equation for calculating the downlink frequency is:

$$f_{down} = 581.398 - f_{up}$$

The transponder accepts input signals from 435.423 to 435.573 MHz. Plugging these numbers into the equation results in a downlink band of 145.975 to 145.825. Note that the downlink is *inverted*, meaning that the higher the frequency appears in the uplink passband, the lower it appears in the downlink passband. This is shown graphically in Fig 13-4.

Two important results occur when inverted transponders are used. First, the total Doppler shift is minimized since the shift experienced on the downlink partially compensates for the (inverted) shift experienced on the uplink. (The uplink shifts more than the downlink if it's on a higher frequency.) Second, SSB signals are inverted. Although there isn't an established rule, most stations transmit on USB, resulting in LSB downlink signals.

An informal band plan has evolved for linear transponder operation. Generally, the lower third of the passband is used for CW operation, the upper third for SSB, and the middle third for mixed CW/SSB and special modes. AO-10 had a more complete band plan that included reserved channels for special functions. The difficulties now being encountered by AO-10 have rendered this band plan moot.

Don't Be a Transponder Hog: A linear transponder retransmits a faithful reproduction of the signals received in the uplink passband. This means that the loudest signal received will be the loudest signal retransmitted. If a received signal is so strong that retransmitting it would cause an overload, automatic gain control (AGC) within the transponder reduces the transponder's amplification. However, this reduces all of the signals passing through it!

Strong signals don't receive any benefit from being loud since the downlink is "maxed out" anyway. Instead, all of the other users are disrupted since their signals are reduced. Avoiding this condition is just common sense—and good operating practice. It's easy, too: just note the signal level of the satellite's beacon transmission and adjust your transmit power to a level just sufficient to make your downlink appear at the same level as the beacon. Any additional power is too much!

Digital Transponders

The subject of digital transponders is a bit more complex than that of analog transponders. The first amateur digital satellite communications were performed through analog transponders. While it was successful, it wasn't particularly efficient. Analog transponders are optimized for low-duty-cycle modulations such as CW and SSB. Digital transmissions are usually constant-power signals that take a disproportionate fraction of the transponder power. This isn't a problem with digital transponders since they demodulate received signals and transmit using an independent digital transmitter. More importantly, digital transponders receive transmissions from the ground, process them using an onboard computer and respond appropriately. This process lets the digital transponder screen the incoming data for errors and enables one of the most important and useful results of this type of system: *store-and-forward* message relay.

Describing digital satellite operation is a little like shooting at a moving target. The modulation schemes are fixed by hardware. What the satellite does with the received data is wholly dependent on the programming of the on-board computer—programming that can be changed from the ground command stations. As you might guess, the operational specifics of a given digital satellite may change over time.

The "file server" satellites can be accessed using the AMSAT PACSAT software. The IBM PC software package (software is also available for the Apple Macintosh) consists of several programs: *PB* receives broadcasts from the satellites, *PG* lets you transfer files to and from the satellite, and *PHS* and *PFHADD* process files for satellite transfer. *WISP* is a complete easy-to-use software package designed to work in Microsoft *Windows*. These programs or their equivalent must be used to access the PACSATs. Both software packages are available from AMSAT,

Transponder Schedules

We've seen how tracking the satellite's location is important for scheduling satellite operating. But most satellites have a schedule of their own. Those that have several transponders may use different modes at different times. And some may not have sufficient power to operate their transponders all the time, especially during periods of extended eclipse.[2] Using the satellite requires that you know when the transponder will be operating and what mode will be available.

AO-13 Operating Schedule

In the case of AO-13, the directional antennas on the spacecraft are the primary determinant of which mode is used. The orientation of AO-13 in space is fixed. This means that if

Table 13-2
Digital Transponder Frequencies and Modes

Satellite	Uplink(s) (MHz)	Downlink(s) (MHz)	Data Format
AMSAT-OSCAR-16	145.90 145.92 145.94 145.96	437.025 437.05 2401.10	1200 bit/s PSK AX.25
DOVE-OSCAR-17	None	145.825 2401.22	Digitized voice with 1200 bit/s AFSK AX.25 (FM) telemetry
WEBERSAT-OSCAR-18	None	437.075 437.10	1200 bit/s PSK AX .25
LUSAT-OSCAR-19	145.84 145.86 145.88 145.90	437.125 437.15	1200 bit/s PSK AX .25
UoSAT-OSCAR 22	145.90 145.975	435.120	9600 bit/s FSK (FM)
KITSAT-OSCAR 23	145.85 145.90	435.175	9600 bit/s FSK (FM)
KITSAT-OSCAR 25	145.87 145.98	436.50	9600 bit/s FM
ITAMSAT-OSCAR 26	145.875 145.900 145.925 145.950	435.870	1200 bit/s PSK

the antennas point directly at Earth at one point in the orbit, they'll point directly *away* from the Earth when the satellite is at the opposite point in the orbit. Normally, AO-13 is oriented so that the antennas point at the Earth when the satellite is at apogee. As the satellite moves away from apogee, the antennas begin to point off into space, making the satellite unusable in modes that require high-gain antennas—Mode S. AO-13 is then switched by the on-board control computer to Mode B, which has wider beamwidth antennas. As the satellite nears perigee, the computer selects omnidirectional antennas for Mode B.

Unfortunately, antenna direction is not the only consideration that affects the positioning of AO-13. The solar panels on the satellite are affixed to the sides of the structure. To ensure sufficient sunlight to keep AO-13 in operation, it must be oriented so that sunlight strikes the panels directly. Depending on the particulars of the orbital plane and the sun's location, this sometimes requires that the satellite be tilted so that the antennas don't point directly at the Earth at apogee. Usually the pointing difference is fairly small, but it does result in a change in transponder operating schedules from time to time. The operating schedule isn't changed frequently—every few months is typical—but you will need to know it. The schedule can be found in AMSAT bulletins and in *OSCAR Satellite Report*, among other places.

The schedule is based on the satellite's location in its orbit using a parameter called *mean anomaly* or "MA." (This number is also called "phase" by some references to distinguish it from a Keplerian element of the same name.) MA is simply a number from 0 to 255 that specifies the time since perigee. Since the orbit is 256 time slices around, apogee occurs at MA 128. Ideally, the Mode L and S antennas would be pointed directly at the Earth at MA 128. The angle of the sun with respect to the satellite usually requires some alterations in its position. Occasionally the satellite may be oriented at an angle that doesn't allow these modes. In this case, the satellite will stay in Mode B. If a particular orbit places the satellite in a shadow for a long period of time, the transponders are shut off altogether.

Modern amateur satellite tracking programs are set up to predict the operating modes. You simply enter the transponder schedule and let the computer determine which mode the satellite will select at any given time. Even the older programs generally will compute the MA value for you, making mode determination fairly easy. If you aren't using a computer program to determine the mode, it still isn't difficult to figure it out.

The orbital period of OSCAR 13 is about 687 minutes. That means each "slice" of MA is 687/256 or 2.6836 minutes. Suppose you know that perigee (MA 0) occurs at 8:20 AM and the transponder schedule shows Mode B to be active from MA 40 to MA 165. MA 40 occurs at 40 × 2.6836 or 107 minutes after perigee. With a perigee at 8:20 AM, MA 40 is at 10:07 AM. Likewise, MA 165 is at 8:20 AM plus 165 × 2.6836 minutes, or 3:43 PM. Therefore, the AO-13 transponder will be in Mode B from 10:07 AM to 3:43 PM. (Of course, you still have to use your tracking technique to see if the satellite will be in view of your station during that period.)

But how do you know when perigee occurs? One way is to monitor the ARRL satellite bulletins from W1AW. The reference orbits for Phase 3 satellites—including AO-13—specify the first apogee of the day for that satellite. Knowing that AO-13's orbital period is about 687 minutes, we can reason that perigee must occur 687/2 = 343.5 minutes before apogee. With this knowledge we can easily calculate perigee time and go on to compute the transponder schedule.

Other Satellite Schedules

Other satellites have operating limitations for various reasons. AO-10 is operable only when its power system is adequate to run the transponder. Since AO-10 is no longer under effective control due to its computer failure, operability is a matter of luck. You'll know the batteries are being depleted when the frequency of the transponder becomes unstable and the downlink signals start to warble (known as FMing). If you hear this, *don't use the transponder*. AMSAT issues bulletins defining when the transponder seems to be usable.

Manned Missions

One of the more exciting operations available to today's amateur is communicating with men and women in space. At present, a semi-permanent 2-meter FM voice and packet station is operational from the Russian Mir space station. While operation is generally unscheduled, numerous voice and packet contacts with Mir have occurred on 145.55 MHz. The packet system aboard Mir uses standard 1200-baud AFSK packet. It often runs in the "mailbox" mode, allowing ground stations to exchange messages with the cosmonauts aboard Mir.

SAREX, the Shuttle Amateur Radio EXperiment, is a

Astronaut Linda Godwin, N5RAX, Pauses during her mission duties to operate the SAREX station aboard the space shuttle *Endeavour*. (Photo courtesy of NASA)

For more information...
Publications

The Satellite Experimenter's Handbook, 2nd edition, ARRL (Order #3185, $20). This is *the* book to have. It's a complete guide to amateur satellites.

The *AMSAT Journal* comes with your membership in AMSAT ($30 in the United States, $36 Mexico and Canada and $45 in all other areas). Contains news, technical articles and commentary. Available from AMSAT, PO Box 27, Washington, DC 20044.

OSCAR Satellite Report, a semimonthly newsletter, contains late-breaking satellite news, current Keplerian elements and commentary. *Satellite Operator*, a monthly newsletter, contains feature articles and news reports, plus Keplerian elements for nonamateur satellites. Both are available from R. Myers Communications, PO Box 17108, Fountain Hills, AZ 85269-7108.

A Beginner's Guide to OSCAR 13 ($7) and *Digital Satellite Guide* ($10) walk you through the process of setting up your station for OSCAR 13 or the PACSATs, respectively, then show you how to operate these satellites. Available from AMSAT.

Computer Bulletin Boards

DRIG Datalink BBS	214-394-7438
Celestial RCP/M BBS	513-253-9767
N8EMR BBS	614-895-2553

AMSAT Information Nets

Area	Day	Time	Frequency
International	Sunday	1800 UTC	14.282 MHz
US East Coast	Tuesday	2100 EST/EDT	3.840 MHz
US Central	Tuesday	2100 CST/CDT	3.840 MHz
US West Coast	Tuesday	2000 PST/PDT	3.840 MHz

continuing series of Amateur Radio operations from US space shuttle missions. While the focus of these missions has evolved toward space-to-classroom use of Amateur Radio, there are still many opportunities for the average ham to communicate with shuttle astronauts. The current SAREX suite comprises several equipment configurations: a simple battery-operated hand-held transceiver for FM voice-only QSOs, a voice-plus-packet setup (using 1200-baud AFSK packet), and a voice/packet/slow-scan televi-

sion system with fast-scan uplink capability. Which configuration can be flown on a shuttle mission depends mostly on the available stowage space, which in turn depends on the duration of the flight (because longer flights need to devote more space to storage of food and other consumables).

SAREX uses split-frequency operation. A 600-kHz split is used to accommodate fixed-offset 2-meter FM radios. It is very important *not* to transmit on the SAREX downlink frequency. Aside from the futility of doing so—the astronaut isn't listening to that frequency—you'll likely interfere with other stations that are trying to receive the SAREX transmissions.

Because SAREX is of secondary importance to the shuttle mission as a whole, SAREX operations are at the mercy of the needs of the primary payload. Open operations are also scheduled around space-to-classroom links, which complicates the picture somewhat. If you plan to operate via SAREX, it's wise to monitor W1AW and WA3NAN (the Goddard Space Center Amateur Radio Club station) for late-breaking schedule changes.

Try It Yourself

As you can see, it's easier than ever to add satellite communications to your list of operating modes. With new technolo-gies on the horizon, it promises to become even easier in the future. Just like CW, RTTY, DXing, contesting and so on, satellites can become an integral part of your enjoyment of Amateur Radio. Don't be fooled into thinking that satellite operating is only for wealthy hams with lots of equipment and real estate. It's for you too!

Notes
[1] Available for $7 ($8 Canada and Mexico, $10 elsewhere) from AMSAT, PO Box 27, Washington, DC 20044.
[2] All amateur satellites operate from battery systems that require recharging via the satellite's solar panels. As a satellite orbits the Earth, it passes from sunlight into the Earth's shadow and back again. Occasionally, due to the geometry of the orbit and/or the orientation of the satellite to the sun, the panels may not receive enough sunlight to recharge the batteries and operate the transponder. Under these circumstances, the satellite control stations have no option but to turn off the transmitters lest the battery system be permanently damaged.

Emergency Communications

People in desperate need of help can depend on Amateur Radio to serve them. Since 1913, ham communicators have been dedicated volunteers for the public interest, convenience and necessity by handling free and reliable communications for people in disaster-stricken areas, until normal communications are restored.

RICHARD R. REGENT, K9GDF
5003 SOUTH 26TH ST
MILWAUKEE, WI 53221

a particular response and careful planning. Large cities usually have capable relief efforts handled by paid professionals, and there always seems to be some equipment and facilities that remain operable. Even though damage may be more concentrated outside a city, it can be remote from fire-fighting or public-works equipment and law enforcement authorities. The rural public then, with

Emergency communication is an Amateur Radio communication directly relating to the immediate safety of human life or the immediate protection of property, and usually concerns disasters, severe weather or vehicular accidents. The ability of amateurs to respond effectively to these situations with emergency communications depends on practical plans, formalized procedures and trained operators.

PLANS

More than any other facet of Amateur Radio, emergency communication requires a plan—an orderly arrangement of time, talent and activities that ensures that performance is smooth and objectives are met. Basically, a plan is a method of achieving a goal. Lack of an emergency-communications plan could hamper urgent operations, defer crucial decisions or delay critical supplies. Be sure to analyze what emergencies are likely to occur in your area, develop guidelines for providing communications after a disaster, know the proper contact people and inform local authorities of your group's capabilities.

Start with a small plan, such as developing a community awareness program for severe-weather emergencies. Next, test the plan a piece at a time, but redefine the plan if it is unsatisfactory. Testing with simulated-emergency drills teaches communicators what to do in a real emergency, without a great deal of risk. Finally, prepare a few contingency plans just in case the original plan fails.

One of the best ways to learn how to plan for emergencies is to join your local Amateur Radio Emergency Service (ARES) group, where members train and prepare constantly in an organized way. Before an emergency occurs, register with your ARRL Emergency Coordinator (EC). The EC will explain to civic and relief agencies in your community what the Amateur Service can offer in time of disaster. ARES and the EC are explained later in this chapter.

Communications for city or rural emergencies each require

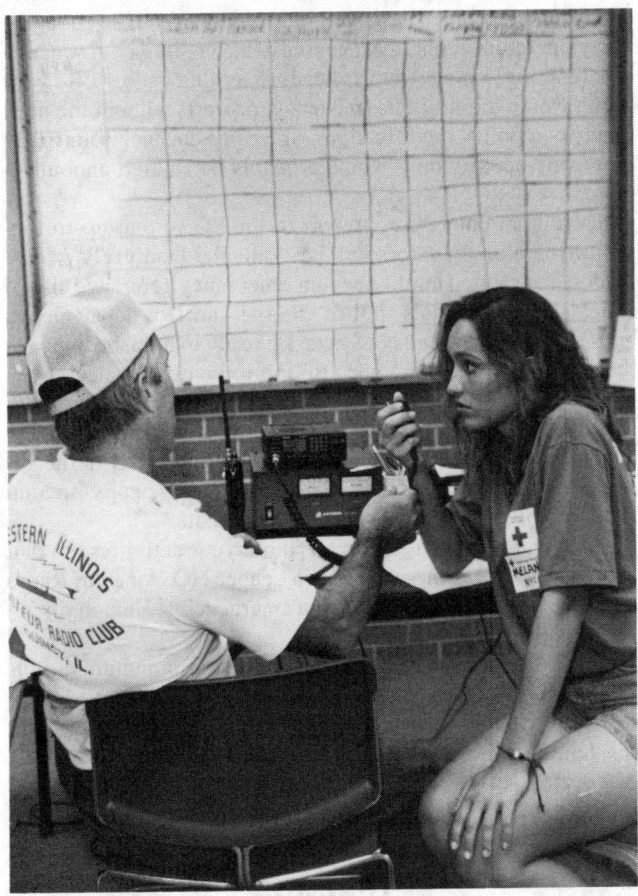

Jim Funk, N9JF, and his daughter Melanie, N9IQV, operate from the Adams County American Red Cross Chapter House in Quincy, Illinois, during the Great Flood of 1993. (*photo by Sandy Martin*)

few volunteers spread over a wide area, may be isolated, unable to call for help or incapable of reporting all of the damage.

It's futile to look back regretfully at past emergencies and wish you had been better prepared. Prepare yourself now for emergency communications by maintaining a dependable transmitter-receiver setup and an emergency-power source. Have a plan ready and learn proper procedures.

PROCEDURES

Besides having plans, it is also necessary to have procedures—the best methods or ways to do a job. Procedures become habits, independent of a plan, when everyone knows what happens next and can tell others what to do. Actually, the size of a disaster affects the size of the response, but not the procedures.

Before disasters occur, there are many existing procedures; for example, how to correctly coordinate or deploy people, equipment and supplies. There are procedures to use a repeater and an autopatch, to check into a net and to format or handle traffic. Because it takes time to learn activities that are not normally used every day, excessively detailed procedures will confuse people and should be avoided.

Specific Procedures

Your EC will have developed a procedure to activate the ARES group, but will need your help to make it work. A telephone alerting "tree" call-up, even if based on a current list of phone numbers, might fail if there are gaps in the calling sequence, members are not near a phone or there is no phone service.

Consider alternative procedures, use alerting tones and frequent announcements on a well-monitored repeater to round up many operators at once. An unused 2-meter simplex frequency can function for alerting; instead of turning radios off, your group would monitor this frequency for alerts without the need of any equipment modifications. Since this channel is normally quiet, any activity on it would probably be an alert announcement.

During an emergency, report to the EC so that up-to-the-minute data on operators will be available. Don't rely on one leader; everyone should keep an emergency reference list of relief-agency officials, police, sheriff and fire departments, ambulance service and NTS nets. Be ready to help, but stay off the air unless there is a specific job to be done that you can efficiently handle. Always listen before you transmit. Work and cooperate with the local civic and relief agencies as the EC suggests; offer these agencies your services directly in the absence of an EC. During a major emergency, copy special W1AW bulletins for the latest developments.

Afterward, let your EC know about your activities, so that a timely report can be submitted to League HQ. Amateur Radio has won glowing public tribute in emergencies. Help maintain this record.

For a comprehensive study of emergency communications procedures, see the ARRL *Emergency Coordinator's Manual* (FSD-9). Contact the ARRL HQ Field Services Department for details on how to obtain a copy. "Experience is the worst teacher: it gives the test before presenting the lesson" (Vernon's Law). Train and drill now so you can be prepared for an emergency.

TRAINING

Amateurs need training in operating procedures and communications skills. In an emergency, radios don't communicate, but people do. Because amateurs with all sorts of

Southeastern Floods

Macon (Georgia) Amateur Radio Club members, on July 4, 1994, begin communicating reports of damage caused by spring flooding and Tropical Storm Alberto. Torrential rains pour down more than 12 inches of water on ground already saturated by several previous days' precipitation. Amateurs report bridges down, roads out, homes under water, and youth camps cut off from roads. Red Cross officials open shelters where hams relay instructions and messages. Meanwhile mobile and home-based operators assist the Georgia Emergency Management Association to keep in touch with officials in Atlanta.

The ARES works with Red Cross and National Weather Service staff. The Red Cross deploys damage assessment teams who need communications in order to decide the structural safety of bridges and other public facilities.

As the Ocmulgee River nears flood stage, the situation becomes alarming. But on the scene are hams, often the first to see the need for evacuation or support. A raging river breaks over its banks and levees to inundate a water treatment plant. Fresh water is dispersed in trucks to the most needy sites, coordinated with Amateur Radio communications. This frees law enforcement officers to handle a backlog of other duties.

After the weather abates on July 8, net operators relay Health-and-Welfare messages. In the southeastern floods disaster Amateur Radio communication is found to be flexible and adaptable by providing trained operators and hardware to others. It is effective at providing a total communications network.

varied interests participate, many of those who offer to help may not have experience in public-service activities. Rarely are there enough trained operators, especially if a crisis persists for a long time.

Proper disaster training replaces chaotic pleas with smooth organized communications. The ARRL recognizes the need for emergency preparedness and emergency-communications training through sponsorship of the ARES. Well-trained communicators respond during drills or actual emergencies with quick, effective and efficient communications. Each understands his or her role in the plan and sets a good example by knowing the proper procedures to use.

Whether training takes place with programs at club meetings, on-the-air or with a personal approach, the basic subjects should cover emergency communications, traffic handling, net or repeater operation and technical knowledge. Get as many people involved as possible to learn how emergency communications should be handled. Explain what's going on and assign each participant a useful role.

Practical on-the-air activities, such as the ARRL's Field Day and Simulated Emergency Test offer training opportunities on a nationwide basis for individuals and groups. Participation in such events reveals weak areas where discussions and more training are needed. In addition, drills and tests can be designed specifically to check dependability of emergency equipment, or to rate training in the local area.

Field Day

The League's Field Day (FD) gets more amateurs out of their cozy shacks and into tents on hilltops than any other event. You may not be operating from a tent after a disaster but the training you will get from FD is invaluable.

In the League Field Day event, a premium is placed on sharp operating skills, adapting equipment that can meet challenges of emergency preparedness and flexible logistics. Amateurs assemble portable stations capable of long-range communications at almost any place and under varying conditions. Alternatives to commercial power in the form of generators, car batteries, windmills or solar power are used to power equipment to make as many contacts as possible. FD is held on the fourth full weekend of June, but enthusiasts get the most out of their training by keeping preparedness programs alive during the rest of the year.

Simulated Emergency Test

The ARRL Simulated Emergency Test (SET) builds emergency-communications character.

The purposes of SET are to:

- Help amateurs gain experience in communicating using standard procedures under simulated emergency conditions, and to experiment with some new concepts.
- Determine strong points, capabilities and limitations in providing emergency communications to improve the response to a real emergency.
- Provide a demonstration, to served agencies and the public through the news media, of the value of Amateur Radio, particularly in time of need.

The goals of SET are to:

- Strengthen VHF-to-HF links at the local level, ensuring that ARES and NTS work in concert.
- Encourage greater use of digital modes for handling high-volume traffic and point-to-point Welfare messages of the affected simulated-disaster area.
- Implement the Memoranda of Understanding between the League, the users and cooperative agencies.
- Focus energies on ARES communications at the local level. Increase use and recognition of tactical communication on behalf of served agencies; using less amateur-to-amateur formal radiogram traffic.

Help promote the SET on nets and repeaters with announcements or bulletins, or at club meetings and publicize it in club newsletters. SET is conducted on the third full weekend of October. However, some groups have their SETs any time during the period of September 1 through November 30, especially if an alternate date coincides more favorably with a planned communications activity and provides greater publicity. Specific SET rules are published in *QST*.

Drills and Tests

A drill or test that includes interest and practical value makes a group glad to participate because it seems worthy of their efforts. Formulate training around a simulated disaster such as weather-caused disasters or vehicle accidents. Elaborate on the situation to develop a realistic scenario or have the drill in conjunction with a local event. Many ARRL Section Emergency Coordinators (SECs) have developed training activities that are specifically designed for your state, section or local area.

During a drill:

1) Announce the simulated emergency situation, activate the emergency net and dispatch mobiles to served agencies.

2) Originate messages and requests for supplies on behalf of served agencies by using tactical communications.

3) Use emergency-powered repeaters and employ digital modes.

4) As warranted by traffic loads, assign liaison stations to receive traffic on the local net and relay to your section net. Be

Chris Sundlee, KC6UIQ, at one of the stations in the mobile unit at the Pepperdine University forward fire base. The efforts of Chris and many other hams played a vital role in the fires that plagued California in 1993. (*KD6NEA photo*)

sure there is a representative on each session of the section nets to receive traffic coming to your area.

After a drill:

1) Determine the results of the emergency communications.

2) Critique the drill.

3) Report your efforts, including any photos, clippings and other items of interest, to your SEC or ARRL HQ.

ARRL Emergency Coordinator and Certification Course

The League has a certification program to provide training and formal recognition of amateur achievement in the field of emergency communications. This course is administered to ECs (or potential ECs) by the Section Emergency Coordinator for each section. The ARRL also offers the public-service-oriented awards program of certificates for Public Service Honor Roll, Emergency Communications Commendation and Public Service Commendation.

Net Operator Training

Network discipline and message-handling procedures are fundamental concepts of Amateur Radio operation. Training should involve as many different operators as possible in Net Control Station and liaison functions; don't have the same operator performing the same functions repeatedly. There should be plenty of work for everyone. Good liaison and cooperation at all levels of NTS requires versatile operators who can operate either phone or CW. Even though phone operators may not feel comfortable on CW and vice versa, encourage net operators to gain familiarity on both modes by giving them proper training.

The liaison duties to serve between different NTS region net cycles as well as between section nets are examples of the need for versatile operators.

Ask your ARRL Section Traffic Manager (STM) to visit your club to conduct a training seminar and to provide any operating tips pertaining to traffic nets in your section. If no local traffic net exists, your club should consider initiating a net on an available 2-meter repeater. Coordinate these efforts with the STM and the trustee(s) of the repeater you'll use. Encourage club members to participate in traffic-handling activities, either from their home stations or as a group activity from a message center. For more information on NTS and traffic handling, see Chapter 15.

Los Angeles Area Earthquake

The earthquake tremor jolts Los Angeles area residents from their sleep on January 17, 1994, at 4:31 AM. Without warning, 300 schools are seriously damaged, thousands of small business crumble in ruins and power lines tear apart. The quake severed three aqueducts, cutting off water service to thousands of Angelenos. Older, unreinforced masonry buildings and modern wood-frame apartments built on top of ground floor garages sustain some of the heaviest damage. The tremor destroys thousands of homes. Eleven major roads are blocked from debris or damaged. The Santa Monica Freeway, hardest hit, was scheduled to be steel retrofit in just one month. Gas lines and water mains burst sending geysers of flames and water into the air.

SEC K6TMJ activates the ARES Emergency Communications Van at the San Fernando Valley area hospital to handle emergency traffic. He moves to recovery efforts the next day near the Devonshire Police Station in Northridge. Officials use fiber-optics cameras and dogs to find victims. Medical facilities need more trained personnel to cope with massive casualties. A Red Cross shelter is set up in the Birmingham High School.

AA6RK starts his generator running since there is no electricity in the San Fernando Valley where he lives. He relays more than 300 messages on 20 meters both in and out of the epicenter, the 818 telephone area code, in two days.

Amateurs comply with rules not to release names of survivors for the first 48 hours after the disaster and the dusk to dawn curfew. Other hams send traffic into the Valley using repeaters and to the Filmore Memorial building shelter operations. Some packet radio bulletin boards handle over 4000 messages for the San Fernando and Santa Clarita Valleys. In the Los Angeles Section alone, ARES had more than 100 hams volunteer for hospital and agency emergency communications services.

It is LA's worst earthquake since the 1971 Sylmar quake. Although 6.6 on the Richter scale is violent, it is not The Big One. Yet 55 people died due to this quake. It will be remembered that Amateur Radio definitely provided significant reliable communications to many relief agencies in a time of crisis.

METHODS OF HANDLING INFORMATION

Emergency Operations Center

Amateur Radio emergency communications frequently use the combined concepts of a Command Post (CP) and an Emergency Operations Center (EOC). See Fig 14-1.

Although a Command Post controls initial activities in emergency and disaster situations, the CP may be unapparent because it is self-starting and automatic. The CP's general procedure is to assess the situation, report to a dispatcher and ask for equipment and people.

Consider an automobile accident where a citizen or an amateur, first on the scene, becomes a temporary Command Post to call or radio for help. A law-enforcement officer is dispatched to the accident scene in a squad car and, upon arriving, takes over the CP tasks. Incidentally, a Command Post may expand into multiple CPs or move to accommodate the situation. Relief efforts, like those in this simple example of an automobile accident, begin when someone takes charge, makes a decision and directs the efforts of others.

The Emergency Operations Center responds to a Command Post by dispatching equipment and helpers, anticipating needs to supply support and assistance, and may send more equipment and people to a staging area to be stored where they can be available almost instantly.

If the status of an accident changes (a car hits a utility pole, which later causes a fire), the CP gives the EOC an updated report then keeps control until the support agencies arrive and take over their specific responsibilities: Injuries—medical; fires—fire department; disabled vehicles—law enforcement or tow truck; and utility poles—utility company. By being outside the perimeter of dangerous activities, the EOC can use the proper type of radio communications, concentrate on gathering data from other agencies and then provide the right response.

As an analogy, think of the CPs, who request action and provide information, being similar to net participants checking into an amateur net with emergency or priority traffic. The EOC, who coordinates relief efforts, then functions as a Net Control Station.

Whether there is a minor vehicle accident or a major disaster operation, the effectiveness of the amateur effort in an emergency depends mainly on handling information.

Incident Command System

The Incident Command System (ICS) is a management tool

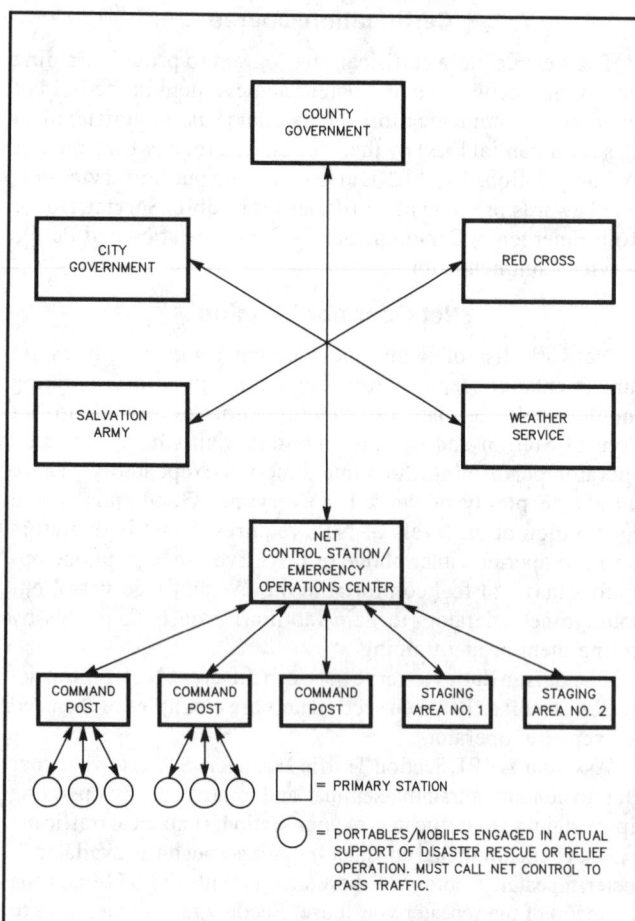

Fig 14-1— The interaction between the EOC/NCS and the command post(s) in a local emergency.

that provides a coordinated system of command, communications, organization and accountability in managing emergency events and is rapidly being adopted by professional emergency responders throughout the country. Amateurs should become familiar with ICS to work with agencies in a variety of multiple jurisdictions and political boundaries situations.

Incident Command Systems use:
- Clear text and common terms. Participants are expected to be familiar with ICS terminology. When the Incident Commander orders "a strike team of Type 2 trucks," everyone affiliated with filling the order knows exactly what is being requested.
- Unified Command. The Incident Commander is the only boss and is responsible for the overall operation.
- Flexibility. Functions such as planning, logistics, operations, finance and working with the press are described in detail so the organization size can change to match the particular incident's requirements. The IC can contract to a single individual for a small incident or expand to a Command Staff for a large incident.
- Concise Span of Control. Since management works well with a small number of people, the ICS typically uses a "strike team": a group of about five people having similar resources, plus a leader, who use common communications.

In some areas the ICS evaluates and determines what resources will be needed to start recovery. Amateur communicators are within the Logistics Section, Service Branch and Communications Unit of an ICS (Fig 14-2).

Tactical Traffic

Whether traffic is tactical, by formal message, packet radio or amateur television, success depends on knowing which to use.

Tactical traffic is first-response communications in an emergency situation involving a few operators in a small area. It may be urgent instructions or inquiries such as "send an ambulance" or "who has the medical supplies?" Tactical traffic, even though unformatted and seldom written, is particularly important in localized communications when working with government and law-enforcement agencies.

The 146.52 MHz FM calling frequency—or VHF and UHF

repeaters and net frequencies (see Fig 14-3)—are typically used for tactical communications. This is a natural choice because FM mobile, portable and fixed-station equipment is so plentiful and popular.

One way to make tactical net operation clear is to use tactical call signs—words that describe a function, location or agency. Their use prevents confusing listeners or agencies who are monitoring. When operators change shifts or locations, the set of tactical calls remains the same. Amateurs may use tactical call signs like "parade headquarters," "finish line," "Red Cross," "Net Control" or "Weather Center" to promote efficiency and coordination in public-service communication activities. However, amateurs must identify their station operation with its FCC-assigned call sign at the end of a transmission or series of transmissions and at intervals not to exceed 10 minutes.

Another tip is to use the 12-hour local-time system for time and dates when working with relief agencies, unless they understand the 24-hour or UTC systems.

Taking part in a tactical net as an ARES team member requires some discipline and following a few rules:

1) Report to the Net Control Station (NCS) promptly as soon as you arrive at your station.

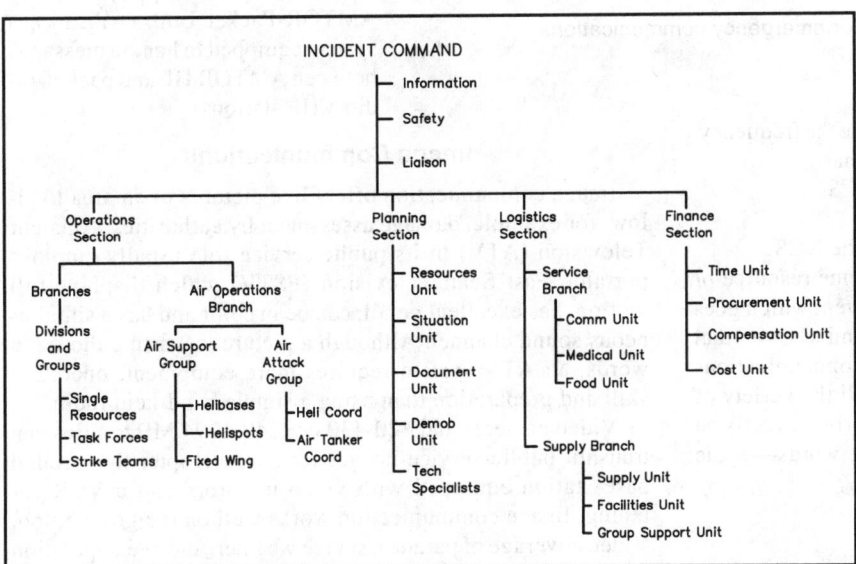

Fig 14-2—The Incident Command System structure.

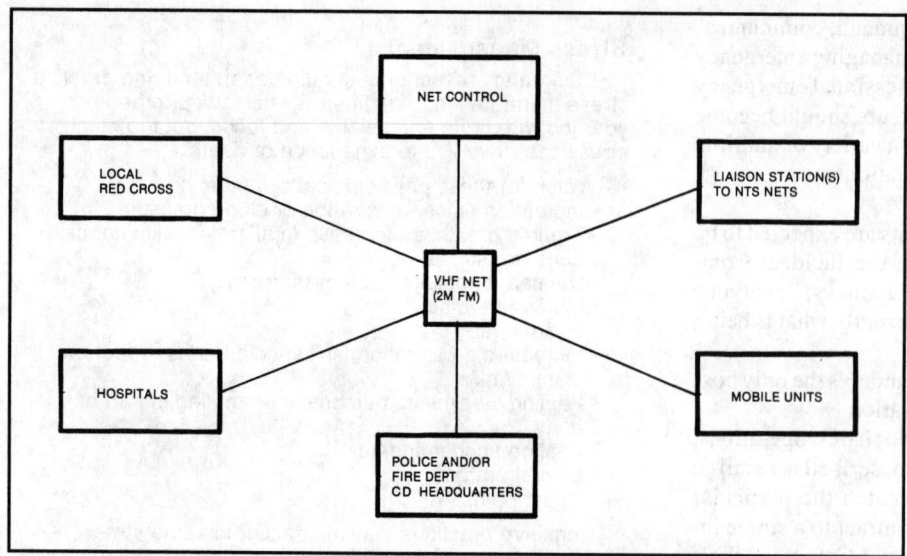

Fig 14-3—Typical station deployment for local ARES net coverage in an emergency.

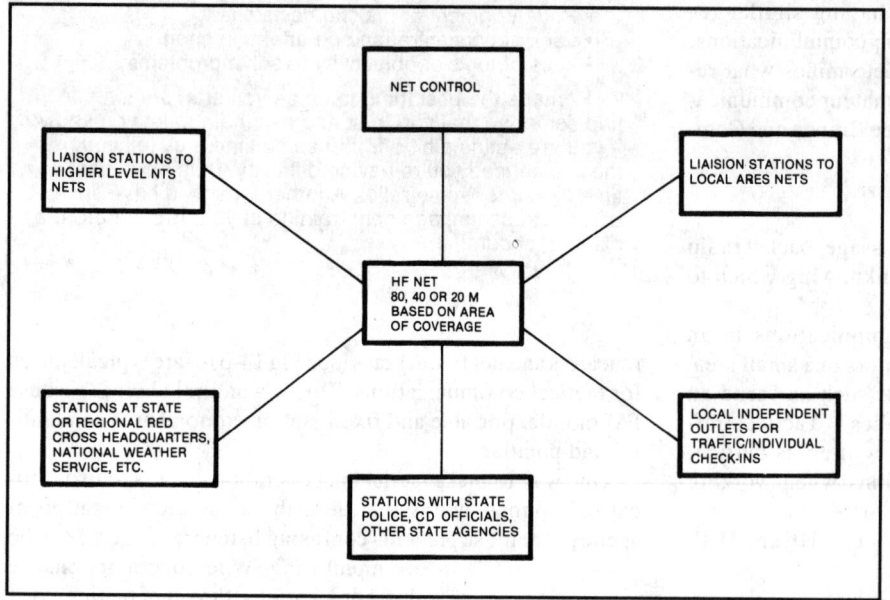

Fig 14-4—Typical structure of an HF network for emergency communications.

2) Ask the NCS for permission before you use the frequency.

3) Use the frequency for traffic, not chit-chat.

4) Answer promptly when called by the NCS.

5) Use tactical call signs.

6) Follow the net protocol established by the NCS.

In some relief activities, tactical nets become resource or command nets. A resource net is used for an event which goes beyond the boundaries of a single jurisdiction and when mutual aid is needed. A command net is used for communications between EOCs and ARES leaders. Yet with all the variety of nets, sometimes the act of simply putting the parties directly on the radio—instead of trying to interpret their words—is the best approach.

Formal Message Traffic

Formal message traffic is long-term communications that involve many people over a large area. It's generally cast in standard ARRL-message format and handled on well-established National Traffic System (NTS) nets, primarily on 75-meter SSB, 80-meter CW or 2-meter FM (see Fig 14-4). [In addition, there is a regular liaison to the International Assistance and Traffic Net, IATN, now officially designated the NTS Atlantic Region Net (ARN). The net meets on 14.303 MHz daily at 1130 UTC (1100 during the summer), to provide international traffic outlets, as suggested in Fig 14-5.]

Formal messages can be used for severe weather and disaster reports. These radiograms, already familiar to many agency officials and to the public, avoid message duplication while ensuring accuracy. Messages should be read to the originators before sending them, since the originators are responsible for their content. When accuracy is more important than speed, getting the message on paper before it is transmitted is an inherent advantage of formal traffic.

Packet Radio

Packet radio is a powerful tool for traffic handling, especially with detailed or lengthy text (see Chapter 10). Prepare and edit messages off line as text files. These can then be sent error free in just seconds, an important time-saver for busy traffic channels. Public service agencies are impressed by fast and accurate printed messages. Packet radio stations can even be mobile or portable. Relaying might be supplemented by AMTOR-Packet Link (APLink), a system equipped to handle messages between AMTOR HF and packet-radio VHF stations.

Image Communications

Image communication offers live pictures of an area to allow, for example, damage assessment by authorities. Amateur Television (ATV) in its public-service role usually employs portable Fast Scan Television (FSTV) which displays full motion, has excellent detail, can be in color and has a simultaneous sound channel. Although a picture is worth a thousand words, an ATV system requires more equipment, operating skill and preparation than using a simple hand-held radio.

Video cameras and 420-430 or 1240-1294 MHz radios can transmit public-service images from a helicopter to a ground base station equipped with video monitors and a VCR for taping. Image communication works well on the ground, too. Video coverage of parades, severe weather, and even operation Santa Claus in hospitals adds another dimension to your information.

AMATEUR RADIO GROUPS

The Amateur Radio Emergency Service

In 1935, the League developed what is now called the Amateur Radio Emergency Service (ARES), an organization of radio amateurs who have voluntarily registered their capabilities and equipment for emergency communications (see Fig 14-6). They are groups of trained operators ready to serve the public when disaster strikes and regular communications fail. ARES often recruits members from existing clubs, and includes amateurs outside the club area since emergencies do not recognize boundary lines.

Are you interested in public-service activities, or preparing for emergency communications? Join ARES in your area and exchange ideas, ask questions and help with message centers, sports events or weather spotting. Any licensed amateur with a sincere desire to serve is eligible for ARES membership. The possession of emergency-powered equipment is recommended, but it is not a requirement. An ARES group needs to refresh its training with meetings, scheduled nets, drills or real emergencies. An effective ARES group is to our benefit, as well as to the benefit of the entire community. Information about ARES may be obtained from your ARRL Section Manager (listed on page 8 of *QST*) or League Headquarters.

The Amateur Radio Emergency Service has responded

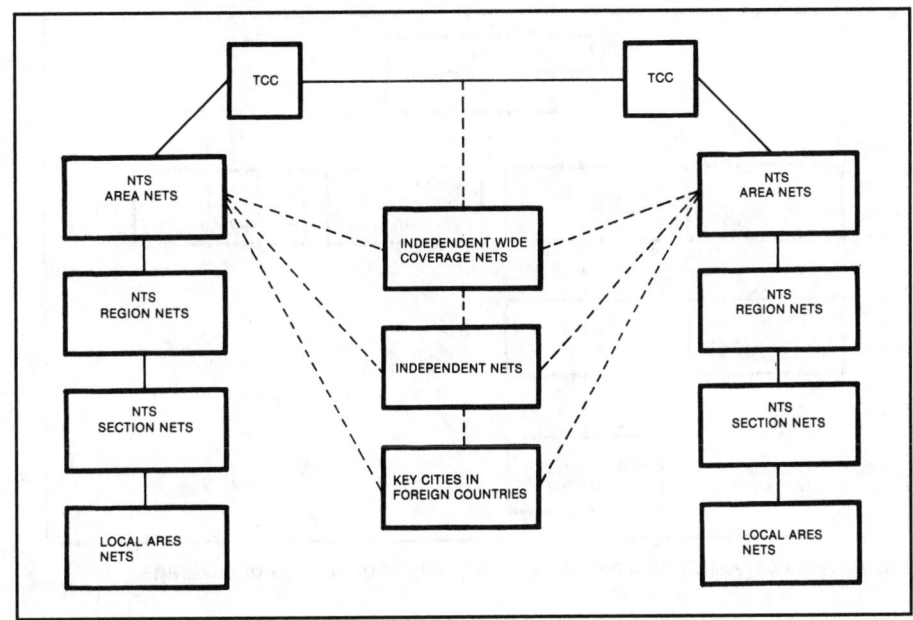

Fig 14-5—Emergency communications through liaison to NTS for handling formal message traffic.

countless times to communications emergencies. ARES also introduces Amateur Radio to the ever changing stream of agency officials. Experience has proven that radio amateurs react and work together more capably in time of emergency when practice has been conducted in an organized group. There is no substitute for experience gained—before the need arises.

Official Emergency Station

After you get some ARES training and practice, you might want to refine your skills in emergency communications. If you

Los Angeles Section Emergency Coordinator Hank Magid, K6YMJ, surveys and reports from the scene at the epicenter of the 1994 Los Angeles earthquake.

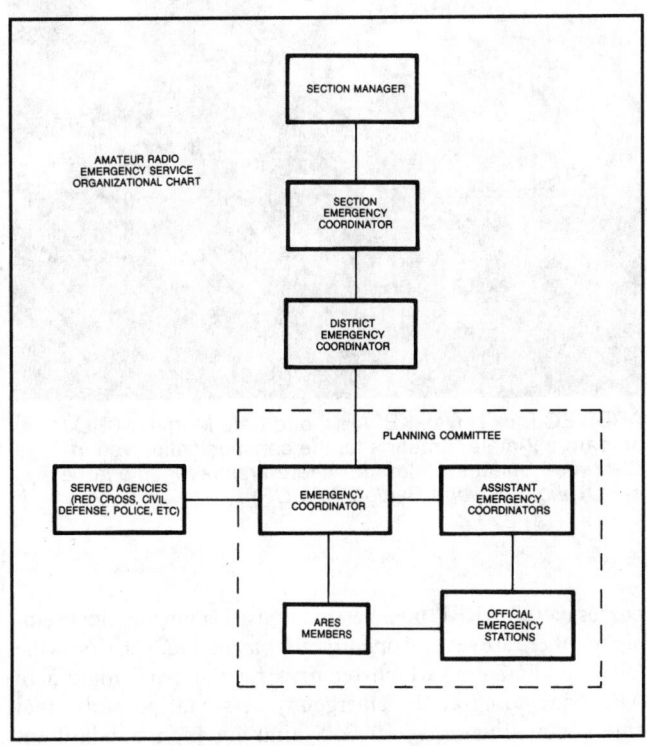

Fig 14-6—Section structure for ARES.

Fig 14-7—Local ARES structure for a county, city or other area of coverage.

ARES EC Dick Doyal, KBØAHR, and Dale Morrie, NØROJ, prepare a 2-meter antenna for the communication van in Glenwood Springs, Colorado, where fires ravaged a large area. (*photo by John Radloff, NØMOR*)

possess a full ARRL membership and a Technician license or higher, there are several opportunities available. The first is the OES appointment, which requires regular participation in ARES including drills, emergency nets and, possibly, real emergency situations. An OES aims for high standards of activity, emergency-preparedness and operating skills.

Emergency Coordinator

Next, when you feel qualified enough to become a team leader of your local ARES group, consider the EC appointment, if that position is vacant in your area. An EC, usually responsible for a county, is the person who can plan, organize, maintain response-readiness and coordinate for emergency communications (see Fig 14-7).

Much of the work involves promoting a working relationship with local government and agencies. The busy EC can hold meetings, train members, keep records, encourage newcomers, determine equipment availability, lead others in drills or be first on the scene in an actual disaster. Some highly populated or emergency-prone areas may also need one or more Assistant Emergency Coordinators (AEC) to help the EC. The AEC is an unofficial appointment made by an EC and, as a specific exception, can hold any class license and need not be an ARRL member.

District Emergency Coordinator

If there are many ARES groups in an area, a DEC may be appointed. Usually responsible for several counties, the DEC coordinates emergency plans between local ARES groups, encourages activity on ARES nets, directs the overall communication needs of a large area or can be a backup for an EC. As a model emergency communicator, the DEC trains clubs in tactical traffic, formal traffic, disaster communications and operating skills.

Section Emergency Coordinator

Finally, there is one rare individual who can qualify as top leader of each ARRL Section emergency structure, the SEC. Only the Section Manager can appoint a candidate to become the SEC. The SEC does some fairly hefty work on a section-wide level: making policies and plans and establishing goals, selecting the DECs and ECs, promoting ARES membership and keeping tabs on emergency preparedness. During an actual emergency, the SEC follows activities from behind the scenes, making sure that plans work and section communications are effective.

Section Manager

The overall leader of the ARRL Field Organization in each

section is the SM, who is elected by the League membership in that section. The SM not only appoints the SEC to handle details of ARES and emergency-preparedness activities, but also appoints a Section Traffic Manager (STM) to handle details of the NTS and formal message traffic operation. Response to emergency and public service needs combines the ARES and the NTS. [For further details on the ARRL Field Organization, see Chapter 17.]

RADIO AMATEUR CIVIL EMERGENCY SERVICE

The Radio Amateur Civil Emergency Service (RACES) was set up in 1952 as a special phase of the Amateur Radio service conducted by volunteer licensed amateurs. It is designed to provide emergency communications to local or state civil-preparedness agencies.

RACES operation is authorized during periods of local, regional or national civil emergencies by the FCC upon request of a state or federal official. While RACES was originally based on potential use during wartime, it has evolved over the years (as has the meaning of "civil defense" which is now called "civil preparedness"), to encompass all types of emergencies and natural disasters. RACES is sponsored by Federal Emergency Management Agency.

Amateurs operating in a local RACES organization must be officially enrolled in that local civil-preparedness group. RACES operation is conducted by amateurs using their own primary station licenses, and by existing RACES stations. The FCC no longer issues new RACES (WC prefix) station call signs. Operator privileges in RACES are dependent upon, and identical to, those for the class of license held in the Amateur Radio service. All of the authorized frequencies and emissions allocated to the Amateur Radio service are also available to RACES on a shared basis. But in the event that the President invokes his War Emergency Powers, amateurs involved with RACES would be limited to the certain frequencies, while all other amateur operation would be silenced.

When operating in a RACES capacity, RACES stations and amateurs registered in the local RACES organization may not communicate with amateurs not operating in a similar capacity. See FCC regulations for further information.

Although RACES and ARES are separate entities, the League advocates dual membership and cooperative efforts between both groups whenever possible. The RACES regulations make it simple for an ARES group whose members are all enrolled in and certified by RACES to operate in an emergency with great flexibility. Using the same operators and the same frequencies, an ARES group also enrolled as RACES can "switch hats" from ARES to RACES or RACES to ARES to meet the requirements of the situation as it develops. For example, during a "non-declared emergency," ARES can operate under ARES, but when an emergency or disaster is officially declared by a state or federal authority, the operation can become RACES with no change in personnel or frequencies.

Where there currently is no RACES, it would be a simple matter for an ARES group to enroll in that capacity, after a presentation to the civil-preparedness authorities. For more information on RACES, contact your State Emergency Management, Civil-Preparedness Office, or FEMA.

THE NATIONAL TRAFFIC SYSTEM

In 1949 the League created the National Traffic System to handle medium- and long-haul formal message traffic through networks whose operations can be expedited to meet the needs of an emergency situation.

The main function of NTS in an emergency is to link various local activities and to allow traffic destined outside of a local area to be systematically relayed to the addressee. In a few rare cases, a message can be handled by taking it directly to a net in the state where the addressee lives for rapid delivery by an amateur there within toll-free calling distance. However, NTS is set up on the basis of being able to relay large amounts of traffic systematically, efficiently and according to an established flow pattern. This proven and dependable scheme is what makes NTS so vital to emergency communications. See Chapter 15 for more information on NTS.

Additional details on ARES and NTS can be found in the *Public Service Communications Manual*, the *Emergency Communicator's Manual* and *The ARRL Net Directory*, all published by the ARRL (send an SASE for ordering information). Information on emergency communications and traffic handling also appears regularly in *QST*.

GOVERNMENT AND RELIEF AGENCIES

Government and relief agencies provide effective emergency management to help communities in disasters. They must rely on normal communications, but even reliable communications systems may fail, be unavailable or become overloaded in an emergency. In disaster situations, agency-to-agency radio systems may be incompatible.

Fortunately, Amateur Radio communicators can serve and support in these situations. We can bridge communications gaps with mobile, portable and fixed stations. We also supply trained volunteers needed by the agencies for collection and exchange of critical emergency information.

As government and relief agencies are restrained by budget cuts, the general public is becoming more dependent on volunteer programs. They quickly recognize the value of efforts of radio amateurs who serve the public interest. This public recognition is important support for the continued existence and justification of Amateur Radio.

By using Amateur Radio operators in the amateur frequency bands, the ARRL has been and continues to be in the forefront of supplying emergency communications, either directly to the general public or through various agencies. In fact, there are several organizations that have signed official Memorandas of Understanding with the League:

• The American National Red Cross
• The Salvation Army
• The Federal Emergency Management Agency
• The National Communications System
• The Associated Public Safety Communications Officers, Inc.
• National Weather Service

The American National Red Cross and the Salvation Army extend assistance to individuals and families in times of disaster. Red Cross chapters, for example, establish, coordinate and maintain continuity of communications during disaster-relief operations (both national and international). These agencies have long recognized that the Amateur Radio service, because of its excellent geographical station coverage, can render valuable aid.

The Federal Emergency Management Agency (FEMA) is a federal agency that supports state and local civil-preparedness and emergency-management agencies. With headquarters in Washington and 10 regional offices throughout the country, it can provide technical assistance, guidance and financial aid to state and local governments wishing to upgrade their emergency communications and warning systems. Since FEMA is in charge of the RACES program, its recognition of ARRL-sponsored emergency preparedness programs can be a powerful tool in selling ARES capability to local emergency-management officials.

The National Communication System (NCS) is a confederation of government agencies and departments established by a Presidential Memorandum to ensure that the critical telecommunication needs of the Federal Government can be met in an emergency. The ARRL Field Organization continues to participate in communications tests sponsored by NCS to study telecommunications readiness in any conceivable national emergency. Through this participation, radio amateurs have received recognition at the highest levels of government.

The Associated Public Safety Communications Officers (APCO) and amateurs share common bonds of communications in the public interest. APCO is made up of law-enforcement, fire and public-safety communications personnel. These officials have primary responsibility for the management, maintenance and operation of communications facilities in the public domain. They also establish international standards for public-safety communications, professionalism and continuity of communications through education, standardization and the exchange of information.

The National Weather Service (NWS) consists of a national headquarters in Washington, DC, six regional offices and over 200 local offices throughout the US. The ARRL Field Organization cooperates with NWS in establishing SKYWARN networks for weather spotting and communications. SKYWARN is a plan sponsored by NWS to report and track destructive storms or other severe weather conditions.

An increased awareness of radio amateur capabilities has also been fostered by ARRL's active participation as a member of the National Volunteer Organizations Active in Disaster (NVOAD). NVOAD coordinates the volunteer efforts of its 23 member-agencies (the Red Cross and Salvation Army are among its members).

ARRL and radio amateurs continue to accelerate their presence with these agencies This enhanced image and recognition of Amateur Radio attracts more "customers" for amateurs and our communications skills.

PUBLIC SERVICE EVENTS

Amateur Radio has a unique responsibility to render emer-

gency communications in times of disaster when normal communications are not available. Progressive experience can be gained by helping with public-service events such as message centers, parades and sports events. Besides serving the community, participation teaches skills that can be applied during an emergency.

Amateurs, according to FCC regulations, are responsible to aid during emergencies, but cannot receive compensation for this aid. This requirement does not remove public events sponsored by for-profit organizations from the list of events at which you can help protect the public welfare.

Message Centers

Message centers are Amateur Radio showcase stations set up and operated in conspicuous places to send free radiogram messages for the public. Many clubs provide such radiogram message services at shopping malls, fairs, information booths, festivals, exhibits, conferences and at special local events to demonstrate Amateur Radio while handling traffic. Message centers also afford opportunities to train operators, to practice handling messages, which are similar to disaster Welfare traffic, and to show the public that amateurs are capable, serious and responsible communicators. A plan for a message center should include a display, equipment, public relations and operators.

Successful centers include eye-catching organized displays and attractive stations that promote Amateur Radio and lure newcomers to the hobby. Cover a table with bright-colored cloth. Put out a few radio magazines and hang up some QSLs, maps, posters and operating aids. ARRL Headquarters can supply free literature, pamphlets, press handouts and various informative videotapes to help make your club display a success.

Equipment should be simple, safe and uncomplicated. Try only a few of the following: HF stations with CW, SSB, or digital capability and VHF stations with FM, packet radio, satellite or ATV activities. Keep expensive radios secure from possible theft, out of reach from visitors or stored under a protected table. Decide on whether to use power from batteries, commercial power or generators. Check prior to the event for receiver hash or public-address system interference. To instill interest, have an easy-to-operate shortwave receiver accessible for spectators to tune and listen. To promote a good public-relations image, consider the visitors; use an extension loudspeaker, desk microphone or computer readout of code. Clear, local contacts impress people more than long-distance, scratchy QSOs.

It may be necessary to form a barrier to separate visitors and busy operators. Even so, have friendly and knowledgeable club members available. They should clearly describe to the passerby what Amateur Radio can do, answer questions, explain the message process, assist in preparing messages and so on. Avoid ham jargon. Keep explanations simple and informative; because of the varying degrees of interest the public has, only a few will be interested in specific details. Discuss the variety of Amateur Radio, even if it is not demonstrated. Most visitors will want to know about the role of Amateur Radio in the community, about FCC exams and the license process.

Well-planned message centers can handle thousands of messages at one event. Some visitors bring their address books and excitedly write out bunches of messages. Others inquisitively watch or wait for their turns. Handling messages for the local community is a valuable way amateurs can enhance public recognition. It also keeps you in practice and your equipment ready for emergencies. Before getting involved in message centers, you should become familiar with the traffic-handling procedures as described in Chapter 15 (which also contains additional hints on exhibit-station operation).

Parades

From a communications viewpoint, parades simply relocate people and equipment, with a few requests for supplies or medical aid along the way. Therefore, parades simulate evacuations.

The first step in handling parade communications is to select a group representative who will contact the parade organizers and other officials during the planning stages. This representative can then instruct the ham group and describe what must be done.

As for most public-service events, arrive early. Wear proper identification, whether it is a special jacket, cap or badge. The basic order for the day is to provide communications. Leave parade decisions to officials and first-aid or medical care to Red Cross personnel.

Certain operators should be assigned to find parade officials and help them exchange information. Others work in pairs or teams with the parade marshall and parade specialists like the band, float, vehicle and marcher officials. These officials will be busy once the bus loads of uniformed marchers arrive at the staging area. Set up net controls at parade headquarters where a telephone is available, at an announcement or information booth or at the staging area near the starting line. Fill vacant positions with standby operators. Establish a procedure for helping lost children.

Next, after initial parade staging is underway, many other operators have specific on-route assignments. Position hand-held operators at intervals along the parade route where assignments can be carried out immediately. Operators can walk, march or ride along with the parade to report any specific problems. If possible, it's a good idea to place one amateur on top of a high building to spot general problems. Communicators near loud marching bands will need earphones. Hot weather can keep first-aid volunteers busy with victims of dehydration, sunstroke and heat exhaustion. In some cases, a request for an ambulance, paramedic or water will be needed. Parade floats may need repairs, fuel or towing.

Then, at the de-staging area, more officials rely on communications to prevent clogging the area and stopping the parade. Large parades can require over 75 amateurs to handle communications, and they must plan and practice many months before the event.

As a result, amateurs provide reliable communications to resolve unanticipated problems and handle last-minute changes with immediate action. For more information on public service events, see the *ARRL Special Events Communications Manual*. Contact the ARRL Headquarters Field Services Department for details on how to obtain a copy.

Sports Events

Sports events include all types of outdoor athletic activities where there is competition or movement, or when participants are timed. Again, it is important to assess the communication needs and plan with event officials, law-enforcement officers, medical and first-aid personnel. Know the route and preview the course. Everyone should understand their functions.

Like parades, there are strategic locations where sport-event amateur communicators could be positioned to be most useful:

• Stationed with or shadowing event officials and organizers or law-enforcement officers.

• At check-in points, staging areas or starting lines.
• At water stops, aid stations and checkpoints around the course or route.
• Near road intersections or sharp turns where safety of crowds or participants is crucial.
• In the course preview car, pace car, follow-up vehicle or roving sheriff's squad.
• At the net control station.
• Alongside first aid or Red Cross stations, medical personnel or ambulance.
• At message centers.
• Near a telephone or at a home station with a telephone.
• At the de-staging area or finish line.

There are many types of sports events in which amateurs serve. The list is nearly endless: air shows, balloon races, soap-box derbies, road rallies, solar car races, Boy Scout hikes, bicycle tours, regattas, golf tournaments, kayak and river-raft races, football games, mini-olympics, Special Olympics, World Olympics, ski-lift operations, slalom courses, cross-country meets, and so on.

A popular event is the *A-Thon*: the walk-athon, marathon, sport-athon, bike-athon or even a triathlon (usually running, bicycling and swimming or canoeing). Long-distance marathons require a large, disciplined force of communicators. This is the case with the New York City Marathon, which is served by well over 200 ham radio operators. Good coverage allows operators to radio the identifying numbers and locations of marathon participants who drop out of the course. This information is posted at a base station to help friends and relatives locate their favorite runners, or to help account for all participants.

Physical endurance and fitness is not only for the participants; hams use roller skates, bicycles, golf-carts, motorcycles or cars around the routes of sport events. Some communicators cover a race at a particular point until all participants have passed, then they become mobile and leapfrog up the route to repeat the procedure. Roving mobiles, sometimes called *roamers*, can keep track of everyone's location, determine race miles remaining, warn of potential traffic hazards, find lost people or items, report harassing spectators or post signs for riders or runners. Yet, even with this variety, keep in mind that an amateur is not to be used as a parking-lot attendant or crowd-control police. Take only those tasks you're capable of handling, like being a communicator.

Portable operators, those not on wheels, can report medical emergencies, give weather reports and handle messages for participants or spectators. Separate UHF channels for administrative purposes are useful for getting supplies like drinking cups, water or bandages, and to report malfunctioning facilities or request more volunteers. Telephones may be crowded in the area, so first-aid stations or checkpoints with bulletin boards can double as NTS message centers. It is then easy for the NCS to give assignments and tell the officials that the route is ready and participants are set to go.

During major sports events, the Red Cross can dispatch their own first-aid vehicles and keep them near crowds or hazardous areas for fast accident response. A specific medical-radio frequency is often assigned. Be prepared to report anything from minor medical situations to serious problems where an ambulance might be called to take the victim to a hospital for examination. If you do call for the ambulance, assign someone to flag it down and give the driver final directions to the victim's location.

Public-service events are great teachers. Once you gain skill in working with the public in hectic and near-emergency conditions, you're well prepared to tackle communications for more difficult natural-disaster assignments.

NATURAL DISASTERS AND CALAMITIES

Nature relentlessly concocts severe weather and natural calamities that can cause human suffering and create needs which the victims cannot alleviate without assistance. Despite the spectrum of requirements desperately needed to help people in a disaster, it is generally understood and agreed that amateurs will neither seek nor accept any duties other than Amateur Radio communications. Volunteer communicators do not, for example, enforce local laws, make major decisions, work as common laborers, rent generators, tents or lights to the public and so on. Instead, amateurs simply handle *radio communications*.

Here are several typical categories of amateur disaster and calamity communications:

• Severe-weather spotting and reporting
• Supporting evacuation of people to safe areas
• Shelter operations
• Assisting government groups and agencies
• Victim rescue operations
• Medical help requests
• Critical supplies requests
• Health-and-Welfare traffic
• Property damage surveys and cleanup

Severe Weather Spotting and Reporting

Nasty weather hits somewhere every day. Long ago, amateurs exchanged simple information among themselves about the approach and progress of storms. Next, concerned hams phoned the weather offices to share a few reports they thought might be of particular interest to the public. National Weather Service forecasters were relying on spotters: police, sheriff, highway patrol, emergency government and trained individuals who reported weather information by telephone. But when severe weather strikes, professional spotters may be burdened with law-enforcement tasks, phone lines may become overloaded, special communication circuits might go out of service or, worse yet, there can be a loss of electrical power. Because of these uncertainties, weather-center officials welcomed amateur operators and encouraged them to install their battery-powered radio equipment on-site so forecasters could monitor the weather nets, request specific area observations and maintain communications in a serious emergency.

Those first, informal weather nets had great potential to access perhaps hundreds of observers in a wide area. Many Meteorologists-in-Charge eagerly began to instruct hams in the types of information needed during severe-weather emergencies, including radar interpreting. Eventually, Amateur Radio SKYWARN operations developed as an important part of community disaster preparedness programs. Accurate observations and rapid communications during extreme weather situations now proves to be fundamental to the NWS. Amateur Radio operators nationwide are a first-response group invaluable to the success of an early storm-warning effort. Weather spotting is popular because the procedures are easy to learn and reports can be given from the relative safety and convenience of a home or an auto.

For example, during a severe-storm episode in Wisconsin in 1994, there was danger of local flooding. A quick check of conditions at the homes of hams operating in an ARES net rapidly revealed trouble spots over a 40 square mile area. This information proved invaluable to emergency government and weather officials, yet no ham had to leave home to participate.

Weather reports on a severe-weather net are limited to drastic weather data, unless specifically requested by the net-control operator. So, most amateurs monitor net operations and transmit only when they can help.

Weather forecasters, depending on their geographical location, need certain information.

During the summer or thunderstorm season report:
• Tornadoes, funnels or wall clouds.
• Hail.
• Damaging winds, usually 50 miles per hour or greater.
• Flash flooding.
• Heavy rains, rate of 2 inches per hour or more.

During the winter or snow season report:
• High winds.
• Heavy, drifting snow.
• Freezing precipitation.
• Sleet.
• New snow accumulation of 2 or more inches.

Here's a four-step method to describe the weather you spot:
1) *What*: Tornadoes, funnels, heavy rain and so on.
2) *Where*: Direction and distance from a known location; for example, 3 miles south of Newington.
3) *When:* Time of observation.
4) *How:* Storm's direction, speed of travel, size, intensity and destructiveness. Include uncertainty as needed. ("Funnel cloud, but too far away to be certain it is on the ground.")

Alerting the Weather Net

The Net Control Station, using a VHF repeater, directs and maintains control over traffic being passed on the Weather Net. The station also collates reports, relates pertinent material to the Weather Service and organizes liaison with other area repeaters. Priority Stations, those that are assigned tactical call signs, may call any other station without going through net control. The NCS might start the net upon hearing a National Oceanic and Atmospheric Administration (NOAA) radio alert, or upon request by NWS or the EC. The NCS should keep in mind that the general public or government officials might be listening to net operations with scanners.

Here are some guidelines an NCS might use to initiate and handle a severe-weather net on a repeater:
1) Activate alert tone on repeater.
2) Read weather-net activation format.
3) Appoint a backup NCS to copy and log all traffic—and to take over in the event the NCS goes off-the-air or needs relief.
4) Ask NWS for the current weather status.
5) Check in all available operators.
6) Assign operators to priority stations and liaisons.
7) Give severe-weather report outline and updates.
8) Be apprised of situations and assignments by EC.
9) Periodically read instructions on net procedures and types of severe weather to report.
10) Acknowledge and respond to all calls immediately.
11) Require that net stations request permission to leave the net.
12) During periods of inactivity and to keep the frequency open, make periodic announcements that a net is in progress.
13) Close the net after operations conclude.

ARES Personal Checklist

The following represents recommendations of equipment and supplies ARES members should consider having available for use during an emergency or public-service activity.

Forms of Identification
- ARES Identification Card
- FCC Amateur Radio license
- driver's license

Radio Gear
- rig (VHF)
- microphone
- headphones
- power supply/extra batteries
- antennas with mounts
- spare fuses
- patch cords/adapters (BNC to PL259/RCA phono to PL259)
- SWR meter
- extra coax

Writing Gear
- pen/pencil/eraser
- clipboard
- message forms
- logbook
- note paper

Personal Gear (short duration)
- snacks
- liquid refreshments
- throat lozenges
- personal medicine
- aspirin
- extra pair of prescription glasses
- sweater/jacket

Personal Gear (72-hour duration)
- foul-weather gear
- three-day supply of drinking water
- cooler with three-day supply of food
- mess kit with cleaning supplies
- first-aid kit
- personal medicine
- aspirin
- throat lozenges
- sleeping bag
- toilet articles
- mechanical or battery powered alarm clock
- flashlight with batteries/lantern
- candles
- waterproof matches
- extra pair of prescription glasses

Tool Box (72-hour duration)
- screwdrivers
- pliers
- socket wrenches
- electrical tape
- 12/120-V soldering iron
- solder
- volt-ohm meter

Other (72-hour duration)
- rig (HF)
- hatchet/ax
- saw
- pick
- shovel
- siphon
- jumper cables
- generator, spare plugs and oil
- kerosene lights, camping lantern or candles
- 3/8-inch hemp rope
- highway flares
- extra gasoline and oil

At the Weather Service

The NCS position at the Weather Service, when practical, can be handled by the EC and other ARES personnel. Operators assigned there must have 2-meter hand-helds with fully charged batteries. The station located at the NWS office may also be connected to other positions with an off-the-air intercom system. This allows some traffic handling without loading up the repeater. Designate a supplementary radio channel in anticipation of an overload or loss of primary communications circuits.

If traffic is flowing faster than you can easily copy and relay, NWS personnel may request that a hand-held radio be placed at the severe-weather desk. This arrangement allows them to monitor incoming traffic directly. Nevertheless, all traffic should be written on report forms. If a disaster should occur during a severe-weather net, shift to disaster-relief operations.

Repeater Liaisons

Assign properly equipped and located stations to act as liaisons with other repeaters. Two stations should be appointed to each liaison assignment. One monitors the weather repeater at all times and switches to the assigned repeater just long enough to pass traffic. The second monitors the assigned repeater and switches to the weather repeater just long enough to pass traffic. If there aren't enough qualified liaison stations, one station can be given both assignments.

Weather Warnings

NWS policy is to issue warnings only when there is absolute certainty, for fear of the "cry wolf" syndrome (premature warnings cause the public to ignore later warnings). Public confidence increases with reliable weather warnings. When NWS calls a weather alert, it will contact the local EC by phone or voice-message pager, or the EC may call NWS to check on a weather situation.

Hurricanes

A hurricane is declared when a storm's winds reach 75 miles per hour or more. These strong winds may cause storm-surge waves along shores and flooding inland.

A Hurricane *Watch* means a hurricane may threaten coastal and inland areas. Storm landfall is a possibility, but it is not necessarily imminent. Listen for further advisories and be prepared to act promptly if a warning is issued.

A Hurricane *Warning* is issued when a hurricane is expected to strike within 24 hours. It may include an assessment of flood danger, small-craft or gale warnings, estimated storm effects and recommended emergency procedures.

Amateurs, in the 4th and 5th call areas in particular, can spot and report the approach of hurricanes well ahead of any news service. In fact, their information is sometimes edited and then broadcast on the local radio or TV to keep citizens informed.

The Hurricane Watch Net on 14.325 MHz, for example, serves either the Atlantic or Pacific during a watch or warning period and keeps in touch with the National Hurricane Center. Frequent, detailed information is issued on nets when storms pose a threat to the US mainland. In addition to hurricane spotting, local communicators may announce that residents have evacuated from low-lying flood areas and coastal shore. Other amateurs across the country can help by relaying information, keeping the net frequency clear and by listening.

Tornadoes

A tornado is an intensely destructive whirlwind formed from strongly rising air currents. With winds of up to 300 miles per hour, tornadoes appear as rotating, funnel-shaped clouds from gray to black in color. They extend toward the ground from the base of a thundercloud. Tornadoes may sound like the roaring of an airplane or locomotive. Even though they are short lived over a small area, tornadoes are the most violent of all atmospheric phenomena. Tornadoes that don't touch the ground are called *funnels*.

A Tornado *Watch* is issued when a tornado may occur near your area. Carefully watch the sky.

A Tornado *Warning* means take shelter immediately, a tornado has actually been sighted or indicated by radar. Protect yourself from being blown away, struck by falling objects or injured by flying debris.

"Tornado alley" runs in the 5th, 9th and 10th US call areas. Amateurs in these areas often receive Tornado Spotter's Training and refresher courses presented by NWS personnel.

Amateur Radio's quick-response capability has reduced injuries and fatalities by giving early warnings to residents. Veteran operators know exactly how serious a tornado can be. They've seen tornadoes knock out telephone and electrical services just as quickly as they flip over trucks or destroy homes. Traffic lights and gas pumps won't work without electricity, creating problems for motorists and fuel shortages for electric generators.

After a tornado strikes, amateurs provide communications in cooperation with local government and relief agencies. Welfare messages are sent from shelters where survivors receive assistance. Teams of amateurs and officials also survey and report property damage.

Floods, Mud Slides and Tidal Waves

Floods occur when excessive rainfall causes rivers to overflow their banks, when heavy rains and warmer-than-usual temperatures melt excessive quantities of snow, or when dams break. Floods can be minor, moderate or major. Floods or volcanic eruptions may melt mountain snow, causing mud slides. A tidal wave or tsunami is actually a series of waves caused by a disturbance which may be associated with earthquakes, volcanoes or sometimes hurricanes.

Don't wait for the water level to rise or for officials to ask for help; sound the alarm and activate a weather net immediately. Besides handling weather data for NWS, enact the response plans necessary to relay tactical flood information to local officials. Assist their decision making by answering the following questions:

1) Which rivers and streams are affected and what are their conditions?

2) When will flooding probably begin, and where are the flood plain areas?

3) What are river-level or depth-gauge readings for comparison to flood levels?

Mobile operators may find roads flooded and bridges washed out. Flood-rescue operations then may be handled by marine police boats with an amateur aboard. If power and telephones are out, portable radio operators can help with relief operations to evacuate families to care facilities where a fixed station should be set up. The officials will need to know the number and location of evacuees. As the river recedes, the water level drops in some areas, but it may rise elsewhere to threaten residents. Liaisons to repeaters downstream can warn others, possibly through the Emergency Broadcast System, of impending flooding. Property-damage reports and welfare traffic will usually be followed by disaster relief and clean-up operations.

Winter Storms

A Winter Storm *Watch* indicates there is a threat of severe winter weather in a particular area. A Winter Storm *Warning* is issued when heavy snow (6 inches or more in a 12-hour period, 8 inches or more in a 24-hour period), freezing rain, sleet or a substantial layer of ice is expected to accumulate.

Freezing rain or freezing drizzle is forecast when expected rain is likely to freeze when it strikes the ground. Freezing rain or ice storms can bring down wires, causing telephone and power outages.

Sleet consists of small particles of ice, usually mixed with rain. If enough sleet accumulates on the ground, it will make roads slippery.

Blizzards are the most dangerous of all winter storms. They are a combination of cold air, heavy snow and strong winds. Blizzards can isolate communities. A Blizzard *Warning* is issued when there is considerable snow and winds of 35 miles per hour or more. A *Severe Blizzard Warning* means that a heavy snow is expected, with winds of at least 45 miles per hour and temperatures 10° F or lower.

Travelers advisories are issued when ice and snow are expected to hinder travel, but not seriously enough to require warnings. Blizzards and snowstorms can create vehicle-traffic problems by making roads impassable, stranding motorists or drastically delaying their progress.

In some areas, local snowmobile club members or 4-wheel-drive-vehicle enthusiasts cooperate with hams to coordinate and transport key medical personnel when snowdrifts render roads almost impassable. They may also assist search teams looking for motorists.

Brush and Forest Fires

A prolonged period of hot, dry weather parches shrubs, brush and trees. This dry vegetation, ignited by lightning, arson or even a helicopter crash, can start a forest fire. The fires quickly become worse when winds spread the burning material.

An amateur who spots a blaze that has the potential of growing into a forest fire can radio the Park Service and ask them to dispatch a district ranger and fire-fighting equipment. When fires are out of control, hams help with communications to evacuate people, report the fires' movement or radio requests for supplies and volunteers. ARES groups over a wide geographical area can set up portable repeaters, digital links or ATV stations.

Amateurs, in the 6th and 7th US call areas in particular, should get adequate safety training, including fire line safety and fire shelter deployment. Even then, only travel to a fire operation upon receiving clear dispatch instructions from a competent authority. If you don't have proper training, inform whoever is in charge of your lack of training to prevent being given an inappropriate or dangerous assignment.

Fire safety rules are of special importance in an emergency, but also should be observed every day to prevent disaster. Since most fires occur in the home, even an alert urban amateur can spot and report a building on fire. Fires are extinguished by taking away the fuel or air (smother it), or by cooling it with water and fire-extinguishing chemicals. A radio call to the fire department will bring the needed control. The Red Cross generally helps find shelter for the homeless after fires make large buildings, such as apartments, uninhabitable. So even in cities, communications for routing people during and after fires may be needed.

Earthquakes and Volcanic Eruptions

Earthquakes are caused when underground forces break and shift rock beneath the surface, causing the Earth's crust to shake or tremble. The actual movement of the earth is seldom a direct cause of death or injury, but can cause buildings and other structures to collapse and may annihilate communications, telephone service and electrical power. Most casualties result from falling objects and debris, splintering glass and fires.

Earthquakes strike without warning, but everyone knows that the quake happened. Amateurs should first ensure that the immediate surroundings and loved ones are safe, and then begin monitoring the ARES frequencies. Amateurs are often the first to alert communities immediately after earthquakes and volcanic eruptions occur. Their warnings are definitely credited with lessening personal injuries in the area.

Amateurs may assist with rescue operations, getting medical help or critical supplies and helping with damage appraisals. There will be communications, usually on VHF or UHF, for Red Cross logistics and government agencies. After vital communications is handled, the rest of the world will be trying to get information through Health-and-Welfare traffic.

Shelter Operations

A shelter or relief center is a temporary place of protection where rescuers can bring disaster victims and where supplies can be dispensed. Many displaced people can stay at the homes of friends or relatives, but those searching for family members or in need are housed in shelters.

Whether a shelter is for a few stranded motorists during a snowstorm, or a whole community of homeless residents after a disaster, it is an ideal location to set up an Amateur Radio communication base station. An alternate station location would be an ARES mobile communications van, if available, near the shelter.

Once officials determine the locations for shelters, radio operators can be assigned to set up equipment at the sites. In fact, the amateur station operators could share a table with shelter registration workers. Make sure you obtain permission for access to the shelter to assist and do not upset the evacuees. Use of repeaters and autopatches allows Welfare phone calls for those inside the shelters to inform and reassure friends, families and relatives. Use of a computer may help to control and distribute information. *ARESdata* is a packet-radio database program that works well with emergency and public-service communications (see December 1990 *QST*, page 75).

Health-and-Welfare Traffic

There can be a tremendous amount of radio traffic to handle during a disaster. This is due in part because phone lines that remain in working order should be reserved for emergency use by those people in peril.

Shortly after a major disaster, Emergency messages within the disaster area often have life-and-death urgency. Of course, they receive primary emphasis. Much of their local traffic will be on VHF or UHF. Next, Priority traffic, messages of an emergency-related nature but not of the utmost urgency, are handled. Then, Welfare traffic is originated by evacuees at shelters or by the injured at hospitals and relayed by Amateur Radio. It flows one way and results in timely advisories to those waiting outside the disaster area.

Incoming Health-and-Welfare traffic should be handled only after all emergency and priority traffic is cleared. Don't solicit traffic going to an emergency because it can severely overload an already busy system. Welfare inquiries can take time to discover hard-to-find answers. An advisory to the inquirer uses even more time. Meanwhile, some questions might have already been answered through restored circuits.

Shelter stations, acting as net control stations, can exchange information on the HF bands directly with destination areas as propagation permits. Or, they can handle formal traffic through a few outside operators on VHF who, in turn, can link to NTS stations. By having many NTS-trained amateurs, it's easy to adapt to whatever communications are required.

Property Damage Surveys

Damage caused by natural disasters can be sudden and extensive. Responsible officials near the disaster area, paralyzed without communications, will need help to contact appropriate officials outside to give damage reports. Such data will be used to initiate and coordinate disaster relief. Amateur Radio operators offer to help but often are unable to cross roadblocks established to limit access by sightseers and potential looters. Proper emergency responder identification will be required to gain access into these areas. In some instances, call-letter license plates on the front of the car or placards inside windshields may help. It's important for amateurs to keep complete and accurate logs for use by officials to survey damage, or to use as a guide for replacement operators.

Accidents and Hazards

The most difficult scenarios to prepare for are accidents and hazardous situations. They are unpredictable and can happen anywhere. Generally, an emergency *autopatch* is used only to report incidents that pose threats to life or personal safety, such as vehicle accidents, disabled vehicles or debris in traffic, injured persons, criminal activities and fires.

Using the keypad featured on most modern VHF hand-held and mobile radios, the operator activates a repeater autopatch by sending a particular code. The repeater connects to a telephone line and routes the incoming and outgoing audio accordingly. By dialing 9-1-1 (or another emergency number), the operator has direct access to law-enforcement agencies.

Vehicle Accidents

Vehicle-accident reports, by far the most common public-service activity on repeaters, can involve anything from bikes, motorcycles and automobiles, to buses, trucks, trains and airplanes. Law-enforcement offices usually accept reports of such incidents anywhere in their county and will relay information to the proper agency when it pertains to adjacent areas.

Here's a typical autopatch procedure:

1) Give your call and say "emergency patch."
2) Drop your carrier momentarily.
3) Key in the access code.
4) Wait for police or fire operator to answer.
5) Answer the questions that the operator asks.
6) After operator acknowledges, dump the patch by keying in the dump code.
7) Give your call and say "patch clear."

When you report a vehicle accident, remain calm and get as much information as you can. This is one time you certainly have the right to break into a conversation on a repeater. Use plain language, say exactly what you mean, and be brief and to the point. Do not guess about injuries; if you don't know, say so. Some accidents may look worse than they really are; requesting an ambulance to be sent needlessly could divert it away from a bona-fide accident injury occurring at the same time elsewhere. And besides, police cruisers are generally only minutes away in an urban area.

Here's what you should report for a vehicle accident:

1) Highway number (eg, I-43, I-94, US-45).
2) Direction of travel (north, south, east, west).
3) Address or street intersection, if on city streets, closest exit on highway.
4) Traffic blocked, or if accident is out of traffic.
5) Apparent injuries, number and extent.
6) Vehicles on fire, smoking or a fuel spill.

Example: "This is WB8IMY, reporting a two-car accident, I-94 at Edgerton, northbound, blocking lane number two, property damage only."

The first activities handled by experts at a vehicle-accident scene are keyed to rescue, stabilize and transport the victims. Then they ensure security, develop a perimeter, handle vehicle traffic and control or prevent fires from gasoline spills. Finally, routine operations restore the area with towing, wrecking and salvage.

The ability to call the police or for an ambulance, without depending on another amateur to monitor the frequency, saves precious minutes. Quick reaction and minimum delay is what makes an autopatch useful in emergencies. The autopatch, when used responsibly, is a valuable asset to the community.

Freeway Warning

Many public-safety agencies recommend that you do not stop on freeways or expressways to render assistance at an accident scene unless you are involved, are a witness or have sufficient medical training. Freeways are extremely dangerous because of the heavy flow of high-speed traffic. Even under ideal conditions, driver fatigue or inattentiveness, high speeds and short distances between vehicles often make it impossible to stop a vehicle from striking stationary objects. If you must stop on a freeway, pull out of traffic and onto distress lanes. Exercise extreme caution to protect yourself. Don't add to the traffic problem. Instead, radio for help.

WORKING WITH PUBLIC-SAFETY AGENCIES

Amateur Radio affords public-safety agencies, such as local police and fire officials, with an extremely valuable resource in times of emergency. Once initial acceptance by the authorities is achieved, an ongoing working relationship between amateurs and safety agencies is based on the efficiency of our performance. Officials tend to be very cautious and skeptical about those who are not members of the public-safety professions. At times, officials may have trouble separating problem solvers from problem makers, but being understaffed and on a limited budget, they often accept communications help if it is offered in the proper spirit.

Here are several image-building rules for working with safety agencies:

• Maintain group unity. Work within ARES, RACES or local club groups. Position your EC as the direct link with the agencies.
• Be honest. If your group cannot handle a request, say so and explain why. Safety personnel often risk their lives based on a fellow disaster-worker's promise to perform.
• Equip conservatively. Do not have more flashing lights, signs, decals and antennas than used by the average police or fire vehicle. Safety professionals are trained against overkill and use the minimum resources necessary to get the job done.
• Look professional. In the field, wear a simple jump-suit or jacket with an ARES patch to give a professional image and to help officials identify radio operators.
• Respect authority. Only assume the level of authority and responsibility that has been given to you.
• Publicize Amateur Radio. If contacted by members of the press, restrict comments solely to the amateurs' role in the situation. Emergency status and names of victims should only come from a press information officer, or the government agency concerned.

The public-service lifeline provided by Amateur Radio must be understood by public-safety agencies before the next disaster occurs. Have an amateur representative meet with public-safety officials in advance of major emergencies so each group will know the capabilities of the other. The representative should appear professional with a calm, businesslike manner and wear conservative attire. It is up to you to invite the local agencies to observe or cooperatively participate with your group.

Assisting the Police

Act as a communicator and radio for the police when you spot criminal activities. Memorize a description of the suspects for apprehension, or information about a vehicle for later recovery. Use caution—uniformed personnel arriving on the scene do not know who you are. Be an observer; let the officers act. For instance, if you see a vehicle traveling at high speed at night without lights, report the location and direction of travel to the law enforcement agency. Don't follow the vehicle.

Some Amateur Radio groups provide police communication assistance involving free taxi service to citizens who do not wish to drive home on New Year's Eve. They also act as lookouts for vandals on city/town streets and on freeway overpasses on Halloween. Hams have even reported sighting a person who is about to make a life-threatening jump into a river. Assisting the police in small ways can make a big difference in the community.

Search and Rescue

Amateurs helping search for an injured climber use repeaters to coordinate the rescue. A small airplane crashes, and amateurs direct the search by tracking from its Emergency Locator Transmitter. No matter what the situation, it's reassuring to team up with local search-and-rescue organizations who have familiarity with the area. Once a victim is found, the hams can radio the status, autopatch for medical information, guide further help to the area and plan for a return transportation. If the victim is found in good condition, Amateur Radio can bolster the hopes of base-camp personnel and the family of the victim with direct communications.

Even in cities, searches are occasionally necessary. An elderly person out for a walk gets lost and doesn't return home. After a reasonable time, a local search team plans and coordinates a search. Amateurs take part by providing communications, a valuable part of any search. When the missing person is discovered, there may be a need to radio for an ambulance for transportation to a nearby hospital.

Hospital Communications

Hospital phones can fail. For example, when a construction crew using a backhoe may accidentally cut though the main trunk line supplying telephone service to several hundred users, including the hospital. Such major hospital telephone outages can block incoming emergency phone calls. In addition, the hospital staff cannot telephone to discuss medical treatment with outside specialists.

Several hand-held equipped amateurs can first handle emergency calls from nursing homes, fire departments and police stations. They can also provide communications to temporarily replace a defective hospital paging system. Next, they can help restore critical interdepartmental hospital communications and, finally, communications with nearby hospitals.

Preparation for hospital communications begins by cooperating with administrators and public-relations personnel. You'll need their permission to perform inside signal checks, install outside antennas or set up net control stations in the hospital.

Working with local hospitals doesn't always involve extreme situations. You may simply be asked to relay information from the poison control center to a campsite victim. Or you may participate in an emergency exercise where reports of the "victims'" conditions are sent to the hospital from a disaster site. One typical drill involved a simulated crop-duster plane crash on an elementary school playground during recess. The doctors and officials depended on communications to find casualties who were contaminated by crop-dusting chemicals. Again, never give the names of victims or fatalities over the air since this information must be handled only by the proper agency.

Toxic-Chemical Spills and Hazardous Materials

A toxic-chemical spill suddenly appears when gasoline pours from a ruptured bulk-storage tank. A water supply is unexpectedly contaminated, or a fire causes chlorine gas to escape at an apartment swimming pool. On a highway, a faulty shutoff valve lets chemicals leak from a truck, or drums of chemicals fall onto the highway and rupture. Amateur communications have helped in all these situations.

Caution: don't rush into a Hazardous Materials (HAZMAT) incident area without knowing what's involved. Vehicles carrying 1000 pounds or more of a HAZMAT are required by Federal regulations to display a placard bearing a four-digit identification number. Radio the placard number to the authorities, and they will decide whether to send HAZMAT experts to contain the spills.

Follow directions from those in command. Provide communications to help them evacuate residents in the immediate area and coordinate between the spill site and the shelter buildings. Hams also assist public-service agencies by setting flares for traffic control, helping reroute motorists and so on. We're occasionally asked to make autopatch calls for police or fire-department workers on the scene as they try to determine the nature of the chemicals.

The National Transportation Safety Board, the Environmental Protection Agency and many local police, fire, and emergency government departments continue to praise ARES volunteers in their assistance with toxic spills.

Traffic Handling

MARIA L. EVANS, KT5Y
8461 BROWNS STATION DR
COLUMBIA, MO 65202

F or just pennies a day, you can protect you and your family from all sorts of catastrophic illnesses with the Mutual of Podunk healthcare policy....

Yes, at the amazing low price of $9.95, you can turn boring old potatoes, carrots and okra into culinary masterpieces with the Super Veggie-Whatchamadoodler—a modest investment for your family's mealtime happiness....

Tired of all that ugly fat on your otherwise-beautiful body? Our Exer-Torture home gym will trim those thunder thighs in 30 days or your money back....

Ah, those pitches. Everybody is trying to sell something—even the participants in the specialty modes of Amateur Radio. Our eyes light up thinking about all those wondrous gizmos that digitize, packetize and equalize. Those kinds of specialty modes are easily remembered and can quickly gain popularity. Most people, though, forget about the oldest specialty mode in Amateur Radio—traffic handling.

Admittedly, it's hard for traffic handlers to compete with other specialty modes because it just doesn't look exciting. After all, tossing messages from Great-Uncle Levi and Grandma Strauss sure doesn't put you on the leading edge of our high-tech hobby, does it? Yet, over 500 nets are members of the ARRL National Traffic System (NTS), probably one of the most highly organized special interests of Amateur Radio. Today's traffic handlers are at this very moment setting new standards for traffic handling via AMTOR and packet—two of the hottest specialty modes going—as well as using traditional modes. If you enjoy emergency-communications preparation, traffic handling is for you. Sure, it's true that over 90% of all messages handled via Amateur Radio are of the "at the state fair, wish you were here" or the "happy holidays" variety—certainly not life-and-death stuff. But consider this: Your local fire department often conducts drills without ever putting out a real fire, your civil defense goes out regularly and looks for "pink tornadoes," and many department stores hire people to come in and pretend to be shoplifters to check the alertness of their employees. Similarly, when real emergencies rear their ugly heads, traffic handlers just take it all in stride and churn messages out like they always do.

Traffic handling is also an excellent way to paint a friendly picture of Amateur Radio to the nonamateur public. What sometimes seems to be unimportant to us is usually never unimportant to that person in the address block of a message. Almost every traffic handler can relate stories of delivering a Christmas message from some long-lost friend or relative that touched the heart of the recipient such that they could hear the tears welling up at the other end of the phone. Those happy recipients will always

mentally connect Amateur Radio traffic handling with good and happy things, and often this is the most satisfying part of this hobby within a hobby.

Next time you go to a hamfest, see if you can spot the traffic handlers. They are almost always reveling in a big social cluster, sharing stories and enjoying a unique camaraderie. Traffickers always enthusiastically look forward to the next hamfest, because it means another chance to spend time with their special friends of the airwaves. These friendships often last a lifetime, transcending barriers of age, geographical distance, background, gender or physical ability.

Young or old, rural or urban, there's a place reserved for you on the traffic nets. Young people often can gain respect among a much older peer group and obtain high levels of responsibility through the traffic nets, good preparation for job opportunities and scholarships. "Nine to fivers" on a tight schedule can still manage to get a regular dose of Amateur Radio in just 15 to 30 minutes of net operation—a lot of hamming in a little time. Retired people can stay active in an important activity and provide a service to the general public.

Even if you live in the sticks, where you rarely get a delivery, you can still perform a vital function in NTS as a net control station (NCS) or as a representative to the upper echelons of NTS. You don't have to check in every night (a popular myth about traffic handling—if you can donate time just once a week you are certainly welcome on NTS). You don't need fancy antennas or huge amplifiers, and you don't even need those ARRL message pads. For the cost of a pad or paper and a pencil, you can interface with a system that covers thousands of miles and consists of tens of thousands of users. A world of fun and friendship is waiting for you in traffic handling. You only have to check into a net to become part of it.

MAKING THE BIG STEP: CHOOSING A NET AND CHECKING INTO IT

Checking into a net for the first time is a lot like making your first dive off the high board. It's not usually very pretty, but it's a start. Once you've gotten over the initial shock of hitting the water that hard, the next one comes a lot easier. But, like a beginning diver, you can do a little advance preparation that will get you emotionally prepared for your first plunge.

Before you attempt to interface with the world of NTS, you need the right "software." Get a copy of *The ARRL Net Directory*, available from ARRL ($1) and two free ARRL operating aids—FSD-3, the list of ARRL numbered radiograms, and FSD-218, the pink "Q-signals for nets and message form" card. Read all the articles in the *Net Directory*, and keep the two operating aids between the covers of your logbook for quick reference. After you have a basic understanding of the materials, you're ready to pick out a net in your ARRL section/state that suits you.

Go through the *Net Directory* and match up your time schedule with the nets in your section/state. If you're a Novice, your section/state slow-speed net is a good choice (or your neighboring section, if yours doesn't have one). But if you aren't a Novice, and still don't have the confidence to try the "big" CW section net, don't feel embarrassed about checking into a slow-speed net. Slow-speed nets are chock-full of veterans that help with NCS and NTS duties, and they're willing to help you, too. Perhaps you'd like to try the section phone net or weather net or a local 2-meter net. Some sections have also established RTTY nets. At any rate, you are the sole judge of what you want to try first.

Once you've chosen a net, it's a good idea to listen to it for a few days before you check in. Although this chapter will deal with a generalized format for net operation, each net is "fingerprinted" with its own special style of operation, and it's best to become acquainted with it before you jump in. If you have a friend who checks into the net, let your friend "take you by the hand" and tell you about the ins and outs of the net.

When the big day arrives, keep in mind that everybody on that net had to check into the net for the first time once. You aren't doing anything different than the rest of them, and this, like your first QSO as a Novice, is just another "rite of passage" in the ham world. You will discover that it doesn't hurt, and that many other folks will be pleased, even happy, that you checked in with them. (See accompanying sidebar for general recommendations on how to make your NCS love you!)

On CW

First, we'll pay a visit to a session of the Mowegia CW Net, not so long ago, on an 80-meter frequency not so far away. KØSI is calling tonight's session, using that peculiar CW net shorthand that we aren't too used to yet.

KØSI: MCWN MCWN MCWN DE KØSI KØSI KØSI GE AND PSE QNZ V V V MCWN QNI QTC?

(Translation: Calling the Mowegia CW Net, calling the Mowegia CW Net, this is KØSI. Good evening, this is a directed net, please zero beat with me. [Then, he sent some Vs to aid stations in zero beating him.] Calling the Mowegia CW Net...any checkins, any traffic?)

In the meantime, NDØN, K2ONP and WØOUD are waiting in the wings to check in. Since each is an experienced traffic handler, each listens carefully before jumping in, so as to not step on anyone.

NDØN: N

KØSI: N

(Notice that NDØN just sent the first letter of his suffix, and the NCS acknowledged it. This is common practice, but if you have the letter E or T or K as the first letter of your suffix, or if it is the same as another net operator's first letter, you might use another letter. It's not a hard-and-fast rule.)

NDØN: DE NDØN GE PETE QTC1 SOUTH CORNER 1 AIØO K

(NDØN has one piece of traffic for South Corner, and one for AIØO.)

KØSI: NDØN DE KØSI GE JOHN QSL TU $\overline{\text{AS}}$

(Good evening, John, I acknowledge your traffic. Thank you, please stand by.)

As a net operator, you have a duty to be self-disciplined. A net is only as good as its worst operator. You can be an exemplary net operator by following a few easy guidelines.

1) *Zero beat the NCS.* The NCS doesn't have time to chase all over the band for you. Make sure you're on frequency, and you will never be known at the annual net picnic as "old so-and-so who's always off frequency."

2) *Don't be late.* There's no such thing as "fashionably late" on a net. Liaison stations are on a tight timetable. Don't hold them up by checking in 10 minutes late with three pieces of traffic.

3) *Speak only when spoken to by the NCS.* Unless it is a bona fide emergency situation, you don't need to "help" the NCS unless asked. If you need to contact the NCS, make it brief. Resist the urge to help clear the frequency for the NCS or to "advise" the NCS. The NCS, not you, is boss.

4) Unless otherwise instructed by the NCS, *transmit only to the NCS.* Side comments to another station in the net are out of order.

5) *Stay until you are excused.* If the NCS calls you and you don't respond because you're getting a "cold one" from the fridge, the NCS may assume you've left the net, and net business may be stymied. If you need to leave the net prematurely, contact the NCS and simply ask to be excused (QNX PSE on CW).

6) *Be brief when transmitting to the NCS.* A simple "yes" (c) or "no" (N) will usually suffice. Shaggy dog tales only waste valuable net time.

7) *Know how the net runs.* The NCS doesn't have time to explain procedure to you. After you have been on the net for a while, you should already know these things.

K2ONP: M

KØSI: M

K2ONP: DE K2ONP GE PETE TEN REP QRU K

(K2ONP is representative to the NTS Tenth Region Net tonight, and he has no traffic.)

KØSI: K2ONP DE KØSI HI GEO TU $\overline{\text{AS}}$

WØOUD: BK

KØSI: BK

WØOUD: DE WØOUD GE PETE QRU K

KØSI: WØOUD DE KØSI GE LETHA QNU TU $\overline{\text{AS}}$

(Good evening, Letha, the net has traffic for you. Please stand by.)

(Since Letha lives in South Corner, the NCS is going to move WØOUD and NDØN off frequency to pass the South Corner traffic.)

KØSI: ØN?

NDØN: HR

(HR [here], or c [yes] are both acceptable ways to answer the NCS, who will usually use only your suffix from here on out to address questions to you.)

KØSI: OUD?

WØOUD: C

KØSI: NDØN ES WØOUD PSE UP 4 1 SOUTH CORNER K

(Please go up 4 kHz and pass the South Corner traffic.)

NDØN: GG (going)

WØOUD: GG

(When two stations go off frequency, the receiving station always calls the transmitting station. If the NCS had said WØOUD PSE QNV NDØN UP 4 GET 1 SOUTH CORNER, WØOUD would have called NDØN first on frequency to see if she copied him. This is done often when conditions are bad. If they don't make connection, they will return to net frequency. If they

ARRL QN Signals for CW Net Use

QNA*	Answer in prearranged order.
QNB*	Act as relay Between _____ and _____.
QNC	All net stations Copy.
	I have a message for all net stations.
QND*	Net is Directed (controlled by net control station.)
QNE*	Entire net stand by.
QNF	Net is Free (not controlled.)
QNG	Take over as net control station.
QNH	Your net frequency is High.
QNI	Net stations report In.
	I am reporting into the net. (Follow with a list of traffic or QRU.)
QNJ	Can you copy me?
QNK*	Transmit messages for _____ to _____.
QNL	Your net frequency is Low.
QNM*	You are QRMing the net. Stand by.
QNN	Net control station is _____.
	What station has net control?
QNO	Station is leaving the net.
QNP	Unable to copy you.
	Unable to copy _____.
QNQ*	Move frequency to _____ and wait for _____ to finish handling traffic. Then send him traffic for _____.
QNR*	Answer _____ and Receive traffic.
QNS	Following Stations are in the net.*
	(Follow with list.)
	Request list of stations in the net.
QNT	I request permission to leave the net for _____ minutes.
QNU*	The net has traffic for you. Stand by.
QNV*	Establish contact with _____ on this frequency. If successful, move to _____ and send him traffic for _____.
QNW	How do I route messages for _____?
QNX*	You are excused from the net.*
	Request to be excused from the net.
QNY*	Shift to another frequency (or to _____ kHz) to clear traffic with _____.
QNZ	Zero beat your signal with mine.

*For use only by the Net Control Station.

Notes on Use of QN Signals

The QN signals listed above are special ARRL signals for use in amateur CW nets *only*. They are not for use in casual amateur conversation. Other meanings that may be used in other services do not apply. Do not use QN signals on phone nets. *Say it with words.* QN signals need not be followed by a question mark, even though the meaning may be interrogatory.

These "Special QN Signals for New Use" originated in the late 1940s in the Michigan QMN Net, and were first known to Headquarters through the then head traffic honcho W8FX. Ev Battey, W1UE, then ARRL assistant communications manager, thought enough of them to print them in *QST* and later to make them standard for ARRL nets, with a few modifications. The original list was designed to make them easy to remember by association. For example, QNA meant "Answer in *A*lphabetical order," QNB meant "Act as relay *Between*...," QNC meant "All *Net* Copy," QND meant "*Net* is *Directed*," etc. Subsequent modifications have tended away from this very principle, however, in order that some of the less-used signals could be changed to another, more needed, use.

Since the QN signals started being used by amateurs, international QN signals having entirely different meanings have been adopted. Concerned that this might make our use of QN signals with our own meanings at best obsolete, at worst illegal, ARRL informally queried FCC's legal branch. The opinion then was that no difficulty was foreseen as long as we continued to use them only in amateur nets.

Occasionally, a purist will insist that our use of QN signals for net purposes is illegal. Should anyone feel so strongly about it as to make a legal test, we might find ourselves deprived of their use. Until or unless we reach such an eventuality, however, let's continue to use them, and use them right. After all, we were using them first.

do make connection, and pass the traffic, they will return as soon as they are done.)

AIØO: O

KØSI: O

AIØO: DE AIØO GE PETE QRU K

KØSI: AIØO DE KØSI GE ROB QNU PSE UP 4 1 AIØO WID NDØN AFTER WØOUD THEN BOTH QNX 73 K

(Good evening, Rob, the net has traffic for you. Please go up 4 and get one for you from NDØN [WID means "with"] after he finishes with WØOUD. Then, when you are both finished, you and NDØN are both excused from the net. 73!)

AIØO: 73 GG

As you can see, it doesn't take much to say a lot on a CW net. Now that all the net business is taken care of, The NCS will start excusing other stations. Since K2ONP has a schedule to make with the Tenth Region Net, he will be excused first.

KØSI: K2ONP DE KØSI TU GEO FER QNI NW QRU QNX TU 73 K

K2ONP: TU PETE CUL 73 DE K2ONP S̄K̄

WØOUD: OUD

(Letha is back from receiving her traffic.)

KØSI: OUD TU LETHA NW QRU QNX 88 K

WØOUD: GN PETE CUL 88 DE WØOUD S̄K̄

Now, the NCS will close the net.

KØSI: MCWN NW QNF [the net is free] GN DE KØSI CL

When you check into a CW net for the first time, don't worry about speed. The NCS will answer you at about the speed you check into the net. You will discover that everyone on a CW net checks in with a different speed, just as everyone has a different voice on SSB. It's nothing to be self-conscious about. As the saying goes, "We're all in this together." The goal is to pass the traffic correctly, with 100 percent accuracy, not burn up the ether with our spiffy fists. Likewise, don't hesitate to slow down for someone else. Remember, when handling traffic, 100 percent is the minimum acceptable performance level!

On SSB

Now, let's journey through a world with plenty of shadow and substance, the Mowegia Single Sideband Net.

As we look in on KØPCK calling tonight's session, keep in mind these few pointers:

1) The net preamble, given at the beginning of each session, will usually give you all the information you need to survive on the net. Method of checking in varies greatly from net to net. For instance, some section nets have a prearranged net roll, some take checkins by alphabetical order, or some even take checkins by city, county or geographical area. Don't feel intimidated by a prearranged net roll. Those nets will always stand by near the end of the session to take stations not on the net roster.

2) As you listen to a net, you will find that on phone, formality also varies. Some SSB nets are strictly business, while others are "chattier." However, don't always confuse lack of formality with looseness on a net. There is still a

definite net procedure to adhere to.

3) Once, on a close play, the catcher asked umpire Bill Klem, "Well, what is it?"

"It ain't nothin' till I call it," he growled.

By the same token, you need to keep in mind that the NCS is the absolute boss when the net is in session. On CW nets, this doesn't seem to be much of a problem to the tightness of operation, but on phone nets, sometimes a group of "well-meaners" can really slow down the net. So don't "help" unless NCS tells you to.

Since net time is upon us, let's get back to the beginning of the Mowegia Single Sideband Net.

"Calling the Mowegia Single Sideband Net, calling the Mowegia Single Sideband Net. This is KØPCK, net control. The Mowegia Single Sideband Net meets on 3988 kHz nightly at 6 PM for the purpose of handling traffic in Mowegia and to provide a link for out-of-section traffic through the ARRL National Traffic System. My name is Ben, Bravo Echo November, located in Toad Lick, Mowegia. When I call for the letter corresponding to the first letter of the suffix of your call, please give your call sign only."

"Is there any emergency, priority or time value traffic for the net?"
(wait 5 seconds)
"Any relays?"
(wait 5 seconds)
"Any low-power, mobile or portable stations wishing to check in?"
(wait 5 seconds)
"Any relays?"
(wait 5 seconds)
"This is KØPCK for the Mowegia Sideband Net. Do we have any formal written traffic?"
"KØORB."
"KØORB, Good evening Bill. List your traffic, please."
"Good evening, Ben. I have two pieces to go out-of-state."
"Very good. Who is our Tenth Region Rep tonight?"
"Good evening, Ben. This is NIØR, Ten Rep."
"NIØR, this is KØPCK. Hi, Roger. Please call KØORB, move him to 3973 and get Bill's out-of-state traffic."
"KØORB, this NIØR. See you on 73."
"Going. KØORB."

Now, let's sit back and analyze this. As you can see, the format is pretty much identical except that it takes more words. Oh, yes, one other difference...you may have noticed that not one single Q-signal was used! Q-signals should not be used on phone nets; using them detracts from the "professionalism" of a net. Work hard at avoiding them, and you will reduce the "lingo barrier," making it a little less intimidating for a potential new checkin.

Two Special Cases

RTTY nets are almost identical to CW nets, except for the method of transmission, but you need to remember two extra tips:

1) It is always important to zero-beat the NCS, but on RTTY it is crucial! It doesn't take much to turn QTC1 Corn Shuck Holler to QTHGCRRX ULUSK HXLBELZZ4.

2) Stay far away from the key that sends RYRYRYRYRYRY. If you follow rule number one, RYing isn't necessary.

The other "exception to the rule" is the local 2-meter FM net. Many 2-meter nets are designed for ragchewing or weather spotting, so often if you bring traffic to the net, be prepared to coach someone in the nuances of traffic handling.

Who Owns the Frequency?

Traffic nets sometimes have difficulties when it comes time for the call-up, and a ragchew is taking place on the published net frequency. What to do? Well, you could break in on the ragchew and ask politely if the participants would mind relinquishing the frequency. This usually works, but what if it doesn't? The net has no more right to the frequency than the stations occupying it at net time, and the ragchew stations would be perfectly within their rights to decline to relinquish it.

The best thing to do in such a case is to call the net near, but not directly on, the normal frequency—far enough (hopefully) to avoid causing QRM, but not so far that net stations can't find the net. Usually, the ragchewers will hear the net and move a bit farther away—or even if they don't, the net can usually live with the situation until the ragchew is over.

It is possible to conceive of a situation, especially on 75-meter phone, in which the net frequency is occupied and the entire segment from 3850 to 4000 kHz is wall-to-wall stations. In such a case, the net will just have to do the best it can and take its chances on QRMing and being QRMed. In any case, it is *not* productive to argue about who has the most right to the frequency. Common courtesy says that the first occupants have, but there are many extenuating circumstances. Avoid such controversies, especially on the air.

Accordingly, net frequencies should be considered "approximate," inasmuch as it may be necessary for nets to vary their frequencies according to band conditions at the time. Further, no amateur or organization has any preemptory right to any specific amateur frequency.

A few other things to remember:

1) Unlike a "double" on CW or SSB, where the NCS might even still get both calls, a "double" on an FM repeater either captures only one station, or makes an ear-splitting squeally heterodyne. Drag your feet a little before you check in, so you are less likely to double.

2) Be especially aware to wait for the squelch tail or courtesy beep. More people seem to time out the repeater on net night than any other!

3) Remember that a lot of people have scanners, many with the local repeater programmed in on one of the channels. Design your behavior in such a way that it attracts nonhams to Amateur Radio. In other words, don't do anything you wouldn't do in front of the whole town! [For more information on RTTY operating, see Chapter 9; see Chapter 11 for additional details about FM.]

Oh, yes. One other thing will insure your success and build your reputation as a good traffic handler. It's a simple thing, really, but you'd be surprised how much mileage you can get out of saying "please" and "thank you."

MAKING IT, TAKING IT AND GIVING IT AWAY: MESSAGE HANDLING AND MESSAGE FORM

By this stage, you have probably been checking into the net for a while, and things have started to move along quite smoothly on your journey as a rookie traffic handler. But in the life of any new traffic op, the fateful day comes along when the NCS points RF at you and says those words that strike fear in almost every newcomer: "Pse up 4 get one QTC."

Now what? You could suddenly have "rig trouble" or a

"power outage" or a "telephone call," or just bump your dial and disappear. After all, it has been done before, and everyone that didn't do it sure thought of it the first time they were asked to take traffic! Of course, there is a more honorable route—go ahead and take it! Chances are you will be no worse than anyone else your first time out. The late Bob Peavler, WØBV, a well-known and respected trafficker, used to say, "Don't be afraid of making a mistake. Chances are we have already made all of them before you."

To ease the shock of your first piece of traffic, maybe it would be a good idea to go over message form "by the book"—ARRL message form, that is.

ARRL Message Form—the Right Way, and the Right Way

A common line of non-traffickers is, "Aw, why do they have to go through that ARRL message-form stuff? It just confuses people and besides, my message is just a few words or so. It's silly to go through all that rigamarole."

Well, then, let's imagine that you are going to write a letter to your best friend. What do you think would happen to your letter if you decided that the standardized method the post office used was "silly," so you signed the front, put the addressee's address where the return address is supposed to go, and stamped the inside of the letter? It would probably end up in the Dead Letter Office.

Amateur message form is standardized so it will reach its destination speedily and correctly. It is very important for every amateur to understand correct message form, because you never know when you will be called upon in a emergency. Most nonhams think all hams know how to handle messages, and it's troublesome to discover how few do. You can completely change the meaning of a piece of traffic by accident if you don't know the ARRL message form, and as you will see later, this can be a real "disaster." Learn it the right way, and this will never happen.

If you will examine the sample message in this section, you will notice that the message is essentially broken into four parts: the preamble, the address block, the text and the signature. The preamble is analogous to the return address in a letter and contains the following:

1) The number denotes the message number of the originating station. Most traffic handlers begin with number 1 on January 1, but some stations with heavy volumes of traffic begin the numbering sequence every quarter or every month.

2) The precedence indicates the relative importance of the message. Most messages are Routine (R) precedence—in fact, about 99 out of 100 are in this category. You might ask, then why use any precedence on routine messages? The reason is that operators should get used to having a precedence on messages so they will be accustomed to it and be alerted in case a message shows up with a different precedence. A Routine message is one that has no urgency aspect of any kind, such as a greeting. And that's what most amateur messages are—just greetings.

The Welfare (W) precedence refers to either an inquiry as to the health and welfare of an individual in the disaster area or an advisory from the disaster area that indicates all is well. Welfare traffic is handled only after all emergency and priority traffic is cleared. The Red Cross equivalent to an incoming Welfare message is DWI (Disaster Welfare Inquiry).

The Priority (P) precedence is getting into the category of high importance and is applicable in a number of circumstances: (1) important messages having a specific time limit, (2) official messages not covered in the emergency category, (3) press dispatches and emergency-related traffic not of the utmost urgency, and (4) notice of death or injury in a disaster area, personal or official.

The highest order of precedence is EMERGENCY (always spelled out, regardless of mode). This indicates any message having life-and-death urgency to any person or group of persons, which is transmitted by Amateur Radio in the absence of regular commercial facilities. This includes official messages of welfare agencies during emergencies requesting supplies, materials or instructions vital to relief of stricken populace in emergency areas. During normal times, it will be very rare.

3) Handling Instructions are optional cues to handle a message in a specific way. For instance, HXG tells us to cancel delivery if it requires a toll call or mail delivery, and to service it back instead. Most messages will not contain handling instructions.

4) Although the station of origin block seems self-explanatory, many new traffic handlers make the common

Handling Instructions

HXA—(Followed by number.) Collect landline delivery authorized by addressee within _____ miles. (If no number, authorization is unlimited.)

HXB—(Followed by number.) Cancel message if not delivered within _____ hours of filing time; service originating station.

HXC—Report date and time of delivery (TOD) to originating station.

HXD—Report to originating station the identity of station from which received, plus date, time and method of delivery.

HXE—Delivering station get reply from addressee, originate message back.

HXF—(Followed by number.) Hold delivery until _____ (date).

HXG—Delivery by mail or landline toll call not required. If toll or other expense involved, cancel message and service originating station.

An HX prosign (when used) will be inserted in the message preamble before the station of origin, thus: NR 207 R HXA50 W1AW 12 . . . (etc). If more than one HX prosign is used, they can be combined if no numbers are to be inserted; otherwise the HX should be repeated, thus: NR 207 R HXAC W1AW . . . (etc), but: NR 207 R HXA50 HXC W1AW . . . (etc). On phone, use phonetics for the letter or letters following the HX, to insure accuracy.

mistake of exchanging their call sign for the station of origin after handling it. The station of origin never changes. That call serves as the return route should the message encounter trouble, and replacing it with your call will eliminate that route. A good rule of thumb is *never* to change any part of a message.

5) The check is merely the word count of the text of the message. The signature is not counted in the check. If you discover that the check is wrong, you may not change it, but you may amend it by putting a slash bar and the amended count after the original count. [See below for additional information on the message check.]

Another common mistake of new traffickers involves "ARL" checks. A check of ARL 8 merely means the text has an ARL numbered radiogram message text in it, and a word count of 8. It does not mean ARRL numbered message no. 8. This confusion has happened before, with unpleasant results. For instance, an amateur with limited traffic experience once received a message with a check of ARL 13. The message itself was an innocuous little greeting from some sort of fair, but the amateur receiving it thought the message was ARRL numbered message 13—"Medical emergency exists here." Consequently, he unknowingly put a family through a great deal of unnecessary stress. When the smoke cleared, the family was on the verge of bringing legal action against the ham, who himself developed an intense hatred for traffic of any sort and refused to ever handle another message. These kinds of episodes certainly don't help the "white hat" image of Amateur Radio!

When counting messages, don't forget that each "X-ray" (instead of periods), "Query" and initial group counts as a word. Ten-digit telephone numbers count as three words; the ARRL-recommended procedure for counting the telephone number in the text of a radiogram message is to separate the telephone number into groups, with the area code (if any) counting as one word, the three-digit exchange counting as one word, and the last four digits counting as one word. Separating the telephone number into separate groups also helps to minimize garbling. Also remember that closings such as "love" or "sincerely" (that would be in the signature of a letter) are considered part of the text in a piece of amateur traffic.

6) The place of origin can either be the location (City/State or City/Province) of the originating station or the location of the third party wishing to initiate a message through the originating station. Use standard abbreviations for state or province. ZIP or postal codes are not necessary. For messages from outside the US and Canada, city and country is usually used. If a message came from the MARS (Military Affiliate Radio System) system, place of origin may read something like "Korea via Mars."

7) The filing time is another option, usually used if speed of delivery is of significant importance. Filing times should be in UTC or "Zulu" time.

8) The final part of the preamble, the date, is the month and day the message was filed—year isn't necessary.

Next in the message is the address. Although things like ZIP code and phone number aren't entirely necessary, the more items included in the address, the better its chances of reaching its destination. To experienced traffic handlers, ZIP codes and telephone area codes can be tip-offs to what area of the state the traffic goes, and can serve as a method of verification in case of garbling. For example, all ZIP codes in Minnesota start with a 5. Therefore, if a piece of traffic

Checking Your Message

Traffic handlers don't have to dine out to fight over the check! Even good ops find much confusion when counting up the text of a message. You can eliminate some of this confusion by remembering these basic rules:

1) Punctuation ("X-rays," "Querys") count separately as a word.

2) Mixed letter-number groups (1700Z, for instance) count as one word.

3) Initial or number groups count as one word if sent together, two if sent separately.

4) The signature does not count as part of the text, but any closing lines, such as "Love" or "Best wishes" do.

Here are some examples:
- Charles J McClain—3 words
- W B Stewart—3 words
- St Louis—2 words
- 3 PM—2 words
- SASE—1 word
- ARL FORTY SIX—3 words
- 2N1601—1 word
- Seventy three—2 words
- 73—1 word

Telephone numbers count as 3 words (area code, prefix, number), and ZIP codes count as one. ZIP + 4 codes count as two words. Canadian postal codes count as two words (first three characters, last three characters.)

Although it is improper to change the text of a message, you may change the check. Always do this by following the original check with a slash bar, then the corrected check. On phone, use the words "corrected to."

Book Messages

When sending book traffic, always send the common parts first, followed by the "uncommon" parts. For example:

R NØFQW ARL 8 BETHEL MO SEP 7 B̄T̄
ARL FIFTY ONE BETHEL SHEEP FESTIVAL X LOVE B̄T̄
PHIL AND JANE B̄T̄

NR 107 TONY AND LYN CALHOUN ĀĀ
160 NORTH DOUGLAS ĀĀ SPRINGFIELD IL 62702 B̄T̄

NR 108 JOE WOOD AJØX ĀĀ
84 MAIN STREET ĀĀ
LAUREL MS 39440 B̄T̄

NR 109 JEAN WILCOX ĀĀ
1243 EDGEWOOD DRIVE ĀĀ
LODI CA 95240 ĀR̄ N

Before sending the book traffic to another operator, announce beforehand that it is book traffic. Say "Follows book traffic." Then use the above format. On CW, a simple HR BK TFC will do.

sent as St Joseph, MO, with a ZIP of 56374 has been garbled along the way, it conceivably can be rerouted. So, when it comes to addresses, the adage "the more, the better" applies.

The text, of course, is the message itself. You can expedite the counting of the check by following this simple rule—when copying by hand, write five words to a line. When copying with a typewriter, or when sending a message via RTTY, type the message 10 words to a line. You will discover that this is a quick way to see if your message count agrees with the check. If you don't agree, nine times out of 10 you have dropped or added an "X-ray," so copy carefully. Another important thing to remember is that you never end a text with a "X-ray"—it just wastes space and makes the word count longer.

Finally, the signature. Remember, complimentary closing words like "sincerely" belong in the text, not the signature. In addition, signatures like "Dody, Vanessa, Jeremy,

MARS

Most modern-day traffic handlers don't realize that our standard message preamble is largely fashioned after the form used in the Army Amateur Radio System, which had its heyday in the '30s. The ARRL standard preamble prior to that time was quite different, but the AARS form was adopted because it had advantages and was so widely used. AARS nets were numerous in the '30s, using WL calls on two frequencies (3497.5 and 6990 kc.) outside the amateur bands and many frequencies, using amateur calls and amateur participants, inside the bands.

MARS, the post-World War II successor to AARS, encompasses all three of the US armed services. Although it operates numerous nets, all of which are outside the amateur bands on military frequencies, thousands of amateurs participate in MARS. This service performs some traffic coverage that we amateurs are not permitted to perform. MARS, which stands for "Military Affiliate Radio System," is conducted in three different organizations under the direction of the Army, Navy and Air Force. Therefore, MARS is not in the strictest sense Amateur Radio.

Nevertheless, many amateur messages find their way into MARS circuits, and MARS messages find their way into amateur nets. In fact, NTS has a semiformal liaison with MARS to handle the many messages from families in the States to their sons and daughters serving overseas.

Traffic for some points overseas at which US military personnel are stationed can be handled via MARS, provided a complete military address is given, even though some of these points cannot be covered by Amateur Radio or NTS. The traffic is originated in standard ARRL form and refiled into MARS form (now quite a bit different from ours) when it is introduced into a MARS circuit for transmission overseas. In this manner, traffic may be exchanged with military personnel in West Germany,

Japan and a few other countries that do not otherwise permit the handling of international third-party traffic.

Traffic coming from MARS circuits into amateur nets for delivery by Amateur Radio are converted from MARS to amateur form and handled as any other amateur message. The *exception* occurs when traffic originates overseas. In this case, the name of the country in which it originates, followed by "via MARS," should appear as the place of origin, so it does not appear that such messages were handled illegally by Amateur Radio. There are places where US military personnel are stationed that even MARS cannot handle traffic with, presumably because of objections by the host country. This information is in the hands of MARS "gateway" stations and changes from time to time.

The amount of MARS traffic appearing on amateur nets is not great, since MARS has a pretty good system of handling it on MARS frequencies, but it is important that we maintain liaison as closely as possible since all civilian MARS members are US licensed amateurs.

Information concerning MARS may be obtained directly from the individual branches at these addresses:

Air Force MARS
Chief, Air Force MARS
HQ, AFCC/DOOCC
Scott AFB, IL 62225-6001

Navy-Marine Corps MARS
Director, Navy-Marine Corps MARS
Naval Communications Unit
Washington, DC 20390-5161

HQ Army MARS
US Army Information Systems Command
AS-OPS-OA
Ft Huachuca, AZ 85613

Ashleigh, and Uncle Porter," no matter how long, go entirely on the signature line.

At the bottom of our sample message you will see call signs next to the blanks marked "sent" and "received." These are not sent as the message, but are just bookkeeping notes for your own files. If necessary, you could help the originating station trace the path of the message.

Keeping It Legal

In the FCC rules under the "Prohibited transmissions" heading (§97.113), it states that "no amateur station shall transmit any communication the purpose of which is to facilitate the business or commercial affairs of any party...except as necessary to providing emergency communications." Under the same heading, it further states that "no station shall transmit messages for hire or for material compensation, direct or indirect, paid or promised."

The FCC rules also have a section directly addressing traffic handling—§97.115, "Third party communications," which reads as follows:

(a) an amateur station may transmit messages for a third party to:

(1) Any station within the jurisdiction of the United States.

(2) Any station within the jurisdiction of any foreign government whose administration has made arrangements with the United States to allow amateur stations to be used for transmitting international communications on behalf of third parties. No station shall transmit messages for a third party to any station within the jurisdiction of any foreign government whose administration has not made such an arrangement. This prohibition does not apply to a message for any third party who is eligible to be a control operator of the station.

(b) The third party may participate in stating the message

where:

(1) The control operator is present at the control point and is continuously monitoring and supervising the third party's participation; and

(2) The third party is not a prior amateur service licensee whose license was revoked; suspended for less than the balance of the license term and the suspension is still in effect; suspended for the balance of the license term and relicensing has not taken place; or surrendered for cancellation following notice of revocation suspension or monetary forfeiture proceedings. The third party may not be the subject of a cease and desist order which relates to amateur service operation and which is still in effect.

(c) At the end of an exchange of international third party communications, the station must also transmit in the station identification procedure the call sign of the station with which a third party message was exchanged.

Note that emergency communications is defined in §97.403 as providing "essential communication needs in connection with the immediate safety of human life and immediate protection of property when normal communications systems are not available."

It's self-explanatory. Every amateur should be familiar with these rules. Also, while third-party traffic is permitted in the US and Canada, this is not so for most other nations. A special legal agreement is required in each country to make such traffic permissible, both internally and externally (except if the message is addressed to another amateur). A list of third-party agreements in effect when this book was published appears in Chapter 5. Check this list before agreeing to handle *any kind* of international traffic.

Such agreements specify that only unimportant, personal, non-business communications be handled—things that ordinarily would not utilize commercial facilities. (In an emergency situation, amateurs generally handle traffic *first* and face the possible consequences *later*. It is not unusual for a special limited-duration third-party agreement to be instituted by the affected country during an overseas disaster.) The key point here is, particularly under routine day-to-day nonemergency conditions, if we value our privileges, we must take care not to abuse any regulations, whether it be on the national or international level.

Some Helpful Hints about Receiving Traffic

1) Once you have committed the format of ARRL message form to memory, there's no need to use the "official" message pads from ARRL except for deliveries. Traffic handlers have many varied materials on hand for message handling. Some just use scrap paper. Many buy inexpensive 200-sheet 6- × 9-inch plain tablets available at stationery stores. Those who like to copy with a typewriter often use roll paper or fan-fold computer paper to provide a continuous stream of paper, separating each message as needed. As long as you can keep track of it, anything goes in the way of writing material.

2) Don't say "QSL" or "I roger number..." unless you mean it! It's not "Roger" unless you've received the contents of the message 100 percent. It's no shame to ask for fills (ie, repeats of parts of the message). Make sure you have received the traffic correctly before going on to the next one.

3) Full- (QSK) or semi-breakin can be very useful in handling traffic on CW (or VOX on SSB). If you get behind, saying "break" or sending a string of dits will alert the other op that you need a fill.

4) You can get a fill by asking for "word before" (WB), "word after" (WA), "all before" (AB), "all after" (AA) or "between" (BN).

Sending the Traffic

Just because you've taken a few messages, don't get the notion that being good at receiving traffic makes you a good sender, too. In the "dark ages" of hamdom, when one had to journey to the FCC office to take exams, one not only had to receive code in the code test, but send it as well. It used to be amazing how many folks could copy perfect 20-WPM code only to be inadequate on the sending portion. A lot of good ops became quite amazed that they could be so ham-fisted on a straight key!

Good traffic operators know they have to learn the nuances of sending messages as well as getting them. Your ability to send can "make or break" the other operator's ability to receive traffic in poor conditions. Imagine yourself and the other operator as a pair of computers interfaced over the telephone. Your computer (your brain) must successfully transfer data through your modem (your rig) over the lines of communication (the amateur frequencies) to the other modem, and ultimately, to the other computer.

Of course, for this transfer to be successful, two major items must be just right. The modems must operate at the same baud rate, and they must operate on matching protocol. Likewise, you must be careful to send your traffic at a comfortable speed for the receiving op, and use standardized protocol (standard ARRL message form). As you will see, these are slightly different for phone and CW, with even a couple of other deviations for RTTY or FM.

Sending the Traffic by CW

Someone once remarked, "The nice thing about CW traffic handling is that you have to spell it as you go along, so

you don't usually have to spell words over." Also, the other main difference in CW traffic handling is that you tell someone when to go to the next line or section by use of the prosigns $\overline{AA}$ or $\overline{BT}$. Keeping this in mind, let's show how our sample message would be sent:

NR 133 R HXG WØMME ARL8 MOUNT PLEASANT IA 1700Z SEP 1
MR MRS JEFF HOLTZCLAW $\overline{AA}$
ROUTE 1 BOX 127 $\overline{AA}$
TONGANOXIE KS 66086 $\overline{AA}$
913 555 1212 $\overline{BT}$
ARL FIFTY ONE OLD THRESHERS REUNION X LOVE $\overline{BT}$
UNCLE CHUCKIE $\overline{AR}$ N (if you have no more messages) or
$\overline{AR}$ B (if you have further messages)

Now, let's examine a few points of interest:

1) You don't need to send "preamble words" such as precedence and check. The other operator is probably as familiar with standard ARRL form as you are (maybe more!).

2) The first three letters of the month are sufficient when sending the date.

3) In the address, always spell out words like "route" and "street."

4) Do not send dashes in the body of telephone numbers; it just wastes time.

5) Always, always, always spell out each word in the text! For example, "ur" for "your" could be misconstrued as the first two letters of the next word. Abbreviations are great for ragchewing, but not for the text of a radiogram.

6) Sometimes, if you have sent a number of messages, when you get to the next-to-last message, it's a good idea to send $\overline{AR}$ 1 instead of $\overline{AR}$ B to alert the other station that you have just one more.

7) If the other operator "breaks" you with a string of dits, stop sending and wait for the last word received by the other side. Then, when you resume sending, start up with that word, and continue through the message.

Becoming a proficient CW traffic sender is tough at first, but once you've mastered the basics, it will become second nature—no kidding!

Sending the Traffic by Phone

Phone traffic handling is a lot like the infield fly rule in baseball—everyone thinks they know the rule, but in truth few really do. Correct message handling via phone can be just as efficient as via CW if and only if the two operators follow these basic rules:

1) If it's not an actual part of the message, don't say it.

2) Unless it's a very weird spelling, don't spell it.

3) Don't spell it phonetically unless it's a letter group or mixed group, or the receiving station didn't get it when you spelled it alphabetically.

Keeping these key points in mind, let's waltz through our sample message. This is how an efficient phone traffic handler would send the message:

"Number one hundred thirty three, routine, Hotel X-ray Golf, WØ Mike Mike Echo, ARL EIGHT, Mount Pleasant, Iowa, seventeen hundred Zulu, September one."

"Mr and Mrs Jeff Holtzclaw, route one, box twenty seven, Tonganoxie, T-O-N, G-A-N, O-X, I-E, Kansas, six six zero eight six. Nine one three, five five five, one two, one two. Break." You would then let up on the PTT switch (or pause if using VOX) and give the operator any fills needed in the first half of the message.

"ARL FIFTY ONE Old Threshers Reunion X-ray Love

Break Uncle Chuckie. End, no more'' (if you have no more messages), ''more'' (if you have more messages).

Notice that in phone traffic handling, a pause is the counterpart for $\overline{AA}$. Also notice the lack of extraneous words. You don't need to say ''check,'' or ''signature,'' or ''Jones, common spelling.'' (If it's common spelling, why tell someone?) You only spell the uncommon. Most importantly, you speak at about half reading speed to give the other person time to write. If the receiving operator types, or if you have worked with the other op a long time and know his capabilities, you can speak faster. Always remember that any fill slows down the message more than if you had sent the message slowly to begin with!

Oh, Yes... Those Exceptions!

Once again, RTTY and VHF FM provide the exceptions to the rules. RTTY traffic is very much like CW traffic. You include the prosigns $\overline{AA}$, $\overline{BT}$, and $\overline{AR}$, as a typed AA, BT or AR. Single-space your message (you don't need to waste paper) and send ''BBBB'' or ''NNNN'' as the last line of your message depending on whether you have more traffic or not. Use three or four lines between messages. This allows you to get four or five average messages on a standard 8½ × 11-inch sheet of paper. For further details, see Chapter 9.

When sending a message over your local repeater, remember that you often will be working with someone who isn't a traffic handler. It may be necessary to break more often (between the preamble and the address, for instance). Also, always make sure they understand about ARL numbered radiogram texts, and if they don't have a list, tell them what the message means. [The complete list of ARRL numbered radiogram texts appears in Chapter 17.] FM is a quiet mode, so you can get away with less spelling than you do on SSB.

If you yourself are already into traffic, don't try to force-feed correct traffic procedure in the case of someone just starting out; ease the person into it a little at a time. You will give a nontrafficker a more positive impression of traffic handling and may even make someone more receptive to joining a net. After all, our goal as traffic handlers is to make it look fun!

Delivering a Message

Up to now, all our traffic work has been carried out in the safe, secure world of Amateur Radio. All of this changes, though, when we get a piece of traffic for delivery. Suddenly, we're thrust out among the general public in the real world with this little message and expected to give it to someone. Unfortunately, many hams don't realize the importance of this little action and miss an opportunity to engrave a favorable impression of Amateur Radio on nonhams. It's ironic that many hams can chat for hours on CW or SSB, but can't pick up the telephone and deliver a 15-word message without mumbling, stuttering or acting embarrassed. Delivering messages should be a treat, not a chore.

Let's go through a few guidelines for delivery, and if you keep these tips in mind, you and the party on the other end of the phone will enjoy the delivery.

1) Ask for the person named in the message. If he or she is not home, ask the person on the phone if they would take a message for that person.

2) Introduce yourself. Don't you hate phone calls from people you don't know and don't bother to give a name? Chances are they're trying to sell you something, and you brush them off. Most people have no idea what Amateur

Radio is about, and it's up to you to make a good first impression.

3) Tell who the message is from before you give the message. Since the signature appears at the end of the message, most hams give it last, but you will hold the deliveree's attention longer if you give it first. When you get letters in the mail, you check out the return addresses first, don't you? Then you open them in some sort of order of importance. Likewise, the party on the phone will want to know the sender of the message first.

A good way to start off a delivery is to say something like, ''Hi, Mr/Mrs/Miss so-and-so, my name is whatever, and I received a greeting message via Amateur Radio for you from wherever from such-and-such person.'' This usually gives you some credibility with your listener, because you mentioned someone they know. They will usually respond by telling such-and-such is their relative, college friend, etc. At that point, you have become less of a stranger in their eyes, and now they don't have to worry about you trying to sell them some aluminum siding or a lake lot at Casa Burrito Estates. Make sure you say it's a *greeting* message, too, to allay any fears of the addressee that some bad news is imminent.

4) When delivering the message, skip the preamble and just give the text, avoiding ARL text abbreviations. Chances are, Grandma Ollie doesn't give two hoots about the check of a message, and thinks ARL FORTY SIX is an all-purpose cleaner. Always give the ''translation'' of an ARL numbered text, even if the message is going to another ham.

5) Unless it is a ''mass origination'' message from a fair or other special event, ask the party if they would like to send a return message. Explain that it's absolutely free, and that you would be happy to send a reply if they wish. Experienced traffickers can vouch that it's easy to get a lot of return and repeat business once you've opened the door to someone. It's not uncommon for strangers to ask for your name or phone number once they discover Amateur Radio is a handy way to tell relatives ''Arriving 3 PM on Flight 202 next Wednesday.''

To Mail or Not to Mail

Suppose you get a message that doesn't have a phone number, or the message would require a toll call. Then what? If you don't know anyone on 2 meters that could deliver it, or Directory Assistance is of no help, you are faced with the decision whether to mail it or not. There is no hard and fast rule on this (unless, of course, the message has an HXG attached). Always remember that since this is a free service, you are under no obligation to shell out a stamp just because you accepted the message.

Many factors may influence your decision. If you live in a large urban area, you probably have more deliveries than most folks, and mail delivery could be a big out-of-pocket expense that you're not willing to accept. If you live out in the wide open spaces, you may be the only ham for miles around, and probably consider mail delivery more often than most. Are you a big softie on Christmas or Mother's Day? If so, you may be willing to brunt the expense of a few stamps during those times of the year when you wouldn't otherwise. At any rate, the decision is entirely up to you.

Although you may be absolved from the responsibility of mailing a message, you don't just chuck the message in the trash. You do have a duty to inform the originating station that the message could not be delivered. A simple ARL SIXTY SEVEN followed by a brief reason (no listing, no one home for three days, mail delivery returned by post office, and so forth) will suffice. This message always goes to the station of origin, not the person in the signature. The originating station will appreciate your courtesy.

Now That You're Moving Up in the World

By now, you are starting to get a grasp of the traffic world. You've been checking in to a net on a regular basis, and you're pretty good at message form. Maybe you've even delivered a few messages. Now you are ready to graduate from Basic Traffic 101 and enroll in Intermediate Traffic 102. Good for you! You have now surpassed 80 percent of your peers in a skillful specialty area of Amateur Radio. However, there's still a lot to learn, so let's move on.

Book Messages

Over the years, book messages have caused a lot of needless headaches and consternation among even the best traffic handlers. Many hams avoid booking anything just because they think it's too confusing. Truthfully, book messages are fairly simple to understand, but folks tend to make them harder than they actually are.

So, just what are book messages? Book messages are merely messages with the same text and different addresses. They come in two categories—ones with different signatures, and ones with the same signatures. Elsewhere in the chapter are examples. Often you will see book messages around holiday times and during fairs or other public events.

Oh, yes . . . one other thing about book messages. When you check into a net with a bunch of book messages, give the regular message count only. Don't say, "I have a book of seven for Outer Baldonia." (The NCS has enough to keep track of without having to break his train of thought to divide by three.) Say instead, "Seven Outer Baldonia." Then, when you and the station from Outer Baldonia go off frequency to pass the traffic, tell him that it is book traffic. When he tells you to begin sending, give common parts first, then the "uncommon" parts (addresses and possibly signatures.) By following this procedure, you will avoid a lot of confusion.

Suppose you get a book of traffic on the NTS Region net bound for your state, but to different towns. When you take them to your section net, you will not be able to send them as a book, since they must be sent to different stations. Now what? Simply "unbook" them, and send them as individual messages. For instance, let's say you get a book of three messages for the Mowegia section from the region net. Two are for Mowegia City, and one is for Swan Valley. Simply list your traffic as Mowegia City 2 and Swan Valley 1, for a total count of 3. Books aren't ironclad chunks of traffic, but a step-saver that can be used to your advantage. They can be un-booked at any time. Use them whenever you can, and don't be afraid of them.

BECOMING NCS AND LIVING TO TELL ABOUT IT

Some momentous evening in your traffic career, you may be called upon to take the net. Perhaps the NCS had a power failure, or is on vacation, or perhaps a vacancy occurred in the daily NCS rotation on your favorite net. Should this be the case, consider yourself lucky. Net Managers entrust few members with net-control duties.

Of course, you probably won't be thinking how lucky you are when the Net Manager says "QNG" and sticks your call after it. Once again, just like your first checkin or your first piece of traffic, you will just have to grit your teeth and live through it. However, you can make the jump easier by following these hints *long before* you are asked to be a net control:

1) Become familiar with the other stations on the net. Even if you never become NCS, it pays to know who you work with and where they live.

2) Pay close attention to the stations that go off frequency to pass traffic. What frequencies does the net use to move traffic? Which stations are off frequency at the moment? You will gain a feel for the net control job just by keeping track of the action.

3) Try to guess what the NCS will do next. You will discover many dilemmas when you try to second-guess the NCS. Often different amounts of traffic with equal precedence appear on the net, and a skillful NCS must rank them in order of importance. For instance, if you follow the NCS closely, you will discover that traffic for the NTS rep, such as out-of-state traffic, gets higher priority than one for the NTS rep's city. Situations like these are fun to second-guess when you are standing by on the net and will better prepare you for the day *you* might get to run the net.

Should that day arrive, just keep your cool and try to implement the techniques used by your favorite net-control stations. After a few rounds of NCS duties, you will develop your own style, and who knows? Perhaps some new hopeful for NCS will try to emulate *you* some day! See accompanying sidebars for further hints on developing a "type-NCS" personality and on proper net-control methods.

HANDLING TRAFFIC AT FAIRS OR OTHER PUBLIC EVENTS

A very special and important aspect of message handling is that of how to handle traffic at public events. If the event is of any size, like a state fair, it doesn't take long to swamp a group of operators with traffic. Only by efficient, tight organization can a handful of amateurs keep a lid on the backlog.

No matter what size your public event, the following points need to be considered for any traffic station accessible to the public:

1) Often, more nontraffic handlers than traffic handlers will be working in the booth. This means your group will have to lay out a standard operating procedure to help the non-traffickers assist the experienced ops.

Jobs such as meeting the public, filling out the message blanks, sorting the "in" and "out" piles, and keeping the booth tidied up, can all be performed by people with little or no traffic skill, and is a good way to introduce those people into the world of traffic.

Are You a Type-NCS Personality?

As net control station (NCS), it pays to remember that the net regulars are the net. Your function is to preside over the net in the most efficient, businesslike way possible so that the net participants can promptly finish their duties and go on to other ones. You must be tolerant and calm, yet confident and quick in your decisions. An ability to "take things as they come" is a must. Remember that you were appointed NCS because your Net Manager believes in you and your abilities.

1) *Be the boss, but don't be bossy.* It's your job to teach net discipline and train new net operators (and retrain some old ones!). You are the absolute boss when the net is in session, even over your Net Manager. However, you must be a "benevolent monarch" rather than a tyrant. Nets lose participation quickly one night a week when it's Captain Bligh's turn to call the net. If the net has a good turnout every night but one, that tells something about its NCS.

2) *Be punctual.* Many of the net participants have other commitments or nets to attend to; liaison stations are often on a tight schedule to make the NTS region or area net. If you, as NCS, don't care when the net starts, others will think it's okay for them to be late, too. Then traffic doesn't get passed in time, and someone may miss his NTS liaison. In short, the system is close to breaking down.

3) *Know your territory.* Your members have names—use them. They also live somewhere—by knowing their locations, you can quickly ascertain who needs to get the traffic. As NCS, it's your responsibility to know the geography of your net. You also need to understand where your net fits into the scheme of NTS.

4) *Take extra care to keep your antennas in good shape,* because an NCS can't run a net with a "wimpy" signal. Although you don't have to be the loudest one on the net, you do have to be heard. You will discover that the best way to do this is to have a good antenna system. A linear amplifier alone won't help you hear those weak checkins!

5) *The NCS establishes the net frequency.* Just because the *Net Directory* lists a certain frequency doesn't give you squatters rights to it if a QSO is already in progress there. Move to a nearby clear frequency, close enough for the net to find you. QRM is a fact of life on HF, especially on 75/80 meters, so live with it.

6) *Keep a log of every net session.* Just because the FCC dropped the logging requirements doesn't mean that you have to drop it. It's a personal decision. The Net Manager may need information about a checkin or a piece of traffic, and your log details can be helpful to him in determining what happened on a particular night.

7) *Don't hamstring the net by waiting to move the traffic.* Your duty is to get traffic moving as quickly as possible. As soon as you can get two stations moving, send them off to clear the traffic. If you have more than one station holding traffic for the same city, let the "singles" (stations with only one piece for that city) go before the ones with more than one piece for that city. The quicker the net gets the traffic moved, the sooner the net can be finished and the net operators can be free to do whatever they want.

---

2) If you plan to handle fairly large amounts of traffic, the incoming traffic needs to be sorted. A good system is to have an "in-state" pile, an "in-region" pile, and three piles for the three areas of NTS. After the traffic has been sent, it needs to be stacked in numerical order in the "out" pile. Keeping it in numerical order makes it easier to find should it need to be referred to.

Since your station will be on a number of hours, plan to check into your region and area, as well as your section, net. Another good idea is to have "helpers" on 2 meters who can also take some of your traffic to the region and/or area net. These arrangements need to be worked out in advance.

3) Make up your radiogram blanks so that most of the preamble is already on them, and all you need to fill in is the number, check and date. In the message portion, put only about 20 word lines to discourage lengthy messages. Try to convince your "customer" to use a standard ARL text so you can book your messages.

The Net Control Sheet

Net controlling is no easy task, requiring much talent on the part of the operator. A useful prop is a set system for keeping track of net operation so that you don't get mixed up and start to lose control. This can best be effected by a sheet of paper on which you record who has what traffic, who covers what and who is on what side frequency. Trying to keep this information in your head is a losing battle unless you have a remarkable memory.

There are many methods for doing this, depending to a great extent on what net is being controlled and the exact procedure used. In general, however, the best method is to list the calls of stations reporting in vertically down the page, followed by their coverage, if known. The coverage may be unnecessary if the NCS knows his stations, but it is a good idea to leave space for it in case an unfamiliar station reports in and you have to ask his location.

Next, list horizontally across the page the traffic reported by each station coming into the net, using destination (abbreviated) followed by the number of messages. From this you can see at a glance when traffic flow can start. In most nets, it is best to start it right away, and not wait until all stations have reported in. As traffic is passed, it can be crossed off. Whenever you get a station who has no traffic and for whom there is none, that station can be excused (QNX) and crossed off the list. As stations clear traffic and there is none for them, they also can be excused.

If side frequency (QNY) procedure is used, net controlling is a bit more complicated, but the use of such frequencies vastly speeds up the process. In this case, you will need to keep track of who is on which side frequency clearing which traffic, and you will be kept busy dispatching them, sending stations up or down to meet stations already on side frequencies, checking stations back into the net as they return, etc. Both your fingers and your key or mike button will be kept going, and it can be a nightmare if not handled properly.

Probably the best method is to keep the side frequencies on a separate sheet, each side frequency utilizing a separate column, labeled up 5, up 10, down 5, down 10, or whatever spacing intervals you find practical. As two stations are dispatched to a side frequency, enter the suffixes of their calls in the appropriate column, at the same time crossing out the traffic they are sent there to handle. When they return, cross them out of the side-frequency column; this side frequency can then accommodate two other stations. Of course, if you dispatch a third station to the side frequency to wait to clear traffic to one of the stations already there, enter him in the column also, and then only one of the two originally dispatched will return, so just cross off that one. When all your listed traffic is crossed off, the net is ready to secure (QNX QNF).

There are a number of refinements to this method, but the above is basic and a good way to start. Experience will soon indicate better ways to do it. For instance, K2KIR's novel approach follows.

Net Controlling EAN

At the NTS area and region levels, the complexity of the net control task often suggests the desirability of using a matrix form of log sheet. Many of the Eastern Area Net (EAN) Net Control Stations use the form here. A similar form for region nets can be made by replacing the RN column headings with the section net designations and the CAN/PAN columns with single column titled THRU (or EAN, CAN or PAN). Although this form can be used with no other accessories than a pencil, the use of moveable objects such as 6-32 hex nuts, buttons or push pins materially aids the visualization of which stations are on the net frequency and hence which stations are available for pairing off and passing traffic.

As stations are sent off frequency, their hex nuts are moved from the net column to the appropriate side frequency and a single diagonal line is drawn through the traffic being cleared. If, for any reason, the traffic is not cleared, a circle is drawn around the traffic total as a reminder to the NCS that it still must be cleared. (In other words, the circle overrides the slash.) Assuming the traffic is cleared on the second attempt, an opposite diagonal is drawn, thus totally crossing out the traffic quantity.

For the example shown, the following notes should explain how the various features of this NCS sheet are utilized:

Halfway through the Eastern Area Net session:

• W1EFW and VE3GOL are both clear and have been QNXed at 0055 and 0057, respectively.

• W8PMJ has cleared his CAN and PAN traffic, as well as 8RN traffic from W1TN. He has yet to clear his 3RN, plus receive 8RN from W3YQ. He is presently DOWN TEN with W2CS.

• W2CS was previously sent off to clear his 4RN with W4NTO, but they were unable to complete the pairing. The NCS will try again later, perhaps by using a relay station.

• WB4PNY has a net report for K2KIR, to be sent direct to him if he QNIs.

• N2AKZ and KW1U are UP TEN, clearing 2RN traffic.

• KQ3T, W4NTO and K4ZK are standing by on the net frequency, waiting for new assignments.

There are many techniques for determining in what order to clear the traffic listed, but a couple of fairly common ones are worth mentioning.

1) Assign highest priorities to stations having the largest totals. Thus, W1NJM (PAN RX), W8PMJ (8RN TX and RX) and W3YQ (3RN TX) should be kept waiting as little as possible. If conditions are good, QNQing other stations to these three will speed things up immensely.

2) Tackle the smaller individual destination totals first. This allows early excusing of stations having small traffic totals to clear, and avoids a last-minute panic near the end of the net.

3) Clear short-haul pairings first. In the winter near a sunspot minimum, short-haul communications are most

							30				EASTERN	AREA	NET								
+30	+25	+20	+15	+10	+5	IN	STATION	OUT	1RN	2RN	3RN	4RN	8RN	ECN	CAN	PAN	OTR	−5	−10	−15	−20	−25
						55	W1EFW	55	⧄													
					⬡	31	N2AKZ			⧄												
					㉝	KQ3T				⧄											
					㉜	W4NTO					⧄										
					33	W8PMJ				2		⧄		✗	7		⊙				
					36	VE3GOL	57	✗			⧄	✗								
				⬡	34	WB4PNY						⧄			K1R					
					30	W1NJM									⧄		⊙			
				⬡	30	㊉W3YQ		✗	2		✗	4	✗	✗	2✗					
					31	㊉W1TN		✗	✗		✗			✗		⊙			
					32	㊉W2CS				⊗✗			3		⊙			
					㉝	K4ZK			✗				7					
				⬡	34	㊉KW1U		✗	✗	1								

FIG. 1

likely to be successful early in the net session before the skip has gone out. At all other times, especially in the summer with nets on Daylight Savings Time, it delays long-haul pairings until band conditions have improved for those paths.

Obviously, there are times when these (or any other algorithms) will conflict with each other. In the final analysis, a good NCS bases his decisions on the specifics of a given net session: the band conditions, the operators, the traffic distribution and the time remaining.—*Bud Hippisley, K2KIR*

A real time- and headache-saver in this department is to fill out the message blank for the sender. This way, you can write in the X-rays and other jargon that the sender is unaware of.

4) Most importantly, realize that you sometimes have to work at getting "customers" as much as if you were selling something! Most people have no concept of Amateur Radio at all, and don't understand how message handling works. ("How can they get it? They don't have a radio like that" is a very common question!)

Use posters to make your booth appealing to the eye. Make sure one of the posters is of the "How your message gets to its destination" variety, such as the one shown in this section. Don't go over someone's head when answering a question—explain it simply and succinctly.

Finally, don't be afraid to "solicit" business. Get up in front of the booth and say "hi" to folks. If they say "hi" back, ask them if they would like to send a free greeting to a friend or relative anywhere in the US (grandparents and grandchildren are the easiest to convince!). Even if they decide not to send a message, your friendliness will help keep our image of "good guys in white hats" viable among the general public, which is every bit as great a service as message handling. For further hints, see accompanying sidebar.

THE ARRL NATIONAL TRAFFIC SYSTEM— MESSAGE HANDLING'S "ROAD MAP"

Although you probably never think about it, when you check into your local net or section net, you are participating in one of the most cleverly designed game plans ever written—the National Traffic System (NTS). Even though the ARRL conceived NTS way back in 1949, and it has grown from one

Handy Hints for Handling Traffic at Fairs or Other Public Events

1) Although you may only be there a day or two, don't compromise your station too much. Try to put up the most you can for an antenna system because band conditions on traffic nets in the summer can really be the pits! Usually, you will be surrounded by electrical lines at fairs, so a line filter is a must. An inboard SSB or CW filter in your rig is a definite plus, too, and may save you many headaches.

2) Don't huddle around the rigs or seat yourself in the back of the booth. Get up front and meet the people. After all, your purpose is to "show off" Amateur Radio to perk the interest of nonhams.

3) Most people will not volunteer to send a piece of traffic, nor will they believe a message is really "free." It's up to you to solicit "business." Be cheerful.

4) Always use "layman's language" when explaining about Amateur Radio to nonhams. Say "message," not "traffic." Don't ramble about the workings

2. AN AMATEUR RADIO OPERATOR TAKES YOUR MESSAGE AND RELAYS IT VIA AMATEUR RADIO TO THE CITY OF DESTINATION.

3. THE RECEIVING STATION GETS YOUR MESSAGE AND CALLS YOUR AUNT PATTY ON THE TELEPHONE.

4. AUNT PATTY GETS YOUR MESSAGE AND IS GLAD TO HEAR FROM YOU.

FREE RADIOGRAMS

1. YOU SEND A MESSAGE FROM THE FAIR TO YOUR AUNT PATTY, VIA AMATEUR RADIO.

of NTS or repeaters; your listener just wants to know how Aunt Patty will get the message. "We take the message and send it via Amateur Radio to Aunt Patty's town, and the ham there will call her on the phone and deliver it to her," will do.

5) Make sure your pencils or pens are attached to the booth with a long string, or you will be out of writing utensils in the first hour!

6) Make sure there are plenty of instructions around for hams not familiar with traffic handling to help them "get the hang of" the situation.

7) Make sure your booth is colorful and attractive. You will catch the public's eye better if you give them something to notice, such as this suggested poster idea.

NATIONAL TRAFFIC SYSTEM AREA & REGION MAP

regular cycle to two to three, NTS hasn't outgrown itself and remains the most streamlined method of traffic handling in the world. (During this discussion, please refer to the accompanying Section/Region/Area map, the NTS Routing Guide and the NTS Flow Chart.)

Actually, the National Traffic System can trace its roots to the railroad's adoption of Standard Time back in 1883,

when radio was still only a wild dream. Three of the Standard Time Zones are the basis for the three NTS areas—Eastern Area (Eastern Time Zone), Central Area (Central Time Zone) and Pacific area (Mountain and Pacific Time Zones). Within these areas are a total of 12 regions. You may wonder why NTS has 12 regions—why not just break it up into 10 regions, one for each area? Ah, but check the map. You will discover that NTS not only covers the US, but our Canadian neighbors as well. Then, of course, the region nets are linked to section/local nets.

The interconnecting lines between the boxes on the flow chart represent liaison stations to and from each level of NTS. The liaisons from area net to area net have a special name, the Transcontinental Corps (TCC). In addition to the functions shown, TCC stations also link the various cycles of NTS to each other.

The clever part about the NTS setup, though, is that in any given cycle of NTS, all nets in the same level commence at approximately the same local time. This allows time for liaisons to the next level to pick up any outgoing traffic and meet the next net. In addition, this gives the TCC stations at least an hour before their duties commence on another area net or their schedule begins with another TCC station.

The original NTS plan calls for four cycles of traffic nets, but usually two cycles are sufficient to handle a normal load of traffic on the system. However, during the holiday season, or in times of emergency, many more messages are dumped into the system, forcing NTS to expand to four cycles temporarily. The cycles of normal operation are Cycle Two, the daytime cycle, which consists primarily of phone nets, and

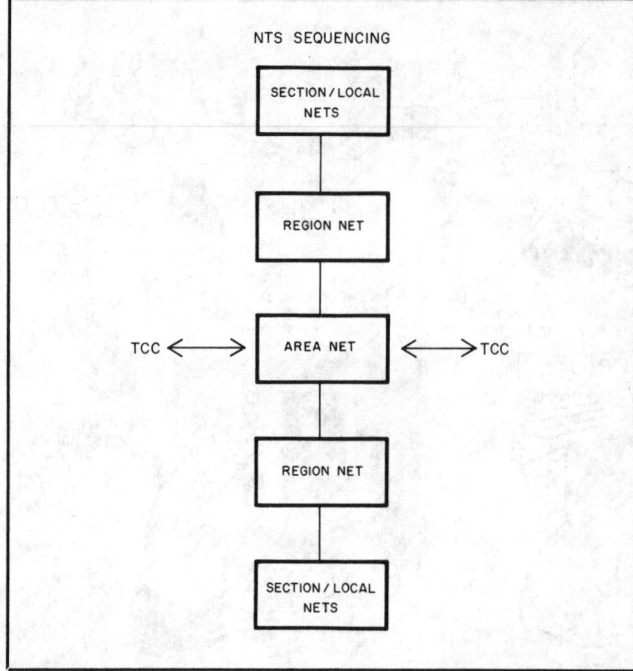

NTS SEQUENCING

SECTION / LOCAL NETS

REGION NET

AREA NET

TCC ←→ AREA NET ←→ TCC

REGION NET

SECTION / LOCAL NETS

Table 1
National Traffic System Routing Guide

State/Province	Abbrev.	Region	Area
Alaska	AK	7	PAN
Alabama	AL	5	CAN
Alberta	AB	7	PAN
Arizona	AZ	12	PAN
Arkansas	AR	5	CAN
British Columbia	BC	7	PAN
California	CA	6	PAN
Colorado	CO	12	PAN
Connecticut	CT	1	EAN
Delaware	DE	3	EAN
District of Columbia	DC	3	EAN
Florida	FL	4	EAN
Georgia	GA	4	EAN
Guam	GU	6	PAN
Hawaii	HI	6	PAN
Idaho	ID	7	PAN
Illinois	IL	9	CAN
Indiana	IN	9	CAN
Iowa	IA	10	CAN
Kansas	KS	10	CAN
Kentucky	KY	9	CAN
Labrador	LB	11	EAN
Louisiana	LA	5	CAN
Maine	ME	1	EAN
Manitoba	MB	10	CAN
Maryland	MD	3	EAN
Massachusetts	MA	1	EAN
Michigan	MI	8	EAN
Minnesota	MN	10	CAN
Mississippi	MS	5	CAN
Missouri	MO	10	CAN
Montana	MT	7	PAN
Nebraska	NE	10	CAN
Nevada	NV	6	PAN
New Brunswick	NB	11	EAN
New Hampshire	NH	1	EAN
New Jersey	NJ	2	EAN
New Mexico	NM	12	PAN
New York	NY	2	EAN
Newfoundland	NF	11	EAN
North Carolina	NC	4	EAN
North Dakota	ND	10	CAN
Nova Scotia	NS	11	EAN
Ohio	OH	8	EAN
Oklahoma	OK	5	CAN
Ontario	ON	11	EAN
Oregon	OR	7	PAN
Pennsylvania	PA	3	EAN
Prince Edward Island	PEI	11	EAN
Puerto Rico	PR	4	EAN
Quebec	PQ	11	EAN
Rhode Island	RI	1	EAN
Saskatchewan	SK	10	CAN
South Carolina	SC	4	EAN
South Dakota	SD	10	CAN
Tennessee	TN	5	CAN
Texas	TX	5	CAN
Utah	UT	12	PAN
Vermont	VT	1	EAN
Virginia	VA	4	EAN
Washington	WA	7	PAN
West Virginia	WV	8	EAN
Wisconsin	WI	9	CAN
Wyoming	WY	12	PAN
Virgin Islands	VI	4	EAN
APO New York	APO NY	2	EAN
APO San Francisco	APO SF	6	PAN

Cycle Four, the nighttime cycle, made up mostly of CW nets. In addition, Cycle One has been implemented in the Pacific and Eastern areas. Now that the rudiments of NTS have been covered, let's see where you fit in.

NTS and You

Before the adoption of NTS, upper-level traffic handlers worked a system called the "trunk line" system, where a handful of stations carried the burden of cross-country traffic, day in day out. Nowadays, no one has to be an "iron man" or "iron woman" within NTS if they choose not to. If each liaison slot and TCC slot were filled by one person, one day a week, this would allow over 1000 hams to participate in NTS! Unfortunately, many hams have to double- and triple-up duties, so there is plenty of room for any interested amateur.

An NTS liaison spot one day (or night) a week is a great way to stay active in the traffic circuit. Many hams who don't have time to make the section nets get satisfaction in the traffic world by holding a TCC slot or area liasion once a week. If you would enjoy such a post, drop a note to your STM, Net Manager or TCC Director. They will be happy to add another to their fold.

However, remember that the area and region net are very different from your section or local net in one aspect. The function of the section or local net is to "saturate" its jurisdiction, so the more checkins, the better. On the upper-level nets, though, the name of the game is to move the traffic as quickly and efficiently as possible. Therefore, additional checkins— other than specified liaisons and stations holding traffic— only slow down the net. [However, if you are a station holding traffic to be moved, you can enter NTS at any level to pass your traffic, even if you've never been on an upper-level net before. Entering the system at the section or local level is preferred.]

If you are interested in finding out more about the "brass tacks" of the workings to NTS, get a copy of the *Public Service Communications Manual* (FSD-235), available free from ARRL HQ. Every aspect of NTS is explained, as well as information about local net operating procedure, RACES and ARES operation. The *PSCM* will also orient you with net procedure of region and area NTS nets. It takes a little more skill and savvy to become a regular part of NTS, but the rewards are worth the effort. If you have the chance, go for it!

So There You Have It

Although this chapter is by no means a complete guide to

Fame and Glory: Your Traffic Total, PSHR Report, and Traffic Awards and Appointments

Even if you handle only one message in a month's time, you should send a message to your ARRL Section Traffic Manager (STM) reporting your activity. Your report should include your total originations, messages received, messages sent and deliveries.

An *origination* is any message obtained from a third party for sending from your station. If you send a message to Uncle Filbert on his birthday, you don't get an origination. However, if your mom or your neighbor wants you to send him one with her signature, it qualifies (it counts as one originated and one sent). The origination category is essentially an "extra" credit for an off-the-air function. This is because of the critical value of contact with the general public and to motivate traffickers to be somewhat more aggressive in making their message-handling services known to the general public.

Any formal piece of traffic you get via Amateur Radio counts as a message *received*. Any message you send via Amateur Radio, even if you originated it, counts as a message *sent*. Therefore, any time you relay a message, you get two points: one received and one sent.

Any time you take a message and give it to the party it's addressed to, on a mode other than Amateur Radio, you are credited with *a delivery*. (It's okay if the addressee is a ham.) As long as you deliver it off the air (eg, telephone, mail, in person), you get a delivery point.

Your monthly report to your STM, if sent in radiogram *format,* should look something like this:

```
NR 111 R NIØR 15 ST JOSEPH MO NOV 2
BLAIR CARMICHAEL WBØPLY
MISSOURI STM
FULTON MO 65251

OCTOBER TRAFFIC ORIG 2 RCVD 5 SENT 6 DLVD 1
TOTAL 14 X 73
ROGER NIØR
```

If you have a traffic total of 500 or more in any month, or have over 100 originations-plus-deliveries in a month, you are eligible for the Brass Pounders League (BPL) certificate, even if you did all that traffic on SSB! If you make BPL three times, you receive a handsome medallion for your shack.

Another mark of distinction is the Public Service Honor Roll (PSHR). You don't have to handle a single message to get PSHR, so it's a favorite among traffickers in rural areas. The categories have traditionally been listed in the Public Service column of *QST,* and they include checking into public service nets, acting as net control, handling an emergency message, and so on. See the Public Service column for particulars.

Almost any station, with regular participation, can get an ARRL Net Certificate through the Net Manager. Once you have participated regularly in a net for three months, you are eligible. If you are a League member, you can also become eligible for an Official Relay Station appointment in the ARRL Field Organization (for details concerning the ARRL Field Organization, see Chapter 17). Either certificate makes a handsome addition to your shack.

traffic handling, it should serve as a good reference for verterans and newcomers alike. If you've never been involved in message handling, perhaps your interest has been piqued. Should that be the case, don't put it off. Find a net that's custom-made for you and check into it! You'll find plenty of fine folks that will soon become close friends as you begin to work with them, learn from them, and yes, even chat with them when the net is over.

The roots of traffic handling run deep into the history of Amateur Radio, yet its branches reach out toward many tomorrows. Our future lies in proving our worth to the nonham public, and what better way to ensure the continuance of our hobby than by uniting family and friends via Amateur Radio?

Sure, it takes some effort, but any trafficker will tell you he stays with it because of the satisfaction he gets from hearing those voices on the other end of the phone say, "Oh, isn't that nice!" We've plenty of room for you—come grow with us!

Tips on Handling NTS Traffic By Packet Radio

Listing Messages

* After logging on to your local NTS-supported bulletin board, type the command LT, meaning List Traffic. The BBS will sort and display an index of all NTSXX traffic awaiting delivery. The index will contain information that looks something like this:

MSG#	TR	SIZE	TO	FROM	@BBS	DATE	Title
200	TN	282	NTSCT	KY1T		870225	QTC 1 Hartford
198	TN	302	NTSRI	K1CE		870224	QTC 1 Cranston
192	TN	215	NTSIN	WF4R		870224	QTC 1 Indianapolis
190	TN	200	NTSCT	AJ6F		870224	QTC 1 Waterbury
188	TN	315	NTSCT	KH6WZ		870224	QTC 1 Newington
187	TN	320	NTSCT	K6TP		870224	QTC 1 New Haven
186	TN	300	NTSAL	WA4STO		870224	QTC 1 Birmingham
184	TN	295	NTSCA	WB4FDT		870224	QTC 1 Fresno

Receiving Messages

* To take a message off the Bulletin Board for telephone delivery to the third party, or for relay to a NTS Local or Section Net, type the R or RT command, meaning Read Traffic, and the message number. R 188 will cause the BBS to find and send the message text file containing the RADIOGRAM for the third party in Newington. The RADIOGRAM will look like any other, with preamble, address, text and signature, only some additional packet-related message header information is added. This information includes the routing path of the message for auditing purposes; e.g., to discern any excessive delays in the system.

* After the message is saved to the printer or disk, the message should be KILLED by using the KT command, meaning Kill Traffic, and the message number. In the above case, at the BBS prompt, type KT 188. This prevents the message from being delivered twice.

* At the time the message is killed, many BBS's will automatically send a message back to the station in the FROM field with information on who took the traffic, and when it was taken!

Delivering or Relaying A Message

* A downloaded RADIOGRAM should, of course, be handled expeditiously in the traditional way: telephone delivery, or relay to a phone or cw net.

Sending Messages

* To send a RADIOGRAM, use the ST command, meaning Send Traffic. The BBS will prompt you for the NTSXX address (NTSOH, for example), the message title which should contain the city in the address of the RADIOGRAM (QTC 1 Dayton), and the text of the message in RADIOGRAM format. The BBS, usually within the hour, will check its outgoing mailpouch, find the NTSOH message and automatically forward it to the next packet station in line to the NTSOH node. Note: Some states have more than one ARRL Section. If you do not know the destination ARRL Section ("Is San Angelo in the ARRL Northern, Southern or West Texas Section?"), then simply use the state designator NTSTX.

*Note: While NTS/packet radio message forwarding is evolving rapidly, there are still some gaps. When uploading an NTS message destined for a distant state, use handling instruction "HXC" to ask the delivering station to report back to you the date and time of delivery.

* Unbundle your messages please: one NTS message per BBS message. Please remember that traffic eventually will have to be broken down to the individual addressee somewhere down the line for ultimate delivery. When you place two or more NTS messages destined for different addressees within one packet message, eventually the routing will require the messages to be broken up by either the BBS SYSOP or the relay station, placing an additional, unreasonable burden on them. Therefore it is good practice for the originator to expend the extra word processing in the first place and create individual messages per city regardless if there are common parts of other messages. This means that book messages are not suitable in packet at this time unless they are going to the same city. Bottom line: Messages should be sent unbundled. (Tnx NI6A for this tip)

We Want You!

Local and Section BBS's need to be checked daily for NTS traffic. SYSOPs and STMs can't do it alone. They need your help to clear NTS RADIOGRAMS every day, seven days a week, for delivery and relay. If you are a traffic handler/packeteer, contact your Section Traffic Manager or Section Manager for information on existing NTS/packet procedures in your Section.

If you are a packeteer, and know nothing of NTS traffic handling, contact ARRL HQ, your Section Manager or Section Traffic Manager for information on how you can put your packet radio gear to use in serving the public in routine times, but especially in times of emergency!

And, if you enjoy phone/cw traffic handling, but aren't on packet yet, discover the incredible speed, and accuracy of packet radio traffic handling. You probably already have a small computer and 2-meter rig; all you need is a packet radio "black box" to connect between your 2-meter rig and computer. Cost: about $130. For more information on packet radio, contact ARRL HQ.

Image Communications

BRUCE BROWN, WA9GVK
3422 SILVER MAPLE PLACE
FALLS CHURCH, VA 22042

and

RALPH TAGGART, WB8DQT
602 SOUTH JEFFERSON
MASON, MI 48854

Amateur Radio operators have always had a strong propensity and well-deserved reputation for resourcefulness, whether it be fashioning an envelope detector out of a safety pin, razor blade and pencil lead, or making antennas out of coat hangers and toy springs. On a larger scale, hams have successfully transformed piles of surplus parts and metal scraps into radio transmitters capable of worldwide communications. Amateur image communicators continue to uphold this venerable tradition, but now with a slightly different leaning.

This time, hams have been successfully incorporating elements of the home video and personal computer revolutions into their image systems. Components such as camcorders, low-cost color video cameras and monitors, videotape recorders, computer processors, printers and large capacity solid-state memories are having a profound beneficial effect upon the operation, performance and costs related to the transmission and reception of both still and moving pictures over the airwaves. Most hams are surprised to learn that some of their home video and computer equipment can indeed serve as the foundation for building a reasonably priced, yet highly effective, image communication system. Once the virtual private domain of the experimenter, image communication is now being enjoyed by an ever-expanding segment of the amateur population.

In this chapter, the goal is to improve your image—quite literally—and you won't even have to attend charm school! The focus will be on the three main image communications systems: fast-scan amateur television (FSTV), also referred to as ATV; slow-scan television (SSTV); and facsimile (fax). Each mode has something unique and fascinating to offer. (The technical aspects of these systems are fully covered in the periodicals listed at the end of this chapter.) Regulatory requirements along with beneficial operating suggestions and procedures for achieving more effective communications will be discussed. Armed with this information and aided by some old-fashioned common sense and on-the-air courtesy, you'll be all set to see and be seen. After all, hams should be seen as well as heard!

FAST-SCAN AMATEUR TELEVISION

"If it looks like a duck, sounds like a duck, walks like a duck, then it must be a...," or so the saying goes. Well, fast-scan amateur television (FSTV), also referred to simply as amateur television (ATV), certainly looks and sounds like commercial-broadcast-TV quality, and the pictures move around on the screen like commercial-broadcast TV. But alas, ATV is still just "amateur." The similarity between the systems is by no means accidental since ATVers use the same basic transmission standards as does the commercial world. Amateurs are prohibited, however, from engaging in any overt forms of broadcasting or entertainment. Also, amateurs operate at considerably lower power levels that must be compensated through the use of high-gain antennas,

Fig 16-1—Richard Logan, WB3EPX, is an active ATV operator. His station is designed with space limitations in mind.

sensitive preamps and low-loss cable at the receiver. Nevertheless, there is no greater compliment that an ATVer can be paid than hearing from a receiving station that your picture is coming across "just like commercial broadcast quality." That's like getting a 5-by-9 report on the phone bands! Most folks viewing ATV for the first time are surprised by the performance levels attainable by the relatively simple and low-cost amateur TV equipment.

Because of its large bandwidth, ATV may not be operated below the 70-cm band. This, in turn, makes ATV largely a line-of-sight or local-coverage system. Exceptions do occur, though, especially when inversion layers in the atmosphere are present and ducting occurs. Under these conditions, DX of several hundred miles is possible.

Television is not exactly new to Amateur Radio; enterprising amateurs have been involved in this branch of the electronics art for more than 60 years. *QST* began carrying articles on TV as early as 1925. Although the early experiments used crude mechanical methods, the advent of the moderately priced iconoscope in 1940 made electronic television a practical reality for the amateur. As reported in the November 1940 issue of *Radio* magazine, "Two-way television was demonstrated for the first time by Amateur Radio men of the W2USA Radio Club, between their station in New York City and their glass-enclosed booth in the Communications Building at the World's Fair, New York." By 1960, amateurs were beginning to experiment with color TV and in the '70s hams learned how to modify inexpensive surplus UHF taxicab radios for ATV application.

In recent years, ATV has undergone some dramatic changes, most notably in performance improvements and expanded portable applications. The most evident is in the widespread use of color because of the availability of low-cost color cameras and color monitors originally aimed at the home-video entertainment market. Color transmission has also been facilitated by the expanded use of broadband solid-state transmitters and stripline amplifiers, which provide the necessary frequency response and phase characteristics to faithfully transmit color. The effectiveness of the low-power solid-state equipment has been further enhanced through the expanded use of ATV repeaters throughout the country, and low-noise GaAsFET receiver preamps. Finally, the small-size ATV transmitters are making their way out of the shack and into the field for use in scientific and public service applications.

License Requirements and Operating Frequencies

ATV can be used by any ham holding a Technician or higher-class license. Most operations can be found in the 420 to 440-MHz segment of the 70-cm band, with operation in the 910-928-MHz portion of the 33-cm band and the 1240-1294-MHz portion of the 23-cm band gaining in popularity. Novices are permitted to use the 1270 to 1295 MHz segment of the 23-cm band with a 5-watt power restriction. See the ARRL band plans in Chapter 2 for more information. Some amateurs are also experimenting with video transmission in the 2300-2450 band using modified equipment originally intended for commercial microwave TV distribution systems, and some testing can even be found at 10 GHz using Gunnplexers and surplus microwave burglar-alarm systems. It is important that you first check with your ATV club, VHF/UHF local frequency coordination organization, or the *ARRL Repeater Directory* to determine the specific ATV frequencies for your area.

Most FSTV activity occurs in the 70-cm band due to lower equipment costs and generally favorable propagation conditions. Picture carrier frequencies of 439.25, 426.25 and

421.25 MHz are the most popular. Increased levels of interference from other services sharing this band have motivated interest in the relatively clear higher frequencies. In the 33-cm band, 911.25 and 923.25 MHz are fairly standard, while for 23 cm suggested channels are 1241.25, 1253.25 and 1277.25 MHz. Many population centers have FSTV repeaters, which may be either in-band (426.25 MHz input/439.25 MHz output, for example) or cross-band (426.25 MHz input/923.25 or 1253.25 MHz output). ATVers always try to avoid interfering with weak signal work (DX/EME) on 432 MHz and 1296 MHz. Amateurs in the cities of Seattle, Detroit, Cleveland and Buffalo are prohibited by the FCC from operating in the 421-430 MHz spectrum. Additionally, 50-watt PEP output power limitations apply to certain amateurs operating adjacent to designated military installations in the US.

Antenna Polarization

Antenna polarization also varies throughout the country. Most nonrepeater areas tend to use horizontal, as do the DXers. Most repeaters still use vertical, although the trend is toward horizontal to increase propagation range and reduce interference from other services. It's highly important that you are using the same polarization as those with whom you intend to communicate. Using a different polarization can mean the difference between receiving a good picture and none at all. Again, check with your local ATV club or frequency coordinators.

Sound Format

Another interoperability requirement that will affect your equipment selection and operation, is the method of transmitting and receiving sound. As with frequencies and antenna polarization, voice format varies in different regions. There are three possibilities:

1) *4.5-MHz FM subcarrier audio;* audio can be heard on a standard TV set along with the picture. This is also the format used by commercial-broadcast TV.

2) *On-carrier audio;* requires a separate audio receiver to demodulate. This mode was an outgrowth of modifying amateur or surplus business transmitters having an existing "on carrier" FM voice capability and modifying the unit for AM video using collector, plate or grid modulation of the final. Some ATV repeaters will convert the on-carrier audio at its input to the subcarrier format on output so that standard TV receivers can be used in the shack to receive the audio.

3) *2 meters;* although this mode requires a separate rig, it has the advantage of letting other local hams know what is happening on ATV to publicize your activities. It allows nonATVers to participate in ATV nets and can be instrumental in picking up many new video converts. In sum, before setting up your station and attempting to operate, be sure to check:

1) simplex and repeater frequencies
2) antenna polarization
3) sound format

Equipment

Primary components of an ATV station are the camera, transmitter, receive-downconverter, standard commercial UHF/VHF television receiver, low-loss transmission line and high-gain antenna. The camcorder you purchased to chronicle the "wonder years" of your children is an ideal starting point. If you don't already have one, used black-and-white cameras can be found at hamfests in the $25 range, while color units with built-in electronic viewfinders—less recorders—are typically priced under $100. Some of the fancy, and more expensive, cameras have built-in character generators to compose on-screen messages and IDs. A video recorder (VCR) or color graphics generated on a personal computer can be used in place of the camera to show a variety of preplanned, prerecorded material. This video is fed into an ATV transmitter or transceiver connected to a high-gain antenna via lowloss cable. Transceivers, which are commercially available from several sources, are especially popular due to their ease of use and compact size. A linear power amplifier also may be used to increase transmission range.

To receive, a downconverter (or transceiver) translates the incoming signal to an unused VHF channel, typically Channel 2 or 3, for reception on an unmodified home TV. Even general purpose commercial UHF/VHF receivers such as the ICOM R7000, equipped with a special adapter, can be used to receive ATV. Add a mike and provide good lighting, and you'll be able to send and receive color television pictures that rival the quality seen on commercial broadcast stations. A typical ATV setup is shown in Fig 16-1.

Lighting

Good lighting is one of the most important and underestimated aspects of ATV. If it's a choice between good picture contrast generated by good lighting or higher transmitted power, contrast wins most every time. In other words, a high-contrast picture from a low-powered transmitter will look much better through the "snow" than a low-contrast picture from a high-powered transmitter. Fig 16-2 shows one possible lighting configuration to eliminate harsh shadows. For black-and-white work, incandescent 100-watt bulbs with reflectors will do a good job. For color, tungsten halogen lights, available at your local camera store, will give you excellent contrast while maintaining the proper color hue for home cameras. Fluorescent lights also can be effective, particularly from an overhead position, to soften background shadows. Though these different light sources appear to be "white," they all contain a degree of color as perceived by the color television camera. This amount of color, referred to as color temperature, is measured in kelvins. Typical values are:

100 W incandescent	2900
Tungsten halogen	3200
Photoflood	3400
Fluorescent	4500
Daylight	6300

The lower temperatures "redden" the picture; the higher temperatures tend to produce "bluer" pictures. Color television cameras typically have built-in filters or make automatic electronic adjustments to compensate for the different color temperatures. When using multiple light sources, try to use lamps that are nearly equivalent in color temperature to maintain consistent color accuracy throughout the scene.

Illumination is not only helpful to improve contrast but also allows cameras to operate with a smaller lens aperture (higher f stop), improving the depth of field. This means that objects both near and far can be viewed in sharp focus. You can experiment by monitoring your camera output on a closed-circuit local video monitor prior to transmitting. Once the lighting looks good, set the video gain control as high as possible before the picture smears or "whites out" as seen on a video monitor driven by the demodulated RF output of your transmitter, or by reports from remote receivers.

DXing

Besides using mast-mounted preamplifiers and low-loss

Fig 16-2—A popular lighting arrangement for ATV or SSTV pictures. Using two lights helps eliminate harsh shadows that result from a single light source. Frequently one light is set brighter or closer than the other to provide a little shading.

Hardline, there are a few tips to increase your chances for ATV DX. The best times for optimum tropospheric conditions tend to be around sunrise and right after sunset. A clue that 70-cm propagation may be especially good is to monitor commercial UHF TV. If you begin to see distant commercial stations, especially on Channels 14 through 30, there may be a real opportunity for some good DX at 70 cm. Horizontal is the preferred polarization of DXers due to better demonstrated propagation characteristics, especially through vegetation. To punch through noise, it is helpful to use pictures with very high contrast, especially big, bold, black call letters against a white background. These pictures will stay in sync despite very weak signal conditions.

Identifying

Fast-scan TV transmissions may be identified in video, or by CW or voice on the audio on-carrier or subcarrier portion. A picture of your QSL card or vanity ham-license plate makes an ideal video identification.

Signal Reporting

Signal reporting on ATV is markedly different than in all other aspects of Amateur Radio. Instead of the RST system used for voice and CW, image communicators use the "P" system which describes the amount of "snow" or noise observed on the screen (see Fig 16-3) and described below:

P5—Excellent; no visible noise. "Closed-circuit" or commercial-broadcast quality.

P4—Very good; some noise visible.

P3—Fairly good; noticeable or moderate noise present but picture is very usable.

P2—Passable; high noise; picture can still be seen but small details are lost.

Pl—Weak picture, limited use; very high, objectionable noise.

P0—Not usable; picture is totally lost in noise; can detect sync bars or barely detect picture presence but is totally unusable.

In some cases, you can split hairs with the system. If you receive an excellent picture with very slight noise, you may want to say it is a "four point five" or P4.5.

What to Show

There are many uses for ATV around the shack to show off your equipment, prize QSLs, family, pets, friends, photographs, home videotapes and vacation movies. How about showing a videotape on how you erected your 100-foot antenna tower? (You'll be surprised how few hams really know how it's done!) How about an informative presentation showing the progress you're making on a new electronics project? You can use multiple cameras and even special-effects generators to superimpose lettering over your pictures or perform split-screen effects. If you also operate SSTV, you can retransmit

UNITED STATES ATV SOCIETY
AMATEUR RADIO FAST SCAN TELEVISION
VIDEO PICTURE STANDARDS

P0—Total noise visible. No picture at all or detectable Video Sync Bars.

P1—High noise visible. Weak picture.

P2—High noise visible. Fair picture. Fair detail.

P3—Noise visible. Strong picture. Recognizable detail.

P4—Slight noise visible. Very strong picture. Good detail.

P5—No noise visible. Closed circuit picture. Excellent detail.

Fig 16-3—ATV picture quality reporting system: P5—Excellent, P4—Very Good, P3—Fairly Good, P2—Passable, P1—Weak Picture, P0—No picture at all. *(Photos courtesy of* SPEC-COM *and Dave Williams, WB0ZJP)*

Fig 16-4—During the National Disaster Medical System drill, ATV was used to send pictures of simulated injury victims arriving at a Virginia hospital to drill coordinators at a USAF base command post in Maryland. *(WA9GVK photo)*

(A)

the SSTV images you've received from overseas to all the local amateurs via ATV. Amateur television has been successfully transmitted from private aircraft, model rockets, and even radio-controlled airplanes and robots. The FCC has given permission to amateurs to retransmit the Space Shuttle video provided over NASA's public information channel on Satcom 1 R. As part of the Shuttle Amateur Radio Experiment (SAREX) program, ATVers successfully uplinked with 70-cm fast-scan TV directly to the STS-37 and STS-50 manned spacecraft in 1991 and 1992 respectively. Videotape exchanges with amateurs of other countries also makes interesting viewing. You can also show off the video output of your home computer to display animated color graphics or repeat W1AW RTTY and ARRL bulletin-board messages.

Thanks to the availability of compact and portable gear, ATV has been making tremendous inroads in the public service arena nationwide. Hams have used ATV during national disaster drills to send video from hospitals back to emergency command posts; see Fig 16-4. In Oklahoma, a camera and ATV transmitter mounted atop a tower searches the skies for tornadoes and relays the pictures directly to the National Weather Service, providing a valuable early warning capability. Amateur TV is used annually to help monitor and coordinate the Tournament of Roses Parade, marathons and yacht races; see Fig 16-5. During the Christmas season, amateurs have taken their TV systems into hospitals to enable children to see and talk to Santa Claus over live TV. Hams have used their equipment to send Voyager 2-received pictures of Jupiter and Saturn from the Jet Propulsion Laboratory in Pasadena, California, to remote locations such as an observatory and convention center to enable more widespread viewing. Amateur TV is also always an attention-getter for Amateur Radio demonstrations held at shopping malls and Boy Scout Jamborees.

Of all the specialized applications of amateur video transmission, none have captured the excitement of the ATV community in recent years more than high-altitude ballooning. Groups throughout the US launch 5 to 8 foot diameter helium balloons equipped with 1-watt ATV transmitters, cameras, 2-meter beacons and environmental sensors. Spectacular pictures from altitudes of 100,000 feet have been received at ranges of 300 miles. Schedules and frequencies for the events are published in periodicals listed at the end of this chapter. Just before launch, ground and chaseplane voice coordination is usually conducted on 7.232 MHz.

(B)

Fig 16-5—Doug McKinney, KC3RL, checks out his portable 8-watt ATV station powered by a 10 ampere-hour motorcycle battery. This arrangement is ideal for public service applications, such as emergency drills, parades and marathon coordination. *(WA9GVK photo)*

Maintaining Strong ATV Activity

Based upon the experiences of ATV organizations throughout the country, several recommendations have emerged as key factors for maintaining enthusiasm and steady growth of on-the-air activity and club support:

1) *Weekly on-the-air meeting.* Set aside a specific weekly time for an on-the-air get-together. This is the time when newcomers can be guaranteed that there will be a signal available for receiver testing and someone available to check the quality of their transmitted picture. It's a time to compare notes and help each other out by exchanging suggestions. (By the way, it's also the best time to show off the videotapes of your recent vacation trip to ensure it will be watched by a large audience!) The meeting is run by a control station who asks all "video rangers" to check in and then queries each one in a disciplined roundtable fashion so each person has a chance to transmit. The weekly get-together usually turns out to be one of the most busy, lively and enlightening aspects of ATV operation.

2) *Use 2 meters for audio.* Even though your ATV operations may use an on-carrier or subcarrier audio signal, 2 meters is recommended for voice to publicize your operations to the rest of the amateur world who may not be fully aware of all the ATV activity in the 70-cm and higher regions of the spectrum. ATV repeaters can be adapted to receive and transmit the audio portion on 2 meters.

3) *Promote a club project.* Public service efforts are ideal because they not only perform a valuable community service but they also generate a sense of purpose and enthusiasm among club members. It's a good feeling to work as a team and use your gear productively.

4) *Vary the picture content.* It's great to see your pretty face all the time on ATV but try to be creative in what you show. Variety is the spice of ATV life!

Self-Monitoring

Although you can usually depend on the received station to give you on-the-air signal reports, the quickest and most accurate method is to use an RF demodulator driving a video monitor to check the quality of your transmitted signal. You'll be able to determine contrast, focusing, framing, color level and color hue, while making sure you are not bending and rolling by excessive sync compression. Commercial ATV transceivers generally have this monitoring function as a built in capability. Video samplers, which convert RF into a video signal that can be fed directly into a TV monitor, are available from a variety of sources including PC Electronics, 2522 Paxson Ln, Arcadia, CA 91007-8537; Wyman Research, 9145 South, 850 West, Waldron, IN 46182; and Pauldon Associates, 210 Utica St, Tonawanda, NY 14150.

An inherent advantage of working through a crossband repeater is the ability to operate in full-duplex to monitor your signal from the repeater while simultaneously transmitting. This is possible since receiver desense—a normal problem with in-band transmit/receive operation—is virtually eliminated.

Why not adjust the transmitter for best picture on the receiver in the shack? The problem is that receiver overloading from the transmitter and downconverter will usually give misleading results. The preferable methods are to employ the sampler or crossband repeater. Both approaches allow an operator to make picture adjustments without relying on descriptions from other stations. Making effective use of these techniques will save you much on-the-air time in getting your station properly adjusted.

TV Audio Noise Squelch

One of the more annoying, but easily solvable, aspects of ATV operation is the audio noise generated by your TV when no signal is being received. When using your commercial TV set to receive video and audio subcarrier, you may notice that when you switch from receive to transmit, the TV set will emit an undesirable hiss because of the absence of a receive signal. That's the same sound that may have awakened you after you dozed off during the late, late show after the station went off the air! Hams have resorted to laboriously turning down the TV audio each time they transmit. However, there are two simple automated and effective alternatives for solving the problem:

1) *Noise squelch* . Communications Concepts, 508 Millstone Dr, Beavercreek, OH 45434-5840, markets a very inexpensive noise squelch that connects to the earphone jack of your TV. An external speaker is then connected to the unit's output. Without touching any knobs or controls, the squelch will automatically turn off the TV audio to the speaker if only noise is present on the channel. When the TV receives a signal, the squelch will sense the quieting of noise and will allow the audio to pass through to the speaker.

2) *VHF TV relay.* Another technique is to use an RFD relay connected to the TV's VHF antenna terminals along with an RF modulator capable of upconverting camera video to the same VHF channel utilized by the receive converter. The relay is activated using your ATV transmitter PTT line. When you are in the transmit mode, the up-converted video from your camera is fed via a relay into the VHF terminals of the TV. On receive, the relay connects the ATV receiver downconverter output to the TV's VHF antenna input. This not only quiets the TV on transmit but allows the TV to perform a dual function of both ATV receiver display and local camera monitor. This technique is fully described in a construction article in August 1985 *QST*, pp 32-37.

SLOW-SCAN TELEVISION

Nestled among the CW, phone and RTTY operators in the Amateur Radio bands is a sizable following of hams who regularly exchange still pictures in a matter of seconds virtually anywhere on Earth. They are using a system called slow-scan television (SSTV), which was originally designed by an amateur in the early 1960s. Over the years, the amateur community has been continually refining and improving the quality of SSTV. Amateur success with SSTV for almost three decades has led to its application by the military and commercial users as a reliable long-range, narrow-bandwidth transmission system. The worldwide appeal of SSTV is evident by the many DX stations that are now equipped for this type of picture transmission. Several amateurs have even worked over 100 DXCC countries on SSTV!

Just as the name implies, SSTV is the transmission of a picture by very slowly transmitting the picture elements, while a television monitor at the receiving end reproduces it in step. The most basic SSTV signal for black and white transmission consists of a variable frequency audio tone from 1500 Hz for black to 2300 Hz for white, with 1200 Hz used for synchronization pulses. Unlike fast-scan television, which uses 30 frames per second, a single SSTV frame takes at least eight to fill the screen. Additionally, the vertical resolution of SSTV is only 120 lines (or 128 for some digital systems) compared with 525 lines for fast-scan. These disadvantages are offset by the fact that SSTV requires less than 1/2000 of a fast-scan TV's bandwidth. Thus, the FCC permits it in any amateur phone band.

The basic SSTV format represents a trade-off among bandwidth, picture rate and resolution. To achieve practical HF

Fig 16-6—SSTV picture as seen on a standard TV set using a digitial scan converter.

long-distance communications, the SSTV spectrum was designed to fit into a standard 3-kHz voice bandwidth through a reduction in picture resolution and frame rate. Thus, SSTV resolution is lower than FSTV and is displayed in the form of still pictures. A sample SSTV picture is shown in Fig 16-6.

Extraordinary new developments are emerging from the amalgamation of consumer color cameras, digital techniques and personal computers. The greatest advancements currently being made in the realm of color SSTV are increased resolution and improved noise immunity. That's the good news. The bad news is that these efforts have spawned over 40 SSTV modes that are not interoperable. Table 16-1 lists many of the formats. Until firm standards can be agreed upon in the SSTV community, interface and software developers are designing products with a multiple mode capability. For example, the Amiga AVT is capable of supporting 56 modes.

License Requirements and Operating Frequencies

Since SSTV operation is restricted to the phone bands, license requirements are identical to those for voice operations as listed in Chapter 2. While most activity occurs on HF using SSB, it is also well suited to FM at VHF or UHF frequencies. In the US, slow-scan TV using double-sideband AM or FM on the HF bands is not permitted.

The common accepted SSTV calling frequencies are 3.845 MHz (Advanced), 7.171 MHz (Advanced), 14.230 MHz (General) 21.340 MHz (General) and 28.680 MHz (General). Traditionally, 20 meters has been the most popular band for SSTV operations, especially on Saturday and Sunday mornings. A weekly international SSTV net is held each Saturday at 1800 UTC on 14.230 MHz. Many years ago, when SSTV was first authorized, the FCC recommended that SSTVers not spread out across the band even though it was legal to do so. A "gentlemen's agreement" has remained to this day that SSTVers operate as close as possible to the above calling frequencies to maintain problem-free operation.

Identifying

On SSTV, the legal identification must be made by voice or CW. Sending "This is W9NTP" on the screen is not sufficient. Most stations intersperse the picture with comments anyway, so voice ID is not much of a problem. Otherwise SSTV operating procedures are quite similar to those used on SSB.

Equipment

All you need to get started is an SSB station (or FM station for VHF/UHF), a monitor, scan converter and a video source. Like

Table 16-1
Various SSTV Formats

Frame Time (sec)	Resolution*	Type	Origin/Remark
12	128 × 120	comp	Robot
24	128 × 120	comp	Robot
36	256 × 240	comp	Robot (most popular color mode)
72	256 × 240	comp	Robot (QRM resistive)
12	128 × 120	B&W	Robot
24	256 × 240	B&W	Robot
36	256 × 240	B&W	Robot
17	128 × 256	B&W	WA7WOD
25.5	128 × 128	LS	WA7WOD
51	128 × 256	LS	WA7WOD
102	256 × 256	LS	WA7WOD
17	256 × 128	B&W	Microcraft
34	256 × 256	B&W	Microcraft
16	128 × 240	B&W	Wraase (Germany)
24	128 × 120	LS	Wraase
48	128 × 240	LS	Wraase
96	256 × 240	LS	Wraase
8	120 × 120	B&W	Original format
8.5	128 × 128	B&W	Digital version of above (most popular B&W mode)
7	128 × 128	B&W	European version of above
5	96 × 96	B&W	Visitel (Mitsubishi)
24	128 × 120	LS	AVT Amiga
90	256 × 240	LS	AVT Amiga
94	320 × 200	LS	AVT Amiga
125	640 × 400	B&W	AVT Amiga

*horizontal resolution (pixels) × vertical resolution (lines/frame)

Notes

B&W—Black and white

LS—Line sequential color: Red line is sent, followed by Green line, then Blue. The three lines are integrated to form one new color line. The process repeats for the remaining lines until the frame is completed.

Comp—Composite color: Color picture is compatible with B&W standards. Brightness and color information are transmitted separately on different lines or line segments. Line and frame synchronization, audio frequencies and formats may vary for each mode.

RTTY, SSTV is a 100%-duty-cycle transmission. Most sideband rigs will have to run considerably below their voice power ratings to avoid ruining the final amplifier or power supply. Early SSTV monitors used long-persistence CRTs much like classical radar displays. In a darkened environment, the image remained visible for a few seconds while the frame was completed. This type of reception is unusual today and has been replaced with digital scan converters which convert SSTV to FSTV to place a bright image on a conventional television monitor.

Observing SSTV pictures on scan converters is a result of recent advances in digital techniques. In receive mode, the device converts the incoming audio to a fast-scan video signal that is usable by a conventional fast-scan TV monitor. Similarly, on transmit the converter changes the fast-scan camera output to a standard slow-scan signal. Examples of commercial scan converters include the Robot 1200C and the Wraase SC-I. A personal computer equipped with the proper software and interface also makes a highly versatile slow-scan converter. Computers successfully used for SSTV include the Tandy/Radio Shack TRS-80, Commodore C64/128, Amiga 500/2000, Apple, Atari ST and IBM PC and compatible computers. A comprehensive listing of software and associated interfaces is obtainable from SPEC-COM Magazine, PO Box 1002, Dubuque, IA 52004-1002.

The video source can be a black-and-white or color camera, video recorder or audio tape recorder, or the video output from a personal color-graphics computer. Since the signal emanating from the transmit-side of the scan converter is audio, it can be recorded on a cassette machine. The frequency response is not critical, but some consideration should be given for the wow and flutter specifications to minimize picture skew. You can have a friend pre-record some pictures from his/her camera onto your recorder and use them for on-the-air transmissions. The combinations of computer, camera and tapes (both audio and video) are endless. The superposition of graphics and images both generated by the operator and those received over the air can be recorded on either audio or video tapes for later use.

To send SSTV over the air, just couple the audio output of the scan converter directly into the microphone input of any single-sideband transmitter or FM transmitter on VHF or UHF. To receive, you tune in the signals on an SSB receiver and feed the tones into the scan converter which is connected to the monitor.

Setting Up Your Studio

Standard photographic lighting arrangements work well for TV. The set-up in Fig 16-2 is good because it reduces background shadows and harsh shadows on the subject. A pair of 100-watt bulbs with reflectors would be a minimum requirement for the typical camera. Outdoors, of course, the light level is pretty much predetermined. Care must be taken, however, to prevent direct sunlight from shining into the lens, since most TV camera pickup tubes can be damaged. Adjust the brightness and contrast controls on the camera so that the darkest screen area comes out black while the brightest areas reach maximum brightness without "blooming" on the monitor screen. This assumes that the monitor brightness and contrast controls are set correctly. (Many of the newer cameras are fully automatic and do not need any adjustment.) It is also highly useful to run a test audio tape consisting of a test pattern with various gray shades through your system to properly set up your equipment. Such a tape may be obtained by an on-the-air recording from another amateur or obtained commercially. Many computer programs and special modes of commercial scan converters also provide the test signals to adjust your equipment properly.

Operating

Prior to sending a picture, it is important to voice the specific SSTV mode that will be used. It is customary to send two or three frames of each subject to ensure the receiving station gets at least one good-quality picture. Many operators then follow with descriptive verbal comments about the picture.

An SSTV signal must be tuned in properly so the picture will come out with the proper brightness and synchronization. If the signal is not "in sync," the picture will appear wildly skewed. The easiest way to tune SSTV is to wait for the transmitting operator to say something on voice and then fine tune for proper pitch. With experience, you may find you are able to zero in on an SSTV signal by listening to the sync pulses and by watching for proper synchronization on the screen. Many SSTV monitors are equipped with automatic or manual tuning aids.

If you want to record slow-scan pictures off the air, there are several ways of doing it. One is to tape record the audio signal for later playback. The other is to take a photographic picture of the image right from the SSTV screen. An instant print camera equipped with a closeup lens enables you to see the results shortly after the picture is taken. If you want to do this without darkening the room lights, you'll have to fabricate a light-tight hood to fit between the camera and monitor screen. It is also possible to feed the converter's fast-scan output to a videotape recorder for later viewing.

Picture Subjects

The selection of things to show is endless, but you will find that high contrast black-and-white or high-saturation color pictures, that are not too cluttered, work best. A "live" close-in shot of you in front of the operating position is probably the most desirable picture. Don't forget a couple of frames with your call, name and QTH. A closeup of your QSL card is ideal, and how about showing off your prized QSLs? Illustrating a technical conversation with a simple schematic diagram can often make clear a point that would be very difficult to get across on voice alone. There is a wealth of good SSTV material in newspapers. Home-brew cartoons are also very popular subjects.

Some SSTVers find it convenient to mount the camera in front of an easel onto which various prepared cards can be inserted. These cards can be photographs, drawings or lettered signs. A kitchen-type noteboard or menu board with press-on removable letters is handy for making headings. Today it is more convenient to use the keyboard on a computer or other specialized captioning devices to produce these letters.

It is fun to make up taped "programs" of related pictures. Some operators have dozens of programs on different cassette tapes that can be selected and put on the air in seconds. People might enjoy seeing a pictorial tour of your shack and QTH. Do you have another hobby? Maybe that person you are working would like to learn about your model train collection. A closeup of your 1862 Cannonball Express would be a lot more interesting than a few bland comments about the weather.

Spectacular pictures of Saturn and Jupiter taken by the Voyager 2 spacecraft and received at the Jet Propulsion

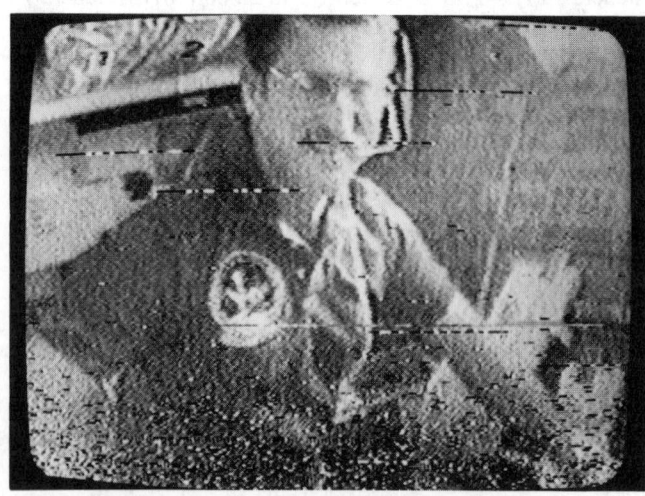

Fig 16-7—SSTV picture using the ROBOT 1200C format transmitted from the space shuttle during the STS-51F mission.

Laboratory in California have been relayed via SSTV instantaneously to amateurs around the world. Color SSTV using the ROBOT 1200C scan-converter format came into its own when Space Shuttle *Challenger* carried aloft SSTV equipment on the STS-51F mission and exchanged pictures with Amateur Radio clubs throughout the world (Fig 16-7).

Of course, there are more down-to-earth applications of SSTV. Slow-scan television has good potential for rendering service to the public by means of third-party traffic. For example, the scientists working on the Antarctic ice pack often don't see their families for months—except by amateur SSTV. This was one of the demonstrations that was used by slow-scan pioneers in 1968 to convince the FCC that slow-scan television should be permanently authorized in the ham bands. Anyone can write or call almost anywhere in the world these days, but grandma and grandpa would certainly enjoy seeing pictures of the kids from 3000 miles away by SSTV.

In whatever you show, use your imagination. But remember—never to pass up the opportunity to transmit your own image occasionally to let everyone on the frequency know with whom they are communicating!

FACSIMILE

Facsimile (fax) is a method for transmitting very high resolution still pictures using voice-bandwidth radio circuits. The narrow bandwidth of the fax signal, equivalent to SSTV, provides the potential for worldwide communications on the HF bands. Fax is the oldest of the image-transmitting technologies and has been the primary method of transmitting newspaper photos and weather charts. Facsimile is also used to transmit high-resolution cloud images from both polar-orbit and geostationary satellites. Many of these images are retransmitted using fax on the HF bands.

The resolution of typical fax images greatly exceeds what can be obtained using SSTV or even conventional television (typical images will be made up of 800 to 1600 scanning lines). This high resolution is achieved by slowing down the rate at which the lines are transmitted, resulting in image transmission times in the 4- to 10-minute range. Prior to the advent of digital technology, the only practical way to display such images was to print each line directly to paper as it arrived. The mechanical systems for accomplishing this are known as *facsimile recorders* and are based on either photographic media (a modulated light source exposing film or paper) or various types of direct-printing technologies including electrostatic and electrolytic papers.

Fig 16-8—Surplus Western Union Telefax Transceiver. The drum accommodates a sheet of paper approximately 6 × 4 in.

Modern desktop computers have virtually eliminated bulky fax recorders from most amateur installations. Now the incoming image can be stored in computer memory and viewed on a standard TV monitor or a high-resolution computer graphics display. The use of a color display system makes it entirely practical to transmit color fax images when band conditions permit.

The same computer-based system that handles fax images is often capable of SSTV operation as well, blurring what was once a clear distinction between the two modes. The advent of the personal computer has provided amateurs with a wide range of options within a single imaging installation. SSTV images of low or moderate resolution can be transmitted when crowded band conditions favor short frame transmission times. When band conditions are stable and interference levels are low, the ability to transmit very-high-resolution fax images is just a few keystrokes away!

Equipment

The simplest fax transmitter is basically mechanical in nature and consists of a drum rotated at constant speed by a precision motor/gear train combination. The photograph to be transmitted is wrapped around the drum and the image is scanned by a high-intensity light source as the drum spins, providing the equivalent of horizontal scanning in a television system. The light source is moved slowly along the spinning drum at a constant rate to provide the "vertical" scanning function. At the receiving end a similar drum/motor combination operates at the same speed as the transmitter. A modulated light source can be used to expose film or photographic paper on the receiving drum. As mentioned previously, various kinds of electrosensitive paper can also be used in conjunction with a wire stylus to recreate the image without the need for photographic processing.

There are a number of systems available on the surplus market designed for different transmission formats. All of these are analog systems that are quite distinct from the digital telephone fax units that are so common today. In fact, telephone fax units are not suitable for high-resolution grayscale images and are not compatible with common amateur formats.

One inexpensive surplus unit that lends itself to ready conversion is the Western Union desk fax transceiver shown in Fig 16-8. Conversion of a surplus unit typically involves modifying the speed of the motor drive trains, either electronically or mechanically, and possible modification of the video or printing circuits. While the Western Union transceivers are inexpensive (typically $50 or so), other systems may range in price up to $1000 or more. Prior to purchasing a more expensive unit, you need to know if conversion is necessary and practical, and whether spare parts and any specialized recording papers will continue to be available.

The steady increase in the memory capacity of personal computers and the evolution of high-resolution computer graphics has resulted in a major upsurge in fax operation. Modifying a mechanical fax system may not always be practical, but computer-based systems, whether controlled by a personal computer or an internal microprocessor, are extremely adaptable. Adding additional modes typically involves modifications to software with little or no hardware alterations.

Fax display systems for simple home computers such as the Radio Shack Color Computer and VIC-20 (see August 1985 *QST*, pp 25-33), tend to use external video and timing circuits connected to the computer via the game or accessory ports. Composite television monitors or standard TV sets are used for image display. More complex computers, such as the IBM PC (see April 1990 *QST*, pp 15-24 and May 1990 *QST*, pp 17-21) and the Amiga, use plug-in hardware cards and rely on the computer's high-resolution graphics capability for image display.

Depending on the computer and the software, images can be saved as disk files, and a variety of image-processing functions including "zooming" (to reveal increased detail), contrast enhancement, noise filtering, image inversions and other special effects are possible. A hard copy of selected images can be obtained by using graphics printers, or video printers—or by simply photographing the computer display.

Frequencies and Operational Procedures

Amateur fax operators have established the following frequencies for nets and general operating:

7.245 MHz
14.245 MHz
21.345 MHz (International Facsimile Net, 1600 UTC
            every Sunday)
28.945 MHz

Anyone with an interest in fax is invited to check into the International Facsimile Net, but all call-ins should be made on voice. The net control station (NCS) coordinates net operations and selects stations for fax transmission based on propagation and other band conditions.

Calling CQ on facsimile (preceded by voice or CW identification per FCC regulations) is poor operating practice since it could take several minutes for the receiving operator to identify the transmitting station. Reliable voice contact should precede any attempt at image transmission. This allows the two stations to coordinate on procedures such as image format. It requires very good conditions to transmit a high-resolution fax image, and if band conditions are marginal the use of SSTV may be a desirable option.

With a general-coverage receiver, it is also possible to receive press wirephotos as well as weather charts and rebroadcasts of satellite images from stations throughout the world.

Formats and Standards

All facsimile transmissions using voice transmitters are accomplished by feeding a video modulated tone into the transmitter's audio input. On frequencies below 30 MHz this tone or *subcarrier* is frequency modulated between 1500 (black) and 2300 Hz (white). This permits the use of audio limiters at the receiving end, making reception relatively immune to fading. Accurate tuning of the SSB receiver is critical to obtain faithful image reproduction. Many fax adapters incorporate circuits to indicate when the signal is properly tuned.

On the VHF and microwave frequencies used by weather satellites, the subcarrier is amplitude modulated (maximum amplitude = white, minimum amplitude = black). This is practical because the transmissions are made using FM modulation of the carrier, avoiding the effect of Doppler frequency shifts caused by the rapid movement of the spacecraft in orbit.

Unlike conventional television and SSTV, fax transmissions do not employ line synchronization pulses. Proper synchronization of both the transmitter and receiver in a mechanical system is maintained by operating the drums using precision synchronous motors locked to a crystal oscillator frequency standard. Computer-based systems derive their timing from crystal-locked counter chains.

One of the primary standards in facsimile is the rate at which lines are transmitted, expressed as the number of lines per minute (LPM). Successful reception of signals requires that the same line rate be used at both ends of the circuit. Commonly encountered speeds and the associated services include 90 LPM (wirephotos), 120 LPM (HF weather charts and satellite images and polar orbit satellite transmissions), 180 LPM (wirephotos) and 240 LPM (polar orbit and geostationary WEFAX satellites). Some old wirephoto equipment can still be found operating at 60 LPM, but this is too slow for practical amateur work where transmissions need to be kept shorter than 10 minutes. Some 360- and 480-LPM systems are also in use. At those speeds the video bandwidth becomes excessive for use on amateur HF bands, although they would be usable on VHF and UHF links. In general, 120 to 240 LPM systems are probably optimum for amateur use, providing the best trade-off between resolution and image transmission time.

Selection of a suitable surplus fax recorder or conversion of such a system requires dealing with the transmission rate (LPM), the scanning density (lines per inch) relative to the size of the image and the modulation format. These subjects are discussed at length, complete with conversion data, in the "Image Communications" chapter of any recent editions of the *ARRL Handbook*.

Color Fax

Amateur fax systems based on the use of high-resolution computer displays are capable of displaying color images using techniques similar to those in use for color SSTV. In such cases, the original color image must be scanned as separate red, green and blue images and then stored in memory for transmission.

Transmission is accomplished in one of two ways. Each red, green and blue version of the complete image can be transmitted **in its entirety** in sequence. This is known as *frame sequential* transmission. Assuming that the images are formatted correctly at the receiving end, a reproduction of the original color image will be displayed on the color graphics monitor. Alternatively, the red, green and blue versions of each **line** of the original image can be transmitted in sequence (*line sequential* transmission) such that the image will appear, line by line, in full color as it is received.

At present, two factors limit the extensive use of color on amateur facsimile. First, high-resolution fax images use a great deal of memory. (A color image requires three times the memory capacity of a grayscale image of the same resolution.) The second factor is time. A grayscale fax image requires from 4 to 10 minutes for transmission while an equivalent color image would require three times as long. HF band conditions and interference levels are not suited for such extended transmissions. Experimentation on VHF and UHF is quite practical, however, and the use of digital image compression techniques can be expected to make color more practical in the years to come.

Weather-Satellite Fax Reception

At present, the area of greatest amateur fax activity involves the reception of images transmitted by weather satellites. All of these spacecraft use AM video subcarrier modulation with line rates of 120 and 240 LPM. There are two major categories of weather satellites: those in near-polar orbits at relatively

Fig 16-9—A two-pass NOAA-11 mosaic image covering almost all of North America.

low altitudes (600-800 miles) and geostationary spacecraft located 22,000 miles above the equator.

Most polar-orbiting spacecraft are capable of transmitting both visible light and infrared (IR) images. The US NOAA spacecraft transmit simultaneous visible and IR views during daylight passes with two channels of IR data available at night. The Soviet METEOR satellites transmit visible light images in daylight and IR pictures at night. Transmissions from these spacecraft occur in the 137-138 MHz range (137.50 and 137.62 MHz in the case of the NOAA satellites) using frequency modulation of the carrier.

Reception requires the use of a receiver with an IF bandwidth of 30-40 kHz and receivers designed for this service, such as the Vanguard Labs WEPIX 2000, are among the most popular options. Modern low-noise RF preamplifiers make it possible to use omnidirectional turnstile of helix antennas without the need to track the spacecraft. Optimum passes, 10-15 minutes in duration, occur twice a day and an orbital prediction program can be a great help in determining when to listen for a specific spacecraft. AMSAT (the Radio Amateur Satellite Corporation) provides such programs on disk for a number of different personal computers. Contact AMSAT at PO Box 27, Washington, DC 20044.

Image data are transmitted continuously from polar orbit spacecraft. If the entire pass is recorded, you can expect images to cover an area about 1500 miles north to south and 800 miles east to west in the case of an optimum pass. Not many years ago, home-built or surplus mechanical fax recorders were the primary display systems, but most stations today are using dedicated scan converters or systems using computers with high-resolution graphics capabilities.

Geostationary spacecraft obtain images of the entire globe and transmit them to earth in a series of quadrants. In addition, many of these spacecraft also relay weather charts and mosaics prepared from polar-orbit spacecraft data. The US GOES satellite system ideally consists of two primary spacecraft, one over South America and the other over the eastern Pacific, often with a third spacecraft devoted entirely to image relay functions. (Premature failure of spacecraft may reduce the number of available satellites.) A consortium of European nations operates the METEOSAT spacecraft stationed over Africa while the Japanese operate their GMS satellite over the western Pacific.

All of the geostationary satellites have FM transmission formats similar to those of the polar-orbiters, but transmissions are made at 1691 MHz. Reception typically involves the

Fig 16-10—A northeast quadrant infrared image obtained from the GOES satellite. High, cold clouds are visible in white in the infrared format.

use of a small dish antenna (4-foot models will provide a reasonable gain margin) operating into a low-noise RF pre-amplifier/downconverter designed to convert the signal to the 137-138 MHz range. This permits the VHF receiver to serve as an IF for the geostationary system as well as provide reception of polar-orbit transmissions.

The *Weather Satellite Handbook,* published by the ARRL, provides comprehensive coverage of the various aspects of amateur weather satellite activity, including chapters on antennas, receivers, digital display systems and tracking programs.

In addition to the *ARRL Handbook* and *QST*, the following periodicals regularly contain information on image communications:

Amateur Television Quarterly
3 North Court St
Crown Point, IN 46307
(219) 662-6396

SPEC-COM
PO Box 1002
Dubuque, IA 52004-1002
tel (319) 557-8791
fax (319) 583-6462

Acknowledgment: Frame and resolution data for SSTV mode chart furnished by Advanced Electronic Applications.

CHAPTER 17

References

INTRODUCTION

"...the number of countries worked is increasingly becoming the criterion of excellence among outstanding DX stations."

Clinton B. DeSoto, W1CBD, October 1935 *QST*

From its simple beginnings, culminating in the announcement of the new DX award, **The DX Century Club**, in September 1937 *QST* (which was itself based on the "ARRL List of Countries" published in January 1937 *QST*), membership in the ARRL DX Century Club (DXCC) has been *the* mark of distinction among radio amateurs the world over. That it is regarded with such prestige by DXers is a testament to its integrity and level of achievement. The high standards of DXCC are intensely defended and supported by its membership. The rules established by the founders of DXCC were consistent with the art of Amateur Radio as it existed at the time. As technology improved the ability to communicate, the rules were progressively changed to maintain a competitive environment and complement the gaining popularity of DXCC.

Because of the vast changes in the international scene brought about by World War II, it logically followed that DXCC needed to be recast, as indicated in December 1945 *QST*. Ultimately, after a great deal of study, the first postwar DXCC Countries List emerged as published in February 1947 *QST*. The new DXCC Rules appeared in March 1947 *QST*. Contacts were valid from November 15, 1945, the date US amateurs were authorized by the FCC to return to the air.

The DXCC rules today represent the aggregate of experience gained from administering postwar DXCC. Some countries on the DXCC Countries List do not, of course, meet the present criteria. This includes countries "grandfathered" from the WWII era or those that met the criteria as it existed at the time and are not subject to deletion (see Section III for the appropriate grounds for deletion). Changes are announced under DXCC Notes in *QST*.

SECTION I. BASIC RULES

1) The DX Century Club Award, with certificate and lapel pin (there is a nominal fee of $5 for the DXCC lapel pin) is available to Amateur Radio operators throughout the world (see #15 below for the DXCC Award Fee Schedule). ARRL membership is required for DXCC applicants in the US and possessions, and Puerto Rico. ARRL membership is not required for foreign applicants. All DXCCs are endorsable (see Rule 5). There are 12 separate DXCC awards available, plus the DXCC Honor Roll:

(a) **Mixed** (general type): Contacts may be made using any mode since November 15, 1945.

(b) **Phone:** Contacts must be made using radiotelephone since November 15, 1945. Confirmations for cross-mode contacts for this award must be dated September 30, 1981, or earlier.

(c) **CW:** Contacts must be made using CW since January 1, 1975. Confirmations for cross-mode contacts for this award must

be dated September 30, 1981, or earlier.

(d) **RTTY:** Contacts must be made using radioteletype since November 15, 1945. (Baudot, ASCII, AMTOR and packet count as RTTY.) Confirmations for cross-mode contacts for this award must be dated September 30, 1981, or earlier.

(e) **160 Meter:** Contacts must be made on 160 meters since November 15, 1945.

(f) **80 Meter:** Contacts must be made on 80 meters since November 15, 1945.

(g) **40 Meter:** Contacts must be made on 40 meters since November 15, 1945.

(h) **10 Meter:** Contacts must be made on 10 meters since November 15, 1945.

(i) **6 Meter:** Contacts must be made on 6 meters since November 15, 1945

(j) **2 Meter:** Contacts must be made on 2 meters since November 15, 1945.

(k) **Satellite:** Contacts must be made using satellites since March 1, 1965. Confirmations must indicate satellite QSO.

(l) **Five-Band DXCC (5BDXCC):** The 5BDXCC certificate is available for working and confirming 100 current DXCC countries (deleted countries don't count for this award) on each of the following five bands: 80, 40, 20, 15, and 10 Meters. Contacts are valid from November 15, 1945. The 5BDXCC is endorsable for these additional bands: 160, 17, 12, 6, and 2 meters. 5BDXCC qualifiers are eligible for an individually engraved plaque (at a charge of $25.00 US plus shipping).

(m) **Honor Roll:** Attaining the DXCC Honor Roll represents the pinnacle of DX achievement:

*__Mixed__—To qualify, you must have a total confirmed country count that places you among the numerical top ten DXCC countries total on the current DXCC Countries List (example: if there are 326 current DXCC countries, you must have at least 317 countries confirmed).

*__Phone__—same as Mixed.

*__CW__—same as Mixed.

To establish the number of DXCC country credits needed to qualify for the Honor Roll, the maximum possible number of current countries available for credit is published monthly in *QST*. First-time Honor Roll members are recognized monthly in *QST*. Complete Honor Roll standings are published annually in *QST* usually in the July issue. See DXCC Notes in *QST* for specific information on qualifying for this Honor Roll standings list. Once recognized on this list or in a subsequent monthly update of new members, you retain your Honor Roll standing until the next standings list is published. In addition, Honor Roll members are recognized in the DXCC Annual List for those who have been listed in the previous Honor Roll listings or have gained Honor Roll status in a subsequent monthly listing. Honor Roll qualifiers receive an Honor Roll endorsement sticker for their DXCC certificate and are eligible for an Honor Roll lapel pin ($5) and an Honor Roll plaque ($25 plus shipping). Write the DXCC Desk for details.

#1 Honor Roll: To qualify for a Mixed, Phone or CW Number One plaque, you must have worked every country on the current DXCC Countries List. Write the DXCC Desk for details.

2) Written proof (confirmations, ie, QSL cards) of having made two-way communication must be submitted directly to ARRL Headquarters for all DXCC countries claimed. Photocopies and electronically transmitted confirmations (including, but not limited to FAX, telex and telegram) are not acceptable for DXCC purposes. Applicants for their first DXCC award may have the cards checked by ARRL DXCC Field Representatives—see Section V for details. The use of the official DXCC application forms or an approved facsimile (eg, produced by a computer program) is required. Complete application materials are available from ARRL Headquarters. Confirmations for a total of 100 or more countries must be included with your first application. By ARRL Board of Directors action, 10-MHz confirmations are creditable to the Mixed, CW and RTTY awards only.

3) The ARRL DXCC Countries List criteria will be used in determining what constitutes a DXCC country.

4) Confirmation data for two-way communications (ie, contacts) must include the call signs of both stations, the country, mode, and date, time and frequency band.

5) Endorsement stickers for affixing to certificates or pins will be awarded as additional DXCC credits are granted. For the Mixed, Phone, CW, RTTY and 10-Meter DXCC, these stickers are in exact multiples of 25, ie 125, 150, etc, between 100 and 250 DXCC countries; in multiples of 10 between 250 and 300, and in multiples of 5 above 300 DXCC countries. For 160-Meter, 80-Meter, 40-Meter, 6-Meter, 2-Meter and Satellite DXCC, the stickers are in exact multiples of 10 starting at 100 and multiples of 5 above 200. Confirmations for DXCC countries may be submitted for credits in any increment. (See #15 for applicable fees, if any.)

6) All contacts must be made with amateur stations working in the authorized amateur bands or with other stations licensed or authorized to work amateurs. Contacts made through "repeater" devices or any other power relay method (aside from Satellite DXCC) are invalid for DXCC credit.

7) Any Amateur Radio operation should take place only with the complete approval and understanding of appropriate administration officials. In countries where amateurs are licensed in the normal manner, credit may be claimed only for stations using regular government-assigned call signs or portable call signs where reciprocal agreements exist or the host government has so authorized portable operation. No credit may be claimed for contacts with stations in any country that has temporarily or permanently closed down Amateur Radio operations by special government edict where amateur licenses were formerly issued in the normal manner. Some countries, in spite of such prohibitions, issue authorizations which are acceptable.

8) All stations contacted must be "land stations." Contacts with ships and boats, anchored or under way, and airborne aircraft, cannot be counted.

9) All stations must be contacted from the same DXCC country.

10) All contacts must be made by the same station licensee. However, contacts may have been made under different call signs in the same country if the licensee for all was the same. That is, you may simultaneously feed one DXCC from several call signs held, as long as the provisions of Rule 9 are met.

11) Any altered, forged, or otherwise invalid confirmations submitted by an applicant for DXCC credit may result in disqualification of the applicant. Any holder of a DXCC award submitting altered, forged or otherwise invalid confirmations may forfeit the right to continued DXCC membership. The ARRL Awards Committee shall rule in these matters and may also determine the eligibility of any DXCC applicant who was ever barred from DXCC to reapply and the conditions of such application.

12) Operations Ethics:

(a) Fair play and good sportsmanship in operating are required of all DXCC members. In the event of specific objections relative to continued poor operating ethics, an individual may be disqualified from DXCC by action on the ARRL Awards Committee.

(b) Credit for contacts with individuals who have displayed continued poor operating ethics may be disallowed by action of the ARRL Awards Committee.

(c) For (a) and (b) above, "operating" includes confirmation procedures and/or documentation submitted for DXCC accreditation.

13) Each DXCC applicant must stipulate that he/she has observed all DXCC rules as well as all pertinent governmental regulations established for Amateur Radio in the country or countries concerned, and agrees to be bound by the decisions of the ARRL Awards Committee. Decisions of the ARRL Awards Committee regarding interpretations of the rules here printed or later amended shall be final.

14) All DXCC applications (both new and endorsements) must include sufficient funds to cover the cost of returning all confirmations (QSL cards) via the method chosen. Funds must be in US dollars, utilizing US currency, check or money order made payable to the ARRL, or International Reply Coupons (IRCs). Address all correspondence and inquiries relating to the various DXCC awards and all applications to: ARRL Headquarters, DXCC Desk, 225 Main St, Newington, CT 06111, USA.

15) Effective January 1, 1994, all amateurs applying for their very first DXCC Award will be charged a one-time Registration Fee of $10.00. This same fee applies to both ARRL members and foreign non-members, and both will receive one DXCC certificate and a DXCC pin. Applicants must provide funds for Postage charges for QSL return.

(b) A $5.00 Shipping and handling fee will be charged for each additional DXCC certificate issued, whether new or replacement. A DXCC pin will be included with each certificate.

(c) Endorsements and new applications may be presented at ARRL HQ, and at certain ARRL conventions. When presented in this manner, such applications shall be limited to 120 cards maximum, and a $2.00 handling charge will apply.

(d) Each ARRL member will be allowed one submission in each calender year at no cost (except as in (c) above, or return postage). This annual submission may include up to 120 QSL cards for any number of DXCC Awards, and may be a combination of new and endorsement applications. Fees as in (b) above will apply for additional new DXCC Awards.

(e) A $0.10 fee will be charged for each QSO credited beyond the limits described in (c) and (g).

(f) Foreign non-ARRL members will be allowed the same annual submission as ARRL members, however, they will be charged a $10.00 DXCC Award fee, in addition to return postage charges. Fees in (b), (c) and (e) may also apply.

(g) DXCC participants who wish to submit more than once per year will be charged a DXCC fee for each additional submission made during the remainder of the calendar year. These fees are dependent upon membership status: ARRL Members: $10.00 (for the first 100 cards) Foreign non-members: $20.00 (for the first 100 cards). Additionally, return postage must be provided by applicant, and charges from (b), (c) and (e) above may be applied.

16) The ARRL DX Advisory Committee (DXAC) requests your comments and suggestions for improving DXCC. Address correspondence, including petitions for new country consideration, to ARRL Headquarters, DXAC, 225 Main St, Newington, CT 06111, USA. You may send e-mail to: dxac@arrl.org.

SECTION II. COUNTRIES LIST CRITERIA

The ARRL DXCC Countries List is the result of progressive changes in DXing since 1945. The full list will not necessarily conform completely with current criteria since some of the listings were recognized from pre-WWII or were accredited from earlier versions of the criteria. While the general policy has remained the same, specific mileages in Point 2(a) and Point 3, mentioned in the following criteria, have been used in considerations made April 1960 and after. The specific mileage in Point 2(b) has been used in considerations made April 1963 and after.

When an area in question meets *at least one* of the following three points, it is eligible as a separate country listing for the DXCC Countries List. These criteria address considerations by virtue of Government [Point 1] or geographical separation [Points 2 and 3], while Point 4 addresses ineligible areas. All distances are given in statute miles.

Point 1, GOVERNMENT

An independent country or nation-state having *sovereignty* (that is, a body politic or society united together, occupying a definite territory and having a definite population, politically organized and controlled under one exclusive regime, and engaging in foreign relations—including the capacity to carry out obligations of international law and applicable international agreements) constitutes a separate DXCC country by reason of **Government.** This *may* be indicated by membership in the United Nations (UN). However, some nations that possess the attributes of sovereignty are *not* members of the UN, although these nations may have been *recognized* by a number of UN-member nations. Recognition is the formal act of one nation committing itself to treat an entity as a sovereign state. There are some entities that have been admitted to the UN that lack the requisite attributes of sovereignty and, as a result, are *not* recognized by a number of UN-member nations.

Other entities which are not totally independent may also be considered for separate DXCC country status by reason of Government. Included are Territories, Protectorates, Dependencies, Associated States, and so on. Such an entity may delegate to another country or international organization a measure of its authority (such as the conduct of its foreign relations in whole or in part, or other functions such as customs, communications or diplomatic protection) *without* surrendering its sovereign status. DXCC country status for such an entity is individually considered, based on all the available facts in the particular case. In making a reasonable determination as to whether a sufficient degree of sovereignty exists for DXCC purposes, the following characteristics (list not necessarily all-inclusive) are taken into consideration:

(a) Membership in specialized agencies of the UN, such as the International Telecommunications Union (ITU).

(b) Authorized use of ITU-assigned call sign prefixes.

(c) Diplomatic relations (entering into international agreements and/or supporting embassies and consulates), and maintaining a standing army.

(d) Regulation of foreign trade and commerce, customs, immigration and licensing (including landing and operating permits), and the issuance of currency and stamps.

An entity that qualifies under Point 1, but consists of two or more separate land areas, will be considered a single DXCC country (since none of these areas alone retains an independent capacity to carry out the obligations of sovereignty), *unless* the areas can qualify under Points 2 or 3.

Point 2, SEPARATION BY WATER

An island or a group of islands which is part of a DXCC country established by reason of **Government,** Point 1, is considered as a separate DXCC country under the following conditions:

(a) The island or islands are situated off shore, geographically separated by a minimum of 225 miles of open water from a continent, another island or group of islands that make up *any* part of the "parent" DXCC country.

For any *additional* island or islands to qualify as an additional separate DXCC country or countries, such must qualify under Point 2(b).

(b) This point applies to the "second" island or island grouping geographically separated from the "first" DXCC country created under Point 2(a). For the second island or island grouping to qualify, at least a 500-mile separation of open water from the first is required, as well as meeting the 225-mile requirement of (a) from the "parent". For any subsequent island(s) to qualify, the 500-mile separation would again have to be met. This precludes, for example, using the 225-mile measurement *for each* of several islands from the parent country to make several DXCC countries.

(c) An island is defined as a naturally formed area of land surrounded by water, the surface of which is above water at high tide. Rocks which cannot sustain human habitation shall not be considered for DXCC country status.

(d) An island must meet or exceed size standards. To be eligible for consideration, the island must be visible, and named, on a chart with a resolution no greater than 1:1,000,000. Charts used must be from recognized national mapping agencies. The island must consist of a single unbroken piece of land not less than 10,000 square feet in area, which is above water at high tide. The area requirements should be demonstrated by a chart with appropriate resolution.

Point 3, SEPARATION BY ANOTHER DXCC COUNTRY

(a) Where a Point 1 DXCC country, composed of one or more continental land areas or of continental land areas and islands, is totally separated by an intervening DXCC country into two land areas which are at least 75 miles apart, two DXCC countries result. This distance is measured along the great circle between the two closest points of the two areas divided. The measured distance may include inland lakes and seas which are part of the intervening DXCC country. The test for total separation into two areas requires that a great circle cannot be drawn from any point on the continental land and/or islands of one area to any point on the continental land and/or islands of the other area without intersecting any land of the intervening DXCC country.

(b) Where a Point 1 DXCC country, composed entirely of islands, is totally separated by an intervening DXCC country into two areas, then two DXCC countries result. No minimal distance is required for the separation. The test for total separation into two areas requires that a great circle cannot be drawn from any point on any island of one area to any point on any island of the other area without intersecting any land of the intervening DXCC country.

Point 4, INELIGIBLE AREAS

(a) Any area which is unclaimed or unowned by any recognized government does not count as a separate DXCC country.

(b) Any area which is classified as a Demilitarized Zone, Neutral Zone or Buffer Zone does not count as a separate DXCC country.

(c) The following do not count as a separate DXCC country from the host country: Embassies, consulates and extra-territorial legal entities of any nature, including, but not limited to, monuments, offices of the United Nations agencies or related organizations, other inter-governmental organizations or diplomatic missions.

SECTION III. DELETION CRITERIA

A DXCC country is subject to deletion from the ARRL DXCC Countries List if political change causes it to cease to meet Point 1 of the Countries List Criteria (a derivative of such change may cause it to cease to meet Points 2 or 3) or if it falls into Point 4 of the criteria. Additions to and deletions from the DXCC Countries List come about as a result of a myriad of such political changes. Reviewing the nature of the changes which have occurred since 1945 as they affect DXCC, these changes can be grouped into categories as follows:

(a) **Annexation.** When an area that has been recognized as a separate country under Point 1 is annexed or absorbed by an adjacent Point 1 country, the annexed area becomes a deleted country. Examples: India annexed Sikkim (AC3); China annexed Tibet (AC4); Indonesia annexed Portuguese Timor (CR8).

(b) **Unification.** When two or more entities that have been separate DXCC countries under Point 1 unite or combine into a single entity under a common administration, one new DXCC country is created and two or more DXCC countries have been deleted. Example: Italian Somaliland (I5) plus British Somaliland (VQ6) became Somalia (6O/T5).

(c) **Partition.** When one country is divided or partitioned into two or more countries, one DXCC country is deleted and two or more DXCC countries are created. Example: French Equatorial Africa (FQ) was deleted and replaced by Central Africa (TL), Congo (TN), Gabon (TR) and Chad (TT). The partition category is *not* employed when the original political entity continues in some form. That is, if part of country A splits off to form country B, the original DXCC country (A) is retained and one new DXCC country (B) is added. Examples: the British Sovereign Bases on Cyprus (ZC4); Aruba (P4).

(d) **Independence**. Mere independence does not result in a Countries List deletion. Examples: the Tonga Islands, then a British protectorate (VR5), is the same country as the present listing of the Kingdom of Tonga (A3). Further, an entity already recognized as a separate DXCC country is *not* deleted because of a change in its independent status. Bangladesh (S2) is the same listing as East Pakistan (AP), which was already separate from West Pakistan by virtue of Point 3. Also, a country that merely changes its name (such as when Upper Volta became Burkina Faso) does not change its basic status as a DXCC country on the DXCC Countries List.

SECTION IV. ACCREDITATION CRITERIA

1) The many vagaries of how each nation manages its telecommunications matters does not lend itself to a hard set of rules that can be applied across the board in accrediting all Amateur Radio DX operations. However, during the course of more than 40 years of DXCC administration, basic standards have evolved in determining whether a DX operation meets the test of legitimate operation. The intent is to assure that DXCC credit is given only for contacts with operations that are conducted appropriately in two respects: (1) proper licensing; and (2) physical presence in the country to be credited.

2) The following points should be of particular interest to those seeking accreditation for a DX operation:

(a) The vast majority of operations are accredited routinely without any requirement for submitting authenticating documentation.

(b) In countries where Amateur Radio operation has not been permitted or has been suspended or where some reluctance to license amateur stations has been evidenced, authenticating documents *may* be required prior to accrediting an operation.

(c) Some DXCC countries, even though part of a country with no Amateur Radio restrictions, nevertheless require the permission of a governmental agency or private party prior to conducting Amateur Radio operations on territory within their jurisdiction. Examples: Desecheo I. (KP5); Palmyra I. (KH5); Kingman Reef (KH5K).

3) In those cases where supporting documentation is required, the following should be used as a guide as to what information may be necessary to make a reasonable determination of the validity of the operation:

(a) Photocopy of license or operating authorization.

(b) Photocopy of passport entry and exit stamps.

(c) For islands, a landing permit and/or signed statement of the transporting ship's, boat's, or aircraft's captain, showing all pertinent data, such as date, place of landing, etc.

(d) For some locations where special permission is known to be required to gain access, evidence of this permission having been given is required.

4) These accreditation requirements are intended to preserve the DXCC program's integrity and to ensure that the program does not encourage amateurs to "bend the rules" in their enthusiasm, thus jeopardizing the future development of Amateur Radio. Every effort will be made to apply these criteria in a uniform manner in conformity with these objectives.

SECTION V. FIELD CHECKING OF QSL CARDS

QSL cards for new DXCC awards may be checked by two DXCC Field Representatives. This program applies to any DXCC award for an individual or station. Specifically excluded from this program are 5BDXCC, 6-meter, 2-meter and Satellite DXCC, and all 160-meter QSLs.

1) Countries Eligible for Field Checking:

(a) Eligible countries will be indicated in the ARRL DXCC Countries List, and are subject to change. Only cards from these eligible countries may be checked by DXCC Field Representatives. QSLs for other DXCC countries must be submitted directly to ARRL Headquarters.

(b) The ARRL Awards Committee determines which countries are eligible for Field Checking.

2) DXCC Field Representatives:

(a) DXCC Field Representatives must be ARRL members who have a DXCC award endorsed for at least 300 countries.

(b) To become a DXCC Field Representative, a person must be nominated by a DX club. (A DX club is an ARRL-affiliated club with at least 25 members who are DXCC members and which has, as its primary interest, DX. If there are any questions regarding the validity of a DX club, the issue shall be determined by the Division Director where the DX club is located.) A person does not have to be a member to be nominated by a DX club.

(c) DXCC Field Representatives are approved by the Director of the ARRL Division in which they reside and appointed by the President of the ARRL.

(d) DXCC Field Representative appointments must be renewed annually.

3) Card Checking Process:

(a) Only cards from the list of eligible countries can be checked by DXCC Field Representatives. An application for a new award shall contain a minimum of 100 QSL confirmations from the list and shall not contain any QSLs from countries that are not on the list of eligible countries. Additional cards should not be sent to HQ with field checked applications. The application may contain the maximum number of countries that appear on the list of eligible countries. That is, if there are 245 countries on the list, field-checked applications could contain 245 countries.

(b) It is the applicant's responsibility to get cards to and from the DXCC Field Representatives.

(c) Field Representatives may, at their own discretion, handle members' cards by mail.

(d) The ARRL is not responsible for cards handled by DXCC Field Representatives and will not honor any claims.

(e) The QSL cards must be checked by two DXCC Field Representatives.

(f) The applicant and both DXCC Field Representatives must sign the application form. (See SECTION I no. 11 regarding altered, forged or otherwise invalid confirmations.)

(g) The applicant shall provide a stamped no. 10 envelope (business size) addressed to ARRL HQ to the DXCC Field Representatives. The applicant shall also provide the applicable fees (check or money order payable to ARRL—no cash).

(h) The DXCC Field Representatives will forward completed applications and appropriate fee(s) to ARRL HQ.

4) ARRL HQ involvement in the card checking process:

a) ARRL HQ staff will receive field-checked applications, enter application data into DXCC records and issue DXCC credits and awards as appropriate.

(b) ARRL HQ staff will perform random audits of applications. Applicants or members may be requested to forward cards to HQ for checking before or after credit is issued.

c) The applicant and both DXCC Field Representatives will be advised of any errors or discrepancies encountered by ARRL staff.

(d) ARRL HQ staff provides instructions and guidelines to DXCC Field Representatives.

5) Applicants and DXCC members may send cards to

ARRL Headquarters at any time for review or recheck if the individual feels that an incorrect determination has been made.

Recent changes to the DXCC List

Added to the DXCC Country List:

Prefix	Country	*Published DXCC List Changed* *
7O	Yemen	Mar 91
ZSØ, 1	Penguin Is.	Sep 91
9A, YU2	Croatia	Jan 93
S5, YU3	Slovenia	Jan 93
T9, 4N4, 4O4, YU4	Bosnia-Herzegovina	Jan 93
Z3, 4N5, YU5	Macedonia (former Yugoslav Rep.)	Jun 93
OK-OL	Czech Rep.	Jun 93
OM	Slovak Rep.	Jun 93
P5	North Korea	Oct 95
BV9P	Pratas I.	Apr 96
BS7	Scarborough Reef	Apr 96

Moved from deleted list to active status:

E3	Eritrea	Feb 94

Moved to deleted list:

DM, Y2-9	German Democratic Rep	Mar 91
4W	Yemen Arab Rep (North)	Mar 91
7O	PDR of Yemen (South)	Mar 91
—	Abu Ail	Jun 93
OK-OM	Czechoslovakia	Jun 93
ZSØ, 1	Penguin Is.	Jul 94
ZS9	Walvis Bay	Jul 94

* DXCC List edition date when the change first appeared. See notes in the countries list for effective dates.

Rules changes:

SECTION II, Point 3	Sep 91
SECTION V (added)	Sep 91
SECTION V, 2), (d) new wording	Jan 92
SECTION I, 1), (l) 5BDXCC Start Date	Jun 93
SECTION I, 15) Fees change	Feb 94
SECTION I, 15), (c) "Walk in" card limit	Feb 94
SECTION V, field checking for any new DXCCs expanded	Feb 94
SECTION V, 160-m QSLs ineligible for field checking	Jul 94
SECTION I, 15), new fees	Apr 96
SECTION V, endorsements may be field checked	Apr 96

Note: There were several changes to countries eligible for field checking made effective January 1993.

In the February 1994 list, several "U" block prefixes changed. See Prefix Cross References section that follows the Countries List.

ARRL DXCC COUNTRIES LIST

NOTE: • INDICATES COUNTRIES ELIGIBLE FOR FIELD CHECKING.

NOTE: * INDICATES CURRENT LIST OF COUNTRIES FOR WHICH QSLs MAY BE FORWARDED BY THE ARRL MEMBERSHIP OUTGOING QSL SERVICE.

NOTE: † INDICATES COUNTRIES WITH WHICH U.S. AMATEURS MAY LEGALLY HANDLE THIRD-PARTY MESSAGE TRAFFIC.

Prefix	Country	CONTINENT	ZONE ITU	ZONE CQ	MIXED	PHONE	CW	RTTY	SAT	160	80	40	20	17	15	12	10
1AØ[1]	Sov. Mil. Order of Malta	EU	28	15													
1S[1]	Spratly Is.	AS	50	26													
• 3A*	Monaco	EU	27	14													
3B6, 7*	Agalega & St. Brandon	AF	53	39													
• 3B8*	Mauritius	AF	53	39													
• 3B9*	Rodriguez I.	AF	53	39													
• 3C	Equatorial Guinea	AF	47	36													
3CØ	Annobon I.	AF	52	36													
• 3D2*	Fiji	OC	56	32													
3D2*	Conway Reef	OC	56	32													
• 3D2*	Rotuma I.	OC	56	32													
• 3DA†*	Swaziland	AF	57	38													
3V	Tunisia	AF	37	33													
3W, XV	Vietnam	AS	49	26													
3X	Guinea	AF	46	35													
3Y*	Bouvet	AF	67	38													
3Y*	Peter I I.	AN	72	12													
4J, 4K*	Azerbaijan	AS	29	21													
4L*	Georgia	AS	29	21													
• 4P-4S*	Sri Lanka	AS	41	22													
• 4U_ITU†*	ITU HQ	EU	28	14													
• 4U_UN*	United Nations HQ	NA	08	05													
• 4X, 4Z†*	Israel	AS	39	20													
5A	Libya	AF	38	34													
• 5B*	Cyprus	AS	39	20													
• 5H-5I	Tanzania	AF	53	37													
• 5N-5O*	Nigeria	AF	46	35													
5R-5S	Madagascar	AF	53	39													
• 5T[2]	Mauritania	AF	46	35													
5U[3]	Niger	AF	46	35													
• 5V*	Togo	AF	46	35													
• 5W*	Western Samoa	OC	62	32													

Prefix	Country	CONTINENT	ZONE		MIXED	PHONE	CW	RTTY	SAT	160	80	40	20	17	15	12	10	
			ITU	CQ														
5X	Uganda	AF	48	37														
● 5Y-5Z*	Kenya	AF	48	37														
● 6V-6W4*	Senegal	AF	46	35														
● 6Y†*	Jamaica	NA	11	08														
7O5	Yemen	AS	39	21, 37														
● 7P*	Lesotho	AF	57	38														
● 7Q	Malawi	AF	53	37														
● 7T-7Y*	Algeria	AF	37	33														
● 8P*	Barbados	NA	11	08														
● 8Q	Maldives	AS/AF	41	22														
● 8R†*	Guyana	SA	12	09														
9A6*	Croatia	EU	28	15														
9G7†	Ghana	AF	46	35														
● 9H*	Malta	EU	28	15														
● 9I-9J*	Zambia	AF	53	36														
● 9K*	Kuwait	AS	39	21														
● 9L†*	Sierra Leone	AF	46	35														
● 9M2, 48*	West Malaysia	AS	54	28														
● 9M6, 88*	East Malaysia	OC	54	28														
● 9N	Nepal	AS	42	22														
● 9Q-9T	Zaire	AF	52	36														
9U9	Burundi	AF	52	36														
● 9V10*	Singapore	AS	54	28														
● 9X9	Rwanda	AF	52	36														
● 9Y-9Z†*	Trinidad & Tobago	SA	11	09														
● A2*	Botswana	AF	57	38														
● A3*	Tonga	OC	62	32														
● A4*	Oman	AS	39	21														
A5	Bhutan	AS	41	22														
● A6	United Arab Emirates	AS	39	21														
● A7*	Qatar	AS	39	21														
● A9*	Bahrain	AS	39	21														
● AP-AS*	Pakistan	AS	41	21														
BS711	Scarborough Reef	AS	50	27														
● BV*	Taiwan	AS	44	24														
BV9P12	Pratas I.	AS	44	24														
● BY,BT*	China	AS	(A)	23,24														
● C2*	Nauru	OC	65	31														
● C3*	Andorra	EU	27	14														
● C5†*	The Gambia	AF	46	35														

Prefix	Country	Continent	ZONE ITU	ZONE CQ	MIXED	PHONE	CW	RTTY	SAT	160	80	40	20	17	15	12	10		
C6*	Bahamas	NA	11	08															
C8-9*	Mozambique	AF	53	37															
CA-CE†*	Chile	SA	14,16	12															
CE0†*	Easter I.	SA	63	12															
CE0†*	Juan Fernandez Is.	SA	14	12															
CE0†*	San Felix & San Ambrosio	SA	14	12															
CE9/KC4▲*	Antarctica	AN	(B)	(C)															
CM, CO†*	Cuba	NA	11	08															
CN*	Morocco	AF	37	33															
CP†*	Bolivia	SA	12,14	10															
CT*	Portugal	EU	37	14															
CT3*	Madeira Is.	AF	36	33															
CU*	Azores	EU	36	14															
CV-CX†*	Uruguay	SA	14	13															
CY0*	Sable I.	NA	09	05															
CY9*	St. Paul I.	NA	09	05															
D2-3	Angola	AF	52	36															
D4*	Cape Verde	AF	46	35															
D6†13*	Comoros	AF	53	39															
DA-DL14*	Fed. Rep. of Germany	EU	28	14															
DU-DZ†*	Philippines	OC	50	27															
E315	Eritrea	AF	48	37															
EA-EH*	Spain	EU	37	14															
EA6-EH6*	Balearic Is.	EU	37	14															
EA8-EH8*	Canary Is.	AF	36	33															
EA9-EH9*	Ceuta & Melilla	AF	37	33															
EI-EJ*	Ireland	EU	27	14															
EK*	Armenia	AS	29	21															
EL†*	Liberia	AF	46	35															
EP-EQ	Iran	AS	40	21															
ER*	Moldova	EU	29	16															
ES*	Estonia	EU	29	15															
ET*	Ethiopia	AF	48	37															
EU, EV, EW*	Belarus	EU	29	16															
EX*	Kyrgyzstan	AS	30,31	17															
EY*	Tajikistan	AS	30	17															
EZ*	Turkmenistan	AS	30	17															
F*	France	EU	27	14															
FG*	Guadeloupe	NA	11	08															
FJ, FS1*	Saint Martin	NA	11	08															

Prefix	Country	CONTINENT	ZONE ITU	ZONE CQ	MIXED	PHONE	CW	RTTY	SAT	160	80	40	20	17	15	12	10
FH11*	Mayotte	AF	53	39													
FK*	New Caledonia	OC	56	32													
FM*	Martinique	NA	11	08													
FO*	Clipperton I.	NA	10	07													
FO*	French Polynesia	OC	63	32													
FP*	St. Pierre & Miquelon	NA	09	05													
FR/G16*	Glorioso Is.	AF	53	39													
FR/J, E16*	Juan de Nova, Europa	AF	53	39													
FR*	Reunion	AF	53	39													
FR/T*	Tromelin I.	AF	53	39													
FT8W*	Crozet I.	AF	68	39													
FT8X*	Kerguelen Is.	AF	68	39													
FT8Z*	Amsterdam & St. Paul Is.	AF	68	39													
FW*	Wallis & Futuna Is.	OC	62	32													
FY*	French Guiana	SA	12	09													
G, GX*#	England	EU	27	14													
GD, GT*	Isle of Man	EU	27	14													
GI, GN*	Northern Ireland	EU	27	14													
GJ, GH*	Jersey	EU	27	14													
GM, GS*	Scotland	EU	27	14													
GU, GP*	Guernsey	EU	27	14													
GW, GC*	Wales	EU	27	14													
H4*	Solomon Is.	OC	51	28													
HA, HG*	Hungary	EU	28	15													
HB*	Switzerland	EU	28	14													
HBØ*	Liechtenstein	EU	28	14													
HC-HD†*	Ecuador	SA	12	10													
HC8-HD8†*	Galapagos Is.	SA	12	10													
HH†*	Haiti	NA	11	08													
HI†*	Dominican Republic	NA	11	08													
HJ-HK†*	Colombia	SA	12	09													
HKØ†*	Malpelo I.	SA	12	09													
HKØ†*	San Andres & Providencia	NA	11	07													
HL*	South Korea	AS	44	25													
HO-HP†*	Panama	NA	11	07													
HQ-HR†*	Honduras	NA	11	07													
HS*	Thailand	AS	49	26													
HV*	Vatican	EU	28	15													
HZ*	Saudi Arabia	AS	39	21													
I*	Italy	EU	28	15, 33													

#Third-party traffic permitted with special-events stations in the United Kingdom having the prefix GB *only*, with the exception that GB3 stations are not included in this agreement.

Prefix	Country	Continent	ZONE ITU	ZONE CQ	MIXED	PHONE	CW	RTTY	SAT	160	80	40	20	17	15	12	10	
ISØ, IMØ *	Sardinia	EU	28	15														
J2*	Djibouti	AF	48	37														
J3†*	Grenada	NA	11	08														
J5	Guinea-Bissau	AF	46	35														
J6†*	St. Lucia	NA	11	08														
J7†*	Dominica	NA	11	08														
J8†*	St. Vincent	NA	11	08														
JA-JS*	Japan	AS	45	25														
JD1[17]*	Minami Torishima	OC	90	27														
JD1[18]*	Ogasawara	AS	45	27														
JT- JV*	Mongolia	AS	32,33	23														
JW*	Svalbard	EU	18	40														
JX*	Jan Mayen	EU	18	40														
JY†*	Jordan	AS	39	20														
K,W, N, AA-AK†	United States of America	NA	6,7,8	3,4,5														
KC6[19]	Belau (W. Caroline Is.)	OC	64	27														
KG4†*	Guantanamo Bay	NA	11	08														
KHØ†	Mariana Is.	OC	64	27														
KH1†	Baker & Howland Is.	OC	61	31														
KH2†*	Guam	OC	64	27														
KH3†*	Johnston I.	OC	61	31														
KH4†	Midway I.	OC	61	31														
KH5†	Palmyra & Jarvis Is.	OC	61,62	31														
KH5K†	Kingman Reef	OC	61	31														
KH6†*	Hawaii	OC	61	31														
KH7†	Kure I.	OC	61	31														
KH8†	American Samoa	OC	62	32														
KH9†	Wake I.	OC	65	31														
KL7†*	Alaska	NA	1, 2	1														
KP1†	Navassa I.	NA	11	08														
KP2†*	Virgin Is.	NA	11	08														
KP4†*	Puerto Rico	NA	11	08														
KP5[20]†	Desecheo I.	NA	11	08														
LA-LN*	Norway	EU	18	14														
LO-LW†*	Argentina	SA	14,16	13														
LX*	Luxembourg	EU	27	14														
LY*	Lithuania	EU	29	15														
LZ*	Bulgaria	EU	28	20														
OA-OC†*	Peru	SA	12	10														
OD*	Lebanon	AS	39	20														

Prefix	Country	CONTINENT	ZONE ITU	ZONE CQ	MIXED	PHONE	CW	RTTY	SAT	160	80	40	20	17	15	12	10
● OE*	Austria	EU	28	15													
● OF-OI*	Finland	EU	18	15													
● OHØ*	Aland Is.	EU	18	15													
● OJØ*	Market Reef	EU	18	15													
OK-OL21*	Czech Rep.	EU	28	15													
OM21*	Slovak Rep.	EU	28	15													
● ON-OT*	Belgium	EU	27	14													
● OX*	Greenland	NA	5, 75	40													
● OY*	Faroe Is.	EU	18	14													
● OZ*	Denmark	EU	18	14													
● P222*	Papua New Guinea	OC	51	28													
● P423*	Aruba	SA	11	09													
P524	North Korea	AS	44	25													
● PA-PI*	Netherlands	EU	27	14													
● PJ2, 4, 9*	Bonaire,Curacao(Neth. Antilles)	SA	11	09													
● PJ5-8*	St.Maarten,Saba,St.Eustatius	NA	11	08													
● PP-PY†*	Brazil	SA	(D)	11													
● PPØ-PYØ†*	Fernando de Noronha	SA	13	11													
PPØ-PYØ†*	St. Peter & St. Paul Rocks	SA	13	11													
● PPØ-PYØ†*	Trindade & Martim Vaz Is.	SA	15	11													
● PZ*	Suriname	SA	12	09													
● R1FJ	Franz Josef Land	EU	75	40													
R1MV	Malyj Vysotskij I	EU	29	16													
SØ1,25*	Western Sahara	AF	46	33													
S2*	Bangladesh	AS	41	22													
S56*	Slovenia	EU	28	15													
● S7	Seychelles	AF	53	39													
● S9*	Sao Tome & Principe	AF	47	36													
● SA-SM*	Sweden	EU	18	14													
● SN-SR*	Poland	EU	28	15													
● ST*	Sudan	AF	47, 48	34													
● STØ*	Southern Sudan	AF	47, 48	34													
● SU*	Egypt	AF	38	34													
● SV-SZ*	Greece	EU	28	20													
SV/A*	Mount Athos	EU	28	20													
● SV5*	Dodecanese	EU	28	20													
● SV9*	Crete	EU	28	20													
● T226	Tuvalu	OC	65	31													
● T3Ø	W. Kiribati (Gilbert Is.)	OC	65	31													
● T31	C. Kiribati (Brit. Phoenix Is.)	OC	62	31													

Prefix	Country	Continent	ZONE ITU	ZONE CQ	MIXED	PHONE	CW	RTTY	SAT	160	80	40	20	17	15	12	10
T32	E. Kiribati (Line Is.)	OC	61,63	31													
T33	Banaba I. (Ocean I.)	OC	65	31													
T5	Somalia	AF	48	37													
T7*	San Marino	EU	28	15													
T9 27*	Bosnia-Herzegovina	EU	28	15													
TA-TC*	Turkey	EU/AS	39	20													
TF*	Iceland	EU	17	40													
TG, TD†*	Guatemala	NA	11	07													
TI, TE†*	Costa Rica	NA	11	07													
TI9†*	Cocos I.	NA	11	07													
TJ	Cameroon	AF	47	36													
TK*	Corsica	EU	28	15													
TL 28	Central Africa	AF	47	36													
TN 29	Congo	AF	52	36													
TR 30*	Gabon	AF	52	36													
TT 31	Chad	AF	47	36													
TU 32*	Côte d'Ivoire	AF	46	35													
TY 33	Benin	AF	46	35													
TZ 34	Mali	AF	46	35													
UA-UI1,3,4,6 RA-RZ*	European Russia	EU	(E)	16													
UA2*	Kaliningrad	EU	29	15													
UA-UI8, 9, Ø RA-RZ*	Asiatic Russia	AS	(F)	(G)													
UJ-UM*	Uzbekistan	AS	30	17													
UN-UQ*	Kazakhstan	AS	29-31	17													
UR-UZ, EM-EO*	Ukraine	EU	29	16													
V2†*	Antigua & Barbuda	NA	11	08													
V3†*	Belize	NA	11	07													
V4 35†*	St. Kitts & Nevis	NA	11	08													
V5*	Namibia	AF	57	38													
V6 36†	Micronesia (E. Caroline Is.)	OC	65	27													
V7†*	Marshall Is.	OC	65	31													
V8*	Brunei	OC	54	28													
VE, VO, VY†*	Canada	NA	(H)	1-5													
VK†*	Australia	OC	(I)	29,30													
VKØ†*	Heard I.	AF	68	39													
VKØ†*	Macquarie I.	OC	60	30													
VK9C†*	Cocos-Keeling Is.	OC	54	29													
VK9L†*	Lord Howe I.	OC	60	30													
VK9M†*	Mellish Reef	OC	56	30													
VK9N†*	Norfolk I.	OC	60	32													

Prefix	Country	CONTINENT	ZONE ITU	ZONE CQ	MIXED	PHONE	CW	RTTY	SAT	160	80	40	20	17	15	12	10
● VK9W†*	Willis I.	OC	55	30													
● VK9X†*	Christmas I.	OC	54	29													
● VP2E35*	Anguilla	NA	11	08													
● VP2M35	Montserrat	NA	11	08													
● VP2V35*	British Virgin Is.	NA	11	08													
● VP5*	Turks & Caicos Is.	NA	11	08													
● VP8*	Falkland Is.	SA	16	13													
● VP8, LU*	South Georgia I.	SA	73	13													
● VP8, LU*	South Orkney Is.	SA	73	13													
● VP8, LU*	South Sandwich Is.	SA	73	13													
VP8, LU, CE9, HFØ, 4K1*	South Shetland Is.	SA	73	13													
● VP9*	Bermuda	NA	11	05													
● VQ9*	Chagos Is.	AF	41	39													
● VR6†*	Pitcairn I.	OC	63	32													
● VS6, VR2*	Hong Kong	AS	44	24													
● VU*	India	AS	41	22													
● VU*	Andaman & Nicobar Is.	AS	49	26													
● VU*	Laccadive Is.	AS	41	22													
● XA-XI†*	Mexico	NA	10	06													
XA4-XI4†*	Revilla Gigedo	NA	10	06													
● XT37*	Burkina Faso	AF	46	35													
XU	Cambodia	AS	49	26													
XW	Laos	AS	49	26													
● XX9	Macao	AS	44	24													
XY-XZ	Myanmar	AS	49	26													
YA	Afghanistan	AS	40	21													
● YB-YH38*	Indonesia	OC	51,54	28													
YI*	Iraq	AS	39	21													
● YJ*	Vanuatu	OC	56	32													
YK*	Syria	AS	39	20													
● YL*	Latvia	EU	29	15													
● YN†*	Nicaragua	NA	11	07													
● YO-YR*	Romania	EU	28	20													
● YS†*	El Salvador	NA	11	07													
● YT-YU, YZ*	Yugoslavia	EU	28	15													
● YV-YY†*	Venezuela	SA	12	09													
YVØ†*	Aves I.	NA	11	08													
● Z2*	Zimbabwe	AF	53	38													
Z339*	Macedonia (former Yugoslav Rep.)	EU	28	15													
ZA*	Albania	EU	28	15													

Prefix	Country	Continent	ZONE		MIXED	PHONE	CW	RTTY	SAT	160	80	40	20	17	15	12	10		
			ITU	CQ															
ZB2*	Gibraltar	EU	37	14															
ZC4⁴⁰ *	UK Sov. Base Areas on Cyprus	AS	39	20															
ZD7*	St. Helena	AF	66	36															
ZD8*	Ascension I.	AF	66	36															
ZD9	Tristan da Cunha & Gough I.	AF	66	38															
ZF*	Cayman Is.	NA	11	08															
ZK1*	N. Cook Is.	OC	62	32															
ZK1*	S. Cook Is.	OC	62	32															
ZK2*	Niue	OC	62	32															
ZK3*	Tokelau Is.	OC	62	31															
ZL-ZM*	New Zealand	OC	60	32															
ZL7*	Chatham Is.	OC	60	32															
ZL8*	Kermadec Is.	OC	60	32															
ZL9*	Auckland & Campbell Is.	OC	60	32															
ZP†*	Paraguay	SA	14	11															
ZR-ZU*	South Africa	AF	57	38															
ZS8*	Prince Edward & Marion Is.	AF	57	38															

1 Unofficial prefix.

2 (5T) Only contacts made June 20, 1960, and after, count for this country.

3 (5U) Only contacts made August 3, 1960, and after, count for this country.

4 (6W) Only contacts made June 20, 1960, and after, count for this country.

5 (7O) Only contacts made May 22, 1990, and after, count for this country.

6 (9A, S5) Only contacts made June 26, 1991, and after, count for this country.

7 (9G) Only contacts made March 5, 1957, and after, count for this country.

8 (9M2, 4, 6, 8) Only contacts made September 16, 1963, and after, count for this country.

9 (9U, 9X) Only contacts made July 1, 1962, and after, count for this country.

10 (9V) Contacts made from September 16, 1963 to August 8, 1965, count for West Malaysia.

11 (BS7) Only contacts made January 1, 1995, and after, count for this country.

12 (BV9P) Only contacts made January 1, 1994, and after, count for this country.

13 (D6, FH8) Only contacts made July 5, 1975, and after, count for this country.

14 (DA-DL) Only contacts made with DA-DL stations September 17, 1973, and after, and contacts made with Y2-Y9 stations October 3, 1990 and after, count for this country.

15 (E3) Only contacts made November 14, 1962, and before, or May 24, 1991, and after, count for this country.

16 (FR) Only contacts made June 25, 1960, and after, count for this country.

17 (JD) Formerly Marcus Island.

18 (JD) Formerly Bonin and Volcano Islands.

19 (KC6) Includes Yap Islands December 31, 1980, and before.

20 (KP5, KP4) Only contacts made March 1, 1979, and after, count for this country.

21 (OK-OL, OM) Only contacts made January 1, 1993, and after, count for this country.

22 (P2) Only contacts made September 16, 1975, and after, count for this country.

23 (P4) Only contacts made January 1, 1986, and after, count for this country.

24 (P5) Only contacts made May 14, 1995, and after count for this country.

25 (SØ) Contacts with Rio de Oro (Spanish Sahara), EA9, also count for this country.

26 (T2) Only contacts made January 1, 1976, and after, count for this country.

27 (T9) Only contacts made October 15, 1991 and after count for this country.

28 (TL) Only contacts made August 13, 1960, and after, count for this country.

29 (TN) Only contacts made August 15, 1960, and after, count for this country.

30 (TR) Only contacts made August 17, 1960, and after, count for this country.

31 (TT) Only contacts made August 11, 1960, and after count for this country.

32 (TU) Only contacts made August 7, 1960, and after, count for this country.

33 (TY) Only contacts made August 1, 1960, and after, count for this country.

34 (TZ) Only contacts made June 20, 1960, and after, count for this country.

35 (V4, VP2) For DXCC credit for contacts made May 31, 1958 and before, see page 97, June 1958 QST.

36 (V6) Includes Yap Islands January 1, 1981, and after.

37 (XT) Only contacts made August 5, 1960, and after, count for this country.

38 (YB) Only contacts made May 1, 1963, and after, count for this country.

39 (Z3) Only contacts made September 8, 1991, and after, count for this country.

40 (ZC4) Only contacts made August 16, 1960, and after, count for this country.

▲Also 3Y, 8J1, ATØ, DPØ, FT8Y, LU, OR4, R1AN, VKØ, VP8, ZL5, ZS1, ZXØ, etc. QSL via country under whose auspices the particular station is operating. The availability of a third-party traffic agreement and a QSL Bureau applies to the country under whose auspices the particular station is operating.

Zone Notes can be found with Prefix Cross References.

DELETED COUNTRIES

Credit for any of these countries can be given if the date of contact in question agrees with the date(s) shown in the corresponding footnote.

Prefix	Country	CONTINENT	ZONE ITU	ZONE CQ	MIXED	PHONE	CW	RTTY	SAT	160	80	40	20	17	15	12	10	
2	Blenheim Reef	AF	41	39														
3	Geyser Reef	AF	53	39														
4	Abu Ail Is.	AS	39	21														
1M[1, 5]	Minerva Reef	OC	62	32														
4W[6]	Yemen Arab Rep.	AS	39	21														
7J1[7]	Okino Tori-shima	AS	45	27														
8Z4[8]	Saudi Arabia/Iraq Neut. Zone	AS	39	21														
8Z5, 9K3[9]	Kuwait/Saudi Arabia Neut. Zone	AS	39	21														
9S4[10]	Saar	EU	28	14														
9U5[11]	Ruanda-Urundi	AF	52	36														
AC3[1, 12]	Sikkim	AS	41	22														
AC4[1, 13]	Tibet	AS	41	23														
C9[14]	Manchuria	AS	33	24														
CN2[15]	Tangier	AF	37	33														
CR8[16]	Damao, Diu	AS	41	22														
CR8[16]	Goa	AS	41	22														
CR8, CR10[17]	Portuguese Timor	OC	54	28														
DA-DM[18]	Germany	EU	28	14														
DM, Y2-9[19]	German Dem. Rep.	EU	28	14														
EA9[20]	Ifni	AF	37	33														
FF[21]	French West Africa	AF	46	35														
FH, FB8[22]	Comoros	AF	53	39														
FI8[23]	French Indo-China	AS	49	26														
FN8[24]	French India	AS	41	22														
FQ8[25]	Fr. Equatorial Africa	AF	47,52	36														
HKØ[26]	Bajo Nuevo	NA	11	08														
HKØ,KP3,KS4[26]	Serrana Bank & Roncador Cay	NA	11	07														
I1[27]	Trieste	EU	28	15														
I5[28]	Italian Somaliland	AF	48	37														
JZØ[29]	Netherlands N. Guinea	OC	51	28														
KR6,8,JR6,KA6[30]	Okinawa (Ryukyu Is.)	AS	45	25														
KS4[31]	Swan Is.	NA	11	07														
KZ5[32]	Canal Zone	NA	11	07														
OK-OM[33]	Czechoslovakia	EU	28	15														
P2, VK9[34]	Papua Territory	OC	51	28														
P2, VK9[34]	Terr. New Guinea	OC	51	28														
PK1-3[35]	Java	OC	54	28														
PK4[35]	Sumatra	OC	54	28														
PK5[35]	Netherlands Borneo	OC	54	28														

Prefix	Country	Continent	ZONE ITU	ZONE CQ	MIXED	PHONE	CW	RTTY	SAT	160	80	40	20	17	15	12	10
PK6[35]	Celebe & Molucca Is.	OC	54	28													
UN1[36]	Karelo-Finnish Rep.	EU	19	16													
VO[37]	Newfoundland, Labrador	NA	09	02,05													
VQ1, 5H1[38]	Zanzibar	AF	53	37													
VQ6[39]	British Somaliland	AF	48	37													
VQ9[40]	Aldabra	AF	53	39													
VQ9[40]	Desroches	AF	53	39													
VQ9[40]	Farquhar	AF	53	39													
VS2, 9M2[41]	Malaya	AS	54	28													
VS4[41]	Sarawak	OC	54	28													
VS9A, P, S[42]	People's Dem. Rep. of Yemen	AS	39	21													
VS9H[43]	Kuria Muria I.	AS	39	21													
VS9K[44]	Kamaran Is.	AS	39	21													
ZC5[41]	British North Borneo	OC	54	28													
ZC6, 4X1[45]	Palestine	AS	39	20													
ZD4[46]	Gold Coast, Togoland	AF	46	35													
ZSØ, 1[47]	Penguin Is.	AF	57	38													
ZS9[48]	Walvis Bay	AF	57	38													

NOTES:

[1] Unofficial prefix.

[2] (Blenheim Reef) Only contacts made from May 4, 1967, to June 30, 1975, count for this counrty. Contacts made July 1, 1975, and after, count as Chagos (VQ9).

[3] (Geyser Reef) Only contacts made from May 4, 1967, to February 28, 1978, count for this country.

[4] (Abu Ail Is.) Only contacts made March 30, 1991, and before, count for this country.

[5] (1M) Only contacts made from July 15, 1972, and before, count for this country. Contacts made July 16, 1972, and after, count as Tonga (A3).

[6] (4W) Only contacts made May 21, 1990, and before, count for this country.

[7] (7J1) Only contacts made from May 30, 1976, to November 30, 1980, count for this country. Contacts made December 1, 1980, and after, count as Ogasawara (JD1).

[8] (8Z4) Only contacts made December 25, 1981, and before, count for this country.

[9] (8Z5, 9K3) Only contacts made December 14, 1969, and before, count for this country.

[10] (9S4) Only contacts made March 31, 1957, and before, count for this country.

[11] (9U5) Only contacts made from July 1, 1960, to June 30, 1962, count for this country. Contacts made July 1, 1962, and after, count as Burundi (9U) or Rwanda (9X).

[12] (AC3) Only contacts made April 30, 1975, and before, count for this country. Contacts made May 1, 1975, and after, count as India (VU).

[13] (AC4) Only contacts made May 30, 1974, and before, count for this country. Contacts made May 31, 1974 and after, count as China (BY).

[14] (C9) Only contacts made September 16, 1963, and before, count for this country. Contacts made September 15, 1963, and after, count as China (BY).

[15] (CN2) Only contacts made June 30, 1960, and before, count for this country. Contacts made July 1, 1960, and after, count as Morocco (CN).

[16] (CR8) Only contacts made December 31, 1961, and before, count for this country.

[17] (CR8, CR10) Only contacts made September 14, 1976, and before, count for this country.

[18] (DA-DM) Only contacts made September 16, 1973, and before, count for this country. Contacts made September 17, 1973, and after, count as either FRG (DA-DL) or GDR (Y2-Y9).

[19] (DM, Y2-Y9) Only contacts made from September 17, 1973 and October 2, 1990 count for this country. On October 3, 1990 the GDR became part of the FRG.

[20] (EA9) Only contacts made May 13, 1969, and before, count for this country.

[21] (FF) Only contacts made August 6, 1960, and before, count for this country.

[22] (FH, FB8) Only contacts made July 5, 1975, and before, count for this country. Contacts made July 6, 1975, and after, count as Comoros (D6) or Mayotte (FH).

[23] (FI8) Only contacts made December 20, 1950, and before, count for this country.

[24] (FN8) Only contacts made October 31, 1954, and before count for this country.

[25] (FQ8) Only contacts made August 16, 1960, and before, will count for this country.

[26] (HKØ, KP3, KS4) Only contacts made September 16, 1981, and before, count for this country. Contacts made September 17, 1981, and after, count as San Andres (HKØ).

[27] (I1) Only contacts made March 31, 1957, and before, count for this country. Contacts made April 1, 1957, and after, count as Italy (I).

[28] (I5) Only contacts made June 30, 1960, and before, count for this country.

[29] (JZØ) Only contacts made April 30, 1963, and before, count for this country.

[30] (KR6, 8, JR6, KA6) Only contacts made May 14, 1972, and before, count for this country. Contacts made May 15, 1972, and after, count as Japan (JA).

[31] (KS4) Only contacts made August 31, 1972, and before, count for this country. Contacts made September 1, 1972, and after, count as Honduras (HR).

[32] (KZ5) Only contacts made September 30, 1979, and before, count for this country.

[33] (OK-OM) Only contacts made December 31, 1992, and before, count for this country.

[34] (P2, VK9) Only contacts made September 15, 1975, and before, count for this country. Contacts made September 16, 1975, and after, count as Papua New Guinea (P2).

[35] (PK1-6) Only contacts made April 30, 1963, and before, count for this country. Contacts made May 1, 1963, and after, count as Indonesia.

[36] (UN1) Only contacts made June 30, 1960, and before, count for this country. Contacts made July 1, 1960, and after, count as European RSFSR (UA).

[37] (VO) Only contacts made March 31, 1949, and before, count for this country. Contacts made April 1, 1949, and after, count as Canada (VE).

[38] (VQ1, 5H1) Only contacts made May 31, 1974, and before, count for this country. Contacts made June 1, 1974, and after, count as Tanzania (5H).

[39] (VQ6) Only contacts made June 30, 1960, and before, count for this country.

[40] (VQ9) Only contacts made June 28, 1976, and before, count for this country. Contacts made June 29, 1976, and after, count as Seychelles (S7).

[41] (VS2, VS4, ZC5, 9M2) Only contacts made September 15, 1963, and before, count for this country. Contacts made September 16, 1963, and after, count as West Malaysia (9M2) or East Malaysia (9M6, 8).

[42] (VS9A, P,S) Only contacts made May 21, 1990, and before, count for this country.

[43] (VS9H) Only contacts made November 29, 1967, and before, count for this country.

[44] (VS9K) Only contacts made on March 10, 1982, and before, count for this country.

[45] (ZC6, 4X1) Only contacts made June 30, 1968, and before, count for this country. Contacts made July 1, 1968, and after, count as Israel (4X).

[46] (ZD4) Only contacts made March 5, 1957, and before, count for this country.

[47] (ZSØ, 1) Only contacts made February 28, 1994, and before, count for this country.

[48] (ZS9) Only contacts made from September 1, 1977, to February 28, 1994, count for this country.

PREFIX CROSS REFERENCES

A8 = EL
AC (before 1972) = A5
AH = KH
AL7 = KL7
AM-AO = EA
AT-AW = VU
AX = VK
AY-AZ = LU
CF-CK = VE
CL = CO
CQ-CS = CT
CR3 (before 1974) = J5
CR4 (before 1976) = D4
CR5 (before 1976) = S9
CR6 (before 1976) = D2
CR7 (before 1976) = C9
CR9 (before 1985) = XX9
CT2 (before 1986) = CU
CXØ = CE9/VP8
CY-CZ = VE
CYØ (before 1985) = CY9
D7 = HL
DM-DT (before 1980) = Y2-9
E2 = HS
EAØ (before 1969) = 3C
EK, EM-EO, ER-ES, EU-EZ = U
ER (after 1992) = UO
EU (after 1991) = UC
FA-FF (after 1983) =F
FA (before 1963) = 7X
FB8 (before 1961) = 5R
FB8 (before 1985) = FT
FC (before 1985) = TK
FD8 (before 1961) = 5V
FE8 (before 1961) =TJ
FL (before 1978) = J2
FU8 (before 1982) = YJ
GB = G
GC (before 1977) = GJ or GU
H2 = 5B
H3 = HP
H5 (Bophutatswana) = ZS
H7 = YN
HE = HB
HM (before 1982) = HL
HT = YN
HU = YS

HW-HY = F
J4 = SV
KA1 = JD1
KA2AA-KA9ZZ = JA
KC6 (before 1990) = V6
KB6 (before 1979) = KH1
KC4 (Navassa) = KP1
KG6 (before 1979) = KH2
KG6I (before 1970) = JD1
KG6R, S, T (before 1979) = KHØ
KJ6 (before 1979) = KH3
KM6 (before 1979) = KH4
KP4 (Desecheo) = KP5
KP6 (before 1979) = KH5
KS6 (before 1979) = KH8
KV4 (before 1979) = KP2
KW6 (before 1979) = KH9
KX6 (before 1990) = V7
L2-9 = LU
M1 (before 1984) = T7
MP4B (before 1972) = A9
MP4M (before 1972) = A4
MP4Q (before 1972) = A7
MP4T, D (before 1972) = A6
NH = KH
NL7 = KL7
NP = KP
OQ (before 1961) = 9Q
P3 = 5B
P4 (before 1986) = PJ
PX (before 1970) = C3
RA, RN = UA
RB-RR = UB-UR
RS = U
RT = UB
RU-RX = U
S4 (Ciskei) = ZS
S8 (Transkei) = ZS
T4 = CO
T4 (Venda) = ZS
TH, TM, TO-TQ, TV-TX = F
UB (before 1994) = UZ
UC (before 1991) = EU
UD (before 1994) = 4J
UF (before 1994) = 4L
UG (before 1994) = EK
UH (before 1993) = EZ

UI (before 1994) = UJ
UJ (before 1993) = EY
UL (before 1994) = UN
UM (before 1993) = EX
UO (before 1994) = ER
UP (before 1991) = LY
UQ (before 1992) = YL
UR (before 1991) = ES
V9 (Venda) = ZS
VA-VG = VE
VH-VN = VK
VK9 (Nauru) = C2
VP1 (before 1982) = V3
VP2A (before 1982) = V2
VP2D (before 1979) = J7
VP2G (before 1975) = J3
VP2K (before 1984) = V4 or VP2E
VP2L (before 1980) = J6
VP2S (before 1980) = J8
VP3 (before 1967) = 8R
VP4 (before 1963) = 9Y
VP5 (Jamaica) = 6Y
VP5E = VP2E
VP6 (before 1967) = 8P
VP7 (before 1974) = C6
VQ2 (before 1965) = 9J
VQ3 (before 1962) = 5H
VQ4 (before 1964) = 5Z
VQ5 (before 1963) = 5X
VQ8 (before 1969) = 3B
VQ8 (Chagos) = VQ9
VQ9 (Seychelles) = S7
VR1 (before 1980) = T3
VR2 (before 1971) = 3D2
VR2 (after 1991) = VS6
VR3 (before 1980) = T32
VR4 (before 1979) = H4
VR5 (before 1971) = A3
VR8 (before 1979) = T2
VS1 (before 1966) = 9V
VS5 (before 1985) = V8
VS7 (before 1949) = 4S
VS9M = 8Q
VS9O (before 1961) = A4
VX-VY = CYØ/VE
WH = KH

WL7 = KL7
WP = KP
XJ-XO = VE
XP = OX
XQ-XR = CE
XV = 3W
XX7 (before 1976) = C9
YU2 (before 1992) = 9A
YU3 (before1992) = S5
YU4 (before 1992) = T9
YU5 (before 1992) = Z3
ZB1 (before 1965) = 9H
ZD1 (before 1962) = 9L
ZD2 (before 1961) = 5N
ZD3 (before 1966) = C5
ZD5 (before 1969) = 3DA
ZD6 (before 1965) = 7Q
ZE (before 1981) = Z2-9
ZK9 (1983) = ZK2
ZM6 (before 1963) = 5W
ZM7 (before 1984) = ZK3
ZS3 (before 1991) = V5
ZS7 (before 1969) = 3D6
ZS8 (before 1967) = 7P
ZS9 (before 1967) = A2
ZV-ZZ = PY
2D = GD
2E = G
2I = GI
2J = GJ
2M = GM
2U = GU
2W = GW
3B-3C (before 1968) = VE
3D6 (before 1988) = 3DA
3G = CE
3Z = SP
4A-4C = XE
4D-4I = DU
4J-4L = U
4J (after 1991) = EK
4J1F (before 1994) = R1MV
4K (before 1994) = UA
4K1 (before 1994) = CE9/KC4
4K2 (before 1994) = R1FJ
4K3 (before 1994) = UA
4K4 (before 1994) = UAØ

4L (after 1991) = UF
4M = YV
4N-4O = YU
4T = OA
4U1VIC = OE
4V = HH
5J-5K = HK
5L-5M = EL
6C = YK
6D-6J = XE
6O = T5
6T-6U = ST
7A-7I = YB
7G (before 1967) = 3X
7J-7N = JA
7JI = JA1 or JD1
7S = SM
7Z = HZ
8A-8I = YB
8J-8N = JA
8O = A2
8S = SM
9A (before 1984) = T7
9B-9D = EP
9E-9F = ET

CONTINENT
AF = AFRICA
AN = ANTARCTICA
AS = ASIA
EU = EUROPE
NA = NORTH AMERICA
OC = OCEANIA
SA = SOUTH AMERICA

ZONE NOTES
(A) 33, 42, 43, 44
(B) 67, 69-74
(C) 12, 13, 29, 30, 32, 38, 39
(D) 12, 13, 15
(E) 19, 20, 29, 30
(F) 20-26, 30-35, 75
(G) 16, 17, 18, 19, 23
(H) 2, 3, 4, 9, 75
(I) 55, 58, 59

DX ADVISORY COMMITTEE MEMBERS
(Listed by ARRL Division)

Atlantic - Tony Gargano, N2SS
26 Winchester Drive, Sewell, NJ 08080

Central - Jim O'Connell, W9WU
512 West Elm Ave, La Grange, IL 60525

Dakota - Thomas Lutz, WØKZV
3677 Steele St, Minnetonka, MN 55345

Delta - Michael W. Mayer, W5ZPA
5836 Marcia Ave, New Orleans, LA 70124

Great Lakes - Theodore Mirgliotta, KB8NW
16806 W 130th Street, Strongville, OH 44136

Hudson - Bill Hellman, NA2M
3713 Valleyview Street, Mohegan Lake, NY 10547

Midwest - James Spencer, WØSR
3712 Tanger Drive NE, Cedar Rapids, IA 52402

New England - James Dionne, K1MEM
31 DeMarco Road, Sudbury, MA 01776

Northwestern - Dick Moen, N7RO
2935 Plymouth Drive, Bellingham, WA 98225

Pacific - John G. Troster, W6ISQ
82 Belbrook Way, Atherton, CA 94025

Roanoke - Gary Dixon, K4MQG
1606 Crescent Ridge, Fort Mill, SC 29715

Rocky Mountain - Wayne A. Mills, N7NG
PO Box 1945, Jackson, WY 83001

Southeastern - Bart Fay, K4CEF
113 Malor Circle, Madison, AL 35758

Southwestern - John Alexander, K6SVL
28403 Covecrest Dr, Rancho Palos Verdes, CA 90274

West Gulf - Jim Lane, N5DC
1602 Chestnut Ridge Road, Kingwood, TX 77339

RAC - Garth Hamilton, VE3HO, 1309 Pelham St., *Chairman*
Box 1156, Fonthill Ontario L0S 1E0 CANADA

Board Liaison - Lew Gordon, K4VX
PO Box 105, Hannibal, MO 63401

Administrative Liaison - Lisa DeLude, ARRL HQ
225 Main Street, Newington, CT 06111

THE ARRL INCOMING QSL BUREAU SYSTEM

Purpose

Within the US and Canada, the ARRL DX QSL Bureau System is made up of numerous call area bureaus that act as central clearing houses for QSLs arriving from foreign countries. These "incoming" bureaus are staffed by volunteers. The service is free and ARRL membership is not required.

How it Works

Most countries have "outgoing" QSL bureaus that operate in much the same manner as the ARRL Outgoing QSL Service. The member sends his cards to his outgoing bureau where they are packaged and shipped to the appropriate countries.

A majority of the DX QSLs are shipped directly to the individual incoming bureaus where volunteers sort the incoming QSLs by the first letter of the call sign suffix. One individual may be assigned the responsibility of handling from one or more letters of the alphabet. Operating costs are funded from ARRL membership dues.

Claiming your QSLs

Send a 5 x 7.5 or 6 x 9 inch self-addressed, stamped envelope (SASE) to the bureau serving your call-sign district. Neatly print your call sign in the upper-left corner and place your mailing address on the front of the envelope. A suggested way to send envelopes is to affix a First Class stamp and clip extra postage to the envelope. Then, if you receive more than 1 oz. of cards, they can be sent in the single package.

Some incoming bureaus sell envelopes or postage credits in addition to the normal SASE handling. They provide the proper envelope and postage upon the prepayment of a certain fee. The exact arrangements can be obtained by sending your inquiry with an SASE to your area bureau. A list of bureaus appears below.

Helpful Hints

Good cooperation between the DXer and the bureau is important to ensure a smooth flow of cards. Remember that the people who work in the area bureaus are volunteers. They are providing you with a valuable service. With that thought in mind, please pay close attention to the following DOs and DON'Ts.

DOs

- DO keep self-addressed 5 x 7.5 or 6 x 9 inch envelopes on file at your bureau, **with your call in the upper-left corner,** and affix at least one unit of First-Class postage.
- DO send the bureau enough postage to cover SASEs on file and enough to take care of possible postage rate increases.
- DO respond quickly to any bureau request for SASEs, stamps or money. Unclaimed card backlogs are the bureau's biggest problem.
- DO notify the bureau of your new call as you upgrade. Please send SASEs with new call, in addition to SASEs with old call.
- DO include an SASE with any information request to the bureau.
- DO notify the bureau in writing if you don't want your cards.

DON'Ts

- DON'T send domestic US to US cards to your call-area bureau.
- DON'T expect DX cards to arrive for several months after the QSO. Overseas delivery is very slow. Many cards coming from overseas bureaus are over a year old.
- DON'T send your outgoing DX cards to your call-area bureau.
- DON'T send SASEs to your "portable" bureau. For example, AA2Z/1 sends SASEs to the W2 bureau, not the W1 bureau.
- DON'T send SASEs to the ARRL Outgoing QSL Service.
- DON'T send SASEs larger than 6 x 9 inches. SASEs larger than 6 x 9 inches require additional postage surcharges.

ARRL INCOMING DX QSL BUREAU ADDRESSES

First Call Area: All calls* —
W1 QSL Bureau
Y.C.C.C.
P.O. Box 80216
Springfield, MA 01138-0216

Second Call Area: All calls* —
ARRL 2nd District QSL Bureau
N.J.D.X.A.
P.O. Box 599
Morris Plains, NJ 07950

Third Call Area: All calls —
Pennsylvania DX Association
P.O. Box 100
York Haven, PA 17370-0100

Fourth Call Area: All single-letter
prefixes (K4, N4, W4) —
Mecklenburg Amateur Radio Club
P.O. Box DX
Charlotte, NC 28220

Fourth Call Area: All two-letter
prefixes (AA4, KB4, NC4, WD4, etc.) —
Sterling Park Amateur Radio Club
Call Box 599
Sterling, VA 20167

Fifth Call Area: All calls* —
ARRL W5 Incoming QSL Bureau
P.O. Box 50625
Midland, TX 79710

Sixth Call Area: All calls*† —
ARRL Sixth (6th) District DX QSL
Bureau
P.O. Box 1460
Sun Valley, CA 91352

Seventh Call Area: All calls* —
Willamette Valley DX Club, Inc.
P.O. Box 555
Portland, OR 97207

Eighth Call Area: All calls —
8th Area QSL Bureau
P.O. Box 182165
Columbus, OH 43218-2165

Ninth Call Area: All calls* —
Northern Illinois DX Assn
Box 519
Elmhurst, IL 60126

Zero Call Area: All calls* —
WØQSL Bureau
P.O. Box 4798
Overland Park, KS 66204

Puerto Rico: All calls* —
KP4 QSL Bureau
P.O. Box 1061
San Juan, PR 00902

U.S. Virgin Islands: All calls —
Virgin Islands ARC
GPO Box 11360
Charlotte, Amalie
Virgin Islands 00801

Hawaiian Islands: All calls* —
Wayne Jones, NH6GJ
P.O. Box 788
Wahiawa, HI 96786

Alaska: All calls* —
Alaska QSL Bureau
4304 Garfield St.
Anchorage, AK 99503

Guam:
MARC
Box 445
Agana, Guam 96910

SWL:Mike Witkowski, WDX9JFT
4206 Nebel St.
Stevens Point, WI 54481

QSL Cards for Canada may be sent to:
RAC Incoming Bureau
Kennebcasis Valley Amateur
   Radio Club
Box 51
St. John, NB E2L 3X1

QSL cards may also be sent to the individual bureaus:

VE1, VE9, VE0, VY2* —
Brit Fader Memorial QSL Bureau
P.O. Box 8895
Halifax, NS B3K 5M5

RAC National Incoming QSL Bureau
Loyalist City Amateur Radio Club
P.O. Box 51
Saint John, NB E2L 3X1

VE2 —
J. Dube, VE2QK
875 St. Severe St
Troi-Rivieres, PQ G9A 4G4

VE3 —
Ontario Trilliums
Box 157
Downsview, ON M3M 3A3

VE4 —
Adam Romanchuk, VE4SN
26 Morrison St.
Winnipeg, MB R2V 3B4

VE5* —
Bj. Madsen, VE5FX
739 Washington Dr.
Weyburn, SK S4H 3C7

VE6*—
Larry Langston, VE6LL
P.O. Box 3364
Ft Saskatchewan AB T8L 2T3

VE7* —
Dennis Livesey, VE7DK
8309 112th St.
Delta BC V4C 4W7

VE8* —
Rolf Ziemann, VE8RZ
2 Taylor Rd.
Yellowknife, NWT X1A 2K9

VY1 —
W. L. Champagne, VY1AU
P.O. Box 4597
Whitehorse, YU Y1A 2RB

VO1, VO2 —
Roland Peddle, VO1BD
P.O. Box 6
St. John's, NF A1C 5H5

* These bureaus sell envelopes or postage credits. Send an SASE to the bureau for further information.

†These bureaus can only accept specific sized envelopes. Send an SASE to the bureau for further information.

THE ARRL OUTGOING QSL SERVICE

Note: The ARRL QSL Service should not be used to exchange QSL cards within the 48 contiguous states.

One of the greatest bargains of League membership is being able to use the ARRL Outgoing QSL Service to conveniently send your DX QSL cards overseas to foreign QSL Bureaus. Your ticket for using this service is proof of ARRL membership and just $3.00 per pound. For those not quite so DX active (sending 10 cards or less), enclose $1.00. You can't even get a deal like that at your local warehouse supermarket! And the potential savings over the substantial cost of individual QSLing is equal to many times the price of your annual dues. Your cards are sorted promptly by the Outgoing Service staff, and cards are on their way overseas usually within a week of arrival at ARRL HQ. Approximately two million cards are handled by the Service each year!

QSL cards are shipped to QSL Bureaus throughout the world, which are typically maintained by the national Amateur Radio Society of each country. While no cards are sent to individuals or individual QSL managers, keep in mind that what you might lose in speed is more than made up in the convenience and savings of not having to address and mail QSL cards separately. (In the case of DXpeditions and/or active DX stations that use *US QSL managers,* a better approach is to QSL directly to the QSL manager. The various DX newsletters, the W6GO QSL manager directory, and other publications, are good sources of up-to-date QSL manager information.)

As postage costs become increasingly prohibitive, don't go

broke before you're even halfway towards making DXCC. There's a better and cheaper way —"QSL VIA BURO" through the ARRL Outgoing QSL Service!

How To Use The ARRL Outgoing QSL Service
1) Presort your DX QSLs alphabetically by parent call-sign prefix (AP, C6, CE, DL, ES, F, G, JA, LY, PY, U, 5N, 9Y and so on). NOTE: Some countries have a parent prefix and use additional prefixes, ie, CE (parent prefix) = XQ, 3G,... When sorting countries that have multiple prefixes, keep that country's prefixes grouped together in your alphabetical stack. **Addresses are not required. DO NOT separate the country prefix by use of paper clips, rubber bands, slips of paper or envelopes.**

2) Enclose proof of current ARRL Membership. This can be in the form of a photocopy of the white address label from your current copy of QST. You can also write on a slip of paper the information from the label, and use that as proof of Membership. A copy of your current Membership card is also acceptable.

3) Members (including foreign, QSL Managers, or managers for DXpeditions) should enclose payment of $3.00 per each pound of cards or portion thereof—approximately 150 cards weigh one pound. A package of only ten (10) cards or fewer sent in a single shipment costs only $1.00. Please pay by check (or money order) and write your callsign on the check. Send "green stamps" (cash) at your own risk. DO NOT send postage stamps

or IRCs. (DXCC credit CANNOT be used towards the QSL Service fee.)

4) Include only the cards, proof of Membership, and fee in the package. Wrap the package securely and address it to the ARRL Outgoing QSL Service, 225 Main Street, Newington CT 06111.

5) Family members may also use the service by enclosing their QSLs with those of the primary member. Include the appropriate fee with each individual's cards and indicate "family membership" on the primary member's proof of Membership.

6) Blind members who do not receive QST need only include the appropriate fee along with a note indicating the cards are from a blind member.

7) ARRL affiliated-club stations may use the service when submitting club QSLs by indicating the club name. Club secretaries should check affiliation papers to ensure that affiliation is current. In addition to sending club station QSLs through this service, affiliated clubs may also "pool" their members' individual QSL cards to effect an even greater savings. Each club member using this service must also be a League member. Cards should be sorted "en masse" by prefix, and proof of Membership enclosed for each ARRL member.

Recommended QSL-Card Dimensions

The efficient operation of the worldwide system of QSL Bureaus requires that cards be easy to handle and sort. Cards of unusual dimensions, either much larger or much smaller than normal, slow the work of the Bureaus, most of which is done by unpaid volunteers. A review of the cards received by the ARRL Outgoing QSL Service indicates that most fall in the following range: Height = 2.75 to 4.25 in. (70 to 110 mm), Width = 4.75 to 6.25 in. (120 to 160 mm). Cards in this range can be easily sorted, stacked and packaged. Cards outside this range create problems; in particular, the larger cards often cannot be handled without folding or otherwise damaging them. In the interest of efficient operation of the worldwide QSL Bureau system, it is recommended that cards entering the system be limited to the range of dimensions given. [Note: IARU Region 2 has suggested the following dimensions as optimum: Height 3.50 in. (90 mm), Width 5.50 in. (140 mm).]

Countries Not Served By The Outgoing QSL Service

Approximately 260 DXCC countries are served by the ARRL Outgoing QSL Service, as detailed in the ARRL DXCC Countries List. This includes nearly every active country. As noted previously, cards are forwarded from the ARRL Outgoing Service to a counterpart bureau in each of these countries. In some cases, there is no incoming bureau in a particular country and cards therefore cannot be forwarded. However, QSL cards can be forwarded to a QSL manager, ie; 3C1MB via (EA7KF). For this reason, the ARRL Bureaus Service cannot forward cards to the following countries:

COUNTRIES NOT SERVED BY THE OUTGOING QSL SERVICE ———

A5	Bhutan	VP2E	Anguilla	
A6	United Arab Emirates	VP2M	Montserrat	
D2	Angola	XT	Burkina Faso	
EP	Iran	XU	Cambodia	
ET	Ethiopia	XW	Laos	
J5	Guinea-Bissau	XX9	Macao	
KC6	Belau	XZ (1Z)	Myanmar (Burma)	
KHØ	Mariana Is.	YA	Afghanistan	
KH1	Baker and Howland Is.	ZD9	Tristan da Cunha	
KH4	Midway I.	3C	Equatorial Guinea	
KH5	Palmyra and Jarvis Is.	3CØ	Pagalu I.	
KH7	Kure I.	3V	Tunisia	
KH8	Am. Samoa	3W, XV	Vietnam	
KH9	Wake I.	3X	Guinea	
KP1	Navassa I.	5A	Libya	
KP5	Desecheo I.	5H	Tanzania	
P5	North Korea	5R	Madagascar	
S2	Bangladesh	5T	Mauritania	
S7	Seychelles	5U	Niger	
T2	Tuvalu	5X	Uganda	
T3	Kiribati	7O, 4W	Yemen	
T5	Somalia	7Q	Malawi	
TJ	Cameroon	8Q	Maldives	
TL	Central Africa	9G	Ghana	
TN	Congo	9N	Nepal	
TT	Chad	9Q	Zaire	
TY	Benin	9U	Burundi	
TZ	Mali	9X	Rwanda	
V6 (KC6)	Micronesia			

"The countries that currently restrict forwarding of QSL cards to members of that country's national radio organization include the following:"

Egypt	Monaco
France	Morocco
Germany	Portugal
Japan	

NOTE: SWL cards can be forwarded through the QSL Service.

NOTE: We no longer hold cards for countries with no Incoming Bureau. Only cards indicating a QSL manager for a station in these particular countries will be forwarded.

NOTE: The Outgoing QSL Service CANNOT forward stamps, IRCs or "green stamps" (cash) to the foreign QSL bureaus.

Please direct any questions or comments to the ARRL Outgoing QSL Service, 225 Main Street, Newington CT., 06111. Inquires via email may be sent to buro@arrl.org

ALLOCATION OF INTERNATIONAL CALL SIGN SERIES

Call Sign Series	Allocated to	Call Sign Series	Allocated to
AAA-ALZ	United States of America	H2A-H2Z	Cyprus
AMA-AOZ	Spain	H3A-H3Z	Panama
APA-ASZ	Pakistan	H4A-H4Z	Solomon Islands
ATA-AWZ	India	H6A-H7Z	Nicaragua
AXA-AXZ	Australia	H8A-H9Z	Panama
AYA-AZZ	Argentina	IAA-IZZ	Italy
A2A-A2Z	Botswana	JAA-JSZ	Japan
A3A-A3Z	Tonga	JTA-JVZ	Mongolia
A4A-A4Z	Oman	JWA-JXZ	Norway
A5A-A5Z	Bhutan	JYA-JYZ	Jordan
A6A-A6Z	United Arab Emirates	JZA-JZZ	Indonesia
A7A-A7Z	Qatar	J2A-J2Z	Djibouti
A8A-A8Z	Liberia	J3A-J3Z	Grenada
A9A-A9Z	Bahrain	J4A-J4Z	Greece
BAA-BZZ	China	J5A-J5Z	Guinea-Bissau
CAA-CEZ	Chile	J6A-J6Z	Saint Lucia
CFA-CKZ	Canada	J7A-J7Z	Dominica
CLA-CMZ	Cuba	†J8A-J8Z	St. Vincent and the Grenadines
CNA-CNZ	Morocco	KAA-KZZ	United States of America
COA-COZ	Cuba	LAA-LNZ	Norway
CPA-CPZ	Bolivia	LOA-LWZ	Argentina
CQA-CUZ	Portugal	LXA-LXZ	Luxembourg
CVA-CXZ	Uruguay	†LYA-LYZ	Lithuania
CYA-CZZ	Canada	LZA-LZZ	Bulgaria
C2A-C2Z	Nauru	L2A-L9Z	Argentina
C3A-C3Z	Andorra	MAA-MZZ	United Kingdom of Great Britain and Northern Ireland
C4A-C4Z	Cyprus	NAA-NZZ	United States of America
C5A-C5Z	Gambia	OAA-OCZ	Peru
C6A-C6Z	Bahamas	ODA-ODZ	Lebanon
*C7A-C7Z	World Meteorological Organization	OEA-OEZ	Austria
C8A-C9Z	Mozambique	OFA-OJZ	Finland
DAA-DRZ	Germany	OKA-OLZ	Czech Republic
DSA-DTZ	South Korea	OMA-OMZ	Slovak Republic
DUA-DZZ	Philippines	ONA-OTZ	Belgium
D2A-D3Z	Angola	OUA-OZZ	Denmark
D4A-D4Z	Cape Verde	PAA-PIZ	Netherlands
D5A-D5Z	Liberia	PJA-PJZ	Netherlands Antilles
D6A-D6Z	Comoros	PKA-POZ	Indonesia
D7A-D9Z	South Korea	PPA-PYZ	Brazil
EAA-EHZ	Spain	PZA-PZZ	Suriname
EIA-EJZ	Ireland	P2A-P2Z	Papua New Guinea
†EKA-EKZ	Armenia	P3A-P3Z	Cyprus
ELA-ELZ	Liberia	†P4A-P4Z	Aruba
†EMA-EOZ	Ukraine	P5A-P9Z	North Korea
EPA-EQZ	Iran	RAA-RZZ	Russian Federation
†ERA-ERZ	Moldova	SAA-SMZ	Sweden
†ESA-ESZ	Estonia	SNA-SRZ	Poland
ETA-ETZ	Ethiopia	• SSA-SSM	Egypt
EUA-EWZ	Belarus	• SSN-STZ	Sudan
†EXA-EXZ	Kyrgyzstan	SUA-SUZ	Egypt
†EYA-EYZ	Tajikistan	SVA-SZZ	Bangladesh
†EZA-EZZ	Turkmenistan	†S5A-S5Z	Slovenia
†E2A-E2Z	Thailand	S6A-S6Z	Singapore
E3A-E3Z	Eritrea	S7A-S7Z	Seychelles
FAA-FZZ	France	S8A-S8Z	South Africa
GAA-GZZ	United Kingdom of Great Britain and Northern Ireland	S9A-S9Z	Sao Tome and Principe
HAA-HAZ	Hungary	TAA-TCZ	Turkey
HBA-HBZ	Switzerland	TDA-TDZ	Guatemala
HCA-HDZ	Ecuador	TEA-TEZ	Costa Rica
HEA-HEZ	Switzerland	TFA-TFZ	Iceland
HFA-HFZ	Poland	TGA-TGZ	Guatemala
HGA-HGZ	Hungary	THA-THZ	France
HHA-HHZ	Haiti	TIA-TIZ	Costa Rica
HIA-HIZ	Dominican Republic	TJA-TJZ	Cameroon
HJA-HKZ	Colombia	TKA-TKZ	France
HLA-HLZ	South Korea	TLA-TLZ	Central Africa
HMA-HMZ	North Korea	TMA-TMZ	France
HNA-HNZ	Iraq	TNA-TNZ	Congo
HOA-HPZ	Panama	TOA-TQZ	France
HQA-HRZ	Honduras	TRA-TRZ	Gabon
HSA-HSZ	Thailand	TSA-TSZ	Tunisia
HTA-HTZ	Nicaragua	TTA-TTZ	Chad
HUA-HUZ	El Salvador	TUA-TUZ	Ivory Coast
HVA-HVZ	Vatican City	TVA-TXZ	France
HWA-HYZ	France	TYA-TYZ	Benin
HZA-HZZ	Saudi Arabia	TZA-TZZ	Mali

Call Sign Series	Allocated to
T2A-T2Z	Tuvalu
T3A-T3Z	Kiribati
T4A-T4Z	Cuba
T5A-T5Z	Somalia
T6A-T6Z	Afghanistan
† T7A-T7Z	San Marino
T8A-T8Z	Palau
† T9A-T9Z	Bosnia and Herzegovina
UAA-UIZ	Russian Federation
† UJA-UMZ	Uzbekistan
† UNA-UQZ	Kazakhstan
URA-UTZ	Ukraine
† UUA-UZZ	Ukraine
VAA-VGZ	Canada
VHA-VNZ	Australia
VOA-VOZ	Canada
VPA-VSZ	United Kingdom of Great Britain and Northern Ireland
VTA-VWZ	India
VXA-VYZ	Canada
VZA-VZZ	Australia
† V2A-V2Z	Antigua and Barbuda
† V3A-V3Z	Belize
† V4A-V4Z	Saint Kitts and Nevis
† V5A-V5Z	Namibia
† V6A-V6Z	Micronesia
† V7A-V7Z	Marshall Islands
† V8A-V8Z	Brunei
WAA-WZZ	United States of America
XAA-XIZ	Mexico
XJA-XOZ	Canada
XPA-XPZ	Denmark
XQA-XRZ	Chile
XSA-XSZ	China
XTA-XTZ	Burkina Faso
XUA-XUZ	Cambodia
XVA-XVZ	Viet Nam
XWA-XWZ	Laos
XXA-XXZ	Portugal
XYA-XZZ	Myanmar
YAA-YAZ	Afghanistan
YBA-YHZ	Indonesia
YIA-YIZ	Iraq
YJA-YJZ	Vanuatu
YKA-YKZ	Syria
† YLA-YLZ	Latvia
YMA-YMZ	Turkey
YNA-YNZ	Nicaragua
YOA-YRZ	Romania
YSA-YSZ	El Salvador
YTA-YUZ	Yugoslavia
YVA-YYZ	Venezuela
YZA-YZZ	Yugoslavia
Y2A-Y9Z	Germany
ZAA-ZAZ	Albania
ZBA-ZJZ	United Kingdom of Great Britain and Northern Ireland
ZKA-ZMZ	New Zealand
ZNA-ZOZ	United Kingdom of Great Britain and Northern Ireland
ZPA-ZPZ	Paraguay
ZQA-ZQZ	United Kingdom of Great Britain and Northern Ireland
ZRA-ZUZ	South Africa
ZVA-ZZZ	Brazil
† Z2A-Z2Z	Zimbabwe
† Z3A-Z3Z	Macedonia (Former Yugoslav Republic)
2AA-2ZZ	United Kingdom of Great Britain and Northern Ireland
3AA-3AZ	Monaco
3BA-3BZ	Mauritius
3CA-3CZ	Equatorial Guinea
• 3DA-3DM	Swaziland
• 3DN-3DZ	Fiji
3EA-3FZ	Panama
3GA-3GZ	Chile
3HA-3UZ	China
3VA-3VZ	Tunisia
3WA-3WZ	Viet Nam
3XA-3XZ	Guinea
3YA-3YZ	Norway
3ZA-3ZZ	Poland

Call Sign Series	Allocated to
4AA-4CZ	Mexico
4DA-4IZ	Philippines
† 4JA-4KZ	Azerbaijan
† 4LA-4LZ	Georgia
4MA-4MZ	Venezuela
4NA-4OZ	Yugoslavia
4PA-4SZ	Sri Lanka
4TA-4TZ	Peru
* 4UA-4UZ	United Nations
4VA-4VZ	Haiti
4XA-4XZ	Israel
* 4YA-4YZ	International Civil Aviation Organization
4ZA-4ZZ	Israel
5AA-5AZ	Libya
5BA-5BZ	Cyprus
5CA-5GZ	Morocco
5HA-5IZ	Tanzania
5JA-5KZ	Colombia
5LA-5MZ	Liberia
5NA-5OZ	Nigeria
5PA-5QZ	Denmark
5RA-5SZ	Madagascar
5TA-5TZ	Mauritania
5UA-5UZ	Niger
5VA-5VZ	Togo
5WA-5WZ	Western Samoa
5XA-5XZ	Uganda
5YA-5ZZ	Kenya
6AA-6BZ	Egypt
6CA-6CZ	Syria
6DA-6JZ	Mexico
6KA-6NZ	South Korea
6OA-6OZ	Somalia
6PA-6SZ	Pakistan
6TA-6UZ	Sudan
6VA-6WZ	Senegal
6XA-6XZ	Madagascar
6YA-6YZ	Jamaica
6ZA-6ZZ	Liberia
7AA-7IZ	Indonesia
7JA-7NZ	Japan
7OA-7OZ	Yemen
7PA-7PZ	Lesotho
7QA-7QZ	Malawi
7RA-7RZ	Algeria
7SA-7SZ	Sweden
7TA-7YZ	Algeria
7ZA-7ZZ	Saudi Arabia
8AA-8IZ	Indonesia
8JA-8NZ	Japan
8OA-8OZ	Botswana
8PA-8PZ	Barbados
8QA-8QZ	Maldives
8RA-8RZ	Guyana
8SA-8SZ	Sweden
8TA-8YZ	India
8ZA-8ZZ	Saudi Arabia
† 9AA-9AZ	Croatia
9BA-9DZ	Iran
9EA-9FZ	Ethiopia
9GA-9GZ	Ghana
9HA-9HZ	Malta
9IA-9JZ	Zambia
9KA-9KZ	Kuwait
9LA-9LZ	Sierra Leone
9MA-9MZ	Malaysia
9NA-9NZ	Nepal
9OA-9TZ	Zaire
9UA-9UZ	Burundi
9VA-9VZ	Singapore
9WA-9WZ	Malaysia
9XA-9XZ	Rwanda
9YZ-9ZZ	Trinidad and Tobago

• Half-series
* Series allocated to an international organization
† Provisional allocation in accordance with RR2088

ARRL DXCC FIELD REPRESENTATIVES (For information about this program, see Section V of the DXCC rules.)

Atlantic Division

EPA

M. Stokley Benson, AJ3H
1192 Divot Drive
Wescosville, PA 18106

Charles E. Moraller, K2CM
RD 1, Box 99-C
New Ringgold, PA 17960

Norman Zoltack, K3NZ
4333 Locust Drive
Schnecksville, PA 18078

MDC

Murray Green, K3BEQ
5730 Lockwood Road
Cheverly, MD 20785

Everett C. Bollin, WA3DVO
8000 Ray Leonard Court
Palmer Park, MD 20785

Michael C. Cizek, KO7V
Box 227
Oxon Hill, MD 20750

SNJ

John M. Fisher, K2JF
538 Wesley Avenue
Pitman, NJ 08071-1924

James M. Mollica, Sr. K2OWE
22 Wright Loop
Williamstown, NJ 08094

Steve Branca, K2SB
202 Minnetonka Road
Somerdale, NJ 08083

John D. Imhof, N2VW
3 Seeley Drive
Mt. Holly, NJ 08060

WNY

John C. Yodis, K2VV
PO Box 460
Hagaman, NY 12086

Eugene W. Nadolny, W2FXA
21 Hidden Valley Drive
Elma, NY 14059

Bob Rossi, W2HG
75 Olde Erie Trail
Rochester, NY 14626

Elmer Wagner, WB2BNJ
23 Red Barn Circle
Pittsford, NY 14534

Robert Dow, WB2CJL
115 Puritan Road
Tonawanda, NY 14150

James R. Ciurczak, WB2IVO
10404 Cayuga Drive
Niagara Falls, NY 14304

Robert E. Nadolny, WB2YQH
135 Wetherstone Drive
West Seneca, NY 14224

Stephen A. Licht, WF2S
3268 West Main St Rd
Batavia, 14020-9109

Central Division

IL

Tim Wright, KØBFR
1131 Warren
Alton, IL 62002

Jerry Brunning, K9BG
15307 Shamrock Lane
Woodstock, IL 60098

Andrew White, K9CW
1725 Counrty Rd, 2500 N
Thomasboro, IL 61878

Merv Schweigert, K9FD
Rt 2, Box 138A
Red Bud, IL 62278

Robert E. Nielsen, K9RN
1372 Skyridge Drive
Crystal Lake, IL 60014

James J. Coleman, KA6A
250 County Road 700N
Ivesdale, IL 61851

Douglas E. Williams, KD9Q
1717 S. 28th Street
Quincy, IL 62301-6305

Greg Wilson, N4CC
3865 Staunton Road
Edwardsville, IL 62025

Michael L. Nowack, NA9Q
2011 N. Sheridan Dr
Quincy, IL 62301-8929

Ed Doubek, N9RF
25W063 Wood Court
Naperville, IL 60563

Jim Dawson, WW9Q
257 S. 9th Street
East Alton, IL 62024

Gary H. Hilker, K9LJN
804 Otto Road
Machesney Park, IL 61115

IN

Joseph L. Pontek, Sr., K8JP
PO Box 80262
Indianapolis, IN 46280-0262

David Bunte, K9FN
129 Ivy Hill Drive
West Lafayette, IN 47906

Victor Keller, N9GK
4011 Daner Drive
Fort Wayne, IN 46815

John C. Goller, K9UWA
4836 Ranch Road
Leo, IN 46765

James R. Weigand, N9BW
1422 Buckskin Drive
Fort Wayne, IN 46804

Mark Reese, NB9F
3511 Windlass Court
Fort Wayne, IN 46815

Vernon Seitz, W9HLY
45 Homestead
Decatur, IN 46733

Bill Gibbons, W9KBV
535 Grapevine Lane
Fort Wayne, IN 46825

Robert C. Webb, W9OCL
8208 Valley Estates Drive
Indianapolis, IN 46227

William D. DeGeer, W9TY
3601 Tyler Street
Gary, IN 46408

Mark E. Musick WB9CIF
P.O. Box 575
Plainfield, IN 46168

David L Zeph W9ZRX
16310 Springmill Rd
Westfield, IN 46074

Michael Goode N9NS
13040 Broadway St
Indianapolis, IN 46280

WI

James Hugo, KE9ET
713 Vernon Avenue
Madison, WI 53714

Peter Byfield, K9VAL
4534 S. Hill Ct.
Deforest, WI 53532

Ron Gorski, N9AU
3241 S. Clement Avenue
Milwaukee, WI 53207

Gerald Scherkenbach, N9AW
5952 So. Elaine Avenue
Cudahy, WI 53110

Herb Jordan, W9LA
6318 Putnam Road
Madison, WI 53711

Richard F. Roll, W9TA
14880 W. Maple Ridge Court
New Berlin, WI 53151

Frank Holliday, WB9NOV
23 Walworth Ct.
Madison, WI 53705

Ed Toal, WI9L
5141 Sunrise Ridge Trail
Middleton, WI 53562

Jim Promis, WY9E
3745 Rolling Heights
Oneida, WI 54155

Robert Gorsiski, WA9BDX
1123 David Street
Racine, WI 53404

Ken Tate, WB9OBX
728 South Twenty Second St
Manitowoc, WI 54220

David Jaeger, WN9Q
11913 Sunnyslope Road
Whitelaw, WI 54247

George R. Croy, W9MDP
2113 Twin Willows Drive
Appleton, WI 54915

John Meyer, NZ9Z
PO Box 146
Kellnersville, WI 54215

Gary Banks, N9ER
1692 Shangra La
Neenah, WI 54956

Skip Caswell, KE9LK
HC2 Box 1038
Florence, WI 54121

Gary Hoehne, KB9AIT
W8244 Winnegamie Drive
Neenah, WI 54956

Marian McCone, KB9EKO
6899 K
Rhinelander, WI 54501

Myron McCone, K9JJR
6899 Cty Hwy K
Rhinelander, WI 54501

Dakota Division

MN

Curt Swenson, KØCVD
4821 Westminster Road
Minnetonka, MN 55345

Dave Wester, KØIEA
10205 217th Street North
Forest Lake, MN 55205

Gary Reichow, KNØV
15615 Harmony Way
Apple Valley, MN 55124

Keith H. Gilbertson, KSØZ
HC10, Box 44
Detroit Lakes, MN 56501

John Hill, NJØM
3353-34th Avenue South
Minneapolis, MN 55406

Jack Falker, W8KR
5716 View Lane
Edina, MN 55436

ND

Ron L. Roche, KØALL
PO Box 5301
Fargo, ND 58105

Delta Division

AR

Paul E. Wynne, AF5M
5306 Marion Street
N. Little Rock, AR 72118

Stanley W. Krueger, K5AS
3169 Skillern Road
Fayetteville, AR 72703

J. Sherwood Charlton, K5GOE
306 N. Willow Avenue
Fayetteville, AR 72701

Leonard Mendel, K5OVC
309 Rolling Acres Drive
Pearcy, AR 71964

Oliver A. Gade, K9GPN
156 Wildcat Estates Lane
Hot Springs, AR 71913

Earl F. Smith, KD5ZM
818 Green Oak Lane
White Hall, AR 71602

Stewart W. Long, KE5PO
28 Nob View Circle
Little Rock, AR 72205

F. Martin Hankins, W5HTY
406 East Cedar Street
Warren, AR 71671

Andrew Toth, W5LQN
117 Camellia Drive
Hot Springs, AR 71913

Ben Lockerd, W5NF
3231 S. Cliff Drive
Fort Smith, AR 72903

Mark Manes, WC5I
123 Maple Shade Road
Alma, AR 72921

LA

Howard N. Schmidt, K5BLV
PO Box 675-Elm Street
Clinton, LA 70722

John Wondergem, K5KR
600 Smith Drive
Metairie, LA 70005

Jim O'Brien, K5NV
PO Box 403
Harvey, LA 70059

Silvano Amenta, KB5GL
5028 Hearst Street
Metairie, LA 70001

Paul Azar, N5AN
412 Live Oak
Lafayette, LA 70503

Shirley Wondergem, N5GGO
600 Smith Drive
Metairie, LA 70005

Wes Attaway, N5WA
2048 Pepper Ridge
Shreveport, LA 71115

Willis J. Robert, W5RRK
Rt 2 Box 538
Erath, LA 70533

Michael Mayer III, W5ZPA
5836 Marcia Avenue
New Orleans, LA 70124

Roy Bonvillian, WD5DBV
PO Box 350
Cade, LA 70519

Cornell C. Bodensteiner, WD5GJB
129 Patricia Ann Place
Lafayette, LA 70508

Randy Hollier, WX5L
1421 North Atlantic
Metaire, LA 70003

MS

Louis J. Raymond, N4CSF
1720 Thomas Jefferson Drive
Biloxi, MS 39531

C. T. Green, KB5WQ
128 Gandy Circle
Long Beach, MS 39560

Steven Alexander, KX5V
22 Yorkshire Pkwy
Gulfport, MS 39501

Floyd Gerald, N5FG
4706 Washington Avenue
Gulfport, MS 39507-4043

Gary E. Jones, W5VSZ
23 Pirate Drive,
Hattiesburg, MS 39402-9557

W. Paul Howard, WA5TUD
Rt 1, Box 59E
Water Valley, MS 38965

Vic West, WA5SUE
113 Ben Drive
Gulfport, MS 39503

Michael J. Parisey, WDØGML
251 Eisenhower Drive, Apt 337
Biloxi, MS 39531

George G. Bethea, WN5IJZ
609 Woodward Avenue
Gulfport, MS 39501

TN

Gary S. Pitts, AA4DO
2023 Kimberly Drive
Mt. Juliet, TN 37122

John L. Anderson, AG4M
1041 West Outer Drive
Oak Ridge, TN 37830

Robert M. May II, K4SE
PO Box 453
Jonesboro, TN 37659

Don Moore, KB4HU
309 Brookhaven Trail
Smyrna, TN 37167

Vernon T. Underwood, WA4N1E
PO Box 3771
Knoxville, TN 37927

Lawrence D. Strader, WA4ZWH
11915 Turpin Lane
Knoxville, TN 37932

Don Payne, K4ID
117 Sam Davis Drive
Springfield, TN 37172

William B. Jones, WB4PUD
5081-HY 49 West
Springfield, TN 37172

Great Lakes Division

KY

James C. Vaughan, K4TXJ
5504 Datura Lane
Louisville, KY 40258

Robert K. Ledford, KC4MK
4533 Highway 155
Fisherville, KY 40023

Timothy B. Totten, KJ4VH
10400 Broad Run Road
Louisville, KY 40299

Ralph E. Wettle, N4NTQ
884 Clarks Lane
Louisville, KY 40217

Roy M. Dobbs, W4KHL
516 Dorsey Way
Louisville, KY 40223

Gary Hext, WB4FLB
4953 Westgate Drive
Bowling Green, KY 42101

John W. Reasoner WA4QMQ
125 Clearview Ave.
Bowling Green, KY 42101

Larry R. Smith WA4QQV
623 Shady Ln.
Alvaton, KY 42122

William E. Gann N4HID
445 Elrod Rd.
Bowling Green, KY 42104

Paul D Schrader N4XM
7001 Briscoe Kn
Louisville, KY 40228

MI

Wayne W. Wiltse, K8BTH
14468 Bassett Avenue
Livonia, MI 48154

Merlin D. Anderson, K8EFS
4300 South Cochran
Charlotte, MI 48813-9109

Donald L. Defeyter, KC8CY
5061 East Clark Road
Bath, MI 48808

Edmund Gurney, Jr., WD8Y
18712 Westbrook Drive
Livonia, MI 48152

John C. Kroll, K8LJG
3528 Craig Drive
Flint, MI 48506

OH

Pete Michaelis, N8ATR
12224 East River Road
Columbia Station, OH 44028

John M. Hugentober, Sr., N8FU
4441 Andreas Avenue
Cincinnati, OH 45211

Ron Moorefield, W8ILC
8847 Cajun Ct
Huber Heights, OH 45424-7014

Harry T. Flasher, W8KKF
7425 Barr Circle
Dayton, OH 45459

Alfred V. Altomari, W8QWI
2009 W. 36 Street
Lorain, OH 44053

Robert A. Frey, WA6EZV
8729 Sarah's Bend Drive
Cincinnati, OH 45251

Jerome Kurucz, WB8LFO
5338 Edgewood Drive
Lorain, OH 44053

John Udvari, WB8VPA
2896 Pine Lake Road
Uniontown, OH 44685

Reno M. Tonsi, WT8C
9248 Woodvale Court
Mentor, OH 44060

Hudson Division

ENY

Jay Musikar, AF2C
6 Cherry Lane
Putnam Valley, NJ 10579

Saul M. Abrams, K2XA
RD #1, Maple Road
Slingerlands, NY 12159

G. William Hellman, NA2M
3713 Valleyview Street
Mohegan Lake, NY 10547

Steven L. Weinstein, NI2C
45 Esterwood Avenue
Dobbs Ferry, NY 10522

Walter J. Bieber KB2RA
374 N. Little Tor Rd.
New City, NY 10956

David Fisher, KA2CYN
124 Bellows Lane
New City, NY 10956

NLI

Richard J. Tygar, AC2P
5 Clelmsford Drive
Wheatley Heights, NY
11798-1504

Arthur M. Albert, K2ENT
2476 Fortesque Avenue
Oceanside, NY 11572

Larry Strasser, K2LS
1404 Amend Drive
North Merrick, NY 11566-1301

Frank Kiefer, K2PWG
1-Sherrill Lane
Port Jefferson Station, NY 11776

Leonard Zuckerman, KB2HK
2444 Seebode Court
Bellmore, NY 11710-4522

Edward J. Manheimer, KD2OD
14 Seneca Trail
Ridge, NY 11961

Martin P. Miller, NN2C
24 Earl Road
Melville, NY 11747

L. E. Dietrich, NN2G
32 Beach Road
Massapequa, NY 11758

Frank E. Gut, NY2A
129 North Side Road
Wading River, NY 11792

Jim Stiles, W2NJN
21 Branch Drive
Smithtown, NY 11787

Jean Chittenden, WA2BGE
RR2, Box 2105
Syosset, NY 11791

NNJ

William Inkrote, K2NJ
911 Route 579
Flemington, NJ 08822

Robert B. Larson, K2TK
PO Box 53
Stockholm, NJ 07460

Warren Hager, K2UFM
31 Forest Drive
Hillsdale, NJ 07642

Joyce A. Birmingham, KA2ANF
235 Van Emburgh Ave
Ridgewood, NJ 07450

Ronald Loneker, KA2BZS
2092 Nicholl Avenue
Scotch Plains, NJ 07076

Derry Galbreath, KB2HZ
207 Gordon Place
Neshanic Station, NJ 08853

Guy Glaser, KE2CG
240 Grant Avenue
Piscataway, NJ 08854

Ronald Hauser, KE2JR
59 Deltart Drive
Belle Mead, NJ 08502

John Hults, KF2BH
714 Mountain View Road
Asbury, NJ 08802-1026

Harry Johnson, KS2O
424 Jefferson Avenue
Hasbrouck Heights, NJ 07604

Eugene Ingraham, N2BIM
79 Stillwater Road
Newton, NJ 07860

Michael Pagan, N2GBH
1 Woodview Avenue
Long Beach, NJ 07740

Harry Westervelt, NA2K
72 Kuhlthau Avenue
Milltown, NJ 08850

John Sawina, NA2R
61 Gallmeier Road
Frenchtown, NJ 08825-3719

Richard Petermann, NF2K
79 Pompton Avenue
West Paterson, NJ 07424

Peter Pellack, NO2R
77 Money Street
Lodi, NJ 07644

Joseph H. Painter, W2BHM
38 Beech Avenue
Berkeley Heights, NJ 07922

Ted Marks, W2FG
81 Oakey Drive
Kendall Park, NJ 08824

Orion Arnold, W2HN
2 Sleepy Hollow Drive
Ho-Ho-Kus, NJ 07423

T. Edward Berzin, W2MIG
47 Palisade Road
Elizabeth, NJ 07208

Stan Dicks, W8YA
9 Settlers Court
Neshanic Station, NJ 08855

William W. Hudzik, WA2UDT
111 Preston Drive
Gillette, NJ 07933

Angel M. Garcia, WA2VUY
7 Markham Drive
Long Valley, NJ 07853

Mario Karcich, WB2CZB
311 Liberty Avenue
Hillsdale, NJ 07642

Andrew Birmingham, WB2RQX
235 Van Emburgh Avenue
Ridgewood, NJ 07450

Robert C. Shelton, Jr., N2EDF
11 North Clarke Street
Ogdensburg, NJ 07439

Edwin E. Lauster WA2MXW
31 Layne Rd.
Somerset, NJ 08873

Midwest Division

IA

Thomas White, KØVZR
2027 Carter Avenue
Jessup, IA 50648

James R. Pitts, NØAMI
1221 East 35th Street
Des Moines, IA 50317

Robert W. Watson, WØEJ
7431 Macon Drive NE
Cedar Rapids, IA 52402

Thomas Noel Vinson, NYØV
10211 Hall Road
Cedar Rapids, IA 52400

Dale E. Repp, WØIZ
1618 Texas Avenue, NE
Cedar Rapids, IA 52402

Donald L. Schmidt, WØANZ
2161 NW 80th Place
Des Moines, IA 50325-5626

James L. Spencer, WØSR
3712 Tanager Drive, NE
Cedar Rapids, IA 52402

Thomas G. Vavra, WB8ZRL
682 Palisades Access Road
Ely, IA 52227

KS

Rick Barnett, KBØU
PO Box 4798, 8909 W. 81
Overland Park, KS 66204

Richard G. Tucker, WØRT
PO Box 875
Parsons, KS 67357

Dean Lewis, WAØTKJ
609 Otto Avenue
Salina, KS 67401-7261

John Shoultys, WDØBNC
2157 Edward
Salina, KS 67401

MO

Joseph F. Nemecek, KØJN
1208 N.E. 77th Street
Kansas City, MO 64118

Thomas L. Bishop, KØTLM
4936 N. Kansas Avenue
Kansas City, MO 64119

Clifford H. Ahrens, KIØW
65 Pioneer Trail
Hannibal, MO 63401-2744

Ken Kreski, KWØA
6645 St. Johns Lane
House Springs, MO 63051

Jim Higgins, NDØF
219 Birchwood Drive
St. Peters, MO 63376

Udo Heinze, NIØG
#2 Wildwood Circle
Crestwood, MO 63126

Jim Glasscock, WØFF
3416 Manhattan Avenue
St. Louis, MO 63143

Bill Wiese, WØHBH
9520 Old Bonhomme
Olivette, MO 63132

John L. Chass, WØJLC
12167 N.W. Hwy. 45
Parkville, MO 64152

Don Gaikins, WDØCHW
344 Hillside
Webster Groves, MO 63119

NE

Robert W. Staub, Jr., WBØYWO
PO Box 113
Hoskins, NE 68740

New England Division

CT

Joel Wilks, AK1N
27 Champion Hill Road
East Hampton, CT 06424

Al Rousseau, W1FJ
180 Den Quarry Road
Lynn, MA 01904

William C. Stapleford, K1DII
Bird Road, PO Box 1093
Bristol, CT 06010

Ralph M. Hirsch, K1RH
172 Newton Road
Woodbridge, CT 06525

Peter Budnik, KB1HY
17 Atwood Street
Plainville, CT 06062

John P. Larson, NQ1K
34 Donna Drive
Burlington, CT 06013

EMA

William G. Bithell, Jr., K1JKS
96 Greenwood Avenue
Swampscott, MA 01907

Gary Young, K2AJY
1 Sutton Place
Swampscott, MA 01907

Clifford O. Thomson, N1EOA
5 Lakeshore Drive
Beverly, MA 01915

Glenn Scanlon, N1EUO
52 Sunset Dr
Beverly, MA 01915

Ray Sylvester, Jr., NR1R
20 Gardner Road
Reading, MA 01867

Melrose R. Cole, WZ1Q
PO Box 8
Prides Crossing, MA 01965

NH

W. David Gerns, KA1CB
3 Rolling Hill Avenue
Plaistow, NH 03865

Carl Huether, KM1H
169 Jeremy Hill Road
Pelham, NH 03076

Paul I. Cleveland, W1OHA
PO Box 89, Craighill
North Sutton, NH 03260-0089

William A. Dodge, WA1PEL
78 Littleworth Road
Dover, NH 03820

Ann M. Santos, WA1S
245 Colburn Road
Milford, NH 03055

VT

David A. Wilson, N4DW
Rt 1, Box 11A
East Burke, VT 05832

WMA

Gareth B. Gaudette, W1GG
21 Westview Road
Lanesboro, MA 01237

Edward Landry, WA1ZAM
140 Cliff Street
North Adams, MA 01247

Robert Tublitz WT2Q
P.O. Box 847
Stockbridge, MA 01262

Northwestern Division

EWA

Earl M. Ringle, N7ER
N. 11310 Parksmith
Mead, WA 99021

William Martinek, K7EFB
N 522 Moore Road
Veradale, WA 99037

Michael J. Klem, K7ZBV
804 Cedar Avenue
Richland, WA 99352

Herb Rode, KE7PB
P.O. Box 4173
West Richland, WA 99352

Lester S. Morgan, NR7B
N 6811 Monroe Street
Spokane, WA 99208-4129

Donald W. E. Calbick, W7GB
447 Knolls Vista Drive
Moses Lake, WA 98837

Marion Lundrigan, W7OIH
2801 Road 92
Pasco, WA 99301

Warren L. Triebwasser, W7YEM
4624 NW Blvd
Spokane, WA 99205

Jay W. Townsend, WS7I
PO Box 644
Spokane, WA 99210

MT

George E. Martin, K7ABV
3608 5th Avenue S.
Great Falls, MT 59405

Robert E. Leo, W7LR
6790 S. 3rd Road
Bozeman, MT 59715

Doug Laubach, WQ7B
Box 19
Carter, MT 59420

OR

Gerald D. Branson, AA6BB
93787 Dorsey Lane
Junction City, OR 97448

Larry R. Johnson, K7LJ
10036 SW 50th Avenue
Portland, OR 97219

Vincent J. Varnas, K8REG
3229 SW Luradel Street
Portland, OR 97219

Hugh C. Dean, KF7SH
420 W. Rose
Lebanon, OR 97355

Benjamin F. Shearer, Jr.,
N7GMN
15364 S. Clackamas River Dr
Oregon City, OR 97045

Sanford T. Weinstein, NK7Y
2945 SW 4th Avenue
Portland, OR 97201

Robert A. Moore, W7JNC
PO Box 33197
Portland, OR 97233

Michael J. Huey, W7ZI
12385 SW Stillwell Lane
Beaverton, OR 97005

Nels H. Nelson, W7EYE
39105 SW Hartley Road
Gaston, OR 97119

Richard A. Zalewski, W7ZR
10735 SW 175th Avenue
Beaverton, OR 97007

Ronald E. Vincent, WJ7R
3715 University St
Eugene, OR, 97405-4347

WWA

David C. Norton, AB9O
2612 NW 18th Avenue
Camas, WA 98607

Hillar Raamat, N6HR
PO Box 213
Greenbank, WA 98253

Jim Fenstermaker, K9JF
10312 NE 161st Avenue
Vancouver, WA 98662

Herbert Anderson, K7GEX
20148 6th NE
Seattle, WA 98155

Thomas H. Renfro, N7OT
8703 NE 94th Avenue
Vancouver, WA 98662

Dick Moen, N7RO
2935 Plymouth Drive
Bellingham, WA 98225

John Gohndrone, N7TT
PO Box 863
Bothell, WA 98041

Brenda G. Murphy, NS7F
2612 NW 18th Avenue
Camas, WA 98607

Kermit W. Raaen, W7BG
15509 E. Mill Plain Blvd 42
Vancouver, WA 98664

Al Johnson, W7EKM
8186 Stein Road
Custer, WA 98240

Randy Stegemeyer, W7HR
PO Box 1590
Port Orchard, WA 98366

Leonard A. Westbo, Jr., W7MCU
10528 S.E. 323nd Street
Auburn, WA 98002

Alan N. Rovner, WA2TMP
18809 N.E. 21st Street
Vancouver, WA 98684

ID

Don M. Clower, KA7T
5103 W. Cherry Lane
Meridian, ID 83642

Pacific Division

EB

Thomas F. Jones, K6TS
1239 Hillview Drive
Livermore, CA 94550

James W. Sheley, KA6DXY
5849 Hansen Drive
Pleasanton, CA 94566

Charles E. McHenry, W6BSY
1612 Via Escondido
San Lorenzo, CA 94580

Rubin Hughes, WA6AHF
17494 Via Alamitos
San Lorenzo, CA 94580

Randy Wright, WB6CUA
18432 Milmar Road
Castro Valley, CA 94546

NV

Frank Dziurda, K7SFN
225 W. Coyote Drive
Carson City, NV 89704

JB Coats, KB7YX
1839 Deep Creek Drive
Sparks, NV 89434

James R. Frye, NW7O
4120 Oakhill Avenue
Las Vegas, NV 89121

William L. Dawson, W7TVF
HCR 65 Box 71623
Pahrump, NV 89041-9678

PAC

Harry K. Nishiyama, KH6FKG
1990 Hale Hooko Street
Hilo, HI 96720

Richard I. Senones, KH6JEB
95-161 Kauopae Place
Mililani Town, HI 96789

Charles S. Y. Yee, KH6WU
27 Moe Moe Place
Wahiawa, HI 96786

SCV

Stan Goldstein, N6ULU
791 Calabasas
Watsonville, CA 95076

Edward Muns, W0YK
PO Box 1877
Los Gatos, CA 95031-1877

James A. Maxwell, W6CF
PO Box 473
Redwood Estates, CA 95044

John G. Troster, W6ISQ
82 Belbrook Way
Atherton, CA 94025

William J. Stevens, W6ZM
2074 Foxworthy Avenue
San Jose, CA 95124

Gerald D. Griffin, W8MEP
123 Forest Avenue
Grove, CA 93950

Richard G. Whisler, WA6SLO
716 Hill Avenue
South San Francisco, CA 94080

Richard M. Letrich, WB6WKM
3686 Kirk Road
San Jose, CA 95124-3816

SF

Chuck Ternes, N6OJ
535 Cherry Street
Petaluma, CA 94952

Dave Stocham, W6NPY
106 Locust Avenue
Mill Valley, CA 94941

Al Lotze, W6RQ
46 Cragmont Avenue
San Francisco, CA 94116-1308

Jerry Foster, WA6BXV
16 San Ramon
Novato, CA 94947

Leonard R Geraldi K6ANP
9705 Old Redwood Hwy
Penngrove, CA 94951

SJV

Perry Foster, K6XJ
10575 E. Bullard
Clovis, CA 93612

Charles McConnell, W6DPD
1658 W. Mesa
Fresno, CA 93711-1944

Robert Craft, W6FAH
8136 Grenoble Way
Stockton, CA 95210

Charles Allessi, W6IEG
PO Box 1244
Oakhurst, CA 93644

Dennis J. DuGal, WG6P
2008 Sharilyn Drive
Modesto, CA 95355

Robert W. Smith, W6GR
13447 Road 23
Madera, CA 93637

Stanley R. Ostrom, W6XP
4921 E. Townsend
Fresno, CA 93727

Jules Wenglare, W6YO
1416 Seventh Avenue
Delano,CA 93215

Carl Boone, WB6VIN
1642 Seventh Place
Delano, CA 93215

Frederick L Moore W6KUS
2929 Jewetta Ave.
Bakerfield, CA 93312

Fred K. Stenger N6AWD
6000 Hesketh Dr.
Bakerfield, CA 93309

Kenneth H. Day W6YK
5211 Crystal Fall Ln.
Bakerfield, CA 93313

SV

Larry Murdoch, K6AAW
14370 Brian Road
Red Bluff, CA 96080

Ken Anderson, K6PU
PO Box 853
Pine Grove, CA 95665

John D. Brand, K6WC
9655 Tanglewood Circle
Orangevale, CA 95662

Phil Sanders, KF6A
8580 Krogh Ct.
Orangevale, CA 95662

Ted Davis, W6BJH
PO Box 494243
Redding, CA 96049

Jettie B. Hill, W6RFF
306 St. Charles Ct.
Roseville, CA 95661

Jerry Fuller, W7JRY
PO Box 363
Forest Ranch, CA 95942

Roanoke Division

NC

Bill Parris, AA4R
16741 100 Norman Place
Huntersville, NC 28078

Ron Oates, AA4VK
9908 Waterview Road
Raleigh, NC 27615

Bruce M. Gragg, AG4L
Route 2, Box 329
Conover, NC 28613

Frank Dowd, Jr., K4BVQ
PO Box 35430
Charlotte, NC 28235

Larry Sossoman, K4CEB
4383 Ponderosa Lane
Concord, NC 28025

Bill McDowell, K4CIA
2317 Norwood Road
Raleigh, NC 27614

Les Murphy, K4DY
PO Box 626
Hickory, NC 28603-0626

Norman Pendleton, Jr., K4NYV
3617 Country Cove Lane
Raleigh, NC 27606

Mike Jackson, N4AYO
2568 Devon Drive
Dallas, NC 28034

Joe Young, N4DAZ
1912 Miles Chapel Road
Mebane, NC 27302

Roger Burt, N4ZC
RFD 1 Box 246
Mt. Holly, NC 28120

Les Murphy, K4DY
PO Box 626
Hickory, NC 28603-0626

Robert H. McNeill, II, W4MBD
PO Box 843
Morehead City, NC 28557

Mark McIntyre, WA4FFW
2903 Maple Avenue
Burlington, NC 27215

Ewell M. Brown, W4CZU
PO Box 447
Taylorsville, NC 28681

Robert C. Cranford, WA4SSI
772 Fernwood Road
Lincolnton, NC 28092

Joe Simplins, WD4R
2400 Flintwood Lane
Charlotte, NC 28226

Gene Turner, WJ4T
159 Windsor Drive
Graham, NC 27253

Stephen D Budensiek K0SD
225 Beulah Ln
Salisbury, NC 28146

SC

Rick Porter, AA4SC
Box 2731
Rock Hill, SC 29731

Gary Dixon, K4MQG
1606 Crescent Ridge
Ford Mill, SC 29715

Jack Jackson, N4JJ
PO Box 12612
Florence, SC 29504

Charles Johnson, N4TJ
109 Kimberly Lane
Florence, SC 29505

Bill Jennings, W4UNP
630 Whitepine Drive
Catawba, SC 29704

Murphy Ratterree, W4WMQ
264 Wayland Drive
Rock Hill, SC 29732

C. W. Eldridge, WA4PLR
220 Pinewood Lane
Rock Hill, SC 29730

Ted F. Goldthorpe, WA4VCC
209 Swamp Fox Drive
Fort Mill, SC 29715

VA

M. David Gaskins, AA4AV
228 Unser Drive
Chesapeake, VA 23320

Ralph King, K1KOB
3409 Mornington Drive
Chesapeake, VA 23321

John N. Kirkham, KC4B
10920 Byrd Drive
Fairfax, VA 22030

Gordon Garrett, KC4DY
4528 North Fork Road
Elliston, VA 24087

Emmett Stine, KD4OS
1269 Parkside Place
Virginia Beach, VA 23454

Fran Sledge, N4CRU
3004 Oakley Hall Road
Portsmouth, VA 23702

Clay Partin, N4VG
6642 Shingle Ridge Road, SW
Roanoke, VA 24018

Karl H. Oyster, Jr., NQ1W
524 Vicksdell Cres.
Chesapeake, VA 23320-3551

Nap Perry, W4DHZ
9317 Marlow Avenue
Norfolk, VA 23503

Art Westmont, W4EEU
4717 Little John Road
Virginia Beach, VA 23455

Grover Brewer, W4FPW
1359 Eagle Avenue
Norfolk, VA 23518

Howell G. Gwaltney, W4IF
209 Pennington
Portsworkt, VA 23701

Pradyumna Rana, WB4NFO
29 East Chapman Street
Alexandria, VA 22301

Richard T. Williams, WE6H
7902 Viola Street
Springfield, VA 22152

Vance B. Williams KN4JD
824 Moffat Ln.
Virginia Beach, VA 23464

Harold E. Sager KO4OG
308 Plaza Trail Ct.
Virginia Beach,VA 23452

James L. Young K4JDJ
556 Babbtown Rd.
Suffolk, VA 23434

Donald F. Lynch Jr. W4ZYT
1517 W. Little Neck Rd.
Virginia Beach, VA 23452

David R. Klimaj W4JVN
5637 Heming Ave.
Springfield, VA 22151

Rocky Mountain Division

CO

Steve P. Gecewicz, K0CS
PO Box 1048
Elizabeth, CO 80107

John D. Peters, K1ER
2439 Sherman Street
Longmont, CO 80501

Karen Schultz, KA0CDN
15643 East 35th Place
Aurora, CO 80011

Glenn Schultz, W0IJR
15643 East 35th Place
Aurora, CO 80011

Jim Spaulding, W0UO
6277 S. Niagara Court
Englewood, CO 80111

NM

Robert L. Norton, N5EPA
116 Pinon Heights Road
Sandia Park, NM 87047

Arne Gjerning, N7KA
PO Box 1485
Corrales, NM 87048

Fred F. Seifert, W5FS
3106 Florida Street NE
Albuquerque, NM 87110-2617

Eugene F. Carter, W5OLN
320 Camino Tres SW
Albuquerque, NM 87105

Paul Rubinfield, WF5T
PO Box 4909
Sante Fe, NM 87502

UT

Steven C. Salmon, K7OXB
2184 East 6070 South
Salt Lake City, UT 84121

WY

Wayne Mills, N7NG
PO Box 1945
Jackson, WY 83001

Southeastern Division

AL

Bill Christian, K4IKR
2800 Cave Avenue
Huntsville, AL 35810

Tim Pearson, KU4J
6214 Rime Village Drive #205
Huntsville, AL 35806

GA

Frank G. Deak, AF4Y
2566 Glen Circle
Lawrenceville, GA 30244

Randy Tudor, K4ODL
1032 Rockcrest Drive
Marietta, GA 30062

Samuel I. Silverman, KB4NJ
3960 Green Forest Parkway
Smyrna, GA 30082

Neil R. Foster, KC4MJ
3185 Friar Tuck Way
Atlanta, GA 30340

John Smith, KI4XO
450 Chaffin Road
Roswell, GA 30075

Roy E. Conley, N4ONI
1429 Dallas Circle
Marietta, GA 30064-2913

Verne Fowler, W8BLA
113215 Stroup Road
Roswell, GA 30075-2225

Robert Varone, WA4ETN
5177 Holly Springs Dr
Douglasville, GA 30135

R.L. "Gad" Williamson, WA4CUG
5877 Brookside Drive
Mapleton, GA 30059

NFL

William H. Bosley, K3NN
201 Highland Street
Valparaiso, FL 32580-1222

William R. Hicks, K4UTE
7002 Deauville Road
Jacksonville, FL 33205

Jim Buth, K9CJK
Route 4, Box 7216-1
Crawford, FL 32327

H. J. Huddleston, KF4BU
925 forest Avenue
Fort Walton Beach, FL 32547

Leslie E. Smallwood, KW4V
24 Apple Hill Hollow
Casselberry, FL 32707

Ronald E. Blake, N4KE
258 Wesley Road
Green Cove Springs, FL 32043

Richard A. Knox, WR4K
5172 Pine Avenue
Orange Park, FL 32072

Gary L Letchford K0LUZ
214 Tollgate Trail
Longwood, FL 32750

PR

Telesforo Figueroa, KP4P
PO Box 1651
Yabucoa, PR 00767-1651

Eduardo Negron, KP4EQF
Urb. Valle Real AG-78
Ponce, PR 00731

SFL

Harry D. Belock, AA2X
8108 N. W. 102 Terr.
Tamarac, FL 33321

Homer J. Cumm, AB4NS
3599 Peoria Rd
Orange Park, FL 32065

Leon A. Katz, K2EWB
4136 Lakespur Circle North
Palm Beach Gardens, FL 33410

Edwin L. Chinnock, W2FZY
5641 Bayview Drive
Fort Lauderdale, FL 33308

David Beckwith, W2QM
3115 NW 13th Street
Delray Beach, FL 33445

Stephen Sarasohn, WU2W
3288 NW 60 Street
Boca Raton, FL 33496

Michael M. Raskin, K4KUZ
561 West Tropical Way
Plantation, FL 33317

Gary Fowks, K4MF
635 W. 64 Drive
Hialeah, FL 33012

Robert R. Wilson, KI4LP
2323 Treetop Ct
Melbourne, FL 32934

Ernie Greeson, K4RD
1500 E. Terra Mar Drive
Pompano Beach, FL 33062

David Novoa, KP4AM
1994 SW 142nd Avenue
Miami, FL 33175

Peter Rimmel, K8UNP
4209 Madison Street
Hollywood, FL 33021

Marv Westerdahl, KC2KU
1934 Hawaii Avenue N.E.
St. Petersburg, FL 33703

Lefty Boggess, KE4VU
11636 Grove Street North
Seminole, FL 34642

Ray Riker, NY2E
360 Ponte Vedra Road
Palm Springs, FL 33461

Donald B. Search, W3AZD
10550 State Road 84
#147 Park City West
Ft. Lauderdale, FL 33324

Gene Sykes, W4OO
6510 Carambola Circle
West Palm Beach, FL 33406

Mark Horowitz, WA2YMX
6831 SW 16th Street
Plantation, FL 33317

Joseph L. Picior, WB4OSN
61 Eden Lane
Lake Placid, FL 33852

Harry T. Henley, WK9Z
2013 Massachusetts Avenue NE
St. Petersburg, FL 33703

Edward Grogan, WW1N
902 Bay Point Drive
Madeira Beach, FL 33708

Bruce Paul Phegley W4OVU
3940 N.W. 4 Ct.
Coconut Creek, FL 33066

Ralph E. Small K1ZLA
969 Waialae Cir. Ne.
Palm Bay, FL 32905

Leslie E. Smallwood KW4V
24 Apple Hill Hollow
Casselberry, FL 32707

John A. Mann Sr. N4OLE
4400 Sw 3rd St.
Miami, FL 33134

Stephen C Taber W4ITD
2665 NE 26th Ave
Lighthouse Pt, FL 33064

Douglas A McDuff WB4MAI
10380 SW 112th St
Miami, FL 33176

Southwestern Division

AZ

Ned Stearns, AA7A
7038 E. Aster Drive
Scattsdale, AZ 85254

Mike Fulcher, KC7V
6545 E. Montgomery Road
Cave Creek, AZ 85331

Lee Finkel, KY7M
6928 E. Ludlow
Scottsdale, AZ 85254

Hardy Landskov, N7RT
15814 N. 44th Street
Phoenix, AZ 85032

Jim McDonald, N7US
7233 N. 16th Ave
Phoenix, AZ 85021

Steve Towne, NN7X
12002 N. Oakhurst Way
Scottsdale, AZ 85254

Frank Schottke, W2UE
5049 E. Laurel Lane
Scottsdale, AZ 85254

Stuart Greene, WA2MOE
7537 N. 28th Avenue
Phoenix, AZ 85051

William G. Snyder KF8N
8930 N. Camino De La Tierra
Tuscon, AZ 85742

Bernie J. Sasek W0YOY
8925 N. Morning View Dr.
Tuscon, AZ 85737

William J. Stange W9DDP
P.O. Box 927
Patagonia, AZ 85624

Ronald G. Swann WA7BOD
5656 W. Utah St.
Tuscon, AZ 85746

Douglas C. Schaber KG7IZ
2611 N. Camino De Oeste
Tuscon, AZ 85745

LAX

John Alexander, K6SVL
28403 Covecrest Drive
Palos Verdes, CA 90274

William A. Angenent, KN6DV
43150 6th St East
Lancaster, CA 93535

Leonard Svidor, W6AUG
17760 Alonzo Place
Encino, CA 91316

Sheldon C. Shallon, W6EL
11058 Queensland Street
Los Angeles, CA 90034

ORG

Rick Samoian, WB6OKK
5302 Cedarlawn Drive
Placentia, CA 92670

SB

Milt Bramer, N6MB
4161 Shadyglade Drive
Santa Maria, CA 93455

John O. Norback, W6KFV
133 Pino Solo Court
Nipomo, CA 93444

Neil B. Taylor, W6MUS
PO Box 392
Atascadero, CA 93423

Jim Robb, W6OUL
501 N. Poppy
Lompoc, CA 93436

Steve Nolan, WA6ARG
904 N. Western Avenue
Santa Maria, CA 93454

SDG

Bruce Clark, K6JYO
508 Washingtonia Drive
San Marcos, CA 92069

Cliff Smith, K6TWU
10565 Rancho Road
La Mesa, CA 91941

Art Charette, K6XT
16604 Adrienne Way
Ramona, CA 92065

Rick Craig, N6ND
PO Box 741
Ramona, CA 92065

Ed Andress, W6KUT
12821 Corte Dorotea
Poway, CA 92064

George Pugsley, W6ZZ
1362 Via Rancho Prkway
Escondido, CA 92029

Pat Bunsold, WA6MHZ
14291 Rios Canyon Road #33
El Cajon, CA 92021

Charles L. Roy, WS6F
6231 Lake Shore Drive
San Diego, CA 92110

West Gulf Division

NTX

Phil Clements, K5PC
1313 Applegate Lane
Lewisville, TX 75069

Mike Krzystyniak, K9MK
249 Bayne Road
Haslet, TX 76052

Joel Rubenstein, KA5W
3601 Larkin Lane
Rowlett, TX 75088

Vernon Brunson, KD5HO
2717 Whispering Trail
Arlington, TX 76013

Mickey Heimlich, N5AJW
9207 Mill Hollow
Dallas, TX 75243

Orin Helvey, N5ORT
969 Tupelo
Coppell, TX 75019

John Clifford, NK5K
125 Alta Mesa Drive
Fort Worth, TX 76108

David Jaksa, W0VX
626 Torrey Pines Lane
Garland, TX 75044

Thomas Little, W5GVP
2109 Miriam Lane
Arlington, TX 76010

Ken Knudson, W5PLN
5725 Wales Avenue
Fort Worth, TX 76133

Richard Pruitt, WB5TED
2232 Jensen
Fort Worth, TX 76112

OK

George W. Adkins, AD1S
PO Box 88
Wellston, OK 74881

Ed Murta, K5LIL
4408 NW 47th Street
Oklahoma City, OK 73112

Ross Hunt, K5RH
17910 E. 15 Street
Tulsa, OK 74108-5130

Jeff Martin, K5WE
5337 S. 71 E. Avenue
Tulsa, OK 74145

Elton Abbott, KE5JE
216 N. Pecan Street
Pauls Valley, OK 73075

Kenneth L. Adams, KM5H
5201 Philson Farm Road
Bartlesville, OK 74006

Mark Byard, N5OGP
504 Foster Avenue
Ponca City, OK 74601

Coy C. Day, N5OK
Rt 1, Skyline View Estates
Claremore, OK 74017-9801

Bob See, N5PC
1708 E. Woodland Road
Ponca City, OK 74604

Charles E. Calhoun, W0RRY
13914 N. 90 E. Avenue
Collinsville, OK 74021

W. Clegg Hahn, W5CKT
2816 SE Hampden Road
Bartlesville, OK 74006

Gil Wood, W5NUT
36806 Old Hwy 270
Shawnee, OK 74801-8701

Larry Watson, W5VHP
4212 Beacon Court
Bartlesville, OK 74006

Gary N. Gompf, W7FG
3300 Wayside Drive
Bartlesville, OK 74006

Jim Hood, WV5S
11623 Smoking Oaks Drive
Oklahoma City, OK 73150

STX

Robert C. Walworth, AK5B
3210 Chaparral Way
Spring, TX 77380

Don Hall, K5AQ
4870 Enchanted Oaks Drive
College Station, TX 77845

William E. Barnes, K5GE
434 Windcrest Drive
San Antonio, TX 78239

Charles R. Crutchfield, KC5SC
3031 Rose Lane
La Marque, TX 77568

Linda G. Walworth, KE5TF
3210 Chaparral Way
Spring, TX 77380

Evonne M. Lane, KF5MY
1602 Chestnut Ridge
Kingwood, TX 77339

G.B. "Jim" Lane II, N5DC
1602 Chestnut Ridge
Kingwood, TX 77339

Edward L. Linde, N5DDT
3900 Sorrell Cove
Austin, TX 78730

Joseph A. Castorina, N5FJY
4437 F. M. 646 N
Sante Fe, TX 77510

Harley L. Dillon, N5HB
4902 Lambeth
San Antonio, TX 78228

John P. Warren, NT5C
3517 Gattis School Road
Round Rock, TX 78664

Robert A. Wood, WB5CRG
1013 Lewis Drive
Kemah, TX 77565

Wayne Wyatt, WB5QBV
7114 Quail Landing
San Antonio, TX 78250

Randy Light, WC5Q
2100 Bent Oak Street
College Station, TX 77840

WTX

David Zulawski, KA5TQF
2808 Catnip Street
El Paso, TX 79925

Joe T. Milam, NI5F
432 W. Dengar
Midland, TX 79707

Leslie A. Bannon, WF5E
3400 Bedford
Midland, TX 79703

DXCC AWARD APPLICATION

DIRECTIONS: PLEASE PRINT. Use this form for both new and endorsement applications. Complete both sides of this form. (The reverse side is optional if you are sending QSL cards to ARRL HQ.) **Do NOT include other correspondence with this application.**

Please complete all sections of this side.

● I am applying for the following DXCC awards (check ALL as appropriate)

Award Type	New app	Endorsement	*Required Last submission date	*Required Cert No.
Mixed				
Phone				
CW				
RTTY				
SAT.				
160				
80				
40				
10				
6				
2				
* 5BDXCC		160-17-12-6-2	◄— Circle Band(s)	

* 5BDXCC Award Number _____

Needed for (Endorsements)

● FEES: Effective 1 January, 1994:All Amateurs applying for their first DXCC Award - $10 (includes certificate and DXCC pin) Each additional DXCC certificate (new or replacement) - $5 shipping and handling fee (includes certificate and DXCC pin)

    Endorsements and new application presented at ARRL HQ or conventions. (These applications will be limited to 120 QSOs) - $2 handling fee.

    ARRL members may make one free submission of up to 120 QSLs each year, subject to charges above, and return postage. These may be a combination of new awards and endorsements.

    Additional member submissions within a calendar year of up to 100 QSLs will be $10, subject to charges above.

    A $0.10 fee applies for each QSL over the above limits.

    Foreign non-ARRL members will be charged $10 for the first submission in a calendar year, subject to charges above.

    Foreign non-ARRL members will be charged $20 for additional submissions in a calendar year, subject to charges above.

    Return postage or SASE must be provided by the applicant for any cards or requests for information.

● ALL Applicants in the US and possessions, Puerto Rico, and foreign members MUST complete the following statement:

"I currently hold ARRL membership expiring _____." OR attach label from *QST* wrapper;

(month, year)

● ALL applicants must complete the following statement:
"I affirm that I have observed all DXCC rules (see attached) as well as all pertinent governmental regulations established for Amateur Radio in my country. I agree to be bound by the decisions of the ARRL Awards Committee. (Decisions of the ARRL Awards Committee shall be final.)"

Signature _____ Call _____ Date _____

● Send application forms, QSL cards, fees (if applicable) and return postage to : DXCC Desk, ARRL HQ, 225 Main St, Newington, CT 06111, USA. **To confirm receipt include an SASE or post card with your application.**

● For any questions or clarifications, please write to the DXCC Desk separately at the above address. The DXCC Desk can also be contacted as follows:

Telephone 860-594-0200
Fax 860-594-0259 (24-hour direct line to ARRL HQ)

Telex 640215-5052 MCI
Internet dxcc@arrl.org

● Thank you for your cooperation, and **Good DX!**

● Call Sign _____

   Ex Calls _____

● Name _____

        Last (*Spanish* Apellido),   First

   Address _____

        (Number and street)

   _____

        (City, State/Country, ZIP)

   ☐ Check here if new address

● Fill this line out ONLY if you will receive a certificate(s) from this submission. Otherwise, leave blank.

   Name _____

        (Print exactly as you want it to appear on certificate)

● Number of QSL cards enclosed _____

● Number of (QSOs) on QSL cards _____

   (Confirmations)

● Postage = $ _____

   Awards Fee = $ _____   TOTAL = $ _____

   ☐ US Currency _____   ☐ Stamps

   ☐ Check or money order_____   ☐ IRCs

● Return my QSL cards via:

   ☐ Registered mail (**Recommended**)   ☐ First class mail

   ☐ Certified (US only)   ☐ Airmail

   ☐ United Parcel Service (US only)   ☐ Other _____

Note: IRCs valued at $0.60 US. Check with your local Post Office for current mailing rates.

ARRL DXCC FIELD REPRESENTATIVE VERIFICATION

I hereby verify that I have personally inspected the confirmations and verify that this application is correct and true. (DXCC Field Reps must forward paper work to HQ.)

Signature _____ Call_____ Date _____

Signature _____ Call_____ Date _____

MSD-505(396)
Printed in USA

_____
          Your Call

DIRECTIONS: (1) Sort cards and list below first by band, e.g. all the 80-meter cards together, then the 40-meter cards, etc. (2) Within each band, sort and list below by mode, e.g. all the 80-meter phone cards, then the 80-meter CW, 40-meter phone, etc. (3) Make one entry below for each QSO credit. (4) Cards indicating multiple contacts should be placed at the end, listing each contact on a separate line below. (5) QSO Date = Day, Month, Year. (6) Bands =160, 80, 40, 30, 20, 17, 15, 12, 10, 6, 2. (7) Modes = PHONE, CW, RTTY, SAT.

	CALL	QSO DATE (DD/MM/YY)	BAND	MODE	COUNTRY
1		/ /			
2		/ /			
3		/ /			
4		/ /			
5		/ /			
6		/ /			
7		/ /			
8		/ /			
9		/ /			
10		/ /			
11		/ /			
12		/ /			
13		/ /			
14		/ /			
15		/ /			
16		/ /			
17		/ /			
18		/ /			
19		/ /			
20		/ /			
21		/ /			
22		/ /			
23		/ /			
24		/ /			
25		/ /			
26		/ /			
27		/ /			
28		/ /			
29		/ /			
30		/ /			

This side of form may be photocopied if more pages are needed.

Russian Oblasts

Updated January 1996

Call sign number and first letter of suffix	Russian	English	Designator
1A, 1B	Sankt-Peterburg[1]	St. Petersburg	SP
Also 1D, 1F, 1G, 1H, 1I, 1J, 1L, 1M			
1C,	Leningradskaya obl.[1]	Leningradskaya Oblast	LO
1N	Karel'skaya ASSR	Karel'skaya Autonomous Soviet Socialist Republic	KL
1O	Arkhangel'skaya obl.	Arkhangel'skaya Oblast	AR
1P	Nenetskiy AO	Nenetskiy Autonomous District (Okrug)	NO
1Q,1R			
1S	Vologodskaya obl.	Vologodskaya Oblast	VO
1T,1U	Novgorodskaya obl.	Novgorodskaya Oblast	NV
1W, 1X	Pskovskaya obl.	Pskovskaya Oblast	PS
1Y, 1Z	Murmanskaya obl.	Murmanskaya Oblast	MU
2F	Kaliningradskaya obl.[2]	Kaliningradskaya Oblast	KA
3A, 3B	g. Moskva[3]	Moscow City	MA
Also 3C, 3F, 3H			
3D	Moskovskaya obl.[3]	Moskovskaya Oblast	MO
3E	Orlovskaya obl.	Orlovskaya Oblast	OR
3G	Lipetskaya obl.	Lipetskaya Oblast	LP
3I, 3J	Tverskaya obl.	Tverskaya Oblast	TV
3L	Smolenskaya obl.	Smolenskaya Oblast	SM
3M	Yaroslavskaya obl.	Yaroslavskaya Oblast	JA
3N, 3O	Kostromskaya obl.	Kostromskaya Oblast	KS
3P	Tul'skaya obl.	Tul'skaya Oblast	TL
3Q	Voronezhskaya obl.	Voronezhskaya Oblast	VH
3R	Tambovskaya obl.	Tambovskaya Oblast	TB
3S	Ryazanskaya obl.	Ryazanskaya Oblast	RA
3T	Nizhegorodskaya obl.	Nizhegorodskaya Oblast	NN
3U	Ivanovskay aobl.	Ivanovskaya Oblast	IV
3V	Vladimirskaya obl.	Vladimirskaya Oblast	VL
3W	Kurskaya obl.	Kurskaya Oblast	KU
3X	Kaluzhskaya obl.	Kaluzhskaya Oblast	KG
3Y	Bryanskaya obl.	Bryanskaya Oblast	BR
3Z	Belgorodskaya obl.	Belgorodskaya Oblast	BO
4A, 4B	Volgogradskaya obl.	Volgogradskaya Oblast	VG
4C, 4D	Saratovskaya obl.	Saratovskaya Oblast	SA
4F	Penzenskaya obl.	Penzenskaya Oblast	PE
4H. 4I	Samarskaya obl.	Samarskaya Oblast	SR
4L, 4M	Ul'yanovskaya obl.	Ul'yanovskaya Oblast	UL
4N, 4O	Kirovskaya obl.	Kirovskaya Oblast	KI
4P, 4Q	Respublika Tatarstan	Republic of Tatarstan	TA
Also 4R			
4S, 4T	Respublika Mariy-El	Republic of Mariy-El	MR
4U	Mordovskaya SSR	Mordovskaya Soviet Socialist Republic	MD
4W	Respublika Udmurtiya	Republic of Udmurtiya	UD
4Y, 4Z	Respublika Chuvashiya	Republic of Chuvashiya	CU
6A, 6B	Krasnodarskiy kray	Krasnodarskiy Kray	KR
Also 6C, 6D			
6E	Karachayevo-Cherkasskaya	Karachayevo-Cherkasskaya Respublika Republic	KC
6H, 6F	Stavropol'skiy kray	Stavropol'skiy Kray	ST
Also 6H			
6I	Respublika Kalmykiya	Republic of Kalmykiya	KM
6J	Severo-Osetinskaya SSR	North-Osetinskaya Soviet Socialist Republic	SO
6L, 6M	Rostovskaya obl.	Rostovskaya Oblast	RO
Also 6N, 6O			
6P	Chechenskaya Respublika	Chechenskaya Republic	CN
Also 6R			
6U, 6V	Astrakhanskaya obl.	Astrakhanskaya Oblast	AO
6W	Respublika Dagestan	Republic of Dagestan	DA
6X	Respublik Kabardino-Balkariya	Republic of Kabardino-Balkariya	KB
6Y	Respublika Adygeya	Republic of Adygeya	AD
8T	Ust'-Ordynskiy Buryatskiy AO	Ust'-Ordynskiy Buryatskiy Autonomous District (Okrug)	UO
8V	Aginskiy Buryatskiy AO	Aginskiy Buryatskiy Autonomous District (Okrug)	AB
9A, 9B	Chelyabinskaya obl.	Chelyanbinskaya Oblast	CB
9C, 9D	Sverdlovskaya obl.	Sverdlovskaya Oblast	SV
Also 9E			
9F	Permskaya obl.	Permskaya Oblast	PM
9G	Komi-Permyatskiy AO	Komi-Permyatskiy Autonomous District (Okrug)	KP
9H, 9I	Tomskaya obl.	Tomskaya Oblast	TO
9J	Khanty-Mansiyskiy AO	Khanty-Mansiyskiy Autonomous District (Okrug	HM
9K	Yamalo-Nenetskiy AO	Yamalo-Nenetskiy Autonomous District (Okrug)	JN
9L	Tyumenskaya obl.	Tyumenskaya Oblast	TN
9M, 9N	Omskaya obl.	Omskaya Oblast	OM
9O, 9P	Novosibirskaya obl.	Novosibirskaya Oblast	NS
9Q, 9R	Kurganskaya obl.	Kurganskaya Oblast	KN
9S, 9T	Orenburgskaya obl.	Orenburgskaya Oblast	OB
9U, 9V	Kemerovskaya obl.	Kemerovskaya Oblast	KE
9W	Respublika Bashkortostan	Republic of Bashkortostan	BA
9X	Respublika Komi	Republic of Komi	KO
9Y	Altayskiy kray	Altayskiy Kray	AL
9Z	Gorno-Altayskiy AO	Gorno-Altayskiy Autonomous District (Okrug)	GA
0A	Krasnoyarskiy kray	Krasnoyarskiy Kray	KK
0B	Taymyrskiy AO	Taymyrskiy Autonomous District (Okrug)	TM
0C	Khabarovskiy kray	Khabarovskiy Kray	HK
0D	Yevreyskaya AO	Jewish Autonomous Oblast[4]	EA
0E, 0F	Sakhalinskaya obl.	Sakhalinskaya Oblast	SL
Also 0G			
0H	Evenkiyskiy AO	Evenkiyskiy Autonomous District (Okrug)	EW
0I	Magadanskaya obl	Magadanskaya Oblast	MG
0J	Amurskaya obl.	Amurskaya Oblast	AM
0K	Chukotskiy AO	Chukotskiy Autonomous District (Okrug)	CK
0L, 0M	Primorskiy kray	Primorskiy Kray	PK
Also 0N			
0O, 0P	Respublika Buryatiya	Republic of Buryatiya	BU
0Q, 0R	Respublika Sakha	Republic of Sakha	YA
0S, 0T	Irkutskaya obl.	Irkutskaya Oblast	IR
0U, 0V	Chitinskaya obl.	Chitinskaya Oblast	CT
0W	Respublika Khakassiya	Republic of Khakassiya	HA
0X	Koryakskiy AO	Koryakskiy Autonomous District (Okrug)	KJ
0Y	Respublika Tuva	Republic of Tuva[5]	TU
0Z	Kamchatskaya obl.	Kamchatskaya Oblast	KT

[1]St. Petersburg is within, and separate from, Leningradskaya Oblast.

[2]Kaliningradskaya Oblast covers the northern part of the former East Prussia. Annexed by the USSR at the end of World War II and made part of the RSFSR, the area is separated from the rest of Russia by Lithuanian and Belarussian territory. The administrative center of the Oblast is Kaliningrad, formerly Konigsberg.

[3]Moscow city is within, and separate from, Moskovskaya Oblast.

[4]According to a U.S. reference dating back to 1961, the Jewish Autonomous Oblast (Yevreyskaya avtonomnaya oblast or YeAO) was established in 1934 in the Soviet Far East but attracted few Jews; its population was 165,000, including no more than 30,000 Jews.

[5]This is the former Tannu Tuva, which was part of the Chinese Empire until 1911, became a Russian Protectorate in 1914, and proclaimed its independence in 1921. It was annexed by the USSR in 1944 and given the status of an autonomous oblast within the RSFSR. In 1961, it was elevated to the status of an autonomous SSR, still within the RSFSR.

Morse Code for Other Languages

Code	Japanese	Korean	Arabic	Hebrew	Russian	Greek
•	へ he	ㅏ a	ت ta	' vav	Е,Э E	E epsilon
—	ム mu	ㅓ ŏ	ث ta	תת tav	Т T	Т tau
••	˙ nigori	ㅑ ya	ش ya	' yod	И I	I iota
•—	イ i	ㅗ o	ا alif	א aleph	А A	A alpha
—•	タ ta	ㅛ yo	ن noon	נ nun	Н N	N nu
——	ヨ yo	ㅁ m	م meem	מ mem	М M	M mu
•••	ラ ra	ㅕ yŏ	س seen	שׁ shin	С S	Σ sigma
••—	ウ u	ㅌ t	ط ta	ט tet	У U	ΟΥ omicron ypsilon
•—•	ナ na	ㅠ yu	ر ra	ר reish	Р R	P rho
•——	ヤ ya	ㅂ p(b)	ز waw	צ tzadi	В V	Ω omega
—••	ホ ho		د dal	ד dalet	Д D	Δ delta
—•—	ワ wa	ㅇ -ng	ك kaf	ךכ chaf	Н K	K kappa
—··•	リ ri	ㅅ s	غ ghain	ג gimmel	Г G	Γ gamma
—··—	レ re	ㅍ p'	خ kha	ה heh	О O	O omicron
••••	ヌ nu	ㅜ u	ح ha	ח chet	Х H	H eta
•••—	ク ku	ㄹ r-(-l)	ض dad		Ж J	ΗΥ eta ypsilon
••—•	チ ti	ㄴ n	ف fa		Ф F	Φ phi
••——	ノ no				Ю yu	ΑΥ alpha ypsilon
•—••	カ ka	ㄱ k(g)	ل lam	ל lamed	Л L	Λ lambda
•—•—	ロ ro	ㄴ ain			Я ya	ΑΙ alpha iota
•——•	ツ tu	ㅈ ch(j)		ﬦ peh	П P	Π pi
•———	ヲ wo	ㅎ h	ج jeem	ע ayen	Й Y	ΤΙ ypsilon iota
—•••	ハ ha	ㄷ t(d)	ب ba	בב bet	Б B	B beta
—••—	マ ma	ㅋ k'	ص sad		Ь,Ъ mute	Ξ xi
—•—•	ニ ni	ㅊ ch'	ث tha	ס samech	Ц TS	Θ theta
—•——	ケ ke	ㅖ e	ظ za		Ы I	Υ ypsilon
——••	フ hu	ㅌ t'	ذ dhal	‎ zain	З Z	Z zeta
——•—	ネ ne	ㅒ ae	ق qaf	ק kof	Щ SHCH	Ψ psi
———•	ソ so		ز zay		Ч CH	ΕΥ epsilon ypsilon
————	コ ko		ش sheen		Ш SH	X khi
•—•—•—	ト to		ه he			
••——	ミ mi					
●	han-nigori					
•—••	オ o					
•——•	ヰ (w)i					
———•—	ン n					
•—•——	テ te					
•——••	ヱ (w)e					
•—•—•—	- hyphen					
•———•	セ se					
—•••—	メ me					
—••—•	モ mo					
—•———	ユ yu					
—•—•—	キ ki					
—•—••	サ sa					
—•—•—	ル ru					
—•——•	エ e					
——•—•	ヒ hi					
——•——	シ si					
———•—	ア a					
————•	ス su					
			ﻻ lam-alif			

Spanish Phonetics

America	ah-MAIR-ika
Brasil	brah-SIL
Canada	cana-DAH
Dinamarca	dina-MAR-ka
Espana	es-PAHN-yah
Francia	FRAHN-seeah
Grenada	gre-NAH-dah
Holanda	oh-LONN-dah
Italia	i-TAL-eeah
Japon	hop-OWN
Kilowatio	kilo-WAT-eeoh
Lima	LIMA
Mejico	MEH-heeco
Norvega	nor-WAY-gah
Ontario	on-TAR-eeoh
Portugal	portu-GAL
Quito	KEY-toe
Roma	ROW-mah
Santiago	santee-AH-go
Toronto	tor-ON-toe
Uniforme	oonee-FORM-eh
Victoria	vic-TOR-eeah
Washington, Wisky	washingtone, wisky
Xilofono	see-LOW-phono
Yucatan	yuca-TAN
Zelandia	see-LAND-eeah
W	DOE-bleh-vay
Ø	cero	SEH-roe
1	uno	OO-no
2	dos	DOS
3	tres	TRAYCE
4	cuatro	KWAT-roe
5	cinco	SINK-oh
6	seis	SAYCE
7	siete	see-AY-teh
8	ocho	OCH-oh
9	nueve	new-AY-veh

—John Mason Jr., EA4AXW

DX Operating Code

For W/VE Amateurs

Some amateur DXers have caused considerable confusion and interference in their efforts to work DX stations. The points below, if observed by all W/VE amateurs, will help make DX more enjoyable for all.

1) *Call* DX only after he calls CQ, QRZ? or signs S̄K̄, or voice equivalents thereof. Make your calls short.

2) Do not call a DX station:

  a) On the frequency of the station he is calling until you are sure the QSO is over (S̄K̄).

  b) Because you hear someone else calling him.

  c) When he signs K̄N̄, ĀR̄ or C̄L̄.

  d) Exactly on his frequency.

  e) After he calls a directional CQ, unless of course you are in the right direction or area.

3) Keep within frequency band limits. Some DX stations can get away with working outside, but you cannot.

4) Observe calling instructions given by DX stations. Example: 15U means "call 15 kHz *up* from my frequency." 15D means *down*, etc.

5) Give honest reports. Many DX stations *depend* on W/VE reports for adjustment of station and equipment.

6) Keep your signal clean. Key clicks, ripple, feedback or splatter gives you a bad reputation and may get you a citation from FCC.

7) *Listen* and call the station you want. Calling CQ DX is not the best assurance that the rare DX will reply.

8) When there are several W or VE stations waiting, avoid asking DX to "listen for a friend." Also avoid engaging him in a ragchew against his wishes.

For Overseas Amateurs

To all overseas amateur stations:

In their eagerness to work you, many W and VE amateurs resort to practices that cause confusion and QRM. Most of this is good-intentional but ill-advised; some of it is intentional and selfish. The key to the cessation of unethical DX operating practices is in your hands. We believe that your adoption of certain operating habits will increase your enjoyment of Amateur Radio and that of amateurs on this side who are eager to work you. We recommend your adoption of the following principles:

1) Do not answer calls on your own frequency.

2) Answer calls from W/VE stations only when their signals are of good quality.

3) Refuse to answer calls from other stations when you are already in contact with someone, and do not acknowledge calls from amateurs who indicate they wish to be "next."

4) Give *everybody* a break. When many W/VE amateurs are patiently and quietly waiting to work you, avoid complying with requests to "listen for a friend."

5) Tell listeners where to call you by indicating how many kilohertz up (U) or down (D) from your frequency you are listening.

6) Use the ARRL-recommended ending signals, especially K̄N̄ to indicate to impatient listeners the status of the QSO. K̄N̄ means "Go ahead (specific station); all others keep out."

7) Let it be known that you avoid working amateurs who are constant violators of these principles.

The Origin of "73"

The traditional expression "73" goes right back to the beginning of the landline telegraph days. It is found in some of the earliest editions of the numerical codes, each with a different definition, but each with the same idea in mind—it indicated that the end, or signature, was coming up. But there are no data to prove that any of these were used.

The first authentic use of 73 is in the publication *The National Telegraphic Review and Operators' Guide*, first published in April 1857. At that time, 73 meant "My love to you"! Succeeding issues of this publication continued to use this definition of the term. Curiously enough, some of the other numerals used then had the same definition as they have now, but within a short time, the use of 73 began to change.

In the National Telegraph Convention, the numeral was changed from the Valentine-type sentiment to a vague sign of fraternalism. Here, 73 was a greeting, a friendly "word" between operators and it was so used on all wires.

In 1859, the Western Union Company set up the standard "92 Code." A list of numerals from one to 92 was compiled to indicate a series of prepared phrases for use by the operators on the wires. Here, in the 92 Code, 73 changes from a fraternal sign to a very flowery "accept my compliments," which was in keeping with the florid language of that era.

Over the years from 1859 to 1900, the many manuals of telegraphy show variations of this meaning. Dodge's *The Telegraph Instructor* shows it merely as "compliments." The *Twentieth Century Manual of Railway and Commercial Telegraphy* defines it two ways, one listing as "my compliments to you"; but in the glossary of abbreviations it is merely "compliments." Theodore A. Edison's *Telegraphy Self-Taught* shows a return to "accept my compliments." By 1908, however, a later edition of the Dodge Manual gives us today's definition of "best regards" with a backward look at the older meaning in another part of the work where it also lists it as "compliments."

"Best regards" has remained ever since as the "put-it-down-in-black-and-white" meaning of 73 but it has acquired overtones of much warmer meaning. Today, amateurs use it more in the manner that James Reid had intended that it be used—a "friendly word between operators."—*Louise Ramsey Moreau, W3WRE*

The RST System

READABILITY
1—Unreadable.
2—Barely readable, occasional words distinguishable.
3—Readable with considerable difficulty.
4—Readable with practically no difficulty.
5—Perfectly readable.

SIGNAL STRENGTH
1—Faint signals barely perceptible.
2—Very weak signals.
3—Weak signals.
4—Fair signals.
5—Fairly good signals.
6—Good signals.
7—Moderately strong signals.
8—Strong signals.
9—Extremely strong signals.

TONE
1—Sixty-cycle ac or less, very rough and broad.
2—Very rough ac, very harsh and broad.
3—Rough ac tone, rectified but not filtered.
4—Rough note, some trace of filtering.
5—Filtered rectified ac but strongly ripple-modulated.
6—Filtered tone, definite trace of ripple modulation.
7—Near pure tone, trace of ripple modulation.
8—Near perfect tone, slight trace of modulation.
9—Perfect tone, no trace of ripple or modulation of any kind.

The "tone" report refers only to the purity of the signal, and has no connection with its stability or freedom from clicks or chirps. If the signal has the characteristic steadiness of crystal control, add X to the report (e.g., RST 469X). If it has a chirp or "tail" (either on "make" or "break") add C (e.g., 469C). If it has clicks or noticeable other keying transients, add K (e.g., 469K). Of course a signal could have both chirps and clicks, in which case both C and K could be used (e.g., RST 469CK).

Q Signals

Given below are a number of Q signals whose meanings most often need to be expressed with brevity and clarity in amateur work. (Q abbreviations take the form of questions only when each is sent followed by a question mark.)

QRG Will you tell me my exact frequency (or that of ____)? Your exact frequency (or that of ____) is ____ kHz.
QRH Does my frequency vary? Your frequency varies.
QRI How is the tone of my transmission? The tone of your transmission is ____ (1. Good; 2. Variable; 3. Bad).
QRJ Are you receiving me badly? I cannot receive you. Your signals are too weak.
QRK What is the intelligibility of my signals (or those of ____)? The intelligibility of your signals (or those of ____) is ____ (1. Bad; 2. Poor; 3. Fair; 4. Good; 5. Excellent).
QRL Are you busy? I am busy (or I am busy with ____). Please do not interfere.
QRM Is my transmission being interfered with? Your transmission is being interfered with ____ (1. Nil; 2. Slightly; 3. Moderately; 4. Severely; 5. Extremely.)
QRN Are you troubled by static? I am troubled by static ____ (1-5 as under QRM).
QRO Shall I increase power? Increase power.
QRP Shall I decrease power? Decrease power.
QRQ Shall I send faster? Send faster (____ WPM).
QRS Shall I send more slowly? Send more slowly (____ WPM).
QRT Shall I stop sending? Stop sending.
QRU Have you anything for me? I have nothing for you.
QRV Are you ready? I am ready.
QRW Shall I inform ____ that you are calling on ____ kHz? Please inform ____ that I am calling on ____ kHz.
QRX When will you call me again? I will call you again at ____ hours (on ____ kHz).
QRY What is my turn? Your turn is numbered ____
QRZ Who is calling me? You are being called by ____ (on ____ kHz).
QSA What is the strength of my signals (or those of ____)? The strength of your signals (or those of ____) is ____ (1. Scarcely perceptible; 2. Weak; 3. Fairly good; 4. Good; 5. Very good).
QSB Are my signals fading? Your signals are fading.
QSD Is my keying defective? Your keying is defective.

QSG Shall I send ____ messages at a time? Send ____ messages at a time.
QSK Can you hear me between your signals and if so can I break in on your transmission? I can hear you between my signals; break in on my transmission.
QSL Can you acknowledge receipt? I am acknowledging receipt
QSM Shall I repeat the last message which I sent you, or some previous message? Repeat the last message which you sent me [or message(s) number(s) ____].
QSN Did you hear me (or ____) on ____ kHz? I did hear you (or ____) on ____ kHz.
QSO Can you communicate with ____ direct or by relay? I can communicate with ____ direct (or by relay through ____).
QSP Will you relay to ____? I will relay to ____
QST General call preceding a message addressed to all amateurs and ARRL members. This is in effect "CQ ARRL."
QSU Shall I send or reply on this frequency (or on ____ kHz)? Send a series of Vs on this frequency (or ____ kHz).
QSW Will you send on this frequency (or on ____ kHz)? I am going to send on this frequency (or on ____ kHz).
QSX Will you listen to ____ on ____ kHz? I am listening to ____ on ____ kHz.
QSY Shall I change to transmission on another frequency? Change to transmission on another frequency (or on ____ kHz).
QSZ Shall I send each word or group more than once? Send each word or group twice (or ____ times).
QTA Shall I cancel message number ____? Cancel message number ____
QTB Do you agree with my counting of words? I do not agree with your counting of words. I will repeat the first letter or digit of each word or group.
QTC How many messages have you to send? I have ____ messages for you (or for ____).
QTH What is your location? My location is ____
QTR What is the correct time? The time is ____

HAM OUTLINE MAPS

The following Ham Outline Maps are reprinted with the permission of FM-RTTY Publishing. For ordering information regarding these or other Outline Maps, send an SASE to Jim Labo, KØOST, FM-RTTY Publishing, PO Box 35742, Dallas, TX 75235-0742.

ARRL FIELD ORGANIZATION

CALIFORNIA COUNTIES/ARRL SECTIONS
HAM OUTLINE MAP No 27

**ASIAN COUNTRIES
HAM OUTLINE MAP NO 1**

12/95   KØOST

1 - C Honshu	
10	Tokyo
11	Kanagawa
12	Chiba
13	Saitama
14	Ibaraki
15	Tochigi
16	Gunma
17	Yamanashi

2 - C Honshu	
18	Shizuoka
19	Gifu
20	Aichi
21	Mie

3 - C Honshu	
22	Kyoto
23	Shiga
24	Nara
25	Osaka
26	Wakayama
27	Hyogo

4 - W Honshu	
31	Okayama
32	Shimane
33	Yamaguchi
34	Tottori
35	Hiroshima

5 - Shikoku	
36	Kagawa
37	Tokushima
38	Ehime
39	Kochi

6 - Kyushu	
40	Fukuoka
41	Saga
42	Nagasaki
43	Kumamoto
44	Oita
45	Miyazaki
46	Kagoshima
47	Okinawa

7 - N Honshu	
02	Aomori
03	Iwate
04	Akita
05	Yamagata
06	Miyagi
07	Fukushima

8 - Hokkaido	
01	Hokkaido

9 - C Honshu	
28	Toyama
29	Fukui
30	Ishikawa

Ø - C Honshu	
08	Niigata
09	Nagano

JAPANESE PREFECTURES (KENS)
PREFIX: JA-JS 7J-7N 8J-8N
HAM OUTLINE MAP No 2

CHINESE PROVINCES

PREFIX: BA-BZ 3H-3U XS

HAM OUTLINE MAP No 3

1 - Central Mexico

COL	Colima
DF	Federal District
GTO	Guanajuato
HGO	Hidalgo
JAL	Jalisco
MEX	Mexico
MICH	Michoacan
MOR	Morelos
NAY	Nayarit
PUE	Puebla
TLX	Tlaxcala
QRO	Queretago
VER	Veracruz

2 - Northern Mexico

AGS	Aguascalientes
BC	Baja California Norte
BC-T	Baja California Sur
CHIH	Chihuahua
COAH	Coahuila
DGO	Durango
NL	Nuevo Leon
SIN	Sinaloa
SLP	San Luis Potosi
SON	Sonora
TAMPS	Tamaulipas
ZAC	Zacatecas

3 - Southern Mexico

CAMP	Campeche
CHIS	Chiapas
GRO	Guerrero
OAX	Oaxaca
QROO	Quintana Rod
TABS	Tabasco
YUC	Yucatan

MEXICAN STATES
   PREFIX: XA-XI 4A-4C 6D-6J
HAM OUTLINE MAP No 4

EUROPEAN COUNTRIES
HAM OUTLINE MAP No 6
FM-RTTY PUBLISHING • PO BOX 35742 • DALLAS, TEXAS 75235

12/95   KØOST

N°	DEPARTEMENT	PROVINCE DPT
01	AIN	RHONE-ALPES
02	AISNE	PICARDIE
03	ALLIER	AUVERGNE
04	ALPES HAUTE PROVENCE	PROVENCE-COTE D'AZUR
05	ALPES (HAUTES)	PROVENCE-COTE D'AZUR
06	ALPES-MARITIMES	PROVENCE-COTE D'AZUR
07	ARDECHE	RHONE-ALPES
08	ARDENNES	CHAMPAGNE
09	ARIEGE	MIDI-PYRENEES
10	AUBE	CHAMPAGNE
11	AUDE	LANGUEDOC-ROUSSILLON
12	AVEYRON	MIDI-PYRENEES
13	BOUCHES-DU-RHONE	PROVENCE-COTE D'AZUR
14	CALVADOS	BASSE-NORMANDIE
15	CANTAL	AUVERGNE
16	CHARENTE	POITOU-CHARENTE
17	CHARENTE-MARITIME	POITOU-CHARENTE
18	CHER	CENTRE
19	CORREZE	LIMOUSIN
2A	CORSE SUD	CORSE (HAUTE)
2B	CORSE (HAUTE)	CORSE }TK/FC
21	COTE D'OR	BOURGOGNE
22	COTES DU NORD	BRETAGNE
23	CREUSE	LIMOUSIN
24	DORDOGNE	AQUITAINE
25	DOUBS	FRANCHE-COMTE
26	DROME	RHONE-ALPES
27	EURE	HAUTE-NORMANDIE
28	EURE-ET-LOIR	CENTRE
29	FINISTERE	BRETAGNE
30	GARD	LANGUEDOC-ROUSSILLON
31	HAUTE-GARONNE	MIDI-PYRENEES
32	GERS	MIDI-PYRENEES
33	GIRONDE	AQUITAINE
34	HERAULT	LANGUEDOC-ROUSSILLON
35	ILLE-ET-VILAINE	BRETAGNE
36	INDRE	CENTRE
37	INDRE-ET-LOIRE	CENTRE
38	ISERE	RHONE-ALPES
39	JURA	FRANCHE-COMTE
40	LANDES	AQUITAINE
41	LOIR-ET-CHER	CENTRE
42	LOIRE	RHONE-ALPES
43	HAUTE-LOIRE	AUVERGNE
44	LOIRE-ATLANTIQUE	PAYS DE LOIRE
45	LOIRET	CENTRE
46	LOT	MIDI-PYRENEES
47	LOT-ET-GARONNE	AQUITAINE
48	LOZERE	LANGUEDOC-ROUSSILLON
49	MAINE-ET-LOIRE	PAYS DE LOIRE
50	MANCHE	BASSE-NORMANDIE
51	MARNE	CHAMPAGNE
52	MARNE (HAUTE)	CHAMPAGNE
53	MAYENNE	PAYS DE LOIRE
54	MEURTHE-ET-MOSELLE	LORRAINE
55	MEUSE	LORRAINE
56	MORBIHAN	BRETAGNE
57	MOSELLE	LORRAINE
58	NIEVRE	BOURGOGNE
59	NORD	NORD
60	OISE	PICARDIE
61	ORNE	BASSE-NORMANDIE
62	PAS-DE-CALAIS	NORD
63	PUY-DE-DOME	AUVERGNE
64	PYRENEES-ATLANTIQUE	AQUITAINE
65	PYRENEES (HAUTES)	MIDI-PYRENEES
66	PYRENEES-ORIENTALES	LANGUEDOC-ROUSSILLON
67	BAS-RHIN	ALSACE
68	HAUT-RHIN	ALSACE
69	RHONE	RHONE-ALPES
70	SAONE (HAUTE)	FRANCHE-COMTE
71	SAONE-ET-LOIRE	BOURGOGNE
72	SARTHE	PAYS DE LOIRE
73	SAVOIE	RHONE-ALPES
74	HAUTE-SAVOIE	RHONE-ALPES
75	VILLE DE PARIS	ILE-DE-FRANCE
76	SEINE-MARITIME	HAUTE-NORMANDIE
77	SEINE-ET-MARNE	ILE-DE-FRANCE
78	YVELINES	ILE-DE-FRANCE
79	DEUX-SEVRES	POITOU-CHARENTES
80	SOMME	PICARDIE
81	TARN	MIDI-PYRENEES
82	TARN-ET-GARONNE	MIDI-PYRENEES
83	VAR	PROVENCE-COTE D'AZUR
84	VAUCLUSE	PROVENCE-COTE D'AZUR
85	VENDEE	PAYS DE LOIRE
86	VIENNE	POITOU-CHARENTES
87	HAUTE-VIENNE	LIMOUSIN
88	VOSGES	LORRAINE
89	YONNE	BOURGOGNE
90	TERRITOIRE DE BELFORT	FRANCHE-COMTE
91	ESSONNE	ILE-DE-FRANCE
92	HAUTS-DE-SEINE	ILE-DE-FRANCE
93	SEINE-SAINT-DENIS	ILE-DE-FRANCE
94	VAL-DE-MARNE	ILE-DE-FRANCE
95	VAL-D'OISE	ILE-DE-FRANCE

FRENCH DEPARTMENTS
PREFIX: F HW-HY TH TK TM
HAM OUTLINE MAP No 7

AG AARGAU
AI APPENZELL INNER RHODEN
AR APPENZELL OUTER RHODEN
BE BERNE
BL BASLE - COUNTY
BS BASLE - CITY
FR FRIBOURG
GE GENEVA
GL GLARIS
GR GRISONS
JU JURA
LU LUCERNE
NE NEUCHATEL
NW NIDWALDEN
OW OBWALDEN
SG ST GALL
SH SCHAFFHAUSEN
SO SOLOTHURN
SZ SCHWYZ
TG THURGAU
TI TICINO
UR URI
VD VAUD
VS VALAIS
ZG ZUG
ZH ZURICH

SWISS CANTONS
PREFIX: HB HE
HAM OUTLINE MAP No 44

4/91 KØOST

1 - Gotland

| I | Gotland |

2 - Northern

| AC | Vasterbotten |
| BD | Norrbotten |

3 - Northern

X	Gavleborg
Y	Vasternorrland
Z	Jamtland

4 - West Central

S	Varmland
T	Orebro
W	Kopparberg

5 - East Central

A	Stockholm City
B	Stockholm County
C	Uppsala
D	Sodermanland
E	Ostergotland
U	Vastmanland

6 - Southwestern

N	Halland
O	Goteborg och Bohus
P	Alvsborg
R	Skaraborg

7 - Southern

F	Jonkoping
G	Kronoberg
H	Kalmar
K	Blekinge
L	Kristianstad
M	Malmohus

Ø - Stockholm Area

| A | Stockholm City |
| B | Stockholm County |

| SM8 | Maritime Mobile |
| SJ9 | Morokulien |

SK	Club
SL	Military
SM	Individual

SWEDISH COUNTIES (LAENS)
PREFIX: SA-SM 7S 8S
HAM OUTLINE MAP No 8

ITALIAN PROVINCES
 PREFIX: I IA-IZ
 HAM OUTLINE MAP No 9

PROVINCE ABBREVIATIONS:

AG - AGRIGENTO	MO - MODENA
AL - ALESSANDRIA	MS - MASSA
AN - ANCONA	MT - MATERA
AO - AOSTA	NA - NAPOLI
AP - ASCOLI PICENO	NO - NOVARA
AQ - L'AQUILA	NU - NUORO
AR - AREZZO	OR - ORISTANO
AT - ASTI	PA - PALERMO
AV - AVELLINO	PC - PIACENZA
BA - BARI	PD - PADOVA
BG - BERGAMO	PE - PESCARA
BL - BELLUNO	PG - PERUGIA
BN - BENEVENTO	PI - PISA
BO - BOLOGNA	PN - PORDENONE
BR - BRINDISI	PR - PARMA
BS - BRESCIA	PS - PESARO
BZ - BOLZANO	PT - PISTOIA
CA - CAGLIARI	PV - PAVIA
CB - CAMPOBASSO	PZ - POTENZA
CE - CASERTA	RA - RAVENNA
CH - CHIETI	RC - REGGIO CALABRIA
CL - CALTANISSETTA	RE - REGGIO EMILIA
CN - CUNEO	RG - RAGUSA
CO - COMO	RI - RIETI
CR - CREMONA	RO - ROVIGO
CS - COSENZA	SA - SALERNO
CT - CATANIA	SI - SIENA
CZ - CATANZARO	SO - SONDRIO
EN - ENNA	SP - LA SPEZIA
FE - FERRARA	SR - SIRACUSA
FG - FOGGIA	SS - SASSARI
FI - FIRENZE	SV - SAVONA
FO - FORLI	TA - TARANTO
FR - FROSINONE	TE - TERAMO
GE - GENOVA	TN - TRENTO
GO - GORIZIA	TO - TORINO
GR - GROSSETO	TP - TRAPANI
IM - IMPERIA	TR - TERNI
IS - ISERNIA	TS - TRIESTE
LE - LECCE	TV - TREVISO
LI - LIVORNO	UD - UDINE
LT - LATINA	VA - VARESE
LU - LUCCA	VC - VERCELLI
MC - MACERATA	VE - VENEZIA
ME - MESSINA	VI - VICENZA
MI - MILANO	VR - VERONA
MN - MANTOVA	VT - VITERBO

P4 PJ

9Y

YV

HKØ

HK

8R

PZ FY

PYØS

HC8

HC

PYØF

OA

PY

CP

PYØT

ZP

CEØX

CEØA

CEØZ

CX

LU

CE

VP8

VP8, LU-Z

VP8
LU-Z

SOUTH AMERICAN COUNTRIES
HAM OUTLINE MAP No 16

BRAZILIAN STATES
PREFIX: PP-PY ZV-ZZ
HAM OUTLINE MAP No 17

PP1—Espirito Santo ES
PP2—Goias GO
PP5—Santa Catarina SC
PP6—Sergipe SE
PP7—Alagoas AL
PP8—Amazonas AM
PR7—Paraiba PB
PR8—Maranhao MA
PS7—Rio Grande do Norte
PS8—Piaui PI
PT2—Distrite Federal DF (Brasilia)
PT7—Ceara CE
PT8—Acre AC
PT9—Mato Grosso do Sul MS
PU8—Amapa AP
PV8—Roraima RR
PW8—Rondonia RO
PY1—Rio de Janeiro RJ
PY2—Sao Paulo SP
PY3—Rio Gran de do Sul RS
PY4—Minas Gerais MG
PY5—Parana PR
PY6—Bahia BA
PY7—Pernambuco PE
PY8—Para PA
PY9—Mato Grosso MT
PY0T—Trinidade Island
PY0F—Fernando de Noronha Is.
PY0S—Sao Pedro/S. Paulo Rocks

Province is indicated by letter(s)
after number in call sign

A B C	Buenos Aires City
D E	Buenos Aires Province
F	Santa Fe
GA-GO	Chaco
GP-GZ	Formosa
H	Cordoba
I	Misiones
J	Entre Rios
K	Tucuman
L	Corrientes
M	Mendoza
N	Santiago Del Estero
O	Salta
P	San Juan
Q	San Luis
R	Catamarga
S	La Rioja
T	Jujuy
U	La Pampas
V	Rio Negro
W	Chubut
XA-XO	Santa Cruz
XP-XZ	Tierra Del Fuego
Y	Neuquen
Z	Antarctica

ARGENTINE PROVINCES
 PREFIX: LO-LW L2-L9 AY-AZ
HAM OUTLINE MAP No 18

ANTARCTICA & OFFSHORE ISLANDS

PREFIX: ATØ CE9 DFØ EDØ FT-Y HFØ IAØ KC4 LU-Z
OR4 VKØ VP8 ZL5 ZS7 ZXØ 4K1 7S8 8J1

HAM OUTLINE MAP No 20

☒ CLOSED STATION

AFRICAN COUNTRIES
HAM OUTLINE MAP No 22

4/91 KØOST

Some Abbreviations for CW Work

Although abbreviations help to cut down unnecessary transmission, make it a rule not to abbreviate unnecessarily when working an operator of unknown experience.

AA	All after	GND	Ground	SASE	Self-addressed, stamped envelope
AB	All before	GUD	Good	SED	Said
ABT	About	HI	The telegraphic laugh; high	SIG	Signature; signal
ADR	Address	HR	Here, hear	SINE	Operator's personal initials or nickname
AGN	Again	HV	Have		
ANT	Antenna	HW	How	SKED	Schedule
BCI	Broadcast interference	LID	A poor operator	SRI	Sorry
BCL	Broadcast listener	MA, MILS	Milliamperes	SSB	Single sideband
BK	Break; break me; break in	MSG	Message; prefix to radiogram	SVC	Service; prefix to service message
BN	All between; been	N	No	T	Zero
BUG	Semi-automatic key	NCS	Net control station	TFC	Traffic
B4	Before	ND	Nothing doing	TMW	Tomorrow
C	Yes	NIL	Nothing; I have nothing for you	TNX-TKS	Thanks
CFM	Confirm; I confirm	NM	No more	TT	That
CK	Check	NR	Number	TU	Thank you
CL	I am closing my station; call	NW	Now; I resume transmission	TVI	Television interference
CLD-CLG	Called; calling	OB	Old boy	TX	Transmitter
CQ	Calling any station	OC	Old chap	TXT	Text
CUD	Could	OM	Old man	UR-URS	Your; you're; yours
CUL	See you later	OP-OPR	Operator	VFO	Variable-frequency oscillator
CW	Continuous wave (i.e., radiotelegraph)	OT	Old timer; old top	VY	Very
		PBL	Preamble	WA	Word after
DLD-DLVD	Delivered	PSE	Please	WB	Word before
DR	Dear	PWR	Power	WD-WDS	Word; words
DX	Distance, foreign countries	PX	Press	WKD-WKG	Worked; working
ES	And, &	R	Received as transmitted; are	WL	Well; will
FB	Fine business, excellent	RCD	Received	WUD	Would
FM	Frequency modulation	RCVR (RX)	Receiver	WX	Weather
GA	Go ahead (or resume sending)	REF	Refer to; referring to; reference	XCVR	Transceiver
GB	Good-by	RFI	Radio frequency interference	XMTR (TX)	Transmitter
GBA	Give better address	RIG	Station equipment	XTAL	Crystal
GE	Good evening	RPT	Repeat; I repeat	XYL (YF)	Wife
GG	Going	RTTY	Radioteletype	YL	Young lady
GM	Good morning	RX	Receiver	73	Best regards
GN	Good night			88	Love and kisses

RETURN POSTAGE CHART FOR WAS QSL CARDS

(Please include sufficient funds, as follows, for the safe return of your cards)

Foreign Small-Packet Air Mail

	Canada	Mexico	Central America, South America, Caribbean Islands	Europe European Russia	Pacific Rim, Asia, Africa, Asiatic Russia
WAS	$1.75	$2.00	$2.25	$3.25	$4.00
5BWAS	$7.00	$13.00	$10.00	$16.75	$23.25

United States

	Reg.	Cert.	First Class
WAS	$6.75	$2.75	$1.50
5BWAS	$9.00	$5.50	$4.50

All of the above are US funds
For Registered Small-Packet Air Mail add $4.95 to the above fees.
IRCs are valued at $0.50 US
These rates are subject to change at any time.

RETURN POSTAGE CHART FOR DXCC QSL CARDS

(Please include sufficient funds, as follows, for the safe return of your cards)

Award	Registered	Certified	First Class
DXCC (100 card new application)	$8.50 ($14.25[1])	$5.00	$4.00 ($9.25[1])
DXCC (10/25 endorsement)	$6.75 ($8.25[1])	$3.25	$2.25 ($3.75[1])
5BDXCC	$10.25 ($23.75[2])	$6.75	$5.75 ($19.25[2])

[1]Foreign small-packet Air Mail [2]Foreign surface Mail

All of the above are US funds.   **Registered or Certified Mail is recommended.**
IRCs are valued at $0.50 US.   These rates are subject to change at any time.

DXCC Fee Schedule

ALL amateurs initial first-time registration—$10. DXCC participants before Oct. 1, 1990 are exempt from this fee.
ALL League members 1st submission in a calendar year—FREE.
Non-League members outside of USA/Canada 1st submission in a calendar year—$10.
ALL League members 2nd submission in a calendar year—$10.
Non-League members outside of USA/Canada 2nd submission in a calendar year—$20.

MCS-16

The Amateur's Code

The Radio Amateur is:

CONSIDERATE...never knowingly operates in such a way as to lessen the pleasure of others.

LOYAL...offers loyalty, encouragement and support to other amateurs, local clubs, and the American Radio Relay League, through which Amateur Radio in the United States is represented nationally and internationally.

PROGRESSIVE...with knowledge abreast of science, a well-built and efficient station and operation above reproach.

FRIENDLY...slow and patient operating when requested; friendly advice and counsel to the beginner; kindly assistance, cooperation and consideration for the interests of others. These are the hallmarks of the amateur spirit.

BALANCED...radio is an avocation, never interfering with duties owed to family, job, school, or community.

PATRIOTIC...station and skill always ready for service to country and community.

—The original Amateur's Code was written by
Paul M. Segal, W9EEA, in 1928.—

ARRL WAS AWARD APPLICATION

Please PRINT.

Call: _____

Name _____
(Print exactly as you want it to appear on certificate)

List any ex-calls used on
any cards submitted:

Address _____
(number and street)

_____

_____
(City, State/Country, ZIP/PC)

FOR RETURN POSTAGE
Cash/Check/Stamps/IRCs
enclosed.

$ _____

Return my cards via:
☐ Registered
☐ Certified • US only
☐ First-class mail

W/VEs: Membership control number
(from your **QST** wrapper)   :_____

Is this your first WAS application? ☐ YES   ☐ NO

If NO, my original WAS number held is _____

I was first licensed in (year): _____

I am applying for ONE of the following WAS awards (each is separately numbered):
Please Check: (Use MCS-225 for 5BWAS)

☐ Basic Award (mixed bands and/or modes)

☐ SSTV   ☐ 144 MHz   ☐ 50 MHz
☐ OSCAR Satellite   ☐ 160 Meters   ☐ 220 MHz
☐ RTTY   ☐ 75 Meters 2-Letter Extra Class   ☐ 432 MHz

I am applying for the following ENDORSEMENTS (Check **ALL** that apply):

☐ SSB   ☐ Novice   ☐ Packet

        ☐ Single Band: _____
                            (Meters)

☐ CW   ☐ QRP   ☐ EME

The undersigned affirms that he/she has read, understood, and abided by the rules of the WAS award (MCS-264).

Signature _____   Date _____ 19 _____

HF AWARDS MANAGER VERIFICATION

I hereby verify that I have personally inspected the confirmations with all 50 states and verify that this application is correct and true. In my own handwriting, I confirm this application is for the following SPECIALTY awards or

ENDORSEMENTS: _____ (write NONE if none)

_____   _____ H- _____   _____ 19 _____
(signature)   (call)   (Mgr. 4-digit code #)   (date)

DIRECTIONS TO APPLICANT

1. Read the WAS Rules (MCS-264) carefully.

2. Fill out BOTH sides of this application.

3. Sort cards alphabetically by state.

4. Present application and cards to your ARRL Special Service Club HF Awards Manager for verification. **Applications from DX stations may be certified by the Awards Manager of your IARU member-society.**

5. Send application to ARRL HQ. Send cards also ONLY if there is no local HF Awards Manager for verification.

6. If mailing cards, enclose postage sufficient for return of cards (see attached CD-16). MAIL TO:

   ARRL WAS AWARD
   225 Main Street
   Newington, CT 06111 USA

MCS-217 (9/88)

Applicant's call sign _____

STATE	CALL	DATE	BAND	MODE	
Alabama					
Alaska					
Arizona					
Arkansas					
California					
Colorado					
Connecticut					
Delaware					
Florida					
Georgia					
Hawaii					
Idaho					
Illinois					
Indiana					
Iowa					
Kansas					
Kentucky					
Louisiana					
Maine					
Maryland (D.C.)					
Massachusetts					
Michigan					
Minnesota					
Mississippi					
Missouri					
Montana					
Nebraska					
Nevada					
New Hampshire					
New Jersey					
New Mexico					
New York					
North Carolina					
North Dakota					
Ohio					
Oklahoma					
Oregon					
Pennsylvania					
Rhode Island					
South Carolina					
South Dakota					
Tennessee					
Texas					
Utah					
Vermont					
Virginia					
Washington					
West Virginia					
Wisconsin					
Wyoming					

ARRL 5BWAS AWARD APPLICATION

Please PRINT.

Call: _____

List any ex-calls used on
any cards submitted:

_____

Name _____
(Print exactly as you want it to appear on certificate)

Address _____
(number and street)

_____
(City, State/Country, ZIP/PC)

FOR RETURN POSTAGE
Cash/Check/Stamps/IRCs
enclosed.

$ _____

Return my cards via:
☐ Registered
☐ Certified • U.S. only
☐ First-class mail

W/VEs: Membership control number
(from your **QST** wrapper)   : _____

Is this your first WAS application?   ☐ YES   ☐ NO

If NO, my original WAS number held is _____

I was first licensed in (year): _____

☐ $25 is enclosed for a plaque

The undersigned affirms that he/she has read, understood, and abided by the
rules of the 5BWAS award printed below and the WAS Rules (MCS-264).

Signature _____ Date _____ 19 _____

HF AWARDS MANAGER VERIFICATION

I hereby verify that I have personally inspected the confirmations with all 50 states on five different bands and verify that
this application is correct and true.

_____   _____ H-_____ _____ 19 _____
(signature)   (call)   (Mgr. 4-digit code #)   (date)

5BWAS RULES

1. The 5BWAS certificate and plaque (see below) will be issued for having submitted confirmations with each of the 50
   United States for contacts **dated January 1, 1970,** or after, on five amateur bands (10, 18 and 24 MHz excluded).
   Phone and CW segments of a band do not count as separate bands.
2. WAS Rules (MCS-264), that do not conflict with these 5BWAS rules, also apply to the 5BWAS award.
3. There are no specialty 5Band awards or endorsements.
4. The 5BWAS certificate is offered free of charge. A handsome 9 × 12 " personalized walnut plaque is available for a fee of
   $25 US (check or money order).

DIRECTIONS TO APPLICANT

1. Read 5BWAS rules above and WAS rules (MCS-264)
   carefully.

2. Fill out BOTH sides of this application.

3. Sort cards alphabetically by state and band.

4. Present application and cards to your ARRL Special
   Service Club HF Awards Manager for verification.
   Applications from DX stations may be certified by the
   Awards Manager of your IARU member-society.

5. Send application to ARRL HQ. Send cards also ONLY
   if there is no local HF Awards Manager.

6. If mailing cards, enclose postage sufficient for return of
   cards (see attached CD-16). MAIL TO:

   ARRL 5BWAS Award
   225 Main Street
   Newington, CT 06111 USA

7. Enclose $25 (check) if you want the plaque.

MCS-225(589)

5B-WAS RECORD SHEET

DIRECTIONS: Enter callsigns in the appropriate boxes by state and band.
QSLs must be for contacts **on or after January 1, 1970.**

STATE	80	40	20	15	10
Alabama					
Alaska					
Arizona					
Arkansas					
California					
Colorado					
Connecticut					
Delaware					
Florida					
Georgia					
Hawaii					
Idaho					
Illinois					
Indiana					
Iowa					
Kansas					
Kentucky					
Louisiana					
Maine					
Maryland (D.C.)					
Massachusetts					
Michigan					
Minnesota					
Mississippi					
Missouri					
Montana					
Nebraska					
Nevada					
New Hampshire					
New Jersey					
New Mexico					
New York					
North Carolina					
North Dakota					
Ohio					
Oklahoma					
Oregon					
Pennsylvania					
Rhode Island					
South Carolina					
South Dakota					
Tennessee					
Texas					
Utah					
Vermont					
Virginia					
Washington					
West Virginia					
Wisconsin					
Wyoming					

MCS-225(589)

ARRL VUCC AWARD APPLICATION

1. Please PRINT. 2. Use separate application for EACH band.

Call:_____ Name:_____

(Print exactly as you want it on certificate)

List any ex-calls used   Address:_____
on any cards submitted:

(number and street)

_____   _____

(city, state, ZIP)

Membership control number (from your QST wrapper):_____

FOR RETURN POSTAGE
Money/Stamps enclosed
$_____
Include $1 for certificate
or SASE (with appropriate
postage for the weight of
your field sheets) for
endorsement.
Return my cards via:
☐Registered
☐Certified US only
☐United Parcel Service
☐First-class mail

☐Initial application (check one)   ☐Endorsement

Band (check **one** only): ☐50MHz ☐144MHz ☐222MHz
☐432MHz ☐902MHz ☐1296MHz ☐2.3GHz ☐3.4GHz
☐5.7GHz ☐10GHz ☐24GHz ☐47GHz ☐75 GHZ
☐119GHz ☐142GHz ☐241GHz ☐Laser ☐Satellite

INITIAL applicants:_____Number of grid squares claimed.

ENDORSEMENTS

_____ + _____ = _____
Number   Number additional   NEW TOTAL
previously credited in this application

The undersigned affirms he/she has read, understood, and abided by the rules of the VUCC (MSD-261).

Signature_____ Date_____19____

***************************VHF AWARDS MANAGER VERIFICATION*************************
Grid square credits claimed on this application have been verified as correct and true.
TOTAL number of grid squares credited   =_____.

_____ _____ V-_____ _____
(signature)   (call)   (Mgr. 4-digit code #)   (date)

************************DIRECTIONS TO APPLICANT************************

1. Fill out this application and MSD-259. Read the VUCC Rules (MSD-261) carefully.
2. Enclose postage sufficient for return of cards (if mailed), and return address label.
3. For initial applicants, enclose $1 U.S. for mailing certificate (W/VEs only).
4. For endorsements, enclose SASE with appropriate postage for the weight of your field sheets.
5. Sort QSLs alphabetically by field, and numerically within the field, from 00 to 99.
6. Forward cards to nearest VHF Awards Manager (see attached information). DO NOT SEND
   CARDS TO ARRL HQ.

MSD-260(894)

Field:
(first 2 letters of locator)

SQ.	Callsign	SQ.	Callsign
00		50	
01		51	
02		52	
03		53	
04		54	
05		55	
06		56	
07		57	
08		58	
09		59	
10		60	
11		61	
12		62	
13		63	
14		64	
15		65	
16		66	
17		67	
18		68	
19		69	
20		70	
21		71	
22		72	
23		73	
24		74	
25		75	
26		76	
27		77	
28		78	
29		79	
30		80	
31		81	
32		82	
33		83	
34		84	
35		85	
36		86	
37		87	
38		88	
39		89	
40		90	
41		91	
42		92	
43		93	
44		94	
45		95	
46		96	
47		97	
48		98	
49		99	

TOTAL number of grid squares worked in this field =

MSD-259(392)

ARRL VUCC AWARD CALLSIGN:

☐ 50 ☐ 144 ☐ 220 ☐ 432 ☐ 902 ☐ 1296 ☐ 2.3 ☐ 3.4
☐ 5.7 ☐ 10 ☐ 24 ☐ 47 ☐ Laser ☐ Satellite

DIRECTIONS: This side of tally sheet good for two fields. 1. Enter your callsign and check band. 2. Enter two-letter field. 3. Enter callsign of stations worked in appropriate grid square. 4. Total grid squares for each field.

Field:
(first 2 letters of locator)

SQ.	Callsign	SQ.	Callsign
00		50	
01		51	
02		52	
03		53	
04		54	
05		55	
06		56	
07		57	
08		58	
09		59	
10		60	
11		61	
12		62	
13		63	
14		64	
15		65	
16		66	
17		67	
18		68	
19		69	
20		70	
21		71	
22		72	
23		73	
24		74	
25		75	
26		76	
27		77	
28		78	
29		79	
30		80	
31		81	
32		82	
33		83	
34		84	
35		85	
36		86	
37		87	
38		88	
39		89	
40		90	
41		91	
42		92	
43		93	
44		94	
45		95	
46		96	
47		97	
48		98	
49		99	

TOTAL number of grid squares worked in this field =

WAC AWARD APPLICATION

Name_____Callsign_____

Mailing Address_____

City/Town_____State_____ZIP Code_____

This application is for:

☐ Basic certificate (mixed mode)   ☐ QRP endorsement
☐ CW certificate   (5 watts output or less)
☐ Phone certificate   ☐ 1.8-MHz endorsement
☐ SSTV certificate   ☐ 3.5-MHz endorsement
☐ RTTY certificate   ☐ 50-MHz endorsement
☐ FAX certificate   ☐ 144-MHz endorsement
☐ Satellite certificate   ☐ 432-MHz endorsement
☐ 5-band certificate   ☐ 6-band endorsement

☐ My 5-band certificate is dated_____

Enclosed are QSL cards from the following stations:

(Enter band(s) and callsigns)

	MHz	MHz	MHz	MHz	MHz
N. America					
S. America					
Oceania					
Asia					
Europe					
Africa					

The undersigned has abided by all rules set forth for this award.
My ARRL membership does not expire until_____

Signature_____Callsign_____

Date_____   11/94

ITU ZONES FOR US AND CANADA

ITU Zone 1

KL7 Alaska (West of 141 Deg. W.)

ITU Zone 2

KL7 Alaska (East of 141 Deg. W.) VE6, 7, VY1 Canada (South of 80 Deg. N. and West of 110 Deg. W.)

ITU Zone 3

VE3, 4, 5, 8 Canada (South of 80 Deg. N. and between 90 Deg. and 110 Deg. W.)

ITU Zone 4

VE2, 3, 4, 8 Canada (South of 80 Deg. N. and between 70 Deg. and 90 Deg. W. including Baffin Isl.)

ITU Zone 6

W-K-N-A U.S.A. (Washington, Oregon, California, Nevada, Idaho & that part of Montana, Utah & Arizona west of 110 Deg. W.)

ITU Zone 7

W-K-N-A U.S.A. (N. Dakota, S. Dakota, Nebraska, Wyoming, Colorado, New Mexico, Texas, Oklahoma, Kansas, Iowa, Minnesota, that part of Utah & Arizona East of 110 Deg. W., & that part of Michigan, Illinois, Missouri, Arkansas, Mississippi, Tennessee, Louisiana & Wisconsin west of 90 Deg. W.)

ITU Zone 8

W-K-N-A U.S.A. (Indiana, Alabama, Georgia, Florida, Virginia, Kentucky, N. Carolina, S. Carolina, West Virginia, Maryland, Deleware, Ohio, Pennsylvania, New Jersey, New York, Maine, Connecticut, Rhode Island, Massachusetts, New Hampshire, Vermont & that part of Michigan, Illinois, Missouri, Arkansas, Mississippi, Tennessee, Louisiana & Wisconsin east of 90 Deg. W.)

ITU Zone 9

VE1, 2, 8, VY2, VO2 Canada (South of 80 Deg. N. & East of 70 Deg. W., including Labrador, Newfoundland, Nova Scotia, but excluding Baffin Isl.)

ITU Zone 61

KH1 Baker & Howland Isl.
KH3 Johnston Isl.
KH4 Midway Isl.
KH5 Palmyra Isl.
KH6 Hawaiian Isl.
KH7 Kure Isl.

United States Counties

(information courtesy of *CQ*'s Counties Award Record Book; for information on county hunting, see p 8-10)

Alabama (67 counties)
Autauga
Balwin
Barbour
Bibb
Blount
Bullock
Butler
Calhoun
Chambers
Cherokee
Chilton
Choctaw
Clarke
Clay
Cleburne
Coffee
Colbert
Conecuh
Coosa
Covington
Crenshaw
Cullman
Dale
Dallas
Dekalb
Elmore
Escambia
Etowah
Fayette
Franklin
Geneva
Greene
Hale
Henry
Houston
Jackson
Jefferson
Lamar
Lauderdale
Lawrence
Lee
Limestone
Lowndes
Macon
Madison
Marengo
Marion
Marshall
Mobile
Monroe
Montgomery
Morgan
Perry
Pickens
Pike
Randolph
Russel
Saint Clair
Shelby
Sumter
Talladega
Tallapoosa
Tuscaloosa
Walker
Washington
Wilcox
Winston

Alaska (4 counties)
Southeastern
Northwestern
South Central
Central

Arizona (15 counties)
Apache
Cochise
Coconino
Gila
Graham
Greenlee
Lez Paz
Maricopa
Mohave
Navajo
Pima
Pinal
Santa Cruz
Yavapai
Yuma

Arkansas (75 counties)
Arkansas
Ashley
Baxter
Benton
Boone
Bradley
Calhoun
Carroll
Chicot
Clark
Clay
Cleburne
Cleveland
Columbia
Conway
Craighead
Crawford
Crittenden
Cross
Dallas
Desha
Drew
Faulkner
Franklin
Fulton
Garland
Grant
Greene
Hempstead
Hot Spring
Howard
Independence
Izard
Jackson
Jefferson
Johnson
Lafayette
Lawrence
Lee
Lincoln
Little River
Logan
Lonoke
Madison
Marion
Miller
Mississippi
Monroe
Montgomery
Nevada
Newton
Ouachita
Perry
Phillips
Pike
Poinsett

Polk
Pope
Prairie
Pulaski
Randolph
St. Francis
Saline
Scott
Searcy
Sebastian
Sevier
Sharp
Stone
Union
Van Buren
Washington
White Woodruff
Yell

California (58 counties)
Alameda
Alpine
Amador
Butte
Calaveras
Colusa
Contra Costa
Del Norte
El Dorado
Fresno
Glenn
Humboldt
Imperial
Inyo
Kern
Kings
Lake
Lassen
Los Angeles
Madera
Marin
Mariposa
Mendocino
Merced
Modoc
Mono
Monterey
Napa
Nevada
Orange
Placer
Plumas
Riverside
Sacramento
San Benito
San Bernardino
San Diego
San Francisco
San Joaquin
San Luis Obispo
San Mateo
Santa Barbara
Santa Clara
Santa Cruz
Shasta
Sierra
Siskiyou
Solano
Sonoma
Stanislaus
Sutter
Tehama
Trinity
Tulare

Tuolumne
Ventura
Yolo
Yuba

Colorado (63 counties)
Adams
Alamosa
Arapahoe
Archuleta
Baca
Bent
Boulder
Chaffee
Cheyenne
Clear Creek
Conejos
Costilla
Crowley
Custer
Delta
Denver
Dolores
Douglas
Eagle
Elbert
El Paso
Fremont
Garfield
Gilpin
Grand
Gunnison
Hinsdale
Huerfano
Jackson
Jefferson
Kiowa
Kit Carson
Lake
La Plata
Larimer
Las Animas
Lincoln
Logan
Mesa
Mineral
Moffat
Montezuma
Montrose
Morgan
Otero
Ouray
Park
Phillips
Pitkin
Prowers
Pueblo
Rio Blanco
Rio Grande
Routt
Saguache
San Juan
San Miguel
Sedgwick
Summit
Teller
Washington
Weld
Yuma

Connecticut (8 counties)
Fairfield
Hartford
Litchfield

Middlesex
New Haven
New London
Tolland
Windham

Delaware (3 counties)

Kent
New Castle
Sussex

Florida (67 counties)

Alachua
Baker
Bay
Bradford
Brevard
Broward
Calhoun
Charlotte
Citrus
Clay
Collier
Columbia
Dade
De Soto
Dixie
Duval
Escambia
Flagler
Franklin
Gadsden
Gilchrist
Glades
Gulf
Hamilton
Hardee
Hendry
Hernando
Highlands
Hillsborough
Holmes
Indian River
Jackson
Jefferson
Lafayette
Lake
Lee
Leon
Levy
Liberty
Madison
Manatee
Marion
Martin
Monroe
Nassau
Okaloosa
Okeechobee
Orange
Osceola
Palm Beach
Pasco
Pinellas
Polk
Putnam
St. Johns
St. Lucie
Santa Rosa
Sarasota
Seminole
Sumter
Suwannee
Taylor
Union
Volusia
Wakulla
Walton
Washington

Georgia (159 counties)

Appling
Atkinson
Bacon
Baker
Baldwin
Banks
Barrow
Bartow
Ben Hill
Berrien
Bibb
Bleckley
Brantley
Brooks
Bryan
Bulloch
Burke
Butts
Calhoun
Camden
Candler
Carroll
Catoosa
Charlton
Chatham
Chattahoochee
Chattooga
Cherokee
Clarke
Clay
Clayton
Clinch
Cobb
Coffee
Colquitt
Columbia
Cook
Coweta
Crawford
Crisp
Dade
Dawson
Decatur
De Kalb
Dodge
Dooly
Dougherty
Douglas
Early
Echols
Effingham
Elbert
Emanuel
Evans
Fannin
Fayette
Floyd
Forsyth
Franklin
Fulton
Gilmer
Glascock
Glynn
Gordon
Grady
Greene
Gwinnett
Habersham
Hall
Hancock
Haralson
Harris
Hart Heard
Henry
Houston
Irwin
Jackson

Jasper
Jeff Davis
Jefferson
Jenkins
Johnson
Jones
Lamar
Lanier
Laurens
Lee
Liberty
Lincoln
Long
Lowndes
Lumpkin
McDuffie
McIntosh
Macon
Madison
Marion
Meriwether
Miller
Mitchell
Monroe
Montgomery
Morgan
Murray
Muscogee
Newton
Oconee
Oglethorpe
Paulding
Peach
Pickens
Pierce
Pike
Polk
Pulaski
Putnam
Quitman
Rabun
Randolph
Richmond
Rockdale
Schley
Screven
Seminole
Spalding
Stephens
Stewart
Sumter
Talbot
Taliaferro
Tattnall
Taylor
Telfair
Terrell
Thomas
Tift
Toombs
Towns
Treutlen
Troup
Turner
Twiggs
Union
Upson
Walker
Walton
Ware
Warren
Washington
Wayne
Webster
Wheeler
White
Whitfield
Wilcox

Wilkes
Wilkinson
Worth

Hawaii (5 counties)

Hawaii
Honolulu
Kalawao
Kauai
Maui

Idaho (44 counties)

Ada
Adams
Bannock
Bear Lake
Benewah
Bingham
Blaine
Boise
Bonner
Bonneville
Boundary
Butte
Camas
Canyon
Caribou
Cassia
Clark
Clearwater
Custer
Elmore
Franklin
Fremont
Gem
Gooding
Idaho
Jefferson
Jerome
Kootenai
Latah
Lemhi
Lewis
Lincoln
Madison
Minidoka
Nez Perce
Oneida
Owyhee
Payette
Power
Shoshone
Teton
Twin Falls
Valley
Washington

Illinois (102 counties)

Adams
Alexander
Bond
Boone
Brown
Bureau
Calhoun
Carroll
Cass
Champaign
Christian
Clark
Clay
Clinton
Coles
Cook
Crawford
Cumberland
De Kalb
De Witt

Douglas	Winnebago	Steuben	Madison
Du Page	Woodford	Sullivan	Mahaska
Edgar		Switzerland	Marion
Edwards	**Indiana** (92 counties)	Tippecanoe	Marshall
Effingham	Adams	Tipton	Mills
Fayette	Allen	Union	Mitchell
Ford	Bartholomew	Vanderburgh	Monona
Franklin	Benton	Vermillion	Monroe
Fulton	Blackford	Vigo	Montgomery
Gallatin	Boone	Wabash	Muscatine
Greene	Brown	Warren	O'Brien
Grundy	Carroll	Warrick	Osceola
Hamilton	Cass	Washington	Page
Hancock	Clark	Wayne	Palo Alto
Hardin Henderson	Clay	Wells	Plymouth
Henry	Clinton	White	Pocahontas
Iroquois	Crawford	Whitley	Polk
Jackson	Daviess		Pottawattamie
Jasper	Dearborn	**Iowa** (99 counties)	Poweshiek
Jefferson	Decatur	Adair	Ringgold
Jersey	De Kalb	Adams	Sac
Jo Daviess	Delaware	Allamakee	Scott
Johnson	Dubois	Appanoose	Shelby
Kane	Elkhart	Audubon	Sioux
Kankakee	Fayette	Benton	Story
Kendall	Floyd	Black Hawk	Tama
Knox	Fountain	Boone	Taylor
Lake	Franklin	Bremer	Union
La Salle	Fulton	Buchanan	Van Buren
Lawrence	Gibson	Buena Vista	Wapello
Lee	Grant	Butler	Warren
Livingston	Greene	Calhoun	Washington
Logan	Hamilton	Carroll	Wayne
McDonough	Hancock	Cass	Webster
McHenry	Harrison	Cedar	Winnebago
McLean	Hendricks	Cerro Gordo	Winneshiek
Macon	Henry	Cherokee	Woodbury
Macoupin	Howard	Chickasaw	Worth
Madison	Huntington	Clarke	Wright
Marion	Jackson	Clay	
Marshall	Jasper	Clayton	**Kansas** (105 counties)
Mason	Jay	Clinton	Allen
Massac	Jefferson	Crawford	Anderson
Menard	Jennings	Dallas	Atchison
Mercer	Johnson	Davis	Barber
Monroe	Knox	Decatur	Barton
Montgomery	Kosciusko	Delaware	Bourbon
Morgan	Lagrange	Des Moines	Brown
Moultrie	Lake	Dickinson	Butler
Ogle	La Porte	Dubuque	Chase
Peoria	Lawrence	Emmet	Chautauqua
Perry	Madison	Fayette	Cherokee
Piatt	Marion	Floyd	Cheyenne
Pike	Marshall	Franklin	Clark
Pope	Martin	Fremont	Clay
Pulaski	Miami	Greene	Cloud
Putnam	Monroe	Grundy	Coffey
Randolph	Montgomery	Guthrie	Comanche
Richland	Morgan	Hamilton	Cowley
Rock Island	Newton	Hancock	Crawford
St. Clair	Noble	Hardin	Decatur
Saline	Ohio	Harrison	Dickinson
Sangamon	Orange	Henry	Doniphan
Schuyler	Owen	Howard	Douglas
Scott	Parke	Humboldt	Edwards
Shelby	Perry	Ida	Elk
Stark	Pike	Iowa	Ellis
Stephenson	Porter	Jackson	Ellsworth
Tazewell	Posey	Jasper	Finney
Union	Pulaski	Jefferson	Ford
Vermilion	Putnam	Johnson	Franklin
Wabash	Randolph	Jones	Geary
Warren	Ripley	Keokuk	Gove
Washington	Rush	Kossuth	Graham
Wayne	St. Joseph	Lee	Grant
White	Scott	Linn	Gray
Whiteside	Shelby	Louisa	Greeley
Will	Spencer	Lucas	Greenwood
Williamson	Starke	Lyon	Hamilton

Harper
Harvey
Haskell
Hodgeman
Jackson
Jefferson
Jewell
Johnson
Kearny
Kingman
Kiowa
Labette
Lane
Leavenworth
Lincoln
Linn
Logan
Lyon
McPherson
Marion
Marshall
Meade
Miami
Mitchell
Montgomery
Morris
Morton
Nemaha
Neosho
Ness
Norton
Osage
Osborne
Ottawa
Pawnee
Phillips
Pottawatomie
Pratt
Rawlins
Reno
Republic
Rice
Riley
Rooks
Rush
Russell
Saline
Scott
Sedgwick
Seward
Shawnee
Sheridan
Sherman
Smith
Stafford
Stanton
Stevens
Sumner
Thomas
Trego
Wabaunsee
Wallace
Washington
Wichita
Wilson
Woodson
Wyandotte

Kentucky (120 counties)

Adair
Allen
Anderson
Ballard
Barren
Bath
Bell
Boone
Bourbon
Boyd
Boyle
Bracken
Breathitt
Breckenridge
Bullitt
Butler
Caldwell
Calloway
Campbell
Carlisle
Carroll
Carter
Casey
Christian
Clark
Clay
Clinton
Crittenden
Cumberland
Daviess
Edmonson
Elliott
Estill
Fayette
Fleming
Floyd
Franklin
Fulton
Gallatin
Garrard
Grant
Graves
Grayson
Green
Greenup
Hancock
Hardin
Harlan
Harrison
Hart
Henderson
Henry
Hickman
Hopkins
Jackson
Jefferson
Jessamine
Johnson
Kenton
Knott
Knox
Larue
Laurel
Lawrence
Lee
Leslie
Letcher
Lewis
Lincoln
Livingston
Logan
Lyon
McCracken
McCreary
McLean
Madison
Magoffin
Marion
Marshall
Martin
Mason
Meade
Menifee
Mercer
Metcalfe
Monroe
Montgomery
Morgan
Muhlenberg
Nelson
Nicholas
Ohio
Oldham
Owen
Owsley
Pendleton
Perry
Pike
Powell
Pulaski
Robertson
Rockcastle
Rowan
Russell
Scott
Shelby
Simpson
Spencer
Taylor
Todd
Trigg
Trimble
Union
Warren
Washington
Wayne
Webster
Whitley
Wolfe
Woodford

Louisiana (64 parishes)

Acadia
Allen
Ascension
Assumption
Avoyelles
Beauregard
Bienville
Bossier
Caddo
Calcasieu
Caldwell
Cameron
Catahoula
Claiborne
Concordia
De Soto
E. Baton Rouge
East Carroll
East Feliciana
Evangeline
Franklin
Grant
Iberia
Iberville
Jackson
Jefferson
Jefferson Davis
Lafayette
Lafourche
La Salle
Lincoln
Livingston
Madison
Morehouse
Natchitoches
Orleans
Ouachita
Plaquemines
Pointe Coupee
Rapides
Red River
Richland
Sabine
St. Bernard
St. Charles
St. Helena
St. James
St. John the Baptist
St. Landry
St. Martin
St. Mary
St. Tammany
Tangipahoa
Tensas
Terrebone
Union
Vermilion
Vernon
Washington
Webster
W. Baton Rouge
West Carroll
West Feliciana
Winn

Maine (16 counties)

Androscoggin
Aroostook
Cumberland
Franklin
Hancock
Kennebec
Knox
Lincoln
Oxford
Penobscot
Piscataquis
Sagadahoc
Somerset
Waldo
Washington
York

Maryland (24 counties)

Allegany
Anne Arundel
Baltimore
Baltimore City
Calvert
Caroline
Carroll
Cecil
Charles
Dorchester
Frederick
Garrett
Harford
Howard
Kent
Montgomery
Prince Georges
Queen Annes
St. Marys
Somerset
Talbot
Washington
Wicomico
Worcester

Massachusetts (14 counties)

Barnstable
Berkshire
Bristol
Dukes
Essex
Franklin
Hampden
Hampshire
Middlesex
Nantucket
Norfolk
Plymouth
Suffolk
Worcester

Michigan (83 counties)

Alcona
Alger
Allegan
Alpena
Antrim
Arenac
Baraga
Barry
Bay
Benzie
Berrien
Branch
Calhoun
Cass
Charlevoix
Cheboygan
Chippewa
Clare
Clinton
Crawford
Delta
Dickinson
Eaton
Emmet
Genesee
Gladwin
Gogebic
Grand Traverse
Gratiot
Hillsdale
Houghton
Huron
Ingham
Ionia
Iosco
Iron
Isabella
Jackson
Kalamazoo
Kalkaska
Kent
Keweenaw
Lake
Lapeer
Leelanau
Lenawee
Livingston
Luce
Mackinac
Macomb
Manistee
Marquette
Mason
Mecosta
Menominee
Midland
Missaukee
Monroe
Montcalm
Montmorency
Muskegon
Newaygo
Oakland
Oceana
Ogemaw
Ontonagon
Osceola
Oscoda
Otsego
Ottawa
Presque Isle
Roscommon
Saginaw
St. Clair
St. Joseph
Sanilac
Schoolcraft
Shiawassee
Tuscola
Van Buren
Washtenaw
Wayne
Wexford

Minnesota (87 counties)

Aitkin
Anoka
Becker
Beltrami
Benton
Big Stone
Blue Earth
Brown
Carlton
Carver
Cass
Chippewa
Chisago
Clay
Clearwater
Cook
Cottonwood
Crow Wing
Dakota
Dodge
Douglas
Faribault
Fillmore
Freeborn
Goodhue
Grant
Hennepin
Houston
Hubbard
Isanti
Itasa
Jackson
Kanabec
Kandiyohi
Kittson
Koochiching
Lac Qui Parle
Lake
L. of the Woods
Le Sueur
Lincoln
Lyon
McLeod
Mahnomen
Marshall
Martin
Meeker
Mille Lacs
Morrison
Mower
Murray
Nicollet
Nobles
Norman
Olmsted
Otter Trail
Pennington
Pine
Pipestone
Polk
Pope
Ramsey
Red Lake
Redwood
Renville
Rice
Rock
Roseau
St. Louis
Scott
Sherburne
Sibley
Stearns
Steele
Stevens
Swift
Todd
Traverse
Wabasha
Wadena
Waseca
Washington
Watonwan
Wilkin
Winona
Wright
Yellow Medicine

Mississippi (82 counties)

Adams
Alcorn
Amite
Attala
Benton
Bolivar
Calhoun
Carroll
Chickasaw
Choctaw
Claiborne
Clarke
Clay
Coahoma
Copiah
Covington
De Soto
Forrest
Franklin
George
Greene
Grenada
Hancock
Harrison
Hinds
Holmes
Humphreys
Issaquena
Itawamba
Jackson
Jasper
Jefferson
Jefferson Davis
Jones
Kemper
Lafayatte
Lamar
Lauderdale
Lawrence
Leake
Lee
Leflore
Lincoln
Lowndes
Madison
Marion
Marshall
Monroe
Montgomery
Neshoba
Newton
Noxubee
Oktibbeha
Panola
Pearl River
Perry
Pike
Pontotoc
Prentiss
Quitman
Rankin
Scott
Sharkey
Simpson
Smith
Stone
Sunflower
Tallahatchie
Tate
Tippah
Tishomingo
Tunica
Union
Walthall
Warren
Washington
Wayne
Webster
Wilkinson
Winston
Yalobusha
Yazoo

Missouri (115 counties)

Adair
Andrew
Atchison
Audrain
Barry
Barton
Bates
Benton
Bollinger
Boone
Buchanan
Butler
Caldwell
Callaway
Camden
Cape Girardeau
Carroll
Carter
Cass
Cedar
Chariton
Christian
Clark
Clay
Clinton
Cole
Cooper
Crawford
Dade
Dallas
Daviess
De Kalb
Dent
Douglas
Dunklin
Franklin
Gasconade
Gentry
Greene
Grundy
Harrison
Henry
Hickory
Holt
Howard
Howell
Iron
Jackson
Jasper
Jefferson
Johnson
Knox
Laclede
Lafayette
Lawrence
Lewis
Lincoln

Linn
Livingston
McDonald
Macon
Madison
Maries
Marion
Mercer
Miller
Mississippi
Moniteau
Monroe
Montgomery
Morgan
New Madrid
Madrid
Newton
Nodaway
Oregon
Osage
Ozark
Pemiscot
Perry
Pettis
Phelps
Pike
Platte
Polk
Pulaski
Putnam
Ralls
Randolph
Ray
Reynolds
Ripley
St. Charles
St. Clair
St. Francois
St. Louis
St. Louis City
Ste. Genevieve
Saline
Schuyler
Scotland
Scott
Shannon
Shelby
Stoddard
Stone
Sullivan
Taney
Texas
Vernon
Warren
Washington
Wayne
Webster
Worth
Wright

Montana (56 counties)

Beaverhead
Big Horn
Blaine
Broadwater
Carbon
Carter
Cascade
Chouteau
Custer
Daniels
Dawson
Deer Lodge
Fallon
Fergus
Flathead
Gallatin
Garfield
Glacier

Golden Valley
Granite
Hill
Jefferson
Judith Basin
Lake
Lewis and Clark
Liberty
Lincoln
McCone
Madison
Meagher
Mineral
Missoula
Musselshell
Park
Petroleum
Phillips
Pondera
Powder River
Powell Prairie
Ravalli
Richland
Roosevelt
Rosebud
Sanders
Sheridan
Silver Bow
Stillwater
Sweet Grass
Teton
Toole
Treasure
Valley
Wheatland
Wibaux
Yellowstone

Nebraska (93 counties)
Adams
Antelope
Arthur
Banner
Blaine
Boone
Box Butte
Boyd
Brown
Buffalo
Burt
Butler
Cass
Cedar
Chase
Cherry
Cheyenne
Clay
Colfax
Cuming
Custer
Dakota
Dawes
Dawson
Deuel
Dixon
Dodge
Douglas
Dundy
Fillmore
Franklin
Frontier
Furnas
Gage
Garden
Garfield
Gosper
Grant
Greeley
Hall

Hamilton
Harlan
Hayes
Hitchcock
Holt
Hooker
Howard
Jefferson
Johnson
Kearney
Keith
Keya Paha
Kimball
Knox
Lancaster
Lincoln
Logan
Loup
McPherson
Madison
Merrick
Morrill
Nance
Nemaha
Nuckolls
Otoe
Pawnee
Perkins
Phelps
Pierce
Platte
Polk
Red Willow
Richardson
Rock
Saline
Sarpy
Saunders
Scotts Bluff
Seward
Sheridan
Sherman
Sioux
Stanton
Thayer
Thomas
Thurston
Valley
Washington
Wayne
Webster
Wheeler
York

Nevada (16 counties)
Churchill
Clark
Douglas
Elko
Esmeralda
Eureka
Humboldt
Lander
Lincoln
Lyon
Mineral
Nye
Pershing
Storey
Washoe
White Pine

New Hampshire (10 counties)

Belknap
Carroll
Cheshire
Coos
Grafton

Hillsboro
Merrimack
Rockingham
Strafford
Sullivan

New Jersey (21 counties)
Atlantic
Bergen
Burlington
Camden
Cape May
Cumberland
Essex
Gloucester
Hudson
Hunterdon
Mercer
Middlesex
Monmouth
Morris
Ocean
Passaic
Salem
Somerset
Sussex
Union
Warren

New Mexico (33 counties)
Bernalillo
Catron
Chaves
Cibola
Colfax
Curry
De Baca
Dona Ana
Eddy
Grant
Guadalupe
Harding
Hidalgo
Lea
Lincoln
Los Alamos
Luna
McKinley
Mora
Otero
Quay
Rio Arriba
Roosevelt
Sandoval
San Juan
San Miguel
Santa Fe
Sierra
Socorro
Taos
Torrance
Union
Valencia

New York (62 counties)
Albany
Allegany
Bronx
Broome
Cattaraugus
Cayuga
Chautauqua
Chemung
Chenango
Clinton
Columbia
Cortland
Delaware
Dutchess

Erie
Essex
Franklin
Fulton
Genesee
Greene
Hamilton
Herkimer
Jefferson
Kings
Lewis
Livingston
Madison
Monroe
Montgomery
Nassau
New York
Niagara
Oneida
Onondaga
Ontario
Orange
Orleans
Oswego
Otsego
Putnam
Queens
Rensselaer
Richmond
Rockland
St. Lawrence
Saratoga
Schenectady
Schoharie
Schuyler
Seneca
Steuben
Suffolk
Sullivan
Tioga
Tompkins
Ulster
Warren
Washington
Wayne
Westchester
Wyoming
Yates

North Carolina (100 counties)

Alamance
Alexander
Alleghany
Anson
Ashe
Avery
Beaufort
Bertie
Bladen
Brunswick
Buncombe
Burke
Cabarrus
Caldwell
Camden
Carteret
Caswell
Catawba
Chatham
Cherokee
Chowan
Clay
Cleveland
Columbus
Craven
Cumberland
Currituck
Dare

Davidson
Davie
Duplin
Durham
Edgecombe
Forsyth
Franklin
Gaston
Gates
Graham
Granville
Greene
Guilford
Halifax
Harnett
Haywood
Henderson
Hertford
Hoke
Hyde
Iredell
Jackson
Johnston
Jones
Lee
Lenoir
Lincoln
McDowell
Macon
Madison
Martin
Mecklenburg
Mitchell
Montgomery
Moore
Nash
New Hanover
Northampton
Onslow
Orange
Pamlico
Pasquotank
Pender
Perquimans
Person
Pitt
Polk
Randolph
Richmond
Robeson
Rockingham
Rowan
Rutherford
Sampson
Scotland
Stanly
Stokes
Surry
Swain
Transylvania
Tyrrell
Union
Vance
Wake
Warren
Washington
Watauga
Wayne
Wilkes
Wilson
Yadkin
Yancey

North Dakota (53 counties)

Adams
Barnes
Benson
Billings
Bottineau

Bowman
Burke
Burleigh
Cass
Cavalier
Dickey
Divide
Dunn
Eddy
Emmons
Foster
Golden Valley
Grand Forks
Grant
Griggs
Hettinger
Kidder
La Moure
Logan
McHenry
McIntosh
McKenzie
McLean
Mercer
Morton
Mountrail
Nelson
Oliver
Pembina
Pierce
Ramsey
Ranson
Renville
Richland
Rolette
Sargent
Sheridan
Sioux
Slope
Stark
Steele
Stutsman
Towner
Traill
Walsh
Ward
Wells
Williams

Ohio (88 counties)

Adams
Allen
Ashland
Ashtabula
Athens
Auglaize
Belmont
Brown
Butler
Carroll
Champaign
Clark
Clermont
Clinton
Columbiana
Coshocton
Crawford
Cuyahoga
Darke
Defiance
Delaware
Erie
Fairfield
Fayette
Franklin
Fulton
Gallia
Geauga
Greene

Guernsey
Hamilton
Hancock
Hardin
Harrison
Henry
Highland
Hocking
Holmes
Huron
Jackson
Jefferson
Knox
Lake
Lawrence
Licking
Logan
Lorain
Lucas
Madison
Mahoning
Marion
Medina
Meigs
Mercer
Miami
Monroe
Montgomery
Morgan
Morrow
Muskingum
Noble
Ottawa
Paulding
Perry
Pickaway
Pike
Portage
Preble
Putnam
Richland
Ross
Sandusky
Scioto
Seneca
Shelby
Stark
Summit
Trumbull
Tuscarawas
Union
Van Wert
Vinton
Warren
Washington
Wayne
Williams
Wood
Wyandot

Oklahoma (77 counties)

Adair
Alfalfa
Atoka
Beaver
Beckham
Blaine
Bryan
Caddo
Canadian
Carter
Cherokee
Choctaw
Cimarron
Cleveland
Coal
Comanche
Cotton
Craig

Creek
Custer
Delaware
Dewey
Ellis
Garfield
Garvin
Grady
Grant
Greer
Harmon
Harper
Haskell
Hughes
Jackson
Jefferson
Johnston
Kay
Kingfisher
Kiowa
Latimer
Le Flore
Lincoln
Logan
Love
McClain
McCurtain
McIntosh
Major
Marshall
Mayes
Murray
Muskogee
Noble
Nowata
Okfuskee
Oklahoma
Okmulgee
Osage
Ottawa
Pawnee
Payne
Pittsburg
Pontotoc
Pottawatomie
Pushmataha
Roger Mills
Rogers
Seminole
Sequoyah
Stephens
Texas
Tillman
Tulsa
Wagoner
Washington
Washita
Woods
Woodward

Oregon (36 counties)
Baker
Benton
Clackamas
Clatsop
Columbia
Coos
Crook
Curry
Deschutes
Douglas
Gilliam
Grant
Harney
Hood River
Jackson
Jefferson
Josephine
Klamath

Lake
Lane
Lincoln
Linn
Malheur
Marion
Morrow
Multnomah
Polk
Sherman
Tillamook
Umatilla
Union
Wallowa
Wasco
Washington
Wheeler
Yamhill

Pennsylvania (67 counties)
Adams
Allegheny
Armstrong
Beaver
Bedford
Berks
Blair
Bradford
Bucks
Butler
Cambria
Cameron
Carbon
Centre
Chester
Clarion
Clearfield
Clinton
Columbia
Crawford
Cumberland
Dauphin
Delaware
Elk
Erie
Fayette
Forest
Franklin
Fulton
Greene
Huntingdon
Indiana
Jefferson
Juniata
Lackawanna
Lancaster
Lawrence
Lebanon
Lehigh
Luzerne
Lycoming
McKean
Mercer
Mifflin
Monroe
Montgomery
Montour
Northampton
Northumberland
Perry
Philadelphia
Pike Potter
Schuylkill
Snyder
Somerset
Sullivan
Susquehanna
Tioga
Union

Venango
Warren
Washington
Wayne
Westmoreland
Wyoming
York

Rhode Island (5 counties)
Bristol
Kent
Newport
Providence
Washington

South Carolina
  (46 counties)
Abbeville
Aiken
Allendale
Anderson
Bamberg
Barnwell
Beaufort
Berkeley
Calhoun
Charleston
Cherokee
Chester
Chesterfield
Clarendon
Colleton
Darlington
Dillon
Dorchester
Edgefield
Fairfield
Florence
Georgetown
Greenville
Greenwood
Hampton
Horry
Jasper
Kershaw
Lancaster
Laurens
Lee
Lexington
McCormick
Marion
Marlboro
Newberry
Oconee
Orangeburg
Pickens
Richland
Saluda
Spartanburg
Sumter
Union
Williamsburg
York

South Dakota (66 counties)
Aurora
Beadle
Bennett
Bon Homme
Brookings
Brown
Brule
Buffalo
Butte
Campbell
Charles Mix
Clark
Clay
Codington

Corson
Custer
Davison
Day
Deuel
Dewey
Douglas
Edmunds
Fall River
Faulk
Grant
Gregory
Haakon
Hamlin
Hand
Hanson
Harding
Hughes
Hutchinson
Hyde
Jackson
Jerauld
Jones
Kingsbury
Lake
Lawrence
Lincoln
Lyman
McCook
McPherson
Marshall
Meade
Mellette
Miner
Minnehaha
Moody
Pennington
Perkins
Potter
Roberts
Sanborn
Shannon
Spink
Stanley
Sully
Todd
Tripp
Turner
Union
Walworth
Yankton
Ziebach

Tennessee (95 counties)

Anderson
Bedford
Benton
Bledsoe
Blount
Bradley
Campbell
Cannon
Carroll
Carter
Cheatham
Chester
Claiborne
Clay
Cocke
Coffee
Crockett
Cumberland
Davidson
Decatur
DeKalb
Dickson
Dyer
Fayette
Fentress

Franklin
Gibson
Giles
Grainger
Greene
Grundy
Hamblen
Hamilton
Hancock
Hardeman
Hardin
Hawkins
Haywood
Henderson
Henry
Hickman
Houston
Humphreys
Jackson
Jefferson
Johnson
Knox
Lake
Lauderdale
Lawrence
Lewis
Lincoln
Loudon
McMinn
McNairy
Macon
Madison
Marion
Marshall
Maury
Meigs
Monroe
Montgomery
Moore
Morgan
Obion
Overton
Perry
Pickett
Polk
Putnam
Rhea
Roane
Robertson
Rutherford
Scott
Sequatchie
Sevier
Shelby
Smith
Stewart
Sullivan
Sumner
Tipton
Trousdale
Unicoi
Union
Van Buren
Warren
Washington
Wayne
Weakley
White
Williamson
Wilson

Texas (254 counties)

Anderson
Andrews
Angelina
Aransas
Archer
Armstrong
Atascosa

Austin
Bailey
Bandera
Bastrop
Baylor
Bee
Bell
Bexar
Blanco
Borden
Bosque
Bowie
Brazoria
Brazos
Brewster
Briscoe
Brooks
Brown
Burleson
Burnet
Caldwell
Calhoun
Callahan
Cameron
Camp
Carson
Cass
Castro
Chambers
Cherokee
Childress
Clay
Cochran
Coke
Coleman
Collin
Collingsworth
Colorado
Comal
Comanche
Concho
Cooke
Coryell
Cottle
Crane
Crockett
Crosby
Culberson
Dallam
Dallas
Dawson
Deaf Smith
Delta
Denton
De Witt
Dickens
Dimmit
Donley
Duval
Eastland
Ector
Edwards
Ellis
El Paso
Erath
Falls
Fannin
Fayette
Fisher
Floyd
Foard
Fort Bend
Franklin
Freestone
Frio
Gaines
Galveston
Garza
Gillespie

Glasscock
Goliad
Gonzales
Gray
Grayson
Gregg
Grimes
Guadelupe
Hale
Hall
Hamilton
Hansford
Hardeman
Hardin
Harris
Harrison
Hartley
Haskell
Hays
Hemphill
Henderson
Hidalgo
Hill
Hockley
Hood
Hopkins
Houston
Howard
Hudspeth
Hunt
Hutchinson
Irion
Jack
Jackson
Jasper
Jeff Davis
Jefferson
Jim Hogg
Jim Wells
Johnson
Jones
Karnes
Kaufman
Kendall
Kenedy
Kent
Kerr
Kimble
King
Kinney
Kleberg
Knox
Lamar
Lamb
Lampasas
La Salle
Lavaca
Lee
Leon
Liberty
Limestone
Lipscomb
Live Oak
Llano
Loving
Lubbock
Lynn
McCulloch
McLennan
McMullen
Madison
Marion
Martin
Mason
Matagorda
Maverick
Medina
Menard
Midland

Milam
Mills
Mitchell
Montague
Montgomery
Moore
Morris
Motley
Nacogdoches
Navarro
Newton
Nolan
Nueces
Ochiltree
Oldham
Orange
Palo Pinto
Panola
Parker
Parmer
Pecos
Polk
Potter
Presidio
Rains
Randall
Reagan
Real
Red River
Reeves
Refugio
Roberts
Robertson
Rockwall
Runnels
Rusk
Sabine
San Augustine
San Jacinto
San Patricio
San Saba
Schleicher
Scurry
Shackelford
Shelby
Sherman
Smith
Somervell
Starr
Stephens
Sterling
Stonewall
Sutton
Swisher
Tarrant
Taylor
Terrell
Throckmorton
Titus
Tom Green
Travis
Trinity
Tyler
Upshur
Upton
Uvalde
Val Verde
Van Zandt
Victoria
Walker
Waller
Ward
Washington
Webb
Wharton
Wheeler
Wichita
Wilbarger
Willacy

Williamson
Wilson
Winkler
Wise
Wood
Yoakum
Young
Zapata
Zavala

Utah (29 counties)

Beaver
Box Elder
Cache
Carbon
Daggett
Davis
Duchesne
Emery
Garfield
Grand
Iron
Juab
Kane
Millard
Morgan
Piute
Rich
Salt Lake
San Juan
Sanpete
Sevier
Summit
Tooele
Uintah
Utah
Wasatch
Washington
Wayne
Weber

Vermont (14 counties)

Addison
Bennington
Caledonia
Chittendon
Essex
Franklin
Grand Isle
Lamoille
Orange
Orleans
Rutland
Washington
Windham
Windsor

Virginia (95 counties)

Accomack
Albemarle
Alleghany
Amelia
Amherst
Appomattox
Arlington
Augusta
Bath
Bedford
Bland
Botetourt
Brunswick
Buchanan
Buckingham
Campbell
Caroline
Carroll
Charles City
Charlotte
Chesterfield

Clarke
Craig
Culpeper
Cumberland
Dickenson
Dinwiddie
Essex
Fairfax
Fauquier
Floyd
Fluvanna
Franklin
Frederick
Giles
Gloucester
Goochland
Grayson
Greene
Greensville
Halifax
Hanover
Henrico
Henry
Highland
Isle of Wight
James City
King and Queen
King George
King William
Lancaster
Lee
Loudoun
Louisa
Lunenberg
Madison
Mathews
Mecklenburg
Middlesex
Montgomery
Nelson
New Kent
Northampton
Northumberland
Nottoway
Orange
Page
Patrick
Pittsylvania
Powhatan
Prince Edward
Prince George
Prince William
Pulaski
Rappahannock
Richmond
Roanoke
Rockbridge
Rockingham
Russell
Scott
Shenandoah
Smyth
Southampton
Spotsylvania
Stafford
Surry
Sussex
Tazewell
Warren
Washington
Westmoreland
Wise
Wythe
York

Washington (39 counties)

Adams
Asotin
Benton

Chelan
Clallam
Clark
Columbia
Cowlitz
Douglas
Ferry
Franklin
Garfield
Grant
Grays Harbor
Island
Jefferson
King
Kitsap
Kittitas
Klickitat
Lewis
Lincoln
Mason
Okanogan
Pacific
Pend Oreille
Pierce
San Juan
Skagit
Skamania
Snohomish
Spokane
Stevens
Thurston
Wahkiakum
Walla Walla
Whatcom
Whitman
Yakima

West Virginia (55 counties)

Barbour
Berkeley
Boone
Braxton
Brooke
Cabell
Calhoun
Clay
Doddridge
Fayette
Gilmer
Grant
Greenbrier
Hampshire
Hancock
Hardy
Harrison
Jackson
Jefferson
Kanawha
Lewis
Lincoln
Logan
McDowell
Marion
Marshall
Mason
Mercer
Mineral
Mingo
Monongalia
Monroe
Morgan
Nicholas
Ohio
Pendleton
Pleasants
Pocahontas
Preston
Putnam
Raleigh

Randolph
Ritchie
Roane
Summers
Taylor
Tucker
Tyler
Upshur
Wayne
Webster
Wetzel
Wirt
Wood
Wyoming

Wisconsin (72 counties)

Adams
Ashland
Barron
Bayfield
Brown
Buffalo
Burnett
Calumet
Chippewa
Clark
Columbia
Crawford
Dane
Dodge
Door
Douglas
Dunn
Eau Claire
Florence
Fond du Lac
Forest
Grant
Green
Green Lake
Iowa
Iron
Jackson
Jefferson
Juneau
Kenosha
Kewaunee
La Crosse
Lafayette
Langlade
Lincoln
Manitowoc
Marathon
Marinette
Marquette
Menominee
Milwaukee
Monroe
Oconto
Oneida
Outagamie
Ozaukee
Pepin
Pierce
Polk
Portage
Price
Racine
Richland
Rock
Rusk
St. Croix
Sauk
Sawyer
Shawano
Sheboygan
Taylor
Trempealeau
Vernon

Vilas	**Wyoming** (23 counties)	Hot Springs	Sublette
Walworth	Albany	Johnson	Sweetwater
Washburn	Big Horn	Laramie	Teton
Washington	Campbell	Lincoln	Uinta
Waukesha	Carbon	Natrona	Washakie
Waupaca	Converse	Niobrara	Weston
Waushara	Crook	Park	
Winnebago	Fremont	Platte	
Wood	Goshen	Sheridan	**Total US Counties = 3076**

Notes: Since USA-CA started, counties that have been absorbed include Princess Ann, Norfolk and Nansemond in Virginia; Ormsby in Nevada; Washabaugh in South Dakota. Carson City, Nevada, is now considered an independent city. Richmond County in New York is now called Staten Island County. A new county has been added to New Mexico—Cibola (used to be part of Valencia County).

The following is a list of independent cities and the county for which each can be used.

Independent City   Counts as_____ County

[VIRGINIA]

Alexandria	Arlington or Fairfax
Bedford	Bedford
Bristol	Washington
Buena Vista	Rockbridge
Charlottesville	Albermarle
Chesapeake	Isle of Wight
Clifton Forge	Alleghany
Colonial Heights	Chesterfield or Prince George
Covington	Alleghany
Danville	Pittsylvania
Emporia	Greensville
Fairfax	Fairfax
Falls Church	Fairfax
Fort Monroe	York
Franklin	Southampton
Fredericksburg	Spotsylvania
Galax	Carroll or Grayson
Hampton	York
Harrisonburg	Rockingham

Hopewell	Prince George
Lexington	Rockbridge
Lynchburg	Amherst or Bedford or Campbell
Manassas	Prince William
Manassas Park	Prince William
Martinsville	Henry
Newport News	York
Norfolk	Isle of Wight
Norton	Wise
Petersburg	Chesterfield or Dinwiddie or Prince George
Poquoson	York
Portsmouth	Isle of Wight
Radford	Montgomery
Richmond	Chesterfield or Henrico
Roanoke	Roanoke
Salem	Roanoke
South Boston	Halifax
Staunton	Augusta
Suffolk	Isle of Wight or Southampton
Virginia Beach	Isle of Wight
Waynesboro	Augusta
Williamsburg	James City
Winchester	Frederick

Carson City, Nevada	Douglas or Lyon or Story or Washoe
Washington, DC	Montgomery or Prince Georges, Maryland

Propagation and the Proper Use of Sunrise, Sunset and Twilight Tables

*This introduction to the **Worldwide UTC Sunrise, Sunset and Twilight Times** reference section was written by ARRL Technical Advisor Thomas Russell, N4KG. Tom lives near Huntsville, Alabama, USA and is well known for contest and DXploits, especially on the lower HF bands. He has worked 300 countries on 80 meters using a TS-830, an SB-220, a single vertical and various dipoles. His antenna farm also features eight highly directional Beverage antennas for reception. We have included this explanation of propagation to help you use, and benefit from, the tables that follow.*

The transition from dark to daylight and from daylight to dark causes drastic changes to occur in the ionosphere, affecting all high frequency (HF) propagation. The all-band DXer needs to understand these changes to improve his chances of long-distance communication, especially on long or difficult paths.

There are several factors which affect high frequency propagation: path, frequency, solar activity and time. Of all of these factors, frequency and time are the only ones over which we have much control. Understanding the effects of the sun will help the DXer select the best frequency for a given time, or perhaps more importantly, the best time for a given frequency, for each selected path and level of sunspot activity.

Time is a complex factor in propagation, encompassing time in the (11-year) sunspot cycle, time of year (season and tilt of the earth), and time of day (daylight, dark, or twilight).

From October through March, the north pole is in partial to total darkness, permitting over-the-pole, northern-hemisphere low and high frequency propagation between opposite sides of the earth. This is also true for the south pole and the southern hemisphere from March through October. Summertime propagation is totally different from the rest of the year, with HF openings and closings occurring much later in the day. The polar paths are primarily restricted to 40 through 15 meters in the summer, while transequatorial paths predominate on the low bands and on 10 meters.

Generally, the low bands (160, 80, 40 and 30 meters) exhibit best long distance propagation from before sunset, through darkness, to after sunrise. The high bands (20, 17, 15, 12 and 10 meters) tend to open from near sunrise, stay open throughout the day, and close after darkness. These openings and closings are, of course, dependent on solar activity, frequency and season. The exact behavior at sunrise and sunset is highly dependent on frequency. This behavior will next be examined in detail for the northern-hemisphere winter months of October through March.

The sunrise/sunset line, also known as the terminator, grayline or twilight zone, is a circle about the earth whose position depends on the time of year (season) and time of day.

Maximum usable frequencies (MUFs) are highest on the illuminated face of the earth. So is the absorption caused by the effects of the energized D region of the ionosphere. Propagation losses are lowest near the maximum usable frequency.

Maximum usable frequencies drop on the dark side of the earth, to a minimum value of approximately one third the daytime MUF, just before sunrise. Absorption also quickly subsides as the D region is no longer illuminated.

Propagation along the grayline is especially interesting. MUFs are rising rapidly on the sunrise side and are still high on the sunset side of the earth. The D region has not yet been energized on the sunrise side, and is rapidly dissipating on the sunset side of the earth, resulting in low absorption losses. The net result is that for a period ranging from a few minutes at low and high frequencies, to one or two hours at intermediate frequencies (20 meters), with suitable ionospheric conditions, it is possible to communicate with any location within the twilight zone on any HF band!

To better understand these effects, let us examine what happens on a band-by-band basis as the sun begins to rise in the northern hemisphere (eastern US) during the winter months. Before sunrise, while still totally dark, the MUF is at its lowest level of the day. This is usually the least productive time of the day unless 20 meters is already open to Europe, which occasionally happens during high sunspot activity and quiet geomagnetic conditions. As the first rays of light approach overhead, the MUF begins to rise in the direction of the sun (SE) and the high bands will open first in that direction before swinging northward toward Europe.

As the sun continues to rise, it proceeds to illuminate the higher altitudes first and the lower altitudes later, forming a westward-moving wedge of light that cuts into the darkness. This sloping wedge has an especially beneficial effect on low frequency propagation to the west. It tends to act as a reflector, similar to a parabolic reflector, focusing energy (signals coming from the west) along the terminator or grayline.

This focusing effect is seen as a sudden (and often large) increase in signal strength from the west that continues to build to a peak and then fades back down as the D region absorption increases with illumination. The lower the frequency, the earlier the peak and the shorter the duration.

The 160-meter band has the shortest sunrise and sunset peaks of all the amateur bands. These may vary from only a few minutes to 20 or 30 minutes maximum. On October 25, 1987 at 1148 UTC I received the surprise of my life as a voice rose from the noise to S5 or 6 in less than a minute and was completely gone five minutes later. The call was from VU4GDG in the Andaman Islands!

The 160-meter band is also the most unpredictable of all the amateur bands. There may be no peak whatsoever at either sunset or sunrise, yet tremendous propagation may arise a few hours after sunset and drop out after an hour or two. I have also heard signals from Europe and Africa peak and disappear a full hour before their sunrise. As I said, 160 meters is the least predictable of the amateur bands.

The 80-meter band exhibits a consistent pattern of signal enhancement (from the west) beginning with first light, building to a maximum at official sunrise (90° solar zenith angle) which may last from ten to twenty minutes, and slowly dropping from D region absorption as the sun rises. With simple antennas, 30 minutes past sunrise usually marks the end of the opening.

Long-path openings do indeed occur on 80 meters, with shorter peaks than for signals to the west, typically exhibiting a pronounced peak for 5 to 7 minutes and a total duration of 15 to 20 minutes maximum. (The term "long-path" refers to the longer arc of the great circle path between one station and another.) I have observed long-path openings to Asia as far west as the sunset terminator. On January 31, 1989, UJ8JMM heard my long-path signals for a total of 3 minutes at 1232 UTC. My sunrise and his sunset were coincident at 1245 UTC on that date.

These long-path openings are predominantly observed during high sunspot activity and quiet conditions. Some of my better 80-meter morning long-path DX includes HS, UA9 (Zones 17, 18), UI, UJ, UL, UM, VU, 4S7 and 8Q7.

The 40-meter band is the premier low-frequency band, with long-path openings to central Asia from October through March that are more reliable than 20 meters! Contrary to some published propagation material, 40 meter propagation does NOT peak to the west *before* sunrise but about 30 minutes *after* sunrise. Propagation just before sunrise can be quite poor in fact. Peak propagation spans about 30 minutes and drops off slowly for an hour or more. Long-path fans benefit from a convenient beacon located in Kazakhstan (UL7) that transmits the letter "V" at 10 WPM on 7002 kHz. This signal can also be heard in the evenings around 0200 to 0300 UTC.

For polar paths, the 10-meter band is the most difficult of the high-frequency bands, requiring a quite high maximum usable frequency and quiet (solar) conditions. In the eastern US, for example, this band will open to the southeast from just before to just after sunrise, depending on solar flux levels, and swing northward into Europe and Western Asia in another 15 to 20 minutes. From the eastern half of the US, the best times to work central Asia are usually within the first two hours following sunrise on this band. In the spring and fall, there is often a long-path (southeast) opening to the Far East and the South China Sea region in this same time period.

In the evenings, as the sun begins to set, the MUF begins to drop and the D region begins to dissipate, reducing absorption on all frequencies. During high sunspot activity, the higher bands can remain open to the west for 2 or 3 hours past sunset.

In the late afternoon, the low bands will begin to open in a direction across the approaching terminator (northeast in the northern hemisphere winter) beginning with 40 meters as much as 2 hours before sunset. As sunset approaches, signals can be heard more and more to the southeast. From just before sunset until total darkness, signals will peak along the terminator (southeast) on all bands from 160 through 20 meters. Long-path to Southeast Asia is especially good on 20 and 40 meters during this period, and occasionally on 80 and even 160 meters.

The northeastern states have perfect grayline alignment with JA, DU, 9M6, 9M8, V85 and YB in midwinter. The southeastern and midwestern states have perfect grayline alignment with HL, BY4, BY1, XU, XV, XW, HS, 9V1, 9M2 and YB.

On 80 meters, DU, HS, JA, V85, VK9Y, XW, YB, 9M2, 9M6 and 9V1 are known to have been worked from the southeast on the sunset long-path. 9M2AX has been worked by a lucky few on 160-meter long-path. Several problems combine to make low-band, long-path contacts difficult and rare. Low activity, brief openings, atmospheric noise, E-region MUF, absorption and station capabilities are other reasons. Yet with the right timing, persistence can be rewarded with some rare and exotic DX indeed! Although these openings are not always there on a daily basis, they always occur in the time periods indicated. These sunset/sunrise tables should help you plan when to look and listen for the rare DX you need. Good luck!

Worldwide UTC Sunrise, Sunset and Twilight Times

The following pages were prepared by Chuck Hutchinson, K8CH. They contain computer-generated tables of sunrise, sunset and twilight times at various locations throughout the world for the 7th (first half) and 21st day (second half) of each month. Times are in Coordinated Universal Time (UTC). The calculations are based on the latitude and longitude coordinates in Chapter 4 and should be accurate to a minute or two.

These calculations are based on the formulas provided in *DXing by Computer*, by Van Brollini, NS6N, and Walter Buchanan, Published in *Ham Radio*, August 1984.

January 7 — Sunrise and Sunset, in UTC

Pref	State/Province/Country/City	Tlight	Sunrise	Sunset	Tlight
VE1	New Brunswick, Saint John	11:29	12:03	20:56	21:30
	Nova Scotia, Halifax	11:18	11:51	20:49	21:22
	Prince Edward Island, Charlottetown	11:20	11:54	20:41	21:15
VE2	Quebec, Montreal	12:00	12:34	21:26	21:59
	Quebec City	11:54	12:29	21:11	21:46
VE3	Ontario, London	12:24	12:56	22:05	22:37
	Ottawa	12:08	12:42	21:34	22:08
	Sudbury	12:32	13:07	21:52	22:26
	Toronto	12:18	12:51	21:55	22:27
VE4	Manitoba, Winnipeg	13:47	14:25	22:42	23:20
VE5	Saskatchewan, Regina	14:19	14:58	23:10	23:48
	Saskatoon	14:34	15:14	23:10	23:51
VE6	Alberta, Calgary	14:59	15:38	23:45	00:24
	Edmonton	15:06	15:48	23:30	00:13
VE7	British Columbia, Prince George	15:45	16:28	00:05	00:48
	Prince Rupert	16:16	17:00	00:33	01:16
	Vancouver	15:29	16:07	00:29	01:06
VE8	Northwest Territories, Yellowknife	15:57	17:01	22:24	23:29
	Resolute	——	——	——	——
VY1	Yukon, Whitehorse	17:08	18:05	00:06	01:04
VO1	Newfoundland, St. John's	10:43	11:18	19:54	20:30
VO2	Labrador, Goose Bay	11:33	12:15	19:59	20:41
W1	Connecticut, Hartford	11:47	12:18	21:34	22:06
	Maine, Bangor	11:39	12:12	21:09	21:42
	Portland	11:42	12:14	21:19	21:51
	Massachusetts, Boston	11:42	12:13	21:26	21:58
	New Hampshire, Concord	11:45	12:18	21:25	21:57
	Rhode Island, Providence	11:41	12:13	21:29	22:00
	Vermont, Montpelier	11:53	12:26	21:26	21:59
W2	New Jersey, Atlantic City	11:48	12:18	21:48	22:18
	New York, Albany	11:53	12:25	21:36	22:08
	Buffalo	12:14	12:46	21:56	22:28
	New York City	11:49	12:20	21:43	22:13
	Syracuse	12:04	12:36	21:44	22:16
W3	Delaware, Wilmington	11:53	12:23	21:52	22:22
	District of Columbia, Washington	11:57	12:27	22:00	22:30
	Maryland, Baltimore	11:56	12:26	21:57	22:27
	Pennsylvania, Harrisburg	12:00	12:30	21:56	22:26
	Philadelphia	11:52	12:22	21:50	22:20
	Pittsburgh	12:12	12:43	22:08	22:38
	Scranton	11:58	12:29	21:48	22:19
W4	Alabama, Montgomery	12:20	12:47	22:54	23:21
	Florida, Jacksonville	11:58	12:24	22:40	23:07
	Miami	11:43	12:08	22:44	23:09
	Pensacola	12:20	12:46	23:02	23:28
	Georgia, Atlanta	12:15	12:43	22:43	23:10
	Savannah	11:59	12:26	22:34	23:01
	Kentucky, Lexington	12:25	12:54	22:33	23:02
	Louisville	12:30	13:00	22:37	23:07
	North Carolina, Charlotte	12:04	12:32	22:25	22:53
	Raleigh	11:56	12:25	22:15	22:43
	Wilmington	11:50	12:18	22:16	22:44
	South Carolina, Columbia	12:02	12:30	22:29	22:56
	Tennessee, Knoxville	12:18	12:46	22:35	23:04
	Memphis	12:41	13:09	23:03	23:31
	Nashville	12:30	12:58	22:47	23:15
	Virginia, Norfolk	11:50	12:18	22:03	22:31
	Richmond	11:55	12:24	22:06	22:35
W5	Arkansas, Little Rock	12:49	13:17	23:12	23:40
	Louisiana, New Orleans	12:30	12:57	23:15	23:41
	Shreveport	12:50	13:17	23:23	23:50
	Mississippi, Jackson	12:36	13:02	23:10	23:37
	New Mexico, Albuquerque	13:47	14:15	00:09	00:37
	Oklahoma, Oklahoma City	13:11	13:39	23:31	23:59
	Texas, Abilene	13:14	13:41	23:47	00:14
	Amarillo	13:28	13:56	23:49	00:17
	Dallas	13:03	13:30	23:35	00:02
	El Paso	13:40	14:06	00:16	00:43
	San Antonio	13:03	13:29	23:50	00:16
W6	California, Los Angeles	14:31	14:59	00:57	01:25
	San Francisco	14:56	15:25	01:05	01:34
W7	Arizona, Flagstaff	14:07	14:35	00:29	00:57
	Phoenix	14:06	14:33	00:35	01:02
	Idaho, Boise	14:45	15:18	00:23	00:55
	Pocatello	14:29	15:01	00:10	00:42
	Montana, Billings	14:20	14:54	23:44	00:18
	Butte	14:37	15:11	23:59	00:34
	Great Falls	14:37	15:12	23:49	00:24
	Nevada, Las Vegas	14:23	14:52	00:40	01:08
	Reno	14:49	15:19	00:50	01:20
	Oregon, Portland	15:16	15:50	00:42	01:16
	Utah, Salt Lake City	14:21	14:52	00:14	00:45
	Washington, Seattle	15:21	15:56	00:33	01:08
	Spokane	15:02	15:37	00:13	00:48
	Wyoming, Cheyenne	13:53	14:24	23:45	00:16
	Sheridan	14:12	14:45	23:42	00:15
W8	Michigan, Detroit	12:29	13:01	22:14	22:46
	Grand Rapids	12:42	13:14	22:23	22:55
	Sault Ste. Marie	12:46	13:21	22:05	22:40
	Traverse City	12:46	13:19	22:16	22:49
	Ohio, Cincinnati	12:27	12:57	22:30	22:59
	Cleveland	12:22	12:53	22:11	22:42
	Columbus	12:23	12:54	22:21	22:51
	West Virginia, Charleston	12:14	12:44	22:20	22:49
W9	Illinois, Chicago	12:46	13:18	22:34	23:05
	Indiana, Indianapolis	12:36	13:06	22:34	23:04
	Wisconsin, Green Bay	12:55	13:28	22:27	23:00
	Milwaukee	12:50	13:22	22:31	23:03
W0	Colorado, Denver	13:51	14:21	23:50	00:20
	Grand Junction	14:04	14:33	00:06	00:36
	Iowa, Des Moines	13:10	13:41	22:59	23:30
	Kansas, Pratt	13:21	13:50	23:30	23:59
	Wichita	13:15	13:44	23:25	23:54
	Minnesota, Duluth	13:18	13:53	22:35	23:10
	Minneapolis	13:17	13:51	22:46	23:20
	Missouri, Columbia	12:58	13:28	23:01	23:31
	Kansas City	13:08	13:37	23:10	23:40
	St. Louis	12:49	13:18	22:54	23:23
	Nebraska, North Platte	13:37	14:08	23:29	00:00
	Omaha	13:18	13:49	23:09	23:40
	North Dakota, Fargo	13:37	14:12	22:53	23:28
	South Dakota, Rapid City	13:55	14:27	23:29	00:02
-		05:39	06:02	18:09	18:31
1A0	S.M.O.M	06:06	06:38	15:54	16:25
1S	Spratly Is.	22:26	22:49	10:27	10:49
3A	Monaco	06:31	07:04	16:08	16:40
3B6	Agalega	01:35	01:58	14:40	15:03
3B7	St. Brandon	01:11	01:34	14:38	15:01
3B8	Mauritius	01:12	01:36	14:54	15:19
3B9	Rodriguez Is.	00:49	01:14	14:30	14:54
3C	Equatorial Guinea, Bata	05:03	05:25	17:27	17:49
	Malabo	05:10	05:33	17:27	17:50
3C0	Annobon Is.	05:14	05:37	17:49	18:11
3D2	Fiji Is., Suva	17:13	17:37	06:47	07:11
3D2	Conway Reef	17:22	17:46	07:09	07:34
3D2	Rotuma Is.	17:27	17:50	06:39	07:02
3DA	Swaziland, Mbabane	02:44	03:10	16:52	17:18
3V	Tunisia, Tunis	06:03	06:32	16:17	16:46
3W, XV	Vietnam, Ho Chi Minh City (Saigon)	22:50	23:13	10:44	11:07
	Hanoi	23:11	23:35	10:29	10:53
3X	Guinea, Conakry	06:50	07:12	18:48	19:10
3Y	Bouvet	02:28	03:22	20:21	21:15
3Y	Peter I Is.	——	——	——	——
4J, 4K	Azerbaijan, Baku	03:33	04:03	13:28	13:59
4L	Georgia, Tbilisi	03:56	04:28	13:45	14:16
4S	Sri Lanka, Colombo	00:31	00:54	12:38	13:00
4U	ITU Geneva	06:43	07:17	16:04	16:38
4U	United Nations Hq.	11:49	12:20	21:43	22:13
4X, 4Z	Israel, Jerusalem	04:13	04:40	14:49	15:16
5A	Libya, Tripoli	05:45	06:12	16:18	16:45
	Benghazi	05:14	05:41	15:49	16:16
5B	Cyprus, Nicosia	04:27	04:55	14:48	15:16
5H	Tanzania, Dar es Salaam	02:49	03:12	15:43	16:05
5N	Nigeria, Lagos	05:36	05:59	17:45	18:07
5R	Madagascar, Antananarivo	01:55	02:19	15:32	15:56
5T	Mauritania, Nouakchott	07:13	07:37	18:42	19:06
5U	Niger, Niamey	05:53	06:16	17:38	18:01
5V	Togo, Lome	05:44	06:07	17:54	18:17
5W	Western Samoa, Apia	16:43	17:06	05:59	06:22
5X	Uganda, Kampala	03:30	03:52	15:58	16:21
5Z	Kenya, Nairobi	04:30	04:52	17:04	17:26
6W	Senegal, Dakar	07:13	07:37	18:54	19:17
6Y	Jamaica, Kingston	11:16	11:40	22:45	23:09
7O	Yemen, Aden	03:00	03:23	14:47	15:10
	Sanaa	03:08	03:31	14:46	15:09
7P	Lesotho, Maseru	02:50	03:17	17:13	17:40
7Q	Malawi, Lilongwe	02:59	03:23	16:18	16:41
	Blantyre	02:51	03:15	16:16	16:40
7X	Algeria, Algiers	06:32	07:01	16:46	17:15

Pref	State/Province/Country/City	Tlight	Sunrise	Sunset	Tlight
8P	Barbados, Bridgetown	09:59	10:22	21:45	22:08
8Q	Maldive Is.	00:53	01:15	13:08	13:30
8R	Guyana, Georgetown	09:43	10:06	21:50	22:13
9A	Croatia, Zagreb	06:02	06:36	15:26	16:00
9G	Ghana, Accra	05:49	06:12	18:01	18:23
9H	Malta	05:45	06:13	16:02	16:31
9J	Zambia, Lusaka	03:19	03:42	16:42	17:06
9K	Kuwait	03:18	03:44	14:04	14:30
9L	Sierra Leone, Freetown	06:46	07:09	18:48	19:10
9M2	West Malaysia, Kuala Lumpur	22:58	23:21	11:17	11:40
9M6,8	East Malaysia, Sabah, Sandakan(9M6)	21:56	22:19	10:07	10:29
	Sarawak, Kuching (9M8)	22:21	22:43	10:45	11:07
9N	Nepal, Kathmandu	00:45	01:10	11:38	12:03
9Q	Zaire, Kinshasa	04:31	04:53	17:15	17:37
	Kisangani	03:59	04:22	16:27	16:50
	Lubumbashi	03:29	03:52	16:39	17:02
9U	Burundi, Bujumbura	03:37	03:59	16:17	16:40
9V	Singapore	22:46	23:09	11:12	11:34
9X	Rwanda, Kigali	03:36	03:58	16:12	16:34
9Y	Trinidad & Tobago , Port of Spain	10:02	10:24	21:57	22:19
A2	Botswana, Gaborone	03:08	03:33	17:10	17:35
A3	Tonga, Nukualofa	16:41	17:05	06:27	06:51
A4	Oman, Musqat	02:24	02:49	13:33	13:58
A5	Bhutan, Thimbu	00:28	00:53	11:22	11:48
A6	United Arab Emirates, Abu Dhabi	02:43	03:08	13:49	14:14
A7	Qatar, Ad-Dawhah	02:56	03:20	13:58	14:23
A9	Bahrein, Al-Manamah	03:01	03:26	14:00	14:25
AP	Pakistan, Karachi	01:52	02:17	12:57	13:21
AP	Islamabad	01:45	02:12	12:13	12:40
BV	Taiwan, Taipei	22:15	22:40	09:19	09:43
BY	Peoples Rep. of China, Beijing	23:06	23:36	09:03	09:34
	Harbin	22:40	23:14	08:03	08:37
	Shanghai	22:27	22:53	09:06	09:32
	Fuzhou	22:26	22:51	09:25	09:50
	Xian	23:23	23:51	09:49	10:16
	Chongqing	23:24	23:50	10:09	10:35
	Chengdu	23:35	00:02	10:16	10:43
	Lhasa	00:25	00:51	11:10	11:36
	Urumqi	01:11	01:43	10:47	11:19
	Kashi	01:46	02:16	11:46	12:16
C2	Nauru	18:31	18:53	07:02	07:25
C3	Andorra	06:52	07:23	16:35	17:07
C5	The Gambia, Banjul	07:08	07:31	18:53	19:16
C6	Bahamas, Nassau	11:31	11:56	22:34	22:59
C9	Mozambique, Maputo	02:38	03:04	16:46	17:11
	Mozambique	02:30	02:53	15:52	16:15
CE	Chile, Santiago	09:12	09:40	23:57	00:25
CE0	Easter Island	12:04	12:30	02:16	02:42
CE0	Juan Fernandez	09:43	10:12	00:29	00:58
CE0	San Felix	10:08	10:34	00:17	00:43
CM,CO	Cuba, Havana	11:47	12:12	22:58	23:23
CN	Morocco, Casablanca	07:07	07:35	17:36	18:03
CP	Bolivia, La Paz	09:43	10:07	23:11	23:35
CT	Portugal, Lisbon	07:25	07:55	17:29	17:59
CT3	Madeira Islands, Funchal	07:43	08:10	18:16	18:43
CU	Azores , Ponta Delgada	08:29	08:58	18:38	19:07
CX	Uruguay, Montevideo	08:09	08:38	23:02	23:31
CY0	Sable Is.	11:01	11:34	20:37	21:10
CY9	St. Paul Is.	11:11	11:46	20:25	21:00
D2	Angola, Luanda	04:31	04:54	17:31	17:54
D4	Cape Verde, Praia	07:38	08:01	19:18	19:41
D6	Comoros, Moroni	02:24	02:47	15:34	15:57
DA-DL,Y2-Y9	Fed. Rep. of Germany, Bonn	06:54	07:32	15:42	16:21
	Berlin	06:35	07:15	15:08	15:49
DU	Philippines, Manilla	21:59	22:22	09:40	10:03
E3	Eritrea, Asmara	03:29	03:52	15:07	15:31
EA	Spain, Madrid	07:07	07:38	17:03	17:33
EA6	Balearic Is., Palma	06:40	07:10	16:40	17:10
EA8	Canary Is., Las Palmas	07:24	07:50	18:15	18:41
EA9	Ceuta & Melilla, Ceuta	07:03	07:32	17:21	17:50
	Melilla	06:53	07:21	17:14	17:42
EI	Ireland, Dublin	07:56	08:38	16:23	17:04
EK	Armenia, Yerevan	03:54	04:25	13:50	14:20
EL	Liberia, Monrovia	06:33	06:55	18:42	19:04
EP	Iran, Tehran	03:15	03:43	13:33	14:01
ER	Moldova, Kishinev	05:15	05:50	14:30	15:05
ES	Estonia, Tallinn	06:20	07:14	13:38	14:32
ET	Ethiopia, Addis Ababa	03:19	03:42	15:19	15:42
EU,EV,EW	Belarus, Minsk	05:43	06:26	14:04	14:47
EX	Kyrgyzstan, Frunze	02:00	02:32	11:42	12:14
EY	Tajikistan, Samarkand	02:24	02:54	12:23	12:53
	Dushanbe	02:14	02:44	12:16	12:46
EZ	Turkmenistan, Ashkhabad	02:53	03:22	13:01	13:30
F	France, Paris	07:06	07:43	16:09	16:46
FG	Guadeloupe	10:12	10:36	21:49	22:12
FJ, FS	St. Martin	10:21	10:45	21:50	22:14
FH	Mayotte	02:15	02:38	15:30	15:53
FK	New Caledonia, Noumia	17:51	18:16	07:42	08:07
FM	Martinique	10:07	10:30	21:48	22:11
FO	Clipperton Is.	13:13	13:36	01:08	01:31
FO	Fr. Polynesia, Tahiti	15:45	16:09	05:17	05:41
	Rururu, (Austral Is.)	18:52	19:17	08:44	09:09
	Hiva Oa, (Marquesas Is.)	14:38	15:01	03:42	04:04
FP	St. Pierre & Miquelon, St. Pierre	10:53	11:28	20:11	20:46
FR/G	Glorioso	02:10	02:33	15:19	15:42
FR/J,E	Juan de Nova	02:17	02:41	15:47	16:11
	Europa	02:16	02:41	16:07	16:32
FR	Reunion	01:18	01:42	15:04	15:28
FR/T	Tromelin	01:33	01:57	14:59	15:22
FT8W	Crozet	00:12	00:50	16:25	17:02
FT8X	Kerguelen	22:45	23:27	15:30	16:12
FT8Y	Antarctica, Dumont D'Urville	----	15:03	14:28	----
FT8Z	Amsterdam & St. Paul Is., Amsterdam	23:05	23:35	14:14	14:45
FW	Wallis & Futuna Is., Wallis	17:01	17:24	06:17	06:40
FY	Fr. Guiana, Cayenne	09:17	09:39	21:30	21:52
G	England, London	07:25	08:05	16:07	16:46
GD	Isle of Man	07:53	08:37	16:10	16:53
GI	Northern Ireland, Belfast	08:00	08:44	16:14	16:58
GJ	Jersey	07:26	08:03	16:25	17:02
GM	Scotland, Glasgow	07:59	08:45	16:00	16:46
	Aberdeen	07:57	08:45	15:42	16:31
GU	Guernsey	07:29	08:06	16:26	17:04
GW	Wales, Cardiff	07:37	08:17	16:19	16:59
H4	Solomon Islands, Honiara	18:43	19:06	07:45	08:08
HA	Hungary, Budapest	05:55	06:30	15:07	15:43
HB	Switzerland, Bern	06:40	07:15	15:56	16:31
HB0	Liechtenstein	06:32	07:07	15:46	16:22
HC	Ecuador, Quito	10:51	11:13	23:21	23:44
HC8	Galapagos Is.	11:40	12:03	00:12	00:34
HH	Haiti, Port-Au-Prince	10:59	11:23	22:26	22:50
HI	Dominican Republic, Santo Domingo	10:50	11:13	22:17	22:41
HK	Colombia, Bogota	10:43	11:06	22:58	23:20
HK0	Malpelo Is.	11:10	11:33	23:27	23:49
HK0	San Adreas	11:27	11:49	23:15	23:38
HL	Korea, Seoul	22:18	22:47	08:28	08:57
HP	Panama, Panama	11:12	11:35	23:12	23:35
HR	Honduras, Tegucigalpa	11:51	12:14	23:34	23:57
HS	Thailand, Bangkok	23:20	23:43	11:04	11:27
HV	Vatican City	06:06	06:37	15:53	16:25
HZ, 7Z	Saudi Arabia, Dharan	03:03	03:28	14:02	14:27
	Mecca	03:36	04:00	14:52	15:17
	Riyadh	03:13	03:38	14:19	14:44
I	Italy, Rome	06:06	06:37	15:53	16:25
	Trieste	06:11	06:45	15:35	16:09
	Sicily	05:50	06:19	16:00	16:29
IS	Sardinia, Cagliari	06:13	06:43	16:15	16:45
J2	Djibouti, Djibouti	03:06	03:28	14:57	15:19
J3	Grenada	10:06	10:29	21:56	22:19
J5	Guinea-Bissau, Bissau	07:01	07:24	18:51	19:14
J6	St. Lucia	10:06	10:29	21:49	22:13
J7	Dominica	10:10	10:33	21:48	22:11
J8	St. Vincent	10:06	10:29	21:52	22:15
JA-JS	Japan, Tokyo	21:23	21:51	07:41	08:10
	Nagasaki	21:56	22:23	08:28	08:55
	Sapporo	21:34	22:06	07:14	07:46
JD1	Minami Torishima	20:04	20:28	07:10	07:35
JD1	Ogasawara, Kazan Is.	21:02	21:27	07:56	08:21
JT	Mongolia, Ulan Bator	00:05	00:41	09:15	09:50
JW	Svalbard, Spitsbergen	----	----	----	----
JX	Jan Mayen	10:34	----	----	14:43
JY	Jordan, Amman	04:11	04:37	14:46	15:13
KC4	Antarctica, Byrd Station	----	----	----	----
	McMurdo Sound	----	----	----	----
	Palmer Station	----	05:55	02:48	----
KC6	Belau, Yap	20:42	21:05	08:40	09:03
	Koror	20:53	21:16	08:59	09:21
KG4	Guantanamo Bay	11:13	11:37	22:35	22:59
KH0	Mariana Is., Saipan	20:21	20:44	08:00	08:23
KH1	Baker, Howland Is.	17:24	17:47	05:52	06:14
KH2	Guam, Agana	20:22	20:45	08:07	08:30
KH3	Johnston Is.	17:21	17:45	04:54	05:17
KH4	Midway Is.	18:16	18:42	05:08	05:33
KH5	Palmyra Is.	16:37	17:00	04:48	05:10
KH5K	Kingman Reef	16:43	17:05	04:48	05:10
KH6	Hawaii, Hilo	16:32	16:56	03:55	04:19
	Honolulu	16:46	17:10	04:04	04:28
KH7	Kure Is.	18:21	18:46	05:11	05:37
KH8	American Samoa, Pago Pago	16:37	17:00	05:56	06:20
KH9	Wake Is.	19:05	19:28	06:29	06:53
KL7	Alaska, Adak	19:12	19:52	03:51	04:31
	Anchorage	18:10	19:10	01:01	02:00
	Fairbanks	18:29	19:47	00:07	01:25
	Juneau	16:51	17:42	00:23	01:14
	Nome	19:36	20:52	01:22	02:38
KP1	Navassa Is.	11:10	11:33	22:37	23:01
KP2	Virgin Islands, Charlotte Amalie	10:29	10:53	21:57	22:21
KP4	Puerto Rico, San Juan	10:35	10:58	22:02	22:26
KP5	Desecheo Is.	10:39	11:03	22:08	22:31
LA-LJ	Norway, Oslo	07:20	08:16	14:29	15:25
LU	Argentina, Buenos Aires	08:19	08:48	23:10	23:39
LX	Luxembourg	06:53	07:31	15:50	16:28
LY	Lithuania, Vilna	05:54	06:38	14:09	14:52
LZ	Bulgaria, Sofia	05:25	05:57	15:07	15:39
OA	Peru, Lima	10:26	10:50	23:38	00:01
OD	Lebanon, Beirut	04:16	04:43	14:43	15:11
OE	Austria, Vienna	06:08	06:44	15:16	15:52
OH	Finland, Helsinki	06:24	07:20	13:30	14:26
OH0	Aland Is.	06:44	07:40	13:50	14:46
OJ0	Market Reef	06:49	07:45	13:53	14:50
OK,OL	Czech Rep., Prague	06:22	07:00	15:15	15:53
OM	Slovak, Rep., Bratislava	06:05	06:41	15:14	15:50
ON	Belgium, Brussels	07:05	07:44	15:52	16:31
OX	Greenland, Godthaab	11:59	13:13	17:51	19:05
	Thule	----	----	----	----

Pref	State/Province/Country/City	Tlight	Sunrise	Sunset	Tlight
OY	Faroe Islands, Torshavn	08:43	09:45	15:20	16:22
OZ	Denmark, Copenhagen	06:51	07:37	14:53	15:39
P2	Papua New Guinea, Madang	19:48	20:11	08:35	08:58
	Port Moresby	19:35	19:57	08:36	08:59
P4	Aruba, Oranjestad	10:35	10:58	22:25	22:47
PA-PI	Netherlands, Amsterdam	07:08	07:49	15:42	16:23
PJ2,4,9	Netherlands Antilles, Willemstad	10:40	11:03	22:28	22:51
PJ5-8	St. Maarten and Saba, St. Maarten	10:21	10:45	21:52	22:15
PY	Brazil, Brasilia	08:23	08:46	21:48	22:11
	Rio De Janeiro	07:49	08:14	21:43	22:08
	Natal	07:50	08:12	20:40	21:02
	Manaus	09:35	09:57	22:15	22:37
	Porto Alegre	08:03	08:30	22:30	22:57
PY0	Fernando De Noronha	07:42	08:05	20:25	20:47
PY0	St. Peter & St. Paul Rocks	07:39	08:01	20:05	20:27
PY0	Trindade & Martin Vaz Is., Trindade	06:58	07:23	20:42	21:07
PZ	Suriname, Paramaribo	09:30	09:52	21:40	22:03
R1FJ_	Franz Josef Land	——	——	——	——
R1MV_	Malyj Vysotskij Is.	06:12	07:10	13:12	14:10
S0	Western Sahara, Smara	07:09	07:34	18:08	18:33
S2	Bangladesh, Dacca	00:17	00:42	11:26	11:50
S5	Slovenia, Ljubljana	06:09	06:43	15:31	16:06
S7	Seychelles, Victoria	01:50	02:12	14:35	14:57
S9	Sao Tome	05:13	05:35	17:42	18:04
SM	Sweden, Stockholm	06:47	07:40	14:06	14:59
SP	Poland, Krakow	05:59	06:37	14:53	15:31
	Warsaw	06:03	06:44	14:39	15:19
ST	Sudan, Khartoum	03:55	04:18	15:32	15:56
ST0	Southern Sudan, Juba	03:41	04:04	15:54	16:17
SU	Egypt, Cairo	04:25	04:51	15:09	15:35
SV	Greece, Athens	05:12	05:41	15:20	15:49
SV/A	Mount Athos	05:15	05:45	15:11	15:41
SV5	Dodecanese, Rhodes	04:50	05:19	15:06	15:35
SV9	Crete	05:00	05:28	15:21	15:49
T2	Tuvalu, Funafuti	17:30	17:53	06:29	06:52
T30	West Kiribati, Bonriki	18:09	18:31	06:34	06:56
T31	Central Kiribati, Kanton	17:01	17:24	05:40	06:03
T32	East Kiribati, Christmas Is.	16:12	16:34	04:35	04:58
T33	Banaba Is.	18:21	18:43	06:52	07:15
T5	Somalia, Mogadishu	02:41	03:04	15:04	15:26
T7	San Marino	06:12	06:45	15:48	16:20
T9	Bosnia-Herzegovina, Sarajevo	05:48	06:20	15:23	15:56
TA	Turkey, Ankara	04:40	05:10	14:38	15:08
	Istanbul	04:58	05:29	14:50	15:20
TF	Iceland, Reykjavik	09:59	11:13	15:54	17:07
TG	Guatemala, Guatemala City	12:05	12:28	23:46	00:09
TI	Costa Rica, San Jose	11:32	11:54	23:28	23:51
TI9	Cocos Is.	11:37	11:59	23:48	00:10
TJ	Cameroon, Yaounde	05:00	05:22	17:16	17:39
TK	Corsica	06:20	06:52	16:07	16:38
TL	Central African Republic, Bangui	04:32	04:55	16:47	17:10
TN	Congo, Brazzaville	04:31	04:53	17:15	17:37
TR	Gabon, Libreville	05:02	05:24	17:30	17:53
TT	Chad, N'Djamena	04:59	05:22	16:49	17:11
TU	Cote d'Ivoire, Abidjan	06:04	06:26	18:16	18:39
TY	Benin, Porto Novo	05:40	06:02	17:48	18:10
TZ	Mali, Bamako	06:32	06:55	18:20	18:42
UA	Russia, European, Leningrad (UA1)	06:01	06:57	13:12	14:07
	Archangel (UA1)	05:53	07:10	11:36	12:53
	Murmansk (UA1)	07:12	——	——	12:34
	Moscow (UA3)	05:11	05:57	13:13	13:58
	Kuybyshev (UA4)	04:11	04:52	12:38	13:19
	Rostov (UA6)	04:33	05:09	13:46	14:21
UA2	Kaliningrad	06:16	07:01	14:26	15:10
UA9,0	Russia, Asiatic, Novosibirsk (UA9)	02:07	02:51	10:16	11:01
	Perm (UA9)	04:07	04:57	11:43	12:33
	Omsk (UA9)	02:45	03:29	10:54	11:39
	Norilsk (UA0)	03:37	——	——	08:49
	Irkutsk (UA0)	00:30	01:11	09:05	09:46
	Vladivostok (UA0)	22:12	22:44	07:51	08:24
	Petropavlovsk (UA0)	20:55	21:37	05:24	06:06
	Khabarovsk (UA0)	22:14	22:50	07:19	07:56
	Krasnoyarsk (UA0)	01:31	02:18	09:31	10:17
	Yakutsk (UA0)	23:37	00:39	06:14	07:16
	Wrangel Island (UA0)	21:59	——	——	02:08
	Kyzyl (UA0Y)	01:07	01:47	09:47	10:27
UJ-UM	Uzbekistan, Bukhoro	02:33	03:03	12:32	13:02
	Tashkent	02:17	02:48	12:08	12:39
UN-UQ	Kazakhstan, Alma-Ata	01:52	02:24	11:31	12:03
UR-UZ,EM-EO	Ukraine, Kiev	05:19	05:57	14:10	14:48
V2	Antigua & Barbuda, St. Johns	10:15	10:38	21:47	22:10
V3	Belize, Belmopan	12:03	12:26	23:35	23:58
V4	St. Kitts & Nevis	10:18	10:42	21:50	22:13
V5	Namibia, Windhoek	03:48	04:13	17:41	18:06
V6	Micronesia, Ponape	19:17	19:40	07:24	07:47
V7	Marshall Islands, Kwajalein	18:45	19:08	06:45	07:07
V8	Brunei, Bandar Seri Begawan	22:08	22:30	10:21	10:44
VK	Australia, Canberra (VK1)	18:27	18:56	09:22	09:51

Pref	State/Province/Country/City	Tlight	Sunrise	Sunset	Tlight
	Sydney (VK2)	18:22	18:51	09:10	09:39
	Melbourne (VK3)	18:35	19:05	09:45	10:16
	Brisbane (VK4)	18:33	18:59	08:47	09:14
	Adelaide (VK5)	19:10	19:39	10:03	10:32
	Perth (VK6)	20:50	21:18	11:26	11:54
	Hobart, Tasmania (VK7)	18:06	18:40	09:52	10:27
	Darwin (VK8)	20:34	20:57	09:46	10:10
VK0	Heard Is.	22:02	22:51	15:32	16:22
VK0	Macquarie Is.	16:05	16:59	10:01	10:55
VK9C	Cocos-Keeling Is.	22:51	23:14	12:02	12:25
VK9L	Lord Howe Is.	17:58	18:25	08:32	09:00
VK9M	Mellish Reef	18:44	19:08	08:16	08:40
VK9N	Norfolk Is.	17:29	17:56	07:51	08:17
VK9W	Willis Is.	19:12	19:36	08:39	09:03
VK9X	Christmas Is.	22:18	22:41	11:24	11:47
VP2E	Anguilla	10:21	10:45	21:50	22:13
VP2M	Montserrat	10:16	10:39	21:49	22:13
VP2V	British Virgin Is., Tortola	10:28	10:52	21:56	22:19
VP5	Turks & Caicos Islands, Grand Turk	10:59	11:24	22:17	22:41
VP8	Falkland Islands, Stanley	06:57	07:43	00:10	00:57
VP8	So. Georgia Is.	05:11	06:04	23:01	23:54
VP8	So. Orkney Is.	04:07	05:47	00:28	02:07
VP8	So. Sandwich Islands, Saunders Is.	03:50	04:58	22:46	23:54
VP8	So. Shetland Is., King George Is.	——	06:21	01:36	——
VP9	Bermuda	10:54	11:20	21:28	21:55
VQ9	Chagos, Diego Garcia	00:37	01:00	13:32	13:54
VR2,VS6	Hong Kong	22:39	23:03	09:53	10:17
VR6	Pitcairn Is.	13:31	13:57	03:35	04:00
VU	India, Bombay	01:19	01:43	12:45	13:09
	Calcutta	00:23	00:47	11:36	12:00
	New Delhi	01:19	01:45	12:08	12:34
	Bangalore	00:50	01:13	12:37	13:00
VU	Andaman Islands, Port Blair	23:47	00:10	11:38	12:01
VU	Laccadive Is.	01:04	01:26	13:00	13:23
XE	Mexico, Mexico City (XE1)	12:48	13:11	00:12	00:36
	Chihuahua (XE2)	13:32	13:58	00:21	00:47
	Merida (XE3)	12:13	12:37	23:31	23:55
XF4	Revilla Gigedo	13:37	14:00	01:02	01:26
XT	Burkina Faso, Ouagadougou	06:06	06:29	17:54	18:17
XU	Cambodia, Phnom Penh	22:59	23:22	10:50	11:13
XW	Laos, Viangchan	23:18	23:42	10:48	11:11
XX9	Macao	22:42	23:06	09:56	10:20
XZ	Myanmar, Yangon	23:43	00:06	11:16	11:40
YA	Afghanistan, Kandahar	02:09	02:35	12:49	13:15
	Kabul	02:02	02:30	12:27	12:55
YB-YD	Indonesia, Jakarta	22:22	22:44	11:12	11:35
	Medan	23:10	23:33	11:28	11:51
	Pontianak	22:22	22:44	10:52	11:14
	Jayapura	20:12	20:35	08:50	09:13
YI	Iraq, Baghdad	03:38	04:05	14:09	14:36
YJ	Vanuatu, Port Vila	17:54	18:18	07:26	07:50
YK	Syria, Damascus	04:12	04:39	14:41	15:08
YL	Latvia, Riga	06:11	06:59	13:59	14:47
YN	Nicaragua, Managua	11:43	12:06	23:33	23:56
YO	Romania, Bucharest	05:18	05:51	14:51	15:24
YS	El Salvador, San Salvador	11:59	12:22	23:43	00:06
YU	Yugoslavia, Belgrade	05:42	06:15	15:11	15:45
YV	Venezuela, Caracas	10:25	10:47	22:19	22:42
YV0	Aves Is.	10:20	10:43	21:57	22:20
Z2	Zimbabwe, Harare	03:03	03:27	16:36	17:00
Z3	Macedonia, (ex Yugoslav), Skopje	05:31	06:02	15:17	15:49
ZA	Albania, Tirane	05:35	06:06	15:26	15:57
ZB2	Gibraltar	07:04	07:33	17:21	17:50
ZC4	British Cyprus	04:27	04:55	14:51	15:19
ZD7	St. Helena	05:34	05:58	19:00	19:24
ZD8	Ascension Is.	06:23	06:46	19:20	19:43
ZD9	Tristan da Cunha	05:06	05:37	20:12	20:43
ZF	Cayman Is.	11:36	12:00	23:00	23:24
ZK1	No. Cook Is., Manihiki	16:05	16:28	05:10	05:33
ZK1	So. Cook Is., Rarotonga	15:39	16:04	05:25	05:50
ZK2	Niue	16:20	16:44	05:58	06:22
ZK3	Tokelaus, Atafu	16:54	17:18	05:54	06:17
ZL	New Zealand, Auckland (ZL1)	16:38	17:09	07:43	08:14
	Wellington (ZL2)	16:22	16:56	07:57	08:30
	Christchurch (ZL3)	16:22	16:57	08:13	08:48
	Dunedin (ZL4)	16:19	16:56	08:30	09:08
ZL5	Antarctica, Scott Base	——	——	——	——
ZL7	Chatham Is.	15:36	16:11	07:31	08:07
ZL8	Kermadec Is.	16:32	16:59	06:55	07:22
ZL9	Auckland & Campbell Is., Auckland	16:06	16:51	09:08	09:52
	Campbell Is.	15:43	16:31	09:07	09:55
ZP	Paraguay, Asuncion	08:41	09:07	22:45	23:11
ZS	South Africa, Cape Town (ZS1)	03:14	03:42	18:01	18:30
	Port Elizabeth (ZS2)	02:44	03:13	17:32	18:01
	Bloemfontein (ZS4)	02:56	03:23	17:19	17:46
	Durban (ZS5)	02:35	03:02	17:01	17:28
	Johannesburg (ZS6)	02:56	03:22	17:04	17:30
ZS8	Marion Is.	01:05	01:43	17:25	18:03

Pref	State/Province/Country/City	Tlight	Sunrise	Sunset	Tlight
VE1	New Brunswick, Saint John	11:24	11:57	21:13	21:46
	Nova Scotia, Halifax	11:12	11:44	21:05	21:38
	Prince Edward Island, Charlottetown	11:14	11:47	20:58	21:32
VE2	Quebec, Montreal	11:54	12:27	21:43	22:15
	Quebec City	11:48	12:22	21:29	22:03
VE3	Ontario, London	12:19	12:51	22:21	22:52
	Ottawa	12:03	12:35	21:51	22:24
	Sudbury	12:26	13:00	22:09	22:43
	Toronto	12:13	12:45	22:11	22:43
VE4	Manitoba, Winnipeg	13:40	14:16	23:02	23:38
VE5	Saskatchewan, Regina	14:12	14:48	23:29	00:06
	Saskatoon	14:25	15:03	23:31	00:10
VE6	Alberta, Calgary	14:51	15:28	00:05	00:43
	Edmonton	14:56	15:37	23:52	00:33
VE7	British Columbia, Prince George	15:35	16:16	00:28	01:09
	Prince Rupert	16:06	16:47	00:56	01:37
	Vancouver	15:22	15:58	00:48	01:24
VE8	Northwest Territories, Yellowknife	15:39	16:37	23:00	23:58
	Resolute	16:57	——	——	20:04
VY1	Yukon, Whitehorse	16:52	17:45	00:37	01:30
VO1	Newfoundland, St. John's	10:36	11:10	20:12	20:47
VO2	Labrador, Goose Bay	11:23	12:03	20:21	21:01
W1	Connecticut, Hartford	11:42	12:13	21:50	22:20
	Maine, Bangor	11:34	12:06	21:26	21:58
	Portland	11:37	12:09	21:35	22:06
	Massachusetts, Boston	11:37	12:08	21:42	22:12
	New Hampshire, Concord	11:41	12:12	21:41	22:12
	Rhode Island, Providence	11:37	12:08	21:45	22:15
	Vermont, Montpelier	11:48	12:20	21:42	22:14
W2	New Jersey, Atlantic City	11:44	12:13	22:03	22:32
	New York, Albany	11:49	12:20	21:52	22:23
	Buffalo	12:10	12:41	22:12	22:43
	New York City	11:46	12:15	21:58	22:27
	Syracuse	11:59	12:30	22:00	22:31
W3	Delaware, Wilmington	11:49	12:19	22:06	22:36
	District of Columbia, Washington	11:54	12:23	22:14	22:43
	Maryland, Baltimore	11:53	12:22	22:12	22:41
	Pennsylvania, Harrisburg	11:56	12:26	22:11	22:40
	Philadelphia	11:49	12:18	22:05	22:34
	Pittsburgh	12:09	12:38	22:23	22:52
	Scranton	11:54	12:24	22:03	22:33
W4	Alabama, Montgomery	12:19	12:45	23:06	23:33
	Florida, Jacksonville	11:57	12:23	22:52	23:18
	Miami	11:44	12:08	22:54	23:19
	Pensacola	12:19	12:45	23:14	23:40
	Georgia, Atlanta	12:14	12:41	22:56	23:23
	Savannah	11:58	12:24	22:46	23:12
	Kentucky, Lexington	12:22	12:50	22:47	23:15
	Louisville	12:28	12:56	22:51	23:20
	North Carolina, Charlotte	12:02	12:29	22:38	23:06
	Raleigh	11:54	12:22	22:28	22:56
	Wilmington	11:48	12:15	22:29	22:56
	South Carolina, Columbia	12:00	12:27	22:42	23:09
	Tennessee, Knoxville	12:16	12:43	22:49	23:17
	Memphis	12:39	13:06	23:16	23:43
	Nashville	12:28	12:55	23:00	23:28
	Virginia, Norfolk	11:47	12:15	22:16	22:44
	Richmond	11:53	12:21	22:19	22:48
W5	Arkansas, Little Rock	12:47	13:14	23:25	23:52
	Louisiana, New Orleans	12:30	12:55	23:26	23:52
	Shreveport	12:49	13:15	23:36	00:02
	Mississippi, Jackson	12:34	13:01	23:22	23:48
	New Mexico, Albuquerque	13:45	14:13	00:22	00:49
	Oklahoma, Oklahoma City	13:09	13:37	23:44	00:12
	Texas, Abilene	13:13	13:39	00:00	00:26
	Amarillo	13:26	13:53	00:02	00:30
	Dallas	13:02	13:28	23:47	00:14
	El Paso	13:39	14:05	00:28	00:54
	San Antonio	13:03	13:28	00:01	00:26
W6	California, Los Angeles	14:29	14:56	01:10	01:37
	San Francisco	14:53	15:22	01:19	01:47
W7	Arizona, Flagstaff	14:05	14:33	00:42	01:09
	Phoenix	14:04	14:31	00:47	01:14
	Idaho, Boise	14:40	15:12	00:39	01:10
	Pocatello	14:24	14:55	00:26	00:57
	Montana, Billings	14:15	14:48	00:01	00:34
	Butte	14:31	15:04	00:17	00:50
	Great Falls	14:30	15:04	00:07	00:41
	Nevada, Las Vegas	14:21	14:49	00:53	01:21
	Reno	14:46	15:15	01:04	01:33
	Oregon, Portland	15:11	15:44	00:59	01:32
	Utah, Salt Lake City	14:17	14:47	00:29	00:59
	Washington, Seattle	15:14	15:49	00:51	01:25
	Spokane	14:55	15:29	00:31	01:05
	Wyoming, Cheyenne	13:49	14:19	00:00	00:30
	Sheridan	14:06	14:39	23:58	00:31
W8	Michigan, Detroit	12:25	12:55	22:30	23:00
	Grand Rapids	12:37	13:08	22:38	23:10
	Sault Ste. Marie	12:40	13:14	22:23	22:56
	Traverse City	12:41	13:13	22:33	23:05
	Ohio, Cincinnati	12:24	12:53	22:44	23:13
	Cleveland	12:18	12:48	22:27	22:57
	Columbus	12:20	12:49	22:36	23:05
	West Virginia, Charleston	12:11	12:40	22:34	23:03
W9	Illinois, Chicago	12:42	13:13	22:49	23:20
	Indiana, Indianapolis	12:32	13:02	22:49	23:18
	Wisconsin, Green Bay	12:50	13:22	22:43	23:15

Pref	State/Province/Country/City	Tlight	Sunrise	Sunset	Tlight
	Milwaukee	12:46	13:17	22:47	23:18
W0	Colorado, Denver	13:47	14:17	00:04	00:34
	Grand Junction	14:00	14:30	00:20	00:49
	Iowa, Des Moines	13:06	13:36	23:14	23:44
	Kansas, Pratt	13:18	13:47	23:44	00:12
	Wichita	13:13	13:41	23:39	00:07
	Minnesota, Duluth	13:12	13:45	22:53	23:26
	Minneapolis	13:12	13:44	23:03	23:35
	Missouri, Columbia	12:55	13:24	23:15	23:44
	Kansas City	13:04	13:34	23:24	23:53
	St. Louis	12:46	13:15	23:08	23:37
	Nebraska, North Platte	13:33	14:03	23:44	00:14
	Omaha	13:14	13:44	23:24	23:54
	North Dakota, Fargo	13:31	14:04	23:11	23:45
	South Dakota, Rapid City	13:50	14:21	23:45	00:17
-		05:45	06:07	18:14	18:36
1A0	S.M.O.M	06:02	06:33	16:09	16:40
1S	Spratly Is.	22:30	22:52	10:33	10:56
3A	Monaco	06:26	06:58	16:24	16:56
3B6	Agalega	01:43	02:05	14:43	15:06
3B7	St. Brandon	01:20	01:43	14:40	15:03
3B8	Mauritius	01:22	01:46	14:56	15:19
3B9	Rodriguez Is.	00:59	01:23	14:31	14:55
3C	Equatorial Guinea, Bata	05:08	05:30	17:32	17:54
	Malabo	05:15	05:37	17:33	17:55
3C0	Annobon Is.	05:20	05:42	17:54	18:16
3D2	Fiji Is., Suva	17:22	17:46	06:48	07:12
3D2	Conway Reef	17:32	17:56	07:10	07:34
3D2	Rotuma Is.	17:35	17:58	06:42	07:04
3DA	Swaziland, Mbabane	02:55	03:20	16:52	17:17
3V	Tunisia, Tunis	06:01	06:29	16:31	16:59
3W, XV	Vietnam, Ho Chi Minh City (Saigon)	22:54	23:16	10:51	11:14
	Hanoi	23:12	23:36	10:39	11:02
3X	Guinea, Conakry	06:54	07:16	18:55	19:17
3Y	Bouvet	02:57	03:46	20:08	20:57
3Y	Peter I Is.	——	06:48	05:38	——
4J, 4K	Azerbaijan, Baku	03:29	03:59	13:43	14:13
4L	Georgia, Tbilisi	03:52	04:23	14:00	14:30
4S	Sri Lanka, Colombo	00:36	00:58	12:44	13:06
4U	ITU Geneva	06:37	07:10	16:21	16:55
4U	United Nations Hq.	11:46	12:15	21:58	22:27
4X, 4Z	Israel, Jerusalem	04:12	04:38	15:01	15:28
5A	Libya, Tripoli	05:44	06:10	16:31	16:57
	Benghazi	05:13	05:39	16:02	16:28
5B	Cyprus, Nicosia	04:25	04:52	15:01	15:29
5H	Tanzania, Dar es Salaam	02:56	03:19	15:46	16:09
5N	Nigeria, Lagos	05:41	06:03	17:51	18:13
5R	Madagascar, Antananarivo	02:04	02:28	15:33	15:57
5T	Mauritania, Nouakchott	07:15	07:38	18:51	19:14
5U	Niger, Niamey	05:56	06:19	17:46	18:09
5V	Togo, Lome	05:49	06:11	18:01	18:23
5W	Western Samoa, Apia	16:51	17:14	06:02	06:24
5X	Uganda, Kampala	03:35	03:57	16:04	16:26
5Z	Kenya, Nairobi	04:36	04:58	17:09	17:31
6W	Senegal, Dakar	07:16	07:39	19:02	19:25
6Y	Jamaica, Kingston	11:18	11:41	22:54	23:17
7O	Yemen, Aden	03:03	03:26	14:55	15:18
	Sanaa	03:10	03:33	14:54	15:17
7P	Lesotho, Maseru	03:03	03:29	17:12	17:38
7Q	Malawi, Lilongwe	03:08	03:31	16:20	16:43
	Blantyre	03:00	03:23	16:18	16:41
7X	Algeria, Algiers	06:29	06:57	17:00	17:28
8P	Barbados, Bridgetown	10:02	10:25	21:53	22:16
8Q	Maldive Is.	00:58	01:20	13:14	13:36
8R	Guyana, Georgetown	09:48	10:10	21:57	22:19
9A	Croatia, Zagreb	05:57	06:30	15:43	16:16
9G	Ghana, Accra	05:54	06:16	18:07	18:29
9H	Malta	05:43	06:10	16:16	16:43
9J	Zambia, Lusaka	03:27	03:50	16:44	17:07
9K	Kuwait	03:18	03:43	14:16	14:41
9L	Sierra Leone, Freetown	06:50	07:12	18:54	19:17
9M2	West Malaysia, Kuala Lumpur	23:03	23:25	11:23	11:45
9M6,8	East Malaysia, Sabah, Sandakan(9M6)	22:01	22:23	10:13	10:35
	Sarawak, Kuching (9M8)	22:26	22:48	10:51	11:13
9N	Nepal, Kathmandu	00:45	01:10	11:49	12:14
9Q	Zaire, Kinshasa	04:37	04:59	17:19	17:41
	Kisangani	04:05	04:27	16:33	16:55
	Lubumbashi	03:37	03:59	16:42	17:04
9U	Burundi, Bujumbura	03:43	04:05	16:22	16:44
9V	Singapore	22:52	23:14	11:17	11:39
9X	Rwanda, Kigali	03:42	04:04	16:17	16:39
9Y	Trinidad & Tobago , Port of Spain	10:05	10:28	22:04	22:26
A2	Botswana, Gaborone	03:19	03:44	17:10	17:35
A3	Tonga, Nukualofa	16:51	17:15	06:28	06:52
A4	Oman, Musqat	02:25	02:49	13:43	14:07
A5	Bhutan, Thimbu	00:28	00:53	11:33	11:58
A6	United Arab Emirates, Abu Dhabi	02:44	03:08	13:59	14:23
A7	Qatar, Ad-Dawhah	02:56	03:20	14:09	14:33
A9	Bahrein, Al-Manamah	03:01	03:26	14:11	14:35
AP	Pakistan, Karachi	01:53	02:17	13:07	13:31
AP	Islamabad	01:43	02:10	12:25	12:52
BV	Taiwan, Taipei	22:16	22:40	09:29	09:53
BY	Peoples Rep. of China, Beijing	23:02	23:32	09:18	09:48
	Harbin	22:34	23:07	08:21	08:53
	Shanghai	22:26	22:52	09:17	09:43
	Fuzhou	22:26	22:51	09:36	10:01
	Xian	23:21	23:48	10:01	10:28

Pref	State/Province/Country/City	Tlight	Sunrise	Sunset	Tlight
	Chongqing	23:23	23:49	10:20	10:46
	Chengdu	23:34	00:00	10:28	10:54
	Lhasa	00:24	00:50	11:22	11:47
	Urumqi	01:06	01:37	11:03	11:35
	Kashi	01:43	02:12	12:01	12:30
C2	Nauru	18:37	18:59	07:07	07:29
C3	Andorra	06:47	07:18	16:51	17:22
C5	The Gambia, Banjul	07:11	07:34	19:01	19:23
C6	Bahamas, Nassau	11:31	11:56	22:45	23:09
C9	Mozambique, Maputo	02:50	03:15	16:45	17:11
	Mozambique	02:38	03:01	15:54	16:17
CE	Chile, Santiago	09:25	09:53	23:54	00:22
CE0	Easter Island	12:15	12:41	02:15	02:41
CE0	Juan Fernandez	09:57	10:25	00:27	00:54
CE0	San Felix	10:20	10:45	00:17	00:42
CM,CO	Cuba, Havana	11:48	12:12	23:08	23:32
CN	Morocco, Casablanca	07:06	07:33	17:48	18:15
CP	Bolivia, La Paz	09:52	10:15	23:13	23:36
CT	Portugal, Lisbon	07:22	07:51	17:44	18:13
CT3	Madeira Islands, Funchal	07:42	08:08	18:28	18:55
CU	Azores , Ponta Delgada	08:26	08:55	18:52	19:20
CX	Uruguay, Montevideo	08:23	08:52	22:59	23:28
CY0	Sable Is.	10:56	11:28	20:53	21:25
CY9	St. Paul Is.	11:05	11:39	20:43	21:17
D2	Angola, Luanda	04:39	05:01	17:34	17:57
D4	Cape Verde, Praia	07:40	08:03	19:26	19:49
D6	Comoros, Moroni	02:32	02:55	15:37	16:00
DA-DL,Y2-Y9	Fed. Rep. of Germany, Bonn	06:46	07:23	16:02	16:39
	Berlin	06:26	07:05	15:29	16:08
DU	Philippines, Manilla	22:02	22:25	09:48	10:11
E3	Eritrea, Asmara	03:31	03:54	15:16	15:38
EA	Spain, Madrid	07:04	07:33	17:17	17:47
EA6	Balearic Is., Palma	06:36	07:06	16:55	17:24
EA8	Canary Is., Las Palmas	07:24	07:49	18:26	18:51
EA9	Ceuta & Melilla, Ceuta	07:01	07:29	17:35	18:02
	Melilla	06:51	07:18	17:27	17:54
EI	Ireland, Dublin	07:47	08:27	16:45	17:25
EK	Armenia, Yerevan	03:51	04:20	14:05	14:35
EL	Liberia, Monrovia	06:37	06:59	18:48	19:10
EP	Iran, Tehran	03:13	03:40	13:46	14:14
ER	Moldova, Kishinev	05:09	05:42	14:48	15:22
ES	Estonia, Tallinn	06:06	06:56	14:07	14:57
ET	Ethiopia, Addis Ababa	03:23	03:45	15:26	15:48
EU,EV,EW	Belarus, Minsk	05:33	06:14	14:26	15:07
EX	Kyrgyzstan, Frunze	01:56	02:27	11:58	12:29
EY	Tajikistan, Samarkand	02:20	02:49	12:37	13:07
	Dushanbe	02:11	02:40	12:31	13:00
EZ	Turkmenistan, Ashkhabad	02:50	03:19	13:15	13:44
F	France, Paris	06:59	07:34	16:28	17:03
FG	Guadeloupe	10:15	10:38	21:57	22:20
FJ, FS	St. Martin	10:24	10:47	21:59	22:22
FH	Mayotte	02:23	02:46	15:32	15:55
FK	New Caledonia, Noumia	18:02	18:26	07:43	08:07
FM	Martinique	10:10	10:33	21:56	22:19
FO	Clipperton Is.	13:17	13:39	01:16	01:38
FO	Fr. Polynesia, Tahiti	15:55	16:18	05:19	05:42
	Rurutu, (Austral Is.)	19:02	19:26	08:44	09:09
	Hiva Oa, (Marquesas Is.)	14:46	15:08	03:45	04:07
FP	St. Pierre & Miquelon, St. Pierre	10:47	11:21	20:28	21:02
FR/G	Glorioso	02:18	02:41	15:22	15:45
FR/J,E	Juan de Nova	02:26	02:50	15:49	16:12
	Europa	02:26	02:50	16:07	16:32
FR	Reunion	01:28	01:52	15:05	15:29
FR/T	Tromelin	01:42	02:05	15:01	15:24
FT8W	Crozet	00:32	01:08	16:17	16:53
FT8X	Kerguelen	23:07	23:47	15:21	16:00
FT8Y	Antarctica, Dumont D'Urville	—	16:41	13:00	—
FT8Z	Amsterdam & St. Paul Is., Amsterdam	23:20	23:50	14:11	14:41
FW	Wallis & Futuna Is., Wallis	17:09	17:32	06:19	06:42
FY	Fr. Guiana, Cayenne	09:21	09:43	21:36	21:58
G	England, London	07:16	07:54	16:27	17:05
GD	Isle of Man	07:43	08:24	16:33	17:14
GI	Northern Ireland, Belfast	07:50	08:31	16:37	17:19
GJ	Jersey	07:19	07:54	16:44	17:20
GM	Scotland, Glasgow	07:48	08:31	16:24	17:08
	Aberdeen	07:44	08:30	16:08	16:54
GU	Guernsey	07:21	07:57	16:46	17:22
GW	Wales, Cardiff	07:29	08:07	16:40	17:18
H4	Solomon Islands, Honiara	18:51	19:13	07:48	08:11
HA	Hungary, Budapest	05:49	06:23	15:25	16:00
HB	Switzerland, Bern	06:34	07:08	16:14	16:47
HB0	Liechtenstein	06:26	07:00	16:04	16:38
HC	Ecuador, Quito	10:57	11:19	23:26	23:48
HC8	Galapagos Is.	11:46	12:08	00:17	00:39
HH	Haiti, Port-Au-Prince	11:01	11:24	22:35	22:59
HI	Dominican Republic, Santo Domingo	10:52	11:15	22:26	22:49
HK	Colombia, Bogota	10:48	11:10	23:04	23:26
HK0	Malpelo Is.	11:15	11:37	23:33	23:55
HK0	San Adreas	11:30	11:52	23:22	23:45
HL	Korea, Seoul	22:15	22:43	08:42	09:10
HP	Panama, Panama	11:16	11:38	23:19	23:41
HR	Honduras, Tegucigalpa	11:54	12:17	23:42	00:05
HS	Thailand, Bangkok	23:23	23:46	11:12	11:34
HV	Vatican City	06:02	06:32	16:09	16:39
HZ, 7Z	Saudi Arabia, Dharan	03:04	03:28	14:13	14:37
	Mecca	03:37	04:01	15:02	15:26
	Riyadh	03:14	03:38	14:29	14:53

Pref	State/Province/Country/City	Tlight	Sunrise	Sunset	Tlight
I	Italy, Rome	06:02	06:32	16:09	16:39
	Trieste	06:05	06:38	15:52	16:25
	Sicily	05:47	06:15	16:14	16:42
IS	Sardinia, Cagliari	06:10	06:39	16:29	16:58
J2	Djibouti, Djibouti	03:09	03:31	15:04	15:27
J3	Grenada	10:09	10:32	22:03	22:26
J5	Guinea-Bissau, Bissau	07:05	07:27	18:59	19:21
J6	St. Lucia	10:09	10:32	21:57	22:20
J7	Dominica	10:12	10:35	21:56	22:19
J8	St. Vincent	10:09	10:32	21:59	22:22
JA-JS	Japan, Tokyo	21:20	21:48	07:55	08:22
	Nagasaki	21:55	22:21	08:41	09:07
	Sapporo	21:29	22:00	07:30	08:01
JD1	Minami Torishima	20:04	20:29	07:20	07:45
JD1	Ogasawara, Kazan Is.	21:02	21:26	08:07	08:32
JT	Mongolia, Ulan Bator	23:58	00:33	09:33	10:07
JW	Svalbard, Spitsbergen	—	—	—	—
JX	Jan Mayen	09:49	—	—	15:38
JY	Jordan, Amman	04:09	04:36	14:58	15:24
KC4	Antarctica, Byrd Station	—	—	—	—
	McMurdo Sound	—	—	—	—
	Palmer Station	—	06:48	02:05	—
KC6	Belau, Yap	20:46	21:08	08:47	09:10
	Koror	20:58	21:20	09:05	09:27
KG4	Guantanamo Bay	11:15	11:38	22:45	23:08
KH0	Mariana Is., Saipan	20:24	20:47	08:08	08:31
KH1	Baker, Howland Is.	17:30	17:52	05:57	06:19
KH2	Guam, Agana	20:25	20:48	08:15	08:37
KH3	Johnston Is.	17:24	17:47	05:02	05:26
KH4	Midway Is.	18:16	18:41	05:19	05:44
KH5	Palmyra Is.	16:42	17:04	04:54	05:16
KH5K	Kingman Reef	16:47	17:09	04:54	05:16
KH6	Hawaii, Hilo	16:34	16:57	04:05	04:28
	Honolulu	16:48	17:11	04:13	04:37
KH7	Kure Is.	18:21	18:46	05:22	05:48
KH8	American Samoa, Pago Pago	16:46	17:09	05:59	06:22
KH9	Wake Is.	19:07	19:30	06:38	07:02
KL7	Alaska, Adak	19:03	19:42	04:12	04:50
	Anchorage	17:54	18:48	01:33	02:27
	Fairbanks	18:07	19:14	00:50	01:57
	Juneau	16:38	17:26	00:51	01:38
	Nome	19:15	20:21	02:04	03:09
KP1	Navassa Is.	11:12	11:35	22:46	23:09
KP2	Virgin Islands, Charlotte Amalie	10:31	10:54	22:06	22:29
KP4	Puerto Rico, San Juan	10:37	11:00	22:11	22:34
KP5	Desecheo Is.	10:41	11:05	22:16	22:40
LA-LJ	Norway, Oslo	07:05	07:56	14:59	15:50
LU	Argentina, Buenos Aires	08:33	09:01	23:07	23:36
LX	Luxembourg	06:46	07:22	16:10	16:46
LY	Lithuania, Vilna	05:44	06:25	14:32	15:13
LZ	Bulgaria, Sofia	05:20	05:51	15:23	15:54
OA	Peru, Lima	10:35	10:57	23:41	00:03
OD	Lebanon, Beirut	04:14	04:41	14:56	15:23
OE	Austria, Vienna	06:02	06:36	15:34	16:09
OH	Finland, Helsinki	06:09	07:01	14:00	14:52
OH0	Aland Is.	06:29	07:21	14:20	15:12
OJ0	Market Reef	06:33	07:25	14:24	15:16
OK,OL	Czech Rep., Prague	06:14	06:51	15:35	16:11
OM	Slovak, Rep., Bratislava	05:58	06:33	15:32	16:07
ON	Belgium, Brussels	06:57	07:34	16:12	16:49
OX	Greenland, Godthaab	11:38	12:43	18:32	19:37
	Thule	—	—	—	—
OY	Faroe Islands, Torshavn	08:25	09:22	15:53	16:50
OZ	Denmark, Copenhagen	06:40	07:23	15:17	16:01
P2	Papua New Guinea, Madang	19:55	20:17	08:39	09:02
	Port Moresby	19:42	20:05	08:40	09:02
P4	Aruba, Oranjestad	10:38	11:01	22:32	22:55
PA-PI	Netherlands, Amsterdam	06:59	07:38	16:04	16:43
PJ2,4,9	Netherlands Antilles, Willemstad	10:44	11:06	22:36	22:58
PJ5-8	St. Maarten and Saba, St. Maarten	10:23	10:46	22:00	22:23
PY	Brazil, Brasilia	08:31	08:55	21:50	22:13
	Rio De Janeiro	07:59	08:24	21:43	22:08
	Natal	07:57	08:19	20:44	21:06
	Manaus	09:41	10:03	22:19	22:42
	Porto Alegre	08:15	08:42	22:29	22:55
PY0	Fernando De Noronha	07:49	08:11	20:29	20:52
PY0	St. Peter & St. Paul Rocks	07:44	08:06	20:10	20:32
PY0	Trindade & Martin Vaz Is., Trindade	07:08	07:32	20:43	21:07
PZ	Suriname, Paramaribo	09:34	09:56	21:46	22:08
R1FJ_	Franz Josef Land	—	—	—	—
R1MV_	Malyj Vysotskij Is.	05:56	06:49	13:43	14:36
S0	Western Sahara, Smara	07:09	07:34	18:18	18:43
S2	Bangladesh, Dacca	00:18	00:42	11:36	12:00
S5	Slovenia, Ljubljana	06:03	06:36	15:49	16:22
S7	Seychelles, Victoria	01:56	02:18	14:39	15:01
S9	Sao Tome	05:19	05:41	17:47	18:09
SM	Sweden, Stockholm	06:32	07:22	14:34	15:24
SP	Poland, Krakow	05:52	06:28	15:13	15:49
	Warsaw	05:54	06:33	15:00	15:39
ST	Sudan, Khartoum	03:57	04:20	15:41	16:04
ST0	Southern Sudan, Juba	03:46	04:08	16:00	16:22
SU	Egypt, Cairo	04:24	04:50	15:20	15:46
SV	Greece, Athens	05:09	05:38	15:34	16:02
SV/A	Mount Athos	05:11	05:41	15:26	15:56
SV5	Dodecanese, Rhodes	04:48	05:16	15:20	15:47
SV9	Crete	04:58	05:26	15:34	16:01
T2	Tuvalu, Funafuti	17:37	18:00	06:33	06:55

Pref	State/Province/Country/City	Tlight	Sunrise	Sunset	Tlight
T30	West Kiribati, Bonriki	18:14	18:36	06:39	07:01
T31	Central Kiribati, Kanton	17:08	17:30	05:45	06:07
T32	East Kiribati, Christmas Is.	16:17	16:39	04:41	05:03
T33	Banaba Is.	18:27	18:49	06:57	07:19
T5	Somalia, Mogadishu	02:46	03:08	15:09	15:31
T7	San Marino	06:07	06:39	16:04	16:36
T9	Bosnia-Herzegovina, Sarajevo	05:43	06:14	15:39	16:11
TA	Turkey, Ankara	04:36	05:06	14:52	15:22
	Istanbul	04:55	05:25	15:05	15:35
TF	Iceland, Reykjavik	09:39	10:43	16:34	17:38
TG	Guatemala, Guatemala City	12:08	12:31	23:54	00:17
TI	Costa Rica, San Jose	11:35	11:58	23:35	23:58
TI9	Cocos Is.	11:41	12:03	23:54	00:16
TJ	Cameroon, Yaounde	05:05	05:27	17:22	17:44
TK	Corsica	06:16	06:47	16:22	16:53
TL	Central African Republic, Bangui	04:37	04:59	16:53	17:15
TN	Congo, Brazzaville	04:37	04:59	17:19	17:41
TR	Gabon, Libreville	05:08	05:30	17:35	17:57
TT	Chad, N'Djamena	05:02	05:25	16:56	17:19
TU	Cote d'Ivoire, Abidjan	06:09	06:31	18:22	18:44
TY	Benin, Porto Novo	05:44	06:06	17:54	18:16
TZ	Mali, Bamako	06:35	06:58	18:27	18:50
UA	Russia, European, Leningrad (UA1)	05:46	06:37	13:41	14:32
	Archangel (UA1)	05:32	06:38	12:19	13:25
	Murmansk (UA1)	06:38	08:21	11:36	13:18
	Moscow (UA3)	05:00	05:43	13:37	14:20
	Kuybyshev (UA4)	04:01	04:41	12:59	13:39
	Rostov (UA6)	04:27	05:01	14:04	14:38
UA2	Kaliningrad	06:06	06:48	14:49	15:32
UA9,0	Russia, Asiatic, Novosibirsk (UA9)	01:56	02:38	10:40	11:22
	Perm (UA9)	03:54	04:41	12:10	12:57
	Omsk (UA9)	02:34	03:16	11:18	12:00
	Norilsk (UA0)	03:02	04:50	07:46	09:34
	Irkutsk (UA0)	00:21	01:00	09:27	10:05
	Vladivostok (UA0)	22:07	22:38	08:07	08:39
	Petropavlovsk (UA0)	20:46	21:26	05:46	06:25
	Khabarovsk (UA0)	22:07	22:42	07:38	08:13
	Krasnoyarsk (UA0)	01:20	02:04	09:55	10:39
	Yakutsk (UA0)	23:19	00:16	06:47	07:44
	Wrangel Island (UA0)	21:14	———	———	03:03
	Kyzyl (UA0Y)	00:59	01:37	10:08	10:46
UJ-UM	Uzbekistan, Bukhoro	02:30	02:59	12:47	13:16
	Tashkent	02:13	02:43	12:23	12:53
UN-UQ	Kazakhstan, Alma-Ata	01:47	02:19	11:47	12:18
UR-UZ,EM-EO	Ukraine, Kiev	05:11	05:48	14:29	15:06
V2	Antigua & Barbuda, St. Johns	10:17	10:40	21:56	22:19
V3	Belize, Belmopan	12:05	12:28	23:43	00:06
V4	St. Kitts & Nevis	10:20	10:43	21:58	22:21
V5	Namibia, Windhoek	03:59	04:23	17:41	18:06
V6	Micronesia, Ponape	19:22	19:44	07:31	07:53
V7	Marshall Islands, Kwajalein	18:49	19:11	06:51	07:14
V8	Brunei, Bandar Seri Begawan	22:12	22:35	10:27	10:49
VK	Australia, Canberra (VK1)	18:41	19:09	09:19	09:47
	Sydney (VK2)	18:36	19:04	09:07	09:35
	Melbourne (VK3)	18:50	19:20	09:41	10:11
	Brisbane (VK4)	18:45	19:10	08:47	09:12
	Adelaide (VK5)	19:24	19:52	10:00	10:28
	Perth (VK6)	21:03	21:30	11:24	11:52
	Hobart, Tasmania (VK7)	18:23	18:56	09:46	10:19
	Darwin (VK8)	20:42	21:05	09:49	10:12
VK0	Heard Is.	22:28	23:14	15:20	16:06
VK0	Macquarie Is.	16:34	17:23	09:47	10:36
VK9C	Cocos-Keeling Is.	22:59	23:21	12:05	12:28
VK9L	Lord Howe Is.	18:11	18:38	08:31	08:58
VK9M	Mellish Reef	18:53	19:17	08:18	08:41
VK9N	Norfolk Is.	17:41	18:07	07:50	08:16
VK9W	Willis Is.	19:21	19:44	08:41	09:04
VK9X	Christmas Is.	22:26	22:49	11:27	11:49
VP2E	Anguilla	10:23	10:47	21:58	22:22
VP2M	Montserrat	10:18	10:41	21:58	22:21
VP2V	British Virgin Is., Tortola	10:30	10:53	22:05	22:28

January 21   Sunrise and Sunset, in UTC

Pref	State/Province/Country/City	Tlight	Sunrise	Sunset	Tlight
VP5	Turks & Caicos Islands, Grand Turk	11:01	11:25	22:26	22:50
VP8	Falkland Islands, Stanley	07:22	08:05	23:59	00:42
VP8	So. Georgia Is.	05:40	06:28	22:48	23:36
VP8	So. Orkney Is.	05:07	06:20	00:05	01:18
VP8	So. Sandwich Islands, Saunders Is.	04:28	05:27	22:28	23:27
VP8	So. Shetland Is., King George Is.	05:32	06:59	01:09	02:36
VP9	Bermuda	10:52	11:19	21:40	22:06
VQ9	Chagos, Diego Garcia	00:44	01:07	13:35	13:58
VR2,VS6	Hong Kong	22:40	23:04	10:03	10:26
VR6	Pitcairn Is.	13:42	14:07	03:34	03:59
VU	India, Bombay	01:21	01:45	12:54	13:17
	Calcutta	00:24	00:48	11:46	12:10
	New Delhi	01:19	01:44	12:20	12:45
	Bangalore	00:53	01:16	12:44	13:07
VU	Andaman Islands, Port Blair	23:51	00:13	11:46	12:08
VU	Laccadive Is.	01:07	01:30	13:07	13:30
XE	Mexico, Mexico City (XE1)	12:49	13:13	00:21	00:44
	Chihuahua (XE2)	13:31	13:57	00:32	00:58
	Merida (XE3)	12:14	12:38	23:41	00:04
XF4	Revilla Gigedo	13:38	14:02	01:11	01:35
XT	Burkina Faso, Ouagadougou	06:09	06:32	18:02	18:25
XU	Cambodia, Phnom Penh	23:03	23:25	10:58	11:20
XW	Laos, Viangchan	23:21	23:44	10:56	11:20
XX9	Macao	22:43	23:07	10:06	10:29
XZ	Myanmar, Yangon	23:45	00:08	11:25	11:48
YA	Afghanistan, Kandahar	02:08	02:34	13:01	13:27
	Kabul	02:00	02:27	12:40	13:07
YB-YD	Indonesia, Jakarta	22:28	22:51	11:16	11:38
	Medan	23:15	23:37	11:34	11:56
	Pontianak	22:28	22:50	10:57	11:19
	Jayapura	20:18	20:40	08:55	09:17
YI	Iraq, Baghdad	03:37	04:03	14:22	14:48
YJ	Vanuatu, Port Vila	18:03	18:27	07:28	07:51
YK	Syria, Damascus	04:10	04:37	14:53	15:20
YL	Latvia, Riga	05:59	06:44	14:24	15:10
YN	Nicaragua, Managua	11:46	12:09	23:40	00:03
YO	Romania, Bucharest	05:13	05:45	15:07	15:39
YS	El Salvador, San Salvador	12:02	12:24	23:50	00:13
YU	Yugoslavia, Belgrade	05:37	06:09	15:28	16:00
YV	Venezuela, Caracas	10:28	10:50	22:27	22:49
YV0	Aves Is.	10:22	10:45	22:05	22:28
Z2	Zimbabwe, Harare	03:12	03:36	16:37	17:01
Z3	Macedonia, (ex Yugoslav), Skopje	05:26	05:57	15:33	16:03
ZA	Albania, Tirane	05:31	06:02	15:41	16:11
ZB2	Gibraltar	07:02	07:30	17:35	18:02
ZC4	British Cyprus	04:26	04:53	15:04	15:31
ZD7	St. Helena	05:43	06:06	19:02	19:25
ZD8	Ascension Is.	06:30	06:53	19:24	19:46
ZD9	Tristan da Cunha	05:21	05:51	20:09	20:38
ZF	Cayman Is.	11:38	12:01	23:09	23:33
ZK1	No. Cook Is., Manihiki	16:13	16:35	05:14	05:36
ZK1	So. Cook Is., Rarotonga	15:49	16:13	05:26	05:50
ZK2	Niue	16:30	16:53	05:59	06:23
ZK3	Tokelaus, Atafu	17:03	17:25	05:57	06:20
ZL	New Zealand, Auckland (ZL1)	16:53	17:23	07:40	08:09
	Wellington (ZL2)	16:39	17:11	07:52	08:24
	Christchurch (ZL3)	16:40	17:13	08:07	08:41
	Dunedin (ZL4)	16:38	17:14	08:23	08:59
ZL5	Antarctica, Scott Base	———	———	———	———
ZL7	Chatham Is.	15:54	16:28	07:25	07:59
ZL8	Kermadec Is.	16:44	17:10	06:54	07:20
ZL9	Auckland & Campbell Is., Auckland	16:30	17:12	08:57	09:39
	Campbell Is.	16:09	16:53	08:55	09:40
ZP	Paraguay, Asuncion	08:52	09:17	22:45	23:10
ZS	South Africa, Cape Town (ZS1)	03:27	03:55	17:58	18:26
	Port Elizabeth (ZS2)	02:58	03:26	17:29	17:57
	Bloemfontein (ZS4)	03:08	03:35	17:18	17:44
	Durban (ZS5)	02:48	03:14	17:00	17:26
	Johannesburg (ZS6)	03:07	03:33	17:04	17:29
ZS8	Marion Is.	01:25	02:02	17:17	17:54

Pref	State/Province/Country/City	Tlight	Sunrise	Sunset	Tlight
VE1	New Brunswick, Saint John	11:08	11:39	21:37	22:08
	Nova Scotia, Halifax	10:56	11:27	21:29	22:00
	Prince Edward Island, Charlottetown	10:57	11:29	21:23	21:55
VE2	Quebec, Montreal	11:38	12:09	22:07	22:38
	Quebec City	11:31	12:03	21:54	22:26
VE3	Ontario, London	12:05	12:35	22:43	23:13
	Ottawa	11:46	12:17	22:16	22:47
	Sudbury	12:09	12:41	22:34	23:06
	Toronto	11:58	12:28	22:34	23:04
VE4	Manitoba, Winnipeg	13:20	13:54	23:30	00:04
VE5	Saskatchewan, Regina	13:51	14:26	23:58	00:33
	Saskatoon	14:03	14:39	00:02	00:38
VE6	Alberta, Calgary	14:30	15:05	00:35	01:10
	Edmonton	14:33	15:10	00:25	01:02
VE7	British Columbia, Prince George	15:11	15:49	01:01	01:39
	Prince Rupert	15:42	16:20	01:29	02:08
	Vancouver	15:03	15:36	01:16	01:49
VE8	Northwest Territories, Yellowknife	15:02	15:53	23:50	00:41
	Resolute	15:05	17:34	19:33	22:03
VY1	Yukon, Whitehorse	16:18	17:06	01:22	02:10
VO1	Newfoundland, St. John's	10:18	10:51	20:38	21:11
VO2	Labrador, Goose Bay	11:00	11:37	20:53	21:30
W1	Connecticut, Hartford	11:28	11:58	22:11	22:40
	Maine, Bangor	11:18	11:48	21:49	22:20
	Portland	11:22	11:52	21:58	22:28
	Massachusetts, Boston	11:23	11:52	22:04	22:33
	New Hampshire, Concord	11:26	11:56	22:04	22:33
	Rhode Island, Providence	11:23	11:52	22:06	22:35
	Vermont, Montpelier	11:32	12:02	22:06	22:36
W2	New Jersey, Atlantic City	11:32	12:00	22:23	22:51
	New York, Albany	11:34	12:04	22:14	22:43
	Buffalo	11:55	12:25	22:34	23:03
	New York City	11:32	12:01	22:18	22:47
	Syracuse	11:44	12:14	22:23	22:52
W3	Delaware, Wilmington	11:37	12:05	22:26	22:55
	District of Columbia, Washington	11:41	12:09	22:34	23:02
	Maryland, Baltimore	11:40	12:08	22:32	23:00
	Pennsylvania, Harrisburg	11:43	12:12	22:31	22:59
	Philadelphia	11:36	12:04	22:25	22:53
	Pittsburgh	11:56	12:24	22:43	23:12
	Scranton	11:40	12:09	22:24	22:53
W4	Alabama, Montgomery	12:10	12:36	23:22	23:48
	Florida, Jacksonville	11:49	12:14	23:07	23:32
	Miami	11:38	12:02	23:07	23:31
	Pensacola	12:11	12:36	23:29	23:54
	Georgia, Atlanta	12:04	12:30	23:12	23:38
	Savannah	11:49	12:14	23:02	23:27
	Kentucky, Lexington	12:10	12:38	23:06	23:33
	Louisville	12:16	12:43	23:10	23:38
	North Carolina, Charlotte	11:52	12:18	22:56	23:22
	Raleigh	11:44	12:10	22:46	23:12
	Wilmington	11:39	12:05	22:46	23:12
	South Carolina, Columbia	11:51	12:17	22:58	23:24
	Tennessee, Knoxville	12:05	12:32	23:07	23:33
	Memphis	12:29	12:55	23:33	23:59
	Nashville	12:17	12:44	23:18	23:45
	Virginia, Norfolk	11:36	12:03	22:35	23:02
	Richmond	11:41	12:08	22:38	23:05
W5	Arkansas, Little Rock	12:37	13:03	23:42	00:09
	Louisiana, New Orleans	12:22	12:47	23:41	00:06
	Shreveport	12:40	13:05	23:51	00:17
	Mississippi, Jackson	12:26	12:51	23:38	00:03
	New Mexico, Albuquerque	13:35	14:02	00:39	01:06
	Oklahoma, Oklahoma City	12:59	13:25	00:02	00:28
	Texas, Abilene	13:04	13:29	00:15	00:41
	Amarillo	13:16	13:42	00:20	00:46
	Dallas	12:53	13:18	00:03	00:29
	El Paso	13:30	13:56	00:44	01:09
	San Antonio	12:55	13:20	00:15	00:40
W6	California, Los Angeles	14:20	14:46	01:27	01:53
	San Francisco	14:42	15:09	01:38	02:05
W7	Arizona, Flagstaff	13:55	14:22	00:59	01:26
	Phoenix	13:55	14:21	01:04	01:29
	Idaho, Boise	14:25	14:55	01:01	01:32
	Pocatello	14:09	14:39	00:48	01:18
	Montana, Billings	13:58	14:29	00:26	00:57
	Butte	14:14	14:46	00:41	01:13
	Great Falls	14:12	14:45	00:33	01:05
	Nevada, Las Vegas	14:10	14:37	01:11	01:38
	Reno	14:34	15:02	01:24	01:52
	Oregon, Portland	14:54	15:26	01:23	01:55
	Utah, Salt Lake City	14:04	14:32	00:50	01:19
	Washington, Seattle	14:56	15:29	01:17	01:49
	Spokane	14:37	15:10	00:57	01:30
	Wyoming, Cheyenne	13:36	14:05	00:21	00:50
	Sheridan	13:50	14:21	00:22	00:53
W8	Michigan, Detroit	12:10	12:40	22:51	23:21
	Grand Rapids	12:22	12:52	23:01	23:31
	Sault Ste. Marie	12:23	12:55	22:48	23:20
	Traverse City	12:25	12:56	22:56	23:27
	Ohio, Cincinnati	12:12	12:40	23:04	23:31
	Cleveland	12:04	12:33	22:48	23:17
	Columbus	12:07	12:35	22:56	23:24
	West Virginia, Charleston	11:59	12:27	22:53	23:21
W9	Illinois, Chicago	12:28	12:57	23:11	23:40
	Indiana, Indianapolis	12:20	12:48	23:09	23:37
	Wisconsin, Green Bay	12:34	13:05	23:07	23:37

Pref	State/Province/Country/City	Tlight	Sunrise	Sunset	Tlight
	Milwaukee	12:31	13:01	23:10	23:39
W0	Colorado, Denver	13:35	14:03	00:24	00:53
	Grand Junction	13:48	14:16	00:40	01:08
	Iowa, Des Moines	12:52	13:21	23:35	00:04
	Kansas, Pratt	13:07	13:34	00:03	00:30
	Wichita	13:01	13:28	23:57	00:25
	Minnesota, Duluth	12:54	13:26	23:18	23:50
	Minneapolis	12:56	13:27	23:27	23:58
	Missouri, Columbia	12:43	13:11	23:35	00:03
	Kansas City	12:52	13:20	23:44	00:12
	St. Louis	12:34	13:02	23:27	23:55
	Nebraska, North Platte	13:20	13:49	00:05	00:34
	Omaha	13:01	13:29	23:45	00:14
	North Dakota, Fargo	13:13	13:45	23:36	00:09
	South Dakota, Rapid City	13:34	14:04	00:08	00:39
-	-	05:49	06:10	18:17	18:39
1A0	S.M.O.M	05:48	06:17	16:31	17:00
1S	Spratly Is.	22:31	22:52	10:40	11:01
3A	Monaco	06:11	06:41	16:47	17:17
3B6	Agalega	01:50	02:12	14:43	15:04
3B7	St. Brandon	01:29	01:52	14:37	14:59
3B8	Mauritius	01:33	01:56	14:51	15:14
3B9	Rodriguez Is.	01:10	01:33	14:27	14:50
3C	Equatorial Guinea, Bata	05:12	05:33	17:36	17:57
	Malabo	05:18	05:39	17:38	17:59
3C0	Annobon Is.	05:25	05:46	17:56	18:18
3D2	Fiji Is., Suva	17:33	17:55	06:45	07:08
3D2	Conway Reef	17:44	18:07	07:05	07:28
3D2	Rotuma Is.	17:43	18:05	06:40	07:02
3DA	Swaziland, Mbabane	03:09	03:33	16:45	17:09
3V	Tunisia, Tunis	05:50	06:17	16:49	17:16
3W, XV	Vietnam, Ho Chi Minh City (Saigon)	22:54	23:16	10:58	11:20
	Hanoi	23:08	23:31	10:49	11:12
3X	Guinea, Conakry	06:54	07:16	19:01	19:23
3Y	Bouvet	03:38	04:21	19:39	20:22
3Y	Peter I Is.	——	08:57	03:35	——
4J, 4K	Azerbaijan, Baku	03:16	03:45	14:04	14:32
4L	Georgia, Tbilisi	03:38	04:07	14:21	14:51
4S	Sri Lanka, Colombo	00:37	00:58	12:50	13:11
4U	ITU Geneva	06:20	06:52	16:46	17:18
4U	United Nations Hq.	11:32	12:01	22:18	22:47
4X, 4Z	Israel, Jerusalem	04:03	04:29	15:17	15:42
5A	Libya, Tripoli	05:35	06:01	16:47	17:12
	Benghazi	05:05	05:30	16:17	16:43
5B	Cyprus, Nicosia	04:15	04:41	15:19	15:45
5H	Tanzania, Dar es Salaam	03:03	03:24	15:47	16:09
5N	Nigeria, Lagos	05:42	06:04	17:56	18:18
5R	Madagascar, Antananarivo	02:15	02:38	15:29	15:52
5T	Mauritania, Nouakchott	07:12	07:35	19:00	19:23
5U	Niger, Niamey	05:55	06:17	17:54	18:16
5V	Togo, Lome	05:50	06:12	18:06	18:27
5W	Western Samoa, Apia	17:00	17:22	06:00	06:22
5X	Uganda, Kampala	03:39	04:01	16:07	16:28
5Z	Kenya, Nairobi	04:40	05:02	17:11	17:33
6W	Senegal, Dakar	07:15	07:37	19:10	19:33
6Y	Jamaica, Kingston	11:16	11:38	23:04	23:26
7O	Yemen, Aden	03:03	03:25	15:03	15:25
	Sanaa	03:09	03:31	15:03	15:25
7P	Lesotho, Maseru	03:18	03:43	17:04	17:29
7Q	Malawi, Lilongwe	03:17	03:39	16:18	16:40
	Blantyre	03:09	03:32	16:15	16:38
7X	Algeria, Algiers	06:18	06:45	17:18	17:45
8P	Barbados, Bridgetown	10:01	10:23	22:01	22:23
8Q	Maldive Is.	01:00	01:21	13:19	13:40
8R	Guyana, Georgetown	09:49	10:11	22:02	22:24
9A	Croatia, Zagreb	05:40	06:11	16:08	16:39
9G	Ghana, Accra	05:56	06:17	18:12	18:33
9H	Malta	05:32	05:59	16:33	17:00
9J	Zambia, Lusaka	03:37	03:59	16:42	17:04
9K	Kuwait	03:10	03:35	14:30	14:55
9L	Sierra Leone, Freetown	06:51	07:13	19:00	19:22
9M2	West Malaysia, Kuala Lumpur	23:06	23:27	11:27	11:49
9M6,8	East Malaysia, Sabah, Sandakan(9M6)	22:03	22:24	10:18	10:40
	Sarawak, Kuching (9M8)	22:29	22:51	10:54	11:16
9N	Nepal, Kathmandu	00:38	01:02	12:03	12:27
9Q	Zaire, Kinshasa	04:43	05:04	17:21	17:42
	Kisangani	04:08	04:30	16:36	16:57
	Lubumbashi	03:45	04:07	16:40	17:03
9U	Burundi, Bujumbura	03:48	04:09	16:24	16:45
9V	Singapore	22:55	23:16	11:20	11:42
9X	Rwanda, Kigali	03:46	04:08	16:19	16:40
9Y	Trinidad & Tobago , Port of Spain	10:05	10:27	22:10	22:32
A2	Botswana, Gaborone	03:32	03:56	17:04	17:28
A3	Tonga, Nukualofa	17:03	17:26	06:23	06:46
A4	Oman, Musqat	02:20	02:44	13:55	14:18
A5	Bhutan, Thimbu	00:21	00:45	11:47	12:11
A6	United Arab Emirates, Abu Dhabi	02:39	03:02	14:11	14:35
A7	Qatar, Ad-Dawhah	02:50	03:14	14:21	14:45
A9	Bahrein, Al-Manamah	02:55	03:19	14:23	14:47
AP	Pakistan, Karachi	01:48	02:11	13:19	13:43
AP	Islamabad	01:34	02:00	12:42	13:08
BV	Taiwan, Taipei	22:10	22:34	09:41	10:05
BY	Peoples Rep. of China, Beijing	22:49	23:18	09:38	10:07
	Harbin	22:17	22:49	08:45	09:16
	Shanghai	22:17	22:43	09:33	09:58
	Fuzhou	22:20	22:44	09:49	10:13
	Xian	23:12	23:38	10:18	10:44

Pref	State/Province/Country/City	Tlight	Sunrise	Sunset	Tlight
	Chongqing	23:16	23:41	10:35	11:00
	Chengdu	23:26	23:51	10:43	11:08
	Lhasa	00:17	00:42	11:36	12:01
	Urumqi	00:50	01:21	11:26	11:56
	Kashi	01:30	01:58	12:21	12:49
C2	Nauru	18:41	19:02	07:10	07:32
C3	Andorra	06:33	07:02	17:13	17:43
C5	The Gambia, Banjul	07:10	07:32	19:09	19:31
C6	Bahamas, Nassau	11:26	11:50	22:57	23:21
C9	Mozambique, Maputo	03:04	03:28	16:39	17:03
	Mozambique	02:47	03:10	15:52	16:14
CE	Chile, Santiago	09:43	10:10	23:44	00:10
CE0	Easter Island	12:30	12:54	02:08	02:33
CE0	Juan Fernandez	10:15	10:42	00:16	00:43
CE0	San Felix	10:34	10:58	00:10	00:34
CM,CO	Cuba, Havana	11:44	12:07	23:20	23:43
CN	Morocco, Casablanca	06:56	07:22	18:05	18:31
CP	Bolivia, La Paz	10:02	10:25	23:10	23:32
CT	Portugal, Lisbon	07:10	07:38	18:03	18:31
CT3	Madeira Islands, Funchal	07:33	07:58	18:44	19:10
CU	Azores , Ponta Delgada	08:15	08:42	19:11	19:38
CX	Uruguay, Montevideo	08:42	09:09	22:48	23:15
CY0	Sable Is.	10:41	11:11	21:16	21:46
CY9	St. Paul Is.	10:47	11:19	21:09	21:41
D2	Angola, Luanda	04:46	05:07	17:34	17:56
D4	Cape Verde, Praia	07:39	08:01	19:34	19:56
D6	Comoros, Moroni	02:40	03:02	15:36	15:58
DA-DL,Y2-Y9	Fed. Rep. of Germany, Bonn	06:25	07:00	16:31	17:06
	Berlin	06:03	06:40	16:00	16:37
DU	Philippines, Manilla	22:01	22:23	09:56	10:19
E3	Eritrea, Asmara	03:30	03:52	15:24	15:46
EA	Spain, Madrid	06:50	07:19	17:38	18:06
EA6	Balearic Is., Palma	06:24	06:52	17:14	17:43
EA8	Canary Is., Las Palmas	07:17	07:42	18:40	19:04
EA9	Ceuta & Melilla, Ceuta	06:51	07:17	17:52	18:19
	Melilla	06:41	07:07	17:44	18:11
EI	Ireland, Dublin	07:24	08:01	17:17	17:54
EK	Armenia, Yerevan	03:37	04:06	14:25	14:54
EL	Liberia, Monrovia	06:39	07:00	18:53	19:15
EP	Iran, Tehran	03:02	03:29	14:04	14:31
ER	Moldova, Kishinev	04:51	05:23	15:14	15:46
ES	Estonia, Tallinn	05:35	06:20	14:49	15:34
ET	Ethiopia, Addis Ababa	03:24	03:45	15:32	15:54
EU,EV,EW	Belarus, Minsk	05:09	05:47	14:59	15:37
EX	Kyrgyzstan, Frunze	01:41	02:11	12:20	12:49
EY	Tajikistan, Samarkand	02:07	02:36	12:57	13:25
	Dushanbe	01:59	02:26	12:50	13:18
EZ	Turkmenistan, Ashkhabad	02:39	03:06	13:34	14:01
F	France, Paris	06:40	07:14	16:55	17:29
FG	Guadeloupe	10:13	10:35	22:06	22:28
FJ, FS	St. Martin	10:21	10:43	22:09	22:31
FH	Mayotte	02:32	02:54	15:31	15:53
FK	New Caledonia, Noumia	18:14	18:37	07:38	08:01
FM	Martinique	10:09	10:31	22:04	22:27
FO	Clipperton Is.	13:17	13:39	01:22	01:44
FO	Fr. Polynesia, Tahiti	16:05	16:28	05:16	05:38
	Rurutu, (Austral Is.)	19:14	19:38	08:39	09:02
	Hiva Oa, (Marquesas Is.)	14:53	15:15	03:44	04:06
FP	St. Pierre & Miquelon, St. Pierre	10:30	11:02	20:54	21:26
FR/G	Glorioso	02:26	02:48	15:21	15:43
FR/J,E	Juan de Nova	02:36	02:59	15:46	16:08
	Europa	02:38	03:02	16:02	16:26
FR	Reunion	01:39	02:03	15:00	15:23
FR/T	Tromelin	01:52	02:14	14:58	15:20
FT8W	Crozet	01:00	01:33	15:58	16:32
FT8X	Kerguelen	23:39	00:15	14:58	15:35
FT8Y	Antarctica, Dumont D'Urville	16:26	18:01	11:47	13:22
FT8Z	Amsterdam & St. Paul Is., Amsterdam	23:40	00:09	13:58	14:26
FW	Wallis & Futuna Is., Wallis	17:18	17:40	06:18	06:40
FY	Fr. Guiana, Cayenne	09:23	09:45	21:41	22:02
G	England, London	06:55	07:31	16:57	17:33
GD	Isle of Man	07:19	07:57	17:06	17:45
GI	Northern Ireland, Belfast	07:25	08:04	17:11	17:50
GJ	Jersey	06:59	07:33	17:12	17:46
GM	Scotland, Glasgow	07:21	08:01	17:00	17:40
	Aberdeen	07:16	07:58	16:46	17:28
GU	Guernsey	07:01	07:35	17:13	17:47
GW	Wales, Cardiff	07:07	07:43	17:10	17:45
H4	Solomon Islands, Honiara	18:58	19:19	07:48	08:10
HA	Hungary, Budapest	05:31	06:03	15:51	16:24
HB	Switzerland, Bern	06:16	06:48	16:39	17:11
HB0	Liechtenstein	06:08	06:40	16:30	17:02
HC	Ecuador, Quito	11:00	11:22	23:29	23:51
HC8	Galapagos Is.	11:50	12:12	00:20	00:41
HH	Haiti, Port-Au-Prince	10:58	11:21	22:45	23:08
HI	Dominican Republic, Santo Domingo	10:49	11:11	22:36	22:58
HK	Colombia, Bogota	10:50	11:12	23:08	23:30
HK0	Malpelo Is.	11:18	11:39	23:37	23:59
HK0	San Adreas	11:29	11:51	23:30	23:52
HL	Korea, Seoul	22:04	22:31	09:00	09:28
HP	Panama, Panama	11:17	11:38	23:25	23:47
HR	Honduras, Tegucigalpa	11:53	12:15	23:50	00:12
HS	Thailand, Bangkok	23:22	23:44	11:19	11:41
HV	Vatican City	05:48	06:17	16:30	16:59
HZ, 7Z	Saudi Arabia, Dharan	02:58	03:22	14:26	14:50
	Mecca	03:33	03:56	15:13	15:36
	Riyadh	03:09	03:32	14:41	15:05

February 7   Sunrise and Sunset, in UTC

Pref	State/Province/Country/City	Tlight	Sunrise	Sunset	Tlight
I	Italy, Rome	05:48	06:17	16:30	16:59
	Trieste	05:49	06:20	16:17	16:48
	Sicily	05:36	06:03	16:32	17:00
IS	Sardinia, Cagliari	05:58	06:25	16:49	17:17
J2	Djibouti, Djibouti	03:09	03:30	15:11	15:33
J3	Grenada	10:09	10:31	22:11	22:33
J5	Guinea-Bissau, Bissau	07:04	07:26	19:06	19:28
J6	St. Lucia	10:08	10:30	22:05	22:27
J7	Dominica	10:11	10:33	22:05	22:27
J8	St. Vincent	10:08	10:30	22:07	22:29
JA-JS	Japan, Tokyo	21:10	21:37	08:12	08:39
	Nagasaki	21:46	22:11	08:57	09:22
	Sapporo	21:14	21:44	07:52	08:22
JD1	Minami Torishima	19:59	20:23	07:32	07:56
JD1	Ogasawara, Kazan Is.	20:55	21:19	08:20	08:44
JT	Mongolia, Ulan Bator	23:40	00:13	09:59	10:32
JW	Svalbard, Spitsbergen	09:05	—	—	13:14
JX	Jan Mayen	08:42	10:07	15:26	16:52
JY	Jordan, Amman	04:01	04:26	15:14	15:39
KC4	Antarctica, Byrd Station	—	—	—	—
	McMurdo Sound	—	—	—	—
	Palmer Station	06:37	07:53	01:06	02:23
KC6	Belau, Yap	20:46	21:08	08:54	09:15
	Koror	20:59	21:20	09:11	09:33
KG4	Guantanamo Bay	11:11	11:34	22:55	23:18
KH0	Mariana Is., Saipan	20:22	20:44	08:17	08:39
KH1	Baker, Howland Is.	17:33	17:55	06:01	06:22
KH2	Guam, Agana	20:24	20:46	08:23	08:45
KH3	Johnston Is.	17:21	17:44	05:12	05:34
KH4	Midway Is.	18:09	18:34	05:33	05:57
KH5	Palmyra Is.	16:44	17:05	04:59	05:20
KH5K	Kingman Reef	16:48	17:10	05:00	05:21
KH6	Hawaii, Hilo	16:31	16:53	04:15	04:37
	Honolulu	16:44	17:07	04:24	04:47
KH7	Kure Is.	18:14	18:38	05:36	06:01
KH8	American Samoa, Pago Pago	16:55	17:17	05:57	06:19
KH9	Wake Is.	19:03	19:26	06:48	07:11
KL7	Alaska, Adak	18:42	19:17	04:43	05:18
	Anchorage	17:20	18:08	02:19	03:08
	Fairbanks	17:25	18:21	01:49	02:46
	Juneau	16:08	16:52	01:31	02:14
	Nome	18:34	19:29	03:01	03:57
KP1	Navassa Is.	11:09	11:31	22:56	23:18
KP2	Virgin Islands, Charlotte Amalie	10:28	10:51	22:16	22:38
KP4	Puerto Rico, San Juan	10:34	10:56	22:21	22:43
KP5	Desecheo Is.	10:39	11:01	22:26	22:49
LA-LJ	Norway, Oslo	06:33	07:19	15:43	16:29
LU	Argentina, Buenos Aires	08:51	09:18	22:56	23:23
LX	Luxembourg	06:26	07:00	16:38	17:12
LY	Lithuania, Vilna	05:19	05:58	15:06	15:44
LZ	Bulgaria, Sofia	05:06	05:35	15:45	16:15
OA	Peru, Lima	10:43	11:05	23:39	00:01
OD	Lebanon, Beirut	04:05	04:31	15:12	15:38
OE	Austria, Vienna	05:43	06:16	16:01	16:34
OH	Finland, Helsinki	05:36	06:23	14:44	15:31
OH0	Aland Is.	05:56	06:43	15:04	15:51
OJ0	Market Reef	06:01	06:47	15:08	15:55
OK,OL	Czech Rep., Prague	05:54	06:29	16:03	16:38
OM	Slovak, Rep., Bratislava	05:40	06:13	15:59	16:31
ON	Belgium, Brussels	06:36	07:11	16:41	17:16
OX	Greenland, Godthaab	10:58	11:52	19:28	20:23
	Thule	13:49	—	—	19:49
OY	Faroe Islands, Torshavn	07:50	08:39	16:42	17:32
OZ	Denmark, Copenhagen	06:14	06:54	15:53	16:33
P2	Papua New Guinea, Madang	20:00	20:22	08:41	09:02
	Port Moresby	19:49	20:11	08:39	09:01
P4	Aruba, Oranjestad	10:38	11:00	22:39	23:01
PA-PI	Netherlands, Amsterdam	06:37	07:13	16:35	17:11
PJ2,4,9	Netherlands Antilles, Willemstad	10:43	11:05	22:43	23:05
PJ5-8	St. Maarten and Saba, St. Maarten	10:21	10:43	22:10	22:32
PY	Brazil, Brasilia	08:41	09:03	21:47	22:10
	Rio De Janeiro	08:12	08:35	21:38	22:01
	Natal	08:03	08:24	20:45	21:06
	Manaus	09:46	10:07	22:21	22:43
	Porto Alegre	08:32	08:57	22:20	22:45
PY0	Fernando De Noronha	07:54	08:15	20:31	20:53
PY0	St. Peter & St. Paul Rocks	07:47	08:09	20:14	20:35
PY0	Trindade & Martin Vaz Is., Trindade	07:20	07:43	20:39	21:02
PZ	Suriname, Paramaribo	09:36	09:57	21:51	22:13
R1FJ_	Franz Josef Land	07:39	—	—	09:44
R1MV_	Malyj Vysotskij Is.	05:23	06:10	14:28	15:15
S0	Western Sahara, Smara	07:03	07:27	18:31	18:55
S2	Bangladesh, Dacca	00:13	00:36	11:48	12:11
S5	Slovenia, Ljubljana	05:46	06:18	16:13	16:45
S7	Seychelles, Victoria	02:01	02:23	14:40	15:02
S9	Sao Tome	05:22	05:44	17:50	18:11
SM	Sweden, Stockholm	06:01	06:46	15:16	16:01
SP	Poland, Krakow	05:32	06:06	15:41	16:16
	Warsaw	05:32	06:08	15:31	16:07
ST	Sudan, Khartoum	03:56	04:18	15:49	16:12
ST0	Southern Sudan, Juba	03:48	04:09	16:05	16:27
SU	Egypt, Cairo	04:16	04:41	15:35	16:00
SV	Greece, Athens	04:57	05:25	15:53	16:20
SV/A	Mount Athos	04:58	05:27	15:46	16:15
SV5	Dodecanese, Rhodes	04:37	05:04	15:38	16:04
SV9	Crete	04:48	05:14	15:51	16:18
T2	Tuvalu, Funafuti	17:44	18:06	06:33	06:54

Pref	State/Province/Country/City	Tlight	Sunrise	Sunset	Tlight
T30	West Kiribati, Bonriki	18:17	18:39	06:43	07:04
T31	Central Kiribati, Kanton	17:12	17:34	05:47	06:09
T32	East Kiribati, Christmas Is.	16:20	16:42	04:45	05:06
T33	Banaba Is.	18:31	18:52	07:00	07:22
T5	Somalia, Mogadishu	02:49	03:11	15:13	15:35
T7	San Marino	05:52	06:22	16:27	16:57
T9	Bosnia-Herzegovina, Sarajevo	05:27	05:58	16:02	16:33
TA	Turkey, Ankara	04:23	04:52	15:12	15:41
	Istanbul	04:41	05:10	15:26	15:55
TF	Iceland, Reykjavik	08:58	09:53	17:30	18:25
TG	Guatemala, Guatemala City	12:07	12:29	00:02	00:25
TI	Costa Rica, San Jose	11:36	11:57	23:42	00:04
TI9	Cocos Is.	11:43	12:04	23:59	00:20
TJ	Cameroon, Yaounde	05:07	05:29	17:27	17:48
TK	Corsica	06:02	06:31	16:44	17:13
TL	Central African Republic, Bangui	04:39	05:01	16:58	17:19
TN	Congo, Brazzaville	04:43	05:04	17:21	17:42
TR	Gabon, Libreville	05:11	05:33	17:39	18:00
TT	Chad, N'Djamena	05:02	05:24	17:03	17:25
TU	Cote d'Ivoire, Abidjan	06:11	06:32	18:27	18:49
TY	Benin, Porto Novo	05:45	06:07	17:59	18:21
TZ	Mali, Bamako	06:35	06:57	18:35	18:57
UA	Russia, European, Leningrad (UA1)	05:14	06:00	14:25	15:11
	Archangel (UA1)	04:50	05:46	13:17	14:13
	Murmansk (UA1)	05:42	06:55	13:08	14:20
	Moscow (UA3)	04:34	05:14	14:13	14:53
	Kuybyshev (UA4)	03:38	04:15	13:31	14:09
	Rostov (UA6)	04:09	04:41	14:30	15:02
UA2	Kaliningrad	05:40	06:19	15:24	16:03
UA9,0	Russia, Asiatic, Novosibirsk (UA9)	01:31	02:10	11:14	11:53
	Perm (UA9)	03:25	04:08	12:49	13:32
	Omsk (UA9)	02:09	02:48	11:52	12:31
	Norilsk (UA0)	02:04	03:18	09:24	10:38
	Irkutsk (UA0)	23:59	00:35	09:57	10:34
	Vladivostok (UA0)	21:52	22:22	08:30	09:00
	Petropavlovsk (UA0)	20:23	21:00	06:18	06:55
	Khabarovsk (UA0)	21:48	22:22	08:05	08:38
	Krasnoyarsk (UA0)	00:54	01:34	10:31	11:11
	Yakutsk (UA0)	22:44	23:33	07:36	08:26
	Wrangel Island (UA0)	20:06	21:32	02:51	04:17
	Kyzyl (UA0Y)	00:37	01:13	10:38	11:14
UJ-UM	Uzbekistan, Bukhoro	02:17	02:45	13:07	13:35
	Tashkent	02:00	02:28	12:44	13:13
UN-UQ	Kazakhstan, Alma-Ata	01:32	02:02	12:10	12:40
UR-UZ,EM-EO	Ukraine, Kiev	04:50	05:25	14:58	15:33
V2	Antigua & Barbuda, St. Johns	10:15	10:37	22:05	22:27
V3	Belize, Belmopan	12:03	12:25	23:52	00:15
V4	St. Kitts & Nevis	10:18	10:40	22:08	22:30
V5	Namibia, Windhoek	04:11	04:35	17:36	17:59
V6	Micronesia, Ponape	19:23	19:45	07:36	07:58
V7	Marshall Islands, Kwajalein	18:50	19:11	06:58	07:19
V8	Brunei, Bandar Seri Begawan	22:15	22:36	10:32	10:54
VK	Australia, Canberra (VK1)	19:00	19:27	09:07	09:35
	Sydney (VK2)	18:54	19:21	08:56	09:23
	Melbourne (VK3)	19:11	19:39	09:28	09:57
	Brisbane (VK4)	18:59	19:24	08:39	09:04
	Adelaide (VK5)	19:43	20:10	09:49	10:16
	Perth (VK6)	21:20	21:46	11:15	11:41
	Hobart, Tasmania (VK7)	18:48	19:19	09:30	10:01
	Darwin (VK8)	20:50	21:12	09:48	10:10
VK0	Heard Is.	23:06	23:47	14:54	15:34
VK0	Macquarie Is.	17:16	17:59	09:18	10:01
VK9C	Cocos-Keeling Is.	23:07	23:29	12:04	12:26
VK9L	Lord Howe Is.	18:27	18:53	08:21	08:47
VK9M	Mellish Reef	19:04	19:26	08:14	08:37
VK9N	Norfolk Is.	17:57	18:22	07:41	08:07
VK9W	Willis Is.	19:31	19:53	08:38	09:01
VK9X	Christmas Is.	22:33	22:55	11:26	11:48
VP2E	Anguilla	10:21	10:43	22:08	22:31
VP2M	Montserrat	10:16	10:38	22:07	22:29
VP2V	British Virgin Is., Tortola	10:27	10:50	22:14	22:37
VP5	Turks & Caicos Islands, Grand Turk	10:57	11:20	22:37	23:00
VP8	Falkland Islands, Stanley	07:57	08:36	23:34	00:13
VP8	So. Georgia Is.	06:20	07:03	22:19	23:01
VP8	So. Orkney Is.	06:12	07:08	23:23	00:19
VP8	So. Sandwich Islands, Saunders Is.	05:19	06:07	21:53	22:42
VP8	So. Shetland Is., King George Is.	06:50	07:51	00:23	01:24
VP9	Bermuda	10:44	11:09	21:56	22:21
VQ9	Chagos, Diego Garcia	00:51	01:12	13:36	13:57
VR2,VS6	Hong Kong	22:36	22:59	10:14	10:37
VR6	Pitcairn Is.	13:56	14:20	03:28	03:52
VU	India, Bombay	01:18	01:41	13:04	13:27
	Calcutta	00:20	00:43	11:57	12:20
	New Delhi	01:11	01:36	12:34	12:58
	Bangalore	00:53	01:15	12:52	13:14
VU	Andaman Islands, Port Blair	23:50	00:12	11:53	12:15
VU	Laccadive Is.	01:08	01:29	13:14	13:36
XE	Mexico, Mexico City (XE1)	12:46	13:09	00:31	00:54
	Chihuahua (XE2)	13:24	13:49	00:46	01:11
	Merida (XE3)	12:10	12:33	23:51	00:14
XF4	Revilla Gigedo	13:35	13:58	01:21	01:44
XT	Burkina Faso, Ouagadougou	06:09	06:31	18:09	18:31
XU	Cambodia, Phnom Penh	23:02	23:24	11:05	11:27
XW	Laos, Viangchan	23:18	23:40	11:06	11:29
XX9	Macao	22:39	23:02	10:17	10:40
XZ	Myanmar, Yangon	23:43	00:05	11:34	11:56
YA	Afghanistan, Kandahar	02:00	02:25	13:16	13:41
	Kabul	01:51	02:17	12:57	13:23
YB-YD	Indonesia, Jakarta	22:34	22:56	11:17	11:39
	Medan	23:18	23:39	11:38	12:00
	Pontianak	22:32	22:53	11:00	11:21
	Jayapura	20:23	20:44	08:57	09:19
YI	Iraq, Baghdad	03:28	03:53	14:38	15:04
YJ	Vanuatu, Port Vila	18:14	18:36	07:25	07:47
YK	Syria, Damascus	04:01	04:27	15:10	15:36
YL	Latvia, Riga	05:31	06:12	15:02	15:44
YN	Nicaragua, Managua	11:46	12:08	23:48	00:09
YO	Romania, Bucharest	04:57	05:28	15:31	16:01
YS	El Salvador, San Salvador	12:00	12:23	23:58	00:20
YU	Yugoslavia, Belgrade	05:21	05:51	15:52	16:23
YV	Venezuela, Caracas	10:28	10:50	22:33	22:55
YV0	Aves Is.	10:21	10:43	22:14	22:36
Z2	Zimbabwe, Harare	03:23	03:45	16:34	16:57
Z3	Macedonia, (ex Yugoslav), Skopje	05:12	05:42	15:54	16:24
ZA	Albania, Tirane	05:18	05:47	16:02	16:31
ZB2	Gibraltar	06:51	07:18	17:52	18:19
ZC4	British Cyprus	04:16	04:42	15:21	15:48
ZD7	St. Helena	05:53	06:15	18:59	19:22
ZD8	Ascension Is.	06:37	06:59	19:24	19:46
ZD9	Tristan da Cunha	05:41	06:09	19:56	20:24
ZF	Cayman Is.	11:35	11:58	23:19	23:42
ZK1	No. Cook Is., Manihiki	16:20	16:42	05:13	05:35
ZK1	So. Cook Is., Rarotonga	16:01	16:24	05:22	05:45
ZK2	Niue	16:40	17:03	05:55	06:18
ZK3	Tokelaus, Atafu	17:10	17:31	05:57	06:19
ZL	New Zealand, Auckland (ZL1)	17:13	17:41	07:27	07:55
	Wellington (ZL2)	17:02	17:33	07:36	08:06
	Christchurch (ZL3)	17:05	17:37	07:50	08:21
	Dunedin (ZL4)	17:06	17:39	08:04	08:37
ZL5	Antarctica, Scott Base	—	—	—	—
ZL7	Chatham Is.	16:20	16:52	07:07	07:39
ZL8	Kermadec Is.	17:00	17:25	06:45	07:11
ZL9	Auckland & Campbell Is., Auckland	17:04	17:42	08:34	09:11
	Campbell Is.	16:46	17:25	08:29	09:09
ZP	Paraguay, Asuncion	09:06	09:30	22:39	23:03
ZS	South Africa, Cape Town (ZS1)	03:46	04:12	17:48	18:14
	Port Elizabeth (ZS2)	03:16	03:43	17:19	17:45
	Bloemfontein (ZS4)	03:24	03:49	17:09	17:34
	Durban (ZS5)	03:03	03:29	16:51	17:17
	Johannesburg (ZS6)	03:21	03:46	16:57	17:21
ZS8	Marion Is.	01:54	02:28	16:57	17:31

Pref	State/Province/Country/City	Tlight	Sunrise	Sunset	Tlight
VE1	New Brunswick, Saint John	10:48	11:18	21:58	22:28
	Nova Scotia, Halifax	10:37	11:07	21:49	22:18
	Prince Edward Island, Charlottetown	10:37	11:07	21:44	22:15
VE2	Quebec, Montreal	11:18	11:48	22:27	22:57
	Quebec City	11:10	11:41	22:16	22:47
VE3	Ontario, London	11:47	12:15	23:02	23:31
	Ottawa	11:26	11:56	22:36	23:06
	Sudbury	11:49	12:19	22:55	23:26
	Toronto	11:40	12:09	22:53	23:22
VE4	Manitoba, Winnipeg	12:57	13:29	23:54	00:27
VE5	Saskatchewan, Regina	13:27	14:01	00:23	00:56
	Saskatoon	13:37	14:12	00:28	01:03
VE6	Alberta, Calgary	14:06	14:39	01:00	01:34
	Edmonton	14:06	14:42	00:53	01:28
VE7	British Columbia, Prince George	14:44	15:20	01:29	02:05
	Prince Rupert	15:14	15:51	01:58	02:35
	Vancouver	14:40	15:12	01:39	02:12
VE8	Northwest Territories, Yellowknife	14:23	15:10	00:31	01:18
	Resolute	13:44	15:16	21:51	23:23
VY1	Yukon, Whitehorse	15:43	16:27	02:00	02:44
VO1	Newfoundland, St. John's	09:57	10:28	21:00	21:32
VO2	Labrador, Goose Bay	10:34	11:09	21:21	21:56
W1	Connecticut, Hartford	11:11	11:39	22:29	22:57
	Maine, Bangor	10:58	11:28	22:09	22:39
	Portland	11:03	11:32	22:17	22:46
	Massachusetts, Boston	11:05	11:34	22:22	22:50
	New Hampshire, Concord	11:08	11:36	22:22	22:51
	Rhode Island, Providence	11:06	11:34	22:24	22:52
	Vermont, Montpelier	11:13	11:42	22:25	22:55
W2	New Jersey, Atlantic City	11:16	11:43	22:39	23:06
	New York, Albany	11:16	11:45	22:32	23:01
	Buffalo	11:37	12:06	22:52	23:21
	New York City	11:15	11:43	22:35	23:03
	Syracuse	11:26	11:55	22:41	23:10
W3	Delaware, Wilmington	11:20	11:48	22:43	23:10
	District of Columbia, Washington	11:26	11:53	22:50	23:17
	Maryland, Baltimore	11:24	11:52	22:48	23:15
	Pennsylvania, Harrisburg	11:27	11:54	22:48	23:15
	Philadelphia	11:19	11:47	22:41	23:09
	Pittsburgh	11:39	12:07	23:00	23:28
	Scranton	11:23	11:51	22:41	23:10
W4	Alabama, Montgomery	11:58	12:23	23:34	23:59
	Florida, Jacksonville	11:38	12:02	23:18	23:42
	Miami	11:28	11:52	23:16	23:40
	Pensacola	12:00	12:24	23:40	00:04
	Georgia, Atlanta	11:51	12:17	23:25	23:51
	Savannah	11:37	12:02	23:14	23:39
	Kentucky, Lexington	11:55	12:22	23:21	23:48
	Louisville	12:00	12:27	23:26	23:53
	North Carolina, Charlotte	11:38	12:04	23:09	23:35
	Raleigh	11:30	11:56	23:00	23:26
	Wilmington	11:26	11:51	22:59	23:24
	South Carolina, Columbia	11:38	12:03	23:11	23:37
	Tennessee, Knoxville	11:51	12:17	23:21	23:47
	Memphis	12:15	12:41	23:47	00:12
	Nashville	12:03	12:29	23:32	23:58
	Virginia, Norfolk	11:21	11:48	22:49	23:16
	Richmond	11:26	11:53	22:53	23:20
W5	Arkansas, Little Rock	12:24	12:49	23:56	00:21
	Louisiana, New Orleans	12:11	12:35	23:52	00:16
	Shreveport	12:27	12:52	00:04	00:29
	Mississippi, Jackson	12:13	12:38	23:50	00:15
	New Mexico, Albuquerque	13:21	13:47	00:53	01:19
	Oklahoma, Oklahoma City	12:45	13:11	00:16	00:42
	Texas, Abilene	12:51	13:16	00:28	00:53
	Amarillo	13:02	13:28	00:33	00:59
	Dallas	12:40	13:05	00:16	00:41
	El Paso	13:18	13:43	00:56	01:21
	San Antonio	12:44	13:08	00:26	00:50
W6	California, Los Angeles	14:07	14:32	01:40	02:06
	San Francisco	14:26	14:53	01:53	02:19
W7	Arizona, Flagstaff	13:42	14:07	01:13	01:39
	Phoenix	13:42	14:07	01:16	01:42
	Idaho, Boise	14:07	14:36	01:20	01:50
	Pocatello	13:51	14:20	01:07	01:35
	Montana, Billings	13:38	14:08	00:46	01:17
	Butte	13:54	14:25	01:02	01:33
	Great Falls	13:51	14:22	00:55	01:26
	Nevada, Las Vegas	13:56	14:22	01:25	01:51
	Reno	14:17	14:45	01:40	02:08
	Oregon, Portland	14:34	15:05	01:44	02:14
	Utah, Salt Lake City	13:47	14:15	01:07	01:35
	Washington, Seattle	14:35	15:06	01:39	02:10
	Spokane	14:15	14:47	01:19	01:50
	Wyoming, Cheyenne	13:19	13:47	00:38	01:06
	Sheridan	13:31	14:01	00:42	01:12
W8	Michigan, Detroit	11:53	12:21	23:09	23:38
	Grand Rapids	12:04	12:33	23:19	23:48
	Sault Ste. Marie	12:02	12:33	23:09	23:40
	Traverse City	12:05	12:35	23:16	23:46
	Ohio, Cincinnati	11:56	12:23	23:20	23:47
	Cleveland	11:47	12:15	23:05	23:33
	Columbus	11:51	12:18	23:12	23:40
	West Virginia, Charleston	11:44	12:11	23:09	23:36
W9	Illinois, Chicago	12:11	12:39	23:28	23:57
	Indiana, Indianapolis	12:03	12:31	23:25	23:53
	Wisconsin, Green Bay	12:15	12:44	23:26	23:56

Pref	State/Province/Country/City	Tlight	Sunrise	Sunset	Tlight
	Milwaukee	12:13	12:42	23:28	23:57
W0	Colorado, Denver	13:18	13:46	00:41	01:08
	Grand Junction	13:32	13:59	00:56	01:23
	Iowa, Des Moines	12:34	13:03	23:53	00:21
	Kansas, Pratt	12:52	13:18	00:18	00:45
	Wichita	12:46	13:13	00:12	00:39
	Minnesota, Duluth	12:33	13:04	23:39	00:10
	Minneapolis	12:36	13:06	23:47	00:17
	Missouri, Columbia	12:27	12:54	23:51	00:18
	Kansas City	12:36	13:03	00:00	00:27
	St. Louis	12:18	12:45	23:43	00:10
	Nebraska, North Platte	13:03	13:31	00:22	00:50
	Omaha	12:43	13:12	00:02	00:30
	North Dakota, Fargo	12:52	13:23	23:58	00:29
	South Dakota, Rapid City	13:15	13:45	00:28	00:57
-		05:49	06:10	18:17	18:38
1A0	S.M.O.M	05:31	05:59	16:48	17:17
1S	Spratly Is.	22:28	22:49	10:42	11:04
3A	Monaco	05:52	06:22	17:06	17:35
3B6	Agalega	01:54	02:15	14:39	15:00
3B7	St. Brandon	01:36	01:58	14:31	14:53
3B8	Mauritius	01:41	02:03	14:43	15:06
3B9	Rodriguez Is.	01:18	01:40	14:19	14:42
3C	Equatorial Guinea, Bata	05:11	05:32	17:36	17:57
	Malabo	05:17	05:38	17:39	18:00
3C0	Annobon Is.	05:25	05:46	17:55	18:17
3D2	Fiji Is., Suva	17:39	18:02	06:38	07:00
3D2	Conway Reef	17:52	18:15	06:57	07:20
3D2	Rotuma Is.	17:48	18:09	06:36	06:57
3DA	Swaziland, Mbabane	03:19	03:43	16:35	16:58
3V	Tunisia, Tunis	05:35	06:02	17:03	17:30
3W, XV	Vietnam, Ho Chi Minh City (Saigon)	22:50	23:12	11:02	11:23
	Hanoi	23:01	23:24	10:57	11:19
3X	Guinea, Conakry	06:51	07:12	19:04	19:25
3Y	Bouvet	04:12	04:51	19:09	19:48
3Y	Peter I Is.	08:46	10:05	02:26	03:46
4J, 4K	Azerbaijan, Baku	02:59	03:27	14:20	14:48
4L	Georgia, Tbilisi	03:21	03:49	14:39	15:07
4S	Sri Lanka, Colombo	00:35	00:56	12:52	13:13
4U	ITU Geneva	06:00	06:30	17:07	17:38
4U	United Nations Hq.	11:15	11:43	22:35	23:03
4X, 4Z	Israel, Jerusalem	03:51	04:16	15:29	15:54
5A	Libya, Tripoli	05:23	05:48	16:59	17:24
	Benghazi	04:52	05:17	16:29	16:54
5B	Cyprus, Nicosia	04:01	04:27	15:33	15:58
5H	Tanzania, Dar es Salaam	03:05	03:26	15:44	16:05
5N	Nigeria, Lagos	05:40	06:01	17:58	18:19
5R	Madagascar, Antananarivo	02:22	02:45	15:22	15:45
5T	Mauritania, Nouakchott	07:06	07:28	19:06	19:28
5U	Niger, Niamey	05:51	06:12	17:58	18:20
5V	Togo, Lome	05:48	06:10	18:07	18:29
5W	Western Samoa, Apia	17:05	17:26	05:55	06:16
5X	Uganda, Kampala	03:39	04:00	16:06	16:28
5Z	Kenya, Nairobi	04:41	05:02	17:11	17:32
6W	Senegal, Dakar	07:10	07:31	19:15	19:37
6Y	Jamaica, Kingston	11:09	11:31	23:10	23:32
7O	Yemen, Aden	02:58	03:20	15:07	15:28
	Sanaa	03:03	03:25	15:08	15:30
7P	Lesotho, Maseru	03:30	03:54	16:52	17:17
7Q	Malawi, Lilongwe	03:22	03:44	16:13	16:34
	Blantyre	03:15	03:37	16:09	16:31
7X	Algeria, Algiers	06:04	06:30	17:32	17:59
8P	Barbados, Bridgetown	09:57	10:19	22:05	22:26
8Q	Maldive Is.	00:59	01:20	13:20	13:41
8R	Guyana, Georgetown	09:47	10:08	22:04	22:25
9A	Croatia, Zagreb	05:20	05:50	16:28	16:59
9G	Ghana, Accra	05:54	06:15	18:13	18:34
9H	Malta	05:18	05:44	16:48	17:14
9J	Zambia, Lusaka	03:42	04:04	16:36	16:58
9K	Kuwait	02:59	03:23	14:41	15:05
9L	Sierra Leone, Freetown	06:48	07:09	19:03	19:24
9M2	West Malaysia, Kuala Lumpur	23:05	23:26	11:28	11:49
9M6,8	East Malaysia, Sabah, Sandakan(9M6)	22:01	22:22	10:20	10:41
	Sarawak, Kuching (9M8)	22:29	22:50	10:54	11:15
9N	Nepal, Kathmandu	00:28	00:52	12:13	12:36
9Q	Zaire, Kinshasa	04:44	05:05	17:19	17:40
	Kisangani	04:08	04:30	16:36	16:57
	Lubumbashi	03:49	04:11	16:36	16:58
9U	Burundi, Bujumbura	03:49	04:10	16:22	16:43
9V	Singapore	22:55	23:16	11:21	11:42
9X	Rwanda, Kigali	03:47	04:08	16:18	16:39
9Y	Trinidad & Tobago , Port of Spain	10:02	10:23	22:14	22:35
A2	Botswana, Gaborone	03:42	04:05	16:54	17:17
A3	Tonga, Nukualofa	17:11	17:33	06:15	06:38
A4	Oman, Musqat	02:12	02:35	14:03	14:26
A5	Bhutan, Thimbu	00:11	00:35	11:57	12:20
A6	United Arab Emirates, Abu Dhabi	02:30	02:53	14:20	14:43
A7	Qatar, Ad-Dawhah	02:41	03:05	14:30	14:53
A9	Bahrein, Al-Manamah	02:46	03:09	14:33	14:56
AP	Pakistan, Karachi	01:39	02:02	13:28	13:51
AP	Islamabad	01:21	01:46	12:55	13:20
BV	Taiwan, Taipei	22:01	22:24	09:50	10:13
BY	Peoples Rep. of China, Beijing	22:33	23:01	09:55	10:22
	Harbin	21:57	22:27	09:06	09:36
	Shanghai	22:06	22:30	09:44	10:09
	Fuzhou	22:11	22:34	09:58	10:22
	Xian	22:58	23:24	10:32	10:57

Pref	State/Province/Country/City	Tlight	Sunrise	Sunset	Tlight
	Chongqing	23:05	23:29	10:46	11:10
	Chengdu	23:15	23:39	10:54	11:19
	Lhasa	00:06	00:30	11:47	12:11
	Urumqi	00:32	01:01	11:45	12:14
	Kashi	01:14	01:42	12:37	13:04
C2	Nauru	18:41	19:02	07:10	07:31
C3	Andorra	06:15	06:44	17:31	18:00
C5	The Gambia, Banjul	07:06	07:27	19:13	19:35
C6	Bahamas, Nassau	11:17	11:40	23:06	23:29
C9	Mozambique, Maputo	03:14	03:37	16:28	16:52
	Mozambique	02:53	03:15	15:46	16:08
CE	Chile, Santiago	09:57	10:23	23:30	23:56
CE0	Easter Island	12:40	13:04	01:57	02:21
CE0	Juan Fernandez	10:29	10:55	00:02	00:28
CE0	San Felix	10:44	11:08	23:59	00:23
CM,CO	Cuba, Havana	11:35	11:58	23:28	23:51
CN	Morocco, Casablanca	06:43	07:09	18:18	18:43
CP	Bolivia, La Paz	10:08	10:30	23:04	23:26
CT	Portugal, Lisbon	06:54	07:21	18:19	18:46
CT3	Madeira Islands, Funchal	07:20	07:45	18:57	19:22
CU	Azores , Ponta Delgada	08:00	08:26	19:26	19:53
CX	Uruguay, Montevideo	08:57	09:23	22:33	23:00
CY0	Sable Is.	10:22	10:51	21:35	22:05
CY9	St. Paul Is.	10:26	10:57	21:31	22:02
D2	Angola, Luanda	04:49	05:10	17:31	17:52
D4	Cape Verde, Praia	07:34	07:56	19:39	20:01
D6	Comoros, Moroni	02:44	03:06	15:31	15:53
DA-DL,Y2-Y9	Fed. Rep. of Germany, Bonn	06:01	06:35	16:56	17:30
	Berlin	05:37	06:12	16:27	17:02
DU	Philippines, Manilla	21:56	22:17	10:01	10:23
E3	Eritrea, Asmara	03:25	03:46	15:29	15:51
EA	Spain, Madrid	06:34	07:02	17:55	18:22
EA6	Balearic Is., Palma	06:08	06:35	17:31	17:58
EA8	Canary Is., Las Palmas	07:07	07:31	18:50	19:14
EA9	Ceuta & Melilla, Ceuta	06:37	07:03	18:07	18:33
	Melilla	06:27	06:53	17:58	18:24
EI	Ireland, Dublin	06:57	07:33	17:44	18:20
EK	Armenia, Yerevan	03:21	03:49	14:42	15:10
EL	Liberia, Monrovia	06:37	06:58	18:55	19:16
EP	Iran, Tehran	02:48	03:14	14:18	14:44
ER	Moldova, Kishinev	04:30	05:01	15:35	16:06
ES	Estonia, Tallinn	05:01	05:44	15:25	16:07
ET	Ethiopia, Addis Ababa	03:21	03:42	15:35	15:56
EU,EV,EW	Belarus, Minsk	04:42	05:19	15:27	16:03
EX	Kyrgyzstan, Frunze	01:23	01:52	12:38	13:07
EY	Tajikistan, Samarkand	01:51	02:19	13:14	13:41
	Dushanbe	01:43	02:10	13:06	13:34
EZ	Turkmenistan, Ashkhabad	02:23	02:50	13:49	14:16
F	France, Paris	06:18	06:50	17:18	17:50
FG	Guadeloupe	10:07	10:29	22:11	22:33
FJ, FS	St. Martin	10:15	10:37	22:15	22:37
FH	Mayotte	02:37	02:58	15:26	15:48
FK	New Caledonia, Noumia	18:22	18:45	07:29	07:52
FM	Martinique	10:04	10:25	22:09	22:31
FO	Clipperton Is.	13:13	13:35	01:26	01:47
FO	Fr. Polynesia, Tahiti	16:11	16:34	05:09	05:31
	Rurutu, (Austral Is.)	19:23	19:46	08:30	08:53
	Hiva Oa, (Marquesas Is.)	14:57	15:18	03:41	04:02
FP	St. Pierre & Miquelon, St. Pierre	10:09	10:40	21:15	21:46
FR/G	Glorioso	02:30	02:52	15:17	15:38
FR/J,E	Juan de Nova	02:43	03:05	15:39	16:01
	Europa	02:47	03:10	15:54	16:17
FR	Reunion	01:47	02:10	14:52	15:14
FR/T	Tromelin	01:58	02:20	14:52	15:14
FT8W	Crozet	01:22	01:54	15:37	16:08
FT8X	Kerguelen	00:05	00:39	14:34	15:08
FT8Y	Antarctica, Dumont D'Urville	17:51	18:57	10:49	11:56
FT8Z	Amsterdam & St. Paul Is., Amsterdam	23:57	00:24	13:42	14:09
FW	Wallis & Futuna Is., Wallis	17:23	17:45	06:13	06:34
FY	Fr. Guiana, Cayenne	09:22	09:43	21:42	22:03
G	England, London	06:30	07:04	17:23	17:57
GD	Isle of Man	06:51	07:28	17:35	18:11
GI	Northern Ireland, Belfast	06:57	07:34	17:40	18:17
GJ	Jersey	06:36	07:09	17:35	18:08
GM	Scotland, Glasgow	06:52	07:30	17:31	18:09
	Aberdeen	06:46	07:25	17:18	17:58
GU	Guernsey	06:38	07:11	17:37	18:10
GW	Wales, Cardiff	06:43	07:17	17:35	18:10
H4	Solomon Islands, Honiara	19:01	19:23	07:44	08:06
HA	Hungary, Budapest	05:09	05:41	16:13	16:45
HB	Switzerland, Bern	05:55	06:26	17:00	17:32
HB0	Liechtenstein	05:47	06:18	16:52	17:23
HC	Ecuador, Quito	11:01	11:22	23:29	23:50
HC8	Galapagos Is.	11:50	12:12	00:19	00:40
HH	Haiti, Port-Au-Prince	10:52	11:14	22:51	23:13
HI	Dominican Republic, Santo Domingo	10:42	11:05	22:42	23:04
HK	Colombia, Bogota	10:49	11:10	23:10	23:31
HK0	Malpelo Is.	11:16	11:37	23:38	23:59
HK0	San Adreas	11:25	11:46	23:34	23:55
HL	Korea, Seoul	21:49	22:15	09:15	09:42
HP	Panama, Panama	11:14	11:35	23:28	23:49
HR	Honduras, Tegucigalpa	11:48	12:10	23:54	00:16
HS	Thailand, Bangkok	23:17	23:39	11:24	11:46
HV	Vatican City	05:30	05:59	16:48	17:16
HZ, 7Z	Saudi Arabia, Dharan	02:48	03:11	14:35	14:59
	Mecca	03:25	03:48	15:20	15:43
	Riyadh	03:00	03:23	14:50	15:13
I	Italy, Rome	05:30	05:59	16:48	17:16
	Trieste	05:29	05:59	16:37	17:08
	Sicily	05:21	05:47	16:47	17:14
IS	Sardinia, Cagliari	05:42	06:09	17:05	17:32
J2	Djibouti, Djibouti	03:05	03:26	15:15	15:36
J3	Grenada	10:05	10:26	22:15	22:36
J5	Guinea-Bissau, Bissau	07:00	07:22	19:10	19:31
J6	St. Lucia	10:03	10:25	22:10	22:31
J7	Dominica	10:05	10:27	22:10	22:32
J8	St. Vincent	10:04	10:26	22:12	22:33
JA-JS	Japan, Tokyo	20:56	21:22	08:26	08:52
	Nagasaki	21:33	21:58	09:09	09:34
	Sapporo	20:56	21:25	08:11	08:40
JD1	Minami Torishima	19:51	20:14	07:41	08:04
JD1	Ogasawara, Kazan Is.	20:45	21:09	08:30	08:54
JT	Mongolia, Ulan Bator	23:18	23:50	10:21	10:53
JW	Svalbard, Spitsbergen	06:51	09:27	12:52	15:28
JX	Jan Mayen	07:41	08:51	16:42	17:52
JY	Jordan, Amman	03:49	04:13	15:26	15:51
KC4	Antarctica, Byrd Station	——	——	——	——
	McMurdo Sound	——	14:14	11:59	——
	Palmer Station	07:44	08:43	00:15	01:15
KC6	Belau, Yap	20:43	21:05	08:57	09:18
	Koror	20:56	21:18	09:13	09:34
KG4	Guantanamo Bay	11:04	11:27	23:02	23:24
KH0	Mariana Is., Saipan	20:17	20:39	08:22	08:43
KH1	Baker, Howland Is.	17:33	17:54	06:00	06:21
KH2	Guam, Agana	20:20	20:41	08:27	08:49
KH3	Johnston Is.	17:15	17:37	05:17	05:39
KH4	Midway Is.	17:59	18:23	05:43	06:07
KH5	Palmyra Is.	16:42	17:03	05:01	05:22
KH5K	Kingman Reef	16:46	17:07	05:02	05:23
KH6	Hawaii, Hilo	16:24	16:46	04:21	04:44
	Honolulu	16:36	16:59	04:31	04:54
KH7	Kure Is.	18:03	18:27	05:47	06:11
KH8	American Samoa, Pago Pago	17:00	17:22	05:51	06:13
KH9	Wake Is.	18:57	19:19	06:55	07:17
KL7	Alaska, Adak	18:17	18:51	05:08	05:43
	Anchorage	16:43	17:28	02:58	03:43
	Fairbanks	16:42	17:33	02:36	03:28
	Juneau	15:36	16:17	02:05	02:45
	Nome	17:51	18:42	03:48	04:38
KP1	Navassa Is.	11:02	11:25	23:02	23:24
KP2	Virgin Islands, Charlotte Amalie	10:22	10:44	22:22	22:44
KP4	Puerto Rico, San Juan	10:27	10:49	22:27	22:49
KP5	Desecheo Is.	10:32	10:54	22:32	22:54
LA-LJ	Norway, Oslo	05:59	06:42	16:19	17:02
LU	Argentina, Buenos Aires	09:06	09:32	22:42	23:08
LX	Luxembourg	06:03	06:36	17:01	17:34
LY	Lithuania, Vilna	04:52	05:28	15:34	16:11
LZ	Bulgaria, Sofia	04:48	05:17	16:04	16:32
OA	Peru, Lima	10:47	11:09	23:35	23:56
OD	Lebanon, Beirut	03:52	04:17	15:26	15:51
OE	Austria, Vienna	05:21	05:53	16:23	16:55
OH	Finland, Helsinki	05:02	05:45	15:22	16:05
OH0	Aland Is.	05:22	06:05	15:42	16:25
OJ0	Market Reef	05:26	06:09	15:45	16:29
OK,OL	Czech Rep., Prague	05:31	06:04	16:28	17:01
OM	Slovak, Rep., Bratislava	05:18	05:50	16:21	16:53
ON	Belgium, Brussels	06:12	06:45	17:06	17:40
OX	Greenland, Godthaab	10:16	11:06	20:14	21:04
	Thule	12:11	14:02	19:35	21:26
OY	Faroe Islands, Torshavn	07:12	07:58	17:23	18:09
OZ	Denmark, Copenhagen	05:45	06:23	16:23	17:01
P2	Papua New Guinea, Madang	20:02	20:23	08:38	09:00
	Port Moresby	19:53	20:14	08:36	08:57
P4	Aruba, Oranjestad	10:34	10:55	22:43	23:05
PA-PI	Netherlands, Amsterdam	06:11	06:46	17:01	17:36
PJ2,4,9	Netherlands Antilles, Willemstad	10:39	11:00	22:47	23:09
PJ5-8	St. Maarten and Saba, St. Maarten	10:15	10:37	22:16	22:38
PY	Brazil, Brasilia	08:47	09:09	21:41	22:03
	Rio De Janeiro	08:20	08:44	21:29	21:52
	Natal	08:05	08:26	20:42	21:03
	Manaus	09:47	10:08	22:20	22:41
	Porto Alegre	08:44	09:08	22:08	22:33
PY0	Fernando De Noronha	07:55	08:16	20:29	20:50
PY0	St. Peter & St. Paul Rocks	07:47	08:08	20:14	20:35
PY0	Trindade & Martin Vaz Is., Trindade	07:28	07:50	20:31	20:53
PZ	Suriname, Paramaribo	09:34	09:55	21:53	22:14
R1FJ_	Franz Josef Land	04:37	——	——	12:45
R1MV_	Malyj Vysotskij Is.	04:48	05:32	15:06	15:50
S0	Western Sahara, Smara	06:54	07:17	18:41	19:04
S2	Bangladesh, Dacca	00:05	00:27	11:56	12:19
S5	Slovenia, Ljubljana	05:26	05:57	16:34	17:05
S7	Seychelles, Victoria	02:03	02:24	14:38	14:59
S9	Sao Tome	05:22	05:43	17:50	18:11
SM	Sweden, Stockholm	05:28	06:10	15:52	16:34
SP	Poland, Krakow	05:08	05:41	16:05	16:38
	Warsaw	05:07	05:41	15:57	16:32
ST	Sudan, Khartoum	03:50	04:12	15:54	16:16
ST0	Southern Sudan, Juba	03:46	04:07	16:06	16:28
SU	Egypt, Cairo	04:05	04:29	15:46	16:10
SV	Greece, Athens	04:42	05:09	16:03	16:31
SV/A	Mount Athos	04:42	05:09	16:03	16:31
SV5	Dodecanese, Rhodes	04:23	04:49	15:52	16:18
SV9	Crete	04:34	05:00	16:05	16:31
T2	Tuvalu, Funafuti	17:47	18:09	06:29	06:50

Pref	State/Province/Country/City	Tlight	Sunrise	Sunset	Tlight
T30	West Kiribati, Bonriki	18:17	18:38	06:43	07:04
T31	Central Kiribati, Kanton	17:13	17:35	05:46	06:07
T32	East Kiribati, Christmas Is.	16:20	16:41	04:45	05:06
T33	Banaba Is.	18:31	18:52	07:00	07:21
T5	Somalia, Mogadishu	02:49	03:10	15:13	15:35
T7	San Marino	05:33	06:02	16:46	17:15
T9	Bosnia-Herzegovina, Sarajevo	05:09	05:38	16:22	16:51
TA	Turkey, Ankara	04:07	04:34	15:29	15:56
	Istanbul	04:24	04:52	15:43	16:11
TF	Iceland, Reykjavik	08:17	09:07	18:16	19:06
TG	Guatemala, Guatemala City	12:02	12:23	00:07	00:29
TI	Costa Rica, San Jose	11:32	11:54	23:45	00:06
TI9	Cocos Is.	11:41	12:02	00:00	00:22
TJ	Cameroon, Yaounde	05:06	05:27	17:28	17:49
TK	Corsica	05:44	06:13	17:02	17:30
TL	Central African Republic, Bangui	04:38	04:59	16:59	17:20
TN	Congo, Brazzaville	04:44	05:05	17:19	17:40
TR	Gabon, Libreville	05:11	05:32	17:38	17:59
TT	Chad, N'Djamena	04:58	05:19	17:07	17:29
TU	Cote d'Ivoire, Abidjan	06:09	06:30	18:29	18:50
TY	Benin, Porto Novo	05:43	06:05	18:01	18:22
TZ	Mali, Bamako	06:30	06:52	18:39	19:00
UA	Russia, European, Leningrad (UA1)	04:40	05:23	15:01	15:44
	Archangel (UA1)	04:08	04:59	14:04	14:55
	Murmansk (UA1)	04:49	05:51	14:11	15:13
	Moscow (UA3)	04:05	04:43	14:43	15:21
	Kuybyshev (UA4)	03:11	03:47	13:59	14:34
	Rostov (UA6)	03:48	04:19	14:52	15:23
UA2	Kaliningrad	05:12	05:49	15:53	16:30
UA9,0	Russia, Asiatic, Novosibirsk (UA9)	01:03	01:40	11:44	12:21
	Perm (UA9)	02:53	03:34	13:23	14:03
	Omsk (UA9)	01:41	02:18	12:22	12:59
	Norilsk (UA0)	01:10	02:13	10:29	11:32
	Irkutsk (UA0)	23:34	00:08	10:24	10:59
	Vladivostok (UA0)	21:34	22:03	08:49	09:18
	Petropavlovsk (UA0)	19:57	20:32	06:45	07:20
	Khabarovsk (UA0)	21:26	21:58	08:28	09:00
	Krasnoyarsk (UA0)	00:24	01:03	11:02	11:40
	Yakutsk (UA0)	22:06	22:52	08:17	09:03
	Wrangel Island (UA0)	19:06	20:16	04:07	05:17
	Kyzyl (UA0Y)	00:12	00:46	11:04	11:39
UJ-UM	Uzbekistan, Bukhoro	02:01	02:28	13:23	13:51
	Tashkent	01:43	02:11	13:02	13:30
UN-UQ	Kazakhstan, Alma-Ata	01:14	01:43	12:28	12:57
UR-UZ,EM-EO	Ukraine, Kiev	04:27	05:00	15:23	15:56
V2	Antigua & Barbuda, St. Johns	10:09	10:31	22:10	22:32
V3	Belize, Belmopan	11:57	12:19	23:58	00:20
V4	St. Kitts & Nevis	10:12	10:34	22:13	22:35
V5	Namibia, Windhoek	04:20	04:43	17:27	17:50
V6	Micronesia, Ponape	19:21	19:42	07:38	07:59
V7	Marshall Islands, Kwajalein	18:46	19:08	07:00	07:22
V8	Brunei, Bandar Seri Begawan	22:13	22:34	10:33	10:54
VK	Australia, Canberra (VK1)	19:15	19:41	08:53	09:19
	Sydney (VK2)	19:09	19:34	08:43	09:08
	Melbourne (VK3)	19:27	19:54	09:12	09:40
	Brisbane (VK4)	19:10	19:34	08:28	08:52
	Adelaide (VK5)	19:57	20:24	09:34	10:00
	Perth (VK6)	21:33	21:58	11:02	11:27
	Hobart, Tasmania (VK7)	19:08	19:38	09:10	09:40
	Darwin (VK8)	20:55	21:16	09:43	10:05
VK0	Heard Is.	23:37	00:14	14:25	15:03
VK0	Macquarie Is.	17:49	18:29	08:48	09:27
VK9C	Cocos-Keeling Is.	23:11	23:33	11:59	12:21
VK9L	Lord Howe Is.	18:40	19:06	08:08	08:33
VK9M	Mellish Reef	19:10	19:32	08:08	08:30
VK9N	Norfolk Is.	18:08	18:33	07:30	07:54
VK9W	Willis Is.	19:37	19:59	08:32	08:54
VK9X	Christmas Is.	22:37	22:59	11:22	11:44
VP2E	Anguilla	10:14	10:36	22:14	22:36
VP2M	Montserrat	10:10	10:32	22:12	22:34
VP2V	British Virgin Is., Tortola	10:21	10:43	22:21	22:43

February 21　　　**Sunrise and Sunset, in UTC**

Pref	State/Province/Country/City	Tlight	Sunrise	Sunset	Tlight
VP5	Turks & Caicos Islands, Grand Turk	10:49	11:12	22:44	23:07
VP8	Falkland Islands, Stanley	08:26	09:03	23:07	23:43
VP8	So. Georgia Is.	06:53	07:32	21:49	22:28
VP8	So. Orkney Is.	06:59	07:48	22:43	23:32
VP8	So. Sandwich Islands, Saunders Is.	05:58	06:42	21:18	22:02
VP8	So. Shetland Is., King George Is.	07:42	08:33	23:40	00:31
VP9	Bermuda	10:31	10:56	22:08	22:33
VQ9	Chagos, Diego Garcia	00:53	01:15	13:33	13:54
VR2,VS6	Hong Kong	22:28	22:51	10:22	10:44
VR6	Pitcairn Is.	14:06	14:29	03:18	03:42
VU	India, Bombay	01:12	01:34	13:10	13:33
	Calcutta	00:12	00:35	12:05	12:28
	New Delhi	01:01	01:25	12:44	13:08
	Bangalore	00:48	01:10	12:56	13:18
VU	Andaman Islands, Port Blair	23:46	00:08	11:56	12:18
VU	Laccadive Is.	01:04	01:26	13:17	13:38
XE	Mexico, Mexico City (XE1)	12:39	13:02	00:38	01:00
	Chihuahua (XE2)	13:14	13:38	00:57	01:21
	Merida (XE3)	12:03	12:26	23:59	00:21
XF4	Revilla Gigedo	13:29	13:51	01:28	01:50
XT	Burkina Faso, Ouagadougou	06:04	06:26	18:13	18:35
XU	Cambodia, Phnom Penh	22:58	23:20	11:08	11:30
XW	Laos, Viangchan	23:12	23:34	11:12	11:34
XX9	Macao	22:31	22:53	10:24	10:47
XZ	Myanmar, Yangon	23:37	23:59	11:39	12:01
YA	Afghanistan, Kandahar	01:48	02:13	13:27	13:52
	Kabul	01:37	02:03	13:10	13:36
YB-YD	Indonesia, Jakarta	22:37	22:58	11:14	11:36
	Medan	23:17	23:38	11:39	12:00
	Pontianak	22:32	22:53	11:00	11:21
	Jayapura	20:24	20:45	08:56	09:17
YI	Iraq, Baghdad	03:15	03:40	14:51	15:16
YJ	Vanuatu, Port Vila	18:20	18:42	07:18	07:40
YK	Syria, Damascus	03:48	04:13	15:23	15:48
YL	Latvia, Riga	05:00	05:40	15:34	16:13
YN	Nicaragua, Managua	11:42	12:03	23:51	00:13
YO	Romania, Bucharest	04:38	05:08	15:50	16:20
YS	El Salvador, San Salvador	11:56	12:17	00:03	00:24
YU	Yugoslavia, Belgrade	05:01	05:31	16:12	16:42
YV	Venezuela, Caracas	10:25	10:46	22:37	22:58
YV0	Aves Is.	10:15	10:37	22:19	22:41
Z2	Zimbabwe, Harare	03:29	03:51	16:27	16:49
Z3	Macedonia, (ex Yugoslav), Skopje	04:55	05:23	16:12	16:41
ZA	Albania, Tirane	05:01	05:29	16:20	16:48
ZB2	Gibraltar	06:37	07:03	18:07	18:33
ZC4	British Cyprus	04:02	04:28	15:35	16:00
ZD7	St. Helena	05:59	06:21	18:53	19:15
ZD8	Ascension Is.	06:40	07:01	19:21	19:42
ZD9	Tristan da Cunha	05:58	06:25	19:40	20:08
ZF	Cayman Is.	11:28	11:50	23:26	23:48
ZK1	No. Cook Is., Manihiki	16:24	16:46	05:09	05:30
ZK1	So. Cook Is., Rarotonga	16:09	16:32	05:13	05:36
ZK2	Niue	16:48	17:10	05:48	06:10
ZK3	Tokelaus, Atafu	17:13	17:34	05:54	06:15
ZL	New Zealand, Auckland (ZL1)	17:29	17:56	07:12	07:39
	Wellington (ZL2)	17:21	17:50	07:18	07:47
	Christchurch (ZL3)	17:26	17:56	07:30	08:00
	Dunedin (ZL4)	17:29	18:00	07:42	08:14
ZL5	Antarctica, Scott Base	——	14:01	12:15	——
ZL7	Chatham Is.	16:41	17:11	06:47	07:18
ZL8	Kermadec Is.	17:12	17:36	06:34	06:58
ZL9	Auckland & Campbell Is., Auckland	17:32	18:07	08:08	08:43
	Campbell Is.	17:16	17:53	08:01	08:38
ZP	Paraguay, Asuncion	09:16	09:39	22:29	22:52
ZS	South Africa, Cape Town (ZS1)	04:00	04:26	17:34	18:00
	Port Elizabeth (ZS2)	03:30	03:56	17:05	17:31
	Bloemfontein (ZS4)	03:36	04:00	16:58	17:22
	Durban (ZS5)	03:16	03:40	16:39	17:04
	Johannesburg (ZS6)	03:32	03:55	16:47	17:10
ZS8	Marion Is.	02:17	02:49	16:35	17:07

Pref	State/Province/Country/City	Tlight	Sunrise	Sunset	Tlight
VE1	New Brunswick, Saint John	10:22	10:52	22:18	22:48
	Nova Scotia, Halifax	10:12	10:41	22:09	22:38
	Prince Edward Island, Charlottetown	10:11	10:40	22:06	22:36
VE2	Quebec, Montreal	10:52	11:22	22:48	23:18
	Quebec City	10:43	11:13	22:38	23:08
VE3	Ontario, London	11:23	11:51	23:21	23:49
	Ottawa	11:01	11:30	22:57	23:26
	Sudbury	11:22	11:52	23:17	23:47
	Toronto	11:15	11:44	23:12	23:41
VE4	Manitoba, Winnipeg	12:27	12:59	00:19	00:51
VE5	Saskatchewan, Regina	12:57	13:30	00:48	01:21
	Saskatoon	13:06	13:39	00:55	01:29
VE6	Alberta, Calgary	13:35	14:08	01:26	01:59
	Edmonton	13:33	14:08	01:21	01:56
VE7	British Columbia, Prince George	14:10	14:45	01:58	02:33
	Prince Rupert	14:40	15:16	02:28	03:03
	Vancouver	14:11	14:43	02:03	02:35
VE8	Northwest Territories, Yellowknife	13:37	14:22	01:14	01:59
	Resolute	12:17	13:37	23:25	00:44
VY1	Yukon, Whitehorse	15:00	15:42	02:40	03:22
VO1	Newfoundland, St. John's	09:29	10:00	21:23	21:54
VO2	Labrador, Goose Bay	10:01	10:35	21:49	22:24
W1	Connecticut, Hartford	10:48	11:16	22:47	23:15
	Maine, Bangor	10:33	11:02	22:29	22:59
	Portland	10:39	11:08	22:36	23:05
	Massachusetts, Boston	10:42	11:10	22:40	23:08
	New Hampshire, Concord	10:44	11:12	22:41	23:10
	Rhode Island, Providence	10:43	11:11	22:42	23:09
	Vermont, Montpelier	10:48	11:17	22:45	23:14
W2	New Jersey, Atlantic City	10:55	11:21	22:55	23:22
	New York, Albany	10:53	11:21	22:51	23:19
	Buffalo	11:13	11:41	23:11	23:39
	New York City	10:53	11:21	22:53	23:20
	Syracuse	11:02	11:31	23:00	23:28
W3	Delaware, Wilmington	10:59	11:26	22:59	23:26
	District of Columbia, Washington	11:05	11:32	23:06	23:32
	Maryland, Baltimore	11:03	11:30	23:04	23:31
	Pennsylvania, Harrisburg	11:05	11:32	23:05	23:32
	Philadelphia	10:58	11:25	22:58	23:25
	Pittsburgh	11:17	11:44	23:17	23:44
	Scranton	11:00	11:28	22:59	23:27
W4	Albama, Montgomery	11:41	12:05	23:46	00:11
	Florida, Jacksonville	11:22	11:46	23:29	23:53
	Miami	11:15	11:38	23:25	23:48
	Pensacola	11:44	12:08	23:51	00:15
	Georgia, Atlanta	11:34	11:58	23:38	00:03
	Savannah	11:20	11:44	23:26	23:50
	Kentucky, Lexington	11:35	12:01	23:36	00:03
	Louisville	11:40	12:06	23:41	00:08
	North Carolina, Charlotte	11:19	11:45	23:23	23:48
	Raleigh	11:11	11:36	23:14	23:39
	Wilmington	11:08	11:33	23:12	23:37
	South Carolina, Columbia	11:20	11:45	23:24	23:49
	Tennessee, Knoxville	11:32	11:58	23:35	00:01
	Memphis	11:57	12:22	00:00	00:25
	Nashville	11:44	12:09	23:46	00:12
	Virginia, Norfolk	11:02	11:28	23:04	23:30
	Richmond	11:06	11:32	23:08	23:34
W5	Arkansas, Little Rock	12:05	12:31	00:09	00:34
	Louisiana, New Orleans	11:56	12:19	00:03	00:27
	Shreveport	12:10	12:35	00:16	00:40
	Mississippi, Jackson	11:56	12:21	00:02	00:26
	New Mexico, Albuquerque	13:03	13:28	01:07	01:32
	Oklahoma, Oklahoma City	12:26	12:52	00:30	00:55
	Texas, Abilene	12:34	12:59	00:40	01:04
	Amarillo	12:43	13:09	00:47	01:12
	Dallas	12:23	12:48	00:28	00:53
	El Paso	13:02	13:26	01:07	01:32
	San Antonio	12:29	12:53	00:36	01:00
W6	California, Los Angeles	13:49	14:14	01:53	02:18
	San Francisco	14:06	14:33	02:08	02:34
W7	Arizona, Flagstaff	13:23	13:48	01:27	01:52
	Phoenix	13:24	13:49	01:29	01:54
	Idaho, Boise	13:43	14:11	01:40	02:08
	Pocatello	13:28	13:56	01:25	01:54
	Montana, Billings	13:12	13:42	01:08	01:37
	Butte	13:28	13:58	01:23	01:53
	Great Falls	13:23	13:54	01:18	01:48
	Nevada, Las Vegas	13:37	14:03	01:40	02:05
	Reno	13:56	14:23	01:57	02:23
	Oregon, Portland	14:09	14:38	02:05	02:34
	Utah, Salt Lake City	13:25	13:52	01:24	01:52
	Washington, Seattle	14:07	14:38	02:01	02:32
	Spokane	13:48	14:19	01:42	02:13
	Wyoming, Cheyenne	12:57	13:24	00:56	01:23
	Sheridan	13:06	13:35	01:02	01:31
W8	Michigan, Detroit	11:30	11:58	23:28	23:56
	Grand Rapids	11:40	12:09	23:38	00:06
	Sault Ste. Marie	11:36	12:06	23:31	00:01
	Traverse City	11:40	12:10	23:37	00:06
	Ohio, Cincinnati	11:35	12:02	23:36	00:02
	Cleveland	11:24	11:52	23:23	23:51
	Columbus	11:29	11:56	23:29	23:56
	West Virginia, Charleston	11:23	11:50	23:24	23:51
W9	Illinois, Chicago	11:48	12:16	23:46	00:14
	Indiana, Indianapolis	11:42	12:09	23:42	00:09
	Wisconsin, Green Bay	11:50	12:19	23:46	00:15

March 7   **Sunrise and Sunset, in UTC**

Pref	State/Province/Country/City	Tlight	Sunrise	Sunset	Tlight
	Milwaukee	11:49	12:18	23:47	00:15
W0	Colorado, Denver	12:57	13:24	00:57	01:24
	Grand Junction	13:11	13:38	01:12	01:39
	Iowa, Des Moines	12:12	12:40	00:11	00:38
	Kansas, Pratt	12:32	12:58	00:33	00:59
	Wichita	12:26	12:52	00:28	00:54
	Minnesota, Duluth	12:07	12:37	00:01	00:32
	Minneapolis	12:11	12:40	00:07	00:37
	Missouri, Columbia	12:06	12:33	00:07	00:34
	Kansas City	12:15	12:42	00:16	00:43
	St. Louis	11:58	12:24	23:59	00:25
	Nebraska, North Platte	12:41	13:08	00:40	01:07
	Omaha	12:21	12:49	00:20	00:48
	North Dakota, Fargo	12:25	12:56	00:20	00:50
	South Dakota, Rapid City	12:51	13:19	00:47	01:16
-	-	05:47	06:07	18:14	18:35
1A0	S.M.O.M	05:08	05:36	17:06	17:34
1S	Spratly Is.	22:22	22:43	10:43	11:04
3A	Monaco	05:28	05:57	17:25	17:54
3B6	Agalega	01:56	02:17	14:32	14:53
3B7	St. Brandon	01:40	02:02	14:21	14:43
3B8	Mauritius	01:47	02:09	14:32	14:54
3B9	Rodriguez Is.	01:24	01:46	14:08	14:30
3C	Equatorial Guinea, Bata	05:08	05:29	17:34	17:55
	Malabo	05:13	05:34	17:37	17:58
3C0	Annobon Is.	05:24	05:44	17:52	18:13
3D2	Fiji Is., Suva	17:45	18:07	06:28	06:49
3D2	Conway Reef	17:59	18:21	06:45	07:07
3D2	Rotuma Is.	17:50	18:12	06:28	06:49
3DA	Swaziland, Mbabane	03:29	03:52	16:21	16:44
3V	Tunisia, Tunis	05:16	05:42	17:18	17:44
3W, XV	Vietnam, Ho Chi Minh City (Saigon)	22:43	23:05	11:03	11:24
	Hanoi	22:50	23:12	11:03	11:25
3X	Guinea, Conakry	06:45	07:06	19:05	19:26
3Y	Bouvet	04:45	05:21	18:33	19:10
3Y	Peter I Is.	10:05	11:08	01:18	02:21
4J, 4K	Azerbaijan, Baku	02:38	03:05	14:37	15:05
4L	Georgia, Tbilisi	02:58	03:26	14:57	15:25
4S	Sri Lanka, Colombo	00:29	00:50	12:52	13:13
4U	ITU Geneva	05:33	06:03	17:28	17:58
4U	United Nations Hq.	10:53	11:21	22:53	23:20
4X, 4Z	Israel, Jerusalem	03:35	03:59	15:41	16:05
5A	Libya, Tripoli	05:06	05:30	17:11	17:36
	Benghazi	04:36	05:00	16:41	17:06
5B	Cyprus, Nicosia	03:43	04:08	15:46	16:11
5H	Tanzania, Dar es Salaam	03:06	03:27	15:39	16:00
5N	Nigeria, Lagos	05:35	05:56	17:58	18:19
5R	Madagascar, Antananarivo	02:28	02:50	15:11	15:33
5T	Mauritania, Nouakchott	06:56	07:18	19:11	19:33
5U	Niger, Niamey	05:43	06:04	18:01	18:22
5V	Togo, Lome	05:44	06:05	18:07	18:28
5W	Western Samoa, Apia	17:08	17:29	05:46	06:08
5X	Uganda, Kampala	03:37	03:57	16:04	16:25
5Z	Kenya, Nairobi	04:39	05:00	17:07	17:28
6W	Senegal, Dakar	07:01	07:23	19:19	19:40
6Y	Jamaica, Kingston	10:59	11:21	23:14	23:36
7O	Yemen, Aden	02:51	03:12	15:09	15:31
	Sanaa	02:55	03:16	15:11	15:33
7P	Lesotho, Maseru	03:41	04:05	16:36	17:00
7Q	Malawi, Lilongwe	03:25	03:47	16:04	16:26
	Blantyre	03:20	03:41	16:00	16:22
7X	Algeria, Algiers	05:45	06:10	17:47	18:13
8P	Barbados, Bridgetown	09:49	10:11	22:08	22:29
8Q	Maldive Is.	00:55	01:15	13:19	13:40
8R	Guyana, Georgetown	09:42	10:03	22:04	22:25
9A	Croatia, Zagreb	04:54	05:24	16:50	17:19
9G	Ghana, Accra	05:49	06:10	18:13	18:34
9H	Malta	04:59	05:24	17:02	17:27
9J	Zambia, Lusaka	03:47	04:08	16:27	16:48
9K	Kuwait	02:44	03:08	14:51	15:15
9L	Sierra Leone, Freetown	06:42	07:03	19:04	19:25
9M2	West Malaysia, Kuala Lumpur	23:01	23:22	11:26	11:47
9M6,8	East Malaysia, Sabah, Sandakan(9M6)	21:56	22:17	10:19	10:40
	Sarawak, Kuching (9M8)	22:26	22:47	10:52	11:13
9N	Nepal, Kathmandu	00:13	00:37	12:22	12:45
9Q	Zaire, Kinshasa	04:44	05:04	17:14	17:35
	Kisangani	04:06	04:27	16:33	16:54
	Lubumbashi	03:52	04:13	16:28	16:50
9U	Burundi, Bujumbura	03:48	04:09	16:18	16:39
9V	Singapore	22:52	23:13	11:18	11:39
9X	Rwanda, Kigali	03:45	04:06	16:14	16:35
9Y	Trinidad & Tobago , Port of Spain	09:55	10:16	22:15	22:36
A2	Botswana, Gaborone	03:51	04:14	16:41	17:04
A3	Tonga, Nukualofa	17:17	17:40	06:03	06:26
A4	Oman, Musqat	01:59	02:22	14:11	14:33
A5	Bhutan, Thimbu	23:57	00:20	12:06	12:29
A6	United Arab Emirates, Abu Dhabi	02:17	02:40	14:28	14:51
A7	Qatar, Ad-Dawhah	02:28	02:51	14:38	15:01
A9	Bahrein, Al-Manamah	02:32	02:55	14:41	15:05
AP	Pakistan, Karachi	01:26	01:48	13:36	13:59
AP	Islamabad	01:03	01:28	13:08	13:33
BV	Taiwan, Taipei	21:48	22:11	09:58	10:21
BY	Peoples Rep. of China, Beijing	22:12	22:39	10:12	10:39
	Harbin	21:31	22:01	09:27	09:56
	Shanghai	21:49	22:14	09:56	10:20
	Fuzhou	21:57	22:20	10:07	10:30
	Xian	22:40	23:06	10:45	11:10

Pref	State/Province/Country/City	Tlight	Sunrise	Sunset	Tlight
	Chongqing	22:49	23:13	10:56	11:20
	Chengdu	22:59	23:23	11:05	11:30
	Lhasa	23:50	00:14	11:58	12:21
	Urumqi	00:07	00:36	12:04	12:33
	Kashi	00:53	01:20	12:53	13:20
C2	Nauru	18:39	19:00	07:07	07:27
C3	Andorra	05:52	06:20	17:50	18:18
C5	The Gambia, Banjul	06:58	07:19	19:16	19:37
C6	Bahamas, Nassau	11:04	11:27	23:14	23:37
C9	Mozambique, Maputo	03:23	03:46	16:14	16:38
	Mozambique	02:57	03:19	15:37	15:59
CE	Chile, Santiago	10:11	10:36	23:12	23:37
CE0	Easter Island	12:50	13:14	01:43	02:06
CE0	Juan Fernandez	10:43	11:08	23:44	00:09
CE0	San Felix	10:54	11:17	23:45	00:09
CM,CO	Cuba, Havana	11:23	11:46	23:35	23:57
CN	Morocco, Casablanca	06:26	06:51	18:31	18:55
CP	Bolivia, La Paz	10:13	10:34	22:54	23:16
CT	Portugal, Lisbon	06:34	07:00	18:35	19:01
CT3	Madeira Islands, Funchal	07:03	07:28	19:09	19:33
CU	Azores , Ponta Delgada	07:40	08:06	19:41	20:07
CX	Uruguay, Montevideo	09:11	09:36	22:14	22:40
CY0	Sable Is.	09:58	10:26	21:55	22:24
CY9	St. Paul Is.	09:59	10:29	21:53	22:23
D2	Angola, Luanda	04:50	05:11	17:25	17:46
D4	Cape Verde, Praia	07:25	07:47	19:42	20:04
D6	Comoros, Moroni	02:47	03:08	15:24	15:45
DA-DL,Y2-Y9	Fed. Rep. of Germany, Bonn	05:31	06:03	17:22	17:55
	Berlin	05:05	05:39	16:55	17:29
DU	Philippines, Manilla	21:47	22:09	10:05	10:26
E3	Eritrea, Asmara	03:16	03:37	15:33	15:54
EA	Spain, Madrid	06:12	06:39	18:12	18:39
EA6	Balearic Is., Palma	05:47	06:13	17:47	18:14
EA8	Canary Is., Las Palmas	06:52	07:16	19:00	19:24
EA9	Ceuta & Melilla, Ceuta	06:18	06:43	18:21	18:46
	Melilla	06:08	06:34	18:12	18:37
EI	Ireland, Dublin	06:24	06:59	18:13	18:48
EK	Armenia, Yerevan	02:59	03:26	14:59	15:26
EL	Liberia, Monrovia	06:32	06:53	18:55	19:16
EP	Iran, Tehran	02:29	02:55	14:32	14:58
ER	Moldova, Kishinev	04:03	04:33	15:57	16:28
ES	Estonia, Tallinn	04:20	05:01	16:02	16:43
ET	Ethiopia, Addis Ababa	03:15	03:36	15:36	15:57
EU,EV,EW	Belarus, Minsk	04:09	04:44	15:57	16:32
EX	Kyrgyzstan, Frunze	00:59	01:27	12:57	13:25
EY	Tajikistan, Samarkand	01:30	01:57	13:30	13:57
	Dushanbe	01:22	01:48	13:22	13:49
EZ	Turkmenistan, Ashkhabad	02:03	02:29	14:05	14:31
F	France, Paris	05:49	06:21	17:42	18:14
FG	Guadeloupe	09:59	10:20	22:15	22:36
FJ, FS	St. Martin	10:05	10:26	22:20	22:41
FH	Mayotte	02:40	03:01	15:18	15:39
FK	New Caledonia, Noumia	18:30	18:52	07:17	07:40
FM	Martinique	09:55	10:17	22:13	22:34
FO	Clipperton Is.	13:07	13:28	01:27	01:48
FO	Fr. Polynesia, Tahiti	16:17	16:38	04:59	05:21
	Rurutu, (Austral Is.)	19:30	19:53	08:18	08:40
	Hiva Oa, (Marquesas Is.)	14:58	15:20	03:34	03:55
FP	St. Pierre & Miquelon, St. Pierre	09:42	10:12	21:37	22:07
FR/G	Glorioso	02:32	02:54	15:09	15:30
FR/J,E	Juan de Nova	02:48	03:09	15:29	15:51
	Europa	02:54	03:17	15:41	16:04
FR	Reunion	01:54	02:16	14:40	15:02
FR/T	Tromelin	02:02	02:23	14:43	15:04
FT8W	Crozet	01:45	02:16	15:10	15:40
FT8X	Kerguelen	00:31	01:04	14:04	14:36
FT8Y	Antarctica, Dumont D'Urville	18:56	19:52	09:49	10:45
FT8Z	Amsterdam & St. Paul Is., Amsterdam	00:13	00:40	13:21	13:47
FW	Wallis & Futuna Is., Wallis	17:26	17:47	06:04	06:26
FY	Fr. Guiana, Cayenne	09:17	09:38	21:41	22:02
G	England, London	05:59	06:32	17:50	18:23
GD	Isle of Man	06:17	06:53	18:05	18:40
GI	Northern Ireland, Belfast	06:23	06:59	18:10	18:46
GJ	Jersey	06:07	06:39	18:00	18:32
GM	Scotland, Glasgow	06:16	06:53	18:02	18:39
	Aberdeen	06:08	06:46	17:52	18:30
GU	Guernsey	06:09	06:41	18:02	18:34
GW	Wales, Cardiff	06:12	06:45	18:02	18:35
H4	Solomon Islands, Honiara	19:03	19:24	07:38	07:59
HA	Hungary, Budapest	04:42	05:13	16:36	17:07
HB	Switzerland, Bern	05:28	05:59	17:23	17:53
HB0	Liechtenstein	05:20	05:50	17:14	17:45
HC	Ecuador, Quito	10:58	11:19	23:26	23:47
HC8	Galapagos Is.	11:48	12:09	00:16	00:37
HH	Haiti, Port-Au-Prince	10:42	11:03	22:56	23:18
HI	Dominican Republic, Santo Domingo	10:32	10:54	22:47	23:09
HK	Colombia, Bogota	10:45	11:05	23:09	23:30
HK0	Malpelo Is.	11:12	11:33	23:37	23:58
HK0	San Adreas	11:18	11:39	23:36	23:57
HL	Korea, Seoul	21:29	21:55	09:31	09:57
HP	Panama, Panama	11:08	11:29	23:29	23:50
HR	Honduras, Tegucigalpa	11:40	12:01	23:58	00:19
HS	Thailand, Bangkok	23:09	23:30	11:27	11:48
HV	Vatican City	05:07	05:35	17:06	17:34
HZ, 7Z	Saudi Arabia, Dharan	02:34	02:57	14:44	15:07
	Mecca	03:14	03:36	15:27	15:49
	Riyadh	02:47	03:10	14:58	15:21

Pref	State/Province/Country/City	Tlight	Sunrise	Sunset	Tlight
I	Italy, Rome	05:07	05:35	17:06	17:34
	Trieste	05:03	05:33	16:58	17:28
	Sicily	05:01	05:27	17:03	17:29
IS	Sardinia, Cagliari	05:21	05:47	17:21	17:48
J2	Djibouti, Djibouti	02:58	03:19	15:17	15:38
J3	Grenada	09:58	10:19	22:17	22:38
J5	Guinea-Bissau, Bissau	06:53	07:14	19:12	19:33
J6	St. Lucia	09:55	10:16	22:13	22:34
J7	Dominica	09:57	10:18	22:13	22:35
J8	St. Vincent	09:56	10:17	22:14	22:36
JA-JS	Japan, Tokyo	20:37	21:03	08:40	09:06
	Nagasaki	21:16	21:41	09:21	09:46
	Sapporo	20:32	21:00	08:30	08:58
JD1	Minami Torishima	19:38	20:01	07:49	08:11
JD1	Ogasawara, Kazan Is.	20:31	20:54	08:39	09:03
JT	Mongolia, Ulan Bator	22:51	23:22	10:44	11:15
JW	Svalbard, Spitsbergen	04:50	06:40	15:34	17:23
JX	Jan Mayen	06:32	07:36	17:52	18:56
JY	Jordan, Amman	03:32	03:56	15:38	16:02
KC4	Antarctica, Byrd Station	—	11:40	04:41	—
	McMurdo Sound	14:37	17:05	09:03	11:31
	Palmer Station	08:41	09:33	23:20	00:12
KC6	Belau, Yap	20:37	20:58	08:58	09:19
	Koror	20:51	21:12	09:13	09:34
KG4	Guantanamo Bay	10:54	11:16	23:07	23:29
KH0	Mariana Is., Saipan	20:08	20:30	08:25	08:47
KH1	Baker, Howland Is.	17:31	17:51	05:58	06:19
KH2	Guam, Agana	20:12	20:33	08:30	08:51
KH3	Johnston Is.	17:06	17:28	05:22	05:43
KH4	Midway Is.	17:44	18:08	05:53	06:16
KH5	Palmyra Is.	16:37	16:58	05:00	05:21
KH5K	Kingman Reef	16:40	17:01	05:02	05:23
KH6	Hawaii, Hilo	16:13	16:35	04:27	04:49
	Honolulu	16:25	16:47	04:38	05:00
KH7	Kure Is.	17:48	18:12	05:57	06:20
KH8	American Samoa, Pago Pago	17:03	17:25	05:43	06:04
KH9	Wake Is.	18:46	19:08	07:00	07:22
KL7	Alaska, Adak	17:45	18:19	05:35	06:09
	Anchorage	15:59	16:43	03:39	04:22
	Fairbanks	15:51	16:40	03:25	04:13
	Juneau	14:57	15:36	02:40	03:20
	Nome	17:01	17:49	04:35	05:23
KP1	Navassa Is.	10:52	11:14	23:07	23:29
KP2	Virgin Islands, Charlotte Amalie	10:12	10:34	22:27	22:49
KP4	Puerto Rico, San Juan	10:17	10:39	22:32	22:54
KP5	Desecheo Is.	10:22	10:44	22:37	22:59
LA-LJ	Norway, Oslo	05:17	05:58	16:58	17:39
LU	Argentina, Buenos Aires	09:20	09:45	22:23	22:48
LX	Luxembourg	05:34	06:06	17:26	17:58
LY	Lithuania, Vilna	04:17	04:53	16:04	16:40
LZ	Bulgaria, Sofia	04:24	04:53	16:22	16:51
OA	Peru, Lima	10:50	11:11	23:27	23:48
OD	Lebanon, Beirut	03:34	03:59	15:38	16:03
OE	Austria, Vienna	04:53	05:24	16:47	17:18
OH	Finland, Helsinki	04:19	05:01	16:00	16:42
OH0	Aland Is.	04:39	05:21	16:20	17:02
OJ0	Market Reef	04:43	05:25	16:24	17:06
OK,OL	Czech Rep., Prague	05:01	05:33	16:53	17:25
OM	Slovak, Rep., Bratislava	04:50	05:21	16:44	17:15
ON	Belgium, Brussels	05:41	06:14	17:32	18:05
OX	Greenland, Godthaab	09:26	10:14	21:01	21:49
	Thule	10:31	12:02	21:29	23:00
OY	Faroe Islands, Torshavn	06:27	07:11	18:05	18:49
OZ	Denmark, Copenhagen	05:09	05:46	16:55	17:32
P2	Papua New Guinea, Madang	20:02	20:23	08:34	08:54
	Port Moresby	19:54	20:15	08:29	08:50
P4	Aruba, Oranjestad	10:27	10:48	22:46	23:07
PA-PI	Netherlands, Amsterdam	05:39	06:13	17:29	18:03
PJ2,4,9	Netherlands Antilles, Willemstad	10:31	10:52	22:50	23:11
PJ5-8	St. Maarten and Saba, St. Maarten	10:05	10:27	22:20	22:42
PY	Brazil, Brasilia	08:51	09:13	21:32	21:53
	Rio De Janeiro	08:28	08:51	21:16	21:39
	Natal	08:05	08:26	20:37	20:58
	Manaus	09:46	10:07	22:16	22:37
	Porto Alegre	08:55	09:19	21:52	22:16
PY0	Fernando De Noronha	07:55	08:15	20:25	20:46
PY0	St. Peter & St. Paul Rocks	07:45	08:05	20:11	20:32
PY0	Trindade & Martin Vaz Is., Trindade	07:34	07:56	20:19	20:42
PZ	Suriname, Paramaribo	09:29	09:50	21:53	22:14
R1FJ_	Franz Josef Land	02:21	04:24	12:53	14:56
R1MV_	Malyj Vysotskij Is.	04:05	04:47	15:45	16:27
S0	Western Sahara, Smara	06:40	07:03	18:49	19:13
S2	Bangladesh, Dacca	23:52	00:15	12:03	12:26
S5	Slovenia, Ljubljana	05:00	05:30	16:55	17:25
S7	Seychelles, Victoria	02:03	02:24	14:34	14:55
S9	Sao Tome	05:20	05:41	17:47	18:08
SM	Sweden, Stockholm	04:47	05:28	16:29	17:09
SP	Poland, Krakow	04:39	05:11	16:30	17:03
	Warsaw	04:35	05:09	16:25	16:58
ST	Sudan, Khartoum	03:42	04:03	15:58	16:20
ST0	Southern Sudan, Juba	03:42	04:03	16:06	16:27
SU	Egypt, Cairo	03:50	04:13	15:57	16:21
SV	Greece, Athens	04:22	04:48	16:23	16:50
SV/A	Mount Athos	04:20	04:47	16:20	16:47
SV5	Dodecanese, Rhodes	04:04	04:29	16:06	16:32
SV9	Crete	04:15	04:41	16:19	16:44
T2	Tuvalu, Funafuti	17:49	18:10	06:23	06:44

Pref	State/Province/Country/City	Tlight	Sunrise	Sunset	Tlight
T30	West Kiribati, Bonriki	18:14	18:35	06:41	07:01
T31	Central Kiribati, Kanton	17:12	17:33	05:42	06:03
T32	East Kiribati, Christmas Is.	16:17	16:38	04:43	05:04
T33	Banaba Is.	18:29	18:50	06:57	07:17
T5	Somalia, Mogadishu	02:46	03:06	15:12	15:32
T7	San Marino	05:09	05:37	17:06	17:34
T9	Bosnia-Herzegovina, Sarajevo	04:44	05:13	16:41	17:10
TA	Turkey, Ankara	03:45	04:13	15:46	16:13
	Istanbul	04:01	04:29	16:01	16:28
TF	Iceland, Reykjavik	07:27	08:15	19:02	19:50
TG	Guatemala, Guatemala City	11:53	12:15	00:11	00:32
TI	Costa Rica, San Jose	11:26	11:47	23:46	00:07
TI9	Cocos Is.	11:37	11:57	00:00	00:21
TJ	Cameroon, Yaounde	05:02	05:23	17:27	17:47
TK	Corsica	05:21	05:49	17:20	17:48
TL	Central African Republic, Bangui	04:34	04:55	16:58	17:19
TN	Congo, Brazzaville	04:44	05:04	17:14	17:35
TR	Gabon, Libreville	05:09	05:29	17:36	17:57
TT	Chad, N'Djamena	04:51	05:12	17:10	17:31
TU	Cote d'Ivoire, Abidjan	06:04	06:25	18:28	18:49
TY	Benin, Porto Novo	05:38	05:59	18:01	18:22
TZ	Mali, Bamako	06:23	06:44	18:41	19:03
UA	Russia, European, Leningrad (UA1)	03:58	04:40	15:39	16:21
	Archangel (UA1)	03:17	04:06	14:51	15:40
	Murmansk (UA1)	03:47	04:45	15:12	16:10
	Moscow (UA3)	03:29	04:06	15:15	15:52
	Kuybyshev (UA4)	02:39	03:13	14:27	15:02
	Rostov (UA6)	03:20	03:51	15:14	15:45
UA2	Kaliningrad	04:37	05:13	16:24	17:00
UA9,0	Russia, Asiatic, Novosibirsk (UA9)	00:28	01:04	12:14	12:51
	Perm (UA9)	02:14	02:53	13:58	14:37
	Omsk (UA9)	01:06	01:42	12:52	13:29
	Norilsk (UA0)	00:07	01:06	11:31	12:30
	Irkutsk (UA0)	23:02	23:36	10:51	11:25
	Vladivostok (UA0)	21:10	21:38	09:08	09:36
	Petropavlovsk (UA0)	19:24	19:59	07:13	07:48
	Khabarovsk (UA0)	20:58	21:29	08:51	09:22
	Krasnoyarsk (UA0)	23:48	00:25	11:34	12:11
	Yakutsk (UA0)	21:21	22:05	08:59	09:43
	Wrangel Island (UA0)	17:57	19:01	05:17	06:21
	Kyzyl (UA0Y)	23:41	00:14	11:31	12:04
UJ-UM	Uzbekistan, Bukhoro	01:39	02:06	13:40	14:07
	Tashkent	01:20	01:48	13:19	13:47
UN-UQ	Kazakhstan, Alma-Ata	00:50	01:19	12:48	13:16
UR-UZ,EM-EO	Ukraine, Kiev	03:57	04:29	15:48	16:21
V2	Antigua & Barbuda, St. Johns	09:59	10:21	22:15	22:36
V3	Belize, Belmopan	11:47	12:09	00:03	00:24
V4	St. Kitts & Nevis	10:02	10:24	22:18	22:40
V5	Namibia, Windhoek	04:27	04:50	17:15	17:37
V6	Micronesia, Ponape	19:16	19:37	07:38	07:59
V7	Marshall Islands, Kwajalein	18:40	19:01	07:01	07:22
V8	Brunei, Bandar Seri Begawan	22:09	22:30	10:33	10:53
VK	Australia, Canberra (VK1)	19:29	19:55	08:33	08:59
	Sydney (VK2)	19:22	19:47	08:24	08:49
	Melbourne (VK3)	19:43	20:10	08:51	09:18
	Brisbane (VK4)	19:20	19:44	08:14	08:37
	Adelaide (VK5)	20:12	20:37	09:15	09:41
	Perth (VK6)	21:46	22:10	10:45	11:09
	Hobart, Tasmania (VK7)	19:28	19:57	08:46	09:15
	Darwin (VK8)	20:58	21:19	09:35	09:56
VK0	Heard Is.	00:08	00:43	13:51	14:26
VK0	Macquarie Is.	18:22	18:59	08:11	08:48
VK9C	Cocos-Keeling Is.	23:14	23:35	11:51	12:13
VK9L	Lord Howe Is.	18:53	19:17	07:51	08:16
VK9M	Mellish Reef	19:15	19:37	07:58	08:20
VK9N	Norfolk Is.	18:19	18:43	07:14	07:38
VK9W	Willis Is.	19:41	20:03	08:22	08:44
VK9X	Christmas Is.	22:39	23:00	11:15	11:36
VP2E	Anguilla	10:04	10:26	22:19	22:41
VP2M	Montserrat	10:01	10:22	22:17	22:38
VP2V	British Virgin Is., Tortola	10:11	10:33	22:25	22:47

March 7   **Sunrise and Sunset, in UTC**

Pref	State/Province/Country/City	Tlight	Sunrise	Sunset	Tlight
VP5	Turks & Caicos Islands, Grand Turk	10:38	11:00	22:51	23:13
VP8	Falkland Islands, Stanley	08:56	09:30	22:35	23:09
VP8	So. Georgia Is.	07:26	08:02	21:13	21:50
VP8	So. Orkney Is.	07:44	08:28	21:58	22:42
VP8	So. Sandwich Islands, Saunders Is.	06:37	07:17	20:38	21:18
VP8	So. Shetland Is., King George Is.	08:30	09:16	22:51	23:38
VP9	Bermuda	10:14	10:39	22:20	22:44
VQ9	Chagos, Diego Garcia	00:54	01:15	13:27	13:48
VR2,VS6	Hong Kong	22:16	22:39	10:28	10:51
VR6	Pitcairn Is.	14:14	14:37	03:05	03:28
VU	India, Bombay	01:01	01:23	13:16	13:38
	Calcutta	00:00	00:22	12:12	12:34
	New Delhi	00:46	01:10	12:54	13:18
	Bangalore	00:40	01:02	12:59	13:20
VU	Andaman Islands, Port Blair	23:39	00:00	11:58	12:20
VU	Laccadive Is.	00:58	01:19	13:18	13:39
XE	Mexico, Mexico City (XE1)	12:29	12:51	00:43	01:05
	Chihuahua (XE2)	12:59	13:23	01:07	01:30
	Merida (XE3)	11:52	12:14	00:05	00:27
XF4	Revilla Gigedo	13:19	13:40	01:33	01:55
XT	Burkina Faso, Ouagadougou	05:57	06:18	18:16	18:37
XU	Cambodia, Phnom Penh	22:51	23:12	11:10	11:32
XW	Laos, Viangchan	23:02	23:24	11:17	11:39
XX9	Macao	22:19	22:41	10:31	10:54
XZ	Myanmar, Yangon	23:28	23:50	11:44	12:05
YA	Afghanistan, Kandahar	01:32	01:56	13:39	14:03
	Kabul	01:19	01:44	13:23	13:48
YB-YD	Indonesia, Jakarta	22:37	22:58	11:09	11:30
	Medan	23:13	23:34	11:38	11:59
	Pontianak	22:29	22:50	10:57	11:18
	Jayapura	20:23	20:44	08:52	09:13
YI	Iraq, Baghdad	02:58	03:22	15:03	15:28
YJ	Vanuatu, Port Vila	18:25	18:47	07:08	07:30
YK	Syria, Damascus	03:31	03:56	15:35	16:00
YL	Latvia, Riga	04:23	05:01	16:08	16:46
YN	Nicaragua, Managua	11:35	11:56	23:54	00:15
YO	Romania, Bucharest	04:13	04:42	16:10	16:39
YS	El Salvador, San Salvador	11:48	12:09	00:06	00:27
YU	Yugoslavia, Belgrade	04:36	05:05	16:32	17:01
YV	Venezuela, Caracas	10:18	10:39	22:38	22:59
YV0	Aves Is.	10:06	10:28	22:23	22:44
Z2	Zimbabwe, Harare	03:34	03:56	16:17	16:39
Z3	Macedonia, (ex Yugoslav), Skopje	04:32	05:00	16:30	16:58
ZA	Albania, Tirane	04:38	05:06	16:37	17:05
ZB2	Gibraltar	06:18	06:44	18:21	18:46
ZC4	British Cyprus	03:44	04:09	15:48	16:13
ZD7	St. Helena	06:03	06:25	18:44	19:05
ZD8	Ascension Is.	06:41	07:02	19:15	19:36
ZD9	Tristan da Cunha	06:13	06:40	19:20	19:46
ZF	Cayman Is.	11:17	11:39	23:31	23:53
ZK1	No. Cook Is., Manihiki	16:26	16:47	05:02	05:23
ZK1	So. Cook Is., Rarotonga	16:16	16:38	05:02	05:24
ZK2	Niue	16:53	17:15	05:37	05:59
ZK3	Tokelaus, Atafu	17:14	17:35	05:48	06:09
ZL	New Zealand, Auckland (ZL1)	17:45	18:11	06:52	07:18
	Wellington (ZL2)	17:40	18:08	06:55	07:23
	Christchurch (ZL3)	17:46	18:15	07:05	07:34
	Dunedin (ZL4)	17:51	18:22	07:16	07:46
ZL5	Antarctica, Scott Base	14:28	17:04	09:06	11:42
ZL7	Chatham Is.	17:02	17:31	06:22	06:51
ZL8	Kermadec Is.	17:22	17:46	06:18	06:42
ZL9	Auckland & Campbell Is., Auckland	18:00	18:33	07:36	08:10
	Campbell Is.	17:46	18:21	07:28	08:03
ZP	Paraguay, Asuncion	09:25	09:48	22:15	22:38
ZS	South Africa, Cape Town (ZS1)	04:13	04:39	17:15	17:41
	Port Elizabeth (ZS2)	03:44	04:09	16:46	17:12
	Bloemfontein (ZS4)	03:47	04:10	16:42	17:06
	Durban (ZS5)	03:27	03:51	16:23	16:47
	Johannesburg (ZS6)	03:41	04:04	16:32	16:56
ZS8	Marion Is.	02:41	03:12	16:07	16:38

Pref	State/Province/Country/City	Tlight	Sunrise	Sunset	Tlight
VE1	New Brunswick, Saint John	09:56	10:26	22:37	23:06
	Nova Scotia, Halifax	09:47	10:16	22:27	22:56
	Prince Edward Island, Charlottetown	09:44	10:14	22:25	22:55
VE2	Quebec, Montreal	10:26	10:56	23:07	23:36
	Quebec City	10:16	10:46	22:57	23:28
VE3	Ontario, London	10:58	11:27	23:37	00:06
	Ottawa	10:35	11:04	23:15	23:45
	Sudbury	10:55	11:25	23:36	00:07
	Toronto	10:50	11:19	23:30	23:58
VE4	Manitoba, Winnipeg	11:57	12:29	00:41	01:14
VE5	Saskatchewan, Regina	12:26	12:59	01:11	01:44
	Saskatoon	12:33	13:07	01:20	01:54
VE6	Alberta, Calgary	13:04	13:37	01:49	02:22
	Edmonton	12:59	13:34	01:47	02:22
VE7	British Columbia, Prince George	13:36	14:11	02:25	03:00
	Prince Rupert	14:06	14:41	02:55	03:30
	Vancouver	13:41	14:13	02:25	02:57
VE8	Northwest Territories, Yellowknife	12:51	13:36	01:53	02:38
	Resolute	10:50	12:11	00:42	02:03
VY1	Yukon, Whitehorse	14:16	14:59	03:15	03:58
VO1	Newfoundland, St. John's	09:01	09:32	21:43	22:14
VO2	Labrador, Goose Bay	09:27	10:02	22:15	22:50
W1	Connecticut, Hartford	10:25	10:52	23:03	23:31
	Maine, Bangor	10:07	10:37	22:47	23:17
	Portland	10:14	10:43	22:53	23:22
	Massachusetts, Boston	10:18	10:46	22:56	23:24
	New Hampshire, Concord	10:19	10:47	22:58	23:27
	Rhode Island, Providence	10:19	10:47	22:58	23:25
	Vermont, Montpelier	10:23	10:52	23:03	23:32
W2	New Jersey, Atlantic City	10:33	10:59	23:09	23:36
	New York, Albany	10:29	10:57	23:07	23:35
	Buffalo	10:49	11:17	23:28	23:56
	New York City	10:30	10:58	23:08	23:35
	Syracuse	10:38	11:06	23:17	23:45
W3	Delaware, Wilmington	10:37	11:04	23:14	23:41
	District of Columbia, Washington	10:43	11:10	23:20	23:46
	Maryland, Baltimore	10:42	11:08	23:18	23:45
	Pennsylvania, Harrisburg	10:42	11:09	23:19	23:47
	Philadelphia	10:36	11:03	23:13	23:40
	Pittsburgh	10:55	11:22	23:32	23:59
	Scranton	10:37	11:04	23:15	23:42
W4	Albama, Montgomery	11:23	11:48	23:56	00:21
	Florida, Jacksonville	11:05	11:29	23:38	00:02
	Miami	11:01	11:24	23:32	23:55
	Pensacola	11:27	11:51	00:00	00:24
	Georgia, Atlanta	11:15	11:40	23:49	00:14
	Savannah	11:02	11:27	23:36	00:00
	Kentucky, Lexington	11:14	11:40	23:50	00:16
	Louisville	11:19	11:45	23:55	00:21
	North Carolina, Charlotte	11:00	11:25	23:35	00:00
	Raleigh	10:51	11:17	23:26	23:51
	Wilmington	10:49	11:14	23:23	23:48
	South Carolina, Columbia	11:01	11:26	23:35	00:00
	Tennessee, Knoxville	11:12	11:38	23:47	00:13
	Memphis	11:37	12:03	00:12	00:37
	Nashville	11:24	11:49	23:59	00:24
	Virginia, Norfolk	10:41	11:07	23:17	23:43
	Richmond	10:46	11:12	23:21	23:47
W5	Arkansas, Little Rock	11:46	12:11	00:21	00:46
	Louisiana, New Orleans	11:39	12:03	00:12	00:35
	Shreveport	11:53	12:17	00:26	00:51
	Mississippi, Jackson	11:39	12:03	00:12	00:37
	New Mexico, Albuquerque	12:44	13:09	01:18	01:44
	Oklahoma, Oklahoma City	12:07	12:32	00:41	01:07
	Texas, Abilene	12:17	12:41	00:50	01:15
	Amarillo	12:24	12:49	00:59	01:24
	Dallas	12:05	12:30	00:38	01:03
	El Paso	12:44	13:08	01:17	01:42
	San Antonio	12:13	12:37	00:45	01:09
W6	California, Los Angeles	13:30	13:55	02:04	02:29
	San Francisco	13:45	14:12	02:21	02:47
W7	Arizona, Flagstaff	13:04	13:29	01:38	02:04
	Phoenix	13:06	13:31	01:40	02:05
	Idaho, Boise	13:18	13:46	01:57	02:26
	Pocatello	13:03	13:32	01:42	02:10
	Montana, Billings	12:45	13:15	01:26	01:56
	Butte	13:01	13:31	01:42	02:12
	Great Falls	12:56	13:26	01:38	02:08
	Nevada, Las Vegas	13:17	13:43	01:52	02:18
	Reno	13:34	14:01	02:11	02:38
	Oregon, Portland	13:42	14:12	02:23	02:53
	Utah, Salt Lake City	13:02	13:29	01:40	02:07
	Washington, Seattle	13:39	14:10	02:22	02:53
	Spokane	13:20	13:51	02:02	02:33
	Wyoming, Cheyenne	12:33	13:01	01:11	01:39
	Sheridan	12:40	13:09	01:20	01:49
W8	Michigan, Detroit	11:06	11:34	23:44	00:12
	Grand Rapids	11:16	11:44	23:55	00:23
	Sault Ste. Marie	11:09	11:39	23:50	00:20
	Traverse City	11:15	11:44	23:55	00:24
	Ohio, Cincinnati	11:13	11:40	23:50	00:16
	Cleveland	11:01	11:28	23:39	00:06
	Columbus	11:07	11:34	23:44	00:11
	West Virginia, Charleston	11:02	11:28	23:38	00:05
W9	Illinois, Chicago	11:24	11:52	00:02	00:30
	Indiana, Indianapolis	11:20	11:47	23:57	00:24
	Wisconsin, Green Bay	11:24	11:53	00:04	00:33

Pref	State/Province/Country/City	Tlight	Sunrise	Sunset	Tlight
	Milwaukee	11:25	11:53	00:04	00:32
W0	Colorado, Denver	12:35	13:02	01:12	01:39
	Grand Junction	12:50	13:16	01:26	01:53
	Iowa, Des Moines	11:48	12:16	00:26	00:54
	Kansas, Pratt	12:11	12:37	00:46	01:13
	Wichita	12:05	12:31	00:41	01:07
	Minnesota, Duluth	11:39	12:09	00:21	00:51
	Minneapolis	11:45	12:15	00:26	00:55
	Missouri, Columbia	11:44	12:11	00:21	00:48
	Kansas City	11:54	12:20	00:30	00:57
	St. Louis	11:36	12:03	00:13	00:39
	Nebraska, North Platte	12:17	12:45	00:55	01:23
	Omaha	11:58	12:25	00:36	01:03
	North Dakota, Fargo	11:58	12:28	00:40	01:10
	South Dakota, Rapid City	12:25	12:54	01:05	01:34
-	-	05:43	06:03	18:10	18:31
1A0	S.M.O.M.	04:44	05:12	17:22	17:50
1S	Spratly Is.	22:15	22:36	10:43	11:04
3A	Monaco	05:03	05:32	17:43	18:11
3B6	Agalega	01:56	02:17	14:24	14:45
3B7	St. Brandon	01:43	02:04	14:11	14:32
3B8	Mauritius	01:52	02:14	14:20	14:42
3B9	Rodriguez Is.	01:28	01:50	13:56	14:18
3C	Equatorial Guinea, Bata	05:04	05:24	17:31	17:52
	Malabo	05:08	05:28	17:35	17:56
3C0	Annobon Is.	05:20	05:41	17:48	18:08
3D2	Fiji Is., Suva	17:48	18:16	06:16	06:38
3D2	Conway Reef	18:04	18:26	06:32	06:55
3D2	Rotuma Is.	17:52	18:13	06:19	06:40
3DA	Swaziland, Mbabane	03:36	03:59	16:06	16:29
3V	Tunisia, Tunis	04:55	05:21	17:31	17:57
3W, XV	Vietnam, Ho Chi Minh City (Saigon)	22:35	22:56	11:04	11:25
	Hanoi	22:38	23:00	11:07	11:30
3X	Guinea, Conakry	06:37	06:58	19:05	19:26
3Y	Bouvet	05:13	05:49	17:58	18:34
3Y	Peter I Is.	11:04	12:02	00:16	01:14
4J, 4K	Azerbaijan, Baku	02:15	02:42	14:52	15:19
4L	Georgia, Tbilisi	02:35	03:02	15:13	15:41
4S	Sri Lanka, Colombo	00:23	00:44	12:51	13:11
4U	ITU Geneva	05:06	05:36	17:48	18:18
4U	United Nations Hq.	10:30	10:58	23:08	23:35
4X, 4Z	Israel, Jerusalem	03:17	03:42	15:50	16:15
5A	Libya, Tripoli	04:48	05:12	17:21	17:46
	Benghazi	04:18	04:42	16:51	17:16
5B	Cyprus, Nicosia	03:23	03:49	15:58	16:23
5H	Tanzania, Dar es Salaam	03:05	03:26	15:32	15:53
5N	Nigeria, Lagos	05:29	05:50	17:57	18:17
5R	Madagascar, Antananarivo	02:32	02:54	15:00	15:22
5T	Mauritania, Nouakchott	06:45	07:07	19:15	19:36
5U	Niger, Niamey	05:34	05:55	18:02	18:24
5V	Togo, Lome	05:38	05:59	18:05	18:26
5W	Western Samoa, Apia	17:10	17:31	05:37	05:59
5X	Uganda, Kampala	03:33	03:53	16:00	16:21
5Z	Kenya, Nairobi	04:36	04:56	17:03	17:24
6W	Senegal, Dakar	06:52	07:13	19:20	19:42
6Y	Jamaica, Kingston	10:49	11:10	23:18	23:40
7O	Yemen, Aden	02:42	03:03	15:10	15:32
	Sanaa	02:45	03:06	15:14	15:35
7P	Lesotho, Maseru	03:50	04:13	16:20	16:44
7Q	Malawi, Lilongwe	03:27	03:48	15:55	16:16
	Blantyre	03:22	03:44	15:50	16:12
7X	Algeria, Algiers	05:24	05:50	18:00	18:25
8P	Barbados, Bridgetown	09:40	10:02	22:09	22:30
8Q	Maldive Is.	00:49	01:10	13:17	13:37
8R	Guyana, Georgetown	09:35	09:56	22:03	22:24
9A	Croatia, Zagreb	04:27	04:57	17:08	17:38
9G	Ghana, Accra	05:43	06:04	18:11	18:32
9H	Malta	04:39	05:05	17:14	17:39
9J	Zambia, Lusaka	03:49	04:10	16:17	16:38
9K	Kuwait	02:28	02:51	15:00	15:24
9M2	West Malaysia, Kuala Lumpur	22:56	23:17	11:24	11:44
9M6,8	East Malaysia, Sabah, Sandakan (9M6)	21:50	22:11	10:18	10:39
	Sarawak, Kuching (9M8)	22:22	22:42	10:49	11:10
9N	Nepal, Kathmandu	23:58	00:21	12:30	12:53
9Q	Zaire, Kinshasa	04:42	05:02	17:09	17:30
	Kisangani	04:02	04:23	16:29	16:50
	Lubumbashi	03:52	04:14	16:20	16:41
9U	Burundi, Bujumbura	03:46	04:06	16:13	16:34
9V	Singapore	22:48	23:08	11:15	11:36
9X	Rwanda, Kigali	03:43	04:03	16:10	16:30
9Y	Trinidad & Tobago , Port of Spain	09:47	10:08	22:16	22:37
A2	Botswana, Gaborone	03:57	04:20	16:27	16:49
A3	Tonga, Nukualofa	17:22	17:44	05:51	06:13
A4	Oman, Musqat	01:46	02:08	14:16	14:39
A5	Bhutan, Thimbu	23:42	00:05	12:13	12:37
A6	United Arab Emirates, Abu Dhabi	02:03	02:26	14:34	14:57
A7	Qatar, Ad-Dawhah	02:14	02:37	14:45	15:08
A9	Bahrein, Al-Manamah	02:17	02:40	14:49	15:12
AP	Pakistan, Karachi	01:12	01:34	13:42	14:05
AP	Islamabad	00:45	01:09	13:19	13:43
BV	Taiwan, Taipei	21:34	21:57	10:05	10:28
BY	Peoples Rep. of China, Beijing	21:49	22:16	10:26	10:53
	Harbin	21:05	21:34	09:46	10:15
	Shanghai	21:32	21:56	10:05	10:29
	Fuzhou	21:42	22:06	10:14	10:37
	Xian	22:22	22:47	10:56	11:21

Pref	State/Province/Country/City	Tlight	Sunrise	Sunset	Tlight
	Chongqing	22:33	22:57	11:05	11:29
	Chengdu	22:42	23:06	11:15	11:39
	Lhasa	23:34	23:58	12:06	12:30
	Urumqi	23:42	00:11	12:22	12:50
	Kashi	00:31	00:58	13:08	13:35
C2	Nauru	18:35	18:56	07:03	07:23
C3	Andorra	05:27	05:56	18:06	18:34
C5	The Gambia, Banjul	06:49	07:10	19:17	19:38
C6	Bahamas, Nassau	10:50	11:12	23:20	23:43
C9	Mozambique, Maputo	03:30	03:53	16:00	16:23
	Mozambique	02:59	03:21	15:27	15:49
CE	Chile, Santiago	10:22	10:47	22:53	23:18
CE0	Easter Island	12:58	13:21	01:28	01:51
CE0	Juan Fernandez	10:54	11:19	23:25	23:50
CE0	San Felix	11:01	11:24	23:31	23:54
CM,CO	Cuba, Havana	11:10	11:32	23:40	00:03
CN	Morocco, Casablanca	06:07	06:32	18:41	19:06
CP	Bolivia, La Paz	10:16	10:37	22:44	23:05
CT	Portugal, Lisbon	06:12	06:39	18:49	19:15
CT3	Madeira Islands, Funchal	06:45	07:10	19:19	19:43
CU	Azores , Ponta Delgada	07:19	07:45	19:54	20:21
CX	Uruguay, Montevideo	09:23	09:48	21:55	22:20
CY0	Sable Is.	09:33	10:01	22:12	22:41
CY9	St. Paul Is.	09:31	10:01	22:13	22:43
D2	Angola, Luanda	04:50	05:11	17:17	17:38
D4	Cape Verde, Praia	07:16	07:37	19:44	20:06
D6	Comoros, Moroni	02:48	03:09	15:15	15:36
DA-DL,Y2-Y9	Fed. Rep. of Germany, Bonn	05:00	05:33	17:45	18:18
	Berlin	04:33	05:07	17:20	17:54
DU	Philippines, Manilla	21:38	21:59	10:06	10:28
E3	Eritrea, Asmara	03:06	03:28	15:35	15:56
EA	Spain, Madrid	05:49	06:17	18:27	18:54
EA6	Balearic Is., Palma	05:25	05:51	18:01	18:28
EA8	Canary Is., Las Palmas	06:36	07:00	19:08	19:32
EA9	Ceuta & Melilla, Ceuta	05:58	06:23	18:33	18:58
	Melilla	05:49	06:14	18:23	18:49
EI	Ireland, Dublin	05:51	06:25	18:39	19:13
EK	Armenia, Yerevan	02:37	03:04	15:14	15:41
EL	Liberia, Monrovia	06:26	06:47	18:53	19:14
EP	Iran, Tehran	02:09	02:35	14:44	15:10
ER	Moldova, Kishinev	03:35	04:06	16:17	16:48
ES	Estonia, Tallinn	03:39	04:20	16:36	17:16
ET	Ethiopia, Addis Ababa	03:08	03:29	15:36	15:56
EU,EV,EW	Belarus, Minsk	03:34	04:10	16:23	16:58
EX	Kyrgyzstan, Frunze	00:35	01:03	13:14	13:42
EY	Tajikistan, Samarkand	01:08	01:35	13:45	14:12
	Dushanbe	01:00	01:27	13:37	14:03
EZ	Turkmenistan, Ashkhabad	01:42	02:08	14:18	14:44
F	France, Paris	05:20	05:52	18:04	18:35
FG	Guadeloupe	09:48	10:10	22:17	22:39
FJ, FS	St. Martin	09:54	10:15	22:23	22:45
FH	Mayotte	02:41	03:02	15:09	15:30
FK	New Caledonia, Noumia	18:35	18:58	07:04	07:26
FM	Martinique	09:46	10:07	22:14	22:36
FO	Clipperton Is.	12:59	13:20	01:27	01:48
FO	Fr. Polynesia, Tahiti	16:20	16:42	04:48	05:10
	Rurutu, (Austral Is.)	19:36	19:58	08:05	08:27
	Hiva Oa, (Marquesas Is.)	14:59	15:20	03:26	03:47
FP	St. Pierre & Miquelon, St. Pierre	09:15	09:45	21:57	22:27
FR/G	Glorioso	02:33	02:54	15:01	15:22
FR/J,E	Juan de Nova	02:51	03:12	15:19	15:40
	Europa	03:00	03:22	15:28	15:51
FR	Reunion	01:59	02:21	14:28	14:50
FR/T	Tromelin	02:04	02:26	14:32	14:54
FT8W	Crozet	02:05	02:35	14:43	15:13
FT8X	Kerguelen	00:54	01:26	13:34	14:06
FT8Y	Antarctica, Dumont D'Urville	19:48	20:40	08:53	09:46
FT8Z	Amsterdam & St. Paul Is., Amsterdam	00:27	00:53	13:00	13:26
FW	Wallis & Futuna Is., Wallis	17:28	17:49	05:55	06:16
FY	Fr. Guiana, Cayenne	09:12	09:33	21:39	22:00
G	England, London	05:28	06:01	18:14	18:47
GD	Isle of Man	05:42	06:18	18:32	19:07
GI	Northern Ireland, Belfast	05:48	06:24	18:37	19:13
GJ	Jersey	05:38	06:10	18:22	18:53
GM	Scotland, Glasgow	05:40	06:17	18:31	19:08
	Aberdeen	05:29	06:08	18:23	19:01
GU	Guernsey	05:40	06:12	18:24	18:56
GW	Wales, Cardiff	05:40	06:13	18:26	18:59
H4	Solomon Islands, Honiara	19:03	19:24	07:30	07:51
HA	Hungary, Budapest	04:14	04:45	16:56	17:27
HB	Switzerland, Bern	05:01	05:31	17:43	18:13
HB0	Liechtenstein	04:52	05:23	17:34	18:05
HC	Ecuador, Quito	10:55	11:15	23:22	23:43
HC8	Galapagos Is.	11:45	12:05	00:12	00:33
HH	Haiti, Port-Au-Prince	10:30	10:52	23:00	23:22
HI	Dominican Republic, Santo Domingo	10:21	10:43	22:51	23:12
HK	Colombia, Bogota	10:39	11:00	23:07	23:27
HK0	Malpelo Is.	11:07	11:28	23:35	23:55
HK0	San Adreas	11:09	11:30	23:37	23:58
HL	Korea, Seoul	21:08	21:34	09:44	10:10
HP	Panama, Panama	11:00	11:21	23:28	23:49
HR	Honduras, Tegucigalpa	11:31	11:52	23:59	00:21
HS	Thailand, Bangkok	23:00	23:21	11:28	11:50
HV	Vatican City	04:44	05:12	17:22	17:50
HZ, 7Z	Saudi Arabia, Dharan	02:20	02:43	14:51	15:14
	Mecca	03:01	03:24	15:32	15:54
	Riyadh	02:33	02:56	15:04	15:27

Pref	State/Province/Country/City	Tlight	Sunrise	Sunset	Tlight
I	Italy, Rome	04:44	05:12	17:22	17:50
	Trieste	04:36	05:06	17:17	17:47
	Sicily	04:40	05:06	17:16	17:42
IS	Sardinia, Cagliari	04:59	05:25	17:35	18:02
J2	Djibouti, Djibouti	02:49	03:10	15:18	15:39
J3	Grenada	09:49	10:10	22:18	22:39
J5	Guinea-Bissau, Bissau	06:45	07:06	19:13	19:34
J6	St. Lucia	09:46	10:07	22:14	22:36
J7	Dominica	09:47	10:08	22:16	22:37
J8	St. Vincent	09:47	10:08	22:16	22:37
JA-JS	Japan, Tokyo	20:17	20:43	08:52	09:18
	Nagasaki	20:58	21:23	09:32	09:56
	Sapporo	20:08	20:36	08:47	09:15
JD1	Minami Torishima	19:24	19:47	07:55	08:18
JD1	Ogasawara, Kazan Is.	20:15	20:39	08:47	09:10
JT	Mongolia, Ulan Bator	22:22	22:53	11:05	11:36
JW	Svalbard, Spitsbergen	02:48	04:42	17:24	19:18
JX	Jan Mayen	05:23	06:27	18:53	19:57
JY	Jordan, Amman	03:14	03:39	15:48	16:12
KC4	Antarctica, Byrd Station	11:44	13:52	02:22	04:29
	McMurdo Sound	17:07	18:48	07:12	08:53
	Palmer Station	09:28	10:17	22:29	23:18
KC6	Belau, Yap	20:30	20:51	08:58	09:18
	Koror	20:45	21:05	09:12	09:33
KG4	Guantanamo Bay	10:42	11:04	23:11	23:33
KH0	Mariana Is., Saipan	19:59	20:20	08:27	08:49
KH1	Baker, Howland Is.	17:27	17:47	05:54	06:15
KH2	Guam, Agana	20:03	20:24	08:31	08:52
KH3	Johnston Is.	16:55	17:17	05:25	05:46
KH4	Midway Is.	17:29	17:52	06:01	06:24
KH5	Palmyra Is.	16:31	16:52	04:59	05:19
KH5K	Kingman Reef	16:34	16:55	05:01	05:22
KH6	Hawaii, Hilo	16:01	16:23	04:31	04:53
	Honolulu	16:12	16:35	04:42	05:04
KH7	Kure Is.	17:33	17:56	06:05	06:28
KH8	American Samoa, Pago Pago	17:05	17:27	05:33	05:55
KH9	Wake Is.	18:35	18:57	07:04	07:26
KL7	Alaska, Adak	17:13	17:47	06:00	06:33
	Anchorage	15:15	15:58	04:15	04:58
	Fairbanks	15:00	15:49	04:08	04:57
	Juneau	14:17	14:57	03:12	03:52
	Nome	16:10	16:59	05:18	06:06
KP1	Navassa Is.	10:41	11:03	23:11	23:32
KP2	Virgin Islands, Charlotte Amalie	10:01	10:23	22:30	22:52
KP4	Puerto Rico, San Juan	10:06	10:28	22:35	22:57
KP5	Desecheo Is.	10:11	10:33	22:41	23:02
LA-LJ	Norway, Oslo	04:34	05:16	17:32	18:14
LU	Argentina, Buenos Aires	09:32	09:57	22:04	22:29
LX	Luxembourg	05:04	05:36	17:48	18:20
LY	Lithuania, Vilna	03:42	04:18	16:32	17:07
LZ	Bulgaria, Sofia	04:00	04:28	16:39	17:07
OA	Peru, Lima	10:51	11:12	23:18	23:40
OD	Lebanon, Beirut	03:15	03:40	15:49	16:14
OE	Austria, Vienna	04:25	04:56	17:08	17:39
OH	Finland, Helsinki	03:37	04:19	16:35	17:17
OH0	Aland Is.	03:57	04:39	16:55	17:37
OJ0	Market Reef	04:01	04:43	16:59	17:41
OK,OL	Czech Rep., Prague	04:31	05:03	17:15	17:48
OM	Slovak, Rep., Bratislava	04:22	04:53	17:05	17:36
ON	Belgium, Brussels	05:10	05:43	17:55	18:28
OX	Greenland, Godthaab	08:36	09:24	21:43	22:31
	Thule	08:51	10:24	23:00	00:33
OY	Faroe Islands, Torshavn	05:41	06:25	18:43	19:27
OZ	Denmark, Copenhagen	04:32	05:09	17:23	18:00
P2	Papua New Guinea, Madang	20:00	20:21	08:28	08:48
	Port Moresby	19:54	20:15	08:22	08:43
P4	Aruba, Oranjestad	10:18	10:39	22:46	23:08
PA-PI	Netherlands, Amsterdam	05:07	05:41	17:54	18:28
PJ2,4,9	Netherlands Antilles, Willemstad	10:22	10:44	22:51	23:12
PJ5-8	St. Maarten and Saba, St. Maarten	09:54	10:16	22:23	22:45
PY	Brazil, Brasilia	08:54	09:15	21:22	21:43
	Rio De Janeiro	08:34	08:56	21:03	21:25
	Natal	08:04	08:24	20:31	20:52
	Manaus	09:44	10:04	22:11	22:32
	Porto Alegre	09:04	09:28	21:35	21:59
PY0	Fernando De Noronha	07:52	08:13	20:20	20:40
PY0	St. Peter & St. Paul Rocks	07:40	08:01	20:08	20:28
PY0	Trindade & Martin Vaz Is., Trindade	07:39	08:01	20:07	20:29
PZ	Suriname, Paramaribo	09:23	09:44	21:51	22:12
R1FJ_	Franz Josef Land	00:01	02:11	14:59	17:09
R1MV_	Malyj Vysotskij Is.	03:22	04:04	16:21	17:03
S0	Western Sahara, Smara	06:25	06:48	18:57	19:20
S2	Bangladesh, Dacca	23:39	00:01	12:09	12:32
S5	Slovenia, Ljubljana	04:33	05:03	17:14	17:44
S7	Seychelles, Victoria	02:01	02:22	14:28	14:49
S9	Sao Tome	05:16	05:37	17:43	18:04
SM	Sweden, Stockholm	04:06	04:47	17:02	17:43
SP	Poland, Krakow	04:08	04:41	16:53	17:25
	Warsaw	04:02	04:36	16:49	17:23
ST	Sudan, Khartoum	03:32	03:53	16:00	16:22
ST0	Southern Sudan, Juba	03:36	03:57	16:04	16:25
SU	Egypt, Cairo	03:33	03:57	16:06	16:29
SV	Greece, Athens	04:01	04:27	16:37	17:03
SV/A	Mount Athos	03:57	04:25	16:35	17:02
SV5	Dodecanese, Rhodes	03:44	04:09	16:19	16:44
SV9	Crete	03:56	04:21	16:31	16:56
T2	Tuvalu, Funafuti	17:48	18:09	06:16	06:37

Pref	State/Province/Country/City	Tlight	Sunrise	Sunset	Tlight
T30	West Kiribati, Bonriki	18:10	18:31	06:37	06:58
T31	Central Kiribati, Kanton	17:10	17:30	05:37	05:58
T32	East Kiribati, Christmas Is.	16:12	16:33	04:40	05:00
T33	Banaba Is.	18:25	18:46	06:53	07:13
T5	Somalia, Mogadishu	02:41	03:02	15:09	15:29
T7	San Marino	04:43	05:12	17:23	17:52
T9	Bosnia-Herzegovina, Sarajevo	04:19	04:48	16:59	17:27
TA	Turkey, Ankara	03:23	03:50	16:00	16:27
	Istanbul	03:38	04:06	16:16	16:44
TF	Iceland, Reykjavik	06:38	07:26	19:44	20:32
TG	Guatemala, Guatemala City	11:44	12:05	00:12	00:34
TI	Costa Rica, San Jose	11:18	11:39	23:46	00:07
TI9	Cocos Is.	11:31	11:51	23:58	00:19
TJ	Cameroon, Yaounde	04:57	05:17	17:24	17:45
TK	Corsica	04:58	05:26	17:36	18:04
TL	Central African Republic, Bangui	04:28	04:49	16:56	17:17
TN	Congo, Brazzaville	04:42	05:02	17:09	17:30
TR	Gabon, Libreville	05:05	05:25	17:32	17:53
TT	Chad, N'Djamena	04:42	05:03	17:10	17:32
TU	Cote d'Ivoire, Abidjan	05:59	06:19	18:26	18:47
TY	Benin, Porto Novo	05:32	05:53	18:00	18:21
TZ	Mali, Bamako	06:14	06:35	18:42	19:04
UA	Russia, European, Leningrad (UA1)	03:16	03:58	16:14	16:55
	Archangel (UA1)	02:27	03:15	15:34	16:23
	Murmansk (UA1)	02:45	03:43	16:06	17:04
	Moscow (UA3)	02:52	03:29	15:44	16:20
	Kuybyshev (UA4)	02:05	02:40	14:53	15:28
	Rostov (UA6)	02:52	03:23	15:35	16:05
UA2	Kaliningrad	04:02	04:38	16:52	17:28
UA9,0	Russia, Asiatic, Novosibirsk (UA9)	23:52	00:28	12:42	13:18
	Perm (UA9)	01:35	02:14	14:29	15:08
	Omsk (UA9)	00:30	01:06	13:20	13:56
	Norilsk (UA0)	23:03	00:03	12:26	13:25
	Irkutsk (UA0)	22:29	23:03	11:16	11:50
	Vladivostok (UA0)	20:45	21:14	09:25	09:53
	Petropavlovsk (UA0)	18:51	19:25	07:39	08:13
	Khabarovsk (UA0)	20:29	21:00	09:12	09:44
	Krasnoyarsk (UA0)	23:11	23:48	12:03	12:40
	Yakutsk (UA0)	20:35	21:19	09:37	10:21
	Wrangel Island (UA0)	16:47	17:52	06:17	07:22
	Kyzyl (UA0Y)	23:09	23:42	11:55	12:29
UJ-UM	Uzbekistan, Bukhoro	01:17	01:44	13:54	14:21
	Tashkent	00:57	01:24	13:35	14:02
UN-UQ	Kazakhstan, Alma-Ata	00:25	00:54	13:05	13:33
UR-UZ,EM-EO	Ukraine, Kiev	03:26	03:59	16:11	16:44
V2	Antigua & Barbuda, St. Johns	09:49	10:10	22:18	22:39
V3	Belize, Belmopan	11:37	11:58	00:06	00:27
V4	St. Kitts & Nevis	09:52	10:13	22:21	22:43
V5	Namibia, Windhoek	04:33	04:55	17:02	17:24
V6	Micronesia, Ponape	19:09	19:30	07:37	07:58
V7	Marshall Islands, Kwajalein	18:33	18:54	07:01	07:22
V8	Brunei, Bandar Seri Begawan	22:03	22:24	10:31	10:51
VK	Australia, Canberra (VK1)	19:42	20:07	08:14	08:39
	Sydney (VK2)	19:34	19:59	08:05	08:30
	Melbourne (VK3)	19:57	20:23	08:30	08:57
	Brisbane (VK4)	19:28	19:51	07:58	08:21
	Adelaide (VK5)	20:24	20:49	08:56	09:21
	Perth (VK6)	21:56	22:20	10:27	10:51
	Hobart, Tasmania (VK7)	19:46	20:14	08:21	08:50
	Darwin (VK8)	20:59	21:20	09:26	09:48
VK0	Heard Is.	00:34	01:09	13:18	13:52
VK0	Macquarie Is.	18:51	19:27	07:36	08:12
VK9C	Cocos-Keeling Is.	23:15	23:36	11:43	12:04
VK9L	Lord Howe Is.	19:03	19:27	07:34	07:58
VK9M	Mellish Reef	19:19	19:40	07:47	08:09
VK9N	Norfolk Is.	18:28	18:51	06:58	07:22
VK9W	Willis Is.	19:44	20:06	08:12	08:34
VK9X	Christmas Is.	22:40	23:01	11:07	11:28
VP2E	Anguilla	09:53	10:15	22:23	22:44
VP2M	Montserrat	09:50	10:12	22:19	22:41
VP2V	British Virgin Is., Tortola	10:00	10:21	22:29	22:51

Pref	State/Province/Country/City	Tlight	Sunrise	Sunset	Tlight
VP5	Turks & Caicos Islands, Grand Turk	10:26	10:48	22:56	23:18
VP8	Falkland Islands, Stanley	09:21	09:54	22:03	22:36
VP8	So. Georgia Is.	07:54	08:29	20:39	21:14
VP8	So. Orkney Is.	08:21	09:03	21:14	21:57
VP8	So. Sandwich Islands, Saunders Is.	07:10	07:49	19:59	20:38
VP8	So. Shetland Is., King George Is.	09:10	09:54	22:06	22:50
VP9	Bermuda	09:57	10:21	22:30	22:55
VQ9	Chagos, Diego Garcia	00:53	01:14	13:20	13:41
VR2,VS6	Hong Kong	22:03	22:26	10:34	10:56
VR6	Pitcairn Is.	14:21	14:44	02:51	03:13
VU	India, Bombay	00:50	01:12	13:19	13:41
	Calcutta	23:47	00:09	12:17	12:40
	New Delhi	00:30	00:54	13:02	13:26
	Bangalore	00:32	00:53	13:00	13:21
VU	Andaman Islands, Port Blair	23:31	23:52	11:59	12:20
VU	Laccadive Is.	00:50	01:11	13:18	13:39
XE	Mexico, Mexico City (XE1)	12:17	12:39	00:47	01:09
	Chihuahua (XE2)	12:43	13:07	01:15	01:39
	Merida (XE3)	11:40	12:02	00:09	00:32
XF4	Revilla Gigedo	13:07	13:29	01:37	01:58
XT	Burkina Faso, Ouagadougou	05:48	06:10	18:17	18:38
XU	Cambodia, Phnom Penh	22:43	23:04	11:11	11:32
XW	Laos, Viangchan	22:51	23:13	11:20	11:42
XX9	Macao	22:06	22:28	10:36	10:59
XZ	Myanmar, Yangon	23:17	23:39	11:47	12:08
YA	Afghanistan, Kandahar	01:15	01:39	13:48	14:12
	Kabul	01:00	01:25	13:35	14:00
YB-YD	Indonesia, Jakarta	22:36	22:56	11:03	11:24
	Medan	23:08	23:29	11:35	11:56
	Pontianak	22:26	22:46	10:53	11:14
	Jayapura	20:20	20:41	08:47	09:08
YI	Iraq, Baghdad	02:40	03:04	15:13	15:38
YJ	Vanuatu, Port Vila	18:29	18:50	06:57	07:19
YK	Syria, Damascus	03:12	03:37	15:46	16:11
YL	Latvia, Riga	03:45	04:23	16:38	17:16
YN	Nicaragua, Managua	11:26	11:47	23:54	00:16
YO	Romania, Bucharest	03:48	04:17	16:28	16:57
YS	El Salvador, San Salvador	11:39	12:00	00:07	00:29
YU	Yugoslavia, Belgrade	04:10	04:39	16:50	17:20
YV	Venezuela, Caracas	10:10	10:31	22:38	22:59
YV0	Aves Is.	09:56	10:18	22:25	22:47
Z2	Zimbabwe, Harare	03:38	04:00	16:06	16:28
Z3	Macedonia, (ex Yugoslav), Skopje	04:08	04:36	16:46	17:14
ZA	Albania, Tirane	04:15	04:42	16:53	17:20
ZB2	Gibraltar	05:58	06:24	18:33	18:59
ZC4	British Cyprus	03:25	03:50	15:59	16:25
ZD7	St. Helena	06:06	06:27	18:34	18:55
ZD8	Ascension Is.	06:40	07:01	19:08	19:29
ZD9	Tristan da Cunha	06:26	06:52	19:00	19:26
ZF	Cayman Is.	11:06	11:28	23:35	23:57
ZK1	No. Cook Is., Manihiki	16:27	16:48	04:54	05:15
ZK1	So. Cook Is., Rarotonga	16:21	16:43	04:49	05:11
ZK2	Niue	16:57	17:19	05:26	05:48
ZK3	Tokelaus, Atafu	17:13	17:34	05:41	06:02
ZL	New Zealand, Auckland (ZL1)	17:58	18:24	06:31	06:57
	Wellington (ZL2)	17:56	18:24	06:31	06:59
	Christchurch (ZL3)	18:04	18:33	06:40	07:09
	Dunedin (ZL4)	18:11	18:41	06:49	07:19
ZL5	Antarctica, Scott Base	17:06	18:49	07:13	08:57
ZL7	Chatham Is.	17:20	17:49	05:57	06:25
ZL8	Kermadec Is.	17:31	17:55	06:02	06:26
ZL9	Auckland & Campbell Is., Auckland	18:24	18:57	07:05	07:38
	Campbell Is.	18:12	18:46	06:55	07:29
ZP	Paraguay, Asuncion	09:31	09:54	22:01	22:24
ZS	South Africa, Cape Town (ZS1)	04:25	04:50	16:57	17:22
	Port Elizabeth (ZS2)	03:56	04:21	16:27	16:52
	Bloemfontein (ZS4)	03:55	04:19	16:26	16:49
	Durban (ZS5)	03:36	04:00	16:07	16:30
	Johannesburg (ZS6)	03:48	04:11	16:18	16:41
ZS8	Marion Is.	03:01	03:32	15:40	16:10

Pref	State/Province/Country/City	Tlight	Sunrise	Sunset	Tlight
VE1	New Brunswick, Saint John	09:23	09:54	22:59	23:29
	Nova Scotia, Halifax	09:15	09:44	22:48	23:18
	Prince Edward Island, Charlottetown	09:10	09:41	22:48	23:18
VE2	Quebec, Montreal	09:53	10:23	23:29	23:59
	Quebec City	09:41	10:12	23:21	23:52
VE3	Ontario, London	10:28	10:57	23:57	00:26
	Ottawa	10:02	10:32	23:37	00:07
	Sudbury	10:21	10:52	00:00	00:30
	Toronto	10:19	10:49	23:50	00:19
VE4	Manitoba, Winnipeg	11:19	11:52	01:08	01:41
VE5	Saskatchewan, Regina	11:48	12:22	01:39	02:12
	Saskatoon	11:53	12:28	01:49	02:24
VE6	Alberta, Calgary	12:25	12:59	02:17	02:51
	Edmonton	12:17	12:53	02:19	02:55
VE7	British Columbia, Prince George	12:53	13:30	02:56	03:33
	Prince Rupert	13:22	13:59	03:27	04:04
	Vancouver	13:04	13:37	02:51	03:24
VE8	Northwest Territories, Yellowknife	11:51	12:39	02:40	03:28
	Resolute	08:36	10:24	04:08	04:08
VY1	Yukon, Whitehorse	13:21	14:06	03:58	04:43
VO1	Newfoundland, St. John's	08:26	08:58	22:08	22:39
VO2	Labrador, Goose Bay	08:45	09:21	22:46	23:22
W1	Connecticut, Hartford	09:55	10:24	23:22	23:50
	Maine, Bangor	09:35	10:05	23:09	23:39
	Portland	09:43	10:12	23:14	23:43
	Massachusetts, Boston	09:48	10:17	23:16	23:44
	New Hampshire, Concord	09:48	10:17	23:18	23:47
	Rhode Island, Providence	09:50	10:18	23:16	23:45
	Vermont, Montpelier	09:51	10:21	23:24	23:53
W2	New Jersey, Atlantic City	10:05	10:33	23:26	23:53
	New York, Albany	09:58	10:27	23:27	23:56
	Buffalo	10:18	10:47	23:47	00:16
	New York City	10:02	10:30	23:26	23:54
	Syracuse	10:07	10:36	23:37	00:06
W3	Delaware, Wilmington	10:09	10:37	23:31	23:58
	District of Columbia, Washington	10:16	10:43	23:36	00:03
	Maryland, Baltimore	10:14	10:41	23:35	00:02
	Pennsylvania, Harrisburg	10:14	10:42	23:37	00:05
	Philadelphia	10:08	10:35	23:30	23:57
	Pittsburgh	10:26	10:54	23:49	00:17
	Scranton	10:08	10:36	23:33	00:01
W4	Alabama, Montgomery	11:01	11:26	00:08	00:33
	Florida, Jacksonville	10:44	11:09	23:48	00:13
	Miami	10:42	11:06	23:39	00:03
	Pensacola	11:06	11:31	00:10	00:35
	Georgia, Atlanta	10:52	11:17	00:02	00:27
	Savannah	10:40	11:05	23:47	00:12
	Kentucky, Lexington	10:47	11:14	00:05	00:32
	Louisville	10:52	11:19	00:11	00:38
	North Carolina, Charlotte	10:36	11:02	23:48	00:14
	Raleigh	10:26	10:52	23:40	00:06
	Wilmington	10:25	10:51	23:36	00:01
	South Carolina, Columbia	10:38	11:03	23:48	00:14
	Tennessee, Knoxville	10:47	11:13	00:01	00:27
	Memphis	11:13	11:39	00:25	00:51
	Nashville	10:59	11:25	00:13	00:39
	Virginia, Norfolk	10:16	10:42	23:32	23:58
	Richmond	10:20	10:46	23:37	00:03
W5	Arkansas, Little Rock	11:22	11:48	00:34	01:00
	Louisiana, New Orleans	11:18	11:43	00:22	00:46
	Shreveport	11:30	11:55	00:38	01:03
	Mississippi, Jackson	11:16	11:41	00:24	00:49
	New Mexico, Albuquerque	12:20	12:45	01:32	01:58
	Oklahoma, Oklahoma City	11:42	12:08	00:55	01:21
	Texas, Abilene	11:54	12:19	01:02	01:27
	Amarillo	12:00	12:26	01:12	01:38
	Dallas	11:42	12:07	00:51	01:16
	El Paso	12:22	12:47	01:29	01:53
	San Antonio	11:52	12:16	00:55	01:19
W6	California, Los Angeles	13:07	13:32	02:17	02:43
	San Francisco	13:19	13:46	02:37	03:03
W7	Arizona, Flagstaff	12:39	13:05	01:52	02:18
	Phoenix	12:43	13:08	01:52	02:17
	Idaho, Boise	12:47	13:16	02:17	02:47
	Pocatello	12:33	13:02	02:02	02:31
	Montana, Billings	12:12	12:43	01:49	02:19
	Butte	12:28	12:59	02:05	02:36
	Great Falls	12:21	12:52	02:02	02:33
	Nevada, Las Vegas	12:52	13:18	02:06	02:32
	Reno	13:07	13:34	02:28	02:55
	Oregon, Portland	13:10	13:40	02:45	03:16
	Utah, Salt Lake City	12:33	13:01	01:57	02:25
	Washington, Seattle	13:04	13:36	02:46	03:18
	Spokane	12:45	13:16	02:26	02:58
	Wyoming, Cheyenne	12:05	12:33	01:29	01:57
	Sheridan	12:08	12:38	01:42	02:12
W8	Michigan, Detroit	10:36	11:04	00:03	00:32
	Grand Rapids	10:45	11:14	00:15	00:44
	Sault Ste. Marie	10:35	11:06	00:13	00:44
	Traverse City	10:42	11:12	00:16	00:46
	Ohio, Cincinnati	10:46	11:13	00:06	00:34
	Cleveland	10:32	11:00	23:57	00:25
	Columbus	10:39	11:06	00:01	00:29
	West Virginia, Charleston	10:35	11:02	23:54	00:21
W9	Illinois, Chicago	10:55	11:23	00:21	00:50
	Indiana, Indianapolis	10:52	11:19	00:14	00:41
	Wisconsin, Green Bay	10:52	11:22	00:25	00:55

Pref	State/Province/Country/City	Tlight	Sunrise	Sunset	Tlight
	Milwaukee	10:54	11:23	00:23	00:52
W0	Colorado, Denver	12:07	12:35	01:29	01:56
	Grand Junction	12:22	12:50	01:43	02:10
	Iowa, Des Moines	11:19	11:47	00:45	01:13
	Kansas, Pratt	11:45	12:11	01:02	01:29
	Wichita	11:39	12:06	00:56	01:23
	Minnesota, Duluth	11:05	11:36	00:44	01:15
	Minneapolis	11:13	11:43	00:47	01:17
	Missouri, Columbia	11:17	11:44	00:37	01:05
	Kansas City	11:26	11:54	00:47	01:14
	St. Louis	11:09	11:36	00:29	00:56
	Nebraska, North Platte	11:49	12:17	01:13	01:41
	Omaha	11:29	11:57	00:54	01:22
	North Dakota, Fargo	11:24	11:55	01:03	01:34
	South Dakota, Rapid City	11:54	12:23	01:26	01:55
-		05:58	05:58	18:05	18:26
1A0	S.M.O.M	04:15	04:43	17:41	18:10
1S	Spratly Is.	22:05	22:27	10:42	11:03
3A	Monaco	04:32	05:01	18:03	18:32
3B6	Agalega	01:56	02:17	14:14	14:35
3B7	St. Brandon	01:45	02:07	13:58	14:20
3B8	Mauritius	01:56	02:18	14:05	14:27
3B9	Rodriguez Is.	01:32	01:54	13:42	14:04
3C	Equatorial Guinea, Bata	04:58	05:18	17:27	17:48
	Malabo	05:01	05:21	17:32	17:53
3C0	Annobon Is.	05:16	05:37	17:42	18:03
3D2	Fiji Is., Suva	17:52	18:14	06:03	06:25
3D2	Conway Reef	18:09	18:31	06:17	06:39
3D2	Rotuma Is.	17:52	18:13	06:08	06:30
3DA	Swaziland, Mbabane	03:44	04:07	15:48	16:11
3V	Tunisia, Tunis	04:30	04:56	17:46	18:12
3W, XV	Vietnam, Ho Chi Minh City (Saigon)	22:25	22:46	11:04	11:25
	Hanoi	22:22	22:44	11:13	11:35
3X	Guinea, Conakry	06:27	06:49	19:05	19:26
3Y	Bouvet	05:45	06:21	17:16	17:51
3Y	Peter I Is.	12:08	13:06	23:02	00:01
4J, 4K	Azerbaijan, Baku	01:47	02:14	15:10	15:38
4L	Georgia, Tbilisi	02:05	02:34	15:31	16:00
4S	Sri Lanka, Colombo	00:14	00:35	12:49	13:10
4U	ITU Geneva	04:33	05:04	18:10	18:41
4U	United Nations Hq.	10:02	10:30	23:26	23:54
4X, 4Z	Israel, Jerusalem	02:55	03:20	16:02	16:27
5A	Libya, Tripoli	04:25	04:50	17:33	17:58
	Benghazi	03:56	04:21	17:03	17:28
5B	Cyprus, Nicosia	02:59	03:25	16:12	16:37
5H	Tanzania, Dar es Salaam	03:03	03:24	15:24	15:45
5N	Nigeria, Lagos	05:21	05:42	17:55	18:16
5R	Madagascar, Antananarivo	02:36	02:58	14:46	15:08
5T	Mauritania, Nouakchott	06:31	06:53	19:18	19:40
5U	Niger, Niamey	05:22	05:44	18:04	18:25
5V	Togo, Lome	05:30	05:51	18:03	18:24
5W	Western Samoa, Apia	17:11	17:32	05:26	05:47
5X	Uganda, Kampala	03:27	03:48	15:55	16:16
5Z	Kenya, Nairobi	04:31	04:52	16:57	17:18
6W	Senegal, Dakar	06:40	07:01	19:22	19:44
6Y	Jamaica, Kingston	10:35	10:57	23:21	23:43
7O	Yemen, Aden	02:31	02:52	15:11	15:33
	Sanaa	02:32	02:54	15:16	15:38
7P	Lesotho, Maseru	03:59	04:23	16:00	16:24
7Q	Malawi, Lilongwe	03:28	03:50	15:43	16:05
	Blantyre	03:24	03:46	15:38	15:59
7X	Algeria, Algiers	04:59	05:25	18:14	18:41
8P	Barbados, Bridgetown	09:29	09:50	22:10	22:31
8Q	Maldive Is.	00:42	01:03	13:14	13:35
8R	Guyana, Georgetown	09:27	09:48	22:01	22:22
9A	Croatia, Zagreb	03:54	04:25	17:31	18:01
9G	Ghana, Accra	05:36	05:57	18:09	18:29
9H	Malta	04:14	04:40	17:28	17:54
9J	Zambia, Lusaka	03:51	04:13	16:05	16:26
9K	Kuwait	02:07	02:31	15:10	15:34
9L	Sierra Leone, Freetown	06:26	06:47	19:02	19:23
9M2	West Malaysia, Kuala Lumpur	22:50	23:10	11:20	11:41
9M6,8	East Malaysia, Sabah, Sandakan(9M6)	21:42	22:03	10:15	10:36
	Sarawak, Kuching (9M8)	22:16	22:36	10:45	11:05
9N	Nepal, Kathmandu	23:39	00:02	12:39	13:02
9Q	Zaire, Kinshasa	04:38	04:59	17:02	17:23
	Kisangani	03:57	04:17	16:25	16:45
	Lubumbashi	03:53	04:14	16:10	16:31
9U	Burundi, Bujumbura	03:42	04:03	16:06	16:27
9V	Singapore	22:42	23:03	11:11	11:31
9X	Rwanda, Kigali	03:38	03:59	16:04	16:25
9Y	Trinidad & Tobago , Port of Spain	09:37	09:58	22:15	22:37
A2	Botswana, Gaborone	04:04	04:27	16:09	16:32
A3	Tonga, Nukualofa	17:27	17:49	05:36	05:58
A4	Oman, Musqat	01:29	01:52	14:23	14:46
A5	Bhutan, Thimbu	23:23	23:46	12:22	12:46
A6	United Arab Emirates, Abu Dhabi	01:46	02:09	14:41	15:04
A7	Qatar, Ad-Dawhah	01:56	02:19	14:52	15:16
A9	Bahrein, Al-Manamah	01:59	02:22	14:57	15:20
AP	Pakistan, Karachi	00:54	01:17	13:50	14:13
AP	Islamabad	00:21	00:47	13:31	13:56
BV	Taiwan, Taipei	21:16	21:39	10:12	10:35
BY	Peoples Rep. of China, Beijing	21:21	21:49	10:43	11:11
	Harbin	20:31	21:02	10:08	10:38
	Shanghai	21:11	21:35	10:16	10:41
	Fuzhou	21:24	21:47	10:22	10:45
	Xian	21:58	22:23	11:09	11:34

Pref	State/Province/Country/City	Tlight	Sunrise	Sunset	Tlight
	Chongqing	22:12	22:36	11:15	11:39
	Chengdu	22:21	22:45	11:25	11:50
	Lhasa	23:13	23:37	12:16	12:41
	Urumqi	23:11	23:40	12:42	13:12
	Kashi	00:03	00:31	13:25	13:52
C2	Nauru	18:30	18:51	06:57	07:18
C3	Andorra	04:57	05:26	18:25	18:54
C5	The Gambia, Banjul	06:37	06:59	19:19	19:40
C6	Bahamas, Nassau	10:32	10:55	23:28	23:51
C9	Mozambique, Maputo	03:38	04:01	15:42	16:05
	Mozambique	03:01	03:23	15:15	15:37
CE	Chile, Santiago	10:34	10:59	22:31	22:56
CE0	Easter Island	13:06	13:30	01:09	01:32
CE0	Juan Fernandez	11:06	11:31	23:03	23:28
CE0	San Felix	11:09	11:32	23:12	23:36
CM,CO	Cuba, Havana	10:53	11:16	23:47	00:09
CN	Morocco, Casablanca	05:44	06:10	18:54	19:19
CP	Bolivia, La Paz	10:18	10:40	22:31	22:52
CT	Portugal, Lisbon	05:45	06:12	19:05	19:32
CT3	Madeira Islands, Funchal	06:23	06:48	19:31	19:56
CU	Azores , Ponta Delgada	06:53	07:19	20:10	20:37
CX	Uruguay, Montevideo	09:36	10:02	21:32	21:57
CY0	Sable Is.	09:01	09:31	22:33	23:02
CY9	St. Paul Is.	08:56	09:28	22:37	23:08
D2	Angola, Luanda	04:49	05:10	17:08	17:29
D4	Cape Verde, Praia	07:03	07:25	19:46	20:08
D6	Comoros, Moroni	02:48	03:09	15:05	15:26
DA-DL,Y2-Y9	Fed. Rep. of Germany, Bonn	04:21	04:55	18:13	18:46
	Berlin	03:52	04:27	17:49	18:25
DU	Philippines, Manilla	21:26	21:47	10:08	10:30
E3	Eritrea, Asmara	02:54	03:15	15:37	15:59
EA	Spain, Madrid	05:21	05:49	18:44	19:12
EA6	Balearic Is., Palma	04:57	05:24	18:18	18:46
EA8	Canary Is., Las Palmas	06:17	06:40	19:18	19:41
EA9	Ceuta & Melilla, Ceuta	05:33	05:59	18:47	19:13
	Melilla	05:24	05:50	18:37	19:03
EI	Ireland, Dublin	05:08	05:44	19:09	19:45
EK	Armenia, Yerevan	02:08	02:36	15:31	15:59
EL	Liberia, Monrovia	06:18	06:39	18:51	19:12
EP	Iran, Tehran	01:45	02:11	14:58	15:24
ER	Moldova, Kishinev	03:01	03:32	16:41	17:12
ES	Estonia, Tallinn	02:46	03:29	17:16	17:59
ET	Ethiopia, Addis Ababa	02:58	03:19	15:35	15:56
EU,EV,EW	Belarus, Minsk	02:51	03:28	16:55	17:31
EX	Kyrgyzstan, Frunze	00:04	00:33	13:33	14:02
EY	Tajikistan, Samarkand	00:40	01:07	14:02	14:29
	Dushanbe	00:33	01:00	13:53	14:20
EZ	Turkmenistan, Ashkhabad	01:16	01:43	14:34	15:01
F	France, Paris	04:44	05:16	18:29	19:01
FG	Guadeloupe	09:36	09:57	22:20	22:42
FJ, FS	St. Martin	09:40	10:02	22:27	22:49
FH	Mayotte	02:42	03:03	14:58	15:19
FK	New Caledonia, Noumia	18:41	19:03	06:48	07:11
FM	Martinique	09:34	09:55	22:16	22:38
FO	Clipperton Is.	12:49	13:10	01:27	01:48
FO	Fr. Polynesia, Tahiti	16:23	16:45	04:35	04:56
	Rurutu, (Austral Is.)	19:42	20:04	07:49	08:11
	Hiva Oa, (Marquesas Is.)	14:58	15:19	03:16	03:38
FP	St. Pierre & Miquelon, St. Pierre	08:41	09:12	22:20	22:51
FR/G	Glorioso	02:33	02:55	14:50	15:12
FR/J,E	Juan de Nova	02:54	03:15	15:06	15:27
	Europa	03:05	03:28	15:13	15:35
FR	Reunion	02:04	02:26	14:12	14:35
FR/T	Tromelin	02:07	02:28	14:20	14:41
FT8W	Crozet	02:27	02:57	14:10	14:40
FT8X	Kerguelen	01:20	01:52	12:58	13:30
FT8Y	Antarctica, Dumont D'Urville	20:44	21:37	07:47	08:39
FT8Z	Amsterdam & St. Paul Is., Amsterdam	00:42	01:08	12:34	13:01
FW	Wallis & Futuna Is., Wallis	17:29	17:50	05:44	06:05
FY	Fr. Guiana, Cayenne	09:04	09:25	21:37	21:58
G	England, London	04:48	05:22	18:42	19:17
GD	Isle of Man	04:59	05:36	19:04	19:41
GI	Northern Ireland, Belfast	05:04	05:41	19:10	19:47
GJ	Jersey	05:01	05:34	18:48	19:20
GM	Scotland, Glasgow	04:54	05:33	19:05	19:44
	Aberdeen	04:41	05:21	18:59	19:39
GU	Guernsey	05:02	05:35	18:50	19:23
GW	Wales, Cardiff	05:00	05:35	18:54	19:29
H4	Solomon Islands, Honiara	19:02	19:23	07:21	07:42
HA	Hungary, Budapest	03:39	04:10	17:20	17:52
HB	Switzerland, Bern	04:26	04:57	18:06	18:37
HB0	Liechtenstein	04:17	04:49	17:58	18:29
HC	Ecuador, Quito	10:50	11:10	23:17	23:38
HC8	Galapagos Is.	11:40	12:01	00:07	00:28
HH	Haiti, Port-Au-Prince	10:16	10:38	23:04	23:26
HI	Dominican Republic, Santo Domingo	10:07	10:29	22:54	23:16
HK	Colombia, Bogota	10:32	10:53	23:04	23:25
HK0	Malpelo Is.	11:00	11:21	23:31	23:52
HK0	San Adreas	10:58	11:19	23:38	23:59
HL	Korea, Seoul	20:42	21:09	09:59	10:26
HP	Panama, Panama	10:51	11:12	23:27	23:49
HR	Honduras, Tegucigalpa	11:19	11:40	00:01	00:22
HS	Thailand, Bangkok	22:48	23:10	11:30	11:51
HV	Vatican City	04:14	04:43	17:41	18:09
HZ, 7Z	Saudi Arabia, Dharan	02:01	02:25	14:59	15:22
	Mecca	02:46	03:08	15:37	15:59
	Riyadh	02:16	02:39	15:11	15:34

Pref	State/Province/Country/City	Tlight	Sunrise	Sunset	Tlight
I	Italy, Rome	04:14	04:43	17:41	18:09
	Trieste	04:03	04:34	17:39	18:10
	Sicily	04:14	04:41	17:31	17:58
IS	Sardinia, Cagliari	04:31	04:59	17:52	18:19
J2	Djibouti, Djibouti	02:39	03:00	15:18	15:39
J3	Grenada	09:38	10:00	22:18	22:40
J5	Guinea-Bissau, Bissau	06:34	06:55	19:13	19:35
J6	St. Lucia	09:34	09:56	22:16	22:37
J7	Dominica	09:34	09:56	22:18	22:40
J8	St. Vincent	09:36	09:57	22:17	22:38
JA-JS	Japan, Tokyo	19:53	20:19	09:06	09:32
	Nagasaki	20:36	21:01	09:44	10:09
	Sapporo	19:37	20:06	09:06	09:35
JD1	Minami Torishima	19:07	19:30	08:02	08:25
JD1	Ogasawara, Kazan Is.	19:56	20:20	08:56	09:19
JT	Mongolia, Ulan Bator	21:47	22:19	11:29	12:01
JW	Svalbard, Spitsbergen	——	02:07	19:48	——
JX	Jan Mayen	03:47	05:03	20:07	21:23
JY	Jordan, Amman	02:52	03:17	15:59	16:24
KC4	Antarctica, Byrd Station	14:18	16:27	23:37	01:46
	McMurdo Sound	19:07	20:49	05:01	06:43
	Palmer Station	10:19	11:08	21:27	22:16
KC6	Belau, Yap	20:20	20:41	08:57	09:18
	Koror	20:36	20:57	09:11	09:32
KG4	Guantanamo Bay	10:27	10:49	23:16	23:38
KH0	Mariana Is., Saipan	19:46	20:08	08:29	08:51
KH1	Baker, Howland Is.	17:21	17:42	05:49	06:10
KH2	Guam, Agana	19:51	20:13	08:33	08:54
KH3	Johnston Is.	16:42	17:04	05:28	05:49
KH4	Midway Is.	17:09	17:33	06:10	06:34
KH5	Palmyra Is.	16:23	16:44	04:56	05:17
KH5K	Kingman Reef	16:25	16:46	05:00	05:21
KH6	Hawaii, Hilo	15:47	16:09	04:35	04:58
	Honolulu	15:57	16:19	04:48	05:10
KH7	Kure Is.	17:13	17:37	06:14	06:38
KH8	American Samoa, Pago Pago	17:07	17:28	05:21	05:43
KH9	Wake Is.	18:20	18:42	07:08	07:31
KL7	Alaska, Adak	16:33	17:08	06:28	07:03
	Anchorage	14:19	15:05	04:59	05:45
	Fairbanks	13:53	14:47	05:00	05:53
	Juneau	13:27	14:08	03:50	04:32
	Nome	15:05	15:57	06:09	07:02
KP1	Navassa Is.	10:27	10:49	23:14	23:36
KP2	Virgin Islands, Charlotte Amalie	09:47	10:09	22:34	22:56
KP4	Puerto Rico, San Juan	09:52	10:14	22:39	23:01
KP5	Desecheo Is.	09:57	10:19	22:44	23:06
LA-LJ	Norway, Oslo	03:40	04:24	18:13	18:57
LU	Argentina, Buenos Aires	09:45	10:10	21:41	22:06
LX	Luxembourg	04:27	05:00	18:14	18:47
LY	Lithuania, Vilna	02:58	03:35	17:04	17:41
LZ	Bulgaria, Sofia	03:30	03:59	16:58	17:27
OA	Peru, Lima	10:51	11:13	23:08	23:29
OD	Lebanon, Beirut	02:52	03:17	16:02	16:27
OE	Austria, Vienna	03:49	04:21	17:32	18:04
OH	Finland, Helsinki	02:42	03:27	17:17	18:01
OH0	Aland Is.	03:02	03:47	17:37	18:21
OJ0	Market Reef	03:06	03:51	17:41	18:25
OK,OL	Czech Rep., Prague	03:53	04:26	17:42	18:16
OM	Slovak, Rep., Bratislava	03:46	04:18	17:29	18:01
ON	Belgium, Brussels	04:31	05:05	18:23	18:57
OX	Greenland, Godthaab	07:32	08:24	22:34	23:25
	Thule	05:47	08:20	00:54	03:27
OY	Faroe Islands, Torshavn	04:42	05:30	19:28	20:15
OZ	Denmark, Copenhagen	03:47	04:25	17:58	18:36
P2	Papua New Guinea, Madang	19:58	20:18	08:20	08:41
	Port Moresby	19:53	20:14	08:12	08:33
P4	Aruba, Oranjestad	10:07	10:28	22:47	23:08
PA-PI	Netherlands, Amsterdam	04:26	05:01	18:23	18:59
PJ2,4,9	Netherlands Antilles, Willemstad	10:11	10:33	22:52	23:13
PJ5-8	St. Maarten and Saba, St. Maarten	09:40	10:02	22:27	22:49
PY	Brazil, Brasilia	08:56	09:18	21:09	21:31
	Rio De Janeiro	08:40	09:02	20:47	21:09
	Natal	08:01	08:22	20:23	20:44
	Manaus	09:40	10:01	22:04	22:25
	Porto Alegre	09:14	09:38	21:15	21:39
PY0	Fernando de Noronha	07:49	08:10	20:13	20:34
PY0	St. Peter & St. Paul Rocks	07:35	07:56	20:03	20:24
PY0	Trindade & Martin Vaz Is., Trindade	07:43	08:06	19:52	20:15
PZ	Suriname, Paramaribo	09:15	09:36	21:49	22:10
R1FJ_	Franz Josef Land	——	23:12	17:47	——
R1MV_	Malyj Vysotskij Is.	02:27	03:12	17:03	17:48
S0	Western Sahara, Smara	06:07	06:30	19:05	19:28
S2	Bangladesh, Dacca	23:22	23:44	12:16	12:39
S5	Slovenia, Ljubljana	04:00	04:31	17:37	18:08
S7	Seychelles, Victoria	01:58	02:19	14:21	14:42
S9	Sao Tome	05:11	05:31	17:38	17:59
SM	Sweden, Stockholm	03:13	03:56	17:42	18:25
SP	Poland, Krakow	03:31	04:04	17:20	17:53
	Warsaw	03:22	03:57	17:19	17:54
ST	Sudan, Khartoum	03:19	03:41	16:03	16:25
ST0	Southern Sudan, Juba	03:29	03:50	16:01	16:22
SU	Egypt, Cairo	03:12	03:36	16:16	16:40
SV	Greece, Athens	03:35	04:01	16:53	17:19
SV/A	Mount Athos	03:29	03:57	16:52	17:20
SV5	Dodecanese, Rhodes	03:18	03:45	16:33	16:59
SV9	Crete	03:32	03:57	16:45	17:10
T2	Tuvalu, Funafuti	17:47	18:08	06:07	06:28

Pref	State/Province/Country/City	Tlight	Sunrise	Sunset	Tlight
T30	West Kiribati, Bonriki	18:04	18:25	06:33	06:54
T31	Central Kiribati, Kanton	17:06	17:27	05:31	05:51
T32	East Kiribati, Christmas Is.	16:06	16:27	04:36	04:56
T33	Banaba Is.	18:20	18:41	06:47	07:08
T5	Somalia, Mogadishu	02:35	02:56	15:05	15:25
T7	San Marino	04:12	04:42	17:44	18:13
T9	Bosnia-Herzegovina, Sarajevo	03:48	04:17	17:19	17:49
TA	Turkey, Ankara	02:55	03:23	16:17	16:45
	Istanbul	03:09	03:37	16:34	17:02
TF	Iceland, Reykjavik	05:33	06:25	20:34	21:26
TG	Guatemala, Guatemala City	11:32	11:53	00:14	00:36
TI	Costa Rica, San Jose	11:08	11:30	23:46	00:07
TI9	Cocos Is.	11:23	11:44	23:56	00:17
TJ	Cameroon, Yaounde	04:50	05:11	17:21	17:42
TK	Corsica	04:28	04:57	17:55	18:23
TL	Central African Republic, Bangui	04:21	04:42	16:53	17:14
TN	Congo, Brazzaville	04:38	04:59	17:02	17:23
TR	Gabon, Libreville	04:59	05:20	17:27	17:48
TT	Chad, N'Djamena	04:31	04:52	17:11	17:32
TU	Cote d'Ivoire, Abidjan	05:51	06:12	18:24	18:45
TY	Benin, Porto Novo	05:24	05:45	17:58	18:19
TZ	Mali, Bamako	06:03	06:24	18:43	19:05
UA	Russia, European, Leningrad (UA1)	02:22	03:06	16:55	17:39
	Archangel (UA1)	01:21	02:14	16:26	17:19
	Murmansk (UA1)	01:21	02:27	17:11	18:18
	Moscow (UA3)	02:06	02:45	16:18	16:56
	Kuybyshev (UA4)	01:23	01:59	15:24	16:00
	Rostov (UA6)	02:17	02:49	15:59	16:30
UA2	Kaliningrad	03:17	03:55	17:25	18:03
UA9,0	Russia, Asiatic, Novosibirsk (UA9)	23:07	23:45	13:15	13:53
	Perm (UA9)	00:45	01:26	15:07	15:48
	Omsk (UA9)	23:45	00:23	13:53	14:31
	Norilsk (UA0)	21:38	22:46	13:33	14:41
	Irkutsk (UA0)	21:48	22:24	11:46	12:21
	Vladivostok (UA0)	20:15	20:44	09:44	10:14
	Petropavlovsk (UA0)	18:09	18:45	08:09	08:45
	Khabarovsk (UA0)	19:53	20:25	09:37	10:10
	Krasnoyarsk (UA0)	22:25	23:04	12:37	13:16
	Yakutsk (UA0)	19:36	20:24	10:22	11:09
	Wrangel Island (UA0)	15:12	16:28	07:32	08:48
	Kyzyl (UA0Y)	22:29	23:04	12:24	12:59
UJ-UM	Uzbekistan, Bukhoro	00:49	01:17	14:11	14:39
	Tashkent	00:28	00:56	13:53	14:21
UN-UQ	Kazakhstan, Alma-Ata	23:55	00:24	13:25	13:54
UR-UZ,EM-EO	Ukraine, Kiev	02:48	03:21	16:38	17:12
V2	Antigua & Barbuda, St. Johns	09:35	09:57	22:21	22:43
V3	Belize, Belmopan	11:23	11:45	00:09	00:31
V4	St. Kitts & Nevis	09:38	10:00	22:24	22:46
V5	Namibia, Windhoek	04:39	05:01	16:46	17:08
V6	Micronesia, Ponape	19:01	19:22	07:35	07:56
V7	Marshall Islands, Kwajalein	18:24	18:45	07:00	07:21
V8	Brunei, Bandar Seri Begawan	21:56	22:16	10:28	10:49
VK	Australia, Canberra (VK1)	19:55	20:21	07:50	08:16
	Sydney (VK2)	19:46	20:11	07:43	08:08
	Melbourne (VK3)	20:12	20:39	08:05	08:31
	Brisbane (VK4)	19:37	20:00	07:39	08:03
	Adelaide (VK5)	20:37	21:02	08:32	08:58
	Perth (VK6)	22:07	22:32	10:06	10:30
	Hobart, Tasmania (VK7)	20:05	20:33	07:52	08:20
	Darwin (VK8)	20:59	21:21	09:16	09:37
VK0	Heard Is.	01:04	01:39	12:37	13:12
VK0	Macquarie Is.	19:23	19:59	06:54	07:30
VK9C	Cocos-Keeling Is.	23:16	23:37	11:32	11:53
VK9L	Lord Howe Is.	19:14	19:38	07:12	07:37
VK9M	Mellish Reef	19:22	19:44	07:33	07:55
VK9N	Norfolk Is.	18:37	19:01	06:38	07:02
VK9W	Willis Is.	19:47	20:08	07:59	08:21
VK9X	Christmas Is.	22:39	23:01	10:57	11:18
VP2E	Anguilla	09:39	10:01	22:26	22:48
VP2M	Montserrat	09:37	09:59	22:22	22:44
VP2V	British Virgin Is., Tortola	09:46	10:08	22:33	22:55

Pref	State/Province/Country/City	Tlight	Sunrise	Sunset	Tlight
VP5	Turks & Caicos Islands, Grand Turk	10:10	10:32	23:01	23:23
VP8	Falkland Islands, Stanley	09:49	10:23	21:24	21:58
VP8	So. Georgia Is.	08:26	09:01	19:57	20:32
VP8	So. Orkney Is.	09:03	09:46	20:22	21:04
VP8	So. Sandwich Islands, Saunders Is.	07:47	08:26	19:11	19:50
VP8	So. Shetland Is., King George Is.	09:55	10:39	21:10	21:55
VP9	Bermuda	09:34	09:59	22:42	23:07
VQ9	Chagos, Diego Garcia	00:51	01:12	13:12	13:33
VR2,VS6	Hong Kong	21:47	22:10	10:39	11:02
VR6	Pitcairn Is.	14:28	14:51	02:33	02:56
VU	India, Bombay	00:35	00:58	13:24	13:46
	Calcutta	23:30	23:53	12:23	12:46
	New Delhi	00:10	00:34	13:12	13:36
	Bangalore	00:20	00:42	13:01	13:23
VU	Andaman Islands, Port Blair	23:20	23:41	12:00	12:21
VU	Laccadive Is.	00:40	01:02	13:18	13:39
XE	Mexico, Mexico City (XE1)	12:03	12:25	00:51	01:13
	Chihuahua (XE2)	12:23	12:47	01:25	01:48
	Merida (XE3)	11:24	11:46	00:15	00:37
XF4	Revilla Gigedo	12:53	13:15	01:41	02:03
XT	Burkina Faso, Ouagadougou	05:37	05:59	18:18	18:39
XU	Cambodia, Phnom Penh	22:32	22:53	11:12	11:33
XW	Laos, Viangchan	22:37	22:59	11:24	11:46
XX9	Macao	21:50	22:13	10:42	11:05
XZ	Myanmar, Yangon	23:04	23:26	11:49	12:11
YA	Afghanistan, Kandahar	00:54	01:18	13:59	14:23
	Kabul	00:37	01:02	13:48	14:13
YB-YD	Indonesia, Jakarta	22:33	22:54	10:55	11:16
	Medan	23:01	23:22	11:32	11:53
	Pontianak	22:20	22:41	10:48	11:09
	Jayapura	20:16	20:37	08:41	09:02
YI	Iraq, Baghdad	02:17	02:42	15:26	15:51
YJ	Vanuatu, Port Vila	18:32	18:54	06:43	07:05
YK	Syria, Damascus	02:49	03:14	15:59	16:24
YL	Latvia, Riga	02:57	03:37	17:14	17:54
YN	Nicaragua, Managua	11:15	11:37	23:55	00:16
YO	Romania, Bucharest	03:16	03:46	16:49	17:19
YS	El Salvador, San Salvador	11:27	11:48	00:09	00:30
YU	Yugoslavia, Belgrade	03:38	04:08	17:12	17:42
YV	Venezuela, Caracas	10:00	10:21	22:38	22:59
YV0	Aves Is.	09:44	10:05	22:28	22:49
Z2	Zimbabwe, Harare	03:41	04:03	15:53	16:14
Z3	Macedonia, (ex Yugoslav), Skopje	03:39	04:07	17:05	17:34
ZA	Albania, Tirane	03:46	04:14	17:11	17:39
ZB2	Gibraltar	05:33	05:59	18:47	19:14
ZC4	British Cyprus	03:01	03:27	16:13	16:38
ZD7	St. Helena	06:08	06:30	18:21	18:43
ZD8	Ascension Is.	06:39	07:00	18:59	19:20
ZD9	Tristan da Cunha	06:41	07:07	18:35	19:01
ZF	Cayman Is.	10:51	11:13	23:40	00:02
ZK1	No. Cook Is., Manihiki	16:26	16:47	04:44	05:05
ZK1	So. Cook Is., Rarotonga	16:26	16:48	04:34	04:56
ZK2	Niue	17:01	17:23	05:12	05:33
ZK3	Tokelaus, Atafu	17:12	17:33	05:32	05:53
ZL	New Zealand, Auckland (ZL1)	18:13	18:39	06:06	06:32
	Wellington (ZL2)	18:14	18:42	06:03	06:31
	Christchurch (ZL3)	18:24	18:53	06:10	06:39
	Dunedin (ZL4)	18:33	19:03	06:16	06:46
ZL5	Antarctica, Scott Base	19:08	20:52	05:00	06:44
ZL7	Chatham Is.	17:41	18:09	05:26	05:55
ZL8	Kermadec Is.	17:41	18:05	05:42	06:06
ZL9	Auckland & Campbell Is., Auckland	18:51	19:24	06:28	07:00
	Campbell Is.	18:41	19:15	06:15	06:49
ZP	Paraguay, Asuncion	09:39	10:02	21:43	22:06
ZS	South Africa, Cape Town (ZS1)	04:37	05:02	16:34	16:59
	Port Elizabeth (ZS2)	04:08	04:33	16:05	16:30
	Bloemfontein (ZS4)	04:05	04:29	16:06	16:30
	Durban (ZS5)	03:46	04:10	15:46	16:10
	Johannesburg (ZS6)	03:56	04:19	16:00	16:23
ZS8	Marion Is.	03:24	03:55	15:06	15:37

Pref	State/Province/Country/City	Tlight	Sunrise	Sunset	Tlight
VE1	New Brunswick, Saint John	08:57	09:29	23:16	23:48
	Nova Scotia, Halifax	08:49	09:20	23:05	23:36
	Prince Edward Island, Charlottetown	08:43	09:15	23:06	23:38
VE2	Quebec, Montreal	09:27	09:58	23:47	00:18
	Quebec City	09:14	09:46	23:40	00:12
VE3	Ontario, London	10:04	10:34	00:13	00:43
	Ottawa	09:36	10:07	23:55	00:27
	Sudbury	09:54	10:26	00:18	00:51
	Toronto	09:54	10:25	00:07	00:37
VE4	Manitoba, Winnipeg	10:49	11:24	01:30	02:05
VE5	Saskatchewan, Regina	11:17	11:52	02:01	02:36
	Saskatoon	11:20	11:57	02:13	02:50
VE6	Alberta, Calgary	11:53	12:29	02:40	03:16
	Edmonton	11:42	12:21	02:44	03:22
VE7	British Columbia, Prince George	12:18	12:57	03:22	04:01
	Prince Rupert	12:46	13:26	03:53	04:33
	Vancouver	12:35	13:09	03:12	03:47
VE8	Northwest Territories, Yellowknife	10:59	11:53	03:19	04:13
	Resolute	——	08:37	03:59	——
VY1	Yukon, Whitehorse	12:34	13:24	04:34	05:24
VO1	Newfoundland, St. John's	07:58	08:31	22:27	23:00
VO2	Labrador, Goose Bay	08:10	08:49	23:11	23:49
W1	Connecticut, Hartford	09:32	10:01	23:37	00:06
	Maine, Bangor	09:09	09:41	23:26	23:57
	Portland	09:18	09:48	23:30	00:01
	Massachusetts, Boston	09:24	09:54	23:31	00:01
	New Hampshire, Concord	09:24	09:54	23:34	00:05
	Rhode Island, Providence	09:27	09:56	23:32	00:01
	Vermont, Montpelier	09:26	09:57	23:41	00:11
W2	New Jersey, Atlantic City	09:44	10:12	23:40	00:08
	New York, Albany	09:34	10:04	23:43	00:13
	Buffalo	09:54	10:24	00:03	00:33
	New York City	09:39	10:08	23:40	00:09
	Syracuse	09:43	10:13	23:53	00:23
W3	Delaware, Wilmington	09:47	10:16	23:45	00:13
	District of Columbia, Washington	09:55	10:23	23:50	00:18
	Maryland, Baltimore	09:53	10:21	23:49	00:17
	Pennsylvania, Harrisburg	09:52	10:20	23:51	00:20
	Philadelphia	09:46	10:14	23:44	00:12
	Pittsburgh	10:04	10:33	00:04	00:33
	Scranton	09:45	10:14	23:48	00:17
W4	Albama, Montgomery	10:43	11:09	00:18	00:44
	Florida, Jacksonville	10:28	10:53	23:57	00:22
	Miami	10:28	10:52	23:46	00:10
	Pensacola	10:50	11:15	00:19	00:44
	Georgia, Atlanta	10:34	11:00	00:12	00:38
	Savannah	10:23	10:48	23:57	00:22
	Kentucky, Lexington	10:27	10:54	00:18	00:46
	Louisville	10:31	10:59	00:24	00:51
	North Carolina, Charlotte	10:17	10:43	00:00	00:26
	Raleigh	10:07	10:34	23:52	00:18
	Wilmington	10:07	10:33	23:47	00:13
	South Carolina, Columbia	10:20	10:46	23:59	00:25
	Tennessee, Knoxville	10:28	10:55	00:13	00:40
	Memphis	10:54	11:21	00:37	01:03
	Nashville	10:39	11:06	00:25	00:52
	Virginia, Norfolk	09:56	10:23	23:44	00:11
	Richmond	09:59	10:27	23:49	00:17
W5	Arkansas, Little Rock	11:04	11:30	00:45	01:11
	Louisiana, New Orleans	11:02	11:27	00:30	00:55
	Shreveport	11:13	11:38	00:48	01:13
	Mississippi, Jackson	10:59	11:25	00:34	00:59
	New Mexico, Albuquerque	12:01	12:27	01:43	02:10
	Oklahoma, Oklahoma City	11:23	11:50	01:07	01:33
	Texas, Abilene	11:37	12:02	01:12	01:37
	Amarillo	11:41	12:07	01:24	01:50
	Dallas	11:25	11:50	01:01	01:26
	El Paso	12:05	12:30	01:38	02:03
	San Antonio	11:37	12:01	01:03	01:28
W6	California, Los Angeles	12:48	13:14	02:28	02:54
	San Francisco	12:59	13:26	02:50	03:17
W7	Arizona, Flagstaff	12:20	12:47	02:03	02:30
	Phoenix	12:25	12:51	02:03	02:29
	Idaho, Boise	12:22	12:52	02:34	03:04
	Pocatello	12:09	12:39	02:18	02:48
	Montana, Billings	11:46	12:18	02:07	02:39
	Butte	12:01	12:33	02:23	02:55
	Great Falls	11:53	12:25	02:21	02:54
	Nevada, Las Vegas	12:32	12:59	02:18	02:45
	Reno	12:45	13:13	02:42	03:10
	Oregon, Portland	12:43	13:15	03:03	03:35
	Utah, Salt Lake City	12:11	12:40	02:12	02:41
	Washington, Seattle	12:36	13:09	03:06	03:39
	Spokane	12:16	12:49	02:46	03:19
	Wyoming, Cheyenne	11:42	12:11	01:44	02:13
	Sheridan	11:42	12:13	01:59	02:30
W8	Michigan, Detroit	10:12	10:42	00:19	00:49
	Grand Rapids	10:21	10:51	00:31	01:01
	Sault Ste. Marie	10:08	10:40	00:32	01:04
	Traverse City	10:17	10:48	00:34	01:05
	Ohio, Cincinnati	10:25	10:53	00:20	00:48
	Cleveland	10:09	10:38	00:12	00:42
	Columbus	10:17	10:45	00:15	00:44
	West Virginia, Charleston	10:14	10:42	00:07	00:35
W9	Illinois, Chicago	10:31	11:01	00:37	01:06
	Indiana, Indianapolis	10:30	10:58	00:28	00:56
	Wisconsin, Green Bay	10:27	10:58	00:43	01:14

Pref	State/Province/Country/City	Tlight	Sunrise	Sunset	Tlight
	Milwaukee	10:30	11:00	00:40	01:10
W0	Colorado, Denver	11:45	12:14	01:43	02:11
	Grand Junction	12:01	12:29	01:56	02:24
	Iowa, Des Moines	10:56	11:25	01:00	01:29
	Kansas, Pratt	11:24	11:51	01:15	01:42
	Wichita	11:18	11:46	01:09	01:37
	Minnesota, Duluth	10:38	11:10	01:03	01:36
	Minneapolis	10:47	11:18	01:05	01:36
	Missouri, Columbia	10:56	11:24	00:51	01:19
	Kansas City	11:05	11:33	01:00	01:28
	St. Louis	10:48	11:16	00:42	01:10
	Nebraska, North Platte	11:26	11:55	01:28	01:57
	Omaha	11:06	11:35	01:09	01:38
	North Dakota, Fargo	10:56	11:29	01:22	01:55
	South Dakota, Rapid City	11:29	11:59	01:43	02:13
-	-	05:34	05:55	18:02	18:23
1A0	S.M.O.M	03:51	04:21	17:57	18:26
1S	Spratly Is.	21:58	22:20	10:42	11:03
3A	Monaco	04:07	04:38	18:20	18:50
3B6	Agalega	01:56	02:17	14:07	14:28
3B7	St. Brandon	01:48	02:10	13:49	14:11
3B8	Mauritius	02:00	02:22	13:54	14:17
3B9	Rodriguez Is.	01:36	01:58	13:31	13:53
3C	Equatorial Guinea, Bata	04:53	05:14	17:24	17:45
	Malabo	04:55	05:16	17:30	17:51
3C0	Annobon Is.	05:13	05:34	17:38	17:59
3D2	Fiji Is., Suva	17:55	18:17	05:53	06:15
3D2	Conway Reef	18:13	18:36	06:06	06:28
3D2	Rotuma Is.	17:53	18:14	06:00	06:22
3DA	Swaziland, Mbabane	03:50	04:14	15:34	15:57
3V	Tunisia, Tunis	04:10	04:37	17:58	18:25
3W, XV	Vietnam, Ho Chi Minh City (Saigon)	22:17	22:39	11:04	11:26
	Hanoi	22:10	22:33	11:17	11:40
3X	Guinea, Conakry	06:20	06:42	19:05	19:26
3Y	Bouvet	06:10	06:47	16:43	17:20
3Y	Peter I Is.	12:57	14:00	22:01	23:04
4J, 4K	Azerbaijan, Baku	01:24	01:53	15:24	15:53
4L	Georgia, Tbilisi	01:42	02:11	15:47	16:16
4S	Sri Lanka, Colombo	00:08	00:29	12:48	13:09
4U	ITU Geneva	04:06	04:38	18:29	19:01
4U	United Nations Hq.	09:39	10:08	23:40	00:09
4X, 4Z	Israel, Jerusalem	02:38	03:04	16:11	16:37
5A	Libya, Tripoli	04:08	04:34	17:43	18:09
	Benghazi	03:39	04:04	17:13	17:38
5B	Cyprus, Nicosia	02:40	03:06	16:23	16:49
5H	Tanzania, Dar es Salaam	03:01	03:23	15:18	15:39
5N	Nigeria, Lagos	05:14	05:36	17:54	18:15
5R	Madagascar, Antananarivo	02:39	03:01	14:35	14:58
5T	Mauritania, Nouakchott	06:21	06:43	19:22	19:44
5U	Niger, Niamey	05:13	05:35	18:05	18:27
5V	Togo, Lome	05:24	05:45	18:02	18:23
5W	Western Samoa, Apia	17:12	17:33	05:17	05:39
5X	Uganda, Kampala	03:23	03:45	15:52	16:13
5Z	Kenya, Nairobi	04:28	04:49	16:53	17:15
6W	Senegal, Dakar	06:30	06:52	19:24	19:46
6Y	Jamaica, Kingston	10:24	10:46	23:25	23:47
7O	Yemen, Aden	02:22	02:44	15:13	15:34
	Sanaa	02:23	02:45	15:18	15:40
7P	Lesotho, Maseru	04:07	04:31	15:45	16:09
7Q	Malawi, Lilongwe	03:30	03:52	15:35	15:56
	Blantyre	03:26	03:48	15:28	15:50
7X	Algeria, Algiers	04:39	05:06	18:26	18:54
8P	Barbados, Bridgetown	09:20	09:42	22:11	22:33
8Q	Maldive Is.	00:36	00:58	13:12	13:33
8R	Guyana, Georgetown	09:21	09:42	22:00	22:22
9A	Croatia, Zagreb	03:28	04:00	17:49	18:21
9G	Ghana, Accra	05:30	05:51	18:07	18:28
9H	Malta	03:55	04:21	17:40	18:07
9J	Zambia, Lusaka	03:53	04:15	15:55	16:17
9K	Kuwait	01:51	02:16	15:18	15:43
9L	Sierra Leone, Freetown	06:19	06:40	19:02	19:23
9M2	West Malaysia, Kuala Lumpur	22:45	23:06	11:18	11:39
9M6,8 10:35	East Malaysia, Sabah, Sandakan(9M6)	21:36	21:58	10:14	
	Sarawak, Kuching (9M8)	22:11	22:32	10:42	11:03
9N	Nepal, Kathmandu	23:24	23:48	12:46	13:10
9Q	Zaire, Kinshasa	04:36	04:57	16:57	17:18
	Kisangani	03:53	04:14	16:21	16:42
	Lubumbashi	03:53	04:15	16:02	16:23
9U	Burundi, Bujumbura	03:39	04:00	16:02	16:23
9V	Singapore	22:37	22:59	11:08	11:29
9X	Rwanda, Kigali	03:35	03:56	16:00	16:21
9Y	Trinidad & Tobago , Port of Spain	09:30	09:51	22:16	22:37
A2	Botswana, Gaborone	04:10	04:33	15:56	16:19
A3	Tonga, Nukualofa	17:31	17:54	05:24	05:47
A4	Oman, Musqat	01:16	01:39	14:29	14:52
A5	Bhutan, Thimbu	23:08	23:32	12:29	12:53
A6	United Arab Emirates, Abu Dhabi	01:32	01:56	14:47	15:11
A7	Qatar, Ad-Dawhah	01:42	02:06	14:59	15:22
A9	Bahrein, Al-Manamah	01:45	02:08	15:03	15:27
AP	Pakistan, Karachi	00:40	01:04	13:56	14:20
AP	Islamabad	00:03	00:29	13:42	14:08
BV	Taiwan, Taipei	21:02	21:26	10:19	10:42
BY	Peoples Rep. of China, Beijing	20:59	21:28	10:58	11:26
	Harbin	20:05	20:37	10:26	10:58
	Shanghai	20:54	21:19	10:25	10:51
	Fuzhou	21:10	21:34	10:28	10:52

Pref	State/Province/Country/City	Tlight	Sunrise	Sunset	Tlight
	Xian	21:40	22:06	11:20	11:46
	Chongqing	21:56	22:21	11:24	11:49
	Chengdu	22:04	22:29	11:34	11:59
	Lhasa	22:57	23:22	12:25	12:50
	Urumqi	22:46	23:17	12:59	13:30
	Kashi	23:42	00:10	13:39	14:07
C2	Nauru	18:27	18:48	06:54	07:15
C3	Andorra	04:34	05:03	18:41	19:11
C5	The Gambia, Banjul	06:28	06:50	19:20	19:42
C6	Bahamas, Nassau	10:18	10:42	23:34	23:58
C9	Mozambique, Maputo	03:44	04:08	15:28	15:52
	Mozambique	03:03	03:25	15:06	15:28
CE	Chile, Santiago	10:44	11:09	22:14	22:39
CE0	Easter Island	13:13	13:37	00:55	01:19
CE0	Juan Fernandez	11:16	11:41	22:46	23:11
CE0	San Felix	11:15	11:39	22:59	23:22
CM,CO	Cuba, Havana	10:40	11:04	23:52	00:15
CN	Morocco, Casablanca	05:26	05:52	19:04	19:30
CP	Bolivia, La Paz	10:21	10:43	22:21	22:43
CT	Portugal, Lisbon	05:24	05:52	19:18	19:46
CT3	Madeira Islands, Funchal	06:05	06:31	19:41	20:06
CU	Azores , Ponta Delgada	06:32	06:59	20:23	20:50
CX	Uruguay, Montevideo	09:47	10:12	21:14	21:39
CY0	Sable Is.	08:37	09:07	22:49	23:20
CY9	St. Paul Is.	08:29	09:01	22:56	23:29
D2	Angola, Luanda	04:48	05:09	17:01	17:23
D4	Cape Verde, Praia	06:54	07:16	19:49	20:11
D6	Comoros, Moroni	02:48	03:10	14:57	15:18
DA-DL,Y2-Y9	Fed. Rep. of Germany, Bonn	03:50	04:25	18:35	19:11
	Berlin	03:18	03:56	18:14	18:51
DU	Philippines, Manilla	21:16	21:38	10:10	10:32
E3	Eritrea, Asmara	02:44	03:06	15:39	16:01
EA	Spain, Madrid	04:59	05:27	18:59	19:27
EA6	Balearic Is., Palma	04:35	05:04	18:32	19:00
EA8	Canary Is., Las Palmas	06:01	06:26	19:25	19:50
EA9	Ceuta & Melilla, Ceuta	05:14	05:40	18:59	19:25
	Melilla	05:05	05:32	18:49	19:15
EI	Ireland, Dublin	04:34	05:12	19:35	20:13
EK	Armenia, Yerevan	01:46	02:15	15:46	16:14
EL	Liberia, Monrovia	06:11	06:33	18:50	19:11
EP	Iran, Tehran	01:25	01:52	15:10	15:37
ER	Moldova, Kishinev	02:33	03:06	17:00	17:33
ES	Estonia, Tallinn	02:02	02:49	17:49	18:36
ET	Ethiopia, Addis Ababa	02:51	03:12	15:35	15:56
EU,EV,EW	Belarus, Minsk	02:16	02:55	17:21	18:00
EX	Kyrgyzstan, Frunze	23:40	00:10	13:49	14:19
EY	Tajikistan, Samarkand	00:18	00:47	14:16	14:44
	Dushanbe	00:11	00:39	14:07	14:35
EZ	Turkmenistan, Ashkhabad	00:55	01:23	14:47	15:14
F	France, Paris	04:14	04:48	18:50	19:24
FG	Guadeloupe	09:26	09:48	22:22	22:44
FJ, FS	St. Martin	09:29	09:51	22:30	22:52
FH	Mayotte	02:43	03:05	14:49	15:11
FK	New Caledonia, Noumia	18:45	19:08	06:36	06:59
FM	Martinique	09:24	09:46	22:18	22:40
FO	Clipperton Is.	12:41	13:03	01:27	01:49
FO	Fr. Polynesia, Tahiti	16:26	16:48	04:25	04:47
	Rurutu, (Austral Is.)	19:46	20:09	07:37	08:00
	Hiva Oa, (Marquesas Is.)	14:58	15:19	03:09	03:31
FP	St. Pierre & Miquelon, St. Pierre	08:13	08:46	22:39	23:11
FR/G	Glorioso	02:34	02:55	14:43	15:04
FR/J,E	Juan de Nova	02:56	03:18	14:56	15:18
	Europa	03:10	03:33	15:01	15:23
FR	Reunion	02:08	02:31	14:01	14:24
FR/T	Tromelin	02:09	02:31	14:11	14:32
FT8W	Crozet	02:45	03:15	13:45	14:16
FT8X	Kerguelen	01:40	02:12	12:31	13:03
FT8Y	Antarctica, Dumont D'Urville	21:28	22:24	06:52	07:49
FT8Z	Amsterdam & St. Paul Is., Amsterdam	00:54	01:21	12:15	12:42
FW	Wallis & Futuna Is., Wallis	17:30	17:51	05:36	05:57
FY	Fr. Guiana, Cayenne	08:59	09:20	21:35	21:56
G	England, London	04:16	04:52	19:05	19:42
GD	Isle of Man	04:23	05:02	19:30	20:10
GI	Northern Ireland, Belfast	04:27	05:07	19:37	20:16
GJ	Jersey	04:31	05:05	19:09	19:43
GM	Scotland, Glasgow	04:16	04:57	19:34	20:15
	Aberdeen	04:01	04:44	19:29	20:13
GU	Guernsey	04:32	05:07	19:11	19:46
GW	Wales, Cardiff	04:28	05:04	19:18	19:54
H4	Solomon Islands, Honiara	19:01	19:23	07:14	07:35
HA	Hungary, Budapest	03:11	03:44	17:40	18:13
HB	Switzerland, Bern	03:59	04:31	18:25	18:58
HB0	Liechtenstein	03:50	04:22	18:17	18:50
HC	Ecuador, Quito	10:46	11:07	23:14	23:35
HC8	Galapagos Is.	11:36	11:57	00:03	00:24
HH	Haiti, Port-Au-Prince	10:05	10:28	23:07	23:30
HI	Dominican Republic, Santo Domingo	09:56	10:19	22:58	23:20
HK	Colombia, Bogota	10:26	10:47	23:02	23:23
HK0	Malpelo Is.	10:55	11:16	23:29	23:51
HK0	San Adreas	10:49	11:11	23:39	00:01
HL	Korea, Seoul	20:22	20:49	10:12	10:39
HP	Panama, Panama	10:44	11:05	23:27	23:49
HR	Honduras, Tegucigalpa	11:10	11:31	00:03	00:25
HS	Thailand, Bangkok	22:39	23:01	11:32	11:53
HV	Vatican City	03:51	04:20	17:56	18:26
HZ, 7Z	Saudi Arabia, Dharan	01:47	02:11	15:06	15:30
	Mecca	02:34	02:56	15:42	16:05

Pref	State/Province/Country/City	Tlight	Sunrise	Sunset	Tlight
	Riyadh	02:02	02:26	15:17	15:41
I	Italy, Rome	03:51	04:20	17:56	18:26
	Trieste	03:37	04:09	17:58	18:29
	Sicily	03:54	04:21	17:44	18:11
IS	Sardinia, Cagliari	04:10	04:38	18:06	18:34
J2	Djibouti, Djibouti	02:31	02:52	15:19	15:40
J3	Grenada	09:30	09:52	22:19	22:41
J5	Guinea-Bissau, Bissau	06:25	06:47	19:14	19:36
J6	St. Lucia	09:25	09:47	22:18	22:40
J7	Dominica	09:25	09:47	22:20	22:42
J8	St. Vincent	09:27	09:49	22:18	22:40
JA-JS	Japan, Tokyo	19:34	20:00	09:18	09:45
	Nagasaki	20:18	20:44	09:54	10:19
	Sapporo	19:13	19:43	09:23	09:53
JD1	Minami Torishima	18:53	19:17	08:08	08:31
JD1	Ogasawara, Kazan Is.	19:41	20:05	09:03	09:27
JT	Mongolia, Ulan Bator	21:19	21:52	11:49	12:23
JW	Svalbard, Spitsbergen	—	—	—	—
JX	Jan Mayen	01:54	03:48	21:15	23:09
JY	Jordan, Amman	02:35	03:01	16:09	16:34
KC4	Antarctica, Byrd Station	18:16	—	—	23:39
	McMurdo Sound	20:41	23:03	02:40	05:02
	Palmer Station	10:59	11:51	20:37	21:30
KC6	Belau, Yap	20:13	20:34	08:57	09:18
	Koror	20:29	20:51	09:10	09:31
KG4	Guantanamo Bay	10:15	10:38	23:20	23:43
KH0	Mariana Is., Saipan	19:36	19:58	08:32	08:54
KH1	Baker, Howland Is.	17:17	17:38	05:46	06:07
KH2	Guam, Agana	19:42	20:04	08:34	08:56
KH3	Johnston Is.	16:32	16:54	05:31	05:53
KH4	Midway Is.	16:54	17:18	06:18	06:42
KH5	Palmyra Is.	16:17	16:38	04:55	05:16
KH5K	Kingman Reef	16:18	16:40	04:59	05:21
KH6	Hawaii, Hilo	15:35	15:58	04:39	05:02
	Honolulu	15:45	16:07	04:52	05:15
KH7	Kure Is.	16:58	17:22	06:22	06:46
KH8	American Samoa, Pago Pago	17:08	17:30	05:13	05:35
KH9	Wake Is.	18:09	18:31	07:12	07:35
KL7	Alaska, Adak	16:01	16:37	06:52	07:29
	Anchorage	13:30	14:21	05:35	06:26
	Fairbanks	12:54	13:55	05:45	06:46
	Juneau	12:45	13:40	04:22	05:07
	Nome	14:06	15:07	06:53	07:53
KP1	Navassa Is.	10:16	10:39	23:18	23:40
KP2	Virgin Islands, Charlotte Amalie	09:36	09:58	22:37	23:00
KP4	Puerto Rico, San Juan	09:41	10:03	22:43	23:05
KP5	Desecheo Is.	09:46	10:09	22:48	23:10
LA-LJ	Norway, Oslo	02:55	03:43	18:48	19:36
LU	Argentina, Buenos Aires	09:55	10:21	21:23	21:49
LX	Luxembourg	03:56	04:31	18:36	19:10
LY	Lithuania, Vilna	02:22	03:02	17:31	18:10
LZ	Bulgaria, Sofia	03:06	03:36	17:14	17:44
OA	Peru, Lima	10:52	11:13	23:00	23:21
OD	Lebanon, Beirut	02:34	03:00	16:13	16:39
OE	Austria, Vienna	03:20	03:54	17:52	18:26
OH	Finland, Helsinki	01:56	02:45	17:51	18:40
OH0	Aland Is.	02:16	03:05	18:11	19:00
OJ0	Market Reef	02:20	03:09	18:16	19:05
OK,OL	Czech Rep., Prague	03:22	03:57	18:04	18:39
OM	Slovak, Rep., Bratislava	03:18	03:51	17:49	18:23
ON	Belgium, Brussels	04:00	04:35	18:46	19:22
OX	Greenland, Godthaab	06:34	07:34	23:16	00:16
	Thule	—	05:51	03:16	—
OY	Faroe Islands, Torshavn	03:52	04:45	20:06	20:59
OZ	Denmark, Copenhagen	03:09	03:50	18:26	19:07
P2	Papua New Guinea, Madang	19:56	20:17	08:15	08:36
	Port Moresby	19:53	20:14	08:05	08:27
P4	Aruba, Oranjestad	09:59	10:20	22:48	23:10
PA-PI	Netherlands, Amsterdam	03:53	04:30	18:48	19:25
PJ2,4,9	Netherlands Antilles, Willemstad	10:03	10:24	22:53	23:15
PJ5-8	St. Maarten and Saba, St. Maarten	09:30	09:52	22:30	22:52
PY	Brazil, Brasilia	08:58	09:20	21:00	21:22
	Rio De Janeiro	08:45	09:08	20:34	20:57
	Natal	07:59	08:21	20:17	20:39
	Manaus	09:37	09:58	22:00	22:21
	Porto Alegre	09:23	09:47	20:59	21:24
PY0	Fernando De Noronha	07:47	08:08	20:08	20:29
PY0	St. Peter & St. Paul Rocks	07:31	07:52	20:00	20:21
PY0	Trindade & Martin Vaz Is., Trindade	07:47	08:10	19:41	20:04
PZ	Suriname, Paramaribo	09:10	09:31	21:47	22:09
R1FJ_	Franz Josef Land	—	—	—	—
R1MV_	Malyj Vysotskij Is.	01:40	02:29	17:39	18:28
S0	Western Sahara, Smara	05:52	06:16	19:12	19:35
S2	Bangladesh, Dacca	23:09	23:32	12:21	12:45
S5	Slovenia, Ljubljana	03:33	04:05	17:55	18:27
S7	Seychelles, Victoria	01:56	02:17	14:16	14:37
S9	Sao Tome	05:07	05:28	17:35	17:56
SM	Sweden, Stockholm	02:29	03:16	18:16	19:03
SP	Poland, Krakow	03:00	03:35	17:42	18:17
	Warsaw	02:49	03:26	17:43	18:20
ST	Sudan, Khartoum	03:09	03:31	16:05	16:27
ST0	Southern Sudan, Juba	03:23	03:44	15:59	16:21
SU	Egypt, Cairo	02:56	03:21	16:24	16:49
SV	Greece, Athens	03:14	03:41	17:06	17:33
SV/A	Mount Athos	03:07	03:36	17:06	17:35
SV5	Dodecanese, Rhodes	02:59	03:26	16:45	17:12
SV9	Crete	03:12	03:39	16:56	17:22

Pref	State/Province/Country/City	Tlight	Sunrise	Sunset	Tlight
T2	Tuvalu, Funafuti	17:46	18:08	06:00	06:21
T30	West Kiribati, Bonriki	18:00	18:21	06:30	06:51
T31	Central Kiribati, Kanton	17:03	17:24	05:26	05:47
T32	East Kiribati, Christmas Is.	16:02	16:23	04:33	04:54
T33	Banaba Is.	18:17	18:38	06:44	07:05
T5	Somalia, Mogadishu	02:30	02:52	15:02	15:23
T7	San Marino	03:47	04:18	18:00	18:31
T9	Bosnia-Herzegovina, Sarajevo	03:23	03:53	17:36	18:07
TA	Turkey, Ankara	02:33	03:02	16:32	17:00
	Istanbul	02:47	03:16	16:49	17:18
TF	Iceland, Reykjavik	04:36	05:35	21:17	22:16
TG	Guatemala, Guatemala City	11:22	11:44	00:16	00:38
TI	Costa Rica, San Jose	11:01	11:22	23:46	00:08
TI9	Cocos Is.	11:17	11:38	23:54	00:16
TJ	Cameroon, Yaounde	04:44	05:06	17:19	17:40
TK	Corsica	04:05	04:34	18:10	18:40
TL	Central African Republic, Bangui	04:16	04:37	16:51	17:12
TN	Congo, Brazzaville	04:36	04:57	16:57	17:18
TR	Gabon, Libreville	04:55	05:17	17:24	17:45
TT	Chad, N'Djamena	04:23	04:44	17:12	17:34
TU	Cote d'Ivoire, Abidjan	05:45	06:06	18:22	18:43
TY	Benin, Porto Novo	05:18	05:39	17:57	18:18
TZ	Mali, Bamako	05:54	06:16	18:45	19:06
UA	Russia, European, Leningrad (UA1)	01:37	02:25	17:29	18:17
	Archangel (UA1)	00:22	01:23	17:10	18:11
	Murmansk (UA1)	23:56	01:22	18:09	19:35
	Moscow (UA3)	01:28	02:10	16:46	17:27
	Kuybyshev (UA4)	00:49	01:27	15:49	16:27
	Rostov (UA6)	01:49	02:22	16:18	16:51
UA2	Kaliningrad	02:40	03:20	17:52	18:32
UA9,0	Russia, Asiatic, Novosibirsk (UA9)	22:31	23:11	13:42	14:23
	Perm (UA9)	00:03	00:48	15:38	16:23
	Omsk (UA9)	23:09	23:49	14:20	15:01
	Norilsk (UA0)	20:10	21:40	14:32	16:01
	Irkutsk (UA0)	21:15	21:52	12:10	12:47
	Vladivostok (UA0)	19:50	20:21	10:01	10:31
	Petropavlovsk (UA0)	17:35	18:13	08:34	09:12
	Khabarovsk (UA0)	19:24	19:58	09:58	10:32
	Krasnoyarsk (UA0)	21:47	22:28	13:06	13:47
	Yakutsk (UA0)	18:46	19:39	11:00	11:53
	Wrangel Island (UA0)	13:19	15:13	08:40	10:34
	Kyzyl (UA0Y)	21:57	22:33	12:47	13:24
UJ-UM	Uzbekistan, Bukhoro	00:28	00:56	14:25	14:54
	Tashkent	00:05	00:34	14:08	14:37
UN-UQ	Kazakhstan, Alma-Ata	23:30	00:00	13:41	14:11
UR-UZ,EM-EO	Ukraine, Kiev	02:17	02:52	17:00	17:36
V2	Antigua & Barbuda, St. Johns	09:25	09:47	22:24	22:46
V3	Belize, Belmopan	11:13	11:35	00:12	00:34
V4	St. Kitts & Nevis	09:28	09:50	22:27	22:49
V5	Namibia, Windhoek	04:43	05:06	16:34	16:56
V6	Micronesia, Ponape	18:55	19:16	07:34	07:56
V7	Marshall Islands, Kwajalein	18:16	18:38	07:00	07:22
V8	Brunei, Bandar Seri Begawan	21:50	22:11	10:26	10:47
VK	Australia, Canberra (VK1)	20:06	20:32	07:32	07:58
	Sydney (VK2)	19:56	20:22	07:25	07:51
	Melbourne (VK3)	20:25	20:51	07:45	08:12
	Brisbane (VK4)	19:44	20:07	07:25	07:49
	Adelaide (VK5)	20:47	21:13	08:14	08:40
	Perth (VK6)	22:16	22:41	09:49	10:14
	Hobart, Tasmania (VK7)	20:20	20:49	07:29	07:58
	Darwin (VK8)	21:00	21:22	09:08	09:29
VK0	Heard Is.	01:28	02:03	12:06	12:42
VK0	Macquarie Is.	19:49	20:26	06:20	06:58
VK9C	Cocos-Keeling Is.	23:16	23:38	11:24	11:46
VK9L	Lord Howe Is.	19:23	19:47	06:56	07:21
VK9M	Mellish Reef	19:25	19:47	07:23	07:46
VK9N	Norfolk Is.	18:45	19:09	06:24	06:48
VK9W	Willis Is.	19:49	20:11	07:50	08:12
VK9X	Christmas Is.	22:39	23:01	10:50	11:11
VP2E	Anguilla	09:28	09:51	22:30	22:52
VP2M	Montserrat	09:27	09:49	22:25	22:47

Pref	State/Province/Country/City	Tlight	Sunrise	Sunset	Tlight
VP2V	British Virgin Is., Tortola	09:35	09:57	22:36	22:59
VP5	Turks & Caicos Islands, Grand Turk	09:58	10:21	23:06	23:28
VP8	Falkland Islands, Stanley	10:11	10:46	20:54	21:29
VP8	So. Georgia Is.	08:50	09:27	19:24	20:01
VP8	So. Orkney Is.	09:36	10:20	19:40	20:25
VP8	So. Sandwich Islands, Saunders Is.	08:15	08:56	18:34	19:15
VP8	So. Shetland Is., King George Is.	10:30	11:16	20:27	21:13
VP9	Bermuda	09:17	09:43	22:52	23:17
VQ9	Chagos, Diego Garcia	00:50	01:11	13:06	13:27
VR2,VS6	Hong Kong	21:35	21:58	10:44	11:08
VR6	Pitcairn Is.	14:34	14:57	02:20	02:43
VU	India, Bombay	00:24	00:47	13:27	13:50
	Calcutta	23:18	23:41	12:28	12:51
	New Delhi	23:55	00:19	13:20	13:44
	Bangalore	00:12	00:33	13:02	13:24
VU	Andaman Islands, Port Blair	23:12	23:34	12:00	12:22
VU	Laccadive Is.	00:33	00:54	13:18	13:40
XE	Mexico, Mexico City (XE1)	11:52	12:14	00:55	01:18
	Chihuahua (XE2)	12:08	12:32	01:33	01:57
	Merida (XE3)	11:12	11:35	00:19	00:42
XF4	Revilla Gigedo	12:42	13:04	01:44	02:07
XT	Burkina Faso, Ouagadougou	05:29	05:51	18:19	18:40
XU	Cambodia, Phnom Penh	22:24	22:46	11:12	11:34
XW	Laos, Viangchan	22:26	22:49	11:27	11:49
XX9	Macao	21:38	22:01	10:47	11:10
XZ	Myanmar, Yangon	22:54	23:16	11:52	12:15
YA	Afghanistan, Kandahar	00:37	01:02	14:08	14:33
	Kabul	00:18	00:44	13:59	14:25
YB-YD	Indonesia, Jakarta	22:32	22:53	10:49	11:11
	Medan	22:56	23:17	11:30	11:51
	Pontianak	22:17	22:38	10:44	11:06
	Jayapura	20:13	20:34	08:37	08:58
YI	Iraq, Baghdad	01:59	02:25	15:36	16:01
YJ	Vanuatu, Port Vila	18:35	18:57	06:33	06:55
YK	Syria, Damascus	02:31	02:57	16:09	16:35
YL	Latvia, Riga	02:17	03:00	17:44	18:27
YN	Nicaragua, Managua	11:07	11:29	23:56	00:18
YO	Romania, Bucharest	02:51	03:22	17:06	17:37
YS	El Salvador, San Salvador	11:18	11:40	00:10	00:32
YU	Yugoslavia, Belgrade	03:12	03:43	17:29	18:01
YV	Venezuela, Caracas	09:52	10:14	22:39	23:00
YV0	Aves Is.	09:34	09:56	22:30	22:52
Z2	Zimbabwe, Harare	03:44	04:06	15:42	16:05
Z3	Macedonia, (ex Yugoslav), Skopje	03:15	03:45	17:21	17:50
ZA	Albania, Tirane	03:23	03:52	17:26	17:55
ZB2	Gibraltar	05:14	05:40	18:59	19:26
ZC4	British Cyprus	02:43	03:09	16:24	16:50
ZD7	St. Helena	06:10	06:32	18:12	18:34
ZD8	Ascension Is.	06:38	06:59	18:53	19:14
ZD9	Tristan da Cunha	06:53	07:20	18:15	18:42
ZF	Cayman Is.	10:40	11:02	23:44	00:06
ZK1	No. Cook Is., Manihiki	16:26	16:48	04:37	04:58
ZK1	So. Cook Is., Rarotonga	16:30	16:52	04:23	04:45
ZK2	Niue	17:05	17:27	05:01	05:23
ZK3	Tokelaus, Atafu	17:11	17:33	05:25	05:47
ZL	New Zealand, Auckland (ZL1)	18:24	18:51	05:47	06:14
	Wellington (ZL2)	18:29	18:57	05:41	06:10
	Christchurch (ZL3)	18:40	19:09	05:47	06:16
	Dunedin (ZL4)	18:51	19:21	05:51	06:22
ZL5	Antarctica, Scott Base	20:44	23:12	02:33	05:01
ZL7	Chatham Is.	17:57	18:26	05:02	05:32
ZL8	Kermadec Is.	17:49	18:13	05:27	05:51
ZL9	Auckland & Campbell Is., Auckland	19:12	19:46	05:59	06:32
	Campbell Is.	19:04	19:39	05:44	06:20
ZP	Paraguay, Asuncion	09:45	10:08	21:30	21:53
ZS	South Africa, Cape Town (ZS1)	04:47	05:13	16:16	16:42
	Port Elizabeth (ZS2)	04:18	04:44	15:47	16:13
	Bloemfontein (ZS4)	04:13	04:37	15:51	16:15
	Durban (ZS5)	03:54	04:18	15:31	15:55
	Johannesburg (ZS6)	04:02	04:26	15:46	16:10
ZS8	Marion Is.	03:42	04:14	14:41	15:12

Pref	State/Province/Country/City	Tlight	Sunrise	Sunset	Tlight
VE1	New Brunswick, Saint John	08:31	09:04	23:37	00:10
	Nova Scotia, Halifax	08:23	08:56	23:25	23:58
	Prince Edward Island, Charlottetown	08:16	08:50	23:27	00:01
VE2	Quebec, Montreal	09:00	09:34	00:07	00:41
	Quebec City	08:46	09:20	00:01	00:36
VE3	Ontario, London	09:39	10:11	00:32	01:03
	Ottawa	09:09	09:42	00:15	00:49
	Sudbury	09:26	10:01	00:40	01:14
	Toronto	09:29	10:02	00:26	00:58
VE4	Manitoba, Winnipeg	10:17	10:55	01:54	02:32
VE5	Saskatchewan, Regina	10:44	11:23	02:26	03:05
	Saskatoon	10:45	11:25	02:40	03:21
VE6	Alberta, Calgary	11:20	11:59	03:06	03:45
	Edmonton	11:05	11:47	03:13	03:55
VE7	British Columbia, Prince George	11:40	12:23	03:52	04:34
	Prince Rupert	12:08	12:51	04:23	05:07
	Vancouver	12:04	12:41	03:36	04:13
VE8	Northwest Territories, Yellowknife	09:55	11:03	04:05	05:12
	Resolute	----	----	----	----
VY1	Yukon, Whitehorse	11:38	12:38	05:15	06:15
VO1	Newfoundland, St. John's	07:29	08:04	22:49	23:25
VO2	Labrador, Goose Bay	07:34	08:16	23:40	00:22
W1	Connecticut, Hartford	09:08	09:39	23:54	00:25
	Maine, Bangor	08:43	09:16	23:46	00:19
	Portland	08:53	09:25	23:49	00:21
	Massachusetts, Boston	09:00	09:32	23:49	00:21
	New Hampshire, Concord	08:59	09:31	23:53	00:25
	Rhode Island, Providence	09:03	09:34	23:49	00:20
	Vermont, Montpelier	09:00	09:33	00:00	00:33
W2	New Jersey, Atlantic City	09:22	09:52	23:56	00:25
	New York, Albany	09:10	09:42	00:01	00:32
	Buffalo	09:30	10:02	00:22	00:53
	New York City	09:17	09:47	23:57	00:28
	Syracuse	09:19	09:50	00:11	00:43
W3	Delaware, Wilmington	09:26	09:55	00:01	00:30
	District of Columbia, Washington	09:34	10:03	00:05	00:34
	Maryland, Baltimore	09:31	10:01	00:04	00:34
	Pennsylvania, Harrisburg	09:30	10:00	00:08	00:38
	Philadelphia	09:24	09:54	00:00	00:30
	Pittsburgh	09:42	10:12	00:20	00:50
	Scranton	09:22	09:52	00:05	00:36
W4	Alabama, Montgomery	10:26	10:53	00:30	00:56
	Florida, Jacksonville	10:12	10:38	00:08	00:33
	Miami	10:15	10:39	23:54	00:19
	Pensacola	10:34	11:00	00:30	00:56
	Georgia, Atlanta	10:16	10:43	00:25	00:52
	Savannah	10:06	10:33	00:08	00:35
	Kentucky, Lexington	10:06	10:35	00:33	01:02
	Louisville	10:11	10:40	00:39	01:08
	North Carolina, Charlotte	09:58	10:26	00:13	00:40
	Raleigh	09:48	10:16	00:05	00:33
	Wilmington	09:49	10:16	23:59	00:26
	South Carolina, Columbia	10:02	10:29	00:11	00:38
	Tennessee, Knoxville	10:09	10:37	00:27	00:55
	Memphis	10:36	11:03	00:50	01:17
	Nashville	10:20	10:48	00:39	01:07
	Virginia, Norfolk	09:36	10:04	23:58	00:26
	Richmond	09:39	10:08	00:04	00:32
W5	Arkansas, Little Rock	10:45	11:13	00:58	01:25
	Louisiana, New Orleans	10:47	11:12	00:40	01:06
	Shreveport	10:56	11:22	00:59	01:26
	Mississippi, Jackson	10:42	11:09	00:45	01:11
	New Mexico, Albuquerque	11:42	12:10	01:56	02:24
	Oklahoma, Oklahoma City	11:04	11:32	01:20	01:48
	Texas, Abilene	11:20	11:46	01:23	01:50
	Amarillo	11:22	11:50	01:37	02:04
	Dallas	11:08	11:34	01:12	01:39
	El Paso	11:48	12:15	01:49	02:16
	San Antonio	11:21	11:47	01:13	01:39
W6	California, Los Angeles	12:30	12:57	02:40	03:07
	San Francisco	12:38	13:07	03:04	03:33
W7	Arizona, Flagstaff	12:02	12:29	02:16	02:44
	Phoenix	12:07	12:34	02:15	02:42
	Idaho, Boise	11:57	12:29	02:53	03:25
	Pocatello	11:44	12:16	02:36	03:08
	Montana, Billings	11:19	11:52	02:28	03:01
	Butte	11:34	12:08	02:44	03:18
	Great Falls	11:24	11:59	02:44	03:19
	Nevada, Las Vegas	12:13	12:41	02:32	03:00
	Reno	12:23	12:53	02:57	03:27
	Oregon, Portland	12:17	12:50	03:24	03:57
	Utah, Salt Lake City	11:48	12:29	02:29	02:59
	Washington, Seattle	12:07	12:43	03:28	04:03
	Spokane	11:47	12:23	03:09	03:44
	Wyoming, Cheyenne	11:19	11:49	02:01	02:32
	Sheridan	11:16	11:49	02:19	02:52
W8	Michigan, Detroit	09:48	10:19	00:37	01:08
	Grand Rapids	09:57	10:29	00:49	01:21
	Sault Ste. Marie	09:40	10:14	00:53	01:27
	Traverse City	09:51	10:24	00:53	01:26
	Ohio, Cincinnati	10:03	10:33	00:35	01:05
	Cleveland	09:45	10:16	00:30	01:00
	Columbus	09:55	10:25	00:31	01:01
	West Virginia, Charleston	09:53	10:23	00:22	00:51
W9	Illinois, Chicago	10:08	10:39	00:54	01:25
	Indiana, Indianapolis	10:08	10:38	00:44	01:14
	Wisconsin, Green Bay	10:01	10:34	01:02	01:35
	Milwaukee	10:06	10:37	00:58	01:30
W0	Colorado, Denver	11:24	11:53	01:59	02:28
	Grand Junction	11:40	12:09	02:12	02:41
	Iowa, Des Moines	10:33	11:03	01:17	01:48
	Kansas, Pratt	11:04	11:32	01:29	01:58
	Wichita	10:58	11:27	01:24	01:52
	Minnesota, Duluth	10:10	10:44	01:25	01:59
	Minneapolis	10:21	10:54	01:25	01:58
	Missouri, Columbia	10:35	11:04	01:06	01:36
	Kansas City	10:44	11:13	01:16	01:45
	St. Louis	10:27	10:57	00:57 .	01:26
	Nebraska, North Platte	11:03	11:33	01:45	02:16
	Omaha	10:43	11:13	01:26	01:57
	North Dakota, Fargo	10:28	11:03	01:44	02:19
	South Dakota, Rapid City	11:03	11:36	02:02	02:34
-		05:31	05:53	18:00	18:21
1A0	S.M.O.M	03:28	03:59	18:14	18:45
1S	Spratly Is.	21:52	22:14	10:43	11:05
3A	Monaco	03:42	04:14	18:38	19:11
3B6	Agalega	01:57	02:19	14:00	14:22
3B7	St. Brandon	01:51	02:13	13:40	14:03
3B8	Mauritius	02:05	02:28	13:44	14:07
3B9	Rodriguez Is.	01:41	02:04	13:21	13:44
3C	Equatorial Guinea, Bata	04:50	05:11	17:22	17:44
	Malabo	04:51	05:13	17:29	17:51
3C0	Annobon Is.	05:10	05:32	17:35	17:57
3D2	Fiji Is., Suva	17:59	18:21	05:43	06:06
3D2	Conway Reef	18:19	18:42	05:55	06:18
3D2	Rotuma Is.	17:55	18:17	05:54	06:16
3DA	Swaziland, Mbabane	03:58	04:22	15:21	15:45
3V	Tunisia, Tunis	03:50	04:19	18:12	18:40
3W, XV	Vietnam, Ho Chi Minh City (Saigon)	22:10	22:32	11:06	11:28
	Hanoi	21:59	22:22	11:23	11:47
3X	Guinea, Conakry	06:14	06:36	19:06	19:28
3Y	Bouvet	06:37	07:16	16:09	16:48
3Y	Peter I Is.	13:52	15:08	20:49	22:05
4J, 4K	Azerbaijan, Baku	01:02	01:32	15:41	16:11
4L	Georgia, Tbilisi	01:19	01:50	16:04	16:35
4S	Sri Lanka, Colombo	00:03	00:24	12:48	13:10
4U	ITU Geneva	03:39	04:13	18:50	19:24
4U	United Nations Hq.	09:17	09:47	23:57	00:28
4X, 4Z	Israel, Jerusalem	02:22	02:48	16:23	16:49
5A	Libya, Tripoli	03:51	04:18	17:55	18:21
	Benghazi	03:22	03:48	17:24	17:50
5B	Cyprus, Nicosia	02:21	02:49	16:36	17:04
5H	Tanzania, Dar es Salaam	03:01	03:23	15:13	15:35
5N	Nigeria, Lagos	05:09	05:31	17:54	18:16
5R	Madagascar, Antananarivo	02:43	03:06	14:26	14:49
5T	Mauritania, Nouakchott	06:11	06:34	19:26	19:49
5U	Niger, Niamey	05:05	05:28	18:08	18:31
5V	Togo, Lome	05:19	05:41	18:02	18:24
5W	Western Samoa, Apia	17:14	17:36	05:10	05:32
5X	Uganda, Kampala	03:21	03:42	15:50	16:11
5Z	Kenya, Nairobi	04:25	04:47	16:51	17:12
6W	Senegal, Dakar	06:22	06:44	19:28	19:50
6Y	Jamaica, Kingston	10:14	10:37	23:29	23:52
7O	Yemen, Aden	02:14	02:37	15:15	15:38
	Sanaa	02:14	02:37	15:22	15:45
7P	Lesotho, Maseru	04:16	04:41	15:31	15:56
7Q	Malawi, Lilongwe	03:32	03:54	15:27	15:49
	Blantyre	03:30	03:52	15:20	15:42
7X	Algeria, Algiers	04:19	04:48	18:40	19:09
8P	Barbados, Bridgetown	09:12	09:35	22:14	22:36
8Q	Maldive Is.	00:32	00:54	13:11	13:33
8R	Guyana, Georgetown	09:15	09:37	22:01	22:22
9A	Croatia, Zagreb	03:01	03:34	18:10	18:43
9G	Ghana, Accra	05:25	05:47	18:07	18:29
9H	Malta	03:35	04:03	17:53	18:21
9J	Zambia, Lusaka	03:56	04:18	15:47	16:10
9K	Kuwait	01:36	02:01	15:28	15:54
9L	Sierra Leone, Freetown	06:13	06:35	19:03	19:25
9M2	West Malaysia, Kuala Lumpur	22:41	23:02	11:17	11:39
9M6,8	East Malaysia, Sabah, Sandakan(9M6)	21:31	21:53	10:14	10:36
	Sarawak, Kuching (9M8)	22:08	22:29	10:40	11:02
9N	Nepal, Kathmandu	23:09	23:34	12:55	13:20
9Q	Zaire, Kinshasa	04:35	04:57	16:53	17:15
	Kisangani	03:50	04:11	16:19	16:41
	Lubumbashi	03:55	04:17	15:55	16:17
9U	Burundi, Bujumbura	03:38	03:59	15:58	16:20
9V	Singapore	22:34	22:56	11:06	11:27
9X	Rwanda, Kigali	03:33	03:55	15:57	16:18
9Y	Trinidad & Tobago , Port of Spain	09:23	09:45	22:18	22:40
A2	Botswana, Gaborone	04:17	04:41	15:44	16:08
A3	Tonga, Nukualofa	17:37	18:00	05:14	05:37
A4	Oman, Musqat	01:04	01:28	14:36	15:00
A5	Bhutan, Thimbu	22:54	23:19	12:38	13:03
A6	United Arab Emirates, Abu Dhabi	01:19	01:44	14:55	15:19
A7	Qatar, Ad-Dawhah	01:29	01:53	15:07	15:31
A9	Bahrein, Al-Manamah	01:31	01:56	15:12	15:36
AP	Pakistan, Karachi	00:27	00:52	14:04	14:28
AP	Islamabad	23:46	00:13	13:54	14:21
BV	Taiwan, Taipei	20:49	21:14	10:26	10:51
BY	Peoples Rep. of China, Beijing	20:37	21:07	11:14	11:44
	Harbin	19:38	20:12	10:47	11:21
	Shanghai	20:38	21:04	10:36	11:02
	Fuzhou	20:56	21:21	10:37	11:01
	Xian	21:21	21:49	11:32	11:59

Pref	State/Province/Country/City	Tlight	Sunrise	Sunset	Tlight
	Chongqing	21:41	22:06	11:34	12:00
	Chengdu	21:48	22:14	11:45	12:11
	Lhasa	22:42	23:08	12:35	13:01
	Urumqi	22:21	22:53	13:18	13:50
	Kashi	23:20	23:50	13:54	14:24
C2	Nauru	18:24	18:46	06:51	07:13
C3	Andorra	04:10	04:41	18:59	19:31
C5	The Gambia, Banjul	06:20	06:43	19:23	19:45
C6	Bahamas, Nassau	10:05	10:29	23:42	00:06
C9	Mozambique, Maputo	03:52	04:16	15:16	15:40
	Mozambique	03:06	03:28	14:58	15:21
CE	Chile, Santiago	10:55	11:21	21:57	22:23
CE0	Easter Island	13:21	13:45	00:42	01:06
CE0	Juan Fernandez	11:27	11:53	22:29	22:55
CE0	San Felix	11:23	11:47	22:46	23:10
CM,CO	Cuba, Havana	10:28	10:52	23:59	00:23
CN	Morocco, Casablanca	05:09	05:36	19:17	19:43
CP	Bolivia, La Paz	10:24	10:46	22:13	22:35
CT	Portugal, Lisbon	05:03	05:32	19:33	20:03
CT3	Madeira Islands, Funchal	05:48	06:15	19:52	20:19
CU	Azores , Ponta Delgada	06:12	06:40	20:37	21:06
CX	Uruguay, Montevideo	09:59	10:25	20:57	21:23
CY0	Sable Is.	08:11	08:44	23:08	23:41
CY9	St. Paul Is.	08:00	08:35	23:18	23:53
D2	Angola, Luanda	04:49	05:10	16:56	17:18
D4	Cape Verde, Praia	06:45	07:08	19:52	20:15
D6	Comoros, Moroni	02:50	03:12	14:50	15:12
DA-DL,Y2-Y9	Fed. Rep. of Germany, Bonn	03:17	03:56	19:00	19:39
	Berlin	02:43	03:24	18:41	19:22
DU	Philippines, Manilla	21:08	21:30	10:14	10:36
E3	Eritrea, Asmara	02:35	02:58	15:43	16:06
EA	Spain, Madrid	04:37	05:07	19:15	19:45
EA6	Balearic Is., Palma	04:14	04:43	18:48	19:17
EA8	Canary Is., Las Palmas	05:46	06:12	19:35	20:00
EA9	Ceuta & Melilla, Ceuta	04:55	05:22	19:12	19:40
	Melilla	04:47	05:14	19:02	19:29
EI	Ireland, Dublin	03:57	04:39	20:03	20:45
EK	Armenia, Yerevan	01:24	01:54	16:02	16:32
EL	Liberia, Monrovia	06:06	06:28	18:50	19:12
EP	Iran, Tehran	01:06	01:34	15:24	15:51
ER	Moldova, Kishinev	02:05	02:40	17:22	17:57
ES	Estonia, Tallinn	01:11	02:06	18:28	19:23
ET	Ethiopia, Addis Ababa	02:45	03:07	15:36	15:58
EU,EV,EW	Belarus, Minsk	01:38	02:21	17:50	18:33
EX	Kyrgyzstan, Frunze	23:16	23:48	14:08	14:39
EY	Tajikistan, Samarkand	23:56	00:26	14:32	15:01
	Dushanbe	23:50	00:19	14:22	14:52
EZ	Turkmenistan, Ashkhabad	00:35	01:03	15:02	15:30
F	France, Paris	03:44	04:21	19:13	19:50
FG	Guadeloupe	09:17	09:39	22:26	22:49
FJ, FS	St. Martin	09:19	09:42	22:35	22:58
FH	Mayotte	02:45	03:07	14:42	15:05
FK	New Caledonia, Noumia	18:51	19:15	06:25	06:49
FM	Martinique	09:16	09:38	22:22	22:44
FO	Clipperton Is.	12:35	12:57	01:29	01:51
FO	Fr. Polynesia, Tahiti	16:30	16:52	04:16	04:38
	Rurutu, (Austral Is.)	19:52	20:16	07:26	07:49
	Hiva Oa, (Marquesas Is.)	14:59	15:21	03:04	03:25
FP	St. Pierre & Miquelon, St. Pierre	07:45	08:20	23:00	23:35
FR/G	Glorioso	02:35	02:57	14:36	14:58
FR/J,E	Juan de Nova	03:00	03:22	14:47	15:10
	Europa	03:16	03:39	14:50	15:13
FR	Reunion	02:13	02:37	13:51	14:14
FR/T	Tromelin	02:12	02:35	14:02	14:25
FT8W	Crozet	03:04	03:36	13:20	13:52
FT8X	Kerguelen	02:01	02:36	12:03	12:37
FT8Y	Antarctica, Dumont D'Urville	22:16	23:21	05:51	06:56
FT8Z	Amsterdam & St. Paul Is., Amsterdam	01:07	01:35	11:56	12:24
FW	Wallis & Futuna Is., Wallis	17:32	17:54	05:28	05:51
FY	Fr. Guiana, Cayenne	08:54	09:16	21:35	21:56
G	England, London	03:42	04:21	19:32	20:11
GD	Isle of Man	03:45	04:28	20:00	20:43
GI	Northern Ireland, Belfast	03:48	04:32	20:07	20:51
GJ	Jersey	04:00	04:37	19:33	20:10
GM	Scotland, Glasgow	03:35	04:21	20:06	20:52
	Aberdeen	03:16	04:05	20:03	20:53
GU	Guernsey	04:01	04:38	19:35	20:13
GW	Wales, Cardiff	03:54	04:34	19:44	20:24
H4	Solomon Islands, Honiara	19:02	19:24	07:08	07:30
HA	Hungary, Budapest	02:42	03:17	18:02	18:37
HB	Switzerland, Bern	03:30	04:05	18:47	19:22
HB0	Liechtenstein	03:21	03:56	18:39	19:14
HC	Ecuador, Quito	10:43	11:05	23:11	23:33
HC8	Galapagos Is.	11:34	11:55	00:01	00:23
HH	Haiti, Port-Au-Prince	09:55	10:18	23:12	23:35
HI	Dominican Republic, Santo Domingo	09:46	10:09	23:03	23:26
HK	Colombia, Bogota	10:22	10:43	23:01	23:23
HK0	Malpelo Is.	10:50	11:12	23:29	23:50
HK0	San Adreas	10:42	11:04	23:42	00:04
HL	Korea, Seoul	20:01	20:30	10:26	10:55
HP	Panama, Panama	10:38	11:00	23:29	23:50
HR	Honduras, Tegucigalpa	11:01	11:24	00:06	00:28
HS	Thailand, Bangkok	22:31	22:53	11:35	11:57
HV	Vatican City	03:27	03:58	18:14	18:45
HZ, 7Z	Saudi Arabia, Dharan	01:33	01:58	15:14	15:39
	Mecca	02:22	02:46	15:48	16:11
	Riyadh	01:49	02:14	15:25	15:49

Pref	State/Province/Country/City	Tlight	Sunrise	Sunset	Tlight
I	Italy, Rome	03:27	03:58	18:14	18:45
	Trieste	03:10	03:44	18:18	18:52
	Sicily	03:33	04:02	17:58	18:27
IS	Sardinia, Cagliari	03:49	04:18	18:21	18:51
J2	Djibouti, Djibouti	02:23	02:45	15:21	15:43
J3	Grenada	09:23	09:45	22:22	22:44
J5	Guinea-Bissau, Bissau	06:18	06:40	19:17	19:39
J6	St. Lucia	09:17	09:39	22:21	22:43
J7	Dominica	09:16	09:39	22:24	22:47
J8	St. Vincent	09:19	09:41	22:21	22:44
JA-JS	Japan, Tokyo	19:15	19:42	09:31	09:59
	Nagasaki	20:01	20:27	10:06	10:32
	Sapporo	18:48	19:20	09:41	10:13
JD1	Minami Torishima	18:41	19:05	08:15	08:39
JD1	Ogasawara, Kazan Is.	19:27	19:52	09:12	09:37
JT	Mongolia, Ulan Bator	20:49	21:25	12:12	12:48
JW	Svalbard, Spitsbergen	——	——	——	——
JX	Jan Mayen	——	02:00	22:59	——
JY	Jordan, Amman	02:18	02:45	16:20	16:46
KC4	Antarctica, Byrd Station	——	——	——	——
	McMurdo Sound	22:52	——	——	02:47
	Palmer Station	11:43	12:41	19:43	20:42
KC6	Belau, Yap	20:06	20:28	08:58	09:20
	Koror	20:24	20:46	09:10	09:32
KG4	Guantanamo Bay	10:05	10:28	23:26	23:49
KH0	Mariana Is., Saipan	19:28	19:50	08:35	08:58
KH1	Baker, Howland Is.	17:14	17:36	05:44	06:06
KH2	Guam, Agana	19:34	19:57	08:37	08:59
KH3	Johnston Is.	16:22	16:45	05:35	05:58
KH4	Midway Is.	16:39	17:04	06:27	06:52
KH5	Palmyra Is.	16:12	16:34	04:55	05:17
KH5K	Kingman Reef	16:13	16:35	05:00	05:22
KH6	Hawaii, Hilo	15:25	15:48	04:45	05:08
	Honolulu	15:33	15:57	04:58	05:22
KH7	Kure Is.	16:43	17:08	06:31	06:56
KH8	American Samoa, Pago Pago	17:11	17:33	05:05	05:27
KH9	Wake Is.	17:58	18:22	07:18	07:41
KL7	Alaska, Adak	15:26	16:06	07:19	07:59
	Anchorage	12:33	13:34	06:18	07:19
	Fairbanks	11:32	12:57	06:38	08:03
	Juneau	11:57	12:49	04:58	05:50
	Nome	12:48	14:10	07:45	09:07
KP1	Navassa Is.	10:06	10:29	23:23	23:46
KP2	Virgin Islands, Charlotte Amalie	09:26	09:49	22:42	23:05
KP4	Puerto Rico, San Juan	09:31	09:54	22:48	23:11
KP5	Desecheo Is.	09:36	09:59	22:53	23:16
LA-LJ	Norway, Oslo	02:02	02:59	19:27	20:24
LU	Argentina, Buenos Aires	10:07	10:33	21:06	21:32
LX	Luxembourg	03:25	04:02	19:00	19:37
LY	Lithuania, Vilna	01:43	02:27	18:01	18:45
LZ	Bulgaria, Sofia	02:42	03:13	17:32	18:04
OA	Peru, Lima	10:54	11:16	22:53	23:15
OD	Lebanon, Beirut	02:16	02:43	16:25	16:52
OE	Austria, Vienna	02:51	03:26	18:15	18:51
OH	Finland, Helsinki	01:03	02:01	18:32	19:29
OH0	Aland Is.	01:23	02:21	18:52	19:49
OJ0	Market Reef	01:26	02:24	18:56	19:54
OK,OL	Czech Rep., Prague	02:50	03:28	18:29	19:07
OM	Slovak, Rep., Bratislava	02:49	03:24	18:12	18:48
ON	Belgium, Brussels	03:27	04:05	19:12	19:50
OX	Greenland, Godthaab	05:19	06:38	00:08	01:27
	Thule	——	——	——	——
OY	Faroe Islands, Torshavn	02:51	03:56	20:51	21:56
OZ	Denmark, Copenhagen	02:28	03:14	18:58	19:44
P2	Papua New Guinea, Madang	19:55	20:16	08:11	08:32
	Port Moresby	19:54	20:16	08:00	08:22
P4	Aruba, Oranjestad	09:51	10:14	22:50	23:13
PA-PI	Netherlands, Amsterdam	03:17	03:58	19:15	19:56
PJ2,4,9	Netherlands Antilles, Willemstad	09:55	10:17	22:55	23:18
PJ5-8	St. Maarten and Saba, St. Maarten	09:20	09:43	22:35	22:58
PY	Brazil, Brasilia	09:01	09:24	20:52	21:14
	Rio De Janeiro	08:51	09:14	20:23	20:47
	Natal	07:59	08:21	20:13	20:35
	Manaus	09:35	09:57	21:57	22:18
	Porto Alegre	09:32	09:57	20:45	21:10
PY0	Fernando De Noronha	07:45	08:07	20:04	20:26
PY0	St. Peter & St. Paul Rocks	07:27	07:49	19:58	20:20
PY0	Trindade & Martin Vaz Is., Trindade	07:52	08:15	19:31	19:54
PZ	Suriname, Paramaribo	09:05	09:26	21:47	22:09
R1FJ_	Franz Josef Land	——	——	——	——
R1MV_	Malyj Vysotskij Is.	00:45	01:44	18:20	19:19
S0	Western Sahara, Smara	05:39	06:03	19:20	19:45
S2	Bangladesh, Dacca	22:56	23:20	12:29	12:53
S5	Slovenia, Ljubljana	03:06	03:40	18:16	18:50
S7	Seychelles, Victoria	01:54	02:16	14:12	14:34
S9	Sao Tome	05:04	05:25	17:33	17:55
SM	Sweden, Stockholm	01:39	02:34	18:54	19:48
SP	Poland, Krakow	02:28	03:06	18:06	18:44
	Warsaw	02:14	02:54	18:10	18:50
ST	Sudan, Khartoum	03:00	03:23	16:09	16:32
ST0	Southern Sudan, Juba	03:18	03:40	15:59	16:21
SU	Egypt, Cairo	02:41	03:06	16:35	17:00
SV	Greece, Athens	02:53	03:22	17:20	17:49
SV/A	Mount Athos	02:45	03:15	17:23	17:53
SV5	Dodecanese, Rhodes	02:39	03:07	16:59	17:27
SV9	Crete	02:54	03:21	17:09	17:37
T2	Tuvalu, Funafuti	17:47	18:09	05:55	06:16

Pref	State/Province/Country/City	Tlight	Sunrise	Sunset	Tlight
T30	West Kiribati, Bonriki	17:56	18:18	06:28	06:50
T31	Central Kiribati, Kanton	17:01	17:23	05:23	05:45
T32	East Kiribati, Christmas Is.	15:58	16:20	04:31	04:53
T33	Banaba Is.	18:14	18:36	06:41	07:03
T5	Somalia, Mogadishu	02:27	02:48	15:00	15:22
T7	San Marino	03:22	03:54	18:19	18:52
T9	Bosnia-Herzegovina, Sarajevo	02:58	03:30	17:55	18:27
TA	Turkey, Ankara	02:11	02:41	16:48	17:17
	Istanbul	02:24	02:54	17:06	17:37
TF	Iceland, Reykjavik	03:22	04:40	22:08	23:26
TG	Guatemala, Guatemala City	11:14	11:36	00:20	00:42
TI	Costa Rica, San Jose	10:54	11:16	23:48	00:10
TI9	Cocos Is.	11:12	11:34	23:54	00:16
TJ	Cameroon, Yaounde	04:40	05:02	17:18	17:40
TK	Corsica	03:41	04:12	18:28	18:59
TL	Central African Republic, Bangui	04:11	04:33	16:50	17:12
TN	Congo, Brazzaville	04:35	04:57	16:53	17:15
TR	Gabon, Libreville	04:52	05:14	17:22	17:44
TT	Chad, N'Djamena	04:15	04:38	17:14	17:37
TU	Cote d'Ivoire, Abidjan	05:40	06:02	18:22	18:44
TY	Benin, Porto Novo	05:12	05:34	17:57	18:19
TZ	Mali, Bamako	05:47	06:09	18:47	19:09
UA	Russia, European, Leningrad (UA1)	00:45	01:41	18:08	19:05
	Archangel (UA1)	23:03	00:26	18:03	19:25
	Murmansk (UA1)	——	00:01	19:27	——
	Moscow (UA3)	00:47	01:33	17:18	18:04
	Kuybyshev (UA4)	00:12	00:54	16:17	16:59
	Rostov (UA6)	01:20	01:56	16:40	17:16
UA2	Kaliningrad	02:00	02:45	18:23	19:08
UA9,0	Russia, Asiatic, Novosibirsk (UA9)	21:51	22:36	14:13	14:58
	Perm (UA9)	23:17	00:08	16:14	17:05
	Omsk (UA9)	22:29	23:14	14:51	15:36
	Norilsk (UA0)	——	20:15	15:53	——
	Irkutsk (UA0)	20:40	21:21	12:37	13:18
	Vladivostok (UA0)	19:26	19:58	10:19	10:51
	Petropavlovsk (UA0)	16:59	17:40	09:02	09:44
	Khabarovsk (UA0)	18:54	19:30	10:21	10:57
	Krasnoyarsk (UA0)	21:05	21:52	13:38	14:25
	Yakutsk (UA0)	17:45	18:50	11:45	12:50
	Wrangel Island (UA0)	——	13:25	10:23	——
	Kyzyl (UA0Y)	21:22	22:02	13:14	13:54
UJ-UM	Uzbekistan, Bukhoro	00:06	00:36	14:41	15:11
	Tashkent	23:42	00:13	14:25	14:55
UN-UQ	Kazakhstan, Alma-Ata	23:05	23:37	13:59	14:31
UR-UZ,EM-EO	Ukraine, Kiev	01:44	02:23	17:25	18:04
V2	Antigua & Barbuda, St. Johns	09:15	09:38	22:28	22:51
V3	Belize, Belmopan	11:03	11:26	00:16	00:39
V4	St. Kitts & Nevis	09:18	09:41	22:32	22:55
V5	Namibia, Windhoek	04:49	05:13	16:23	16:46
V6	Micronesia, Ponape	18:49	19:11	07:35	07:57
V7	Marshall Islands, Kwajalein	18:10	18:32	07:01	07:23
V8	Brunei, Bandar Seri Begawan	21:45	22:07	10:26	10:48
VK	Australia, Canberra (VK1)	20:18	20:45	07:15	07:41
	Sydney (VK2)	20:08	20:34	07:09	07:35
	Melbourne (VK3)	20:38	21:06	07:26	07:54
	Brisbane (VK4)	19:52	20:16	07:12	07:36
	Adelaide (VK5)	20:59	21:26	07:57	08:24
	Perth (VK6)	22:26	22:52	09:34	09:59
	Hobart, Tasmania (VK7)	20:37	21:07	07:07	07:37
	Darwin (VK8)	21:02	21:24	09:01	09:23
VK0	Heard Is.	01:53	02:31	11:34	12:12
VK0	Macquarie Is.	20:16	20:55	05:47	06:26
VK9C	Cocos-Keeling Is.	23:18	23:40	11:17	11:39
VK9L	Lord Howe Is.	19:33	19:58	06:41	07:06
VK9M	Mellish Reef	19:29	19:51	07:15	07:37
VK9N	Norfolk Is.	18:54	19:19	06:10	06:34
VK9W	Willis Is.	19:52	20:15	07:41	08:04
VK9X	Christmas Is.	22:41	23:03	10:44	11:06
VP2E	Anguilla	09:18	09:41	22:35	22:58
VP2M	Montserrat	09:17	09:40	22:29	22:52
VP2V	British Virgin Is., Tortola	09:25	09:48	22:41	23:04

Pref	State/Province/Country/City	Tlight	Sunrise	Sunset	Tlight
VP5	Turks & Caicos Islands, Grand Turk	09:46	10:10	23:12	23:35
VP8	Falkland Islands, Stanley	10:35	11:11	20:24	21:00
VP8	So. Georgia Is.	09:17	09:56	18:50	19:29
VP8	So. Orkney Is.	10:11	10:59	18:57	19:45
VP8	So. Sandwich Islands, Saunders Is.	08:46	09:30	17:56	18:39
VP8	So. Shetland Is., King George Is.	11:07	11:59	19:40	20:31
VP9	Bermuda	09:00	09:27	23:03	23:29
VQ9	Chagos, Diego Garcia	00:50	01:12	13:01	13:23
VR2,VS6	Hong Kong	21:23	21:47	10:51	11:15
VR6	Pitcairn Is.	14:41	15:05	02:08	02:32
VU	India, Bombay	00:14	00:37	13:32	13:56
	Calcutta	23:06	23:30	12:35	12:59
	New Delhi	23:40	00:05	13:29	13:54
	Bangalore	00:04	00:26	13:05	13:28
VU	Andaman Islands, Port Blair	23:05	23:27	12:03	12:25
VU	Laccadive Is.	00:26	00:48	13:20	13:42
XE	Mexico, Mexico City (XE1)	11:41	12:04	01:01	01:24
	Chihuahua (XE2)	11:53	12:18	01:42	02:07
	Merida (XE3)	11:01	11:24	00:25	00:49
XF4	Revilla Gigedo	12:31	12:54	01:50	02:13
XT	Burkina Faso, Ouagadougou	05:21	05:44	18:21	18:43
XU	Cambodia, Phnom Penh	22:17	22:39	11:15	11:37
XW	Laos, Viangchan	22:16	22:39	11:32	11:55
XX9	Macao	21:26	21:50	10:54	11:17
XZ	Myanmar, Yangon	22:45	23:07	11:57	12:19
YA	Afghanistan, Kandahar	00:21	00:47	14:19	14:45
	Kabul	00:00	00:27	14:11	14:38
YB-YD	Indonesia, Jakarta	22:31	22:53	10:45	11:07
	Medan	22:52	23:13	11:29	11:51
	Pontianak	22:14	22:35	10:42	11:04
	Jayapura	20:11	20:33	08:34	08:55
YI	Iraq, Baghdad	01:42	02:09	15:47	16:14
YJ	Vanuatu, Port Vila	18:39	19:01	06:24	06:47
YK	Syria, Damascus	02:14	02:41	16:21	16:48
YL	Latvia, Riga	01:33	02:22	18:18	19:06
YN	Nicaragua, Managua	11:00	11:22	23:58	00:21
YO	Romania, Bucharest	02:25	02:58	17:25	17:58
YS	El Salvador, San Salvador	11:10	11:32	00:13	00:36
YU	Yugoslavia, Belgrade	02:46	03:19	17:49	18:22
YV	Venezuela, Caracas	09:46	10:08	22:40	23:02
YV0	Aves Is.	09:25	09:48	22:34	22:57
Z2	Zimbabwe, Harare	03:48	04:11	15:33	15:56
Z3	Macedonia, (ex Yugoslav), Skopje	02:51	03:23	17:38	18:09
ZA	Albania, Tirane	03:00	03:31	17:43	18:14
ZB2	Gibraltar	04:54	05:22	19:13	19:41
ZC4	British Cyprus	02:24	02:52	16:36	17:04
ZD7	St. Helena	06:13	06:36	18:03	18:26
ZD8	Ascension Is.	06:38	07:00	18:47	19:09
ZD9	Tristan da Cunha	07:06	07:34	17:57	18:24
ZF	Cayman Is.	10:29	10:53	23:49	00:12
ZK1	No. Cook Is., Manihiki	16:27	16:49	04:31	04:53
ZK1	So. Cook Is., Rarotonga	16:35	16:58	04:12	04:35
ZK2	Niue	17:09	17:32	04:51	05:14
ZK3	Tokelaus, Atafu	17:12	17:34	05:20	05:42
ZL	New Zealand, Auckland (ZL1)	18:37	19:05	05:29	05:56
	Wellington (ZL2)	18:44	19:14	05:20	05:49
	Christchurch (ZL3)	18:57	19:27	05:24	05:54
	Dunedin (ZL4)	19:10	19:41	05:27	05:59
ZL5	Antarctica, Scott Base	23:00	——	——	02:41
ZL7	Chatham Is.	18:14	18:45	04:39	05:10
ZL8	Kermadec Is.	17:58	18:23	05:13	05:38
ZL9	Auckland & Campbell Is., Auckland	19:35	20:11	05:29	06:05
	Campbell Is.	19:29	20:06	05:13	05:50
ZP	Paraguay, Asuncion	09:52	10:16	21:18	21:42
ZS	South Africa, Cape Town (ZS1)	04:59	05:25	16:00	16:26
	Port Elizabeth (ZS2)	04:30	04:56	15:31	15:57
	Bloemfontein (ZS4)	04:22	04:46	15:37	16:02
	Durban (ZS5)	04:03	04:28	15:17	15:42
	Johannesburg (ZS6)	04:10	04:34	15:33	15:58
ZS8	Marion Is.	04:02	04:34	14:15	14:48

Pref	State/Province/Country/City	Tlight	Sunrise	Sunset	Tlight
VE1	New Brunswick, Saint John	08:13	08:48	23:53	00:28
	Nova Scotia, Halifax	08:05	08:40	23:41	00:15
	Prince Edward Island, Charlottetown	07:57	08:33	23:44	00:20
VE2	Quebec, Montreal	08:42	09:17	00:24	00:59
	Quebec City	08:26	09:03	00:19	00:55
VE3	Ontario, London	09:23	09:56	00:47	01:20
	Ottawa	08:51	09:26	00:32	01:07
	Sudbury	09:07	09:43	00:57	01:33
	Toronto	09:12	09:46	00:41	01:15
VE4	Manitoba, Winnipeg	09:54	10:35	02:14	02:54
VE5	Saskatchewan, Regina	10:21	11:02	02:47	03:28
	Saskatoon	10:20	11:04	03:02	03:46
VE6	Alberta, Calgary	10:56	11:38	03:27	04:09
	Edmonton	10:38	11:24	03:36	04:23
VE7	British Columbia, Prince George	11:12	11:59	04:15	05:03
	Prince Rupert	11:39	12:27	04:47	05:36
	Vancouver	11:42	12:21	03:56	04:35
VE8	Northwest Territories, Yellowknife	08:49	10:23	04:44	06:18
	Resolute	——	——	——	——
VY1	Yukon, Whitehorse	10:49	12:03	05:50	07:04
VO1	Newfoundland, St. John's	07:09	07:46	23:07	23:45
VO2	Labrador, Goose Bay	07:07	07:53	00:03	00:49
W1	Connecticut, Hartford	08:53	09:25	00:09	00:41
	Maine, Bangor	08:25	09:00	00:02	00:37
	Portland	08:36	09:10	00:05	00:39
	Massachusetts, Boston	08:44	09:17	00:04	00:37
	New Hampshire, Concord	08:43	09:16	00:08	00:41
	Rhode Island, Providence	08:47	09:20	00:03	00:36
	Vermont, Montpelier	08:43	09:17	00:16	00:50
W2	New Jersey, Atlantic City	09:08	09:39	00:09	00:40
	New York, Albany	08:54	09:27	00:16	00:49
	Buffalo	09:13	09:47	00:37	01:10
	New York City	09:01	09:33	00:11	00:43
	Syracuse	09:02	09:35	00:26	01:00
W3	Delaware, Wilmington	09:11	09:42	00:14	00:45
	District of Columbia, Washington	09:20	09:50	00:18	00:48
	Maryland, Baltimore	09:17	09:48	00:17	00:48
	Pennsylvania, Harrisburg	09:15	09:46	00:21	00:53
	Philadelphia	09:09	09:40	00:13	00:44
	Pittsburgh	09:27	09:58	00:34	01:05
	Scranton	09:06	09:38	00:19	00:52
W4	Alabama, Montgomery	10:16	10:43	00:39	01:07
	Florida, Jacksonville	10:02	10:29	00:17	00:43
	Miami	10:07	10:32	00:02	00:27
	Pensacola	10:24	10:51	00:39	01:05
	Georgia, Atlanta	10:04	10:32	00:35	01:03
	Savannah	09:56	10:23	00:18	00:45
	Kentucky, Lexington	09:53	10:23	00:45	01:15
	Louisville	09:57	10:27	00:51	01:21
	North Carolina, Charlotte	09:46	10:15	00:24	00:52
	Raleigh	09:36	10:04	00:16	00:46
	Wilmington	09:37	10:05	00:10	00:38
	South Carolina, Columbia	09:50	10:18	00:22	00:50
	Tennessee, Knoxville	09:56	10:25	00:38	01:07
	Memphis	10:24	10:52	01:01	01:29
	Nashville	10:07	10:36	00:50	01:19
	Virginia, Norfolk	09:23	09:53	00:10	00:39
	Richmond	09:26	09:55	00:16	00:46
W5	Arkansas, Little Rock	10:33	11:02	01:09	01:37
	Louisiana, New Orleans	10:37	11:03	00:49	01:16
	Shreveport	10:45	11:12	01:09	01:37
	Mississippi, Jackson	10:31	10:59	00:55	01:22
	New Mexico, Albuquerque	11:30	11:58	02:07	02:36
	Oklahoma, Oklahoma City	10:52	11:21	01:31	02:00
	Texas, Abilene	11:09	11:36	01:33	02:01
	Amarillo	11:10	11:39	01:48	02:16
	Dallas	10:57	11:24	01:22	01:50
	El Paso	11:38	12:05	01:59	02:26
	San Antonio	11:12	11:38	01:22	01:48
W6	California, Los Angeles	12:19	12:47	02:51	03:19
	San Francisco	12:25	12:55	03:17	03:46
W7	Arizona, Flagstaff	11:50	12:18	02:27	02:56
	Phoenix	11:56	12:24	02:25	02:53
	Idaho, Boise	11:40	12:14	03:08	03:42
	Pocatello	11:28	12:01	02:51	03:24
	Montana, Billings	11:00	11:36	02:44	03:20
	Butte	11:15	11:51	03:01	03:37
	Great Falls	11:03	11:41	03:01	03:39
	Nevada, Las Vegas	12:00	12:29	02:43	03:12
	Reno	12:09	12:40	03:11	03:41
	Oregon, Portland	11:58	12:34	03:40	04:16
	Utah, Salt Lake City	11:33	12:05	02:43	03:14
	Washington, Seattle	11:47	12:25	03:46	04:23
	Spokane	11:27	12:05	03:27	04:04
	Wyoming, Cheyenne	11:04	11:35	02:15	02:47
	Sheridan	10:58	11:33	02:35	03:10
W8	Michigan, Detroit	09:32	10:05	00:51	01:24
	Grand Rapids	09:40	10:14	01:04	01:37
	Sault Ste. Marie	09:20	09:57	01:10	01:47
	Traverse City	09:33	10:07	01:09	01:44
	Ohio, Cincinnati	09:49	10:20	00:48	01:19
	Cleveland	09:30	10:02	00:44	01:16
	Columbus	09:40	10:11	00:45	01:16
	West Virginia, Charleston	09:40	10:10	00:35	01:05
W9	Illinois, Chicago	09:52	10:24	01:08	01:41
	Indiana, Indianapolis	09:54	10:25	00:57	01:28
	Wisconsin, Green Bay	09:43	10:18	01:18	01:53
	Milwaukee	09:49	10:22	01:13	01:46
W0	Colorado, Denver	11:09	11:40	02:12	02:43
	Grand Junction	11:26	11:56	02:25	02:55
	Iowa, Des Moines	10:17	10:49	01:32	02:04
	Kansas, Pratt	10:50	11:20	01:41	02:11
	Wichita	10:45	11:15	01:36	02:06
	Minnesota, Duluth	09:50	10:27	01:42	02:19
	Minneapolis	10:03	10:38	01:41	02:16
	Missouri, Columbia	10:21	10:51	01:19	01:50
	Kansas City	10:30	11:00	01:29	01:59
	St. Louis	10:14	10:44	01:10	01:40
	Nebraska, North Platte	10:48	11:19	01:59	02:31
	Omaha	10:27	10:59	01:40	02:12
	North Dakota, Fargo	10:08	10:45	02:01	02:38
	South Dakota, Rapid City	10:46	11:20	02:18	02:52
-		05:30	05:52	18:00	18:22
1A0	S.M.O.M	03:12	03:44	18:28	19:01
1S	Spratly Is.	21:49	22:12	10:45	11:07
3A	Monaco	03:25	03:59	18:54	19:28
3B6	Agalega	01:59	02:21	13:58	14:20
3B7	St. Brandon	01:55	02:18	13:36	13:59
3B8	Mauritius	02:10	02:33	13:39	14:02
3B9	Rodriguez Is.	01:45	02:09	13:16	13:39
3C	Equatorial Guinea, Bata	04:49	05:11	17:23	17:45
	Malabo	04:50	05:12	17:30	17:52
3C0	Annobon Is.	05:10	05:32	17:35	17:57
3D2	Fiji Is., Suva	18:03	18:26	05:39	06:02
3D2	Conway Reef	18:24	18:48	05:49	06:13
3D2	Rotuma Is.	17:57	18:20	05:50	06:13
3DA	Swaziland, Mbabane	04:05	04:29	15:14	15:38
3V	Tunisia, Tunis	03:37	04:07	18:24	18:53
3W, XV	Vietnam, Ho Chi Minh City (Saigon)	22:07	22:30	11:09	11:31
	Hanoi	21:52	22:17	11:29	11:53
3X	Guinea, Conakry	06:11	06:33	19:09	19:31
3Y	Bouvet	06:58	07:39	15:46	16:27
3Y	Peter I Is.	14:37	16:17	19:40	21:20
4J, 4K	Azerbaijan, Baku	00:47	01:19	15:54	16:26
4L	Georgia, Tbilisi	01:03	01:35	16:18	16:51
4S	Sri Lanka, Colombo	00:00	00:22	12:50	13:13
4U	ITU Geneva	03:19	03:56	19:07	19:43
4U	United Nations Hq.	09:01	09:33	00:11	00:43
4X, 4Z	Israel, Jerusalem	02:11	02:38	16:32	16:59
5A	Libya, Tripoli	03:40	04:08	18:04	18:32
	Benghazi	03:11	03:38	17:34	18:01
5B	Cyprus, Nicosia	02:09	02:38	16:47	17:16
5H	Tanzania, Dar es Salaam	03:03	03:25	15:11	15:33
5N	Nigeria, Lagos	05:07	05:29	17:56	18:18
5R	Madagascar, Antananarivo	02:48	03:11	14:21	14:44
5T	Mauritania, Nouakchott	06:05	06:29	19:31	19:55
5U	Niger, Niamey	05:01	05:24	18:12	18:35
5V	Togo, Lome	05:17	05:39	18:03	18:26
5W	Western Samoa, Apia	17:17	17:40	05:07	05:29
5X	Uganda, Kampala	03:20	03:42	15:50	16:12
5Z	Kenya, Nairobi	04:25	04:47	16:50	17:12
6W	Senegal, Dakar	06:17	06:40	19:32	19:55
6Y	Jamaica, Kingston	10:09	10:32	23:34	23:58
7O	Yemen, Aden	02:10	02:33	15:19	15:42
	Sanaa	02:09	02:32	15:26	15:49
7P	Lesotho, Maseru	04:24	04:49	15:23	15:48
7Q	Malawi, Lilongwe	03:36	03:58	15:23	15:46
	Blantyre	03:33	03:56	15:16	15:39
7X	Algeria, Algiers	04:07	04:36	18:52	19:22
8P	Barbados, Bridgetown	09:08	09:31	22:18	22:41
8Q	Maldive Is.	00:30	00:52	13:12	13:35
8R	Guyana, Georgetown	09:13	09:35	22:02	22:25
9A	Croatia, Zagreb	02:42	03:18	18:26	19:02
9G	Ghana, Accra	05:23	05:45	18:08	18:31
9H	Malta	03:23	03:52	18:05	18:34
9J	Zambia, Lusaka	04:00	04:22	15:43	16:06
9K	Kuwait	01:26	01:53	15:37	16:03
9L	Sierra Leone, Freetown	06:10	06:33	19:05	19:27
9M2	West Malaysia, Kuala Lumpur	22:39	23:01	11:18	11:40
9M6,8	East Malaysia, Sabah, Sandakan(9M6)	21:29	21:51	10:16	10:38
	Sarawak, Kuching (9M8)	22:07	22:29	10:41	11:03
9N	Nepal, Kathmandu	23:01	23:26	13:03	13:29
9Q	Zaire, Kinshasa	04:36	04:58	16:52	17:14
	Kisangani	03:49	04:11	16:20	16:42
	Lubumbashi	03:57	04:20	15:52	16:15
9U	Burundi, Bujumbura	03:38	04:00	15:58	16:20
9V	Singapore	22:33	22:55	11:06	11:28
9X	Rwanda, Kigali	03:33	03:55	15:56	16:18
9Y	Trinidad & Tobago , Port of Spain	09:19	09:42	22:20	22:43
A2	Botswana, Gaborone	04:23	04:48	15:37	16:02
A3	Tonga, Nukualofa	17:42	18:06	05:08	05:32
A4	Oman, Musqat	00:56	01:21	14:42	15:07
A5	Bhutan, Thimbu	22:45	23:11	12:46	13:12
A6	United Arab Emirates, Abu Dhabi	01:12	01:37	15:02	15:27
A7	Qatar, Ad-Dawhah	01:21	01:46	15:14	15:39
A9	Bahrein, Al-Manamah	01:23	01:48	15:19	15:45
AP	Pakistan, Karachi	00:19	00:44	14:11	14:36
AP	Islamabad	23:34	00:02	14:04	14:32
BV	Taiwan, Taipei	20:41	21:06	10:34	10:59
BY	Peoples Rep. of China, Beijing	20:23	20:54	11:27	11:58
	Harbin	19:19	19:55	11:04	11:39
	Shanghai	20:27	20:54	10:46	11:13
	Fuzhou	20:48	21:13	10:44	11:10
	Xian	21:10	21:38	11:43	12:11

Pref	State/Province/Country/City	Tlight	Sunrise	Sunset	Tlight
	Chongqing	21:31	21:57	11:43	12:09
	Chengdu	21:38	22:05	11:54	12:21
	Lhasa	22:32	22:59	12:44	13:10
	Urumqi	22:04	22:38	13:33	14:07
	Kashi	23:06	23:37	14:07	14:38
C2	Nauru	18:24	18:46	06:51	07:13
C3	Andorra	03:53	04:26	19:14	19:47
C5	The Gambia, Banjul	06:16	06:39	19:27	19:50
C6	Bahamas, Nassau	09:57	10:22	23:49	00:14
C9	Mozambique, Maputo	03:58	04:23	15:08	15:33
	Mozambique	03:09	03:32	14:54	15:17
CE	Chile, Santiago	11:04	11:31	21:47	22:14
CE0	Easter Island	13:28	13:53	00:34	00:59
CE0	Juan Fernandez	11:37	12:03	22:19	22:46
CE0	San Felix	11:30	11:54	22:39	23:03
CM,CO	Cuba, Havana	10:21	10:46	00:06	00:30
CN	Morocco, Casablanca	04:57	05:25	19:27	19:55
CP	Bolivia, La Paz	10:28	10:51	22:08	22:31
CT	Portugal, Lisbon	04:49	05:20	19:46	20:16
CT3	Madeira Islands, Funchal	05:37	06:05	20:02	20:30
CU	Azores , Ponta Delgada	05:58	06:28	20:49	21:19
CX	Uruguay, Montevideo	10:08	10:36	20:46	21:13
CY0	Sable Is.	07:54	08:28	23:24	23:58
CY9	St. Paul Is.	07:40	08:17	23:36	00:13
D2	Angola, Luanda	04:50	05:13	16:54	17:16
D4	Cape Verde, Praia	06:41	07:04	19:56	20:19
D6	Comoros, Moroni	02:53	03:15	14:47	15:10
DA-DL,Y2-Y9	Fed. Rep. of Germany, Bonn	02:53	03:35	19:21	20:03
	Berlin	02:17	03:01	19:03	19:48
DU	Philippines, Manilla	21:03	21:26	10:18	10:41
E3	Eritrea, Asmara	02:31	02:54	15:47	16:10
EA	Spain, Madrid	04:22	04:53	19:29	20:00
EA6	Balearic Is., Palma	03:59	04:30	19:01	19:32
EA8	Canary Is., Las Palmas	05:37	06:03	19:43	20:09
EA9	Ceuta & Melilla, Ceuta	04:42	05:11	19:23	19:52
	Melilla	04:35	05:13	19:13	19:41
EI	Ireland, Dublin	03:30	04:16	20:26	21:12
EK	Armenia, Yerevan	01:09	01:41	16:16	16:47
EL	Liberia, Monrovia	06:04	06:26	18:52	19:14
EP	Iran, Tehran	00:54	01:23	15:35	16:04
ER	Moldova, Kishinev	01:45	02:22	17:39	18:16
ES	Estonia, Tallinn	00:28	01:34	18:59	20:05
ET	Ethiopia, Addis Ababa	02:42	03:04	15:38	16:01
EU,EV,EW	Belarus, Minsk	01:10	01:58	18:14	19:01
EX	Kyrgyzstan, Frunze	22:59	23:33	14:23	14:56
EY	Tajikistan, Samarkand	23:42	00:13	14:45	15:16
	Dushanbe	23:36	00:07	14:35	15:06
EZ	Turkmenistan, Ashkhabad	00:21	00:51	15:14	15:44
F	France, Paris	03:23	04:02	19:32	20:11
FG	Guadeloupe	09:12	09:35	22:31	22:54
FJ, FS	St. Martin	09:14	09:37	22:40	23:03
FH	Mayotte	02:48	03:11	14:39	15:02
FK	New Caledonia, Noumia	18:57	19:21	06:19	06:43
FM	Martinique	09:11	09:34	22:26	22:49
FO	Clipperton Is.	12:31	12:54	01:32	01:54
FO	Fr. Polynesia, Tahiti	16:34	16:57	04:11	04:34
	Rurutu, (Austral Is.)	19:58	20:22	07:20	07:44
	Hiva Oa, (Marquesas Is.)	15:01	15:23	03:01	03:23
FP	St. Pierre & Miquelon, St. Pierre	07:26	08:03	23:17	23:54
FR/G	Glorioso	02:38	03:00	14:33	14:56
FR/J,E	Juan de Nova	03:04	03:27	14:43	15:06
	Europa	03:21	03:45	14:44	15:07
FR	Reunion	02:19	02:42	13:45	14:08
FR/T	Tromelin	02:16	02:39	13:58	14:21
FT8W	Crozet	03:19	03:52	13:04	13:37
FT8X	Kerguelen	02:19	02:54	11:44	12:20
FT8Y	Antarctica, Dumont D'Urville	22:55	00:12	05:00	06:17
FT8Z	Amsterdam & St. Paul Is., Amsterdam	01:19	01:47	11:44	12:13
FW	Wallis & Futuna Is., Wallis	17:35	17:57	05:25	05:48
FY	Fr. Guiana, Cayenne	08:52	09:14	21:36	21:58
G	England, London	03:17	04:00	19:53	20:36
GD	Isle of Man	03:16	04:04	20:24	21:12
GI	Northern Ireland, Belfast	03:19	04:08	20:31	21:20
GJ	Jersey	03:38	04:18	19:52	20:32
GM	Scotland, Glasgow	03:03	03:55	20:32	21:24
	Aberdeen	02:41	03:37	20:31	21:28
GU	Guernsey	03:39	04:19	19:55	20:35
GW	Wales, Cardiff	03:29	04:12	20:05	20:48
H4	Solomon Islands, Honiara	19:04	19:26	07:06	07:28
HA	Hungary, Budapest	02:22	02:59	18:20	18:57
HB	Switzerland, Bern	03:11	03:48	19:04	19:41
HB0	Liechtenstein	03:01	03:38	18:57	19:34
HC	Ecuador, Quito	10:43	11:05	23:11	23:33
HC8	Galapagos Is.	11:33	11:55	00:01	00:23
HH	Haiti, Port-Au-Prince	09:50	10:13	23:17	23:41
HI	Dominican Republic, Santo Domingo	09:40	10:04	23:08	23:32
HK	Colombia, Bogota	10:20	10:42	23:03	23:25
HK0	Malpelo Is.	10:49	11:11	23:30	23:52
HK0	San Adreas	10:38	11:00	23:45	00:08
HL	Korea, Seoul	19:48	20:18	10:38	11:08
HP	Panama, Panama	10:35	10:57	23:31	23:53
HR	Honduras, Tegucigalpa	10:57	11:20	00:10	00:33
HS	Thailand, Bangkok	22:27	22:50	11:38	12:01
HV	Vatican City	03:12	03:44	18:28	19:01
HZ, 7Z	Saudi Arabia, Dharan	01:25	01:50	15:22	15:47
	Mecca	02:16	02:40	15:54	16:18
	Riyadh	01:42	02:07	15:32	15:57

Pref	State/Province/Country/City	Tlight	Sunrise	Sunset	Tlight
I	Italy, Rome	03:12	03:44	18:28	19:01
	Trieste	02:51	03:27	18:35	19:10
	Sicily	03:20	03:50	18:10	18:40
IS	Sardinia, Cagliari	03:34	04:05	18:34	19:05
J2	Djibouti, Djibouti	02:20	02:42	15:24	15:47
J3	Grenada	09:19	09:42	22:25	22:47
J5	Guinea-Bissau, Bissau	06:14	06:37	19:20	19:43
J6	St. Lucia	09:13	09:35	22:25	22:47
J7	Dominica	09:11	09:34	22:28	22:51
J8	St. Vincent	09:15	09:38	22:25	22:48
JA-JS	Japan, Tokyo	19:02	19:31	09:43	10:11
	Nagasaki	19:50	20:17	10:16	10:43
	Sapporo	18:31	19:05	09:56	10:29
JD1	Minami Torishima	18:33	18:58	08:22	08:47
JD1	Ogasawara, Kazan Is.	19:18	19:44	09:20	09:46
JT	Mongolia, Ulan Bator	20:29	21:07	12:30	13:08
JW	Svalbard, Spitsbergen	—	—	—	—
JX	Jan Mayen	—	—	—	—
JY	Jordan, Amman	02:08	02:35	16:30	16:57
KC4	Antarctica, Byrd Station	—	—	—	—
	McMurdo Sound	—	—	—	—
	Palmer Station	12:17	13:24	19:00	20:07
KC6	Belau, Yap	20:03	20:26	09:01	09:23
	Koror	20:21	20:44	09:12	09:35
KG4	Guantanamo Bay	09:59	10:22	23:31	23:55
KH0	Mariana Is., Saipan	19:23	19:46	08:39	09:03
KH1	Baker, Howland Is.	17:14	17:36	05:44	06:06
KH2	Guam, Agana	19:30	19:53	08:41	09:04
KH3	Johnston Is.	16:17	16:40	05:40	06:03
KH4	Midway Is.	16:30	16:56	06:35	07:01
KH5	Palmyra Is.	16:10	16:32	04:57	05:19
KH5K	Kingman Reef	16:10	16:33	05:02	05:24
KH6	Hawaii, Hilo	15:19	15:42	04:50	05:14
	Honolulu	15:27	15:51	05:04	05:29
KH7	Kure Is.	16:34	17:00	06:40	07:06
KH8	American Samoa, Pago Pago	17:14	17:37	05:01	05:24
KH9	Wake Is.	17:53	18:16	07:23	07:47
KL7	Alaska, Adak	15:01	15:45	07:40	08:24
	Anchorage	11:40	12:58	06:54	08:12
	Fairbanks	—	12:09	07:27	—
	Juneau	11:19	12:19	05:28	06:29
	Nome	—	13:23	08:32	—
KP1	Navassa Is.	10:01	10:24	23:28	23:51
KP2	Virgin Islands, Charlotte Amalie	09:20	09:44	22:47	23:11
KP4	Puerto Rico, San Juan	09:25	09:49	22:53	23:16
KP5	Desecheo Is.	09:31	09:54	22:58	23:21
LA-LJ	Norway, Oslo	01:16	02:26	20:01	21:10
LU	Argentina, Buenos Aires	10:17	10:44	20:55	21:22
LX	Luxembourg	03:03	03:43	19:20	20:00
LY	Lithuania, Vilna	01:14	02:03	18:25	19:14
LZ	Bulgaria, Sofia	02:25	02:58	17:47	18:20
OA	Peru, Lima	10:56	11:19	22:50	23:12
OD	Lebanon, Beirut	02:04	02:32	16:36	17:04
OE	Austria, Vienna	02:30	03:08	18:34	19:12
OH	Finland, Helsinki	00:16	01:27	19:05	20:16
OH0	Aland Is.	00:36	01:47	19:25	20:36
OJ0	Market Reef	00:39	01:50	19:30	20:41
OK,OL	Czech Rep., Prague	02:27	03:08	18:49	19:30
OM	Slovak, Rep., Bratislava	02:28	03:06	18:30	19:08
ON	Belgium, Brussels	03:03	03:45	19:32	20:14
OX	Greenland, Godthaab	—	05:52	00:54	—
	Thule	—	—	—	—
OY	Faroe Islands, Torshavn	01:51	03:18	21:29	22:56
OZ	Denmark, Copenhagen	01:56	02:48	19:23	20:15
P2	Papua New Guinea, Madang	19:56	20:18	08:10	08:32
	Port Moresby	19:56	20:18	07:57	08:20
P4	Aruba, Oranjestad	09:48	10:10	22:54	23:16
PA-PI	Netherlands, Amsterdam	02:51	03:36	19:37	20:21
PJ2,4,9	Netherlands Antilles, Willemstad	09:51	10:14	22:59	23:22
PJ5-8	St. Maarten and Saba, St. Maarten	09:15	09:38	22:39	23:03
PY	Brazil, Brasilia	09:05	09:28	20:47	21:10
	Rio De Janeiro	08:57	09:21	20:17	20:41
	Natal	08:00	08:22	20:12	20:34
	Manaus	09:36	09:58	21:56	22:18
	Porto Alegre	09:40	10:06	20:36	21:02
PY0	Fernando De Noronha	07:46	08:08	20:03	20:25
PY0	St. Peter & St. Paul Rocks	07:27	07:49	19:59	20:21
PY0	Trindade & Martin Vaz Is., Trindade	07:57	08:21	19:25	19:49
PZ	Suriname, Paramaribo	09:02	09:25	21:49	22:11
R1FJ_	Franz Josef Land	—	—	—	—
R1MV_	Malyj Vysotskij Is.	23:56	01:09	18:54	20:08
S0	Western Sahara, Smara	05:30	05:56	19:28	19:53
S2	Bangladesh, Dacca	22:49	23:13	12:35	13:00
S5	Slovenia, Ljubljana	02:47	03:23	18:33	19:09
S7	Seychelles, Victoria	01:55	02:17	14:11	14:33
S9	Sao Tome	05:03	05:25	17:33	17:55
SM	Sweden, Stockholm	00:56	02:02	19:25	20:31
SP	Poland, Krakow	02:05	02:46	18:26	19:07
	Warsaw	01:48	02:32	18:32	19:16
ST	Sudan, Khartoum	02:56	03:19	16:13	16:36
ST0	Southern Sudan, Juba	03:17	03:39	16:01	16:23
SU	Egypt, Cairo	02:31	02:57	16:44	17:10
SV	Greece, Athens	02:40	03:10	17:33	18:03
SV/A	Mount Athos	02:30	03:02	17:36	18:07
SV5	Dodecanese, Rhodes	02:27	02:56	17:11	17:40
SV9	Crete	02:42	03:10	17:20	17:49
T2	Tuvalu, Funafuti	17:49	18:11	05:52	06:15

Pref	State/Province/Country/City	Tlight	Sunrise	Sunset	Tlight
T30	West Kiribati, Bonriki	17:56	18:18	06:29	06:51
T31	Central Kiribati, Kanton	17:01	17:23	05:22	05:44
T32	East Kiribati, Christmas Is.	15:57	16:19	04:32	04:54
T33	Banaba Is.	18:14	18:36	06:41	07:03
T5	Somalia, Mogadishu	02:26	02:48	15:01	15:23
T7	San Marino	03:05	03:39	18:35	19:09
T9	Bosnia-Herzegovina, Sarajevo	02:40	03:14	18:10	18:45
TA	Turkey, Ankara	01:57	02:28	17:01	17:32
	Istanbul	02:08	02:40	17:20	17:52
TF	Iceland, Reykjavik	———	03:55	22:53	———
TG	Guatemala, Guatemala City	11:09	11:32	00:24	00:47
TI	Costa Rica, San Jose	10:51	11:14	23:50	00:13
TI9	Cocos Is.	11:10	11:32	23:56	00:18
TJ	Cameroon, Yaounde	04:39	05:01	17:19	17:41
TK	Corsica	03:25	03:58	18:42	19:15
TL	Central African Republic, Bangui	04:09	04:32	16:52	17:14
TN	Congo, Brazzaville	04:36	04:58	16:52	17:14
TR	Gabon, Libreville	04:52	05:14	17:22	17:44
TT	Chad, N'Djamena	04:12	04:34	17:18	17:40
TU	Cote d'Ivoire, Abidjan	05:38	06:01	18:23	18:46
TY	Benin, Porto Novo	05:10	05:32	17:59	18:21
TZ	Mali, Bamako	05:43	06:05	18:51	19:13
UA	Russia, European, Leningrad (UA1)	00:00	01:08	18:41	19:50
	Archangel (UA1)	———	23:38	18:50	———
	Murmansk (UA1)	———	21:57	21:31	———
	Moscow (UA3)	00:15	01:07	17:44	18:36
	Kuybyshev (UA4)	23:45	00:31	16:40	17:26
	Rostov (UA6)	01:00	01:38	16:58	17:36
UA2	Kaliningrad	01:30	02:20	18:48	19:38
UA9,0	Russia, Asiatic, Novosibirsk (UA9)	22:11	22:49	14:38	15:28
	Perm (UA9)	22:39	23:38	16:43	17:43
	Omsk (UA9)	21:59	22:49	15:16	16:06
	Norilsk (UA0)	———	———	———	———
	Irkutsk (UA0)	20:14	20:59	12:59	13:43
	Vladivostok (UA0)	19:09	19:43	10:34	11:08
	Petropavlovsk (UA0)	16:32	17:18	09:25	10:10
	Khabarovsk (UA0)	18:33	19:12	10:40	11:18
	Krasnoyarsk (UA0)	20:33	21:25	14:04	14:57
	Yakutsk (UA0)	16:45	18:12	12:23	13:50
	Wrangel Island (UA0)	———	———	———	———
	Kyzyl (UA0Y)	20:57	21:41	13:35	14:19
UJ-UM	Uzbekistan, Bukhoro	23:51	00:22	14:55	15:26
	Tashkent	23:27	23:59	14:39	15:11
UN-UQ	Kazakhstan, Alma-Ata	22:49	23:22	14:15	14:48
UR-UZ,EM-EO	Ukraine, Kiev	01:21	02:02	17:46	18:27
V2	Antigua & Barbuda, St. Johns	09:10	09:34	22:33	22:56
V3	Belize, Belmopan	10:58	11:21	00:21	00:45
V4	St. Kitts & Nevis	09:13	09:36	22:36	23:00
V5	Namibia, Windhoek	04:55	05:19	16:16	16:40
V6	Micronesia, Ponape	18:47	19:09	07:37	07:59
V7	Marshall Islands, Kwajalein	18:07	18:30	07:04	07:26
V8	Brunei, Bandar Seri Begawan	21:43	22:06	10:27	10:49
VK	Australia, Canberra (VK1)	20:28	20:55	07:04	07:31
	Sydney (VK2)	20:17	20:44	06:58	07:25
	Melbourne (VK3)	20:49	21:18	07:14	07:43
	Brisbane (VK4)	19:59	20:24	07:04	07:29
	Adelaide (VK5)	21:09	21:37	07:47	08:14
	Perth (VK6)	22:35	23:01	09:24	09:50
	Hobart, Tasmania (VK7)	20:50	21:22	06:52	07:23
	Darwin (VK8)	21:05	21:27	08:57	09:20
VK0	Heard Is.	02:13	02:52	11:12	11:52
VK0	Macquarie Is.	20:37	21:19	05:23	06:05
VK9C	Cocos-Keeling Is.	23:21	23:43	11:14	11:37
VK9L	Lord Howe Is.	19:41	20:08	06:32	06:58
VK9M	Mellish Reef	19:33	19:56	07:10	07:33
VK9N	Norfolk Is.	19:02	19:27	06:01	06:27
VK9W	Willis Is.	19:56	20:19	07:37	08:00
VK9X	Christmas Is.	22:43	23:05	10:41	11:04
VP2E	Anguilla	09:13	09:36	22:40	23:03
VP2M	Montserrat	09:12	09:36	22:34	22:57
VP2V	British Virgin Is., Tortola	09:19	09:43	22:46	23:10

Pref	State/Province/Country/City	Tlight	Sunrise	Sunset	Tlight
VP5	Turks & Caicos Islands, Grand Turk	09:40	10:04	23:18	23:42
VP8	Falkland Islands, Stanley	10:54	11:32	20:03	20:41
VP8	So. Georgia Is.	09:38	10:19	18:27	19:09
VP8	So. Orkney Is.	10:39	11:31	18:25	19:17
VP8	So. Sandwich Islands, Saunders Is.	09:11	09:57	17:28	18:15
VP8	So. Shetland Is., King George Is.	11:37	12:34	19:05	20:01
VP9	Bermuda	08:49	09:17	23:13	23:40
VQ9	Chagos, Diego Garcia	00:51	01:14	12:59	13:21
VR2,VS6	Hong Kong	21:16	21:40	10:57	11:22
VR6	Pitcairn Is.	14:48	15:12	02:01	02:25
VU	India, Bombay	00:08	00:32	13:38	14:01
	Calcutta	22:59	23:23	12:41	13:06
	New Delhi	23:31	23:57	13:38	14:04
	Bangalore	00:00	00:23	13:09	13:32
VU	Andaman Islands, Port Blair	23:01	23:24	12:06	12:29
VU	Laccadive Is.	00:23	00:46	13:22	13:45
XE	Mexico, Mexico City (XE1)	11:35	11:59	01:06	01:30
	Chihuahua (XE2)	11:43	12:09	01:51	02:17
	Merida (XE3)	10:54	11:19	00:31	00:55
XF4	Revilla Gigedo	12:25	12:49	01:55	02:19
XT	Burkina Faso, Ouagadougou	05:17	05:40	18:25	18:47
XU	Cambodia, Phnom Penh	22:13	22:36	11:18	11:41
XW	Laos, Viangchan	22:11	22:34	11:37	12:00
XX9	Macao	21:19	21:43	11:00	11:24
XZ	Myanmar, Yangon	22:39	23:03	12:01	12:25
YA	Afghanistan, Kandahar	00:11	00:38	14:28	14:55
	Kabul	23:48	00:17	14:22	14:50
YB-YD	Indonesia, Jakarta	22:32	22:54	10:43	11:05
	Medan	22:50	23:12	11:30	11:52
	Pontianak	22:13	22:35	10:42	11:04
	Jayapura	20:11	20:33	08:33	08:55
YI	Iraq, Baghdad	01:31	01:58	15:58	16:25
YJ	Vanuatu, Port Vila	18:43	19:06	06:20	06:43
YK	Syria, Damascus	02:02	02:30	16:31	16:59
YL	Latvia, Riga	00:58	01:54	18:45	19:41
YN	Nicaragua, Managua	10:56	11:18	00:02	00:24
YO	Romania, Bucharest	02:07	02:42	17:41	18:16
YS	El Salvador, San Salvador	11:06	11:29	00:17	00:40
YU	Yugoslavia, Belgrade	02:28	03:03	18:05	18:40
YV	Venezuela, Caracas	09:42	10:05	22:43	23:06
YV0	Aves Is.	09:20	09:43	22:38	23:01
Z2	Zimbabwe, Harare	04:32	04:15	15:29	15:52
Z3	Macedonia, (ex Yugoslav), Skopje	02:36	03:08	17:53	18:25
ZA	Albania, Tirane	02:44	03:17	17:57	18:29
ZB2	Gibraltar	04:42	05:11	19:24	19:53
ZC4	British Cyprus	02:13	02:41	16:47	17:15
ZD7	St. Helena	06:17	06:40	17:59	18:22
ZD8	Ascension Is.	06:40	07:02	18:45	19:08
ZD9	Tristan da Cunha	07:17	07:45	17:45	18:13
ZF	Cayman Is.	10:23	10:47	23:55	00:18
ZK1	No. Cook Is., Manihiki	16:30	16:52	04:28	04:51
ZK1	So. Cook Is., Rarotonga	16:40	17:04	04:06	04:30
ZK2	Niue	17:14	17:37	04:46	05:10
ZK3	Tokelaus, Atafu	17:13	17:36	05:18	05:40
ZL	New Zealand, Auckland (ZL1)	18:48	19:16	05:17	05:45
	Wellington (ZL2)	18:57	19:27	05:07	05:37
	Christchurch (ZL3)	19:11	19:42	05:09	05:41
	Dunedin (ZL4)	19:25	19:58	05:10	05:43
ZL5	Antarctica, Scott Base	———	———	———	———
ZL7	Chatham Is.	18:28	19:00	04:24	04:56
ZL8	Kermadec Is.	18:06	18:31	05:04	05:30
ZL9	Auckland & Campbell Is., Auckland	19:53	20:30	05:10	05:47
	Campbell Is.	19:48	20:27	04:52	05:31
ZP	Paraguay, Asuncion	09:58	10:23	21:11	21:35
ZS	South Africa, Cape Town (ZS1)	05:08	05:35	15:50	16:17
	Port Elizabeth (ZS2)	04:39	05:06	15:20	15:47
	Bloemfontein (ZS4)	04:29	04:55	15:28	15:54
	Durban (ZS5)	04:11	04:37	15:08	15:33
	Johannesburg (ZS6)	04:17	04:41	15:26	15:51
ZS8	Marion Is.	04:18	04:51	13:58	14:32

Pref	State/Province/Country/City	Tlight	Sunrise	Sunset	Tlight
VE1	New Brunswick, Saint John	07:59	08:37	00:09	00:46
	Nova Scotia, Halifax	07:53	08:29	23:56	00:32
	Prince Edward Island, Charlottetown	07:43	08:21	00:00	00:38
VE2	Quebec, Montreal	08:28	09:06	00:39	01:17
	Quebec City	08:12	08:51	00:35	01:14
VE3	Ontario, London	09:11	09:46	01:01	01:36
	Ottawa	08:37	09:15	00:47	01:25
	Sudbury	08:53	09:31	01:13	01:52
	Toronto	09:00	09:36	00:56	01:31
VE4	Manitoba, Winnipeg	09:37	10:21	02:33	03:16
VE5	Saskatchewan, Regina	10:03	10:48	03:06	03:50
	Saskatoon	10:00	10:48	03:22	04:10
VE6	Alberta, Calgary	10:38	11:23	03:46	04:32
	Edmonton	10:15	11:07	03:58	04:49
VE7	British Columbia, Prince George	10:49	11:41	04:37	05:30
	Prince Rupert	11:15	12:09	05:10	06:04
	Vancouver	11:25	12:08	04:14	04:56
VE8	Northwest Territories, Yellowknife	——	09:47	05:24	——
	Resolute	——	——	——	——
VY1	Yukon, Whitehorse	09:46	11:34	06:23	08:11
VO1	Newfoundland, St. John's	06:54	07:34	23:24	00:04
VO2	Labrador, Goose Bay	06:45	07:36	00:24	01:15
W1	Connecticut, Hartford	08:42	09:16	00:22	00:56
	Maine, Bangor	08:13	08:49	00:17	00:54
	Portland	08:24	08:59	00:19	00:55
	Massachusetts, Boston	08:33	09:07	00:18	00:52
	New Hampshire, Concord	08:31	09:06	00:22	00:57
	Rhode Island, Providence	08:37	09:11	00:17	00:51
	Vermont, Montpelier	08:30	09:06	00:31	01:07
W2	New Jersey, Atlantic City	08:58	09:30	00:21	00:53
	New York, Albany	08:42	09:17	00:30	01:04
	Buffalo	09:02	09:37	00:51	01:26
	New York City	08:51	09:24	00:24	00:57
	Syracuse	08:50	09:25	00:41	01:16
W3	Delaware, Wilmington	09:01	09:34	00:27	00:59
	District of Columbia, Washington	09:10	09:42	00:30	01:02
	Maryland, Baltimore	09:07	09:39	00:30	01:02
	Pennsylvania, Harrisburg	09:05	09:37	00:34	01:07
	Philadelphia	08:59	09:32	00:26	00:59
	Pittsburgh	09:17	09:50	00:47	01:20
	Scranton	08:56	09:29	00:33	01:07
W4	Alabama, Montgomery	10:09	10:37	00:50	01:18
	Florida, Jacksonville	09:57	10:24	00:26	00:53
	Miami	10:02	10:28	00:10	00:36
	Pensacola	10:18	10:46	00:48	01:16
	Georgia, Atlanta	09:57	10:26	00:46	01:14
	Savannah	09:49	10:17	00:28	00:56
	Kentucky, Lexington	09:44	10:15	00:58	01:29
	Louisville	09:48	10:19	01:03	01:35
	North Carolina, Charlotte	09:38	10:08	00:35	01:04
	Raleigh	09:28	09:58	00:28	00:58
	Wilmington	09:30	09:59	00:21	00:50
	South Carolina, Columbia	09:43	10:12	00:33	01:01
	Tennessee, Knoxville	09:48	10:18	00:49	01:19
	Memphis	10:16	10:45	01:12	01:41
	Nashville	09:59	10:29	01:02	01:32
	Virginia, Norfolk	09:15	09:45	00:22	00:52
	Richmond	09:17	09:48	00:28	00:59
W5	Arkansas, Little Rock	10:26	10:55	01:20	01:49
	Louisiana, New Orleans	10:31	10:58	00:59	01:26
	Shreveport	10:38	11:07	01:19	01:48
	Mississippi, Jackson	10:25	10:53	01:05	01:33
	New Mexico, Albuquerque	11:22	11:52	02:18	02:48
	Oklahoma, Oklahoma City	10:44	11:14	01:43	02:12
	Texas, Abilene	11:02	11:31	01:43	02:12
	Amarillo	11:02	11:32	01:59	02:28
	Dallas	10:50	11:18	01:33	02:01
	El Paso	11:32	12:00	02:09	02:37
	San Antonio	11:06	11:33	01:31	01:58
W6	California, Los Angeles	12:12	12:41	03:02	03:31
	San Francisco	12:16	12:47	03:29	04:00
W7	Arizona, Flagstaff	11:42	12:12	02:38	03:08
	Phoenix	11:49	12:18	02:36	03:04
	Idaho, Boise	11:28	12:03	03:23	03:58
	Pocatello	11:16	11:51	03:05	03:40
	Montana, Billings	10:47	11:24	03:00	03:38
	Butte	11:02	11:40	03:17	03:55
	Great Falls	10:49	11:28	03:18	03:58
	Nevada, Las Vegas	11:52	12:22	02:55	03:25
	Reno	11:59	12:32	03:23	03:55
	Oregon, Portland	11:45	12:22	03:56	04:33
	Utah, Salt Lake City	11:23	11:56	02:56	03:29
	Washington, Seattle	11:32	12:12	04:03	04:43
	Spokane	11:12	11:52	03:44	04:24
	Wyoming, Cheyenne	10:53	11:26	02:28	03:02
	Sheridan	10:46	11:22	02:50	03:27
W8	Michigan, Detroit	09:21	09:55	01:05	01:40
	Grand Rapids	09:29	10:04	01:18	01:53
	Sault Ste. Marie	09:07	09:45	01:27	02:05
	Traverse City	09:20	09:57	01:25	02:01
	Ohio, Cincinnati	09:40	10:12	01:01	01:33
	Cleveland	09:19	09:53	00:57	01:31
	Columbus	09:30	10:03	00:58	01:30
	West Virginia, Charleston	09:31	10:02	00:47	01:19
W9	Illinois, Chicago	09:41	10:15	01:22	01:56
	Indiana, Indianapolis	09:44	10:16	01:10	01:42
	Wisconsin, Green Bay	09:31	10:07	01:33	02:09

Pref	State/Province/Country/City	Tlight	Sunrise	Sunset	Tlight
	Milwaukee	09:37	10:12	01:27	02:02
W0	Colorado, Denver	10:59	11:32	02:25	02:57
	Grand Junction	11:16	11:48	02:37	03:09
	Iowa, Des Moines	10:06	10:40	01:45	02:19
	Kansas, Pratt	10:42	11:13	01:54	02:24
	Wichita	10:36	11:07	01:48	02:19
	Minnesota, Duluth	09:36	10:15	01:59	02:38
	Minneapolis	09:50	10:27	01:56	02:33
	Missouri, Columbia	10:11	10:43	01:32	02:04
	Kansas City	10:20	10:52	01:41	02:13
	St. Louis	10:04	10:36	01:22	01:54
	Nebraska, North Platte	10:37	11:10	02:12	02:46
	Omaha	10:17	10:50	01:53	02:27
	North Dakota, Fargo	09:54	10:33	02:18	02:57
	South Dakota, Rapid City	10:34	11:10	02:32	03:08
-		05:32	05:55	18:02	18:24
1A0	S.M.O.M	03:01	03:35	18:42	19:16
1S	Spratly Is.	21:49	22:12	10:49	11:12
3A	Monaco	03:13	03:49	19:09	19:44
3B6	Agalega	02:03	02:26	13:58	14:21
3B7	St. Brandon	02:00	02:23	13:35	13:58
3B8	Mauritius	02:16	02:40	13:36	14:00
3B9	Rodriguez Is.	01:51	02:15	13:14	13:38
3C	Equatorial Guinea, Bata	04:50	05:12	17:26	17:48
	Malabo	04:51	05:13	17:33	17:56
3C0	Annobon Is.	05:12	05:35	17:37	17:59
3D2	Fiji Is., Suva	18:09	18:32	05:37	06:01
3D2	Conway Reef	18:30	18:55	05:47	06:11
3D2	Rotuma Is.	18:02	18:25	05:50	06:13
3DA	Swaziland, Mbabane	04:12	04:38	15:10	15:35
3V	Tunisia, Tunis	03:29	04:00	18:35	19:06
3W, XV	Vietnam, Ho Chi Minh City (Saigon)	22:06	22:29	11:13	11:37
	Hanoi	21:49	22:14	11:36	12:01
3X	Guinea, Conakry	06:10	06:33	19:13	19:36
3Y	Bouvet	07:17	08:01	15:28	16:12
3Y	Peter I Is.	15:21	——	——	20:40
4J, 4K	Azerbaijan, Baku	00:37	01:10	16:07	16:40
4L	Georgia, Tbilisi	00:52	01:26	16:32	17:06
4S	Sri Lanka, Colombo	00:00	00:23	12:54	13:17
4U	ITU Geneva	03:06	03:44	19:23	20:01
4U	United Nations Hq.	08:51	09:24	00:24	00:57
4X, 4Z	Israel, Jerusalem	02:05	02:33	16:42	17:10
5A	Libya, Tripoli	03:34	04:02	18:15	18:43
	Benghazi	03:05	03:33	17:44	18:12
5B	Cyprus, Nicosia	02:02	02:31	16:58	17:28
5H	Tanzania, Dar es Salaam	03:06	03:28	15:12	15:35
5N	Nigeria, Lagos	05:07	05:30	17:59	18:22
5R	Madagascar, Antananarivo	02:54	03:17	14:19	14:43
5T	Mauritania, Nouakchott	06:03	06:27	19:38	20:02
5U	Niger, Niamey	05:00	05:23	18:17	18:40
5V	Togo, Lomé	05:17	05:40	18:07	18:29
5W	Western Samoa, Apia	17:22	17:45	05:06	05:29
5X	Uganda, Kampala	03:22	03:44	15:52	16:15
5Z	Kenya, Nairobi	04:27	04:50	16:52	17:15
6W	Senegal, Dakar	06:16	06:39	19:37	20:01
6Y	Jamaica, Kingston	10:06	10:30	23:41	00:05
7O	Yemen, Aden	02:09	02:33	15:24	15:47
	Sanaa	02:08	02:31	15:32	15:55
7P	Lesotho, Maseru	04:32	04:58	15:18	15:44
7Q	Malawi, Lilongwe	03:40	04:03	15:23	15:46
	Blantyre	03:38	04:02	15:15	15:38
7X	Algeria, Algiers	03:58	04:29	19:04	19:34
8P	Barbados, Bridgetown	09:07	09:31	22:23	22:46
8Q	Maldive Is.	00:31	00:54	13:16	13:38
8R	Guyana, Georgetown	09:13	09:36	22:06	22:29
9A	Croatia, Zagreb	02:29	03:06	18:42	19:20
9G	Ghana, Accra	05:24	05:46	18:12	18:35
9H	Malta	03:15	03:45	18:16	18:46
9J	Zambia, Lusaka	04:05	04:28	15:42	16:06
9K	Kuwait	01:21	01:48	15:46	16:13
9L	Sierra Leone, Freetown	06:10	06:33	19:09	19:32
9M2	West Malaysia, Kuala Lumpur	22:40	23:03	11:21	11:43
9M6,8	East Malaysia, Sabah, Sandakan(9M6)	21:30	21:52	10:19	10:42
	Sarawak, Kuching (9M8)	22:08	22:31	10:43	11:06
9N	Nepal, Kathmandu	22:56	23:22	13:12	13:39
9Q	Zaire, Kinshasa	04:38	05:01	16:53	17:16
	Kisangani	03:51	04:13	16:22	16:44
	Lubumbashi	04:02	04:24	15:52	16:15
9U	Burundi, Bujumbura	03:41	04:03	15:59	16:22
9V	Singapore	22:35	22:57	11:09	11:31
9X	Rwanda, Kigali	03:35	03:58	15:58	16:21
9Y	Trinidad & Tobago , Port of Spain	09:19	09:42	22:25	22:48
A2	Botswana, Gaborone	04:30	04:55	15:34	15:59
A3	Tonga, Nukualofa	17:48	18:12	05:06	05:30
A4	Oman, Musqat	00:52	01:18	14:50	15:15
A5	Bhutan, Thimbu	22:40	23:06	12:55	13:21
A6	United Arab Emirates, Abu Dhabi	01:08	01:33	15:09	15:35
A7	Qatar, Ad-Dawhah	01:17	01:42	15:22	15:48
A9	Bahrein, Al-Manamah	01:18	01:44	15:28	15:54
AP	Pakistan, Karachi	00:15	00:41	14:19	14:44
AP	Islamabad	23:27	23:56	14:15	14:44
BV	Taiwan, Taipei	20:37	21:03	10:42	11:07
BY	Peoples Rep. of China, Beijing	20:13	20:45	11:40	12:13
	Harbin	19:06	19:44	11:19	11:57
	Shanghai	20:21	20:49	10:55	11:23
	Fuzhou	20:44	21:10	10:53	11:19
	Xian	21:02	21:32	11:54	12:23

Pref	State/Province/Country/City	Tlight	Sunrise	Sunset	Tlight
	Chongqing	21:25	21:52	11:52	12:19
	Chengdu	21:32	22:00	12:04	12:31
	Lhasa	22:27	22:54	12:53	13:20
	Urumqi	21:52	22:28	13:48	14:24
	Kashi	22:56	23:28	14:20	14:52
C2	Nauru	18:25	18:48	06:53	07:16
C3	Andorra	03:42	04:17	19:28	20:02
C5	The Gambia, Banjul	06:15	06:38	19:32	19:55
C6	Bahamas, Nassau	09:53	10:18	23:57	00:23
C9	Mozambique, Maputo	04:06	04:31	15:05	15:30
	Mozambique	03:14	03:38	14:53	15:16
CE	Chile, Santiago	11:14	11:41	21:41	22:09
CE0	Easter Island	13:36	14:01	00:30	00:56
CE0	Juan Fernandez	11:46	12:14	22:13	22:41
CE0	San Felix	11:37	12:02	22:35	23:00
CM,CO	Cuba, Havana	10:18	10:43	00:13	00:38
CN	Morocco, Casablanca	04:50	05:19	19:37	20:06
CP	Bolivia, La Paz	10:33	10:57	22:07	22:30
CT	Portugal, Lisbon	04:40	05:12	19:58	20:30
CT3	Madeira Islands, Funchal	05:31	05:59	20:13	20:41
CU	Azores , Ponta Delgada	05:50	06:21	21:02	21:32
CX	Uruguay, Montevideo	10:19	10:46	20:40	21:08
CY0	Sable Is.	07:42	08:18	23:39	00:14
CY9	St. Paul Is.	07:26	08:05	23:52	00:32
D2	Angola, Luanda	04:54	05:17	16:54	17:17
D4	Cape Verde, Praia	06:39	07:03	20:02	20:25
D6	Comoros, Moroni	02:57	03:20	14:47	15:10
DA-DL,Y2-Y9	Fed. Rep. of Germany, Bonn	02:35	03:20	19:40	20:25
	Berlin	01:56	02:45	19:24	20:13
DU	Philippines, Manilla	21:02	21:25	10:23	10:47
E3	Eritrea, Asmara	02:29	02:53	15:53	16:16
EA	Spain, Madrid	04:12	04:44	19:42	20:15
EA6	Balearic Is., Palma	03:50	04:22	19:14	19:46
EA8	Canary Is., Las Palmas	05:32	05:59	19:52	20:19
EA9	Ceuta & Melilla, Ceuta	04:34	05:04	19:35	20:05
	Melilla	04:27	04:57	19:24	19:54
EI	Ireland, Dublin	03:08	03:59	20:48	21:39
EK	Armenia, Yerevan	00:59	01:32	16:29	17:01
EL	Liberia, Monrovia	06:05	06:27	18:56	19:18
EP	Iran, Tehran	00:46	01:16	15:46	16:16
ER	Moldova, Kishinev	01:31	02:10	17:56	18:35
ES	Estonia, Tallinn	23:44	01:08	19:30	20:54
ET	Ethiopia, Addis Ababa	02:42	03:05	15:42	16:05
EU,EV,EW	Belarus, Minsk	00:47	01:40	18:36	19:28
EX	Kyrgyzstan, Frunze	22:48	23:23	14:37	15:12
EY	Tajikistan, Samarkand	23:32	00:05	14:58	15:30
	Dushanbe	23:27	23:58	14:48	15:20
EZ	Turkmenistan, Ashkhabad	00:12	00:43	15:26	15:57
F	France, Paris	03:07	03:48	19:50	20:31
FG	Guadeloupe	09:10	09:34	22:36	23:00
FJ, FS	St. Martin	09:11	09:35	22:46	23:10
FH	Mayotte	02:53	03:16	14:39	15:02
FK	New Caledonia, Noumia	19:04	19:28	06:17	06:41
FM	Martinique	09:10	09:33	22:31	22:55
FO	Clipperton Is.	12:31	12:54	01:36	01:59
FO	Fr. Polynesia, Tahiti	16:39	17:03	04:09	04:33
	Rurutu, (Austral Is.)	20:05	20:29	07:17	07:41
	Hiva Oa, (Marquesas Is.)	15:05	15:27	03:01	03:24
FP	St. Pierre & Miquelon, St. Pierre	07:12	07:51	23:34	00:13
FR/G	Glorioso	02:42	03:05	14:33	14:56
FR/J,E	Juan de Nova	03:09	03:33	14:41	15:05
	Europa	03:28	03:52	14:41	15:05
FR	Reunion	02:25	02:49	13:43	14:07
FR/T	Tromelin	02:21	02:44	13:57	14:20
FT8W	Crozet	03:33	04:07	12:53	13:27
FT8X	Kerguelen	02:35	03:12	11:31	12:08
FT8Y	Antarctica, Dumont D'Urville	23:31	01:11	04:05	05:46
FT8Z	Amsterdam & St. Paul Is., Amsterdam	01:29	01:59	11:37	12:06
FW	Wallis & Futuna Is., Wallis	17:39	18:02	05:24	05:47
FY	Fr. Guiana, Cayenne	08:53	09:16	21:39	22:02
G	England, London	02:58	03:45	20:13	20:59
GD	Isle of Man	02:52	03:46	20:47	21:40
GI	Northern Ireland, Belfast	02:55	03:49	20:54	21:49
GJ	Jersey	03:21	04:04	20:10	20:53
GM	Scotland, Glasgow	02:36	03:35	20:56	21:55
	Aberdeen	02:09	03:15	20:58	22:04
GU	Guernsey	03:22	04:05	20:13	20:56
GW	Wales, Cardiff	03:10	03:57	20:25	21:12
H4	Solomon Islands, Honiara	19:08	19:30	07:06	07:29
HA	Hungary, Budapest	02:07	02:47	18:37	19:17
HB	Switzerland, Bern	02:56	03:35	19:21	20:00
HB0	Liechtenstein	02:47	03:26	19:14	19:53
HC	Ecuador, Quito	10:45	11:07	23:14	23:36
HC8	Galapagos Is.	11:35	11:57	00:03	00:25
HH	Haiti, Port-Au-Prince	09:47	10:11	23:24	23:48
HI	Dominican Republic, Santo Domingo	09:38	10:02	23:14	23:38
HK	Colombia, Bogota	10:21	10:43	23:06	23:29
HK0	Malpelo Is.	10:50	11:12	23:33	23:56
HK0	San Adreas	10:37	11:00	23:50	00:13
HL	Korea, Seoul	19:40	20:10	10:50	11:21
HP	Panama, Panama	10:34	10:57	23:35	23:59
HR	Honduras, Tegucigalpa	10:56	11:19	00:15	00:38
HS	Thailand, Bangkok	22:25	22:49	11:44	12:07
HV	Vatican City	03:01	03:35	18:42	19:16
HZ, 7Z	Saudi Arabia, Dharan	01:20	01:46	15:30	15:56
	Mecca	02:12	02:37	16:01	16:26
	Riyadh	01:38	02:03	15:40	16:05

Pref	State/Province/Country/City	Tlight	Sunrise	Sunset	Tlight
I	Italy, Rome	03:01	03:35	18:42	19:16
	Trieste	02:38	03:16	18:51	19:28
	Sicily	03:12	03:42	18:22	18:53
IS	Sardinia, Cagliari	03:25	03:57	18:47	19:19
J2	Djibouti, Djibouti	02:19	02:42	15:29	15:52
J3	Grenada	09:18	09:41	22:30	22:53
J5	Guinea-Bissau, Bissau	06:14	06:37	19:25	19:48
J6	St. Lucia	09:11	09:35	22:30	22:53
J7	Dominica	09:10	09:33	22:34	22:57
J8	St. Vincent	09:14	09:37	22:30	22:53
JA-JS	Japan, Tokyo	18:54	19:24	09:54	10:24
	Nagasaki	19:43	20:11	10:26	10:54
	Sapporo	18:20	18:55	10:10	10:45
JD1	Minami Torishima	18:29	18:55	08:30	08:55
JD1	Ogasawara, Kazan Is.	19:13	19:40	09:29	09:55
JT	Mongolia, Ulan Bator	20:14	20:54	12:47	13:28
JW	Svalbard, Spitsbergen	—	—	—	—
JX	Jan Mayen	—	—	—	—
JY	Jordan, Amman	02:01	02:29	16:40	17:08
KC4	Antarctica, Byrd Station	—	—	—	—
	McMurdo Sound	—	—	—	—
	Palmer Station	12:49	14:09	18:20	19:40
KC6	Belau, Yap	20:03	20:26	09:05	09:28
	Koror	20:22	20:44	09:16	09:39
KG4	Guantanamo Bay	09:56	10:20	23:38	00:02
KH0	Mariana Is., Saipan	19:22	19:45	08:45	09:09
KH1	Baker, Howland Is.	17:15	17:38	05:47	06:09
KH2	Guam, Agana	19:29	19:52	08:46	09:09
KH3	Johnston Is.	16:15	16:39	05:46	06:09
KH4	Midway Is.	16:25	16:52	06:44	07:11
KH5	Palmyra Is.	16:10	16:33	05:00	05:23
KH5K	Kingman Reef	16:10	16:33	05:06	05:28
KH6	Hawaii, Hilo	15:16	15:40	04:57	05:21
	Honolulu	15:24	15:48	05:11	05:36
KH7	Kure Is.	16:29	16:55	06:48	07:15
KH8	American Samoa, Pago Pago	17:19	17:42	05:01	05:24
KH9	Wake Is.	17:50	18:14	07:29	07:54
KL7	Alaska, Adak	14:42	15:29	08:00	08:48
	Anchorage	10:13	12:28	07:29	09:43
	Fairbanks	—	11:17	08:22	—
	Juneau	10:42	11:55	05:56	07:10
	Nome	—	12:35	09:25	—
KP1	Navassa Is.	09:58	10:22	23:34	23:58
KP2	Virgin Islands, Charlotte Amalie	09:18	09:42	22:54	23:18
KP4	Puerto Rico, San Juan	09:23	09:47	22:59	23:23
KP5	Desecheo Is.	09:28	09:53	23:04	23:28
LA-LJ	Norway, Oslo	00:26	01:58	20:32	22:05
LU	Argentina, Buenos Aires	10:27	10:55	20:49	21:17
LX	Luxembourg	02:46	03:29	19:38	20:21
LY	Lithuania, Vilna	00:50	01:44	18:48	19:42
LZ	Bulgaria, Sofia	02:14	02:49	18:01	18:36
OA	Peru, Lima	11:01	11:24	22:50	23:13
OD	Lebanon, Beirut	01:57	02:26	16:46	17:15
OE	Austria, Vienna	02:14	02:55	18:51	19:32
OH	Finland, Helsinki	23:23	00:59	19:37	21:13
OH0	Aland Is.	23:43	01:19	19:57	21:33
OJ0	Market Reef	23:44	01:22	20:03	21:40
OK,OL	Czech Rep., Prague	02:10	02:54	19:08	19:51
OM	Slovak, Rep., Bratislava	02:13	02:53	18:47	19:28
ON	Belgium, Brussels	02:44	03:30	19:52	20:37
OX	Greenland, Godthaab	—	05:07	01:43	—
	Thule	—	—	—	—
OY	Faroe Islands, Torshavn	—	02:44	22:07	—
OZ	Denmark, Copenhagen	01:29	02:28	19:48	20:46
P2	Papua New Guinea, Madang	19:58	20:21	08:11	08:33
	Port Moresby	19:59	20:22	07:58	08:20
P4	Aruba, Oranjestad	09:47	10:10	22:59	23:22
PA-PI	Netherlands, Amsterdam	02:31	03:20	19:58	20:46
PJ2,4,9	Netherlands Antilles, Willemstad	09:50	10:13	23:04	23:27
PJ5-8	St. Maarten and Saba, St. Maarten	09:12	09:36	22:46	23:10
PY	Brazil, Brasilia	09:10	09:33	20:46	21:10
	Rio De Janeiro	09:04	09:28	20:14	20:38
	Natal	08:03	08:26	20:13	20:35
	Manaus	09:38	10:01	21:57	22:20
	Porto Alegre	09:49	10:15	20:31	20:57
PY0	Fernando De Noronha	07:48	08:11	20:05	20:27
PY0	St. Peter & St. Paul Rocks	07:28	07:51	20:01	20:24
PY0	Trindade & Martin Vaz Is., Trindade	08:04	08:28	19:23	19:47
PZ	Suriname, Paramaribo	09:03	09:26	21:52	22:15
R1FJ_	Franz Josef Land	—	—	—	—
R1MV_	Malyj Vysotskij Is.	22:55	00:40	19:27	21:12
S0	Western Sahara, Smara	05:26	05:52	19:36	20:02
S2	Bangladesh, Dacca	22:45	23:10	12:43	13:08
S5	Slovenia, Ljubljana	02:34	03:12	18:49	19:27
S7	Seychelles, Victoria	01:58	02:20	14:12	14:35
S9	Sao Tome	05:05	05:27	17:36	17:58
SM	Sweden, Stockholm	00:13	01:36	19:56	21:18
SP	Poland, Krakow	01:48	02:32	18:45	19:28
	Warsaw	01:28	02:16	18:52	19:40
ST	Sudan, Khartoum	02:54	03:18	16:19	16:43
ST0	Southern Sudan, Juba	03:17	03:40	16:04	16:26
SU	Egypt, Cairo	02:25	02:52	16:53	17:20
SV	Greece, Athens	02:31	03:02	17:45	18:16
SV/A	Mount Athos	02:20	02:53	17:49	18:22
SV5	Dodecanese, Rhodes	02:18	02:49	17:22	17:52
SV9	Crete	02:34	03:03	17:31	18:01
T2	Tuvalu, Funafuti	17:52	18:15	05:53	06:15

Pref	State/Province/Country/City	Tlight	Sunrise	Sunset	Tlight
T30	West Kiribati, Bonriki	17:57	18:19	06:31	06:54
T31	Central Kiribati, Kanton	17:04	17:26	05:24	05:46
T32	East Kiribati, Christmas Is.	15:59	16:21	04:35	04:57
T33	Banaba Is.	18:15	18:38	06:43	07:06
T5	Somalia, Mogadishu	02:27	02:49	15:04	15:26
T7	San Marino	02:53	03:28	18:50	19:25
T9	Bosnia-Herzegovina, Sarajevo	02:28	03:04	18:25	19:01
TA	Turkey, Ankara	01:47	02:20	17:14	17:46
	Istanbul	01:58	02:31	17:34	18:07
TF	Iceland, Reykjavik	——	03:10	23:43	——
TG	Guatemala, Guatemala City	11:08	11:31	00:29	00:53
TI	Costa Rica, San Jose	10:51	11:14	23:55	00:18
TI9	Cocos Is.	11:11	11:33	23:59	00:22
TJ	Cameroon, Yaounde	04:40	05:02	17:22	17:45
TK	Corsica	03:14	03:48	18:56	19:30
TL	Central African Republic, Bangui	04:10	04:33	16:55	17:17
TN	Congo, Brazzaville	04:38	05:01	16:53	17:16
TR	Gabon, Libreville	04:54	05:16	17:25	17:47
TT	Chad, N'Djamena	04:11	04:34	17:23	17:46
TU	Cote d'Ivoire, Abidjan	05:39	06:02	18:27	18:49
TY	Benin, Porto Novo	05:11	05:33	18:02	18:25
TZ	Mali, Bamako	05:42	06:05	18:56	19:19
UA	Russia, European, Leningrad (UA1)	23:10	00:41	19:13	20:44
	Archangel (UA1)	——	22:49	19:44	——
	Murmansk (UA1)	——	——	——	——
	Moscow (UA3)	23:48	00:47	18:08	19:08
	Kuybyshev (UA4)	23:24	00:14	17:02	17:52
	Rostov (UA6)	00:45	01:25	17:15	17:55
UA2	Kaliningrad	01:05	02:01	19:11	20:07
UA9,0	Russia, Asiatic, Novosibirsk (UA9)	20:56	21:52	15:02	15:58
	Perm (UA9)	22:04	23:15	17:11	18:22
	Omsk (UA9)	21:34	22:30	15:40	16:36
	Norilsk (UA0)	——	——	——	——
	Irkutsk (UA0)	19:54	20:43	13:20	14:08
	Vladivostok (UA0)	18:57	19:32	10:49	11:24
	Petropavlovsk (UA0)	16:11	17:01	09:46	10:36
	Khabarovsk (UA0)	18:17	18:58	10:57	11:38
	Krasnoyarsk (UA0)	20:05	21:05	14:29	15:29
	Yakutsk (UA0)	——	17:38	13:01	——
	Wrangel Island (UA0)	——	——	——	——
	Kyzyl (UA0Y)	20:38	21:25	13:55	14:42
UJ-UM	Uzbekistan, Bukhoro	23:41	00:14	15:07	15:40
	Tashkent	23:16	23:50	14:52	15:26
UN-UQ	Kazakhstan, Alma-Ata	22:37	23:12	14:29	15:04
UR-UZ,EM-EO	Ukraine, Kiev	01:03	01:48	18:05	18:49
V2	Antigua & Barbuda, St. Johns	09:08	09:32	22:39	23:03
V3	Belize, Belmopan	10:56	11:20	00:27	00:51
V4	St. Kitts & Nevis	09:11	09:35	22:42	23:06
V5	Namibia, Windhoek	05:02	05:26	16:14	16:38
V6	Micronesia, Ponape	18:47	19:10	07:40	08:03
V7	Marshall Islands, Kwajalein	18:07	18:30	07:08	07:31
V8	Brunei, Bandar Seri Begawan	21:44	22:07	10:31	10:53
VK	Australia, Canberra (VK1)	20:38	21:06	06:57	07:25
	Sydney (VK2)	20:27	20:54	06:53	07:20
	Melbourne (VK3)	21:00	21:29	07:07	07:36
	Brisbane (VK4)	20:07	20:32	07:00	07:25
	Adelaide (VK5)	21:19	21:47	07:40	08:08
	Perth (VK6)	22:44	23:11	09:19	09:46
	Hobart, Tasmania (VK7)	21:03	21:35	06:43	07:15
	Darwin (VK8)	21:09	21:32	08:57	09:20
VK0	Heard Is.	02:31	03:13	10:57	11:38
VK0	Macquarie Is.	20:56	21:41	05:05	05:50
VK9C	Cocos-Keeling Is.	23:25	23:48	11:14	11:37
VK9L	Lord Howe Is.	19:51	20:17	06:26	06:53
VK9M	Mellish Reef	19:38	20:02	07:08	07:32
VK9N	Norfolk Is.	19:10	19:36	05:57	06:23
VK9W	Willis Is.	20:01	20:25	07:36	07:59
VK9X	Christmas Is.	22:47	23:10	10:41	11:04
VP2E	Anguilla	09:10	09:35	22:46	23:10
VP2M	Montserrat	09:10	09:34	22:40	23:04
VP2V	British Virgin Is., Tortola	09:17	09:41	22:53	23:17
VP5	Turks & Caicos Islands, Grand Turk	09:37	10:01	23:25	23:50
VP8	Falkland Islands, Stanley	11:11	11:51	19:48	20:28
VP8	So. Georgia Is.	09:57	10:41	18:10	18:54
VP8	So. Orkney Is.	11:04	12:02	17:59	18:57
VP8	So. Sandwich Islands, Saunders Is.	09:33	10:23	17:07	17:57
VP8	So. Shetland Is., King George Is.	12:04	13:07	18:35	19:39
VP9	Bermuda	08:43	09:11	23:23	23:51
VQ9	Chagos, Diego Garcia	00:55	01:17	13:00	13:23
VR2,VS6	Hong Kong	21:13	21:37	11:05	11:30
VR6	Pitcairn Is.	14:55	15:20	01:57	02:22
VU	India, Bombay	00:06	00:30	13:44	14:08
	Calcutta	22:55	23:20	12:49	13:14
	New Delhi	23:26	23:52	13:47	14:13
	Bangalore	23:59	00:22	13:14	13:37
VU	Andaman Islands, Port Blair	23:00	23:23	12:11	12:34
VU	Laccadive Is.	00:23	00:46	13:27	13:50
XE	Mexico, Mexico City (XE1)	11:33	11:57	01:12	01:37
	Chihuahua (XE2)	11:38	12:05	02:00	02:26
	Merida (XE3)	10:51	11:16	00:38	01:03
XF4	Revilla Gigedo	12:23	12:47	02:01	02:26
XT	Burkina Faso, Ouagadougou	05:17	05:40	18:30	18:53
XU	Cambodia, Phnom Penh	22:12	22:35	11:23	11:46
XW	Laos, Viangchan	22:09	22:33	11:43	12:07
XX9	Macao	21:16	21:40	11:07	11:32
XZ	Myanmar, Yangon	22:38	23:01	12:07	12:31
YA	Afghanistan, Kandahar	00:05	00:32	14:38	15:05
	Kabul	23:41	00:10	14:33	15:02
YB-YD	Indonesia, Jakarta	22:35	22:58	10:44	11:07
	Medan	22:51	23:14	11:33	11:56
	Pontianak	22:15	22:37	10:45	11:07
	Jayapura	20:14	20:36	08:35	08:57
YI	Iraq, Baghdad	01:24	01:53	16:08	16:36
YJ	Vanuatu, Port Vila	18:48	19:12	06:18	06:42
YK	Syria, Damascus	01:55	02:24	16:42	17:11
YL	Latvia, Riga	00:27	01:32	19:12	20:16
YN	Nicaragua, Managua	10:55	11:18	00:06	00:30
YO	Romania, Bucharest	01:55	02:31	17:56	18:33
YS	El Salvador, San Salvador	11:04	11:28	00:22	00:46
YU	Yugoslavia, Belgrade	02:15	02:52	18:21	18:57
YV	Venezuela, Caracas	09:42	10:05	22:48	23:11
YV0	Aves Is.	09:19	09:42	22:44	23:08
Z2	Zimbabwe, Harare	03:58	04:21	15:27	15:51
Z3	Macedonia, (ex Yugoslav), Skopje	02:25	02:59	18:07	18:41
ZA	Albania, Tirane	02:34	03:07	18:11	18:44
ZB2	Gibraltar	04:34	05:04	19:36	20:06
ZC4	British Cyprus	02:05	02:34	16:58	17:27
ZD7	St. Helena	06:22	06:46	17:58	18:21
ZD8	Ascension Is.	06:43	07:06	18:46	19:09
ZD9	Tristan da Cunha	07:28	07:57	17:38	18:07
ZF	Cayman Is.	10:21	10:45	00:01	00:25
ZK1	No. Cook Is., Manihiki	16:33	16:56	04:28	04:51
ZK1	So. Cook Is., Rarotonga	16:47	17:11	04:04	04:28
ZK2	Niue	17:19	17:43	04:44	05:08
ZK3	Tokelaus, Atafu	17:17	17:40	05:19	05:41
ZL	New Zealand, Auckland (ZL1)	18:59	19:28	05:10	05:39
	Wellington (ZL2)	19:09	19:40	04:58	05:29
	Christchurch (ZL3)	19:24	19:56	04:59	05:32
	Dunedin (ZL4)	19:39	20:13	04:59	05:34
ZL5	Antarctica, Scott Base	——	——	——	——
ZL7	Chatham Is.	18:42	19:14	04:14	04:47
ZL8	Kermadec Is.	18:14	18:40	05:00	05:26
ZL9	Auckland & Campbell Is., Auckland	20:10	20:49	04:56	05:34
	Campbell Is.	20:06	20:47	04:36	05:17
ZP	Paraguay, Asuncion	10:06	10:31	21:07	21:32
ZS	South Africa, Cape Town (ZS1)	05:18	05:46	15:44	16:11
	Port Elizabeth (ZS2)	04:49	05:17	15:14	15:42
	Bloemfontein (ZS4)	04:38	05:04	15:24	15:50
	Durban (ZS5)	04:20	04:46	15:03	15:29
	Johannesburg (ZS6)	04:24	04:49	15:22	15:47
ZS8	Marion Is.	04:32	05:07	13:47	14:22

Pref	State/Province/Country/City	Tlight	Sunrise	Sunset	Tlight
VE1	New Brunswick, Saint John	07:58	08:36	00:15	00:53
	Nova Scotia, Halifax	07:52	08:29	00:02	00:39
	Prince Edward Island, Charlottetown	07:41	08:20	00:07	00:46
VE2	Quebec, Montreal	08:27	09:05	00:46	01:24
	Quebec City	08:10	08:50	00:42	01:22
VE3	Ontario, London	09:10	09:45	01:07	01:43
	Ottawa	08:36	09:14	00:54	01:32
	Sudbury	08:51	09:30	01:20	01:59
	Toronto	08:59	09:35	01:02	01:38
VE4	Manitoba, Winnipeg	09:34	10:19	02:40	03:25
VE5	Saskatchewan, Regina	10:00	10:46	03:13	03:59
	Saskatoon	09:56	10:45	03:31	04:20
VE6	Alberta, Calgary	10:34	11:21	03:54	04:41
	Edmonton	10:11	11:04	04:06	04:59
VE7	British Columbia, Prince George	10:44	11:39	04:46	05:40
	Prince Rupert	11:10	12:06	05:19	06:14
	Vancouver	11:23	12:06	04:21	05:05
VE8	Northwest Territories, Yellowknife	——	09:38	05:39	——
	Resolute	——	——	——	——
VY1	Yukon, Whitehorse	——	11:27	06:36	——
VO1	Newfoundland, St. John's	06:52	07:32	23:32	00:12
VO2	Labrador, Goose Bay	06:40	07:33	00:33	01:25
W1	Connecticut, Hartford	08:41	09:15	00:28	01:03
	Maine, Bangor	08:11	08:49	00:24	01:01
	Portland	08:23	08:59	00:26	01:02
	Massachusetts, Boston	08:32	09:07	00:24	00:59
	New Hampshire, Concord	08:30	09:06	00:29	01:04
	Rhode Island, Providence	08:36	09:10	00:23	00:58
	Vermont, Montpelier	08:29	09:06	00:37	01:14
W2	New Jersey, Atlantic City	08:58	09:30	00:27	01:00
	New York, Albany	08:41	09:17	00:36	01:11
	Buffalo	09:01	09:36	00:57	01:33
	New York City	08:51	09:24	00:30	01:04
	Syracuse	08:49	09:25	00:47	01:23
W3	Delaware, Wilmington	09:01	09:34	00:33	01:05
	District of Columbia, Washington	09:10	09:42	00:36	01:08
	Maryland, Baltimore	09:07	09:39	00:36	01:08
	Pennsylvania, Harrisburg	09:04	09:37	00:40	01:13
	Philadelphia	08:59	09:32	00:32	01:05
	Pittsburgh	09:16	09:49	00:53	01:26
	Scranton	08:55	09:29	00:39	01:13
W4	Alabama, Montgomery	10:09	10:38	00:55	01:23
	Florida, Jacksonville	09:57	10:25	00:31	00:59
	Miami	10:03	10:29	00:14	00:41
	Pensacola	10:19	10:47	00:53	01:21
	Georgia, Atlanta	09:58	10:27	00:51	01:20
	Savannah	09:50	10:18	00:33	01:01
	Kentucky, Lexington	09:44	10:15	01:03	01:35
	Louisville	09:48	10:20	01:09	01:41
	North Carolina, Charlotte	09:39	10:08	00:40	01:10
	Raleigh	09:28	09:58	00:33	01:03
	Wilmington	09:30	10:00	00:26	00:55
	South Carolina, Columbia	09:43	10:13	00:38	01:07
	Tennessee, Knoxville	09:48	10:19	00:55	01:25
	Memphis	10:16	10:46	01:17	01:47
	Nashville	09:59	10:30	01:07	01:38
	Virginia, Norfolk	09:15	09:46	00:27	00:58
	Richmond	09:17	09:48	00:33	01:05
W5	Arkansas, Little Rock	10:26	10:56	01:25	01:54
	Louisiana, New Orleans	10:32	10:59	01:04	01:31
	Shreveport	10:39	11:07	01:25	01:53
	Mississippi, Jackson	10:25	10:54	01:10	01:38
	New Mexico, Albuquerque	11:23	11:52	02:24	02:53
	Oklahoma, Oklahoma City	10:44	11:14	01:48	02:18
	Texas, Abilene	11:03	11:31	01:49	02:17
	Amarillo	11:03	11:32	02:04	02:34
	Dallas	10:50	11:19	01:38	02:06
	El Paso	11:32	12:00	02:14	02:42
	San Antonio	11:07	11:34	01:36	02:03
W6	California, Los Angeles	12:12	12:41	03:07	03:36
	San Francisco	12:16	12:47	03:34	04:06
W7	Arizona, Flagstaff	11:42	12:12	02:44	03:14
	Phoenix	11:49	12:18	02:41	03:10
	Idaho, Boise	11:27	12:03	03:29	04:05
	Pocatello	11:15	11:51	03:12	03:47
	Montana, Billings	10:45	11:23	03:07	03:45
	Butte	11:00	11:39	03:24	04:02
	Great Falls	10:47	11:27	03:25	04:06
	Nevada, Las Vegas	11:52	12:23	03:00	03:31
	Reno	11:59	12:32	03:29	04:02
	Oregon, Portland	11:43	12:21	04:02	04:41
	Utah, Salt Lake City	11:22	11:56	03:02	03:35
	Washington, Seattle	11:30	12:11	04:10	04:51
	Spokane	11:10	11:51	03:51	04:32
	Wyoming, Cheyenne	10:52	11:26	02:34	03:08
	Sheridan	10:44	11:21	02:57	03:34
W8	Michigan, Detroit	09:20	09:55	01:11	01:46
	Grand Rapids	09:28	10:03	01:25	02:00
	Sault Ste. Marie	09:05	09:44	01:33	02:13
	Traverse City	09:19	09:56	01:31	02:08
	Ohio, Cincinnati	09:39	10:12	01:07	01:39
	Cleveland	09:18	09:52	01:03	01:38
	Columbus	09:30	10:03	01:04	01:36
	West Virginia, Charleston	09:30	10:02	00:53	01:25
W9	Illinois, Chicago	09:40	10:15	01:28	02:03
	Indiana, Indianapolis	09:43	10:16	01:16	01:48
	Wisconsin, Green Bay	09:30	10:07	01:40	02:17

Pref	State/Province/Country/City	Tlight	Sunrise	Sunset	Tlight
	Milwaukee	09:36	10:12	01:34	02:09
W0	Colorado, Denver	10:59	11:32	02:31	03:03
	Grand Junction	11:16	11:48	02:43	03:15
	Iowa, Des Moines	10:06	10:40	01:51	02:26
	Kansas, Pratt	10:41	11:13	01:59	02:30
	Wichita	10:36	11:07	01:54	02:25
	Minnesota, Duluth	09:34	10:14	02:06	02:45
	Minneapolis	09:48	10:26	02:03	02:40
	Missouri, Columbia	10:11	10:43	01:38	02:10
	Kansas City	10:20	10:52	01:47	02:19
	St. Louis	10:04	10:36	01:28	02:00
	Nebraska, North Platte	10:36	11:10	02:18	02:52
	Omaha	10:16	10:50	02:00	02:33
	North Dakota, Fargo	09:52	10:32	02:25	03:05
	South Dakota, Rapid City	10:32	11:09	02:39	03:15
-		05:35	05:58	18:05	18:27
1A0	S.M.O.M	03:00	03:35	18:48	19:23
1S	Spratly Is.	21:52	22:14	10:53	11:16
3A	Monaco	03:12	03:48	19:15	19:51
3B6	Agalega	02:06	02:29	14:00	14:23
3B7	St. Brandon	02:04	02:27	13:37	14:00
3B8	Mauritius	02:20	02:44	13:38	14:02
3B9	Rodriguez Is.	01:55	02:19	13:16	13:40
3C	Equatorial Guinea, Bata	04:53	05:15	17:29	17:51
	Malabo	04:53	05:16	17:36	17:59
3C0	Annobon Is.	05:15	05:38	17:40	18:02
3D2	Fiji Is., Suva	18:13	18:36	05:39	06:03
3D2	Conway Reef	18:34	18:59	05:48	06:13
3D2	Rotuma Is.	18:05	18:28	05:52	06:15
3DA	Swaziland, Mbabane	04:17	04:42	15:11	15:37
3V	Tunisia, Tunis	03:29	04:00	18:41	19:12
3W, XV	Vietnam, Ho Chi Minh City (Saigon)	22:08	22:32	11:17	11:40
	Hanoi	21:51	22:16	11:40	12:05
3X	Guinea, Conakry	06:13	06:36	19:16	19:39
3Y	Bouvet	07:25	08:10	15:26	16:10
3Y	Peter I Is.	15:37	——	——	20:30
4J, 4K	Azerbaijan, Baku	00:37	01:10	16:13	16:47
4L	Georgia, Tbilisi	00:52	01:26	16:38	17:12
4S	Sri Lanka, Colombo	00:03	00:26	12:57	13:20
4U	ITU Geneva	03:04	03:43	19:30	20:09
4U	United Nations Hq.	08:51	09:24	00:30	01:04
4X, 4Z	Israel, Jerusalem	02:05	02:34	16:47	17:15
5A	Libya, Tripoli	03:34	04:03	18:20	18:48
	Benghazi	03:05	03:34	17:49	18:17
5B	Cyprus, Nicosia	02:02	02:32	17:03	17:33
5H	Tanzania, Dar es Salaam	03:09	03:32	15:15	15:37
5N	Nigeria, Lagos	05:10	05:33	18:03	18:25
5R	Madagascar, Antananarivo	02:58	03:21	14:21	14:45
5T	Mauritania, Nouakchott	06:04	06:29	19:42	20:06
5U	Niger, Niamey	05:02	05:25	18:21	18:44
5V	Togo, Lome	05:20	05:43	18:10	18:33
5W	Western Samoa, Apia	17:25	17:48	05:08	05:31
5X	Uganda, Kampala	03:24	03:47	15:55	16:18
5Z	Kenya, Nairobi	04:30	04:53	16:55	17:18
6W	Senegal, Dakar	06:18	06:41	19:41	20:05
6Y	Jamaica, Kingston	10:08	10:32	23:45	00:09
7O	Yemen, Aden	02:11	02:35	15:28	15:51
	Sanaa	02:09	02:33	15:36	15:59
7P	Lesotho, Maseru	04:37	05:03	15:19	15:45
7Q	Malawi, Lilongwe	03:44	04:07	15:25	15:48
	Blantyre	03:42	04:06	15:17	15:40
7X	Algeria, Algiers	03:58	04:29	19:09	19:40
8P	Barbados, Bridgetown	09:09	09:33	22:26	22:50
8Q	Maldive Is.	00:34	00:56	13:19	13:42
8R	Guyana, Georgetown	09:16	09:38	22:09	22:32
9A	Croatia, Zagreb	02:27	03:05	18:49	19:27
9G	Ghana, Accra	05:26	05:49	18:15	18:38
9H	Malta	03:15	03:45	18:22	18:52
9J	Zambia, Lusaka	04:08	04:32	15:44	16:08
9K	Kuwait	01:22	01:49	15:51	16:18
9L	Sierra Leone, Freetown	06:12	06:35	19:12	19:35
9M2	West Malaysia, Kuala Lumpur	22:43	23:06	11:24	11:47
9M6,8	East Malaysia, Sabah, Sandakan(9M6)	21:32	21:55	10:22	10:45
	Sarawak, Kuching (9M8)	22:11	22:34	10:46	11:09
9N	Nepal, Kathmandu	22:56	23:23	13:17	13:43
9Q	Zaire, Kinshasa	04:41	05:04	16:56	17:19
	Kisangani	03:53	04:16	16:25	16:47
	Lubumbashi	04:05	04:28	15:54	16:17
9U	Burundi, Bujumbura	03:43	04:06	16:02	16:24
9V	Singapore	22:37	23:00	11:12	11:34
9X	Rwanda, Kigali	03:38	04:01	16:01	16:23
9Y	Trinidad & Tobago , Port of Spain	09:21	09:44	22:28	22:52
A2	Botswana, Gaborone	04:35	05:00	15:35	16:00
A3	Tonga, Nukualofa	17:52	18:17	05:07	05:32
A4	Oman, Musqat	00:54	01:19	14:54	15:20
A5	Bhutan, Thimbu	22:41	23:08	12:59	13:26
A6	United Arab Emirates, Abu Dhabi	01:09	01:35	15:14	15:40
A7	Qatar, Ad-Dawhah	01:18	01:44	15:27	15:52
A9	Bahrein, Al-Manamah	01:19	01:45	15:32	15:58
AP	Pakistan, Karachi	00:16	00:42	14:23	14:49
AP	Islamabad	23:27	23:57	14:20	14:49
BV	Taiwan, Taipei	20:38	21:04	10:46	11:12
BY	Peoples Rep. of China, Beijing	20:12	20:45	11:46	12:19
	Harbin	19:04	19:43	11:26	12:04
	Shanghai	20:22	20:50	11:00	11:28
	Fuzhou	20:45	21:11	10:57	11:23
	Xian	21:03	21:32	11:59	12:28

Pref	State/Province/Country/City	Tlight	Sunrise	Sunset	Tlight
	Chongqing	21:26	21:53	11:57	12:24
	Chengdu	21:33	22:01	12:09	12:37
	Lhasa	22:27	22:55	12:58	13:25
	Urumqi	21:51	22:27	13:55	14:31
	Kashi	22:56	23:28	14:26	14:58
C2	Nauru	18:28	18:51	06:56	07:19
C3	Andorra	03:41	04:16	19:34	20:09
C5	The Gambia, Banjul	06:17	06:40	19:36	19:59
C6	Bahamas, Nassau	09:54	10:20	00:02	00:28
C9	Mozambique, Maputo	04:10	04:35	15:06	15:31
	Mozambique	03:18	03:41	14:55	15:19
CE	Chile, Santiago	11:19	11:47	21:42	22:10
CE0	Easter Island	13:40	14:06	00:32	00:57
CE0	Juan Fernandez	11:51	12:19	22:14	22:42
CE0	San Felix	11:42	12:07	22:36	23:02
CM,CO	Cuba, Havana	10:19	10:44	00:17	00:43
CN	Morocco, Casablanca	04:51	05:20	19:43	20:12
CP	Bolivia, La Paz	10:37	11:00	22:09	22:33
CT	Portugal, Lisbon	04:40	05:12	20:04	20:36
CT3	Madeira Islands, Funchal	05:31	06:00	20:18	20:46
CU	Azores , Ponta Delgada	05:49	06:21	21:07	21:38
CX	Uruguay, Montevideo	10:24	10:52	20:40	21:08
CY0	Sable Is.	07:41	08:17	23:45	00:21
CY9	St. Paul Is.	07:24	08:04	23:59	00:39
D2	Angola, Luanda	04:57	05:20	16:57	17:19
D4	Cape Verde, Praia	06:41	07:05	20:05	20:29
D6	Comoros, Moroni	03:00	03:23	14:49	15:12
DA-DL,Y2-Y9	Fed. Rep. of Germany, Bonn	02:32	03:18	19:48	20:34
	Berlin	01:52	02:43	19:32	20:23
DU	Philippines, Manilla	21:04	21:27	10:27	10:50
E3	Eritrea, Asmara	02:31	02:54	15:57	16:20
EA	Spain, Madrid	04:11	04:44	19:48	20:21
EA6	Balearic Is., Palma	03:49	04:22	19:20	19:52
EA8	Canary Is., Las Palmas	05:33	06:00	19:57	20:24
EA9	Ceuta & Melilla, Ceuta	04:34	05:04	19:40	20:10
	Melilla	04:27	04:57	19:29	19:59
EI	Ireland, Dublin	03:04	03:56	20:56	21:49
EK	Armenia, Yerevan	00:59	01:32	16:35	17:08
EL	Liberia, Monrovia	06:07	06:30	18:59	19:22
EP	Iran, Tehran	00:46	01:16	15:52	16:22
ER	Moldova, Kishinev	01:29	02:09	18:03	18:43
ES	Estonia, Tallinn	23:29	01:03	19:41	21:14
ET	Ethiopia, Addis Ababa	02:44	03:07	15:46	16:09
EU,EV,EW	Belarus, Minsk	00:43	01:37	18:44	19:39
EX	Kyrgyzstan, Frunze	22:47	23:22	14:43	15:19
EY	Tajikistan, Samarkand	23:32	00:04	15:03	15:36
	Dushanbe	23:26	23:58	14:53	15:26
EZ	Turkmenistan, Ashkhabad	00:12	00:43	15:32	16:03
F	France, Paris	03:04	03:47	19:57	20:40
FG	Guadeloupe	09:12	09:36	22:40	23:04
FJ, FS	St. Martin	09:13	09:37	22:50	23:14
FH	Mayotte	02:56	03:19	14:41	15:04
FK	New Caledonia, Noumia	19:08	19:32	06:18	06:43
FM	Martinique	09:12	09:35	22:35	22:58
FO	Clipperton Is.	12:33	12:56	01:40	02:03
FO	Fr. Polynesia, Tahiti	16:43	17:07	04:11	04:35
	Rurutu, (Austral Is.)	20:09	20:33	07:19	07:43
	Hiva Oa, (Marquesas Is.)	15:08	15:31	03:03	03:26
FP	St. Pierre & Miquelon, St. Pierre	07:10	07:50	23:41	00:20
FR/G	Glorioso	02:46	03:08	14:35	14:58
FR/J,E	Juan de Nova	03:13	03:37	14:43	15:07
	Europa	03:32	03:57	14:43	15:07
FR	Reunion	02:29	02:53	13:44	14:08
FR/T	Tromelin	02:25	02:48	13:59	14:22
FT8W	Crozet	03:39	04:14	12:52	13:27
FT8X	Kerguelen	02:41	03:19	11:30	12:07
FT8Y	Antarctica, Dumont D'Urville	23:44	01:38	03:44	05:39
FT8Z	Amsterdam & St. Paul Is., Amsterdam	01:35	02:04	11:37	12:07
FW	Wallis & Futuna Is., Wallis	17:43	18:06	05:27	05:50
FY	Fr. Guiana, Cayenne	08:55	09:18	21:43	22:06
G	England, London	02:55	03:42	20:21	21:08
GD	Isle of Man	02:47	03:43	20:56	21:51
GI	Northern Ireland, Belfast	02:50	03:46	21:03	22:00
GJ	Jersey	03:19	04:02	20:18	21:01
GM	Scotland, Glasgow	02:29	03:31	21:05	22:07
	Aberdeen	02:01	03:11	21:08	22:18
GU	Guernsey	03:20	04:03	20:21	21:04
GW	Wales, Cardiff	03:07	03:55	20:33	21:21
H4	Solomon Islands, Honiara	19:11	19:34	07:08	07:31
HA	Hungary, Budapest	02:05	02:46	18:44	19:24
HB	Switzerland, Bern	02:54	03:34	19:28	20:08
HB0	Liechtenstein	02:45	03:25	19:21	20:01
HC	Ecuador, Quito	10:47	11:10	23:16	23:39
HC8	Galapagos Is.	11:38	12:00	00:06	00:28
HH	Haiti, Port-Au-Prince	09:49	10:13	23:28	23:52
HI	Dominican Republic, Santo Domingo	09:40	10:04	23:18	23:43
HK	Colombia, Bogota	10:23	10:46	23:09	23:32
HK0	Malpelo Is.	10:52	11:15	23:36	23:59
HK0	San Adreas	10:39	11:02	23:54	00:17
HL	Korea, Seoul	19:39	20:11	10:56	11:27
HP	Panama, Panama	10:37	11:00	23:39	00:02
HR	Honduras, Tegucigalpa	10:58	11:21	00:19	00:42
HS	Thailand, Bangkok	22:27	22:51	11:47	12:11
HV	Vatican City	03:00	03:34	18:48	19:22
HZ, 7Z	Saudi Arabia, Dharan	01:21	01:47	15:35	16:01
	Mecca	02:14	02:39	16:05	16:30
	Riyadh	01:39	02:04	15:44	16:10

Pref	State/Province/Country/City	Tlight	Sunrise	Sunset	Tlight
I	Italy, Rome	03:00	03:34	18:48	19:22
	Trieste	02:36	03:15	18:57	19:36
	Sicily	03:11	03:43	18:28	18:59
IS	Sardinia, Cagliari	03:25	03:57	18:53	19:25
J2	Djibouti, Djibouti	02:21	02:44	15:32	15:56
J3	Grenada	09:20	09:43	22:33	22:57
J5	Guinea-Bissau, Bissau	06:16	06:39	19:28	19:52
J6	St. Lucia	09:13	09:37	22:34	22:57
J7	Dominica	09:11	09:35	22:38	23:01
J8	St. Vincent	09:16	09:39	22:34	22:57
JA-JS	Japan, Tokyo	18:54	19:25	09:59	10:29
	Nagasaki	19:43	20:12	10:31	11:00
	Sapporo	18:19	18:54	10:17	10:52
JD1	Minami Torishima	18:30	18:56	08:34	09:00
JD1	Ogasawara, Kazan Is.	19:14	19:41	09:34	10:00
JT	Mongolia, Ulan Bator	20:11	20:53	12:54	13:36
JW	Svalbard, Spitsbergen	——	——	——	——
JX	Jan Mayen	——	——	——	——
JY	Jordan, Amman	02:02	02:30	16:45	17:13
KC4	Antarctica, Byrd Station	——	——	——	——
	McMurdo Sound	——	——	——	——
	Palmer Station	13:00	14:26	18:09	19:34
KC6	Belau, Yap	20:05	20:28	09:09	09:32
	Koror	20:24	20:47	09:20	09:42
KG4	Guantanamo Bay	09:57	10:22	23:42	00:07
KH0	Mariana Is., Saipan	19:23	19:47	08:49	09:13
KH1	Baker, Howland Is.	17:18	17:41	05:50	06:12
KH2	Guam, Agana	19:31	19:54	08:50	09:13
KH3	Johnston Is.	16:17	16:41	05:49	06:13
KH4	Midway Is.	16:26	16:53	06:49	07:16
KH5	Palmyra Is.	16:13	16:36	05:03	05:26
KH5K	Kingman Reef	16:13	16:36	05:09	05:32
KH6	Hawaii, Hilo	15:17	15:42	05:01	05:26
	Honolulu	15:25	15:50	05:16	05:40
KH7	Kure Is.	16:29	16:56	06:53	07:20
KH8	American Samoa, Pago Pago	17:23	17:46	05:03	05:26
KH9	Wake Is.	17:51	18:16	07:34	07:58
KL7	Alaska, Adak	14:38	15:27	08:08	08:57
	Anchorage	——	12:20	07:42	——
	Fairbanks	——	10:59	08:47	——
	Juneau	10:32	11:50	06:07	07:26
	Nome	——	12:19	09:47	——
KP1	Navassa Is.	10:00	10:24	23:38	00:02
KP2	Virgin Islands, Charlotte Amalie	09:20	09:44	22:58	23:22
KP4	Puerto Rico, San Juan	09:24	09:49	23:03	23:27
KP5	Desecheo Is.	09:30	09:54	23:08	23:32
LA-LJ	Norway, Oslo	00:06	01:52	20:44	22:31
LU	Argentina, Buenos Aires	10:32	11:00	20:50	21:18
LX	Luxembourg	02:43	03:27	19:45	20:29
LY	Lithuania, Vilna	00:45	01:41	18:57	19:53
LZ	Bulgaria, Sofia	02:13	02:48	18:08	18:43
OA	Peru, Lima	11:04	11:27	22:52	23:15
OD	Lebanon, Beirut	01:58	02:27	16:51	17:21
OE	Austria, Vienna	02:12	02:54	18:58	19:40
OH	Finland, Helsinki	22:59	00:53	19:50	21:43
OH0	Aland Is.	23:19	01:13	20:10	22:03
OJ0	Market Reef	23:17	01:16	20:15	22:13
OK,OL	Czech Rep., Prague	02:07	02:52	19:15	20:00
OM	Slovak, Rep., Bratislava	02:10	02:52	18:55	19:36
ON	Belgium, Brussels	02:41	03:28	19:59	20:46
OX	Greenland, Godthaab	——	04:52	02:04	——
	Thule	——	——	——	——
OY	Faroe Islands, Torshavn	——	02:36	22:21	——
OZ	Denmark, Copenhagen	01:23	02:24	19:57	20:58
P2	Papua New Guinea, Madang	20:02	20:24	08:13	08:36
	Port Moresby	20:03	20:26	08:00	08:23
P4	Aruba, Oranjestad	09:49	10:12	23:02	23:26
PA-PI	Netherlands, Amsterdam	02:27	03:17	20:06	20:56
PJ2,4,9	Netherlands Antilles, Willemstad	09:52	10:16	23:08	23:31
PJ5-8	St. Maarten and Saba, St. Maarten	09:14	09:38	22:50	23:14
PY	Brazil, Brasilia	09:14	09:37	20:48	21:12
	Rio De Janeiro	09:08	09:32	20:16	20:40
	Natal	08:06	08:29	20:15	20:38
	Manaus	09:41	10:04	22:00	22:23
	Porto Alegre	09:53	10:20	20:32	20:58
PY0	Fernando de Noronha	07:51	08:14	20:08	20:30
PY0	St. Peter & St. Paul Rocks	07:31	07:53	20:04	20:27
PY0	Trindade & Martin Vaz Is., Trindade	08:08	08:32	19:25	19:49
PZ	Suriname, Paramaribo	09:05	09:28	21:56	22:18
R1FJ_	Franz Josef Land	——	——	——	——
R1MV_	Malyj Vysotskij Is.	——	00:34	19:40	——
S0	Western Sahara, Smara	05:27	05:53	19:41	20:07
S2	Bangladesh, Dacca	22:46	23:12	12:47	13:13
S5	Slovenia, Ljubljana	02:32	03:11	18:56	19:34
S7	Seychelles, Victoria	02:01	02:23	14:15	14:37
S9	Sao Tome	05:08	05:30	17:38	18:01
SM	Sweden, Stockholm	23:59	01:30	20:07	21:39
SP	Poland, Krakow	01:45	02:30	18:52	19:37
	Warsaw	01:25	02:14	19:00	19:50
ST	Sudan, Khartoum	02:56	03:20	16:23	16:47
ST0	Southern Sudan, Juba	03:20	03:42	16:07	16:30
SU	Egypt, Cairo	02:26	02:53	16:58	17:25
SV	Greece, Athens	02:31	03:02	17:50	18:22
SV/A	Mount Athos	02:20	02:53	17:55	18:28
SV5	Dodecanese, Rhodes	02:18	02:49	17:28	17:58
SV9	Crete	02:34	03:04	17:37	18:07
T2	Tuvalu, Funafuti	17:56	18:18	05:55	06:18

Pref	State/Province/Country/City	Tlight	Sunrise	Sunset	Tlight
T30	West Kiribati, Bonriki	18:00	18:22	06:34	06:57
T31	Central Kiribati, Kanton	17:07	17:29	05:27	05:49
T32	East Kiribati, Christmas Is.	16:01	16:24	04:38	05:00
T33	Banaba Is.	18:18	18:41	06:46	07:09
T5	Somalia, Mogadishu	02:30	02:52	15:07	15:29
T7	San Marino	02:51	03:28	18:56	19:32
T9	Bosnia-Herzegovina, Sarajevo	02:27	03:03	18:32	19:08
TA	Turkey, Ankara	01:47	02:19	17:20	17:52
	Istanbul	01:57	02:31	17:40	18:14
TF	Iceland, Reykjavik	———	02:56	00:02	———
TG	Guatemala, Guatemala City	11:10	11:33	00:33	00:56
TI	Costa Rica, San Jose	10:53	11:16	23:58	00:21
TI9	Cocos Is.	11:13	11:36	00:03	00:25
TJ	Cameroon, Yaounde	04:42	05:05	17:26	17:48
TK	Corsica	03:13	03:48	19:02	19:37
TL	Central African Republic, Bangui	04:13	04:35	16:58	17:21
TN	Congo, Brazzaville	04:41	05:04	16:56	17:19
TR	Gabon, Libreville	04:56	05:19	17:27	17:50
TT	Chad, N'Djamena	04:13	04:36	17:26	17:50
TU	Cote d'Ivoire, Abidjan	05:42	06:04	18:30	18:53
TY	Benin, Porto Novo	05:13	05:36	18:06	18:29
TZ	Mali, Bamako	05:44	06:07	18:59	19:23
UA	Russia, European, Leningrad (UA1)	22:51	00:35	19:25	21:09
	Archangel (UA1)	———	22:32	20:06	———
	Murmansk (UA1)	———	———	———	———
	Moscow (UA3)	23:42	00:44	18:18	19:20
	Kuybyshev (UA4)	23:19	00:12	17:10	18:02
	Rostov (UA6)	00:43	01:24	17:22	18:03
UA2	Kaliningrad	01:00	01:58	19:20	20:19
UA9,0	Russia, Asiatic, Novosibirsk (UA9)	20:50	21:48	15:11	16:09
	Perm (UA9)	21:55	23:10	17:21	18:37
	Omsk (UA9)	21:28	22:26	15:49	16:47
	Norilsk (UA0)	———	———	———	———
	Irkutsk (UA0)	19:51	20:40	13:28	14:17
	Vladivostok (UA0)	18:56	19:32	10:55	11:31
	Petropavlovsk (UA0)	16:07	16:58	09:54	10:46
	Khabarovsk (UA0)	18:15	18:57	11:04	11:47
	Krasnoyarsk (UA0)	19:59	21:01	14:39	15:41
	Yakutsk (UA0)	———	17:30	13:15	———
	Wrangel Island (UA0)	———	———	———	———
	Kyzyl (UA0Y)	20:35	21:23	14:03	14:52
UJ-UM	Uzbekistan, Bukhoro	23:41	00:14	15:13	15:46
	Tashkent	23:16	23:49	14:58	15:32
UN-UQ	Kazakhstan, Alma-Ata	22:36	23:12	14:36	15:11
UR-UZ,EM-EO	Ukraine, Kiev	01:01	01:46	18:12	18:58
V2	Antigua & Barbuda, St. Johns	09:10	09:34	22:43	23:07
V3	Belize, Belmopan	10:57	11:21	00:31	00:55
V4	St. Kitts & Nevis	09:13	09:37	22:46	23:10
V5	Namibia, Windhoek	05:06	05:30	16:15	16:40
V6	Micronesia, Ponape	18:49	19:12	07:44	08:06
V7	Marshall Islands, Kwajalein	18:09	18:32	07:12	07:35
V8	Brunei, Bandar Seri Begawan	21:47	22:09	10:34	10:56
VK	Australia, Canberra (VK1)	20:43	21:12	06:58	07:26
	Sydney (VK2)	20:32	21:00	06:53	07:21
	Melbourne (VK3)	21:05	21:35	07:07	07:37
	Brisbane (VK4)	20:11	20:37	07:01	07:27
	Adelaide (VK5)	21:24	21:53	07:41	08:09
	Perth (VK6)	22:49	23:16	09:20	09:47
	Hobart, Tasmania (VK7)	21:09	21:42	06:42	07:15
	Darwin (VK8)	21:13	21:36	08:59	09:22
VK0	Heard Is.	02:38	03:21	10:54	11:37
VK0	Macquarie Is.	21:04	21:49	05:03	05:48
VK9C	Cocos-Keeling Is.	23:29	23:52	11:16	11:39
VK9L	Lord Howe Is.	19:55	20:22	06:27	06:54
VK9M	Mellish Reef	19:42	20:06	07:10	07:34
VK9N	Norfolk Is.	19:14	19:41	05:58	06:24
VK9W	Willis Is.	20:05	20:28	07:38	08:01
VK9X	Christmas Is.	22:50	23:13	10:44	11:06
VP2E	Anguilla	09:12	09:36	22:50	23:14
VP2M	Montserrat	09:12	09:36	22:44	23:08
VP2V	British Virgin Is., Tortola	09:18	09:42	22:57	23:21

June 21   Sunrise and Sunset, in UTC

Pref	State/Province/Country/City	Tlight	Sunrise	Sunset	Tlight
VP5	Turks & Caicos Islands, Grand Turk	09:38	10:03	23:29	23:54
VP8	Falkland Islands, Stanley	11:18	11:59	19:46	20:27
VP8	So. Georgia Is.	10:05	10:49	18:08	18:52
VP8	So. Orkney Is.	11:13	12:13	17:53	18:53
VP8	So. Sandwich Islands, Saunders Is.	09:41	10:33	17:03	17:55
VP8	So. Shetland Is., King George Is.	12:14	13:20	18:29	19:35
VP9	Bermuda	08:43	09:12	23:28	23:56
VQ9	Chagos, Diego Garcia	00:58	01:21	13:02	13:25
VR2,VS6	Hong Kong	21:14	21:39	11:09	11:34
VR6	Pitcairn Is.	14:59	15:24	01:59	02:24
VU	India, Bombay	00:07	00:32	13:48	14:13
	Calcutta	22:57	23:22	12:53	13:18
	New Delhi	23:26	23:53	13:51	14:18
	Bangalore	00:01	00:24	13:17	13:41
VU	Andaman Islands, Port Blair	23:02	23:26	12:14	12:38
VU	Laccadive Is.	00:25	00:48	13:30	13:53
XE	Mexico, Mexico City (XE1)	11:34	11:59	01:17	01:41
	Chihuahua (XE2)	11:39	12:06	02:04	02:31
	Merida (XE3)	10:53	11:18	00:42	01:07
XF4	Revilla Gigedo	12:25	12:49	02:05	02:30
XT	Burkina Faso, Ouagadougou	05:19	05:42	18:33	18:56
XU	Cambodia, Phnom Penh	22:14	22:38	11:26	11:50
XW	Laos, Viangchan	22:10	22:34	11:47	12:11
XX9	Macao	21:17	21:42	11:12	11:37
XZ	Myanmar, Yangon	22:39	23:03	12:11	12:35
YA	Afghanistan, Kandahar	00:05	00:33	14:43	15:11
	Kabul	23:41	00:11	14:38	15:07
YB-YD	Indonesia, Jakarta	22:38	23:01	10:47	11:09
	Medan	22:54	23:16	11:36	11:59
	Pontianak	22:18	22:40	10:48	11:10
	Jayapura	20:17	20:39	08:37	09:00
YI	Iraq, Baghdad	01:24	01:53	16:13	16:42
YJ	Vanuatu, Port Vila	18:52	19:16	06:20	06:44
YK	Syria, Damascus	01:56	02:25	16:47	17:16
YL	Latvia, Riga	00:20	01:28	19:21	20:30
YN	Nicaragua, Managua	10:57	11:20	00:10	00:33
YO	Romania, Bucharest	01:54	02:31	18:03	18:40
YS	El Salvador, San Salvador	11:06	11:30	00:26	00:49
YU	Yugoslavia, Belgrade	02:14	02:51	18:27	19:05
YV	Venezuela, Caracas	09:44	10:07	22:51	23:14
YV0	Aves Is.	09:20	09:44	22:48	23:12
Z2	Zimbabwe, Harare	04:02	04:25	15:29	15:53
Z3	Macedonia, (ex Yugoslav), Skopje	02:24	02:58	18:13	18:47
ZA	Albania, Tirane	02:33	03:07	18:17	18:51
ZB2	Gibraltar	04:34	05:04	19:41	20:12
ZC4	British Cyprus	02:05	02:35	17:03	17:33
ZD7	St. Helena	06:26	06:50	18:00	18:23
ZD8	Ascension Is.	06:46	07:09	18:48	19:11
ZD9	Tristan da Cunha	07:33	08:02	17:39	18:08
ZF	Cayman Is.	10:22	10:47	00:05	00:30
ZK1	No. Cook Is., Manihiki	16:37	17:00	04:31	04:53
ZK1	So. Cook Is., Rarotonga	16:51	17:15	04:06	04:30
ZK2	Niue	17:23	17:47	04:46	05:10
ZK3	Tokelaus, Atafu	17:20	17:43	05:21	05:44
ZL	New Zealand, Auckland (ZL1)	19:04	19:33	05:11	05:40
	Wellington (ZL2)	19:15	19:46	04:58	05:29
	Christchurch (ZL3)	19:30	20:02	04:59	05:32
	Dunedin (ZL4)	19:45	20:20	04:59	05:33
ZL5	Antarctica, Scott Base	———	———	———	———
ZL7	Chatham Is.	18:47	19:21	04:14	04:47
ZL8	Kermadec Is.	18:19	18:45	05:01	05:27
ZL9	Auckland & Campbell Is., Auckland	20:17	20:56	04:54	05:33
	Campbell Is.	20:14	20:55	04:34	05:16
ZP	Paraguay, Asuncion	10:10	10:35	21:09	21:34
ZS	South Africa, Cape Town (ZS1)	05:23	05:51	15:44	16:12
	Port Elizabeth (ZS2)	04:54	05:22	15:15	15:43
	Bloemfontein (ZS4)	04:42	05:09	15:25	15:51
	Durban (ZS5)	04:25	04:51	15:04	15:30
	Johannesburg (ZS6)	04:29	04:54	15:24	15:49
ZS8	Marion Is.	04:39	05:14	13:46	14:21

Pref	State/Province/Country/City	Tlight	Sunrise	Sunset	Tlight
VE1	New Brunswick, Saint John	08:06	08:44	00:14	00:51
	Nova Scotia, Halifax	08:00	08:36	00:01	00:38
	Prince Edward Island, Charlottetown	07:50	08:28	00:05	00:43
VE2	Quebec, Montreal	08:35	09:13	00:45	01:22
	Quebec City	08:19	08:58	00:40	01:19
VE3	Ontario, London	09:18	09:53	01:06	01:41
	Ottawa	08:44	09:22	00:53	01:30
	Sudbury	09:00	09:38	01:18	01:57
	Toronto	09:07	09:43	01:01	01:37
VE4	Manitoba, Winnipeg	09:44	10:28	02:38	03:21
VE5	Saskatchewan, Regina	10:10	10:55	03:11	03:55
	Saskatoon	10:07	10:55	03:27	04:15
VE6	Alberta, Calgary	10:45	11:30	03:51	04:36
	Edmonton	10:23	11:14	04:03	04:54
VE7	British Columbia, Prince George	10:57	11:49	04:42	05:34
	Prince Rupert	11:23	12:16	05:15	06:08
	Vancouver	11:32	12:15	04:19	05:01
VE8	Northwest Territories, Yellowknife	——	09:56	05:28	——
	Resolute	——	——	——	——
VY1	Yukon, Whitehorse	09:58	11:42	06:28	08:11
VO1	Newfoundland, St. John's	07:01	07:41	23:30	00:09
VO2	Labrador, Goose Bay	06:52	07:43	00:29	01:20
W1	Connecticut, Hartford	08:49	09:22	00:28	01:02
	Maine, Bangor	08:20	08:56	00:23	00:59
	Portland	08:31	09:06	00:25	01:00
	Massachusetts, Boston	08:40	09:14	00:23	00:58
	New Hampshire, Concord	08:38	09:13	00:28	01:03
	Rhode Island, Providence	08:44	09:17	00:23	00:56
	Vermont, Montpelier	08:37	09:13	00:36	01:12
W2	New Jersey, Atlantic City	09:05	09:37	00:27	00:59
	New York, Albany	08:49	09:24	00:35	01:10
	Buffalo	09:09	09:44	00:56	01:31
	New York City	08:58	09:31	00:30	01:03
	Syracuse	08:57	09:32	00:46	01:21
W3	Delaware, Wilmington	09:08	09:40	00:32	01:04
	District of Columbia, Washington	09:17	09:49	00:36	01:07
	Maryland, Baltimore	09:14	09:46	00:35	01:07
	Pennsylvania, Harrisburg	09:12	09:44	00:40	01:12
	Philadelphia	09:06	09:39	00:32	01:04
	Pittsburgh	09:24	09:56	00:52	01:25
	Scranton	09:02	09:36	00:38	01:12
W4	Albama, Montgomery	10:16	10:44	00:55	01:23
	Florida, Jacksonville	10:03	10:30	00:32	00:59
	Miami	10:09	10:35	00:16	00:41
	Pensacola	10:25	10:52	00:54	01:21
	Georgia, Atlanta	10:04	10:33	00:51	01:20
	Savannah	09:56	10:24	00:34	01:02
	Kentucky, Lexington	09:50	10:22	01:03	01:34
	Louisville	09:55	10:26	01:09	01:40
	North Carolina, Charlotte	09:45	10:15	00:40	01:10
	Raleigh	09:34	10:04	00:33	01:03
	Wilmington	09:37	10:06	00:26	00:55
	South Carolina, Columbia	09:50	10:19	00:38	01:07
	Tennessee, Knoxville	09:55	10:25	00:55	01:25
	Memphis	10:23	10:52	01:17	01:47
	Nashville	10:06	10:36	01:07	01:37
	Virginia, Norfolk	09:21	09:52	00:27	00:58
	Richmond	09:24	09:55	00:33	01:04
W5	Arkansas, Little Rock	10:33	11:02	01:25	01:54
	Louisiana, New Orleans	10:38	11:05	01:04	01:32
	Shreveport	10:45	11:13	01:25	01:53
	Mississippi, Jackson	10:32	11:00	01:11	01:39
	New Mexico, Albuquerque	11:29	11:58	02:24	02:53
	Oklahoma, Oklahoma City	10:51	11:21	01:48	02:18
	Texas, Abilene	11:09	11:37	01:49	02:17
	Amarillo	11:09	11:39	02:04	02:34
	Dallas	10:56	11:25	01:38	02:07
	El Paso	11:38	12:06	02:15	02:42
	San Antonio	11:13	11:40	01:37	02:04
W6	California, Los Angeles	12:18	12:47	03:07	03:36
	San Francisco	12:23	12:54	03:34	04:05
W7	Arizona, Flagstaff	11:49	12:18	02:44	03:14
	Phoenix	11:56	12:24	02:41	03:10
	Idaho, Boise	11:35	12:10	03:28	04:03
	Pocatello	11:23	11:58	03:11	03:45
	Montana, Billings	10:54	11:31	03:05	03:43
	Butte	11:09	11:46	03:22	04:00
	Great Falls	10:56	11:35	03:24	04:03
	Nevada, Las Vegas	11:59	12:29	03:00	03:30
	Reno	12:06	12:38	03:29	04:01
	Oregon, Portland	11:52	12:29	04:01	04:38
	Utah, Salt Lake City	11:30	12:03	03:01	03:34
	Washington, Seattle	11:39	12:19	04:08	04:48
	Spokane	11:19	11:59	03:49	04:29
	Wyoming, Cheyenne	11:00	11:33	02:34	03:07
	Sheridan	10:53	11:29	02:56	03:32
W8	Michigan, Detroit	09:28	10:02	01:11	01:45
	Grand Rapids	09:36	10:10	01:24	01:59
	Sault Ste. Marie	09:14	09:52	01:32	02:10
	Traverse City	09:27	10:03	01:30	02:07
	Ohio, Cincinnati	09:46	10:18	01:06	01:38
	Cleveland	09:26	09:59	01:03	01:36
	Columbus	09:37	10:09	01:03	01:36
	West Virginia, Charleston	09:37	10:09	00:53	01:24
W9	Illinois, Chicago	09:48	10:22	01:28	02:02
	Indiana, Indianapolis	09:51	10:23	01:15	01:48
	Wisconsin, Green Bay	09:38	10:14	01:38	02:15

Pref	State/Province/Country/City	Tlight	Sunrise	Sunset	Tlight
	Milwaukee	09:44	10:19	01:33	02:07
W0	Colorado, Denver	11:06	11:38	02:30	03:02
	Grand Junction	11:23	11:55	02:43	03:15
	Iowa, Des Moines	10:13	10:47	01:51	02:24
	Kansas, Pratt	10:48	11:19	01:59	02:30
	Wichita	10:43	11:14	01:53	02:24
	Minnesota, Duluth	09:43	10:22	02:04	02:43
	Minneapolis	09:57	10:33	02:02	02:38
	Missouri, Columbia	10:18	10:50	01:37	02:09
	Kansas City	10:27	10:59	01:47	02:19
	St. Louis	10:11	10:43	01:28	01:59
	Nebraska, North Platte	10:44	11:17	02:18	02:51
	Omaha	10:24	10:57	01:59	02:32
	North Dakota, Fargo	10:01	10:40	02:23	03:02
	South Dakota, Rapid City	10:41	11:16	02:38	03:14
-	-	05:38	06:01	18:08	18:30
1A0	S.M.O.M	03:08	03:42	18:48	19:22
1S	Spratly Is.	21:55	22:18	10:55	11:18
3A	Monaco	03:20	03:55	19:14	19:49
3B6	Agalega	02:09	02:32	14:04	14:27
3B7	St. Brandon	02:06	02:29	13:41	14:04
3B8	Mauritius	02:22	02:46	13:43	14:07
3B9	Rodriguez Is.	01:57	02:21	13:20	13:44
3C	Equatorial Guinea, Bata	04:56	05:18	17:32	17:54
	Malabo	04:57	05:19	17:39	18:02
3C0	Annobon Is.	05:18	05:41	17:43	18:05
3D2	Fiji Is., Suva	18:15	18:38	05:43	06:07
3D2	Conway Reef	18:36	19:00	05:53	06:17
3D2	Rotuma Is.	18:08	18:31	05:56	06:19
3DA	Swaziland, Mbabane	04:18	04:43	15:17	15:42
3V	Tunisia, Tunis	03:36	04:06	18:41	19:11
3W, XV	Vietnam, Ho Chi Minh City (Saigon)	22:13	22:36	11:19	11:42
	Hanoi	21:56	22:20	11:42	12:06
3X	Guinea, Conakry	06:17	06:39	19:19	19:42
3Y	Bouvet	07:22	08:06	15:35	16:19
3Y	Peter I Is.	15:25	——	——	20:49
4J, 4K	Azerbaijan, Baku	00:44	01:17	16:13	16:45
4L	Georgia, Tbilisi	00:59	01:33	16:37	17:11
4S	Sri Lanka, Colombo	00:07	00:29	13:00	13:23
4U	ITU Geneva	03:13	03:51	19:28	20:06
4U	United Nations Hq.	08:58	09:31	00:30	01:03
4X, 4Z	Israel, Jerusalem	02:11	02:39	16:48	17:16
5A	Libya, Tripoli	03:40	04:08	18:20	18:48
	Benghazi	03:11	03:39	17:49	18:17
5B	Cyprus, Nicosia	02:08	02:38	17:04	17:33
5H	Tanzania, Dar es Salaam	03:12	03:34	15:18	15:41
5N	Nigeria, Lagos	05:14	05:36	18:05	18:28
5R	Madagascar, Antananarivo	03:00	03:23	14:25	14:49
5T	Mauritania, Nouakchott	06:09	06:33	19:43	20:07
5U	Niger, Niamey	05:06	05:30	18:23	18:46
5V	Togo, Lome	05:24	05:46	18:13	18:35
5W	Western Samoa, Apia	17:28	17:51	05:12	05:35
5X	Uganda, Kampala	03:28	03:50	15:58	16:21
5Z	Kenya, Nairobi	04:33	04:56	16:59	17:21
6W	Senegal, Dakar	06:22	06:45	19:43	20:07
6Y	Jamaica, Kingston	10:13	10:37	23:46	00:10
7O	Yemen, Aden	02:16	02:39	15:30	15:53
	Sanaa	02:14	02:37	15:38	16:01
7P	Lesotho, Maseru	04:38	05:04	15:25	15:50
7Q	Malawi, Lilongwe	03:46	04:09	15:29	15:52
	Blantyre	03:44	04:08	15:21	15:44
7X	Algeria, Algiers	04:05	04:35	19:09	19:40
8P	Barbados, Bridgetown	09:13	09:37	22:29	22:52
8Q	Maldive Is.	00:37	01:00	13:22	13:44
8R	Guyana, Georgetown	09:19	09:42	22:12	22:35
9A	Croatia, Zagreb	02:36	03:13	18:47	19:25
9G	Ghana, Accra	05:30	05:52	18:18	18:40
9H	Malta	03:22	03:52	18:22	18:52
9J	Zambia, Lusaka	04:10	04:34	15:49	16:12
9K	Kuwait	01:27	01:54	15:52	16:19
9L	Sierra Leone, Freetown	06:16	06:39	19:15	19:38
9M2	West Malaysia, Kuala Lumpur	22:46	23:09	11:27	11:49
9M6,8	East Malaysia, Sabah, Sandakan(9M6)	21:36	21:59	10:25	10:48
	Sarawak, Kuching (9M8)	22:14	22:37	10:49	11:12
9N	Nepal, Kathmandu	23:02	23:28	13:18	13:44
9Q	Zaire, Kinshasa	04:44	05:07	17:00	17:22
	Kisangani	03:57	04:19	16:28	16:50
	Lubumbashi	04:08	04:30	15:58	16:21
9U	Burundi, Bujumbura	03:47	04:09	16:05	16:28
9V	Singapore	22:41	23:03	11:15	11:37
9X	Rwanda, Kigali	03:41	04:04	16:04	16:27
9Y	Trinidad & Tobago , Port of Spain	09:25	09:48	22:31	22:54
A2	Botswana, Gaborone	04:36	05:01	15:40	16:05
A3	Tonga, Nukualofa	17:54	18:18	05:12	05:36
A4	Oman, Musqat	00:59	01:24	14:56	15:21
A5	Bhutan, Thimbu	22:47	23:13	13:00	13:27
A6	United Arab Emirates, Abu Dhabi	01:14	01:40	15:15	15:41
A7	Qatar, Ad-Dawhah	01:23	01:49	15:28	15:53
A9	Bahrein, Al-Manamah	01:25	01:51	15:33	15:59
AP	Pakistan, Karachi	00:22	00:47	14:25	14:50
AP	Islamabad	23:34	00:03	14:20	14:49
BV	Taiwan, Taipei	20:44	21:09	10:47	11:13
BY	Peoples Rep. of China, Beijing	20:19	20:52	11:46	12:18
	Harbin	19:13	19:50	11:25	12:02
	Shanghai	20:28	20:56	11:01	11:29
	Fuzhou	20:50	21:16	10:58	11:24
	Xian	21:09	21:38	11:59	12:28

Pref	State/Province/Country/City	Tlight	Sunrise	Sunset	Tlight
	Chongqing	21:32	21:59	11:58	12:25
	Chengdu	21:39	22:06	12:09	12:37
	Lhasa	22:33	23:00	12:59	13:26
	Urumqi	21:59	22:34	13:53	14:29
	Kashi	23:03	23:35	14:26	14:58
C2	Nauru	18:32	18:54	06:59	07:22
C3	Andorra	03:49	04:23	19:33	20:08
C5	The Gambia, Banjul	06:21	06:44	19:38	20:01
C6	Bahamas, Nassau	09:59	10:25	00:03	00:29
C9	Mozambique, Maputo	04:12	04:37	15:11	15:36
	Mozambique	03:20	03:44	14:59	15:23
CE	Chile, Santiago	11:20	11:47	21:48	22:15
CE0	Easter Island	13:42	14:07	00:37	01:02
CE0	Juan Fernandez	11:52	12:19	22:20	22:47
CE0	San Felix	11:43	12:08	22:41	23:06
CM,CO	Cuba, Havana	10:24	10:49	00:19	00:44
CN	Morocco, Casablanca	04:57	05:26	19:43	20:12
CP	Bolivia, La Paz	10:39	11:02	22:13	22:37
CT	Portugal, Lisbon	04:47	05:18	20:04	20:35
CT3	Madeira Islands, Funchal	05:37	06:06	20:18	20:46
CU	Azores , Ponta Delgada	05:56	06:27	21:07	21:38
CX	Uruguay, Montevideo	10:24	10:52	20:46	21:14
CY0	Sable Is.	07:49	08:25	23:44	00:19
CY9	St. Paul Is.	07:33	08:12	23:58	00:37
D2	Angola, Luanda	05:00	05:23	17:00	17:23
D4	Cape Verde, Praia	06:46	07:09	20:07	20:31
D6	Comoros, Moroni	03:03	03:26	14:53	15:16
DA-DL,Y2-Y9	Fed. Rep. of Germany, Bonn	02:43	03:27	19:45	20:30
	Berlin	02:04	02:52	19:29	20:18
DU	Philippines, Manilla	21:08	21:32	10:29	10:52
E3	Eritrea, Asmara	02:35	02:59	15:59	16:22
EA	Spain, Madrid	04:18	04:51	19:47	20:20
EA6	Balearic Is., Palma	03:57	04:29	19:19	19:51
EA8	Canary Is., Las Palmas	05:39	06:05	19:58	20:24
EA9	Ceuta & Melilla, Ceuta	04:41	05:11	19:40	20:10
	Melilla	04:34	05:03	19:30	19:59
EI	Ireland, Dublin	03:16	04:06	20:53	21:43
EK	Armenia, Yerevan	01:06	01:39	16:34	17:07
EL	Liberia, Monrovia	06:11	06:33	19:02	19:24
EP	Iran, Tehran	00:53	01:23	15:52	16:21
ER	Moldova, Kishinev	01:38	02:17	18:01	18:40
ES	Estonia, Tallinn	23:54	01:16	19:34	20:56
ET	Ethiopia, Addis Ababa	02:48	03:11	15:48	16:11
EU,EV,EW	Belarus, Minsk	00:55	01:47	18:41	19:33
EX	Kyrgyzstan, Frunze	22:55	23:30	14:42	15:17
EY	Tajikistan, Samarkand	23:39	00:11	15:03	15:35
	Dushanbe	23:33	00:05	14:53	15:25
EZ	Turkmenistan, Ashkhabad	00:19	00:50	15:31	16:03
F	France, Paris	03:14	03:55	19:55	20:36
FG	Guadeloupe	09:16	09:40	22:42	23:06
FJ, FS	St. Martin	09:18	09:42	22:52	23:16
FH	Mayotte	02:58	03:21	14:45	15:08
FK	New Caledonia, Noumia	19:09	19:34	06:23	06:47
FM	Martinique	09:16	09:40	22:37	23:00
FO	Clipperton Is.	12:37	13:00	01:42	02:05
FO	Fr. Polynesia, Tahiti	16:45	17:09	04:16	04:39
	Rurutu, (Austral Is.)	20:11	20:35	07:23	07:48
	Hiva Oa, (Marquesas Is.)	15:11	15:33	03:07	03:30
FP	St. Pierre & Miquelon, St. Pierre	07:19	07:58	23:39	00:18
FR/G	Glorioso	02:48	03:11	14:39	15:02
FR/J,E	Juan de Nova	03:15	03:39	14:48	15:11
	Europa	03:34	03:58	14:47	15:12
FR	Reunion	02:31	02:55	13:49	14:13
FR/T	Tromelin	02:27	02:50	14:03	14:27
FT8W	Crozet	03:38	04:13	13:00	13:34
FT8X	Kerguelen	02:40	03:17	11:38	12:15
FT8Y	Antarctica, Dumont D'Urville	23:35	01:13	04:15	05:54
FT8Z	Amsterdam & St. Paul Is., Amsterdam	01:35	02:04	11:44	12:13
FW	Wallis & Futuna Is., Wallis	17:45	18:08	05:31	05:54
FY	Fr. Guiana, Cayenne	08:59	09:22	21:45	22:08
G	England, London	03:05	03:52	20:18	21:04
GD	Isle of Man	03:00	03:53	20:52	21:45
GI	Northern Ireland, Belfast	03:02	03:57	20:59	21:53
GJ	Jersey	03:29	04:11	20:15	20:58
GM	Scotland, Glasgow	02:44	03:42	21:01	21:59
	Aberdeen	02:18	03:23	21:02	22:08
GU	Guernsey	03:29	04:12	20:18	21:01
GW	Wales, Cardiff	03:18	04:04	20:30	21:16
H4	Solomon Islands, Honiara	19:14	19:36	07:12	07:35
HA	Hungary, Budapest	02:14	02:54	18:42	19:22
HB	Switzerland, Bern	03:03	03:42	19:26	20:05
HB0	Liechtenstein	02:54	03:33	19:19	19:58
HC	Ecuador, Quito	10:51	11:13	23:20	23:42
HC8	Galapagos Is.	11:41	12:04	00:09	00:31
HH	Haiti, Port-Au-Prince	09:54	10:18	23:29	23:53
HI	Dominican Republic, Santo Domingo	09:44	10:08	23:20	23:44
HK	Colombia, Bogota	10:27	10:49	23:12	23:35
HK0	Malpelo Is.	10:56	11:18	23:39	00:01
HK0	San Adreas	10:43	11:06	23:56	00:19
HL	Korea, Seoul	19:46	20:17	10:56	11:26
HP	Panama, Panama	10:41	11:04	23:41	00:04
HR	Honduras, Tegucigalpa	11:02	11:25	00:21	00:44
HS	Thailand, Bangkok	22:32	22:55	11:50	12:13
HV	Vatican City	03:07	03:41	18:47	19:21
HZ, 7Z	Saudi Arabia, Dharan	01:27	01:53	15:36	16:02
	Mecca	02:19	02:43	16:07	16:31
	Riyadh	01:44	02:10	15:45	16:11

Pref	State/Province/Country/City	Tlight	Sunrise	Sunset	Tlight
I	Italy, Rome	03:07	03:41	18:47	19:21
	Trieste	02:45	03:22	18:56	19:33
	Sicily	03:18	03:49	18:28	18:58
IS	Sardinia, Cagliari	03:32	04:04	18:52	19:24
J2	Djibouti, Djibouti	02:25	02:48	15:35	15:58
J3	Grenada	09:24	09:47	22:36	22:59
J5	Guinea-Bissau, Bissau	06:20	06:43	19:31	19:54
J6	St. Lucia	09:18	09:41	22:36	22:59
J7	Dominica	09:16	09:39	22:40	23:03
J8	St. Vincent	09:20	09:43	22:36	22:59
JA-JS	Japan, Tokyo	19:01	19:31	09:59	10:29
	Nagasaki	19:50	20:18	10:31	11:00
	Sapporo	18:27	19:02	10:16	10:51
JD1	Minami Torishima	18:36	19:01	08:36	09:01
JD1	Ogasawara, Kazan Is.	19:20	19:46	09:35	10:01
JT	Mongolia, Ulan Bator	20:21	21:01	12:53	13:33
JW	Svalbard, Spitsbergen	——	——	——	——
JX	Jan Mayen	——	——	——	——
JY	Jordan, Amman	02:08	02:36	16:45	17:13
KC4	Antarctica, Byrd Station	——	——	——	——
	McMurdo Sound	——	——	——	——
	Palmer Station	12:53	14:12	18:29	19:47
KC6	Belau, Yap	20:09	20:32	09:11	09:34
	Koror	20:28	20:50	09:22	09:45
KG4	Guantanamo Bay	10:02	10:27	23:44	00:08
KH0	Mariana Is., Saipan	19:28	19:51	08:51	09:14
KH1	Baker, Howland Is.	17:21	17:44	05:53	06:15
KH2	Guam, Agana	19:35	19:58	08:52	09:15
KH3	Johnston Is.	16:21	16:45	05:51	06:15
KH4	Midway Is.	16:32	16:58	06:50	07:16
KH5	Palmyra Is.	16:17	16:39	05:06	05:29
KH5K	Kingman Reef	16:17	16:39	05:12	05:34
KH6	Hawaii, Hilo	15:22	15:47	05:03	05:27
	Honolulu	15:30	15:55	05:17	05:42
KH7	Kure Is.	16:35	17:02	06:54	07:21
KH8	American Samoa, Pago Pago	17:25	17:48	05:07	05:30
KH9	Wake Is.	17:56	18:21	07:35	08:00
KL7	Alaska, Adak	14:49	15:36	08:05	08:52
	Anchorage	10:33	12:36	07:33	09:35
	Fairbanks	——	11:27	08:24	——
	Juneau	10:51	12:03	06:01	07:13
	Nome	——	12:44	09:27	——
KP1	Navassa Is.	10:05	10:29	23:40	00:04
KP2	Virgin Islands, Charlotte Amalie	09:24	09:48	22:59	23:23
KP4	Puerto Rico, San Juan	09:29	09:53	23:05	23:29
KP5	Desecheo Is.	09:35	09:59	23:10	23:34
LA-LJ	Norway, Oslo	00:36	02:06	20:37	22:07
LU	Argentina, Buenos Aires	10:32	11:00	20:56	21:23
LX	Luxembourg	02:53	03:36	19:43	20:26
LY	Lithuania, Vilna	00:58	01:52	18:53	19:47
LZ	Bulgaria, Sofia	02:21	02:55	18:07	18:41
OA	Peru, Lima	11:07	11:29	22:56	23:19
OD	Lebanon, Beirut	02:04	02:33	16:52	17:21
OE	Austria, Vienna	02:21	03:02	18:56	19:37
OH	Finland, Helsinki	23:34	01:07	19:42	21:15
OH0	Aland Is.	23:54	01:27	20:02	21:35
OJ0	Market Reef	23:55	01:30	20:07	21:42
OK,OL	Czech Rep., Prague	02:17	03:01	19:13	19:56
OM	Slovak, Rep., Bratislava	02:20	03:00	18:53	19:33
ON	Belgium, Brussels	02:52	03:37	19:57	20:42
OX	Greenland, Godthaab	——	05:16	01:46	——
	Thule	——	——	——	——
OY	Faroe Islands, Torshavn	——	02:53	22:10	——
OZ	Denmark, Copenhagen	01:37	02:35	19:52	20:51
P2	Papua New Guinea, Madang	20:05	20:27	08:17	08:39
	Port Moresby	20:05	20:28	08:04	08:26
P4	Aruba, Oranjestad	09:53	10:16	23:04	23:28
PA-PI	Netherlands, Amsterdam	02:38	03:27	20:03	20:51
PJ2,4,9	Netherlands Antilles, Willemstad	09:56	10:20	23:10	23:33
PJ5-8	St. Maarten and Saba, St. Maarten	09:19	09:43	22:51	23:15
PY	Brazil, Brasilia	09:16	09:39	20:53	21:16
	Rio De Janeiro	09:09	09:34	20:20	20:45
	Natal	08:09	08:32	20:19	20:41
	Manaus	09:44	10:07	22:04	22:26
	Porto Alegre	09:54	10:21	20:38	21:04
PY0	Fernando De Noronha	07:54	08:17	20:11	20:33
PY0	St. Peter & St. Paul Rocks	07:34	07:57	20:07	20:30
PY0	Trindade & Martin Vaz Is., Trindade	08:09	08:33	19:30	19:54
PZ	Suriname, Paramaribo	09:09	09:32	21:58	22:21
R1FJ_	Franz Josef Land	——	——	——	——
R1MV_	Malyj Vysotskij Is.	23:07	00:48	19:32	21:13
S0	Western Sahara, Smara	05:32	05:58	19:42	20:08
S2	Bangladesh, Dacca	22:51	23:17	12:49	13:14
S5	Slovenia, Ljubljana	02:41	03:18	18:54	19:32
S7	Seychelles, Victoria	02:04	02:26	14:18	14:41
S9	Sao Tome	05:11	05:33	17:42	18:04
SM	Sweden, Stockholm	00:23	01:44	20:00	21:21
SP	Poland, Krakow	01:55	02:39	18:50	19:33
	Warsaw	01:36	02:23	18:57	19:45
ST	Sudan, Khartoum	03:00	03:24	16:25	16:48
ST0	Southern Sudan, Juba	03:23	03:46	16:10	16:32
SU	Egypt, Cairo	02:32	02:59	16:59	17:26
SV	Greece, Athens	02:38	03:09	17:50	18:21
SV/A	Mount Athos	02:27	03:00	17:55	18:27
SV5	Dodecanese, Rhodes	02:25	02:55	17:28	17:58
SV9	Crete	02:40	03:10	17:37	18:07
T2	Tuvalu, Funafuti	17:58	18:21	05:59	06:22

Pref	State/Province/Country/City	Tlight	Sunrise	Sunset	Tlight
T30	West Kiribati, Bonriki	18:03	18:26	06:37	07:00
T31	Central Kiribati, Kanton	17:10	17:32	05:30	05:52
T32	East Kiribati, Christmas Is.	16:05	16:27	04:41	05:03
T33	Banaba Is.	18:22	18:44	06:49	07:12
T5	Somalia, Mogadishu	02:33	02:56	15:10	15:32
T7	San Marino	03:00	03:35	18:55	19:31
T9	Bosnia-Herzegovina, Sarajevo	02:35	03:11	18:31	19:06
TA	Turkey, Ankara	01:54	02:26	17:19	17:52
	Istanbul	02:05	02:38	17:39	18:12
TF	Iceland, Reykjavik	–––	03:19	23:45	–––
TG	Guatemala, Guatemala City	11:14	11:38	00:35	00:58
TI	Costa Rica, San Jose	10:57	11:20	00:01	00:24
TI9	Cocos Is.	11:17	11:39	00:05	00:28
TJ	Cameroon, Yaounde	04:46	05:08	17:28	17:51
TK	Corsica	03:21	03:55	19:02	19:36
TL	Central African Republic, Bangui	04:16	04:39	17:01	17:23
TN	Congo, Brazzaville	04:44	05:07	17:00	17:22
TR	Gabon, Libreville	05:00	05:22	17:31	17:53
TT	Chad, N'Djamena	04:17	04:40	17:28	17:52
TU	Cote d'Ivoire, Abidjan	05:45	06:08	18:33	18:55
TY	Benin, Porto Novo	05:17	05:39	18:08	18:31
TZ	Mali, Bamako	05:48	06:11	19:02	19:25
UA	Russia, European, Leningrad (UA1)	23:20	00:49	19:17	20:46
	Archangel (UA1)	–––	22:58	19:46	–––
	Murmansk (UA1)	–––	–––	–––	–––
	Moscow (UA3)	23:56	00:55	18:13	19:12
	Kuybyshev (UA4)	23:31	00:21	17:06	17:57
	Rostov (UA6)	00:53	01:32	17:20	18:00
UA2	Kaliningrad	01:13	02:09	19:16	20:12
UA9,0	Russia, Asiatic, Novosibirsk (UA9)	21:03	21:59	15:07	16:02
	Perm (UA9)	22:13	23:23	17:15	18:26
	Omsk (UA9)	21:41	22:37	15:45	16:40
	Norilsk (UA0)	–––	–––	–––	–––
	Irkutsk (UA0)	20:02	20:50	13:25	14:13
	Vladivostok (UA0)	19:04	19:39	10:54	11:29
	Petropavlovsk (UA0)	16:18	17:08	09:51	10:41
	Khabarovsk (UA0)	18:24	19:05	11:02	11:43
	Krasnoyarsk (UA0)	20:13	21:12	14:34	15:33
	Yakutsk (UA0)	–––	17:47	13:04	–––
	Wrangel Island (UA0)	–––	–––	–––	–––
	Kyzyl (UA0Y)	20:45	21:32	14:00	14:47
UJ-UM	Uzbekistan, Bukhoro	23:48	00:21	15:13	15:45
	Tashkent	23:23	23:56	14:58	15:31
UN-UQ	Kazakhstan, Alma-Ata	22:44	23:19	14:34	15:10
UR-UZ,EM-EO	Ukraine, Kiev	01:11	01:55	18:10	18:54
V2	Antigua & Barbuda, St. Johns	09:14	09:38	22:45	23:09
V3	Belize, Belmopan	11:02	11:26	00:33	00:57
V4	St. Kitts & Nevis	09:17	09:41	22:48	23:12
V5	Namibia, Windhoek	05:08	05:32	16:20	16:44
V6	Micronesia, Ponape	18:53	19:16	07:46	08:09
V7	Marshall Islands, Kwajalein	18:13	18:36	07:14	07:37
V8	Brunei, Bandar Seri Begawan	21:50	22:13	10:37	10:59
VK	Australia, Canberra (VK1)	20:44	21:12	07:04	07:32
	Sydney (VK2)	20:32	21:00	06:59	07:27
	Melbourne (VK3)	21:06	21:35	07:14	07:43
	Brisbane (VK4)	20:13	20:38	07:06	07:32
	Adelaide (VK5)	21:25	21:53	07:47	08:15
	Perth (VK6)	22:50	23:17	09:25	09:52
	Hobart, Tasmania (VK7)	21:09	21:41	06:49	07:21
	Darwin (VK8)	21:15	21:38	09:03	09:26
VK0	Heard Is.	02:36	03:18	11:04	11:45
VK0	Macquarie Is.	21:01	21:45	05:13	05:57
VK9C	Cocos-Keeling Is.	23:31	23:54	11:20	11:43
VK9L	Lord Howe Is.	19:56	20:23	06:33	07:00
VK9M	Mellish Reef	19:44	20:08	07:15	07:38
VK9N	Norfolk Is.	19:16	19:41	06:03	06:29
VK9W	Willis Is.	20:07	20:30	07:42	08:05
VK9X	Christmas Is.	22:53	23:16	10:47	11:10
VP2E	Anguilla	09:17	09:41	22:52	23:16
VP2M	Montserrat	09:17	09:40	22:46	23:10
VP2V	British Virgin Is., Tortola	09:23	09:47	22:58	23:22

Pref	State/Province/Country/City	Tlight	Sunrise	Sunset	Tlight
VP5	Turks & Caicos Islands, Grand Turk	09:43	10:08	23:31	23:55
VP8	Falkland Islands, Stanley	11:16	11:56	19:55	20:35
VP8	So. Georgia Is.	10:02	10:46	18:17	19:01
VP8	So. Orkney Is.	11:08	12:06	18:06	19:04
VP8	So. Sandwich Islands, Saunders Is.	09:38	10:28	17:15	18:05
VP8	So. Shetland Is., King George Is.	12:09	13:12	18:43	19:46
VP9	Bermuda	08:50	09:18	23:29	23:57
VQ9	Chagos, Diego Garcia	01:01	01:23	13:06	13:29
VR2,VS6	Hong Kong	21:19	21:44	11:10	11:35
VR6	Pitcairn Is.	15:01	15:26	02:04	02:29
VU	India, Bombay	00:12	00:36	13:50	14:14
	Calcutta	23:02	23:27	12:55	13:20
	New Delhi	23:32	23:59	13:52	14:19
	Bangalore	00:05	00:28	13:20	13:43
VU	Andaman Islands, Port Blair	23:07	23:30	12:17	12:40
VU	Laccadive Is.	00:29	00:52	13:33	13:56
XE	Mexico, Mexico City (XE1)	11:39	12:03	01:18	01:43
	Chihuahua (XE2)	11:45	12:11	02:05	02:32
	Merida (XE3)	10:58	11:22	00:44	01:08
XF4	Revilla Gigedo	12:29	12:53	02:07	02:31
XT	Burkina Faso, Ouagadougou	05:23	05:46	18:35	18:59
XU	Cambodia, Phnom Penh	22:19	22:42	11:29	11:52
XW	Laos, Viangchan	22:15	22:39	11:49	12:13
XX9	Macao	21:22	21:47	11:13	11:38
XZ	Myanmar, Yangon	22:44	23:08	12:13	12:37
YA	Afghanistan, Kandahar	00:11	00:39	14:43	15:11
	Kabul	23:48	00:17	14:38	15:07
YB-YD	Indonesia, Jakarta	22:41	23:04	10:50	11:13
	Medan	22:57	23:20	11:39	12:02
	Pontianak	22:21	22:43	10:51	11:13
	Jayapura	20:20	20:42	08:41	09:03
YI	Iraq, Baghdad	01:31	01:59	16:14	16:42
YJ	Vanuatu, Port Vila	18:54	19:18	06:24	06:48
YK	Syria, Damascus	02:02	02:31	16:48	17:16
YL	Latvia, Riga	00:36	01:40	19:16	20:20
YN	Nicaragua, Managua	11:01	11:24	00:12	00:35
YO	Romania, Bucharest	02:02	02:38	18:02	18:38
YS	El Salvador, San Salvador	11:11	11:34	00:28	00:51
YU	Yugoslavia, Belgrade	02:22	02:59	18:26	19:03
YV	Venezuela, Caracas	09:48	10:11	22:54	23:17
YV0	Aves Is.	09:25	09:48	22:50	23:13
Z2	Zimbabwe, Harare	04:04	04:27	15:33	15:57
Z3	Macedonia, (ex Yugoslav), Skopje	02:31	03:05	18:12	18:46
ZA	Albania, Tirane	02:41	03:14	18:16	18:49
ZB2	Gibraltar	04:41	05:11	19:41	20:11
ZC4	British Cyprus	02:12	02:41	17:04	17:33
ZD7	St. Helena	06:28	06:52	18:04	18:28
ZD8	Ascension Is.	06:49	07:12	18:52	19:15
ZD9	Tristan da Cunha	07:33	08:02	17:45	18:14
ZF	Cayman Is.	10:27	10:51	00:07	00:31
ZK1	No. Cook Is., Manihiki	16:39	17:02	04:34	04:57
ZK1	So. Cook Is., Rarotonga	16:53	17:17	04:10	04:34
ZK2	Niue	17:25	17:49	04:51	05:15
ZK3	Tokelaus, Atafu	17:23	17:46	05:25	05:47
ZL	New Zealand, Auckland (ZL1)	19:04	19:33	05:17	05:46
	Wellington (ZL2)	19:15	19:46	05:04	05:35
	Christchurch (ZL3)	19:29	20:02	05:06	05:39
	Dunedin (ZL4)	19:44	20:18	05:06	05:40
ZL5	Antarctica, Scott Base	–––	–––	–––	–––
ZL7	Chatham Is.	18:47	19:20	04:21	04:54
ZL8	Kermadec Is.	18:20	18:46	05:06	05:32
ZL9	Auckland & Campbell Is., Auckland	20:15	20:54	05:03	05:41
	Campbell Is.	20:11	20:52	04:44	05:24
ZP	Paraguay, Asuncion	10:12	10:36	21:14	21:39
ZS	South Africa, Cape Town (ZS1)	05:24	05:51	15:50	16:18
	Port Elizabeth (ZS2)	04:55	05:22	15:21	15:48
	Bloemfontein (ZS4)	04:44	05:09	15:30	15:56
	Durban (ZS5)	04:26	04:52	15:10	15:36
	Johannesburg (ZS6)	04:30	04:55	15:29	15:54
ZS8	Marion Is.	04:38	05:12	13:54	14:29

Pref	State/Province/Country/City	Tlight	Sunrise	Sunset	Tlight
VE1	New Brunswick, Saint John	08:21	08:56	00:05	00:40
	Nova Scotia, Halifax	08:13	08:48	23:52	00:27
	Prince Edward Island, Charlottetown	08:05	08:41	23:56	00:32
VE2	Quebec, Montreal	08:50	09:25	00:35	01:11
	Quebec City	08:34	09:11	00:30	01:07
VE3	Ontario, London	09:31	10:04	00:58	01:31
	Ottawa	08:58	09:34	00:43	01:19
	Sudbury	09:15	09:51	01:08	01:45
	Toronto	09:20	09:54	00:52	01:27
VE4	Manitoba, Winnipeg	10:02	10:43	02:26	03:07
VE5	Saskatchewan, Regina	10:28	11:10	02:58	03:40
	Saskatoon	10:27	11:11	03:14	03:58
VE6	Alberta, Calgary	11:03	11:46	03:39	04:21
	Edmonton	10:44	11:31	03:48	04:35
VE7	British Columbia, Prince George	11:19	12:07	04:27	05:15
	Prince Rupert	11:46	12:34	05:00	05:48
	Vancouver	11:49	12:29	04:07	04:47
VE8	Northwest Territories, Yellowknife	08:49	10:29	04:58	06:38
	Resolute	----	----	----	----
VY1	Yukon, Whitehorse	10:53	12:09	06:03	07:20
VO1	Newfoundland, St. John's	07:16	07:54	23:19	23:57
VO2	Labrador, Goose Bay	07:14	08:00	00:15	01:01
W1	Connecticut, Hartford	09:01	09:33	00:20	00:52
	Maine, Bangor	08:33	09:08	00:14	00:49
	Portland	08:44	09:18	00:16	00:50
	Massachusetts, Boston	08:52	09:25	00:15	00:48
	New Hampshire, Concord	08:51	09:24	00:19	00:53
	Rhode Island, Providence	08:56	09:28	00:15	00:47
	Vermont, Montpelier	08:51	09:25	00:27	01:02
W2	New Jersey, Atlantic City	09:16	09:47	00:20	00:51
	New York, Albany	09:02	09:35	00:27	01:00
	Buffalo	09:22	09:55	00:48	01:21
	New York City	09:10	09:42	00:22	00:54
	Syracuse	09:10	09:44	00:38	01:11
W3	Delaware, Wilmington	09:19	09:51	00:25	00:58
	District of Columbia, Washington	09:28	09:59	00:29	01:00
	Maryland, Baltimore	09:25	09:56	00:28	00:59
	Pennsylvania, Harrisburg	09:23	09:55	00:32	01:04
	Philadelphia	09:18	09:49	00:24	00:56
	Pittsburgh	09:35	10:07	00:45	01:17
	Scranton	09:14	09:47	00:31	01:03
W4	Albama, Montgomery	10:24	10:52	00:50	01:18
	Florida, Jacksonville	10:11	10:38	00:27	00:54
	Miami	10:16	10:41	00:12	00:38
	Pensacola	10:33	11:00	00:50	01:16
	Georgia, Atlanta	10:13	10:41	00:46	01:14
	Savannah	10:04	10:32	00:29	00:56
	Kentucky, Lexington	10:01	10:31	00:57	01:27
	Louisville	10:06	10:36	01:02	01:32
	North Carolina, Charlotte	09:55	10:23	00:35	01:03
	Raleigh	09:44	10:13	00:27	00:56
	Wilmington	09:46	10:14	00:21	00:49
	South Carolina, Columbia	09:59	10:27	00:33	01:01
	Tennessee, Knoxville	10:05	10:34	00:49	01:18
	Memphis	10:32	11:01	01:12	01:40
	Nashville	10:16	10:45	01:01	01:30
	Virginia, Norfolk	09:32	10:01	00:21	00:50
	Richmond	09:34	10:04	00:27	00:57
W5	Arkansas, Little Rock	10:42	11:11	01:20	01:48
	Louisiana, New Orleans	10:46	11:12	01:00	01:27
	Shreveport	10:54	11:21	01:20	01:48
	Mississippi, Jackson	10:40	11:08	01:06	01:33
	New Mexico, Albuquerque	11:39	12:07	02:18	02:47
	Oklahoma, Oklahoma City	11:01	11:29	01:42	02:11
	Texas, Abilene	11:18	11:45	01:44	02:12
	Amarillo	11:19	11:47	01:59	02:27
	Dallas	11:05	11:33	01:33	02:01
	El Paso	11:47	12:14	02:10	02:37
	San Antonio	11:21	11:47	01:33	01:59
W6	California, Los Angeles	12:27	12:56	03:02	03:30
	San Francisco	12:33	13:03	03:28	03:58
W7	Arizona, Flagstaff	11:58	12:27	02:38	03:07
	Phoenix	12:05	12:33	02:36	03:04
	Idaho, Boise	11:48	12:22	03:19	03:53
	Pocatello	11:36	12:09	03:02	03:36
	Montana, Billings	11:08	11:44	02:56	03:32
	Butte	11:23	11:59	03:13	03:49
	Great Falls	11:11	11:49	03:13	03:51
	Nevada, Las Vegas	12:09	12:38	02:54	03:24
	Reno	12:17	12:48	03:23	03:53
	Oregon, Portland	12:06	12:42	03:52	04:27
	Utah, Salt Lake City	11:41	12:13	02:54	03:26
	Washington, Seattle	11:55	12:33	03:58	04:35
	Spokane	11:35	12:13	03:38	04:16
	Wyoming, Cheyenne	11:12	11:44	02:26	02:58
	Sheridan	11:06	11:41	02:46	03:21
W8	Michigan, Detroit	09:40	10:13	01:03	01:35
	Grand Rapids	09:48	10:22	01:15	01:49
	Sault Ste. Marie	09:28	10:05	01:22	01:59
	Traverse City	09:41	10:16	01:21	01:56
	Ohio, Cincinnati	09:58	10:28	00:55	01:30
	Cleveland	09:38	10:10	00:55	01:27
	Columbus	09:48	10:20	00:56	01:27
	West Virginia, Charleston	09:48	10:19	00:46	01:16
W9	Illinois, Chicago	10:00	10:33	01:20	01:52
	Indiana, Indianapolis	10:02	10:33	01:08	01:39
	Wisconsin, Green Bay	09:52	10:26	01:29	02:04

July 21   Sunrise and Sunset, in UTC

Pref	State/Province/Country/City	Tlight	Sunrise	Sunset	Tlight
	Milwaukee	09:57	10:31	01:24	01:58
W0	Colorado, Denver	11:17	11:49	02:23	02:54
	Grand Junction	11:34	12:05	02:36	03:07
	Iowa, Des Moines	10:25	10:58	01:43	02:15
	Kansas, Pratt	10:59	11:29	01:53	02:22
	Wichita	10:53	11:23	01:47	02:17
	Minnesota, Duluth	09:58	10:35	01:54	02:31
	Minneapolis	10:11	10:46	01:52	02:28
	Missouri, Columbia	10:29	11:00	01:30	02:01
	Kansas City	10:38	11:09	01:40	02:11
	St. Louis	10:22	10:52	01:21	01:51
	Nebraska, North Platte	10:56	11:28	02:10	02:42
	Omaha	10:36	11:08	01:51	02:23
	North Dakota, Fargo	10:16	10:53	02:13	02:50
	South Dakota, Rapid City	10:54	11:28	02:29	03:03
-		05:40	06:02	18:09	18:31
1A0	S.M.O.M	03:20	03:53	18:40	19:12
1S	Spratly Is.	21:59	22:21	10:55	11:18
3A	Monaco	03:33	04:07	19:05	19:39
3B6	Agalega	02:09	02:32	14:07	14:30
3B7	St. Brandon	02:05	02:28	13:45	14:08
3B8	Mauritius	02:20	02:44	13:48	14:11
3B9	Rodriguez Is.	01:56	02:19	13:25	13:49
3C	Equatorial Guinea, Bata	04:58	05:20	17:33	17:55
	Malabo	04:59	05:21	17:40	18:02
3C0	Annobon Is.	05:20	05:42	17:45	18:07
3D2	Fiji Is., Suva	18:13	18:37	05:48	06:11
3D2	Conway Reef	18:34	18:58	05:58	06:22
3D2	Rotuma Is.	18:08	18:30	06:00	06:23
3DA	Swaziland, Mbabane	04:15	04:40	15:23	15:48
3V	Tunisia, Tunis	03:46	04:15	18:35	19:04
3W, XV	Vietnam, Ho Chi Minh City (Saigon)	22:16	22:39	11:19	11:42
	Hanoi	22:02	22:26	11:40	12:04
3X	Guinea, Conakry	06:20	06:43	19:19	19:41
3Y	Bouvet	07:10	07:52	15:53	16:35
3Y	Peter I Is.	14:52	16:36	19:40	21:25
4J, 4K	Azerbaijan, Baku	00:55	01:27	16:05	16:37
4L	Georgia, Tbilisi	01:11	01:44	16:30	17:02
4S	Sri Lanka, Colombo	00:10	00:32	13:00	13:23
4U	ITU Geneva	03:27	04:04	19:18	19:55
4U	United Nations Hq.	09:10	09:42	00:22	00:54
4X, 4Z	Israel, Jerusalem	02:20	02:47	16:43	17:10
5A	Libya, Tripoli	03:49	04:16	18:15	18:43
	Benghazi	03:20	03:47	17:44	18:12
5B	Cyprus, Nicosia	02:18	02:47	16:58	17:27
5H	Tanzania, Dar es Salaam	03:13	03:35	15:21	15:43
5N	Nigeria, Lagos	05:17	05:39	18:06	18:28
5R	Madagascar, Antananarivo	02:58	03:21	14:30	14:54
5T	Mauritania, Nouakchott	06:14	06:38	19:42	20:05
5U	Niger, Niamey	05:11	05:34	18:22	18:45
5V	Togo, Lome	05:27	05:49	18:13	18:36
5W	Western Samoa, Apia	17:27	17:50	05:16	05:39
5X	Uganda, Kampala	03:30	03:52	16:00	16:22
5Z	Kenya, Nairobi	04:35	04:57	17:00	17:22
6W	Senegal, Dakar	06:27	06:50	19:42	20:05
6Y	Jamaica, Kingston	10:18	10:41	23:45	00:08
7O	Yemen, Aden	02:20	02:43	15:29	15:52
	Sanaa	02:19	02:42	15:36	16:00
7P	Lesotho, Maseru	04:35	05:00	15:32	15:57
7Q	Malawi, Lilongwe	03:46	04:08	15:33	15:56
	Blantyre	03:44	04:06	15:25	15:48
7X	Algeria, Algiers	04:15	04:45	19:03	19:33
8P	Barbados, Bridgetown	09:18	09:41	22:28	22:51
8Q	Maldive Is.	00:40	01:02	13:22	13:45
8R	Guyana, Georgetown	09:23	09:45	22:12	22:35
9A	Croatia, Zagreb	02:50	03:26	18:38	19:14
9G	Ghana, Accra	05:33	05:55	18:18	18:41
9H	Malta	03:32	04:01	18:16	18:45
9J	Zambia, Lusaka	04:10	04:33	15:53	16:16
9K	Kuwait	01:35	02:02	15:48	16:14
9L	Sierra Leone, Freetown	06:20	06:42	19:15	19:37
9M2	West Malaysia, Kuala Lumpur	22:49	23:11	11:28	11:50
9M6,8	East Malaysia, Sabah, Sandakan(9M6)	21:39	22:01	10:26	10:48
	Sarawak, Kuching (9M8)	22:17	22:39	10:51	11:13
9N	Nepal, Kathmandu	23:09	23:35	13:14	13:40
9Q	Zaire, Kinshasa	04:45	05:08	17:02	17:24
	Kisangani	03:59	04:21	16:29	16:51
	Lubumbashi	04:07	04:30	16:02	16:24
9U	Burundi, Bujumbura	03:48	04:10	16:07	16:29
9V	Singapore	22:43	23:05	11:16	11:38
9X	Rwanda, Kigali	03:43	04:05	16:06	16:28
9Y	Trinidad & Tobago , Port of Spain	09:29	09:52	22:31	22:53
A2	Botswana, Gaborone	04:34	04:58	15:46	16:11
A3	Tonga, Nukualofa	17:52	18:16	05:17	05:41
A4	Oman, Musqat	01:05	01:30	14:53	15:18
A5	Bhutan, Thimbu	22:54	23:20	12:57	13:23
A6	United Arab Emirates, Abu Dhabi	01:21	01:46	15:12	15:37
A7	Qatar, Ad-Dawhah	01:30	01:55	15:25	15:50
A9	Bahrein, Al-Manamah	01:32	01:57	15:30	15:55
AP	Pakistan, Karachi	00:28	00:53	14:21	14:46
AP	Islamabad	23:43	00:11	14:15	14:43
BV	Taiwan, Taipei	20:50	21:16	10:44	11:09
BY	Peoples Rep. of China, Beijing	20:31	21:02	11:38	12:10
	Harbin	19:27	20:03	11:15	11:51
	Shanghai	20:36	21:03	10:56	11:23
	Fuzhou	20:57	21:22	10:55	11:20
	Xian	21:18	21:47	11:54	12:22

Pref	State/Province/Country/City	Tlight	Sunrise	Sunset	Tlight
	Chongqing	21:40	22:06	11:53	12:20
	Chengdu	21:47	22:14	12:05	12:32
	Lhasa	22:41	23:08	12:54	13:21
	Urumqi	22:12	22:46	13:45	14:19
	Kashi	23:14	23:45	14:18	14:50
C2	Nauru	18:33	18:55	07:01	07:23
C3	Andorra	04:01	04:35	19:25	19:58
C5	The Gambia, Banjul	06:25	06:48	19:37	20:00
C6	Bahamas, Nassau	10:06	10:31	00:00	00:25
C9	Mozambique, Maputo	04:09	04:33	15:17	15:42
	Mozambique	03:20	03:43	15:04	15:26
CE	Chile, Santiago	11:15	11:42	21:56	22:23
CE0	Easter Island	13:39	14:03	00:43	01:08
CE0	Juan Fernandez	11:47	12:14	22:28	22:55
CE0	San Felix	11:40	12:05	22:48	23:12
CM,CO	Cuba, Havana	10:30	10:55	00:16	00:41
CN	Morocco, Casablanca	05:06	05:34	19:38	20:06
CP	Bolivia, La Paz	10:38	11:01	22:18	22:41
CT	Portugal, Lisbon	04:58	05:28	19:57	20:28
CT3	Madeira Islands, Funchal	05:46	06:14	20:13	20:41
CU	Azores , Ponta Delgada	06:07	06:37	21:01	21:30
CX	Uruguay, Montevideo	10:19	10:47	20:55	21:22
CY0	Sable Is.	08:02	08:37	23:35	00:09
CY9	St. Paul Is.	07:48	08:25	23:47	00:25
D2	Angola, Luanda	05:00	05:23	17:03	17:26
D4	Cape Verde, Praia	06:50	07:13	20:06	20:29
D6	Comoros, Moroni	03:03	03:25	14:57	15:19
DA-DL,Y2-Y9	Fed. Rep. of Germany, Bonn	03:01	03:43	19:33	20:15
	Berlin	02:24	03:09	19:16	20:01
DU	Philippines, Manilla	21:13	21:36	10:28	10:51
E3	Eritrea, Asmara	02:40	03:03	15:57	16:21
EA	Spain, Madrid	04:30	05:02	19:40	20:11
EA6	Balearic Is., Palma	04:08	04:39	19:12	19:43
EA8	Canary Is., Las Palmas	05:46	06:12	19:54	20:20
EA9	Ceuta & Melilla, Ceuta	04:51	05:20	19:34	20:03
	Melilla	04:43	05:12	19:24	19:53
EI	Ireland, Dublin	03:37	04:24	20:38	21:25
EK	Armenia, Yerevan	01:17	01:49	16:27	16:58
EL	Liberia, Monrovia	06:14	06:36	19:02	19:24
EP	Iran, Tehran	01:03	01:32	15:46	16:15
ER	Moldova, Kishinev	01:53	02:30	17:51	18:28
ES	Estonia, Tallinn	00:33	01:41	19:13	20:20
ET	Ethiopia, Addis Ababa	02:51	03:14	15:48	16:11
EU,EV,EW	Belarus, Minsk	01:17	02:05	18:26	19:14
EX	Kyrgyzstan, Frunze	23:08	23:41	14:34	15:07
EY	Tajikistan, Samarkand	23:50	00:21	14:56	15:27
	Dushanbe	23:44	00:15	14:46	15:17
EZ	Turkmenistan, Ashkhabad	00:29	01:00	15:25	15:55
F	France, Paris	03:30	04:10	19:44	20:23
FG	Guadeloupe	09:21	09:44	22:41	23:04
FJ, FS	St. Martin	09:23	09:46	22:50	23:14
FH	Mayotte	02:58	03:21	14:49	15:11
FK	New Caledonia, Noumia	19:07	19:31	06:28	06:52
FM	Martinique	09:21	09:44	22:36	22:59
FO	Clipperton Is.	12:41	13:03	01:42	02:04
FO	Fr. Polynesia, Tahiti	16:44	17:07	04:20	04:44
	Rurutu, (Austral Is.)	20:08	20:32	07:29	07:53
	Hiva Oa, (Marquesas Is.)	15:11	15:33	03:11	03:33
FP	St. Pierre & Miquelon, St. Pierre	07:34	08:11	23:29	00:06
FR/G	Glorioso	02:48	03:10	14:43	15:05
FR/J,E	Juan de Nova	03:14	03:37	14:52	15:15
	Europa	03:32	03:56	14:53	15:17
FR	Reunion	02:29	02:53	13:54	14:18
FR/T	Tromelin	02:26	02:49	14:08	14:30
FT8W	Crozet	03:30	04:03	13:12	13:46
FT8X	Kerguelen	02:30	03:06	11:52	12:28
FT8Y	Antarctica, Dumont D'Urville	23:08	00:28	05:04	06:23
FT8Z	Amsterdam & St. Paul Is., Amsterdam	01:29	01:58	11:53	12:21
FW	Wallis & Futuna Is., Wallis	17:45	18:08	05:34	05:57
FY	Fr. Guiana, Cayenne	09:02	09:24	21:46	22:08
G	England, London	03:24	04:08	20:05	20:48
GD	Isle of Man	03:22	04:11	20:36	21:25
GI	Northern Ireland, Belfast	03:26	04:15	20:44	21:33
GJ	Jersey	03:46	04:26	20:04	20:44
GM	Scotland, Glasgow	03:09	04:02	20:44	21:37
	Aberdeen	02:47	03:44	20:44	21:41
GU	Guernsey	03:46	04:27	20:07	20:47
GW	Wales, Cardiff	03:37	04:20	20:17	21:01
H4	Solomon Islands, Honiara	19:14	19:36	07:15	07:38
HA	Hungary, Budapest	02:30	03:07	18:32	19:09
HB	Switzerland, Bern	03:18	03:56	19:16	19:53
HB0	Liechtenstein	03:09	03:46	19:08	19:46
HC	Ecuador, Quito	10:53	11:15	23:21	23:43
HC8	Galapagos Is.	11:43	12:05	00:11	00:33
HH	Haiti, Port-Au-Prince	09:59	10:23	23:28	23:51
HI	Dominican Republic, Santo Domingo	09:50	10:13	23:18	23:42
HK	Colombia, Bogota	10:30	10:52	23:13	23:35
HK0	Malpelo Is.	10:59	11:21	23:40	00:02
HK0	San Adreas	10:47	11:10	23:55	00:18
HL	Korea, Seoul	19:57	20:26	10:49	11:19
HP	Panama, Panama	10:44	11:07	23:41	00:04
HR	Honduras, Tegucigalpa	11:06	11:29	00:20	00:43
HS	Thailand, Bangkok	22:36	22:59	11:49	12:12
HV	Vatican City	03:20	03:52	18:39	19:12
HZ, 7Z	Saudi Arabia, Dharan	01:34	01:59	15:32	15:58
	Mecca	02:25	02:49	16:04	16:29
	Riyadh	01:51	02:16	15:42	16:07

Pref	State/Province/Country/City	Tlight	Sunrise	Sunset	Tlight
I	Italy, Rome	03:20	03:52	18:39	19:12
	Trieste	02:59	03:35	18:46	19:22
	Sicily	03:29	03:58	18:21	18:51
IS	Sardinia, Cagliari	03:43	04:14	18:45	19:16
J2	Djibouti, Djibouti	02:29	02:52	15:34	15:57
J3	Grenada	09:28	09:51	22:35	22:58
J5	Guinea-Bissau, Bissau	06:24	06:47	19:30	19:53
J6	St. Lucia	09:22	09:45	22:35	22:58
J7	Dominica	09:21	09:44	22:38	23:02
J8	St. Vincent	09:24	09:47	22:35	22:58
JA-JS	Japan, Tokyo	19:11	19:40	09:54	10:22
	Nagasaki	19:58	20:26	10:26	10:54
	Sapporo	18:40	19:13	10:07	10:41
JD1	Minami Torishima	18:42	19:07	08:33	08:58
JD1	Ogasawara, Kazan Is.	19:27	19:53	09:31	09:57
JT	Mongolia, Ulan Bator	20:36	21:15	12:42	13:20
JW	Svalbard, Spitsbergen	——	——	——	——
JX	Jan Mayen	——	——	——	——
JY	Jordan, Amman	02:17	02:44	16:41	17:08
KC4	Antarctica, Byrd Station	——	——	——	——
	McMurdo Sound	——	——	——	——
	Palmer Station	12:30	13:39	19:05	20:13
KC6	Belau, Yap	20:13	20:35	09:11	09:34
	Koror	20:31	20:53	09:22	09:45
KG4	Guantanamo Bay	10:08	10:32	23:42	00:06
KH0	Mariana Is., Saipan	19:32	19:56	08:50	09:13
KH1	Baker, Howland Is.	17:23	17:46	05:54	06:16
KH2	Guam, Agana	19:39	20:02	08:51	09:14
KH3	Johnston Is.	16:26	16:50	05:50	06:13
KH4	Midway Is.	16:39	17:05	06:46	07:12
KH5	Palmyra Is.	16:20	16:42	05:07	05:29
KH5K	Kingman Reef	16:20	16:42	05:12	05:34
KH6	Hawaii, Hilo	15:28	15:52	05:01	05:25
	Honolulu	15:36	16:00	05:15	05:39
KH7	Kure Is.	16:43	17:09	06:50	07:16
KH8	American Samoa, Pago Pago	17:24	17:47	05:11	05:34
KH9	Wake Is.	18:02	18:26	07:33	07:57
KL7	Alaska, Adak	15:08	15:52	07:52	08:36
	Anchorage	11:43	13:04	07:07	08:29
	Fairbanks	——	12:13	07:42	——
	Juneau	11:24	12:26	05:41	06:43
	Nome	——	13:27	08:48	——
KP1	Navassa Is.	10:10	10:33	23:38	00:02
KP2	Virgin Islands, Charlotte Amalie	09:30	09:53	22:58	23:21
KP4	Puerto Rico, San Juan	09:34	09:58	23:03	23:27
KP5	Desecheo Is.	09:40	10:04	23:08	23:32
LA-LJ	Norway, Oslo	01:21	02:32	20:14	21:25
LU	Argentina, Buenos Aires	10:28	10:55	21:04	21:31
LX	Luxembourg	03:10	03:51	19:31	20:12
LY	Lithuania, Vilna	01:21	02:10	18:38	19:27
LZ	Bulgaria, Sofia	02:34	03:07	17:59	18:32
OA	Peru, Lima	11:06	11:29	23:00	23:22
OD	Lebanon, Beirut	02:13	02:41	16:46	17:15
OE	Austria, Vienna	02:37	03:16	18:45	19:24
OH	Finland, Helsinki	00:21	01:33	19:18	20:31
OH0	Aland Is.	00:41	01:53	19:38	20:51
OJ0	Market Reef	00:43	01:56	19:43	20:57
OK,OL	Czech Rep., Prague	02:35	03:16	19:01	19:42
OM	Slovak, Rep., Bratislava	02:36	03:14	18:42	19:20
ON	Belgium, Brussels	03:10	03:52	19:44	20:26
OX	Greenland, Godthaab	——	05:57	01:09	——
	Thule	——	——	——	——
OY	Faroe Islands, Torshavn	01:52	03:23	21:43	23:14
OZ	Denmark, Copenhagen	02:02	02:55	19:36	20:28
P2	Papua New Guinea, Madang	20:06	20:28	08:19	08:41
	Port Moresby	20:06	20:28	08:07	08:29
P4	Aruba, Oranjestad	09:57	10:20	23:04	23:27
PA-PI	Netherlands, Amsterdam	02:59	03:43	19:49	20:34
PJ2,4,9	Netherlands Antilles, Willemstad	10:01	10:23	23:09	23:32
PJ5-8	St. Maarten and Saba, St. Maarten	09:24	09:47	22:50	23:13
PY	Brazil, Brasilia	09:15	09:38	20:57	21:20
	Rio De Janeiro	09:07	09:31	20:26	20:50
	Natal	08:10	08:32	20:21	20:43
	Manaus	09:46	10:08	22:06	22:28
	Porto Alegre	09:51	10:16	20:45	21:11
PY0	Fernando De Noronha	07:56	08:18	20:13	20:35
PY0	St. Peter & St. Paul Rocks	07:36	07:58	20:09	20:31
PY0	Trindade & Martin Vaz Is., Trindade	08:08	08:31	19:35	19:58
PZ	Suriname, Paramaribo	09:12	09:34	21:59	22:21
R1FJ_	Franz Josef Land	——	——	——	——
R1MV_	Malyj Vysotskij Is.	00:00	01:15	19:08	20:23
S0	Western Sahara, Smara	05:39	06:05	19:38	20:04
S2	Bangladesh, Dacca	22:58	23:23	12:46	13:11
S5	Slovenia, Ljubljana	02:55	03:31	18:45	19:21
S7	Seychelles, Victoria	02:05	02:27	14:21	14:43
S9	Sao Tome	05:13	05:35	17:43	18:05
SM	Sweden, Stockholm	01:01	02:08	19:39	20:45
SP	Poland, Krakow	02:13	02:54	18:38	19:19
	Warsaw	01:55	02:40	18:44	19:28
ST	Sudan, Khartoum	03:05	03:28	16:24	16:47
ST0	Southern Sudan, Juba	03:26	03:48	16:11	16:33
SU	Egypt, Cairo	02:40	03:06	16:54	17:21
SV	Greece, Athens	02:48	03:18	17:44	18:14
SV/A	Mount Athos	02:39	03:10	17:47	18:19
SV5	Dodecanese, Rhodes	02:35	03:04	17:22	17:51
SV9	Crete	02:50	03:19	17:31	18:00
T2	Tuvalu, Funafuti	17:59	18:21	06:02	06:24

Pref	State/Province/Country/City	Tlight	Sunrise	Sunset	Tlight
T30	West Kiribati, Bonriki	18:05	18:27	06:39	07:01
T31	Central Kiribati, Kanton	17:11	17:33	05:32	05:54
T32	East Kiribati, Christmas Is.	16:07	16:29	04:42	05:04
T33	Banaba Is.	18:23	18:45	06:51	07:13
T5	Somalia, Mogadishu	02:35	02:58	15:11	15:33
T7	San Marino	03:13	03:47	18:46	19:21
T9	Bosnia-Herzegovina, Sarajevo	02:48	03:23	18:22	18:56
TA	Turkey, Ankara	02:05	02:37	17:12	17:43
	Istanbul	02:16	02:49	17:31	18:04
TF	Iceland, Reykjavik	——	03:59	23:08	——
TG	Guatemala, Guatemala City	11:19	11:42	00:34	00:57
TI	Costa Rica, San Jose	11:01	11:23	00:00	00:23
TI9	Cocos Is.	11:20	11:42	00:06	00:28
TJ	Cameroon, Yaounde	04:48	05:10	17:29	17:51
TK	Corsica	03:33	04:06	18:54	19:26
TL	Central African Republic, Bangui	04:19	04:41	17:02	17:24
TN	Congo, Brazzaville	04:45	05:08	17:02	17:24
TR	Gabon, Libreville	05:02	05:24	17:32	17:54
TT	Chad, N'Djamena	04:21	04:44	17:28	17:51
TU	Cote d'Ivoire, Abidjan	05:48	06:10	18:33	18:56
TY	Benin, Porto Novo	05:20	05:42	18:09	18:31
TZ	Mali, Bamako	05:52	06:15	19:01	19:24
UA	Russia, European, Leningrad (UA1)	00:04	01:15	18:55	20:05
	Archangel (UA1)	——	23:42	19:06	——
	Murmansk (UA1)	——	——	——	——
	Moscow (UA3)	00:22	01:14	17:57	18:49
	Kuybyshev (UA4)	23:52	00:39	16:52	17:39
	Rostov (UA6)	01:08	01:46	17:10	17:48
UA2	Kaliningrad	01:37	02:27	19:00	19:51
UA9,0	Russia, Asiatic, Novosibirsk (UA9)	21:27	22:18	14:51	15:41
	Perm (UA9)	22:45	23:45	16:56	17:57
	Omsk (UA9)	22:05	22:56	15:29	16:19
	Norilsk (UA0)	——	——	——	——
	Irkutsk (UA0)	20:22	21:06	13:11	13:56
	Vladivostok (UA0)	19:17	19:51	10:46	11:19
	Petropavlovsk (UA0)	16:39	17:25	09:37	10:23
	Khabarovsk (UA0)	18:41	19:20	10:51	11:30
	Krasnoyarsk (UA0)	20:39	21:32	14:17	15:10
	Yakutsk (UA0)	16:46	18:17	12:37	14:08
	Wrangel Island (UA0)	——	——	——	——
	Kyzyl (UA0Y)	21:05	21:48	13:47	14:31
UJ-UM	Uzbekistan, Bukhoro	00:00	00:31	15:06	15:37
	Tashkent	23:35	00:07	14:50	15:22
UN-UQ	Kazakhstan, Alma-Ata	22:57	23:31	14:26	15:00
UR-UZ,EM-EO	Ukraine, Kiev	01:29	02:10	17:58	18:39
V2	Antigua & Barbuda, St. Johns	09:19	09:43	22:43	23:07
V3	Belize, Belmopan	11:07	11:31	00:32	00:55
V4	St. Kitts & Nevis	09:22	09:46	22:47	23:10
V5	Namibia, Windhoek	05:05	05:29	16:26	16:49
V6	Micronesia, Ponape	18:56	19:19	07:47	08:09
V7	Marshall Islands, Kwajalein	18:17	18:39	07:14	07:36
V8	Brunei, Bandar Seri Begawan	21:53	22:15	10:37	10:59
VK	Australia, Canberra (VK1)	20:39	21:06	07:13	07:40
	Sydney (VK2)	20:28	20:55	07:07	07:34
	Melbourne (VK3)	21:00	21:29	07:23	07:52
	Brisbane (VK4)	20:10	20:35	07:13	07:38
	Adelaide (VK5)	21:20	21:47	07:55	08:23
	Perth (VK6)	22:46	23:12	09:33	09:59
	Hobart, Tasmania (VK7)	21:02	21:33	07:01	07:32
	Darwin (VK8)	21:15	21:38	09:07	09:29
VK0	Heard Is.	02:25	03:04	11:20	12:00
VK0	Macquarie Is.	20:49	21:31	05:30	06:12
VK9C	Cocos-Keeling Is.	23:31	23:54	11:24	11:46
VK9L	Lord Howe Is.	19:52	20:18	06:41	07:07
VK9M	Mellish Reef	19:43	20:06	07:19	07:42
VK9N	Norfolk Is.	19:12	19:38	06:10	06:36
VK9W	Willis Is.	20:06	20:29	07:46	08:09
VK9X	Christmas Is.	22:53	23:15	10:51	11:13
VP2E	Anguilla	09:22	09:46	22:50	23:14
VP2M	Montserrat	09:22	09:45	22:44	23:08
VP2V	British Virgin Is., Tortola	09:28	09:52	22:57	23:20

Pref	State/Province/Country/City	Tlight	Sunrise	Sunset	Tlight
VP5	Turks & Caicos Islands, Grand Turk	09:49	10:13	23:28	23:52
VP8	Falkland Islands, Stanley	11:06	11:44	20:11	20:49
VP8	So. Georgia Is.	09:50	10:31	18:35	19:16
VP8	So. Orkney Is.	10:51	11:45	18:31	19:24
VP8	So. Sandwich Islands, Saunders Is.	09:23	10:10	17:35	18:22
VP8	So. Shetland Is., King George Is.	11:50	12:47	19:11	20:08
VP9	Bermuda	08:58	09:26	23:24	23:51
VQ9	Chagos, Diego Garcia	01:01	01:24	13:09	13:31
VR2,VS6	Hong Kong	21:25	21:49	11:08	11:32
VR6	Pitcairn Is.	14:58	15:23	02:10	02:34
VU	India, Bombay	00:18	00:41	13:48	14:12
	Calcutta	23:08	23:33	12:52	13:16
	New Delhi	23:40	00:06	13:48	14:14
	Bangalore	00:09	00:32	13:19	13:42
VU	Andaman Islands, Port Blair	23:11	23:33	12:16	12:39
VU	Laccadive Is.	00:33	00:55	13:33	13:55
XE	Mexico, Mexico City (XE1)	11:44	12:08	01:16	01:40
	Chihuahua (XE2)	11:52	12:18	02:01	02:27
	Merida (XE3)	11:04	11:28	00:42	01:06
XF4	Revilla Gigedo	12:35	12:58	02:05	02:29
XT	Burkina Faso, Ouagadougou	05:27	05:50	18:35	18:58
XU	Cambodia, Phnom Penh	22:23	22:45	11:28	11:51
XW	Laos, Viangchan	22:20	22:44	11:47	12:11
XX9	Macao	21:28	21:52	11:10	11:35
XZ	Myanmar, Yangon	22:49	23:12	12:12	12:35
YA	Afghanistan, Kandahar	00:20	00:46	14:39	15:06
	Kabul	23:57	00:25	14:33	15:01
YB-YD	Indonesia, Jakarta	22:42	23:04	10:53	11:15
	Medan	23:00	23:22	11:40	12:02
	Pontianak	22:23	22:45	10:52	11:14
	Jayapura	20:21	20:43	08:43	09:05
YI	Iraq, Baghdad	01:40	02:07	16:08	16:36
YJ	Vanuatu, Port Vila	18:53	19:16	06:29	06:52
YK	Syria, Damascus	02:11	02:39	16:42	17:10
YL	Latvia, Riga	01:04	02:01	18:58	19:55
YN	Nicaragua, Managua	11:05	11:28	00:12	00:34
YO	Romania, Bucharest	02:16	02:50	17:53	18:27
YS	El Salvador, San Salvador	11:15	11:38	00:27	00:50
YU	Yugoslavia, Belgrade	02:36	03:11	18:17	18:52
YV	Venezuela, Caracas	09:52	10:14	22:53	23:16
YV0	Aves Is.	09:30	09:53	22:48	23:12
Z2	Zimbabwe, Harare	04:02	04:26	15:38	16:01
Z3	Macedonia, (ex Yugoslav), Skopje	02:44	03:16	18:04	18:37
ZA	Albania, Tirane	02:53	03:25	18:08	18:41
ZB2	Gibraltar	04:50	05:20	19:35	20:04
ZC4	British Cyprus	02:21	02:50	16:58	17:27
ZD7	St. Helena	06:27	06:50	18:09	18:31
ZD8	Ascension Is.	06:50	07:12	18:55	19:17
ZD9	Tristan da Cunha	07:28	07:56	17:54	18:22
ZF	Cayman Is.	10:33	10:56	00:05	00:29
ZK1	No. Cook Is., Manihiki	16:40	17:02	04:38	05:00
ZK1	So. Cook Is., Rarotonga	16:51	17:15	04:16	04:39
ZK2	Niue	17:24	17:47	04:56	05:19
ZK3	Tokelaus, Atafu	17:23	17:46	05:28	05:50
ZL	New Zealand, Auckland (ZL1)	18:59	19:27	05:26	05:54
	Wellington (ZL2)	19:08	19:38	05:15	05:45
	Christchurch (ZL3)	19:22	19:53	05:18	05:49
	Dunedin (ZL4)	19:36	20:09	05:19	05:52
ZL5	Antarctica, Scott Base	——	——	——	——
ZL7	Chatham Is.	18:39	19:11	04:32	05:04
ZL8	Kermadec Is.	18:16	18:42	05:13	05:39
ZL9	Auckland & Campbell Is., Auckland	20:05	20:42	05:17	05:55
	Campbell Is.	20:00	20:39	04:59	05:39
ZP	Paraguay, Asuncion	10:09	10:33	21:20	21:44
ZS	South Africa, Cape Town (ZS1)	05:19	05:46	15:58	16:25
	Port Elizabeth (ZS2)	04:50	05:17	15:29	15:56
	Bloemfontein (ZS4)	04:40	05:06	15:37	16:03
	Durban (ZS5)	04:22	04:48	15:17	15:42
	Johannesburg (ZS6)	04:27	04:52	15:35	16:00
ZS8	Marion Is.	04:29	05:03	14:06	14:40

Pref	State/Province/Country/City	Tlight	Sunrise	Sunset	Tlight
VE1	New Brunswick, Saint John	08:42	09:15	23:44	00:18
	Nova Scotia, Halifax	08:34	09:07	23:33	00:05
	Prince Edward Island, Charlottetown	08:27	09:01	23:35	00:09
VE2	Quebec, Montreal	09:11	09:44	00:15	00:48
	Quebec City	08:57	09:31	00:09	00:43
VE3	Ontario, London	09:50	10:22	00:39	01:11
	Ottawa	09:20	09:53	00:23	00:56
	Sudbury	09:37	10:11	00:47	01:21
	Toronto	09:40	10:12	00:33	01:06
VE4	Manitoba, Winnipeg	10:28	11:06	02:02	02:39
VE5	Saskatchewan, Regina	10:56	11:34	02:34	03:12
	Saskatoon	10:57	11:37	02:47	03:27
VE6	Alberta, Calgary	11:31	12:10	03:13	03:52
	Edmonton	11:17	11:59	03:20	04:02
VE7	British Columbia, Prince George	11:52	12:34	03:59	04:41
	Prince Rupert	12:20	13:03	04:30	05:13
	Vancouver	12:15	12:52	03:44	04:20
VE8	Northwest Territories, Yellowknife	10:09	11:15	04:10	05:17
	Resolute	—	—	—	—
VY1	Yukon, Whitehorse	11:52	12:50	05:21	06:20
VO1	Newfoundland, St. John's	07:40	08:15	22:57	23:32
VO2	Labrador, Goose Bay	07:45	08:27	23:47	00:28
W1	Connecticut, Hartford	09:19	09:50	00:02	00:33
	Maine, Bangor	08:54	09:27	23:54	00:27
	Portland	09:04	09:36	23:57	00:29
	Massachusetts, Boston	09:11	09:42	23:57	00:28
	New Hampshire, Concord	09:10	09:42	00:01	00:32
	Rhode Island, Providence	09:14	09:45	23:57	00:28
	Vermont, Montpelier	09:11	09:44	00:08	00:40
W2	New Jersey, Atlantic City	09:33	10:02	00:04	00:33
	New York, Albany	09:21	09:52	00:09	00:40
	Buffalo	09:41	10:12	00:30	01:01
	New York City	09:27	09:57	00:05	00:35
	Syracuse	09:29	10:01	00:19	00:51
W3	Delaware, Wilmington	09:36	10:06	00:09	00:38
	District of Columbia, Washington	09:44	10:14	00:13	00:42
	Maryland, Baltimore	09:42	10:11	00:12	00:42
	Pennsylvania, Harrisburg	09:40	10:10	00:16	00:46
	Philadelphia	09:34	10:04	00:08	00:38
	Pittsburgh	09:52	10:22	00:28	00:58
	Scranton	09:32	10:03	00:13	00:44
W4	Alabama, Montgomery	10:37	11:03	00:38	01:04
	Florida, Jacksonville	10:23	10:48	00:16	00:42
	Miami	10:25	10:49	00:03	00:27
	Pensacola	10:44	11:10	00:38	01:04
	Georgia, Atlanta	10:26	10:53	00:33	01:00
	Savannah	10:16	10:43	00:17	00:43
	Kentucky, Lexington	10:17	10:45	00:41	01:10
	Louisville	10:21	10:50	00:47	01:16
	North Carolina, Charlotte	10:09	10:36	00:21	00:48
	Raleigh	09:58	10:26	00:13	00:41
	Wilmington	09:59	10:26	00:07	00:35
	South Carolina, Columbia	10:12	10:39	00:20	00:47
	Tennessee, Knoxville	10:19	10:47	00:35	01:03
	Memphis	10:46	11:13	00:58	01:25
	Nashville	10:30	10:58	00:47	01:15
	Virginia, Norfolk	09:47	10:15	00:06	00:34
	Richmond	09:49	10:18	00:12	00:40
W5	Arkansas, Little Rock	10:56	11:23	01:06	01:33
	Louisiana, New Orleans	10:57	11:22	00:49	01:14
	Shreveport	11:06	11:32	01:08	01:34
	Mississippi, Jackson	10:52	11:19	00:53	01:20
	New Mexico, Albuquerque	11:52	12:20	02:04	02:32
	Oklahoma, Oklahoma City	11:15	11:42	01:28	01:56
	Texas, Abilene	11:30	11:56	01:32	01:58
	Amarillo	11:33	12:00	01:45	02:12
	Dallas	11:18	11:44	01:21	01:47
	El Paso	11:59	12:25	01:58	02:24
	San Antonio	11:31	11:57	01:22	01:47
W6	California, Los Angeles	12:41	13:08	02:49	03:16
	San Francisco	12:49	13:17	03:12	03:41
W7	Arizona, Flagstaff	12:12	12:40	02:25	02:52
	Phoenix	12:18	12:44	02:23	02:50
	Idaho, Boise	12:08	12:40	03:00	03:32
	Pocatello	11:55	12:27	02:44	03:15
	Montana, Billings	11:30	12:03	02:35	03:09
	Butte	11:45	12:19	02:52	03:26
	Great Falls	11:35	12:10	02:51	03:26
	Nevada, Las Vegas	12:23	12:51	02:40	03:08
	Reno	12:34	13:03	03:06	03:35
	Oregon, Portland	12:28	13:01	03:31	04:05
	Utah, Salt Lake City	11:59	12:29	02:37	03:07
	Washington, Seattle	12:18	12:54	03:35	04:11
	Spokane	11:58	12:34	03:16	03:51
	Wyoming, Cheyenne	11:30	12:00	02:09	02:39
	Sheridan	11:27	12:00	02:27	02:59
W8	Michigan, Detroit	09:59	10:30	00:45	01:16
	Grand Rapids	10:08	10:39	00:57	01:29
	Sault Ste. Marie	09:51	10:25	01:01	01:35
	Traverse City	10:01	10:34	01:01	01:34
	Ohio, Cincinnati	10:14	10:43	00:43	01:13
	Cleveland	09:56	10:27	00:37	01:08
	Columbus	10:05	10:35	00:39	01:09
	West Virginia, Charleston	10:04	10:33	00:30	00:59
W9	Illinois, Chicago	10:18	10:49	01:02	01:33
	Indiana, Indianapolis	10:19	10:48	00:52	01:21
	Wisconsin, Green Bay	10:12	10:45	01:10	01:42

Pref	State/Province/Country/City	Tlight	Sunrise	Sunset	Tlight
	Milwaukee	10:16	10:48	01:06	01:37
W0	Colorado, Denver	11:34	12:04	02:07	02:36
	Grand Junction	11:50	12:19	02:20	02:49
	Iowa, Des Moines	10:43	11:14	01:25	01:56
	Kansas, Pratt	11:14	11:43	01:37	02:06
	Wichita	11:09	11:37	01:32	02:00
	Minnesota, Duluth	10:21	10:55	01:32	02:07
	Minneapolis	10:32	11:05	01:32	02:05
	Missouri, Columbia	10:45	11:14	01:14	01:44
	Kansas City	10:54	11:23	01:24	01:53
	St. Louis	10:38	11:07	01:05	01:34
	Nebraska, North Platte	11:14	11:44	01:53	02:23
	Omaha	10:53	11:24	01:34	02:04
	North Dakota, Fargo	10:39	11:13	01:51	02:26
	South Dakota, Rapid City	11:14	11:46	02:10	02:42
-	-	05:40	06:02	18:09	18:30
1A0	S.M.O.M	03:38	04:09	18:22	18:53
1S	Spratly Is.	22:02	22:24	10:52	11:14
3A	Monaco	03:53	04:25	18:46	19:18
3B6	Agalega	02:06	02:28	14:10	14:32
3B7	St. Brandon	02:00	02:22	13:50	14:12
3B8	Mauritius	02:14	02:37	13:54	14:17
3B9	Rodriguez Is.	01:49	02:12	13:31	13:54
3C	Equatorial Guinea, Bata	04:59	05:20	17:32	17:53
	Malabo	05:00	05:22	17:38	18:00
3C0	Annobon Is.	05:20	05:41	17:45	18:06
3D2	Fiji Is., Suva	18:08	18:30	05:53	06:16
3D2	Conway Reef	18:27	18:51	06:05	06:28
3D2	Rotuma Is.	18:04	18:26	06:03	06:25
3DA	Swaziland, Mbabane	04:06	04:31	15:31	15:55
3V	Tunisia, Tunis	04:01	04:29	18:20	18:48
3W, XV	Vietnam, Ho Chi Minh City (Saigon)	22:20	22:42	11:15	11:37
	Hanoi	22:09	22:32	11:32	11:55
3X	Guinea, Conakry	06:23	06:45	19:15	19:37
3Y	Bouvet	06:44	07:23	16:20	16:59
3Y	Peter I Is.	13:57	15:12	21:04	22:18
4J, 4K	Azerbaijan, Baku	01:13	01:43	15:49	16:19
4L	Georgia, Tbilisi	01:29	02:00	16:12	16:43
4S	Sri Lanka, Colombo	00:12	00:34	12:57	13:19
4U	ITU Geneva	03:49	04:23	18:58	19:31
4U	United Nations Hq.	09:27	09:57	00:05	00:35
4X, 4Z	Israel, Jerusalem	02:32	02:58	16:31	16:57
5A	Libya, Tripoli	04:01	04:28	18:03	18:29
	Benghazi	03:32	03:58	17:32	17:58
5B	Cyprus, Nicosia	02:32	02:59	16:44	17:12
5H	Tanzania, Dar es Salaam	03:10	03:32	15:22	15:44
5N	Nigeria, Lagos	05:19	05:40	18:03	18:25
5R	Madagascar, Antananarivo	02:52	03:15	14:36	14:58
5T	Mauritania, Nouakchott	06:20	06:43	19:35	19:58
5U	Niger, Niamey	05:15	05:37	18:17	18:39
5V	Togo, Lome	05:28	05:50	18:11	18:33
5W	Western Samoa, Apia	17:23	17:45	05:20	05:42
5X	Uganda, Kampala	03:30	03:51	15:59	16:21
5Z	Kenya, Nairobi	04:35	04:56	17:00	17:22
6W	Senegal, Dakar	06:31	06:54	19:37	19:59
6Y	Jamaica, Kingston	10:24	10:47	23:38	00:01
7O	Yemen, Aden	02:24	02:46	15:24	15:46
	Sanaa	02:24	02:46	15:31	15:53
7P	Lesotho, Maseru	04:25	04:49	15:41	16:06
7Q	Malawi, Lilongwe	03:41	04:03	15:37	15:59
	Blantyre	03:38	04:01	15:30	15:52
7X	Algeria, Algiers	04:30	04:58	18:49	19:17
8P	Barbados, Bridgetown	09:22	09:44	22:23	22:45
8Q	Maldive Is.	00:41	01:03	13:20	13:42
8R	Guyana, Georgetown	09:25	09:46	22:10	22:31
9A	Croatia, Zagreb	03:12	03:45	18:17	18:51
9G	Ghana, Accra	05:34	05:56	18:16	18:38
9H	Malta	03:46	04:14	18:02	18:29
9J	Zambia, Lusaka	04:05	04:27	15:57	16:19
9K	Kuwait	01:46	02:12	15:37	16:02
9L	Sierra Leone, Freetown	06:23	06:44	19:12	19:34
9M2	West Malaysia, Kuala Lumpur	22:50	23:12	11:26	11:48
9M6,8	East Malaysia, Sabah, Sandakan(9M6)	21:41	22:03	10:23	10:45
	Sarawak, Kuching (9M8)	22:17	22:39	10:49	11:11
9N	Nepal, Kathmandu	23:19	23:44	13:04	13:29
9Q	Zaire, Kinshasa	04:44	05:06	17:02	17:24
	Kisangani	03:59	04:20	16:29	16:50
	Lubumbashi	04:04	04:26	16:05	16:27
9U	Burundi, Bujumbura	03:47	04:08	16:08	16:29
9V	Singapore	22:43	23:05	11:15	11:37
9X	Rwanda, Kigali	03:42	04:04	16:06	16:28
9Y	Trinidad & Tobago , Port of Spain	09:32	09:54	22:27	22:49
A2	Botswana, Gaborone	04:25	04:49	15:54	16:18
A3	Tonga, Nukualofa	17:45	18:09	05:24	05:47
A4	Oman, Musqat	01:13	01:37	14:44	15:08
A5	Bhutan, Thimbu	23:04	23:29	12:47	13:12
A6	United Arab Emirates, Abu Dhabi	01:29	01:54	15:03	15:27
A7	Qatar, Ad-Dawhah	01:39	02:03	15:15	15:40
A9	Bahrein, Al-Manamah	01:41	02:06	15:20	15:45
AP	Pakistan, Karachi	00:37	01:01	14:12	14:37
AP	Islamabad	23:56	00:23	14:02	14:29
BV	Taiwan, Taipei	20:59	21:24	11:05	11:30
BY	Peoples Rep. of China, Beijing	20:48	21:18	11:22	11:52
	Harbin	19:49	20:22	10:54	11:28
	Shanghai	20:48	21:14	10:45	11:11
	Fuzhou	21:06	21:31	10:45	11:10
	Xian	21:32	21:59	11:40	12:08

Pref	State/Province/Country/City	Tlight	Sunrise	Sunset	Tlight
	Chongqing	21:51	22:16	11:42	12:08
	Chengdu	21:59	22:24	11:53	12:19
	Lhasa	22:52	23:18	12:43	13:09
	Urumqi	22:32	23:04	13:26	13:58
	Kashi	23:31	00:00	14:02	14:32
C2	Nauru	18:33	18:55	07:01	07:22
C3	Andorra	04:20	04:52	19:07	19:38
C5	The Gambia, Banjul	06:30	06:52	19:32	19:54
C6	Bahamas, Nassau	10:15	10:39	23:51	00:15
C9	Mozambique, Maputo	04:00	04:24	15:26	15:50
	Mozambique	03:15	03:37	15:08	15:30
CE	Chile, Santiago	11:03	11:29	22:08	22:34
CE0	Easter Island	13:29	13:54	00:52	01:16
CE0	Juan Fernandez	11:35	12:02	22:39	23:05
CE0	San Felix	11:31	11:55	22:56	23:20
CM,CO	Cuba, Havana	10:38	11:02	00:08	00:31
CN	Morocco, Casablanca	05:19	05:46	19:25	19:52
CP	Bolivia, La Paz	10:33	10:55	22:22	22:45
CT	Portugal, Lisbon	05:14	05:43	19:41	20:10
CT3	Madeira Islands, Funchal	05:59	06:25	20:01	20:27
CU	Azores , Ponta Delgada	06:22	06:51	20:45	21:14
CX	Uruguay, Montevideo	10:07	10:33	21:07	21:33
CY0	Sable Is.	08:22	08:54	23:16	23:48
CY9	St. Paul Is.	08:11	08:46	23:26	00:00
D2	Angola, Luanda	04:58	05:19	17:06	17:27
D4	Cape Verde, Praia	06:55	07:18	20:01	20:23
D6	Comoros, Moroni	02:59	03:21	15:00	15:22
DA-DL,Y2-Y9	Fed. Rep. of Germany, Bonn	03:28	04:07	19:08	19:46
	Berlin	02:54	03:35	18:48	19:29
DU	Philippines, Manilla	21:18	21:40	10:23	10:45
E3	Eritrea, Asmara	02:45	03:08	15:52	16:14
EA	Spain, Madrid	04:47	05:17	19:23	19:53
EA6	Balearic Is., Palma	04:24	04:54	18:56	19:25
EA8	Canary Is., Las Palmas	05:57	06:22	19:43	20:08
EA9	Ceuta & Melilla, Ceuta	05:05	05:33	19:20	19:48
	Melilla	04:57	05:25	19:10	19:37
EI	Ireland, Dublin	04:09	04:51	20:10	20:52
EK	Armenia, Yerevan	01:35	02:05	16:10	16:40
EL	Liberia, Monrovia	06:16	06:37	18:59	19:21
EP	Iran, Tehran	01:17	01:44	15:32	15:59
ER	Moldova, Kishinev	02:16	02:51	17:29	18:04
ES	Estonia, Tallinn	01:24	02:18	18:34	19:28
ET	Ethiopia, Addis Ababa	02:54	03:16	15:45	16:07
EU,EV,EW	Belarus, Minsk	01:50	02:33	17:57	18:40
EX	Kyrgyzstan, Frunze	23:27	23:58	14:16	14:47
EY	Tajikistan, Samarkand	00:07	00:37	14:40	15:09
	Dushanbe	00:01	00:30	14:30	15:00
EZ	Turkmenistan, Ashkhabad	00:45	01:14	15:10	15:38
F	France, Paris	03:55	04:32	19:21	19:57
FG	Guadeloupe	09:26	09:49	22:35	22:58
FJ, FS	St. Martin	09:29	09:52	22:44	23:06
FH	Mayotte	02:54	03:16	14:52	15:14
FK	New Caledonia, Noumia	19:00	19:23	06:35	06:59
FM	Martinique	09:26	09:48	22:31	22:53
FO	Clipperton Is.	12:44	13:06	01:38	02:00
FO	Fr. Polynesia, Tahiti	16:39	17:01	04:25	04:48
	Rurutu, (Austral Is.)	20:01	20:24	07:36	07:59
	Hiva Oa, (Marquesas Is.)	15:08	15:30	03:13	03:35
FP	St. Pierre & Miquelon, St. Pierre	07:57	08:31	23:08	23:42
FR/G	Glorioso	02:44	03:06	14:46	15:08
FR/J,E	Juan de Nova	03:09	03:31	14:57	15:19
	Europa	03:24	03:48	15:00	15:23
FR	Reunion	02:22	02:45	14:00	14:24
FR/T	Tromelin	02:21	02:43	14:12	14:34
FT8W	Crozet	03:12	03:43	13:31	14:03
FT8X	Kerguelen	02:09	02:43	12:14	12:48
FT8Y	Antarctica, Dumont D'Urville	22:22	23:26	06:05	07:09
FT8Z	Amsterdam & St. Paul Is., Amsterdam	01:16	01:43	12:07	12:34
FW	Wallis & Futuna Is., Wallis	17:41	18:03	05:38	06:00
FY	Fr. Guiana, Cayenne	09:04	09:25	21:44	22:05
G	England, London	03:53	04:32	19:39	20:18
GD	Isle of Man	03:56	04:40	20:07	20:50
GI	Northern Ireland, Belfast	04:00	04:44	20:14	20:57
GJ	Jersey	04:11	04:48	19:40	20:17
GM	Scotland, Glasgow	03:47	04:33	20:12	20:58
	Aberdeen	03:29	04:17	20:10	20:59
GU	Guernsey	04:12	04:49	19:43	20:20
GW	Wales, Cardiff	04:06	04:46	19:51	20:31
H4	Solomon Islands, Honiara	19:11	19:33	07:18	07:39
HA	Hungary, Budapest	02:53	03:28	18:10	18:45
HB	Switzerland, Bern	03:41	04:16	18:55	19:29
HB0	Liechtenstein	03:32	04:07	18:47	19:21
HC	Ecuador, Quito	10:52	11:14	23:21	23:42
HC8	Galapagos Is.	11:43	12:04	00:10	00:32
HH	Haiti, Port-Au-Prince	10:05	10:28	23:21	23:44
HI	Dominican Republic, Santo Domingo	09:56	10:19	23:12	23:35
HK	Colombia, Bogota	10:31	10:53	23:11	23:32
HK0	Malpelo Is.	11:00	11:21	23:38	00:00
HK0	San Adreas	10:51	11:13	23:51	00:13
HL	Korea, Seoul	20:12	20:40	10:34	11:03
HP	Panama, Panama	10:47	11:09	23:38	23:59
HR	Honduras, Tegucigalpa	11:11	11:33	00:15	00:37
HS	Thailand, Bangkok	22:41	23:03	11:44	12:06
HV	Vatican City	03:38	04:09	18:22	18:52
HZ, 7Z	Saudi Arabia, Dharan	01:43	02:08	15:23	15:47
	Mecca	02:32	02:56	15:57	16:20
	Riyadh	01:59	02:24	15:33	15:58

Pref	State/Province/Country/City	Tlight	Sunrise	Sunset	Tlight
I	Italy, Rome	03:38	04:09	18:22	18:52
	Trieste	03:21	03:54	18:26	18:59
	Sicily	03:44	04:12	18:06	18:35
IS	Sardinia, Cagliari	03:59	04:28	18:29	18:59
J2	Djibouti, Djibouti	02:33	02:55	15:30	15:52
J3	Grenada	09:32	09:55	22:30	22:53
J5	Guinea-Bissau, Bissau	06:28	06:50	19:26	19:48
J6	St. Lucia	09:27	09:49	22:30	22:52
J7	Dominica	09:26	09:48	22:33	22:55
J8	St. Vincent	09:29	09:51	22:30	22:52
JA-JS	Japan, Tokyo	19:25	19:53	09:39	10:07
	Nagasaki	20:11	20:38	10:14	10:40
	Sapporo	18:59	19:31	09:49	10:20
JD1	Minami Torishima	18:51	19:15	08:24	08:48
JD1	Ogasawara, Kazan Is.	19:37	20:02	09:21	09:46
JT	Mongolia, Ulan Bator	21:00	21:36	12:20	12:55
JW	Svalbard, Spitsbergen	——	——	——	——
JX	Jan Mayen	——	02:19	22:57	——
JY	Jordan, Amman	02:29	02:55	16:28	16:55
KC4	Antarctica, Byrd Station	——	——	——	——
	McMurdo Sound	22:48	——	——	03:09
	Palmer Station	11:49	12:47	19:56	20:54
KC6	Belau, Yap	20:16	20:38	09:07	09:29
	Koror	20:33	20:55	09:19	09:41
KG4	Guantanamo Bay	10:15	10:38	23:34	23:58
KH0	Mariana Is., Saipan	19:38	20:00	08:44	09:07
KH1	Baker, Howland Is.	17:24	17:45	05:53	06:15
KH2	Guam, Agana	19:44	20:06	08:46	09:08
KH3	Johnston Is.	16:32	16:55	05:44	06:06
KH4	Midway Is.	16:49	17:14	06:35	07:00
KH5	Palmyra Is.	16:21	16:43	05:04	05:26
KH5K	Kingman Reef	16:22	16:44	05:09	05:31
KH6	Hawaii, Hilo	15:34	15:58	04:54	05:17
	Honolulu	15:43	16:07	05:07	05:31
KH7	Kure Is.	16:53	17:18	06:40	07:05
KH8	American Samoa, Pago Pago	17:20	17:42	05:15	05:37
KH9	Wake Is.	18:08	18:31	07:26	07:49
KL7	Alaska, Adak	15:38	16:17	07:26	08:06
	Anchorage	12:46	13:47	06:24	07:24
	Fairbanks	11:48	13:11	06:43	08:05
	Juneau	12:10	13:01	05:05	05:56
	Nome	13:04	14:23	07:50	09:10
KP1	Navassa Is.	10:16	10:39	23:32	23:54
KP2	Virgin Islands, Charlotte Amalie	09:36	09:59	22:51	23:14
KP4	Puerto Rico, San Juan	09:41	10:04	22:56	23:19
KP5	Desecheo Is.	09:46	10:09	23:01	23:24
LA-LJ	Norway, Oslo	02:15	03:11	19:34	20:30
LU	Argentina, Buenos Aires	10:15	10:42	21:16	21:43
LX	Luxembourg	03:36	04:13	19:07	19:45
LY	Lithuania, Vilna	01:55	02:39	18:08	18:51
LZ	Bulgaria, Sofia	02:53	03:24	17:40	18:12
OA	Peru, Lima	11:03	11:25	23:03	23:25
OD	Lebanon, Beirut	02:26	02:53	16:33	17:00
OE	Austria, Vienna	03:02	03:37	18:23	18:58
OH	Finland, Helsinki	01:16	02:13	18:38	19:35
OH0	Aland Is.	01:36	02:33	18:58	19:55
OJ0	Market Reef	01:39	02:36	19:02	19:59
OK,OL	Czech Rep., Prague	03:01	03:39	18:36	19:14
OM	Slovak, Rep., Bratislava	03:00	03:35	18:19	18:55
ON	Belgium, Brussels	03:38	04:16	19:19	19:57
OX	Greenland, Godthaab	05:34	06:51	00:13	01:30
	Thule	——	——	——	——
OY	Faroe Islands, Torshavn	03:05	04:08	20:56	22:00
OZ	Denmark, Copenhagen	02:40	03:25	19:04	19:50
P2	Papua New Guinea, Madang	20:04	20:26	08:20	08:42
	Port Moresby	20:03	20:25	08:09	08:31
P4	Aruba, Oranjestad	10:01	10:23	22:59	23:22
PA-PI	Netherlands, Amsterdam	03:29	04:09	19:22	20:03
PJ2,4,9	Netherlands Antilles, Willemstad	10:05	10:27	23:04	23:27
PJ5-8	St. Maarten and Saba, St. Maarten	09:30	09:53	22:43	23:06
PY	Brazil, Brasilia	09:10	09:32	21:01	21:24
	Rio De Janeiro	09:00	09:23	20:33	20:56
	Natal	08:08	08:30	20:22	20:44
	Manaus	09:45	10:06	22:06	22:27
	Porto Alegre	09:40	10:05	20:55	21:20
PY0	Fernando De Noronha	07:54	08:16	20:14	20:35
PY0	St. Peter & St. Paul Rocks	07:37	07:58	20:08	20:29
PY0	Trindade & Martin Vaz Is., Trindade	08:01	08:24	19:41	20:04
PZ	Suriname, Paramaribo	09:14	09:36	21:56	22:18
R1FJ_	Franz Josef Land	——	——	——	——
R1MV_	Malyj Vysotskij Is.	00:58	01:56	18:26	19:24
S0	Western Sahara, Smara	05:49	06:13	19:29	19:53
S2	Bangladesh, Dacca	23:06	23:30	12:37	13:01
S5	Slovenia, Ljubljana	03:17	03:51	18:24	18:58
S7	Seychelles, Victoria	02:04	02:25	14:21	14:43
S9	Sao Tome	05:13	05:35	17:42	18:04
SM	Sweden, Stockholm	01:52	02:46	19:00	19:54
SP	Poland, Krakow	02:40	03:17	18:14	18:51
	Warsaw	02:25	03:06	18:17	18:57
ST	Sudan, Khartoum	03:10	03:33	16:18	16:40
ST0	Southern Sudan, Juba	03:28	03:49	16:08	16:30
SU	Egypt, Cairo	02:51	03:16	16:43	17:09
SV	Greece, Athens	03:04	03:33	17:28	17:57
SV/A	Mount Athos	02:56	03:26	17:31	18:00
SV5	Dodecanese, Rhodes	02:50	03:18	17:07	17:35
SV9	Crete	03:04	03:32	17:17	17:45
T2	Tuvalu, Funafuti	17:56	18:18	06:04	06:26

Pref	State/Province/Country/City	Tlight	Sunrise	Sunset	Tlight
T30	West Kiribati, Bonriki	18:06	18:27	06:38	06:59
T31	Central Kiribati, Kanton	17:10	17:32	05:32	05:54
T32	East Kiribati, Christmas Is.	16:08	16:29	04:41	05:02
T33	Banaba Is.	18:23	18:45	06:51	07:12
T5	Somalia, Mogadishu	02:36	02:58	15:10	15:31
T7	San Marino	03:33	04:05	18:27	18:59
T9	Bosnia-Herzegovina, Sarajevo	03:08	03:41	18:03	18:35
TA	Turkey, Ankara	02:22	02:52	16:56	17:25
	Istanbul	02:34	03:05	17:14	17:45
TF	Iceland, Reykjavik	03:37	04:53	22:13	23:30
TG	Guatemala, Guatemala City	11:24	11:46	00:29	00:51
TI	Costa Rica, San Jose	11:04	11:26	23:57	00:19
TI9	Cocos Is.	11:21	11:43	00:03	00:25
TJ	Cameroon, Yaounde	04:50	05:11	17:27	17:49
TK	Corsica	03:52	04:23	18:36	19:07
TL	Central African Republic, Bangui	04:21	04:42	17:00	17:21
TN	Congo, Brazzaville	04:44	05:06	17:02	17:24
TR	Gabon, Libreville	05:02	05:23	17:31	17:53
TT	Chad, N'Djamena	04:25	04:47	17:23	17:46
TU	Cote d'Ivoire, Abidjan	05:50	06:12	18:31	18:53
TY	Benin, Porto Novo	05:22	05:44	18:06	18:28
TZ	Mali, Bamako	05:56	06:18	18:56	19:18
UA	Russia, European, Leningrad (UA1)	00:58	01:53	18:15	19:11
	Archangel (UA1)	23:19	00:39	18:08	19:28
	Murmansk (UA1)	——	00:16	19:29	——
	Moscow (UA3)	00:59	01:45	17:25	18:11
	Kuybyshev (UA4)	00:24	01:05	16:24	17:06
	Rostov (UA6)	01:32	02:07	16:48	17:23
UA2	Kaliningrad	02:12	02:57	18:30	19:14
UA9,0	Russia, Asiatic, Novosibirsk (UA9)	22:03	22:47	14:20	15:05
	Perm (UA9)	23:29	00:20	16:20	17:11
	Omsk (UA9)	22:41	23:25	14:58	15:43
	Norilsk (UA0)	——	20:31	15:55	——
	Irkutsk (UA0)	20:52	21:32	12:44	13:24
	Vladivostok (UA0)	19:37	20:08	10:27	10:59
	Petropavlovsk (UA0)	17:11	17:52	09:09	09:50
	Khabarovsk (UA0)	19:05	19:41	10:28	11:04
	Krasnoyarsk (UA0)	21:17	22:03	13:45	14:31
	Yakutsk (UA0)	17:59	19:02	11:50	12:54
	Wrangel Island (UA0)	——	13:44	10:22	——
	Kyzyl (UA0Y)	21:34	22:13	13:21	14:01
UJ-UM	Uzbekistan, Bukhoro	00:16	00:46	14:49	15:19
	Tashkent	23:53	00:23	14:33	15:03
UN-UQ	Kazakhstan, Alma-Ata	23:16	23:48	14:07	14:39
UR-UZ,EM-EO	Ukraine, Kiev	01:56	02:34	17:33	18:11
V2	Antigua & Barbuda, St. Johns	09:25	09:48	22:37	23:00
V3	Belize, Belmopan	11:13	11:36	00:25	00:48
V4	St. Kitts & Nevis	09:28	09:51	22:40	23:03
V5	Namibia, Windhoek	04:58	05:21	16:32	16:56
V6	Micronesia, Ponape	18:59	19:20	07:44	08:06
V7	Marshall Islands, Kwajalein	18:20	18:42	07:10	07:32
V8	Brunei, Bandar Seri Begawan	21:55	22:16	10:35	10:57
VK	Australia, Canberra (VK1)	20:26	20:53	07:25	07:52
	Sydney (VK2)	20:16	20:42	07:19	07:45
	Melbourne (VK3)	20:46	21:14	07:37	08:04
	Brisbane (VK4)	20:00	20:25	07:22	07:46
	Adelaide (VK5)	21:08	21:34	08:08	08:34
	Perth (VK6)	22:35	23:00	09:44	10:09
	Hobart, Tasmania (VK7)	20:45	21:15	07:17	07:47
	Darwin (VK8)	21:11	21:33	09:10	09:32
VK0	Heard Is.	02:00	02:38	11:46	12:23
VK0	Macquarie Is.	20:23	21:02	05:58	06:37
VK9C	Cocos-Keeling Is.	23:27	23:49	11:27	11:49
VK9L	Lord Howe Is.	19:41	20:07	06:51	07:17
VK9M	Mellish Reef	19:37	20:00	07:24	07:47
VK9N	Norfolk Is.	19:02	19:27	06:20	06:44
VK9W	Willis Is.	20:01	20:23	07:51	08:14
VK9X	Christmas Is.	22:50	23:11	10:53	11:15
VP2E	Anguilla	09:28	09:51	22:43	23:06
VP2M	Montserrat	09:27	09:50	22:38	23:01
VP2V	British Virgin Is., Tortola	09:34	09:57	22:50	23:13

Pref	State/Province/Country/City	Tlight	Sunrise	Sunset	Tlight
VP5	Turks & Caicos Islands, Grand Turk	09:56	10:20	23:20	23:44
VP8	Falkland Islands, Stanley	10:43	11:19	20:35	21:11
VP8	So. Georgia Is.	09:24	10:03	19:02	19:41
VP8	So. Orkney Is.	10:18	11:06	19:09	19:57
VP8	So. Sandwich Islands, Saunders Is.	08:53	09:37	18:07	18:51
VP8	So. Shetland Is., King George Is.	11:14	12:05	19:52	20:43
VP9	Bermuda	09:10	09:37	23:11	23:38
VQ9	Chagos, Diego Garcia	00:59	01:21	13:11	13:32
VR2,VS6	Hong Kong	21:33	21:56	11:00	11:23
VR6	Pitcairn Is.	14:50	15:14	02:18	02:42
VU	India, Bombay	00:24	00:47	13:41	14:04
	Calcutta	23:16	23:40	12:44	13:07
	New Delhi	23:50	00:15	13:38	14:03
	Bangalore	00:13	00:36	13:14	13:36
VU	Andaman Islands, Port Blair	23:14	23:36	12:12	12:34
VU	Laccadive Is.	00:36	00:58	13:29	13:51
XE	Mexico, Mexico City (XE1)	11:51	12:14	01:09	01:32
	Chihuahua (XE2)	12:03	12:28	01:51	02:16
	Merida (XE3)	11:11	11:34	00:34	00:57
XF4	Revilla Gigedo	12:41	13:04	01:58	02:21
XT	Burkina Faso, Ouagadougou	05:31	05:53	18:30	18:52
XU	Cambodia, Phnom Penh	22:26	22:48	11:24	11:46
XW	Laos, Viangchan	22:26	22:49	11:41	12:03
XX9	Macao	21:36	21:59	11:02	11:26
XZ	Myanmar, Yangon	22:54	23:17	12:05	12:28
YA	Afghanistan, Kandahar	00:31	00:57	14:27	14:53
	Kabul	00:10	00:37	14:19	14:47
YB-YD	Indonesia, Jakarta	22:40	23:02	10:54	11:16
	Medan	23:01	23:23	11:38	12:00
	Pontianak	22:23	22:45	10:52	11:13
	Jayapura	20:21	20:42	08:43	09:04
YI	Iraq, Baghdad	01:52	02:19	15:56	16:22
YJ	Vanuatu, Port Vila	18:47	19:10	06:34	06:57
YK	Syria, Damascus	02:24	02:51	16:29	16:56
YL	Latvia, Riga	01:45	02:33	18:24	19:13
YN	Nicaragua, Managua	11:09	11:31	00:07	00:29
YO	Romania, Bucharest	02:36	03:09	17:33	18:06
YS	El Salvador, San Salvador	11:20	11:42	00:22	00:45
YU	Yugoslavia, Belgrade	02:57	03:30	17:57	18:30
YV	Venezuela, Caracas	09:55	10:17	22:49	23:11
YV0	Aves Is.	09:35	09:57	22:43	23:05
Z2	Zimbabwe, Harare	03:57	04:19	15:43	16:06
Z3	Macedonia, (ex Yugoslav), Skopje	03:02	03:33	17:46	18:17
ZA	Albania, Tirane	03:11	03:41	17:51	18:22
ZB2	Gibraltar	05:05	05:33	19:21	19:49
ZC4	British Cyprus	02:35	03:02	16:45	17:12
ZD7	St. Helena	06:22	06:45	18:13	18:35
ZD8	Ascension Is.	06:47	07:09	18:57	19:19
ZD9	Tristan da Cunha	07:14	07:42	18:07	18:35
ZF	Cayman Is.	10:39	11:02	23:58	00:21
ZK1	No. Cook Is., Manihiki	16:36	16:58	04:40	05:02
ZK1	So. Cook Is., Rarotonga	16:44	17:07	04:22	04:45
ZK2	Niue	17:18	17:41	05:01	05:24
ZK3	Tokelaus, Atafu	17:21	17:43	05:30	05:51
ZL	New Zealand, Auckland (ZL1)	18:46	19:13	05:39	06:06
	Wellington (ZL2)	18:52	19:21	05:31	06:00
	Christchurch (ZL3)	19:05	19:35	05:35	06:05
	Dunedin (ZL4)	19:17	19:49	05:38	06:09
ZL5	Antarctica, Scott Base	22:56	——	——	03:03
ZL7	Chatham Is.	18:22	18:53	04:50	05:21
ZL8	Kermadec Is.	18:06	18:31	05:23	05:48
ZL9	Auckland & Campbell Is., Auckland	19:43	20:18	05:41	06:16
	Campbell Is.	19:36	20:13	05:24	06:01
ZP	Paraguay, Asuncion	10:00	10:24	21:28	21:52
ZS	South Africa, Cape Town (ZS1)	05:07	05:33	16:10	16:36
	Port Elizabeth (ZS2)	04:38	05:04	15:41	16:07
	Bloemfontein (ZS4)	04:30	04:55	15:47	16:12
	Durban (ZS5)	04:12	04:37	15:27	15:52
	Johannesburg (ZS6)	04:18	04:42	15:43	16:07
ZS8	Marion Is.	04:10	04:42	14:26	14:58

Pref	State/Province/Country/City	Tlight	Sunrise	Sunset	Tlight
VE1	New Brunswick, Saint John	09:00	09:32	23:22	23:54
	Nova Scotia, Halifax	08:52	09:23	23:11	23:42
	Prince Edward Island, Charlottetown	08:46	09:18	23:12	23:44
VE2	Quebec, Montreal	09:30	10:01	23:53	00:24
	Quebec City	09:17	09:49	23:46	00:18
VE3	Ontario, London	10:07	10:37	00:19	00:49
	Ottawa	09:38	10:10	00:01	00:33
	Sudbury	09:57	10:29	00:24	00:57
	Toronto	09:57	10:28	00:13	00:43
VE4	Manitoba, Winnipeg	10:51	11:26	01:36	02:11
VE5	Saskatchewan, Regina	11:19	11:55	02:07	02:43
	Saskatoon	11:22	11:59	02:20	02:57
VE6	Alberta, Calgary	11:56	12:32	02:46	03:22
	Edmonton	11:44	12:23	02:50	03:29
VE7	British Columbia, Prince George	12:20	12:59	03:29	04:08
	Prince Rupert	12:48	13:28	04:00	04:39
	Vancouver	12:37	13:12	03:19	03:53
VE8	Northwest Territories, Yellowknife	11:00	11:54	03:26	04:21
	Resolute	———	08:33	04:13	———
VY1	Yukon, Whitehorse	12:35	13:25	04:41	05:31
VO1	Newfoundland, St. John's	08:01	08:34	22:33	23:06
VO2	Labrador, Goose Bay	08:13	08:51	23:17	23:56
W1	Connecticut, Hartford	09:35	10:04	23:43	00:12
	Maine, Bangor	09:12	09:44	23:32	00:03
	Portland	09:21	09:52	23:36	00:07
	Massachusetts, Boston	09:27	09:57	23:37	00:07
	New Hampshire, Concord	09:27	09:57	23:40	00:10
	Rhode Island, Providence	09:30	09:59	23:37	00:07
	Vermont, Montpelier	09:29	10:00	23:47	00:17
W2	New Jersey, Atlantic City	09:47	10:15	23:46	00:14
	New York, Albany	09:37	10:07	23:48	00:18
	Buffalo	09:57	10:27	00:09	00:39
	New York City	09:42	10:11	23:46	00:15
	Syracuse	09:46	10:16	23:59	00:29
W3	Delaware, Wilmington	09:51	10:19	23:50	00:19
	District of Columbia, Washington	09:58	10:26	23:55	00:23
	Maryland, Baltimore	09:56	10:24	23:54	00:22
	Pennsylvania, Harrisburg	09:55	10:24	23:57	00:26
	Philadelphia	09:49	10:17	23:50	00:18
	Pittsburgh	10:07	10:36	00:09	00:38
	Scranton	09:48	10:17	23:54	00:23
W4	Albama, Montgomery	10:47	11:12	00:23	00:49
	Florida, Jacksonville	10:32	10:57	00:02	00:27
	Miami	10:32	10:56	23:51	00:15
	Pensacola	10:53	11:18	00:25	00:49
	Georgia, Atlanta	10:37	11:03	00:18	00:44
	Savannah	10:26	10:52	00:02	00:28
	Kentucky, Lexington	10:30	10:58	00:24	00:51
	Louisville	10:35	11:02	00:29	00:57
	North Carolina, Charlotte	10:20	10:47	00:05	00:32
	Raleigh	10:10	10:37	23:57	00:24
	Wilmington	10:10	10:36	23:52	00:18
	South Carolina, Columbia	10:23	10:49	00:04	00:30
	Tennessee, Knoxville	10:31	10:58	00:19	00:45
	Memphis	10:58	11:24	00:42	01:09
	Nashville	10:42	11:09	00:30	00:57
	Virginia, Norfolk	09:59	10:26	23:49	00:17
	Richmond	10:02	10:30	23:55	00:22
W5	Arkansas, Little Rock	11:07	11:33	00:50	01:17
	Louisiana, New Orleans	11:06	11:31	00:36	01:00
	Shreveport	11:16	11:42	00:53	01:19
	Mississippi, Jackson	11:03	11:28	00:39	01:04
	New Mexico, Albuquerque	12:04	12:30	01:49	02:15
	Oklahoma, Oklahoma City	11:26	11:53	01:12	01:39
	Texas, Abilene	11:40	12:06	01:17	01:43
	Amarillo	11:44	12:11	01:29	01:56
	Dallas	11:28	11:54	01:06	01:32
	El Paso	12:09	12:34	01:43	02:09
	San Antonio	11:40	12:05	01:09	01:33
W6	California, Los Angeles	12:52	13:18	02:33	02:59
	San Francisco	13:02	13:29	02:55	03:23
W7	Arizona, Flagstaff	12:24	12:50	02:09	02:35
	Phoenix	12:28	12:54	02:08	02:34
	Idaho, Boise	12:25	12:55	02:40	03:10
	Pocatello	12:12	12:42	02:24	02:54
	Montana, Billings	11:49	12:20	02:13	02:45
	Butte	12:04	12:36	02:29	03:01
	Great Falls	11:55	12:28	02:27	03:01
	Nevada, Las Vegas	12:36	13:03	02:24	02:51
	Reno	12:48	13:16	02:47	03:16
	Oregon, Portland	12:46	13:18	03:09	03:41
	Utah, Salt Lake City	12:14	12:43	02:18	02:47
	Washington, Seattle	12:39	13:12	03:12	03:45
	Spokane	12:19	12:52	02:52	03:25
	Wyoming, Cheyenne	11:45	12:14	01:50	02:19
	Sheridan	11:45	12:16	02:05	02:36
W8	Michigan, Detroit	10:15	10:45	00:25	00:54
	Grand Rapids	10:24	10:54	00:37	01:07
	Sault Ste. Marie	10:10	10:43	00:38	01:10
	Traverse City	10:20	10:51	00:39	01:11
	Ohio, Cincinnati	10:28	10:56	00:25	00:54
	Cleveland	10:12	10:41	00:18	00:47
	Columbus	10:20	10:49	00:21	00:49
	West Virginia, Charleston	10:18	10:45	00:13	00:41
W9	Illinois, Chicago	10:34	11:04	00:42	01:12
	Indiana, Indianapolis	10:33	11:02	00:33	01:02
	Wisconsin, Green Bay	10:30	11:01	00:48	01:19

Pref	State/Province/Country/City	Tlight	Sunrise	Sunset	Tlight
	Milwaukee	10:33	11:03	00:45	01:16
W0	Colorado, Denver	11:49	12:17	01:48	02:17
	Grand Junction	12:04	12:32	02:02	02:30
	Iowa, Des Moines	10:59	11:28	01:06	01:35
	Kansas, Pratt	11:27	11:55	01:20	01:48
	Wichita	11:22	11:49	01:15	01:42
	Minnesota, Duluth	10:40	11:13	01:09	01:42
	Minneapolis	10:50	11:21	01:11	01:42
	Missouri, Columbia	10:59	11:27	00:57	01:25
	Kansas City	11:08	11:36	01:06	01:34
	St. Louis	10:52	11:19	00:48	01:15
	Nebraska, North Platte	11:29	11:58	01:34	02:03
	Omaha	11:09	11:38	01:15	01:44
	North Dakota, Fargo	10:59	11:32	01:28	02:01
	South Dakota, Rapid City	11:32	12:02	01:49	02:19
-		05:38	05:59	18:06	18:27
1A0	S.M.O.M	03:54	04:24	18:02	18:32
1S	Spratly Is.	22:03	22:24	10:46	11:08
3A	Monaco	04:10	04:41	18:25	18:56
3B6	Agalega	02:00	02:22	14:11	14:32
3B7	St. Brandon	01:52	02:14	13:53	14:15
3B8	Mauritius	02:05	02:27	13:58	14:21
3B9	Rodriguez Is.	01:41	02:03	13:35	13:57
3C	Equatorial Guinea, Bata	04:57	05:19	17:28	17:50
	Malabo	05:00	05:21	17:34	17:55
3C0	Annobon Is.	05:17	05:38	17:42	18:04
3D2	Fiji Is., Suva	17:59	18:22	05:57	06:19
3D2	Conway Reef	18:18	18:41	06:09	06:32
3D2	Rotuma Is.	17:58	18:19	06:05	06:26
3DA	Swaziland, Mbabane	03:55	04:19	15:38	16:01
3V	Tunisia, Tunis	04:13	04:40	18:03	18:30
3W, XV	Vietnam, Ho Chi Minh City (Saigon)	22:21	22:43	11:09	11:30
	Hanoi	22:14	22:37	11:22	11:45
3X	Guinea, Conakry	06:24	06:46	19:09	19:31
3Y	Bouvet	06:16	06:53	16:45	17:22
3Y	Peter I Is.	13:05	14:09	22:01	23:05
4J, 4K	Azerbaijan, Baku	01:28	01:56	15:30	15:59
4L	Georgia, Tbilisi	01:45	02:15	15:52	16:22
4S	Sri Lanka, Colombo	00:12	00:34	12:53	13:14
4U	ITU Geneva	04:09	04:41	18:35	19:07
4U	United Nations Hq.	09:42	10:11	23:46	00:15
4X, 4Z	Israel, Jerusalem	02:42	03:07	16:17	16:42
5A	Libya, Tripoli	04:11	04:37	17:48	18:14
	Benghazi	03:42	04:08	17:18	17:43
5B	Cyprus, Nicosia	02:43	03:10	16:28	16:55
5H	Tanzania, Dar es Salaam	03:06	03:27	15:22	15:43
5N	Nigeria, Lagos	05:19	05:40	17:58	18:19
5R	Madagascar, Antananarivo	02:44	03:06	14:39	15:02
5T	Mauritania, Nouakchott	06:25	06:47	19:26	19:49
5U	Niger, Niamey	05:17	05:39	18:10	18:32
5V	Togo, Lome	05:28	05:49	18:06	18:28
5W	Western Samoa, Apia	17:17	17:38	05:22	05:43
5X	Uganda, Kampala	03:28	03:49	15:56	16:17
5Z	Kenya, Nairobi	04:32	04:53	16:58	17:19
6W	Senegal, Dakar	06:34	06:56	19:29	19:51
6Y	Jamaica, Kingston	10:28	10:50	23:30	23:52
7O	Yemen, Aden	02:26	02:48	15:17	15:39
	Sanaa	02:27	02:49	15:23	15:45
7P	Lesotho, Maseru	04:12	04:37	15:49	16:13
7Q	Malawi, Lilongwe	03:35	03:56	15:39	16:00
	Blantyre	03:31	03:53	15:32	15:54
7X	Algeria, Algiers	04:42	05:09	18:32	18:59
8P	Barbados, Bridgetown	09:24	09:46	22:16	22:38
8Q	Maldive Is.	00:41	01:02	13:16	13:38
8R	Guyana, Georgetown	09:25	09:46	22:05	22:26
9A	Croatia, Zagreb	03:31	04:02	17:55	18:27
9G	Ghana, Accra	05:34	05:55	18:12	18:33
9H	Malta	03:58	04:25	17:45	18:12
9J	Zambia, Lusaka	03:58	04:20	15:59	16:21
9K	Kuwait	01:55	02:20	15:23	15:48
9L	Sierra Leone, Freetown	06:23	06:45	19:06	19:28
9M2	West Malaysia, Kuala Lumpur	22:49	23:10	11:22	11:44
9M6,8	East Malaysia, Sabah, Sandakan(9M6)	21:41	22:02	10:19	10:40
	Sarawak, Kuching (9M8)	22:16	22:37	10:46	11:07
9N	Nepal, Kathmandu	23:27	23:52	12:51	13:16
9Q	Zaire, Kinshasa	04:41	05:02	17:01	17:22
	Kisangani	03:57	04:18	16:26	16:47
	Lubumbashi	03:58	04:19	16:06	16:27
9U	Burundi, Bujumbura	03:44	04:05	16:06	16:27
9V	Singapore	22:42	23:03	11:12	11:33
9X	Rwanda, Kigali	03:39	04:01	16:04	16:25
9Y	Trinidad & Tobago , Port of Spain	09:34	09:55	22:21	22:42
A2	Botswana, Gaborone	04:15	04:38	16:00	16:23
A3	Tonga, Nukualofa	17:36	17:59	05:28	05:51
A4	Oman, Musqat	01:20	01:43	14:34	14:57
A5	Bhutan, Thimbu	23:12	23:36	12:34	12:59
A6	United Arab Emirates, Abu Dhabi	01:36	02:00	14:52	15:16
A7	Qatar, Ad-Dawhah	01:46	02:10	15:04	15:28
A9	Bahrein, Al-Manamah	01:48	02:12	15:08	15:32
AP	Pakistan, Karachi	00:44	01:08	14:01	14:25
AP	Islamabad	00:07	00:33	13:47	14:13
BV	Taiwan, Taipei	21:06	21:30	10:24	10:47
BY	Peoples Rep. of China, Beijing	21:02	21:31	11:03	11:32
	Harbin	20:08	20:40	10:32	11:04
	Shanghai	20:57	21:23	10:31	10:56
	Fuzhou	21:14	21:37	10:34	10:57
	Xian	21:43	22:09	11:25	11:51

Pref	State/Province/Country/City	Tlight	Sunrise	Sunset	Tlight
	Chongqing	22:00	22:24	11:29	11:54
	Chengdu	22:08	22:33	11:40	12:05
	Lhasa	23:01	23:26	12:30	12:55
	Urumqi	22:49	23:20	13:05	13:35
	Kashi	23:45	00:13	13:44	14:12
C2	Nauru	18:31	18:52	06:58	07:19
C3	Andorra	04:37	05:06	18:47	19:17
C5	The Gambia, Banjul	06:32	06:54	19:25	19:47
C6	Bahamas, Nassau	10:22	10:45	23:39	00:03
C9	Mozambique, Maputo	03:49	04:13	15:32	15:55
	Mozambique	03:08	03:30	15:10	15:32
CE	Chile, Santiago	10:49	11:15	22:17	22:43
CE0	Easter Island	13:18	13:42	00:59	01:23
CE0	Juan Fernandez	11:21	11:47	22:49	23:15
CE0	San Felix	11:20	11:44	23:02	23:26
CM,CO	Cuba, Havana	10:44	11:07	23:57	00:20
CN	Morocco, Casablanca	05:30	05:56	19:10	19:36
CP	Bolivia, La Paz	10:25	10:47	22:25	22:47
CT	Portugal, Lisbon	05:27	05:55	19:24	19:52
CT3	Madeira Islands, Funchal	06:09	06:35	19:46	20:12
CU	Azores , Ponta Delgada	06:35	07:03	20:28	20:56
CX	Uruguay, Montevideo	09:52	10:18	21:17	21:43
CY0	Sable Is.	08:40	09:10	22:55	23:26
CY9	St. Paul Is.	08:31	09:04	23:02	23:35
D2	Angola, Luanda	04:53	05:14	17:06	17:27
D4	Cape Verde, Praia	06:58	07:20	19:53	20:15
D6	Comoros, Moroni	02:53	03:15	15:01	15:23
DA-DL,Y2-Y9	Fed. Rep. of Germany, Bonn	03:52	04:28	18:41	19:17
	Berlin	03:20	03:58	18:20	18:58
DU	Philippines, Manilla	21:20	21:42	10:15	10:37
E3	Eritrea, Asmara	02:48	03:10	15:44	16:06
EA	Spain, Madrid	05:02	05:31	19:04	19:33
EA6	Balearic Is., Palma	04:39	05:07	18:38	19:06
EA8	Canary Is., Las Palmas	06:05	06:29	19:31	19:55
EA9	Ceuta & Melilla, Ceuta	05:17	05:44	19:04	19:31
	Melilla	05:09	05:35	18:54	19:21
EI	Ireland, Dublin	04:36	05:15	19:41	20:19
EK	Armenia, Yerevan	01:49	02:18	15:51	16:20
EL	Liberia, Monrovia	06:16	06:37	18:55	19:16
EP	Iran, Tehran	01:29	01:55	15:15	15:42
ER	Moldova, Kishinev	02:36	03:09	17:06	17:39
ES	Estonia, Tallinn	02:03	02:51	17:56	18:44
ET	Ethiopia, Addis Ababa	02:55	03:17	15:39	16:01
EU,EV,EW	Belarus, Minsk	02:18	02:57	17:27	18:06
EX	Kyrgyzstan, Frunze	23:43	00:13	13:55	14:25
EY	Tajikistan, Samarkand	00:21	00:50	14:21	14:50
	Dushanbe	00:15	00:43	14:12	14:40
EZ	Turkmenistan, Ashkhabad	00:58	01:26	14:52	15:20
F	France, Paris	04:17	04:51	18:56	19:30
FG	Guadeloupe	09:30	09:52	22:27	22:49
FJ, FS	St. Martin	09:33	09:55	22:35	22:57
FH	Mayotte	02:48	03:09	14:54	15:15
FK	New Caledonia, Noumia	18:50	19:13	06:40	07:03
FM	Martinique	09:28	09:50	22:23	22:45
FO	Clipperton Is.	12:45	13:07	01:32	01:53
FO	Fr. Polynesia, Tahiti	16:31	16:53	04:29	04:51
	Rurutu, (Austral Is.)	19:51	20:14	07:41	08:04
	Hiva Oa, (Marquesas Is.)	15:02	15:24	03:14	03:35
FP	St. Pierre & Miquelon, St. Pierre	08:16	08:49	22:45	23:17
FR/G	Glorioso	02:39	03:00	14:47	15:08
FR/J,E	Juan de Nova	03:01	03:23	15:00	15:22
	Europa	03:15	03:38	15:05	15:27
FR	Reunion	02:13	02:36	14:05	14:28
FR/T	Tromelin	02:14	02:36	14:15	14:37
FT8W	Crozet	02:51	03:21	13:48	14:19
FT8X	Kerguelen	01:46	02:19	12:33	13:06
FT8Y	Antarctica, Dumont D'Urville	21:36	22:32	06:53	07:50
FT8Z	Amsterdam & St. Paul Is., Amsterdam	00:59	01:26	12:18	12:45
FW	Wallis & Futuna Is., Wallis	17:34	17:56	05:40	06:01
FY	Fr. Guiana, Cayenne	09:03	09:24	21:40	22:01
G	England, London	04:18	04:55	19:12	19:48
GD	Isle of Man	04:25	05:05	19:37	20:16
GI	Northern Ireland, Belfast	04:30	05:10	19:43	20:23
GJ	Jersey	04:34	05:08	19:15	19:49
GM	Scotland, Glasgow	04:18	05:00	19:40	20:22
	Aberdeen	04:03	04:46	19:36	20:20
GU	Guernsey	04:35	05:10	19:17	19:52
GW	Wales, Cardiff	04:30	05:07	19:24	20:01
H4	Solomon Islands, Honiara	19:06	19:27	07:18	07:39
HA	Hungary, Budapest	03:14	03:47	17:46	18:19
HB	Switzerland, Bern	04:01	04:34	18:31	19:04
HB0	Liechtenstein	03:52	04:25	18:23	18:56
HC	Ecuador, Quito	10:50	11:11	23:18	23:39
HC8	Galapagos Is.	11:41	12:02	00:08	00:29
HH	Haiti, Port-Au-Prince	10:09	10:32	23:12	23:34
HI	Dominican Republic, Santo Domingo	10:00	10:23	23:03	23:25
HK	Colombia, Bogota	10:30	10:52	23:06	23:28
HK0	Malpelo Is.	10:59	11:20	23:34	23:55
HK0	San Adreas	10:53	11:15	23:44	00:06
HL	Korea, Seoul	20:25	20:52	10:17	10:45
HP	Panama, Panama	10:48	11:09	23:32	23:53
HR	Honduras, Tegucigalpa	11:14	11:36	00:07	00:29
HS	Thailand, Bangkok	22:43	23:05	11:36	11:58
HV	Vatican City	03:54	04:23	18:02	18:31
HZ, 7Z	Saudi Arabia, Dharan	01:51	02:14	15:11	15:35
	Mecca	02:37	03:00	15:47	16:10
	Riyadh	02:06	02:29	15:22	15:46

Pref	State/Province/Country/City	Tlight	Sunrise	Sunset	Tlight
I	Italy, Rome	03:54	04:23	18:02	18:31
	Trieste	03:40	04:11	18:04	18:35
	Sicily	03:57	04:24	17:49	18:17
IS	Sardinia, Cagliari	04:13	04:41	18:11	18:39
J2	Djibouti, Djibouti	02:35	02:56	15:24	15:45
J3	Grenada	09:34	09:56	22:24	22:46
J5	Guinea-Bissau, Bissau	06:29	06:51	19:19	19:41
J6	St. Lucia	09:29	09:51	22:22	22:44
J7	Dominica	09:29	09:51	22:25	22:47
J8	St. Vincent	09:31	09:53	22:23	22:45
JA-JS	Japan, Tokyo	19:37	20:04	09:23	09:50
	Nagasaki	20:21	20:47	09:59	10:25
	Sapporo	19:16	19:46	09:28	09:59
JD1	Minami Torishima	18:57	19:21	08:13	08:36
JD1	Ogasawara, Kazan Is.	19:45	20:09	09:08	09:33
JT	Mongolia, Ulan Bator	21:21	21:55	11:55	12:29
JW	Svalbard, Spitsbergen	—	—	—	—
JX	Jan Mayen	01:46	03:47	21:25	23:26
JY	Jordan, Amman	02:39	03:04	16:14	16:40
KC4	Antarctica, Byrd Station	16:32	—	—	23:34
	McMurdo Sound	20:52	23:22	02:29	05:00
	Palmer Station	11:06	11:59	20:39	21:31
KC6	Belau, Yap	20:17	20:38	09:02	09:23
	Koror	20:34	20:55	09:14	09:36
KG4	Guantanamo Bay	10:19	10:42	23:25	23:48
KH0	Mariana Is., Saipan	19:40	20:03	08:36	08:58
KH1	Baker, Howland Is.	17:22	17:43	05:51	06:12
KH2	Guam, Agana	19:46	20:08	08:39	09:01
KH3	Johnston Is.	16:36	16:58	05:35	05:58
KH4	Midway Is.	16:57	17:22	06:23	06:47
KH5	Palmyra Is.	16:21	16:43	05:00	05:21
KH5K	Kingman Reef	16:23	16:44	05:04	05:25
KH6	Hawaii, Hilo	15:39	16:02	04:44	05:07
	Honolulu	15:48	16:11	04:57	05:20
KH7	Kure Is.	17:01	17:26	06:27	06:51
KH8	American Samoa, Pago Pago	17:13	17:35	05:17	05:39
KH9	Wake Is.	18:13	18:35	07:17	07:40
KL7	Alaska, Adak	16:03	16:40	06:58	07:35
	Anchorage	13:31	14:23	05:43	06:34
	Fairbanks	12:54	13:56	05:52	06:55
	Juneau	12:46	13:32	04:29	05:14
	Nome	14:07	15:08	07:01	08:02
KP1	Navassa Is.	10:20	10:43	23:23	23:45
KP2	Virgin Islands, Charlotte Amalie	09:40	10:02	22:42	23:05
KP4	Puerto Rico, San Juan	09:45	10:07	22:48	23:10
KP5	Desecheo Is.	09:50	10:13	22:53	23:15
LA-LJ	Norway, Oslo	02:56	03:45	18:55	19:44
LU	Argentina, Buenos Aires	10:01	10:26	21:26	21:52
LX	Luxembourg	03:59	04:34	18:42	19:17
LY	Lithuania, Vilna	02:24	03:04	17:37	18:17
LZ	Bulgaria, Sofia	03:09	03:39	17:20	17:50
OA	Peru, Lima	10:57	11:18	23:04	23:26
OD	Lebanon, Beirut	02:37	03:03	16:18	16:44
OE	Austria, Vienna	03:23	03:56	17:59	18:32
OH	Finland, Helsinki	01:58	02:47	17:59	18:48
OH0	Aland Is.	02:18	03:07	18:19	19:08
OJ0	Market Reef	02:21	03:10	18:23	19:12
OK,OL	Czech Rep., Prague	03:25	04:00	18:10	18:46
OM	Slovak, Rep., Bratislava	03:21	03:54	17:55	18:29
ON	Belgium, Brussels	04:02	04:38	18:52	19:28
OX	Greenland, Godthaab	06:35	07:35	23:24	00:24
	Thule	—	05:36	03:40	—
OY	Faroe Islands, Torshavn	03:53	04:46	20:13	21:07
OZ	Denmark, Copenhagen	03:11	03:52	18:32	19:14
P2	Papua New Guinea, Madang	20:00	20:21	08:19	08:40
	Port Moresby	19:58	20:19	08:10	08:31
P4	Aruba, Oranjestad	10:03	10:25	22:53	23:14
PA-PI	Netherlands, Amsterdam	03:55	04:32	18:54	19:31
PJ2,4,9	Netherlands Antilles, Willemstad	10:07	10:29	22:58	23:19
PJ5-8	St. Maarten and Saba, St. Maarten	09:34	09:56	22:35	22:57
PY	Brazil, Brasilia	09:03	09:25	21:04	21:26
	Rio De Janeiro	08:50	09:13	20:38	21:01
	Natal	08:04	08:25	20:22	20:43
	Manaus	09:42	10:03	22:04	22:25
	Porto Alegre	09:28	09:52	21:03	21:27
PY0	Fernando De Noronha	07:51	08:12	20:12	20:34
PY0	St. Peter & St. Paul Rocks	07:35	07:56	20:05	20:26
PY0	Trindade & Martin Vaz Is., Trindade	07:52	08:15	19:45	20:08
PZ	Suriname, Paramaribo	09:14	09:35	21:52	22:13
R1FJ_	Franz Josef Land	—	—	—	—
R1MV_	Malyj Vysotskij Is.	01:41	02:31	17:46	18:36
S0	Western Sahara, Smara	05:56	06:20	19:17	19:41
S2	Bangladesh, Dacca	23:12	23:36	12:27	12:50
S5	Slovenia, Ljubljana	03:36	04:08	18:01	18:33
S7	Seychelles, Victoria	02:00	02:21	14:20	14:41
S9	Sao Tome	05:11	05:32	17:40	18:01
SM	Sweden, Stockholm	02:31	03:18	18:23	19:10
SP	Poland, Krakow	03:03	03:38	17:48	18:23
	Warsaw	02:51	03:28	17:49	18:26
ST	Sudan, Khartoum	03:13	03:35	16:10	16:32
ST0	Southern Sudan, Juba	03:27	03:49	16:04	16:25
SU	Egypt, Cairo	03:00	03:25	16:30	16:54
SV	Greece, Athens	03:17	03:45	17:11	17:39
SV/A	Mount Athos	03:10	03:39	17:12	17:41
SV5	Dodecanese, Rhodes	03:02	03:29	16:51	17:18
SV9	Crete	03:16	03:42	17:01	17:28
T2	Tuvalu, Funafuti	17:51	18:12	06:04	06:26

Pref	State/Province/Country/City	Tlight	Sunrise	Sunset	Tlight
T30	West Kiribati, Bonriki	18:04	18:25	06:35	06:56
T31	Central Kiribati, Kanton	17:07	17:28	05:30	05:52
T32	East Kiribati, Christmas Is.	16:06	16:27	04:37	04:58
T33	Banaba Is.	18:21	18:42	06:48	07:09
T5	Somalia, Mogadishu	02:35	02:56	15:06	15:27
T7	San Marino	03:50	04:21	18:06	18:37
T9	Bosnia-Herzegovina, Sarajevo	03:26	03:56	17:42	18:12
TA	Turkey, Ankara	02:37	03:05	16:37	17:06
	Istanbul	02:50	03:19	16:55	17:24
TF	Iceland, Reykjavik	04:37	05:37	21:25	22:25
TG	Guatemala, Guatemala City	11:26	11:48	00:21	00:43
TI	Costa Rica, San Jose	11:05	11:27	23:51	00:12
TI9	Cocos Is.	11:21	11:42	23:59	00:20
TJ	Cameroon, Yaounde	04:49	05:10	17:23	17:45
TK	Corsica	04:08	04:37	18:16	18:46
TL	Central African Republic, Bangui	04:20	04:41	16:56	17:17
TN	Congo, Brazzaville	04:41	05:02	17:01	17:22
TR	Gabon, Libreville	05:00	05:21	17:28	17:50
TT	Chad, N'Djamena	04:27	04:49	17:17	17:38
TU	Cote d'Ivoire, Abidjan	05:49	06:11	18:27	18:48
TY	Benin, Porto Novo	05:22	05:43	18:01	18:23
TZ	Mali, Bamako	05:58	06:20	18:49	19:11
UA	Russia, European, Leningrad (UA1)	01:38	02:27	17:36	18:25
	Archangel (UA1)	00:22	01:24	17:18	18:19
	Murmansk (UA1)	23:54	01:22	18:18	19:47
	Moscow (UA3)	01:30	02:12	16:53	17:34
	Kuybyshev (UA4)	00:51	01:29	15:55	16:33
	Rostov (UA6)	01:52	02:25	16:24	16:57
UA2	Kaliningrad	02:42	03:23	17:59	18:39
UA9,0	Russia, Asiatic, Novosibirsk (UA9)	22:33	23:13	13:49	14:30
	Perm (UA9)	00:05	00:50	15:45	16:30
	Omsk (UA9)	23:11	23:51	14:27	15:08
	Norilsk (UA0)	20:07	21:39	14:41	16:13
	Irkutsk (UA0)	21:18	21:55	12:16	12:53
	Vladivostok (UA0)	19:53	20:24	10:07	10:37
	Petropavlovsk (UA0)	17:37	18:16	08:40	09:18
	Khabarovsk (UA0)	19:27	20:01	10:04	10:38
	Krasnoyarsk (UA0)	21:49	22:31	13:12	13:54
	Yakutsk (UA0)	18:47	19:40	11:07	12:01
	Wrangel Island (UA0)	13:11	15:12	08:49	10:51
	Kyzyl (UA0Y)	21:59	22:36	12:54	13:30
UJ-UM	Uzbekistan, Bukhoro	00:31	00:59	14:31	14:59
	Tashkent	00:08	00:37	14:14	14:43
UN-UQ	Kazakhstan, Alma-Ata	23:33	00:03	13:47	14:17
UR-UZ,EM-EO	Ukraine, Kiev	02:19	02:55	17:07	17:42
V2	Antigua & Barbuda, St. Johns	09:29	09:51	22:29	22:51
V3	Belize, Belmopan	11:17	11:39	00:17	00:39
V4	St. Kitts & Nevis	09:32	09:54	22:32	22:54
V5	Namibia, Windhoek	04:48	05:11	16:37	17:00
V6	Micronesia, Ponape	18:59	19:20	07:39	08:00
V7	Marshall Islands, Kwajalein	18:21	18:42	07:05	07:26
V8	Brunei, Bandar Seri Begawan	21:54	22:15	10:31	10:52
VK	Australia, Canberra (VK1)	20:11	20:37	07:35	08:01
	Sydney (VK2)	20:02	20:27	07:29	07:54
	Melbourne (VK3)	20:30	20:57	07:49	08:15
	Brisbane (VK4)	19:49	20:13	07:29	07:53
	Adelaide (VK5)	20:53	21:19	08:18	08:44
	Perth (VK6)	22:21	22:46	09:53	10:18
	Hobart, Tasmania (VK7)	20:26	20:55	07:32	08:01
	Darwin (VK8)	21:05	21:27	09:12	09:33
VK0	Heard Is.	01:34	02:10	12:09	12:44
VK0	Macquarie Is.	19:55	20:32	06:23	07:00
VK9C	Cocos-Keeling Is.	23:21	23:43	11:28	11:50
VK9L	Lord Howe Is.	19:28	19:53	07:00	07:25
VK9M	Mellish Reef	19:29	19:52	07:27	07:50
VK9N	Norfolk Is.	18:50	19:14	06:27	06:51
VK9W	Willis Is.	19:54	20:16	07:54	08:16
VK9X	Christmas Is.	22:44	23:06	10:54	11:16
VP2E	Anguilla	09:32	09:55	22:35	22:57
VP2M	Montserrat	09:31	09:53	22:30	22:52
VP2V	British Virgin Is., Tortola	09:39	10:01	22:41	23:04
VP5	Turks & Caicos Islands, Grand Turk	10:02	10:24	23:11	23:33
VP8	Falkland Islands, Stanley	10:17	10:52	20:57	21:31
VP8	So. Georgia Is.	08:57	09:34	19:26	20:03
VP8	So. Orkney Is.	09:43	10:27	19:42	20:27
VP8	So. Sandwich Islands, Saunders Is.	08:22	09:03	18:36	19:17
VP8	So. Shetland Is., King George Is.	10:37	11:24	20:28	21:15
VP9	Bermuda	09:21	09:46	22:57	23:22
VQ9	Chagos, Diego Garcia	00:55	01:16	13:10	13:31
VR2,VS6	Hong Kong	21:38	22:02	10:49	11:13
VR6	Pitcairn Is.	14:39	15:02	02:24	02:47
VU	India, Bombay	00:28	00:51	13:32	13:55
	Calcutta	23:22	23:45	12:33	12:57
	New Delhi	23:59	00:23	13:25	13:49
	Bangalore	00:16	00:37	13:07	13:29
VU	Andaman Islands, Port Blair	23:16	23:38	12:05	12:27
VU	Laccadive Is.	00:37	00:59	13:23	13:44
XE	Mexico, Mexico City (XE1)	11:55	12:18	01:00	01:23
	Chihuahua (XE2)	12:11	12:36	01:38	02:02
	Merida (XE3)	11:16	11:39	00:24	00:47
XF4	Revilla Gigedo	12:46	13:08	01:49	02:12
XT	Burkina Faso, Ouagadougou	05:33	05:55	18:23	18:45
XU	Cambodia, Phnom Penh	22:28	22:50	11:17	11:39
XW	Laos, Viangchan	22:30	22:53	11:32	11:54
XX9	Macao	21:41	22:04	10:52	11:15
XZ	Myanmar, Yangon	22:58	23:20	11:57	12:19
YA	Afghanistan, Kandahar	00:41	01:06	14:13	14:38
	Kabul	00:22	00:48	14:04	14:30
YB-YD	Indonesia, Jakarta	22:36	22:57	10:54	11:15
	Medan	23:00	23:21	11:34	11:56
	Pontianak	22:21	22:42	10:49	11:10
	Jayapura	20:18	20:39	08:41	09:02
YI	Iraq, Baghdad	02:03	02:28	15:41	16:07
YJ	Vanuatu, Port Vila	18:40	19:02	06:37	06:59
YK	Syria, Damascus	02:35	03:01	16:14	16:40
YL	Latvia, Riga	02:19	03:02	17:50	18:34
YN	Nicaragua, Managua	11:11	11:33	00:01	00:22
YO	Romania, Bucharest	02:54	03:25	17:12	17:43
YS	El Salvador, San Salvador	11:22	11:44	00:15	00:37
YU	Yugoslavia, Belgrade	03:15	03:46	17:35	18:06
YV	Venezuela, Caracas	09:56	10:18	22:43	23:05
YV0	Aves Is.	09:38	10:00	22:35	22:57
Z2	Zimbabwe, Harare	03:49	04:11	15:46	16:09
Z3	Macedonia, (ex Yugoslav), Skopje	03:18	03:48	17:26	17:56
ZA	Albania, Tirane	03:26	03:55	17:32	18:01
ZB2	Gibraltar	05:17	05:44	19:05	19:32
ZC4	British Cyprus	02:46	03:12	16:29	16:55
ZD7	St. Helena	06:15	06:37	18:16	18:38
ZD8	Ascension Is.	06:42	07:04	18:57	19:18
ZD9	Tristan da Cunha	06:58	07:25	18:19	18:45
ZF	Cayman Is.	10:44	11:06	23:49	00:11
ZK1	No. Cook Is., Manihiki	16:31	16:52	04:41	05:03
ZK1	So. Cook Is., Rarotonga	16:35	16:57	04:26	04:49
ZK2	Niue	17:09	17:32	05:05	05:27
ZK3	Tokelaus, Atafu	17:16	17:37	05:30	05:51
ZL	New Zealand, Auckland (ZL1)	18:30	18:56	05:51	06:17
	Wellington (ZL2)	18:34	19:03	05:44	06:13
	Christchurch (ZL3)	18:45	19:15	05:50	06:19
	Dunedin (ZL4)	18:56	19:27	05:54	06:25
ZL5	Antarctica, Scott Base	20:56	23:33	02:21	04:59
ZL7	Chatham Is.	18:02	18:32	05:05	05:35
ZL8	Kermadec Is.	17:54	18:18	05:30	05:55
ZL9	Auckland & Campbell Is., Auckland	19:18	19:52	06:01	06:35
	Campbell Is.	19:10	19:46	05:47	06:22
ZP	Paraguay, Asuncion	09:50	10:13	21:34	21:57
ZS	South Africa, Cape Town (ZS1)	04:53	05:18	16:20	16:45
	Port Elizabeth (ZS2)	04:24	04:49	15:51	16:16
	Bloemfontein (ZS4)	04:18	04:42	15:55	16:19
	Durban (ZS5)	03:59	04:24	15:35	15:59
	Johannesburg (ZS6)	04:07	04:31	15:50	16:13
ZS8	Marion Is.	03:48	04:19	14:44	15:15

Pref	State/Province/Country/City	Tlight	Sunrise	Sunset	Tlight
VE1	New Brunswick, Saint John	09:22	09:52	22:52	23:22
	Nova Scotia, Halifax	09:13	09:43	22:41	23:11
	Prince Edward Island, Charlottetown	09:09	09:39	22:41	23:11
VE2	Quebec, Montreal	09:52	10:22	23:22	23:52
	Quebec City	09:40	10:11	23:14	23:44
VE3	Ontario, London	10:26	10:55	23:50	00:19
	Ottawa	10:01	10:31	23:30	00:00
	Sudbury	10:20	10:51	23:52	00:23
	Toronto	10:18	10:47	23:43	00:13
VE4	Manitoba, Winnipeg	11:18	11:52	01:00	01:33
VE5	Saskatchewan, Regina	11:47	12:21	01:31	02:05
	Saskatoon	11:53	12:27	01:42	02:16
VE6	Alberta, Calgary	12:24	12:58	02:10	02:44
	Edmonton	12:17	12:53	02:11	02:47
VE7	British Columbia, Prince George	12:53	13:29	02:48	03:25
	Prince Rupert	13:22	13:59	03:19	03:56
	Vancouver	13:04	13:36	02:44	03:16
VE8	Northwest Territories, Yellowknife	11:53	12:40	02:30	03:18
	Resolute	08:49	10:32	02:04	03:46
VY1	Yukon, Whitehorse	13:22	14:07	03:49	04:34
VO1	Newfoundland, St. John's	08:25	08:56	22:00	22:32
VO2	Labrador, Goose Bay	08:45	09:21	22:38	23:14
W1	Connecticut, Hartford	09:54	10:22	23:15	23:43
	Maine, Bangor	09:34	10:04	23:02	23:32
	Portland	09:41	10:11	23:07	23:36
	Massachusetts, Boston	09:46	10:15	23:09	23:38
	New Hampshire, Concord	09:47	10:16	23:11	23:40
	Rhode Island, Providence	09:48	10:17	23:10	23:38
	Vermont, Montpelier	09:50	10:19	23:17	23:46
W2	New Jersey, Atlantic City	10:03	10:31	23:20	23:47
	New York, Albany	09:57	10:26	23:20	23:49
	Buffalo	10:17	10:46	23:41	00:09
	New York City	10:00	10:28	23:19	23:47
	Syracuse	10:06	10:35	23:30	23:59
W3	Delaware, Wilmington	10:07	10:35	23:24	23:52
	District of Columbia, Washington	10:14	10:41	23:30	23:57
	Maryland, Baltimore	10:12	10:40	23:28	23:56
	Pennsylvania, Harrisburg	10:12	10:40	23:30	23:58
	Philadelphia	10:06	10:33	23:23	23:51
	Pittsburgh	10:25	10:52	23:43	00:11
	Scranton	10:06	10:34	23:27	23:55
W4	Albama, Montgomery	10:58	11:23	00:02	00:27
	Florida, Jacksonville	10:42	11:06	23:43	00:07
	Miami	10:40	11:03	23:34	23:57
	Pensacola	11:04	11:28	00:05	00:29
	Georgia, Atlanta	10:49	11:15	23:56	00:21
	Savannah	10:38	11:03	23:41	00:06
	Kentucky, Lexington	10:45	11:12	23:59	00:26
	Louisville	10:50	11:17	00:04	00:31
	North Carolina, Charlotte	10:34	10:59	23:42	00:08
	Raleigh	10:24	10:50	23:34	00:00
	Wilmington	10:23	10:48	23:30	23:55
	South Carolina, Columbia	10:36	11:01	23:42	00:08
	Tennessee, Knoxville	10:45	11:11	23:55	00:21
	Memphis	11:11	11:37	00:19	00:45
	Nashville	10:57	11:23	00:07	00:33
	Virginia, Norfolk	10:14	10:40	23:25	23:52
	Richmond	10:18	10:44	23:30	23:57
W5	Arkansas, Little Rock	11:20	11:46	00:28	00:53
	Louisiana, New Orleans	11:16	11:40	00:16	00:40
	Shreveport	11:28	11:53	00:32	00:57
	Mississippi, Jackson	11:14	11:39	00:18	00:43
	New Mexico, Albuquerque	12:17	12:43	01:26	01:51
	Oklahoma, Oklahoma City	11:40	12:06	00:49	01:15
	Texas, Abilene	11:52	12:17	00:56	01:21
	Amarillo	11:58	12:23	01:06	01:32
	Dallas	11:40	12:05	00:45	01:10
	El Paso	12:20	12:44	01:23	01:47
	San Antonio	11:50	12:14	00:49	01:13
W6	California, Los Angeles	13:04	13:30	02:11	02:36
	San Francisco	13:17	13:44	02:31	02:57
W7	Arizona, Flagstaff	12:37	13:03	01:46	02:12
	Phoenix	12:41	13:06	01:46	02:11
	Idaho, Boise	12:45	13:14	02:10	02:40
	Pocatello	12:31	13:00	01:55	02:24
	Montana, Billings	12:11	12:41	01:42	02:12
	Butte	12:27	12:57	01:58	02:28
	Great Falls	12:20	12:51	01:55	02:26
	Nevada, Las Vegas	12:50	13:16	02:00	02:26
	Reno	13:05	13:32	02:21	02:49
	Oregon, Portland	13:08	13:39	02:38	03:08
	Utah, Salt Lake City	12:32	13:00	01:51	02:19
	Washington, Seattle	13:03	13:35	02:39	03:10
	Spokane	12:44	13:15	02:19	02:51
	Wyoming, Cheyenne	12:03	12:31	01:23	01:51
	Sheridan	12:07	12:36	01:35	02:05
W8	Michigan, Detroit	10:34	11:03	23:57	00:25
	Grand Rapids	10:44	11:13	00:08	00:37
	Sault Ste. Marie	10:34	11:04	00:06	00:37
	Traverse City	10:41	11:11	00:09	00:39
	Ohio, Cincinnati	10:44	11:11	00:00	00:27
	Cleveland	10:30	10:58	23:51	00:19
	Columbus	10:37	11:05	23:55	00:22
	West Virginia, Charleston	10:33	11:00	23:48	00:15
W9	Illinois, Chicago	10:53	11:21	00:15	00:43
	Indiana, Indianapolis	10:50	11:18	00:07	00:35
	Wisconsin, Green Bay	10:51	11:21	00:18	00:48

Pref	State/Province/Country/City	Tlight	Sunrise	Sunset	Tlight
	Milwaukee	10:53	11:22	00:17	00:46
W0	Colorado, Denver	12:05	12:33	01:22	01:50
	Grand Junction	12:21	12:48	01:36	02:03
	Iowa, Des Moines	11:17	11:46	00:38	01:07
	Kansas, Pratt	11:43	12:09	00:56	01:22
	Wichita	11:37	12:04	00:50	01:17
	Minnesota, Duluth	11:04	11:35	00:37	01:08
	Minneapolis	11:12	11:41	00:40	01:10
	Missouri, Columbia	11:16	11:43	00:31	00:58
	Kansas City	11:25	11:52	00:40	01:07
	St. Louis	11:08	11:34	00:22	00:49
	Nebraska, North Platte	11:47	12:15	01:07	01:35
	Omaha	11:27	11:55	00:47	01:15
	North Dakota, Fargo	11:23	11:54	00:56	01:27
	South Dakota, Rapid City	11:52	12:22	01:19	01:48
-		05:33	05:54	18:01	18:22
1A0	S.M.O.M	04:13	04:41	17:35	18:03
1S	Spratly Is.	22:02	22:23	10:37	10:58
3A	Monaco	04:31	05:00	17:56	18:25
3B6	Agalega	01:51	02:12	14:10	14:31
3B7	St. Brandon	01:40	02:02	13:55	14:16
3B8	Mauritius	01:51	02:13	14:02	14:24
3B9	Rodriguez Is.	01:27	01:49	13:39	14:01
3C	Equatorial Guinea, Bata	04:53	05:14	17:23	17:43
	Malabo	04:57	05:17	17:27	17:48
3C0	Annobon Is.	05:12	05:32	17:38	17:59
3D2	Fiji Is., Suva	17:47	18:08	06:00	06:21
3D2	Conway Reef	18:04	18:26	06:14	06:36
3D2	Rotuma Is.	17:47	18:09	06:05	06:26
3DA	Swaziland, Mbabane	03:38	04:02	15:45	16:08
3V	Tunisia, Tunis	04:28	04:54	17:39	18:06
3W, XV	Vietnam, Ho Chi Minh City (Saigon)	22:22	22:43	10:59	11:20
	Hanoi	22:19	22:41	11:07	11:30
3X	Guinea, Conakry	06:24	06:45	19:00	19:21
3Y	Bouvet	05:37	06:13	17:15	17:51
3Y	Peter I Is.	11:57	12:54	23:06	00:03
4J, 4K	Azerbaijan, Baku	01:45	02:13	15:03	15:31
4L	Georgia, Tbilisi	02:04	02:32	15:25	15:53
4S	Sri Lanka, Colombo	00:11	00:32	12:44	13:05
4U	ITU Geneva	04:32	05:02	18:03	18:34
4U	United Nations Hq.	10:00	10:28	23:19	23:47
4X, 4Z	Israel, Jerusalem	02:53	03:18	15:56	16:21
5A	Libya, Tripoli	04:23	04:48	17:27	17:52
	Benghazi	03:54	04:18	16:57	17:22
5B	Cyprus, Nicosia	02:57	03:23	16:05	16:31
5H	Tanzania, Dar es Salaam	02:58	03:19	15:20	15:41
5N	Nigeria, Lagos	05:17	05:38	17:50	18:11
5R	Madagascar, Antananarivo	02:31	02:52	14:43	15:05
5T	Mauritania, Nouakchott	06:28	06:50	19:13	19:35
5U	Niger, Niamey	05:19	05:40	17:59	18:20
5V	Togo, Lome	05:26	05:47	17:59	18:20
5W	Western Samoa, Apia	17:06	17:27	05:22	05:44
5X	Uganda, Kampala	03:23	03:44	15:51	16:12
5Z	Kenya, Nairobi	04:27	04:48	16:53	17:14
6W	Senegal, Dakar	06:36	06:58	19:18	19:39
6Y	Jamaica, Kingston	10:31	10:53	23:16	23:38
7O	Yemen, Aden	02:27	02:49	15:07	15:28
	Sanaa	02:29	02:51	15:11	15:33
7P	Lesotho, Maseru	03:54	04:18	15:58	16:21
7Q	Malawi, Lilongwe	03:24	03:45	15:40	16:01
	Blantyre	03:19	03:41	15:34	15:56
7X	Algeria, Algiers	04:57	05:23	18:08	18:34
8P	Barbados, Bridgetown	09:25	09:47	22:05	22:27
8Q	Maldive Is.	00:38	00:59	13:09	13:30
8R	Guyana, Georgetown	09:23	09:44	21:57	22:18
9A	Croatia, Zagreb	03:53	04:23	17:24	17:54
9G	Ghana, Accra	05:32	05:53	18:04	18:25
9H	Malta	04:12	04:38	17:22	17:48
9J	Zambia, Lusaka	03:46	04:08	16:01	16:23
9K	Kuwait	02:05	02:29	15:04	15:28
9L	Sierra Leone, Freetown	06:22	06:43	18:57	19:18
9M2	West Malaysia, Kuala Lumpur	22:46	23:06	11:16	11:37
9M6,8	East Malaysia, Sabah, Sandakan(9M6)	21:38	21:59	10:11	10:32
	Sarawak, Kuching (9M8)	22:12	22:32	10:40	11:01
9N	Nepal, Kathmandu	23:36	00:00	12:33	12:57
9Q	Zaire, Kinshasa	04:34	04:55	16:58	17:19
	Kisangani	03:52	04:13	16:20	16:41
	Lubumbashi	03:48	04:09	16:06	16:27
9U	Burundi, Bujumbura	03:38	03:58	16:02	16:23
9V	Singapore	22:38	22:58	11:06	11:27
9X	Rwanda, Kigali	03:34	03:55	16:00	16:20
9Y	Trinidad & Tobago, Port of Spain	09:34	09:55	22:11	22:32
A2	Botswana, Gaborone	03:59	04:22	16:06	16:29
A3	Tonga, Nukualofa	17:22	17:44	05:33	05:55
A4	Oman, Musqat	01:26	01:49	14:18	14:40
A5	Bhutan, Thimbu	23:20	23:44	12:16	12:40
A6	United Arab Emirates, Abu Dhabi	01:43	02:06	14:36	14:59
A7	Qatar, Ad-Dawhah	01:53	02:16	14:47	15:10
A9	Bahrein, Al-Manamah	01:56	02:19	14:51	15:14
AP	Pakistan, Karachi	00:51	01:14	13:44	14:07
AP	Islamabad	00:19	00:44	13:25	13:50
BV	Taiwan, Taipei	21:13	21:36	10:07	10:30
BY	Peoples Rep. of China, Beijing	21:20	21:47	10:37	11:04
	Harbin	20:30	21:01	10:01	10:31
	Shanghai	21:08	21:33	10:10	10:35
	Fuzhou	21:21	21:45	10:16	10:39
	Xian	21:56	22:21	11:03	11:28

Pref	State/Province/Country/City	Tlight	Sunrise	Sunset	Tlight
	Chongqing	22:10	22:34	11:10	11:34
	Chengdu	22:18	22:43	11:20	11:44
	Lhasa	23:11	23:35	12:11	12:35
	Urumqi	23:10	23:39	12:35	13:05
	Kashi	00:02	00:29	13:18	13:45
C2	Nauru	18:26	18:47	06:53	07:14
C3	Andorra	04:56	05:25	18:19	18:47
C5	The Gambia, Banjul	06:34	06:55	19:14	19:35
C6	Bahamas, Nassau	10:29	10:52	23:22	23:45
C9	Mozambique, Maputo	03:32	03:55	15:39	16:02
	Mozambique	02:56	03:18	15:12	15:33
CE	Chile, Santiago	10:28	10:53	22:29	22:53
CE0	Easter Island	13:01	13:24	01:06	01:30
CE0	Juan Fernandez	11:00	11:25	23:01	23:25
CE0	San Felix	11:03	11:26	23:10	23:33
CM,CO	Cuba, Havana	10:50	11:13	23:41	00:04
CN	Morocco, Casablanca	05:42	06:07	18:48	19:13
CP	Bolivia, La Paz	10:13	10:35	22:27	22:49
CT	Portugal, Lisbon	05:43	06:10	18:58	19:25
CT3	Madeira Islands, Funchal	06:21	06:46	19:25	19:50
CU	Azores , Ponta Delgada	06:51	07:17	20:04	20:30
CX	Uruguay, Montevideo	09:30	09:55	21:29	21:55
CY0	Sable Is.	09:00	09:29	22:26	22:55
CY9	St. Paul Is.	08:55	09:26	22:30	23:01
D2	Angola, Luanda	04:44	05:05	17:04	17:25
D4	Cape Verde, Praia	07:00	07:22	19:42	20:03
D6	Comoros, Moroni	02:43	03:04	15:01	15:22
DA-DL,Y2-Y9	Fed. Rep. of Germany, Bonn	04:21	04:54	18:05	18:39
	Berlin	03:51	04:26	17:42	18:17
DU	Philippines, Manilla	21:22	21:44	10:03	10:25
E3	Eritrea, Asmara	02:50	03:12	15:32	15:54
EA	Spain, Madrid	05:19	05:47	18:38	19:05
EA6	Balearic Is., Palma	04:55	05:23	18:12	18:39
EA8	Canary Is., Las Palmas	06:14	06:38	19:12	19:36
EA9	Ceuta & Melilla, Ceuta	05:31	05:57	18:41	19:07
	Melilla	05:22	05:48	18:31	18:57
EI	Ireland, Dublin	05:08	05:44	19:01	19:37
EK	Armenia, Yerevan	02:07	02:34	15:25	15:52
EL	Liberia, Monrovia	06:14	06:35	18:47	19:08
EP	Iran, Tehran	01:43	02:09	14:52	15:18
ER	Moldova, Kishinev	03:00	03:31	16:34	17:05
ES	Estonia, Tallinn	02:47	03:30	17:07	17:50
ET	Ethiopia, Addis Ababa	02:54	03:16	15:30	15:51
EU,EV,EW	Belarus, Minsk	02:51	03:28	16:47	17:23
EX	Kyrgyzstan, Frunze	00:03	00:32	13:27	13:55
EY	Tajikistan, Samarkand	00:38	01:06	13:55	14:23
	Dushanbe	00:31	00:58	13:47	14:14
EZ	Turkmenistan, Ashkhabad	01:14	01:41	14:27	14:54
F	France, Paris	04:43	05:15	18:22	18:54
FG	Guadeloupe	09:32	09:54	22:15	22:37
FJ, FS	St. Martin	09:37	09:58	22:22	22:43
FH	Mayotte	02:37	02:59	14:54	15:15
FK	New Caledonia, Noumia	18:36	18:58	06:45	07:08
FM	Martinique	09:30	09:52	22:11	22:33
FO	Clipperton Is.	12:45	13:07	01:22	01:43
FO	Fr. Polynesia, Tahiti	16:18	16:40	04:31	04:53
	Rurutu, (Austral Is.)	19:36	19:59	07:46	08:08
	Hiva Oa, (Marquesas Is.)	14:53	15:14	03:13	03:34
FP	St. Pierre & Miquelon, St. Pierre	08:40	09:11	22:13	22:44
FR/G	Glorioso	02:29	02:50	14:47	15:08
FR/J,E	Juan de Nova	02:49	03:10	15:02	15:24
	Europa	03:00	03:22	15:10	15:32
FR	Reunion	01:59	02:21	14:09	14:32
FR/T	Tromelin	02:02	02:24	14:16	14:38
FT8W	Crozet	02:20	02:50	14:09	14:39
FT8X	Kerguelen	01:12	01:44	12:57	13:29
FT8Y	Antarctica, Dumont D'Urville	20:34	21:26	07:49	08:42
FT8Z	Amsterdam & St. Paul Is., Amsterdam	00:36	01:02	12:32	12:59
FW	Wallis & Futuna Is., Wallis	17:24	17:45	05:40	06:02
FY	Fr. Guiana, Cayenne	09:00	09:21	21:32	21:53
G	England, London	04:47	05:22	18:34	19:09
GD	Isle of Man	04:59	05:36	18:56	19:32
GI	Northern Ireland, Belfast	05:04	05:41	19:02	19:39
GJ	Jersey	05:00	05:33	18:40	19:13
GM	Scotland, Glasgow	04:54	05:33	18:57	19:35
	Aberdeen	04:42	05:21	18:51	19:30
GU	Guernsey	05:02	05:34	18:42	19:15
GW	Wales, Cardiff	05:00	05:34	18:47	19:21
H4	Solomon Islands, Honiara	18:57	19:18	07:17	07:38
HA	Hungary, Budapest	03:38	04:09	17:13	17:44
HB	Switzerland, Bern	04:25	04:56	17:59	18:30
HB0	Liechtenstein	04:16	04:48	17:51	18:22
HC	Ecuador, Quito	10:46	11:06	23:13	23:34
HC8	Galapagos Is.	11:36	11:56	00:03	00:24
HH	Haiti, Port-Au-Prince	10:13	10:35	22:59	23:21
HI	Dominican Republic, Santo Domingo	10:04	10:26	22:49	23:11
HK	Colombia, Bogota	10:28	10:49	22:59	23:20
HK0	Malpelo Is.	10:56	11:17	23:27	23:48
HK0	San Adreas	10:54	11:16	23:33	23:55
HL	Korea, Seoul	20:40	21:07	09:53	10:19
HP	Panama, Panama	10:47	11:08	23:23	23:44
HR	Honduras, Tegucigalpa	11:15	11:37	23:56	00:17
HS	Thailand, Bangkok	22:45	23:06	11:25	11:47
HV	Vatican City	04:13	04:41	17:34	18:02
HZ, 7Z	Saudi Arabia, Dharan	01:58	02:22	14:53	15:17
	Mecca	02:43	03:05	15:32	15:54
	Riyadh	02:13	02:36	15:06	15:29

Pref	State/Province/Country/City	Tlight	Sunrise	Sunset	Tlight
I	Italy, Rome	04:13	04:41	17:34	18:02
	Trieste	04:02	04:32	17:32	18:03
	Sicily	04:12	04:39	17:25	17:51
IS	Sardinia, Cagliari	04:30	04:57	17:46	18:13
J2	Djibouti, Djibouti	02:35	02:56	15:13	15:35
J3	Grenada	09:35	09:56	22:13	22:35
J5	Guinea-Bissau, Bissau	06:30	06:51	19:09	19:30
J6	St. Lucia	09:31	09:52	22:11	22:33
J7	Dominica	09:31	09:53	22:13	22:35
J8	St. Vincent	09:32	09:54	22:12	22:33
JA-JS	Japan, Tokyo	19:51	20:17	09:00	09:26
	Nagasaki	20:33	20:58	09:38	10:03
	Sapporo	19:35	20:04	09:00	09:29
JD1	Minami Torishima	19:04	19:27	07:56	08:19
JD1	Ogasawara, Kazan Is.	19:53	20:17	08:50	09:14
JT	Mongolia, Ulan Bator	21:46	22:18	11:22	11:54
JW	Svalbard, Spitsbergen	——	02:22	19:26	——
JX	Jan Mayen	03:54	05:07	19:54	21:08
JY	Jordan, Amman	02:50	03:15	15:53	16:18
KC4	Antarctica, Byrd Station	13:58	16:04	23:52	01:57
	McMurdo Sound	18:50	20:31	05:11	06:51
	Palmer Station	10:09	10:58	21:29	22:18
KC6	Belau, Yap	20:16	20:37	08:52	09:13
	Koror	20:32	20:53	09:06	09:27
KG4	Guantanamo Bay	10:24	10:46	23:11	23:33
KH0	Mariana Is., Saipan	19:43	20:04	08:25	08:46
KH1	Baker, Howland Is.	17:17	17:38	05:45	06:06
KH2	Guam, Agana	19:48	20:09	08:28	08:49
KH3	Johnston Is.	16:39	17:01	05:23	05:44
KH4	Midway Is.	17:06	17:30	06:04	06:28
KH5	Palmyra Is.	16:19	16:40	04:52	05:13
KH5K	Kingman Reef	16:21	16:42	04:55	05:16
KH6	Hawaii, Hilo	15:43	16:06	04:30	04:53
	Honolulu	15:54	16:16	04:42	05:05
KH7	Kure Is.	17:10	17:34	06:08	06:32
KH8	American Samoa, Pago Pago	17:02	17:24	05:18	05:39
KH9	Wake Is.	18:17	18:39	07:03	07:25
KL7	Alaska, Adak	16:33	17:07	06:21	06:55
	Anchorage	14:20	15:06	04:50	05:35
	Fairbanks	13:56	14:49	04:50	05:42
	Juneau	13:28	14:09	03:42	04:23
	Nome	15:07	15:59	05:59	06:51
KP1	Navassa Is.	10:24	10:46	23:09	23:31
KP2	Virgin Islands, Charlotte Amalie	09:44	10:06	22:29	22:51
KP4	Puerto Rico, San Juan	09:49	10:11	22:34	22:56
KP5	Desecheo Is.	09:54	10:16	22:39	23:01
LA-LJ	Norway, Oslo	03:42	04:25	18:04	18:48
LU	Argentina, Buenos Aires	09:39	10:04	21:38	22:04
LX	Luxembourg	04:26	04:59	18:07	18:40
LY	Lithuania, Vilna	02:58	03:35	16:56	17:33
LZ	Bulgaria, Sofia	03:28	03:57	16:52	17:20
OA	Peru, Lima	10:47	11:08	23:04	23:25
OD	Lebanon, Beirut	02:50	03:15	15:56	16:21
OE	Austria, Vienna	03:48	04:20	17:25	17:57
OH	Finland, Helsinki	02:44	03:28	17:08	17:52
OH0	Aland Is.	03:04	03:48	17:28	18:12
OJ0	Market Reef	03:07	03:51	17:32	18:16
OK,OL	Czech Rep., Prague	03:52	04:25	17:35	18:08
OM	Slovak, Rep., Bratislava	03:46	04:17	17:22	17:54
ON	Belgium, Brussels	04:31	05:04	18:16	18:49
OX	Greenland, Godthaab	07:34	08:25	22:23	23:15
	Thule	06:15	08:30	00:36	02:51
OY	Faroe Islands, Torshavn	04:44	05:31	19:19	20:05
OZ	Denmark, Copenhagen	03:47	04:25	17:49	18:28
P2	Papua New Guinea, Madang	19:53	20:14	08:16	08:37
	Port Moresby	19:49	20:10	08:09	08:30
P4	Aruba, Oranjestad	10:04	10:25	22:42	23:04
PA-PI	Netherlands, Amsterdam	04:25	05:00	18:16	18:51
PJ2,4,9	Netherlands Antilles, Willemstad	10:08	10:29	22:47	23:08
PJ5-8	St. Maarten and Saba, St. Maarten	09:37	09:59	22:22	22:44
PY	Brazil, Brasilia	08:51	09:13	21:06	21:27
	Rio De Janeiro	08:35	08:57	20:44	21:06
	Natal	07:57	08:18	20:19	20:40
	Manaus	09:36	09:56	22:00	22:21
	Porto Alegre	09:09	09:33	21:12	21:36
PY0	Fernando de Noronha	07:45	08:06	20:09	20:30
PY0	St. Peter & St. Paul Rocks	07:31	07:51	19:59	20:20
PY0	Trindade & Martin Vaz Is., Trindade	07:38	08:00	19:49	20:11
PZ	Suriname, Paramaribo	09:12	09:33	21:44	22:05
R1FJ_	Franz Josef Land	——	23:31	17:20	——
R1MV_	Malyj Vysotskij Is.	02:28	03:12	16:54	17:39
S0	Western Sahara, Smara	06:04	06:27	18:59	19:22
S2	Bangladesh, Dacca	23:19	23:42	12:10	12:33
S5	Slovenia, Ljubljana	03:59	04:29	17:30	18:00
S7	Seychelles, Victoria	01:53	02:14	14:17	14:38
S9	Sao Tome	05:07	05:27	17:34	17:55
SM	Sweden, Stockholm	03:14	03:57	17:33	18:16
SP	Poland, Krakow	03:30	04:03	17:12	17:45
	Warsaw	03:21	03:56	17:11	17:46
ST	Sudan, Khartoum	03:16	03:37	15:58	16:20
ST0	Southern Sudan, Juba	03:25	03:46	15:57	16:18
SU	Egypt, Cairo	03:10	03:34	16:10	16:34
SV	Greece, Athens	03:33	03:59	16:46	17:13
SV/A	Mount Athos	03:28	03:55	16:46	17:13
SV5	Dodecanese, Rhodes	03:16	03:43	16:27	16:53
SV9	Crete	03:29	03:55	16:38	17:04
T2	Tuvalu, Funafuti	17:43	18:04	06:03	06:24

Pref	State/Province/Country/City	Tlight	Sunrise	Sunset	Tlight
T30	West Kiribati, Bonriki	18:00	18:21	06:29	06:50
T31	Central Kiribati, Kanton	17:01	17:22	05:27	05:47
T32	East Kiribati, Christmas Is.	16:02	16:23	04:31	04:52
T33	Banaba Is.	18:16	18:37	06:43	07:04
T5	Somalia, Mogadishu	02:31	02:52	15:00	15:21
T7	San Marino	04:11	04:40	17:37	18:06
T9	Bosnia-Herzegovina, Sarajevo	03:46	04:16	17:12	17:42
TA	Turkey, Ankara	02:54	03:21	16:11	16:38
	Istanbul	03:08	03:36	16:28	16:56
TF	Iceland, Reykjavik	05:36	06:27	20:24	21:15
TG	Guatemala, Guatemala City	11:28	11:50	00:09	00:31
TI	Costa Rica, San Jose	11:05	11:26	23:41	00:02
TI9	Cocos Is.	11:19	11:40	23:51	00:12
TJ	Cameroon, Yaounde	04:46	05:07	17:17	17:37
TK	Corsica	04:27	04:55	17:48	18:17
TL	Central African Republic, Bangui	04:17	04:38	16:48	17:09
TN	Congo, Brazzaville	04:34	04:55	16:58	17:19
TR	Gabon, Libreville	04:55	05:16	17:23	17:44
TT	Chad, N'Djamena	04:28	04:49	17:06	17:28
TU	Cote d'Ivoire, Abidjan	05:47	06:08	18:19	18:40
TY	Benin, Porto Novo	05:20	05:41	17:53	18:14
TZ	Mali, Bamako	05:59	06:21	18:39	19:00
UA	Russia, European, Leningrad (UA1)	02:24	03:07	16:46	17:29
	Archangel (UA1)	01:23	02:15	16:16	17:08
	Murmansk (UA1)	01:26	02:31	17:00	18:05
	Moscow (UA3)	02:07	02:45	16:09	16:48
	Kuybyshev (UA4)	01:23	01:59	15:16	15:51
	Rostov (UA6)	02:16	02:48	15:51	16:23
UA2	Kaliningrad	03:17	03:55	17:17	17:54
UA9,0	Russia, Asiatic, Novosibirsk (UA9)	23:07	23:45	13:07	13:45
	Perm (UA9)	00:46	01:27	14:58	15:39
	Omsk (UA9)	23:45	00:23	13:45	14:23
	Norilsk (UA0)	21:43	22:49	13:21	14:27
	Irkutsk (UA0)	21:48	22:23	11:38	12:13
	Vladivostok (UA0)	20:13	20:42	09:38	10:07
	Petropavlovsk (UA0)	18:09	18:45	08:01	08:37
	Khabarovsk (UA0)	19:52	20:24	09:30	10:02
	Krasnoyarsk (UA0)	22:25	23:04	12:29	13:08
	Yakutsk (UA0)	19:38	20:25	10:13	10:59
	Wrangel Island (UA0)	15:18	16:32	07:19	08:33
	Kyzyl (UA0Y)	22:29	23:03	12:16	12:51
UJ-UM	Uzbekistan, Bukhoro	00:48	01:15	14:05	14:32
	Tashkent	00:26	00:54	13:46	14:14
UN-UQ	Kazakhstan, Alma-Ata	23:53	00:22	13:18	13:47
UR-UZ,EM-EO	Ukraine, Kiev	02:47	03:21	16:31	17:04
V2	Antigua & Barbuda, St. Johns	09:32	09:54	22:16	22:38
V3	Belize, Belmopan	11:20	11:42	00:04	00:26
V4	St. Kitts & Nevis	09:35	09:57	22:19	22:41
V5	Namibia, Windhoek	04:33	04:56	16:43	17:05
V6	Micronesia, Ponape	18:57	19:18	07:31	07:52
V7	Marshall Islands, Kwajalein	18:20	18:41	06:56	07:17
V8	Brunei, Bandar Seri Begawan	21:52	22:13	10:23	10:44
VK	Australia, Canberra (VK1)	19:49	20:14	07:48	08:13
	Sydney (VK2)	19:40	20:05	07:40	08:05
	Melbourne (VK3)	20:06	20:32	08:03	08:29
	Brisbane (VK4)	19:31	19:55	07:37	08:00
	Adelaide (VK5)	20:31	20:56	08:30	08:55
	Perth (VK6)	22:01	22:26	10:03	10:28
	Hobart, Tasmania (VK7)	19:58	20:27	07:50	08:19
	Darwin (VK8)	20:55	21:16	09:12	09:33
VK0	Heard Is.	00:57	01:31	12:37	13:11
VK0	Macquarie Is.	19:15	19:51	06:53	07:29
VK9C	Cocos-Keeling Is.	23:11	23:32	11:29	11:50
VK9L	Lord Howe Is.	19:08	19:32	07:10	07:34
VK9M	Mellish Reef	19:17	19:39	07:30	07:52
VK9N	Norfolk Is.	18:32	18:55	06:36	07:00
VK9W	Willis Is.	19:42	20:03	07:56	08:18
VK9X	Christmas Is.	22:35	22:56	10:54	11:15
VP2E	Anguilla	09:36	09:58	22:21	22:43
VP2M	Montserrat	09:34	09:55	22:17	22:39
VP2V	British Virgin Is., Tortola	09:42	10:04	22:28	22:50

Pref	State/Province/Country/City	Tlight	Sunrise	Sunset	Tlight
VP5	Turks & Caicos Islands, Grand Turk	10:07	10:29	22:56	23:18
VP8	Falkland Islands, Stanley	09:41	10:15	21:23	21:57
VP8	So. Georgia Is.	08:18	08:53	19:56	20:32
VP8	So. Orkney Is.	08:54	09:36	20:23	21:05
VP8	So. Sandwich Islands, Saunders Is.	07:38	08:17	19:12	19:51
VP8	So. Shetland Is., King George Is.	09:46	10:30	21:12	21:56
VP9	Bermuda	09:32	09:57	22:36	23:01
VQ9	Chagos, Diego Garcia	00:47	01:08	13:08	13:29
VR2,VS6	Hong Kong	21:44	22:07	10:34	10:57
VR6	Pitcairn Is.	14:23	14:46	02:30	02:53
VU	India, Bombay	00:32	00:54	13:18	13:40
	Calcutta	23:28	23:50	12:18	12:40
	New Delhi	00:08	00:32	13:06	13:30
	Bangalore	00:17	00:38	12:56	13:18
VU	Andaman Islands, Port Blair	23:17	23:38	11:55	12:16
VU	Laccadive Is.	00:37	00:58	13:13	13:34
XE	Mexico, Mexico City (XE1)	12:00	12:22	00:46	01:08
	Chihuahua (XE2)	12:20	12:44	01:19	01:43
	Merida (XE3)	11:21	11:43	00:09	00:32
XF4	Revilla Gigedo	12:50	13:12	01:36	01:58
XT	Burkina Faso, Ouagadougou	05:34	05:55	18:13	18:34
XU	Cambodia, Phnom Penh	22:29	22:50	11:07	11:28
XW	Laos, Viangchan	22:34	22:56	11:19	11:41
XX9	Macao	21:47	22:10	10:37	10:59
XZ	Myanmar, Yangon	23:01	23:23	11:45	12:06
YA	Afghanistan, Kandahar	00:51	01:16	13:53	14:17
	Kabul	00:34	01:00	13:42	14:07
YB-YD	Indonesia, Jakarta	22:29	22:50	10:51	11:12
	Medan	22:57	23:18	11:28	11:49
	Pontianak	22:16	22:37	10:44	11:05
	Jayapura	20:12	20:33	08:37	08:58
YI	Iraq, Baghdad	02:15	02:40	15:20	15:45
YJ	Vanuatu, Port Vila	18:27	18:49	06:40	07:02
YK	Syria, Damascus	02:47	03:12	15:53	16:18
YL	Latvia, Riga	02:57	03:37	17:05	17:45
YN	Nicaragua, Managua	11:12	11:33	23:50	00:12
YO	Romania, Bucharest	03:15	03:44	16:42	17:12
YS	El Salvador, San Salvador	11:24	11:45	00:04	00:25
YU	Yugoslavia, Belgrade	03:36	04:06	17:05	17:35
YV	Venezuela, Caracas	09:56	10:18	22:34	22:55
YV0	Aves Is.	09:40	10:02	22:23	22:44
Z2	Zimbabwe, Harare	03:36	03:58	15:49	16:11
Z3	Macedonia, (ex Yugoslav), Skopje	03:37	04:05	16:59	17:27
ZA	Albania, Tirane	03:44	04:12	17:04	17:33
ZB2	Gibraltar	05:31	05:57	18:41	19:07
ZC4	British Cyprus	02:59	03:25	16:07	16:32
ZD7	St. Helena	06:03	06:25	18:18	18:39
ZD8	Ascension Is.	06:34	06:55	18:55	19:16
ZD9	Tristan da Cunha	06:35	07:01	18:32	18:58
ZF	Cayman Is.	.10:48	11:10	23:35	23:57
ZK1	No. Cook Is., Manihiki	16:22	16:43	04:41	05:02
ZK1	So. Cook Is., Rarotonga	16:20	16:43	04:31	04:53
ZK2	Niue	16:56	17:18	05:08	05:30
ZK3	Tokelaus, Atafu	17:08	17:29	05:28	05:49
ZL	New Zealand, Auckland (ZL1)	18:07	18:33	06:04	06:30
	Wellington (ZL2)	18:08	18:35	06:01	06:29
	Christchurch (ZL3)	18:17	18:46	06:09	06:37
	Dunedin (ZL4)	18:26	18:56	06:15	06:45
ZL5	Antarctica, Scott Base	18:51	20:34	05:10	06:53
ZL7	Chatham Is.	17:34	18:03	05:25	05:53
ZL8	Kermadec Is.	17:35	17:59	05:39	06:03
ZL9	Auckland & Campbell Is., Auckland	18:44	19:16	06:27	07:00
	Campbell Is.	18:34	19:08	06:15	06:49
ZP	Paraguay, Asuncion	09:33	09:56	21:41	22:03
ZS	South Africa, Cape Town (ZS1)	04:31	04:56	16:32	16:57
	Port Elizabeth (ZS2)	04:02	04:27	16:02	16:27
	Bloemfontein (ZS4)	03:59	04:23	16:03	16:27
	Durban (ZS5)	03:40	04:04	15:44	16:08
	Johannesburg (ZS6)	03:50	04:14	15:57	16:20
ZS8	Marion Is.	03:17	03:48	15:05	15:35

Pref	State/Province/Country/City	Tlight	Sunrise	Sunset	Tlight
VE1	New Brunswick, Saint John	09:40	10:09	22:25	22:54
	Nova Scotia, Halifax	09:30	09:59	22:15	22:44
	Prince Edward Island, Charlottetown	09:27	09:57	22:13	22:43
VE2	Quebec, Montreal	10:10	10:39	22:55	23:24
	Quebec City	09:59	10:29	22:46	23:16
VE3	Ontario, London	10:42	11:11	23:25	23:54
	Ottawa	10:18	10:48	23:03	23:33
	Sudbury	10:39	11:09	23:25	23:55
	Toronto	10:34	11:03	23:18	23:46
VE4	Manitoba, Winnipeg	11:40	12:12	00:30	01:02
VE5	Saskatchewan, Regina	12:10	12:42	01:00	01:33
	Saskatoon	12:16	12:50	01:09	01:43
VE6	Alberta, Calgary	12:47	13:20	01:38	02:11
	Edmonton	12:42	13:17	01:36	02:11
VE7	British Columbia, Prince George	13:19	13:54	02:14	02:49
	Prince Rupert	13:48	14:24	02:44	03:19
	Vancouver	13:25	13:56	02:14	02:45
VE8	Northwest Territories, Yellowknife	12:32	13:17	01:43	02:29
	Resolute	10:27	11:49	00:36	01:58
VY1	Yukon, Whitehorse	13:58	14:41	03:05	03:48
VO1	Newfoundland, St. John's	08:44	09:15	21:32	22:02
VO2	Labrador, Goose Bay	09:10	09:45	22:04	22:39
W1	Connecticut, Hartford	10:09	10:36	22:51	23:18
	Maine, Bangor	09:51	10:20	22:36	23:05
	Portland	09:58	10:26	22:41	23:10
	Massachusetts, Boston	10:02	10:30	22:44	23:12
	New Hampshire, Concord	10:03	10:31	22:46	23:14
	Rhode Island, Providence	10:03	10:31	22:45	23:13
	Vermont, Montpelier	10:07	10:36	22:51	23:20
W2	New Jersey, Atlantic City	10:17	10:44	22:57	23:24
	New York, Albany	10:12	10:41	22:55	23:23
	Buffalo	10:33	11:01	23:16	23:44
	New York City	10:14	10:42	22:56	23:23
	Syracuse	10:22	10:50	23:05	23:33
W3	Delaware, Wilmington	10:21	10:48	23:01	23:28
	District of Columbia, Washington	10:27	10:54	23:07	23:34
	Maryland, Baltimore	10:26	10:52	23:06	23:33
	Pennsylvania, Harrisburg	10:26	10:53	23:07	23:34
	Philadelphia	10:20	10:47	23:00	23:27
	Pittsburgh	10:39	11:06	23:20	23:47
	Scranton	10:21	10:48	23:03	23:30
W4	Albama, Montgomery	11:08	11:32	23:44	00:08
	Florida, Jacksonville	10:50	11:14	23:25	23:49
	Miami	10:45	11:08	23:19	23:42
	Pensacola	11:12	11:36	23:47	00:11
	Georgia, Atlanta	10:59	11:24	23:36	00:01
	Savannah	10:47	11:11	23:23	23:47
	Kentucky, Lexington	10:58	11:24	23:37	00:04
	Louisville	11:03	11:29	23:42	00:09
	North Carolina, Charlotte	10:44	11:10	23:22	23:47
	Raleigh	10:35	11:01	23:13	23:39
	Wilmington	10:33	10:58	23:10	23:35
	South Carolina, Columbia	10:46	11:11	23:23	23:48
	Tennessee, Knoxville	10:56	11:22	23:35	00:00
	Memphis	11:22	11:47	23:59	00:25
	Nashville	11:08	11:34	23:46	00:12
	Virginia, Norfolk	10:26	10:51	23:04	23:30
	Richmond	10:30	10:56	23:09	23:35
W5	Arkansas, Little Rock	11:31	11:56	00:08	00:33
	Louisiana, New Orleans	11:24	11:48	23:59	00:23
	Shreveport	11:37	12:02	00:13	00:38
	Mississippi, Jackson	11:23	11:48	23:59	00:24
	New Mexico, Albuquerque	12:28	12:53	01:06	01:31
	Oklahoma, Oklahoma City	11:51	12:16	00:29	00:54
	Texas, Abilene	12:01	12:26	00:37	01:02
	Amarillo	12:08	12:34	00:46	01:11
	Dallas	11:49	12:14	00:26	00:50
	El Paso	12:29	12:53	01:04	01:29
	San Antonio	11:57	12:21	00:32	00:56
W6	California, Los Angeles	13:14	13:39	01:52	02:17
	San Francisco	13:30	13:56	02:09	02:35
W7	Arizona, Flagstaff	12:48	13:13	01:26	01:51
	Phoenix	12:50	13:15	01:27	01:52
	Idaho, Boise	13:01	13:30	01:45	02:14
	Pocatello	12:47	13:15	01:30	01:58
	Montana, Billings	12:29	12:59	01:15	01:44
	Butte	12:45	13:15	01:31	02:00
	Great Falls	12:39	13:10	01:26	01:57
	Nevada, Las Vegas	13:01	13:27	01:39	02:05
	Reno	13:18	13:45	01:59	02:26
	Oregon, Portland	13:26	13:56	02:11	02:41
	Utah, Salt Lake City	12:46	13:13	01:27	01:55
	Washington, Seattle	13:23	13:54	02:10	02:41
	Spokane	13:03	13:34	01:51	02:21
	Wyoming, Cheyenne	12:17	12:45	00:59	01:26
	Sheridan	12:24	12:53	01:08	01:38
W8	Michigan, Detroit	10:49	11:17	23:32	00:00
	Grand Rapids	11:00	11:28	23:43	00:11
	Sault Ste. Marie	10:52	11:22	23:38	00:08
	Traverse City	10:58	11:27	23:43	00:12
	Ohio, Cincinnati	10:57	11:24	23:37	00:04
	Cleveland	10:45	11:12	23:27	23:54
	Columbus	10:51	11:18	23:32	23:59
	West Virginia, Charleston	10:46	11:12	23:26	23:52
W9	Illinois, Chicago	11:08	11:36	23:50	00:18
	Indiana, Indianapolis	11:04	11:31	23:44	00:11
	Wisconsin, Green Bay	11:08	11:37	23:52	00:21
	Milwaukee	11:09	11:37	23:52	00:20
W0	Colorado, Denver	12:19	12:46	00:59	01:26
	Grand Junction	12:34	13:00	01:14	01:40
	Iowa, Des Moines	11:32	12:00	00:14	00:42
	Kansas, Pratt	11:55	12:21	00:34	01:00
	Wichita	11:49	12:15	00:28	00:55
	Minnesota, Duluth	11:23	11:53	00:09	00:39
	Minneapolis	11:29	11:58	00:14	00:43
	Missouri, Columbia	11:29	11:55	00:09	00:35
	Kansas City	11:38	12:04	00:18	00:44
	St. Louis	11:20	11:47	00:00	00:27
	Nebraska, North Platte	12:01	12:29	00:43	01:10
	Omaha	11:42	12:09	00:23	00:51
	North Dakota, Fargo	11:41	12:12	00:28	00:58
	South Dakota, Rapid City	12:09	12:38	00:53	01:22
-		05:29	05:49	17:56	18:17
1A0	S.M.O.M	04:28	04:56	17:10	17:38
1S	Spratly Is.	22:00	22:21	10:29	10:50
3A	Monaco	04:47	05:16	17:31	17:59
3B6	Agalega	01:42	02:03	14:09	14:30
3B7	St. Brandon	01:29	01:51	13:56	14:18
3B8	Mauritius	01:38	02:00	14:05	14:27
3B9	Rodriguez Is.	01:15	01:37	13:42	14:04
3C	Equatorial Guinea, Bata	04:49	05:10	17:17	17:38
	Malabo	04:53	05:14	17:21	17:42
3C0	Annobon Is.	05:06	05:27	17:34	17:54
3D2	Fiji Is., Suva	17:35	17:57	06:02	06:23
3D2	Conway Reef	17:50	18:13	06:17	06:40
3D2	Rotuma Is.	17:38	17:59	06:05	06:26
3DA	Swaziland, Mbabane	03:23	03:46	15:51	16:14
3V	Tunisia, Tunis	04:40	05:05	17:18	17:44
3W, XV	Vietnam, Ho Chi Minh City (Saigon)	22:21	22:42	10:50	11:11
	Hanoi	22:23	22:45	10:54	11:16
3X	Guinea, Conakry	06:23	06:44	18:51	19:12
3Y	Bouvet	05:02	05:38	17:41	18:16
3Y	Peter I Is.	10:56	11:54	23:57	00:54
4J, 4K	Azerbaijan, Baku	01:59	02:26	14:40	15:07
4L	Georgia, Tbilisi	02:19	02:46	15:01	15:28
4S	Sri Lanka, Colombo	00:09	00:29	12:37	12:58
4U	ITU Geneva	04:50	05:20	17:36	18:06
4U	United Nations Hq.	10:14	10:42	22:56	23:23
4X, 4Z	Israel, Jerusalem	03:02	03:26	15:38	16:02
5A	Libya, Tripoli	04:32	04:57	17:09	17:33
	Benghazi	04:02	04:27	16:38	17:03
5B	Cyprus, Nicosia	03:08	03:33	15:45	16:11
5H	Tanzania, Dar es Salaam	02:51	03:12	15:18	15:38
5N	Nigeria, Lagos	05:15	05:35	17:43	18:04
5R	Madagascar, Antananarivo	02:18	02:40	14:45	15:07
5T	Mauritania, Nouakchott	06:30	06:52	19:01	19:23
5U	Niger, Niamey	05:19	05:41	17:49	18:10
5V	Togo, Lome	05:23	05:44	17:52	18:12
5W	Western Samoa, Apia	16:56	17:17	05:23	05:44
5X	Uganda, Kampala	03:19	03:39	15:46	16:07
5Z	Kenya, Nairobi	04:22	04:42	16:49	17:09
6W	Senegal, Dakar	06:37	06:58	19:07	19:28
6Y	Jamaica, Kingston	10:34	10:55	23:04	23:26
7O	Yemen, Aden	02:27	02:49	14:57	15:18
	Sanaa	02:30	02:52	15:00	15:22
7P	Lesotho, Maseru	03:37	04:01	16:05	16:29
7Q	Malawi, Lilongwe	03:13	03:35	15:40	16:01
	Blantyre	03:09	03:30	15:35	15:57
7X	Algeria, Algiers	05:08	05:34	17:47	18:13
8P	Barbados, Bridgetown	09:26	09:47	21:55	22:16
8Q	Maldive Is.	00:35	00:56	13:03	13:23
8R	Guyana, Georgetown	09:21	09:42	21:49	22:10
9A	Croatia, Zagreb	04:11	04:41	16:57	17:26
9G	Ghana, Accra	05:29	05:50	17:57	18:18
9H	Malta	04:23	04:49	17:01	17:27
9J	Zambia, Lusaka	03:35	03:57	16:02	16:24
9K	Kuwait	02:12	02:36	14:47	15:11
9L	Sierra Leone, Freetown	06:21	06:42	18:49	19:10
9M2	West Malaysia, Kuala Lumpur	22:42	23:03	11:10	11:30
9M6,8	East Malaysia, Sabah, Sandakan(9M6)	21:36	21:57	10:04	10:25
	Sarawak, Kuching (9M8)	22:07	22:28	10:35	10:56
9N	Nepal, Kathmandu	23:43	00:06	12:17	12:40
9Q	Zaire, Kinshasa	04:28	04:48	16:55	17:15
	Kisangani	03:48	04:09	16:15	16:36
	Lubumbashi	03:39	04:00	16:05	16:27
9U	Burundi, Bujumbura	03:32	03:52	15:59	16:19
9V	Singapore	22:33	22:54	11:01	11:22
9X	Rwanda, Kigali	03:28	03:49	15:56	16:16
9Y	Trinidad & Tobago , Port of Spain	09:33	09:54	22:02	22:23
A2	Botswana, Gaborone	03:44	04:07	16:11	16:34
A3	Tonga, Nukualofa	17:09	17:31	05:36	05:58
A4	Oman, Musqat	01:31	01:53	14:03	14:26
A5	Bhutan, Thimbu	23:27	23:50	12:00	12:24
A6	United Arab Emirates, Abu Dhabi	01:48	02:11	14:21	14:44
A7	Qatar, Ad-Dawhah	01:59	02:22	14:32	14:55
A9	Bahrein, Al-Manamah	02:02	02:25	14:35	14:59
AP	Pakistan, Karachi	00:56	01:19	13:29	13:52
AP	Islamabad	00:29	00:54	13:06	13:31
BV	Taiwan, Taipei	21:19	21:42	09:52	10:15
BY	Peoples Rep. of China, Beijing	21:33	22:00	10:14	10:41
	Harbin	20:48	21:18	09:34	10:03
	Shanghai	21:17	21:41	09:52	10:17
	Fuzhou	21:27	21:50	10:01	10:24
	Xian	22:06	22:31	10:43	11:08

Pref	State/Province/Country/City	Tlight	Sunrise	Sunset	Tlight
	Chongqing	22:17	22:41	10:52	11:16
	Chengdu	22:27	22:51	11:02	11:26
	Lhasa	23:19	23:42	11:53	12:17
	Urumqi	23:26	23:55	12:10	12:38
	Kashi	00:15	00:42	12:55	13:22
C2	Nauru	18:21	18:42	06:48	07:09
C3	Andorra	05:11	05:39	17:54	18:22
C5	The Gambia, Banjul	06:36	06:55	19:04	19:25
C6	Bahamas, Nassau	10:34	10:57	23:07	23:30
C9	Mozambique, Maputo	03:17	03:40	15:45	16:08
	Mozambique	02:46	03:07	15:13	15:34
CE	Chile, Santiago	10:09	10:34	22:38	23:03
CE0	Easter Island	12:45	13:08	01:12	01:36
CE0	Juan Fernandez	10:41	11:06	23:10	23:35
CE0	San Felix	10:48	11:11	23:15	23:38
CM,CO	Cuba, Havana	10:55	11:17	23:27	23:50
CN	Morocco, Casablanca	05:52	06:17	18:29	18:54
CP	Bolivia, La Paz	10:02	10:24	22:29	22:50
CT	Portugal, Lisbon	05:56	06:23	18:36	19:03
CT3	Madeira Islands, Funchal	06:30	06:54	19:06	19:31
CU	Azores , Ponta Delgada	07:03	07:29	19:42	20:08
CX	Uruguay, Montevideo	09:10	09:35	21:39	22:05
CY0	Sable Is.	09:16	09:45	22:00	22:29
CY9	St. Paul Is.	09:14	09:45	22:01	22:32
D2	Angola, Luanda	04:36	04:57	17:03	17:24
D4	Cape Verde, Praia	07:01	07:22	19:31	19:52
D6	Comoros, Moroni	02:34	02:55	15:01	15:22
DA-DL,Y2-Y9	Fed. Rep. of Germany, Bonn	04:43	05:16	17:34	18:06
	Berlin	04:16	04:50	17:08	17:43
DU	Philippines, Manilla	21:23	21:44	09:53	10:14
E3	Eritrea, Asmara	02:51	03:13	15:21	15:43
EA	Spain, Madrid	05:33	06:01	18:14	18:42
EA6	Balearic Is., Palma	05:09	05:35	17:49	18:16
EA8	Canary Is., Las Palmas	06:21	06:44	18:55	19:19
EA9	Ceuta & Melilla, Ceuta	05:42	06:08	18:20	18:46
	Melilla	05:33	05:58	18:11	18:36
EI	Ireland, Dublin	05:34	06:08	18:27	19:02
EK	Armenia, Yerevan	02:21	02:48	15:02	15:29
EL	Liberia, Monrovia	06:11	06:32	18:40	19:00
EP	Iran, Tehran	01:54	02:19	14:32	14:57
ER	Moldova, Kishinev	03:19	03:49	16:06	16:36
ES	Estonia, Tallinn	03:21	04:02	16:25	17:06
ET	Ethiopia, Addis Ababa	02:53	03:14	15:22	15:43
EU,EV,EW	Belarus, Minsk	03:17	03:53	16:12	16:47
EX	Kyrgyzstan, Frunze	00:19	00:47	13:02	13:30
EY	Tajikistan, Samarkand	00:52	01:19	13:32	13:59
	Dushanbe	00:44	01:11	13:24	13:51
EZ	Turkmenistan, Ashkhabad	01:26	01:53	14:06	14:32
F	France, Paris	05:03	05:35	17:52	18:23
FG	Guadeloupe	09:34	09:55	22:04	22:25
FJ, FS	St. Martin	09:39	10:01	22:10	22:31
FH	Mayotte	02:28	02:49	14:54	15:15
FK	New Caledonia, Noumia	18:22	18:44	06:49	07:11
FM	Martinique	09:31	09:52	22:01	22:22
FO	Clipperton Is.	12:45	13:06	01:13	01:34
FO	Fr. Polynesia, Tahiti	16:06	16:28	04:33	04:55
	Rurutu, (Austral Is.)	19:23	19:45	07:50	08:12
	Hiva Oa, (Marquesas Is.)	14:45	15:06	03:12	03:33
FP	St. Pierre & Miquelon, St. Pierre	08:58	09:29	21:45	22:15
FR/G	Glorioso	02:20	02:41	14:46	15:07
FR/J,E	Juan de Nova	02:37	02:59	15:04	15:26
	Europa	02:46	03:09	15:13	15:36
FR	Reunion	01:46	02:08	14:13	14:35
FR/T	Tromelin	01:51	02:12	14:18	14:39
FT8W	Crozet	01:53	02:23	14:26	14:56
FT8X	Kerguelen	00:42	01:14	13:18	13:49
FT8Y	Antarctica, Dumont D'Urville	19:39	20:31	08:34	09:27
FT8Z	Amsterdam & St. Paul Is., Amsterdam	00:14	00:40	12:44	13:10
FW	Wallis & Futuna Is., Wallis	17:14	17:35	05:41	06:02
FY	Fr. Guiana, Cayenne	08:58	09:18	21:25	21:46
G	England, London	05:11	05:44	18:02	18:36
GD	Isle of Man	05:25	06:01	18:21	18:56
GI	Northern Ireland, Belfast	05:30	06:06	18:26	19:02
GJ	Jersey	05:21	05:53	18:10	18:42
GM	Scotland, Glasgow	05:23	06:00	18:20	18:57
	Aberdeen	05:12	05:50	18:12	18:50
GU	Guernsey	05:23	05:55	18:12	18:44
GW	Wales, Cardiff	05:23	05:56	18:15	18:48
H4	Solomon Islands, Honiara	18:49	19:10	07:16	07:37
HA	Hungary, Budapest	03:57	04:28	16:44	17:15
HB	Switzerland, Bern	04:44	05:15	17:31	18:01
HB0	Liechtenstein	04:36	05:06	17:22	17:53
HC	Ecuador, Quito	10:41	11:01	23:08	23:29
HC8	Galapagos Is.	11:31	11:51	23:58	00:19
HH	Haiti, Port-Au-Prince	10:16	10:37	22:46	23:08
HI	Dominican Republic, Santo Domingo	10:06	10:28	22:37	22:59
HK	Colombia, Bogota	10:25	10:45	22:53	23:13
HK0	Malpelo Is.	10:53	11:14	23:21	23:41
HK0	San Adreas	10:54	11:15	23:24	23:45
HL	Korea, Seoul	20:52	21:18	09:31	09:57
HP	Panama, Panama	10:46	11:07	23:15	23:35
HR	Honduras, Tegucigalpa	11:16	11:37	23:46	00:07
HS	Thailand, Bangkok	22:45	23:07	11:15	11:36
HV	Vatican City	04:28	04:56	17:10	17:38
HZ, 7Z	Saudi Arabia, Dharan	02:04	02:27	14:38	15:01
	Mecca	02:46	03:09	15:18	15:40
	Riyadh	02:18	02:41	14:51	15:14
I	Italy, Rome	04:28	04:56	17:10	17:38
	Trieste	04:20	04:50	17:05	17:35
	Sicily	04:24	04:50	17:03	17:29
IS	Sardinia, Cagliari	04:43	05:10	17:23	17:50
J2	Djibouti, Djibouti	02:35	02:56	15:04	15:25
J3	Grenada	09:35	09:56	22:04	22:25
J5	Guinea-Bissau, Bissau	06:30	06:51	18:59	19:20
J6	St. Lucia	09:31	09:53	22:01	22:22
J7	Dominica	09:32	09:54	22:02	22:24
J8	St. Vincent	09:33	09:54	22:02	22:23
JA-JS	Japan, Tokyo	20:02	20:27	08:40	09:05
	Nagasaki	20:43	21:07	09:19	09:44
	Sapporo	19:51	20:20	08:34	09:03
JD1	Minami Torishima	19:09	19:32	07:42	08:04
JD1	Ogasawara, Kazan Is.	20:00	20:23	08:34	08:57
JT	Mongolia, Ulan Bator	22:06	22:37	10:53	11:24
JW	Svalbard, Spitsbergen	02:20	04:16	17:21	19:17
JX	Jan Mayen	05:02	06:07	18:45	19:50
JY	Jordan, Amman	02:59	03:23	15:35	15:59
KC4	Antarctica, Byrd Station	11:45	13:50	01:55	04:00
	McMurdo Sound	17:04	18:44	06:48	08:28
	Palmer Station	09:18	10:07	22:10	22:59
KC6	Belau, Yap	20:15	20:36	08:44	09:05
	Koror	20:30	20:51	08:58	09:19
KG4	Guantanamo Bay	10:27	10:49	22:58	23:20
KH0	Mariana Is., Saipan	19:44	20:05	08:14	08:35
KH1	Baker, Howland Is.	17:13	17:33	05:40	06:01
KH2	Guam, Agana	19:48	20:09	08:18	08:39
KH3	Johnston Is.	16:41	17:02	05:11	05:33
KH4	Midway Is.	17:13	17:37	05:48	06:11
KH5	Palmyra Is.	16:17	16:37	04:45	05:06
KH5K	Kingman Reef	16:19	16:40	04:48	05:08
KH6	Hawaii, Hilo	15:47	16:08	04:18	04:40
	Honolulu	15:57	16:20	04:29	04:51
KH7	Kure Is.	17:17	17:41	05:52	06:15
KH8	American Samoa, Pago Pago	16:52	17:13	05:19	05:40
KH9	Wake Is.	18:20	18:42	06:51	07:13
KL7	Alaska, Adak	16:56	17:30	05:48	06:22
	Anchorage	14:57	15:40	04:05	04:48
	Fairbanks	14:41	15:30	03:58	04:48
	Juneau	13:59	14:39	03:01	03:41
	Nome	15:52	16:40	05:08	05:57
KP1	Navassa Is.	10:26	10:48	22:57	23:19
KP2	Virgin Islands, Charlotte Amalie	09:46	10:08	22:17	22:39
KP4	Puerto Rico, San Juan	09:51	10:13	22:22	22:44
KP5	Desecheo Is.	09:56	10:18	22:27	22:49
LA-LJ	Norway, Oslo	04:16	04:58	17:22	18:03
LU	Argentina, Buenos Aires	09:19	09:44	21:48	22:13
LX	Luxembourg	04:47	05:19	17:37	18:09
LY	Lithuania, Vilna	03:25	04:01	16:21	16:56
LZ	Bulgaria, Sofia	03:44	04:12	16:27	16:55
OA	Peru, Lima	10:37	10:58	23:04	23:25
OD	Lebanon, Beirut	03:00	03:25	15:37	16:02
OE	Austria, Vienna	04:08	04:39	16:56	17:27
OH	Finland, Helsinki	03:19	04:01	16:25	17:07
OH0	Aland Is.	03:39	04:21	16:45	17:27
OJ0	Market Reef	03:43	04:25	16:49	17:31
OK,OL	Czech Rep., Prague	04:14	04:46	17:04	17:36
OM	Slovak, Rep., Bratislava	04:05	04:36	16:53	17:24
ON	Belgium, Brussels	04:53	05:26	17:44	18:17
OX	Greenland, Godthaab	08:18	09:06	21:33	22:21
	Thule	08:26	10:01	22:55	00:29
OY	Faroe Islands, Torshavn	05:23	06:07	18:33	19:17
OZ	Denmark, Copenhagen	04:15	04:52	17:13	17:49
P2	Papua New Guinea, Madang	19:46	20:07	08:13	08:34
	Port Moresby	19:40	20:01	08:07	08:28
P4	Aruba, Oranjestad	10:03	10:25	22:33	22:54
PA-PI	Netherlands, Amsterdam	04:50	05:24	17:42	18:16
PJ2,4,9	Netherlands Antilles, Willemstad	10:08	10:29	22:37	22:58
PJ5-8	St. Maarten and Saba, St. Maarten	09:39	10:01	22:10	22:32
PY	Brazil, Brasilia	08:40	09:02	21:07	21:28
	Rio De Janeiro	08:21	08:43	20:48	21:10
	Natal	07:50	08:10	20:17	20:37
	Manaus	09:30	09:50	21:57	22:17
	Porto Alegre	08:51	09:15	21:20	21:43
PY0	Fernando De Noronha	07:38	07:59	20:05	20:26
PY0	St. Peter & St. Paul Rocks	07:26	07:47	19:54	20:14
PY0	Trindade & Martin Vaz Is., Trindade	07:25	07:47	19:52	20:14
PZ	Suriname, Paramaribo	09:09	09:30	21:37	21:58
R1FJ_	Franz Josef Land	23:31	01:44	14:57	17:11
R1MV_	Malyj Vysotskij Is.	03:04	03:46	16:10	16:53
S0	Western Sahara, Smara	06:10	06:33	18:43	19:07
S2	Bangladesh, Dacca	23:24	23:46	11:56	12:19
S5	Slovenia, Ljubljana	04:17	04:47	17:03	17:32
S7	Seychelles, Victoria	01:47	02:08	14:14	14:35
S9	Sao Tome	05:02	05:23	17:29	17:50
SM	Sweden, Stockholm	03:48	04:29	16:52	17:33
SP	Poland, Krakow	03:52	04:24	16:41	17:14
	Warsaw	03:45	04:19	16:38	17:12
ST	Sudan, Khartoum	03:17	03:38	15:47	16:08
ST0	Southern Sudan, Juba	03:22	03:43	15:50	16:11
SU	Egypt, Cairo	03:18	03:42	15:53	16:17
SV	Greece, Athens	03:45	04:11	16:24	16:51
SV/A	Mount Athos	03:42	04:09	16:22	16:49
SV5	Dodecanese, Rhodes	03:28	03:54	16:06	16:32
SV9	Crete	03:40	04:06	16:18	16:43
T2	Tuvalu, Funafuti	17:34	17:55	06:01	06:22

Pref	State/Province/Country/City	Tlight	Sunrise	Sunset	Tlight
T30	West Kiribati, Bonriki	17:56	18:16	06:23	06:44
T31	Central Kiribati, Kanton	16:56	17:16	05:23	05:43
T32	East Kiribati, Christmas Is.	15:58	16:19	04:26	04:46
T33	Banaba Is.	18:11	18:32	06:38	06:59
T5	Somalia, Mogadishu	02:27	02:48	14:55	15:15
T7	San Marino	04:27	04:56	17:11	17:40
T9	Bosnia-Herzegovina, Sarajevo	04:03	04:32	16:47	17:15
TA	Turkey, Ankara	03:07	03:34	15:48	16:15
	Istanbul	03:22	03:50	16:04	16:31
TF	Iceland, Reykjavik	06:19	07:07	19:34	20:22
TG	Guatemala, Guatemala City	11:29	11:50	23:59	00:20
TI	Costa Rica, San Jose	11:04	11:25	23:33	23:54
TI9	Cocos Is.	11:16	11:37	23:44	00:05
TJ	Cameroon, Yaounde	04:42	05:03	17:10	17:31
TK	Corsica	04:42	05:10	17:24	17:52
TL	Central African Republic, Bangui	04:14	04:35	16:42	17:03
TN	Congo, Brazzaville	04:28	04:48	16:55	17:15
TR	Gabon, Libreville	04:51	05:11	17:18	17:39
TT	Chad, N'Djamena	04:27	04:49	16:57	17:18
TU	Cote d'Ivoire, Abidjan	05:44	06:05	18:12	18:33
TY	Benin, Porto Novo	05:18	05:39	17:46	18:07
TZ	Mali, Bamako	05:59	06:21	18:29	18:50
UA	Russia, European, Leningrad (UA1)	02:58	03:40	16:03	16:45
	Archangel (UA1)	02:08	02:57	15:25	16:14
	Murmansk (UA1)	02:24	03:23	15:57	16:56
	Moscow (UA3)	02:35	03:12	15:33	16:10
	Kuybyshev (UA4)	01:48	02:23	14:42	15:17
	Rostov (UA6)	02:36	03:06	15:23	15:54
UA2	Kaliningrad	03:44	04:21	16:41	17:17
UA9,0	Russia, Asiatic, Novosibirsk (UA9)	23:35	00:11	12:31	13:07
	Perm (UA9)	01:17	01:56	14:19	14:58
	Omsk (UA9)	00:13	00:49	13:09	13:45
	Norilsk (UA0)	22:43	23:43	12:18	13:17
	Irkutsk (UA0)	22:12	22:46	11:05	11:39
	Vladivostok (UA0)	20:29	20:58	09:12	09:41
	Petropavlovsk (UA0)	18:34	19:08	07:27	08:02
	Khabarovsk (UA0)	20:13	20:44	09:01	09:32
	Krasnoyarsk (UA0)	22:54	23:31	11:52	12:29
	Yakutsk (UA0)	20:17	21:01	09:27	10:11
	Wrangel Island (UA0)	16:27	17:32	06:10	07:15
	Kyzyl (UA0Y)	22:52	23:26	11:44	12:17
UJ-UM	Uzbekistan, Bukhoro	01:01	01:28	13:42	14:09
	Tashkent	00:41	01:08	13:23	13:50
UN-UQ	Kazakhstan, Alma-Ata	00:09	00:38	12:52	13:21
UR-UZ,EM-EO	Ukraine, Kiev	03:09	03:42	16:00	16:32
V2	Antigua & Barbuda, St. Johns	09:34	09:55	22:04	22:26
V3	Belize, Belmopan	11:22	11:43	23:52	00:14
V4	St. Kitts & Nevis	09:37	09:59	22:07	22:29
V5	Namibia, Windhoek	04:20	04:42	16:47	17:09
V6	Micronesia, Ponape	18:55	19:16	07:23	07:44
V7	Marshall Islands, Kwajalein	18:19	18:40	06:47	07:08
V8	Brunei, Bandar Seri Begawan	21:49	22:09	10:17	10:37
VK	Australia, Canberra (VK1)	19:29	19:54	07:58	08:24
	Sydney (VK2)	19:21	19:46	07:50	08:15
	Melbourne (VK3)	19:45	20:11	08:15	08:41
	Brisbane (VK4)	19:15	19:38	07:43	08:06
	Adelaide (VK5)	20:11	20:36	08:40	09:05
	Perth (VK6)	21:43	22:07	10:12	10:36
	Hobart, Tasmania (VK7)	19:33	20:02	08:05	08:33
	Darwin (VK8)	20:45	21:06	09:12	09:33
VK0	Heard Is.	00:23	00:57	13:01	13:35
VK0	Macquarie Is.	18:40	19:16	07:19	07:55
VK9C	Cocos-Keeling Is.	23:02	23:23	11:28	11:49
VK9L	Lord Howe Is.	18:50	19:14	07:18	07:43
VK9M	Mellish Reef	19:05	19:27	07:32	07:54
VK9N	Norfolk Is.	18:15	18:39	06:43	07:06
VK9W	Willis Is.	19:31	19:52	07:57	08:19
VK9X	Christmas Is.	22:26	22:47	10:53	11:14
VP2E	Anguilla	09:38	10:00	22:09	22:31
VP2M	Montserrat	09:35	09:57	22:06	22:27
VP2V	British Virgin Is., Tortola	09:45	10:07	22:16	22:37

Pref	State/Province/Country/City	Tlight	Sunrise	Sunset	Tlight
VP5	Turks & Caicos Islands, Grand Turk	10:11	10:33	22:42	23:04
VP8	Falkland Islands, Stanley	09:09	09:43	21:46	22:19
VP8	So. Georgia Is.	07:43	08:18	20:21	20:57
VP8	So. Orkney Is.	08:11	08:53	20:56	21:38
VP8	So. Sandwich Islands, Saunders Is.	06:59	07:38	19:41	20:20
VP8	So. Shetland Is., King George Is.	09:00	09:44	21:47	22:32
VP9	Bermuda	09:41	10:06	22:17	22:42
VQ9	Chagos, Diego Garcia	00:39	01:00	13:06	13:27
VR2,VS6	Hong Kong	21:48	22:11	10:20	10:43
VR6	Pitcairn Is.	14:08	14:31	02:35	02:58
VU	India, Bombay	00:35	00:57	13:06	13:28
	Calcutta	23:32	23:54	12:04	12:26
	New Delhi	00:15	00:38	12:49	13:13
	Bangalore	00:17	00:38	12:46	13:08
VU	Andaman Islands, Port Blair	23:16	23:37	11:45	12:07
VU	Laccadive Is.	00:36	00:57	13:05	13:26
XE	Mexico, Mexico City (XE1)	12:03	12:24	00:34	00:56
	Chihuahua (XE2)	12:28	12:51	01:02	01:26
	Merida (XE3)	11:25	11:47	23:56	00:18
XF4	Revilla Gigedo	12:52	13:14	01:23	01:45
XT	Burkina Faso, Ouagadougou	05:34	05:55	18:03	18:24
XU	Cambodia, Phnom Penh	22:28	22:49	10:57	11:19
XW	Laos, Viangchan	22:36	22:58	11:07	11:28
XX9	Macao	21:51	22:13	10:23	10:45
XZ	Myanmar, Yangon	23:03	23:24	11:33	11:55
YA	Afghanistan, Kandahar	01:00	01:24	13:35	13:59
	Kabul	00:45	01:10	13:22	13:47
YB-YD	Indonesia, Jakarta	22:22	22:42	10:49	11:09
	Medan	22:54	23:14	11:21	11:42
	Pontianak	22:11	22:32	10:39	10:59
	Jayapura	20:06	20:27	08:33	08:54
YI	Iraq, Baghdad	02:24	02:49	15:01	15:25
YJ	Vanuatu, Port Vila	18:15	18:37	06:42	07:04
YK	Syria, Damascus	02:57	03:22	15:33	15:58
YL	Latvia, Riga	03:27	04:06	16:27	17:05
YN	Nicaragua, Managua	11:11	11:33	23:41	00:02
YO	Romania, Bucharest	03:32	04:01	16:16	16:45
YS	El Salvador, San Salvador	11:24	11:45	23:54	00:15
YU	Yugoslavia, Belgrade	03:54	04:23	16:38	17:08
YV	Venezuela, Caracas	09:56	10:17	22:25	22:46
YV0	Aves Is.	09:42	10:03	22:12	22:33
Z2	Zimbabwe, Harare	03:24	03:46	15:51	16:13
Z3	Macedonia, (ex Yugoslav), Skopje	03:52	04:20	16:34	17:02
ZA	Albania, Tirane	03:59	04:26	16:41	17:08
ZB2	Gibraltar	05:42	06:08	18:21	18:46
ZC4	British Cyprus	03:09	03:35	15:47	16:12
ZD7	St. Helena	05:52	06:14	18:19	18:40
ZD8	Ascension Is.	06:26	06:47	18:53	19:14
ZD9	Tristan da Cunha	06:14	06:40	18:44	19:10
ZF	Cayman Is.	10:51	11:13	23:22	23:44
ZK1	No. Cook Is., Manihiki	16:13	16:34	04:40	05:01
ZK1	So. Cook Is., Rarotonga	16:07	16:29	04:34	04:56
ZK2	Niue	16:44	17:06	05:11	05:33
ZK3	Tokelaus, Atafu	17:00	17:21	05:26	05:47
ZL	New Zealand, Auckland (ZL1)	17:46	18:12	06:15	06:41
	Wellington (ZL2)	17:44	18:12	06:15	06:43
	Christchurch (ZL3)	17:52	18:21	06:24	06:53
	Dunedin (ZL4)	17:59	18:29	06:32	07:02
ZL5	Antarctica, Scott Base	17:03	18:45	06:49	08:31
ZL7	Chatham Is.	17:08	17:37	05:40	06:09
ZL8	Kermadec Is.	17:18	17:42	05:46	06:10
ZL9	Auckland & Campbell Is., Auckland	18:12	18:45	06:48	07:21
	Campbell Is.	18:01	18:35	06:38	07:12
ZP	Paraguay, Asuncion	09:18	09:41	21:46	22:09
ZS	South Africa, Cape Town (ZS1)	04:12	04:37	16:41	17:06
	Port Elizabeth (ZS2)	03:43	04:08	16:12	16:37
	Bloemfontein (ZS4)	03:42	04:06	16:10	16:34
	Durban (ZS5)	03:23	03:47	15:51	16:15
	Johannesburg (ZS6)	03:35	03:58	16:03	16:26
ZS8	Marion Is.	02:50	03:20	15:23	15:53

Pref	State/Province/Country/City	Tlight	Sunrise	Sunset	Tlight
VE1	New Brunswick, Saint John	10:00	10:29	21:54	22:24
	Nova Scotia, Halifax	09:50	10:19	21:45	22:14
	Prince Edward Island, Charlottetown	09:48	10:18	21:42	22:12
VE2	Quebec, Montreal	10:30	10:59	22:24	22:54
	Quebec City	10:20	10:51	22:14	22:44
VE3	Ontario, London	11:00	11:28	22:57	23:25
	Ottawa	10:38	11:08	22:33	23:02
	Sudbury	11:00	11:30	22:53	23:23
	Toronto	10:53	11:21	22:49	23:17
VE4	Manitoba, Winnipeg	12:04	12:37	23:55	00:27
VE5	Saskatchewan, Regina	12:35	13:07	00:24	00:57
	Saskatoon	12:43	13:17	00:32	01:05
VE6	Alberta, Calgary	13:13	13:46	01:02	01:35
	Edmonton	13:11	13:45	00:57	01:32
VE7	British Columbia, Prince George	13:48	14:23	01:34	02:09
	Prince Rupert	14:18	14:53	02:04	02:39
	Vancouver	13:48	14:20	01:39	02:11
VE8	Northwest Territories, Yellowknife	13:15	14:00	00:50	01:35
	Resolute	11:56	13:16	22:59	00:19
VY1	Yukon, Whitehorse	14:38	15:20	02:16	02:58
VO1	Newfoundland, St. John's	09:07	09:37	20:59	21:30
VO2	Labrador, Goose Bay	09:38	10:13	21:25	22:00
W1	Connecticut, Hartford	10:26	10:53	22:23	22:51
	Maine, Bangor	10:10	10:40	22:06	22:35
	Portland	10:16	10:45	22:12	22:41
	Massachusetts, Boston	10:19	10:47	22:16	22:44
	New Hampshire, Concord	10:21	10:49	22:17	22:46
	Rhode Island, Providence	10:20	10:48	22:18	22:46
	Vermont, Montpelier	10:26	10:55	22:21	22:50
W2	New Jersey, Atlantic City	10:32	10:59	22:31	22:58
	New York, Albany	10:30	10:58	22:27	22:55
	Buffalo	10:51	11:19	22:47	23:16
	New York City	10:31	10:58	22:29	22:56
	Syracuse	10:40	11:08	22:36	23:05
W3	Delaware, Wilmington	10:36	11:03	22:36	23:03
	District of Columbia, Washington	10:42	11:09	22:42	23:09
	Maryland, Baltimore	10:41	11:07	22:40	23:07
	Pennsylvania, Harrisburg	10:42	11:09	22:41	23:08
	Philadelphia	10:35	11:02	22:34	23:01
	Pittsburgh	10:54	11:22	22:53	23:20
	Scranton	10:37	11:05	22:35	23:03
W4	Alabama, Montgomery	11:18	11:43	23:23	23:47
	Florida, Jacksonville	10:59	11:23	23:05	23:29
	Miami	10:52	11:15	23:01	23:24
	Pensacola	11:21	11:45	23:27	23:51
	Georgia, Atlanta	11:11	11:36	23:14	23:39
	Savannah	10:57	11:22	23:02	23:27
	Kentucky, Lexington	11:12	11:38	23:13	23:39
	Louisville	11:17	11:44	23:18	23:44
	North Carolina, Charlotte	10:57	11:22	22:59	23:25
	Raleigh	10:48	11:14	22:50	23:16
	Wilmington	10:45	11:10	22:48	23:13
	South Carolina, Columbia	10:57	11:22	23:01	23:26
	Tennessee, Knoxville	11:09	11:35	23:11	23:37
	Memphis	11:34	11:59	23:37	00:02
	Nashville	11:21	11:47	23:23	23:48
	Virginia, Norfolk	10:39	11:05	22:40	23:06
	Richmond	10:44	11:10	22:44	23:11
W5	Arkansas, Little Rock	11:43	12:08	23:46	00:11
	Louisiana, New Orleans	11:33	11:57	23:39	00:03
	Shreveport	11:48	12:12	23:52	00:17
	Mississippi, Jackson	11:34	11:58	23:38	00:03
	New Mexico, Albuquerque	12:40	13:06	00:43	01:08
	Oklahoma, Oklahoma City	12:03	12:29	00:06	00:31
	Texas, Abilene	12:12	12:36	00:16	00:41
	Amarillo	12:21	12:46	00:23	00:49
	Dallas	12:00	12:25	00:04	00:29
	El Paso	12:39	13:03	00:44	01:08
	San Antonio	12:06	12:30	00:13	00:37
W6	California, Los Angeles	13:26	13:51	01:29	01:54
	San Francisco	13:44	14:10	01:44	02:11
W7	Arizona, Flagstaff	13:00	13:26	01:03	01:28
	Phoenix	13:01	13:26	01:05	01:30
	Idaho, Boise	13:20	13:48	01:16	01:45
	Pocatello	13:05	13:33	01:02	01:30
	Montana, Billings	12:49	13:19	00:44	01:13
	Butte	13:05	13:35	01:00	01:29
	Great Falls	13:01	13:32	00:54	01:24
	Nevada, Las Vegas	13:14	13:40	01:16	01:42
	Reno	13:33	14:00	01:33	02:00
	Oregon, Portland	13:46	14:16	01:41	02:10
	Utah, Salt Lake City	13:02	13:30	01:01	01:28
	Washington, Seattle	13:45	14:16	01:38	02:08
	Spokane	13:25	13:56	01:18	01:49
	Wyoming, Cheyenne	12:34	13:01	00:32	00:59
	Sheridan	12:43	13:12	00:38	01:08
W8	Michigan, Detroit	11:07	11:35	23:04	23:32
	Grand Rapids	11:18	11:46	23:14	23:43
	Sault Ste. Marie	11:13	11:43	23:07	23:37
	Traverse City	11:18	11:47	23:13	23:42
	Ohio, Cincinnati	11:12	11:39	23:12	23:39
	Cleveland	11:01	11:29	22:59	23:27
	Columbus	11:06	11:33	23:05	23:32
	West Virginia, Charleston	11:00	11:27	23:01	23:27
W9	Illinois, Chicago	11:25	11:53	23:23	23:51
	Indiana, Indianapolis	11:19	11:46	23:18	23:45
	Wisconsin, Green Bay	11:27	11:56	23:23	23:52

Pref	State/Province/Country/City	Tlight	Sunrise	Sunset	Tlight
	Milwaukee	11:27	11:55	23:23	23:52
W0	Colorado, Denver	12:34	13:01	00:34	01:01
	Grand Junction	12:49	13:15	00:48	01:15
	Iowa, Des Moines	11:49	12:17	23:47	00:15
	Kansas, Pratt	12:09	12:35	00:10	00:36
	Wichita	12:03	12:29	00:04	00:30
	Minnesota, Duluth	11:44	12:14	23:37	00:08
	Minneapolis	11:48	12:18	23:43	00:13
	Missouri, Columbia	11:43	12:10	23:43	00:10
	Kansas City	11:53	12:19	23:52	00:19
	St. Louis	11:35	12:01	23:35	00:02
	Nebraska, North Platte	12:18	12:45	00:16	00:43
	Omaha	11:58	12:26	23:56	00:24
	North Dakota, Fargo	12:03	12:33	23:56	00:26
	South Dakota, Rapid City	12:28	12:57	00:24	00:53
-		05:23	05:44	17:51	18:12
1A0	S.M.O.M	04:45	05:13	16:43	17:11
1S	Spratly Is.	21:59	22:20	10:20	10:41
3A	Monaco	05:05	05:34	17:02	17:30
3B6	Agalega	01:32	01:54	14:08	14:30
3B7	St. Brandon	01:17	01:38	13:58	14:20
3B8	Mauritius	01:24	01:46	14:09	14:31
3B9	Rodriguez Is.	01:00	01:22	13:45	14:07
3C	Equatorial Guinea, Bata	04:45	05:06	17:11	17:32
	Malabo	04:50	05:10	17:14	17:35
3C0	Annobon Is.	05:00	05:21	17:29	17:50
3D2	Fiji Is., Suva	17:21	17:43	06:05	06:26
3D2	Conway Reef	17:35	17:58	06:22	06:44
3D2	Rotuma Is.	17:27	17:48	06:05	06:26
3DA	Swaziland, Mbabane	03:05	03:28	15:58	16:21
3V	Tunisia, Tunis	04:53	05:19	16:54	17:20
3W, XV	Vietnam, Ho Chi Minh City (Saigon)	22:20	22:41	10:40	11:01
	Hanoi	22:27	22:49	10:39	11:02
3X	Guinea, Conakry	06:21	06:43	18:42	19:03
3Y	Bouvet	04:20	04:57	18:10	18:47
3Y	Peter I Is.	09:40	10:43	00:56	02:00
4J, 4K	Azerbaijan, Baku	02:15	02:42	14:14	14:41
4L	Georgia, Tbilisi	02:36	03:03	14:33	15:01
4S	Sri Lanka, Colombo	00:06	00:27	12:29	12:49
4U	ITU Geneva	05:11	05:41	17:05	17:35
4U	United Nations Hq.	10:31	10:58	22:29	22:56
4X, 4Z	Israel, Jerusalem	03:12	03:36	15:17	15:41
5A	Libya, Tripoli	04:43	05:07	16:47	17:12
	Benghazi	04:13	04:37	16:18	16:42
5B	Cyprus, Nicosia	03:20	03:45	15:22	15:48
5H	Tanzania, Dar es Salaam	02:42	03:03	15:15	15:36
5N	Nigeria, Lagos	05:12	05:33	17:35	17:56
5R	Madagascar, Antananarivo	02:04	02:26	14:48	15:10
5T	Mauritania, Nouakchott	06:33	06:55	18:48	19:10
5U	Niger, Niamey	05:20	05:41	17:38	17:59
5V	Togo, Lome	05:21	05:41	17:44	18:05
5W	Western Samoa, Apia	16:45	17:06	05:23	05:45
5X	Uganda, Kampala	03:13	03:34	15:41	16:01
5Z	Kenya, Nairobi	04:16	04:36	16:44	17:05
6W	Senegal, Dakar	06:38	07:00	18:55	19:17
6Y	Jamaica, Kingston	10:36	10:58	22:51	23:13
7O	Yemen, Aden	02:28	02:49	14:46	15:07
	Sanaa	02:32	02:53	14:48	15:10
7P	Lesotho, Maseru	03:17	03:41	16:14	16:38
7Q	Malawi, Lilongwe	03:02	03:23	15:41	16:03
	Blantyre	02:56	03:18	15:37	15:59
7X	Algeria, Algiers	05:22	05:48	17:23	17:49
8P	Barbados, Bridgetown	09:26	09:47	21:44	22:05
8Q	Maldive Is.	00:31	00:52	12:56	13:16
8R	Guyana, Georgetown	09:19	09:39	21:41	22:02
9A	Croatia, Zagreb	04:31	05:01	16:26	16:55
9G	Ghana, Accra	05:26	05:47	17:49	18:10
9H	Malta	04:36	05:02	16:38	17:04
9J	Zambia, Lusaka	03:23	03:45	16:04	16:25
9K	Kuwait	02:21	02:45	14:28	14:52
9L	Sierra Leone, Freetown	06:19	06:40	18:40	19:01
9M2	West Malaysia, Kuala Lumpur	22:38	22:59	11:03	11:24
9M6,8	East Malaysia, Sabah, Sandakan(9M6)	21:33	21:54	09:56	10:17
	Sarawak, Kuching (9M8)	22:03	22:23	10:29	10:50
9N	Nepal, Kathmandu	23:51	00:14	11:59	12:22
9Q	Zaire, Kinshasa	04:20	04:41	16:51	17:12
	Kisangani	03:43	04:03	16:10	16:31
	Lubumbashi	03:28	03:49	16:05	16:27
9U	Burundi, Bujumbura	03:25	03:46	15:55	16:16
9V	Singapore	22:29	22:49	10:55	11:16
9X	Rwanda, Kigali	03:22	03:43	15:51	16:12
9Y	Trinidad & Tobago , Port of Spain	09:32	09:53	21:52	22:13
A2	Botswana, Gaborone	03:27	03:50	16:18	16:41
A3	Tonga, Nukualofa	16:54	17:16	05:40	06:03
A4	Oman, Musqat	01:36	01:59	13:47	14:10
A5	Bhutan, Thimbu	23:34	23:57	11:42	12:06
A6	United Arab Emirates, Abu Dhabi	01:54	02:17	14:04	14:27
A7	Qatar, Ad-Dawhah	02:05	02:28	14:15	14:38
A9	Bahrein, Al-Manamah	02:09	02:32	14:18	14:41
AP	Pakistan, Karachi	01:03	01:26	13:13	13:35
AP	Islamabad	00:40	01:05	12:44	13:09
BV	Taiwan, Taipei	21:25	21:48	09:35	09:58
BY	Peoples Rep. of China, Beijing	21:49	22:16	09:48	10:15
	Harbin	21:09	21:38	09:03	09:33
	Shanghai	21:27	21:51	09:32	09:56
	Fuzhou	21:34	21:57	09:43	10:06
	Xian	22:18	22:43	10:21	10:46

Pref	State/Province/Country/City	Tlight	Sunrise	Sunset	Tlight
	Chongqing	22:26	22:50	10:33	10:57
	Chengdu	22:36	23:00	10:42	11:06
	Lhasa	23:27	23:51	11:34	11:58
	Urumqi	23:45	00:13	11:41	12:09
	Kashi	00:30	00:57	12:30	12:57
C2	Nauru	18:16	18:36	06:43	07:04
C3	Andorra	05:29	05:57	17:26	17:54
C5	The Gambia, Banjul	06:35	06:56	18:52	19:14
C6	Bahamas, Nassau	10:41	11:04	22:50	23:13
C9	Mozambique, Maputo	02:59	03:23	15:51	16:15
	Mozambique	02:34	02:55	15:14	15:36
CE	Chile, Santiago	09:47	10:12	22:49	23:14
CE0	Easter Island	12:27	12:50	01:20	01:43
CE0	Juan Fernandez	10:19	10:44	23:21	23:46
CE0	San Felix	10:30	10:53	23:22	23:46
CM,CO	Cuba, Havana	11:00	11:23	23:11	23:34
CN	Morocco, Casablanca	06:03	06:28	18:07	18:32
CP	Bolivia, La Paz	09:49	10:11	22:31	22:53
CT	Portugal, Lisbon	06:11	06:38	18:11	18:38
CT3	Madeira Islands, Funchal	06:40	07:05	18:45	19:10
CU	Azores , Ponta Delgada	07:17	07:43	19:18	19:44
CX	Uruguay, Montevideo	08:47	09:13	21:52	22:17
CY0	Sable Is.	09:35	10:04	21:31	22:00
CY9	St. Paul Is.	09:36	10:07	21:29	22:00
D2	Angola, Luanda	04:27	04:48	17:01	17:22
D4	Cape Verde, Praia	07:02	07:24	19:19	19:41
D6	Comoros, Moroni	02:23	02:45	15:01	15:22
DA-DL,Y2-Y9	Fed. Rep. of Germany, Bonn	05:08	05:41	16:58	17:31
	Berlin	04:43	05:17	16:31	17:05
DU	Philippines, Manilla	21:24	21:46	09:41	10:03
E3	Eritrea, Asmara	02:53	03:14	15:09	15:31
EA	Spain, Madrid	05:49	06:16	17:48	18:15
EA6	Balearic Is., Palma	05:24	05:51	17:23	17:50
EA8	Canary Is., Las Palmas	06:29	06:53	18:37	19:00
EA9	Ceuta & Melilla, Ceuta	05:55	06:20	17:57	18:22
	Melilla	05:45	06:11	17:48	18:13
EI	Ireland, Dublin	06:02	06:36	17:49	18:24
EK	Armenia, Yerevan	02:36	03:04	14:35	15:02
EL	Liberia, Monrovia	06:09	06:30	18:32	18:52
EP	Iran, Tehran	02:06	02:32	14:09	14:34
ER	Moldova, Kishinev	03:40	04:11	15:34	16:04
ES	Estonia, Tallinn	03:58	04:39	15:38	16:18
ET	Ethiopia, Addis Ababa	02:52	03:13	15:13	15:34
EU,EV,EW	Belarus, Minsk	03:46	04:21	15:33	16:08
EX	Kyrgyzstan, Frunze	00:37	01:05	12:33	13:02
EY	Tajikistan, Samarkand	01:07	01:34	13:06	13:33
	Dushanbe	00:59	01:26	12:59	13:25
EZ	Turkmenistan, Ashkhabad	01:40	02:07	13:41	14:07
F	France, Paris	05:27	05:58	17:18	17:50
FG	Guadeloupe	09:35	09:57	21:51	22:13
FJ, FS	St. Martin	09:42	10:03	21:56	22:18
FH	Mayotte	02:16	02:38	14:55	15:16
FK	New Caledonia, Noumia	18:06	18:29	06:54	07:17
FM	Martinique	09:32	09:54	21:49	22:11
FO	Clipperton Is.	12:44	13:05	01:04	01:25
FO	Fr. Polynesia, Tahiti	15:53	16:15	04:36	04:58
	Rurutu, (Austral Is.)	19:07	19:30	07:55	08:17
	Hiva Oa, (Marquesas Is.)	14:35	14:56	03:11	03:32
FP	St. Pierre & Miquelon, St. Pierre	09:20	09:50	21:13	21:43
FR/G	Glorioso	02:09	02:30	14:46	15:07
FR/J,E	Juan de Nova	02:24	02:46	15:06	15:28
	Europa	02:31	02:53	15:18	15:41
FR	Reunion	01:31	01:53	14:17	14:39
FR/T	Tromelin	01:38	02:00	14:20	14:41
FT8W	Crozet	01:21	01:52	14:47	15:17
FT8X	Kerguelen	00:07	00:40	13:41	14:14
FT8Y	Antarctica, Dumont D'Urville	18:31	19:28	09:27	10:24
FT8Z	Amsterdam & St. Paul Is., Amsterdam	23:49	00:16	12:58	13:25
FW	Wallis & Futuna Is., Wallis	17:03	17:24	05:41	06:03
FY	Fr. Guiana, Cayenne	08:54	09:15	21:18	21:39
G	England, London	05:37	06:10	17:26	17:59
GD	Isle of Man	05:55	06:30	17:41	18:16
GI	Northern Ireland, Belfast	06:00	06:36	17:46	18:22
GJ	Jersey	05:45	06:17	17:36	18:08
GM	Scotland, Glasgow	05:54	06:31	17:38	18:15
	Aberdeen	05:45	06:24	17:28	18:06
GU	Guernsey	05:47	06:19	17:38	18:10
GW	Wales, Cardiff	05:49	06:22	17:38	18:11
H4	Solomon Islands, Honiara	18:39	19:00	07:14	07:36
HA	Hungary, Budapest	04:19	04:50	16:12	16:43
HB	Switzerland, Bern	05:06	05:36	16:59	17:29
HB0	Liechtenstein	04:57	05:28	16:50	17:21
HC	Ecuador, Quito	10:35	10:56	23:03	23:24
HC8	Galapagos Is.	11:25	11:46	23:53	00:14
HH	Haiti, Port-Au-Prince	10:19	10:40	22:33	22:55
HI	Dominican Republic, Santo Domingo	10:09	10:31	22:24	22:46
HK	Colombia, Bogota	10:21	10:42	22:45	23:06
HK0	Malpelo Is.	10:49	11:10	23:14	23:34
HK0	San Adreas	10:54	11:16	23:13	23:34
HL	Korea, Seoul	21:06	21:32	09:07	09:33
HP	Panama, Panama	10:45	11:06	23:05	23:26
HR	Honduras, Tegucigalpa	11:17	11:38	23:34	23:56
HS	Thailand, Bangkok	22:46	23:07	11:04	11:25
HV	Vatican City	04:45	05:13	16:42	17:10
HZ, 7Z	Saudi Arabia, Dharan	02:11	02:35	14:20	14:43
	Mecca	02:51	03:13	15:03	15:25
	Riyadh	02:24	02:47	14:34	14:57

Pref	State/Province/Country/City	Tlight	Sunrise	Sunset	Tlight
I	Italy, Rome	04:45	05:13	16:42	17:10
	Trieste	04:40	05:10	16:35	17:04
	Sicily	04:38	05:04	16:39	17:05
IS	Sardinia, Cagliari	04:58	05:25	16:57	17:24
J2	Djibouti, Djibouti	02:35	02:56	14:54	15:15
J3	Grenada	09:35	09:56	21:53	22:15
J5	Guinea-Bissau, Bissau	06:30	06:51	18:49	19:10
J6	St. Lucia	09:32	09:53	21:49	22:11
J7	Dominica	09:34	09:55	21:50	22:12
J8	St. Vincent	09:33	09:54	21:51	22:12
JA-JS	Japan, Tokyo	20:14	20:40	08:17	08:42
	Nagasaki	20:53	21:18	08:58	09:22
	Sapporo	20:09	20:38	08:06	08:34
JD1	Minami Torishima	19:15	19:38	07:25	07:48
JD1	Ogasawara, Kazan Is.	20:08	20:31	08:16	08:39
JT	Mongolia, Ulan Bator	22:28	22:59	10:21	10:51
JW	Svalbard, Spitsbergen	04:30	06:20	15:07	16:57
JX	Jan Mayen	06:10	07:14	17:27	18:31
JY	Jordan, Amman	03:09	03:34	15:14	15:39
KC4	Antarctica, Byrd Station	——	11:12	04:22	——
	McMurdo Sound	14:06	16:39	08:42	11:15
	Palmer Station	08:17	09:09	22:58	23:50
KC6	Belau, Yap	20:14	20:35	08:34	08:55
	Koror	20:28	20:49	08:50	09:11
KG4	Guantanamo Bay	10:31	10:53	22:44	23:06
KH0	Mariana Is., Saipan	19:45	20:07	08:02	08:23
KH1	Baker, Howland Is.	17:08	17:28	05:35	05:55
KH2	Guam, Agana	19:49	20:10	08:06	08:28
KH3	Johnston Is.	16:43	17:05	04:58	05:20
KH4	Midway Is.	17:21	17:45	05:29	05:53
KH5	Palmyra Is.	16:14	16:35	04:37	04:58
KH5K	Kingman Reef	16:17	16:38	04:39	05:00
KH6	Hawaii, Hilo	15:50	16:12	04:04	04:26
	Honolulu	16:02	16:24	04:14	04:36
KH7	Kure Is.	17:26	17:49	05:33	05:57
KH8	American Samoa, Pago Pago	16:40	17:02	05:20	05:41
KH9	Wake Is.	18:23	18:45	06:37	06:59
KL7	Alaska, Adak	17:23	17:56	05:11	05:45
	Anchorage	15:37	16:20	03:14	03:58
	Fairbanks	15:29	16:18	03:00	03:49
	Juneau	14:35	15:14	02:16	02:55
	Nome	16:39	17:27	04:11	04:59
KP1	Navassa Is.	10:29	10:51	22:44	23:06
KP2	Virgin Islands, Charlotte Amalie	09:49	10:11	22:03	22:25
KP4	Puerto Rico, San Juan	09:54	10:16	22:08	22:30
KP5	Desecheo Is.	09:59	10:21	22:14	22:36
LA-LJ	Norway, Oslo	04:54	05:36	16:33	17:15
LU	Argentina, Buenos Aires	08:56	09:22	22:00	22:26
LX	Luxembourg	05:11	05:43	17:02	17:34
LY	Lithuania, Vilna	03:55	04:30	15:40	16:16
LZ	Bulgaria, Sofia	04:02	04:30	15:59	16:27
OA	Peru, Lima	10:26	10:48	23:04	23:25
OD	Lebanon, Beirut	03:11	03:36	15:15	15:40
OE	Austria, Vienna	04:31	05:02	16:23	16:54
OH	Finland, Helsinki	03:57	04:39	15:36	16:18
OH0	Aland Is.	04:17	04:59	15:56	16:38
OJ0	Market Reef	04:21	05:03	16:00	16:42
OK,OL	Czech Rep., Prague	04:39	05:11	16:29	17:01
OM	Slovak, Rep., Bratislava	04:28	04:59	16:20	16:51
ON	Belgium, Brussels	05:19	05:52	17:08	17:41
OX	Greenland, Godthaab	09:04	09:52	20:36	21:24
	Thule	10:11	11:42	21:04	22:35
OY	Faroe Islands, Torshavn	06:05	06:49	17:40	18:25
OZ	Denmark, Copenhagen	04:46	05:23	16:31	17:08
P2	Papua New Guinea, Madang	19:39	20:00	08:10	08:31
	Port Moresby	19:31	19:52	08:06	08:27
P4	Aruba, Oranjestad	10:03	10:25	22:22	22:43
PA-PI	Netherlands, Amsterdam	05:17	05:51	17:05	17:39
PJ2,4,9	Netherlands Antilles, Willemstad	10:08	10:29	22:26	22:48
PJ5-8	St. Maarten and Saba, St. Maarten	09:42	10:04	21:57	22:19
PY	Brazil, Brasilia	08:28	08:49	21:09	21:30
	Rio De Janeiro	08:05	08:27	20:53	21:16
	Natal	07:42	08:03	20:14	20:35
	Manaus	09:23	09:44	21:53	22:14
	Porto Alegre	08:32	08:56	21:29	21:53
PY0	Fernando De Noronha	07:31	07:52	20:02	20:23
PY0	St. Peter & St. Paul Rocks	07:21	07:42	19:48	20:09
PY0	Trindade & Martin Vaz Is., Trindade	07:11	07:33	19:56	20:19
PZ	Suriname, Paramaribo	09:06	09:27	21:29	21:50
R1FJ_	Franz Josef Land	02:01	04:04	12:26	14:30
R1MV_	Malyj Vysotskij Is.	03:43	04:25	15:21	16:03
S0	Western Sahara, Smara	06:17	06:40	18:26	18:49
S2	Bangladesh, Dacca	23:29	23:52	11:40	12:03
S5	Slovenia, Ljubljana	04:37	05:07	16:32	17:01
S7	Seychelles, Victoria	01:39	02:00	14:11	14:31
S9	Sao Tome	04:57	05:17	17:24	17:45
SM	Sweden, Stockholm	04:25	05:05	16:05	16:45
SP	Poland, Krakow	04:16	04:48	16:07	16:39
	Warsaw	04:12	04:46	16:01	16:34
ST	Sudan, Khartoum	03:19	03:40	15:35	15:56
ST0	Southern Sudan, Juba	03:19	03:40	15:42	16:03
SU	Egypt, Cairo	03:27	03:51	15:33	15:57
SV	Greece, Athens	03:59	04:26	16:00	16:26
SV/A	Mount Athos	03:57	04:24	15:56	16:23
SV5	Dodecanese, Rhodes	03:41	04:07	15:43	16:08
SV9	Crete	03:53	04:18	15:55	16:21
T2	Tuvalu, Funafuti	17:25	17:46	06:00	06:21

Pref	State/Province/Country/City	Tlight	Sunrise	Sunset	Tlight
T30	West Kiribati, Bonriki	17:51	18:12	06:17	06:38
T31	Central Kiribati, Kanton	16:49	17:10	05:19	05:39
T32	East Kiribati, Christmas Is.	15:54	16:14	04:20	04:40
T33	Banaba Is.	18:06	18:26	06:33	06:54
T5	Somalia, Mogadishu	02:23	02:43	14:48	15:09
T7	San Marino	04:46	05:15	16:42	17:11
T9	Bosnia-Herzegovina, Sarajevo	04:22	04:50	16:17	16:46
TA	Turkey, Ankara	03:23	03:50	15:22	15:49
	Istanbul	03:39	04:06	15:37	16:04
TF	Iceland, Reykjavik	07:05	07:53	18:38	19:25
TG	Guatemala, Guatemala City	11:30	11:52	23:47	00:09
TI	Costa Rica, San Jose	11:03	11:24	23:23	23:44
TI9	Cocos Is.	11:13	11:34	23:37	23:58
TJ	Cameroon, Yaounde	04:39	05:00	17:03	17:24
TK	Corsica	04:59	05:27	16:56	17:24
TL	Central African Republic, Bangui	04:11	04:31	16:35	16:56
TN	Congo, Brazzaville	04:20	04:41	16:51	17:12
TR	Gabon, Libreville	04:45	05:06	17:13	17:33
TT	Chad, N'Djamena	04:27	04:49	16:46	17:07
TU	Cote d'Ivoire, Abidjan	05:41	06:02	18:05	18:26
TY	Benin, Porto Novo	05:15	05:36	17:38	17:59
TZ	Mali, Bamako	06:00	06:21	18:18	18:39
UA	Russia, European, Leningrad (UA1)	03:36	04:17	15:15	15:56
	Archangel (UA1)	02:55	03:44	14:27	15:15
	Murmansk (UA1)	03:25	04:23	14:47	15:45
	Moscow (UA3)	03:06	03:43	14:51	15:28
	Kuybyshev (UA4)	02:16	02:51	14:03	14:38
	Rostov (UA6)	02:58	03:28	14:50	15:21
UA2	Kaliningrad	04:15	04:51	16:00	16:36
UA9,0	Russia, Asiatic, Novosibirsk (UA9)	00:05	00:41	11:50	12:27
	Perm (UA9)	01:52	02:31	13:33	14:13
	Omsk (UA9)	00:43	01:19	12:28	13:05
	Norilsk (UA0)	23:45	00:44	11:06	12:05
	Irkutsk (UA0)	22:39	23:13	10:27	11:01
	Vladivostok (UA0)	20:47	21:16	08:44	09:12
	Petropavlovsk (UA0)	19:02	19:36	06:49	07:24
	Khabarovsk (UA0)	20:35	21:07	08:27	08:59
	Krasnoyarsk (UA0)	23:26	00:03	11:10	11:47
	Yakutsk (UA0)	20:59	21:43	08:34	09:19
	Wrangel Island (UA0)	17:35	18:39	04:52	05:56
	Kyzyl (UA0Y)	23:18	23:52	11:07	11:41
UJ-UM	Uzbekistan, Bukhoro	01:17	01:44	13:16	13:43
	Tashkent	00:57	01:25	12:56	13:23
UN-UQ	Kazakhstan, Alma-Ata	00:27	00:56	12:24	12:52
UR-UZ,EM-EO	Ukraine, Kiev	03:34	04:07	15:24	15:57
V2	Antigua & Barbuda, St. Johns	09:36	09:58	21:51	22:13
V3	Belize, Belmopan	11:24	11:46	23:39	00:01
V4	St. Kitts & Nevis	09:39	10:01	21:55	22:16
V5	Namibia, Windhoek	04:04	04:26	16:52	17:14
V6	Micronesia, Ponape	18:53	19:14	07:15	07:36
V7	Marshall Islands, Kwajalein	18:17	18:38	06:38	06:59
V8	Brunei, Bandar Seri Begawan	21:46	22:06	10:09	10:30
VK	Australia, Canberra (VK1)	19:06	19:31	08:11	08:36
	Sydney (VK2)	18:59	19:24	08:01	08:27
	Melbourne (VK3)	19:20	19:46	08:29	08:55
	Brisbane (VK4)	18:57	19:20	07:51	08:14
	Adelaide (VK5)	19:48	20:14	08:52	09:18
	Perth (VK6)	21:22	21:47	10:22	10:46
	Hobart, Tasmania (VK7)	19:05	19:33	08:23	08:52
	Darwin (VK8)	20:34	20:56	09:12	09:33
VK0	Heard Is.	23:44	00:19	13:29	14:04
VK0	Macquarie Is.	17:58	18:35	07:49	08:26
VK9C	Cocos-Keeling Is.	22:51	23:12	11:28	11:50
VK9L	Lord Howe Is.	18:29	18:54	07:28	07:53
VK9M	Mellish Reef	18:52	19:14	07:35	07:57
VK9N	Norfolk Is.	17:56	18:19	06:51	07:15
VK9W	Willis Is.	19:18	19:40	07:59	08:21
VK9X	Christmas Is.	22:16	22:37	10:52	11:13
VP2E	Anguilla	09:41	10:03	21:56	22:18
VP2M	Montserrat	09:38	09:59	21:53	22:15
VP2V	British Virgin Is., Tortola	09:48	10:10	22:02	22:24

Pref	State/Province/Country/City	Tlight	Sunrise	Sunset	Tlight
VP5	Turks & Caicos Islands, Grand Turk	10:15	10:37	22:27	22:50
VP8	Falkland Islands, Stanley	08:31	09:06	22:12	22:47
VP8	So. Georgia Is.	07:02	07:38	20:51	21:28
VP8	So. Orkney Is.	07:19	08:03	21:35	22:20
VP8	So. Sandwich Islands, Saunders Is.	06:13	06:53	20:16	20:56
VP8	So. Shetland Is., King George Is.	08:05	08:52	22:29	23:16
VP9	Bermuda	09:52	10:16	21:56	22:21
VQ9	Chagos, Diego Garcia	00:31	00:52	13:04	13:25
VR2,VS6	Hong Kong	21:53	22:16	10:05	10:27
VR6	Pitcairn Is.	13:51	14:14	02:42	03:05
VU	India, Bombay	00:38	01:00	12:52	13:14
	Calcutta	23:37	23:59	11:48	12:11
	New Delhi	00:23	00:47	12:31	12:54
	Bangalore	00:17	00:39	12:35	12:57
VU	Andaman Islands, Port Blair	23:16	23:37	11:35	11:56
VU	Laccadive Is.	00:35	00:56	12:55	13:16
XE	Mexico, Mexico City (XE1)	12:06	12:28	00:20	00:42
	Chihuahua (XE2)	12:36	13:00	00:43	01:07
	Merida (XE3)	11:29	11:51	23:41	00:04
XF4	Revilla Gigedo	12:56	13:17	01:09	01:31
XT	Burkina Faso, Ouagadougou	05:34	05:55	17:52	18:14
XU	Cambodia, Phnom Penh	22:28	22:49	10:47	11:08
XW	Laos, Viangchan	22:39	23:01	10:53	11:15
XX9	Macao	21:56	22:18	10:08	10:30
XZ	Myanmar, Yangon	23:05	23:27	11:20	11:42
YA	Afghanistan, Kandahar	01:09	01:33	13:15	13:39
	Kabul	00:56	01:22	13:00	13:25
YB-YD	Indonesia, Jakarta	22:14	22:34	10:46	11:07
	Medan	22:50	23:11	11:15	11:35
	Pontianak	22:06	22:27	10:34	10:54
	Jayapura	20:00	20:20	08:29	08:50
YI	Iraq, Baghdad	02:35	03:00	14:39	15:04
YJ	Vanuatu, Port Vila	18:02	18:24	06:45	07:07
YK	Syria, Damascus	03:08	03:33	15:12	15:37
YL	Latvia, Riga	04:01	04:39	15:43	16:21
YN	Nicaragua, Managua	11:11	11:33	23:30	23:51
YO	Romania, Bucharest	03:51	04:20	15:46	16:15
YS	El Salvador, San Salvador	11:25	11:46	23:42	00:04
YU	Yugoslavia, Belgrade	04:13	04:43	16:08	16:38
YV	Venezuela, Caracas	09:55	10:16	22:15	22:36
YV0	Aves Is.	09:43	10:05	22:00	22:21
Z2	Zimbabwe, Harare	03:11	03:33	15:54	16:16
Z3	Macedonia, (ex Yugoslav), Skopje	04:09	04:37	16:07	16:35
ZA	Albania, Tirane	04:15	04:43	16:13	16:41
ZB2	Gibraltar	05:55	06:21	17:57	18:23
ZC4	British Cyprus	03:21	03:46	15:25	15:50
ZD7	St. Helena	05:40	06:01	18:21	18:42
ZD8	Ascension Is.	06:18	06:39	18:51	19:12
ZD9	Tristan da Cunha	05:50	06:16	18:57	19:24
ZF	Cayman Is.	10:54	11:16	23:08	23:30
ZK1	No. Cook Is., Manihiki	16:03	16:24	04:39	05:00
ZK1	So. Cook Is., Rarotonga	15:52	16:15	04:39	05:01
ZK2	Niue	16:30	16:52	05:14	05:36
ZK3	Tokelaus, Atafu	16:51	17:12	05:25	05:46
ZL	New Zealand, Auckland (ZL1)	17:21	17:48	06:29	06:55
	Wellington (ZL2)	17:17	17:44	06:32	07:00
	Christchurch (ZL3)	17:23	17:52	06:42	07:12
	Dunedin (ZL4)	17:28	17:58	06:53	07:23
ZL5	Antarctica, Scott Base	13:57	16:38	08:46	11:27
ZL7	Chatham Is.	16:38	17:08	05:59	06:29
ZL8	Kermadec Is.	16:59	17:23	05:55	06:19
ZL9	Auckland & Campbell Is., Auckland	17:36	18:09	07:14	07:47
	Campbell Is.	17:22	17:57	07:05	07:40
ZP	Paraguay, Asuncion	09:01	09:24	21:52	22:15
ZS	South Africa, Cape Town (ZS1)	03:50	04:15	16:53	17:18
	Port Elizabeth (ZS2)	03:21	03:46	16:23	16:49
	Bloemfontein (ZS4)	03:23	03:47	16:19	16:43
	Durban (ZS5)	03:03	03:27	16:00	16:24
	Johannesburg (ZS6)	03:17	03:41	16:10	16:33
ZS8	Marion Is.	02:17	02:48	15:45	16:15

Pref	State/Province/Country/City	Tlight	Sunrise	Sunset	Tlight
VE1	New Brunswick, Saint John	10:18	10:48	21:30	22:00
	Nova Scotia, Halifax	10:07	10:37	21:21	21:50
	Prince Edward Island, Charlottetown	10:06	10:37	21:16	21:47
VE2	Quebec, Montreal	10:48	11:18	21:59	22:29
	Quebec City	10:39	11:10	21:48	22:19
VE3	Ontario, London	11:16	11:45	22:34	23:03
	Ottawa	10:56	11:26	22:08	22:38
	Sudbury	11:18	11:49	22:27	22:58
	Toronto	11:09	11:38	22:25	22:54
VE4	Manitoba, Winnipeg	12:26	12:59	23:26	23:59
VE5	Saskatchewan, Regina	12:57	13:30	23:55	00:29
	Saskatoon	13:07	13:41	00:01	00:35
VE6	Alberta, Calgary	13:35	14:09	00:32	01:06
	Edmonton	13:36	14:11	00:25	01:01
VE7	British Columbia, Prince George	14:13	14:49	01:02	01:38
	Prince Rupert	14:44	15:20	01:31	02:07
	Vancouver	14:09	14:42	01:11	01:44
VE8	Northwest Territories, Yellowknife	13:52	14:39	00:05	00:52
	Resolute	13:10	14:41	21:27	22:58
VY1	Yukon, Whitehorse	15:12	15:56	01:34	02:18
VO1	Newfoundland, St. John's	09:26	09:57	20:33	21:04
VO2	Labrador, Goose Bay	10:03	10:38	20:53	21:29
W1	Connecticut, Hartford	10:41	11:09	22:01	22:29
	Maine, Bangor	10:28	10:58	21:41	22:11
	Portland	10:33	11:02	21:49	22:18
	Massachusetts, Boston	10:35	11:04	21:54	22:22
	New Hampshire, Concord	10:37	11:06	21:54	22:23
	Rhode Island, Providence	10:36	11:04	21:56	22:24
	Vermont, Montpelier	10:43	11:12	21:57	22:27
W2	New Jersey, Atlantic City	10:46	11:13	22:11	22:38
	New York, Albany	10:46	11:15	22:04	22:33
	Buffalo	11:07	11:35	22:24	22:53
	New York City	10:45	11:13	22:07	22:35
	Syracuse	10:56	11:25	22:13	22:42
W3	Delaware, Wilmington	10:50	11:18	22:15	22:42
	District of Columbia, Washington	10:56	11:23	22:22	22:49
	Maryland, Baltimore	10:54	11:22	22:20	22:47
	Pennsylvania, Harrisburg	10:56	11:24	22:20	22:47
	Philadelphia	10:49	11:17	22:13	22:41
	Pittsburgh	11:09	11:37	22:32	23:00
	Scranton	10:53	11:21	22:13	22:41
W4	Alabama, Montgomery	11:28	11:53	23:06	23:31
	Florida, Jacksonville	11:08	11:32	22:50	23:14
	Miami	10:59	11:22	22:48	23:11
	Pensacola	11:30	11:54	23:12	23:36
	Georgia, Atlanta	11:21	11:47	22:57	23:22
	Savannah	11:07	11:32	22:46	23:10
	Kentucky, Lexington	11:25	11:52	22:53	23:19
	Louisville	11:30	11:57	22:58	23:24
	North Carolina, Charlotte	11:08	11:34	22:41	23:07
	Raleigh	11:00	11:26	22:32	22:58
	Wilmington	10:56	11:21	22:31	22:56
	South Carolina, Columbia	11:08	11:33	22:43	23:09
	Tennessee, Knoxville	11:21	11:47	22:53	23:19
	Memphis	11:45	12:11	23:18	23:44
	Nashville	11:33	11:59	23:04	23:30
	Virginia, Norfolk	10:51	11:18	22:21	22:48
	Richmond	10:56	11:23	22:25	22:51
W5	Arkansas, Little Rock	11:54	12:19	23:28	23:53
	Louisiana, New Orleans	11:41	12:06	23:24	23:48
	Shreveport	11:58	12:22	23:36	00:00
	Mississippi, Jackson	11:43	12:08	23:22	23:47
	New Mexico, Albuquerque	12:52	13:17	00:25	00:50
	Oklahoma, Oklahoma City	12:15	12:41	23:48	00:13
	Texas, Abilene	12:22	12:46	00:00	00:24
	Amarillo	12:32	12:58	00:05	00:31
	Dallas	12:10	12:35	23:48	00:13
	El Paso	12:48	13:13	00:27	00:52
	San Antonio	12:15	12:39	23:58	00:22
W6	California, Los Angeles	13:37	14:02	01:12	01:37
	San Francisco	13:56	14:23	01:25	01:51
W7	Arizona, Flagstaff	13:12	13:37	00:45	01:10
	Phoenix	13:12	13:37	00:48	01:13
	Idaho, Boise	13:36	14:06	00:52	01:22
	Pocatello	13:21	13:50	00:39	01:07
	Montana, Billings	13:08	13:38	00:19	00:49
	Butte	13:24	13:54	00:34	01:05
	Great Falls	13:20	13:52	00:27	00:58
	Nevada, Las Vegas	13:26	13:52	00:57	01:23
	Reno	13:47	14:15	01:12	01:39
	Oregon, Portland	14:04	14:34	01:16	01:46
	Utah, Salt Lake City	13:17	13:45	00:39	01:07
	Washington, Seattle	14:04	14:36	01:11	01:42
	Spokane	13:45	14:16	00:51	01:23
	Wyoming, Cheyenne	12:49	13:17	00:10	00:38
	Sheridan	13:01	13:30	00:14	00:44
W8	Michigan, Detroit	11:22	11:51	22:41	23:10
	Grand Rapids	11:34	12:03	22:51	23:20
	Sault Ste. Marie	11:32	12:03	22:41	23:12
	Traverse City	11:35	12:05	22:48	23:18
	Ohio, Cincinnati	11:26	11:53	22:51	23:19
	Cleveland	11:17	11:45	22:37	23:05
	Columbus	11:21	11:48	22:44	23:12
	West Virginia, Charleston	11:14	11:41	22:41	23:08
W9	Illinois, Chicago	11:41	12:09	23:00	23:29
	Indiana, Indianapolis	11:33	12:01	22:57	23:25
	Wisconsin, Green Bay	11:44	12:14	22:58	23:28

Pref	State/Province/Country/City	Tlight	Sunrise	Sunset	Tlight
	Milwaukee	11:43	12:12	23:00	23:29
W0	Colorado, Denver	12:48	13:16	00:13	00:40
	Grand Junction	13:02	13:29	00:28	00:55
	Iowa, Des Moines	12:04	12:32	23:25	23:53
	Kansas, Pratt	12:22	12:48	23:50	00:16
	Wichita	12:16	12:43	23:44	00:11
	Minnesota, Duluth	12:03	12:34	23:11	23:42
	Minneapolis	12:06	12:36	23:19	23:49
	Missouri, Columbia	11:57	12:24	23:23	23:50
	Kansas City	12:06	12:33	23:32	23:59
	St. Louis	11:48	12:15	23:15	23:42
	Nebraska, North Platte	12:33	13:01	23:54	00:22
	Omaha	12:13	12:41	23:34	00:02
	North Dakota, Fargo	12:22	12:53	23:30	00:01
	South Dakota, Rapid City	12:45	13:14	00:00	00:29
-	-	05:20	05:41	17:48	18:09
1A0	S.M.O.M	05:01	05:29	16:20	16:49
1S	Spratly Is.	21:59	22:20	10:13	10:35
3A	Monaco	05:22	05:51	16:38	17:07
3B6	Agalega	01:25	01:46	14:09	14:31
3B7	St. Brandon	01:07	01:29	14:01	14:23
3B8	Mauritius	01:12	01:35	14:14	14:36
3B9	Rodriguez Is.	00:49	01:12	13:50	14:12
3C	Equatorial Guinea, Bata	04:42	05:03	17:07	17:28
	Malabo	04:47	05:08	17:10	17:31
3C0	Annobon Is.	04:56	05:17	17:26	17:47
3D2	Fiji Is., Suva	17:11	17:33	06:08	06:31
3D2	Conway Reef	17:23	17:46	06:27	06:50
3D2	Rotuma Is.	17:19	17:41	06:06	06:28
3DA	Swaziland, Mbabane	02:51	03:15	16:05	16:29
3V	Tunisia, Tunis	05:05	05:32	16:35	17:02
3W, XV	Vietnam, Ho Chi Minh City (Saigon)	22:21	22:42	10:33	10:54
	Hanoi	22:31	22:54	10:28	10:51
3X	Guinea, Conakry	06:21	06:43	18:35	18:57
3Y	Bouvet	03:44	04:23	18:38	19:17
3Y	Peter I Is.	08:22	09:40	01:54	03:11
4J, 4K	Azerbaijan, Baku	02:29	02:57	13:52	14:20
4L	Georgia, Tbilisi	02:51	03:19	14:11	14:39
4S	Sri Lanka, Colombo	00:05	00:26	12:23	12:44
4U	ITU Geneva	05:29	06:00	16:39	17:10
4U	United Nations Hq.	10:45	11:13	22:07	22:35
4X, 4Z	Israel, Jerusalem	03:21	03:46	15:01	15:25
5A	Libya, Tripoli	04:53	05:18	16:31	16:56
	Benghazi	04:22	04:47	16:01	16:26
5B	Cyprus, Nicosia	03:31	03:57	15:04	15:30
5H	Tanzania, Dar es Salaam	02:36	02:57	15:15	15:36
5N	Nigeria, Lagos	05:11	05:32	17:29	17:50
5R	Madagascar, Antananarivo	01:53	02:16	14:53	15:15
5T	Mauritania, Nouakchott	06:37	06:59	18:38	19:00
5U	Niger, Niamey	05:21	05:43	17:29	17:51
5V	Togo, Lome	05:19	05:40	17:38	18:00
5W	Western Samoa, Apia	16:36	16:58	05:25	05:47
5X	Uganda, Kampala	03:10	03:31	15:37	15:58
5Z	Kenya, Nairobi	04:12	04:33	16:41	17:02
6W	Senegal, Dakar	06:40	07:02	18:46	19:08
6Y	Jamaica, Kingston	10:40	11:02	22:41	23:03
7O	Yemen, Aden	02:29	02:51	14:38	14:59
	Sanaa	02:34	02:56	14:39	15:01
7P	Lesotho, Maseru	03:02	03:26	16:22	16:47
7Q	Malawi, Lilongwe	02:53	03:15	15:43	16:05
	Blantyre	02:46	03:08	15:40	16:02
7X	Algeria, Algiers	05:34	06:00	17:04	17:30
8P	Barbados, Bridgetown	09:28	09:49	21:36	21:58
8Q	Maldive Is.	00:29	00:51	12:51	13:12
8R	Guyana, Georgetown	09:18	09:39	21:35	21:56
9A	Croatia, Zagreb	04:50	05:20	16:01	16:31
9G	Ghana, Accra	05:25	05:46	17:44	18:05
9H	Malta	04:48	05:14	16:19	16:45
9J	Zambia, Lusaka	03:14	03:36	16:06	16:28
9K	Kuwait	02:29	02:54	14:12	14:37
9L	Sierra Leone, Freetown	06:19	06:40	18:34	18:55
9M2	West Malaysia, Kuala Lumpur	22:36	22:57	10:59	11:20
9M6,8	East Malaysia, Sabah, Sandakan(9M6)	21:32	21:53	09:51	10:12
	Sarawak, Kuching (9M8)	22:00	22:21	10:25	10:46
9N	Nepal, Kathmandu	23:58	00:22	11:44	12:08
9Q	Zaire, Kinshasa	04:15	04:36	16:50	17:11
	Kisangani	03:39	04:00	16:06	16:27
	Lubumbashi	03:20	03:42	16:07	16:28
9U	Burundi, Bujumbura	03:20	03:41	15:53	16:14
9V	Singapore	22:26	22:47	10:51	11:12
9X	Rwanda, Kigali	03:18	03:39	15:49	16:10
9Y	Trinidad & Tobago , Port of Spain	09:33	09:54	21:45	22:06
A2	Botswana, Gaborone	03:13	03:37	16:24	16:48
A3	Tonga, Nukualofa	16:42	17:05	05:45	06:08
A4	Oman, Musqat	01:42	02:05	13:35	13:58
A5	Bhutan, Thimbu	23:41	00:05	11:28	11:52
A6	United Arab Emirates, Abu Dhabi	02:00	02:23	13:51	14:14
A7	Qatar, Ad-Dawhah	02:12	02:35	14:02	14:25
A9	Bahrein, Al-Manamah	02:16	02:39	14:04	14:28
AP	Pakistan, Karachi	01:09	01:32	13:00	13:23
AP	Islamabad	00:51	01:16	12:27	12:52
BV	Taiwan, Taipei	21:31	21:55	09:22	09:45
BY	Peoples Rep. of China, Beijing	22:03	22:30	09:27	09:54
	Harbin	21:27	21:57	08:38	09:08
	Shanghai	21:36	22:00	09:16	09:41
	Fuzhou	21:41	22:04	09:30	09:53
	Xian	22:29	22:54	10:03	10:29

Pref	State/Province/Country/City	Tlight	Sunrise	Sunset	Tlight
	Chongqing	22:35	22:59	10:17	10:42
	Chengdu	22:45	23:09	10:26	10:51
	Lhasa	23:36	00:00	11:19	11:43
	Urumqi	00:01	00:31	11:17	11:46
	Kashi	00:44	01:11	12:09	12:36
C2	Nauru	18:12	18:33	06:40	07:01
C3	Andorra	05:45	06:13	17:03	17:32
C5	The Gambia, Banjul	06:36	06:58	18:44	19:06
C6	Bahamas, Nassau	10:47	11:10	22:37	23:01
C9	Mozambique, Maputo	02:45	03:09	15:59	16:22
	Mozambique	02:24	02:46	15:17	15:38
CE	Chile, Santiago	09:29	09:55	23:00	23:26
CE0	Easter Island	12:12	12:36	01:28	01:52
CE0	Juan Fernandez	10:01	10:27	23:32	23:58
CE0	San Felix	10:16	10:39	23:30	23:53
CM,CO	Cuba, Havana	11:06	11:29	22:59	23:22
CN	Morocco, Casablanca	06:14	06:39	17:50	18:15
CP	Bolivia, La Paz	09:39	10:01	22:34	22:56
CT	Portugal, Lisbon	06:24	06:51	17:51	18:18
CT3	Madeira Islands, Funchal	06:50	07:15	18:28	18:53
CU	Azores , Ponta Delgada	07:30	07:56	18:58	19:24
CX	Uruguay, Montevideo	08:28	08:55	22:03	22:30
CY0	Sable Is.	09:52	10:21	21:07	21:37
CY9	St. Paul Is.	09:55	10:26	21:03	21:34
D2	Angola, Luanda	04:20	04:41	17:02	17:23
D4	Cape Verde, Praia	07:04	07:26	19:10	19:32
D6	Comoros, Moroni	02:15	02:37	15:02	15:23
DA-DL,Y2-Y9	Fed. Rep. of Germany, Bonn	05:30	06:04	16:29	17:02
	Berlin	05:07	05:42	16:00	16:34
DU	Philippines, Manilla	21:26	21:48	09:32	09:54
E3	Eritrea, Asmara	02:55	03:17	15:00	15:22
EA	Spain, Madrid	06:04	06:31	17:27	17:54
EA6	Balearic Is., Palma	05:38	06:05	17:03	17:30
EA8	Canary Is., Las Palmas	06:37	07:01	18:22	18:46
EA9	Ceuta & Melilla, Ceuta	06:07	06:32	17:38	18:04
	Melilla	05:57	06:23	17:30	17:56
EI	Ireland, Dublin	06:26	07:02	17:17	17:52
EK	Armenia, Yerevan	02:51	03:18	14:14	14:42
EL	Liberia, Monrovia	06:08	06:29	18:26	18:47
EP	Iran, Tehran	02:18	02:44	13:50	14:16
ER	Moldova, Kishinev	04:00	04:31	15:08	15:38
ES	Estonia, Tallinn	04:30	05:12	14:58	15:40
ET	Ethiopia, Addis Ababa	02:52	03:13	15:06	15:27
EU,EV,EW	Belarus, Minsk	04:12	04:48	15:00	15:36
EX	Kyrgyzstan, Frunze	00:53	01:21	12:10	12:39
EY	Tajikistan, Samarkand	01:21	01:49	12:46	13:13
	Dushanbe	01:13	01:40	12:38	13:05
EZ	Turkmenistan, Ashkhabad	01:53	02:20	13:21	13:48
F	France, Paris	05:47	06:19	16:51	17:23
FG	Guadeloupe	09:38	10:00	21:42	22:04
FJ, FS	St. Martin	09:45	10:07	21:46	22:08
FH	Mayotte	02:08	02:30	14:56	15:18
FK	New Caledonia, Noumia	17:54	18:17	07:00	07:23
FM	Martinique	09:34	09:56	21:40	22:02
FO	Clipperton Is.	12:44	13:05	00:57	01:18
FO	Fr. Polynesia, Tahiti	15:43	16:05	04:39	05:02
	Rurutu, (Austral Is.)	18:54	19:17	08:01	08:24
	Hiva Oa, (Marquesas Is.)	14:28	14:49	03:11	03:33
FP	St. Pierre & Miquelon, St. Pierre	09:38	10:09	20:47	21:18
FR/G	Glorioso	02:01	02:23	14:47	15:09
FR/J,E	Juan de Nova	02:14	02:36	15:10	15:32
	Europa	02:18	02:41	15:24	15:47
FR	Reunion	01:19	01:42	14:22	14:45
FR/T	Tromelin	01:29	01:51	14:22	14:44
FT8W	Crozet	00:55	01:26	15:06	15:38
FT8X	Kerguelen	23:38	00:12	14:03	14:37
FT8Y	Antarctica, Dumont D'Urville	17:26	18:32	10:17	11:23
FT8Z	Amsterdam & St. Paul Is., Amsterdam	23:29	23:56	13:12	13:39
FW	Wallis & Futuna Is., Wallis	16:54	17:16	05:43	06:05
FY	Fr. Guiana, Cayenne	08:53	09:14	21:13	21:34
G	England, London	06:00	06:34	16:55	17:30
GD	Isle of Man	06:20	06:57	17:07	17:44
GI	Northern Ireland, Belfast	06:26	07:03	17:12	17:49
GJ	Jersey	06:06	06:38	17:08	17:40
GM	Scotland, Glasgow	06:22	06:59	17:03	17:41
	Aberdeen	06:15	06:54	16:51	17:31
GU	Guernsey	06:08	06:41	17:09	17:42
GW	Wales, Cardiff	06:12	06:46	17:08	17:42
H4	Solomon Islands, Honiara	18:32	18:54	07:15	07:36
HA	Hungary, Budapest	04:39	05:10	15:46	16:17
HB	Switzerland, Bern	05:25	05:56	16:33	17:04
HB0	Liechtenstein	05:17	05:48	16:24	16:55
HC	Ecuador, Quito	10:32	10:53	23:00	23:21
HC8	Galapagos Is.	11:21	11:42	23:50	00:11
HH	Haiti, Port-Au-Prince	10:22	10:44	22:23	22:45
HI	Dominican Republic, Santo Domingo	10:13	10:35	22:13	22:35
HK	Colombia, Bogota	10:20	10:41	22:41	23:02
HK0	Malpelo Is.	10:47	11:08	23:09	23:30
HK0	San Adreas	10:56	11:17	23:05	23:26
HL	Korea, Seoul	21:19	21:45	08:47	09:14
HP	Panama, Panama	10:44	11:06	22:59	23:20
HR	Honduras, Tegucigalpa	11:19	11:40	23:26	23:47
HS	Thailand, Bangkok	22:48	23:09	10:55	11:17
HV	Vatican City	05:00	05:28	16:20	16:48
HZ, 7Z	Saudi Arabia, Dharan	02:18	02:42	14:07	14:30
	Mecca	02:56	03:18	14:52	15:14
	Riyadh	02:30	02:53	14:21	14:44

Pref	State/Province/Country/City	Tlight	Sunrise	Sunset	Tlight
I	Italy, Rome	05:00	05:28	16:20	16:48
	Trieste	04:58	05:29	16:09	16:40
	Sicily	04:51	05:17	16:19	16:46
IS	Sardinia, Cagliari	05:12	05:39	16:37	17:04
J2	Djibouti, Djibouti	02:35	02:57	14:46	15:07
J3	Grenada	09:36	09:57	21:46	22:07
J5	Guinea-Bissau, Bissau	06:31	06:52	18:41	19:02
J6	St. Lucia	09:34	09:55	21:41	22:03
J7	Dominica	09:36	09:58	21:41	22:03
J8	St. Vincent	09:35	09:56	21:43	22:04
JA-JS	Japan, Tokyo	20:26	20:52	07:58	08:24
	Nagasaki	21:03	21:28	08:41	09:06
	Sapporo	20:26	20:54	07:43	08:12
JD1	Minami Torishima	19:21	19:44	07:12	07:35
JD1	Ogasawara, Kazan Is.	20:15	20:39	08:02	08:25
JT	Mongolia, Ulan Bator	22:48	23:20	09:54	10:25
JW	Svalbard, Spitsbergen	06:15	08:44	12:36	15:05
JX	Jan Mayen	07:08	08:18	16:17	17:27
JY	Jordan, Amman	03:19	03:44	14:58	15:22
KC4	Antarctica, Byrd Station	—	—	—	—
	McMurdo Sound	—	14:02	11:13	—
	Palmer Station	07:18	08:17	23:43	00:42
KC6	Belau, Yap	20:14	20:35	08:28	08:49
	Koror	20:27	20:48	08:44	09:05
KG4	Guantanamo Bay	10:35	10:57	22:33	22:55
KH0	Mariana Is., Saipan	19:47	20:09	07:53	08:15
KH1	Baker, Howland Is.	17:04	17:25	05:31	05:52
KH2	Guam, Agana	19:50	20:12	07:58	08:20
KH3	Johnston Is.	16:46	17:08	04:49	05:11
KH4	Midway Is.	17:29	17:53	05:15	05:38
KH5	Palmyra Is.	16:13	16:34	04:32	04:53
KH5K	Kingman Reef	16:16	16:38	04:33	04:54
KH6	Hawaii, Hilo	15:54	16:16	03:53	04:15
	Honolulu	16:06	16:29	04:03	04:25
KH7	Kure Is.	17:33	17:57	05:18	05:42
KH8	American Samoa, Pago Pago	16:31	16:53	05:22	05:44
KH9	Wake Is.	18:27	18:49	06:26	06:49
KL7	Alaska, Adak	17:46	18:20	04:41	05:15
	Anchorage	16:12	16:57	02:32	03:16
	Fairbanks	16:10	17:01	02:10	03:01
	Juneau	15:05	15:46	01:38	02:18
	Nome	17:20	18:10	03:21	04:12
KP1	Navassa Is.	10:33	10:55	22:33	22:56
KP2	Virgin Islands, Charlotte Amalie	09:52	10:14	21:53	22:15
KP4	Puerto Rico, San Juan	09:58	10:20	21:58	22:20
KP5	Desecheo Is.	10:03	10:25	22:04	22:26
LA-LJ	Norway, Oslo	05:27	06:10	15:52	16:35
LU	Argentina, Buenos Aires	08:38	09:04	22:12	22:38
LX	Luxembourg	05:32	06:05	16:34	17:06
LY	Lithuania, Vilna	04:21	04:57	15:07	15:44
LZ	Bulgaria, Sofia	04:18	04:46	15:36	16:04
OA	Peru, Lima	10:18	10:40	23:05	23:27
OD	Lebanon, Beirut	03:22	03:47	14:57	15:23
OE	Austria, Vienna	04:51	05:22	15:56	16:27
OH	Finland, Helsinki	04:31	05:14	14:55	15:38
OH0	Aland Is.	04:51	05:34	15:15	15:58
OJ0	Market Reef	04:55	05:38	15:18	16:02
OK,OL	Czech Rep., Prague	05:00	05:33	16:00	16:33
OM	Slovak, Rep., Bratislava	04:48	05:19	15:53	16:25
ON	Belgium, Brussels	05:41	06:15	16:39	17:12
OX	Greenland, Godthaab	09:44	10:34	19:48	20:38
	Thule	11:37	13:26	19:13	21:02
OY	Faroe Islands, Torshavn	06:41	07:27	16:56	17:42
OZ	Denmark, Copenhagen	05:14	05:52	15:56	16:34
P2	Papua New Guinea, Madang	19:33	19:54	08:09	08:30
	Port Moresby	19:24	19:45	08:06	08:28
P4	Aruba, Oranjestad	10:05	10:26	22:14	22:36
PA-PI	Netherlands, Amsterdam	05:41	06:15	16:34	17:09
PJ2,4,9	Netherlands Antilles, Willemstad	10:09	10:31	22:18	22:40
PJ5-8	St. Maarten and Saba, St. Maarten	09:45	10:07	21:47	22:09
PY	Brazil, Brasilia	08:18	08:40	21:12	21:34
	Rio De Janeiro	07:52	08:15	20:59	21:22
	Natal	07:36	07:57	20:13	20:34
	Manaus	09:18	09:39	21:51	22:12
	Porto Alegre	08:15	08:40	21:38	22:03
PY0	Fernando De Noronha	07:26	07:47	20:00	20:21
PY0	St. Peter & St. Paul Rocks	07:18	07:39	19:44	20:05
PY0	Trindade & Martin Vaz Is., Trindade	09:05	09:22	20:01	20:24
PZ	Suriname, Paramaribo	09:05	09:26	21:24	21:45
R1FJ_	Franz Josef Land	04:00	07:29	08:56	12:24
R1MV_	Malyj Vysotskij Is.	04:17	05:01	14:39	15:23
S0	Western Sahara, Smara	06:24	06:47	18:12	18:36
S2	Bangladesh, Dacca	23:35	23:58	11:27	11:50
S5	Slovenia, Ljubljana	04:56	05:26	16:06	16:37
S7	Seychelles, Victoria	01:34	01:55	14:09	14:30
S9	Sao Tome	04:53	05:14	17:21	17:42
SM	Sweden, Stockholm	04:57	05:39	15:25	16:07
SP	Poland, Krakow	04:38	05:11	15:38	16:11
	Warsaw	04:36	05:11	15:30	16:04
ST	Sudan, Khartoum	03:21	03:43	15:26	15:47
ST0	Southern Sudan, Juba	03:17	03:38	15:37	15:59
SU	Egypt, Cairo	03:35	04:00	15:18	15:42
SV	Greece, Athens	04:12	04:39	15:40	16:07
SV/A	Mount Athos	04:12	04:39	15:35	16:02
SV5	Dodecanese, Rhodes	03:53	04:19	15:24	15:50
SV9	Crete	04:04	04:30	15:37	16:03
T2	Tuvalu, Funafuti	17:18	17:40	06:00	06:21

Pref	State/Province/Country/City	Tlight	Sunrise	Sunset	Tlight
T30	West Kiribati, Bonriki	17:48	18:09	06:14	06:35
T31	Central Kiribati, Kanton	16:44	17:06	05:17	05:38
T32	East Kiribati, Christmas Is.	15:51	16:12	04:16	04:37
T33	Banaba Is.	18:02	18:23	06:30	06:51
T5	Somalia, Mogadishu	02:20	02:41	14:44	15:05
T7	San Marino	05:03	05:32	16:18	16:47
T9	Bosnia-Herzegovina, Sarajevo	04:38	05:08	15:54	16:23
TA	Turkey, Ankara	03:37	04:04	15:01	15:28
	Istanbul	03:54	04:22	15:15	15:43
TF	Iceland, Reykjavik	07:45	08:35	17:49	18:39
TG	Guatemala, Guatemala City	11:32	11:54	23:38	00:00
TI	Costa Rica, San Jose	11:03	11:24	23:16	23:37
TI9	Cocos Is.	11:12	11:33	23:31	23:52
TJ	Cameroon, Yaounde	04:37	04:58	16:59	17:20
TK	Corsica	05:14	05:43	16:34	17:02
TL	Central African Republic, Bangui	04:09	04:30	16:30	16:51
TN	Congo, Brazzaville	04:15	04:36	16:50	17:11
TR	Gabon, Libreville	04:42	05:03	17:09	17:30
TT	Chad, N'Djamena	04:29	04:50	16:38	17:00
TU	Cote d'Ivoire, Abidjan	05:40	06:01	18:00	18:21
TY	Benin, Porto Novo	05:14	05:35	17:32	17:53
TZ	Mali, Bamako	06:01	06:22	18:10	18:31
UA	Russia, European, Leningrad (UA1)	04:09	04:52	14:34	15:17
	Archangel (UA1)	03:36	04:27	13:37	14:28
	Murmansk (UA1)	04:16	05:18	13:45	14:47
	Moscow (UA3)	03:34	04:12	14:16	14:54
	Kuybyshev (UA4)	02:41	03:16	13:31	14:07
	Rostov (UA6)	03:17	03:48	14:24	14:55
UA2	Kaliningrad	04:41	05:18	15:26	16:03
UA9,0	Russia, Asiatic, Novosibirsk (UA9)	00:32	01:09	11:16	11:53
	Perm (UA9)	02:22	03:02	12:56	13:36
	Omsk (UA9)	01:10	01:47	11:54	12:31
	Norilsk (UA0)	00:37	01:40	10:03	11:06
	Irkutsk (UA0)	23:03	23:38	09:56	10:31
	Vladivostok (UA0)	21:04	21:33	08:21	08:50
	Petropavlovsk (UA0)	19:26	20:01	06:17	06:53
	Khabarovsk (UA0)	20:56	21:28	08:00	08:32
	Krasnoyarsk (UA0)	23:53	00:32	10:34	11:13
	Yakutsk (UA0)	21:35	22:21	07:50	08:36
	Wrangel Island (UA0)	18:33	19:42	03:42	04:51
	Kyzyl (UA0Y)	23:41	00:16	10:37	11:11
UJ-UM	Uzbekistan, Bukhoro	01:31	01:58	12:55	13:22
	Tashkent	01:12	01:40	12:34	13:02
UN-UQ	Kazakhstan, Alma-Ata	00:44	01:13	12:00	12:29
UR-UZ,EM-EO	Ukraine, Kiev	03:56	04:29	14:55	15:28
V2	Antigua & Barbuda, St. Johns	09:39	10:01	21:42	22:04
V3	Belize, Belmopan	11:27	11:49	23:30	23:52
V4	St. Kitts & Nevis	09:42	10:04	21:45	22:07
V5	Namibia, Windhoek	03:51	04:14	16:57	17:20
V6	Micronesia, Ponape	18:52	19:13	07:09	07:30
V7	Marshall Islands, Kwajalein	18:17	18:38	06:32	06:53
V8	Brunei, Bandar Seri Begawan	21:44	22:05	10:04	10:25
VK	Australia, Canberra (VK1)	18:47	19:13	08:23	08:49
	Sydney (VK2)	18:40	19:06	08:13	08:38
	Melbourne (VK3)	18:59	19:26	08:42	09:09
	Brisbane (VK4)	18:42	19:06	07:59	08:23
	Adelaide (VK5)	19:29	19:55	09:04	09:30
	Perth (VK6)	21:05	21:30	10:32	10:57
	Hobart, Tasmania (VK7)	18:40	19:10	08:40	09:10
	Darwin (VK8)	20:26	20:48	09:14	09:35
VK0	Heard Is.	23:10	23:47	13:54	14:32
VK0	Macquarie Is.	17:22	18:01	08:17	08:56
VK9C	Cocos-Keeling Is.	22:43	23:04	11:30	11:51
VK9L	Lord Howe Is.	18:12	18:37	07:38	08:04
VK9M	Mellish Reef	18:42	19:04	07:38	08:00
VK9N	Norfolk Is.	17:40	18:04	07:00	07:24
VK9W	Willis Is.	19:08	19:30	08:02	08:24
VK9X	Christmas Is.	22:09	22:30	10:53	11:14
VP2E	Anguilla	09:45	10:07	21:46	22:08
VP2M	Montserrat	09:41	10:02	21:44	22:05
VP2V	British Virgin Is., Tortola	09:51	10:13	21:52	22:14

Pref	State/Province/Country/City	Tlight	Sunrise	Sunset	Tlight
VP5	Turks & Caicos Islands, Grand Turk	10:20	10:42	22:16	22:38
VP8	Falkland Islands, Stanley	07:59	08:35	22:37	23:13
VP8	So. Georgia Is.	06:26	07:05	21:18	21:57
VP8	So. Orkney Is.	06:33	07:21	22:12	23:00
VP8	So. Sandwich Islands, Saunders Is.	05:31	06:15	20:47	21:31
VP8	So. Shetland Is., King George Is.	07:15	08:07	23:08	23:59
VP9	Bermuda	10:01	10:26	21:40	22:05
VQ9	Chagos, Diego Garcia	00:24	00:46	13:04	13:25
VR2,VS6	Hong Kong	21:58	22:21	09:53	10:16
VR6	Pitcairn Is.	13:37	14:01	02:49	03:12
VU	India, Bombay	00:42	01:04	12:42	13:04
	Calcutta	23:42	00:05	11:36	11:59
	New Delhi	00:31	00:55	12:16	12:40
	Bangalore	00:19	00:40	12:27	12:49
VU	Andaman Islands, Port Blair	23:17	23:38	11:28	11:49
VU	Laccadive Is.	00:35	00:56	12:48	13:09
XE	Mexico, Mexico City (XE1)	12:10	12:32	00:09	00:31
	Chihuahua (XE2)	12:44	13:08	00:28	00:52
	Merida (XE3)	11:33	11:56	23:30	23:53
XF4	Revilla Gigedo	12:59	13:21	00:59	01:21
XT	Burkina Faso, Ouagadougou	05:35	05:57	17:45	18:06
XU	Cambodia, Phnom Penh	22:29	22:50	10:40	11:01
XW	Laos, Viangchan	22:42	23:04	10:43	11:05
XX9	Macao	22:01	22:24	09:56	10:19
XZ	Myanmar, Yangon	23:08	23:30	11:11	11:33
YA	Afghanistan, Kandahar	01:18	01:43	12:59	13:24
	Kabul	01:07	01:33	12:42	13:07
YB-YD	Indonesia, Jakarta	22:08	22:29	10:45	11:06
	Medan	22:48	23:09	11:10	11:31
	Pontianak	22:03	22:24	10:30	10:51
	Jayapura	19:55	20:16	08:27	08:48
YI	Iraq, Baghdad	02:45	03:10	14:22	14:47
YJ	Vanuatu, Port Vila	17:51	18:14	06:48	07:11
YK	Syria, Damascus	03:18	03:44	14:54	15:20
YL	Latvia, Riga	04:29	05:09	15:07	15:46
YN	Nicaragua, Managua	11:12	11:34	23:23	23:44
YO	Romania, Bucharest	04:08	04:37	15:22	15:52
YS	El Salvador, San Salvador	11:26	11:48	23:34	23:56
YU	Yugoslavia, Belgrade	04:31	05:01	15:44	16:14
YV	Venezuela, Caracas	09:55	10:17	22:08	22:29
YV0	Aves Is.	09:46	10:08	21:50	22:12
Z2	Zimbabwe, Harare	03:01	03:23	15:58	16:20
Z3	Macedonia, (ex Yugoslav), Skopje	04:25	04:53	15:44	16:13
ZA	Albania, Tirane	04:30	04:58	15:52	16:20
ZB2	Gibraltar	06:07	06:33	17:38	18:05
ZC4	British Cyprus	03:32	03:58	15:07	15:32
ZD7	St. Helena	05:30	05:52	18:24	18:46
ZD8	Ascension Is.	06:11	06:32	18:51	19:13
ZD9	Tristan da Cunha	05:29	05:56	19:10	19:37
ZF	Cayman Is.	10:58	11:21	22:57	23:20
ZK1	No. Cook Is., Manihiki	15:55	16:17	04:40	05:01
ZK1	So. Cook Is., Rarotonga	15:40	16:03	04:44	05:07
ZK2	Niue	16:19	16:41	05:18	05:46
ZK3	Tokelaus, Atafu	16:44	17:05	05:25	05:46
ZL	New Zealand, Auckland (ZL1)	17:01	17:28	06:42	07:09
	Wellington (ZL2)	16:54	17:22	06:48	07:16
	Christchurch (ZL3)	16:58	17:28	07:00	07:30
	Dunedin (ZL4)	17:01	17:32	07:12	07:43
ZL5	Antarctica, Scott Base	——	13:53	11:25	——
ZL7	Chatham Is.	16:13	16:44	06:17	06:47
ZL8	Kermadec Is.	16:43	17:08	06:04	06:28
ZL9	Auckland & Campbell Is., Auckland	17:04	17:40	07:37	08:12
	Campbell Is.	16:48	17:25	07:30	08:07
ZP	Paraguay, Asuncion	08:47	09:11	21:59	22:23
ZS	South Africa, Cape Town (ZS1)	03:32	03:57	17:04	17:30
	Port Elizabeth (ZS2)	03:02	03:28	16:35	17:01
	Bloemfontein (ZS4)	03:07	03:32	16:28	16:52
	Durban (ZS5)	02:47	03:12	16:09	16:34
	Johannesburg (ZS6)	03:03	03:27	16:17	16:41
ZS8	Marion Is.	01:50	02:22	16:04	16:36

Pref	State/Province/Country/City	Tlight	Sunrise	Sunset	Tlight
VE1	New Brunswick, Saint John	10:40	11:11	21:04	21:35
	Nova Scotia, Halifax	10:29	10:59	20:56	21:27
	Prince Edward Island, Charlottetown	10:29	11:01	20:50	21:22
VE2	Quebec, Montreal	11:10	11:41	21:34	22:05
	Quebec City	11:03	11:35	21:21	21:53
VE3	Ontario, London	11:37	12:07	22:10	22:40
	Ottawa	11:18	11:50	21:42	22:14
	Sudbury	11:41	12:13	22:01	22:33
	Toronto	11:30	12:00	22:01	22:31
VE4	Manitoba, Winnipeg	12:52	13:27	22:56	23:31
VE5	Saskatchewan, Regina	13:24	13:59	23:25	00:00
	Saskatoon	13:35	14:12	23:28	00:05
VE6	Alberta, Calgary	14:03	14:38	00:01	00:37
	Edmonton	14:06	14:44	23:51	00:29
VE7	British Columbia, Prince George	14:44	15:22	00:27	01:05
	Prince Rupert	15:15	15:54	00:55	01:34
	Vancouver	14:35	15:09	00:42	01:16
VE8	Northwest Territories, Yellowknife	14:36	15:28	23:14	00:05
	Resolute	14:46	———	———	21:20
VY1	Yukon, Whitehorse	15:53	16:41	00:47	01:35
VO1	Newfoundland, St. John's	09:50	10:23	20:05	20:38
VO2	Labrador, Goose Bay	10:33	11:11	20:19	20:57
W1	Connecticut, Hartford	11:00	11:30	21:38	22:08
	Maine, Bangor	10:50	11:21	21:16	21:47
	Portland	10:54	11:24	21:25	21:55
	Massachusetts, Boston	10:55	11:24	21:31	22:00
	New Hampshire, Concord	10:58	11:28	21:31	22:01
	Rhode Island, Providence	10:55	11:24	21:33	22:02
	Vermont, Montpelier	11:04	11:35	21:33	22:03
W2	New Jersey, Atlantic City	11:03	11:32	21:50	22:18
	New York, Albany	11:06	11:36	21:41	22:11
	Buffalo	11:27	11:57	22:01	22:31
	New York City	11:04	11:33	21:46	22:14
	Syracuse	11:16	11:46	21:50	22:20
W3	Delaware, Wilmington	11:08	11:37	21:54	22:22
	District of Columbia, Washington	11:13	11:41	22:01	22:29
	Maryland, Baltimore	11:12	11:40	21:59	22:27
	Pennsylvania, Harrisburg	11:15	11:43	21:58	22:27
	Philadelphia	11:07	11:36	21:52	22:21
	Pittsburgh	11:27	11:56	22:10	22:39
	Scranton	11:12	11:41	21:51	22:20
W4	Albama, Montgomery	11:41	12:07	22:50	23:16
	Florida, Jacksonville	11:20	11:45	22:35	23:00
	Miami	11:09	11:33	22:35	22:59
	Pensacola	11:42	12:07	22:57	23:22
	Georgia, Atlanta	11:35	12:02	22:40	23:06
	Savannah	11:20	11:46	22:30	22:55
	Kentucky, Lexington	11:42	12:09	22:33	23:01
	Louisville	11:47	12:15	22:38	23:06
	North Carolina, Charlotte	11:23	11:50	22:23	22:50
	Raleigh	11:15	11:42	22:13	22:40
	Wilmington	11:10	11:36	22:13	22:40
	South Carolina, Columbia	11:22	11:48	22:26	22:52
	Tennessee, Knoxville	11:36	12:03	22:34	23:01
	Memphis	12:00	12:27	23:01	23:27
	Nashville	11:48	12:15	22:46	23:12
	Virginia, Norfolk	11:07	11:35	22:02	22:29
	Richmond	11:13	11:40	22:06	22:33
W5	Arkansas, Little Rock	12:08	12:35	23:10	23:36
	Louisiana, New Orleans	11:53	12:18	23:09	23:34
	Shreveport	12:11	12:37	23:19	23:45
	Mississippi, Jackson	11:57	12:22	23:06	23:31
	New Mexico, Albuquerque	13:06	13:33	00:07	00:34
	Oklahoma, Oklahoma City	12:30	12:57	23:30	23:56
	Texas, Abilene	12:35	13:01	23:43	00:09
	Amarillo	12:47	13:14	23:47	00:14
	Dallas	12:24	12:49	23:31	23:57
	El Paso	13:01	13:27	00:12	00:37
	San Antonio	12:26	12:51	23:43	00:08
W6	California, Los Angeles	13:51	14:17	00:55	01:21
	San Francisco	14:13	14:41	01:05	01:33
W7	Arizona, Flagstaff	13:27	13:53	00:27	00:53
	Phoenix	13:26	13:52	00:31	00:57
	Idaho, Boise	13:57	14:27	00:29	00:59
	Pocatello	13:41	14:11	00:15	00:45
	Montana, Billings	13:30	14:02	23:53	00:24
	Butte	13:47	14:18	00:08	00:40
	Great Falls	13:44	14:17	00:00	00:32
	Nevada, Las Vegas	13:42	14:08	00:39	01:06
	Reno	14:05	14:33	00:51	01:20
	Oregon, Portland	14:26	14:58	00:50	01:22
	Utah, Salt Lake City	13:36	14:04	00:17	00:46
	Washington, Seattle	14:29	15:01	00:43	01:16
	Spokane	14:09	14:42	00:23	00:56
	Wyoming, Cheyenne	13:08	13:37	23:48	00:17
	Sheridan	13:22	13:53	23:49	00:20
W8	Michigan, Detroit	11:42	12:12	22:19	22:48
	Grand Rapids	11:54	12:24	22:28	22:58
	Sault Ste. Marie	11:55	12:27	22:15	22:47
	Traverse City	11:57	12:28	22:23	22:54
	Ohio, Cincinnati	11:43	12:11	22:31	22:59
	Cleveland	11:36	12:05	22:15	22:44
	Columbus	11:39	12:07	22:23	22:52
	West Virginia, Charleston	11:31	11:58	22:21	22:48
W9	Illinois, Chicago	12:00	12:29	22:38	23:07
	Indiana, Indianapolis	11:51	12:20	22:36	23:05
	Wisconsin, Green Bay	12:06	12:37	22:34	23:04

Pref	State/Province/Country/City	Tlight	Sunrise	Sunset	Tlight
	Milwaukee	12:03	12:33	22:37	23:07
W0	Colorado, Denver	13:06	13:35	23:52	00:20
	Grand Junction	13:20	13:48	00:07	00:35
	Iowa, Des Moines	12:24	12:53	23:02	23:32
	Kansas, Pratt	12:38	13:06	23:30	23:58
	Wichita	12:33	13:00	23:25	23:52
	Minnesota, Duluth	12:26	12:59	22:45	23:17
	Minneapolis	12:28	12:59	22:54	23:25
	Missouri, Columbia	12:14	12:42	23:02	23:30
	Kansas City	12:24	12:52	23:11	23:39
	St. Louis	12:05	12:33	22:55	23:23
	Nebraska, North Platte	12:52	13:21	23:32	00:01
	Omaha	12:32	13:01	23:12	23:41
	North Dakota, Fargo	12:45	13:18	23:03	23:35
	South Dakota, Rapid City	13:06	13:37	23:35	00:06
-	-	05:18	05:40	17:47	18:08
1A0	S.M.O.M	05:20	05:49	15:58	16:27
1S	Spratly Is.	22:01	22:22	10:09	10:31
3A	Monaco	05:43	06:13	16:14	16:44
3B6	Agalega	01:19	01:41	14:13	14:35
3B7	St. Brandon	00:58	01:21	14:07	14:30
3B8	Mauritius	01:01	01:25	14:22	14:45
3B9	Rodriguez Is.	00:39	01:02	13:57	14:20
3C	Equatorial Guinea, Bata	04:41	05:03	17:05	17:27
	Malabo	04:47	05:09	17:07	17:29
3C0	Annobon Is.	04:54	05:16	17:26	17:48
3D2	Fiji Is., Suva	17:01	17:24	06:15	06:38
3D2	Conway Reef	17:12	17:36	06:36	06:59
3D2	Rotuma Is.	17:12	17:34	06:11	06:33
3DA	Swaziland, Mbabane	02:37	03:02	16:16	16:40
3V	Tunisia, Tunis	05:21	05:48	16:16	16:44
3W, XV	Vietnam, Ho Chi Minh City (Saigon)	22:24	22:46	10:27	10:49
	Hanoi	22:39	23:02	10:18	10:41
3X	Guinea, Conakry	06:24	06:46	18:30	18:52
3Y	Bouvet	03:03	03:47	19:13	19:56
3Y	Peter I Is.	———	08:16	03:16	———
4J, 4K	Azerbaijan, Baku	02:48	03:16	13:31	13:59
4L	Georgia, Tbilisi	03:10	03:39	13:49	14:18
4S	Sri Lanka, Colombo	00:07	00:28	12:19	12:41
4U	ITU Geneva	05:52	06:24	16:13	16:45
4U	United Nations Hq.	11:04	11:33	21:46	22:14
4X, 4Z	Israel, Jerusalem	03:34	04:00	14:45	15:10
5A	Libya, Tripoli	05:06	05:32	16:15	16:40
	Benghazi	04:36	05:01	15:45	16:11
5B	Cyprus, Nicosia	03:46	04:13	14:46	15:13
5H	Tanzania, Dar es Salaam	02:32	02:53	15:17	15:39
5N	Nigeria, Lagos	05:12	05:34	17:25	17:47
5R	Madagascar, Antananarivo	01:44	02:07	15:00	15:23
5T	Mauritania, Nouakchott	06:43	07:05	18:29	18:52
5U	Niger, Niamey	05:26	05:48	17:23	17:45
5V	Togo, Lome	05:20	05:42	17:35	17:57
5W	Western Samoa, Apia	16:28	16:51	05:30	05:52
5X	Uganda, Kampala	03:09	03:30	15:36	15:58
5Z	Kenya, Nairobi	04:09	04:31	16:41	17:03
6W	Senegal, Dakar	06:45	07:07	18:39	19:01
6Y	Jamaica, Kingston	10:46	11:08	22:32	22:55
7O	Yemen, Aden	02:33	02:55	14:32	14:54
	Sanaa	02:39	03:01	14:32	14:54
7P	Lesotho, Maseru	02:46	03:12	16:35	17:00
7Q	Malawi, Lilongwe	02:45	03:08	15:48	16:11
	Blantyre	02:38	03:01	15:46	16:08
7X	Algeria, Algiers	05:50	06:17	16:45	17:13
8P	Barbados, Bridgetown	09:32	09:54	21:30	21:52
8Q	Maldive Is.	00:30	00:51	12:48	13:10
8R	Guyana, Georgetown	09:19	09:40	21:32	21:53
9A	Croatia, Zagreb	05:12	05:44	15:35	16:06
9G	Ghana, Accra	05:25	05:47	17:41	18:03
9H	Malta	05:03	05:30	16:01	16:28
9J	Zambia, Lusaka	03:05	03:28	16:12	16:35
9K	Kuwait	02:41	03:06	13:58	14:23
9L	Sierra Leone, Freetown	06:21	06:42	18:30	18:51
9M2	West Malaysia, Kuala Lumpur	22:36	22:57	10:57	11:18
9M6,8	East Malaysia, Sabah, Sandakan(9M6)	21:32	21:54	09:47	10:09
	Sarawak, Kuching (9M8)	21:59	22:20	10:24	10:45
9N	Nepal, Kathmandu	00:09	00:33	11:31	11:55
9Q	Zaire, Kinshasa	04:12	04:33	16:50	17:12
	Kisangani	03:38	04:00	16:05	16:27
	Lubumbashi	03:14	03:36	16:11	16:33
9U	Burundi, Bujumbura	03:17	03:39	15:53	16:15
9V	Singapore	22:25	22:46	10:50	11:11
9X	Rwanda, Kigali	03:15	03:37	15:49	16:10
9Y	Trinidad & Tobago , Port of Spain	09:35	09:57	21:40	22:01
A2	Botswana, Gaborone	03:01	03:25	16:34	16:59
A3	Tonga, Nukualofa	16:31	16:54	05:54	06:17
A4	Oman, Musqat	01:51	02:14	13:23	13:47
A5	Bhutan, Thimbu	23:52	00:16	11:15	11:39
A6	United Arab Emirates, Abu Dhabi	02:09	02:33	13:40	14:03
A7	Qatar, Ad-Dawhah	02:21	02:45	13:49	14:13
A9	Bahrein, Al-Manamah	02:26	02:50	13:52	14:16
AP	Pakistan, Karachi	01:18	01:42	12:48	13:11
AP	Islamabad	01:05	01:31	12:10	12:36
BV	Taiwan, Taipei	21:41	22:05	09:10	09:33
BY	Peoples Rep. of China, Beijing	22:21	22:50	09:06	09:34
	Harbin	21:49	22:21	08:12	08:43
	Shanghai	21:48	22:14	09:01	09:26
	Fuzhou	21:51	22:15	09:17	09:41
	Xian	22:43	23:09	09:46	10:12

Pref	State/Province/Country/City	Tlight	Sunrise	Sunset	Tlight
	Chongqing	22:47	23:12	10:03	10:28
	Chengdu	22:57	23:23	10:11	10:36
	Lhasa	23:48	00:13	11:04	11:29
	Urumqi	00:22	00:53	10:53	11:23
	Kashi	01:02	01:30	11:48	12:16
C2	Nauru	18:10	18:32	06:40	07:01
C3	Andorra	06:05	06:34	16:40	17:10
C5	The Gambia, Banjul	06:40	07:02	18:38	19:00
C6	Bahamas, Nassau	10:57	11:20	22:25	22:49
C9	Mozambique, Maputo	02:32	02:56	16:09	16:34
	Mozambique	02:16	02:39	15:22	15:45
CE	Chile, Santiago	09:11	09:38	23:15	23:42
CE0	Easter Island	11:58	12:23	01:39	02:04
CE0	Juan Fernandez	09:43	10:09	23:47	00:14
CE0	San Felix	10:02	10:27	23:41	00:05
CM,CO	Cuba, Havana	11:14	11:38	22:48	23:11
CN	Morocco, Casablanca	06:28	06:54	17:33	17:59
CP	Bolivia, La Paz	09:31	09:53	22:40	23:03
CT	Portugal, Lisbon	06:42	07:09	17:31	17:58
CT3	Madeira Islands, Funchal	07:04	07:30	18:12	18:38
CU	Azores , Ponta Delgada	07:46	08:14	18:38	19:06
CX	Uruguay, Montevideo	08:09	08:37	22:19	22:47
CY0	Sable Is.	10:13	10:43	20:43	21:14
CY9	St. Paul Is.	10:19	10:52	20:36	21:08
D2	Angola, Luanda	04:15	04:37	17:04	17:26
D4	Cape Verde, Praia	07:09	07:31	19:03	19:25
D6	Comoros, Moroni	02:09	02:31	15:06	15:28
DA-DL,Y2-Y9	Fed. Rep. of Germany, Bonn	05:58	06:33	15:58	16:33
	Berlin	05:36	06:13	15:26	16:03
DU	Philippines, Manilla	21:31	21:53	09:25	09:48
E3	Eritrea, Asmara	03:00	03:22	14:53	15:15
EA	Spain, Madrid	06:22	06:51	17:05	17:34
EA6	Balearic Is., Palma	05:56	06:24	16:42	17:10
EA8	Canary Is., Las Palmas	06:48	07:13	18:08	18:33
EA9	Ceuta & Melilla, Ceuta	06:22	06:49	17:20	17:47
	Melilla	06:12	06:38	17:12	17:38
EI	Ireland, Dublin	06:57	07:34	16:43	17:20
EK	Armenia, Yerevan	03:09	03:38	13:53	14:21
EL	Liberia, Monrovia	06:09	06:30	18:23	18:44
EP	Iran, Tehran	02:33	03:00	13:32	13:59
ER	Moldova, Kishinev	04:23	04:56	14:40	15:13
ES	Estonia, Tallinn	05:09	05:54	14:14	14:59
ET	Ethiopia, Addis Ababa	02:54	03:15	15:01	15:23
EU,EV,EW	Belarus, Minsk	04:42	05:21	14:25	15:03
EX	Kyrgyzstan, Frunze	01:13	01:43	11:47	12:17
EY	Tajikistan, Samarkand	01:39	02:07	12:25	12:53
	Dushanbe	01:30	01:58	12:18	12:46
EZ	Turkmenistan, Ashkhabad	02:10	02:38	13:01	13:29
F	France, Paris	06:13	06:46	16:22	16:55
FG	Guadeloupe	09:43	10:06	21:34	21:57
FJ, FS	St. Martin	09:51	10:14	21:37	22:00
FH	Mayotte	02:01	02:23	15:01	15:23
FK	New Caledonia, Noumia	17:42	18:06	07:08	07:32
FM	Martinique	09:39	10:01	21:33	21:56
FO	Clipperton Is.	12:47	13:09	00:51	01:13
FO	Fr. Polynesia, Tahiti	15:33	15:56	04:46	05:09
	Rurutu, (Austral Is.)	18:43	19:06	08:10	08:33
	Hiva Oa, (Marquesas Is.)	14:22	14:44	03:14	03:36
FP	St. Pierre & Miquelon, St. Pierre	10:02	10:34	20:20	20:53
FR/G	Glorioso	01:55	02:17	14:51	15:13
FR/J,E	Juan de Nova	02:05	02:28	15:16	15:39
	Europa	02:07	02:30	15:33	15:56
FR	Reunion	01:08	01:31	14:30	14:54
FR/T	Tromelin	01:20	01:43	14:28	14:51
FT8W	Crozet	00:26	01:00	15:31	16:04
FT8X	Kerguelen	23:05	23:41	14:31	15:08
FT8Y	Antarctica, Dumont D'Urville	15:38	17:22	11:25	13:09
FT8Z	Amsterdam & St. Paul Is., Amsterdam	23:08	23:36	13:29	13:58
FW	Wallis & Futuna Is., Wallis	16:47	17:09	05:48	06:10
FY	Fr. Guiana, Cayenne	08:53	09:15	21:10	21:32
G	England, London	06:28	07:04	16:24	16:59
GD	Isle of Man	06:52	07:30	16:32	17:11
GI	Northern Ireland, Belfast	06:58	07:37	16:36	17:15
GJ	Jersey	06:32	07:06	16:38	17:12
GM	Scotland, Glasgow	06:55	07:35	16:26	17:06
	Aberdeen	06:50	07:32	16:11	16:53
GU	Guernsey	06:34	07:08	16:40	17:14
GW	Wales, Cardiff	06:40	07:16	16:36	17:12
H4	Solomon Islands, Honiara	18:27	18:49	07:18	07:40
HA	Hungary, Budapest	05:03	05:36	15:18	15:51
HB	Switzerland, Bern	05:48	06:21	16:06	16:38
HB0	Liechtenstein	05:40	06:13	15:57	16:29
HC	Ecuador, Quito	10:30	10:51	22:59	23:20
HC8	Galapagos Is.	11:20	11:41	23:49	00:11
HH	Haiti, Port-Au-Prince	10:28	10:51	22:14	22:36
HI	Dominican Republic, Santo Domingo	10:19	10:42	22:04	22:27
HK	Colombia, Bogota	10:20	10:41	22:38	22:59
HK0	Malpelo Is.	10:47	11:09	23:06	23:28
HK0	San Adreas	10:59	11:21	22:59	23:21
HL	Korea, Seoul	21:35	22:02	08:28	08:55
HP	Panama, Panama	10:46	11:08	22:54	23:16
HR	Honduras, Tegucigalpa	11:23	11:45	23:19	23:41
HS	Thailand, Bangkok	22:52	23:14	10:48	11:10
HV	Vatican City	05:20	05:49	15:57	16:27
HZ, 7Z	Saudi Arabia, Dharan	02:28	02:52	13:54	14:18
	Mecca	03:03	03:27	14:41	15:05
	Riyadh	02:40	03:03	14:10	14:33
I	Italy, Rome	05:20	05:49	15:57	16:27
	Trieste	05:21	05:52	15:44	16:15
	Sicily	05:07	05:34	16:00	16:27
IS	Sardinia, Cagliari	05:29	05:57	16:16	16:44
J2	Djibouti, Djibouti	02:39	03:01	14:40	15:02
J3	Grenada	09:39	10:01	21:40	22:02
J5	Guinea-Bissau, Bissau	06:34	06:56	18:35	18:57
J6	St. Lucia	09:38	10:00	21:34	21:56
J7	Dominica	09:41	10:03	21:34	21:56
J8	St. Vincent	09:39	10:01	21:36	21:58
JA-JS	Japan, Tokyo	20:41	21:08	07:40	08:07
	Nagasaki	21:17	21:43	08:24	08:50
	Sapporo	20:46	21:16	07:19	07:49
JD1	Minami Torishima	19:30	19:54	07:01	07:24
JD1	Ogasawara, Kazan Is.	20:26	20:50	07:48	08:13
JT	Mongolia, Ulan Bator	23:12	23:45	09:26	09:59
JW	Svalbard, Spitsbergen	09:02	—	—	12:16
JX	Jan Mayen	08:20	09:49	14:44	16:13
JY	Jordan, Amman	03:32	03:57	14:42	15:07
KC4	Antarctica, Byrd Station	—	—	—	—
	McMurdo Sound				
	Palmer Station	05:55	07:16	00:43	02:03
KC6	Belau, Yap	20:16	20:38	08:23	08:45
	Koror	20:29	20:50	08:40	09:02
KG4	Guantanamo Bay	10:42	11:05	22:23	22:46
KH0	Mariana Is., Saipan	19:52	20:15	07:45	08:08
KH1	Baker, Howland Is.	17:03	17:24	05:30	05:52
KH2	Guam, Agana	19:54	20:16	07:52	08:14
KH3	Johnston Is.	16:52	17:14	04:40	05:03
KH4	Midway Is.	17:40	18:05	05:01	05:25
KH5	Palmyra Is.	16:13	16:35	04:28	04:50
KH5K	Kingman Reef	16:18	16:40	04:29	04:51
KH6	Hawaii, Hilo	16:01	16:24	03:43	04:06
	Honolulu	16:14	16:37	03:52	04:16
KH7	Kure Is.	17:45	18:09	05:05	05:29
KH8	American Samoa, Pago Pago	16:23	16:46	05:27	05:49
KH9	Wake Is.	18:34	18:57	06:17	06:40
KL7	Alaska, Adak	18:14	18:51	04:09	04:45
	Anchorage	16:54	17:43	01:44	02:32
	Fairbanks	17:00	17:58	01:12	02:10
	Juneau	15:42	16:26	00:56	01:39
	Nome	18:09	19:05	02:24	03:21
KP1	Navassa Is.	10:39	11:02	22:25	22:47
KP2	Virgin Islands, Charlotte Amalie	09:59	10:21	21:44	22:07
KP4	Puerto Rico, San Juan	10:04	10:27	21:49	22:12
KP5	Desecheo Is.	10:09	10:32	21:55	22:17
LA-LJ	Norway, Oslo	06:07	06:54	15:07	15:54
LU	Argentina, Buenos Aires	08:19	08:46	22:27	22:55
LX	Luxembourg	05:59	06:33	16:04	16:38
LY	Lithuania, Vilna	04:52	05:31	14:31	15:10
LZ	Bulgaria, Sofia	04:38	05:08	15:12	15:42
OA	Peru, Lima	10:12	10:34	23:09	23:32
OD	Lebanon, Beirut	03:36	04:02	14:40	15:06
OE	Austria, Vienna	05:15	05:49	15:27	16:01
OH	Finland, Helsinki	05:11	05:58	14:09	14:56
OH0	Aland Is.	05:31	06:18	14:29	15:16
OJ0	Market Reef	05:35	06:22	14:32	15:19
OK,OL	Czech Rep., Prague	05:27	06:01	15:30	16:04
OM	Slovak, Rep., Bratislava	05:12	05:45	15:25	15:58
ON	Belgium, Brussels	06:08	06:44	16:07	16:43
OX	Greenland, Godthaab	10:33	11:28	18:52	19:47
	Thule	13:34	—	—	19:03
OY	Faroe Islands, Torshavn	07:24	08:15	16:06	16:57
OZ	Denmark, Copenhagen	05:47	06:27	15:18	15:59
P2	Papua New Guinea, Madang	19:30	19:51	08:10	08:32
	Port Moresby	19:18	19:40	08:09	08:31
P4	Aruba, Oranjestad	10:08	10:30	22:08	22:30
PA-PI	Netherlands, Amsterdam	06:10	06:46	16:01	16:37
PJ2,4,9	Netherlands Antilles, Willemstad	10:13	10:35	22:12	22:34
PJ5-8	St. Maarten and Saba, St. Maarten	09:51	10:14	21:38	22:01
PY	Brazil, Brasilia	08:10	08:32	21:17	21:40
	Rio De Janeiro	07:40	08:04	21:08	21:32
	Natal	07:32	07:53	20:15	20:36
	Manaus	09:15	09:37	21:51	22:13
	Porto Alegre	07:59	08:25	21:51	22:17
PY0	Fernando De Noronha	07:23	07:45	20:01	20:22
PY0	St. Peter & St. Paul Rocks	07:17	07:38	19:43	20:05
PY0	Trindade & Martin Vaz Is., Trindade	06:48	07:12	20:09	20:32
PZ	Suriname, Paramaribo	09:06	09:27	21:21	21:42
R1FJ_	Franz Josef Land	—	—	—	—
R1MV_	Malyj Vysotskij Is.	04:58	05:45	13:52	14:40
S0	Western Sahara, Smara	06:34	06:58	17:59	18:23
S2	Bangladesh, Dacca	23:44	00:07	11:16	11:40
S5	Slovenia, Ljubljana	05:19	05:50	15:40	16:12
S7	Seychelles, Victoria	01:31	01:52	14:10	14:32
S9	Sao Tome	04:52	05:13	17:20	17:41
SM	Sweden, Stockholm	05:35	06:21	14:41	15:26
SP	Poland, Krakow	05:04	05:39	15:08	15:42
	Warsaw	05:05	05:41	14:57	15:34
ST	Sudan, Khartoum	03:26	03:48	15:18	15:40
ST0	Southern Sudan, Juba	03:18	03:39	15:34	15:56
SU	Egypt, Cairo	03:47	04:12	15:03	15:28
SV	Greece, Athens	04:29	04:57	15:20	15:48
SV/A	Mount Athos	04:30	04:58	15:14	15:42
SV5	Dodecanese, Rhodes	04:09	04:36	15:05	15:32
SV9	Crete	04:19	04:46	15:19	15:46
T2	Tuvalu, Funafuti	17:13	17:35	06:03	06:24

Pref	State/Province/Country/City	Tlight	Sunrise	Sunset	Tlight
T30	West Kiribati, Bonriki	17:47	18:09	06:12	06:34
T31	Central Kiribati, Kanton	16:42	17:03	05:17	05:38
T32	East Kiribati, Christmas Is.	15:50	16:11	04:14	04:36
T33	Banaba Is.	18:00	18:22	06:30	06:51
T5	Somalia, Mogadishu	02:19	02:41	14:43	15:04
T7	San Marino	05:24	05:54	15:54	16:24
T9	Bosnia-Herzegovina, Sarajevo	04:59	05:30	15:29	16:00
TA	Turkey, Ankara	03:55	04:23	14:40	15:08
	Istanbul	04:13	04:42	14:53	15:22
TF	Iceland, Reykjavik	08:33	09:29	16:53	17:49
TG	Guatemala, Guatemala City	11:37	11:59	23:31	23:54
TI	Costa Rica, San Jose	11:06	11:27	23:11	23:33
TI9	Cocos Is.	11:13	11:34	23:28	23:50
TJ	Cameroon, Yaounde	04:37	04:58	16:56	17:18
TK	Corsica	05:34	06:03	16:11	16:41
TL	Central African Republic, Bangui	04:09	04:30	16:27	16:49
TN	Congo, Brazzaville	04:12	04:33	16:50	17:12
TR	Gabon, Libreville	04:41	05:02	17:08	17:30
TT	Chad, N'Djamena	04:32	04:54	16:32	16:54
TU	Cote d'Ivoire, Abidjan	05:40	06:02	17:56	18:18
TY	Benin, Porto Novo	05:15	05:37	17:29	17:50
TZ	Mali, Bamako	06:05	06:27	18:04	18:26
UA	Russia, European, Leningrad (UA1)	04:48	05:35	13:49	14:36
	Archangel (UA1)	04:26	05:23	12:40	13:37
	Murmansk (UA1)	05:19	06:34	12:28	13:43
	Moscow (UA3)	04:07	04:48	13:38	14:18
	Kuybyshev (UA4)	03:11	03:48	12:57	13:35
	Rostov (UA6)	03:41	04:14	13:56	14:29
UA2	Kaliningrad	05:14	05:53	14:49	15:29
UA9,0	Russia, Asiatic, Novosibirsk (UA9)	01:04	01:43	10:40	11:19
	Perm (UA9)	02:58	03:42	12:14	12:57
	Omsk (UA9)	01:42	02:21	11:18	11:57
	Norilsk (UA0)	01:41	02:58	08:44	10:00
	Irkutsk (UA0)	23:32	00:09	09:23	10:00
	Vladivostok (UA0)	21:24	21:54	07:57	08:27
	Petropavlovsk (UA0)	19:56	20:33	05:44	06:21
	Khabarovsk (UA0)	21:21	21:54	07:31	08:05
	Krasnoyarsk (UA0)	00:27	01:08	09:56	10:37
	Yakutsk (UA0)	22:18	23:09	07:00	07:51
	Wrangel Island (UA0)	19:44	21:14	02:09	03:38
	Kyzyl (UA0Y)	00:10	00:46	10:05	10:41
UJ-UM	Uzbekistan, Bukhoro	01:49	02:17	12:34	13:02
	Tashkent	01:31	02:00	12:12	12:41
UN-UQ	Kazakhstan, Alma-Ata	01:04	01:34	11:37	12:07
UR-UZ,EM-EO	Ukraine, Kiev	04:23	04:58	14:24	14:59
V2	Antigua & Barbuda, St. Johns	09:45	10:07	21:33	21:56
V3	Belize, Belmopan	11:33	11:56	23:21	23:44
V4	St. Kitts & Nevis	09:48	10:11	21:36	21:59
V5	Namibia, Windhoek	03:39	04:03	17:06	17:30
V6	Micronesia, Ponape	18:53	19:15	07:05	07:27
V7	Marshall Islands, Kwajalein	18:19	18:41	06:27	06:49
V8	Brunei, Bandar Seri Begawan	21:44	22:06	10:01	10:23
VK	Australia, Canberra (VK1)	18:27	18:55	08:39	09:06
	Sydney (VK2)	18:22	18:49	08:28	08:55
	Melbourne (VK3)	18:38	19:06	09:00	09:29
	Brisbane (VK4)	18:27	18:52	08:10	08:35
	Adelaide (VK5)	19:10	19:38	09:20	09:47
	Perth (VK6)	20:48	21:14	10:46	11:12
	Hobart, Tasmania (VK7)	18:15	18:46	09:02	09:33
	Darwin (VK8)	20:19	20:41	09:18	09:40
VK0	Heard Is.	22:31	23:12	14:27	15:08
VK0	Macquarie Is.	16:41	17:24	08:52	09:35
VK9C	Cocos-Keeling Is.	22:36	22:58	11:34	11:56
VK9L	Lord Howe Is.	17:55	18:21	07:52	08:18
VK9M	Mellish Reef	18:32	18:55	07:45	08:08
VK9N	Norfolk Is.	17:25	17:50	07:12	07:38
VK9W	Willis Is.	18:59	19:22	08:08	08:31
VK9X	Christmas Is.	22:02	22:24	10:56	11:18
VP2E	Anguilla	09:51	10:14	21:37	21:59
VP2M	Montserrat	09:46	10:09	21:35	21:58
VP2V	British Virgin Is., Tortola	09:58	10:20	21:43	22:06

Pref	State/Province/Country/City	Tlight	Sunrise	Sunset	Tlight
VP5	Turks & Caicos Islands, Grand Turk	10:27	10:50	22:06	22:29
VP8	Falkland Islands, Stanley	07:23	08:02	23:07	23:47
VP8	So. Georgia Is.	05:45	06:28	21:53	22:36
VP8	So. Orkney Is.	05:35	06:32	22:58	23:56
VP8	So. Sandwich Islands, Saunders Is.	04:43	05:32	21:28	22:17
VP8	So. Shetland Is., King George Is.	06:11	07:14	23:58	01:01
VP9	Bermuda	10:15	10:40	21:24	21:49
VQ9	Chagos, Diego Garcia	00:20	00:42	13:06	13:27
VR2,VS6	Hong Kong	22:06	22:30	09:42	10:06
VR6	Pitcairn Is.	13:24	13:48	02:59	03:23
VU	India, Bombay	00:49	01:11	12:33	12:55
	Calcutta	23:50	00:14	11:26	11:49
	New Delhi	00:42	01:07	12:02	12:26
	Bangalore	00:23	00:45	12:21	12:43
VU	Andaman Islands, Port Blair	23:20	23:42	11:22	11:44
VU	Laccadive Is.	00:38	00:59	12:43	13:05
XE	Mexico, Mexico City (XE1)	12:17	12:39	00:00	00:23
	Chihuahua (XE2)	12:55	13:20	00:14	00:39
	Merida (XE3)	11:41	12:04	23:20	23:43
XF4	Revilla Gigedo	13:06	13:29	00:50	01:13
XT	Burkina Faso, Ouagadougou	05:39	06:01	17:38	18:00
XU	Cambodia, Phnom Penh	22:32	22:54	10:34	10:56
XW	Laos, Viangchan	22:48	23:11	10:35	10:57
XX9	Macao	22:09	22:32	09:45	10:08
XZ	Myanmar, Yangon	23:13	23:36	11:03	11:25
YA	Afghanistan, Kandahar	01:31	01:56	12:44	13:09
	Kabul	01:22	01:48	12:25	12:51
YB-YD	Indonesia, Jakarta	22:04	22:25	10:47	11:08
	Medan	22:48	23:09	11:08	11:29
	Pontianak	22:01	22:23	10:29	10:51
	Jayapura	19:52	20:14	08:27	08:48
YI	Iraq, Baghdad	02:59	03:25	14:06	14:32
YJ	Vanuatu, Port Vila	17:42	18:05	06:55	07:18
YK	Syria, Damascus	03:32	03:58	14:38	15:04
YL	Latvia, Riga	05:04	05:46	14:27	15:09
YN	Nicaragua, Managua	11:16	11:38	23:17	23:39
YO	Romania, Bucharest	04:29	05:00	14:58	15:28
YS	El Salvador, San Salvador	11:31	11:53	23:27	23:49
YU	Yugoslavia, Belgrade	04:53	05:24	15:19	15:50
YV	Venezuela, Caracas	09:58	10:20	22:02	22:24
YV0	Aves Is.	09:51	10:13	21:43	22:05
Z2	Zimbabwe, Harare	02:51	03:14	16:04	16:27
Z3	Macedonia, (ex Yugoslav), Skopje	04:44	05:14	15:22	15:51
ZA	Albania, Tirane	04:50	05:19	15:29	15:58
ZB2	Gibraltar	06:23	06:49	17:20	17:47
ZC4	British Cyprus	03:47	04:13	14:49	15:15
ZD7	St. Helena	05:21	05:44	18:30	18:52
ZD8	Ascension Is.	06:06	06:28	18:54	19:16
ZD9	Tristan da Cunha	05:09	05:37	19:28	19:56
ZF	Cayman Is.	11:05	11:28	22:48	23:11
ZK1	No. Cook Is., Manihiki	15:49	16:11	04:43	05:05
ZK1	So. Cook Is., Rarotonga	15:29	15:53	04:52	05:16
ZK2	Niue	16:09	16:32	05:26	05:49
ZK3	Tokelaus, Atafu	16:39	17:01	05:27	05:49
ZL	New Zealand, Auckland (ZL1)	16:41	17:09	06:59	07:27
	Wellington (ZL2)	16:29	17:00	07:08	07:39
	Christchurch (ZL3)	16:32	17:04	07:22	07:54
	Dunedin (ZL4)	16:32	17:06	07:36	08:10
ZL5	Antarctica, Scott Base	——	——	——	——
ZL7	Chatham Is.	15:47	16:19	06:40	07:12
ZL8	Kermadec Is.	16:28	16:53	06:16	06:42
ZL9	Auckland & Campbell Is., Auckland	16:30	17:08	08:07	08:45
	Campbell Is.	16:11	16:51	08:02	08:43
ZP	Paraguay, Asuncion	08:34	08:58	22:10	22:34
ZS	South Africa, Cape Town (ZS1)	03:13	03:40	17:19	17:46
	Port Elizabeth (ZS2)	02:44	03:11	16:50	17:17
	Bloemfontein (ZS4)	02:52	03:17	16:40	17:06
	Durban (ZS5)	02:31	02:57	16:22	16:48
	Johannesburg (ZS6)	02:49	03:14	16:28	16:52
ZS8	Marion Is.	01:20	01:54	16:30	17:04

Pref	State/Province/Country/City	Tlight	Sunrise	Sunset	Tlight
VE1	New Brunswick, Saint John	10:58	11:30	20:49	21:22
	Nova Scotia, Halifax	10:46	11:18	20:41	21:13
	Prince Edward Island, Charlottetown	10:48	11:21	20:35	21:08
VE2	Quebec, Montreal	11:28	12:01	21:19	21:51
	Quebec City	11:22	11:55	21:05	21:39
VE3	Ontario, London	11:53	12:24	21:57	22:28
	Ottawa	11:36	12:09	21:27	22:00
	Sudbury	12:00	12:33	21:45	22:19
	Toronto	11:47	12:19	21:47	22:19
VE4	Manitoba, Winnipeg	13:13	13:49	22:38	23:14
VE5	Saskatchewan, Regina	13:45	14:22	23:06	23:42
	Saskatoon	13:58	14:36	23:08	23:46
VE6	Alberta, Calgary	14:24	15:02	23:42	00:19
	Edmonton	14:30	15:10	23:29	00:09
VE7	British Columbia, Prince George	15:08	15:49	00:04	00:45
	Prince Rupert	15:40	16:21	00:33	01:14
	Vancouver	14:56	15:31	00:24	01:00
VE8	Northwest Territories, Yellowknife	15:11	16:09	22:37	23:35
	Resolute	16:23			19:48
VY1	Yukon, Whitehorse	16:24	17:17	00:14	01:07
VO1	Newfoundland, St. John's	10:10	10:44	19:48	20:23
VO2	Labrador, Goose Bay	10:57	11:36	19:57	20:37
W1	Connecticut, Hartford	11:16	11:47	21:26	21:56
	Maine, Bangor	11:07	11:39	21:02	21:34
	Portland	11:11	11:42	21:11	21:42
	Massachusetts, Boston	11:11	11:42	21:18	21:48
	New Hampshire, Concord	11:15	11:46	21:17	21:48
	Rhode Island, Providence	11:11	11:41	21:20	21:51
	Vermont, Montpelier	11:21	11:53	21:18	21:50
W2	New Jersey, Atlantic City	11:18	11:47	21:39	22:08
	New York, Albany	11:23	11:53	21:28	21:58
	Buffalo	11:43	12:14	21:47	22:18
	New York City	11:19	11:49	21:33	22:03
	Syracuse	11:33	12:04	21:36	22:07
W3	Delaware, Wilmington	11:23	11:52	21:42	22:11
	District of Columbia, Washington	11:28	11:56	21:50	22:19
	Maryland, Baltimore	11:27	11:56	21:48	22:17
	Pennsylvania, Harrisburg	11:30	12:00	21:46	22:16
	Philadelphia	11:22	11:52	21:41	22:10
	Pittsburgh	11:43	12:12	21:58	22:28
	Scranton	11:27	11:58	21:39	22:09
W4	Alabama, Montgomery	11:53	12:19	22:42	23:08
	Florida, Jacksonville	11:31	11:57	22:27	22:53
	Miami	11:18	11:42	22:30	22:54
	Pensacola	11:53	12:19	22:49	23:15
	Georgia, Atlanta	11:48	12:15	22:31	22:58
	Savannah	11:32	11:58	22:22	22:48
	Kentucky, Lexington	11:56	12:24	22:22	22:51
	Louisville	12:01	12:30	22:27	22:56
	North Carolina, Charlotte	11:36	12:03	22:14	22:41
	Raleigh	11:28	11:56	22:04	22:31
	Wilmington	11:23	11:49	22:04	22:31
	South Carolina, Columbia	11:35	12:01	22:17	22:44
	Tennessee, Knoxville	11:50	12:17	22:24	22:52
	Memphis	12:13	12:40	22:51	23:19
	Nashville	12:02	12:29	22:36	23:03
	Virginia, Norfolk	11:21	11:49	21:52	22:20
	Richmond	11:27	11:55	21:55	22:23
W5	Arkansas, Little Rock	12:21	12:48	23:01	23:28
	Louisiana, New Orleans	12:04	12:30	23:02	23:27
	Shreveport	12:23	12:49	23:11	23:37
	Mississippi, Jackson	12:08	12:35	22:58	23:24
	New Mexico, Albuquerque	13:19	13:47	23:58	00:25
	Oklahoma, Oklahoma City	12:43	13:11	23:20	23:47
	Texas, Abilene	12:47	13:13	23:35	00:01
	Amarillo	13:00	13:27	23:38	00:05
	Dallas	12:36	13:02	23:23	23:49
	El Paso	13:13	13:39	00:04	00:30
	San Antonio	12:37	13:02	23:36	00:02
W6	California, Los Angeles	14:04	14:30	00:46	01:13
	San Francisco	14:27	14:55	00:54	01:23
W7	Arizona, Flagstaff	13:40	14:07	00:17	00:45
	Phoenix	13:38	14:05	00:23	00:49
	Idaho, Boise	14:14	14:46	00:15	00:46
	Pocatello	13:58	14:29	00:02	00:33
	Montana, Billings	13:48	14:21	23:37	00:10
	Butte	14:05	14:38	23:53	00:26
	Great Falls	14:04	14:38	23:43	00:17
	Nevada, Las Vegas	13:55	14:23	00:29	00:57
	Reno	14:20	14:49	00:40	01:09
	Oregon, Portland	14:45	15:17	00:35	01:08
	Utah, Salt Lake City	13:51	14:21	00:05	00:35
	Washington, Seattle	14:48	15:22	00:27	01:01
	Spokane	14:29	15:03	00:07	00:41
	Wyoming, Cheyenne	13:23	13:53	23:36	00:06
	Sheridan	13:40	14:12	23:34	00:07
W8	Michigan, Detroit	11:59	12:29	22:05	22:36
	Grand Rapids	12:11	12:42	22:14	22:45
	Sault Ste. Marie	12:14	12:47	21:59	22:32
	Traverse City	12:15	12:47	22:09	22:41
	Ohio, Cincinnati	11:58	12:27	22:20	22:49
	Cleveland	11:52	12:22	22:02	22:33
	Columbus	11:54	12:23	22:11	22:41
	West Virginia, Charleston	11:45	12:14	22:10	22:38
W9	Illinois, Chicago	12:16	12:47	22:25	22:55
	Indiana, Indianapolis	12:06	12:36	22:25	22:54
	Wisconsin, Green Bay	12:23	12:55	22:19	22:51

Pref	State/Province/Country/City	Tlight	Sunrise	Sunset	Tlight
	Milwaukee	12:20	12:51	22:23	22:54
W0	Colorado, Denver	13:21	13:50	23:40	00:09
	Grand Junction	13:34	14:03	23:56	00:25
	Iowa, Des Moines	12:39	13:10	22:50	23:20
	Kansas, Pratt	12:52	13:20	23:20	23:48
	Wichita	12:47	13:15	23:14	23:43
	Minnesota, Duluth	12:45	13:19	22:29	23:02
	Minneapolis	12:46	13:18	22:39	23:11
	Missouri, Columbia	12:29	12:58	22:51	23:20
	Kansas City	12:38	13:07	23:00	23:29
	St. Louis	12:20	12:49	22:44	23:12
	Nebraska, North Platte	13:07	13:37	23:20	23:50
	Omaha	12:48	13:18	23:00	23:30
	North Dakota, Fargo	13:04	13:38	22:47	23:21
	South Dakota, Rapid City	13:23	13:55	23:21	23:53
-		05:20	05:42	17:49	18:11
1A0	S.M.O.M	05:36	06:07	15:45	16:15
1S	Spratly Is.	22:05	22:27	10:08	10:31
3A	Monaco	06:00	06:31	16:00	16:31
3B6	Agalega	01:18	01:40	14:18	14:40
3B7	St. Brandon	00:55	01:18	14:14	14:37
3B8	Mauritius	00:57	01:21	14:30	14:54
3B9	Rodriguez Is.	00:34	00:58	14:05	14:29
3C	Equatorial Guinea, Bata	04:43	05:05	17:07	17:29
	Malabo	04:50	05:12	17:08	17:30
3C0	Annobon Is.	04:55	05:17	17:29	17:51
3D2	Fiji Is., Suva	16:57	17:21	06:23	06:46
3D2	Conway Reef	17:07	17:31	06:44	07:08
3D2	Rotuma Is.	17:10	17:33	06:16	06:39
3DA	Swaziland, Mbabane	02:30	02:56	16:26	16:51
3V	Tunisia, Tunis	05:35	06:03	16:06	16:34
3W, XV	Vietnam, Ho Chi Minh City (Saigon)	22:28	22:51	10:26	10:49
	Hanoi	22:47	23:10	10:14	10:38
3X	Guinea, Conakry	06:28	06:50	18:30	18:52
3Y	Bouvet	02:34	03:23	19:41	20:29
3Y	Peter I Is.	——	06:38	04:57	——
4J, 4K	Azerbaijan, Baku	03:03	03:33	13:19	13:48
4L	Georgia, Tbilisi	03:26	03:56	13:36	14:06
4S	Sri Lanka, Colombo	00:10	00:32	12:19	12:41
4U	ITU Geneva	06:11	06:44	15:57	16:30
4U	United Nations Hq.	11:19	11:49	21:33	22:03
4X, 4Z	Israel, Jerusalem	03:46	04:12	14:37	15:03
5A	Libya, Tripoli	05:18	05:44	16:06	16:33
	Benghazi	04:47	05:14	15:37	16:03
5B	Cyprus, Nicosia	03:59	04:26	14:37	15:04
5H	Tanzania, Dar es Salaam	02:31	02:54	15:21	15:43
5N	Nigeria, Lagos	05:16	05:38	17:26	17:48
5R	Madagascar, Antananarivo	01:39	02:03	15:08	15:31
5T	Mauritania, Nouakchott	06:50	07:13	18:26	18:49
5U	Niger, Niamey	05:31	05:54	17:21	17:44
5V	Togo, Lome	05:23	05:45	17:36	17:58
5W	Western Samoa, Apia	16:26	16:49	05:36	05:59
5X	Uganda, Kampala	03:10	03:32	15:38	16:00
5Z	Kenya, Nairobi	04:11	04:33	16:44	17:06
6W	Senegal, Dakar	06:51	07:13	18:37	19:00
6Y	Jamaica, Kingston	10:53	11:16	22:29	22:52
7O	Yemen, Aden	02:38	03:01	14:30	14:53
	Sanaa	02:45	03:08	14:29	14:52
7P	Lesotho, Maseru	02:38	03:04	16:46	17:13
7Q	Malawi, Lilongwe	02:43	03:06	15:55	16:17
	Blantyre	02:35	02:58	15:53	16:16
7X	Algeria, Algiers	06:03	06:31	16:35	17:03
8P	Barbados, Bridgetown	09:37	09:59	21:28	21:51
8Q	Maldive Is.	00:33	00:55	12:49	13:11
8R	Guyana, Georgetown	09:22	09:44	21:32	21:54
9A	Croatia, Zagreb	05:30	06:03	15:19	15:52
9G	Ghana, Accra	05:29	05:51	17:42	18:04
9H	Malta	05:17	05:44	15:51	16:19
9J	Zambia, Lusaka	03:02	03:26	16:19	16:42
9K	Kuwait	02:52	03:17	13:51	14:16
9L	Sierra Leone, Freetown	06:25	06:47	18:29	18:52
9M2	West Malaysia, Kuala Lumpur	22:38	23:00	10:58	11:20
9M6,8	East Malaysia, Sabah, Sandakan(9M6)	21:36	21:58	09:48	10:10
	Sarawak, Kuching (9M8)	22:01	22:23	10:25	10:47
9N	Nepal, Kathmandu	00:19	00:44	11:24	11:49
9Q	Zaire, Kinshasa	04:12	04:34	16:54	17:16
	Kisangani	03:40	04:02	16:07	16:29
	Lubumbashi	03:12	03:35	16:16	16:39
9U	Burundi, Bujumbura	03:18	03:40	15:56	16:19
9V	Singapore	22:26	22:48	10:52	11:14
9X	Rwanda, Kigali	03:16	03:38	15:51	16:13
9Y	Trinidad & Tobago , Port of Spain	09:40	10:02	21:39	22:01
A2	Botswana, Gaborone	02:54	03:19	16:44	17:09
A3	Tonga, Nukualofa	16:26	16:50	06:02	06:26
A4	Oman, Musqat	01:59	02:23	13:18	13:42
A5	Bhutan, Thimbu	00:02	00:27	11:09	11:34
A6	United Arab Emirates, Abu Dhabi	02:18	02:42	13:34	13:59
A7	Qatar, Ad-Dawhah	02:30	02:55	13:44	14:08
A9	Bahrein, Al-Manamah	02:35	03:00	13:46	14:11
AP	Pakistan, Karachi	01:27	01:52	12:42	13:07
AP	Islamabad	01:17	01:44	12:01	12:28
BV	Taiwan, Taipei	21:50	22:14	09:04	09:29
BY	Peoples Rep. of China, Beijing	22:36	23:06	08:54	09:23
	Harbin	22:08	22:40	07:57	08:29
	Shanghai	22:00	22:26	08:53	09:19
	Fuzhou	22:00	22:25	09:11	09:36
	Xian	22:56	23:22	09:37	10:04

Pref	State/Province/Country/City	Tlight	Sunrise	Sunset	Tlight
	Chongqing	22:57	23:23	09:56	10:21
	Chengdu	23:09	23:34	10:04	10:29
	Lhasa	23:59	00:24	10:57	11:23
	Urumqi	00:39	01:11	10:39	11:10
	Kashi	01:17	01:46	11:37	12:06
C2	Nauru	18:11	18:33	06:42	07:04
C3	Andorra	06:21	06:52	16:27	16:58
C5	The Gambia, Banjul	06:46	07:08	18:36	18:58
C6	Bahamas, Nassau	11:06	11:30	22:20	22:44
C9	Mozambique, Maputo	02:25	02:50	16:20	16:45
	Mozambique	02:13	02:36	15:29	15:52
CE	Chile, Santiago	09:01	09:29	23:28	23:56
CE0	Easter Island	11:51	12:16	01:50	02:15
CE0	Juan Fernandez	09:33	10:00	00:01	00:28
CE0	San Felix	09:55	10:20	23:51	00:16
CM,CO	Cuba, Havana	11:23	11:47	22:43	23:07
CN	Morocco, Casablanca	06:40	07:07	17:24	17:51
CP	Bolivia, La Paz	09:27	09:51	22:47	23:11
CT	Portugal, Lisbon	06:56	07:25	17:19	17:48
CT3	Madeira Islands, Funchal	07:16	07:42	18:04	18:30
CU	Azores , Ponta Delgada	08:00	08:28	18:28	18:56
CX	Uruguay, Montevideo	07:59	08:27	22:33	23:01
CY0	Sable Is.	10:30	11:01	20:29	21:01
CY9	St. Paul Is.	10:38	11:12	20:19	20:53
D2	Angola, Luanda	04:14	04:36	17:09	17:31
D4	Cape Verde, Praia	07:15	07:38	19:01	19:24
D6	Comoros, Moroni	02:07	02:30	15:12	15:34
DA-DL,Y2-Y9	Fed. Rep. of Germany, Bonn	06:19	06:56	15:39	16:15
	Berlin	05:59	06:38	15:06	15:45
DU	Philippines, Manilla	21:37	21:59	09:23	09:46
E3	Eritrea, Asmara	03:06	03:29	14:51	15:13
EA	Spain, Madrid	06:37	07:07	16:53	17:23
EA6	Balearic Is., Palma	06:10	06:40	16:30	16:59
EA8	Canary Is., Las Palmas	06:58	07:24	18:01	18:27
EA9	Ceuta & Melilla, Ceuta	06:35	07:03	17:10	17:38
	Melilla	06:25	06:52	17:02	17:30
EI	Ireland, Dublin	07:20	08:00	16:21	17:01
EK	Armenia, Yerevan	03:24	03:54	13:41	14:10
EL	Liberia, Monrovia	06:12	06:34	18:23	18:45
EP	Iran, Tehran	02:47	03:14	13:22	13:50
ER	Moldova, Kishinev	04:42	05:16	14:24	14:58
ES	Estonia, Tallinn	05:39	06:28	13:44	14:33
ET	Ethiopia, Addis Ababa	02:58	03:20	15:01	15:23
EU,EV,EW	Belarus, Minsk	05:07	05:47	14:03	14:43
EX	Kyrgyzstan, Frunze	01:29	02:00	11:33	12:04
EY	Tajikistan, Samarkand	01:54	02:23	12:13	12:42
	Dushanbe	01:45	02:14	12:06	12:35
EZ	Turkmenistan, Ashkhabad	02:24	02:53	12:51	13:19
F	France, Paris	06:33	07:08	16:04	16:39
FG	Guadeloupe	09:49	10:12	21:32	21:55
FJ, FS	St. Martin	09:58	10:21	21:34	21:57
FH	Mayotte	01:59	02:21	15:07	15:30
FK	New Caledonia, Noumia	17:37	18:01	07:17	07:42
FM	Martinique	09:45	10:07	21:31	21:54
FO	Clipperton Is.	12:51	13:14	00:51	01:13
FO	Fr. Polynesia, Tahiti	15:30	15:53	04:53	05:17
	Rurutu, (Austral Is.)	18:37	19:02	08:19	08:43
	Hiva Oa, (Marquesas Is.)	14:21	14:43	03:19	03:42
FP	St. Pierre & Miquelon, St. Pierre	10:21	10:54	20:05	20:38
FR/G	Glorioso	01:53	02:16	14:57	15:19
FR/J,E	Juan de Nova	02:02	02:25	15:23	15:47
	Europa	02:01	02:26	15:42	16:06
FR	Reunion	01:03	01:27	14:39	15:03
FR/T	Tromelin	01:17	01:40	14:34	14:58
FT8W	Crozet	00:08	00:44	15:51	16:27
FT8X	Kerguelen	22:44	23:23	14:54	15:34
FT8Y	Antarctica, Dumont D'Urville	———	16:21	12:30	———
FT8Z	Amsterdam & St. Paul Is., Amsterdam	22:56	23:25	13:45	14:14
FW	Wallis & Futuna Is., Wallis	16:44	17:07	05:54	06:17
FY	Fr. Guiana, Cayenne	08:56	09:18	21:11	21:33
G	England, London	06:50	07:28	16:04	16:42
GD	Isle of Man	07:16	07:57	16:09	16:50
GI	Northern Ireland, Belfast	07:23	08:04	16:14	16:55
GJ	Jersey	06:52	07:28	16:21	16:56
GM	Scotland, Glasgow	07:21	08:04	16:01	16:44
	Aberdeen	07:17	08:03	15:45	16:30
GU	Guernsey	06:55	07:30	16:22	16:58
GW	Wales, Cardiff	07:02	07:40	16:16	16:54
H4	Solomon Islands, Honiara	18:26	18:48	07:23	07:45
HA	Hungary, Budapest	05:22	05:56	15:02	15:36
HB	Switzerland, Bern	06:07	06:41	15:50	16:23
HB0	Liechtenstein	05:59	06:33	15:41	16:14
HC	Ecuador, Quito	10:32	10:53	23:01	23:23
HC8	Galapagos Is.	11:21	11:43	23:52	00:14
HH	Haiti, Port-Au-Prince	10:35	10:59	22:10	22:34
HI	Dominican Republic, Santo Domingo	10:26	10:49	22:00	22:24
HK	Colombia, Bogota	10:23	10:45	22:39	23:01
HK0	Malpelo Is.	10:50	11:12	23:07	23:29
HK0	San Adreas	11:04	11:27	22:57	23:20
HL	Korea, Seoul	21:49	22:17	08:17	08:46
HP	Panama, Panama	10:51	11:13	22:54	23:16
HR	Honduras, Tegucigalpa	11:29	11:51	23:17	23:40
HS	Thailand, Bangkok	22:57	23:20	10:47	11:09
HV	Vatican City	05:36	06:06	15:45	16:15
HZ, 7Z	Saudi Arabia, Dharan	02:38	03:03	13:48	14:13
	Mecca	03:11	03:35	14:37	15:01
	Riyadh	02:48	03:13	14:04	14:29
I	Italy, Rome	05:36	06:06	15:45	16:15
	Trieste	05:39	06:12	15:29	16:01
	Sicily	05:21	05:49	15:49	16:18
IS	Sardinia, Cagliari	05:44	06:13	16:05	16:34
J2	Djibouti, Djibouti	02:44	03:06	14:39	15:02
J3	Grenada	09:44	10:07	21:39	22:01
J5	Guinea-Bissau, Bissau	06:39	07:02	18:34	18:56
J6	St. Lucia	09:44	10:06	21:32	21:55
J7	Dominica	09:47	10:10	21:31	21:54
J8	St. Vincent	09:44	10:07	21:35	21:57
JA-JS	Japan, Tokyo	20:54	21:22	07:30	07:58
	Nagasaki	21:29	21:55	08:16	08:43
	Sapporo	21:03	21:34	07:06	07:37
JD1	Minami Torishima	19:39	20:03	06:56	07:20
JD1	Ogasawara, Kazan Is.	20:36	21:01	07:42	08:07
JT	Mongolia, Ulan Bator	23:32	00:06	09:09	09:43
JW	Svalbard, Spitsbergen	———	———	———	———
JX	Jan Mayen	09:19	———	———	15:18
JY	Jordan, Amman	03:44	04:10	14:34	15:00
KC4	Antarctica, Byrd Station	———	———	———	———
	McMurdo Sound	———	———	———	———
	Palmer Station	———	06:27	01:36	———
KC6	Belau, Yap	20:21	20:43	08:22	08:44
	Koror	20:32	20:54	08:40	09:02
KG4	Guantanamo Bay	10:49	11:12	22:20	22:43
KH0	Mariana Is., Saipan	19:58	20:21	07:43	08:06
KH1	Baker, Howland Is.	17:05	17:27	05:32	05:54
KH2	Guam, Agana	20:00	20:22	07:50	08:12
KH3	Johnston Is.	16:58	17:21	04:38	05:01
KH4	Midway Is.	17:51	18:16	04:54	05:19
KH5	Palmyra Is.	16:17	16:39	04:29	04:51
KH5K	Kingman Reef	16:22	16:44	04:29	04:51
KH6	Hawaii, Hilo	16:08	16:32	03:40	04:03
	Honolulu	16:22	16:46	03:48	04:12
KH7	Kure Is.	17:55	18:20	04:58	05:23
KH8	American Samoa, Pago Pago	16:21	16:44	05:33	05:56
KH9	Wake Is.	18:41	19:04	06:14	06:37
KL7	Alaska, Adak	18:37	19:15	03:49	04:27
	Anchorage	17:27	18:20	01:10	02:04
	Fairbanks	17:39	18:45	00:28	01:35
	Juneau	16:11	16:58	00:28	01:15
	Nome	18:47	19:52	01:42	02:47
KP1	Navassa Is.	10:46	11:09	22:21	22:45
KP2	Virgin Islands, Charlotte Amalie	10:06	10:29	21:41	22:04
KP4	Puerto Rico, San Juan	10:11	10:34	21:46	22:09
KP5	Desecheo Is.	10:16	10:39	21:52	22:15
LA-LJ	Norway, Oslo	06:38	07:29	14:36	15:27
LU	Argentina, Buenos Aires	08:08	08:37	22:41	23:09
LX	Luxembourg	06:19	06:55	15:46	16:22
LY	Lithuania, Vilna	05:17	05:58	14:08	14:50
LZ	Bulgaria, Sofia	04:54	05:25	14:59	15:30
OA	Peru, Lima	10:10	10:32	23:15	23:38
OD	Lebanon, Beirut	03:48	04:15	14:31	14:58
OE	Austria, Vienna	05:35	06:10	15:10	15:45
OH	Finland, Helsinki	05:42	06:33	13:38	14:29
OH0	Aland Is.	06:02	06:53	13:58	14:49
OJ0	Market Reef	06:06	06:58	14:01	14:52
OK,OL	Czech Rep., Prague	05:48	06:24	15:11	15:48
OM	Slovak. Rep., Bratislava	05:32	06:06	15:08	15:43
ON	Belgium, Brussels	06:30	07:07	15:48	16:25
OX	Greenland, Godthaab	11:10	12:14	18:10	19:14
	Thule	———	———	———	———
OY	Faroe Islands, Torshavn	07:58	08:54	15:31	16:27
OZ	Denmark, Copenhagen	06:13	06:56	14:54	15:37
P2	Papua New Guinea, Madang	19:30	19:52	08:14	08:36
	Port Moresby	19:17	19:40	08:14	08:37
P4	Aruba, Oranjestad	10:13	10:35	22:07	22:30
PA-PI	Netherlands, Amsterdam	06:33	07:11	15:40	16:19
PJ2,4,9	Netherlands Antilles, Willemstad	10:18	10:41	22:11	22:33
PJ5-8	St. Maarten and Saba, St. Maarten	09:58	10:21	21:35	21:58
PY	Brazil, Brasilia	08:07	08:30	21:24	21:47
	Rio De Janeiro	07:34	07:59	21:17	21:42
	Natal	07:32	07:54	20:18	20:41
	Manaus	09:16	09:38	21:54	22:16
	Porto Alegre	07:51	08:17	22:03	22:29
PY0	Fernando De Noronha	07:24	07:46	20:04	20:26
PY0	St. Peter & St. Paul Rocks	07:19	07:41	19:45	20:07
PY0	Trindade & Martin Vaz Is., Trindade	06:44	07:08	20:17	20:41
PZ	Suriname, Paramaribo	09:09	09:31	21:21	21:43
R1FJ_	Franz Josef Land	———	———	———	———
R1MV_	Malyj Vysotskij Is.	05:29	06:22	13:20	14:13
S0	Western Sahara, Smara	06:44	07:08	17:54	18:18
S2	Bangladesh, Dacca	23:52	00:16	11:11	11:35
S5	Slovenia, Ljubljana	05:37	06:10	15:25	15:58
S7	Seychelles, Victoria	01:31	01:53	14:14	14:36
S9	Sao Tome	04:53	05:15	17:22	17:44
SM	Sweden, Stockholm	06:05	06:55	14:11	15:01
SP	Poland, Krakow	05:25	06:01	14:49	15:25
	Warsaw	05:28	06:06	14:37	15:15
ST	Sudan, Khartoum	03:32	03:55	15:16	15:39
ST0	Southern Sudan, Juba	03:21	03:43	15:35	15:57
SU	Egypt, Cairo	03:58	04:24	14:56	15:21
SV	Greece, Athens	04:43	05:12	15:10	15:38
SV/A	Mount Athos	04:45	05:15	15:02	15:31
SV5	Dodecanese, Rhodes	04:22	04:50	14:55	15:23
SV9	Crete	04:32	05:00	15:09	15:37
T2	Tuvalu, Funafuti	17:12	17:35	06:07	06:30

Pref	State/Province/Country/City	Tlight	Sunrise	Sunset	Tlight
T30	West Kiribati, Bonriki	17:49	18:11	06:14	06:36
T31	Central Kiribati, Kanton	16:42	17:05	05:20	05:42
T32	East Kiribati, Christmas Is.	15:52	16:14	04:16	04:38
T33	Banaba Is.	18:01	18:23	06:32	06:54
T5	Somalia, Mogadishu	02:21	02:43	14:44	15:06
T7	San Marino	05:41	06:12	15:40	16:11
T9	Bosnia-Herzegovina, Sarajevo	05:16	05:48	15:15	15:47
TA	Turkey, Ankara	04:10	04:39	14:28	14:57
	Istanbul	04:28	04:58	14:41	15:11
TF	Iceland, Reykjavik	09:11	10:14	16:12	17:16
TG	Guatemala, Guatemala City	11:43	12:05	23:29	23:52
TI	Costa Rica, San Jose	11:10	11:32	23:10	23:33
TI9	Cocos Is.	11:16	11:38	23:29	23:51
TJ	Cameroon, Yaounde	04:39	05:01	16:57	17:19
TK	Corsica	05:50	06:20	15:58	16:29
TL	Central African Republic, Bangui	04:12	04:34	16:28	16:50
TN	Congo, Brazzaville	04:12	04:34	16:54	17:16
TR	Gabon, Libreville	04:42	05:04	17:10	17:32
TT	Chad, N'Djamena	04:37	04:59	16:31	16:54
TU	Cote d'Ivoire, Abidjan	05:43	06:05	17:57	18:19
TY	Benin, Porto Novo	05:19	05:41	17:29	17:51
TZ	Mali, Bamako	06:10	06:32	18:02	18:25
UA	Russia, European, Leningrad (UA1)	05:19	06:10	13:19	14:09
	Archangel (UA1)	05:04	06:10	11:57	13:03
	Murmansk (UA1)	06:10	07:49	11:17	12:56
	Moscow (UA3)	04:33	05:16	13:14	13:57
	Kuybyshev (UA4)	03:34	04:14	12:36	13:16
	Rostov (UA6)	04:01	04:35	13:40	14:14
UA2	Kaliningrad	05:39	06:21	14:26	15:08
UA9,0	Russia, Asiatic, Novosibirsk (UA9)	01:29	02:11	10:16	10:58
	Perm (UA9)	03:27	04:14	11:47	12:33
	Omsk (UA9)	02:07	02:49	10:54	11:36
	Norilsk (UA0)	02:33	04:17	07:29	09:13
	Irkutsk (UA0)	23:55	00:33	09:03	09:42
	Vladivostok (UA0)	21:41	22:12	07:43	08:15
	Petropavlovsk (UA0)	20:19	20:59	05:22	06:02
	Khabarovsk (UA0)	21:41	22:16	07:14	07:49
	Krasnoyarsk (UA0)	00:53	01:36	09:32	10:15
	Yakutsk (UA0)	22:52	23:48	06:25	07:21
	Wrangel Island (UA0)	20:44	—	—	02:43
	Kyzyl (UA0Y)	00:32	01:10	09:45	10:23
UJ-UM	Uzbekistan, Bukhoro	02:04	02:33	12:22	12:52
	Tashkent	01:47	02:17	11:59	12:29
UN-UQ	Kazakhstan, Alma-Ata	01:21	01:52	11:23	11:54
UR-UZ,EM-EO	Ukraine, Kiev	04:44	05:21	14:06	14:42
V2	Antigua & Barbuda, St. Johns	09:51	10:14	21:31	21:54
V3	Belize, Belmopan	11:40	12:03	23:18	23:41
V4	St. Kitts & Nevis	09:55	10:18	21:34	21:57
V5	Namibia, Windhoek	03:34	03:58	17:15	17:40
V6	Micronesia, Ponape	18:56	19:19	07:06	07:28
V7	Marshall Islands, Kwajalein	18:24	18:46	06:26	06:49
V8	Brunei, Bandar Seri Begawan	21:47	22:09	10:02	10:24
VK	Australia, Canberra (VK1)	18:17	18:45	08:53	09:21
	Sydney (VK2)	18:12	18:40	08:41	09:09
	Melbourne (VK3)	18:26	18:55	09:15	09:45
	Brisbane (VK4)	18:20	18:46	08:21	08:46
	Adelaide (VK5)	19:00	19:28	09:34	10:02
	Perth (VK6)	20:39	21:06	10:58	11:26
	Hobart, Tasmania (VK7)	17:59	18:32	09:20	09:53
	Darwin (VK8)	20:17	20:40	09:24	09:46
VK0	Heard Is.	22:05	22:50	14:53	15:38
VK0	Macquarie Is.	16:11	17:00	09:20	10:09
VK9C	Cocos-Keeling Is.	22:34	22:57	11:40	12:02
VK9L	Lord Howe Is.	17:46	18:13	08:05	08:32
VK9M	Mellish Reef	18:29	18:52	07:52	08:16
VK9N	Norfolk Is.	17:17	17:43	07:24	07:50
VK9W	Willis Is.	18:56	19:19	08:15	08:39
VK9X	Christmas Is.	22:01	22:24	11:02	11:24
VP2E	Anguilla	09:58	10:21	21:34	21:57
VP2M	Montserrat	09:53	10:16	21:33	21:56
VP2V	British Virgin Is., Tortola	10:04	10:28	21:40	22:03

Pref	State/Province/Country/City	Tlight	Sunrise	Sunset	Tlight
VP5	Turks & Caicos Islands, Grand Turk	10:35	10:59	22:01	22:25
VP8	Falkland Islands, Stanley	06:58	07:41	23:33	00:15
VP8	So. Georgia Is.	05:17	06:04	22:21	23:08
VP8	So. Orkney Is.	04:46	05:58	23:37	00:49
VP8	So. Sandwich Islands, Saunders Is.	04:06	05:04	22:01	22:58
VP8	So. Shetland Is., King George Is.	05:12	06:36	00:41	02:05
VP9	Bermuda	10:26	10:53	21:16	21:42
VQ9	Chagos, Diego Garcia	00:19	00:42	13:10	13:32
VR2,VS6	Hong Kong	22:15	22:38	09:38	10:02
VR6	Pitcairn Is.	13:18	13:43	03:09	03:34
VU	India, Bombay	00:56	01:19	12:29	12:53
	Calcutta	23:59	00:23	11:21	11:45
	New Delhi	00:53	01:18	11:55	12:20
	Bangalore	00:28	00:50	12:19	12:42
VU	Andaman Islands, Port Blair	23:25	23:48	11:21	11:43
VU	Laccadive Is.	00:42	01:04	12:42	13:05
XE	Mexico, Mexico City (XE1)	12:24	12:47	23:56	00:20
	Chihuahua (XE2)	13:06	13:31	00:08	00:33
	Merida (XE3)	11:49	12:12	23:16	23:40
XF4	Revilla Gigedo	13:13	13:36	00:46	01:10
XT	Burkina Faso, Ouagadougou	05:44	06:06	17:37	18:00
XU	Cambodia, Phnom Penh	22:37	23:00	10:33	10:55
XW	Laos, Viangchan	22:55	23:18	10:32	10:55
XX9	Macao	22:17	22:41	09:41	10:05
XZ	Myanmar, Yangon	23:20	23:43	11:00	11:23
YA	Afghanistan, Kandahar	01:42	02:08	12:36	13:02
	Kabul	01:35	02:01	12:16	12:43
YB-YD	Indonesia, Jakarta	22:03	22:26	10:51	11:13
	Medan	22:50	23:12	11:09	11:31
	Pontianak	22:03	22:25	10:32	10:54
	Jayapura	19:53	20:15	08:30	08:52
YI	Iraq, Baghdad	03:11	03:37	13:57	14:24
YJ	Vanuatu, Port Vila	17:38	18:02	07:02	07:26
YK	Syria, Damascus	03:45	04:11	14:29	14:56
YL	Latvia, Riga	05:32	06:17	14:01	14:46
YN	Nicaragua, Managua	11:21	11:43	23:15	23:38
YO	Romania, Bucharest	04:47	05:19	14:43	15:15
YS	El Salvador, San Salvador	11:36	11:59	23:26	23:48
YU	Yugoslavia, Belgrade	05:10	05:43	15:04	15:36
YV	Venezuela, Caracas	10:03	10:25	22:02	22:24
YV0	Aves Is.	09:57	10:20	21:40	22:03
Z2	Zimbabwe, Harare	02:47	03:11	16:12	16:35
Z3	Macedonia, (ex Yugoslav), Skopje	05:00	05:31	15:09	15:39
ZA	Albania, Tirane	05:05	05:35	15:17	15:47
ZB2	Gibraltar	06:36	07:04	17:10	17:38
ZC4	British Cyprus	04:00	04:27	14:40	15:07
ZD7	St. Helena	05:18	05:41	18:36	19:00
ZD8	Ascension Is.	06:05	06:28	18:58	19:20
ZD9	Tristan da Cunha	04:57	05:26	19:43	20:12
ZF	Cayman Is.	11:12	11:36	22:44	23:08
ZK1	No. Cook Is., Manihiki	15:48	16:10	04:48	05:11
ZK1	So. Cook Is., Rarotonga	15:24	15:48	05:01	05:25
ZK2	Niue	16:05	16:29	05:33	05:57
ZK3	Tokelaus, Atafu	16:38	17:00	05:32	05:54
ZL	New Zealand, Auckland (ZL1)	16:29	16:58	07:14	07:43
	Wellington (ZL2)	16:15	16:47	07:25	07:57
	Christchurch (ZL3)	16:16	16:49	07:40	08:14
	Dunedin (ZL4)	16:14	16:50	07:57	08:32
ZL5	Antarctica, Scott Base	—	—	—	—
ZL7	Chatham Is.	15:30	16:04	06:58	07:32
ZL8	Kermadec Is.	16:20	16:46	06:28	06:54
ZL9	Auckland & Campbell Is., Auckland	16:07	16:48	08:31	09:12
	Campbell Is.	15:45	16:30	08:28	09:12
ZP	Paraguay, Asuncion	08:28	08:53	22:20	22:45
ZS	South Africa, Cape Town (ZS1)	03:03	03:31	17:32	18:00
	Port Elizabeth (ZS2)	02:34	03:02	17:03	17:31
	Bloemfontein (ZS4)	02:44	03:10	16:52	17:18
	Durban (ZS5)	02:23	02:50	16:34	17:00
	Johannesburg (ZS6)	02:43	03:08	16:38	17:03
ZS8	Marion Is.	01:01	01:38	16:50	17:27

Pref	State/Province/Country/City	Tlight	Sunrise	Sunset	Tlight
VE1	New Brunswick, Saint John	11:15	11:49	20:41	21:15
	Nova Scotia, Halifax	11:04	11:37	20:34	21:07
	Prince Edward Island, Charlottetown	11:06	11:40	20:26	21:01
VE2	Quebec, Montreal	11:46	12:20	21:11	21:45
	Quebec City	11:40	12:15	20:56	21:31
VE3	Ontario, London	12:10	12:42	21:50	22:22
	Ottawa	11:54	12:28	21:19	21:53
	Sudbury	12:19	12:53	21:37	22:11
	Toronto	12:04	12:37	21:40	22:13
VE4	Manitoba, Winnipeg	13:34	14:11	22:27	23:05
VE5	Saskatchewan, Regina	14:06	14:44	22:54	23:33
	Saskatoon	14:20	15:00	22:55	23:36
VE6	Alberta, Calgary	14:45	15:24	23:30	00:09
	Edmonton	14:52	15:35	23:15	23:57
VE7	British Columbia, Prince George	15:31	16:14	23:50	00:33
	Prince Rupert	16:03	16:46	00:18	01:01
	Vancouver	15:16	15:53	00:14	00:51
VE8	Northwest Territories, Yellowknife	15:43	16:48	22:09	23:14
	Resolute	—	—	—	—
VY1	Yukon, Whitehorse	16:54	17:52	23:50	00:49
VO1	Newfoundland, St. John's	10:29	11:04	19:39	20:15
VO2	Labrador, Goose Bay	11:19	12:01	19:44	20:26
W1	Connecticut, Hartford	11:33	12:04	21:19	21:51
	Maine, Bangor	11:25	11:58	20:54	21:27
	Portland	11:28	12:00	21:04	21:36
	Massachusetts, Boston	11:28	11:59	21:11	21:43
	New Hampshire, Concord	11:31	12:04	21:10	21:42
	Rhode Island, Providence	11:27	11:59	21:14	21:46
	Vermont, Montpelier	11:39	12:12	21:11	21:44
W2	New Jersey, Atlantic City	11:34	12:04	21:33	22:03
	New York, Albany	11:39	12:11	21:21	21:53
	Buffalo	12:00	12:32	21:41	22:13
	New York City	11:35	12:06	21:28	21:58
	Syracuse	11:50	12:22	21:29	22:01
W3	Delaware, Wilmington	11:39	12:09	21:37	22:07
	District of Columbia, Washington	11:43	12:12	21:45	22:15
	Maryland, Baltimore	11:42	12:12	21:43	22:12
	Pennsylvania, Harrisburg	11:46	12:16	21:41	22:11
	Philadelphia	11:38	12:08	21:35	22:05
	Pittsburgh	11:58	12:29	21:53	22:23
	Scranton	11:44	12:15	21:33	22:04
W4	Albama, Montgomery	12:06	12:33	22:39	23:06
	Florida, Jacksonville	11:43	12:10	22:26	22:52
	Miami	11:29	11:54	22:29	22:54
	Pensacola	12:06	12:32	22:47	23:14
	Georgia, Atlanta	12:01	12:29	22:28	22:56
	Savannah	11:45	12:11	22:19	22:46
	Kentucky, Lexington	12:11	12:40	22:18	22:47
	Louisville	12:16	12:46	22:22	22:52
	North Carolina, Charlotte	11:50	12:18	22:10	22:38
	Raleigh	11:42	12:11	22:00	22:28
	Wilmington	11:36	12:04	22:01	22:29
	South Carolina, Columbia	11:48	12:16	22:14	22:42
	Tennessee, Knoxville	12:04	12:32	22:21	22:49
	Memphis	12:27	12:55	22:48	23:16
	Nashville	12:16	12:44	22:32	23:00
	Virginia, Norfolk	11:35	12:04	21:48	22:17
	Richmond	11:41	12:10	21:51	22:20
W5	Arkansas, Little Rock	12:35	13:03	22:58	23:25
	Louisiana, New Orleans	12:16	12:42	23:00	23:26
	Shreveport	12:36	13:03	23:09	23:36
	Mississippi, Jackson	12:21	12:48	22:55	23:22
	New Mexico, Albuquerque	13:33	14:01	23:54	00:22
	Oklahoma, Oklahoma City	12:57	13:25	23:16	23:45
	Texas, Abilene	13:00	13:27	23:33	00:00
	Amarillo	13:14	13:42	23:34	00:02
	Dallas	12:49	13:16	23:20	23:47
	El Paso	13:26	13:52	00:01	00:28
	San Antonio	12:49	13:15	23:35	00:01
W6	California, Los Angeles	14:17	14:45	00:43	01:10
	San Francisco	14:42	15:11	00:50	01:19
W7	Arizona, Flagstaff	13:53	14:21	00:14	00:42
	Phoenix	13:51	14:19	00:20	00:47
	Idaho, Boise	14:31	15:04	00:08	00:40
	Pocatello	14:15	14:47	23:55	00:27
	Montana, Billings	14:07	14:41	23:29	00:03
	Butte	14:23	14:57	23:44	00:19
	Great Falls	14:23	14:58	23:34	00:09
	Nevada, Las Vegas	14:09	14:38	00:25	00:53
	Reno	14:35	15:05	00:35	01:05
	Oregon, Portland	15:02	15:36	00:27	01:01
	Utah, Salt Lake City	14:07	14:38	23:59	00:30
	Washington, Seattle	15:07	15:43	00:18	00:53
	Spokane	14:48	15:23	23:58	00:33
	Wyoming, Cheyenne	13:39	14:10	23:30	00:01
	Sheridan	13:58	14:31	23:27	00:00
W8	Michigan, Detroit	12:15	12:47	21:59	22:31
	Grand Rapids	12:28	13:00	22:08	22:40
	Sault Ste. Marie	12:32	13:07	21:50	22:25
	Traverse City	12:32	13:05	22:01	22:34
	Ohio, Cincinnati	12:13	12:43	22:15	22:45
	Cleveland	12:08	12:39	21:56	22:28
	Columbus	12:09	12:40	22:06	22:36
	West Virginia, Charleston	12:00	12:29	22:05	22:35
W9	Illinois, Chicago	12:32	13:04	22:19	22:50
	Indiana, Indianapolis	12:22	12:52	22:20	22:50
	Wisconsin, Green Bay	12:41	13:14	22:12	22:45

Pref	State/Province/Country/City	Tlight	Sunrise	Sunset	Tlight
	Milwaukee	12:36	13:09	22:16	22:49
W0	Colorado, Denver	13:37	14:07	23:35	00:05
	Grand Junction	13:50	14:19	23:51	00:21
	Iowa, Des Moines	12:56	13:27	22:44	23:15
	Kansas, Pratt	13:07	13:36	23:15	23:45
	Wichita	13:01	13:30	23:10	23:39
	Minnesota, Duluth	13:04	13:39	22:20	22:55
	Minneapolis	13:03	13:37	22:31	23:05
	Missouri, Columbia	12:44	13:14	22:46	23:16
	Kansas City	12:54	13:23	22:55	23:25
	St. Louis	12:35	13:04	22:39	23:08
	Nebraska, North Platte	13:23	13:54	23:14	23:45
	Omaha	13:04	13:35	22:54	23:25
	North Dakota, Fargo	13:23	13:58	22:38	23:13
	South Dakota, Rapid City	13:41	14:13	23:14	23:47
-	-	05:25	05:47	17:54	18:17
1A0	S.M.O.M	05:52	06:24	15:39	16:10
1S	Spratly Is.	22:12	22:34	10:12	10:35
3A	Monaco	06:17	06:50	15:53	16:25
3B6	Agalega	01:20	01:43	14:26	14:49
3B7	St. Brandon	00:56	01:20	14:23	14:47
3B8	Mauritius	00:57	01:22	14:40	15:05
3B9	Rodriguez Is.	00:35	00:59	14:15	14:40
3C	Equatorial Guinea, Bata	04:49	05:11	17:12	17:35
	Malabo	04:56	05:18	17:13	17:35
3C0	Annobon Is.	05:00	05:22	17:35	17:57
3D2	Fiji Is., Suva	16:58	17:22	06:32	06:56
3D2	Conway Reef	17:07	17:32	06:55	07:19
3D2	Rotuma Is.	17:12	17:36	06:25	06:48
3DA	Swaziland, Mbabane	02:29	02:55	16:38	17:04
3V	Tunisia, Tunis	05:49	06:18	16:02	16:31
3W, XV	Vietnam, Ho Chi Minh City (Saigon)	22:36	22:59	10:30	10:52
	Hanoi	22:57	23:21	10:15	10:39
3X	Guinea, Conakry	06:35	06:58	18:33	18:56
3Y	Bouvet	02:13	03:07	20:08	21:02
3Y	Peter I Is.	—	—	—	—
4J, 4K	Azerbaijan, Baku	03:19	03:49	13:13	13:44
4L	Georgia, Tbilisi	03:42	04:14	13:30	14:01
4S	Sri Lanka, Colombo	00:17	00:39	12:23	12:46
4U	ITU Geneva	06:29	07:03	15:49	16:23
4U	United Nations Hq.	11:35	12:06	21:28	21:58
4X, 4Z	Israel, Jerusalem	03:59	04:26	14:35	15:01
5A	Libya, Tripoli	05:31	05:58	16:04	16:31
	Benghazi	05:00	05:27	15:35	16:02
5B	Cyprus, Nicosia	04:13	04:41	14:34	15:02
5H	Tanzania, Dar es Salaam	02:35	02:58	15:28	15:51
5N	Nigeria, Lagos	05:22	05:45	17:30	17:53
5R	Madagascar, Antananarivo	01:40	02:04	15:18	15:42
5T	Mauritania, Nouakchott	06:59	07:22	18:27	18:51
5U	Niger, Niamey	05:39	06:02	17:24	17:47
5V	Togo, Lome	05:30	05:52	17:40	18:02
5W	Western Samoa, Apia	16:28	16:51	05:45	06:08
5X	Uganda, Kampala	03:15	03:38	15:44	16:06
5Z	Kenya, Nairobi	04:15	04:38	16:49	17:12
6W	Senegal, Dakar	06:59	07:22	18:40	19:03
6Y	Jamaica, Kingston	11:02	11:25	22:31	22:54
7O	Yemen, Aden	02:46	03:09	14:33	14:56
	Sanaa	02:53	03:17	14:32	14:55
7P	Lesotho, Maseru	02:36	03:03	16:59	17:26
7Q	Malawi, Lilongwe	02:45	03:08	16:03	16:27
	Blantyre	02:36	03:00	16:02	16:25
7X	Algeria, Algiers	06:18	06:46	16:31	17:00
8P	Barbados, Bridgetown	09:45	10:08	21:31	21:54
8Q	Maldive Is.	00:39	01:01	12:54	13:16
8R	Guyana, Georgetown	09:29	09:51	21:36	21:58
9A	Croatia, Zagreb	05:49	06:23	15:11	15:45
9G	Ghana, Accra	05:35	05:57	17:46	18:09
9H	Malta	05:31	05:59	15:48	16:16
9J	Zambia, Lusaka	03:04	03:28	16:28	16:51
9K	Kuwait	03:04	03:30	13:49	14:15
9L	Sierra Leone, Freetown	06:32	06:54	18:33	18:56
9M2	West Malaysia, Kuala Lumpur	22:44	23:06	11:03	11:25
9M6,8	East Malaysia, Sabah, Sandakan(9M6)	21:42	22:05	09:52	10:15
	Sarawak, Kuching (9M8)	22:06	22:29	10:31	10:53
9N	Nepal, Kathmandu	00:30	00:56	11:23	11:49
9Q	Zaire, Kinshasa	04:16	04:39	17:00	17:23
	Kisangani	03:45	04:07	16:13	16:35
	Lubumbashi	03:14	03:37	16:24	16:47
9U	Burundi, Bujumbura	03:22	03:45	16:03	16:25
9V	Singapore	22:32	22:54	10:57	11:20
9X	Rwanda, Kigali	03:21	03:44	15:57	16:20
9Y	Trinidad & Tobago , Port of Spain	09:47	10:10	21:42	22:05
A2	Botswana, Gaborone	02:53	03:19	16:56	17:21
A3	Tonga, Nukualofa	16:26	16:51	06:13	06:37
A4	Oman, Musqat	02:10	02:34	13:19	13:43
A5	Bhutan, Thimbu	00:13	00:39	11:08	11:33
A6	United Arab Emirates, Abu Dhabi	02:29	02:54	13:34	13:59
A7	Qatar, Ad-Dawhah	02:41	03:06	13:44	14:08
A9	Bahrein, Al-Manamah	02:47	03:12	13:45	14:10
AP	Pakistan, Karachi	01:38	02:03	12:42	13:07
AP	Islamabad	01:31	01:58	11:58	12:26
BV	Taiwan, Taipei	22:01	22:26	09:04	09:29
BY	Peoples Rep. of China, Beijing	22:52	23:22	08:49	09:19
	Harbin	22:26	23:00	07:48	08:22
	Shanghai	22:12	22:39	08:51	09:17
	Fuzhou	22:12	22:37	09:11	09:36
	Xian	23:09	23:37	09:34	10:01

Pref	State/Province/Country/City	Tlight	Sunrise	Sunset	Tlight
	Chongqing	23:10	23:36	09:54	10:20
	Chengdu	23:21	23:47	10:02	10:28
	Lhasa	00:11	00:37	10:55	11:21
	Urumqi	00:57	01:29	10:32	11:04
	Kashi	01:32	02:02	11:32	12:02
C2	Nauru	18:16	18:39	06:48	07:10
C3	Andorra	06:38	07:09	16:20	16:52
C5	The Gambia, Banjul	06:54	07:17	18:38	19:01
C6	Bahamas, Nassau	11:17	11:41	22:20	22:44
C9	Mozambique, Maputo	02:24	02:50	16:31	16:57
	Mozambique	02:15	02:39	15:38	16:01
CE	Chile, Santiago	08:57	09:26	23:43	00:11
CE0	Easter Island	11:49	12:15	02:02	02:28
CE0	Juan Fernandez	09:29	09:57	00:15	00:44
CE0	San Felix	09:54	10:20	00:03	00:29
CM,CO	Cuba, Havana	11:33	11:57	22:44	23:08
CN	Morocco, Casablanca	06:53	07:21	17:21	17:49
CP	Bolivia, La Paz	09:29	09:52	22:57	23:20
CT	Portugal, Lisbon	07:11	07:41	17:15	17:44
CT3	Madeira Islands, Funchal	07:29	07:56	18:01	18:28
CU	Azores , Ponta Delgada	08:15	08:44	18:23	18:53
CX	Uruguay, Montevideo	07:54	08:24	22:48	23:17
CY0	Sable Is.	10:47	11:20	20:22	20:55
CY9	St. Paul Is.	10:57	11:32	20:10	20:46
D2	Angola, Luanda	04:17	04:40	17:17	17:39
D4	Cape Verde, Praia	07:23	07:47	19:03	19:26
D6	Comoros, Moroni	02:09	02:32	15:20	15:43
DA-DL,Y2-Y9	Fed. Rep. of Germany, Bonn	06:40	07:19	15:27	16:06
	Berlin	06:21	07:02	14:53	15:34
DU	Philippines, Manilla	21:45	22:08	09:26	09:49
E3	Eritrea, Asmara	03:14	03:38	14:53	15:16
EA	Spain, Madrid	06:53	07:24	16:48	17:18
EA6	Balearic Is., Palma	06:26	06:56	16:25	16:55
EA8	Canary Is., Las Palmas	07:10	07:36	18:00	18:26
EA9	Ceuta & Melilla, Ceuta	06:49	07:18	17:07	17:35
	Melilla	06:39	07:07	16:59	17:27
EI	Ireland, Dublin	07:43	08:25	16:08	16:49
EK	Armenia, Yerevan	03:40	04:11	13:35	14:06
EL	Liberia, Monrovia	06:18	06:41	18:27	18:50
EP	Iran, Tehran	03:01	03:29	13:18	13:47
ER	Moldova, Kishinev	05:01	05:36	14:16	14:51
ES	Estonia, Tallinn	06:07	07:01	13:23	14:17
ET	Ethiopia, Addis Ababa	03:05	03:28	15:05	15:27
EU,EV,EW	Belarus, Minsk	05:29	06:12	13:49	14:31
EX	Kyrgyzstan, Frunze	01:46	02:18	11:27	11:59
EY	Tajikistan, Samarkand	02:09	02:40	12:08	12:38
	Dushanbe	02:00	02:30	12:02	12:31
EZ	Turkmenistan, Ashkhabad	02:39	03:08	12:46	13:15
F	France, Paris	06:52	07:29	15:54	16:31
FG	Guadeloupe	09:58	10:21	21:34	21:57
FJ, FS	St. Martin	10:07	10:31	21:36	21:59
FH	Mayotte	02:01	02:24	15:15	15:39
FK	New Caledonia, Noumia	17:37	18:02	07:28	07:53
FM	Martinique	09:53	10:16	21:34	21:57
FO	Clipperton Is.	12:59	13:21	00:54	01:17
FO	Fr. Polynesia, Tahiti	15:31	15:55	05:03	05:27
	Rurutu, (Austral Is.)	18:37	19:02	08:29	08:54
	Hiva Oa, (Marquesas Is.)	14:24	14:47	03:27	03:50
FP	St. Pierre & Miquelon, St. Pierre	10:39	11:14	19:56	20:31
FR/G	Glorioso	01:55	02:19	15:05	15:28
FR/J,E	Juan de Nova	02:03	02:27	15:33	15:57
	Europa	02:01	02:26	15:53	16:17
FR	Reunion	01:03	01:28	14:49	15:14
FR/T	Tromelin	01:19	01:42	14:44	15:08
FT8W	Crozet	23:57	00:35	16:11	16:49
FT8X	Kerguelen	22:30	23:12	15:16	15:59
FT8Y	Antarctica, Dumont D'Urville				
FT8Z	Amsterdam & St. Paul Is., Amsterdam	22:50	23:21	14:00	14:31
FW	Wallis & Futuna Is., Wallis	16:46	17:10	06:02	06:26
FY	Fr. Guiana, Cayenne	09:02	09:25	21:16	21:38
G	England, London	07:11	07:51	15:52	16:31
GD	Isle of Man	07:40	08:23	15:55	16:38
GI	Northern Ireland, Belfast	07:46	08:30	15:59	16:43
GJ	Jersey	07:12	07:49	16:10	16:47
GM	Scotland, Glasgow	07:45	08:31	15:45	16:31
	Aberdeen	07:43	08:32	15:27	16:16
GU	Guernsey	07:15	07:52	16:11	16:49
GW	Wales, Cardiff	07:24	08:03	16:04	16:44
H4	Solomon Islands, Honiara	18:29	18:51	07:30	07:53
HA	Hungary, Budapest	05:41	06:17	14:52	15:28
HB	Switzerland, Bern	06:26	07:01	15:41	16:16
HB0	Liechtenstein	06:18	06:53	15:32	16:07
HC	Ecuador, Quito	10:37	10:59	23:07	23:29
HC8	Galapagos Is.	11:26	11:48	23:57	00:20
HH	Haiti, Port-Au-Prince	10:45	11:08	22:12	22:36
HI	Dominican Republic, Santo Domingo	10:35	10:59	22:03	22:26
HK	Colombia, Bogota	10:29	10:51	22:43	23:06
HK0	Malpelo Is.	10:56	11:18	23:12	23:35
HK0	San Adreas	11:12	11:35	23:00	23:23
HL	Korea, Seoul	22:04	22:33	08:13	08:42
HP	Panama, Panama	10:58	11:20	22:57	23:20
HR	Honduras, Tegucigalpa	11:37	12:00	23:19	23:42
HS	Thailand, Bangkok	23:06	23:29	10:49	11:12
HV	Vatican City	05:52	06:23	15:38	16:10
HZ, 7Z	Saudi Arabia, Dharan	02:49	03:14	13:48	14:13
	Mecca	03:21	03:46	14:38	15:02
	Riyadh	02:59	03:24	14:04	14:29

Pref	State/Province/Country/City	Tlight	Sunrise	Sunset	Tlight
I	Italy, Rome	05:52	06:23	15:38	16:10
	Trieste	05:57	06:31	15:20	15:54
	Sicily	05:36	06:05	15:45	16:14
IS	Sardinia, Cagliari	05:59	06:29	16:00	16:30
J2	Djibouti, Djibouti	02:51	03:14	14:42	15:05
J3	Grenada	09:52	10:15	21:41	22:04
J5	Guinea-Bissau, Bissau	06:47	07:10	18:37	19:00
J6	St. Lucia	09:52	10:15	21:35	21:58
J7	Dominica	09:55	10:19	21:34	21:57
J8	St. Vincent	09:52	10:15	21:37	22:00
JA-JS	Japan, Tokyo	21:08	21:37	07:27	07:55
	Nagasaki	21:42	22:09	08:13	08:41
	Sapporo	21:20	21:52	06:59	07:31
JD1	Minami Torishima	19:49	20:14	06:56	07:20
JD1	Ogasawara, Kazan Is.	20:47	21:13	07:41	08:06
JT	Mongolia, Ulan Bator	23:51	00:27	09:00	09:35
JW	Svalbard, Spitsbergen				
JX	Jan Mayen	10:22	——	——	14:26
JY	Jordan, Amman	03:56	04:23	14:31	14:58
KC4	Antarctica, Byrd Station				
	McMurdo Sound	——	——	——	——
	Palmer Station	——	05:38	02:35	——
KC6	Belau, Yap	20:28	20:50	08:26	08:48
	Koror	20:39	21:01	08:44	09:07
KG4	Guantanamo Bay	10:59	11:22	22:21	22:45
KH0	Mariana Is., Saipan	20:07	20:30	07:45	08:09
KH1	Baker, Howland Is.	17:10	17:32	05:38	06:00
KH2	Guam, Agana	20:08	20:31	07:52	08:15
KH3	Johnston Is.	17:07	17:30	04:39	05:03
KH4	Midway Is.	18:02	18:28	04:53	05:19
KH5	Palmyra Is.	16:23	16:46	04:33	04:56
KH5K	Kingman Reef	16:28	16:51	04:33	04:56
KH6	Hawaii, Hilo	16:18	16:42	03:41	04:05
	Honolulu	16:32	16:56	03:49	04:13
KH7	Kure Is.	18:07	18:32	04:57	05:22
KH8	American Samoa, Pago Pago	16:23	16:46	05:42	06:06
KH9	Wake Is.	18:50	19:14	06:15	06:39
KL7	Alaska, Adak	18:58	19:38	03:36	04:16
	Anchorage	17:57	18:57	00:45	01:45
	Fairbanks	18:15	19:34	23:51	01:10
	Juneau	16:38	17:29	00:08	00:59
	Nome	19:23	20:39	01:06	02:22
KP1	Navassa Is.	10:55	11:19	22:23	22:46
KP2	Virgin Islands, Charlotte Amalie	10:15	10:38	21:43	22:06
KP4	Puerto Rico, San Juan	10:20	10:44	21:48	22:11
KP5	Desecheo Is.	10:25	10:49	21:53	22:17
LA-LJ	Norway, Oslo	07:07	08:02	14:14	15:10
LU	Argentina, Buenos Aires	08:04	08:33	22:56	23:25
LX	Luxembourg	06:39	07:17	15:35	16:13
LY	Lithuania, Vilna	05:40	06:24	13:54	14:37
LZ	Bulgaria, Sofia	05:11	05:43	14:53	15:24
OA	Peru, Lima	10:12	10:35	23:23	23:47
OD	Lebanon, Beirut	04:02	04:29	14:28	14:56
OE	Austria, Vienna	05:54	06:31	15:01	15:37
OH	Finland, Helsinki	06:11	07:07	13:15	14:11
OH0	Aland Is.	06:31	07:27	13:35	14:31
OJ0	Market Reef	06:35	07:32	13:38	14:35
OK,OL	Czech Rep., Prague	06:08	06:46	15:00	15:38
OM	Slovak. Rep., Bratislava	05:51	06:27	14:59	15:35
ON	Belgium, Brussels	06:51	07:30	15:37	16:16
OX	Greenland, Godthaab	11:45	13:00	17:35	18:50
	Thule				
OY	Faroe Islands, Torshavn	08:29	09:32	15:04	16:07
OZ	Denmark, Copenhagen	06:37	07:23	14:38	15:24
P2	Papua New Guinea, Madang	19:34	19:56	08:21	08:43
	Port Moresby	19:20	19:43	08:22	08:45
P4	Aruba, Oranjestad	10:21	10:44	22:10	22:33
PA-PI	Netherlands, Amsterdam	06:54	07:35	15:27	16:08
PJ2,4,9	Netherlands Antilles, Willemstad	10:26	10:49	22:14	22:37
PJ5-8	St. Maarten and Saba, St. Maarten	10:07	10:30	21:37	22:00
PY	Brazil, Brasilia	08:08	08:32	21:33	21:57
	Rio De Janeiro	07:34	07:59	21:28	21:53
	Natal	07:35	07:58	20:25	20:48
	Manaus	09:20	09:43	22:00	22:23
	Porto Alegre	07:48	08:16	22:16	22:43
PY0	Fernando De Noronha	07:28	07:50	20:11	20:33
PY0	St. Peter & St. Paul Rocks	07:24	07:47	19:50	20:13
PY0	Trindade & Martin Vaz Is., Trindade	06:44	07:08	20:28	20:52
PZ	Suriname, Paramaribo	09:15	09:38	21:26	21:48
R1FJ_	Franz Josef Land	——	——	——	——
R1MV_	Malyj Vysotskij Is.	05:59	06:56	12:57	13:54
S0	Western Sahara, Smara	06:55	07:20	17:53	18:18
S2	Bangladesh, Dacca	00:03	00:27	11:11	11:36
S5	Slovenia, Ljubljana	05:55	06:29	15:16	15:51
S7	Seychelles, Victoria	01:35	01:58	14:20	14:43
S9	Sao Tome	04:59	05:21	17:27	17:50
SM	Sweden, Stockholm	06:33	07:27	13:50	14:44
SP	Poland, Krakow	05:46	06:23	14:38	15:16
	Warsaw	05:49	06:30	14:24	15:04
ST	Sudan, Khartoum	03:41	04:04	15:18	15:41
ST0	Southern Sudan, Juba	03:27	03:49	15:40	16:02
SU	Egypt, Cairo	04:10	04:37	14:54	15:20
SV	Greece, Athens	04:58	05:27	15:05	15:34
SV/A	Mount Athos	05:01	05:31	14:56	15:27
SV5	Dodecanese, Rhodes	04:36	05:05	14:51	15:20
SV9	Crete	04:46	05:14	15:06	15:34
T2	Tuvalu, Funafuti	17:15	17:38	06:15	06:38

Pref	State/Province/Country/City	Tlight	Sunrise	Sunset	Tlight
T30	West Kiribati, Bonriki	17:54	18:17	06:19	06:42
T31	Central Kiribati, Kanton	16:47	17:09	05:26	05:48
T32	East Kiribati, Christmas Is.	15:58	16:20	04:21	04:43
T33	Banaba Is.	18:06	18:29	06:38	07:00
T5	Somalia, Mogadishu	02:27	02:49	14:49	15:12
T7	San Marino	05:58	06:31	15:33	16:05
T9	Bosnia-Herzegovina, Sarajevo	05:34	06:06	15:08	15:41
TA	Turkey, Ankara	04:26	04:56	14:23	14:53
	Istanbul	04:44	05:15	14:35	15:06
TF	Iceland, Reykjavik	09:46	11:00	15:38	16:52
TG	Guatemala, Guatemala City	11:51	12:14	23:32	23:55
TI	Costa Rica, San Jose	11:17	11:40	23:14	23:37
TI9	Cocos Is.	11:22	11:45	23:33	23:56
TJ	Cameroon, Yaounde	04:45	05:08	17:02	17:24
TK	Corsica	06:06	06:38	15:52	16:23
TL	Central African Republic, Bangui	04:18	04:40	16:33	16:55
TN	Congo, Brazzaville	04:16	04:39	17:00	17:23
TR	Gabon, Libreville	04:48	05:10	17:16	17:38
TT	Chad, N'Djamena	04:45	05:08	16:34	16:57
TU	Cote d'Ivoire, Abidjan	05:50	06:12	18:02	18:24
TY	Benin, Porto Novo	05:25	05:48	17:33	17:56
TZ	Mali, Bamako	06:18	06:41	18:05	18:28
UA	Russia, European, Leningrad (UA1)	05:48	06:43	12:56	13:52
	Archangel (UA1)	05:40	06:57	11:20	12:38
	Murmansk (UA1)	06:59	——	——	12:18
	Moscow (UA3)	04:58	05:44	12:57	13:43
	Kuybyshev (UA4)	03:57	04:39	12:22	13:04
	Rostov (UA6)	04:20	04:55	13:31	14:06
UA2	Kaliningrad	06:02	06:47	14:11	14:55
UA9,0	Russia, Asiatic, Novosibirsk (UA9)	01:53	02:37	10:01	10:46
	Perm (UA9)	03:53	04:44	11:28	12:18
	Omsk (UA9)	02:31	03:15	10:39	11:24
	Norilsk (UA0)	03:24	——	——	08:33
	Irkutsk (UA0)	00:16	00:57	08:50	09:31
	Vladivostok (UA0)	21:58	22:30	07:37	08:09
	Petropavlovsk (UA0)	20:42	21:23	05:09	05:51
	Khabarovsk (UA0)	22:00	22:37	07:04	07:41
	Krasnoyarsk (UA0)	01:18	02:04	09:15	10:02
	Yakutsk (UA0)	23:23	00:26	05:58	07:01
	Wrangel Island (UA0)	21:47	——	——	01:51
	Kyzyl (UA0Y)	00:53	01:33	09:32	10:12
UJ-UM	Uzbekistan, Bukhoro	02:19	02:49	12:17	12:47
	Tashkent	02:03	02:34	11:53	12:24
UN-UQ	Kazakhstan, Alma-Ata	01:38	02:10	11:16	11:49
UR-UZ,EM-EO	Ukraine, Kiev	05:05	05:43	13:55	14:33
V2	Antigua & Barbuda, St. Johns	10:00	10:24	21:32	21:56
V3	Belize, Belmopan	11:49	12:12	23:20	23:44
V4	St. Kitts & Nevis	10:04	10:27	21:35	21:59
V5	Namibia, Windhoek	03:34	03:59	17:26	17:51
V6	Micronesia, Ponape	19:03	19:26	07:10	07:32
V7	Marshall Islands, Kwajalein	18:31	18:53	06:30	06:53
V8	Brunei, Bandar Seri Begawan	21:53	22:16	10:07	10:29
VK	Australia, Canberra (VK1)	18:12	18:41	09:08	09:37
	Sydney (VK2)	18:08	18:37	08:56	09:24
	Melbourne (VK3)	18:20	18:51	09:31	10:02
	Brisbane (VK4)	18:18	18:45	08:33	08:59
	Adelaide (VK5)	18:55	19:24	09:49	10:18
	Perth (VK6)	20:35	21:03	11:12	11:40
	Hobart, Tasmania (VK7)	17:51	18:25	09:38	10:13
	Darwin (VK8)	20:19	20:42	09:32	09:55
VK0	Heard Is.	21:46	22:36	15:19	16:08
VK0	Macquarie Is.	15:49	16:44	09:48	10:42
VK9C	Cocos-Keeling Is.	22:36	22:59	11:48	12:11
VK9L	Lord Howe Is.	17:43	18:11	08:18	08:46
VK9M	Mellish Reef	18:30	18:54	08:02	08:26
VK9N	Norfolk Is.	17:14	17:41	07:36	08:03
VK9W	Willis Is.	18:57	19:21	08:25	08:48
VK9X	Christmas Is.	22:04	22:27	11:09	11:32
VP2E	Anguilla	10:07	10:31	21:35	21:59
VP2M	Montserrat	10:01	10:25	21:35	21:58
VP2V	British Virgin Is., Tortola	10:14	10:37	21:41	22:05

December 7   **Sunrise and Sunset, in UTC**

Pref	State/Province/Country/City	Tlight	Sunrise	Sunset	Tlight
VP5	Turks & Caicos Islands, Grand Turk	10:45	11:09	22:02	22:26
VP8	Falkland Islands, Stanley	06:42	07:28	23:57	00:43
VP8	So. Georgia Is.	04:55	05:49	22:47	23:41
VP8	So. Orkney Is.	03:50	05:31	00:15	01:56
VP8	So. Sandwich Islands, Saunders Is.	03:34	04:43	22:32	23:41
VP8	So. Shetland Is., King George Is.	——	06:05	01:23	——
VP9	Bermuda	10:39	11:06	21:13	21:40
VQ9	Chagos, Diego Garcia	00:23	00:45	13:17	13:40
VR2,VS6	Hong Kong	22:25	22:49	09:38	10:03
VR6	Pitcairn Is.	13:17	13:42	03:20	03:46
VU	India, Bombay	01:05	01:29	12:31	12:54
	Calcutta	00:09	00:33	11:21	11:46
	New Delhi	01:05	01:30	11:54	12:20
	Bangalore	00:36	00:59	12:22	12:45
VU	Andaman Islands, Port Blair	23:33	23:56	11:24	11:46
VU	Laccadive Is.	00:49	01:12	12:46	13:08
XE	Mexico, Mexico City (XE1)	12:33	12:57	23:57	00:21
	Chihuahua (XE2)	13:18	13:43	00:06	00:32
	Merida (XE3)	11:59	12:23	23:17	23:41
XF4	Revilla Gigedo	13:22	13:46	00:48	01:11
XT	Burkina Faso, Ouagadougou	05:52	06:15	17:40	18:03
XU	Cambodia, Phnom Penh	22:45	23:08	10:36	10:58
XW	Laos, Viangchan	23:04	23:28	10:33	10:57
XX9	Macao	22:27	22:52	09:41	10:06
XZ	Myanmar, Yangon	23:29	23:52	11:02	11:25
YA	Afghanistan, Kandahar	01:55	02:21	12:34	13:01
	Kabul	01:48	02:16	12:12	12:40
YB-YD	Indonesia, Jakarta	22:07	22:30	10:58	11:20
	Medan	22:56	23:18	11:14	11:36
	Pontianak	22:08	22:30	10:37	11:00
	Jayapura	19:58	20:20	08:36	08:58
YI	Iraq, Baghdad	03:24	03:51	13:55	14:22
YJ	Vanuatu, Port Vila	17:39	18:03	07:12	07:36
YK	Syria, Damascus	03:58	04:25	14:26	14:54
YL	Latvia, Riga	05:57	06:45	13:44	14:32
YN	Nicaragua, Managua	11:29	11:51	23:18	23:41
YO	Romania, Bucharest	05:04	05:37	14:36	15:09
YS	El Salvador, San Salvador	11:44	12:07	23:28	23:51
YU	Yugoslavia, Belgrade	05:28	06:01	14:56	15:30
YV	Venezuela, Caracas	10:10	10:33	22:05	22:28
YV0	Aves Is.	10:06	10:29	21:43	22:06
Z2	Zimbabwe, Harare	02:48	03:12	16:21	16:45
Z3	Macedonia, (ex Yugoslav), Skopje	05:17	05:48	15:02	15:34
ZA	Albania, Tirane	05:21	05:52	15:11	15:42
ZB2	Gibraltar	06:50	07:18	17:06	17:35
ZC4	British Cyprus	04:13	04:41	14:37	15:04
ZD7	St. Helena	05:20	05:43	18:46	19:09
ZD8	Ascension Is.	06:09	06:31	19:06	19:28
ZD9	Tristan da Cunha	04:51	05:22	19:58	20:29
ZF	Cayman Is.	11:22	11:46	22:46	23:09
ZK1	No. Cook Is., Manihiki	15:51	16:14	04:56	05:19
ZK1	So. Cook Is., Rarotonga	15:24	15:49	05:11	05:36
ZK2	Niue	16:05	16:30	05:43	06:07
ZK3	Tokelaus, Atafu	16:41	17:04	05:39	06:02
ZL	New Zealand, Auckland (ZL1)	16:24	16:54	07:29	08:00
	Wellington (ZL2)	16:07	16:41	07:43	08:16
	Christchurch (ZL3)	16:07	16:42	07:59	08:34
	Dunedin (ZL4)	16:04	16:41	08:16	08:54
ZL5	Antarctica, Scott Base	——	——	——	——
ZL7	Chatham Is.	15:21	15:57	07:17	07:53
ZL8	Kermadec Is.	16:17	16:44	06:41	07:08
ZL9	Auckland & Campbell Is., Auckland	15:51	16:36	08:54	09:39
	Campbell Is.	15:27	16:16	08:53	09:42
ZP	Paraguay, Asuncion	08:27	08:52	22:31	22:57
ZS	South Africa, Cape Town (ZS1)	02:59	03:28	17:47	18:16
	Port Elizabeth (ZS2)	02:29	02:58	17:18	17:47
	Bloemfontein (ZS4)	02:42	03:08	17:05	17:31
	Durban (ZS5)	02:20	02:48	16:47	17:14
	Johannesburg (ZS6)	02:41	03:07	16:50	17:16
ZS8	Marion Is.	00:50	01:29	17:11	17:50

Pref	State/Province/Country/City	Tlight	Sunrise	Sunset	Tlight
VE1	New Brunswick, Saint John	11:26	12:00	20:44	21:18
	Nova Scotia, Halifax	11:14	11:48	20:36	21:10
	Prince Edward Island, Charlottetown	11:17	11:52	20:28	21:03
VE2	Quebec, Montreal	11:57	12:31	21:13	21:47
	Quebec City	11:51	12:26	20:58	21:34
VE3	Ontario, London	12:20	12:53	21:53	22:25
	Ottawa	12:05	12:39	21:22	21:56
	Sudbury	12:29	13:04	21:39	22:14
	Toronto	12:15	12:47	21:43	22:16
VE4	Manitoba, Winnipeg	13:45	14:23	22:28	23:07
VE5	Saskatchewan, Regina	14:17	14:56	22:56	23:35
	Saskatoon	14:32	15:13	22:56	23:37
VE6	Alberta, Calgary	14:57	15:37	23:31	00:11
	Edmonton	15:04	15:48	23:15	23:59
VE7	British Columbia, Prince George	15:43	16:27	23:50	00:34
	Prince Rupert	16:15	16:59	00:18	01:02
	Vancouver	15:27	16:05	00:15	00:53
VE8	Northwest Territories, Yellowknife	15:59	17:07	22:03	23:12
	Resolute	—	—	—	—
VY1	Yukon, Whitehorse	17:09	18:09	23:47	00:47
VO1	Newfoundland, St. John's	10:40	11:16	19:41	20:17
VO2	Labrador, Goose Bay	11:31	12:14	19:44	20:27
W1	Connecticut, Hartford	11:43	12:14	21:22	21:54
	Maine, Bangor	11:35	12:09	20:56	21:30
	Portland	11:38	12:11	21:06	21:39
	Massachusetts, Boston	11:38	12:10	21:14	21:46
	New Hampshire, Concord	11:42	12:14	21:13	21:45
	Rhode Island, Providence	11:37	12:09	21:17	21:49
	Vermont, Montpelier	11:49	12:22	21:13	21:47
W2	New Jersey, Atlantic City	11:43	12:14	21:37	22:07
	New York, Albany	11:49	12:22	21:24	21:56
	Buffalo	12:10	12:43	21:43	22:16
	New York City	11:45	12:16	21:31	22:02
	Syracuse	12:00	12:33	21:32	22:04
W3	Delaware, Wilmington	11:48	12:19	21:40	22:11
	District of Columbia, Washington	11:52	12:22	21:49	22:19
	Maryland, Baltimore	11:52	12:22	21:46	22:16
	Pennsylvania, Harrisburg	11:56	12:26	21:44	22:15
	Philadelphia	11:48	12:18	21:38	22:09
	Pittsburgh	12:08	12:39	21:56	22:27
	Scranton	11:54	12:25	21:36	22:07
W4	Alabama, Montgomery	12:15	12:42	22:43	23:10
	Florida, Jacksonville	11:52	12:19	22:30	22:56
	Miami	11:37	12:03	22:34	22:59
	Pensacola	12:14	12:41	22:52	23:18
	Georgia, Atlanta	12:10	12:38	22:32	23:00
	Savannah	11:53	12:21	22:23	22:50
	Kentucky, Lexington	12:20	12:50	22:21	22:51
	Louisville	12:26	12:56	22:26	22:55
	North Carolina, Charlotte	11:59	12:27	22:14	22:42
	Raleigh	11:52	12:20	22:04	22:32
	Wilmington	11:45	12:13	22:05	22:33
	South Carolina, Columbia	11:57	12:25	22:18	22:46
	Tennessee, Knoxville	12:13	12:42	22:24	22:53
	Memphis	12:36	13:04	22:52	23:20
	Nashville	12:25	12:54	22:35	23:04
	Virginia, Norfolk	11:45	12:14	21:51	22:21
	Richmond	11:51	12:20	21:54	22:23
W5	Arkansas, Little Rock	12:44	13:12	23:01	23:29
	Louisiana, New Orleans	12:25	12:51	23:04	23:31
	Shreveport	12:45	13:12	23:13	23:40
	Mississippi, Jackson	12:30	12:57	22:59	23:26
	New Mexico, Albuquerque	13:42	14:11	23:58	00:26
	Oklahoma, Oklahoma City	13:06	13:35	23:20	23:49
	Texas, Abilene	13:09	13:36	23:37	00:04
	Amarillo	13:23	13:51	23:38	00:06
	Dallas	12:58	13:25	23:24	23:52
	El Paso	13:34	14:01	00:06	00:33
	San Antonio	12:58	13:24	23:39	00:05
W6	California, Los Angeles	14:26	14:54	00:47	01:14
	San Francisco	14:51	15:21	00:53	01:23
W7	Arizona, Flagstaff	14:03	14:31	00:18	00:46
	Phoenix	14:00	14:28	00:24	00:51
	Idaho, Boise	14:42	15:14	00:10	00:43
	Pocatello	14:25	14:57	23:58	00:30
	Montana, Billings	14:17	14:52	23:31	00:06
	Butte	14:34	15:08	23:47	00:21
	Great Falls	14:34	15:10	23:36	00:12
	Nevada, Las Vegas	14:18	14:47	00:29	00:57
	Reno	14:45	15:15	00:38	01:08
	Oregon, Portland	15:13	15:47	00:29	01:04
	Utah, Salt Lake City	14:17	14:48	00:02	00:33
	Washington, Seattle	15:18	15:54	00:19	00:55
	Spokane	14:59	15:35	23:59	00:36
	Wyoming, Cheyenne	13:49	14:20	23:33	00:04
	Sheridan	14:08	14:42	23:29	00:03
W8	Michigan, Detroit	12:25	12:57	22:02	22:34
	Grand Rapids	12:38	13:10	22:10	22:43
	Sault Ste. Marie	12:43	13:18	21:52	22:27
	Traverse City	12:43	13:16	22:04	22:37
	Ohio, Cincinnati	12:23	12:53	22:18	22:48
	Cleveland	12:18	12:49	21:59	22:31
	Columbus	12:19	12:50	22:09	22:40
	West Virginia, Charleston	12:10	12:39	22:08	22:38
W9	Illinois, Chicago	12:42	13:14	22:22	22:53
	Indiana, Indianapolis	12:31	13:02	22:23	22:53
	Wisconsin, Green Bay	12:51	13:25	22:14	22:48
	Milwaukee	12:47	13:19	22:19	22:52
W0	Colorado, Denver	13:46	14:17	23:38	00:09
	Grand Junction	13:59	14:29	23:54	00:24
	Iowa, Des Moines	13:06	13:37	22:47	23:18
	Kansas, Pratt	13:16	13:46	23:19	23:48
	Wichita	13:11	13:40	23:13	23:43
	Minnesota, Duluth	13:15	13:50	22:22	22:57
	Minneapolis	13:14	13:48	22:34	23:07
	Missouri, Columbia	12:54	13:24	22:49	23:19
	Kansas City	13:03	13:33	22:58	23:28
	St. Louis	12:45	13:14	22:42	23:12
	Nebraska, North Platte	13:33	14:04	23:17	23:48
	Omaha	13:14	13:45	22:57	23:28
	North Dakota, Fargo	13:34	14:09	22:40	23:16
	South Dakota, Rapid City	13:51	14:24	23:16	23:50
-		05:31	05:54	18:01	18:24
1A0	S.M.O.M	06:02	06:34	15:42	16:13
1S	Spratly Is.	22:19	22:42	10:18	10:41
3A	Monaco	06:27	07:00	15:55	16:28
3B6	Agalega	01:26	01:49	14:33	14:56
3B7	St. Brandon	01:01	01:25	14:31	14:55
3B8	Mauritius	01:02	01:27	14:48	15:13
3B9	Rodriguez Is.	00:40	01:04	14:24	14:48
3C	Equatorial Guinea, Bata	04:55	05:18	17:19	17:41
	Malabo	05:03	05:25	17:19	17:42
3C0	Annobon Is.	05:06	05:29	17:41	18:04
3D2	Fiji Is., Suva	17:03	17:27	06:40	07:05
3D2	Conway Reef	17:12	17:37	07:03	07:28
3D2	Rotuma Is.	17:18	17:41	06:32	06:55
3DA	Swaziland, Mbabane	02:33	02:59	16:47	17:13
3V	Tunisia, Tunis	05:59	06:28	16:06	16:35
3W, XV	Vietnam, Ho Chi Minh City (Saigon)	22:43	23:06	10:35	10:58
	Hanoi	23:04	23:29	10:20	10:44
3X	Guinea, Conakry	06:42	07:05	18:39	19:02
3Y	Bouvet	02:10	03:06	20:22	21:18
3Y	Peter I Is.	—	—	—	—
4J, 4K	Azerbaijan, Baku	03:29	03:59	13:16	13:47
4L	Georgia, Tbilisi	03:52	04:24	13:33	14:04
4S	Sri Lanka, Colombo	00:24	00:46	12:29	12:52
4U	ITU Geneva	06:40	07:14	15:51	16:26
4U	United Nations Hq.	11:45	12:16	21:31	22:02
4X, 4Z	Israel, Jerusalem	04:08	04:35	14:39	15:06
5A	Libya, Tripoli	05:40	06:07	16:08	16:35
	Benghazi	05:09	05:36	15:39	16:06
5B	Cyprus, Nicosia	04:22	04:50	14:37	15:06
5H	Tanzania, Dar es Salaam	02:41	03:04	15:35	15:58
5N	Nigeria, Lagos	05:29	05:52	17:36	17:59
5R	Madagascar, Antananarivo	01:45	02:09	15:26	15:50
5T	Mauritania, Nouakchott	07:06	07:30	18:33	18:56
5U	Niger, Niamey	05:46	06:10	17:29	17:52
5V	Togo, Lome	05:37	05:59	17:46	18:09
5W	Western Samoa, Apia	16:34	16:57	05:52	06:16
5X	Uganda, Kampala	03:22	03:44	15:51	16:13
5Z	Kenya, Nairobi	04:22	04:44	16:56	17:19
6W	Senegal, Dakar	07:07	07:30	18:45	19:08
6Y	Jamaica, Kingston	11:10	11:33	22:36	23:00
7O	Yemen, Aden	02:53	03:16	14:39	15:02
	Sanaa	03:01	03:24	14:37	15:00
7P	Lesotho, Maseru	02:40	03:07	17:08	17:35
7Q	Malawi, Lilongwe	02:50	03:14	16:11	16:34
	Blantyre	02:42	03:05	16:09	16:33
7X	Algeria, Algiers	06:27	06:56	16:35	17:04
8P	Barbados, Bridgetown	09:52	10:15	21:36	22:00
8Q	Maldive Is.	00:45	01:08	13:00	13:22
8R	Guyana, Georgetown	09:36	09:58	21:42	22:05
9A	Croatia, Zagreb	05:59	06:34	15:13	15:48
9G	Ghana, Accra	05:42	06:04	17:52	18:15
9H	Malta	05:40	06:09	15:51	16:20
9J	Zambia, Lusaka	03:09	03:33	16:36	16:59
9K	Kuwait	03:13	03:39	13:54	14:20
9L	Sierra Leone, Freetown	06:39	07:01	18:39	19:02
9M2	West Malaysia, Kuala Lumpur	22:50	23:13	11:09	11:32
9M6,8	East Malaysia, Sabah, Sandakan(9M6)	21:49	22:12	09:59	10:21
	Sarawak, Kuching (9M8)	22:13	22:35	10:37	11:00
9N	Nepal, Kathmandu	00:39	01:05	11:28	11:54
9Q	Zaire, Kinshasa	04:23	04:45	17:07	17:30
	Kisangani	03:51	04:14	16:19	16:42
	Lubumbashi	03:20	03:43	16:32	16:55
9U	Burundi, Bujumbura	03:28	03:51	16:10	16:32
9V	Singapore	22:38	23:01	11:04	11:26
9X	Rwanda, Kigali	03:27	03:50	16:04	16:27
9Y	Trinidad & Tobago , Port of Spain	09:55	10:17	21:48	22:11
A2	Botswana, Gaborone	02:58	03:24	17:04	17:30
A3	Tonga, Nukualofa	16:31	16:56	06:21	06:46
A4	Oman, Musqat	02:18	02:43	13:23	13:48
A5	Bhutan, Thimbu	00:22	00:47	11:12	11:38
A6	United Arab Emirates, Abu Dhabi	02:37	03:02	13:39	14:04
A7	Qatar, Ad-Dawhah	02:50	03:15	13:48	14:13
A9	Bahrein, Al-Manamah	02:55	03:20	13:50	14:15
AP	Pakistan, Karachi	01:46	02:11	12:47	13:12
AP	Islamabad	01:40	02:07	12:02	12:30
BV	Taiwan, Taipei	22:09	22:34	09:09	09:34
BY	Peoples Rep. of China, Beijing	23:02	23:32	08:52	09:22
	Harbin	22:36	23:11	07:51	08:25
	Shanghai	22:21	22:48	08:55	09:22
	Fuzhou	22:20	22:45	09:15	09:41
	Xian	23:18	23:46	09:38	10:06

Pref	State/Province/Country/City	Tlight	Sunrise	Sunset	Tlight
	Chongqing	23:18	23:45	09:58	10:25
	Chengdu	23:30	23:56	10:06	10:32
	Lhasa	00:19	00:46	11:00	11:26
	Urumqi	01:07	01:40	10:34	11:07
	Kashi	01:42	02:12	11:35	12:05
C2	Nauru	18:23	18:45	06:54	07:17
C3	Andorra	06:48	07:20	16:23	16:55
C5	The Gambia, Banjul	07:01	07:24	18:44	19:07
C6	Bahamas, Nassau	11:25	11:50	22:24	22:49
C9	Mozambique, Maputo	02:28	02:54	16:40	17:06
	Mozambique	02:20	02:44	15:45	16:09
CE	Chile, Santiago	09:00	09:29	23:52	00:21
CE0	Easter Island	11:53	12:20	02:11	02:37
CE0	Juan Fernandez	09:32	10:01	00:24	00:53
CE0	San Felix	09:58	10:24	00:12	00:38
CM,CO	Cuba, Havana	11:41	12:06	22:48	23:13
CN	Morocco, Casablanca	07:02	07:30	17:25	17:53
CP	Bolivia, La Paz	09:34	09:58	23:04	23:28
CT	Portugal, Lisbon	07:21	07:51	17:18	17:48
CT3	Madeira Islands, Funchal	07:38	08:05	18:05	18:32
CU	Azores , Ponta Delgada	08:24	08:54	18:27	18:56
CX	Uruguay, Montevideo	07:58	08:27	22:57	23:27
CY0	Sable Is.	10:57	11:30	20:25	20:58
CY9	St. Paul Is.	11:08	11:44	20:12	20:48
D2	Angola, Luanda	04:23	04:46	17:24	17:47
D4	Cape Verde, Praia	07:31	07:54	19:09	19:32
D6	Comoros, Moroni	02:15	02:38	15:27	15:50
DA-DL,Y2-Y9	Fed. Rep. of Germany, Bonn	06:51	07:31	15:28	16:08
	Berlin	06:33	07:14	14:53	15:35
DU	Philippines, Manilla	21:52	22:16	09:31	09:55
E3	Eritrea, Asmara	03:22	03:45	14:58	15:22
EA	Spain, Madrid	07:03	07:34	16:51	17:22
EA6	Balearic Is., Palma	06:36	07:06	16:28	16:59
EA8	Canary Is., Las Palmas	07:19	07:45	18:05	18:31
EA9	Ceuta & Melilla, Ceuta	06:59	07:27	17:10	17:39
	Melilla	06:48	07:16	17:03	17:31
EI	Ireland, Dublin	07:55	08:38	16:08	16:51
EK	Armenia, Yerevan	03:50	04:21	13:38	14:09
EL	Liberia, Monrovia	06:25	06:48	18:33	18:56
EP	Iran, Tehran	03:10	03:38	13:22	13:51
ER	Moldova, Kishinev	05:12	05:47	14:17	14:53
ES	Estonia, Tallinn	06:21	07:17	13:20	14:16
ET	Ethiopia, Addis Ababa	03:12	03:35	15:11	15:33
EU,EV,EW	Belarus, Minsk	05:42	06:25	13:49	14:32
EX	Kyrgyzstan, Frunze	01:56	02:29	11:29	12:02
EY	Tajikistan, Samarkand	02:19	02:50	12:11	12:41
	Dushanbe	02:10	02:40	12:05	12:35
EZ	Turkmenistan, Ashkhabad	02:49	03:18	12:50	13:19
F	France, Paris	07:03	07:41	15:56	16:33
FG	Guadeloupe	10:06	10:29	21:40	22:03
FJ, FS	St. Martin	10:15	10:39	21:41	22:05
FH	Mayotte	02:06	02:30	15:23	15:46
FK	New Caledonia, Noumia	17:41	18:07	07:36	08:02
FM	Martinique	10:00	10:24	21:39	22:03
FO	Clipperton Is.	13:06	13:29	01:00	01:23
FO	Fr. Polynesia, Tahiti	15:36	16:00	05:11	05:35
	Rurutu, (Austral Is.)	18:42	19:07	08:38	09:03
	Hiva Oa, (Marquesas Is.)	14:29	14:52	03:35	03:58
FP	St. Pierre & Miquelon, St. Pierre	10:50	11:25	19:58	20:33
FR/G	Glorioso	02:01	02:24	15:12	15:35
FR/J,E	Juan de Nova	02:08	02:32	15:41	16:05
	Europa	02:06	02:31	16:01	16:26
FR	Reunion	01:08	01:33	14:58	15:22
FR/T	Tromelin	01:24	01:48	14:52	15:16
FT8W	Crozet	23:58	00:37	16:22	17:01
FT8X	Kerguelen	22:30	23:13	15:28	16:12
FT8Y	Antarctica, Dumont D'Urville	----	----	----	----
FT8Z	Amsterdam & St. Paul Is., Amsterdam	22:53	23:24	14:10	14:42
FW	Wallis & Futuna Is., Wallis	16:52	17:15	06:10	06:33
FY	Fr. Guiana, Cayenne	09:09	09:32	21:22	21:44
G	England, London	07:23	08:03	15:53	16:33
GD	Isle of Man	07:52	08:36	15:55	16:39
GI	Northern Ireland, Belfast	07:59	08:44	15:59	16:43
GJ	Jersey	07:23	08:01	16:12	16:49
GM	Scotland, Glasgow	07:58	08:45	15:44	16:31
	Aberdeen	07:56	08:46	15:26	16:16
GU	Guernsey	07:26	08:04	16:13	16:51
GW	Wales, Cardiff	07:35	08:16	16:05	16:45
H4	Solomon Islands, Honiara	18:34	18:57	07:38	08:01
HA	Hungary, Budapest	05:52	06:28	14:54	15:30
HB	Switzerland, Bern	06:37	07:12	15:43	16:18
HB0	Liechtenstein	06:29	07:05	15:33	16:09
HC	Ecuador, Quito	10:43	11:06	23:13	23:36
HC8	Galapagos Is.	11:32	11:55	00:04	00:27
HH	Haiti, Port-Au-Prince	10:52	11:16	22:17	22:41
HI	Dominican Republic, Santo Domingo	10:43	11:07	22:08	22:32
HK	Colombia, Bogota	10:36	10:58	22:50	23:12
HK0	Malpelo Is.	11:03	11:25	23:19	23:41
HK0	San Adreas	11:20	11:43	23:06	23:29
HL	Korea, Seoul	22:13	22:42	08:17	08:46
HP	Panama, Panama	11:05	11:28	23:03	23:26
HR	Honduras, Tegucigalpa	11:44	12:08	23:25	23:48
HS	Thailand, Bangkok	23:13	23:36	10:55	11:18
HV	Vatican City	06:02	06:34	15:41	16:13
HZ, 7Z	Saudi Arabia, Dharan	02:57	03:23	13:52	14:17
	Mecca	03:29	03:54	14:43	15:07
	Riyadh	03:07	03:32	14:09	14:34
I	Italy, Rome	06:02	06:34	15:41	16:13
	Trieste	06:08	06:42	15:23	15:57
	Sicily	05:45	06:14	15:49	16:18
IS	Sardinia, Cagliari	06:09	06:39	16:03	16:33
J2	Djibouti, Djibouti	02:58	03:21	14:48	15:11
J3	Grenada	09:59	10:22	21:47	22:10
J5	Guinea-Bissau, Bissau	06:54	07:17	18:43	19:06
J6	St. Lucia	09:59	10:22	21:41	22:04
J7	Dominica	10:03	10:26	21:39	22:02
J8	St. Vincent	09:59	10:22	21:43	22:06
JA-JS	Japan, Tokyo	21:18	21:46	07:30	07:59
	Nagasaki	21:51	22:18	08:17	08:45
	Sapporo	21:30	22:02	07:02	07:34
JD1	Minami Torishima	19:58	20:23	07:00	07:25
JD1	Ogasawara, Kazan Is.	20:56	21:21	07:46	08:11
JT	Mongolia, Ulan Bator	00:02	00:38	09:01	09:38
JW	Svalbard, Spitsbergen	----	----	----	----
JX	Jan Mayen	10:52	----	----	14:09
JY	Jordan, Amman	04:05	04:32	14:35	15:02
KC4	Antarctica, Byrd Station	----	----	----	----
	McMurdo Sound	----	----	----	----
	Palmer Station	----	05:20	03:07	----
KC6	Belau, Yap	20:35	20:58	08:32	08:55
	Koror	20:46	21:09	08:50	09:13
KG4	Guantanamo Bay	11:06	11:30	22:26	22:50
KH0	Mariana Is., Saipan	20:14	20:38	07:51	08:14
KH1	Baker, Howland Is.	17:16	17:39	05:44	06:07
KH2	Guam, Agana	20:15	20:38	07:58	08:21
KH3	Johnston Is.	17:15	17:38	04:45	05:08
KH4	Midway Is.	18:11	18:37	04:58	05:23
KH5	Palmyra Is.	16:30	16:53	04:39	05:02
KH5K	Kingman Reef	16:35	16:58	04:39	05:02
KH6	Hawaii, Hilo	16:26	16:50	03:46	04:10
	Honolulu	16:40	17:04	03:54	04:18
KH7	Kure Is.	18:15	18:41	05:01	05:27
KH8	American Samoa, Pago Pago	16:28	16:52	05:50	06:13
KH9	Wake Is.	18:58	19:22	06:20	06:44
KL7	Alaska, Adak	19:10	19:51	03:37	04:18
	Anchorage	18:12	19:14	00:41	01:43
	Fairbanks	18:32	19:58	23:41	01:06
	Juneau	16:51	17:44	00:06	00:59
	Nome	19:40	21:02	00:56	02:19
KP1	Navassa Is.	11:03	11:27	22:28	22:52
KP2	Virgin Islands, Charlotte Amalie	10:22	10:46	21:48	22:12
KP4	Puerto Rico, San Juan	10:28	10:52	21:53	22:17
KP5	Desecheo Is.	10:33	10:57	21:58	22:22
LA-LJ	Norway, Oslo	07:21	08:19	14:11	15:09
LU	Argentina, Buenos Aires	08:07	08:37	23:05	23:35
LX	Luxembourg	06:51	07:29	15:37	16:15
LY	Lithuania, Vilna	05:53	06:37	13:54	14:38
LZ	Bulgaria, Sofia	05:21	05:53	14:55	15:28
OA	Peru, Lima	10:18	10:41	23:31	23:54
OD	Lebanon, Beirut	04:11	04:39	14:32	15:00
OE	Austria, Vienna	06:05	06:42	15:02	15:39
OH	Finland, Helsinki	06:25	07:23	13:12	14:10
OH0	Aland Is.	06:45	07:43	13:32	14:30
OJ0	Market Reef	06:50	07:48	13:35	14:33
OK,OL	Czech Rep., Prague	06:20	06:58	15:02	15:40
OM	Slovak, Rep., Bratislava	06:02	06:38	15:01	15:37
ON	Belgium, Brussels	07:03	07:42	15:38	16:17
OX	Greenland, Godthaab	12:02	13:22	17:27	18:47
	Thule	----	----	----	----
OY	Faroe Islands, Torshavn	08:44	09:50	14:59	16:05
OZ	Denmark, Copenhagen	06:50	07:37	14:37	15:24
P2	Papua New Guinea, Madang	19:40	20:02	08:28	08:50
	Port Moresby	19:26	19:49	08:29	08:52
P4	Aruba, Oranjestad	10:28	10:51	22:16	22:39
PA-PI	Netherlands, Amsterdam	07:06	07:48	15:28	16:09
PJ2,4,9	Netherlands Antilles, Willemstad	10:33	10:56	22:19	22:42
PJ5-8	St. Maarten and Saba, St. Maarten	10:15	10:38	21:42	22:06
PY	Brazil, Brasilia	08:13	08:37	21:41	22:05
	Rio De Janeiro	07:39	08:04	21:37	22:02
	Natal	07:41	08:04	20:32	20:55
	Manaus	09:27	09:49	22:07	22:30
	Porto Alegre	07:52	08:20	22:25	22:52
PY0	Fernando De Noronha	07:34	07:57	20:18	20:40
PY0	St. Peter & St. Paul Rocks	07:31	07:53	19:57	20:20
PY0	Trindade & Martin Vaz Is., Trindade	06:49	07:13	20:36	21:01
PZ	Suriname, Paramaribo	09:22	09:45	21:32	21:54
R1FJ_	Franz Josef Land	----	----	----	----
R1MV_	Malyj Vysotskij Is.	06:13	07:13	12:53	13:53
S0	Western Sahara, Smara	07:03	07:29	17:58	18:23
S2	Bangladesh, Dacca	00:11	00:36	11:16	11:41
S5	Slovenia, Ljubljana	06:06	06:40	15:19	15:53
S7	Seychelles, Victoria	01:41	02:04	14:27	14:50
S9	Sao Tome	05:05	05:28	17:34	17:56
SM	Sweden, Stockholm	06:47	07:43	13:48	14:43
SP	Poland, Krakow	05:57	06:35	14:40	15:18
	Warsaw	06:01	06:42	14:25	15:06
ST	Sudan, Khartoum	03:48	04:12	15:23	15:47
ST0	Southern Sudan, Juba	03:34	03:56	15:46	16:09
SU	Egypt, Cairo	04:19	04:46	14:58	15:25
SV	Greece, Athens	05:07	05:37	15:08	15:38
SV/A	Mount Athos	05:10	05:41	14:59	15:30
SV5	Dodecanese, Rhodes	04:46	05:15	14:55	15:24
SV9	Crete	04:55	05:24	15:10	15:38
T2	Tuvalu, Funafuti	17:21	17:44	06:22	06:45

Pref	State/Province/Country/City	Tlight	Sunrise	Sunset	Tlight
T30	West Kiribati, Bonriki	18:01	18:23	06:26	06:48
T31	Central Kiribati, Kanton	16:53	17:16	05:33	05:55
T32	East Kiribati, Christmas Is.	16:04	16:27	04:27	04:50
T33	Banaba Is.	18:13	18:35	06:44	07:07
T5	Somalia, Mogadishu	02:33	02:56	14:56	15:18
T7	San Marino	06:08	06:41	15:35	16:08
T9	Bosnia-Herzegovina, Sarajevo	05:44	06:17	15:11	15:44
TA	Turkey, Ankara	04:35	05:06	14:26	14:56
	Istanbul	04:54	05:26	14:38	15:09
TF	Iceland, Reykjavik	10:02	11:21	15:30	16:49
TG	Guatemala, Guatemala City	11:58	12:22	23:37	00:01
TI	Costa Rica, San Jose	11:24	11:47	23:20	23:43
TI9	Cocos Is.	11:29	11:52	23:39	00:02
TJ	Cameroon, Yaounde	04:52	05:15	17:08	17:31
TK	Corsica	06:16	06:48	15:55	16:27
TL	Central African Republic, Bangui	04:25	04:47	16:39	17:02
TN	Congo, Brazzaville	04:23	04:45	17:07	17:30
TR	Gabon, Libreville	04:54	05:17	17:22	17:45
TT	Chad, N'Djamena	04:52	05:15	16:40	17:03
TU	Cote d'Ivoire, Abidjan	05:56	06:19	18:08	18:31
TY	Benin, Porto Novo	05:32	05:55	17:39	18:02
TZ	Mali, Bamako	06:25	06:48	18:11	18:34
UA	Russia, European, Leningrad (UA1)	06:02	06:59	12:53	13:51
	Archangel (UA1)	05:57	07:20	11:11	12:34
	Murmansk (UA1)	07:22	—	—	12:09
	Moscow (UA3)	05:10	05:57	12:57	13:44
	Kuybyshev (UA4)	04:09	04:51	12:23	13:05
	Rostov (UA6)	04:30	05:06	13:33	14:09
UA2	Kaliningrad	06:15	07:01	14:10	14:56
UA9,0	Russia, Asiatic, Novosibirsk (UA9)	02:05	02:51	10:01	10:46
	Perm (UA9)	04:07	04:59	11:26	12:18
	Omsk (UA9)	02:43	03:29	10:39	11:24
	Norilsk (UA0)	03:47	—	—	08:23
	Irkutsk (UA0)	00:28	01:10	08:51	09:32
	Vladivostok (UA0)	22:08	22:41	07:39	08:12
	Petropavlovsk (UA0)	20:54	21:36	05:09	05:52
	Khabarovsk (UA0)	22:11	22:48	07:06	07:43
	Krasnoyarsk (UA0)	01:30	02:18	09:15	10:02
	Yakutsk (UA0)	23:38	00:44	05:53	06:59
	Wrangel Island (UA0)	22:17	—	—	01:34
	Kyzyl (UA0Y)	01:05	01:46	09:33	10:14
UJ-UM	Uzbekistan, Bukhoro	02:29	03:00	12:20	12:51
	Tashkent	02:13	02:44	11:56	12:28
UN-UQ	Kazakhstan, Alma-Ata	01:48	02:21	11:19	11:52
UR-UZ,EM-EO	Ukraine, Kiev	05:16	05:55	13:56	14:35
V2	Antigua & Barbuda, St. Johns	10:08	10:32	21:38	22:01
V3	Belize, Belmopan	11:56	12:20	23:25	23:49
V4	St. Kitts & Nevis	10:11	10:35	21:41	22:04
V5	Namibia, Windhoek	03:38	04:03	17:35	18:00
V6	Micronesia, Ponape	19:10	19:33	07:16	07:39
V7	Marshall Islands, Kwajalein	18:38	19:01	06:36	06:59
V8	Brunei, Bandar Seri Begawan	22:00	22:23	10:13	10:36
VK	Australia, Canberra (VK1)	18:15	18:45	09:17	09:47
	Sydney (VK2)	18:11	18:40	09:05	09:34
	Melbourne (VK3)	18:23	18:54	09:41	10:12
	Brisbane (VK4)	18:23	18:49	08:42	09:08
	Adelaide (VK5)	18:58	19:28	09:58	10:28
	Perth (VK6)	20:39	21:07	11:21	11:50
	Hobart, Tasmania (VK7)	17:53	18:28	09:49	10:24
	Darwin (VK8)	20:25	20:48	09:40	10:03
VK0	Heard Is.	21:44	22:36	15:32	16:23
VK0	Macquarie Is.	15:46	16:43	10:01	10:58
VK9C	Cocos-Keeling Is.	22:42	23:05	11:56	12:19
VK9L	Lord Howe Is.	17:47	18:15	08:27	08:56
VK9M	Mellish Reef	18:35	18:59	08:10	08:34
VK9N	Norfolk Is.	17:18	17:46	07:45	08:12
VK9W	Willis Is.	19:03	19:27	08:32	08:56
VK9X	Christmas Is.	22:09	22:33	11:17	11:40
VP2E	Anguilla	10:15	10:39	21:40	22:04
VP2M	Montserrat	10:09	10:33	21:40	22:04
VP2V	British Virgin Is., Tortola	10:21	10:45	21:47	22:10

Pref	State/Province/Country/City	Tlight	Sunrise	Sunset	Tlight
VP5	Turks & Caicos Islands, Grand Turk	10:53	11:17	22:07	22:31
VP8	Falkland Islands, Stanley	06:41	07:29	00:09	00:58
VP8	So. Georgia Is.	04:53	05:48	23:01	23:57
VP8	So. Orkney Is.	—	05:26	00:33	—
VP8	So. Sandwich Islands, Saunders Is.	03:26	04:41	22:48	00:02
VP8	So. Shetland Is., King George Is.	—	05:58	01:43	—
VP9	Bermuda	10:48	11:15	21:17	21:44
VQ9	Chagos, Diego Garcia	00:29	00:51	13:24	13:47
VR2,VS6	Hong Kong	22:33	22:57	09:43	10:08
VR6	Pitcairn Is.	13:21	13:47	03:29	03:55
VU	India, Bombay	01:13	01:37	12:36	13:00
	Calcutta	00:17	00:41	11:26	11:51
	New Delhi	01:13	01:39	11:58	12:24
	Bangalore	00:43	01:06	12:28	12:51
VU	Andaman Islands, Port Blair	23:40	00:03	11:29	11:52
VU	Laccadive Is.	00:56	01:19	12:52	13:14
XE	Mexico, Mexico City (XE1)	12:41	13:05	00:03	00:27
	Chihuahua (XE2)	13:26	13:52	00:11	00:37
	Merida (XE3)	12:06	12:31	23:22	23:46
XF4	Revilla Gigedo	13:30	13:54	00:53	01:17
XT	Burkina Faso, Ouagadougou	05:59	06:22	17:46	18:09
XU	Cambodia, Phnom Penh	22:52	23:15	10:41	11:04
XW	Laos, Viangchan	23:12	23:36	10:39	11:02
XX9	Macao	22:35	23:00	09:46	10:11
XZ	Myanmar, Yangon	23:36	00:00	11:07	11:31
YA	Afghanistan, Kandahar	02:04	02:30	12:38	13:05
	Kabul	01:57	02:25	12:16	12:44
YB-YD	Indonesia, Jakarta	22:13	22:36	11:05	11:28
	Medan	23:03	23:25	11:20	11:43
	Pontianak	22:14	22:37	10:44	11:06
	Jayapura	20:04	20:27	08:43	09:05
YI	Iraq, Baghdad	03:33	04:00	13:59	14:26
YJ	Vanuatu, Port Vila	17:44	18:09	07:20	07:44
YK	Syria, Damascus	04:07	04:34	14:30	14:58
YL	Latvia, Riga	06:10	07:00	13:42	14:32
YN	Nicaragua, Managua	11:36	11:59	23:24	23:47
YO	Romania, Bucharest	05:15	05:48	14:38	15:12
YS	El Salvador, San Salvador	11:52	12:15	23:34	23:57
YU	Yugoslavia, Belgrade	05:38	06:12	14:59	15:33
YV	Venezuela, Caracas	10:17	10:40	22:11	22:34
YV0	Aves Is.	10:13	10:36	21:48	22:11
Z2	Zimbabwe, Harare	02:53	03:18	16:29	16:53
Z3	Macedonia, (ex Yugoslav), Skopje	05:27	05:59	15:05	15:37
ZA	Albania, Tirane	05:31	06:03	15:14	15:45
ZB2	Gibraltar	06:59	07:28	17:10	17:39
ZC4	British Cyprus	04:22	04:50	14:40	15:09
ZD7	St. Helena	05:25	05:49	18:53	19:17
ZD8	Ascension Is.	06:15	06:37	19:13	19:36
ZD9	Tristan da Cunha	04:54	05:25	20:08	20:39
ZF	Cayman Is.	11:30	11:54	22:51	23:15
ZK1	No. Cook Is., Manihiki	15:56	16:20	05:03	05:27
ZK1	So. Cook Is., Rarotonga	15:29	15:54	05:19	05:44
ZK2	Niue	16:10	16:35	05:51	06:16
ZK3	Tokelaus, Atafu	16:47	17:10	05:47	06:10
ZL	New Zealand, Auckland (ZL1)	16:27	16:57	07:39	08:10
	Wellington (ZL2)	16:10	16:43	07:53	08:27
	Christchurch (ZL3)	16:09	16:44	08:10	08:46
	Dunedin (ZL4)	16:05	16:43	08:28	09:06
ZL5	Antarctica, Scott Base	—	—	—	—
ZL7	Chatham Is.	15:23	15:59	07:28	08:04
ZL8	Kermadec Is.	16:21	16:48	06:50	07:17
ZL9	Auckland & Campbell Is., Auckland	15:51	16:37	09:06	09:52
	Campbell Is.	15:26	16:16	09:06	09:56
ZP	Paraguay, Asuncion	08:31	08:57	22:40	23:06
ZS	South Africa, Cape Town (ZS1)	03:02	03:32	17:56	18:25
	Port Elizabeth (ZS2)	02:33	03:02	17:27	17:57
	Bloemfontein (ZS4)	02:46	03:13	17:13	17:41
	Durban (ZS5)	02:24	02:52	16:56	17:23
	Johannesburg (ZS6)	02:46	03:12	16:59	17:25
ZS8	Marion Is.	00:51	01:30	17:22	18:02

W1AW schedule

Pacific	Mtn	Cent	East	Sun	Mon	Tue	Wed	Thu	Fri	Sat
6 am	7 am	8 am	9 am			Fast Code	Slow Code	Fast Code	Slow Code	
7 am	8 am	9 am	10 am			Code Bulletin				
8 am	9 am	10 am	11 am			Teleprinter Bulletin				
9 am	10 am	11 am	noon							
10 am	11 am	noon	1 pm			**Visiting Operator Time**				
11 am	noon	1 pm	2 pm							
noon	1 pm	2 pm	3 pm							
1 pm	2 pm	3 pm	4 pm	Slow Code	Fast Code	Slow Code	Fast Code	Slow Code	Fast Code	Slow Code
2 pm	3 pm	4 pm	5 pm	Code Bulletin						
3 pm	4 pm	5 pm	6 pm	Teleprinter Bulletin						
4 pm	5 pm	6 pm	7 pm	Fast Code	Slow Code	Fast Code	Slow Code	Fast Code	Slow Code	Fast Code
5 pm	6 pm	7 pm	8 pm	Code Bulletin						
6 pm	7 pm	8 pm	9 pm	Teleprinter Bulletin						
6⁴⁵ pm	7⁴⁵ pm	8⁴⁵ pm	9⁴⁵ pm	Voice Bulletin						
7 pm	8 pm	9 pm	10 pm	Slow Code	Fast Code	Slow Code	Fast Code	Slow Code	Fast Code	Slow Code
8 pm	9 pm	10 pm	11 pm	Code Bulletin						
9 pm	10 pm	11 pm	Mdnte	Teleprinter Bulletin						
9⁴⁵ pm	10⁴⁵ pm	11⁴⁵ pm	12⁴⁵ am	Voice Bulletin						

W1AW's schedule is at the same local time throughout the year. The schedule according to your local time will change if your local time does not have seasonal adjustments that are made at the same time as North American time changes between standard time and daylight time. From the first Sunday in April to the last Sunday in October, UTC = Eastern Time + 4 hours. For the rest of the year, UTC = Eastern Time + 5 hours.

❏ *Morse code transmissions:*

Frequencies are 1.818, 3.5815, 7.0475, 14.0475, 18.0975, 21.0675, 28.0675 and 147.555 MHz.

Slow Code = practice sent at 5, 7¹/₂, 10, 13 and 15 wpm.

Fast Code = practice sent at 35, 30, 25, 20, 15, 13 and 10 wpm.

Code practice text is from the pages of *QST*. The source is given at the beginning of each practice session and alternate speeds within each session. For example, "Text is from July 1992 *QST*, pages 9 and 81," indicates that the plain text is from the article on page 9 and mixed number/letter groups are from page 81.

Code bulletins are sent at 18 wpm.

❏ *Teleprinter transmissions:*

Frequencies are 3.625, 7.095, 14.095, 18.1025, 21.095, 28.095 and 147.555 MHz.

Bulletins are sent at 45.45-baud Baudot and 100-baud AMTOR, FEC Mode B. 110-baud ASCII will be sent only as time allows.

On Tuesdays and Saturdays at 6:30 PM Eastern Time, Keplerian elements for many amateur satellites are sent on the regular teleprinter frequencies.

❏ *Voice transmissions:*

Frequencies are 1.855, 3.99, 7.29, 14.29, 18.16, 21.39, 28.59 and 147.555 MHz.

❏ *Miscellanea:*

On Fridays, UTC, a DX bulletin replaces the regular bulletins.

W1AW is open to visitors during normal operating hours: from 1 PM until 1 AM on Mondays, 9 AM until 1 AM Tuesday through Friday, from 1 PM to 1 AM on Saturdays, and from 3:30 PM to 1 AM on Sundays. FCC licensed amateurs may operate the station from 1 to 4 PM Monday through Saturday. Be sure to bring your current FCC amateur license or a photocopy.

In a communication emergency, monitor W1AW for special bulletins as follows: voice on the hour, teleprinter at 15 minutes past the hour, and CW on the half hour.

Headquarters and W1AW are closed on New Year's Day, President's Day, Good Friday, Memorial Day, Independence Day, Labor Day, Thanksgiving and the following Friday, and Christmas Day. On the first Thursday of September, Headquarters and W1AW will be closed during the afternoon.

ARRL Field Organization

The United States is divided into 15 ARRL Divisions. Every two years the ARRL full members in each of these divisions elect a director and a vice director to represent them on the League's Board of Directors. The Board determines the policies of the League, which are carried out by the Headquarters staff. A director's function is principally policymaking at the highest level, but the Board of Directors is all-powerful in the conduct of League affairs.

The 15 divisions are further broken down into 69 sections, and the ARRL full members in each section elect a Section Manager (SM). The SM is the senior elected ARRL official in the section, and in cooperation with the director, fosters and encourages all ARRL activities within the section. A breakdown of sections within each division (and counties within each split-state section) follows:

ATLANTIC DIVISION: *Delaware, Eastern Pennsylvania* (Adams, Berks, Bradford, Bucks, Carbon, Chester, Columbia, Cumberland, Dauphin, Delaware, Juniata, Lackawanna, Lancaster, Lebanon, Lehigh, Luzerne, Lycoming, Monroe, Montgomery, Montour, Northhampton, Northumberland, Perry, Philadelphia, Pike, Schuylkill, Snyder, Sullivan, Susquehanna, Tioga, Union, Wayne, Wyoming, York); *Maryland-D.C.; Southern New Jersey* (Atlantic, Burlington, Camden, Cape May, Cumberland, Gloucester, Mercer Ocean, Salem); *Western New York* (Allegany, Broome, Cattaraugus, Cayuga, Chautauqua, Chemung, Chenango, Clinton, Cortland, Delaware, Erie, Essex, Franklin, Fulton, Genesee, Hamilton, Herkimer, Jefferson, Lewis, Livingston, Madison, Monroe, Montgomery, Niagara, Oneida, Onondaga, Ontario, Orleans, Oswego, Otsego, St. Lawrence, Schoharie, Schuyler, Seneca, Steuben, Tioga, Tompkins, Wayne, Wyoming, Yates); *Western Pennsylvania* (those counties not listed under Eastern Pennsylvania).

CENTRAL DIVISION: *Illinois; Indiana; Wisconsin.*

DAKOTA DIVISION: *Minnesota; North Dakota; South Dakota.*

DELTA DIVISION: *Arkansas; Louisiana; Mississippi; Tennessee*

GREAT LAKES DIVISION: *Kentucky; Michigan; Ohio.*

HUDSON DIVISION: *Eastern New York* (Albany, Columbia, Dutchess, Greene, Orange, Putnam, Rensselaer, Rockland, Saratoga, Schenectady, Sullivan, Ulster, Warren, Washington, Westchester); *N.Y.C.-L.I.* (Bronx, Kings, Nassau, New York, Queens, Staten Island, Suffolk); *Northern New Jersey* Bergen, Essex, Hudson, Hunterdon, Middlesex, Monmouth, Morris, Passaic, Somerset, Sussex, Union, Warren).

MIDWEST DIVISION: *Iowa; Kansas; Missouri; Nebraska.*

NEW ENGLAND DIVISION: *Connecticut; Maine, Eastern Massachusetts* (Barnstable, Bristol, Dukes, Essex, Middlesex, Nantucket, Norfolk, Plymouth, Suffolk); *New Hampshire; Rhode Island; Vermont; Western Massachusetts* (those counties not listed under Eastern Massachusetts).

NORTHWESTERN DIVISION: *Alaska; Idaho; Montana; Oregon; Eastern Washington* (Adams, Asotin, Benton, Chelan, Columbia, Douglas, Ferry, Franklin, Garfield, Grant, Kittitas, Klickitat, Lincoln, Okangogan, Pend Oreille, Spokane, Stevens, Walla, Walla, Whitman, Yakima); *Western Washington* (Challam, Clark Cowlitz, Grays Harbor Island, Jefferson, King, Kitsap, Lewis, Mason, Pacific, Pierce, San Juan, Skagit, Skamania, Snohomish, Thurston, Wahkiakum, Whatcom).

PACIFIC DIVISION: *East Bay* (Alameda, Contra Costa, Napa, Solano); *Nevada; Pacific* (Hawaii and U.S. possessions in the Pacific); *Sacramento Valley* (Alpine, Amador, Butte, Colusa, El Dorado, Glenn, Lassen, Modoc, Nevada, Placer, Plumas, Sacramento, Shasta, Sierra, Siskiyou, Sutter, Tehama, Trinity, Yolo, Yuba); *San Franciso,* (Del Norte, Humboldt, Lake, Marin, Mendocino, San Franciso, Sonoma); *San Joaquin Valley* (Calaveras, Fresno, Kern, Kings, Madera, Mariposa, Merced, Mono, San Joaquin, Stanislaus, Tulare, Tuolumne); *Santa Clara Valley* (Monterey, San Benito, San Mateo, Santa Clara, Santa Cruz).

ROANOKE DIVISION: *North Carolina; South Carolina; Virginia; West Virginia.*

ROCKY MOUNTAIN DIVISION: *Colorado; Utah; New Mexico; Wyoming.*

SOUTHEASTERN DIVISION: *Alabama; Georgia; Northern Florida* (Alachua, Baker, Bay, Bradford, Calhoun, Citrus, Clay, Columbia, Dixie, Duval, Escambia, Flagler, Franklin, Gadsden, Gilchrist, Gulf, Hamilton, Hernando, Holmes, Jackson, Jefferson, Lafayette, Lake, Leon, Levy, Liberty, Madison, Marion, Nassau, Okaloosa, Orange, Pasco, Putnam, Santa Rosa, Seminole, St. Johns, Sumter, Suwanee, Taylor, Union, Volusia, Wakulla, Walton, Washington); *Southern Florida* (those counties not listed under Northern Florida); *Puerto Rico; U.S. Virgin Islands.*

SOUTHWESTERN DIVISION: *Arizona; Los Angeles; Orange* (Inyo, Orange, Riverside, San Bernardino); *San Diego* (Imperial, San Diego); *Santa Barbara* (San Luis Obispo, Santa Barbara, Ventura).

WEST GULF DIVISION: *North Texas* (Anderson, Archer, Baylor, Bell, Bosque, Bowie, Brown, Camp, Cass, Cherokee, Clay, Collin, Comanche, Cooke, Coryell, Dallas, Delta, Denton, Eastland, Ellis, Erath, Falls, Fannin, Franklin, Freestone, Grayson, Gregg, Hamilton, Harrison, Henderson, Hill, Hood, Hopkins, Hunt,

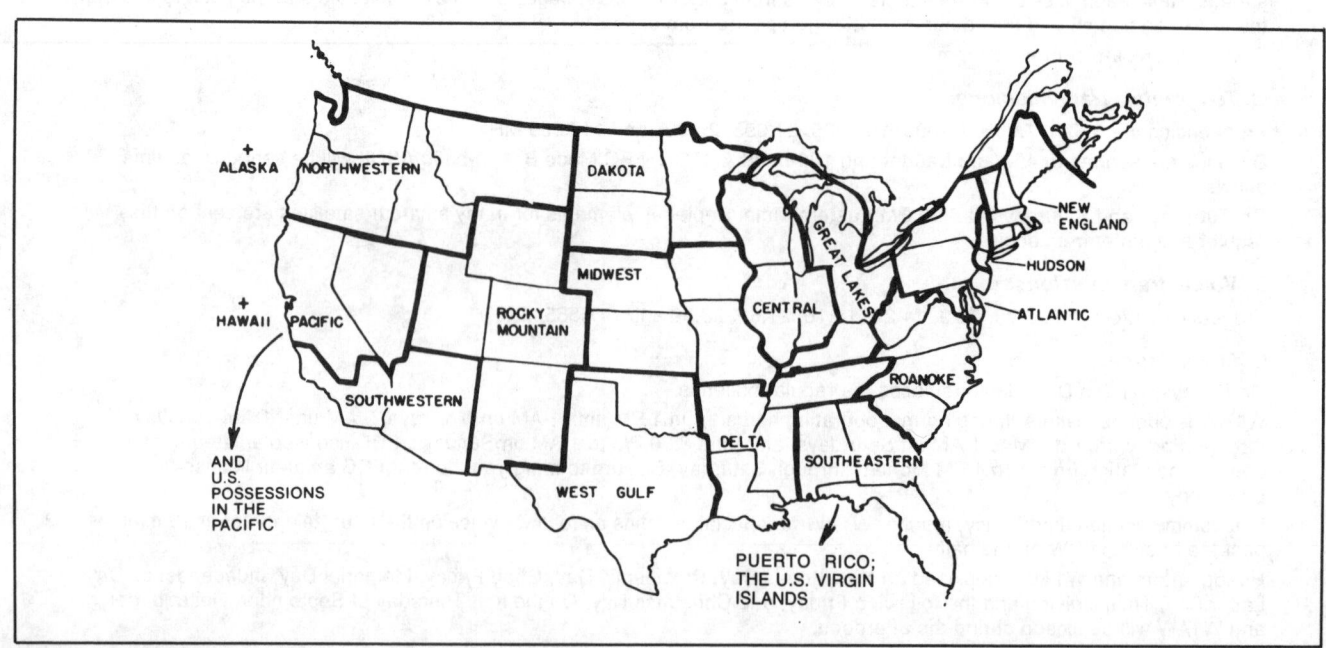

Jack, Johnson, Kaufman, Lamar, Lampasas, Limestone, McLennan, Marion, Mills, Montague, Morris, Nacogdoches, Navarro, Palo Pinto, Panola, Parker, Rains, Red River, Rockwall, Rusk, Shelby, Smith, Somervell, Stephens, Tarrant, Throckmorton, Titus, Upshur, Van Zandt, Wichita, Wilbarger, Wise, Wood, Young); *Oklahoma; South Texas* (Angelina, Aransas, Atacosa, Austin, Bandera, Bastrop, Bee, Bexar, Blanco, Brazoria, Brazos, Brooks, Burleson, Burnet, Caldwell, Calhoun, Cameron, Chambers, Colorado, Comal, Concho, DeWitt, Dimmitt, Duval, Edwards, Fayette, Fort Bend, Frio, Galveston, Gillespie, Goliad, Gonzales, Grimes, Guadalupe, Hardin, Harris, Hays, Hidalgo, Houston, Jackson, Jasper, Jefferson, Jim Hogg, Jim Wells, Karnes, Kendall, Kenedy, Kerr, Kimble, Kinney, Kleberg, LaSalle, Lavaca, Lee, Leon, Liberty, Live Oak, Llano, Madison, Mason, Matagorda, Maverick, McCulloch, McMullen, Medina, Menard, Milam, Montgomery, Newton, Nueces, Orange, Polk, Real, Refugio, Robertson, Sabine, San Augustine, San Jacinto, San Patricio, San Saba, Starr, Travis, Trinity, Tyle, Uvalde, Val Verde, Victoria, Walker, Waller, Washington, Webb, Wharton, Willacy, Williamson, Wilson, Zapata, Zavala); *West Texas* (Andrews, Armstrong, Bailey, Bordon, Brewster, Briscoe, Callahan, Carson, Castro, Childress, Cochran, Coke, Coleman, Collingsworth, Cottle, Crane, Crockett, Crosby, Culberson, Dallam, Dawson, Deaf Smith, Dickens, Donley, Ector, El Paso, Fischer, Floyd, Foard, Gaines, Garza, Glasscock, Gray, Hale, Hall, Hansford, Hardeman, Hartley, Haskell, Hemphill, Hockley, Howard Hudspeth, Hutchinson, Irion, Jeff Davis, Jones, Kent, King, Knox, Lamb, Lipscomb, Loving, Lubbock, Lynn, Martin, Midland, Mitchell, Moore, Motley, Nolan, Ochiltree, Oldham, Parmer, Pecos, Potter, Presidio, Randall, Reagan, Reeves, Roberts, Runnels, Schleicher, Scurry, Shackelford, Sherman, Sterling, Stonewall, Sutton, Swisher, Taylor, Terrell, Terry, Tom Green, Upton, Ward, Wheeler, Winkler, Yoakum).

RAC RADIO AMATEURS OF CANADA-CANADA: *Alberta; British Columbia; Manitoba; Maritime* (Nova Scotia, New Brunswick, Prince Edward Island, Labrador, Newfoundland); *Ontario; Quebec; Saskatchewan.*

SECTION MANAGER

In each ARRL section there is an elected Section Manager (SM) who will have authority over the Field Organization in his or her section, and in cooperation with his Director, foster and encourage ARRL activities and programs within the section. Details regarding the election procedures for SMs are contained in "Rules and Regulations of the ARRL Field Organization," available on request to any ARRL member. Election notices are posted regularly in the Happenings section of *QST.*

Any candidate for the office of Section Manager must be a resident of the section, a licensed amateur of the Technician class or higher, and a full member of the League for a continuous term of at least two years immediately preceding receipt of a petition for nomination at Hq. If elected, he or she must maintain membership throughout the term of office.

The following is a detailed resume of the duties of the Section Manager. In discharging his responsibilities, he:

a) Recruits and appoints nine section-level assistants to serve under his general supervision and to administer the following ARRL programs in the section: emergency communications, message traffic, official observers, affiliated clubs, public information, state government liaison, technical activities and on-the-air bulletins.

b) Supervises the activities of these assistants to ensure continuing progress in accordance with overall ARRL policies and objectives.

c) Appoints qualified ARRL members in the section to volunteer positions of responsibility in support of section programs, or authorizes the respective section-level assistants to make such appointments.

d) Maintains liaison with the Division Director and makes periodic reports to him regarding the status of section activities; receives from him information and guidance pertaining to matters of mutual concern and interest; serves on the Division Cabinet and renders advice as requested by the Division Director; keeps informed on matters of policy that affect section-level programs.

e) Conducts correspondence or other communications, including personal visits to clubs, hamfests and conventions, with ARRL members and affiliated clubs in the section; either responds to their questions or concerns or refers them to the appropriate person or office in the League organization; maintains liaison with, and provides support to, representative repeater frequency-coordinating bodies having jurisdiction in the section.

f) Writes, or supervises preparation of, the monthly Section News column in *QST* to encourage member participation in the ARRL program in the section.

[Note: Move of permanent residence outside the section from which elected will be grounds for declaring the office vacant.]

ARRL LEADERSHIP APPOINTMENTS

Field Organization leadership appointments from the Section Manager are available to qualified ARRL full members in each section. These appointments are as follows: Assistant Section Manager, Section Emergency Coordinator, Section Traffic Manager, Official Observer Coordinator, State Government Liaison, Technical Coordinator, Affiliated Club Coordinator, Public Information Officer, Bulletin Manager, District Emergency Coordinator, Emergency Coordinator and Net Manager. Holders of such appointments may wear the League emblem pin with the distinctive deep-green background. Functions of these leadership officials are described below.

Assistant Section Manager

The ASM is an ARRL section-level official appointed by the Section Manager, in addition to the Section Manager's eight section-level assistants. An ASM may be appointed if the Section Manager believes such an appointment is desirable to meet the goals of the ARRL Field Organization in that section. Thus, the ASM is appointed at the complete discretion of the Section Manager, and serves at the pleasure of the Section Manager.

1) The ASM may serve as a general or as a specialized assistant to the Section Manager. That is, the ASM may assist the Section Manager with general leadership matters as the Section Manager's general understudy, or the ASM may be assigned to handle a specific important function not within the scope of the duties of the Section Manager's eight assistants.

2) At the Section Manager's discretion, the ASM may be designated as the recommended successor to the incumbent Section Manager, in case the Section Manager resigns or is otherwise unable to finish the term of office.

3) The ASM should be familiar with "Guidelines for the ARRL Section Manager," which contains the fundamentals of general section management.

4) The ASM must be an ARRL full member, holding at least a Novice class license.

Section Emergency Coordinator

The SEC must hold a Technician class license or higher and is appointed by the Section Manager to take care of all matters pertaining to emergency communications and the Amateur Radio Emergency Service (ARES) on a sectionwide basis. The duties of the SEC include the following:

1) The encouragement of all groups of community amateurs to establish a local emergency organization.

2) Recommendations to the SM on all section emegency policy and planning, including the development of a section emergency communications plan.

3) Cooperation and Coordination with the Section Traffic Manager so that emergency nets and traffic nets in the section present a unified public sevice front. Cooperation and coordiantion should also be maintained with other section leadership officials as appropriate, particularly the State Governmnent Liaison and the Public Information Coordinator.

4) Recommendations of candidates for Emergency Coordinator and District Emergency Coordinator appointments (and cancellations) to the Section Manager and determinations of areas of jurisdiction of each amateur so appointed. At the SM's discretion, the SEC may be directly in charge of making (and cancelling) such appointments. In the same way, the SEC can handle the Official Emergency Station program.

5) Promotion of ARES membership drives, meetings, activities, tests, procedures, etc., at the section level.

6) Collection and consolidation of Emergency Coordinator (or District Emergency Coordinator) monthly reports and submission of monthly progress summaries to ARRL Hq.

7) Maintenance of contact with other communication services and liaison at the section level with all agencies served in the public interest, particularly in connection with state and local government, civil preparedness, Red Cross, Salvation Army and the National Weather Service. Such contact is maintained in cooperation with the State Government Liaison.

Section Traffic Manager

The STM is appointed by the Section Manager to supervise traffic handling organization at the section level—that is, of coordinating the activities of all traffic nets, both National Traffic System-affiliated and independents, so that routings within the section and connections with other nets to effect orderly and efficient traffic flow are maintained. The STM should be a person at home and familiar with traffic handling on all modes, must have at least a Technician class license, and should possess the willingness and ability to devote equal consideration and time to all section traffic matters. The duties of the STM include the following:

1) Establish, administer, and promote a traffic handling program at the section level, based on, but not restricted to, National Traffic System networks.

2) Develop and implement one or more effective training programs within the section that addresses the needs of both traditional and digital modes of traffic handling. Ensure that Net Managers place particular emphasis on the needs of amateurs new to formal network traffic handling, as well as those who receive, send, and deliver formal traffic on a "casual" basis, via RTTY, Amtor, and Packet based message storage and bulletin board systems.

3) Cooperate and coordinate with the Section Emergency Coordinator so that traffic nets and emergency nets in the section present a unified public service front.

4) Recommend candidates for Net Managers and Official Relay Station appointments to the SM. Issue FSD-211 appointment/cancellation cards and appropriate certificates. At the SM's discretion, the STM may directly make or cancel NM and ORS appointments.

5) Ensure that all traffic nets within the section are properly and adequately staffed, with appropriate direction to Net Managers, as required, which results in coverage of all Net Control liaison functions. Assign liaison coverage adequate to ensure that all digital bulletin boards and message storage systems within the section are polled on a daily basis, to prevent misaddressed, lingering, or duplicated radiogram-formatted message traffic.

6) Maintain familiarity with proper traffic handling and directed net procedures applicable to all normally used modes within the section.

7) Collect and prepare accurate monthly net reports and submit them to ARRL Headquarters, either directly or via the Section Manager, but in any case on or prior to the established deadlines.

Affiliated Club Coordinator

The ACC is the primary contact and resource person for each Amateur Radio club in the section, specializing in providing assistance to clubs. The ACC is appointed by, and reports to, the Section Manager. Duties and qualifications of the ACC include:

1) Volunteer a great deal of time in getting to know the Amateur Radio clubs' members and officers person to person in his or her section. Learn their needs, strengths and interests and work with them to make clubs effective resources in their communities and more enjoyable for their members.

2) Encourage affiliated clubs in the section to become more active and, if the club is already healthy and effective, to apply as a Special Service Club (SSC).

3) Supply interested clubs with SSC application forms.

4) Assist clubs in completing SSC application forms, if requested.

5) Help clubs establish workable programs to use as SSCs.

6) Approve SSC application forms and pass them to the SM.

7) Work with other section leadership officials (Section Emergency Coordinator, Public Information Coordinator, Technical Coordinator, State Government Liason, etc.) to ensure that clubs are involved in the mainstream of ARRL Field Organization activities.

8) Encourage new clubs to become ARRL affiliated.

9) Ensure that annual progress reports (updated officers, liaison mailing addresses, etc.) are forthcoming from all affiliated clubs.

10) Novice Class license; ARRL membership required.

Bulletin Manager

Rapid dissemination of information is the lifeblood of an active, progressive organization. The ARRL Official Bulletin Station network provides a vital communications link for informing the amateur community of the latest developments in Amateur Radio and the ARRL. The ARRL Bulletin Manager is responsible for recruiting and supervising a team of Official Bulletin Stations to disseminate such news and information of interest to amateurs in the section and to provide a means of getting the news and information to all OBS appointees. The bulletins should include the content of ARRL bulletins (transmitted by W1AW), but should also include items of local, section and regional interest from other sources, such as ARRL section leadership officials, as well as information provided by the Division Director.

A special effort should be made to recruit an OBS for each major repeater and packet bulletin board in the section. This is where the greatest "audience" is to be found, many of whom are not sufficiently informed about the latest news of Amateur Radio and the League. Such bulletins should be transmitted regularly, perhaps in conjunction with a repeater net or on a repeater "bulletin board" (tone-accessed recorded announcements for repeater club members).

Although the primary mission of OBS appointees is to copy ARRL bulletins directly from W1AW, in some sections the Bullein Manager may take on the responsibility of retransmitting ARRL bulletins (as well as other information) for the benefit of OBS appointees, on a regularly scheduled day, time and frequency. An agreed-upon schedule should be worked out in advance. Time is of the essence when conveying news; therefore a successful Bulletin Manager will develop ways of communicating with the OBS appointees quickly and efficiently.

Bulletin Managers should be familiar with the position description of the Official Bulletin Station, which appears later. The duties of the Bulletin Manager include the following:

1) The Bulletin Manager must have a Technician class license or higher, and maintain League membership.

2) The Bulletin Manager is appointed by the SM and is required to report regularly to the SM concerning the section's bulletin program.

3) The Bulletin Manager is responsible for recruiting (and, at the discretion of the SM, appointing) and supervising a team of Official Bulletin Stations in the Section. A special effort should be made to recruit OBSs for each major repeater and PBBS in the section.

4) The Bulletin Manager must be capable of copying ARRL bulletins directly from W1AW on the mode(s) necessary. The Bulletin Manager may, in some cases, be required to retransmit ARRL bulletins for OBS appointees who might be unable to copy them directly from W1AW.

5) The Bulletin Manager is also responsible for funneling news and information of a local, section and regional nature to OBS appointees. In so doing, the Bulletin Manager must maintain close contact with other section-level officials, and the Division Director, to maintain an organized and unified information flow within the section.

Official Observer Coordinator

The Official Observer Coordinator is an ARRL section-level leadership official appointed by the Section Manager to supervise the Official Observer program in the section. The OO Coordinator must hold a Technician class (or higher) amateur license and be licensed as a Technician or higher for at least four years.

The Official Observer program has operated for more than half a century, and in that time, OO appointees have assisted thousands of amateurs whose signals, or operating procedures, were not in compliance with the regulations. The function of the

OO is to *listen* for amateurs who might otherwise come to the attention of the FCC and to advise them by mail of the irregularity observed. The OO program is, in essence, for the benefit of amateurs who *want* to be helped. Official Observers must meet high standards of expertise and experience. It is the job of the OO Coordinator to recruit, supervise and direct the efforts of OOs in the section, and to report their activity monthly to the Section Manager and to ARRL Hq.

The OO Coordinator is a key figure in the Amateur Auxiliary to the FCC's Field Operations Bureau, the foundation of which is an enhanced OO program. Jointly created by the FCC and ARRL in response to federal enabling legislation (Public Law 97-259), the Auxiliary permits a close relationship between FCC and ARRL Field Organization volunteers in monitoring the amateur airwaves for potential rules discrepancies/violations. (Contact your Section Manager for further details on the Amateur Auxiliary program.)

Public Information Coordinator

The ARRL Public Information Coordinator is a section-level official appointed by the Section Manager to be the section's expert on public information and public relations matters. The Public Information Coordinator is also responsible for organizing, guiding and coordinating the activities of the Public Information Officers within the section.

The Public Information Coordinator must be a full member of the ARRL and, preferably, have professional public relations or journalism experience or a significantly related background in dealing with the public media.

The purpose of public relations goes beyond column inches and minutes of air time. Those are means to an end—generally, telling a specific story about hams, ham radio or ham-related activities for a specific purpose. Goals may range from recruiting potential hams for a licensing course to improving public awareness of amateurs' service to the community. Likewise, success is measured not in column inches or air time, but in how well that story gets across and how effectively it generates the desired results.

For this reason, public relations are not conducted in a vacuum. Even the best PR is wasted without effective follow-up. To do this best, PR activities must be well-timed and well-coordinated within the amateur community, so that clubs, Elmers, instructors and so on are prepared to deal with the interest the PR generates. Effective PICs will convey this goal-oriented perspective and attitude to their PIOs and help them coordinate public relations efforts with others in their sections.

Recruitment of new hams and League members is an integral part of the job of every League appointee. Appointees should take advantage of every opportunity to recruit a new ham or a member to foster growth of Field Organization programs, and our abilities to serve the public.

Specific Duties of the Public Information Coordinator

1) Advises the Section Manager on building and maintaining a positive public image for Amateur Radio in the section; keeps the SM informed of all significant events which would benefit from the SM's personal involvement and reports regularly to the SM on activities.

2) Counsels the SM in dealing with the media and with government officials, particularly when representing the ARRL and/or Amateur Radio in a public forum.

3) Maintains contact with other section level League officials, particularly the Section Manager and others such as the State Government Liaison, Section Emergency Coordinator and Bulletin Manager on matters appropriate for their attention and to otherwise help to assure and promote a coordinated and cohesive ARRL Field Organization.

4) Works closely with the section Affiliated Club Coordinator and ARRL-affiliated clubs in the section to recruit and train a team of Public Information Officers (PIOs). With the approval of the Section Manager, makes PIO appointments within the section.

5) Works with the SM and other PICs in the division to:

a) develop regional training programs for PIOs and club publicity chairpersons;

b) coordinate public relations efforts for events and activities which may involve more than one section, and

c) provide input on matters before the League's Public Relations Committee for discussion or action.

6) Establishes and coordinates a section-wide Speakers Bureau to provide knowledgeable and effective speakers who are available to address community groups about Amateur Radio, and works with PIOs to promote interest among those groups.

7) Helps local PIOs to recognize and publicize newsworthy stories in their areas. Monitors news releases sent out by the PIOs for stories of broader interest and offers constructive comments for possible improvement. Helps local PIOs in learning to deal with, and attempting to minimize, any negative publicity about Amateur Radio or to correct negative stories incorrectly ascribed to Amateur Radio operators.

8) Working with the PIOs, develops and maintains a comprehensive list of media outlets and contacts in the section for use in section-wide or nationwide mailings.

9) Helps local PIOs prepare emergency response PR kits containing general information on Amateur Radio and on local clubs, which may be distributed in advance to local Emergency Coordinators and District Emergency Coordinators for use in dealing with the media during emergencies.

10) Works with PIOs, SM and ARRL staff to identify and publicize League-related stories of local or regional interest, including election or appointment of ARRL leadership officials, scholarship winners/award winners, *QST* articles by local authors or local achievements noted or featured in *QST*.

11) Familiarize self with ARRL Public Service Announcements (PSAs), brochures and audio-visual materials; assists PIOs in arranging air time for PSAs; helps PIOs and speakers choose and secure appropriate brochures and audio-visual materials for events or presentations.

12) At the request of the Section Manager or Division Director, may assist in preparation of a section or division newsletter.

13) Encourages, organizes and conducts public information/public relations sessions at ARRL hamfests and conventions.

14) Works with PIOs to encourage activities that place Amateur Radio in the public eye, including demonstrations, Field Day activities, etc. and assures that sponsoring organizations are prepared to follow-up on interest generated by these activities.

Most public relations activities are conducted on a local level by affiliated clubs, which generally are established community organizations. PICs should encourage clubs to make public relations a permanent part of their activities.

With the Section Manager's approval, the PIC may appoint club publicity chairpersons or other individuals recommended by affiliated clubs as PIOs. Where the responsibility cannot or will not be assumed by the club, the PIC is encouraged to seek qualified League members who are willing to accept the responsibility of PIO appointments.

Appointees should take advantage of every opportunity to recruit a new ham or member to foster growth of Field Organization programs, and our abilities to serve the public.

State Government Liaison

The State Government Liaison (SGL) shall be an amateur who is aware (at a minimum) of state legislative proposals in the normal course of events and who can watch for those proposals having the potential to affect Amateur Radio without creating a conflict of interest.

The SGL shall collect and promulgate information on state ordinances affecting Amateur Radio and work (with the assistance of other ARRL members) toward assuring that they work to the mutual benefit of society and the Amateur Radio Service.

The SGL shall guide, encourage and support ARRL members in representing the interests of the Amateur Radio Service at all levels. Accordingly, the SGL shall cooperate closely with other section-level League officials, particularly the Section Emergency Coordinator and the Public Information Coordinator.

When monitoring state legislative dockets, SGL's should watch for key words that could lead to potential items affecting Amateur Radio. Antennas (dish, microwave, towers, structures, satellite, television, lighting), mobile radio, radio receivers, radio interference, television interference, scanners, license plates, cable television, ham radio, headphones in automobiles, lightning protection, antenna radiation and biological effects of radio signals are a few of the examples of what to look for.

In those states where there is more than one section, the Section Managers whose territory does not encompass the state capital may simply defer to the SGL appointed by their counterpart in the section where the state capital is located. In this case, the SGL is expected to communicate equally with all Section Managers (and Section Emergency Coordinators and other section-level League officials). In sections where there is more than one government entity, i.e., Maryland, DC, Pacific, there may be a Liaison appointed for each entity.

Technical Coordinator

The ARRL Technical Coordinator (TC) is a section-level official appointed by the Section Manager to coordinate all technical activites within the section. The technical Coordinator must hold a Novice class or higher amateur license. The Technical Coordinator reports to the Section Manager and is expected to maintain contact with other section-level appointees as appropriate to ensure a unified ARRL Field Organization within the section. The duties of the Technical Coordinator are as follows:

1) Supervise and coordinate the work of the section's Technical Specialists (TSs).

2) Encourage amateurs in the section to share their technical achievements with others through the pages of *QST*, and at club meetings, hamfests and conventions.

3) Promote technical advances and experimentation at VHF/UHF and with specialized modes, and work closely with enthusiasts in these fields within the section.

4) Serve as an advisor to radio clubs that sponsor training programs for obtaining amateur licenses or upgraded licenses in cooperation with the ARRL Affiliated Club Coordinator.

5) In times of emergency or disaster, function as the coordinator for establishing an array of equipment for communications use and be available to supply technical expertise to government and relief agencies to set up emergency communication networks, in cooperation with the ARRL Section Emergency Coordinator.

6) Refer amateurs in the section who need technical advice to local TSs.

7) Encourage TSs to serve on RFI and TVI committees in the section for the purpose of rendering technical assistance as needed, in cooperation with the ARRL OO/Coordinator.

8) Be available to assist local technical program committees in arranging suitable programs for ARRL hamfests and conventions.

9) Convey the views of section amateurs and TSs about the technical contents of *QST* and ARRL books to ARRL HQ. Suggestions for improvements should also be called to the attention of the ARRL HQ technical staff.

10) Work with the appointed ARRL TAs (technical advisors) when called upon.

11) Be available to give technical talks at club meetings, hamfests and conventions in the section.

District Emergency Coordinator

The DEC is an ARRL full member of at least Technician class experienced in emergency communications who can assist the SEC by taking charge in the area of jurisdiction especially during an emergency.
The DEC shall:

1) Coordinate the training, organization and emergency participation of Emergency Coordinators in the area of jurisdiction.

2) Make local decisions in the absence of the SEC or through coordination with the SEC concerning the allotment of available amateurs and equipment during an emergency.

3) Coordinate the interrelationship between local emergency plans and between communications networks within the area of jurisdiction.

4) Act as backup for local areas without an Emergency Coordinator and assist in maintaining contact with governmental and other agencies in the area of jurisdiction.

5) Provide direction in the routing and handling of emergency communications of either a formal or tactical nature.

6) Recommend EC appointments to the SEC and advise on OES appointments.

7) Coordinate the reporting and documentation of ARES activities in the area of jurisdiction.

8) Act as a model emergency communicator as evidenced by dedication to purpose, reliability and understanding of emergency communications.

Emergency Coordinator

Appointment of an Emergency Coordinator can be made by the SM, on recommendation of the SEC and DEC (or directly by the SEC), in every community where a qualified amateur of Technician class or higher holding ARRL full membership can be found, since it is on his work and that of the local amateurs working with him that the entire ARES organization is based. While his duties are varied and manifold, they can be generally described as promotion and enhancement of activities of the Amateur Radio Emergency Service for public service at the local level.
The EC shall:

1) Coordinate the training, organization and emergency participation of interested amateurs in the area of jurisdiction.

2) Establish an emergency communications plan for the area that will effectively utilize the group to cover the needs for both tactical and formal message traffic requirements.

3) Establish a viable working relationship with governmental and other agencies who might need the service of ARES (e.g., Red Cross, Salvation Army, Rescue Squads, Weather Bureau, Hospitals, Police Department, Fire Department, etc.).

4) Establish local communications networks run on a regular basis and periodically test these networks by drills.

5) Establish an emergency traffic plan utilizing NTS to the extent possible and coordinate liaison with NTS nets.

6) In times of disaster, evaluate the communications needs of the jurisdiction and respond quickly to those needs.

7) Do all that is possible to further the image of Amateur Radio by dedication to purpose, reliability and a thorough understanding of the mission of Amateur Radio.

Every EC, on appointment, receives a copy of the *Emergency Coordinator's Handbook,* containing full details on how to organize and the basic principles of ARES organization, plus many samples and illustrations. Briefly, a newly appointed EC should undertake the following steps (or review each phase if he is succeeding someone else):

1) A general meeting of local amateurs.

2) Recommend candidates for Official Emergency Station appointment.

3) Designate Assistant ECs to serve specialized areas on a Planning Committee.

4) Have a meeting with the committee to discuss initial plans.

5) Contact local "to-be-served" agencies to determine requirements.

6) Draft detailed plans to serve community requirements.

7) Establish training procedures to ensure that the plans can be carried out in any emergency.

The EC also has certain routine administrative duties as follows:

a) Maintain full and current information on OES appointees and ARES registrants. Keep the DEC and SEC informed on the former.

b) Issue certificates to assistant ECs (not an SM appointment) and assign each a specific subleadership role in the local ARES organization.

c) Report the status of the ARES unit to the DEC each month on the form provided (FSD-212)

d) Coordinate and cooperate with the DEC.

The area of jurisdiction of each EC is determined by the SM and SEC/DEC depending on the ARES organizational plan of the section.

Net Manager

For coordinating and supervising traffic-handling activities in the section, the SM may appoint one or more Net Managers, usually on recommendation of the Section Traffic Manager. The number of NMs appointed may depend on a section's geographical size, the number of nets operating in the section, or other factors having to do with the way the section is organized. In some cases, there may be only one Net Manager in charge of the one section net, or one NM for the phone net, one for the CW net. In larger or more traffic-active sections there may be several, including NMs for the VHF net or nets, for the RTTY net, or NTS local nets not controlled by ECs. All ARRL NMs should work under the STM in a coordinated section traffic plan.

Some nets cover more than one section but operate in NTS at the section level. In this case, the Net Manager is selected by agreement among the STMs concerned and the NM appointment conferred on him by his resident SM.

NMs may conduct any testing of candidates for ORS appointment (see below) that they consider necessary before making appointment recommendations to the STM. Net Managers also have the function of requiring that all traffic handling in ARRL recognized nets is conducted in proper ARRL form.

Remember: All appointees or appointee candidates must be ARRL full members.

ARRL STATION APPOINTMENTS

Field Organization station and individual appointments from the Section Manager are available to qualified ARRL full members in each section. These appointments are as follows: Official Relay Station, Official Emergency Station, Official Bulletin Station, Public Information Officer, Official Observer and Technical Specialist. All appointees receive handsome certificates from the SM and are entitled to wear ARRL membership pins with the distinctive blue background. All appointees are required to submit regular reports to maintain appointments and to remain active in their area of specialty.

The report is the criterion of activity. An appointee who misses three consecutive monthly reports is subject to cancellation by the SM of the appropriate section leadership official, who cannot know what or how much you are doing unless you report. An appointee whose appointment is cancelled for this or other reasons must earn reinstatement by demonstrating activity and adherence to the requirements. Reinstatement of cancelled appointments, and indeed judgment of whether or not a candidate meets the requirements, is at the discretion of the SM and the section leadership.

The detailed qualifications of the six individual "station" appointments are given below. If you are interested, your SM will be glad to receive your application. Use application form FSD-187, reproduced nearby.

Official Relay Station

This is a traffic-handling appointment that is open to all licenses. This appointment applies equally to all modes and all parts of the spectrum. It is for traffic handlers, regardless of how or in what part of the spectrum they do it.

The potential value of the operator who has traffic know-how to his country and community is enhanced by his ability and the readiness of his station to function in the community interest in case of emergency. Traffic awareness and experience are often the signs by which mature amateurs may be distinguished.

Traditionally, there have been considerable differences between procedures for traffic handling by CW, phone, RTTY, ASCII and other modes. Appointment requirements for ORS do not deal with these, but with factors equally applicable to all modes. The appointed ORS may confine activities to one mode or one part of the spectrum if he wishes although versatility does indeed make it possible to perform a more complete public service. The expectation is that the ORS will set the example in traffic handling, however it is done. To the degree that he is deficient in performing traffic functions by any mode, to that extent he does not meet the qualifications for the appointment. Here are the basic requirements:

1) Full ARRL membership and Novice class license or higher.
2) Code and/or voice transmission.
3) Transmission quality, by whatever mode, must be of the highest quality, both technically and operationally. For example, CW signals must be pure, chirpless and clickless, and code sending must be well spaced and properly formed. Voice transmission must be of proper modulation percentage or deviation, precisely enunciated with minimum distortion. RTTY must be clickless, proper shift, etc.
4) All ORSs are expected to follow standard ARRL operating practices (message form, ending signals, abbreviations or prowords, courtecy, etc.).
5) Regular participation in traffic activities, either freelance or ARRL-sponsored. The latter is encouraged, but not required.
6) Handle all record communications speedily and reliably and set the example in efficient operating procedures. All traffic is relayed or delivered promptly after receipt.
7) Report monthly to the STM, including a breakdown of traffic handled during the past calendar month.

Official Emergency Station

Amateur operators of Novice class and above may be appointed OES by their SEC or SM at the recommendation of their ECs or DECs (if no EC) if they are ARRL members and interested in setting high standards of emergency preparedness and operating. In addition to candidates for this appointment operating within their own local EC jurisdictional areas, we want to recruit OESs who can set similar examples in "offshore" emergencies, such as those that frequently occur in foreign countries. Here are the standard qualifications and functions of this appointment.

1) Possession of full ARRL membership and a Technician class license or higher.
2) Regular participation in the local ARES, if any, include all drills and tests, emergency nets and, of course, real emergency situations.
3) Ability to operate independent of commerical mains including at least one-band mobile capability.
4) Must be fully acquainted with standard ARRL message form and capable of using it in handling any third-party messages.
5) Report monthly to the EC/DEC or SEC.

Official Bulletin Station

Rapid dissemination of information is the lifeblood of an active, progressive organization. The ARRL Official Bulletin Station network provides a vital communications link for informing the amateur community of the latest developments in Amateur Radio and the League. ARRL bulletins, containing up-to-the minute news and information of Amateur Radio, are issued by League Hq as soon as such news breaks. These bulletins are transmitted on a regular schedule by ARRL Hq station W1AW.

The primary mission of OBS appointees is to copy these bulletins directly off the air from W1AW—on voice, CW or RTTY/ASCII—and retransmit them locally for the benefit of amateurs in the particular coverage area, many of whom may not be equipped to receive bulletins directly from W1AW.

ARRL bulletins of major importance or of wide-ranging scope are mailed from Hq to each Bulletin Manager and OBS appointee.

**ARRL FIELD ORGANIZATION
APPLICATION FOR STATION APPOINTMENT**

I, _____ licensee of Amateur Radio station _____, hereby submit application to the Section Manager (or designated Section-level official) of the _____ Section, for appointment as:

Official Relay Station _____
Official Emergency Station _____
Official Observer _____   (Must hold minimum Technician Class license for 4 or more years).
Official Bulletin Station _____
Assistant Technical Coordinator _____
Public Information Assistant _____

My qualifications are: _____

(describe appropriate experience and/or interest)

_____

_____

_____

Address: _____
              street   city   state   zip

Phone number: _____
                    work   home

ARRL membership number: _____ (from QST label)

I have read the description of the duties and responsibilities of the appointment, and agree to maintain current League membership, and report my station activity to designated section officials on a monthly basis.

Signed: _____ Date: _____

(FSD-187 7/87)

Use this form to apply for an ARRL station appointment.

However, some bulletins, such as the ARRL DX Bulletin (transmitted on Fridays UTC), are disseminated only by W1AW because of time value. Thus, it is advantageous for each OBS to copy W1AW directly. In some sections, the Bulletin Manager may assume the responsibility of copying the bulletins from W1AW; therefore, individual OBSs should be sure to meet the Bulletin Manager on a regular, agreed-upon schedule to receive the latest bulletins.

Inasmuch as W1AW operates on all bands (160-2 meters), the need for OBSs on HF has lessened somewhat in recent times. However, OBS appointments for HF operation can be conferred by the Section Manager (or the Bulletin Manager, depending on how the SM organizes the section) if the need is apparent. More important, to serve the greatest possible "audience," OBS appointees who can send ARRL bulletins over VHF repeaters are of maximum usefulness and are much in demand. If possible, an OBS who can copy bulletins directly from W1AW (or the Bulletin Manager) should be assigned to each major repeater in the section. Bulletins should be transmitted regularly, perhaps in conjunction with a VHF repeater net, on a repeater bulletin board (tone-accessed recorded announcements for repeater club members), or via a local RTTY (computer) mailbox. Duties and requirements of the OBS include the following:

1) OBS candidates must have Novice class license or higher.

2) Retransmission of ARRL bulletins must be made at least once per week to maintain appointment.

3) OBS candidates are appointed by the Section Manager (or by the Bulletin Manager, if the SM so desires) and must adhere to a schedule that is mutually agreeable, as indicated on appointment application form FSD-187.

4) OBS appointees should send a monthly activity report (such as FSD-210, under "Schedules and Net Affiliations") to the Bulletin Manager, indicating bulletin transmissions made and generally updating the Bulletin Manager to any OBS-related activities. This reporting arrangement may be modified by the Bulletin Manager as he/she sees fit.

5) As directed by the Bulletin Manager, OBSs will include in their bulletin transmissions news of local, section and regional interest.

Public Information Officer

Public Information Officers (PIOs) are appointed by and report to the ARRL section Public Information Coordinator (PIC) generally upon the recommendation of an affiliated club and with the approval of the Section Manager (SM). PIOs are usually club publicity chairpersons and must be full ARRL members. Training for PIOs should be provided regularly on a sectional or regional basis by the PIC and/or other qualified people.

Good "grass roots" public relations activities involve regular and frequent publicizing of amateur activities through local news media plus community activities; school programs; presentations to service clubs and community organizations; exhibits and demonstrations; and other efforts which create a positive public image for Amateur Radio.

Public relations are not conducted in a vacuum. Even the best PR is wasted without effective follow-up. To do this best, PR activities must be well-timed and well-coordinated within the amateur community, so that clubs, Elmers, instructors and so on are prepared to deal with the interest the PR generates.

Recruitment of new hams and League members is an integral part of the job of every League appointee. Appointees should take advantage of every opportunity to recruit a new ham or member to foster growth of Field Organization programs, and our abilities to serve the public.

Specific Duties of the Public Information Officer

1) Establishes and maintains a list of media contacts in the local area; strives to establish and maintain personal contacts with appropriate representatives of those media (editors, news directors, science reporters and so on).

2) Becomes a contact for the local media and assures that editors/reporters who need information about Amateur Radio know where to find it.

3) Works with Local Government Liaisons to establish personal contacts with local government officials where possible and explain to them, briefly and non-technically, about Amateur Radio and how it can help their communities.

4) Keeps informed of activities by local hams and identifies and publicizes those that are newsworthy or carry human interest appeal. (This is usually done through news releases or suggestions for interviews or feature stories.)

5) Attempts to deal with and minimize any negative publicity about Amateur Radio and to correct any negative stories which are incorrectly ascribed to Amateur Radio operators.

6) Generates advance publicity through the local media of scheduled activities of interest to the general public, including licensing classes, hamfests, club meetings, Field Day operations, etc.

7) Works with the section PIC to identify and publicize League-related stories of local news interest, including election and appointment of hams to leadership positions, *QST* articles by local authors or local achievements noted or featured in *QST*.

8) Maintains contact with other League officials in the local area, particularly the Emergency Coordinator and/or District Emergency Coordinator. With the PIC, helps prepare an emergency response PR kit, including general brochures on Amateur Radio and specific information about local clubs. Distributes them to ECs and DECs before an emergency occurs. During emergencies, these kits should be made available to reporters at the scene or at a command post. The PIO should help summarize Amateur Radio activity in an ongoing situation, and follow up any significant emergency communications activities with prompt reporting to media of the extent and nature of Amateur Radio involvement.

9) Assists the section PIC in recruiting hams for the section's Speakers Bureau; promotes interest among community and service organizations in finding out more about Amateur Radio through the bureau and relays requests to the PIC.

10) Helps individual hams and radio clubs to develop and promote good ideas for community projects and special events to display Amateur Radio to the public in a positive light.

11) Attends regional training sessions sponsored by section PICs.

12) Becomes familiar with ARRL Public Service Announcements (PSAs), brochures and audio-visual materials; contacts local radio and TV stations to arrange airing of Amateur Radio PSAs; secures appropriate brochures and audio-visual materials for use in conjunction with planned activities.

13) Keeps the section PIC fully informed on activities and places PIC on news release mailing list.

Official Observer

The Official Observer (OO) program has been sponsored by the League for over 50 years to help amateurs help each other. Official observer appointees have aided thousands of amateurs to maintain their transmitting equipment and operating procedures in compliance with the regulations. The object of the OO program is to notify amateurs by mail of operating/technical irregularities before they come to the attention of the FCC.

The ARRL commitment to volunteer monitoring has been greatly enhanced by the creation of the Amateur Auxiliary to the FCC Field Operations Bureau, designed to enable amateurs to play a more active and direct role in upholding the traditional high standard of conduct on the amateur bands. The OO is the foundation of the Amateur Auxiliary, carrying out the all-important day-to-day maintenance monitoring of the amateur airwaves. Following recommendation by the Section Manager, potential members of the Amateur Auxiliary are provided with training materials, and all applicants must successfully complete a written examination to be enrolled as Official Observers. For further information, please contact your SM.

The OO performs his function by *listening* rather than transmitting, keeping a watchful ear out for such things as frequency instability, harmonics, hum, key clicks, broad signals, distorted audio, overdeviation, out-of-band operation, etc. The OO completes his task once the notification card is sent. Reimbursement for postage expenses are provided for through the SM. The OO:

1) Must be an ARRL full member and have been a licensee of Technician class or higher for at least four years.

2) Must undergo and complete successfully the Amateur Auxiliary training and certification procedure.

3) Must report to the OO Coordinator regularly on FSD-23.

4) Maintain regular activity in sending out notices as observed.

The OO program is one of the most important functions of the League. A sincere dedication to helping our brother and sister amateurs is required for appointment. Only the "very best" are sought.

Technical Specialist

Appointed by the SM, or TC under delegated authority from the SM, the TS supports the TC in two main areas of responsibility: Radio Frequency Interference, and Technical Information. The TS must hold full ARRL membership and at least a Novice class license. TSs can specialize in certain specific technical areas, or can be generalists. Here is a list of specific job duties:

1) Serve as a technical oracle to local hams and clubs. Correspond by telephone and letter on tech topics. Refer correspondents to other sources if specific topic is outside TS's knowledge.

2) Serve as advisor in radio frequency interference issues. RFI can drive a wedge in neighbor and city relations. It will be the TS with a cool head who will resolve problems. Local hams will come to you for guidance in dealing with interference problems.

3) Speak at local clubs on popular tech topics. Let local clubs know you're available and willing.

4) Represent ARRL at technical symposiums in industry; serve on CATV advisory committees; advise municipal governments on technical matters.

5) Work with other ARRL officials and appointees when called upon for technical advice especially in emergency communications situations where technical prowess can mean the difference in getting a communications system up and running, the difference between life and death.

6) Handle other miscellaneous technically related tasks assigned by the Technical Coordinator.

Local Government Liaison

The Local Government Liaison (LGL) is primarily responsible for monitoring proposals and actions by local government bodies and officials which may affect Amateur Radio; for working with the local PIO to alert section leadership officials and area amateurs to any such proposals or actions, and for coordinating local responses. In addition, the LGL serves as a primary contact for amateurs encountering problems dealing with local government agencies, for those who want to avoid problems and for local officials who wish to work with amateurs or simply learn more about Amateur Radio. The most effective LGL will be able to monitor local government dockets consistently, muster local, organized support quickly when necessary, and be well known in the local amateur community as the point person for local government problems. The LGL must be a Full Member of the ARRL. LGLs are appointed by and report to the Section Manager, or State Government Liaison (acting under delegated authority from the SM).

Specific Responsibilities:

1) Monitor proposals and actions of town/city councils zoning appeals boards, and any other legislative or regulatory agencies or officials below the state level whose actions can directly or indirectly affect Amateur Radio.

2) Attend meetings of those bodies when possible, to become familiar with their policies, procedures and members. Assist local amateurs in their dealings with local boards and agencies.

3) Be available to educate elected and appointed officials, formally and informally, about the value of Amateur Radio to their community.

4) Work with the PIO or PIC to inform local amateurs, the SGL and the SM of any proposals of actions which may affect Amateur Radio, and report regularly on the progress or lack thereof.

5) Work with the PIO to organize the necessary local response to any significant proposals or actions, either negative or positive, and coordinate that response.

6) Refer amateurs seeking ARRL Volunteer Counsels to HQ.

7) Register on mailing list for Planning Commission meeting agendas.

8) Work with the PIO and local clubs to build and/or maintain good relations between Amateur Radio and local officials. (For example, invite the mayor to a club dinner or council members to Field Day.)

ARRL Numbered Radiograms

The letters ARL are inserted in the preamble in the check and in the text before spelled out numbers, which represent texts from this list. Note that some ARL texts include insertion of numerals. *Example:* NR 1 R W1AW ARL 5 NEWINGTON CONN DEC 25 DONALD R SMITH A̅A̅ 164 EAST SIXTH AVE A̅A̅ NORTH RIVER CITY MO A̅A̅ PHONE 733 3968 B̅T̅ ARL FIFTY ARL SIXTY ONE B̅T̅ DIANA A̅R̅.

Group One—For possible "Relief Emergency" Use

ONE	Everyone safe here. Please don't worry.
TWO	Coming home as soon as possible.
THREE	Am in ___ hospital. Receiving excellent care and recovering fine.
FOUR	Only slight property damage here. Do not be concerned about disaster reports.
FIVE	Am moving to new location. Send no further mail or communication. Will inform you of new address when relocated.
SIX	Will contact you as soon as possible.
SEVEN	Please reply by Amateur Radio through the amateur delivering this message. This is a free public service.
EIGHT	Need additional ___ mobile or portable equipment for immediate emergency use.
NINE	Additional ___ radio operators needed to assist with emergency at this location.
TEN	Please contact ___. Advise to standby and provide further emergency information, instructions or assistance.
ELEVEN	Establish Amateur Radio emergency communications with ___ on ___ MHz.
TWELVE	Anxious to hear from you. No word in some time. Please contact me as soon as possible.
THIRTEEN	Medical emergency situation exists here.
FOURTEEN	Situation here becoming critical. Losses and damage from ___ increasing.
FIFTEEN	Please advise your condition and what help is needed.
SIXTEEN	Property damage very severe in this area.
SEVENTEEN	REACT communications services also available. Establish REACT communications with ___ on channel ___.
EIGHTEEN	Please contact me as soon as possible at ___.
NINETEEN	Request health and welfare report on ___. (State name, address and telephone number.)
TWENTY	Temporarily stranded. Will need some assistance. Please contact me at ___.
TWENTY ONE	Search and Rescue assistance is needed by local authorities here. Advise availability.
TWENTY TWO	Need accurate information on the extent and type of conditions now existing at your location. Please furnish this information and reply without delay.
TWENTY THREE	Report at once the accessibility and best way to reach your location.
TWENTY FOUR	Evacuation of residents from this area urgently needed. Advise plans for help.
TWENTY FIVE	Furnish as soon as possible the weather conditions at your location.
TWENTY SIX	Help and care for evacuation of sick and injured from this location needed at once.

Emergency/priority messages originating from official sources must carry the signature of the originating official.

Group Two—Routine messages

FORTY SIX	Greetings on your birthday and best wishes for many more to come.
FIFTY	Greetings by Amateur Radio.
FIFTY ONE	Greetings by Amateur Radio. This message is sent as a free public service by ham radio operators here at ___. Am having a wonderful time.
FIFTY TWO	Really enjoyed being with you. Looking forward to getting together again.
FIFTY THREE	Received your ___. It's appreciated; many thanks.
FIFTY FOUR	Many thanks for your good wishes.
FIFTY FIVE	Good news is always welcome. Very delighted to hear about yours.
FIFTY SIX	Congratulations on your ___, a most worthy and deserved achievement.
FIFTY SEVEN	Wish we could be together.
FIFTY EIGHT	Have a wonderful time. Let us know when you return.
FIFTY NINE	Congratulations on the new arrival. Hope mother and child are well.
*SIXTY	Wishing you the best of everything on ___.
SIXTY ONE	Wishing you a very merry Christmas and a happy New Year.
*SIXTY TWO	Greetings and best wishes to you for a pleasant ___ holiday season.
SIXTY THREE	Victory or defeat, our best wishes are with you. Hope you win.
SIXTY FOUR	Arrived safely at ___.
SIXTY FIVE	Arriving ___ on ___. Please arrange to meet me there.
SIXTY SIX	DX QSLs are on hand for you at the ___ QSL Bureau. Send ___ self-addressed envelopes.
SIXTY SEVEN	Your message number ___ undeliverable because of ___. Please advise.
SIXTY EIGHT	Sorry to hear you are ill. Best wishes for a speedy recovery.
SIXTY NINE	Welcome to the _____. We are glad to have you with us and hope you will enjoy the fun and fellowship of the organization.

*Can be used for all holidays.

ARL NUMBERS SHOULD BE SPELLED OUT AT ALL TIMES.

ARRL Recommended Message Precedences

Please observe the following ARRL provisions for PRECEDENCES in connection with written message traffic. These provisions are designed to increase the efficiency of our service both in normal times and in emergency.

Precedences

EMERGENCY	Any message having life and death urgency to any person or group of persons, which is transmitted by Amateur Radio in the absence of regular commercial facilities. This includes official messages of welfare agencies during emergencies requesting supplies, materials or instructions vital to relief of stricken populace in emergency areas. During normal times, it will be *very rare*. On CW/RTTY, this designation will *always* be spelled out. When in doubt, do not use it.
PRIORITY	Use abbreviation P on CW/RTTY. This classification is for (a) important messages having a specific time limit, (b) official messages not covered in the emergency category, (c) press dispatches and emergency-related traffic not of the *utmost* urgency, and (d) notice of death or injury in a disaster area, personal or official.
WELFARE	This classification, abbreviated w on CW/RTTY, refers to either an inquiry as to the health and welfare of an individual in the disaster area or an advisory from the disaster area that indicates all is well. Welfare traffic is handled only after all emergency and priority traffic is cleared. The Red Cross equivalent to an incoming Welfare message is DWI (Disaster Welfare Inquiry).
ROUTINE	Most traffic in normal times will bear this designation. In disaster situations, traffic labeled Routine (R on CW/RTTY) should be handled last, or not at all when circuits are busy with higher precedence traffic.

Note—the precedence always follows the message number. For example, a message number may be 207 R on CW and "Two Zero Seven Routine" on phone.

N5KR AZIMUTHAL EQUIDISTANT MAPS

The following computer-generated azimuthal maps centered on 21 different world locations were produced especially for *The ARRL Operating Manual* by William D. Johnston, N5KR, PO Box 370, White Sands, NM, 88002. Similar maps centered on your QTH or other locations can be ordered from N5KR (please include an SASE with your request for information).

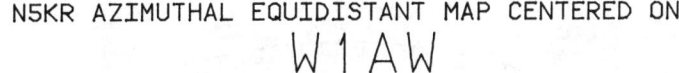

N5KR AZIMUTHAL EQUIDISTANT MAP CENTERED ON
W1AW

SCALE OF DISTANCE FROM CENTER

N5KR AZIMUTHAL EQUIDISTANT MAP CENTERED ON
EASTERN USA

SCALE OF DISTANCE FROM CENTER
KILOMETERS
0   5000   10000   15000   20000

0   2000   4000   6000   8000   10000 12000
MILES

N5KR AZIMUTHAL EQUIDISTANT MAP CENTERED ON
CENTRAL USA

SCALE OF DISTANCE FROM CENTER
KILOMETERS
0   5000   10000   15000   20000
0   2000   4000   6000   8000   10000   12000
MILES

N5KR AZIMUTHAL EQUIDISTANT MAP CENTERED ON
WESTERN USA

SCALE OF DISTANCE FROM CENTER

KILOMETERS

0   5000   10000   15000   20000

0   2000   4000   6000   8000   10000   12000

MILES

N5KR AZIMUTHAL EQUIDISTANT MAP CENTERED ON
ALASKA

SCALE OF DISTANCE FROM CENTER

KILOMETERS

0	5000	10000	15000	20000

0	2000	4000	6000	8000	10000	12000

MILES

HAWAII

SCALE OF DISTANCE FROM CENTER
KILOMETERS

| 0 | 5000 | 10000 | 15000 | 20000 |

| 0 | 2000 | 4000 | 6000 | 8000 | 10000 | 12000 |
MILES

CARIBBEAN

SCALE OF DISTANCE FROM CENTER

KILOMETERS

| 0 | 5000 | 10000 | 15000 | 20000 |

| 0 | 2000 | 4000 | 6000 | 8000 | 10000 | 12000 |

MILES

E. SOUTH AM.

SCALE OF DISTANCE FROM CENTER

KILOMETERS

0	5000	10000	15000	20000

0	2000	4000	6000	8000	10000	12000

MILES

N5KR AZIMUTHAL EQUIDISTANT MAP CENTERED ON
S. SOUTH AM.

SCALE OF DISTANCE FROM CENTER
KILOMETERS

| 0 | 5000 | 10000 | 15000 | 20000 |

| 0 | 2000 | 4000 | 6000 | 8000 | 10000 | 12000 |

MILES

ANTARCTICA

SCALE OF DISTANCE FROM CENTER

KILOMETERS

| 0 | 5000 | 10000 | 15000 | 20000 |

| 0 | 2000 | 4000 | 6000 | 8000 | 10000 | 12000 |

MILES

SCALE OF DISTANCE FROM CENTER
KILOMETERS

0	5000	10000	15000	20000

0	2000	4000	6000	8000	10000	12000

MILES

SCALE OF DISTANCE FROM CENTER

KILOMETERS

0   5000   10000   15000   20000

0   2000   4000   6000   8000   10000   12000

MILES

N5KR AZIMUTHAL EQUIDISTANT MAP CENTERED ON
WEST AFRICA

SCALE OF DISTANCE FROM CENTER
KILOMETERS

EAST AFRICA

SCALE OF DISTANCE FROM CENTER
KILOMETERS
0   5000   10000   15000   20000
0   2000 4000   6000   8000   10000 12000
MILES

N5KR AZIMUTHAL EQUIDISTANT MAP CENTERED ON
SOUTHRN AFRI

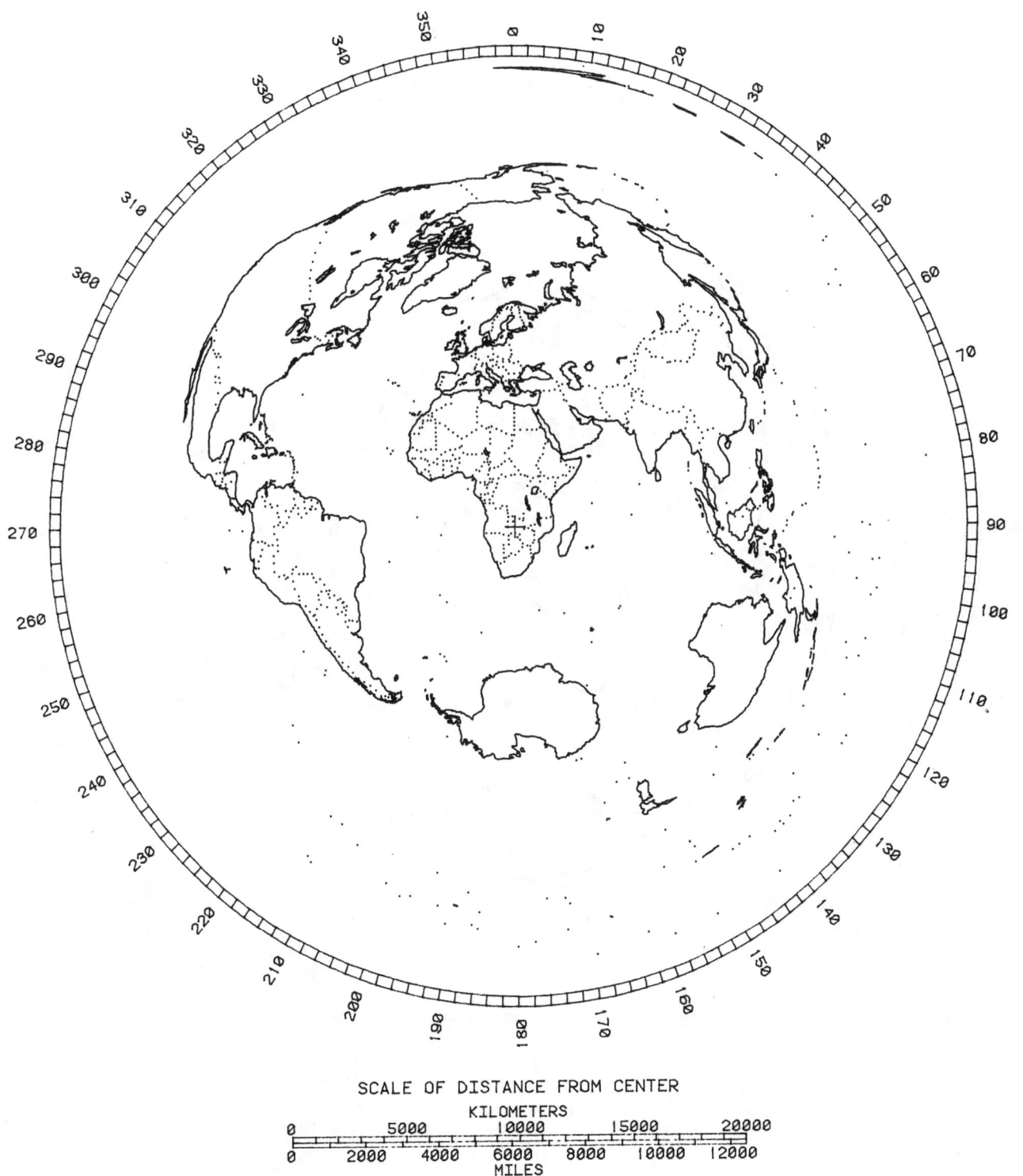

SCALE OF DISTANCE FROM CENTER
KILOMETERS

0	5000	10000	15000	20000

0	2000	4000	6000	8000	10000	12000

MILES

SCALE OF DISTANCE FROM CENTER
KILOMETERS
0   5000   10000   15000   20000
0   2000 4000 6000 8000 10000 12000
MILES

N5KR AZIMUTHAL EQUIDISTANT MAP CENTERED ON
SOUTH ASIA

SCALE OF DISTANCE FROM CENTER
KILOMETERS

0   5000   10000   15000   20000

0   2000   4000   6000   8000   10000   12000
MILES

N5KR AZIMUTHAL EQUIDISTANT MAP CENTERED ON
S.E. ASIA

SCALE OF DISTANCE FROM CENTER
KILOMETERS

| 0 | 5000 | 10000 | 15000 | 20000 |

| 0 | 2000 | 4000 | 6000 | 8000 | 10000 | 12000 |
MILES

N5KR AZIMUTHAL EQUIDISTANT MAP CENTERED ON
FAR EAST

SCALE OF DISTANCE FROM CENTER
KILOMETERS

| 0 | 5000 | 10000 | 15000 | 20000 |

| 0 | 2000 | 4000 | 6000 | 8000 | 10000 | 12000 |
MILES

AUSTRALIA

SCALE OF DISTANCE FROM CENTER
KILOMETERS

0	5000	10000	15000	20000

0	2000	4000	6000	8000	10000	12000
MILES

N5KR AZIMUTHAL EQUIDISTANT MAP CENTERED ON
S. PACIFIC

SCALE OF DISTANCE FROM CENTER
KILOMETERS

| 0 | 5000 | 10000 | 15000 | 20000 |

| 0 | 2000 | 4000 | 6000 | 8000 | 10000 | 12000 |
MILES

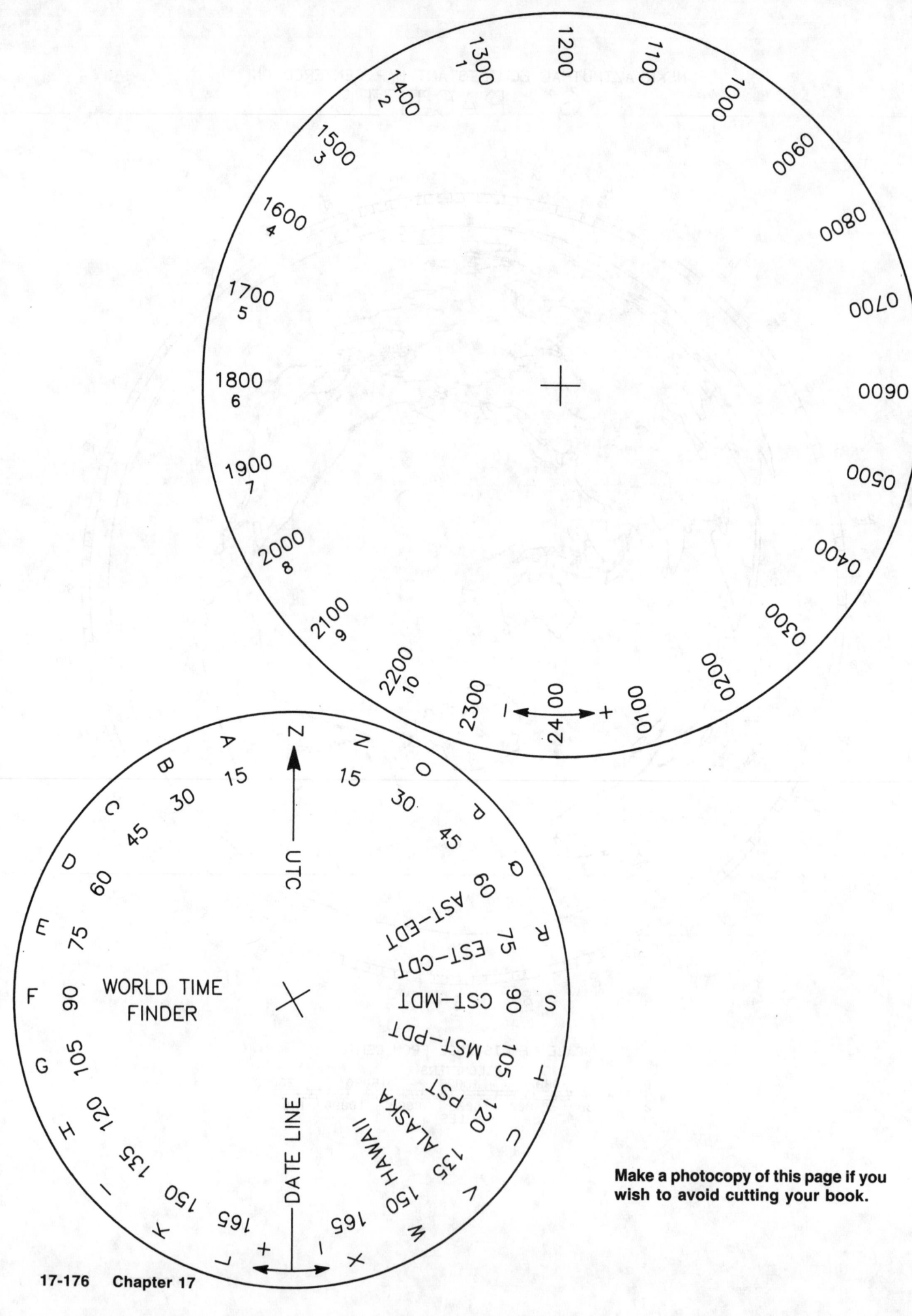

Make a photocopy of this page if you wish to avoid cutting your book.

FCC-Allocated Prefixes for Areas Outside the Continental US

Prefix	*Location*
AH1, KH1, NH1, WH1	Baker, Howland Is
AH2, KH2, NH2, WH2	Guam
AH3, KH3, NH3, WH3	Johnston I
AH4, KH4, NH4, WH4	Midway I
AH5K, KH5K, NH5K, WH5K	Kingman Reef
AH5, KH5, NH5, WH5 (except K suffix)	Palmyra, Jarvis Is
AH6-7, KH6-7, NH6-7, WH6-7	Hawaii
AH7K, KH7K, NH7K, WH7K	Kure I
AH8, KH8, NH8, WH8	American Samoa
AH9, KH9, NH9, WH9	Wake, Wilkes, Peale Is
AHØ, KHØ, NHØ, WHØ	Northern Mariana Is
AL, KL, NL, WL	Alaska
KP1, NP1, WP1	Navassa
KP2, NP2, WP2	Virgin Is
KP3-4, NP3-4, WP3-4	Puerto Rico
KP5, NP5, WP5	Desecheo

About the ARRL

The seed for Amateur Radio was planted in the 1890s, when Guglielmo Marconi began his experiments in wireless telegraphy. Soon he was joined by dozens, then hundreds, of others who were enthusiastic about sending and receiving messages through the air—some with a commercial interest, but others solely out of a love for this new communications medium. The United States government began licensing Amateur Radio operators in 1912.

By 1914, there were thousands of Amateur Radio operators—hams—in the United States. Hiram Percy Maxim, a leading Hartford, Connecticut, inventor and industrialist saw the need for an organization to band together this fledgling group of radio experimenters. In May 1914 he founded the American Radio Relay League (ARRL) to meet that need.

Today ARRL, with more than 170,000 members, is the largest organization of radio amateurs in the United States. The League is a not-for-profit organization that:
- promotes interest in Amateur Radio communications and experimentation
- represents US radio amateurs in legislative matters, and
- maintains fraternalism and a high standard of conduct among Amateur Radio operators.

At League headquarters in the Hartford suburb of Newington, the staff helps serve the needs of members. ARRL is also International Secretariat for the International Amateur Radio Union, which is made up of similar societies in more than 100 countries around the world.

ARRL publishes the monthly journal QST, as well as newsletters and many publications covering all aspects of Amateur Radio. Its headquarters station, W1AW, transmits bulletins of interest to radio amateurs and Morse code practice sessions. The League also coordinates an extensive field organization, which includes volunteers who provide technical information for radio amateurs and public-service activities. ARRL also represents US amateurs with the Federal Communications Commission and other government agencies in the US and abroad.

Membership in ARRL means much more than receiving QST each month. In addition to the services already described, ARRL offers membership services on a personal level, such as the ARRL Volunteer Examiner Coordinator Program and a QSL bureau.

Full ARRL membership (available only to licensed radio amateurs) gives you a voice in how the affairs of the organization are governed. League policy is set by a Board of Directors (one from each of 15 Divisions). Each year, half of the ARRL Board of Directors stands for election by the full members they represent. The day-to-day operation of ARRL HQ is managed by an Executive Vice President and a Chief Financial Officer.

No matter what aspect of Amateur Radio attracts you, ARRL membership is relevant and important. There would be no Amateur Radio as we know it today were it not for the ARRL. We would be happy to welcome you as a member! (An Amateur Radio license is not required for Associate Membership.) For more information about ARRL and answers to any questions you may have about Amateur Radio, write or call:

ARRL Educational Activities Dept
225 Main Street
Newington CT 06111-1494
(860) 594-0301
Prospective new amateurs call:
800-32-NEW HAM (800-326-3942)

Index

PACKET POWER!